Unit Operations

Unit Operations

George Granger Brown
EDWARD DE MILLE CAMPBELL UNIVERSITY PROFESSOR OF CHEMICAL ENGINEERING AND DEAN OF THE COLLEGE OF ENGINEERING, UNIVERSITY OF MICHIGAN

Alan Shivers Foust
PROFESSOR OF CHEMICAL ENGINEERING, UNIVERSITY OF MICHIGAN

George Martin Brown
ASSOCIATE PROFESSOR OF CHEMICAL ENGINEERING, NORTHWESTERN UNIVERSITY

Donald LaVerne Katz
PROFESSOR OF CHEMICAL ENGINEERING AND CHAIRMAN OF THE DEPARTMENT OF CHEMICAL AND METALLURGICAL ENGINEERING, UNIVERSITY OF MICHIGAN

Lloyd Earl Brownell
ASSOCIATE PROFESSOR OF CHEMICAL ENGINEERING, UNIVERSITY OF MICHIGAN

Richard Schneidewind
PROFESSOR OF METALLURGICAL ENGINEERING, UNIVERSITY OF MICHIGAN

Joseph J. Martin
ASSOCIATE PROFESSOR OF CHEMICAL ENGINEERING, UNIVERSITY OF MICHIGAN

Robert Roy White
PROFESSOR OF CHEMICAL ENGINEERING, UNIVERSITY OF MICHIGAN

George Brymer Williams
ASSOCIATE PROFESSOR OF CHEMICAL ENGINEERING, UNIVERSITY OF MICHIGAN

William Platt Wood
PROFESSOR OF METALLURGICAL ENGINEERING, UNIVERSITY OF MICHIGAN

Julius Thomas Banchero
ASSISTANT PROFESSOR OF CHEMICAL ENGINEERING, UNIVERSITY OF MICHIGAN

Jesse Louis York
ASSOCIATE PROFESSOR OF CHEMICAL ENGINEERING, UNIVERSITY OF MICHIGAN

CBS Publishers & Distributors Pvt. Ltd.

New Delhi • Bengaluru • Chennai • Kochi • Kolkata • Mumbai
Hyderabad • Nagpur • Patna • Pune • Vijayawada

ISBN: 81-239-1099-1

First Indian Edition: 1995
Reprint: 2001, 2004, 2005

This edition has been published in India by arrangement with the Author

Published by **Satish Kumar Jain** and produced by **Varun Jain** for
CBS Publishers & Distributors Pvt. Ltd.,
4819/XI Prahlad Street, 24 Ansari Road, Daryaganj, New Delhi - 110002
delhi@cbspd.com, cbspubs@airtelmail.in • www.cbspd.com
Ph.: 23289259, 23266861, 23266867 • Fax: 011-23243014

Corporate Office: 204 FIE, Industrial Area, Patparganj, Delhi - 110 092
Ph: 49344934 • Fax: 011-49344935
E-mail: publishing@cbspd.com • publicity@cbspd.com

Branches:

- ***Bengaluru:*** 2975, 17th Cross, K.R. Road, Bansankari 2nd Stage, Bengaluru - 70 • Ph: +91-80-26771678/79 • Fax: +91-80-26771680 E-mail: cbsbng@gmail.com, bangalore@cbspd.com
- ***Chennai:*** No. 7, Subbaraya Street, Shenoy Nagar, Chennai - 600030 Ph: +91-44-26681266, 26680620 • Fax: +91-44-42032115 E-mail: chennai@cbspd.com
- ***Kochi:*** Ashana House, 39/1904, A.M. Thomas Road, Valanjambalam, Ernakulum, Kochi • Ph: +91-484-4059061-65 Fax: +91-484-4059065 • E-mail: cochin@cbspd.com
- ***Kolkata:*** 6-B, Ground Floor, Rameshwar Shaw Road, Kolkata - 700014 Ph: +91-33-22891126/7/8 • E-mail: kolkata@cbspd.com
- ***Mumbai:*** 83-C, Dr. E. Moses Road, Worli, Mumbai - 400018 Ph: +91-9833017933, 022-24902340/41 • E-mail: mumbai@cbspd.com

Representatives:

- Hyderabad: 0-9885175004
- Patna: 0-9334159340
- Jharkhand: 0-9811541605
- Nagpur: 0-9021734563
- Pune: 0-9623451994
- Uttarakhand: 0-9716462459

Printed at:
J.S. Offset Printers, Delhi (India)

Dedicated to

ALFRED HOLMES WHITE

Preface

This textbook is the first to carry the title *Unit Operations*, but it is not the first to treat the subject.

Modern practice and equipment are emphasized as well as mathematical interpretations, as only by properly designed, constructed, and operated equipment can mathematical treatment yield useful results. The object is to build the student's knowledge and power progressively and continuously until he has a reasonably clear concept of how to approach the problems of design and operation of processing equipment. The unit operations are grouped according to similarities in action or in methods of calculation and presented in sequence according to increasing difficulty.

By grouping similar operations and using a common nomenclature in similar theoretical discussions, we find that the student makes more rapid progress, less effort is required to master nomenclature, and a better understanding is gained of the relationships among the different unit operations. The association and comparison of similar operations from different industrial processes is the essence of unit operations and the major factor in developing chemical, metallurgical, or process engineers capable of successfully designing new plants for conducting new processes. The full advantage of the study of unit operations can be realized only if the unit operations are themselves associated and compared so the engineer may more skillfully select the most suitable operation and equipment desired for each step in the process. The tendency of the specialist to treat each unit operation as a specialty having its own peculiar result, rationalization, and nomenclature is of questionable value in any sustained educational effort and is to be resisted by all means in an undergraduate curriculum.

The arrangement in order of increasing difficulty rather than in order of assumed importance continually presents new advanced intriguing problems to the student, maintains his interest, and encourages him to continue his own development beyond the limitations of the book. The treatment of those operations covering solids in Part I requires little more preparation than is ordinarily given in high school, whereas the treatment of mass transfer in Part IV is suitable for a postgraduate course and is presented with a critical attitude tending to develop the research point of view.

The inductive method is generally followed, relying upon observations from experience rather than upon deductive rationalizations. This method is a powerful tool of the practicing engineer and has been found most satisfactory for undergraduate students. However, kinetic explanations are not neglected and receive increasing emphasis in the last part on energy and mass transfer as an important means to a thorough understanding of the mechanisms involved.

Physics, calculus, and a beginning course in material and energy balances, or thermodynamics, are assumed as prerequisites to unit operations. Even with this background the student may be confused regarding dimensions and energy balances, and these subjects are treated rather fully. It is hoped that all chapters have received sufficiently extensive treatment to meet the requirements of any undergraduate curriculum so that the desired emphasis may be obtained by omission rather than addition. About 180 recitations should be required to cover the entire material in an adequate manner with undergraduate students, allowing 8 to 10 for the first five chapters and 50 to 60 each for Parts II, III, and IV. In a postgraduate course for students who have completed an undergraduate course in unit operations, this time could be reduced by one-third or one-half. With appropriate omissions the text has been used successfully for undergraduate courses of three quarters with a total of 117 class meetings and of two semesters with a total of 105 class meetings, as well as for a single-semester short course of 60 class meetings.

References to the literature are included for the purpose of attracting the student's attention to other sources of information as well as to acknowledge sources. An effort has been made to give credit for all material used, but so many workers have contributed so much that it is impossible to recognize the contributions of everyone. Indebtedness to previous texts and handbooks and to manufacturers of equipment is freely acknowledged. The specific help and suggestions of L. F. Stutzman and George Thodos, Associate Professors, and D. A. Dahlstrom, Assistant Professor of Chemical Engineering, at Northwestern University, F. Charles Moesel and Cedomir Sliepcevich, Assistant Professors of Chemical Engineering at The University of Michigan, Dr. Joseph Allerton, of Sayville, Long Island, and Verne C. Kennedy, Jr., of Chicago, and the frank criticisms of students who have used the material as mimeographed notes have been invaluable. Tolerance and your cooperation in helping to eliminate errors and suggest improvements as they may appear are requested.

THE AUTHORS

Contents

RULE IV OF REASONING

IN EXPERIMENTAL PHILOSOPHY WE ARE TO LOOK UPON PROPOSITIONS INFERRED BY GENERAL INDUCTION FROM PHENOMENA AS ACCURATE OR VERY NEARLY TRUE, NOT-WITHSTANDING ANY CONTRARY HYPOTHESES THAT MAY BE IMAGINED, TILL SUCH TIME AS OTHER PHENOMENA OCCUR, BY WHICH THEY MAY EITHER BE MADE MORE ACCURATE, OR LIABLE TO EXCEPTIONS.

This rule we must follow, that the argument of induction may not be evaded by hypotheses.

J. ISAAC NEWTON, *Principia* (1686)

CHAPTER 1

Introduction to the Unit Operations

In general there are two different approaches to the study of industrial processing. Each particular industry, such as the alcohol, petroleum, plastic, copper, or steel industry, including its characteristic operations, may be studied as a unit; or the different operations common to many industrial processes may be classified, each according to its function without regard to the industry using it, and each such operation studied as a unit operation. Thus heat transfer is a single or unit operation common to practically all industries, and knowledge of the principles of heat transfer is equally useful to an engineer in any industry requiring the transfer of heat.

As industrial processes have become more varied and technical, the fields open to the engineer have widened and it has become increasingly difficult, if not impossible, to cover the various industries in an adequate manner without limiting the students to a few closely related fields. By studying the unit operations themselves and their functions the engineer is trained to recognize these functions in new industrial processes; and by applying his knowledge and skill in the corresponding unit operations he is able to design, construct, and operate a plant for a new process with almost as much confidence as for a proved process. For these reasons the study of unit operations has proved to be the more efficient approach to the study of industrial processing.

Although the importance of these operations that are common to different industries was recognized as early as 1893 by Professor George Lunge,* the concept of unit operations was first crystallized by A. D. Little † in 1915.

> The arts of pulverizing, evaporating, filtering, distilling, and other operations constantly carried on in chemical works have been so thoroughly developed as to amount almost to special sciences.*
>
> Any chemical process, on whatever scale conducted, may be resolved into a coordinate series of what may be termed "Unit Operations," as pulverizing, drying, roasting, crystallizing, filtering, evaporating, electrolyzing, and so on. The number of these basic unit operations is not large and relatively few of them are involved in any particular process. The complexity of chemical engineering results from the variety of conditions as to temperature, pressure, etc., under which the unit operations must be carried out in different processes, and from the limitations as to materials of construction and design of apparatus imposed by the physical and chemical character of the reacting substances.†

A study of the unit operations is just as valuable to the operating engineer as to the designer, since all industrial operations, or plants, are composed physically of a series of unit operations in their proper sequence. The ability or capacity of a plant is no greater than that of its weakest unit. The operator analyzes his complex operations into units for individual improvement, and the designer synthesizes complex operations from a number of unit operations.

UNIT OPERATIONS CLASSIFIED

In this treatment the unit operations are classified or grouped according to their function and the phase

* Professor George Lunge of the Federal Polytechnic School of Zurich, in an address on the "Education of Industrial Chemists" presented at the Congress of Chemists at the Exposition in Chicago, 1893.

† Arthur D. Little as chairman of the Visiting Committee of the Department of Chemistry and Chemical Engineering of the Massachusetts Institute of Technology in a report to the President of the Institute in 1915.

or phases treated. A *phase* is a homogeneous and mechanically distinct or separable mass. Thus sand and water are two mechanically distinct masses, and each represents a separate phase; whether the sand is separate from or suspended in the water makes no difference. An oil phase floating upon water, or emulsified with the water, is a homogeneous mass mechanically distinct from the water whether or not it is continuous; and it is, therefore, a separate phase from the water phase. Similarly, a copper ore contains the mineral chalcopyrite as a separate solid phase from the surrounding gangue or rock, no matter how finely the mineral may be dispersed.

The phases present at any one time may be one or more solid phases, and one or more fluid phases. Sand and water represent one solid and one fluid phase, oil and water are two fluid phases, and the mineral and gangue are two (at least) solid phases. A mixture of solid salt, ice, water, and water vapor contains two solid and two fluid phases. Gases are fluids. Ordinarily there will exist only one gaseous phase.

The order of treatment begins with unit operations that treat solids alone, such as mechanical size separation, size reduction, and conveying of solids. These are followed by operations involving fluids. Since all fluids must be confined to store them or to direct their flow, a solid boundary phase is always involved, whether the solid particles are flowing through the fluid as in classification and flotation, or whether the fluid is flowing through a solid as in fluid transportation or filtration. The operations involving transfer of material from one phase to another are next treated by the method of equilibrium stages or contacts. These include leaching (solid to liquid), extraction (liquid to liquid), gas absorption and distillation (vapor to liquid), and adsorption (fluid to solid). Heat transfer and evaporation follow. Heat transfer deals with the rate of energy transfer and serves as a means of leading directly to the concept of rate of mass transfer as applied in crystallization, drying, absorption, distillation, and the more complicated operations involving catalysts and rates of reaction.

PRACTICAL OPERATIONS

In the study of unit operations, it must always be remembered that a unit operation is simply a unit of a more complex operating plant: a heat exchanger in a sugar plant, a crusher in a cement plant, a distillation column in a petroleum refinery, and that the important requirement in each case is a satisfactory workable overall operation. It makes no difference whether the result is obtained by exact mathematical calculation, by empirical approximation, or by a good guess based on the application of sound judgment, provided it is a satisfactory, workable, economical operation in its entirety.

The unit operations are the best available methods for classifying and formulating the combined experience of engineers as a guide to the operation and design of industrial plants. But these data, although of great help, are inadequate in themselves to insure successful operations. The successful engineer must develop sound judgment by his willingness to try, to recognize failures, and to keep on trying until he arrives at a satisfactory result. Seldom if ever does he have the opportunity to assemble either on paper or in physical form the ideal or perfect operation. Engineering operations require approximations and compromises. If made too nearly perfect, they may cost too much and last too long. Many plants become obsolete before they wear out.

All the information now available started with a single observation. As additional observations were made, the engineering mind began to draw conclusions which could be presented in the form of an empirical tabulation, such as the power required to operate crushing and grinding machines. Frequently these tabulated data could be presented in the form of a graph as a more satisfactory basis for extrapolating and interpolating the results. The next step was to derive an equation for the line representing the plotted data and to indicate means for estimating how the constants in the equation would be affected by different conditions. These equations might then be rationalized or sometimes "derived." However, the student and engineer should always keep in mind that these conclusions are drawn more or less soundly from a series of more or less reliable observations that have been empirically correlated; also, they should remember that the practical operator in the plant who may never have seen the equation or heard the term "unit operation" has probably made more observations himself than all those involved in deriving the equation. But it has taken the practical operator a much longer time to acquire his skill without understanding than it has the modern student of unit operations to acquire his comprehensive understanding.

FUNDAMENTAL CONCEPTS

Certain concepts or conclusions drawn from many observations are regarded as fundamental because, the more carefully the observations are made, the more closely do the data conform to the previous conclusion. Perhaps the most important of these to the engineer is the law of conservation of mass and energy.

Operations involving atomic energy have emphasized the concept that mass and energy are directly related. The quantity of energy equivalent to a unit of mass is so large, about 3×10^{16} ft-lb of energy per pound-mass, or the mass is so small, about 2.6×10^{-14} pound-mass per British thermal unit (Btu), that ordinary means of measurement are incapable of detecting any increase or decrease in mass accompanying a chemical process. In engineering operations, when nuclear changes are not involved, the mass of the products equals the mass of the reactants. This is in accord with engineering experience over many years and simplifies calculations, since material balances can then be made independently of energy balances.

The following four concepts are basic and form the foundation for the calculation of all operations. If nuclear changes are involved, the energy changes become so great that the first and second concepts are not independent and a combined energy and mass balance must be made.

1. The Material Balance

If matter may be neither created nor destroyed, the total mass for all materials entering an operation equals the total mass for all materials leaving that operation, except for any material that may be retained or accumulated in the operation. By the application of this principle, the yields of a chemical reaction or engineering operation are computed.

In continuous operations, material is usually not accumulated in the operation, and a material balance consists simply in charging (or debiting) the operation with all material entering and crediting the operation with all material leaving, in the same manner as used by any accountant. The result must be a balance. The accountant uses dollars as his unit, and the engineer uses pounds, tons, etc. In making a material balance, the engineer should not attempt to use units that may be created or destroyed during the process, such as units of volume or moles, or cubic feet, gallons, barrels, or molecules.

As long as the reaction is chemical and does not destroy or create atoms, it is proper and frequently very convenient to employ atoms as the basis for the material balance. The material balance may be made for the entire plant or for any part of it as a unit, depending upon the problem at hand. It is most conveniently made by adopting as a basis for calculation a fixed quantity of material which passes through the operation unchanged.

2. The Energy Balance

Similarly, an energy balance may be made around any plant or unit operation to determine the energy required to carry on the operation or to maintain the desired operating conditions. The principle is just as important as that of the material balance, and it is used in the same way. The important point to keep in mind is that all energy of all kinds must be included, although it may be converted to a single equivalent form such as Btu's, calories, or foot-pounds for the sake of addition. A balance cannot be made of heat or electrical energy alone, since all energy is convertible and all forms must be included in the balance.

3. The Ideal Contact

Whenever the materials being processed are in contact for any length of time under specified conditions, such as conditions of temperature, pressure, chemical composition, or electrical potential, they tend to approach a definite condition of equilibrium which is determined by the specified conditions. In many cases the rate of approach to these equilibrium conditions is so rapid or the length of time is sufficient that the equilibrium conditions are practically attained at each contact. Such a contact is known as an equilibrium or ideal contact. The calculation of the number of ideal contacts is an important step required in understanding those unit operations involving transfer of material from one phase to another, such as leaching, extraction, absorption, and distillation.

4. Rates of an Operation

In most operations equilibrium is not attained, either because of insufficient time or because it is not desired. As soon as equilibrium is attained no further change can take place and the process stops, but the engineer must keep the process going. For this reason rate operations, such as rate of energy transfer, rate of mass transfer, and rate of chemical reaction, are of the greatest importance and interest. In all such cases the rate and direction depend upon a difference in potential or driving force. The rate usually may be expressed as proportional to a potential drop divided by a resistance. An application of this principle to electrical energy is the familiar Ohm's law for steady or direct current.

$$I = \frac{E_1 - E_2}{R} = \frac{-\Delta E}{R}$$

where I = rate of electron transfer or current of electricity (coulombs/sec. or amp).

E = electrical potential, and ΔE is the increase in potential between points 1 and 2 (volts).

R = resistance (ohms).

In heat transfer under similar conditions for steady flow, the time rate of heat transfer from mass A in contact with mass B is

$$\frac{dQ}{dt} = \frac{T_A - T_B}{R} = \frac{-(T_B - T_A)}{R} = \frac{-\Delta T}{R}$$

where $\frac{dQ}{dt}$ = the instantaneous time rate of heat transfer or the quantity of heat transferred per unit of time from mass A to mass B.

T_A = temperature of mass A.

T_B = temperature of mass B.

$-\Delta T$ = the temperature drop.

R = resistance to heat transfer.

In solving rate problems as in heat transfer or mass transfer with this simple concept, the major difficulty is the evaluation of the resistance term. In practice, the values of the resistance term are generally computed from an empirical correlation of many determinations of transfer rates under different conditions.

The basic concept that rate depends directly upon a potential drop and inversely upon a resistance may be applied to any rate operation, although the rate may be expressed in different ways with particular coefficients for particular cases.

APPLICATION OF CONCEPTS

These principles, used singly or in combination, and the coordinated knowledge of the unit operations as presented in this textbook, the handbooks, and other technical literature constitute the science or theory of the unit operations. Practical engineering consists in applying the understanding of these operations and practical knowledge of the many types of equipment that may be employed to the design and operation of a commercial plant that will show not only a material balance but also a favorable dollar balance.

PART I

Solids

THIS section deals with those operations which treat material in the solid state only: screening, size reduction, and handling of solids. Before discussing these operations the properties of solids should be reviewed.

CHAPTER 2

Properties of Solids

AMONG the many properties of solids, those listed below are of particular significance in engineering operations.

Density, usually expressed by the symbol ρ, is defined as the mass per unit volume. The units are usually pounds per cubic foot, or grams per cubic centimeter.

Specific gravity is the ratio of the density of the material to the density of some reference substance, or ρ/ρ_{ref}. For solids and liquids the reference substance is usually water at 4° C. For most engineering work the specific gravity may be given the same numerical value as density in grams per cubic centimeter, but the specific gravity is a dimensionless ratio.

Bulk (or apparent) density ρ_b is the total mass per unit of total volume. For example, the true density of quartz is 2.65 grams/cc. But a quartz sand of 2.65 grams mass may occupy a total or bulk volume of 2 cc and have a bulk density ρ_b of 1.33 grams/cc. The bulk density is not an intrinsic characteristic of the material since it varies with the size distribution of the particles and their environment. The porosity of the solid itself and the material with which the pores, or voids, are filled also influence the bulk density. For a single nonporous particle the true density ρ equals the bulk density ρ_b.

Hardness[2] * of certain solids such as metals and plastics may be defined as resistance to indentation. The hardness of minerals is usually defined as resistance to scratching and is usually expressed in terms of Mohs' scale, which is based on a series of minerals of increasing hardness numbers as follows:

1	Talc	6	Feldspar
2	Gypsum	7	Quartz
3	Calcite	8	Topaz
4	Fluorite	9	Corundum, sapphire
5	Apatite	10	Diamond

Each mineral in the list will scratch all those of a lower number. A mineral of unknown hardness is rubbed against these test minerals, and its hardness is indicated by the softest material which just scratches it. The approximate hardnesses of some common materials are: dry finger nail, 2.5; copper

* Superior numbers refer to entries in the bibliographies. For this chapter, see p. 8.

TABLE 1. SOME PROPERTIES OF SOLIDS

Material	Density,[1] ρ, lb/cu ft	Bulk Density, lb/cu ft	Mohs Hardness [1]	Crystalline Form [1]
Alumina, Al_2O_3	249		9.0	Hexagonal
Bauxite	159	Crushed 80	Soft	Amorphous
Barites, $BaSO_4$	280	Crushed 180	2.5–3.5	Orthorhombic
Calcite, $CaCO_3$	169	Crushed 90–96	3	Hexagonal
Gypsum, $CaSO_4 \cdot 2H_2O$	145	Crushed 80–100; Powdered 60–80	1.5–2.0	Monoclinic
Hematite, Fe_2O_3	306–330	Crushed 150	5.5–6.5	Triclinic
Pyrites, FeS_2	308–318		6.0–6.5	Cubic
Halite, NaCl	131–162	Crushed 80; Powdered 45	2.5	Cubic
Galena, PbS	460–480		2.5–2.7	Cubic
Quartz, SiO_2	165	Gravel 100–110	7	Hexagonal
Sphalerite, ZnS	255		3.5–4.0	Cubic
Beans, corn, flaxseed, wheat		45–48		
Cottonseed, oats		25–26		
Coal, bituminous	84	44–52	Soft	
Coke		23–34		
Petroleum coke		35–40	Soft	
Portland cement		100		

penny, 3.0; tooth enamel, 5.0; penknife, 5.5; ordinary glass, 5.8.

Brittleness or friability refers to the ease with which a substance may be broken by impact. The hardness of a mineral is not a sure criterion of its brittleness. Horn, some plastics, and gypsum are soft and tough and are not easily broken by impact. Coal is soft and also friable. Friability is the inverse quality to toughness. *Toughness* is the property of metals and alloys called impact resistance.

The crystalline structure and crystal size influence the friability. The structure also determines the shape into which particles naturally break when subjected to a crushing operation. For example, galena, PbS, breaks into cubes; mica into plates; and magnetite into somewhat rounded grains. Those crystalline planes which are easily broken are termed *cleavage planes*. The quantity of work necessary to fracture a unit area of cleavage plane can be determined by experiment. When metals and alloys are stressed beyond their yield points, a similar cleavage takes place in the crystals; but the crystals do not break apart, they simply deform. Wood and asbestos are fibrous, do not possess cleavage planes, and do not crush readily, but must be torn or shredded.

Friction is the resistance to sliding of one material against another material. The coefficient of friction is the ratio of the force parallel to the surface of friction in the direction of motion required to maintain a constant velocity, to the force perpendicular to the surface of friction and normal to the direction of motion.

BIBLIOGRAPHY

1. Kraus, E. H., W. F. Hunt, and L. F. Ramsdell, *Mineralogy*, 3rd ed., McGraw-Hill Book Co. (1936).
2. Williams, S. R., *Hardness and Hardness Measurements*, American Society for Metals, Cleveland (1942).

PROBLEM

A copper tube, 1 in. I.D. and 2 ft long, is filled with steel balls of 1-in. diameter. The space between the balls is filled with water. The specific gravity of steel is 7.8. What is the bulk density of the contents of the tube?

CHAPTER 3

Screening

THE separation of materials on the basis of size is frequently important as a means of preparing a product for sale or for a subsequent operation. It is also a widely used means of analysis, either to control or gage the effectiveness of another operation, such as crushing or grinding, or to determine the value of a product for some specific application.

In the marketing of coal, for example, the size of the particles is the basis of its classification for sale. Certain equipment such as stokers require definite limits of size for successful operation. In the case of sand and gravel for concrete, on the other hand, only a properly blended series of sizes will insure the most dense packing, requiring the minimum of cement and securing the greatest strength and freedom from voids.

It has frequently been observed that the rate of a chemical reaction between a solid and a fluid is roughly proportional to the surface involved. Since the surface areas may be computed from a knowledge of the sizes of the particles, a sizing operation is of particular value in controlling the rates of reactions involving solids. The combustion of powdered coal illustrates the desirability of controlling the grinding operation to produce material of definite size limits in order to control the rate of combustion. Since the setting of Portland cement must take place within a specified time, it has been necessary to specify certain size limits. The hiding power of a paint pigment is indicated by size since it depends upon the projected area of the particles.

Screening is accomplished by passing the material over a surface provided with openings of the desired size. The equipment may take the form of stationary or moving bars, punched metal plate, or woven wire mesh. Screening consists in separating a mixture of various sizes of particles into two or more portions, each of which is more uniform in size of particle than is the original mixture.

Dry screening refers to the treatment of a material containing a natural amount of moisture or a material that has been dried before screening. Wet screening refers to an operation in which water is added to the material being treated for the purpose of washing the fine material through the screen.

The material that fails to pass through the screen is referred to as oversize or plus material, and that which passes through the screen openings is referred to as undersize or minus material. When more than one screen is used and more than two sizes are produced, the various fractions may be designated according to the openings employed in making the separations. For example, Table 2 shows three different ways of indicating sizes.

TABLE 2. THREE METHODS OF INDICATING SIZE FRACTIONS

First	Second	Third
Oversize $\frac{1}{4}$ in.	$+\frac{1}{4}$ in.	$+\frac{1}{4}$ in.
Through $\frac{1}{4}$ in. on $\frac{1}{8}$ in.	$-\frac{1}{4}+\frac{1}{8}$ in.	$\frac{1}{4}/\frac{1}{8}$ in.
Through $\frac{1}{8}$ in. on $\frac{1}{16}$ in.	$-\frac{1}{8}+\frac{1}{16}$ in.	$\frac{1}{8}/\frac{1}{16}$ in.
Undersize	$-\frac{1}{16}$ in.	$\frac{1}{16}/0$ in.

INDUSTRIAL SCREENING EQUIPMENT

Grizzlies are widely used for screening large sizes, particularly of 1 in. and over. They consist simply of a set of parallel bars separated by spacers at the ends. The bars may be laid horizontally or inclined longitudinally 20 to 50 degrees from the horizontal, depending upon the nature of the material treated.

The usual cross section of the bars is trapezoidal with the wide base upward to prevent clogging or wedging of the particles between the bars. Inverted railroad rails are frequently employed. Owing to the wear on the bars, they are frequently made of manganese steel.

Grizzlies are usually constructed about 3 to 4 ft wide with the bars from 8 to 10 ft in length. They are frequently used before material is sent to a crusher to remove the smaller particles from the feed to the crusher.

In some grizzlies, a cam arrangement causes a slight lengthwise reciprocal movement of alternate bars, permitting better flow of material through them and preventing clogging. Endless chains passing over sheaves may replace the bars, constituting the chain grizzly. These more elaborate grizzlies are for somewhat sticky or clay-like material. Figure 1 shows a grizzly mounted on springs with the whole assembly vibrated mechanically.

Fig. 1. Mechanically vibrated bar grizzly. The material enters at the top left and works its way downward to the right. The large or oversize particles are discharged over the lower right end, and the smaller particles pass through the slots between the bars into a hopper directly below. (*Nordberg Mfg. Co.*)

A rough figure for the capacity of grizzlies is approximately 100 to 150 tons of material per square foot of area per 24 hr when the bars are spaced to give about 1 in. of clear opening.

Stationary screens are made of punched metal plate or woven wire mesh, usually set at an angle with the horizontal up to about 60 degrees. They are suitable for intermittent small-scale operations, such as screening sand, gravel, or coal by throwing the material against the screen. When large tonnage is to be handled stationary screens are usually abandoned in favor of the vibrating screens.

Fig. 2. Triple-decked mechanically vibrated screen. (*W. S. Tyler Co.*)

Vibrating screens are used where large tonnages are to be treated. The vibrating motion is imparted to the screen surface by means of cams, eccentric shafts, unbalanced flywheels, or electromagnetic means. A complete screen may have a single screening surface, or it may be double- or triple-decked, as indicated in Fig. 2. This screen is driven by an eccentric shaft, as shown in Fig. 3. The wire screens

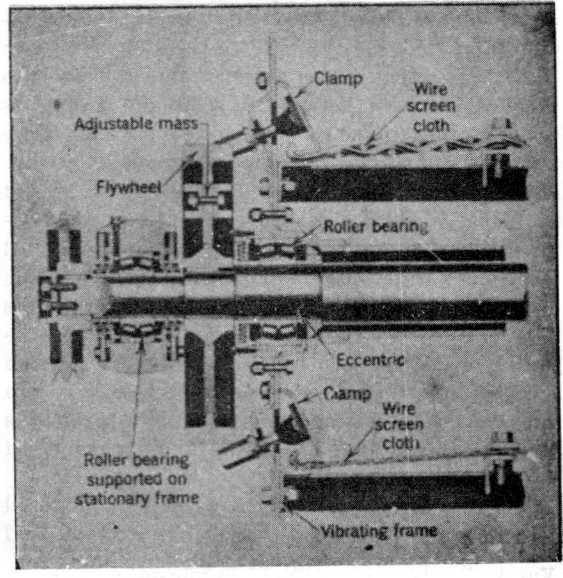

Fig. 3. Sectional drawing of the mechanism of a mechanically vibrated screen. (*W. S. Tyler Co.*)

are held in place under tension by the clamps and supported on the vibrating frame. The rotating shaft is supported on the stationary frame by the outer roller bearings which are fastened to the stationary frame. The eccentric carries the roller bearings which support the vibrating frame. The vibrating frame is positioned by springs, four of which are indicated under covers on each side of the screen shown in Fig. 2. The flywheel is mounted eccentrically on the shaft as a counterbalance to balance the vibrating frame, screens, and load of material being screened. An adjustable mass is provided for controlling the eccentricity of the flywheel. Rotation of the shaft gives the screens a circular motion in the vertical plane. As the screen passes the top of its cycle, the material is thrown clear of the screen surface. The material will be moved along the screen in either direction, depending upon the direction of rotation. Such screens are operated (Fig. 4) with a slope that may vary from the horizontal to about 45 degrees. These screens are operated with an amplitude (diameter of the described circle of motion) up to about ¼ in., depending on the size of the material, and with a frequency of vibration or rotation of about 1200 to 1800 per minute. They are made for heavy duty with openings above 1 in. and are widely used for dry screening of particles from 1 in. down to about 35 mesh * (0.0164 in.) at an angle of about 20 degrees. For wet screening the angle is reduced to about 5 or 10 degrees. Usually the feed enters the top of the screen, but sometimes it is found desirable to feed at the lower end of the screen and discharge at the top.

FIG. 4. Double-decked mechanically vibrated screen operating under test conditions. (*W. S. Tyler Co.*)

FIG. 5. Vibrating mechanism consisting of an adjustable unbalanced flywheel and shaft mounted rigidly on a vibrating screen frame, which is in turn supported on springs. (*Nordberg Mfg. Co.*)

In another type of screen the vibrating frame is mounted on springs, and the belt-driven rotating shaft is mounted only on the vibrating frame. Vibration is caused by creating an unbalance on the rotating shaft by mounting an adjustable unbalanced flywheel (Fig. 5) to give the desired amplitude. Such a drive is mechanically simpler than the counterbalanced eccentric-shaft drive previously described,

* *Mesh* is a term stating the number of openings per linear inch of screen surface. The size of the opening depends on the size of the wire, but for screens of a standard series the mesh is a specific designation of the aperture, as stated in Table 4.

Fig. 6. Vibrating screen employing an electromagnetic vibrating unit attached to the center of the screen. (*W. S. Tyler Co.*)

but it is not capable of as rugged construction. If two unbalanced wheels are rotated in opposite directions at the same number of revolutions per minute (rpm), the vibration may be strictly normal to the plane of the screen.

An electromagnetic vibrator may be attached to the center of the screen, as shown in Fig. 6. The frame is rigid and the screen is vibrated by the solenoid whose core is fastened to the center of the screen. The core of the solenoid works against adjustable spring tension, as shown in Fig. 7. In this way the amplitude of vibration may be varied up to about $\frac{1}{8}$ in. In other types the core is fastened to the frame supporting the screen and vibrates the frame and screen. With such an arrangement the solenoid may be mounted obliquely to the screen surface, giving the screen a motion at an angle to the normal. High amplitude in a screen vibrated

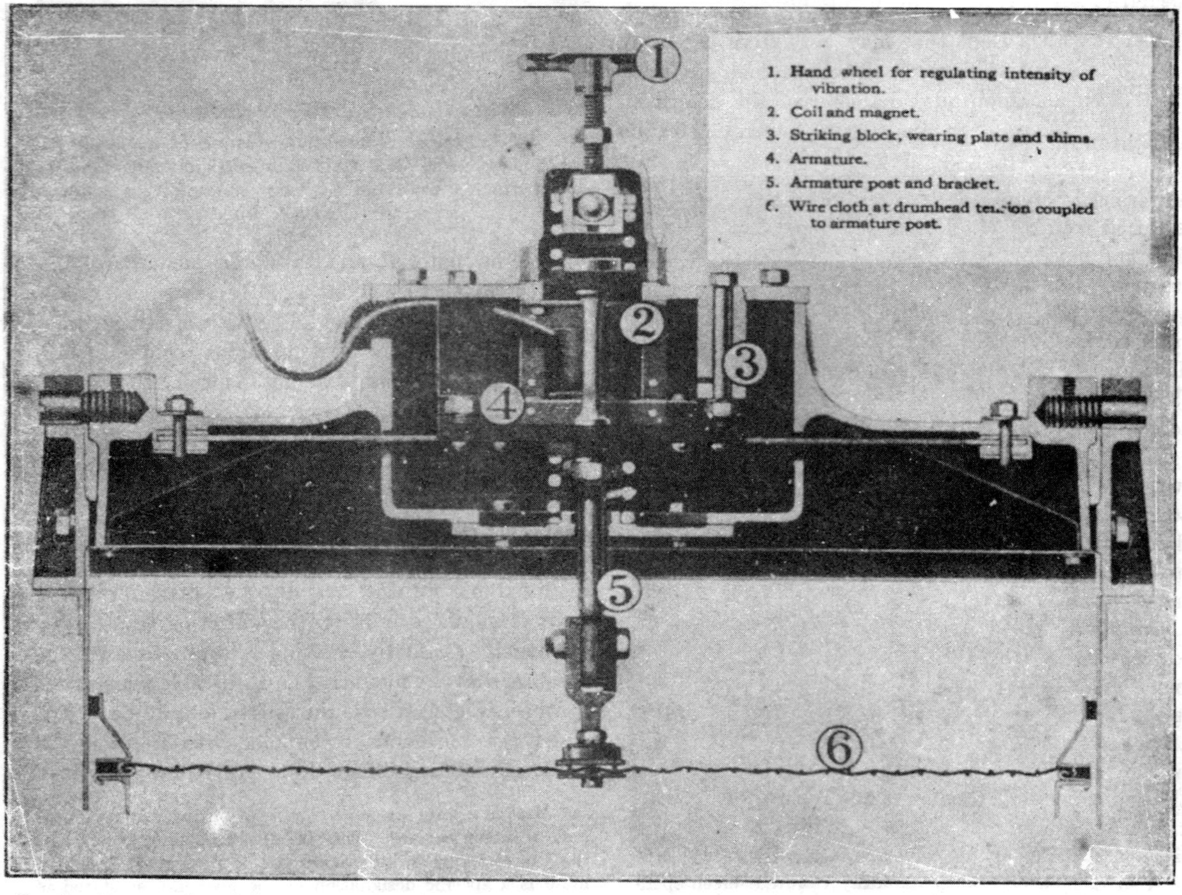

Fig. 7. Cross section of electromagnetic vibrator of Fig. 6. The handwheel, 1, adjusts the spring tension and limits the length of stroke caused by the electromagnet, 2, operating on the armature directly connected to the screen, 6. (*W. S. Tyler Co.*)

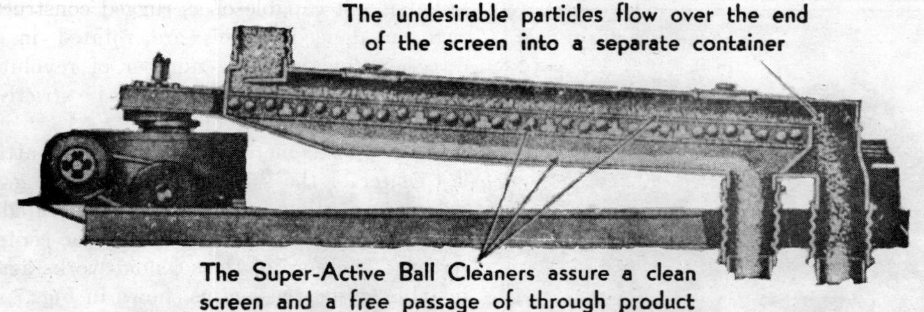

FIG. 8. Reciprocating screen with ball cleaners. The oversize flows over the end of the screen. The balls are confined to limited areas directly beneath the screen, and supported on a coarse wire screen. The balls bounce against this supporting screen and give additional vibration to the screen directly above. The desired fine material passes through the screen. (*J. H. Day Co.*)

in a fixed frame (Fig. 6) may cause fatigue failure of the screen near the clamps. If the screen is supported in a vibrating frame this limitation does not apply but a greater energy is required to overcome the inertia of the mass of the frame. The solenoid is usually caused to strike a block or anvil, thereby suddenly halting the upward motion of the screen and throwing the material clear of the screen. This is particularly desirable in the handling of sticky materials. These screens are normally used for material from about 8 mesh down to 100 mesh or finer and have been successful in wet screening. The frequency of vibration is determined by the frequency of the alternating current used and varies from about 900 to 7200 vibrations per minute. The lower frequency is used for coarser screening (8 mesh) and the higher frequencies for the finer screening.

The capacity of a vibrating screen varies widely with the character of the material treated from 2 tons/sq ft of surface per 24 hr for materials such as damp clay or powdered soap, up to 30 tons for dry material such as coke on screens of about 6 to 8 mesh.

Oscillating screens are characterized by a relatively low speed (300 or 400 oscillations per minute) in a plane essentially parallel to the screen. The riddle is a screen driven in an oscillating path by an eccentric or other mechanism attached to the sole support of the screen, usually a vertical bar extending from the top of the screen box. It is the cheapest form of screen on the market and is used for batch screening.

A sifter is a box-like container holding a number of screen cloths nested on top of one another and oscillated by an eccentric or counterweights in a nearly circular orbit. Many of these devices carry a coarse screen directly below the screening cloths on which rubber balls are confined to limited areas and are caused to bounce against the lower surface of the screen cloths as the device is oscillated.

Reciprocating screens (Fig. 8) are driven by an eccentric under the screen at the feed end. The motion varies from gyratory (about 2 in. in diameter) at the feed end to a reciprocating motion at the discharge end. These screens are usually inclined about 5 degrees, giving the screen a motion normal to the cloths of about 1/10 in. Further vibration may be caused by including rubber balls as shown confined to local areas below the active screen surface. This type of screen is popular and widely used for screening dry chemicals down to about 300 mesh.

Trommels, or revolving screens, consist of a screen cylindrical or conical in form rotated about its axis.

FIG. 9. Compound trommel. (*C. O. Bartlett and Snow Co.*)

Simple trommels may be arranged in series with the undersize of the first passing to the second trommel and the undersize of the second passing to the third, etc. Sometimes the trommels are built with screens of different sizes throughout their length, the feed

Fig. 10. Tandem-type trommel. The feed enters at the left end and passes over successively coarser screens as it works its way to discharge of the oversize at the far right. (*C. O. Bartlett and Snow Co.*)

entering at the end of the finest screen (Fig. 10). In this way it is possible to collect materials of different size ranges from a single trommel. But the operation is not so efficient as that of a series of simple trommels, or a compound trommel.

The compound trommel (Fig. 9) contains two or more concentric screening surfaces mounted on the same shaft. The coarser screening surface is the innermost, and the finest the outermost, with intermediate sizes arranged between the two limits. With provision made for the separate removal of the oversize from each screening surface, the undersize of each screen becomes the feed to the screen of the next smaller aperture.

Conical trommels have the shape of a truncated cone and are generally mounted with their axes horizontal.

Trommels are quite efficient for coarse sizes. The inclination of the trommel varies from about ¾ in. (for wet screening) to 3 in./ft of length, depending on the nature of the material to be processed.

The capacity of the trommel increases with increased speed of rotation up to a point where blinding occurs due to crowding of material through the screen. If the speed of rotation is increased still further to the *critical speed*, the material no longer cascades over the screen surface but is carried around by centrifugal force. The best operating speed is usually about 0.33 to 0.45 times the critical speed.

The critical speed of rotation of a trommel may be computed by equating the force of gravity tending to cause the particle to fall to the centrifugal force tending to carry the particle around.

$$mg = \frac{2mv^2}{D} \quad \text{or} \quad g = \frac{2v^2}{D}$$

where m = mass (lb).
g = acceleration of gravity (ft/sec²).
v = velocity of particle, or of trommel, in circular path (fps).
D = diameter of trommel (ft).

When N = number of revolutions of trommel per minute,

$$v = \frac{\pi DN}{60}$$

$$g = \frac{2}{D}\left(\frac{\pi DN}{60}\right)^2$$

$$N^2 = \frac{60^2 gD}{2(\pi D)^2} = \frac{60^2 g}{2\pi^2 D}$$

$$N = \sqrt{\frac{60^2 g}{2\pi^2 D}}$$

At sea level $g = 32.17$ ft/sec² and $N = 76.65/\sqrt{D}$.

With compound trommels the speed of rotation is naturally governed by the diameter of the outside screen.

Trommels are usually about 3 to 4 ft in diameter, from 5 to 8 ft in length, and driven at 15 to 20 rpm with 2½- to 5-hp motors.

The trommel is best suited for material from ¼ to 2½ in. in size.

Reels are revolving screens driven at relatively high speed. They are used in the flour milling industry and for other light, dry, nonabrasive material. The screening surface consists of silk bolting cloth supported by wire mesh. Speed of rotation is above the critical speed for a trommel and is such as to throw the undersized particles outward through the bolting cloth by centrifugal force. The surface may be cleared by brushes inside the reel. Reels are generally 24 to 40 in. in diameter and 5 to 8 ft long, and they are rotated at speeds that may vary from 100 to 200 rpm.

The effectiveness of screens is based upon both the recovery in the product of the desired material in the feed and the exclusion or rejection from the product of the undesired material in the feed. For example, the specifications for hydraulic hydrated lime (ASTM C141-42) require that the product contain not more than 10 per cent by weight of material coarser than 200 mesh.

If x_P = mass fraction of desired material in product,
x_F = mass fraction of desired material in feed,
x_R = mass fraction of desired material in reject,
P = total mass of product,
F = total mass of feed,
R = total mass of reject,

$$\text{Recovery} = \frac{x_P P}{x_F F}$$

$$\text{Rejection (1 − Effectiveness of recovery of undesired material)} = 1 - \frac{(1 - x_P)P}{(1 - x_F)F}$$

Effectiveness (recovery × rejection)

$$= \frac{x_P P}{x_F F}\left(1 - \frac{(1 - x_F)P}{(1 - x_P)F}\right)$$

Weighing the entire feed and product is not practical, and it is desirable to express the effectiveness from the analyses of samples alone. A material balance around the screening operation gives

$$x_F F = x_P P + x_R R$$

$$F = P + R$$

Substituting for R

$$x_F F = x_P P + x_R F - x_R P$$

Collecting terms

$$F(x_F - x_R) = P(x_P - x_R)$$

$$\frac{P}{F} = \frac{(x_F - x_R)}{(x_P - x_R)}$$

Substituting for P/F

$$\text{Recovery} = \frac{x_P(x_F - x_R)}{x_F(x_P - x_R)}$$

$$\text{Rejection} = 1 - \frac{(1 - x_P)(x_F - x_R)}{(1 - x_F)(x_P - x_R)}$$

Effectiveness * (recovery × rejection)

$$= \frac{x_P(x_F - x_R)}{x_F(x_P - x_R)}\left[1 - \frac{(1 - x_P)(x_F - x_R)}{(1 - x_F)(x_P - x_R)}\right]$$

These equations permit the desired calculation of recovery, rejection, or effectiveness of any sizing operation from size analysis of the streams, without knowledge of the quantities.

Rapid feeding or too steep an angle of the screen gives insufficient time for complete separation of the fine and coarse material. Excessive dampness in the feed may cause cohesion of small particles to form larger masses, or the adhesion of small particles to large particles. Worn screens with enlarged apertures will pass more oversize material into the undersize fraction. Clogged screens (blinded) retain more undersize material in the oversize fraction. The effectiveness as defined above is a numerical expression for the effect of all these factors.

Capacity of screens and effectiveness are closely related. If a low efficiency or effectiveness may be tolerated, the screen may be operated at high capacity. The ability of the device to prevent blinding of the screen surface is probably the most important single factor determining capacity of the screen. In dry screening, the greater the amount of moisture or dampness in any particular material, the

* Other expressions for effectiveness are used, such as the recovery, or rejection, as defined above, or the product of recovery and enrichment.

$$\frac{x_P(x_F - x_R)}{x_F(x_P - x_R)}\left(\frac{x_P - x_F}{1 - x_F}\right)$$

All these expressions give different values, depending on whether the undersize is considered the reject or the product. The expression for effectiveness in the text gives the same value, regardless of whether the undersize is the product or reject.

lower is the capacity of the screen. Because of its greater surface area, finer material can tolerate a greater percentage of moisture. If the feed contains a high proportion of material of a size just slightly smaller than the size of the openings in the screen, called "near mesh," the capacity of the screen will be greatly reduced. For example, if the size of the openings is 1/8 in. and there is a large proportion of 3/32-in. grains in the material to be screened, the screen's capacity, for the same degree of effectiveness, will be much lower than if most of the undersize material is smaller than 1/32 in.

The ratio of the open area of the screen to the total area is an important factor in determining its capacity. Because of the direct dependence of screening capacity upon the area of the screen surface and upon the screen aperture, the capacity is usually expressed [1]* in terms of tons of feed per square foot of screen area per millimeter of screen aperture per 24 hr, as indicated in Table 3. For example, a vibrating screen having 6 sq ft of surface and an aperture of 2 mm may be expected to have an approximate capacity of (5 to 20)(6)(2) = 60 to 240 tons of ore per 24 hr.

TABLE 3. THE APPROXIMATE CAPACITY OF SCREENS FOR DENSE MATERIALS SUCH AS ORES

Type of Screen	Capacity Range, tons/sq ft area/mm aperture/24 hr
Grizzlies	1–6
Stationary screens	1–5
Vibrating screens	5–20
Shaking and oscillating screens *	2–8
Trommels	0.3–2

* If the oscillating screens are also vibrated by means of the rubber balls described, the capacities will be somewhat increased over those of the simple oscillating screens.

DETERMINING PARTICLE SIZE

The size of a particle may be expressed in different ways. If the particle is a sphere, the diameter, the projected area, the volume, or the surface of a particle may be the significant size. If the particle is a cube, the edge length, the projected area, the volume, or the surface may be the significant dimension indicating size.

Various methods are used for measurements of particle size. These depend on the size range, the physical properties, and the condition of dryness or wetness permissible. The following methods are used in laboratory and control work.

Microscope. For very small sizes of the order of a few microns (1 micron equals 0.001 mm), the sample may be placed under a microscope; the size may be determined by simple measurement of a photomicrograph of known magnification, or it may be determined directly by means of a filar micrometer. This device consists of a movable cross hair built into a standard microscope eyepiece. The movement of the cross hair is actuated by a calibrated micrometer screw. The cross hair is moved until it appears in contact with one edge of the particle, and a reading is taken on the micrometer; then the cross hair is moved to the opposite edge of the particle, and another reading is made. The difference in readings is a measure of the particle "diameter." This number divided by the optical magnification of the objective and eyepiece will give the true dimension in inches or other units. The microscopic method is frequently employed to measure particles of dust from the atmosphere and to evaluate the effectiveness of air filters.

Screening. Perhaps the simplest method for laboratory sizing consists in passing the material successively over a series of screens or sieves having progressively smaller openings. The size of a material which has passed through one screen and has been retained on a screen having openings of a smaller size is usually considered to be the arithmetic average of the two screen openings and is called the "average dimension" (or "average diameter") represented by the symbol D_{avg}.

Sedimentation. Sedimentation methods are based on the fact that small particles of a given material fall in a fluid at a rate that is proportional to their size. One method involves shaking a sample of the solid in water; after the mixture stands a definite length of time, portions are removed from different levels by means of a pipette. These portions are evaporated to dryness, and the residues weighed. Other modifications have been developed, such as having one balance pan suspended in the pulp of suspended solids and weighing at intervals as the particles settle on the pan.

Elutriation also depends on the velocity of settling. If the material is placed in a rising stream of fluid having a fixed upward velocity, particles whose normal falling velocity is less than the velocity of the fluid will be carried upward and out of the vessel.

* The bibliography for this chapter appears on p. 22.

If fractions obtained from a series of fluid velocities are collected and weighed, a complete size analysis may be obtained.

Centrifuging. Sedimentation is too slow for particles of diameter under ½ micron. Therefore a centrifugal force is substituted for the normal force of gravity when the size of very small particles is to be determined.

Other Methods. The coercive (magnetic) force of a paramagnetic material such as magnetite is directly proportional to its specific surface, regardless of its shape. This relationship has served as a means of determining the surface, or size, of such particles. The amount of light transmitted through a suspension of a definite quantity of the finely divided solid in kerosene in a tube of specified dimensions depends upon the projected area of the particles and is used as a method of determining particle size.[2] The surface size of quartz particles has been measured in research work by the rate of solution in solutions of hydrofluoric acid. It is assumed that the rate of solution in mass per unit of time is directly proportional to the surface area of quartz.

SCREEN ANALYSES

Screens are generally used for control and analytical work. They are constructed of wire mesh cloth, the diameters of the wire and the spacing of the wires being closely specified. These screens form the bottoms of metal pans about 8 in. in diameter and 2 in. high, whose sides are so fashioned that the bottom of one sieve nests snugly on the top of the next.

Screen Aperture and Screen Interval. The clear space between the individual wires of the screen is termed the screen aperture. Frequently the term *mesh* is applied to the number of apertures per linear inch; for example, a 10-mesh screen will have 10 openings per inch, and the aperture will be 0.1 in. minus the diameter of the wire. Mesh is therefore a nominal figure which does not permit accurate computation of the screen openings or aperture without knowledge of the wire sizes used by the manufacturer.

The screen interval is the relationship between the successive sizes of screen openings in a series. A simple arithmetic series might be used such that the screen openings are 10, 9, 8, 7, 6, 5, 4, 3, 2, and 1 in., for example. The weakness of such a system is that there is a large relative difference between the 1-in. and 2-in. sizes, but the 9-in. and 10-in. sizes are almost alike for practical purposes. All the material under 1 in. down to a micron would be in one fraction.

A more satisfactory series of screens is one in which the opening of each successive member varies from the next by a multiplier such as to give a series having openings of 8, 4, 2, 1, ½, and so forth. These sizes

TABLE 4. TYLER SCREENS[5]

Standard Interval $=\sqrt{2}$, Aperture, in.	Interval $=\sqrt[4]{2}$, for Closer Sizing			
	Aperture, in.	Aperture, mm	Mesh Number	Wire Diameter, in.
1.050	1.050	26.67	....	0.148
	0.883	22.43	....	0.135
0.742	0.742	18.85	....	0.135
	0.624	15.85	....	0.120
0.525	0.525	13.33	....	0.105
	0.441	11.20	....	0.105
0.371	0.371	9.423	....	0.092
	0.312	7.925	2½	0.088
0.263	0.263	6.680	3	0.070
	0.221	5.613	3½	0.065
0.185	0.185	4.699	4	0.065
	0.156	3.962	5	0.044
0.131	0.131	3.327	6	0.036
	0.110	2.794	7	0.0326
0.093	0.093	2.362	8	0.032
	0.078	1.981	9	0.033
0.065	0.065	1.651	10	0.035
	0.055	1.397	12	0.028
0.046	0.046	1.168	14	0.025
	0.0390	0.991	16	0.0235
0.0328	0.0328	0.833	20	0.0172
	0.0276	0.701	24	0.0141
0.0232	0.0232	0.589	28	0.0125
	0.0195	0.495	32	0.0118
0.0164	0.0164	0.417	35	0.0122
	0.0138	0.351	42	0.0100
0.0116	0.0116	0.295	48	0.0092
	0.0097	0.248	60	0.0070
0.0082	0.0082	0.208	65	0.0072
	0.0069	0.175	80	0.0056
0.0058	0.0058	0.147	100	0.0042
	0.0049	0.124	115	0.0038
0.0041	0.0041	0.104	150	0.0026
	0.0035	0.088	170	0.0024
0.0029	0.0029	0.074	200	0.0021
	0.0024	0.061	230	0.0016
0.0021	0.0021	0.053	270	0.0016
	0.0017	0.043	325	0.0014
0.0015	0.0015	0.038	400	0.0010

vary in a geometric progression, and the factor or screen interval is 2. If closer sizing is desired, an additional screen is inserted between each two screens of the previous series and the screen interval becomes $\sqrt{2}$. The standard screens used in the United States employ a screen interval in which the factor is $\sqrt{2}$ although $\sqrt[4]{2}$ is sometimes used for careful work and research.

The first commercial laboratory screens using this system were the *Tyler Standard screens.* This series of screens is based upon a 200-mesh screen with wire 0.0021 in. thick and with an opening of 0.0029 in. (0.0074 cm). The other sizes vary by a fixed ratio of $\sqrt{2}$. A supplementary set can be purchased for intermediate sizes so that the complete set varies by $\sqrt[4]{2}$.

The United States screens introduced by the National Bureau of Standards differ but slightly from the Tyler series, being based on a 1-mm opening (No. 18 mesh) and varying by $\sqrt[4]{2}$.

The British standard screens are similar but have wires of different gage.

Method of Making a Screen Analysis. In making a screen analysis, the individual screens comprising the entire series, varying for example by the ratio $\sqrt{2}$ from 3 mesh to 200 mesh, are cleaned with a brush and tapped free from any adhering particles. They are nested together with the coarsest or 3 mesh at the top and the finest or 200 mesh at the bottom. A bottom pan and top cover are put in place to complete the set. A weighed amount of material is placed upon the top screen, and the cover is replaced. The assembly may be supported and rotated by one hand and bumped against the other hand at intervals to set up a jarring action. After a period of time, the fines, −200 mesh, are removed from the bottom pan. The pan is replaced, and shaking is resumed to see whether any more fines are revealed. When no new material appears in the bottom pan, thus indicating for all practical purposes that the screening operation seems to have been completed, the sieves are disassembled and the individual fractions are weighed. The material, for example, which passed the 100-mesh screen but was retained on the 150-mesh screen is designated the 100/150 or −100 +150 fraction.

Since the shaking of the screens is a tedious process and open to error, mechanical screening is desirable. In one of these machines, the Ro-Tap, the screens are fastened into a vertical framework which is given an elliptical motion in a horizontal plane, a sharp tap or blow being given at the top of the screens for each revolution. Shaking is continued for 15 to 20 min. Other machines employ vibrators or other motions.

Factors which militate against accurate results are overloading of the screens, which may result in blinding which is the wedging of particles in the openings; or electrostatic forces causing small particles to adhere to one another or to large particles. A small amount of moisture may also cause adhesion or cohesion of particles.

Wet-and-dry screening is suitable for very precise screen analyses since it avoids the dangers of adhesion and cohesion. The weighed sample is placed in a beaker and pulped with a nonsolvent, frequently water, and then decanted over the finest screen in the series, for example, 200 mesh. More water is added; stirring and decantation are repeated until no fines are in suspension after stirring. Water from a wash bottle is played on the screen until the drip is clear. The water is decanted from the undersize fraction, and the material is dried. The oversize is also dried and put over the entire series of screens as usual. The new −200 fraction is weighed with the fraction obtained by wet screening. This procedure gives more accurate results since the chance of fine particles clinging to large ones is minimized.

Method of Reporting Screen Analyses. The customary manner of reporting screen analyses is shown in Table 5, in which the mass fractions retained on each of the screens are given.

TABLE 5. TYPICAL SCREEN ANALYSIS

Tyler Screen Mesh	Average Diameter of Particles, D_{avg}		Mass (or Weight) Fraction	Mass Fraction through Each Screen
	cm	in.		
−8 +10	0.2007	0.0791	0.03	1.0
−10 +14	0.1410	0.0555	0.14	0.97
−14 +20	0.1001	0.0394	0.25	0.83
−20 +28	0.0711	0.0280	0.20	0.58
−28 +35	0.0503	0.0198	0.14	0.38
−35 +48	0.0356	0.0140	0.09	0.24
−48 +65	0.0252	0.0099	0.06	0.15
−65 +100	0.0178	0.0070	0.04	0.09
−100 +150	0.0126	0.00496	0.03	0.05
−150 +200	0.0089	0.0035	0.02	0.02
			Total 1.00	

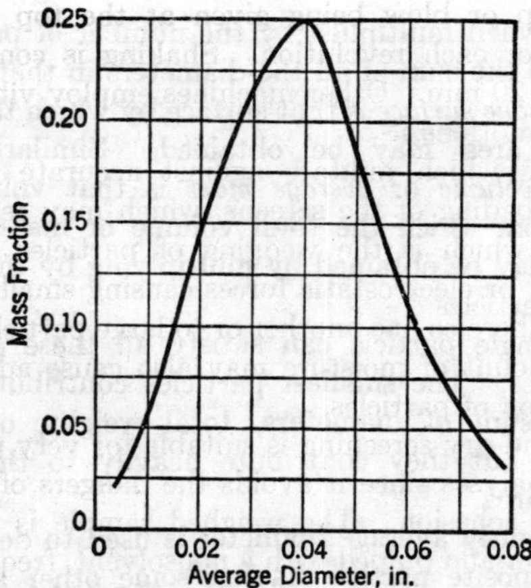

Fig. 11. Fractional plot of screen analysis of Table 5, showing the mass fraction (column 4) retained on screens in the series interval $\sqrt{2}$ as a function of the average particle diameter retained in each fraction (column 3).

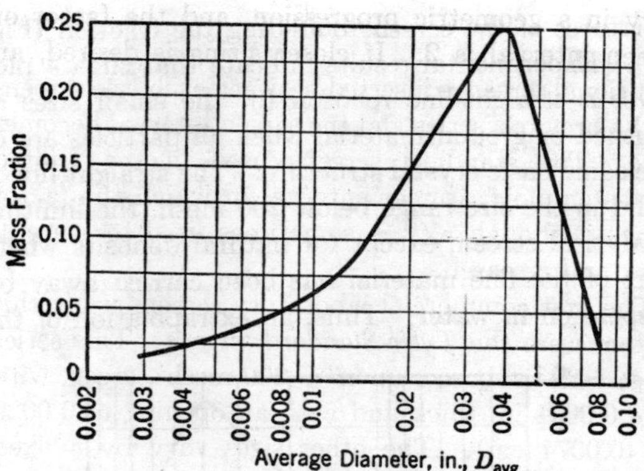

Fig. 13. Fractional plot of screen analysis of Table 5, showing the mass fraction retained on screens in the series interval $\sqrt{2}$ plotted against the logarithm of the average particle diameter retained in each fraction.

These data may be presented graphically by any one of several methods (Figs. 11 to 15). But most of the resulting curves are valuable primarily as pictures of the size distribution of the mixture. Such pictures tell a great deal to an experienced observer but are misleading unless the method of plotting and the materials comprising the mixture are not changed from curve to curve. *Fractional plots* of the mass fraction retained on each screen versus average screen aperture (Fig. 11), or *cumulative plots* of the mass fraction passing each screen versus particular screen aperture (Fig. 12), may be the basis for comparisons of different mixtures of the same materials, indicating changes with time or shipment. The fractional data (Fig. 11) give different curves for different screen intervals and are therefore specific to the particular screen series used as in Table 5. This limitation does not apply to plots of the cumulative data (Figs. 12 and 14) which give the same values regardless of screen intervals. The cumulative plot does not require the computation of average diameter but rather the addition of the fractions passing through the screens.

Ordinary rectangular coordinate plots crowd many points in the small size range into a narrow section of the curve. A better picture is obtained with the logarithm of the average of screen apertures (Figs. 13 and 14), as this spreads the points for the small particles along the dimension scale.

Still further use can be made of a plot of the logarithm of the mass fraction retained on each screen against the logarithm of the arithmetic average of the

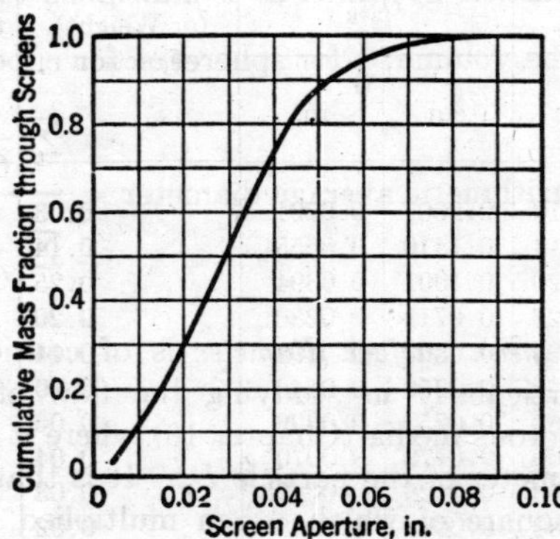

Fig. 12. Cumulative plot of screen analysis of Table 5, showing the mass fraction passing through screens (column 5) as a function of the screen aperture.

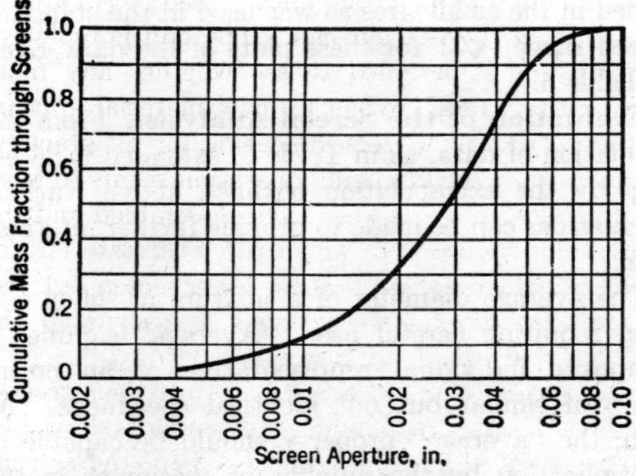

Fig. 14. Cumulative plot of screen analysis of Table 5, showing the mass fraction passing screens as a function of the logarithm of the screen aperture.

apertures of the screens bounding the fraction (Fig. 15). Experimental results indicate that such a plot gives a straight-line relation for the small sizes of crushed or ground material when all particles are of the same basic crystal structure. The straight line is valid in the size range below 200 mesh, the limiting analytical screen, except for natural deposits where part of the fine material has been carried away by suspension in water. Thus, an extrapolation of the

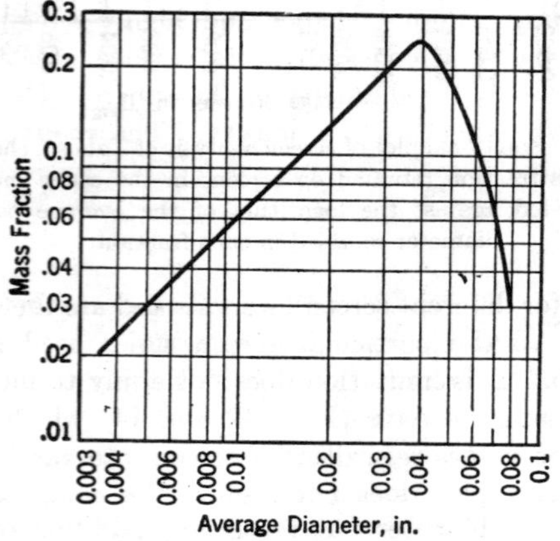

FIG. 15. Log-log fractional plot of screen analysis of Table 5, showing the mass fraction retained on each screen in the series interval $\sqrt{2}$ as a function of the average particle diameter retained in each fraction.

straight line (Fig. 15) will give approximate quantities of the material in each small size range. This extrapolation must be stopped when the total quantity through the 200-mesh screen is obtained by a cumulation of the small fractions. The extrapolation is valid only if the same screen size interval is maintained in the small sizes as was used in the standard screen range, $\sqrt{2}$ for these plots of the data given in Table 5.

Evaluation of the Screen Analysis. From the tabulation of data, as in Table 5, extended if necessary by the extrapolation outlined above, various calculations can be made to provide further information.

The average diameter of a mixture of solids is a term requiring careful use. "Average" signifies a composite individual representative of an entire group of similar but not identical specimens. As such, the "average" property should be capable of multiplication by the number of specimens in the entire group to give a total value for that property. Thus, strictly, *average diameter* is that diameter which, when multiplied by the number of particles, will give the sum of all the diameters in that group. The *average surface* is that surface by which the total surface area may be obtained. Similarly, the *average volume* or *average mass* is that volume or mass from which the total volume or mass of the group may be obtained by multiplying by the number of particles.

No single particle can satisfy all these average properties. The smallest particles contribute little to the sum of diameters, total weight, or total volume, but they contribute heavily to the total surface area.

Frequently average diameter is used to designate the composite particle having some other average property than the average diameter as defined above. Care must be exercised in interpreting average diameter since the term may be so defined as to be useful only for comparison of particles having some other average property. For example, if N_1, N_2, N_3, etc., are the number of particles and x_1, x_2, x_3, etc., are the mass fractions having diameters D_1, D_2, D_3, etc., respectively:

1. The *true arithmetic average diameter* is

$$\frac{N_1D_1 + N_2D_2 + \cdots + N_ND_N}{N_1 + N_2 + \cdots + N_N} = \frac{\Sigma(N_iD_i)}{\Sigma N_i}$$

Since

$$\Sigma N_i = \frac{M}{\rho}\left[\frac{x_1}{C_1D_1^3} + \frac{x_2}{C_2D_2^3} + \cdots + \frac{x_N}{C_ND_N^3}\right] = \frac{M}{\rho}\sum\frac{x}{CD^3}$$

where M = total mass of all particles.

C is a constant depending on the shape of the particle by which D^3 is multiplied to obtain the volume; $\frac{\pi}{6}$ for spheres, 1 for cubes, etc.

$$\text{True arithmetic average diameter} = \frac{\sum\frac{x}{CD^2}}{\sum\frac{x}{CD^3}}$$

2. The *mean surface diameter* is of considerable value, particularly in studying the flow of fluids through porous media (Chapter 16) where it is used as the diameter of the particle D_p. It is that diameter the square of which, when multiplied by the number of particles and also by a suitable constant, B, depending upon the particle shape (π for spheres, 6 for cubes, etc.), gives the total surface of the aggre-

gate number of particles

$$B_1D_1^2N_1 + B_2D_2^2N_2 + \cdots = B(D_{sur})^2\Sigma N_i$$

The mean surface diameter (D_{sur}) is

$$(D_{sur}) = \sqrt{\frac{\Sigma N_iB_iD_i^2}{B\Sigma N_i}} = \sqrt{\frac{\sum \frac{x_iB_i}{C_iD_i}}{B\sum \frac{x_i}{C_iD_i^3}}}$$

3. Similarly the *mean volume diameter* or mean mass diameter equals

$$\sqrt[3]{\frac{\Sigma N_iC_iD_i^3}{C\Sigma N_i}} = \sqrt[3]{\frac{\Sigma x_i}{C\sum \frac{x_i}{C_iD_i^3}}}$$

One of the important properties of solids is the surface area. Since it is impractical to determine the number of particles in a mixture, the usual basis for evaluation of surface is a unit of mass. The *specific surface* or the *surface area per unit of mass* is an important property of solids which varies widely depending upon the condition of the surface as well as the particle size.

The *specific surface* could be computed easily if the particles were of known geometry, but they are of many different shapes and highly irregular. If the sphere is considered, its surface area is πD^2, where D is the diameter of the sphere. Its mass is $\rho\pi D^3/6$, where ρ is its density. The *specific surface of spherical particles* is area divided by mass or $6/\rho D$.

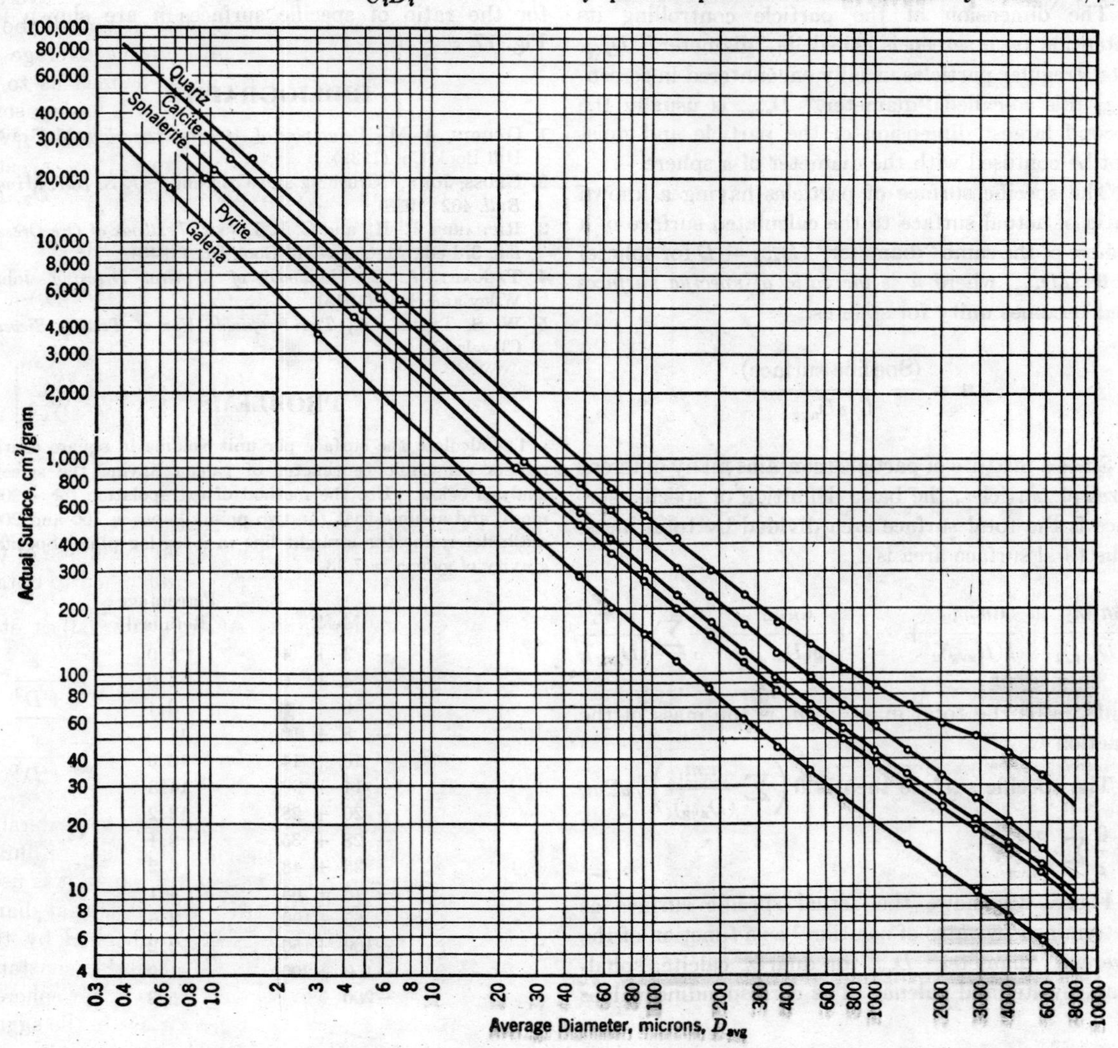

Fig. 16. Actual specific surface as a function of average diameter D_{avg} for quartz, calcite, sphalerite, pyrite, and galena

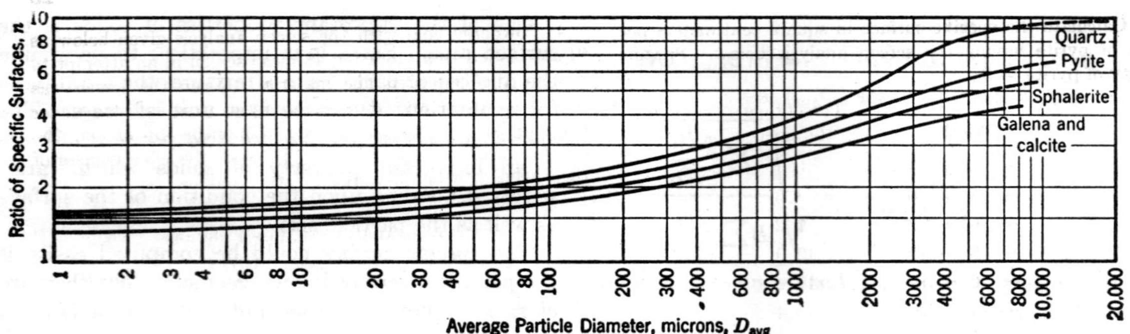

FIG. 17. Ratio n of specific surfaces [2] as a function of average diameter of particles D_{avg} for quartz, pyrite, sphalerite, calcite and galena.

The dimension of the particle controlling its retention on a screen is called its "diameter," D_{avg}. For irregular particles usually encountered in screening, this so-called "diameter," D_{avg}, is usually the second largest dimension of the particle and must not be confused with the diameter of a sphere.

The specific surface of particles having a known ratio of actual surface to the calculated surface of a sphere of the same "diameter" ($D_{avg} = D$ for sphere) is $6n/\rho D_{avg}$, where *n is the ratio of specific surfaces* and becomes unity for spheres.

$$n = \frac{\text{(Specific surface)}}{6/\rho D_{avg}}$$

Since a mixture of particles contains many different sizes of particles, the basic definition of specific surface is the total surface area divided by total mass. The total surface area is

$$\frac{6n_1m_1}{\rho(D_{avg})_1} + \frac{6n_2m_2}{\rho(D_{avg})_2} + \cdots + \frac{6n_Nm_N}{\rho(D_{avg})_N} = \frac{6}{\rho}\sum \frac{n_im_i}{(D_{avg})_i}$$

and Σm_i is the total mass, if m_i is the mass of the fraction i.

The specific surface then is $6\left(\sum \frac{n_im_i}{(D_{avg})_i}\right)/\rho\Sigma m_i$ or $\frac{6}{\rho}\sum \frac{nx}{D_{avg}}$.

Figure 16 shows the actual specific surface as determined by rates of solution [2] as a function of the average "diameter" D_{avg} for quartz, calcite, sphalerite, pyrite, and galena. The corresponding values for the ratio of specific surfaces n are shown in Fig. 17.

BIBLIOGRAPHY

1. GAUDIN, A. M., *Principles of Mineral Dressing*, McGraw-Hill Book Co. (1939).
2. GROSS, JOHN, "Crushing and Grinding," *U. S. Bur. Mines Bull.* 402 (1938).
3. RICHARDS, R. H., and C. E. LOCKE, *Textbook of Ore Dressing*, 3rd ed., McGraw-Hill Book Co. (1940).
4. TAGGART, A. F., *Handbook of Mineral Dressing*, John Wiley and Sons (1945).
5. W. S. TYLER CO., *The Profitable Use of Testing Sieves*, Cleveland.

PROBLEMS

1. Calculate the surface per unit volume in square centimeters per cubic centimeter of galena having the screen analysis below. Use the method of extrapolation for −200 mesh, and assume that the two points between 100 and 200 mesh determine the straight line in a log-log plot. Specific gravity of galena = 7.43.

Mesh	Percentage Retained
− 3 + 4	1.0
− 4 + 6	4.0
− 6 + 8	8.1
− 8 + 10	11.5
− 10 + 14	16.0
− 14 + 20	14.8
− 20 + 28	13.2
− 28 + 35	8.1
− 35 + 48	6.2
− 48 + 65	4.1
− 65 +100	3.6
−100 +150	2.2
−150 +200	1.9
−200	5.3

2. Calculate the specific surface in square centimeters per gram of pyrite having the screen analysis below. Specific gravity of pyrite = 5.0.

Mesh	Percentage Retained
− 3 + 4	0
− 4 + 6	4.0
− 6 + 8	7.2
− 8 + 10	12.0
− 10 + 14	17.6
− 14 + 20	15.4
− 20 + 28	12.0
− 28 + 35	10.0
− 35 + 48	7.2
− 48 + 65	6.0
− 65 +100	3.8
−100 +150	2.8
−150 +200	2.0

3. In the petroleum industry, gas oil is cracked in contact with solid catalysts to yield high-octane blending stocks, with a finely divided clay as the catalyst. The yields obtained are a function of the surface area of the catalyst. The catalyst has a density of 1.20 grams/cc and approximately the same specific surface ratio as quartz. A sample of this material was screened, and the material through the 200-mesh screen was further sized by air elutriation. From the resulting analysis given below, determine the specific surface (square centimeters per gram), the arithmetic average diameter and the mean surface diameter of the catalyst.

The specific surface so computed does not include the surface in the capillaries which may increase the total surface about 3000-fold.

Screened Fraction

Mesh	Mass Fraction
− 48 + 65	0.088
− 65 +100	0.178
−100 +150	0.293
−150 +200	0.194
−200	0.247

Elutriated Fraction

Size Limits, microns	Mass Fraction (of Original Sample)
80–60	0.113
60–40	0.078
40–20	0.042
20–0	0.014

4. Powdered coal with the screen analysis given below as "Feed" is fed to a vibrating 48-mesh screen in an attempt to remove the undesired fine material. When the screen was new the oversize and undersize analyses were as listed under columns headed "New." After 3 months' operation, the analyses are as headed "Old." What is the effectiveness of the screen (*a*) when new and (*b*) when old?

Screen Analyses—Mass Fractions

Mesh	Feed	Oversize New	Oversize Old	Undersize New	Undersize Old
− 3 + 4	0.010	0.012	0.014	.....	.....
− 4 + 6	0.022	0.027	0.031	.....	.....
− 6 + 8	0.063	0.078	0.088	.....	.....
− 8 + 10	0.081	0.100	0.112	.....	.....
− 10 + 14	0.102	0.126	0.142	.....	.....
− 14 + 20	0.165	0.204	0.229	.....	.....
− 20 + 28	0.131	0.162	0.182	.....	.....
− 28 + 35	0.101	0.125	0.104	.....	0.093
− 35 + 48	0.095	0.117	0.065	.....	0.171
− 48 + 65	0.070	0.029	0.025	0.246	0.186
− 65 +100	0.047	0.015	0.008	0.183	0.146
−100 +150	0.031	0.005	.....	0.141	0.111
−150 +200	0.020	.....	.....	0.105	0.071
−200	0.062	.....	.....	0.325	0.222

5. Table salt is being fed to a vibrating screen at the rate of 300 lb/hr. The desired product is the 48/65 mesh fraction. A 48- and a 65-mesh screen are therefore used (double deck), the feed being introduced on the 48-mesh screen, the product being discharged from the 65-mesh screen. During the operation it was observed that the average proportion of oversize:product:undersize was 2:1½:1.

(*a*) Calculate effectiveness of the screener.

(*b*) If screen dimensions were 2 ft by 4 ft, calculate the capacity of the 65-mesh screen on the basis of a perfectly functioning 48-mesh screen and also on the basis of the actual performance of the screen.

Screen Mesh	Feed, mass fraction	Oversize, mass fraction	Product, mass fraction	Undersize, mass fraction
− 10 + 14	0.000356	0.0008	......	....
− 14 + 20	0.00373	0.008	0.0005	0.00003
− 20 + 28	0.089	0.189	0.016	0.00012
− 28 + 35	0.186	0.389	0.039	0.0009
− 35 + 48	0.258	0.337	0.322	0.0036
− 48 + 65	0.281	0.066	0.526	0.344
− 65 +100	0.091	0.005	0.067	0.299
−100 +150	0.062	0.005	0.024	0.237
−150 +200	0.025	0.001	0.002	0.11

6. One ton per hour of dolomite is produced by crushing and then screening through a 14-mesh screen. According to the screen analysis below, calculate (*a*) the total load to crusher and (*b*) the effectiveness of the screen.

Tyler Mesh	Feed to Screen, %	Screen Undersize, Product, %	Screen Oversize, Circulating Load, %
4 on	14.3	....	20
8 on	20.0	....	28
14 on	20.0	0.0	28
28 on	28.5	40.0	24
48 on	8.6	30.0	0 through 28 mesh
100 on	5.7	20.0	
100 through	2.86	10.0	

7. The data below were obtained on the operation of a 6-mesh (square) hummer screen at the tipple of a coal mine. The screening was done to separate a very fine refuse from a fine coal stream so that it could be reprocessed. Calculate (*a*) the recovery and rejection of each size fraction and (*b*) the screen effectiveness.

Feed to Screen, 131 Tons/Hr (Approximately 5% Moisture)

Size	Sample Weight
+¼ in.	3825 grams
¼ × 6 mesh	1006
6 × 14	750
14 × 28	303
28 × 48	219
48 × 0	807

Overflow from Screen

Size	Sample Weight
+¼ in.	2905 grams
¼ × 6 mesh	767
6 × 14	405
14 × 28	117
28 × 48	68
48 × 0	278

Underflow from Screen, 9.8 Tons/Hr (Dry Solids)

Size	Sample, %
¼ × 6 mesh	11.3
6 × 8	7.8
8 × 14	6.9
14 × 28	8.6
28 × 48	3.[illegible]
48 × 0	62.1

CHAPTER 4

Size Reduction of Solids

IN industries that process raw material in the solid state or use solid material in the processing of fluids, reduction in the size of the solid particles is frequently required. In the production of gypsum plaster, the raw gypsum rock is removed from the quarry in large blocks, sometimes 5 ft in diameter. It must be reduced to particles fine enough to pass through a 100-mesh screen in order to provide sufficient specific surface for hydration to take place rapidly. This means a reduction in size from 60 in. to 0.005 in. Pigments in paints must be very fine in order to give good coverage when applied to a surface.

Reduction in size involves the production of smaller mass units from larger mass units of the same material; it therefore follows that the operation must cause fracture to take place in the larger units. This fracturing or shattering of the larger mass units is accomplished by the application of pressure. All *true solid materials* are crystalline in nature; that is, the atoms in the individual crystals are arranged in definite repeating geometric patterns, and there are certain planes in the crystal along which shear takes place more readily. The pressure applied must be sufficient to cause failure by shear along these cleavage planes. If the shear along these planes results in deformation but not rupture, the deformation is called plastic deformation. The segments of the crystal slide along on each other like a pack of cards, the only result being a change in dimensions of the crystal. In order to bring about actual size reduction, it is necessary that the material be actually fractured and that shear movement, once started, results in complete separation of the segments between which the shear failure occurred.

From this, it might appear that the best method of causing rupture to take place in solid material would be the application of shearing loads. However, the orientation of crystals in solid matter is usually so irregular that the direct application of compressive loads is just as effective as shearing loads. All equipment for size reduction of solids uses compression, or shear, or both, as disrupting forces.

OBJECTIVES

The purpose of size reduction is not only to make "little ones out of big ones" when the effectiveness can be measured by the degree of fineness of the product, but also to produce a product of the desired size or size range. The size requirements for various products may vary widely, and hence different machines and procedures are employed. A size range entirely satisfactory for one purpose may be highly undesirable for another, even when the same substance is involved. Powdered coal is widely used for firing industrial furnaces, and lump coal is also fed into furnaces by mechanical stokers. But powdered coal could not be used in the stoker, and lump coal could not be used in the equipment designed for firing pulverized or powdered coal.

In many cases, it is necessary to use a product with rather narrow limits in size variation. It is usually impossible to accomplish this by size reduction only. Screening and classification by various means are required to secure the desired limitation in size range. The two unit operations of size reduction and size separation are further closely associated in that laboratory screen analyses are necessary to evaluate the effectiveness of a given size reduction operation as well as to furnish data for estimating the power or energy required.

Ores of metals consist of varying amounts of valuable minerals associated with undesired gangue minerals. The first step in processing ores for the recovery of metal values is the separation of the values from the gangue, since the ore as taken from the mine contains both types of minerals together in solid masses. Unless the valuable mineral exists in great enough concentration to permit the ore to be reduced to the metal without previous treatment, in which case the gangue is usually separated in the molten state, it is necessary to break up the ore mass mechanically, thus freeing the valuable minerals from the gangue. The minerals are then separated by gravity or flotation methods resulting in concentration of the valuable minerals.

The purposes of size reductions are therefore twofold: (1) To produce solids with desired size ranges or specific surfaces. (2) To break apart minerals or crystals of chemical compounds which are intimately associated in the solid state.

STAGES OF REDUCTION

For successful size reduction, it is necessary that every lump or particle must be broken by contact with other particles or by direct contact with the moving parts of the machine. As the breaking action proceeds, the number of particles increases, requiring more contacts per unit mass. Thus the capacity of a particular machine of fixed dimensions, as in tons per day, is much less for small sizes than for the larger sizes, since it is necessary for the smaller particles to remain in the machine for longer periods of time to sustain the required number of contacts. No device has been developed capable of automatically adjusting itself to the varying requirements of contact. In commercial operations, sufficient capacity in the intermediate and fine ranges of size reduction is obtained either by operating several similar units in parallel or, better, by employing machines which furnish greater numbers of contacts per unit of time.

Machines providing the required large number of contacts, particularly for smaller-size material, have been developed, primarily for the last stages of size reduction.

For commercial reduction in size of masses of solids 1 ft or more in diameter to 200-mesh size, usually at least three stages or steps are followed which are divided according to the types of machines best adapted to each stage. The three steps are:

1. Coarse size reduction: feeds from 2 to 96 in. or more.
2. Intermediate size reduction: feeds from 1 to 3 in.
3. Fine size reduction: feeds from 0.25 to 0.5 in.

OPERATING VARIABLES

The *moisture content* of solids to be reduced in size is important. If it is below 3 or 4 per cent by weight, no particular difficulties are encountered; indeed, it appears that the presence of this amount of moisture is of real benefit in size reduction if for no other reason than for dust control. When moisture content exceeds about 4 per cent, most materials become sticky or pasty with a tendency to clog the machine. This is particularly true in the coarse and intermediate stages.

A large excess of water (50 per cent or more) facilitates the operation by washing the feed into and the product out of the zone of action and by furnishing a means for transporting the solids about the plant as a suspension or slurry. *Wet grinding* is mostly confined to the fine stage of reduction.

The *reduction ratio* is the ratio of the average diameter of the feed to the average diameter of the product. Most machines in the coarser ranges of crushing have a reduction ratio from about 3 to 7. Fine grinders may have a reduction ratio as high as 100.

In *free crushing*, the crushed product with whatever fines have been formed is quickly removed after a relatively short sojourn in the crushing zone. The product may flow out by gravity, be blown out with compressed air, be washed out with water, or be thrown out by centrifugal force. This method of operation prevents the formation of an excessive amount of fines by limiting the number of contacts.

In *choke feeding* (the antithesis of free crushing), the crusher is equipped with a feed hopper and kept filled (or choked) so that it does not freely discharge the crushed product. This increases greatly the proportion of fines produced and decreases the capacity. In some instances choke feeding may result in economy of operation, eliminating one or more reducing stages because of the large quantity of fines produced.

Each stage in size reduction may, and frequently does, have a size-separating unit following it. If the oversize material is returned to the crusher, the operation is termed *closed circuit*. If no material is returned for recrushing, the operation is called *open circuit*. Closed-circuit operation is economical of crushing power, which at best is high, permits smaller units per given tonnage, and produces a material with greater uniformity of size.

Although the size of the feed is an important factor in the selection of a machine, other factors must be considered, such as hardness or structure of the material. From the standpoint of crushing, minerals with a Mohs hardness of 4 or less are classed as soft; others are considered hard. Machines for the coarse preliminary crushing of soft materials do not need to be so sturdily constructed or so elaborate in design as machines for breaking hard materials. In the finer size ranges, similar machines are used for both hard and soft materials.

Machines exerting a tearing action and called *disintegrators* are employed for reducing the size of fibrous materials such as wood and asbestos.

COARSE SIZE REDUCTION

Machines for the coarser stages of size reduction handle feed sizes from 3 to 4 in. and up. For hard materials, either jaw, gyratory, or disk crushers are used. For soft materials where the production of fines is to be limited, as in crushing coal for sale, such devices as hammer mills or toothed rolls are employed.

Coarse Crushers for Hard Materials

Jaw Crushers. Jaw crushers are represented by the Blake and Dodge types and operate by applying a crushing pressure.

The *Blake crusher* (Fig. 18) consists essentially of a cast-steel frame supporting one fixed and one movable jaw. The jaws are made of cast steel lined with a tough abrasion resistant metal, such as manganese steel. The movable jaw is pivoted at the top and operated by the eccentric, pitman, and toggles. The pitman is given a nearly vertical motion by the eccentric, and, since one of the toggles is mounted in rigid journals at one end of the crusher frame, the reciprocating motion of the pitman causes the other toggle to move the jaw back and forth. The jaw is held against the toggle by a tension link and spring. Crushing is accomplished only when the movable jaw moves toward the fixed jaw. This means an intermittent power requirement. In order to equalize this, one or two heavy flywheels are mounted on the main shaft of the crusher. The machine is driven by flat belts or V-belts.

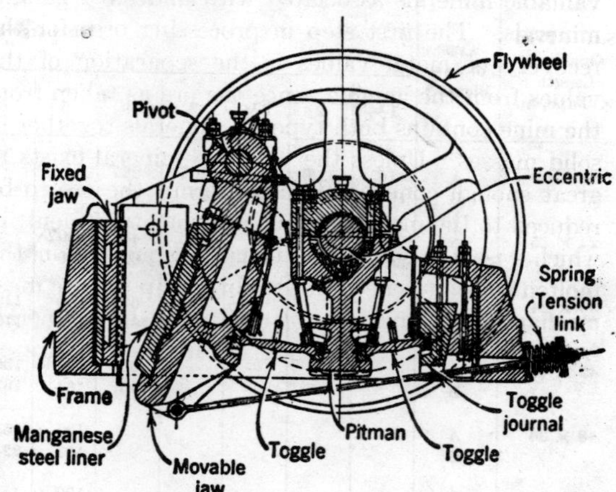

FIG. 18. Sectional drawing of Blake-type jaw crusher. (*Allis-Chalmers Mfg. Co.*)

The *Dodge crusher* (Fig. 19) is subject to uneven stresses inherent in its design and therefore is made only in small sizes. It differs from the Blake crusher in that the movable jaw is pivoted at the bottom and the width of the discharge opening remains practically constant, thereby yielding a more closely sized product. No toggles are required, the jaw being operated through the pitman by the eccentric. If only one size-reducing machine is being employed, the uniformity in size of product may be of advantage, but otherwise the machine is of limited use.

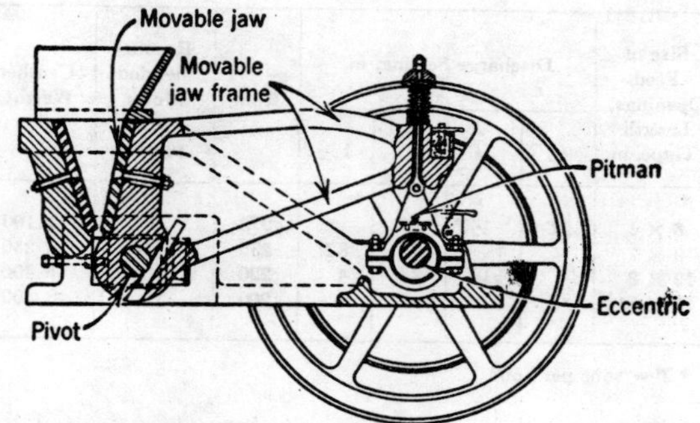

FIG. 19. Sectional drawing of Dodge-type jaw crusher. (*Allis-Chalmers Mfg. Co.*)

The power is applied through a long lever, and if the crusher becomes clogged enormous stresses are set up in the members which become excessive in ma-

TABLE 6. CAPACITIES OF BLAKE JAW CRUSHERS

(Allis-Chalmers Mfg. Co.)

Size of Feed Opening, Length × Gape, in.	Type of Jaw Plates *	Discharge Setting, in.													Rpm	Recommended Motor Horsepower	Crusher Weight, lb
		1½	2	2½	3	4	5	6	7	8	9	10	11	12			
15 × 10	A	7T †	11T	16T	20T	28T									235	15	10,000
	B	16	23	28	35	47											
24 × 15	A		22	28	35	48	60T								210	35	27,000
	B	25	34	43	52	69	86										
36 × 24	A				45	67	88	110T							210	75	70,000
	B				80	102	127	170									
42 × 40	A					90	103	130	155T	190T					190	125	140,000
	B					140	164	197	230	263							
48 × 36	A						120	155	187	225					190	150	145,000
	B						187	224	262	300							
48 × 42	A						120	155	187	225					190	150	160,000
	B						187	224	262	300							
60 × 48	A						150	210	240	265	300T				170	200	215,000
	B						262	314	368	420	472						
84 × 56	A									360	425	480T	565T	630T	90	200	422,000
	B									...	...	...	...	...			
84 × 60	A									360	425	480	565	630	90	250	430,000
	B									...	...	...	...	...			
84 × 66	A									420	480	510	570	630	90	250	460,000
	B									...	...	...	...	...			

* A = standard jaw plates (smooth).
B = "Nonchoking" jaw plates (corrugated).

† T = tons per hour.

TABLE 7. CAPACITIES OF DODGE CRUSHERS

(Allis-Chalmers Mfg. Co.)

Size of Feed Openings, Length × Gape, in.	Discharge Setting, in.				Rpm	Recommended Motor Horsepower	Crusher Weight, lb
	½	¾	1	1½			
6 × 4	¼T *	½T	1T		275	3	1,100
9 × 7		1	2	3T	235	6	3,250
12 × 8		1½	3	4	220	10	5,400
15 × 11		2	4	6	200	15	13,500

* T = tons per hour.

chines with gapes * above 11 in. The constant opening of the jaws at the discharge end gives the Dodge crusher an annoying tendency to clog which is absent in the Blake crusher.

* Gape is the greatest distance between the jaws or crushing surfaces.

There are many different designs of jaw crushers, some of which combine shear with compression. The Universal jaw crusher (Fig. 20) combines the principles of the Dodge and Blake crushers. It gives two crushing strokes per revolution because the pivot is above the bottom end of the jaw, causing the bottom of the jaw to move forward while the upper end of the jaw recedes.

Gyratory Crushers. Gyratory crushers were developed later to supply a machine with greater capacity. Actually, the crushing action of gyratories is similar to the action of jaw crushers in that the moving crushing element approaches to and recedes from a fixed crushing plate.

Figure 21 shows a *suspended-spindle type* of gyratory, consisting of an outer frame carrying an inverted conical surface known as "concaves" and an inner gyrating crushing head. The conical crushing head is supported on a spindle which hangs from a suitable bearing in the upper portion of the machine.

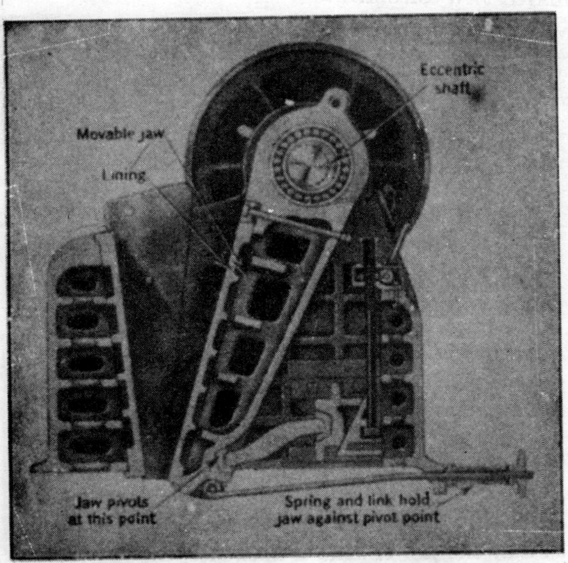

Fig. 20. Sectional drawing of Universal streamlined roller-bearing jaw crusher. (*Universal Engineering Corp.*)

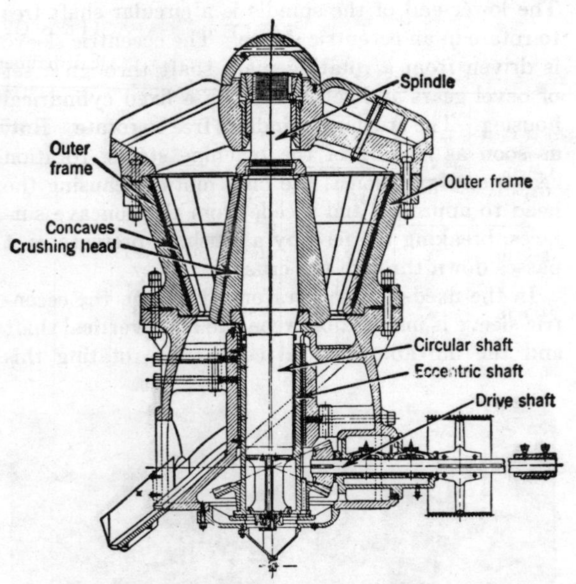

Fig. 21. Sectional drawing of gyratory crusher of suspended-spindle type. (*Allis-Chalmers Mfg. Co.*)

TABLE 8. CAPACITIES OF GYRATORY CRUSHERS

(Allis-Chalmers Mfg. Co.)

Size of Feed Opening, Gape × Length, in.	Finest Setting *		Coarsest Setting †		Driving Pulley, rpm	Recommended Motor, Horsepower	Crusher Weight, lb
	Size of Discharge Opening, in.	Capacity, tons/hr	Size of Discharge Opening, in.	Capacity, tons/hr			
2¼ × 10	⅜	½			700	3	700
8 × 34	1½	25	2½	47	450	15–25	20,000
10 × 40	1¾	39	3½	93	400	25–40	30,000
13 × 45	2	63	3½	128	375	50–75	45,000
16 × 56	3	120	4	176	350	60–100	62,000
20 × 68	3½	152	5	245	330	75–125	94,000
30 × 90	4	235	6½	450	325	125–175	169,000
36 × 126	5	365	6⅝	525	300	175–225	263,000
42 × 132	5½	475	6⅝	615	300	200–275	286,000
50 × 162	6	740	7¼	845	250	225–300	575,000
54 × 162	6¼	875	8	1050	250	225–300	630,000
60 × 174	6¼	990	10	1440	250	225–300	725,000
60 × 182	6½	1420	10½	1900	250	300–500	1,000,000

* Finest permitted for this size gyratory.
† Coarsest permitted for this size gyratory.

The lower end of the spindle is a circular shaft free to rotate in an eccentric sleeve. The eccentric sleeve is driven from a rotating main shaft through a set of bevel gears and rotates within a fixed cylindrical housing. The crushing spindle is free to rotate. But, as soon as feeding of the machine starts, rotation ceases and gyration is the only motion, causing the head to approach and recede from the concave surfaces, breaking the feed by a crushing pressure as it passes down through the crusher.

In the fixed-spindle gyratory (Fig. 22), the eccentric sleeve is inserted between the fixed vertical shaft and the movable vertical cone. By rotating this eccentric sleeve the axis of the cone is given a cylindrical motion with a "parallel pinching" action on the material being crushed.

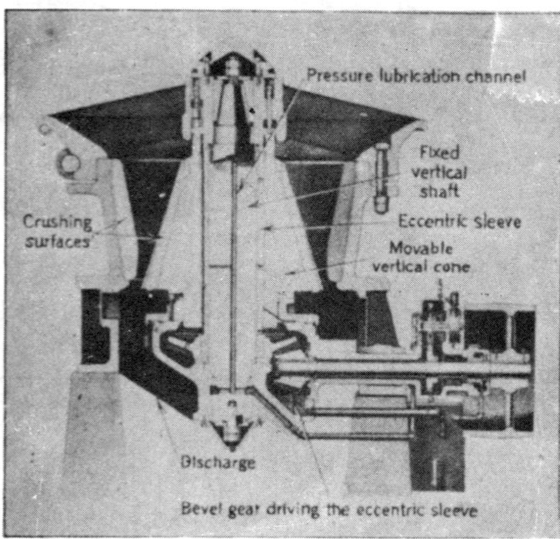

Fig. 22. Sectional drawing of Telsmith parallel pinch crusher. (Smith Engineering Works.)

Gyratory crushers have large capacity because the action is continuous. The capacity is similar to that of a jaw crusher having the same gape and a length L equal to the perimeter of the gyratory. Since all the coarse crushers have greater capacities than the devices for the finer ranges of size reduction, a gyratory of sufficient size to handle the required size of feed may have an excessive capacity. Jaw crushers, therefore, are frequently used for the first coarse breaking operation, followed by gyratories.

Capacities of jaw and gyratory crushers with gapes of 4 in. to 2 ft may be approximated by the Taggart [5] * formula:

$$T = 0.6LS$$

where T = capacity (tons/hr).
L = length of feed opening (in jaw crushers, normal to gape; in gyratories, the perimeter of a circle whose diameter is the arithmetic average of the diameters of the two cones) (in.).
S = greatest width of discharge opening (in.).

Exercise. Compare the capacities as estimated by the Taggart formula with those given in Table 6.

The *power requirements* for jaw and gyratory crushers are about the same, but the gyratory load is somewhat more uniform since it is crushing continuously whereas the jaw crusher works intermittently.

In choosing between a jaw crusher or a gyratory crusher for a given installation, capacity is the criterion. If capacity requirements are small enough so that one jaw crusher is adequate, the jaw crusher is the usual choice because of its lower original cost and upkeep. If capacity requirements are large enough to keep a gyratory in continuous operation, the gyratory is usually preferred. Taggart [5] states an empirical rule that "if the hourly tonnage to be crushed divided by the square of the gape in inches is less than 0.115, use a jaw crusher; otherwise, a gyratory."

Coarse Crushers for Soft Materials

Such materials as coal, gypsum, some types of limestone, ice, fire clay and shales are less hard than 4 on the Mohs scale and do not require the heavy and expensive types of crushers needed for hard materials. Frequently, the size reduction desired for these soft materials excludes the very fine ranges, and most of the crushers designed for such materials produce a small amount of excessively fine material.

The *Bradford breaker* for coal (Fig. 23) combines the two features of breaking and screening. The periphery of the machine is a reinforced screen which allows the coal, when sufficiently reduced in size, to pass through. Breaking is accomplished by rotation of the cylinder. The coal is lifted on interior shelves and broken by falling and striking the coal below as the cylinder is rotated. Harder material such as slate and tramp iron are not broken and gradually pass out from the open end of the breaker as indicated.

* The bibliography for this chapter appears on p. 45.

FIG. 23. Phantom drawing of Bradford breaker. Run-of-mine coal enters through the chute at the far end, is lifted, falls, and is broken by the impact, passing through perforations into the chute below; rock and refuse are plowed out as indicated in the foreground. (*Pennsylvania Crusher Co.*)

FIG. 24. Sectional drawing showing operation of toothed roll crusher. (*Link-Belt Co.*)

A *toothed roll crusher* for coal, gypsum, ice, or other soft materials (Fig. 24) accomplishes breaking by pressure of the teeth against the larger lumps of the material, disintegrating it in much the same manner as ice is broken up manually with an ice pick. Excessive production of fines is thus prevented. Knobbed and smooth rolls (Fig. 30) are also widely used for coarse crushing of soft materials.

FIG. 25. Sectional drawing showing operation of a hammer mill. (*Allis-Chalmers Mfg. Co.*)

A *hammer mill* (Fig. 25) may be used for coal or even fibrous material. Heavy blocks of steel are attached by pins to a disk or disks revolving at high speed within a sturdy housing. The hammers deliver heavy blows to the feed material while it is in suspension, driving it against a breaker plate until it is fine enough to pass through the openings in the cage bars at the bottom of the mill constituting the screen. Some of these mills are built in extremely large sizes, the individual hammers weighing as much as 250 lb. Very sturdy housings are required for such hammer mills. The same type is also adapted to fine pulverizing, the size of the product being controlled by the sizes of the discharge screens. The hammer mill is probably the most versatile type of crushing device currently available. For wet material the cages or screens are replaced with corrugated grinding plates.

A so-called *squirrel-cage disintegrator* (Fig. 26) is useful in tearing apart fibrous material such as wood blocks and asbestos. The device consists of two or more concentric cages rotated in opposite directions. The feed is introduced into the inner cage. Centrifugal force drives the material into the spaces between the rotating cages where it is torn apart, and thence into the outer casing from which it is discharged to a conveyor or storage bin.

INTERMEDIATE SIZE REDUCTION

Cone crushers, developed since the 1920's, have gained such wide acceptance that they may be regarded as standard in the intermediate range. A

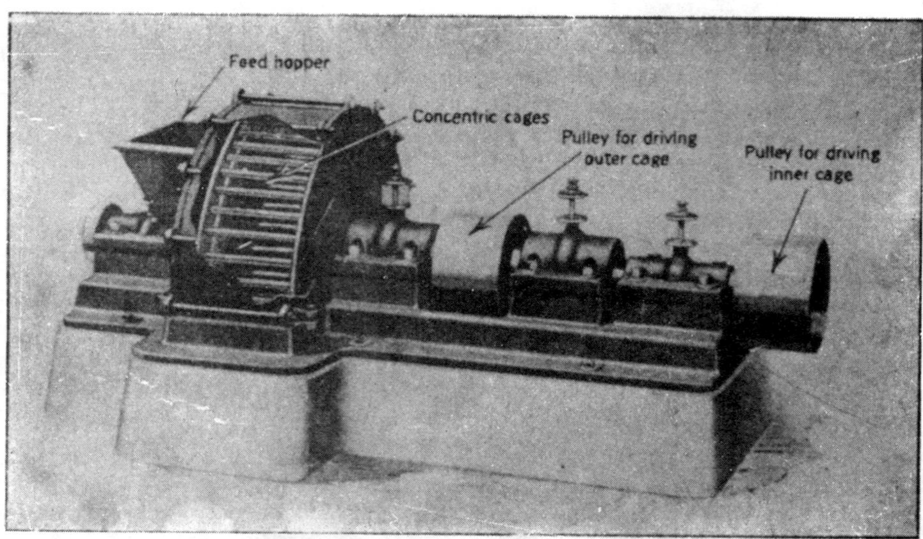

FIG. 26. Cutaway view of squirrel-cage disintegrator. (*C. O. Bartlett and Snow Co.*)

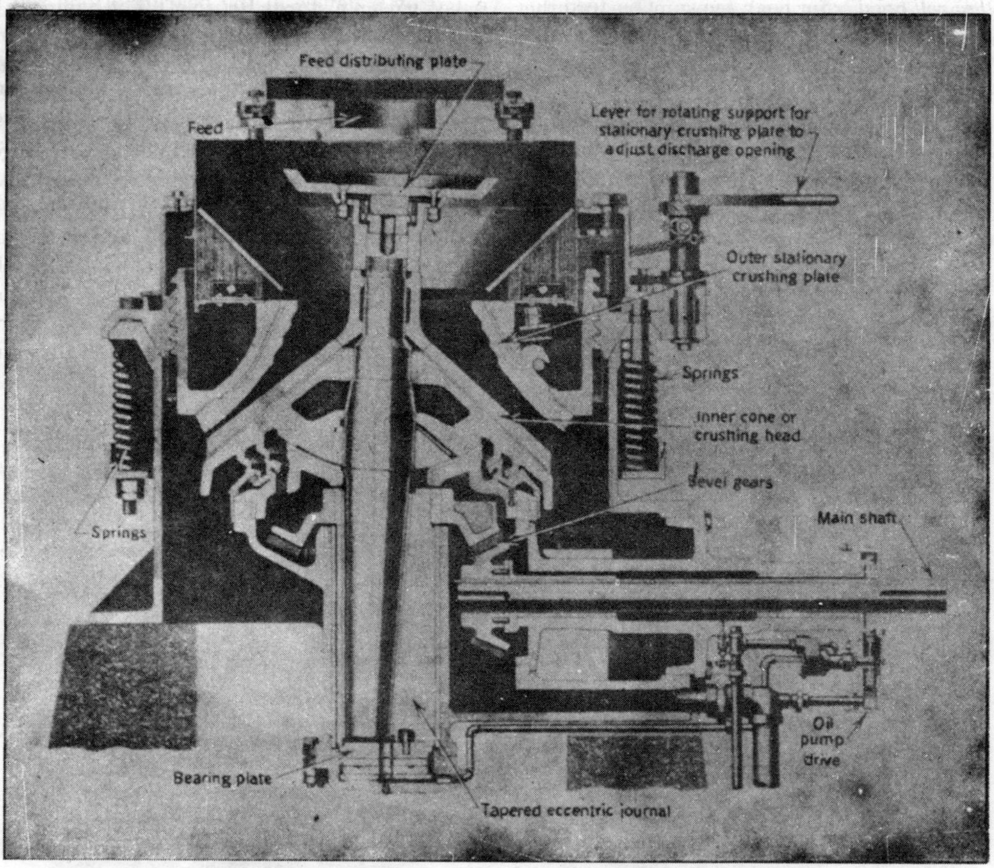

FIG. 27. Sectional drawing of cone crusher. (*Nordberg Mfg. Co.*)

standard cone crusher is shown in Figs. 27 and 28. The drive is similar to that of the gyratory crusher. The inner cone or "crushing head" is supported by the tapered eccentric journal which is rotated by the bevel gears driven by the main shaft. The entire weight of the crushing head and spindle is supported on a bearing plate supplied with oil under pressure. The operation is quite similar to that of the gyratory crusher, but there are two important points of difference. The outer stationary crushing plate flares outward to provide an increasing area of discharge so that the machine can quickly clear itself of the reduced product. This stationary crushing plate is held in position by a nest of heavy helical tension springs so that when tramp iron or other uncrushable objects enter the crushing zone the plate is lifted, preventing fracture of the plate and injury to the machine. These cone crushers are available in two sizes, the standard (Fig. 27) for coarser feed, and a

FIG. 28. Cutaway view showing action of cone crusher. (*Nordberg Mfg. Co.*)

so-called "short head" for finer feed. The feed to cone crushers must be dry and rather uniformly sized. Cone crushers give best results when operating in closed circuit with screens.

The Telsmith Gyrasphere, Fig. 29, is a variation of the cone crusher. The crushing head is spherical in contour, and the crushing plate is held in position by springs under compression instead of tension. The drive and oiling system is similar to that of the cone crusher. The spherical head facilitates discharge of the crushed product.

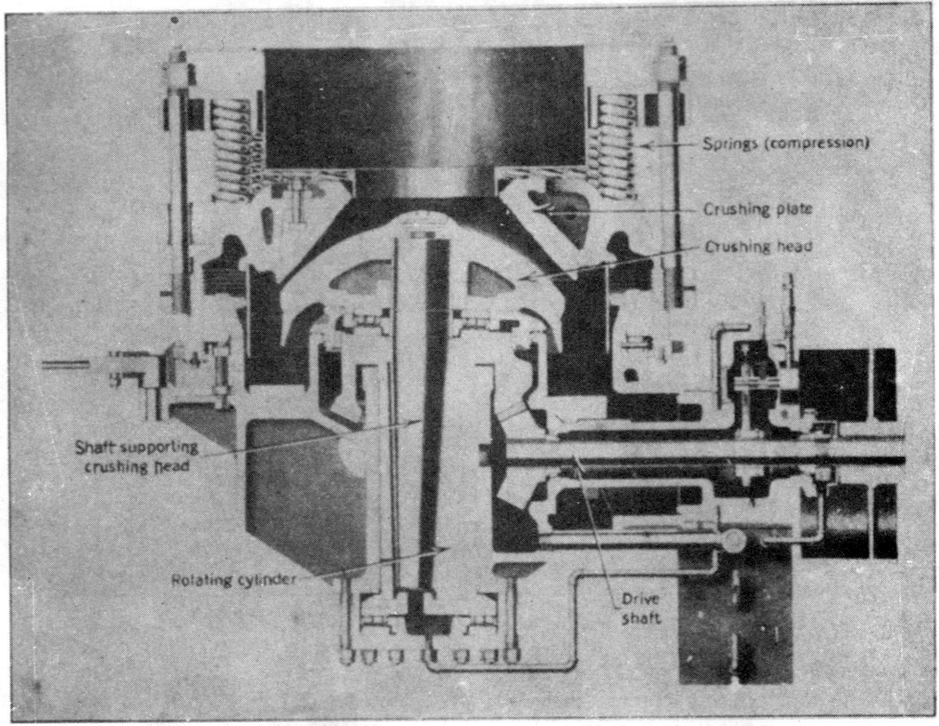

FIG. 29. Sectional drawing of Telsmith Gyrasphere. (*Smith Engineering Works.*)

Crushing rolls consist of two heavy cylinders revolving toward each other, the feed being nipped and pulled downward through the rolls by friction. As shown in Fig. 30, modern crushers drive both rolls positively, breakage being prevented by mounting the bearings of one of the rolls against nests of heavy compression springs. Since there is a considerable amount of wear on the rolls, the crushing surface consists of a tough steel sleeve which is shrunk on to the main cylindrical casting, making possible the replacement of worn crushing surfaces. The wearing of grooves in the surface of the rolls is largely prevented by a device in the bearing of one roll which gives it a limited lateral motion simultaneously with the rotation. The size reduction accomplished by rolls is relatively small, the average diameter of the product being about one-fourth that of the feed.

Cone crushers are replacing rolls for intermediate size reduction of ores because their reduction ratio is two or three times that of rolls and they require less maintenance.

The diameter and spacing of rolls may be varied over rather wide ranges, allowing considerable variations in size of feed and product. This flexibility is a favorable characteristic of crushing rolls, which, combined with the low initial cost, has encouraged the wide adoption of rolls for moderate size reduction of all sizes. The proper diameter and spacing of the rolls, the capacity in tons per hour, and the required horsepower for crushing rolls may be computed as follows.

The coefficient of friction of the mineral against the steel surfaces of the rolls incorporated with a relationship between the dimension of the material to be crushed and the diameter of the rolls determines

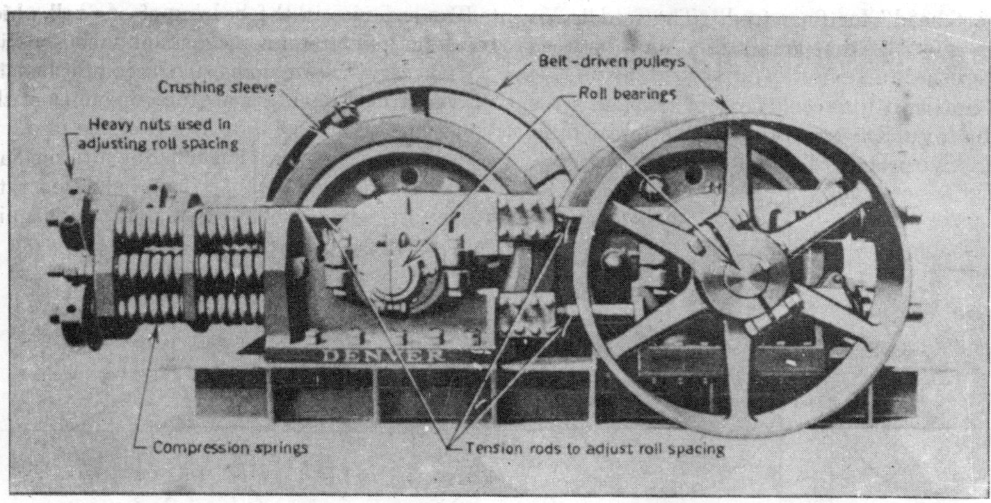

FIG. 30. Crushing rolls. (*Denver Equipment Co.*)

whether or not a particle will be drawn into the rolls and crushed. Figure 31 is a line diagram showing the outline of a spherical particle in position to be crushed between a pair of rolls. The vectors F_T and F_N represent the forces acting on the particle at the point of contact with the roll and may be represented by the resultant force F_R:

A_n = angle of nip (the value for angle A in Fig. 31 corresponding to F_R being horizontal).
D_r = diameter of the rolls.
D_f = diameter of the feed particle.
D_p = maximum dimension of the product (minimum distance between rolls).
F_T = tangential force on the particle.
F_N = normal force on the particle.
F_R = resultant of F_T and F_N.

If F_R is at a negative angle (pointing downward) with the horizontal, as shown in Fig. 31, the particle will be drawn between the rolls. If F_R is at a positive angle with the horizontal, the particle will ride on the rolls or be thrown up and out and will not be crushed. The angle A between the two tangents at the points of contact of the particle with the rolls indicates whether or not the particle will be drawn between the rolls.

The definition of the coefficient of friction is the ratio of the force tangent to the surface to the force normal to the surface. In Fig. 31, this is F_T/F_N.

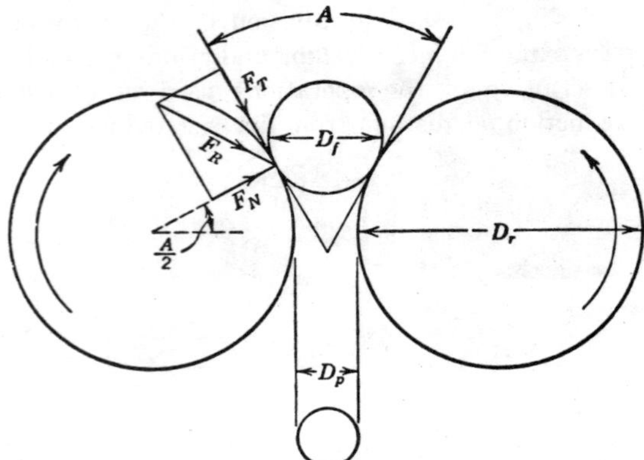

FIG. 31. Forces exerted by crushing rolls for a spherical particle in position to be crushed.

In the limiting case F_R is horizontal and

$$\tan\left(\frac{A}{2}\right) = \frac{F_T}{F_N}$$

which is equal to the coefficient of friction.

If the particle is a sphere,

$$\cos\frac{A}{2} = \frac{\frac{D_r}{2} + \frac{D_p}{2}}{\frac{D_r}{2} + \frac{D_f}{2}} = \frac{D_r + D_p}{D_r + D_f}$$

The value for the angle A corresponding to this limiting case is called the angle of nip, A_n.

For smooth steel rolls the value of the angle of nip A_n is usually about 32 degrees for ordinary rocks. In industrial operations general practice is to determine the theoretical minimum roll diameter D_r, add 1 in. to allow for wear, and select the next larger industrial roll.

If the rolls are operating on a slab of steel (or a particle of similar shape) as indicated in Fig. 32,

$$\cos\frac{A}{2} = \frac{\overline{ae}}{\overline{ac}} = \frac{\frac{D_r}{2} + \frac{D_p}{2}}{\frac{D_r}{2} + bc} = \frac{D_r + D_p}{D_r + \frac{D_f}{\cos(A/2)}}$$

$$D_r\left(1 - \cos\frac{A}{2}\right) = D_f - D_p$$

The limiting value for the angle $A/2$ at which the resulting force is horizontal is called the *angle of bite*

The *theoretical capacity of rolls* is the weight of a ribbon of feed having a width equal to the width of

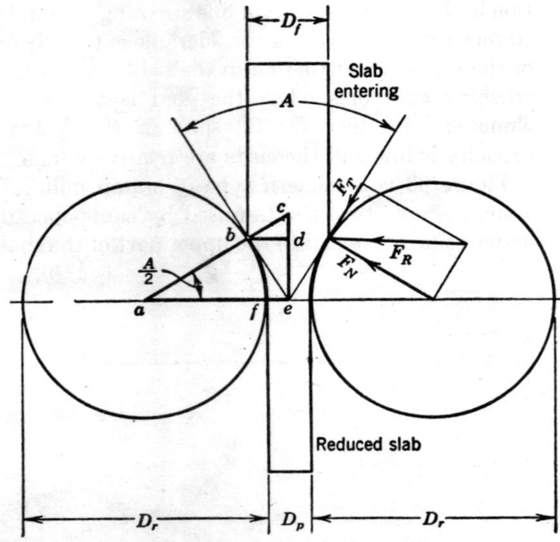

FIG. 32. Forces exerted by crushing rolls on a slab at the approximate angle of bite.

the rolls, a thickness equal to the distance between the rolls, and a length equal to peripheral velocity of the rolls in linear units per interval of time. This may be expressed in tons per hour:

$$T = \frac{60vLD_p\rho}{2000}$$

where T = capacity (tons/hr).
v = peripheral velocity (fpm). For rolls up to 72 in. in diameter, v is usually approximately equal to $300 + 84D_r$.
L = width of rolls (ft).
D_p = distance between rolls (ft).
ρ = density of material (lb/cu ft).

The *actual capacity* is usually from 0.10 to 0.30 of the theoretical.

With the increasing use of cone crushers for intermediate size reduction of ores, the application of rolls in this field is being limited to the size range between cone crushers and fine grinders.

Gravity stamps. The oldest method for size reduction of solids is undoubtedly a husky human being swinging a heavy hammer. When man began to devise mechanical methods for industrial operations, he naturally thought of a rock-crushing device involving a weight to be lifted and dropped on the material to be broken. For this reason the gravity stamp is the oldest recorded method for size reduction in the intermediate and fine size ranges. Gravity stamps are still used to a considerable extent because of the ease of construction in the field, especially for crushing gold ores when the gold is to be amalgamated with mercury, in spite of the fact that capacity is low and the costs are relatively high.

Figure 33 is a modern type of stamp mill. The stamps are vertical shafts raised by cams operating under collars fastened to the upper part of the shafts.

Fig. 33. Gravity stamp mill. (*Allis-Chalmers Mfg. Co.*)

The lower end of each shaft is equipped with a heavy cylindrical metal block which strikes on a stationary hard metal slab. Since a stamp mill has no means of clearing itself of the crushed product, the operation is usually carried out on suspensions of solids in water, which pass slowly through the crushing zone.

The reduction ratios in stamps may be as high as 150, making them one of the most flexible types of machines for size reduction.

FINE SIZE REDUCTION

Size reduction in the finer ranges has usually been termed fine grinding. This is due to the fact that most of the older devices for reduction in this range consisted of two main parts, a stationary surface and a surface rubbed against the stationary surface. The upper and nether millstones used for grinding flour from grain are typical. Such a machine causes disintegration mainly by the application of shear loads. Most recent devices in fine size reduction, such as ball mills, depend more on impact than on shearing forces. The division of the operations of size reduction into crushing and grinding is no longer descriptive of the operations used in coarse size reduction, as distinct from fine size reduction.

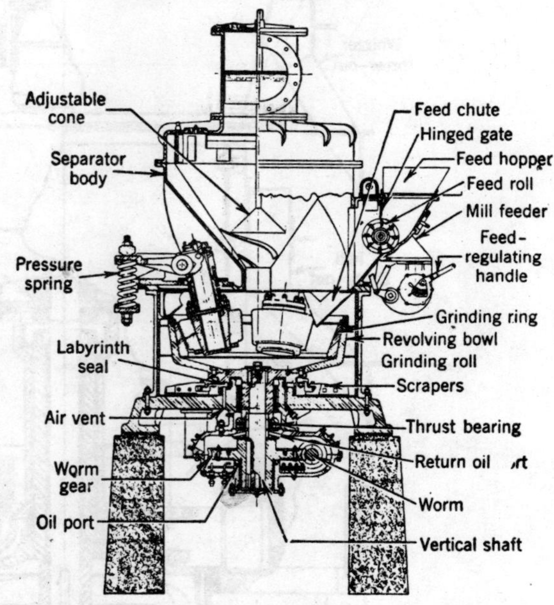

Fig. 34. Cutaway and sectional diagram of bowl mill with air classifier or separator. (*Combustion Engineering Co.*)

In the transition from the old-style shear-grinding devices to the widespread application of ball mills and rod mills, several machines appeared in which the material is reduced in size between rollers, or heavy balls, rolling against a crushing ring. In the Chilean mill, the horizontal axes of the rolls are usually stationary, and the flat pan carrying the crushing ring revolves. The bowl mill (Fig. 34) may be regarded as its modern development.

The Raymond roller mill (Fig. 35) consists of rollers suspended on balanced journals from a rapidly rotating spider mounted on the upper end of the main shaft. The revolving rolls exert pressure on a stationary confining ring by centrifugal force. A plow mounted on the apron or sleeve revolves with the shaft to throw the material into the crushing zone. This mill is usually provided with a sizing feature whereby the material cannot leave the machine until it is fine enough to pass through a screen of given mesh or be lifted by a stream of air of constant velocity. The so-called whizzer consists of vertical vanes rotating rapidly in a horizontal plane to knock oversize particles out of the rising stream.

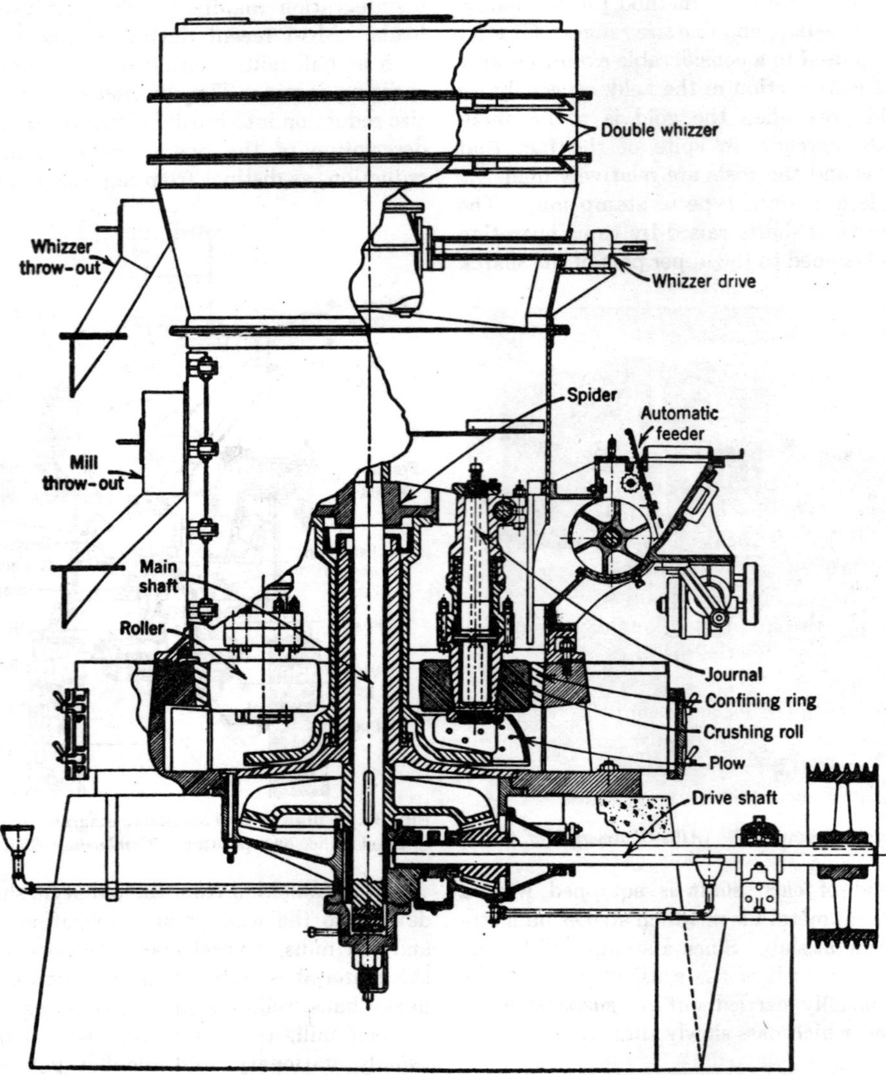

Fig. 35. Cutaway and sectional diagram of Raymond roller mill with air classifier or separator. (*Combustion Engineering Co.*)

Ball mills are horizontal rotating cylindrical or conical steel chambers, approximately half full of steel or iron balls, or flint stones. The size reduction is accomplished by the impact of these balls as they fall back after being lifted by the rotating chamber. The length of the cylinder is usually about equal to the diameter. Most ball mills are continuous in operation, feed entering at one end and discharging through the opposite end or through the periphery. They may be operated either wet or dry.

In cylindrical ball mills the product may be discharged by overflow through a hollow trunnion (Fig. 36). The smaller particles are suspended and carried out by the circulating fluid, such as air or water.

The Hardinge mill (Fig. 37) is typical of cylindro-conical ball mills. The larger balls and larger par-

Fig. 36. Ball mill showing feeder and hollow trunnion. (*Allis-Chalmers Mfg. Co.*)

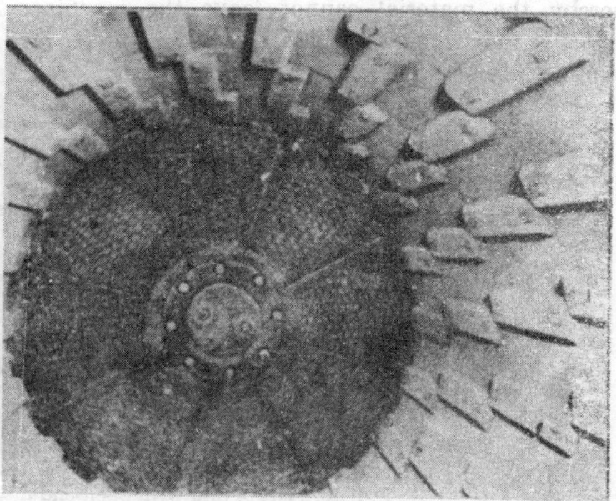

Fig. 38. Interior view of empty ball mill showing grate and rolled steel liners. (*Allis-Chalmers Mfg. Co.*)

ticles of feed are supposed to segregate to a certain extent in the cylindrical portion of the mill with the greatest diameter. Whether or not this supposition is true, there is a definite relationship between size of particles and size of balls required for effective size reduction. In any case the lifting effect on the balls is greatest at the greatest diameter, and the larger balls will be most effective in size reduction at this point.

In "grate mills" the product passes out through the openings in a vertical grate or diaphragm (Fig. 38). In the trunnion mill, the product may be raised by radial plates or scoops on the outside of the grate (Fig. 39), pushed away from the grate by helical vanes on the inner periphery of the cylinder, and discharged from the hollow trunnion by which the mill is supported. If the mill is supported by peripheral tires riding on rollers (Fig. 40), the material simply flows out through the grate and through the open end of the mill.

Compound ball mills consist of two to four cylindrical compartments separated by grates. Each successive compartment is of smaller diameter and contains balls of smaller sizes for finer grinding.

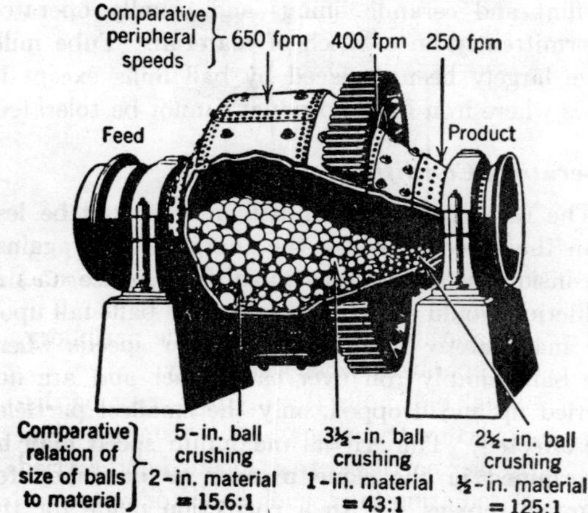

Fig. 37. Cutaway diagram indicating idealized operation of conical ball mill. (*Hardinge Co.*)

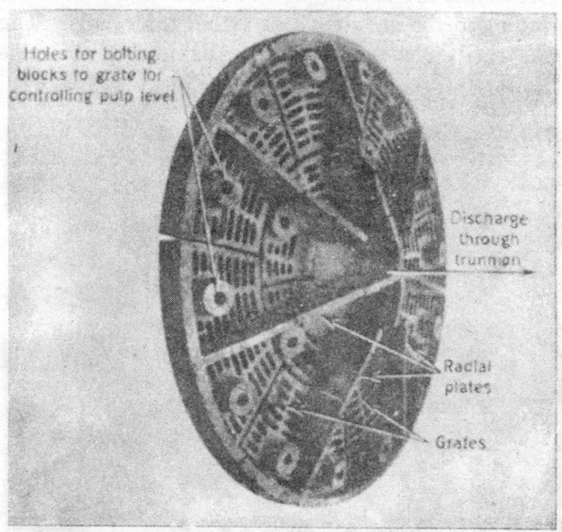

Fig. 39. Outside view of grate showing radial plates which raise the product and cause it to be discharged through the hollow trunnion. (*Allis-Chalmers Mfg. Co.*)

Fig. 40. Grate mill with open-end discharge. (*The Mine and Smelter Supply Co.*)

Such a mill is essentially a series of mills operating continuously.

The liners of ball mills are replaceable and usually made from alloy steel. Other materials such as rubber, cast iron, ceramic, and rock products are sometimes used. The wear on liners is usually from 0.1 to 0.5 lb/ton of product. The balls introduced into the mill vary from 1 to 6 in. in diameter, and the wear is from 1 to 3 lb/ton of product. It is customary to compensate for ball wear by introducing one or more full-sized balls to the mill at least once a day.

Rod mills are similar to ball mills except that the grinding media are steel rods rather than balls. The rods are always longer than the diameter of the mills and therefore lie in the mill parallel to the axis. The impact of the rods is received mainly by the larger particles, causing preferential reduction on the coarsest particles and giving a more closely sized product. Rod mills are more expensive to operate than ball mills, but their use is indicated when a small proportion of fines is desired in the product. Figure 41 shows the inside of a typical rod mill and indicates the wear and replacement of the rods by their different diameters. When the rods become badly worn they must be removed before they bend or break; if they become shorter than the diameter of the mill they may become wedged in such a position as to be held against the lining.

Fig. 41. Interior view of rod mill showing rods in various states of wear from service. (*Allis-Chalmers Mfg. Co.*)

Tube mill is a term used to identify a long cylindrical mill (usually about 22 ft long) utilizing pebbles of flint and ceramic linings and usually operated intermittently on a batch of material. Tube mills have largely been replaced by ball mills except in cases where iron in the product cannot be tolerated.

Operating Conditions

The rate of rotation of ball mills should be less than the speed at which the charge is held against the inside surface by centrifugal force, since no size reduction would take place unless the balls fall upon the material to be crushed. At low speeds where the balls simply roll over each other and are not carried up and dropped, only the smallest particles are affected. The critical maximum speed may be determined in the same manner as described for trommel screens. With a correction made for the diameter of the ball, the critical rate at sea level may be ascertained from the expression

$$N = \frac{76.65}{\sqrt{D - d}}$$

where N = revolutions per minute.
D = diameter of the mill (ft).
d = diameter of the balls (ft).

At low speeds where the contents are simply tumbled or rolled over, the power required to drive the mill varies directly with the speed of rotation. At higher speeds slippage occurs between the contents and the lining, and power requirements increase more slowly with speed of rotation.

Increasing the load (balls and material) in a ball mill will increase the power requirements until the maximum value is reached, after which the power requirement decreases with increasing load as the center of gravity of the load approaches the axis of rotation. For wet grinding the maximum power is required when the weight fraction of solids in the feed is about 0.60 to 0.75. The load may be increased by increasing the weight of balls introduced into the mill, by operating on material (wet pulp) of higher density, or by operating at a higher pulp level. The pulp level or quantity of material being ground in the mill is a major factor in the operation of the mill.

In the simple overflow type of continuous ball mill (no diaphragm), the feed enters at one end and the product flows out through the hollow trunnion

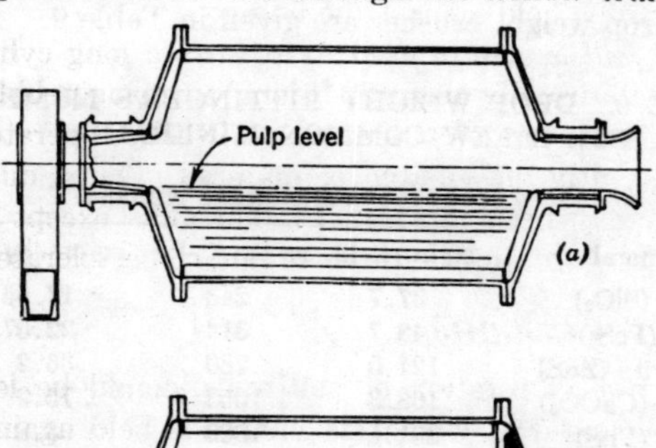

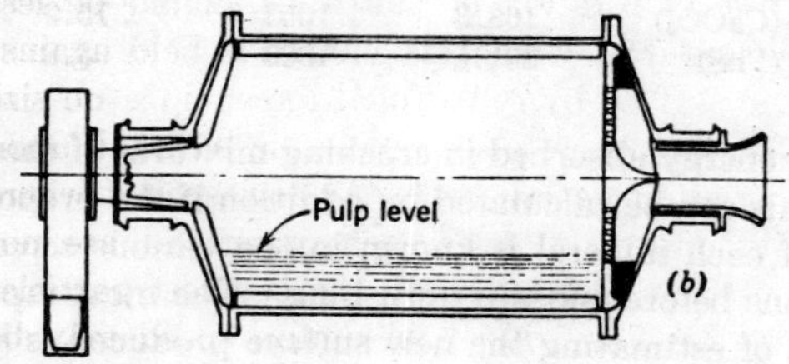

Fig. 42. (*a*) Sectional diagram of overflow ball mill. (*b*) Sectional diagram of ball mill equipped with diaphragm or grate allowing lower pulp levels. (*Allis-Chalmers Mfg. Co.*)

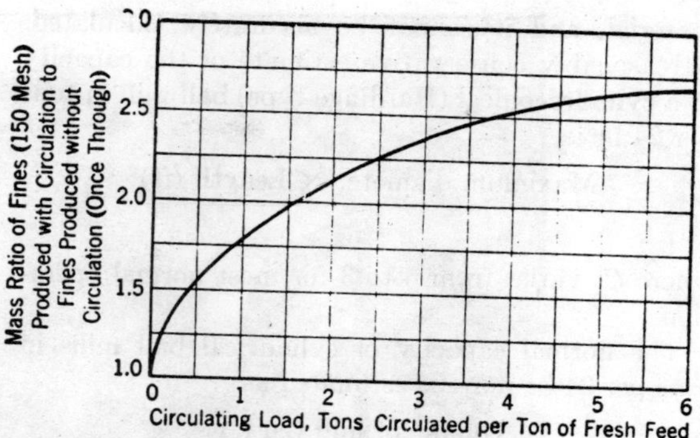

Fig. 43. Relation between circulating load and production of fines in a ball mill being operated in closed circuit.[6]

at the other end, as shown in Fig. 42*a*. The relatively fixed or constant pulp level provided by such a mill means that the effectiveness of grinding can be controlled only by the size and quantity of balls or the rate of feed. With the use of diaphragms the pulp level may be independently controlled at any desired level by making the diaphragm or grate solid for the desired distance from the periphery (Fig. 42*b*).

Lower pulp levels result in greater freedom of movement of the balls with consequent improvement in effectiveness of grinding. In a simple overflow type of mill the balls lose kinetic energy when falling into the dense pulp, and the contact forces between balls under the surface of the pulp is decreased. Mills with diaphragms or grates blocked to maintain the proper pulp level are reported to deliver 25 per cent more product of the correct size range with an increased power requirement of only 20 per cent. Low levels of pulp and decreased time in the mill result also in a decrease of overgrinding.

Closed circuit operation (see diagram accompanying example, p. 44) is usually necessary in ball mill operation since these mills do not have a sizing action on their product. A sizing device, such as a "classifier," is placed in series with the ball mill, and the oversize material from the sizing operation is returned to the mill for further size reduction. In such operations, the circulating load may be the major part of the feed. The present trend is to use high circulating loads. The approximate relationship between the production of fines and circulating load is shown in Fig. 43.

The *capacity of ball mills* depends very largely on the reduction ratio as well as on the hardness of the material, and it cannot be accurately calculated. A reasonably conservative estimate of the capacity of a cylindroconical (Hardinge type) ball mill in tons per 24 hr is

$$\frac{\text{Maximum diameter} \times \text{Length (ft)}}{C_1}$$

where C_1 varies from 6 to 3 for most normal operations.

The normal capacity of cylindrical ball mills in tons per 24 hr may be estimated as

$$\frac{\text{Volume of mill (cu ft)}}{C}$$

where C usually varies from about 1 to 2.

ENERGY REQUIREMENTS

Although most of the power required for driving crushers and grinders is used in overcoming mechanical friction, the actual energy used in size reduction is an important consideration and theoretically is proportional to the new surface produced, as there is no change in the material except size and the

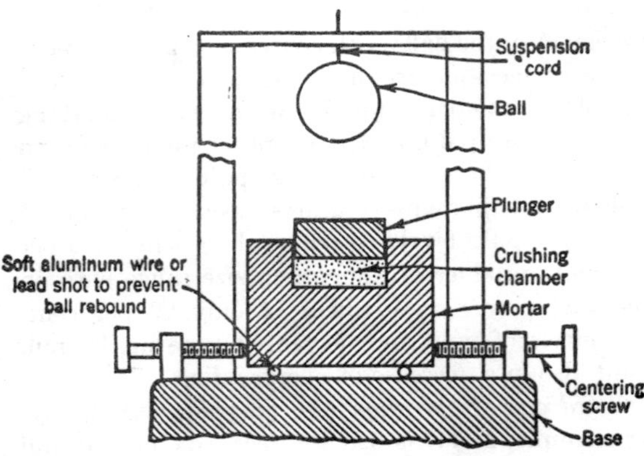

Fig. 44. Diagram of a drop weight crusher.[2]

creation of new surface. This principle was first recognized by Rittinger.[4] Rittinger's law was first confirmed beyond doubt * by the U. S. Bureau of Mines.[2] A drop weight crusher (Fig. 44) was used for accurate determination of the energy expended in crushing, and a rate of solution method for accurate determination of the surface of the particles. The results of their measurements on quartz (SiO_2),

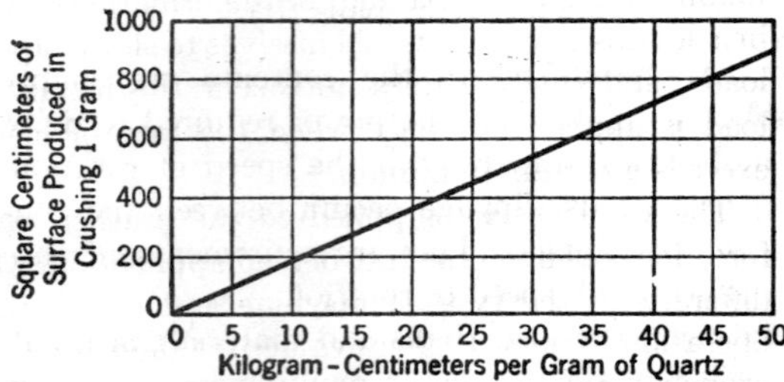

Fig. 45. Relation of energy input to surface produced in crushing quartz with a drop weight crusher.[2]

plotted in Fig. 45, show a constant energy requirement of 1 kg-cm for each 17.56 sq cm of new surface produced for this quartz, or, as usually expressed, 17.56 sq cm of new surface produced by the application of 1 kg-cm of mechanical energy.

Rittinger's number designates the new surface produced per unit of mechanical energy absorbed by the material being crushed. The values vary for different materials, depending on the elastic constants and their relation to the ultimate strength and on the manner or rate of application of the crushing force. A few values of Rittinger's number as determined by a drop weight crusher are given in Table 9.

TABLE 9. DROP WEIGHT RITTINGER'S NUMBER FOR A FEW COMMON MINERALS[2]

Mineral	Rittinger's Number		
	sq in./ft-lb	sq cm/ft-lb	sq cm/kg-cm
Quartz (SiO_2)	37.7	243	17.56
Pyrite (FeS_2)	48.7	314	22.57
Sphalerite (ZnS)	121.0	780	56.2
Calcite ($CaCO_3$)	163.3	1053	75.9
Galena (PbS)	201.5	1300	93.8

The energy absorbed in crushing mixtures of these minerals can be calculated by addition if the proportion of each mineral is known in the various screen fractions before and after crushing. The most rapid means of estimating the new surface produced is by the use of screen analyses as discussed in Chapter 3. Other methods, such as the rate of solution, are more precise but more difficult to execute.

* The principle known as Kick's law, that "the energy required to produce analogous changes of configuration of geometrically similar bodies varies as the volumes or masses of these bodies," was at one time erroneously applied in the theory of crushing. It led to the false conclusion that the energy required in crushing was proportional to the decrease in volume or mass of the particles. This principle is now recognized as applicable only to plastic deformation of particles within the elastic limit and not to crushing.

The mechanical energy supplied to the crusher is always greater than that indicated by Rittinger's number, as friction losses and inertia effects in the equipment require more energy than the actual production of new surface. Also, fracture is accomplished, not by static loading, but by exceeding the minimum rate of loading or deformation. Even brittle substances adjust themselves to slowly applied loads, and fracture does not occur the instant the load is applied but only when the rate of loading exceeds a certain minimum.

The total energy supplied to the crusher, therefore, depends upon the rate of load application, which differs with the type of machine and conditions of operation. Table 10 gives values for the new surface produced per unit of energy supplied to the material being crushed in a laboratory ball mill operated at the same speed but with varying weights of similar balls in the machine while grinding equal weights of quartz.

TABLE 10. EXPERIMENTAL VALUES OF NEW SURFACE PRODUCED PER UNIT OF ENERGY FOR QUARTZ

Calculated by subtracting the energy required to drive the mill containing balls but no material from the total energy required to drive the operating mill for the same length of time.

Total Weight of Ball in Ball Mill, lb	sq in./ft-lb	sq cm/ft-lb	sq cm/kg-cm
36	5.6	36	2.6
71	10.1	65	4.6
142	12.7	82	5.9
178	14.6	94	6.8
249	12.1	78	5.6
Drop weight method	37.6	243	17.56

The new surface produced per unit of energy supplied to the material being crushed in a ball mill is much less than for the drop weight crusher. This may be explained by the high percentage of ineffective blows and other losses in the ball mill. The important practical point is the variation in effectiveness of size reduction with the total weight of balls charged, showing a maximum value at about 175 lb of balls in this particular mill.

Values of the Rittinger number as determined in the drop weight crusher represent maximum effectiveness in size reduction and may be used in calculating the *crushing effectiveness* for any such operation. In the ball mill with 178 lb of balls, the crushing effectiveness is $94/243 = 0.387$. In this manner the performance of various machines, and variations in the same machine, can be compared.

The overall energy effectiveness (or efficiency) of a crusher is always much less than the crushing effectiveness, as the latter does not include the mechanical losses such as friction and inertia. The capacity of ball mills cannot be accurately calculated because of the effects of variables such as the relative grindability of the material and the range in size reduction. An approximate idea of the capacity and power requirements of ball mills, both cylindrical and conical, may be gained by reference to Table 11.

TABLE 11. CAPACITY AND POWER REQUIREMENTS OF BALL MILLS

Size, ft, diameter × length	Approximate Ball Load, lb	Approximate Rpm	Approximate Average Capacity, tons/24 hr			Motor Horsepower
			½ in. to 48 mesh	½ in. to 65 mesh	¼ in. to 100 mesh	
3 × 2	1,000	35	[illegible]	9	5	6–8
3 × 4	2,000	35	24	18	10	12–15
4 × 4	3,300	30	42	30	20	20–25
5 × 4	5,000	29	80	55	30	30–40
5 × 6	7,500	29	120	85	50	40–50
6 × 3	6,000	25	125	85	50	50–60
6 × 5	10,000	25	210	150	90	75–100
6 × 6	12,000	25	250	175	100	90–120
6 × 12	24,000	25	500	340	200	150–200
7 × 6	21,600	23	500	350	200	110–160
8 × 6	28,000	22	620	450	260	150–225
10 × 9	74,000	17	1500	1100	650	550–600
Cylindroconical Mills						
2 × ¾	600	40	4	3	2	2
3 × ¾	1,100	35	12	10	9	5–8
3 × 2	2,000	35	17	15	13	10
5 × 3	9,500	28	100	80	60	40–50
6 × 3	15,000	24	180	120	90	60–75
7 × 4	27,000	23	300	220	150	125
8 × 4	38,000	21	480	350	270	175–200
12 × 6	110,000	16	1800	1400	1000	700–800

Illustrative Example. A ball mill operating in closed circuit with a 100-mesh screen gives the screen analyses below. The ratio of the oversize to the undersize (product) stream is 1.0705 when 200 tons of galena are handled per day.

The ball mill requires 15.0 hp when running empty (with the balls but without galena) and 20.0 hp when delivering 200 tons per day of galena. Find:

1. The effectiveness of crushing based on drop weight crushing as 1.00.
2. The overall energy efficiency.

5. The classifying screen effectiveness.

Mesh	Mill Feed, weight %	Oversize from Screen, weight %	Undersize from Screen, weight %
− 4 + 6	1.0	0	0
− 6 + 8	1.2	0	0
− 8 + 10	2.3	0	0
− 10 + 14	3.5	0	0
− 14 + 20	7.1	0	0
− 20 + 28	15.4	0	0
− 28 + 35	18.5	13.67	0
− 35 + 48	17.2	32.09	0
− 48 + 65	15.6	27.12	0
− 65 +100	10.4	20.70	2.32
−100 +150	6.5	4.35	14.12
−150 +200	1.3	2.07	13.54
−200	0	0	70.02
	100.00	100.00	100.00

Solution.

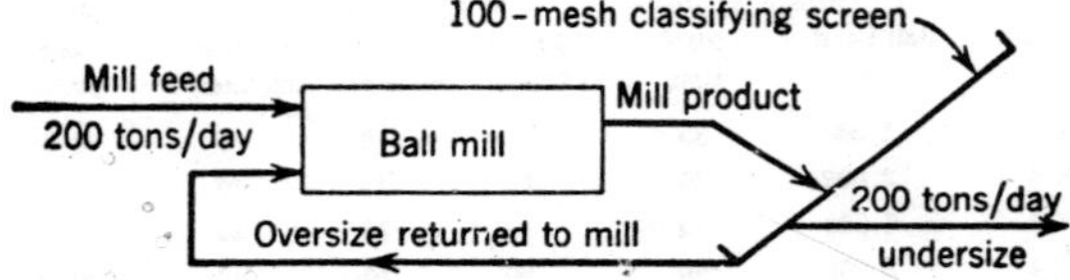

Analysis. The Rittinger number measures the minimum energy required to form new surface. If the new surface created per unit time is calculated, then the minimum energy required for the formation can be calculated.

In order to evaluate the new surface of the product, the minus 200-mesh fractions may be evaluated by the straight-line plot, such as Fig. 15. This method is valid only for the product of a mechanical crushing device and not for the classified product. Therefore the size distribution of the mill product must be computed and extrapolated for the mass fractions retained below 200 mesh. The sum of these fractions must equal the minus 200-mesh fraction.

The fractions of mill product below 200 mesh are then converted to fractions of the undersize stream. The surfaces of the fractions are calculated either from the actual specific surfaces in Fig. 16 or from the relationship (p. 22):

$$\text{Total surface} = \frac{6}{\rho}\sum \frac{n_i m_i}{(D_{avg})_i}$$

n is evaluated from the data of Fig. 17.

Theoretical effectiveness of ball mill

$$= \frac{\text{Minimum power required to create new surface}}{\text{Power increase due to charge}}$$

Overall energy effectiveness of ball mill

$$= \frac{\text{Minimum energy required to create new surface}}{\text{Total energy used}}$$

Effectiveness of classifying screen

$$= \frac{x_P(x_F - x_R)}{x_F(x_P - x_R)}\left[1 - \frac{(1 - x_P)(x_F - x_R)}{(1 - x_F)(x_P - x_R)}\right]$$

Mesh	Calculated Size Distribution: Mass % in Oversize	Mass % in Undersize	Mass in Mill Product per 100 lb of Oversize, Oversize + (Undersize/1.0705), lb	Size Distribution, Mass % in Mill Product
− 28 + 35	13.67	0	13.67	7.07
− 35 + 48	32.09	0	32.09	16.60
− 48 + 65	27.12	0	27.12	14.02
− 65 +100	20.70	2.32	22.86	11.82
−100 +150	4.35	14.12	17.52	9.07
−150 +200	2.07	13.54	14.72	7.62
−200	0	70.02	65.35	33.80
	100.00	100.00	193.33	100.00

Mass % in mill product

$$= \frac{\text{Mass \% oversize} + \dfrac{\text{Mass \% undersize}}{1.0705}}{193.33}$$

The distribution of material smaller than 200 mesh in mill product from an extrapolation on a plot of

log (mass retained, per cent) versus log (D_{avg})

is given below.

Average Diameter (D_{avg}), microns	Mass % in Mill Product	Mass % in Undersize	n	$\dfrac{mn}{D_{avg}}$
63	6.98	14.46	1.65	0.378
45	5.90	12.22	1.60	0.434
31.8	5.00	10.34	1.53	0.497
22.5	4.24	8.78	1.50	0.585
15.9	3.58	7.41	1.45	0.676
11.2	3.03	6.28	1.42	0.795
7.92	2.68	5.58	1.40	0.985
5.59	2.19	4.54	1.375	1.117
3.84	0.20	0.41	1.35	0.144
	33.80	70.02		$5.611 = \sum \dfrac{mn}{D_{avg}}$

Surface area for this fraction of the undersize stream

$$= \frac{5.611 \times 10^4 \times 6}{7.43} = 45{,}250 \text{ sq cm/70.02 grams}$$

Screen Mesh	Mass % in Undersize	Specific Surface, sq cm/gram	Actual Surface, sq cm
− 65 +100	2.32	85.8	199.1
−100 +150	14.12	115.4	1,629.4
−150 +200	13.54	155.1	2,100.0
−200	70.02		45,250.0
	100.00		49,180 sq cm surface area/100 grams of undersize

FEED SURFACE CALCULATIONS

Screen Mesh	Mass % in Feed	Specific Surface, sq cm/gram	Actual Surface, sq cm/100 grams
− 4 + 6	1.0	7.6	7.6
− 6 + 8	1.2	9.9	11.9
− 8 + 10	2.3	12.5	28.8
− 10 + 14	3.5	16.4	57.4
− 14 + 20	7.1	21.1	149.8
− 20 + 28	15.4	26.9	414.5
− 28 + 35	18.5	35.7	660.5
− 35 + 48	17.2	47.2	811.8
− 48 + 65	15.6	63.0	982.8
− 65 +100	10.4	85.8	892.3
−100 +150	6.5	115.4	750.1
−150 +200	1.3	155.1	201.6
			4,968.8 sq cm total surface/100 grams of feed

$$\text{New surface created} = (49{,}180 - 4968)$$

$$= 44{,}212 \text{ sq cm/100 grams of feed}$$

$$\text{Theoretical effectiveness} = \frac{(44{,}212)(9072)(200)}{(6.452)(1.98)(10^6)(24)(201.5)(5)}$$

$$= \frac{1.265 \text{ hp}}{5 \text{ hp}} = 0.253$$

$$\text{Overall energy effectiveness} = \frac{1.265}{20} = 0.0633$$

SCREEN EFFECTIVENESS CALCULATIONS

$$x_P = 1 - 0.0232 = 0.9768$$

$$x_F = 0.0907 + 0.0762 + 0.3380 = 0.5049$$

$$x_R = 0.0642$$

Screen effectiveness

$$= \frac{(0.9768)(0.4407)}{(0.5049)(0.9126)}\left[1 - \frac{(0.0232)(0.4407)}{(0.4951)(0.9126)}\right] = 0.914$$

BIBLIOGRAPHY

1. Gaudin, A. M., *Principles of Mineral Dressing*, McGraw-Hill Book Co. (1939).
2. Gross, John, "Crushing and Grinding," *U. S. Bur. Mines Bull.* 402 (1938). Contains complete bibliography.
3. Richards, R. H., and C. E. Locke, *Textbook of Ore Dressing*, 3rd ed., McGraw-Hill Book Co. (1939).
4. von Rittinger, P. R., *Lehrbuch der Aufbereitungskunde*, Berlin (1867).
5. Taggart, A. F., *Handbook of Mineral Dressing*, John Wiley and Sons (1945).
6. Davis, E. W., "Ball Mill Crushing in Closed Circuit with Screens," *Bull. Univ. Minn.*, **28**, No. 42 (1925), *Bull.* 10, *School of Mines Exp. Sta.*

PROBLEMS

1. A short-head cone crusher is available for crushing 2 tons of pyrites per hour. On similar materials, the overall energy efficiency has been found to be 3.15 per cent. The raw feed is to be crushed by a jaw crusher, whose product constitutes the feed to the cone crusher. The cone crusher operates in closed circuit with a 14-mesh screen. The cone crusher product and recycle stream analyses are given below. On the basis of calculations and of any assumptions which you may find necessary, select a Dodge-type crusher that will do the job.

The surface ratio (n) may be considered to be 6.5 above 3 mesh. The full load energy requirement for the short-head cone crusher is 5 hp. The recycle ratio (recycle stream/product stream) is 1.

Product Mesh	Product Mass %	Recycle Stream Mesh	Recycle Stream Mass %
−14 + 20	29.8	3 + 4	3.8
−20 + 28	30.2	− 4 + 6	10.0
−28 + 35	25.0	− 6 + 8	19.6
−35 + 48	9.6	− 8 +10	26.0
−48 + 65	3.8	−10 +14	36.6
−65 +100	1.6	−14 +20	4.6
	100.0		100.0

2. A ball mill, operated in a closed circuit with a classifier, is used to grind calcite after it has had preliminary crushing in jaw crushers. Screen analyses of the various streams are given below.

The ball mill feed (25 tons/hr) is estimated to have a specific surface of 292 sq cm/gram. When the ball mill is operated with a recycle of 75 tons/hr, 75 kw are required to drive the ball mill. Determine the efficiency of the ball mill.

Tyler Screen Mesh	Ball Mill Feed, mass % retained	Recycle, Classifier Sands, mass % retained	Product, Classifier Overflow, mass % retained
0.525 in. − 0.371 in.	4.7		
0.371 in. − 3 mesh	20.1	6.3	
− 3 mesh + 4 mesh	17.9	7.0	
− 4 + 6	12.1	8.2	
− 6 + 8	8.6	9.3	
− 8 + 10	5.5	3.0	
− 10 + 14	4.7	15.4	
− 14 + 20	2.7	16.9	
− 20 + 28	3.5	20.7	
− 28 + 35	2.9	3.4	4.2
− 35 + 48	1.9	2.8	12.7
− 48 + 65	2.0	1.4	19.3
− 65 +100	1.7	1.2	13.7
−100 +150	1.7	0.8	11.7
−150 +200	1.5	0.6	9.8
−200	8.5	3.0	28.6
Totals	100.0	100.0	100.0

3. A cement plant is grinding 10 tons/hr of a hard rock (specific gravity, 3.8) in a high-speed disk grinder operating in a closed circuit with a 65-mesh screen. Regular checks upon the possibility of oversize particles passing through the screen show that all material in the undersize stream from the screen will pass through a 35-mesh screen.

Drop weight laboratory tests upon the material being crushed indicate that the absorption of 1 ft-lb of energy will result in creation of 110 sq cm of new surface and that the specific surface ratios are identical with those of sphalerite.

(*a*) If the energy efficiency of the grinder is 18 per cent and the known streams have the analyses given below, what is the horsepower required by the grinder?

(*b*) What is the effectiveness of the screen?

Mesh	Raw Feed to Grinder, mass fraction	Discharge from Grinder, mass fraction	Oversize from Screen, mass fraction
− 3 + 4	0.05		
− 4 + 6	0.10		
− 6 + 8	0.20		
− 8 + 10	0.30		
− 10 + 14	0.20	0.04	0.05
− 14 + 20	0.10	0.08	0.10
− 20 + 28	0.05	0.16	0.20
− 28 + 35		0.24	0.30
− 35 + 48		0.17	0.2025
− 48 + 65		0.10	0.0975
− 65 +100		0.08	0.05
−100 +150		0.06	
−150 +200		0.04	
−200 +270		0.02	
−270 +400		0.01	

4. A feed of 150 tons/day of pyrites must be comminuted from the material size given below as feed (the product from a controlling screen) to the size range given below as product (the feed to a reduction process). A ball mill is to be used. It will be loaded with balls to operate at a crushing effectiveness of about 32 per cent.

(*a*) What size cylindrical mill should be selected?

(*b*) What size motor will be needed to drive it?

(*c*) What is the overall energy efficiency?

Mesh	Feed, mass fraction	Product, mass fraction
− 3 + 4	0.036	
− 4 + 6	0.192	
− 6 + 8	0.365	
− 8 + 10	0.284	0.010
− 10 + 14	0.123	0.072
− 14 + 20		0.228
− 20 + 28		0.295
− 28 + 35		0.170
− 35 + 48		0.098
− 48 + 65		0.072
− 65 +100		0.046
−100 +150		0.009
−150 +200		0.002

5. Quartz from the mine is sent over a grizzly with 8-in. spacing and then to a Blake standard jaw crusher with a 40-in. by 42-in. feed opening and a 6-in. discharge setting. The crusher operates at 190 rpm and handles 130 tons/hr of feed. Screen analyses of the feed and product are given below.

(*a*) What are the theoretical power requirements?

(*b*) What size motor is recommended? Why?

(In the size range indicated the average surface ratio may be assumed to be 8.0)

Feed: Screen Aperture, in.	Feed: Mass Fraction	Product: Screen Aperture, in.	Product: Mass Fraction
−34 +28.6	0.181	−6 +4.23	0.123
−28.6 +24.0	0.343	−4.23 +2.99	0.248
−24.0 +20.3	0.220	−2.99 +2.11	0.167
−20.3 +17.0	0.165	−2.11 +1.49	0.105
−17.0 +14.3	0.054	−1.49 +1.05	0.068
−14.3 +12.0	0.037	−1.05 +0.81	0.051
		−0.81 +0.57	0.046
		−0.57 +0.403	0.039
		−0.403 +0.285	0.033
		−0.285 +0.201	0.028
		−0.201 +0.142	0.025
		−0.142 +0.100	0.023
		−0.100 +0.0707	0.018
		−0.0707 +0.0500	0.016
		−0.0500 +0.0353	0.010

6. In an attempt to evaluate the efficiency of a 24-in. by 15-in. Blake jaw crusher, a set of coarse analytical screens was constructed from welded steel rods. The standard Tyler $\sqrt{2}$ relationship between screen apertures was maintained in this series of large screens.

Calcite was fed to the crusher at the rate of 60 tons/hr. The discharge setting of the jaws was 5 in. The crusher was driven by a 35-hp motor. Screen analyses of the feed and the product are given in the table below.

(*a*) Calculate efficiency of the crusher, assuming the motor was operating at an average of ⅕ its rating.

(*b*) How many tons per hour of galena could be fed to the crusher and reduced over the same size range with the same power?

(*c*) What is the capacity according to Taggart's formula?

Aperture of Screen, in.	Feed, mass fraction on screen	Product, mass fraction on screen	Specific Surface Ratio *n* for Average Diameter of Material on Screen
22.6	0.0	0.0	
16.75	0.15	0.0	10.0
11.85	0.35	0.0	9.7
8.40	0.25	0.0	9.5
5.93	0.15	0.0	9.0
4.20	0.10	0.05	8.6
2.97	0.0	0.20	8.0
2.10	0.0	0.45	7.2
1.48	0.0	0.25	6.6
1.05	0.0	0.05	6.2

7. A grinder is to be used to reduce a siliceous ore of the feed size shown below. Laboratory tests on similar equipment indicate that the product size given below will be satisfactory, and that the grinder is approximately 8 per cent efficient in converting input energy into size reduction as evidenced by an increase in surface.

It is estimated that a crusher to handle 10 short tons/hr will cost about $4000. If the crusher operates on a 24-hr basis for 300 days/yr, it is estimated that maintenance costs, overhead, and ordinary replacement costs will be about 50 per cent of power costs. Electric power costs 2 cents/kwhr.

If this machine depreciates on a straight-line basis and its life is estimated at 10 yr, what is the processing cost per ton of ore?

Tyler Mesh	Feed, mass fraction	Product, mass fraction
− 6 + 8	0.143	
− 8 + 10	0.211	
− 10 + 14	0.230	
− 14 + 20	0.186	0.098
− 20 + 28	0.120	0.234
− 28 + 35	0.076	0.277
− 35 + 48	0.034	0.149
− 48 + 65		0.101
− 65 +100		0.068
−100 +150		0.044
−150 +200		0.029

8. A roll crusher is to be used to crush medium hard quartz (specific gravity, 2.65). The product from the crusher is to be fed to a number of rod mills (6 ft by 12 ft) at the rate of 8 tons/hr to each mill. The power consumption of each rod mill is 160 hp, with an overall energy effectiveness or efficiency of 2.0 per cent.

The rod mills operate in closed circuit with a 48-mesh screen. The ratio of recycle to product is 1:1.

If the surface ratio n for quartz is 10 for all sizes above 3 mesh, determine the setting (distance between the rolls) in the roll crusher.

Classifier Product		Recycle Stream	
Mesh	Mass Fraction	Mesh	Mass Fraction
− 35 + 48	0.05	−20 +28	0.05
− 48 + 65	0.80	−28 +35	0.10
− 65 +100	0.10	−35 +48	0.80
−100 +150	0.05	−48 +65	0.05

9. Five tons of a hard rock (specific gravity, 3.8) are fed every hour to a cone crusher in closed circuit with a 48-mesh screen. Regular checks upon the possibility of oversize particles passing through the screen show that all material in the undersize stream from the screen will pass through a 28-mesh screen.

Drop weight laboratory tests upon the material being crushed indicate that the absorption of 1 ft-lb of energy will result in the creation of 110 sq cm of new surface. These tests also indicate that the surface area ratios for the material are identical with those of sphalerite.

If the energy efficiency of the grinder is 18 per cent and the known streams have the analyses given below, what is the horsepower required by the grinder? What is the effectiveness of the screen?

Mesh	Raw Feed to Grinder, mass fraction	Discharge from Grinder, mass fraction	Oversize from Screen, mass fraction
− 3 + 4	0.10		
− 4 + 6	0.20		
− 6 + 8	0.40		
− 8 + 10	0.20		
− 10 + 14	0.10	0.02	0.03
− 14 + 20		0.04	0.06
− 20 + 28		0.06	0.09
− 28 + 35		0.25	0.35
− 35 + 48		0.30	0.30
− 48 + 65		0.20	0.08
− 65 +100		0.06	0.05
−100 +150		0.04	0.04
−150 +200		0.03	

10. Quartz goes through two successive grinders on the same shaft which draws a total of 20 hp. The feed averages 2 in. in diameter and has a surface ratio n of 10. The grinders running empty require 2 hp. Their capacity is 3 tons/hr. The analyses of their products are given below.

(*a*) Calculate the horsepower used in each grinder.

(*b*) Calculate the efficiency of the grinders if Rittinger's number (new surface produced per unit of energy) is 37.6 sq in./ft-lb.

Primary Grinder

Mesh	%
− 4 + 8	20
− 8 + 14	30
−14 + 28	30
−28 + 48	15
−48 +100	5

Final Grinder

Mesh		%
− 28	+ 48	10
− 48	+100	20
−100	+200	30
−200 mesh	+ 0.001 in.	30
− 0.001 in.	+ 0.0003 in.	10

11. A Hardinge mill is grinding cement clinker (specific gravity, 2.2) in an open circuit at the rate of 20 tons/hr. All grinder product must pass 48 mesh, and none is to be wasted. A total of 375 hp is required by the mill, with 5 hp

needed to run the mill empty. Following is a size analysis of the feed and product.

Mesh	Feed, mass %	Product, mass %	n
− ½ in. + ⅜ in.	25.0	....	7.2
− ⅜ in. + 3	27.3	....	6.1
− 3 + 8	19.2	....	4.2
− 8 + 20	19.6	....	3.1
− 20 + 48	8.9	....	2.9
− 48 + 65	....	2.1	2.7
− 65 +100	....	8.2	2.3
−100 +200	....	21.9	1.9
−200 mesh + 0.001 in.	....	67.8	1.6

A closed circuit system is suggested as a means of reducing power costs. A laboratory test indicates the following results could be obtained with a closed circuit unit using the above feed and screening the grinder product on a 48-mesh screen (1 kwhr = 1.341 hp-hr).

Mesh	Screen Feed, mass %	Oversize, mass %	Undersize, mass %
− 3 + 8	4.1	9.1	
− 8 + 20	13.2	29.3	
− 20 + 48	14.1	31.4	Nothing
− 48 + 65	20.2	23.6	coarser
− 65 +100	24.3	6.6	than
−100 +200	16.2	....	48 mesh
−200 mesh + 0.001 in.	7.9	....	

If electricity costs 1 cent/kwhr, how much in power costs could be saved per year by a closed circuit if 480 tons of ground clinker are produced daily, 365 days a year?

What is the energy consumption per hour if the same feed rate (20 tons/hr) is used in the closed circuit?

12. The double-roll toothed crusher of a coal corporation yields a product of the indicated screen analysis when crushing 239 tons/hr using 38 hp.

(*a*) If the surface ratio n is 2, what is Rittinger's number for this particular installation?

(*b*) From this calculation and any data you can find in the handbooks, what is your conclusion as to the efficiency of this crusher?

Screen Size	Feed, mass %	Product, mass %
+2¾ in.	16.88	.....
2¾ × 2	34.62	.....
2 × 1¾	10.82	.....
1¾ × 1½	14.80	0.39
1½ × 1¼	11.12	3.24
1¼ × 1	5.77	12.10
1 × ¾	1.75	16.26
¾ × ⅝	1.44	18.84
⅝ × ⅜	1.04	21.14
⅜ × ¼	0.42	8.02
¼ × 3/16	0.21	5.43
3/16 × 0	1.13	14.10
3/16 × 6 mesh	....	2.83
6 × 8	....	2.42
8 × 14	....	3.35
14 × 28	....	2.09
28 × 48	....	1.28
48 × 0	....	2.13

CHAPTER

5

Handling of Solids

HANDLING of materials deals with their movement over relatively short distances such as from cargo ship and freight car to storage, from plant to plant, or from one piece of apparatus to another, as distinct from the transportation of materials over relatively long distances such as by ship, rail, or motor transport.

Unassisted manpower involving shoveling, pushing, carrying, and lifting is recommended for moving solid materials short distances, usually not exceeding 10 ft, and only when the quantity does not exceed 4 or 5 tons. In exceptional cases, as in an emergency or when the operation is not repetitive, larger quantities may be so moved.

Under normal conditions a man can do work at the rate of about 1500 to 3000 ft-lb/min; for example, an average 150-lb man can lift 20 lb on the end of a shovel a distance of 5 ft within a radius of 5 ft at an average continuous rate of 15 strokes per minute. The handling rate varies considerably with individuals, with the nature of the material being handled, and with the method of handling, but it always decreases rapidly when the material must be lifted higher than 3 to 5 ft or moved a distance greater than 5 to 10 ft. Lifting of more than 100 to 150 lb per man should be avoided as it may result in a strain or rupture.

Assisted manpower with wheelbarrows, two-wheeled hand trucks, four-wheeled floor trucks, or carts is recommended when the radius of work is increased to 100 to 200 ft. Depending on the device, an average man can move 200 to 1000 lb across a horizontal surface at the average rate of 1½ mph, or 130 fpm, excluding loading, unloading, and return. The rolling coefficient of friction usually will be about 0.02 to 0.03. Loading and unloading time varies greatly with the material being handled but runs approximately ¼ to 1 min/100 lb of material. For distances greater than about 200 ft, power trucks or tractor-drawn trucks are more efficient.

To lift loads weighing more than 100 to 150 lb, various mechanical devices are employed to amplify the force exerted by the workman. All these devices, such as chain falls, pulley blocks, hand-power hoists, and jacks, are based on the lever principle. For example, in raising a 1-ton load by means of a chain fall, a man would pull about 50 lb and would pass about 100 ft of chain through his hands per minute, elevating the load about 2 fpm (80% eff.).

The rates given apply to continuous work. It is possible to double or triple the rate of work for short intervals if they are followed by a rest period of equivalent length.

PORTABLE POWER-DRIVEN MACHINES

Electric-storage-battery trucks are available in various capacities, the most common being the 2-ton truck with a speed of about 4 to 6 mph. The general-purpose truck (Fig. 46) has a fixed platform on which the load is carried. The lift-platform truck is a similar unit with a platform that is elevated by power. It is used primarily for handling material on skids. Special types of trucks may be obtained, such as the crane truck, dump body truck, tiering or high-lift trucks. Electric trucks will run about 8 hr without recharging or battery substitution. Ample charging facilities and battery storage must be provided.

Gasoline-powered industrial trucks are similar to the electric trucks but have speeds up to 15 mph.

Fig. 46. Electric-storage-battery truck, fixed-platform type. (*Yale and Towne Mfg. Co.*)

In general, the gasoline trucks are better adapted for longer hauls outdoors, whereas the electric trucks are ideal for short hauls indoors.

Trailers are frequently used in conjunction with industrial trucks. The larger trailers are generally drawn by tractors. Rubber-tired tractors are employed for hauling across relatively smooth and paved surfaces, whereas tractors with caterpillar treads are used over rough terrain. In rough terrain skids often replace trailers.

Tractors, which in themselves carry no load but draw trailers or material on skids, are extremely valuable general-purpose machines and are available in a large range of sizes.

Power shovels are widely used for handling large quantities of solid materials with portable equipment. Open-pit mining, excavation, and the handling of open-air storage piles are examples. Steam shovels are economical to operate, but, because of the extra man hours required to get up steam and other disadvantages of the boiler, the gasoline and diesel-power shovels have gained great favor. Electric shovels are often used where ample electric power is available. Power shovels operate on a turntable and have a boom which may be raised or lowered. The material to be handled is picked up by a thrust or drag motion, elevated with the boom, and rotated to the point of discharge by the turntable. Discharge

Fig. 47. Electric traveling gantry crane, 118-ft span with 50-ft overhang. (*Whiting Corp.*)

is by gravity and occurs when the scoop is inverted or a gate at the back of the scoop is opened. Power shovels are generally used in conjunction with commercial dump trucks but are often employed to load railroad cars directly. The newer shovels have caterpillar treads and can be moved from place to place under their own power. Power shovels are expensive and require skilled operators. Their purchase is warranted only when large quantities of bulk materials are being handled at changing locations. A local contractor can often supply this equipment for short periods or will contract to move a quantity of material.

Gantry or bridge cranes (Fig. 47) are often used for loading and unloading railroad cars or ships. These machines consist of two supporting towers with a bridge between. A trolley moves back and forth across the bridge; the entire crane is on wheels and can be moved along a track. The trolley has a grab bucket or lift. This type of equipment is widely used in the Great Lakes area for handling ore and coal. Electrical drive is the general practice. A skilled operator is required who generally rides in a cab on one of the towers. In the larger and newer designs the operator's cab is placed in the trolley.

PERMANENT INSTALLATIONS FOR HANDLING SOLIDS

Material that is being moved from one fixed location to another fixed location continuously or at frequent intervals usually can be more economically handled by some type of permanent installation. When the material must pass through a series of operations, gravity feed can often be used to advantage, the raw material being brought to the top of the building or structure by a conveyor or elevator and passing downward, operation by operation, until removed as product at the bottom. In ideal situations the only power requirement is for the original elevation of the raw material. Loose material is usually guided by means of a chute or slide. The chutes may be straight or spiral. The angle of the slide with the horizontal must be sufficient to overcome the frictional resistance. The coefficient of friction varies with different materials but is about 0.3 to 0.6 for most dry solids on steel slides. Material often becomes wedged between the sides of the chute, creating additional resistance, so that an angle of 45 degrees or steeper is desirable for the slide.

The *angle of slide* is the angle of minimum slope measured from the horizontal at which any loose solid material will flow. The *angle of repose* is the angle of maximum slope measured from the horizontal at which a heap of loose solid material will stand without sliding, approximately 17 degrees for wet clay earth, 27 degrees for anthracite coal, 31 degrees for fine sand, 35 degrees for bituminous coal, 39 degrees for dry earth, and 39 to 48 degrees for gravel.

Vibrators may be used to keep gravity slides free by giving the slide a vibration in the direction of flow. The most common type is the magnetic vibrator, operating on alternating current with the same frequency of vibration as the current frequency. The displacement of the slide is small, usually less than ⅛ in., but the frequency of vibration keeps the slide in motion relative to the solid material because of the inertia of the solid.

If a heavy spring is employed with the magnetic vibrator, the forces of the spring and magnet become additive in one direction and opposite in the other direction, causing a slower motion of the slide in the direction of the smaller force or acceleration.

The material on the slide tends to ride with it in the direction of slower movement. With this arrangement, gravity may be overcome and a solid material may be conveyed up an incline (Fig. 48).

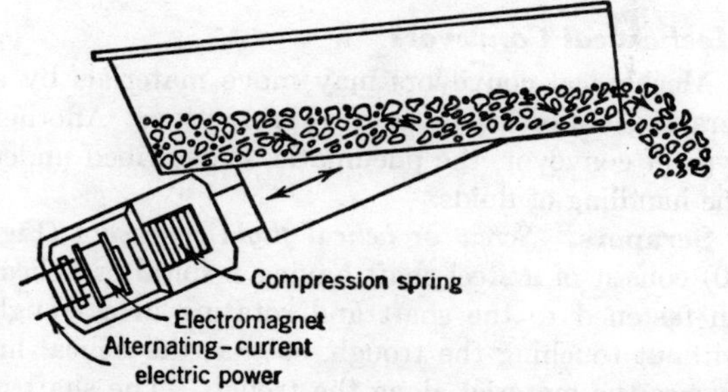

Fig. 48. Diagrammatic representation of magnetic vibrator feeding up an incline to right. The spring and electric magnet work together to give a fast return to the left and a slower movement to the right.

Vibrators are often used to obtain uniform feed from a hopper having an adjustable gate at the bottom. A great advantage of the electric vibrator is that it obtains motion of the slide without any mechanical parts requiring lubrication, resulting in low maintenance costs.

A mechanical feeder placed inside the tank or hopper is illustrated in Fig. 49. Lumps of material

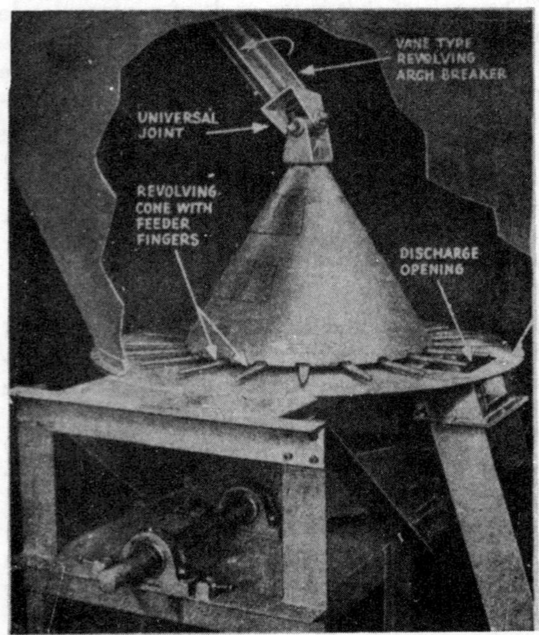

FIG. 49. Mechanical feeder to provide uniform rate of feeding solids from hopper. (*Stephens-Adamson Mfg. Co.*)

are broken by the revolving cone and moved at a uniform rate to the discharge opening by the feeder fingers.

Mechanical Conveyors

Mechanical conveyors may move materials by a scraping action or by a carrying action. Another type of conveyor, the pneumatic, is described under the handling of fluids.

Scrapers. *Screw* or *helical flight conveyors* (Fig. 50) consist of a steel shaft having a spiral or helical fin fastened to the shaft and rotating in a trough without touching the trough, so that the helical fin pushes the material along the trough. The shaft is driven by a motor through gears or a chain. The conveyors are made in sections from 8 to 12 ft in length that may be joined together to obtain the desired length. The torsional stress developed in the shaft usually limits a single drive to about 100 ft. Diameters vary from 3 to 24 in.

Screw conveyors are compact, requiring little headroom and no return mechanism. They are economical in original cost and maintenance. The material is also mixed as it passes through the conveyor. This is often an advantage. The tendency to crush friable materials may or may not be an advantage.

FIG. 50. One section of a screw or helical conveyor. (*Link-Belt Co.*)

For handling abrasive materials, the fin may be made of cast iron or tipped with a hard metal alloy such as manganese steel. The trough may be made considerably larger than the fin so that the material rubs on itself rather than on the trough, thereby reducing wear on the trough. In selection of the diameter of the conveyor, consideration must be given to the size of the lump, as shown in Table 12.

TABLE 12. MAXIMUM SIZE OF LUMPS FOR VARIOUS DIAMETERS OF SCREW CONVEYORS[3]*

Conveyor diameter, in.	3	4	6	9	12	14	16	18	20	24
Lumps 20 to 25 per cent of total, in.	3/8	1/2	3/4	1 1/2	2	2 1/2	3	3	3 1/2	3 1/2
All lumps, in.	1/4	1/4	1/2	3/4	1	1 1/4	1 1/2	2	2	2 1/2

Screw conveyors are used to handle a wide range of materials, such as grain, asphalt, crushed coal, ashes, gravel, and sand. A special type, the ribbon conveyor in which the center portion of the helical fin is absent, is suitable for sticky, gummy liquids, such as molasses, hot tar, and sugar.

The horsepower requirements of screw conveyors and of all conveyors operating with a scraping action may be estimated as the sum of the power to run the equipment alone, the power to overcome the friction of the material being moved, and the power to lift the material any vertical height. For screw conveyors the horsepower to run the equipment alone is essentially that required to overcome bearing and

* The bibliography for this chapter appears on p. 64.

TABLE 13. APPROXIMATE MAXIMUM CAPACITIES OF HORIZONTAL SPIRAL OR SCREW CONVEYORS[2]

Diameter of Screw, in.	Light Nonabrasive Material, e.g., Grain		Heavy Nonabrasive Material, e.g., Coal		Heavy Abrasive Material, e.g., Ash	
	Capacity, cu ft/hr	Maximum Rpm	Capacity, cu ft/hr	Maximum Rpm	Capacity, cu ft/hr	Maximum Rpm
3	74	250	37	125	....	..
4	171	220	86	110	46	90
5	304	210	150	105	85	85
6	500	200	255	100	135	80
7	820	190	410	95	200	75
8	1180	180	590	90	300	75
9	1600	175	780	85	400	70
10	2050	160	1030	80	516	65
12	3300	150	1660	75	820	60
14	4000	140	2000	70	1200	55
16	7000	130	3400	65	1630	50
18	9000	120	4500	60	2100	45
20	12000	115	5800	55	2860	46

gear friction and is proportionally small. For horizontal movement the total horsepower may be estimated as follows.[2]

Horsepower

$$= \frac{(\text{Coefficient})(\text{Capacity, lb/min})(\text{Length, ft})}{33{,}000}$$

where coefficient = 4.0 for ashes.
= 2.5 for coal.
= 1.3 for grain.

Flight conveyors consist of one or two endless chains passing through a trough or set of guides (Fig. 51). The chains have plates of wood or steel called flights attached at regular intervals. The flights are shaped to fit the trough. The chains pull the flights and the material along the trough and pass over sprockets at the end of the run, one of the sprockets acting as the drive. Various designs are used. Some flights scrape on the trough bottom (Fig. 51). Some with two chains support the flights by the chains or rollers or by arms riding on a guide track (Fig. 52). One type has a pair of rollers on each flight and a single connecting chain at the center. Either or both the bottom and upper run may be used in handling material; however, it is most common to employ the lower run. Speeds of 100 fpm are common, but the speed may vary from 25 to 200 fpm.

Flight conveyors may be used on inclines up to 45 degrees, but preferably the incline should be limited

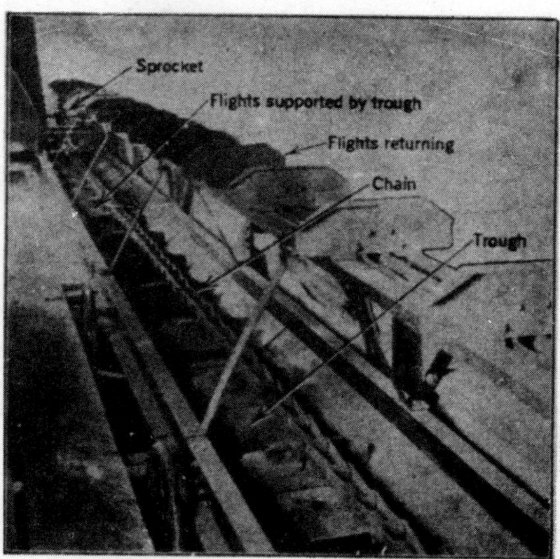

Fig. 51. Flight conveyor with flights supported on the trough. *(Link-Belt Co.)*

to 30 degrees. The conveyors are widely used for loose material that is nonabrasive, such as grain, food waste, garbage, and coal, but they are not suitable for friable abrasive material, such as clinkers, gravel, or crushed ore.

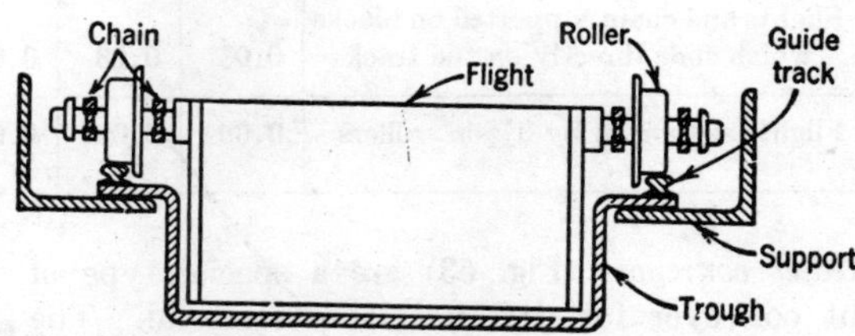

Fig. 52. Sectional diagram of two-chain roller-supported flight conveyor.

Capacities of flight conveyors may be estimated from the following formula.[1] (80% eff.).

$$T = \frac{BDS\rho_b}{6000}$$

where T = tons/hr.
B = width of flight (in.).
D = depth of flight (in.).
S = speed of conveyor (fpm).
ρ_b = bulk density of material handled (lb/cu ft).

This expression applies to horizontal conveyors and to conveyors inclined to an angle not exceeding the angle of repose for the material handled.

The horsepower required to drive flight conveyors may be estimated:

$$hp = \frac{a(T)(L) + b(W)(L)(S) + 10L}{1000}$$

in which hp = total horsepower at conveyor shaft.
T = material handled (tons/hr).
L = length of conveyor from center to center of sprocket (ft).
W = total weight (lb) of chain and flights per foot of distance between centers (both runs). Usually about equal to $T_{max}/4$. See Table 16A.
S = speed (fpm).
a = constant for material (Table 14).
b = constant for conveyor (Table 14).

occupy only a portion of the entire cross-sectional area of the duct. The duct is kept filled or partially filled with material. The movement of the flights conveys the material in the duct because the friction between the particles of the material is greater than the friction between the particles and the smooth walls of the conveyor duct. It is used primarily for handling finely divided or powdered material such as flour, cement, clay, and loose material such as sand, coal, grain, and breakfast foods. One of its major advantages is its ability to elevate vertically (Fig. 54) as well as to convey horizontally. The closed duct may also be important.

Slat or drag conveyors (Fig. 55) consist of one or two chains to which are attached cross bars, usually of wood, which drag on a flat-bottom trough. They are

TABLE 14. POWER CONSTANTS FOR FLIGHT CONVEYORS [2]

	Inclination with Horizontal	0°	5°	10°	15°	20°	25°	30°	35°	40°	45°
	Anthracite	0.343	0.42	0.50	0.586	0.66	0.73	0.79	0.85	0.90	0.945
a	Bituminous	0.60	0.69	0.76	0.83	0.88	0.95	1.02	1.08	1.13	1.15
	Ashes	0.54	0.62	0.72	0.80	0.85	0.90	0.97	1.03	1.06	1.10
b	Flights and chain supported on blocks which slide directly on the track	0.03	0.03	0.03	0.029	0.028	0.027	0.026	0.025	0.023	0.020
	Flights supported by 3½-in. rollers	0.004	0.004	0.004	0.004	0.004	0.004	0.003	0.003	0.003	0.003

Redler conveyors (Fig. 53) are a special type of flight conveyor for dry and loose material. The conveyor (Fig. 53) consists of a metal duct which may be circular, rectangular, or square, through which a chain passes carrying flights. The flights

used for conveying loose material such as chips, sawdust and refuse.

The simplest form of the drag conveyor is the drag chain, consisting of simply one or more endless chains running in a trough, generally of wood. No flights

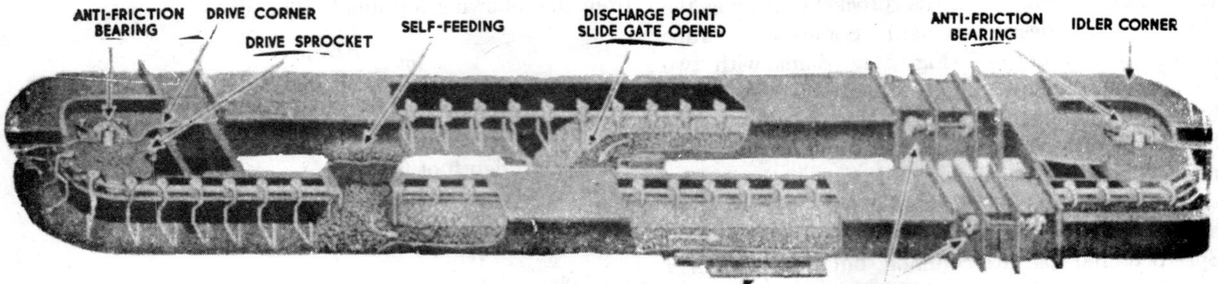

FIG. 53. Phantom drawing of closed-circuit Redler conveyor. (*Stephens-Adamson Mfg. Co.*)

or slats are needed, as contact with the chain is sufficient to move the material. They are primarily for the handling of ashes and similar material. The

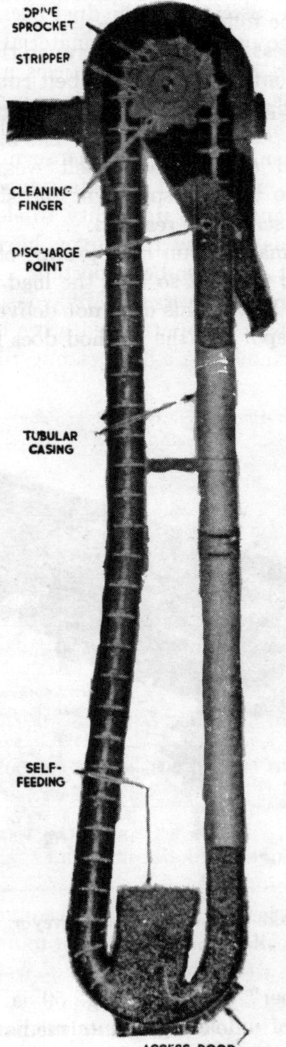

FIG. 54. Phantom drawing of Redler elevator. (*Stephens-Adamson Mfg. Co.*)

depth of material conveyed depends on whether the material is fine or lumpy. The capacities when material is being carried to a depth of 4 in. at a speed of 30 linear fpm vary from 400 to 800 cu ft of material per hour for chains of 46 to 54 in. in width. The capacities are directly proportional to the speed, but speeds should not exceed 30 fpm for abrasive materials.

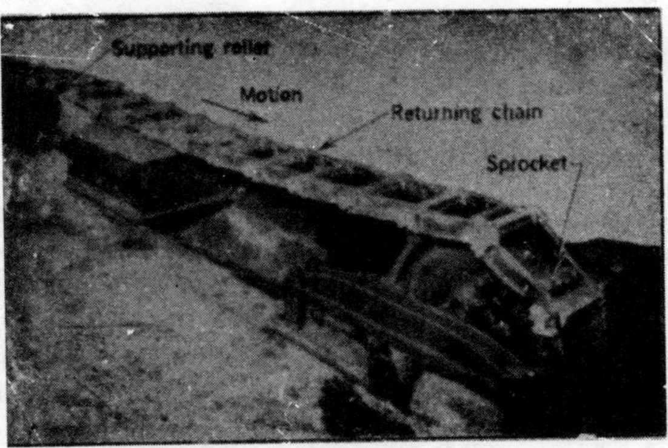

FIG. 55. Drag conveyor for hot cement clinker using cast steel drag chain. (*Chain Belt Co.*)

Carriers. *Belt conveyors*, as the name suggests, consist of endless belts, suitably supported and driven, which carry or transport solids from place to place. Belts are made of canvas, reinforced rubber or balata, and strip steel. Strip steel is also employed for conveying materials through furnaces. Belt conveyors are adapted to wide varieties and quantities of materials, require relatively low power, and can transport solids for long distances. The initial costs are usually higher than for other types of conveyors, but this is not the deciding factor in practice.

FIG. 56. Loaded belt conveyor. (*Chain Belt Co.*)

The loaded belt (Fig. 56) is carried on groups of small idling rollers (Fig. 57), so arranged that the belt forms a trough. The width of the belt varies from 14 to 60 in., and the number of idlers varies correspondingly. The idlers must be spaced so that there will be no sagging of the belt between them

This spacing varies from about 5 ft for narrow belts down to 3 ft for the widest belts. The return idlers are cylindrical rollers, spaced at greater intervals than those carrying the loaded belt. The idlers may be equipped with "antifriction" (ball or roller) bearings or ordinary bushed bearings. Power requirements are considerably lower for antifriction bearings, but original costs are higher.

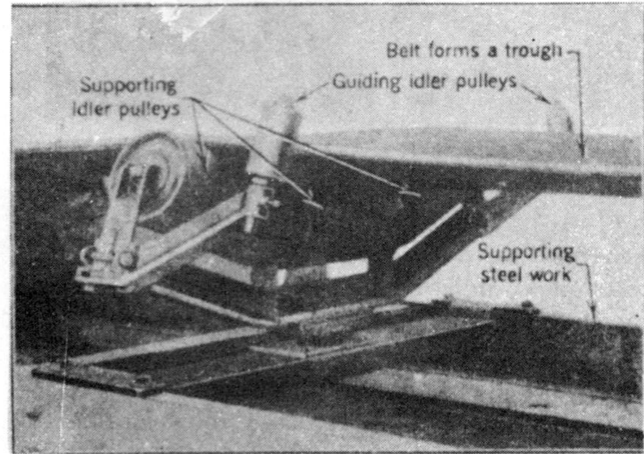

Fig. 57. Empty belt conveyor, showing supporting and guiding idler pulleys. (*Link-Belt Co.*)

Belt conveyors may be driven from any point, although the head or discharge end is the best from the standpoint of stress conditions in the belt. The driving power may be applied through a simple bare steel pulley, a rubber lagged pulley, or a tandem drive consisting of two pulleys connected to the motors through reducing gears.

Changes in loading, temperature, and humidity affect the length of the belt, and some provision must be made for keeping the belt from sagging and becoming loose on the driving pulleys. The simplest form of such "take-up" consists in mounting the shaft of the head or tail pulley in a bearing block which may be moved back and forth. A so-called gravity take-up is shown in Fig. 58. This take-up consists of a pulley mounted in a framework which is free to slide up and down between I-beam guides. The pulley and its framework are supported by the belt between two return idlers. Weights are fastened to the framework to control belt tensions as required. With very long conveyors it is often necessary to remove sections of the belt and replace them as the seasons change.

Belt conveyors may be loaded by any type of feeding device already described.

Unloading the conveyor offers more of a problem unless the load is discharged by gravity where the belt passes around the head pulley. If discharge at an intermediate point is required, one of three methods may be followed.

1. The simplest method is to arrange the supporting idlers horizontally so that the belt runs flat, and to install a *scraper* at an angle of 45 degrees across the belt, at the point of discharge. The bad features of this method are increased belt wear and the tendency of the load to spill over the sides of the belt before the scraper is reached.

2. The belt may be run over groups of idlers set obliquely to tip the belt so that the load spills over the edge of the belt. This does not deliver the load at one definite spot but the method does have some applications.

Fig. 58. Gravity take-up pulley for belt conveyor. (*Stephens-Adamson Mfg. Co.*)

3. The "tripper" shown in Fig. 59 is the most positive means of unloading at intermediate points. The tripper runs on rails which are installed along each side of the belt. It is equipped with two pulleys over which the belt runs. The lower pulley is placed below and behind the upper pulley so that the belt is doubled back, allowing the load to fall over the upper pulley into a chute which may deliver the load all on one side of the belt or on both sides. The tripper may be placed at any desired unloading point, or it may be constructed so that it will move back and forth, distributing the load evenly in a long storage bin. The use of a tripper increases the

Fig. 59. Belt conveyor tripper that may be self-propelled and reversed along tripper track for distributing discharged material. *(Stephens-Adamson Mfg. Co.)*

power required to drive the conveyor, but this is offset by the convenience and flexibility of the unit in unloading the belt.

Capacities of belt conveyors are indicated in Tables 15 and 16 on p. 58.

Power requirements for belt conveyors involve a number of variables and may be estimated by means of empirical formulas such as the following.

1. For *empty* belt conveyor (minimum value):

$$\text{hp} = \frac{F(L + L_0)(0.03WS)}{990}$$

2. For material (excluding belt) conveyed horizontally:

$$\text{hp} = \frac{F(L + L_0)T}{990}$$

3. For elevating material (excluding belt which balances itself):

$$\text{hp} = \frac{T\,\Delta Z}{990} \quad \text{(negative when travel is downward)}$$

4. Total (sum of 1, 2, and 3):

$$\text{hp} = \frac{F(L + L_0)(0.03WS)}{990} + \frac{F(L + L_0)T}{990} + \frac{T\,\Delta Z}{990}$$

$$= \frac{F(L + L_0)(T + 0.03WS) + T\,\Delta Z}{990}$$

where hp = horsepower required.

F = friction factor, 0.05 for plain bearings, 0.03 for antifriction bearings, depending upon installation maintenance.

L = length of conveyor between terminal pulleys (ft).

L_0 = 100 for plain bearings, 150 for antifriction bearings.

S = speed of belt (fpm).

T = material (tons/hr).

ΔZ = increase in elevation of material (ft).

W = mass (lb) of moving parts including belt and idlers per foot of distance between centers of terminal pulleys (both runs).

TABLE 15. MAXIMUM LUMP SIZE AND SPEEDS FOR CONVEYOR BELTS

Belt Width, in.	Maximum Lump Size, in.		Cross-Sectional Area of Load, sq ft	Normal Speed, fpm	Maximum Belt Speeds, fpm		
	Uniform Size	With 90% Fines			Free-Flowing Material *	Average Material †	Abrasive Material ‡
14	2	3	0.11	200	400	300	250
16	2½	4	0.14	200	500	300	250
18	3	5	0.18	250	500	400	300
20	3½	6	0.22	300	600	400	300
24	4½	8	0.33	300	600	500	350
30	6	11	0.53	350	700	500	350
36	8	15	0.78	400	800	600	400
42	10	18	1.09	400	800	600	400
48	12	21	1.46	400	800	600	400
54	14	24	1.90	450	800	600	400
60	16	28	2.40	450	800	600	400

* Free-flowing includes such materials as grain and fine-sized anthracite coal.

† Average includes such materials as coal, crushed stone, sand, and fine ore.

‡ Abrasive includes such materials as coke, screened lump coal, gravel and coarse ore.

TABLE 16. MAXIMUM CAPACITIES * FOR CONVEYOR BELTS

Belt Width, in.	Cu Yd/Hr at 100 Fpm	Maximum Capacity with Materials of Various Bulk Densities, tons/hr at 100 fpm				
		25 Lb/Ft³	50 Lb/Ft³	75 Lb/Ft³	100 Lb/Ft³	150 Lb/Ft³
14	23.6	8	16	24	32	48
16	31.1	10	21	31	42	63
18	39.6	13	27	40	54	81
20	49.3	16	33	49	66	99
24	72.4	24	49	73	98	147
30	116.7	39	79	118	158	237
36	173.3	57	115	172	230	345
42	242.2	82	165	247	330	495
48	324.4	110	220	330	440	660
54	422.2	142	285	427	570	855
60	533.3	180	360	540	720	1080

* Operating capacities of flat belt conveyors are taken at one-half of those listed. Capacities of inclined conveyors are 5 to 10 per cent less than listed. For material weights and speeds other than shown above, use direct proportion for tonnage calculations.

TABLE 16A. APPROXIMATE WEIGHTS OF CONVEYORS

Flight conveyors	
4 × 10 to 6 × 18	0.5 lb/in. of width per running foot
8 × 18 to 10 × 24	1.0 lb/in. of width per running foot
Belt conveyors	1.0 lb/in. of width per running foot

Actual dimensions and weights are available in manufacturers' catalogues.

Power requirements for trippers may be computed as follows.

$$\text{hp} = YS + ZT$$

where S = belt speed (fpm).
T = peak capacity (tons/hr).

Y and Z are constants from following table.[3]

Constant	Width of Belt, in. 14	16	18	20	24	30	36	42	48	54
Y	0.0020	0.0020	0.0026	0.0029	0.0034	0.0047	0.0060	0.0069	0.0082	0.0100
Z	0.0035	0.0035	0.0035	0.0040	0.0040	0.0050	0.0050	0.0055	0.0060	0.0070

An interesting modification of the belt conveyor is the zipper conveyor which is essentially a belt conveyor with the edges zipped together to form an endless tube (Fig. 60). In operation the tube is closed after being loaded and is zipped open at the point of discharge, remaining open until the loading point is reached. Because the tube is flexible and completely encloses the material, it may travel around corners at any angle and in any plane.

Apron conveyors are similar to belt conveyors in that solid materials are carried in a moving trough, but the trough is formed of articulating sections of wood or metal instead of a continuous flexible belt. Apron conveyors are frequently employed when the material to be conveyed is lumpy, abrasive, hot, or otherwise injurious to flexible belts. Their weight restricts apron conveyors to relatively short hauls and much lower speeds, but they are capable of carrying much heavier loads than belt conveyors. The only discharge point is at the head end of the conveyor.

Apron conveyors may appear in several types of construction, but the usual form, shown in Fig. 61, consists of two endless strands of roller chain which are connected by double-beaded steel pans. The idea of the beading is to maintain a continuous trough to prevent leakage in transit, and to prevent material from being wedged between the aprons when the load is being discharged at the head end of the conveyor. The beading also prevents material from slipping backward when the conveyor is

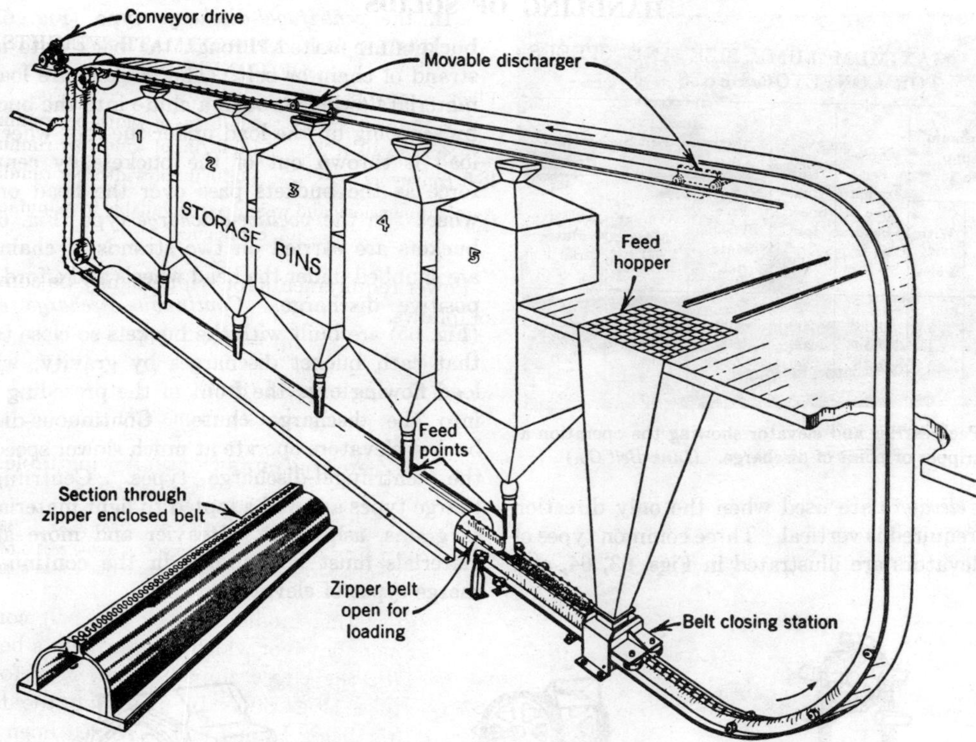

Fig. 60. Drawing indicating the construction and operation of zipper enclosed-belt conveyor-elevator. (*Stephens-Adamson Mfg. Co.*)

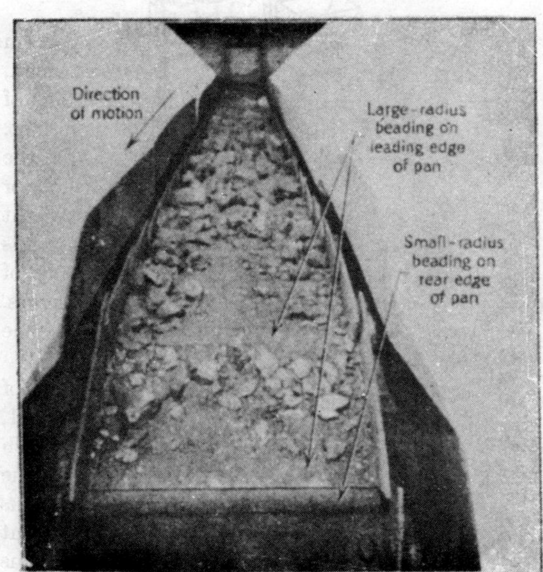

Fig. 61. Apron conveyor viewed from point of discharge. (*Chain Belt Co.*)

inclined. The bead with the larger radius is located on the leading side of the apron or pan and rotates about the bead with the smaller radius, which is located on the rear edge of the apron just ahead. The curvature of the beads is concentric with the line of articulation of the chain joints. Otherwise it would be impossible for the larger bead to rotate smoothly around the smaller bead on the adjacent pan.

The contours of the aprons may be varied until they become a continuous series of buckets pivotally supported between two endless chains. This is a very flexible type of conveyor and is known as the *Peck carrier* (Fig. 62). The buckets maintain their carrying position by gravity and so may convey material horizontally, vertically, and again horizontally, or in any desired path in the same vertical plane. The buckets may be readily discharged at any point by the tripping device indicated, causing each bucket to turn through 90 degrees and dump its load.

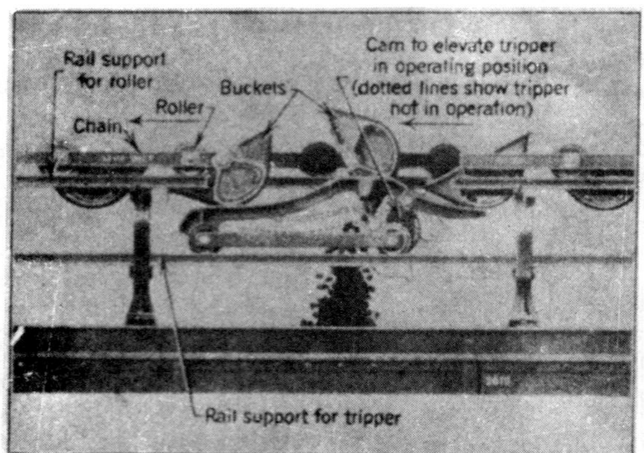

Fig. 62. Peck carrier and elevator showing the operation at the tripper or point of discharge. (*Link-Belt Co.*)

Bucket elevators are used when the only direction of travel required is vertical. Three common types of bucket elevators are illustrated in Figs. 63, 64, and 65.

In the *centrifugal-discharge* type (Fig. 63), the buckets are bolted through the back onto a single strand of chain or belt. The buckets are loaded by material flowing through a chute into the buckets or by scooping up the load under the foot wheel. The load is thrown out of the buckets by centrifugal force as the buckets pass over the head or upper wheel. In the *positive-discharge* type (Fig. 64), the buckets are carried on two strands of chain which are snubbed under the head wheels and afford a more positive discharge. *Continuous-discharge elevators* (Fig. 65) are built with the buckets so close together that each bucket discharges by gravity, with the load flowing over the front of the preceding bucket into the discharge chute. Continuous-discharge bucket elevators operate at much slower speeds than the centrifugal-discharge types. Centrifugal-discharge types are well adapted to light materials such as grains, ashes, etc. Heavier and more abrasive materials must be handled in the continuous-discharge types of elevators.

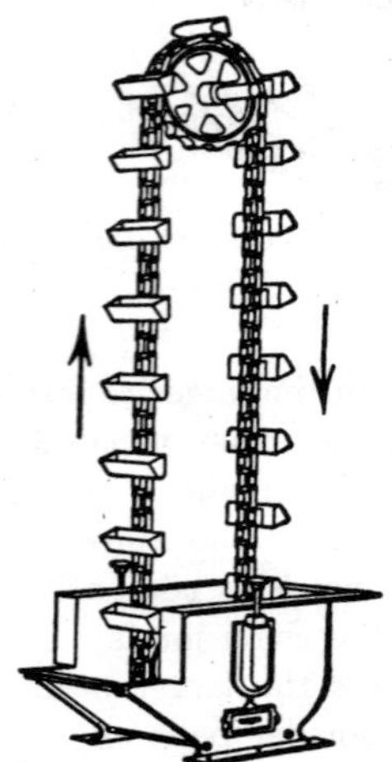

Fig. 63. Drawing illustrating a centrifugal-discharge bucket elevator. The buckets are suitably spaced on a chain or belt which is driven with sufficient speed to discharge the buckets over the head wheel by centrifugal force. It may be used for almost any material that is discharged freely from the buckets. (*Chain Belt Co.*)

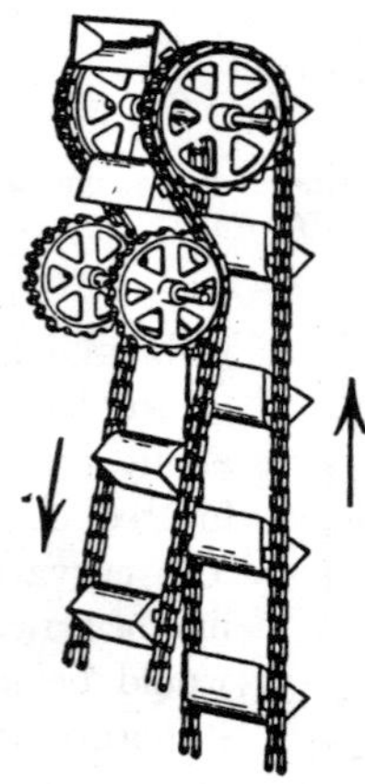

Fig. 64. Drawing illustrating the "positive-discharge" bucket elevator. The buckets are "end-hung" between two strands of chain and are inverted as they pass over the head sprockets and behind the idler wheels, thereby effectively discharging materials not well handled by centrifugal discharge elevators. (*Chain Belt Co.*)

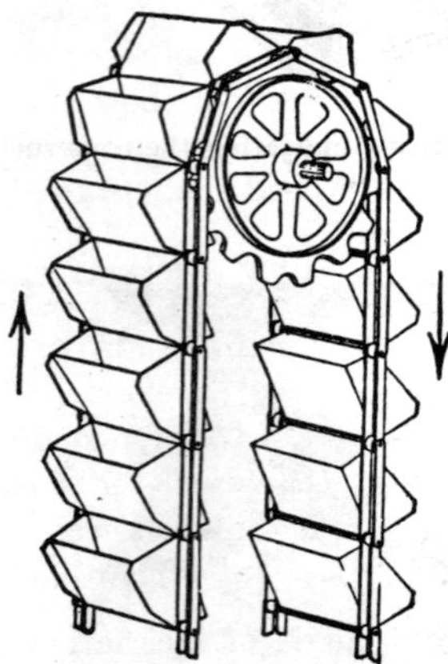

Fig. 65. Drawing illustrating the continuous-discharge bucket elevator. Buckets are mounted close to one another to form a continuous row of buckets. High capacity is obtained by the large number of buckets and not by speed. These elevators may be inclined as well as vertical, and they operate satisfactorily at 45 or more degrees from the horizontal. (*Chain Belt Co.*)

The capacities of apron, pan, and bucket conveyors and bucket elevators vary from a few tons up to about 150 tons/hr for materials having a density of 50 lb/cu ft. The capacity varies approximately directly with the density of the load.

The power required for driving these apron and bucket conveyors may be estimated by the expression [3]

$$\text{hp} = \frac{(\text{Gross turning effort})(\text{Speed, fpm})}{33{,}000}$$

The resistance to turning R_t in pounds at the pitch radius of the head or driving sprockets may be calculated:

For bucket elevators

$$R_t = ML + C$$

For horizontal apron and bucket conveyors

$$R_t = 2(M + W)(LR_f)$$

For inclined apron and bucket conveyors

$$R_t = L(M + W)[(R_f \cos a) + \sin a] + WL[(\cos aR_f) - \sin a]$$

where M = mass (lb) of material conveyed per foot of conveyor or elevator.

W = mass (lb) of chain and aprons or buckets per foot (both runs).

L = length of conveyor or elevator from head shaft to foot shaft (ft).

$R_f = (\text{rolling friction}) = x\left(\frac{d}{D}\right) + \left(\frac{0.06}{D}\right)$

$x = 0.33$ for metal on metal, not greased.

$= 0.20$ for metal on metal, greased.

D = diameter of chain roller (in.).

d = diameter of bushing or pin upon which roller turns.

a = angle of inclination of conveyor or elevator (degrees).

C = force (lb) required to drag the buckets of a centrifugal discharge elevator through the material being loaded. It need not be included in computations when the elevator is loaded by a feeding device. Even when used its value must be estimated.

The turning resistance R_t calculated in this manner should be increased 10 per cent for friction of head and foot shafts, and 15 per cent for each pair of gear reductions to obtain the gross turning effort.

WEIGHT DETERMINATION

The control of material as it flows through a plant is essential for accounting, for interplant records, or for control of the operation or process. Measuring weight is an important method. In the selection of weighing equipment not only must the process or method of manufacture be considered, but various outside agencies, laws, and regulations must frequently be consulted.

A scale is a device which measures the force of gravity on an unknown mass by balancing a known force against the unknown force. Except in small scales the various elements of the scale reduce the applied force, so that a relatively small balancing force may be applied.

If the balancing force is supplied by the force of gravity on any mass, such as a weighbeam or pendulum weights connected to a dial, any change in the gravitational field has a similar and proportional effect on the unknown weight and the balancing weight, and the scale correctly measures the unknown mass in any gravitational field. If the balancing force is supplied by a spring, the scale does not measure the unknown mass but the force of gravity on the unknown mass which varies with the local acceleration of gravity.

Generally scales and other weighing devices are composed of three essential parts; the load-receiving element, the load-bearing members, and the indicating element. The indicator may be a weighbeam, a dial, or an automatic recorder. The selection of the indicating device depends on the application, the operating personnel, and the cost of the installation.

Load-Receiving Element

When an object is weighed it is usually necessary to support the weight on an independent element entirely supported by the load-bearing members. In nearly all large scales this independent element consists of a weighbridge supporting various superstructures such as railroad tracks, a platform for trucks, a hopper, bin, or tank permanently built on the weighbridge. On smaller scales the load-receiving element may be a small platform or pan, or the object may be hung directly on the load-bearing

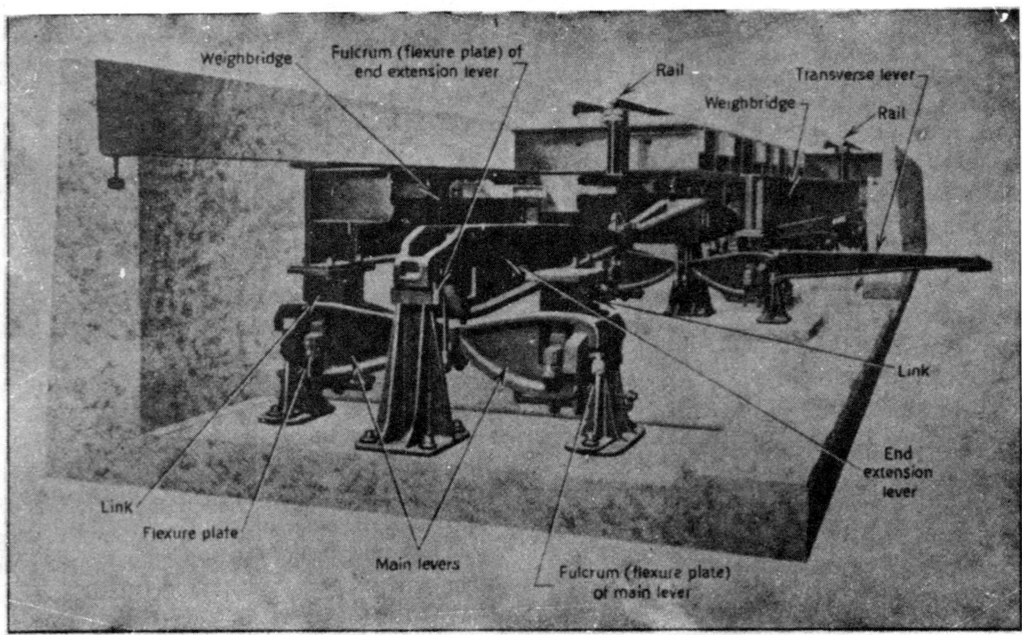

Fig. 65*a*. Cutaway view of typical railroad track scale. Four-section scale, consisting of four sets of main levers. The load is delivered to the indicator by the transverse lever extending out at the right of the picture. (*Fairbanks, Morse & Co.*)

element. Various types of conveyors are frequently placed directly on the load members.

Load-Bearing Members

The load-bearing members generally consist of several interconnected levers. In simple systems only two levers may be required to carry the load and reduce the balancing force required. In heavy-duty railroad scales there may be as many as thirteen or more levers (Fig. 65*a*). The load from the weighbridge is transmitted through links to the main levers, which reduce and transmit the load to the end extension levers, which in turn transmit the load to the transverse lever and finally to the indicating device.

The connection between the weighbridge and the lever system is important for accurate weighing. Only the vertical components of the load must be delivered to the levers. Any horizontal forces due to impact, temperature, or similar factors must be eliminated. This is generally accomplished by suspension links, or by balls or rollers between the bridge and the lever system. It is also necessary to restrain the weighbridge from excessive violent displacement. Various "checking" systems are used to limit horizontal motion. These may be bumpers built into the walls of the scale pit, or they may be horizontal members between the weighbridge and the walls of the scale pit.

Pivots and *bearings* of hardened tool steel are used in most scales to carry the load, as they are practically the only parts of the lever system that are subject to wear or damage. They are frequently inspected and must be continuously protected from corrosion. In scales that are subjected to heavy usage they must be replaced or resharpened periodically. *Flexure plates* are frequently used in place of the more conventional pivots or knife edges. They consist of relatively thin plates of metal restrained in a vertical position so that they are subjected to nearly pure compression. The slight amount of angular movement is provided by the elasticity of the metal. Flexure plates are generally found in the largest types of scales but are also used in laboratory and small capacity scales.

Indicating Elements

The indicating element includes the method of applying a calibrated force to balance the applied load and must indicate by visual or other means the magnitude of the balancing force.

FIG. 65*b*. Belt-conveyor scale. A section of belt is supported by the lever system, and the weight is continuously integrated with belt speed. (*Merrick Scale Mfg. Co.*)

The *weighbeam* consists of a calibrated bar with a movable poise. The poise is positioned along the beam until the load on the lever system is balanced. The position of the poise may then be converted into the amount of load on the scale. The weighbeam is the basis of most indicating elements, and in itself is widely used for scales of all sizes. It is frequently associated with other indicating devices, being used for tare compensation, or as a stand-by unit in case of failure of the other indicator.

Dials may be used on all types of lever systems. They are easy to read, require no manipulation on the part of the operator, and indicate the result rapidly. However, they seldom have more than 1000 graduations around the periphery which limits the readability of dials to $\frac{1}{1000}$ of their total capacity. This limitation is overcome using the dial to indicate only that part of the weight over the weight balanced by drop weights applied to the lever system. For example, a 5000-lb dial with a minimum graduation of 5 lb may be used for loads in excess of 5000 lb. An added weight balancing 5000 lb is added to the beam, and the range of the dial becomes 5000 to 10,000 lb without sacrificing readability.

In pendulum dials the load is balanced by elevating one or more pendulum weights. Although rather complex in design pendulum dials are generally accurate and dependable if properly installed and maintained. Spring-operated dials are simpler in design and maintenance, but not so accurate unless springs are compensated for temperature changes. They are also subject to error because of changes in gravitational force with changes in location.

Automatic recorders may be separate units or associated with either a dial or weighbeam. They are used on railroad scales to allow weighing of the cars as they roll across the scale without stopping. Belt conveyor scales (Fig. 65*b*) continuously integrate the weight of material on the conveyor with the speed of travel of the belt. These are widely used in mines, power plants, or in industrial plants where large quantities of bulk material are handled on belts. Conveyor scales are the only strictly continuous operation in weighing.

Automatic batch controllers are of great importance in the process industries wherever rates of production are high. They may be used to proportion solids (Fig. 65*c*), liquids, or combinations of both. Special indicators on conventional dial or beam scales, such as photoelectric devices, may be used to operate

FIG. 65*c*. Typical batching units. Two different materials are proportioned by weight in the weigh hoppers. (*Streeter-Amel Co.*)

batch controllers. More often special automatic controllers govern the entire process.

Packaging and bagging scales are specialized forms of batch controllers, generally incorporated as integral portions of other machinery.

Counting scales for counting large quantities of small objects have ratio beams that allow rapid determination of count when the identical objects are used as counterbalancing weights.

Electronic scales that eliminate levers are becoming increasingly important in special weighing applications such as large tanks and bins. The load-bearing element consists of a calibrated member with strain gages or other sensitive devices attached. The load is measured by the deflection of the column as the load is applied. The deflection is measured electrically and through the use of servos is indicated by a dial or automatic recorder.

Hydraulic scales also form an increasingly important type in which the levers of a conventional scale are replaced by a hydraulic system that balances the applied load. The force or weight is indicated by a dial or recorder. Hydraulic scales have been useful in small compact crane scales, where the load on the hook of a crane must be determined. The accurate dead-weight tester for measuring high pressure and calibrating pressure gages is a form of hydraulic scale.

BIBLIOGRAPHY

1. Chain Belt Co., Catalog 445 (1945), Milwaukee.
2. Hawley, R. S., *Materials Handling and Factory Transportation*, Ulrich's Book Store, Ann Arbor (1941).
3. Link-Belt Co., Catalog 800, Chicago.

PROBLEMS

1. A pile of crude sulfur must be moved from one plant to another by dump truck. Trucks carrying 5 tons each can make one round trip every hour, exclusive of loading time. The cost of trucking is estimated at $4.00/hr for the truck and driver. A construction company will load the sulfur with a power shovel at $0.40/cu yd and guarantee loading at 15 tons/hr but with a minimum charge of $150. The trucks may be loaded by hand labor at a cost of $1.30/hr per laborer.
   (*a*) What is the maximum volume of the pile for which hand loading would be most economical?
   (*b*) What is the estimated cost to move 10 tons, 100 tons, 1000 tons, by the more economical method?

2. How many man hours are required to load a freight car with 40 tons of product packaged in 100-lb bags if the product must be transported 100 yd by a hand truck which carries 3 bags per trip?

3. One hundred tons/hr of anthracite coal are to be moved horizontally a distance of 100 ft. Select a conveyor of each of the three classes listed, and calculate the power required to operate the system. Choose the smallest conveyor that will do the job.
   (*a*) Screw conveyor.
   (*b*) Flight conveyor.
   (*c*) Belt conveyor.

4. For the conditions of problem 3, specify the type of conveyor and its size that will require the minimum power.

5. A belt conveyor is required to deliver crushed limestone having a bulk density of 75 lb/cu ft at the rate of 200 tons/hr. The conveyor is to be 200 ft between centers of pulleys with a rise of 25 ft. The largest lumps are 4 in. and constitute 15% of the total. The conveyor will discharge over the end. For a belt speed of 200 fpm, what is the minimum width of belt that can be used? Calculate the horsepower for the drive motor.

6. What is the capacity of a flight conveyor of 12 by 24 in. traveling at 100 fpm and handling the crushed limestone of problem 5.

7. A screw conveyor is to be installed to convey 800 bushels of wheat per hour over a distance of 80 ft. Determine the size (diameter), speed (revolutions per minute) and horsepower requirements for the installation.

PART II

Fluids

ALTHOUGH all our activities are completely surrounded by the fluid atmosphere, those operations previously considered in Part I can be calculated and conducted as though in the complete absence of fluids. In the various unit operations to be discussed fluids are involved and the properties of the fluids have an important influence on the operations, particularly as they affect the relative motion between solids and fluids.

CHAPTER

6

Properties of Fluids

DENSITY, specific gravity, and other similar properties have the same significance for fluids as for solids. The definitions of these properties are given in Chapter 2.

VISCOSITY

Viscosity is a unique and most important property of all actual fluids. It is somewhat analogous to resistance to shear in solids. The principal reason for the difference in the flow characteristics of water and of molasses is that molasses has a much higher viscosity than water. The analogy to shear resistance in solids may be referred to in establishing the units of viscosity. Assume two parallel layers in the fluid, as indicated in Fig. 66*a*, each having an area of A square centimeters and with a distance of dy centimeters between them. One layer is considered stationary; the other layer moves at a constant velocity relative to the first layer of dv centimeters per second. A force of F dynes is required to maintain this relative velocity of the moving layer.

For the general case, with a number of layers, when the velocity v is any function of the distance y, as indicated in Fig. 66*b*, at any point

$$\frac{F}{A} = \mu \frac{dv}{dy} \tag{1}$$

where μ is a factor representing the characteristic of the fluid called the viscosity, which is the shear modulus for the fluid. Solving for the viscosity

$$\mu = \frac{F}{A}\frac{dy}{dv} = \frac{d(mv)}{A\,dt}\frac{dy}{dv} \tag{2}$$

Viscosity is the time rate of change of momentum per unit area and velocity gradient. It has the net dimensions of mass/(length)(time) or the dimensions of (force)(time)/(length)^2. These dimensions of viscosity (μ) suggest correctly that it may be evaluated by such means as timing the flow of a given mass of fluid through a standard aperture as well as by measuring the resistance to motion of two parallel surfaces.

In the metric system viscosity is defined as grams mass/(centimeter)(second) and the unit is

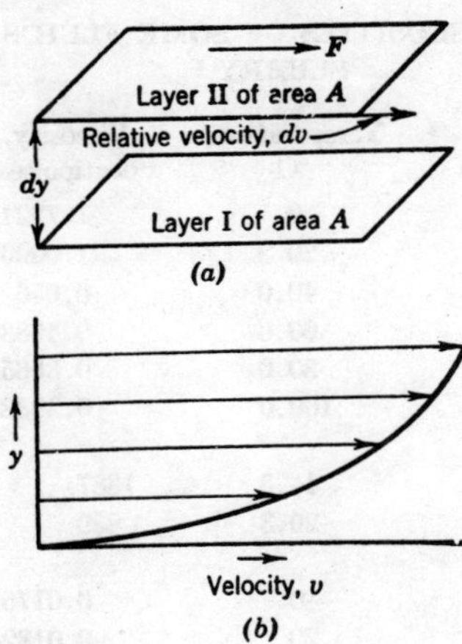

Fig. 66. (*a*) Two layers of fluid in laminar flow; (*b*) velocity of multiple layers of fluid in laminar flow.

called a poise. This unit is rather large for most applications, and viscosities are frequently reported in centipoises. One poise = 100 centipoises.

In English units, the dimensions of viscosity are in pounds mass/(feet)(seconds). This unit has no name as yet, although it is widely used. To convert from poises to the British viscosity unit, which may be abbreviated as Bvu, multiply by 30.48/453.6 or 0.0672.

$$\text{Viscosity in centipoises} \times (6.72 \times 10^{-4}) = \text{Viscosity in British units}$$

Specific viscosity is the ratio of the absolute viscosity in either poises or Bvu to the absolute viscosity of a standard fluid expressed in the same units and taken at the same temperature.

Kinematic viscosity is the absolute viscosity divided by the density (μ/ρ) and has the dimensions of (volume)/(length)(time). It is frequently determined by measuring the time required for a specified volume of fluid to flow through a standard aperture or tube under specified conditions (as in the Saybolt viscosimeter). The unit corresponding to the poise is the stoke, having the dimensions of $(\text{centimeters})^2/(\text{second})$.

An *ideal fluid* is a hypothetical fluid which has a viscosity of zero and interposes no resistance to shear. All actual fluids possess viscosity. The viscosity of a liquid decreases and the viscosity of a gas increases as the temperature increases, the two becoming identical at the critical point.

The viscosities for a few common fluids are listed in Table 17.

TABLE 17. VISCOSITIES OF SOME FLUIDS AND A SLURRY *

Material	Temperature, °C	Viscosity, centipoises
Water	0	1.7921
	20.2	1.0000
	40.0	0.656
	60.0	0.4688
	80.0	0.3565
	100.0	0.2838
Glycerin	14.3	1387
	20.3	830
Air	0	0.0175
	20	0.0182
	50	0.0195
	100	0.0218
	500	0.036

Slurries of $CaCO_3$ in Water (Particles about 5 Microns), % by weight	Temperature, °C	Approximate Bulk Viscosity μ_b, centipoises
0	20	1.0
5	20	1.11
10	20	1.37
15	20	1.81
20	20	2.70
25	20	5.7

* See Appendix for additional viscosity data.

RELATIVE MOTION BETWEEN FLUIDS AND SOLIDS

There are many practical situations in which relative motion exists between fluids and solid boundaries. In some cases these solid boundaries are at rest with respect to the earth while the fluid is moving. In other cases the fluid is more or less at rest while the solid boundaries move with respect to the earth. Engineers encounter problems which fall into either or both of these categories, so it is necessary to develop a generalized treatment for the relative motion between a fluid and a solid.

Geometric Similarity

The different shapes and sizes of solids are almost infinite in number. For convenience, solids may be grouped according to their shapes. When every linear dimension of a given member of a group occurs in a definite or constant ratio to the corresponding linear dimension of any other member of the group, all members of that group are geometrically similar. Geometric similarity is the property of having the same shape and differing only in size and position. The scale models of airplanes or railroads and the corresponding full-size article are excellent examples of geometrically similar systems. Perfect spheres are all geometrically similar, as are cubes, regular tetrahedrons, and a host of other objects. Rectangular parallelepipeds are not necessarily geometrically similar unless multiplying the length of every edge of the smaller one by a constant factor will yield the lengths of the corresponding edges of the larger one.

Dynamic Similarity

Dynamic similarity is the property of having motion of the same form and differing only in size and position. It, therefore, also requires geometric similarity. Two flow systems possess dynamic similarity if the streamlines of fluid flow, or the fluid flow pattern, of one system may be superimposed directly upon the streamlines of the other system with no change other than the equal magnification in all directions of one of the systems. Consider the free passage of air past two spheres of different diameters. If the velocity of the air is the same past both spheres and a cigarette is held in the air stream, two different flow patterns of the smoke around the spheres are apparent. If the velocity past one of the spheres is varied, a particular velocity is found at which the smoke flow patterns appear to be similar. If photographs are then made of the two smoke patterns and the negative of the smaller system is placed in an enlarger, it is possible to adjust the enlarger so that a print can be made which will be identical to the print made from the negative of the larger system. In some flow patterns which oscillate with time, it is necessary to take the pictures at corresponding times.

It is apparent that fluid flow patterns cannot be dynamically similar at points adjacent to solid boundaries that are not geometrically similar, and that *geometric similarity is a prime requisite for dynamic similarity.* If dynamic similarity exists between two systems, the radii of curvature of the flow lines or paths at corresponding points in the two systems occur in a fixed ratio, regardless of the path or point selected. At these corresponding points the velocity ratios are constant, regardless of what particular point is selected; and the ratios of the accelerations at corresponding points are fixed and independent of the position of the points selected. The directions of the velocity and of the acceleration at corresponding points in dynamically similar systems are respectively the same.

Reynolds Number, The Criterion of Dynamic Similarity

Consider the motion or flow of fluid in two geometrically similar systems, with the time-average velocity of the fluid constant at any given point. The paths of the fluids in the two systems are similar, and dynamic similarity is said to exist. The forces acting upon small elements of the fluids at corresponding points in each of the similar systems may be expressed in terms of two forces, one normal to the direction of motion (F_n), and one tangential to the motion (F_t) (Fig. 67).

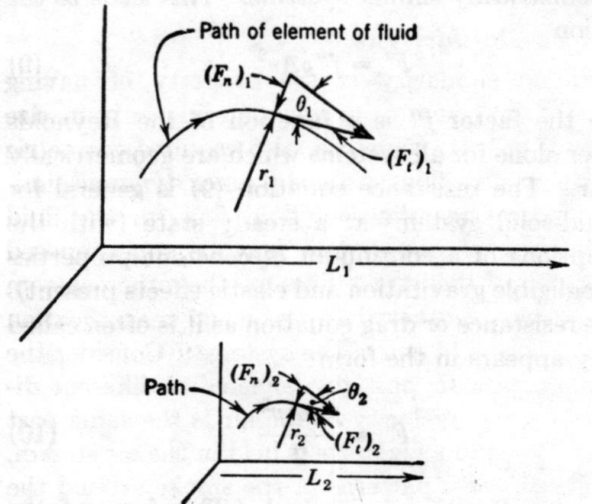

FIG. 67. Dynamically similar fluid motion within geometrically similar boundaries.

The fluid is assumed to be a continuous medium. This assumption is valid except at very low pressures where the mean free paths between the molecules become large and the fluid no longer behaves as a continuous medium. The gravitational and elastic forces are also assumed to be negligible in comparison with inertial and viscous forces. These assumptions are valid when there is no free surface for wave motion and the velocities are considerably less than the velocity of sound in the fluid and lead to the following relations.

NOMENCLATURE

	System 1	System 2
Any linear dimension characterizing the size of the system	L_1	L_2
A linear dimension of the small element	δL_1	δL_2
Mass of the small element	δm_1	δm_2
Radius of path of the small element	r_1	r_2
Velocity of the small element	v_1	v_2
Volume of the small element	$(\delta L_1)^3$	$(\delta L_2)^3$
Representative area of the element	$(\delta L_1)^2$	$(\delta L_2)^2$
Normal force acting on the element	$(F_n)_1$	$(F_n)_2$
Tangential force acting on the element	$(F_t)_1$	$(F_t)_2$
Angle of resultant force on the element	θ_1	θ_2
Density of fluid in a given system	ρ_1	ρ_2
Viscosity of fluid in a given system	μ_1	μ_2

Because of geometric similarity between the two systems,

$$\frac{r_1}{L_1} = \frac{r_2}{L_2} = a \qquad \therefore r = aL \tag{3}$$

$$L_1 = bL_2 \qquad \text{and} \qquad v_1 = cv_2$$

$$dL_1 = \frac{L_1}{L_2}\,dL_2 \qquad dv_1 = \frac{v_1}{v_2}\,dv_2$$

$$\delta L_1 = b\,\delta L_2$$

$$\delta L_1 = \frac{L_1}{L_2}\,\delta L_2$$

The normal force acting on element $F_n = \delta m v^2/r$. If aL is substituted for r and $\rho(\delta L)^3$ for δm

$$F_n = \frac{\rho(\delta L)^3 v^2}{aL} \tag{4}$$

The tangent force acting on element from equation 2

$$F_t = \mu A\,\frac{dv}{dL} = \mu(\delta L)^2\,\frac{dv}{dL}$$

The angle θ of the resultant force can then be expressed in terms of the tangent of the angle which is the same for the two similar systems as

$$\tan\theta = \frac{F_n}{F_t} = \frac{\rho_1(\delta L_1)^3 v_1^2}{aL_1\mu_1(\delta L_1)^2 \dfrac{dv_1}{dL_1}} = \frac{\rho_2(\delta L_2)^3 v_2^2}{aL_2\mu_2(\delta L_2)^2 \dfrac{dv_2}{dL_2}} \tag{5}$$

Substituting $(v_1/v_2)\,dv_2$ for dv_1, $(L_1/L_2)\,dL_2$ for dL_1, $(L_1/L_2)\,\delta L_2$ for δL_1, and simplifying,

$$\frac{L_1\rho_1 v_1}{\mu_1} = \frac{L_2\rho_2 v_2}{\mu_2} \tag{6}$$

In other words, when dynamic similarity exists in two systems, the product of *any* characteristic dimension, *any* velocity, the density, and the reciprocal of the viscosity is the same for both systems when these variables are chosen at corresponding locations. Therefore, these variables (L, v, ρ, μ) in themselves will determine the flow pattern in geometrically similar systems.

The dimensionless expression $Lv\rho/\mu$ is popularly called the Reynolds number after Osborne Reynolds,[2] * who applied it to the problem of flow inside pipes. The derivation, however, is absolutely general for all systems involving relative motion between fluids and solids except in the presence of appreciable gravitational or elastic effects. The application to pipe flow is but one of many such applications. The criterion of dynamic similarity will be found of great utility in all types of fluid flow problems. It is significant only when applied to geometrically similar systems.

The Universal Resistance or Drag Law

Consider the steady passage of an isothermal fluid past the solid boundaries of a given object. The situation may be any system where the solid boundaries may be defined along with the extent of a fluid moving relative to those boundaries, such as an infinite extent of air passing a sphere or a stream of water flowing inside a pipe.

If the solid boundaries are considered to be moving and the fluid preceding the solid is standing still, a problem often encountered is the determination of the force necessary to keep the solid moving with constant velocity through the fluid. The effect of the solid on the fluid will be to accelerate the fluid or to produce momentum in the fluid. Newton's law, that force is equal to the time rate of production of momentum, may be applied to the fluid as follows.

$$F' = \frac{d(mu)}{dt} = m\frac{du}{dt} + u\frac{dm}{dt} \tag{7}$$

where F' = the force acting on the fluid.
m = the mass of the fluid.
u = the maximum velocity to which the fluid is accelerated by the action of the moving solid.

If the solid is moving with a constant velocity v, it follows that the velocity of the fluid due to the action of the solid is a constant independent of time, and equation 7 may be simplified to

$$F' = u\frac{dm}{dt} \tag{8}$$

However, dm/dt is the mass of fluid being acted upon per unit of time and is proportional to a representative area A of the solid and to the distance traveled by the solid in unit time, which is the velocity of the solid. Therefore, with ρ as the density of the fluid,

$$F' \sim \rho A v u$$

The velocity v of the solid and the average velocity u to which the fluid is accelerated are related by a factor which depends only upon the flow pattern for geometrically similar systems. This leads to the equation

$$F' = f''\rho A v^2 \tag{9}$$

where the factor f'' is a function of the Reynolds number alone for all systems which are geometrically similar. The resistance equation (9) is general for all fluid-solid systems at a steady state (with the assumptions of a continuum of constant properties and negligible gravitation and elastic effects present).

The resistance or drag equation as it is often called usually appears in the form

$$F' = \frac{f'\rho A v^2}{2} \tag{10}$$

where one direction is implied. This form of the equation is used because a derivation involving energy yields the term $v^2/2$, and it has been found

* The bibliography for this chapter appears on p. 71.

convenient to use the same type of expression in the resistance term. Fundamentally this causes no difficulty whatsoever, for a factor of 2 can be included in the term f'.

BIBLIOGRAPHY

1. Dodge, R. A., and M. J. Thompson, *Fluid Mechanics*, McGraw-Hill Book Co. (1937).
2. Reynolds, Osborne, "An Experimental Investigation of the Circumstances which Determine Whether the Motion of Water Shall be Direct or Sinuous, and of the Law of Resistance in Parallel Channels," *Phil. Trans. Roy. Soc. (London)*, 174, 935–982 (1883).
3. Rouse, Hunter, *Elementary Mechanics of Fluids*, John Wiley and Sons (1946).

PROBLEMS

1. If a model of a submarine one-fifth the length of its prototype is to be tested in a wind tunnel instead of in water, what air velocity should be used with the model as equivalent to a submerged speed of 6 mph for the prototype?

	Air	Water
Density, lb/cu ft	0.0765	62.3
Viscosity, centipoises	0.0175	0.95

2. If the length is considered the characteristic dimension, what is the Reynolds number for the model which is 14 ft long?

3. If the drag coefficient f' in equation 10 is 1.2 for flat plates at high Reynolds number, determine the force against a signboard 15 ft high and 30 ft long when a 50-mph wind is blowing against it.

At what wind velocity will this force on the sign be doubled?

CHAPTER 7

The Flow of Solids through Fluids

THE direction, upward or downward, of the flow of particles of solids through a fluid depends upon the density of the solid relative to that of the fluid. This principle is the basis for the separation of solid particles according to their densities, an operation known as *sorting*. If a fluid of a density intermediate to the densities of the solids is available, the solid particles may be sorted by simply introducing them into a body of the fluid. The particles of a density less than that of the fluid will rise or float, and the particles of a density greater than that of the fluid will fall or sink. If no simple fluid of the required density range is available, a "complex" fluid or a suspension of very fine (through 325 mesh) solid particles in a fluid having the desired bulk density may be used in a manner similar to that with a simple fluid having a density equal to the bulk density of the suspension.

Instead of a suspension as the means of obtaining a fluid of the desired density, the surface characteristics of the solid and the addition of material adsorbed on the surfaces may be used to cause one material to float and the other material to sink independent of the density relationships of the simple solids and fluid. We have all observed dust particles floating on water, insects walking on water, and possibly the demonstration of a steel needle "floating" on the surface of water. These phenomena are possible because of "surface tension" of the water or the water not "wetting" the solid. If the dust particles or the insect were thoroughly wet, they would sink, and it is only by placing a dry needle carefully on the surface that it can be made to float. A duck floats on the water. But the addition of a "wetting agent" which is adsorbed on the surface of the feathers, causes the water to wet the surface of the feathers, and the duck sinks.

Conversely "flotation agents" may be added. These are selectively adsorbed on one of the solids present and enable these solids to adsorb, or cling to, bubbles of air which give those solid particles an effective density less than that of the surrounding fluid, as the solid particle plus the adsorbed flotation agent and air is the effective particle. Operations of this kind are identified as flotation operations and are of great industrial importance in the sorting and recovery of valuable minerals and in many other separations.

Even if the particles of both solids to be separated are more dense than the fluid so that they both flow in the same direction, downward, when placed in the fluid, separation can be accomplished if the particles flow through the fluid with different velocities. For example, a mixture of galena (PbS) and silica (SiO_2) rock are to be separated. The particles are reduced to a uniform size of about 1 cm. By a laboratory test it may be determined, or by calculation it may be estimated, that galena particles of 1-cm size fall through quiet water at a velocity of 13 fps and that silica particles of the same size fall with a velocity of 7 fps. Then if the mixed solids are introduced into a vertical stream of water rising with a velocity of less than 13 fps and greater than 7 fps the silica particles will be carried upwards and discharged at the overflow, and the galena will sink through the rising fluid and discharge at the bottom. This continuous process, called *elutriation*, uses a cylinder with fluid flowing upward at a constant known velocity. The solid may be introduced as such or as a dispersion in the fluid. The size of the largest particle carried upward by the rising fluid (after flowing for some time) is the size having a maximum settling velocity just less than the upward velocity of the fluid.

Classification is the term used to designate the separation of solid particles on the basis of their velocities of flow through fluids. These velocities depend upon the properties of the solid (density, size, and shape), and the surface or interfacial conditions between the solid and fluid. The relationships between these variables are complex, and the interfacial conditions are generally unpredictable without experimental determination, frequently making calculation impossible.

The experimental determination of the maximum velocity of settling of small particles can be made in a simple manner with a deep beaker or, better, a large graduate. The solid is dispersed in the fluid in the graduate. The dispersion is allowed to stand for a known time interval. The upper part of the dispersion is then siphoned off down to a chosen level. All particles whose settling rates are greater than that settling rate which would carry the particles to the depth of liquid removed in the known time interval will be at the lower depths and will not appear in the dispersion withdrawn. If the largest size of particles in the decanted dispersion is determined, the maximum velocity of that size of particle may be estimated by dividing the depth of fluid decanted by the time interval.

MAXIMUM VELOCITY

Where the shapes of the particles can be defined and surface conditions are of minor importance so that the other variables predominate in determining the flow characteristics, the rates of free settling may be estimated with satisfactory results.

The basic theory of the flow of solids through fluids is derived from the concept of freely moving (or falling) bodies of constant mass under a constant acceleration.

$$v = a\,\Delta t$$

where v represents the velocity of the body relative to the initial position, Δt is the time interval after starting from rest, under the constant acceleration a.

In the case of falling bodies, the acceleration is due to the force of gravity represented by the symbol g. If no other force is acting on the body, g may be substituted for a in the above equation. The presence of any fluid such as air or water introduces two additional forces, the buoyancy effect resulting from displacement of the fluid by the solid, and the frictional resistance from relative motion of the solid and the fluid. Friction increases with increasing velocity until the accelerating and resisting forces are equal. Then the solid continues to move (or fall) at a constant maximum velocity (also called the terminal velocity) unless additional forces upset the balance.

If the following four assumptions are made, the mathematical treatment is relatively simple.

1. The solid is a nonporous, incompressible spherical particle.
2. The fluid is incompressible and of sufficient extent to eliminate the effects of the confining walls.
3. The accelerating force is derived from a uniform gravitational field.
4. The particle is freely moving, that is, other particles are absent or, if present, do not affect the motion of the particle under consideration.

The force causing the particle to move may be expressed in absolute units as the mass of the particle times the acceleration. Since the force is really a summation of several forces the resultant *force tending to move the particle downward* is

$$F' = ma = mg - wg - F_R' \tag{11}$$

where g = acceleration due to gravity.
m = mass of the solid particle.
w = mass of fluid displaced by the particle or having the same volume as the particle.
mg = gravitational force on the particle.
wg = buoyant force on the particle.
F_R' = resisting force due to friction effects or required to accelerate fluid being displaced.

The terms in equation 11 are readily evaluated with the exception of the resisting force F_R'. Newton[4] * developed an expression for the resisting force, already given, as follows.

$$F_R' = (f')A\frac{\rho_l v^2}{2} \tag{10}$$

For a sphere the representative area A may be the projected area $\pi D^2/4$ and

$$F_R' = (f_D)\frac{\pi D^2 \rho_l v^2}{8} \tag{12}$$

* The bibliography for this chapter appears on p. 83.

where f_D is the friction factor for this specific application of equation 10.

If this value of F_R' is substituted in equation 11, the steady-state maximum falling velocity may be computed.

$$m\frac{dv}{dt} = mg - wg - F_R' \tag{11a}$$

$$\left(\frac{\pi D^3}{6}\right)(\rho_s)\frac{dv}{dt} = \frac{\pi D^3}{6}g(\rho_s - \rho_l) - (f_D)\frac{\pi D^2 \rho_l v^2}{8} \tag{11b}$$

$$\frac{dv}{dt} = \frac{(\rho_s - \rho_l)}{\rho_s}g - \frac{3(f_D)\rho_l v^2}{4D\rho_s} \tag{13}$$

At maximum velocity v_m, $dv/dt = 0$.

$$\left(\frac{\rho_s - \rho_l}{\rho_s}\right)g = \frac{3(f_D)\rho_l {v_m}^2}{4D\rho_s}$$

$${v_m}^2 = \frac{4(\rho_s - \rho_l)gD}{3\rho_l(f_D)} \tag{14}$$

$$v_m = \sqrt{\frac{4(\rho_s - \rho_l)gD}{3\rho_l(f_D)}} \tag{14}$$

Equation 14 is frequently termed Newton's law, expressing the *maximum (or terminal) velocity for falling spheres* in terms of a variable factor f_D. Solving equation 14 for f_D, the friction factor

$$f_D = \frac{4(\rho_s - \rho_l)gD}{3{v_m}^2\rho_l} \tag{15}$$

The flow may be laminar or turbulent, as indicated in Fig. 68 which shows photographs of the wake behind a cylinder. Photograph 1 shows laminar flow lines with the fluid flowing in layers around the cylinder. In 2 a small eddy has formed directly behind (to the right) of the cylinder, but the flow is still predominantly laminar. As the rate of flow is increased the eddy becomes larger and more complex as the flow becomes more turbulent, as in photographs 3, 4, 5, and 6. The flow lines around a falling particle are similar to those shown for the cylinder. If the particle is small the flow is more likely to be laminar (Fig. 68, photograph 1), and the viscosity of the fluid is an important factor in determining resistance. If the particle is large, the flow is more likely to be turbulent and accompanied by the formation of eddies and vortices in the fluid behind the moving particle. These eddies introduce large resistances to flow, and the viscosity of the fluid becomes less important, becoming negligible in determining the resistance when the flow conditions become completely turbulent.

For viscous or laminar flow (Fig. 68–1), Stokes[10] showed that the force resisting the motion of a spherical particle is

$$F_R' = 3\pi D\mu v \tag{16}$$

where D = diameter of the spherical particle.
μ = viscosity of the fluid.
v = velocity of the particle relative to the fluid.

Equation 11b for a sphere may then be written

$$\left(\frac{\pi D^3}{6}\rho_s\right)\frac{dv}{dt} = \frac{\pi D^3}{6}g(\rho_s - \rho_l) - 3\pi D\mu v$$

By dividing both sides of the equation by $\pi D^3\rho_s/6$

$$\frac{dv}{dt} = \frac{(\rho_s - \rho_l)g}{\rho_s} - \frac{18\mu v}{D^2\rho_s}$$

At the maximum (or terminal) velocity v_m, $dv/dt = 0$,

$$\frac{(\rho_s - \rho_l)g}{\rho_s} = \frac{18\mu v_m}{D^2\rho_s} \quad \text{and} \quad v_m = \frac{(\rho_s - \rho_l)gD^2}{18\mu} \tag{17}$$

This form of the equation is the usual version of Stokes' law, which is applicable to the fall (or rise) of spherical particles of a nonporous incompressible solid in an incompressible fluid in laminar flow.

If the value for v_m of equation 17 for laminar flow is substituted for v_m in equation 15, the friction factor becomes

$$f_D = \frac{4(\rho_s - \rho_l)gD}{3v_m\rho_l}\,\frac{18\mu}{(\rho_s - \rho_l)gD^2}$$

$$= 24\frac{\mu}{Dv_m\rho_l} = 24\frac{1}{\text{Re}} \tag{18}$$

The expression $\mu/Dv\rho$ was referred to by Newton[4] as the criterion of flow. But it is usually referred to as 1/Re or the reciprocal of the Reynolds number because Osborne Reynolds rediscovered this criterion[6] and applied it to the flow of fluids in pipes.

Whenever laminar flow prevails, equation 18 may be used directly, or the friction drag factor may be expressed as a logarithmic equation

$$\log(f_D) = \log 24 - \log\frac{Dv_m\rho}{\mu}$$

$$= \log 24 - \log \text{Re} \tag{19}$$

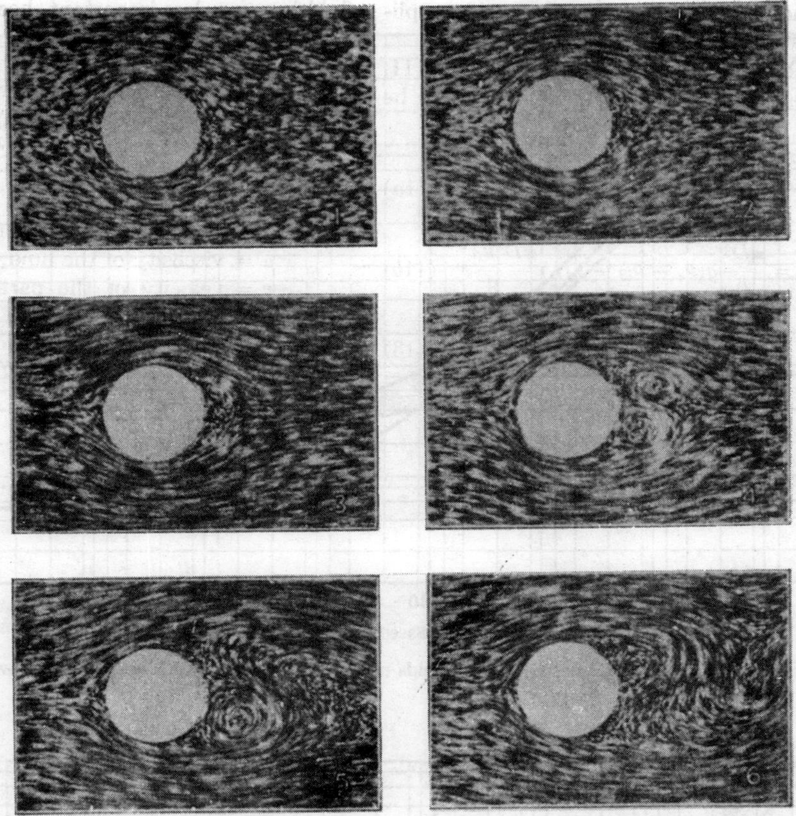

Fig. 68. **Flow lines of fluid passing a cylinder.** (*Hunter Rouse, "Elementary Mechanics of Fluids," John Wiley and Sons, 1946, p. 240, Plate XIV.*)

The friction factor in laminar flow is then represented by a straight line with a slope of −1 on logarithmic coordinates when log f is plotted as a function of log Re, as in Fig. 69 or Fig. 70. This condition exists up to a Reynolds number, Re, of about 3 when the eddies or turbulence begin to have some effect which increases as the values of Re become greater.

In the *turbulent flow* region (Re greater than about 300 or 400), the value for the friction factor becomes nearly constant, independent of Reynolds number as indicated by the horizontal part of the lines on Fig. 69. At a Reynolds number of about 200,000, the friction factor decreases sharply. This is explained by the fluid forming immediately behind the particle an eddy which travels with the particle, thereby constituting a streamlined body composed of sphere and eddy.

If the fluid is water, laminar flow is usually encountered with spheres of most minerals if the diameter is less than 50 microns, and turbulent flow is usually encountered with spheres of 1 mm (1000 microns) or more in diameter. Most of the sizes involved in commercial sedimentation or classification are in the range of 0.05 to 1 mm where the values of the friction factor f_D can be best obtained only from the plot, Fig. 70.

As indicated in Fig. 69, the transition from laminar to turbulent flow is rather indefinite, there being no sharp break from the straight sloping line representing laminar flow (equation 19) to the horizontal line of completely turbulent flow where f_D is practically constant. A particle always tends to move in that manner which offers the greatest resistance. In that region where the friction factor in turbulent flow is not significantly greater than in laminar flow, any local irregularities in conditions would tend to cause a change in type of flow and the flow would

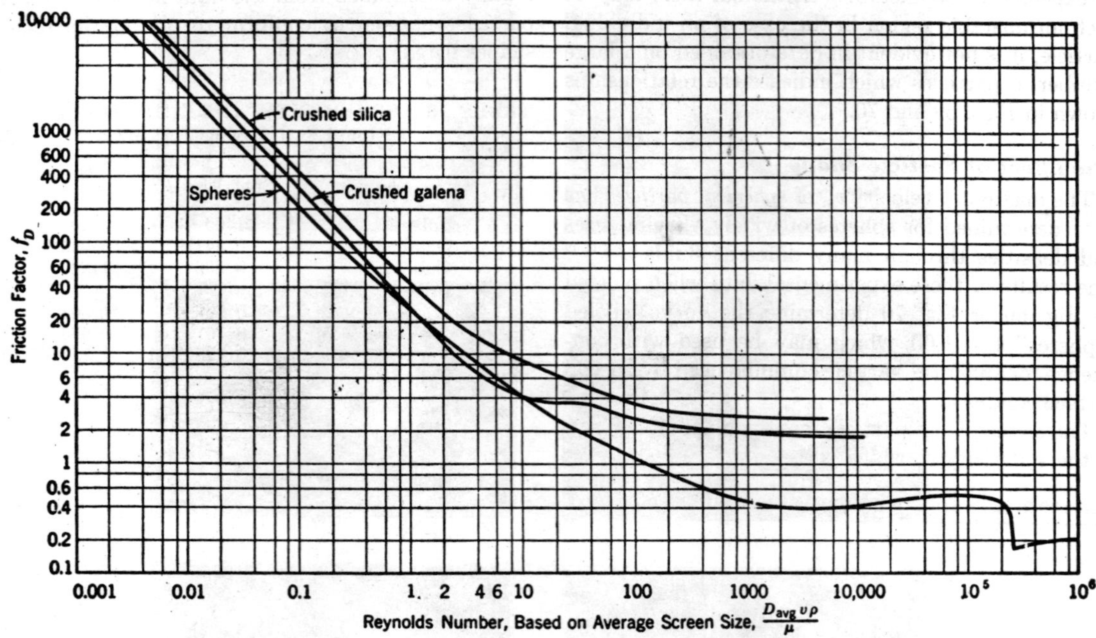

FIG. 69. Friction factor, or drag coefficient, versus Reynolds number for crushed solids and spheres flowing through fluids.[8]

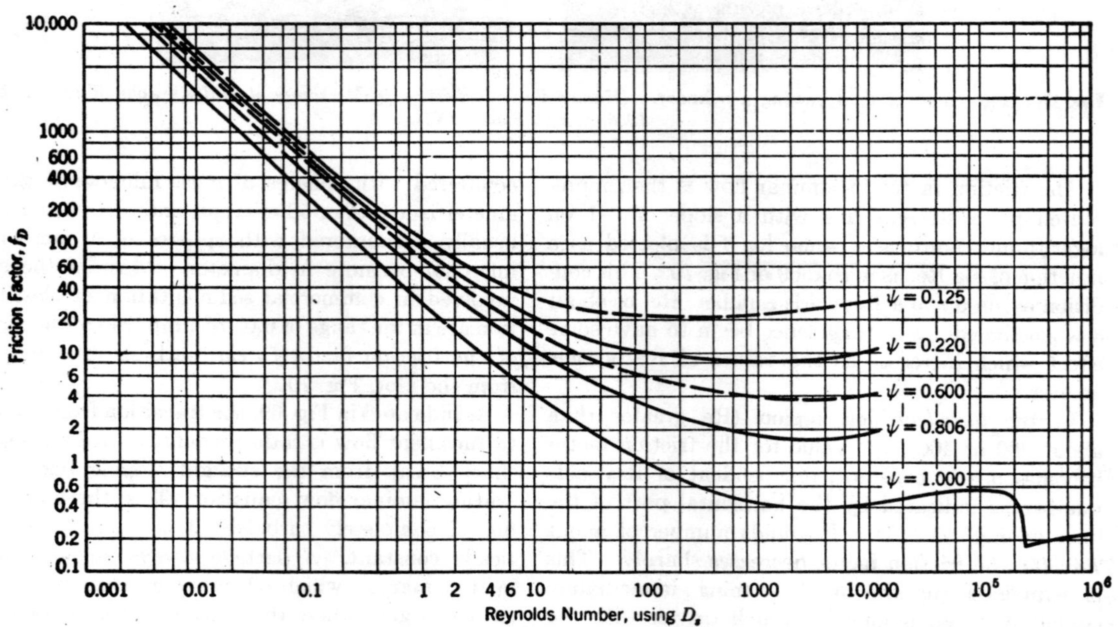

FIG. 70. Friction factor, or drag coefficient, versus Reynolds number for particles of different sphericities.[12]

be expected to be unstable. Although there may be such an unstable region in the case of an individual particle, it is not evident in data obtained on a large number of particles which indicate the relationships shown in Figs. 69 and 70.

The Effect of Particle Shape

The maximum velocity v_m of *spherical particles* has been determined for spheres of widely varying sizes and densities through many different fluids.[5,7,8,11,12] The results of these experimental data when plotted on Fig. 69 or Fig. 70 determine the curve labeled "spheres" $\psi = 1.00$, which may be used with confidence for spherical particles uninfluenced by special surface conditions.

The shape of the particle may be defined in terms of the sphericity ψ, which is

$$\frac{\text{Surface area of sphere having same volume as particle}}{\text{Surface area of particle}} = \psi = \frac{D_{avg}}{D_s}\left(\frac{1}{n}\right)$$

where n = the ratio of specific surfaces, the n of Fig. 17.

D_{avg} = average screen size.

D_s = diameter of sphere having same volume as particle (for a spherical particle $D_s = D$).

The above equation is derived as follows:

$$n = \frac{\text{Specific surface of particle}}{\text{Specific surface of sphere of same diameter}}$$

$$= \frac{\dfrac{\text{Surface of particle}}{\text{Mass of particle}}}{\dfrac{\text{Surface of sphere of same diameter}}{\text{Mass of sphere of same diameter}}}$$

$$= \left(\frac{\pi D_s^2/\psi}{\pi D_s^3 \rho/6}\right)\left(\frac{\pi D_{avg}^3 \rho/6}{\pi D_{avg}^2}\right)$$

$$n = \frac{D_{avg}}{D_s \psi}$$

Therefore,

$$\psi = \frac{D_{avg}}{D_s}\frac{1}{n}$$

The value of D_s is the value to be used for D in all flow equations using sphericity. The friction factor f_D may be estimated as a function of Reynolds number as indicated by the family of curves plotted on Fig. 70, with the sphericity ψ used as a parameter.[11] The sphericity ψ and the value of D_s to be used may be readily determined from the data in Table 18 if the particles happen to conform to one of the regular shapes listed.

TABLE 18. SPHERICITY AND THE VALUE OF D_s RELATED TO SCREEN SIZE

Shape	Sphericity ψ	$\dfrac{D_s}{D_{avg}}$
Sphere	1.00	1.00 *
Octahedron	0.847	0.965
Cube	0.806	1.24
Prisms		
$a \times a \times 2a$	0.767	1.564
$a \times 2a \times 2a$	0.761	0.985
$a \times 2a \times 3a$	0.725	1.127
Cylinders		
$h = 2r$	0.874	1.135
$h = 3r$	0.860	1.31
$h = 10r$	0.691	1.96
$h = 20r$	0.580	2.592
Disks		
$h = 1.33r$	0.858	1.00
$h = r$	0.827	0.909
$h = r/3$	0.594	0.630
$h = r/10$	0.323	0.422
$h = r/15$	0.254	0.368

* Multiply screen size D_{avg} by the factor indicated to get correct value for D_s to be used in the equations.

If the shape is not regular, as would be true for a mutilated cube, the sphericity and the correct value for D_s may be approximated by interpolating between the regular basic crystal form and the sphere, provided extreme care and judgment are used. The actual shapes of particles encountered in industrial operations can usually be estimated only by microscopic examination. The curves of Fig. 70 are based on a "diameter" D_s, which is not the same dimension as the size of a particle determined by a screen analysis (the average of the apertures of the confining screens). The screen size approximates the second largest dimension of the particle. For example a prism a by $2a$ by $3a$ has the effective dimension $2a$ as its "size" in screen analysis, but the diameter of a sphere having the same volume (D_s) is $1.127(2a) = 2.254a$. A cube having an edge of a, which is also the effective dimension in screen analysis, has $1.24a$ as the diameter for a sphere of the same volume (D_s).

In Fig. 69, the curves labeled "crushed silica" and "crushed galena" were plotted from experimental data [8] covering sizes varying from 0.0008 to 1.8 cm, with the size indicated by the screen analysis as the diameter D_{avg}. The anomalous location of these curves results from the use of the dimension obtained by screen analysis (D_{avg}) instead of the diameter of the sphere of equal volume (D_s).

However, the curves plotted in Fig. 69 on the basis of screen sizes (D_{avg}) are convenient as they may be used directly for "crushed galena," and silica without computing the sphericity or D_s.

Flocculation

When many particles are present there is a tendency for individual particles to agglomerate into clusters or flocs. Each floc then behaves much as a single particle of larger size and different shape than the individual particles constituting the floc. For this reason the observed rate of settling is frequently many times greater than that computed for the small individual particle. Flocculation agents which increase the tendency to form flocs are frequently added to increase the speed with which a suspension settles to produce a clear fluid. Deflocculating agents which reduce the tendency to form flocs have the opposite effect.

Hindered Settling

When many particles are present there is mutual interference in the motion of particles, and the velocity of motion or rate of settling is considerably less than that computed by the equations derived on the assumption of free motion of the solid particles. The particle is actually settling through a slurry or suspension of particles in a fluid rather than through the simple fluid itself. Therefore, the bulk density of the slurry ρ_b, calculated by dividing the mass of the fluid and suspended matter by the volume occupied by the slurry, should be used instead of the density ρ of the clear fluid in equation 17. Likewise, the bulk viscosity of the slurry μ_b, as may be determined experimentally, may be used instead of the viscosity μ of the clear fluid.

With these substitutions equation 17 becomes

$$v_H = v_m = \frac{(\rho_s - \rho_b)gD^2}{18\mu_b} \tag{17a}$$

where ρ_b = bulk density of suspension.
μ_b = bulk viscosity of suspension (Table 17).
v_H = hindered settling maximum velocity.

The bulk viscosity is frequently a function of the rate of shear in the case of suspensions and therefore an indefinite and indeterminate value in so far as equation 17a is concerned. However, the interference caused by the presence of many particles is a function of the volume fraction of the slurry (X) occupied by the fluid. Therefore, the viscosity of the fluid may be multiplied by a factor determined as a function of the volume fraction X.

$$\frac{\mu_b}{\mu_l} = \frac{10^{1.82(1-X)}}{Y}$$

This relationship has been developed for spherical particles.[9] Since the bulk density is also a function of the same volume fraction X and the densities of the solid and fluid, a convenient way of estimating the velocity of hindered settling is to apply the factor F_s to the velocity calculated for free settling. This factor F_s is given as a function of the volume fraction of the fluid (X) in Fig. 71 and may be used as a

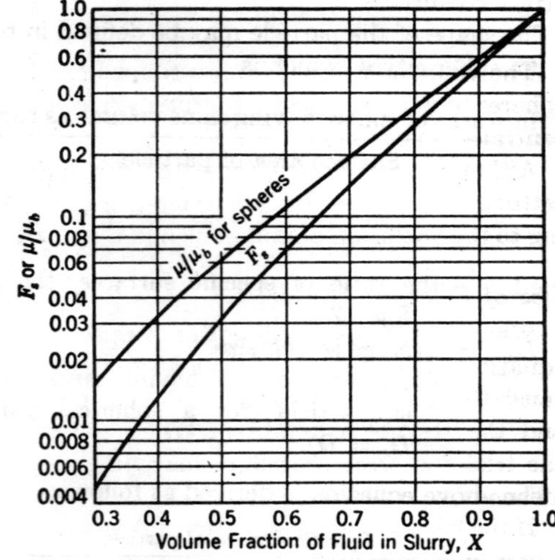

Fig. 71. Settling factor F_s for hindered settling versus volume fraction of fluid in slurry for spherical particles in laminar flow.[9]

multiplying factor[9] in equation 17 for estimating the velocity of settling when many particles are present.

$$v_H = \frac{(\rho_s - \rho_l)gD^2}{18\mu_l} F_s \tag{17H}$$

The value of F_s as given in Fig. 71 for viscous sedimentation of nonflocculated spheres may be obtained from the equation

$$F_s = \frac{v_H}{v_{max}} = \frac{X^2}{10^{1.82(1-X)}}$$

If X is not greater than 0.7, the following simpler relationship may be used.

$$F_s = \frac{v_H}{v_{max}} = 0.123 \frac{X^3}{1 - X}$$

The effect of particle concentration on rate of settling is different for angular particles than for spheres. Apparently the angular particles carry with them a volume of liquid which is proportional to the volume of the solid and a function of particle shape and degree of flocculation. This liquid is not adsorbed on the solid but simply held relatively stagnant by the angularities in the particles. The effect of this stagnant liquid may be incorporated by modifying the above equation as follows.

$$F_s = \frac{v_H}{v_{\max}} = \frac{0.123(1+a)^2}{1-X}\left(X - \frac{a}{1+a}\right)^3$$

where a = volume of immobile liquid per unit volume of solid.

The value of a varies from zero for nonflocculated spheres to unity or greater for highly flocculated concentrated suspensions.

Although flocculation increases the rate in free settling it decreases the rate in hindered settling, owing to this inclusion of immobile liquid with the solid.

Use of Friction Factor

Whenever D, ρ, or v_m is not known, the solution of equation 14 becomes a trial-and-error procedure because these terms appear in both the friction factor and the Reynolds number. When solving for v_m, the trial-and-error procedure may be eliminated by means of equation 22.

If the friction factor (equation 15) is written in the logarithmic form

$$\log (f_D) = \log \frac{4g(\rho_s - \rho_l)D}{3\rho_l} - 2 \log v_m \quad (20)$$

and the Reynolds number

$$\log \text{Re} = \log \frac{D\rho_l}{\mu} + \log v_m \quad (21)$$

eliminating $\log v_m$ between these equations gives

$$\log (f_D) = -2 \log \text{Re} + \log \left(\frac{4g(\rho_s - \rho_l)D^3\rho_l}{3\mu^2}\right) \quad (22)$$

an equation between f and Re in which v_m does not appear and which may be plotted on Fig. 70 as a straight line with a slope of -2 passing through the points

$$(f_D) = \frac{4g(\rho_s - \rho_l)D^3\rho_l}{3\mu^2}, \qquad \text{Re} = 1$$

and

$$(f_D) = 1, \qquad \text{Re} = \sqrt{\frac{4g(\rho_s - \rho_l)D^3\rho_l}{3\mu^2}}$$

The intersection of the straight line of equation 22 when plotted on Fig. 69 or Fig. 70 with the proper line giving f_D as a function of Re is the desired value for Reynolds number incorporating the value of v_m which is the required solution. The value of v_m is then computed directly from the value of Re, since D, ρ, and μ are known.

Exercise. By a similar procedure derive the following equation (23) and show how it can be used to calculate the size of a particle that will have a specified maximum velocity v_m.

$$\log (f_D) = \log \text{Re} + \log \left[\frac{4g(\rho_s - \rho_l)\mu}{3\rho_l^2 v_m^3}\right] \quad (23)$$

Wall Effect on Free Settling

If the fluid is confined in a cylinder or otherwise so that it may not be regarded as infinite in extent, the projected area of the falling particle decreases the effective cross-sectional area of the confining vessel, requiring the fluid to move with a greater velocity than if it were infinite in extent. This action has the effect of increasing the resistance tending to decrease the maximum velocity v_m. If the vessel is cylindrical, the following empirical "correction" factors by which the maximum velocity v_m may be "corrected" or multiplied have been suggested to take into account the wall effect.

For laminar flow [1]

$$\left(1 - \frac{D}{D_c}\right)^{2.25}$$

For turbulent flow [3]

$$\left[1 - \left(\frac{D}{D_c}\right)^{1.5}\right]$$

where D = diameter of the spherical particle.
D_c = diameter of the vessel or container.

Exercise. Draw curves showing the value of the velocity correction factors for wall effect as a function of the ratio of the diameter of the container to the diameter of the spherical particle.

THE GENERAL CASE

In the previous treatment the particle was considered to be moving in one dimension only, downward. But the motion of the particle may have a horizontal component as well as a vertical. In such cases, a balance of forces gives

$$m\frac{dv_h}{dt} = -F' \cos\alpha \tag{24}$$

$$m\frac{dv_v}{dt} = mg\frac{\rho_s - \rho_l}{\rho_s} - F' \sin\alpha \tag{25}$$

where α = angle between direction of motion and the horizontal.
v_h = horizontal component of particle velocity.
v_v = vertical component of particle velocity
$\cos\alpha = v_h/v$.
$\sin\alpha = v_v/v$.
$v^2 = v_h^2 + v_v^2$.

Substituting from equation 10:

$$\frac{dv_h}{dt} = -\frac{f_D\rho_l A v v_h}{2m} \tag{26}$$

$$\frac{dv_v}{dt} = g\left(\frac{\rho_s - \rho_l}{\rho_s}\right) - \frac{f_D\rho_l A v v_v}{2m} \tag{27}$$

Equations 26 and 27 are too complex to solve in general form here.[2a] But for special cases solutions are available.[2]

For laminar flow, equations 26 and 27 may be simplified and integrated directly.

$$f_D = \frac{C}{\text{Re}} = \frac{C\mu}{Dv\rho}$$

$$(f_D)_h = \frac{C}{(\text{Re})_h} = \frac{C\mu}{Dv_h\rho_l}$$

$$(f_D)_v = \frac{C}{(\text{Re})_v} = \frac{C\mu}{Dv_v\rho_l}$$

$$vf_D = v_v(f_D)_v = v_h(f_D)_h$$

For laminar flow

$$\frac{dv_h}{dt} = -\frac{\rho_l A (f_D)_h v_h^2}{2m} \tag{26L}$$

$$\frac{dv_v}{dt} = g\left(\frac{\rho_s - \rho_l}{\rho_s}\right) - \frac{\rho_l A (f_D)_v v_v^2}{2m} \tag{27L}$$

Equations 26L and 27L, which apply only to laminar flow show that the velocity or motion in one direction is independent of the motion in a direction perpendicular to the first motion. Therefore the trajectory of the particle will be that of one motion superimposed on the other, and the equations may be integrated independently.

One-Dimensional Motion in Absence of Gravitational Field

Although such motion is not usually encountered, the consideration of horizontal or any one-dimensional motion alone is an important step in integrating equation 26L.

By substituting $\text{Re}\mu/D\rho$ for v and v_h in equation 26 for the horizontal or one-dimensional component,

$$\frac{d(\text{Re})}{f_D(\text{Re})^2} = -\frac{\mu A}{2Dm}dt \tag{28}$$

which may be integrated between limits $t = 0$ and $t = t$, and $\text{Re} = \text{Re}_0$ at $t = 0$ when $v = v_0$, and $\text{Re} = \text{Re}$ when $v = v$.

$$\frac{\mu A}{2Dm}t = \int_{\text{Re}}^{\text{Re}_0}\frac{d(\text{Re})}{f_D(\text{Re})^2}$$

$$= \int_{\text{Re}}^{\text{Re}_b}\frac{d(\text{Re})}{f_D(\text{Re})^2} - \int_{\text{Re}_0}^{\text{Re}_b}\frac{d(\text{Re})}{f_D(\text{Re})^2} \tag{29}$$

where Re_b = Re at any arbitrary reference base.

In Fig. 72 the value of the following integral

$$\int_{\text{Re}}^{2\times10^5}\frac{d(\text{Re})}{f_D(\text{Re})^2}$$

is plotted against Re for spherical particles. In using Fig. 72, choose a value of v and compute the corresponding value of Reynolds number Re.

The right-hand side of equation 29 is evaluated by adding the respective integrals as determined from Fig. 72 with due regard for the sign. The value of the elapsed time (t) for the chosen one-dimensional velocity (v) can be calculated directly. The time-distance relation is then readily obtainable by graphical integration

$$s = \int_{t=0}^{t=t} v\,dt$$

Graphical integration is readily and accurately accomplished by plotting the values of v as determined as a function of t, as indicated in Fig. 73a, drawing a smooth curve through the points, and determining the area under the curve by the following procedure.

Draw horizontal lines connecting the ordinates, such as lines a–b, c–d, e–f, etc., so that the area above each horizontal line and below the curve equals the

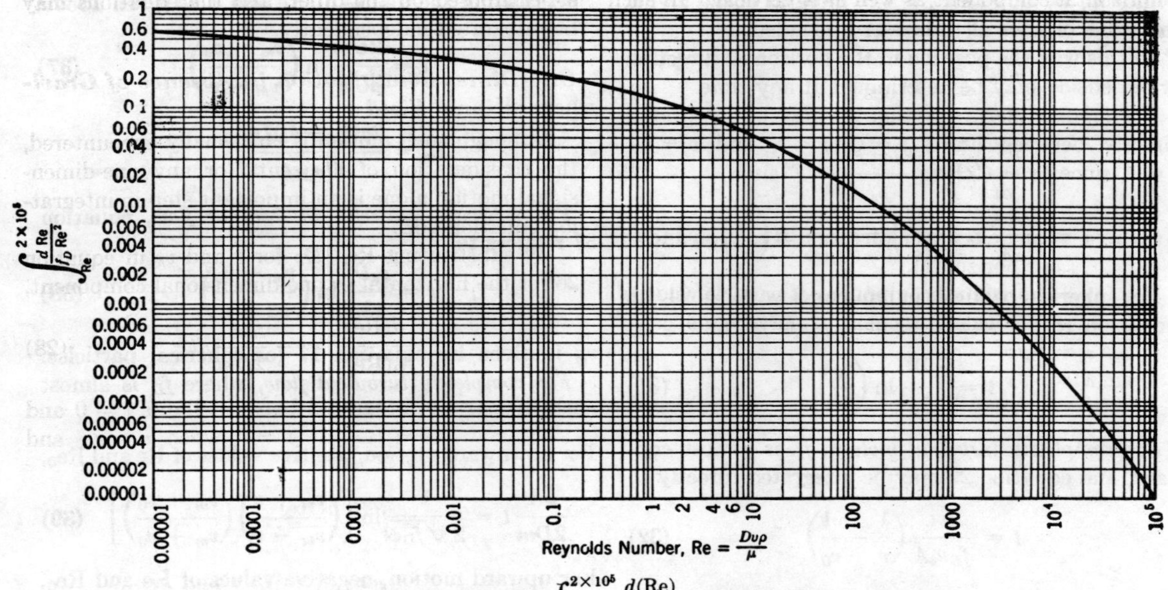

FIG. 72. Values of $\log \int_{Re}^{2\times10^5} \frac{d(\text{Re})}{f_D(\text{Re})^2}$ versus log (Re).

area below the horizontal line and above the curve, that is, the area 1a2 equals 2b3, and the area 3c4 equals the area 4d5, etc. With a little care this can be done quite accurately by eye. The area of the rectangle a,b,10,0 is then equal to the area under the curve 1,2,3,10,0 between the same ordinates. The same is true for each rectangle so constructed.

The integral curve representing the value of the integral

$$\int_{t=0}^{t=t} v\,dt$$

may then be readily drawn as indicated in Fig. 73b, giving

$$s = \int_0^t v\,dt + s_0$$

as a function of t as follows.

Point s_0 at $t = 0$ in Fig. 73b represents the distance or position of the particle at zero time ($t = 0$). The increase in distance Δs between $t = 0$ and $t = 10$ is represented by the area a,b,10,0 of Fig. 73a. The distance s at $t = 10$ is obtained by adding the appropriate value of Δs to s at $t = 0$, that is, laying of AB on Fig. 73b numerically equal to $\Delta s]_{t=0}^{t=10}$ or equal to area a,b,10,0 on Fig. 73a. Likewise, CD equals the area c,d,20,10, and EF equals the area e,f,30,20, etc. A smooth curve drawn through the

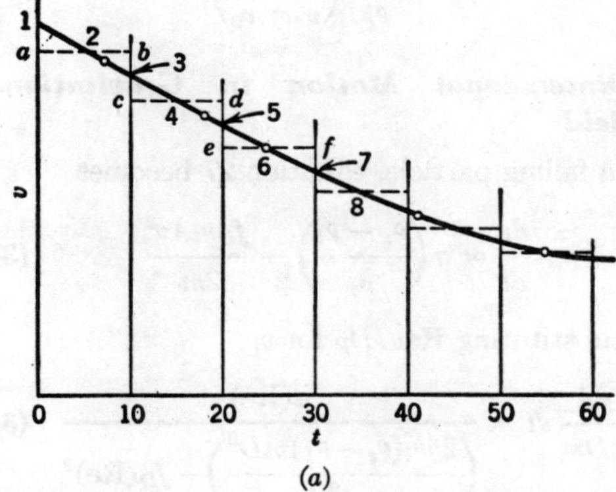

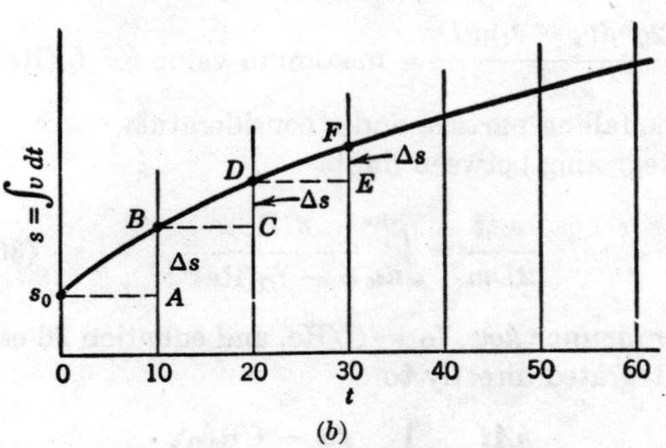

FIG. 73. Graphical integration. (a) The differential curv v versus t; (b) the integral curve, s versus t.

points s_0, B, D, F, etc., is the integral curve giving the value of s for all values of t included.

In this way the position in the horizontal direction or dimension may be determined at any time.

For laminar flow, equation 29 may be integrated directly between limits $t = 0$, $t = t$, and $v = v_0$, $v = v$, since $f_D = C/\text{Re}$.

$$t = \frac{2Dm}{C\mu A} \ln\left(\frac{\text{Re}_0}{\text{Re}}\right) = \frac{2Dm}{C\mu A} \ln\left(\frac{v_0}{v}\right) \tag{30}$$

For spherical particles in laminar flow, $f_D = 24/\text{Re}$ (equation 18); $A = \pi D^2/4$; $m = \rho_s \pi D^3/6$, and

$$t = \frac{\rho_s D^2}{18\mu} \ln\left(\frac{v_0}{v}\right) \tag{31}$$

For completely turbulent motion, f_D is almost constant, and equation 29 may be integrated directly to

$$t = \frac{2m}{f_D \rho_l A}\left(\frac{1}{v} - \frac{1}{v_0}\right) \tag{32}$$

For spherical particles, f_D is about 0.4 and

$$t \simeq \frac{3\rho_s D}{\rho_l}\left(\frac{1}{v} - \frac{1}{v_0}\right) \tag{33}$$

One-Dimensional Motion in Gravitational Field

For a falling particle, equation 27 becomes

$$\frac{dv}{dt} = g\left(\frac{\rho_s - \rho_l}{\rho_s}\right) - \frac{f_D \rho_l A v^2}{2m} \tag{34}$$

By substituting $\text{Re}\mu/D\rho$ for v,

$$\frac{\mu A}{2Dm} dt = \frac{d(\text{Re})}{\left(\dfrac{2g\rho_l(\rho_s - \rho_l)mD^2}{\rho_s \mu^2 A}\right) - f_D(\text{Re})^2} \tag{35}$$

$$\phi = \frac{2g\rho_l(\rho_s - \rho_l)mD^2}{\rho_s \mu^2 A} = \text{maximum value for } f_D(\text{Re})^2$$

for the falling particle under consideration.

Integrating between limits

$$\frac{\mu A t}{2Dm} = \int_{\text{Re}_0}^{\text{Re}} \frac{d(\text{Re})}{\phi - f_D(\text{Re})^2} \tag{36}$$

For laminar flow, $f_D = C/\text{Re}$, and equation 36 can be integrated directly to

$$\frac{\mu A t}{2Dm} = \frac{1}{C} \ln\left(\frac{\phi - C\text{Re}_0}{\phi - C\text{Re}}\right)$$

Substituting the values of Re and ϕ as given above,

$$t = \frac{2Dm}{\mu CA} \ln\left(\frac{v_m - v_0}{v_m - v}\right) \tag{37}$$

where

$$v_m = \frac{2g(\rho_s - \rho_l)Dm}{C\rho_s \mu A}$$

For spherical particles in laminar flow, equation 37 reduces to

$$t = \frac{\rho_s D^2}{18\mu} \ln\left(\frac{v_m - v_0}{v_m - v}\right) \tag{38}$$

v_m is given by equation 17 for spherical particles.

For completely turbulent flow, where f_D is almost constant:

For downward motion, positive values of Re and Re_0,

$$\frac{\mu A}{2Dm} t = \frac{1}{2\sqrt{f_D \phi}} \ln\left[\left(\frac{v_m + v}{v_m - v}\right)\left(\frac{v_m - v_0}{v_m + v_0}\right)\right] \tag{39}$$

For upward motion, negative values of Re and Re_0,

$$\frac{\mu A}{2Dm} t = \frac{1}{\sqrt{f_D \phi}} \tan^{-1}\left[\frac{(v_0 - v)v_m}{v_0 v + v_m^2}\right] \tag{40}$$

where

$$v_m = \sqrt{\frac{2g(\rho_s - \rho_l)m}{\rho_l \rho_s f_D A}}$$

or as given by equation 14 for spherical particles.

When a particle has an initial downward velocity, equation 39 may be used directly. When the initial velocity is upward, equation 40 should be used during the ascent, and equation 39 applied thereafter to the descent with $v_0 = 0$.

For further treatment of this problem and for two-dimensional motion in centrifugal fields, the original reference[2] should be consulted.

The general equations *for motion of a particle in a fluid stream undergoing simple rotational motion* are

$$\frac{dv_r}{dt} = \left(\frac{v_t^2}{r}\right)\left[1 - \left(\frac{\rho_l u_t^2}{\rho_s v_t^2}\right)\right] - \left(\frac{\rho_l f_D A v v_r}{2m}\right) \tag{41}$$

$$\frac{dv_t}{dt} = -\frac{v_t v_r}{r} + \left[\frac{\rho_l f_D A v(u_t - v_t)}{2m}\right] \tag{42}$$

where u_t = tangential velocity of the fluid.
v_t = tangential velocity of the particle.
v_r = radial velocity of the particle.
$v = \sqrt{v_r^2 + (u_t - v_t)^2}$ = velocity of the particle relative to the fluid

For spherical particles in laminar flow

$$\frac{dv_r}{dt} = \frac{v_t^2}{r}\left[1 - \left(\frac{\rho_l}{\rho_s}\right)\left(\frac{u_t}{v_t}\right)^2\right] - \frac{18\mu v_r}{\rho_s D^2} \quad (41L)$$

$$\frac{dv_t}{dt} = -\frac{v_t v_r}{r} + \left[\frac{18\mu(u_t - v_t)}{\rho_s D^2}\right] \quad (42L)$$

When the tangential velocity of the particle v_t is about equal to the tangential velocity of the fluid u_t and v_r is relatively small, equation 41 may be reduced to

$$v_r^2 = \frac{2mu_t(\rho_s - \rho_l)}{\rho_l \rho_s f_D A r} \quad (43)$$

For spherical particles equation 43 becomes

$$v_r = \frac{u_t^2(\rho_s - \rho_l)D^2}{18\mu r} \quad (44)$$

BIBLIOGRAPHY

1. Francis, Alfred W., "Wall Effect in the Falling-Ball Method for Viscosity," *Physics*, **4**, 403–406 (1933).
2. Lapple, C. E., and C. B. Shepherd, "Calculation of Particle Trajectories," *Ind. Eng. Chem.*, **32**, 605 (May 1940).

2a. von Kármán, Theodore, and Maurice A. Biot, *Mathematical Methods in Engineering*, p. 139, McGraw-Hill Boo Co. (1940).

3. Monroe, H. S., "The English vs. the Continental System of Jigging—Is Close Sizing Advantageous?" *Trans. Am. Inst. Mining Met. Engrs.*, **17**, 637–659 (1888–1889).
4. Newton, Isaac, *Mathematical Principles of Natural Philosophy*, Book II (English translation of 1729).
5. Pernolet, V., "A l'étude des préparations mécaniques des minerais, ou expériences propres à établir la théorie des différents systèmes usités ou possibles," *Ann. Mines* [IV], **20**, 379–425 (1851).
6. Reynolds, Osborne, "An Experimental Investigation of the Circumstances which Determine Whether the Motion of Water Shall be Direct or Sinuous, and of the Law of Resistance in Parallel Channels," *Phil. Trans. Roy. Soc. (London)*, **174**, 935–982 (1883).
7. Richards, R. H., and C. E. Locke, *Textbook of Ore Dressing*, Fig. 65, p. 129, 3rd ed., McGraw-Hill Book Co. (1940).
8. Richards, R. H., "Velocity of Galena and Quartz Falling in Water," *Trans. Am. Inst. Mining Met. Engrs.*, **38**, 210–235 (1907).
9. Steinour, H. H., "The Rate of Sedimentation," *Ind. Eng. Chem.*, **36**, 618, 840, 901 (1944).
10. Stokes, G. G., *Mathematical and Physical Papers*, (1901). *Trans. Cambridge Phil. Soc.*, **9**, Part II, pp. 51ff (1851).
11. Waddel, Hakon, "Some New Sedimentation Formulas," *Physics*, **5**, 281–291 (1934).
12. Waddel, Hakon, "The Coefficient of Resistance as a Function of Reynolds' Number for Solids of Various Shapes," *J. Franklin Inst.*, **217**, 459–490 (1934).

PROBLEMS

1. By means of equation 17, compute the maximum velocity at which particles of silica (SiO_2) 0.005 cm in diameter will fall through quiet water:

(*a*) When the slurry is so dilute that free settling prevails.

(*b*) When the mass ratio of water to silica is 2.

2. With data from Tables 1 and 17, compute values for F_s for calcite and compare these values with those given in Fig. 71 for spherical particles.

3. Air is being dried by being bubbled (in very small bubbles) through concentrated sulfuric acid (specific gravity, 1.84; viscosity, 15 centipoises; temperature, 100° F). The sulfuric acid fills a 24-in. tall, 2-in. ID glass tube to a depth of 6 in. The dry air above the acid is at a pressure of 1 atm and at 100° F. If the dry air rate is 3.5 cfm, what is the maximum diameter of a sulfuric acid spray droplet which might be carried out of the apparatus by entrainment in the air stream.

4. A thin water suspension of soil is prepared at 10 A.M. At noon, a sample is drawn from the suspension at a depth of 5 cm. What is the largest particle probably removed by a pipette at this depth (5 cm) if the specific gravity of the soil is 2.5?

5. It is required to classify small particles of charcoal which may be assumed to be disks or cylinders such that $D_s = D_{avg}$, with a specific gravity of 0.8. The charcoal is to be allowed to fall freely through a vertical tower against a rising current of air at 20° C and atmospheric pressure. Calculate the minimum size of charcoal which will settle to the bottom of the tower if the air is rising through the tower with a velocity of 10 fps.

6. A mixture of spherical particles of silica contains particles ranging in size from 14 mesh to 200 mesh. This mixture is to be divided into two fractions by elutriation, utilizing the upward velocity of a stream of water at 55° F rising through a tube 4 in. in diameter.

(*a*) What quantity of water, in gallons per minute, will probably be needed to divide the mixture at a size equal to the aperture of a 48-mesh screen?

(*b*) When the water flow through the tube is 1.5 gpm, what is the smallest size of silica particle which will probably settle through the stream?

7. Crushed silica is to be analyzed by elutriation, with an elutriator whose analyzing zone is a cylinder 3.5 in. in diameter. Water at 55° F is to be used as the analyzing fluid. What rates of flow of that water, measured in gallons per minute, will be necessary to give a sized fraction of the crushed silica corresponding to the −270 +400 mesh fraction obtainable from Tyler standard screens?

8. A falling-ball viscometer operates by timing the fall of a steel ball with a diameter of 0.25 in. (density, 7.9 gm/cc) through the fluid. Oil whose density is 0.88 gm/cc is introduced into the instruments. The steel ball falls a distance of 10 in. through the oil. What is the viscosity of the oil if the time of fall is 6.35 sec?

CHAPTER

8

Classification

SEPARATION of materials into two or more fractions depending upon their rates of flow through fluids is called *classification*. When the different rates of flow are used to separate materials of the same density according to their sizes (and shape), the operation is known as *sizing*. Sizing is a possible substitute for screening. When materials of the same equivalent size are separated according to their densities, the operation is called *sorting*. Since it is impossible in actual operation to obtain a feed of absolutely uniform size, sizing is always involved to a greater or less degree in every sorting operation.

"EQUAL FALLING" PARTICLES

The effects of size and density of particles can be considered in combination to compute that range of sizes of a mixed feed that can be sorted or separated according to density. This is done conveniently by the concept of the "equal falling" particles, the sizes of particles of the two different materials to be sorted which fall through the fluid at equal velocities.

From equation 14, the two different materials, A and B, will have the same maximum velocity v_m when

$$v_m = \sqrt{\frac{4(\rho_A - \rho)gD_A}{3\rho(f_D)_A}} = \sqrt{\frac{4(\rho_B - \rho)gD_B}{3\rho(f_D)_B}}$$

where ρ = density of fluid, and subscripts A and B indicate properties or conditions for materials A and B.

The ratio of the diameters of spherical particles of the two different materials D_A/D_B having the same maximum settling velocity v_m may be called the *spherical settling ratio*. The ratio D/D_{avg} may be obtained from Table 18, p. 77, and used in equation 45.

Solving for the spherical settling ratio,

$$\frac{D_A}{D_B} = \frac{(\rho_B - \rho)}{(\rho_A - \rho)}\left[\frac{(f_D)_A}{(f_D)_B}\right]$$

$$= \left[\frac{(D_{\text{avg}})_A}{(D_{\text{avg}})_B}\frac{D_A}{(D_{\text{avg}})_A}\frac{(D_{\text{avg}})_B}{D_B}\right] \tag{45}$$

Under turbulent conditions, the value of f_D for each solid is constant over the range of conditions involved, and equation 45 may be readily solved. In many cases, as with particles of the same sphericity, the values for f_D are the same for each material and $(f_D)_A/(f_D)_B = 1$.

Under laminar conditions, f_D varies inversely with the Reynolds number Re, from Equation 18.

$$f_D = \frac{24\mu}{v_m\rho}\frac{1}{D}$$

For all materials having the same maximum velocity v_m in the same fluid in laminar flow,* equation 45 may be written

$$\frac{D_A}{D_B} = \frac{(\rho_B - \rho)}{(\rho_A - \rho)}\frac{D_B}{D_A} \quad \text{or} \quad \left(\frac{D_A}{D_B}\right)^2 = \frac{\rho_B - \rho}{\rho_A - \rho} \tag{46}$$

* In the region between laminar and completely turbulent conditions, where many sedimentation problems occur, f varies as a smaller negative power of Re. The following equation, with an exponent greater than 1 but less than 2 for the D_A/D_B term, may be used for particles of the same sphericity or when the value for f may be represented in terms of D and v as the only variables.

$$\frac{(D_A)^n}{(D_B)^n} = \frac{(\rho_B - \rho)}{(\rho_A - \rho)} \tag{47}$$

where n = 1 minus the slope of the curve for the corresponding solid on Fig. 70 over the range covered by operating conditions.

The ratio $(D_{avg})_A/(D_{avg})_B$, as obtained from equation 45, may be called the *settling ratio* and represents the ratio between the maximum and minimum "screen diameters" of particles that may exist in a screened or sized mixture of the materials A and B, if a clean separation between these two materials is to be made by classification, using the fluid in question.

A large settling ratio is desirable to permit separation of material having wide size limits. One way of accomplishing this is the use of a fluid of high density. As the density of the fluid approaches that of one of the solid materials the settling ratio approaches infinity and particles of any size range can be sorted.

If the size ratio $\left(\frac{D_{avg} \text{ of largest particle}}{D_{avg} \text{ of smallest particle}}\right)$ is equal to or less than the settling ratio, a complete separation can be made.

The density of the fluid may be increased by adding solutes such as salt or calcium chloride to water, but this expedient introduces corrosion problems. The effective density of the fluid may be increased by suspending fine particles of a heavy solid such as galena in water, in which case the bulk density of the suspension ρ_b becomes the effective density to be substituted for ρ in the equation. Similarly, as the concentration of the material being separated builds up in the fluid, the bulk density of the fluid approaches the density of the solid material and the settling ratio increases, giving a sharper sorting of the two materials. This principle is used in hindered settling.

EQUIPMENT

The simplest form of classification equipment is the so-called *surface velocity classifier* which, in its elemental form, consists of a tank provided with an

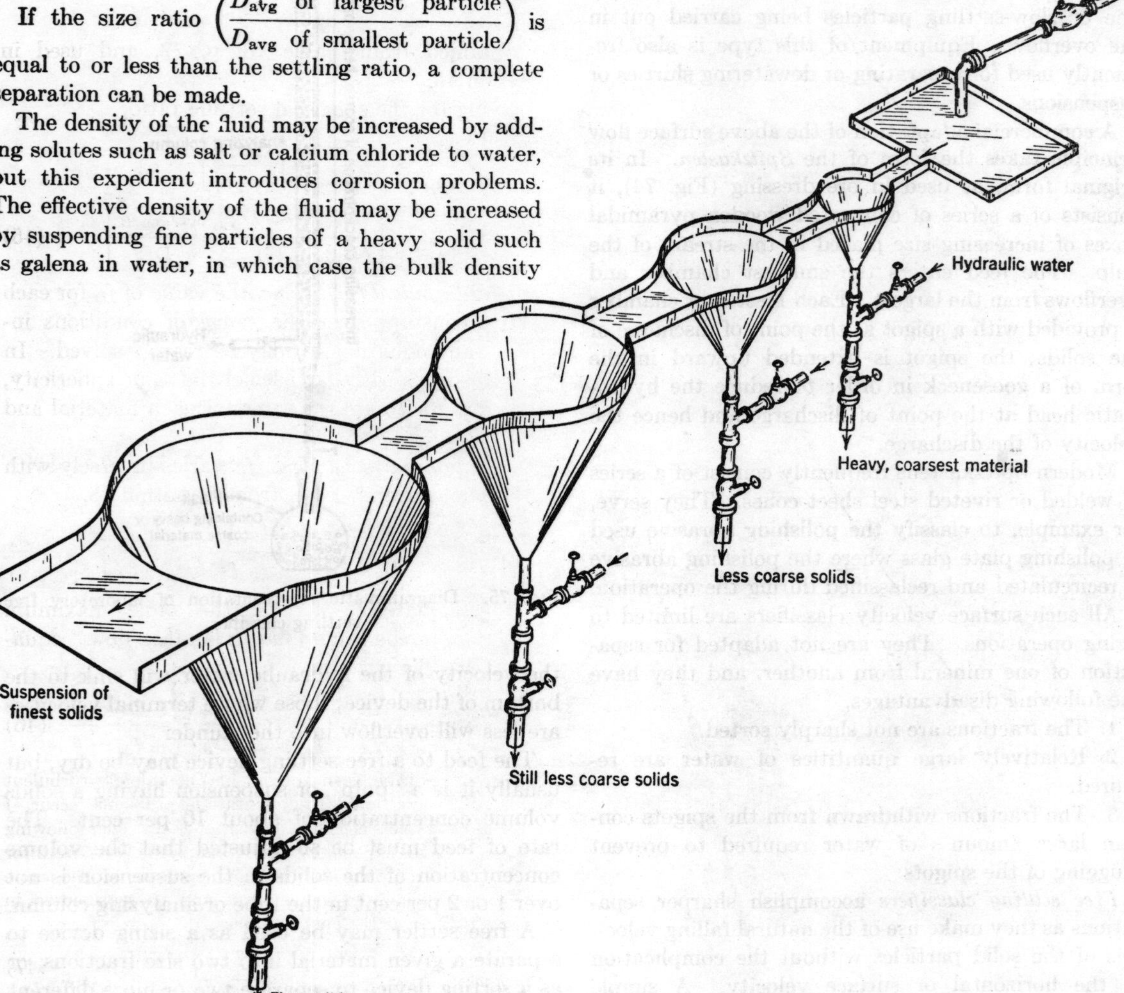

FIG. 74. Diagrammatic representation of a Spitzkasten.

inlet and an outlet opposite each other at or near the top of the tank. The feed, consisting of a pulp of solids suspended in water, enters with a relatively high horizontal velocity. After it enters the enlarged cross section of the tank, its horizontal velocity is decreased and gravity tends to cause the solid particles to sink. The path of each particle is a resultant of the surface velocity in the horizontal direction and the velocity of settling.

If the tank is provided with perpendicular partitions placed normal to the direction of flow, several fractions of solids can be collected, particles having the greatest settling velocity collecting behind the partition nearest the feed inlet, and the exceedingly fine or slow-settling particles being carried out in the overflow. Equipment of this type is also frequently used for separating or dewatering slurries or suspensions.

A commercial adaptation of the above surface flow principle takes the form of the *Spitzkasten*. In its original form, as used in ore dressing (Fig. 74), it consists of a series of connected wooden pyramidal boxes of increasing size placed in the stream of the pulp. The feed enters the smallest chamber and overflows from the largest. Each Kasten or chamber is provided with a spigot at the point of discharge of the solids; the spigot is extended upward in the form of a gooseneck in order to reduce the hydrostatic head at the point of discharge and hence the velocity of the discharge.

Modern Spitzkastens frequently consist of a series of welded or riveted steel sheet cones. They serve, for example, to classify the polishing abrasive used in polishing plate glass where the polishing abrasive is recirculated and reclassified during the operation.

All such surface velocity classifiers are limited to sizing operations. They are not adapted for separation of one mineral from another, and they have the following disadvantages.

1. The fractions are not sharply sorted.

2. Relatively large quantities of water are required.

3. The fractions withdrawn from the spigots contain large amounts of water required to prevent plugging of the spigots.

Free settling classifiers accomplish sharper separations as they make use of the natural falling velocities of the solid particles without the complication of the horizontal or surface velocity. A simple example of a free settling classifier is a vertical tube called the analyzing column provided with a fluid inlet and a solid outlet at the bottom and a feeding device and an overflow launder at the top (Fig. 75). By adjusting the rate of flow of fluid (hydraulic water) into the tube, any predetermined upward velocity of the fluid may be obtained. The feed entering the rising stream of fluid tends to fall; particles whose terminal velocities are greater than

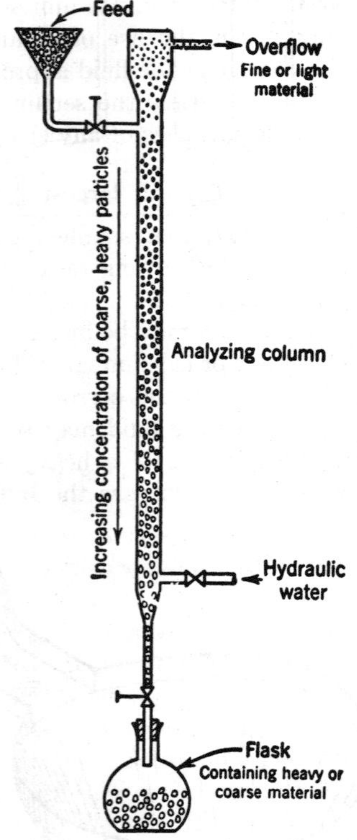

FIG. 75. Diagrammatic representation of laboratory free settling classifier.

the velocity of the hydraulic water will sink to the bottom of the device; those whose terminal velocities are less will overflow into the launder.

The feed to a free settling device may be dry, but usually it is a "pulp" or suspension having a solids volume concentration of about 10 per cent. The rate of feed must be so adjusted that the volume concentration of the solids in the suspension is not over 1 or 2 per cent in the tube or analyzing column.

A free settler may be used as a sizing device to separate a given material into two size fractions, or as a sorting device to separate two or more different materials by virtue of the differences in their densities. The amount of fluid required per ton of solids

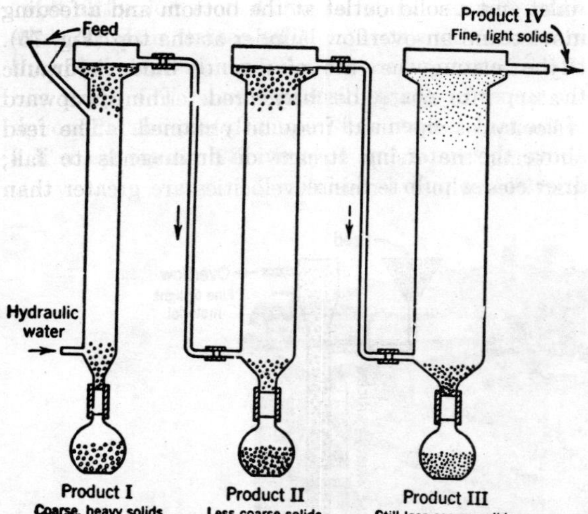

FIG. 76. Diagrammatic representation of elutriation or sizing by free settling.

treated is large although considerably less than with surface velocity classifiers. This item may be a major consideration even when the fluid is water. Many air classifiers are in service where the cost of operating them is less than for competing methods such as screening.

Elutriation is the name given a sizing classification operation conducted in a similar device of laboratory size for the purpose of determining the grain size distribution in the range below the finest screen size. By means of a series of free settling tubes as indicated in Fig. 76, the weighed sample originally placed in the first tube of smallest diameter is classified into fractions according to their settling velocities. The fractions are collected in the successive tubes and may be dried and weighed. Care must be taken that the particles are not agglomerated but well dispersed in the fluid if reliable results are to be obtained.

The *double-cone classifier* (Fig. 77) consists of a fixed external cone provided with a source of fluid (hydraulic water), a spigot at the tip, and a peripheral overflow launder around the open end. A concentric inner truncated cone is so supported that the height of this cone relative to the outer cone can be adjusted as by a handwheel. The feed flows down within the inner cone and out of the opening against a baffle. The hydraulic water or other fluid and the feed are mixed in the vicinity of the baffle and rise in the space between the two cones. Adjustment of the height of the inner cone and control of the amount of hydraulic fluid determines the velocity of fluid in the space between the cones which functions as the analyzing column.

The operating characteristics of a double-cone classifier are determined by the minimum cross-sectional area of the annular space between cones. From this minimum area A and the volumes of feed and hydraulic fluid introduced per unit time, the significant velocity of the analyzing column and the volume fraction of solids Y can be computed. The capacity of the classifier in tons of solid per hour may be estimated as follows:

$$\text{Capacity of double-cone classifier} = 1.8AYv\rho_s$$

where A = minimum cross-sectional area (sq ft).

Y = volume concentration, $\dfrac{\text{volume solid}}{\text{volume total}}$.

v = velocity (fps).

ρ_s = density of solids (lb/cu ft).

The double-cone classifier, as well as the free settling elutriator, is also operated with air or other fluid. Since air is much less viscous than water and also much less dense, the particles have much greater settling velocities, approximately one hundred times as great, in air than in water.

The *Gayco pneumatic classifier* (Fig. 78), although having the appearance of a cone classifier, operates on a different principle. The dry solid feed enters

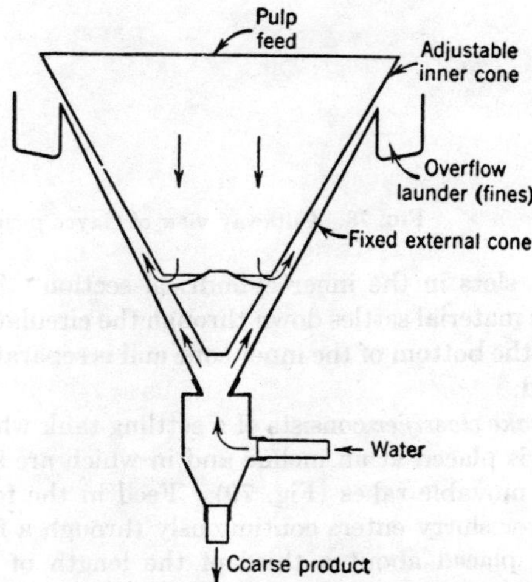

FIG. 77. Diagrammatic representation of double-cone classifier.

through a hollow shaft and falls upon the rotating feeder plate which distributes the particles across the cylindrical classification section. The circulating fan causes air to circulate upward and outward through the inner cylinder, carrying the fines into the outer cone where they settle out and are removed from the bottom. The circulating air returns through slots in the inner cylindrical section. The oversize material settles down through the circulating air into the bottom of the inner cone and is separately removed.

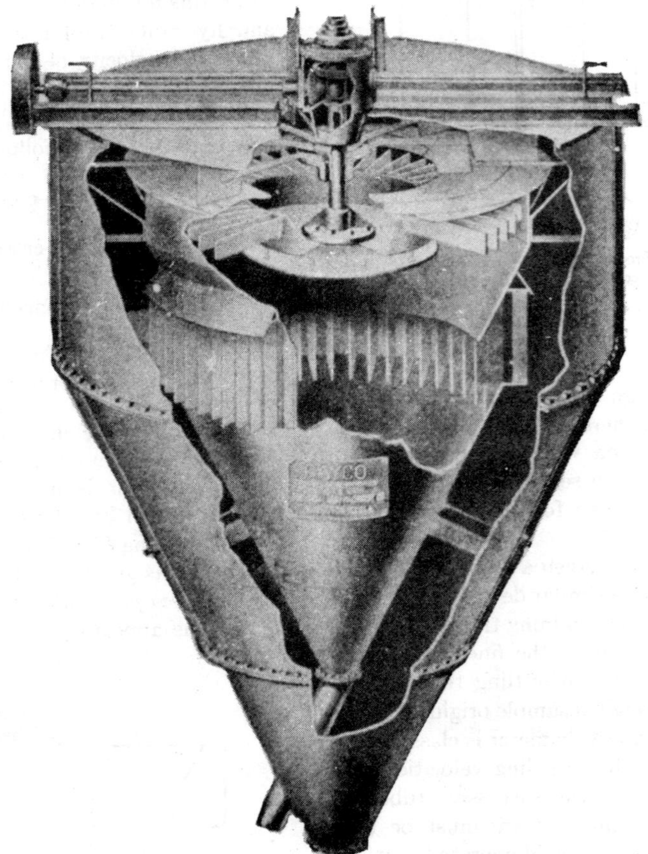

FIG. 78. Cutaway view of Gayco pneumatic classifier. (*Universal Road Machinery Co.*)

The *rake classifier* consists of a settling tank whose bottom is placed at an incline and in which are suspended movable rakes (Fig. 79). Feed in the form of pulp or slurry enters continuously through a feed launder placed about a third of the length of the tank from the lower end. The lower or fine discharge end is provided with an overflow lip over which the fine material which does not settle flows in a continuous stream. The coarser material sinks to the bottom where the rakes gently move it toward the upper or coarse discharge end. The upper end of the tank is open and frequently extends far enough above the water line to provide drainage space if a drier coarse pulp is desired.

At the beginning of the stroke the rakes are dropped to the bottom and slowly drawn toward the upper end. After movement of about a foot or two the rakes are lifted clear of the floor and moved parallel to the bottom toward the fine discharge end. At the completion of this part of their travel the rakes are dropped and the cycle is repeated. The movement of the rakes stirs up the solids. The heavier or faster settling particles collect on the bottom and are carried along with the upward movement of the rakes. The finer particles are

FIG. 79. Cutaway drawing of a rake classifier. (*Dorr Co.*)

thrown above the rakes and travel in suspension toward the fine discharge end.

Classifiers are built with one or more sets of rakes side by side. For closer sizing of the fines a spray of water may be added at the fine discharge end, diluting the pulp and permitting the operation to approach free settling; or the fine discharge may be further classified in a *bowl classifier* built as an extension to the rake classifier (Fig. 80).

The *cross-flow* or *Akins classifier* (Fig. 81) performs in much the same manner as the rake classifier. It consists of a trough of semicylindrical shape, set at an incline of about 2½ in./ft. Revolving within the trough is a shaft carrying a helix. Feed is introduced below the water level. The heavier solids sink and are carried up the slope by the helix. They are discharged at the upper end through an opening in the bottom of the trough above the water level. The fine material overflows at the lower end.

The *Hardinge countercurrent classifier* is a similar device consisting of a slowly revolving drum, the inner surfaces of which carry a helix. Motion is

FIG. 80. Bowl classifier working on overflow of rake classifier. (*Dorr Co.*)

imparted through a trunnion, thereby eliminating any submerged bearing.

Power and operating costs on these classifiers are relatively low. A rake classifier 18 in. wide by 15 ft long requires about 1.5 hp on light service, and 25 hp is sufficient for a classifier 16 ft wide and 30 ft long.

The size or mesh of the separation point, i.e., the largest size in the fines or overflow and the smallest size in the coarse fraction, is controlled by the operating conditions. A finer separation point is obtained by (1) greater dilution of the pulp, (2) lower tilt of the tank, and (3) slower speed of the rakes. For a feed of constant quality the capacity of the classifier is decreased by those changes in operating conditions which give a finer separation point. For this reason other types of classifiers are generally employed when a fine separation point is required.

These sloping classifiers have a practical advantage since they elevate the oversize in the act of classification, thereby frequently eliminating the necessity of a separate elevator when used with a mill in closed circuit operation.

For Classifying Fine Material

The *bowl classifier* (Fig. 80) consists of a short cylinder with a bottom sloping toward the center. The feed enters at the center through a loading well and baffle plate placed just below the surface of the liquid, directing the feed radially outward. The fines overflow into a launder at the periphery of the bowl. The oversize material is raked to the center by the slowly rotating vanes and discharges in an annular opening. When the classifier is operating on the fines from a rake classifier, the oversize from the bowl is fed back to the rake classifier. The bowl classifier is designed for more dilute pulps than the rake classifier. The large overflow permits the handling of large quantities of water with low velocities as required for the more slowly settling small particles.

The *spiral-vane classifier* (Fig. 82) is a device consisting of a cylindrical tank, a central shaft, and spiral or radial arms which sweep the bottom. Feed is pumped in as a slurry. The solids settle and are swept to the center and removed continuously. The fines overflow into a launder as shown.

FIG. 81. Cross-flow classifier. (*Denver Equipment Co.*)

When it is desired to *de-slime sands*, to remove very fine particles such as clay from the coarser sand particles, the feed in the form of a suspension is fed to the center of an inverted cone. The sand settles out, and the water carrying the fines overflows into a peripheral launder. The feed is supplied at a rapid rate, and the vertical velocity of the water is relatively high and will carry in suspension most of the solids of 200 mesh and smaller.

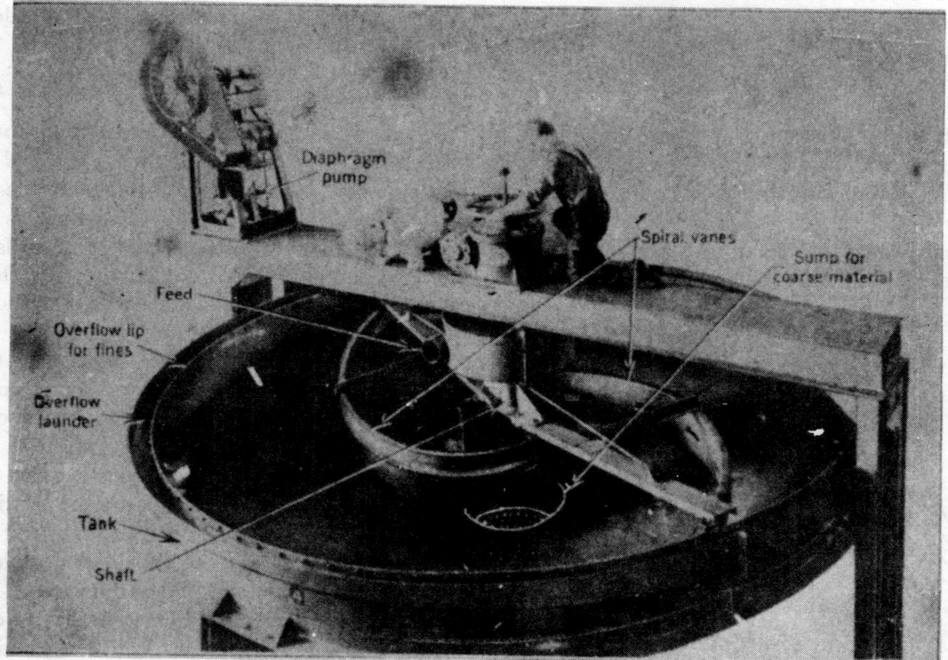

FIG. 82. Spiral vane classifier and adjustable-stroke diaphragm pump. (*Denver Equipment Co.*)

These cones are generally of 60-degree angle, 9 to 10 ft in diameter at the top. The sand is discharged through a spigot at the bottom, either by manual or automatic means, depending on the type of device. A typical cone will handle from 50 to 100 tons of solids per 24 hr. The feed will contain about 90 per cent water, the spigot discharge 30 per cent, and the overflow about 95 per cent.

On the basis of the same standards as in screening, the effectiveness of such a device can be about 70 per cent, with recovery around 90 per cent.

JIGGING

Jigging is a special form of hindered settling which consists of stratification of the particles into layers of different densities by repeatedly affording a very thick suspension of the mixed particles an opportunity to settle or fall for short periods of time and then removing the stratified layers.

In the classification equipment mentioned above, the maximum or terminal velocity of the particle is the significant characteristic which determines the separation of particles. If the particles are allowed only very short settling periods, they will never develop the maximum velocity and the separation is made on the basis of the initial settling velocities of the particles. At the outset of settling the velocity is extremely low and no resisting force due to friction effects has been developed. Therefore, since F_R' is practically zero, equation 11 becomes

$$F' = ma = mg - wg - 0$$

or

$$a = \left(\frac{m - w}{m}\right) g = \left(\frac{\rho_s - \rho}{\rho_s}\right) g = \left(1 - \frac{\rho}{\rho_s}\right) g \quad (48)$$

The initial acceleration a given a particle at the start of settling depends on the force of gravity and the densities of the particle and of the pulp or fluid. It is independent of the size or shape of the particle.

This means that sorting of two materials according to density may be possible, almost regardless of the size distribution of the material, if the settling periods are of extremely short duration.

The relative acceleration or relative initial velocity given particles of two different materials A and B is obtained by dividing equation 48 for material A by equation 48 for material B.

$$\frac{a_A}{a_B} = \frac{(\rho_A - \rho)\ \rho_B}{(\rho_B - \rho)\ \rho_A}$$

The difference between sorting on the basis of initial velocity and on the basis of terminal or maximum velocity is indicated in Fig. 83, which shows the relative velocity of settling as a function of time for different particles. For example, a particle of slate (curve 1) of the same shape and size as a particle of

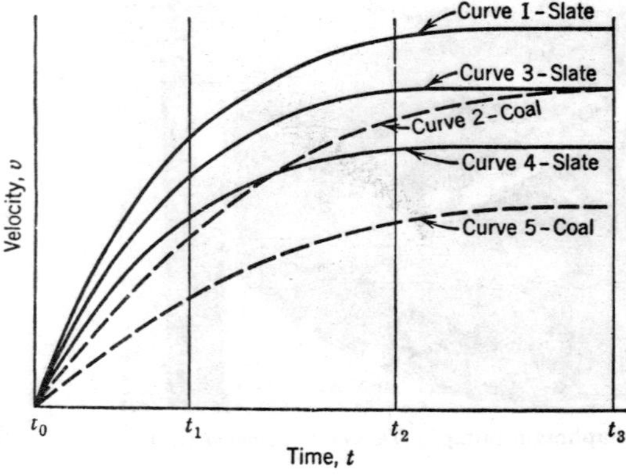

FIG. 83. Relative settling velocities of slate and coal as a function of time from the moment settling was started. Curves 1 and 2 are for particles of the same size. Curve 3 is for smaller particles of slate which have the same terminal settling velocity as the larger particles of coal represented by curve 2. Curves 4 and 5 are for a still smaller size of particle.

coal (curve 2) will settle faster than the particle of coal, owing to its greater density. A smaller particle of slate may be selected (curve 3) which will have a maximum settling rate exactly the same as that of the larger particle of coal (curve 2). A still smaller particle of slate (curve 4) will have a still slower maximum or terminal velocity. Coal of the same size as the smallest slate particle will always settle more slowly (curve 5) than the slate. But the initial velocity or acceleration of all slate particles is greater than that of the coal.

The distance a particle falls in time t equals the integral $\int_{t=0}^{t=t} v\,dt$, or the area under the curve on Fig. 83 from $t = 0$ to $t = t$. If the time is short such as t_1, all slate particles will have settled a greater distance than all coal particles, and the sized feed can be completely sorted or separated into slate-free coal in the tops or "fines" and coal-free slate in the bottoms.

If the time of sorting or settling is allowed to extend to t_2, when the area under curves 2 and 4 ($\int v\,dt$) are about equal, it is impossible to separate the smallest slate particles (curve 4) from the largest coal particles (curve 2) as they are *equal jigging*, that is, they settle equal distances in the allowed settling period. The ratio of the diameters of these particles is called the *jigging ratio* and, similar to the *settling ratio*, indicates the maximum size ratio that can be completely sorted under the conditions indicated.

For settling periods greater than t_3 the terminal velocity controls the separation, and it is impossible to make an effective separation between coal and slate with the size distribution indicated for this feed. The limit to the range of particle size that can be sorted under these conditions is the settling ratio given by equation 45.

As indicated in Fig. 83 the jigging ratio varies enormously with changes in the duration of settling, increasing approximately fourfold as the duration of fall is decreased from 0.50 to 0.10 sec, and more than tenfold as the duration is decreased from 0.5 to 0.05 sec. If jigging is practiced on a feed of a wide size range, a very short settling time must be used for stratification.

Jigs are essentially tanks of rectangular cross section, fitted with a screen placed a short distance below the rim or overflow in a horizontal or slightly sloping position. The screen or the water is given a pulsating or "jigging" motion, which causes alternate upward and downward surges of the fluid through the screen so that the solid particles are lifted free of the screen and allowed to settle for a short time interval, then compacted on the screen in layers, in a series of cycles.

The feed, usually as a pulp, although it may be dry, is introduced over the screen at one side and subjected to a series of short settling periods as it moves across the screen to the overflow. The screen openings are approximately twice the diameter of the largest particle to be passed through the screen. A bedding of larger particles of dense material is usually built up on the screen in a few hours of operation. If this bedding does not develop naturally from the feed, bedding material should be added, made of sized material too large to pass through the screen and of a density approximating that of the heavy material to be concentrated in the lower layer on the screen and in the hutch. Steel shot of twice the diameter of the screen openings is frequently used for this purpose.

Particles of different sizes, of either the same or different densities, do not settle the same distance during one of the short settling periods. A coarse particle may remain suspended, or it may settle for only 0.05 sec out of a cycle of 0.30 sec before it is supported by bridging with other large particles resting on the screen. A small particle may settle for 0.20 sec during the same cycle, part of the time on top of the bed of coarse material and part of the time through the interstices between the large particles. This action, called *consolidation trickling*, represents the settling of fine particles, whereas coarse particles are self-supported and do not settle. The settling of fine particles is much slower during consolidation than during suspension, but the effect may be important if continued for sufficiently long time. It is observed not only in jigging but also in stratification caused solely by lateral vibration without vertical movement and has been called "reverse classification."[1] *

In jigging, the first stratification occurs while the bed is open (solids suspended) and is essentially hindered settling as controlled by the initial velocities or accelerations. This tends to put the coarse heavy grains at the bottom and fine light grains at the top, with the coarse light and fine heavy grains in the middle. The second stratification occurs while the bed is tight (coarse solids self-supported and not suspended) and is essentially consolidation trickling, which tends to put the fine heavy grains at the bottom and coarse light grains at the top with the coarse heavy and fine light grains in the middle. By proper control of the time allowed these two actions an almost perfect stratification according to density can be obtained. The products from a jig are usually, in order of position, (1) the overflow of fine light material; (2) an upper layer above the screen of medium and coarse light material; (3) a lower layer on the screen of medium and coarse heavy material; (4) hutch material of fine heavy material.

* The bibliography for this chapter appears on p. 97.

When *jigging on the screen,* the overflow and the upper layer may be removed over a weir or dam and the lower layer (or concentrate) through a gate or well on the screen. The proper removal of this lower layer is essential for successful jigging and is frequently a difficult problem.

In *hutch jigging* or jigging through the screen, the lower layer is drawn through the screen into the hutch and removed therefrom. A bed is required on the screen of particles of such size as to remain on the screen and support the light product, and the screen openings must be coarse enough for ready passage of heavy product.

Fixed-screen plunger jigs are exemplified by the Harz jig (Fig. 84). The jigging motion is obtained by the plunger reciprocating in the compartment.

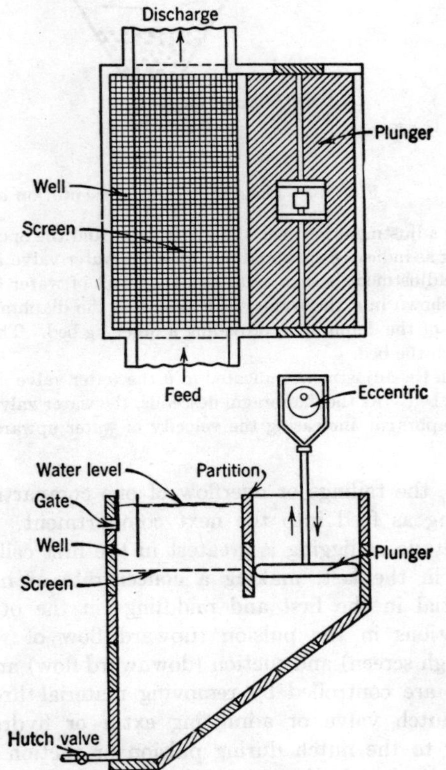

FIG. 84. Fixed-screen plunger jig.

The upper layer is discharged over a weir at the side opposite the feed. The lower layer flows into the well on the screen and is withdrawn through the gate. Such jigs may be constructed of wood, steel, or concrete and are built with several compartments in

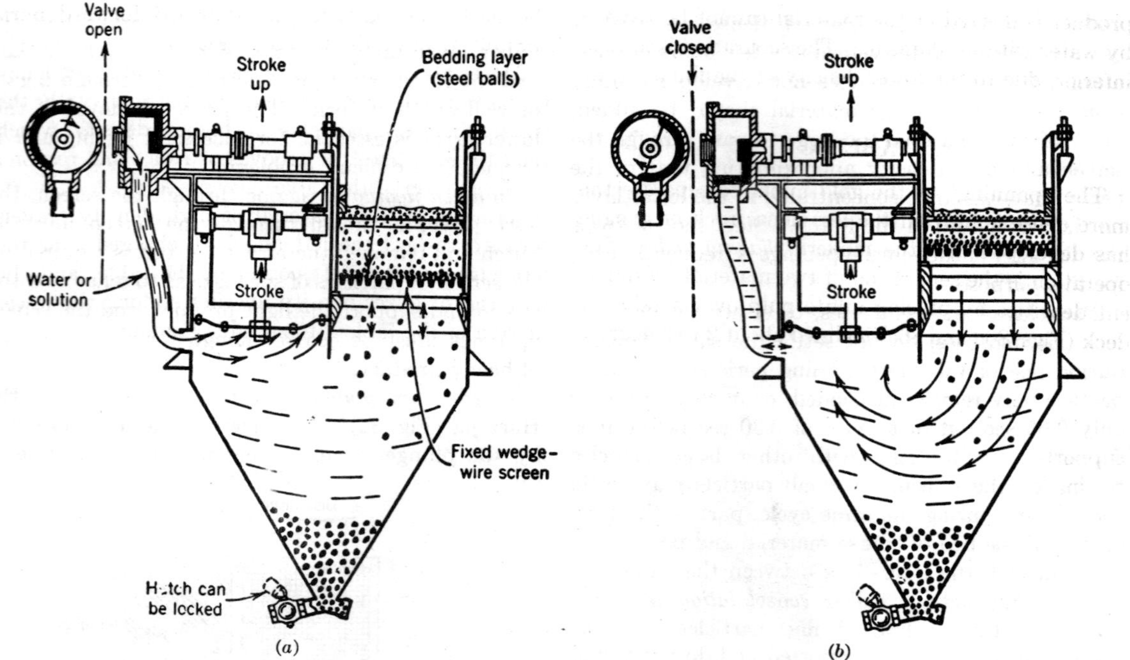

FIG. 85. Diagrammatic representation of a fixed-screen diaphragm jig. (*Denver Equipment Co.*)

The adjustment between the water valve and the operation of the diaphragm may be varied to provide different intensities of jigging as indicated in *a* and *b*. In *a* the water valve is adjusted to work against the diaphragm, giving a mild action, and in *b* the adjustment is such as to add the effect of water flow to that of the diaphragm.

As shown in *a* the water valve opens as the diaphragm begins to rise, and the flow of water into the hutch neutralizes the action of the diaphragm, providing a quiet jig bed. The water valve closes as the diaphragm descends, giving an upward flow through the bed.

With the adjustment indicated in *b* the water valve closes as the diaphragm rises, and water is drawn into the hutch from the jig bed. As the diaphragm descends, the water valve opens and the entering water augments the effect of the jig stroke of the diaphragm, increasing the velocity of water upward through the jig bed.

series, the tailings or overflow of one compartment passing as feed into the next compartment. The amplitude of jigging is greatest in the first cell and least in the last, making a concentrate of heavy material in the first and middlings in the others. Variations in the pulsion (upward flow of water through screen) and suction (downward flow) amplitudes are controlled by removing material through the hutch valve or admitting extra or hydraulic water to the hutch during pulsion or suction. In hutch jigging the dense lower layer is drawn through the screen and discharged through the hutch valve.

Fixed-screen diaphragm jigs are similar to the plunger jigs except that the plunger is sealed to the frame by a rubber diaphragm which prevents leakage of water around the plunger and gives more positive pulsion and suction. The Denver mineral jig (Fig. 85) exemplifies this type and shows how the hydraulic water may be automatically controlled.

Power requirements vary from about 0.1 to 0.15 hp/sq ft of screen area, and capacities range from about 1 to 4 tons/(sq ft)(24 hr).

Movable screen jigs, as the name implies, move the screen in order to get the jigging action. The motion is usually not only reciprocating up and down but also fore and aft, with greater acceleration on the return stroke so as to cause the bed to move forward. The Hancock jig has a capacity up to 300 to 600 tons/24 hr for a screen 25 ft by 4 ft 2 in., or about 3 to 6 tons/sq ft.

Heavy suspensions may be used in place of water when jigging relatively coarse particles with all the advantages of a high density fluid.

Air is also used in pneumatic jigs when a dry

product is desired or the material cannot be treated by water without damage. The separation is always inferior, due to the lower density of the fluid.

TABLING

The "panning" of the gold miner to separate the more dense gold from the less dense rock or gangue has developed into the modern *shaking table.* The operation is the separation of two materials of different densities by passing dilute pulp over a table or deck (Figs. 86*a* and 86*b*) inclined about 2 to 5 degrees from the horizontal. It is given a reciprocating horizontal motion or shake at the rate of 150 to 375 per minute, with a slow forward motion and a fast return, causing the material to move forward in the same manner as for a movable screen jig.

The feed and water are supplied to a launder at the right upper side of the deck. The surface of the table is covered with canvas, wood, linoleum, cement, or otherwise roughened, and usually provided with grooves or cleats to form riffles which may be parallel to the direction of table motion or at an angle thereto. Cleats ¼ to ½ in. wide and ¹⁄₁₆ to ⅛ in. high are usually spaced ½ to 2 in. apart, covering most of the deck as shown.

The net movement of the particles is the resultant of two forces applied to them. Under the influence of the tilt of the table and the velocity and thickness of the film of water flowing over it, the particles of solid tend to be washed straight down the slope over the lower edge but are impeded in this flow by the riffles Under the influence of the jerking motion of the deck, the particles are impelled forward horizontally (from right to left in Fig. 86).

Materials of low density have a greater tendency to be carried in suspension downstream with the water than do particles of higher density which tend to settle into the grooves or riffles and respond to the motion of the table.

As a result, the material of high density travels forward toward the left and is discharged over the left edge or lower left corner of the table, and the material of lower density overflows along the lower margin of the table into suitable launders.

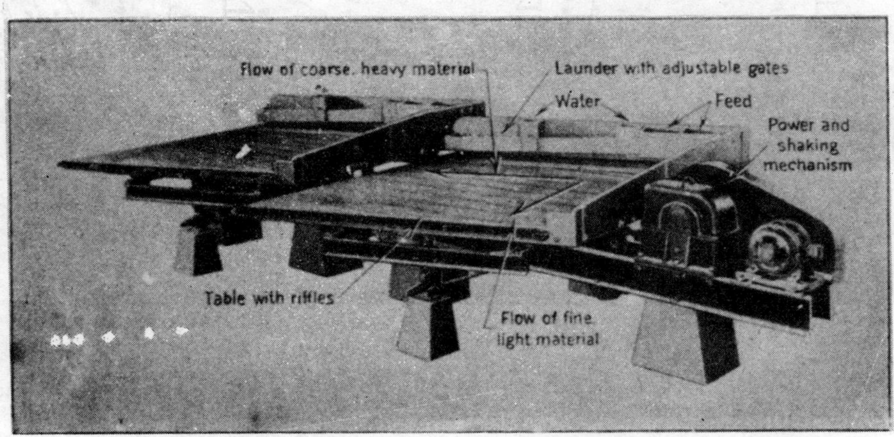

Fig. 86*a*. Duplex diagonal-deck washing table. (*Deister Concentrator Co.*)

Provision is made to alter the degree of tilt of the deck. The drive has a variable speed mechanism, and the coupling to the table provides a means of increasing or decreasing the length of the jerking stroke from a fractional part of an inch to somewhat over 1 in.

Although tables are used for sorting materials from about 6 mesh to 300 mesh, separations of particles of below 48 mesh are usually more economically accomplished by flotation. Successful tabling is dependent upon a wide difference in densities of the minerals fed.

A small table about 15 in. by 30 in. will separate from ½ to 2 tons of solids in 24 hr; a large table about 6 ft by 15 ft will handle 15 to 60 tons of solids in the same period of time. Horsepower requirements are low, in the neighborhood of ½ to 2 hp per table. Water requirements vary from about 3 to 30 gpm per table.

A *dry table* operates by a blast of air passing upward through an inclined perforated table or

FIG. 86*b*. Diagonal-deck concentrating table treating glass sand. (*Deister Concentrator Co.*)

screen. The feed is introduced near the top of the inclined table. The less dense particles are supported by the air above the surface of the table and flow downward to discharge over the lower end. The table is given an oscillating motion similar to that described for Fig. 48 which causes the more dense particles contacting the surface of the table to move upward and discharge over the higher end of the table. Dry tables are used to remove stones and other dense materials from light products such as seeds with capacities of ½ ton/hr for tables about 1½ by 3 ft up to 16 tons/hr for tables about 3½ by 7 ft. The power required is about one horsepower per ton per hour.

Dry tables are also used to remove a relatively large quantity of light material from a dense product.

ELECTROSTATIC CLASSIFICATION

Small particles of different solids may be separated on the basis of their different behaviors in an electric field. These selective forces act in such a manner that the resulting motion causes a decrease in the energy of the system. The forces are due either to an electric charge on the particle or to a difference between the dielectric constants of the particle and the surrounding medium.

Forces due to an electric charge are parallel to the field. Forces due to differences in dielectric constants are parallel to the direction of maximum variation in field intensity and are zero in uniform electric fields.

In conductance separators a conductive particle acquires a charge by coming into contact with a charged surface or screen. Forces created by such charges are relatively large and easily produced. Such separators may be constructed as shown in Fig. 87.

A is the initial material, *B* the less conductive, and *C* the more conductive. The devices in Figs. 87*a* and 87*b* are represented as having rolls beneath hoppers. In some cases these rolls are replaced by inclined chutes. The attracting electrode, indicated as being charged negatively and of cylindrical form, is sometimes of other shape and polarity but is always the opposite of the roll polarity. Instead of a dividing edge to separate the charged and uncharged particles, the former may be collected selectively by adhesion to a surface. Simple charging by conductance is illustrated in *a*, Fig. 87, where in *b* additional charging by ions from a pinpoint electrode is shown. This causes the nonconductors to adhere more firmly to the rolls.

In Fig. 87*c* material is moved mechanically from I to II on a flat surface, such as a grounded vibrating table or a moving belt. A superimposed electrode causes the more conductive material to be levitated and affected less by the mechanical movement. With proper tilting this more conductive material is removed at III.

The applied potential in Fig. 87*c* may be constant in magnitude and polarity, constant in polarity but intermittently reduced to zero, or alternating in polarity. Interruption and alternation of potential cause material to be repeatedly levitated and redeposited on the table and results in a number of successive separations during one passage over the table. The alternations and interruptions may also be synchronized with the table vibrations. The effect of these repetitions is to give higher selectivity and higher efficiency of separation. The selectivity of the separators just described depends not only on conductivity differences of the constituents but also on differences in the density, particle size, and particle shape.

Some separators or precipitators for the removal of suspended solids (smokes) or liquid (mists) are so constructed as to cause the suspension to pass through a screen charged with either positive or negative polarity, which charges the particles in the gas, and then through a maze of metal baffles charged with the opposite polarity. The particles are attracted to the plates from which they are removed periodically. One type of equipment uses a slowly moving endless chain of baffles which are dipped in a liquid bath to clean them. Such equipment is from 85 to 100 per cent effective in separating the suspended particles from the gas.

Contact potential separators act in much the same way but differ in the method of charging the solid particles. This is done by successive contact between the particles and a surface of electrically conductive material whose contact potential lies between the contact potentials of the constituent particles to be separated. The surface or plate is positive to one constituent and negative toward the other. The contacting may be done by allowing the particles to flow over an inclined surface which may be vibrated.

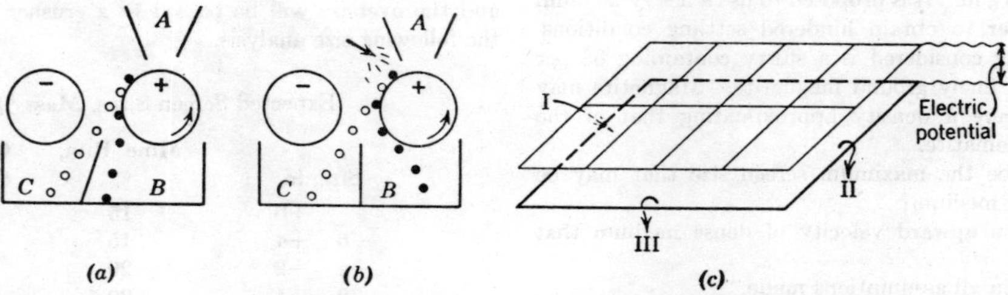

FIG. 87. Diagrammatic representation of conductance separators.[2]

Pyroelectric effects may also be utilized in which the heated material is fed onto a roll, the drop in temperature after deposition on the roll causing electric polarization to appear on some minerals, such as quartz,[3] which then adhere to the roll and may be separated as *B* in Fig. 87*b*.

Electrostatic methods are not so widely used for separating different solids as the other methods described. But they have proved effective in precipitating or separating fine particles from suspensions in gases.

BIBLIOGRAPHY

1. DYER, FRED C., "The Scope for Reverse Classification by Crowded Settling," *Eng. Min. J.*, **127**, 1030–33 (1929).
2. FRAAS, F., and O. C. RALSTON, "Electrostatic Separation of Solids," *Ind. Eng. Chem.*, **32**, 600 (1940).
3. OVERSTROM, G. A., U. S. Patent 1,679,739–40 (Aug. 7, 1928).

PROBLEMS

1. A mixture of silica and galena particles (as crushed) ranging in size from 0.0074 cm to 0.0652 cm is separated by a rising stream of water at 60° F.

(*a*) What velocity of water flow will give an uncontaminated product of galena?

(*b*) What is the maximum size range of this product?

(c) If benzene (specific gravity 0.85, viscosity 0.65 centipoise) is substituted for the water, how will this separation be changed?

2. A mixture of crushed galena and silica ranging from 28 to 200 mesh is to be separated cleanly and completely by screening and classification. Calculate the screen sizes and the water velocities (at 20° C) for vertical classifiers which will accomplish this separation.

3. In the mining of sphalerite (ZnS), a particular concern separates the ore from the quartz gangue by a hydraulic classification on one size fraction in free settling classifiers, using an upward current velocity of water of 50 fpm. Suitable concentration is obtained as long as the size range is held to ⅛ in. by $\frac{1}{16}$ in. It is proposed to use a heavy medium separation in order to obtain hindered settling conditions. The medium being considered is a slurry containing 65 per cent by weight of finely ground magnetite. Magnetite may be assumed to have a density approximating that of the heavier forms of hematite.

(a) What will be the maximum screen size that may be used with the new medium?

(b) Estimate the upward velocity of dense medium that should be used.

State and explain all assumptions made.

4. Soap is to be ground to produce a coarse soap powder which is a granular product with a maximum permissible size of 14 mesh. It is processed in a hammer mill run in closed circuit with a double-cone air classifier using cool air. The analyzing zone of the classifier is the annular zone between the outer and inner cones which have maximum diameters of 48 in. and 36 in., respectively. The air in this zone contains 0.2 per cent of solids by volume. The density of the soap is 1.3 grams/cc.

How much air in cubic feet per minute must be supplied?

5. A gold mining company is installing a plant for the separation of tungsten oxide (density 446 lb/cu ft) from a mixture of 15 per cent by weight silica and 85 per cent tungsten oxide. Fifty tons/day of this mixture (average size is 2 in. in diameter) is to be crushed to the following screen analysis:

Screen Size, in.	Mass Fraction	n (both tungsten oxide and silica)
−0.32 +0.16	0.2	8.7
−0.16 +0.08	0.3	6.7
−0.08 +0.04	0.3	4.5
−0.04 +0.02	0.2	3.4

The discharge from the crusher is to be screened with an 0.08-in. screen, and both the oversize and undersize fractions are to be sent to free settling hydraulic classifiers.

(a) Estimate the rated horsepower of the motor required to drive the crusher.

(b) Will the proposed screening and classification operation effect complete separation of tungsten oxide and silica? If so, what water velocities are required in the classifiers? If not, specify screen sizes and classifier conditions which will.

State and explain any assumptions that are made.

6. A coal company whose reputation has been built upon a low-ash clean product plans to open a new mine. All coal from the new mine is to be washed to eliminate free ash.

The mine-run coal will be first screened over a 6-in. screen, and the oversize will be passed to a crusher and crushed to the following size analysis.

Expected Screen Sizes, Mass %

Size, in.	Mine Run, %	Crushed Coal, %
+6	15	..
−6 +4	15	10
−4 +2	20	20
−2 +1	20	20
−1 +¼	20	40
−¼	10	10

The crusher product is then blended back in with the 6-in. undersize and double-screened to three sizes. The two larger sizes are to be washed by means of hydraulic classification, while the undersize from the finest screen (which has ¼-in. holes) is passed to a fine coal launder.

Float-and-sink tests on the coal indicate that an average of 85 per cent of the mine product is coal of 1.5 specific gravity or lighter, and 15 per cent is free ash of 1.7 specific gravity or heavier. Thus, washing should be done to recover all the 1.5 gravity or lighter and eliminate the 1.7 or heavier.

(a) In the double screening of the −6-in. coal and ash, a ¼-in. fine screen has been specified. What should be the screen size of the coarser screen which will allow complete elimination of free ash in the hydraulic classifier, at the same time permitting as large a weight fraction as possible in the medium-sized higher-priced coal product?

(b) If 300 tons of mine-run coal are processed per hour, how much of each of the final three sized products and free ash are produced per hour?

State and explain all assumptions made.

CHAPTER 9

Flotation

FLOTATION includes any operation in which one solid is separated from another by floating one of them at or on the surface of a fluid. In modern froth flotation the solid particles are continuously agitated in water upon which a thick layer of froth is maintained. Because of differences in the surface properties, one solid more readily adsorbs the water phase, becomes surrounded by water, and sinks. The other solid more readily adsorbs air and becomes at least partially surrounded or covered by air. The average or bulk density of the solid and adsorbed air bubbles is less than that of water, and the whole mass of air and solid floats to the surface to form a mineralized froth which continuously overflows the side of the vessel. Because separation by froth flotation depends upon the surface characteristics of the materials, it is capable of separating materials regardless of their densities.

The equipment and operations for flotation were developed in the mineral industries, and over 80 per cent of mineral concentration is accomplished by this method. The operation is being extended into other fields, such as separating the hulls of wheat from the kernels, printer's ink from reclaimed newspaper pulp and even potassium chloride (KCl) from sodium chloride (NaCl).

Because of the importance of surface conditions and the necessity of the air bubbles supporting the solid in the froth, flotation is performed on finely divided material, usually varying in size from 20 mesh to under 200 mesh. The material, reduced by the ball mill or other fine crusher to the proper size and of the desired pulp density, is fed to a conditioning tank (Fig. 88) which is essentially a cylindrical tank equipped with an efficient agitator. The purpose of the conditioning tank is to bring about the coating of the solid to be floated with the proper flotation reagent. The flotation reagent is fed continuously to the conditioner, sufficient time being allowed to cause complete "filming" of the solid by the reagent. Some filming may be done in the ball mills when a portion of the reagents are added there.

Fig. 88. Sectional drawing showing operation of conditioning tank. (*Denver Equipment Co.*)

The overflow from the conditioner is fed to a flotation cell termed the rougher, where the first or rough separation by flotation is made. The material floated off is called the concentrate as it contains the desired mineral. The other material which sinks in the water and is removed from the bottom is called the tailings.

Since the tailings from the rougher may contain some material desired in the concentrate, they are frequently treated in another cell called the scavenger, as indicated in Fig. 89 under conditions that favor the flotation of the maximum quantity of

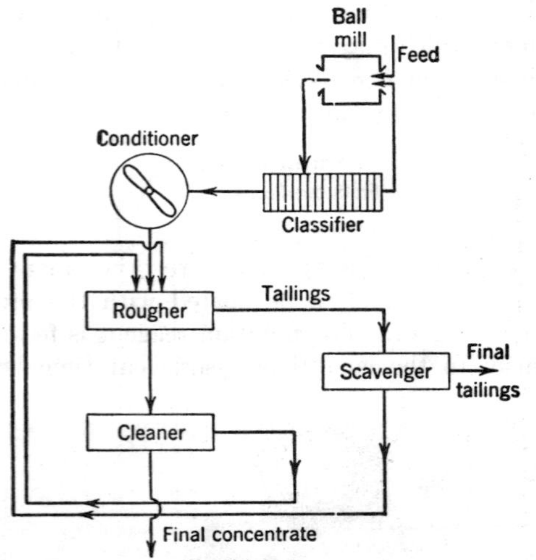

FIG. 89. Simple flow diagram showing separation by flotation using rougher, scavenger, and cleaner (or finisher) cells.

material desired in the concentrate even if a large amount of tailings is also floated. The floated product from the scavenger is returned or cycled to the rougher with the feed. The tailings from the scavenger are the final tailings.

The product floated from the rougher may contain more gangue than is desired. This may be reduced by feeding the overflow from the rougher to a third flotation cell called the cleaner, so operated as to give the desired quality in the concentrate. Under these conditions the tailings from the cleaner contain material desired in the concentrate and are cycled with the feed to the first cell or rougher.

Under special conditions extra scavengers or cleaners may be used in series.

If the ore contains several valuable minerals, a process of consecutive or selective flotation (Fig. 98) may be used to recover the minerals separately. The feed from the first conditioner is treated with the proper reagents to float one of the desired minerals, leaving the others in the tailings with the gangue. These tailings are fed to a second conditioner and treated with the proper reagents for floating the second desired mineral.

The mineral concentrates are generally subjected to sedimentation, filtration, and drying prior to smelting. The tailings are fed to a tailing pond where the solids settle out and the clear water is cycled to the milling and flotation units.

FLOTATION CELLS

A flotation unit or cell is the equipment in which the material is actually separated or floated from the residual tailings. It consists essentially of a vessel or tank provided with a feed at one end, an overflow for froth removal, and a discharge for tailings at the opposite end, with a provision for introducing air for froth formation and agitation. *Pneumatic cells* depend upon compressed air for agitation, give a relatively mild agitation, and produce a clean froth relatively free from gangue. In general, about 50 per cent longer contact time is provided in pneumatic machines, and the pulp must be fully conditioned before flotation. *Mechanical cells* incorporate a mechanical agitator that draws in air and beats it into the pulp. Because of the more violent agitation that can be obtained with mechanical cells, they may give more thorough flotation, and tailings more nearly free from material desired in the concentrate, but the concentrates then contain more gangue than those from cells with less violent agitation. At high altitudes auxiliary air under moderate pressure is frequently supplied to mechanical cells. Mechanical cells have a greater capacity for the same volume and also help condition the pulp, giving greater capacity to the conditioner.

The *Callow cell* is one of the oldest and simplest but is now practically obsolete (Fig. 90). It consists

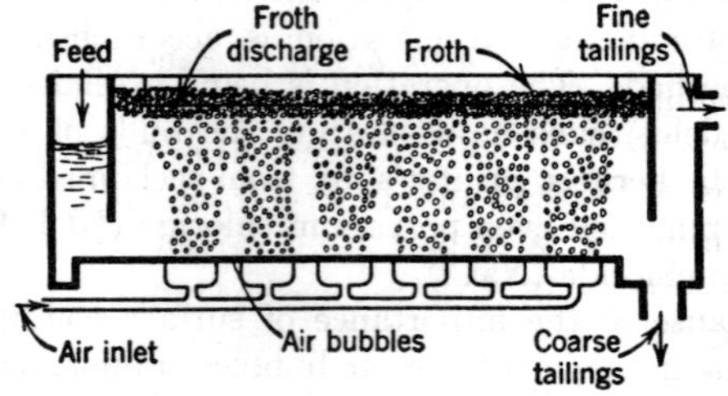

FIG. 90. Diagram representing the simple Callow flotation cell in section.

of a trough 24 to 36 in. wide and 18 to 22 in. high at the overflow level, and as long as required. The feed enters at the high end, the tailings being removed from the other. A "blanket," frequently several layers of canvas fastened to square open-topped cast-iron blanket frames or "pans," covers the bottom of the trough. Compressed air is admitted below the blanket to each frame by pipes projecting through the solid bottom of the trough. The small bubbles of air passing upward from the blanket induce a mild agitation and become adsorbed or attached to the particles to be floated. The froth, usually maintained to a depth of 8 to 10 in. in the tank or trough, overflows into the concentrate launder surrounding the cell.

The air consumption averages 8 to 10 cu ft/(min)(sq ft) at a pressure of about 4 psi until the blanket becomes blinded when higher pressures may be needed. Horizontal canvas blankets have the disadvantage of clogging due to sands and to precipitates of calcium salts in the fibers of the cloth. Carelessness in cleaning with a shovel may rip the cloth. Nevertheless, Callow cells have been satisfactory for many ores, especially those that are easily floated without great agitation.

The *Callow-MacIntosh cell,* Fig. 91, carries a hollow revolving rotor made from a steel tube about 9 in. in diameter and perforated with ¾-in. holes. Canvas or perforated rubber sheet is fastened to the surface by means of steel bands. Two scraper bars made of 2-in. angle iron are bolted the full length of the rotor.

This machine does not blind easily and can handle heavy slurries with a water-to-solid ratio by weight of 1:1. These cells are usually made 10, 15, or 20 ft in length and 24, 30, or 36 in. wide, respectively, but larger sizes up to 30 ft long and 48 in. wide are available, provided with two rotors.

The quantity of air necessary for operation varies from 4 to 8 cu ft/(min)(sq ft of rotor surface) at 2 to 2½ psi. A ½-hp motor is generally capable of turning the rotor at 10 to 15 rpm.

The *"air-lift" cell* is pneumatic but does not employ a blanket and is free from moving parts. It consists of a V-bottomed trough divided into sections by vertical baffles, as shown in cross section in Fig. 92. Air is supplied from a header running the entire length of the trough to the air connections at the top of each cell as shown. The air flows down to the bottom of the vertical air pipe within about 6 in. of the bottom. The air bubbles carry some pulp upward through the perforated apron, thereby agitating the slurry and forming froth which is deflected by the dome against the sloping baffles. The froth overflows over the overflow lips along the sides of the trough into a launder. The feed enters at one end of the trough and receives successive treatment as it passes along the trough. The nonfloatable particles settle in the relatively quiet zone along the sloping sides of the lower part of the cell. The method of feeding and discharging the tailings is similar to that used in the Callow cell.

The air supply is controlled by a main valve to the header, which may be divided into sections about 4 ft long, each one having an individual valve that can control any one section. If the air supply is decreased slightly at the feed end of the tank, the concentrate can be made to overflow in large part near this end. Air is usually supplied at 2 psi at the blower. For roughing operations, 75 to 100 cfm/ft of tank length is used. For cleaning a lower quantity, 45 to 70 cfm/ft will have a gentler action less liable to raise gangue into the froth. The approx-

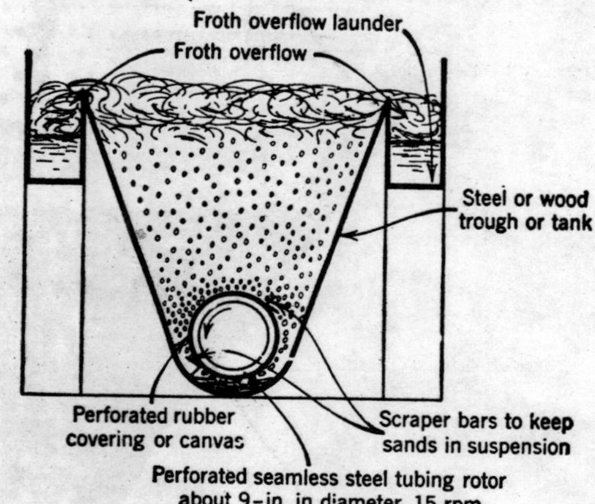

FIG. 91. Cross-sectional diagram of Callow-MacIntosh flotation cell.

TABLE 19. POWER REQUIREMENTS OF LOW-PRESSURE BLOWERS[*]

Air Delivered, cfm	Approximate Horsepower at Pressures of 2 psi	3 psi	4 psi	5 psi
500	6	9.5	13.5	17.5
1000	12	18.5	26	33
2000	24	38	54	70
3000	34	57	76	96
4000	46	76	105	135
5000	58	91	125	160
6000	70	120	168	220
7000	83	135	190	245
8000	96	165	225	310

* The bibliography for this chapter appears on p. 108.

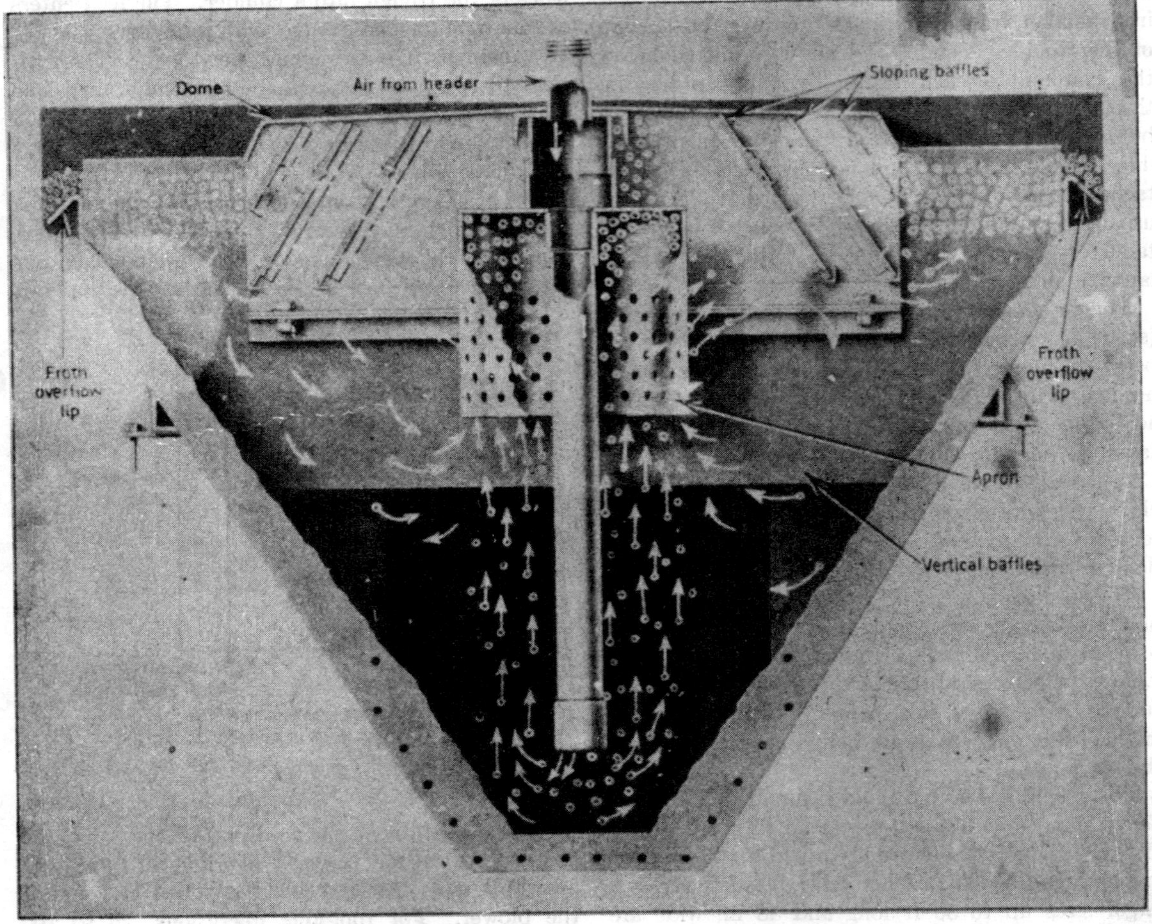

FIG. 92. Cross-sectional drawing of air-lift flotation cell indicating its action. (*Western Machinery Co.*)

imate power required to deliver air at these low pressures is indicated by the data of Table 19.

The *mechanically agitated machine* (Fig. 93) consists of a square cross-sectioned tank provided with an impeller which violently agitates the pulp, effecting some conditioning. The rotation of the impeller sucks air downward through a sleeve surrounding the impeller shaft and breaks it up into fine bubbles. Figure 94 shows the froth overflow and actual appearance of one of these cells in operation. The feed enters at one end of the battery of cells and passes through the desired number of cells to tailings discharge at the opposite end.

At low pressures, elevations above 8000 ft, adequate air is supplied only by aid of a blower.

The *Fagergren machine* (Fig. 95) features a rotor-stator assembly for agitation and aeration of the pulp. The stator consists of cylindrical spacers mounted between two rings rigidly fastened to the tank. The rotor construction is similar to that of the stator, except for upper and lower bladed impellers mounted within the rings. It is suspended on a short drive shaft and rotates within the stator.

The pulp enters directly into the tank through a suitable opening. The pulp is drawn by the impeller blades into the rotor. Rapid pulp displacement creates a partial vacuum which causes air to enter into the rotor through the standpipe. The air is dispersed through the pulp in the form of fine bubbles. In passing between the cylindrical spacers of the rotor and stator, the pulp-water-air mixture is highly agitated, giving effective aeration.

FIG. 93 Cross-sectional drawing of the Denver mechanical flotation machine, indicating its action. *(Denver Equipment Co.)*

FIG. 94. Denver flotation machine in action. *(Denver Equipment Co.)*

As many as six machines may be combined into one unit having a common tank separated by suitable partition plates. In such a case, the pulp flows from one cell to another through openings provided in the partition plates; at the end of the string is a discharge box. The froth is removed from each unit by a rotating skimmer. The power requirements for these mechanical machines may be estimated from Table 20.

TABLE 20. APPROXIMATE POWER REQUIREMENTS FOR MECHANICAL FLOTATION CELLS

Size, cu ft	Horsepower Consumed per Cell	
	Denver	Fagergren
10	1	
12	1.2	1.8–2.0
18	1.4	
24	2.2	3.5–4.0
40	3.2	5
50	4.2	6
70		8
100	9	

FIG. 95. Sectional drawing of Fagergren flotation machine. (*Western Machinery Co.*)

FLOTATION AGENTS

Flotation depends upon the relative adsorption or "wetting" of the solid surfaces by the fluid. This in turn is controlled by surface or interfacial energy of which interfacial tension is the intensive factor. Any surface such as that between water and air resists extension and behaves exactly as if it were in tension. It is this interfacial tension which tends to make small masses of water in air take on a spherical shape or become drops, and small masses of air in water take on a spherical shape and become bubbles, as the sphere has the least surface per unit volume. The interfacial tension can be measured as the force resisting the extension of the interface. Its relative values may be determined by the angle formed between surfaces or interfaces when three or more phases are in contact and at equilibrium.

In general the sum of the components of the interfacial tensions equals zero. If one of the phases is a solid presenting a rigid plane surface, as indicated in Fig. 96, and the other two phases are fluids, the balance of forces parallel to surface of the solid gives

$$\gamma_{SG} = \gamma_{SL} + \gamma_{LG} (\cos \theta)$$

where γ = interfacial tension as indicated.
θ = contact angle.

The contact angle may be determined by placing a polished specimen of solid in the bottom of a flat-

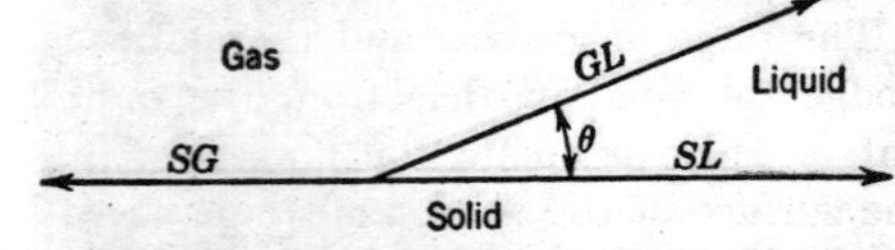

FIG. 96. Diagram of surface tensions involved in three-phase contact.

sided vessel, as indicated in Fig. 97. The surface of the solid may be clean, or it may carry a film of adsorbed material, depending upon the conditions it is desired to measure. The liquid to be investigated is poured into the vessel until the solid is at least 1 in. under the liquid. A flat-ended capillary is introduced perpendicularly over the solid, and air is cautiously blown through the capillary until the bubble makes contact with the solid.

If the solid is easily and completely wetted by the liquid (Fig. 97*a*), the liquid exists as a skin between the solid and air and there is no point of contact of the three phases. The angle formed by a tangent at the apparent point of contact between the three phases is zero.

If the surface of the solid is not completely wetted, the liquid is forced to recede to an equilibrium position, as indicated in Fig. 97*b*, where the forces of the three interfaces are in balance. A tangent to the gas-liquid interface at this point forms the angle θ, defined as the contact angle (always measured through the more dense phase). The angle may be measured by projection of a magnified shadow of the bubble on a screen.

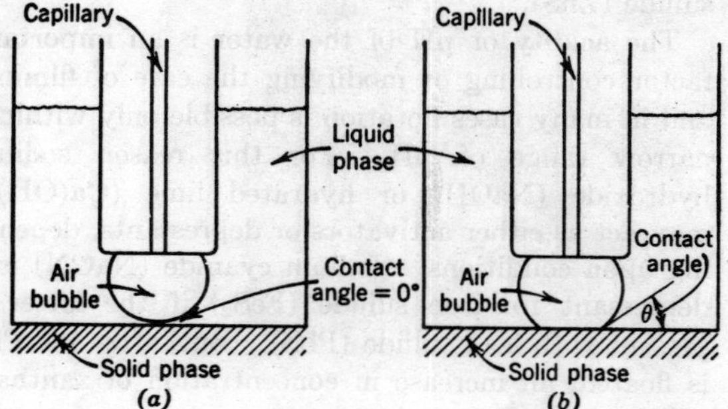

FIG. 97. Diagram representing contact angle between a liquid and a solid.[6] (*a*) Solid completely wet by liquid; (*b*) solid partially wet by liquid.

In flotation a solid particle is attached to a bubble in the same way as the solid and bubble are in contact in Fig. 97. The only difference is that the solid is extremely small and the air bubble is relatively large. The force of gravity and the agitation tend to dislodge the solid particle from the bubble. If the contact angle is small, the liquid will advance over the surface of the solid, and the surface forces holding the solid and bubble together are weak. A large contact angle means easy flotability. When a solid particle attaches itself to a bubble, there is a loss in surface energy $-\Delta E$ per unit area of surface σ, equal to the loss in surface tension or

$$\Delta E = \gamma_{SG}\,\Delta\sigma_{SG} + \gamma_{SL}\,\Delta\sigma_{SL} + \gamma_{LG}\,\Delta\sigma_{LG}$$

$$\Delta\sigma_{SL} = -\Delta\sigma_{SG} = \Delta\sigma_{LG}$$

$$\frac{-\Delta E}{\Delta\sigma_{SG}} = (\gamma_{SL} + \gamma_{LG} - \gamma_{SG})$$

Since

$$\gamma_{SG} = \gamma_{SL} + \gamma_{LG}(\cos\theta)$$

$$\gamma_{SL} - \gamma_{SG} = -\gamma_{LG}(\cos\theta)$$

$$\frac{-\Delta E}{\Delta\sigma_{SG}} = \gamma_{LG}(1 - \cos\theta)$$

This loss in energy $(-\Delta E)$ is a measure of the wettability of the solid phase by the air and therefore an indication of the flotability. It represents the work required to separate the air from a unit surface of solid.

Collectors and promoters are reagents which are adsorbed on the surface of the solid as very thin films and which because of their properties thereby increase the contact angle. The term promoter is particularly applied to agents forming films one molecule thick, such as sodium xanthate, NaS(CS)OR, which is adsorbed on lead sulfide and oriented with the xanthate radical (—SCS—) toward the lead and the hydrocarbon part (R) outward. This gives the surface of the solid a characteristic approaching that of a hydrocarbon which is not wetted by water. If the adsorbed material forms films several molecules thick it is called a collector. Petroleum is an example of a true collector. It has the disadvantage of making a greasy froth containing a sticky mass of bubbles difficult to break down in the subsequent sedimentation operation, and in some cases a trace of oil in the concentrate is undesirable. Pine oil, generally classed as a frothing agent, also acts as a collector and is not so sticky as cresylic acid and petroleum.

The more common promoters used in mineral flotation are the xanthates, the aerofloats, $HS(PS)(OR)_2$, and thiocarbanilide, $HSC(NC_6H_5)(NHC_6H_5)$. The quantity of these reagents used is about 0.05 to 0.15 lb/ton of solids treated. Somewhat larger quantities of collectors, up to about 1 lb/ton, may be used. The greater the length of the nonpolar or hydrocarbon part of the molecule, the more the surface approaches that of paraffin and the greater the contact angle. The R group in the xanthates is frequently the methyl or isoamyl group. In the aerofloats cresyl or phenyl as well as the methyl radical is frequently used and the hydrogen, H, may be replaced by a metal.

Frothing agents are required to prevent coalescence of air bubbles when they reach the surface of the water, thereby maintaining a persistent froth. The agent must be sparingly soluble in water without appreciable ionization and be adsorbed in the interface between water and air tending to reduce the surface tension of water. The heavy alcohols (hexyl alcohol $C_6H_{13}OH$) possess these properties. If two air bubbles collide, the "skins" (interfaces containing water and alcohol) stretch but do not break and the bubbles do not coalesce, because at the point of contact stretching of the film allows water molecules to come into the interface, increasing the surface tension at that point. Alcohols below amyl are too soluble, and those above octyl are too insoluble to make satisfactory frothing agents. Cresylic acid produces an unsatisfactory brittle froth unless used with a trace of petroleum. Pine oil, a mixture of compounds, and eucalyptus oil are good frothing agents requiring no additions. About 0.05 to 0.3 lb of frothing agent is used per ton of solids treated.

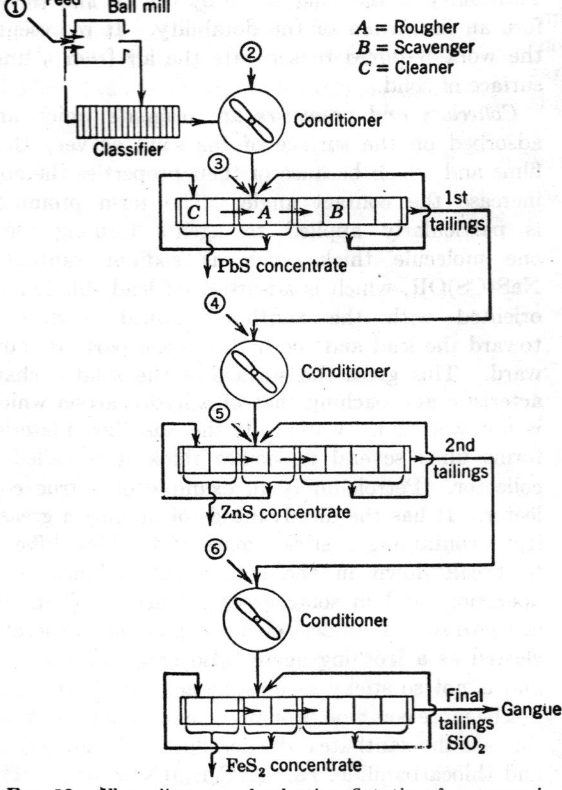

FIG. 98: Flow diagram of selective flotation for recovering PbS, ZnS, and FeS_2.

Modifying agents may be used to "activate" or to "depress" the adsorption of the filming agent (collector or promoter). They react with the surface of the solid, either by chemical action or by adsorption, to change the character of the surface of one or more of the solids, thereby preventing the adsorption of the collector or promoter on the solid. The addition of copper sulfate ($CuSO_4$) activates the surface of zinc sulfide (ZnS) so that the latter is readily floated if a small quantity of copper sulfate ($CuSO_4$) is added in the conditioning tank. This may be due to the formation of copper sulfide (CuS) or perhaps metallic copper (Cu) on the surface of the zinc sulfide (ZnS).

The acidity or pH of the water is an important factor controlling or modifying the ease of filming, and in many cases flotation is possible only within a narrow range of pH. For this reason sodium hydroxide (NaOH) or hydrated lime ($Ca(OH)_2$) may act as either activators or depressants, depending upon conditions. Sodium cyanide (NaCN) is a depressant for iron sulfide (FeS_2) if the latter is present with lead sulfide (PbS). But, after the PbS is floated, an increase in concentration of xanthate will float the FeS_2.

Example. The reagent program for the operation of Fig. 98 is indicated as follows where the numbers correspond to the numbered streams in Fig. 98.

1. To grinding circuit

Na_2CO_3	2–4 lb/ton of solids treated
Xanthate (collector)	0.05–0.15 lb/ton

2. To PbS conditioner (5–15 min)

NaCN (depresses FeS_2)	0.1–0.4 lb/ton
$ZnSO_4$ (depresses ZnS)	0.3–1.2 lb/ton

3. To PbS rougher, $L/S = 2.5/1$ to $1.5/1$ (thick pulp)

Cresylic acid (frother)	0.02–0.15 lb/ton
Xanthate (collector)	0.02–0.10 lb/ton

4. To ZnS conditioner (5–15 min)

$CuSO_4$ (activates ZnS)	0.5–2.0 lb/ton
$Ca(OH)_2$ (depresses FeS_2)	1–4.0 lb/ton

5. To ZnS rougher, pH 8.0–9.0

Pine oil or cresylic acid (frother)	0.05–0.15 lb/ton
Xanthate (collector)	0.10–0.20 lb/ton
Na aerofloat (collector) (if necessary)	0.02–0.10 lb/ton

If it is desired to secure FeS_2, usually unwanted, Na_2CO_3 must be used to raise the pH. Another cycle is then required and there is added

6. To FeS_2 conditioner

Na_2CO_3 (pH control)	0.5–2.0 lb/ton
Xanthate (collector)	0.05–0.15 lb/ton

Dispersants such as sodium silicate, sodium metaphosphate, or soluble starch may be added when necessary to break up agglomerations of mineral and gangue. In such cases the mineral becomes coated with a slime of silicates and becomes unfloatable.

CALCULATIONS FOR A FLOTATION PROCESS

The factors controlling a flotation process are complicated and not readily susceptible of calculation. Sufficient data to serve as a guide in the computations may be obtained by laboratory tests on standardized laboratory equipment, either at the plant or in so-called "ore-testing laboratories" conducted commercially by equipment and chemical companies.

When the best combination of operating conditions has been decided upon, the following pertinent data will be obtainable for each flotation unit.

1. Densities of both minerals.
2. Pulp density of the material in the flotation cell, expressed either as the volume fraction of solid, or as L/S, the water-to-solids ratio by weight.
3. The composition of the feed and of the products.
4. The reagents and quantities of reagents to be used, determined by experiment in each individual problem.
5. The contact time, usually expressed as the average time in minutes that the pulp is in the flotation cell.
6. The type of flotation cell, mechanical or pneumatic, in which the laboratory test was conducted.

To these data must be added:

7. The desired capacity, usually expressed in tons per hour or tons per 24 hr of solids to be handled.
8. The type of flotation equipment to be used.

From this information, one may compute:

1. The capacity of the cell banks and the number of individual machines required.
2. If pneumatic equipment is used, the amount of compressed air required and the horsepower for the compressors.
3. If mechanical equipment is used, the horsepower required.

Illustrative Example. It is desired to recover lead from an ore containing 10 per cent lead sulfide (PbS) and the balance assumed to be silica, 500 tons of ore being treated per 24-hr day.

It is assumed that the concentrate from a single cell is of acceptable purity but the tailings are to be retreated in scavenger cells with return of scavenger concentrate to the rougher.

Laboratory findings indicate that if water-to-solids ratio $L/S = 2$, and the contact time is 8 min in the rougher, and $L/S = 4$ for 15 min in scavenger, with mechanically agitated machines of the Denver type, the following compositions will be found for the various products.

	PbS	SiO_2
Feed, a	10%	90%
Concentrate, b	80	20
Rougher tailings, c	2	98
Scavenger concentrate, d	11	89
Final tailings, e	0.5	99.5

The densities of PbS and SiO_2 are 7.5 and 2.65 gm/cc, respectively.

Solution.

A. Computation of density ρ for all solids.

In feed, PbS: 10 grams = 1.33 cc

SiO_2: 90 grams = 34.00 cc

35.33 cc

$$\text{Average density } \rho_a = \frac{100}{35.33} = 2.83 \text{ grams/cc} \approx 176.5 \text{ lb/cu ft}$$

$$\text{Similarly, } \rho_b = 5.5 \text{ grams/cc} \approx 343 \text{ lb/cu ft}$$
$$\rho_c = 2.682 \text{ grams/cc} \approx 167.3 \text{ lb/cu ft}$$
$$\rho_d = 2.855 \text{ grams/cc} \approx 178 \text{ lb/cu ft}$$
$$\rho_e = 2.679 \text{ grams/cc} \approx 167 \text{ lb/cu ft}$$

B. Computation of the mass of products by material balance.

Overall balance basis: 100 lb of net feed

$$a = b + e = 100$$
$$b = 100 - e$$
$$a = (100 - e) + e$$

An overall PbS balance:

$$(0.1)(100) = (0.8)(100 - e) + (0.005)(e)$$
$$10 = 80 - 0.8e + 0.005e$$
$$0.795e = 70$$
$$e = 88.1 \text{ lb} \quad \text{and} \quad b = 11.9 \text{ lb}$$
$$c = d + e = d + 88.1$$

PbS balance around the scavenger:

$$c = d + e$$
$$(0.02)(d + 88.1) = (0.11)(d) + (0.005)(88.1)$$
$$0.02d + 1.762 = 0.11d + 0.4405$$
$$0.09d = 1.3115$$
$$d = 14.57 \text{ lb} \quad \text{and} \quad c = 102.67 \text{ lb}$$

$$\text{Yield} = \frac{(0.8)(11.9)}{(0.1)(100)} = 95.2\%$$

Purity of concentrate = 80%

C. Computation of tank volumes.

Rougher tank must hold a and d solids:

$$a = 100 \text{ lb} \quad \text{or} \quad \frac{100}{176.5} = 0.5650 \text{ cu ft}$$

$$d = \underline{14.57 \text{ lb}} \quad \text{or} \quad \frac{14.57}{178} = \underline{0.0818 \text{ cu ft}}$$

$$114.57 \text{ lb} \qquad\qquad 0.6468 \text{ cu ft of solids}$$

The average density ρ_{ad} for this mixture of a and d = 176.8 lb/cu ft, with $L/S = 2/1$. 114.57 lb of solid feed require 229.14 lb or 3.675 cu ft of water.

$$\text{Volume fraction of solids in pulp} = \frac{0.6468}{0.6468 + 3.675} = 0.1492$$

Capacity per 1 cu ft of rougher tank volume:

$$\frac{1 \times 0.1492 \times 176.8 \times 60 \times 24}{2000 \times 8} = 2.378 \text{ tons per 24 hr solids handled}$$

The basis for all computations was 100 lb of net feed; if the net feed is 500 tons/24 hr, the rougher tank must handle

$$\frac{114.57}{100} \times 500 = 572.85 \text{ tons/24 hr.}$$

Required capacity for bank of rougher cells:

$$\frac{572.85}{2.378} = 241 \text{ cu ft of tank volume}$$

Scavenger cells:

$$c = 102.67 \text{ lb} \quad \text{or} \quad \frac{102.67}{167.3} = 0.613 \text{ cu ft of solids}$$

$$\text{at } L/S = 4,\ H_2O = 410.68 \text{ lb or } \frac{410.68}{62.11} = 6.590 \text{ cu ft of water}$$

$$\text{Volume fraction of solids} = \frac{0.613}{0.613 + 6.59} = 0.0851$$

Capacity for 1 cu ft of scavenger tank:

$$\frac{1 \times 0.0851 \times 167.3 \times 60 \times 24}{2000 \times 15} = 0.682 \text{ tons solids handled/24 hr}$$

$$\text{Capacity required} = \frac{102.67}{100} \times 500 = 513.35 \text{ tons/24 hr}$$

Volume of tank required for scavenger cells

$$= \frac{513.35}{0.682} = 752 \text{ cu ft}$$

D. Computation of number of cells.

Assume Denver No. 24 machines which have 50 cu ft of volume and require 4.2 hp per cell to operate.

For rougher I:	$\frac{241}{50} = 4.82$	Use 5 cells and 21 hp
For scavenger I:	$\frac{752}{50} = 15.13$	Use 16 cells and 67.2 hp
Total		21 cells and 88.2 hp

Air-Lift Machines as manufactured by the Southwestern Engineering Co. have a standard cross-sectional area of 9.85 sq ft and, owing to the gentler action, usually require about 50 per cent longer contact time for operation.

$$\text{For rougher II: } \frac{241 \times 1.5}{9.85} = 36.7 \text{ ft long}$$

$$\text{For scavenger II: } \frac{752 \times 1.5}{9.85} = 114.2 \text{ ft long, preferably broken down into two units}$$

The amount of air required may be assumed to be 75 cfm/ft in rougher and 100 cfm/ft in scavenger at 2 psi.

Air for rougher:	$75 \times 36.7 =$	2,750 cfm
Air for scavenger:	$100 \times 114.1 =$	11,410 cfm
		14,160 cfm

The power for the air compressor may be read from Table 19.

BIBLIOGRAPHY

1. *Flotation Index*, Great Western Division, Dow Chemical Co., San Francisco.
2. Gaudin, A. M., *Flotation*, McGraw-Hill Book Co. (1932).
3. Gaudin, A. M., *Principles of Mineral Dressing*, McGraw-Hill Book Co. (1939).
4. Rabone, Philip, *Flotation Plant Practice*, Mining Publications Ltd., London (1939).
5. Ralston, Oliver C., "Flotation and Agglomerate Concentration of Non-Metallic Minerals," *U. S. Bur. Mines Rept. Invest.* 3397 (1938).
6. Wark, I. W. and A. B. Cox, *Trans. Am. Inst. Mining Met. Engrs.* **112**, 245 (1934).

1. Ground lead ore is to be concentrated by a single flotation using 1.5 oz of reagent per ton of ore. The feed, concentrate, and tailings have the following compositions by weight on dry basis:

	Feed, %	Concentrate, %	Tailings, %
PbS	30	90	0.9
ZnS	20	5	27.3
$CaCO_3$	40	3	57.9
SiO_2	10	2	13.9

Water is fed to the cell at the rate of 1100 gal/ton of dry concentrate with 99% of the water leaving with the tailings and 1.0% with the concentrate. Calculate:

(*a*) Mass of wet concentrate produced per hour when 10 tons of ore are fed to the cell per 24 hr.

(*b*) Total water required in pounds per hour.

(*c*) The fraction of lead in ore lost in tailings.

(*d*) Ratio of liquid to solid by volume in concentrate.

(*e*) Ratio of liquid to solid by volume in tailings.

2. For a flotation operation, such as that in Fig. 89, the net feed is 1000 tons/24 hr to the rougher. From the following data, compute the capacity and number of flotation cells in each unit and the power consumption.

	Mass % SiO_2	Mass % CuS
Feed	98	2
Tailings from rougher	99	1
Rougher concentrate	60	40
Tailings from scavenger	99.6	0.4
Scavenger concentrate	50	50
Tailings from cleaner	80	20
Final concentrate	1	99

Specific Gravity		Water-to-Solid Ratios		Contact Time, min	
SiO_2	2.65	Rougher	2/1	Rougher	8
CuS	4.60	Scavenger	4/1	Scavenger	12
H_2O	1.0	Cleaner	6/1	Cleaner	10
				Conditioner	8

(*a*) Use Denver No. 24 cells whose cubic capacity is 50 cu ft and whose power consumption is 4.2 hp per cell.

(*b*) Use Air-Lift cells whose cross-sectional area up to froth overflow is 9.85 sq ft and compute the lengths of the troughs and air required at 3 psi.

CHAPTER 10

Sedimentation

THE separation of a suspension into a supernatant clear fluid and a rather dense slurry containing a higher concentration of solid is called sedimentation. Commercial sedimentation of water suspensions is conducted as a continuous process in "thickeners" or large tanks which receive the suspension or dilute slurry at the center or side, permit the overflow of supernatant liquid, and produce a sludge from the bottom of the tank.

LABORATORY BATCH SEDIMENTATION

Before the continuous operation is considered, a general conception of the operation may be gained from simple batch sedimentation, illustrated by suspending some finely divided solid in water in a graduated cylinder and allowing the contents of the cylinder to stand undisturbed. The time rate of decrease in height of the visible interface between supernatant clear liquid and slurry containing the particles is the "sedimentation rate." This small-scale experiment must be conducted at a uniform temperature to avoid movement of fluid or convection due to density differences resulting from differences in temperature. The observed height of the interface is plotted as a function of time in Fig. 99, which also indicates the progress of the sedimentation in the cylinder.

At the start of a batch sedimentation the concentration of solids is uniform throughout the cylinder. Soon after the process begins, all the particles of suspended solid fall through the fluid at their maximum velocities v_m under the existing conditions of hindered settling. For a closely sized solid, all particles fall at about the same velocity and a sharp line of demarcation is observed between the supernatant clear liquid (zone *A*) and the slurry (zone *B*) as the process continues. In a slurry containing particles of different sizes, including very fine solids, the larger particles will settle faster, the line of demarcation is not sharp, and the supernatant liquid may be hazy or milky. In either case, the particles near the bottom of the container begin to pile up on the bottom, building up the concentrated sludge (zone *D*) as indicated in Fig. 99. There may not be a well-defined interface separating zones *B* and *D*, but in all cases the concentrated sludge builds up as the sedimentation proceeds. So long as the two interfaces are relatively far apart, the solid particles in zone *B* continue to fall at their constant maximum velocities and no change in sedimentation rate is observed since the density or concentration of the solids in the suspension near the upper interface remains constant.

The concentrations of zones *B* and *D* are also plotted in Fig. 99. Zone *B* maintains a constant composition until the interface between zones *A* and *B* approaches the interface between zones *B* and *D*. As the upper interface approaches the sludge building up on the bottom of the container, the density and viscosity of the suspension surrounding the falling particles increase, with a corresponding decrease in settling velocity. This velocity continues to decrease during a so-called transition period, after which the slurry appears uniform as a heavy sludge (zone *D*), the settling zone *B* having disappeared. The sedimentation process from then on consists only of the continuation of the slow compaction of the solid in zone *D*. During this compaction of the sludge the fluid may be considered to be flowing through a porous bed of decreasing permeability. An ultimate height of the suspension is reached which represents the maximum compaction of the solid when surrounded by the given fluid.

Equations 14, 17, or 17H, when properly interpreted, using bulk density and bulk viscosity for hindered settling, may serve to express the rate of settling of particles in sedimentation. But other variables are frequently of great importance. Small particles may be dispersed or agglomerated into larger particles. The effective size and density of the particles may also be affected by adsorption of or from the fluid. These changes in the effective size (D) and density (ρ_s) may change the sedimentation rate by a relatively large factor. For example, Table 21 reports the observed data on two batch sedimentations run on slurries of the same material of the same concentration and starting from the same height. The only difference is the addition of a small amount of wetting agent to the second test.

The presence of the small amount of nacconal had a pronounced effect on the velocity of settling, increasing the rate of sedimentation during the first third of the distance and retarding the rate during compaction as well as preventing close compaction. The screen size of the dry solid particles is insufficient information to determine settling velocities. Adsorption on the particles may cause agglomeration of particles or simply increase the effective size by the additional material adsorbed. Apparently some such action was caused by the nacconal in the tests reported in Table 21. In view of the complex nature of sedimentation, reliable calculations can be made only when based on actual measurements of the rate of sedimentation under carefully controlled conditions.

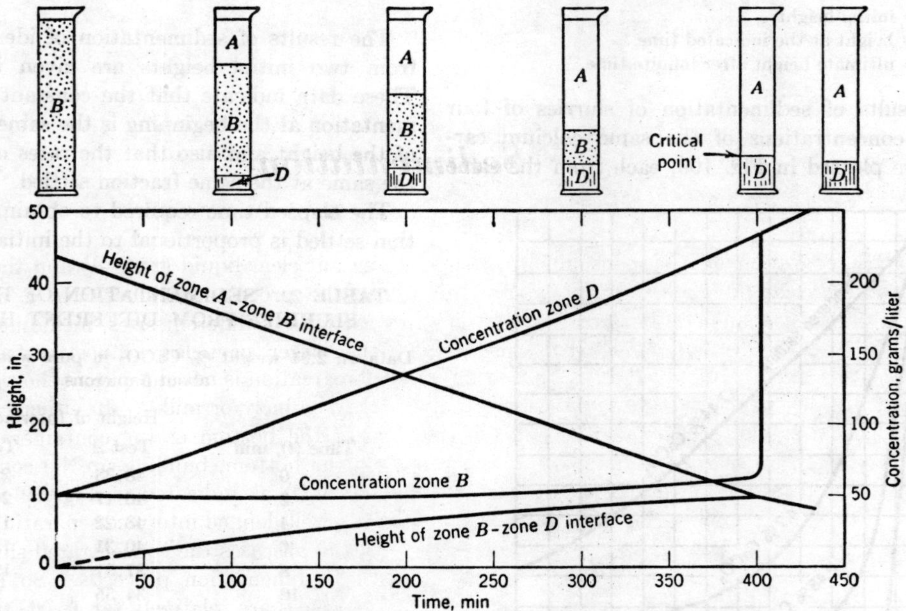

FIG. 99. Height of interfaces and concentration of zones in batch sedimentation.

TABLE 21. EXPERIMENTAL SEDIMENTATION

Data on pulp containing 7.44 per cent of $CaCO_3$ of an average size estimated to be about 5 microns in diameter and 2.26 grams/cc density in H_2O at 20° C.

Time, min	Height of Interface, cm: Pure Water	Height of Interface, cm: 0.1% Nacconal Freshly Added
0	30.84	30.88
4	29.89	27.21
8	29.10	21.65
12	28.30	15.85
16	27.50	14.48
27	25.29	12.97
37	23.34	12.23
52	20.34	11.42
72	16.48	10.74
102	11.48	9.92
117	9.80	9.62
151	7.80	9.13
∞	6.95	8.44

Exercise. From the data of Table 21, prepare a plot showing the velocity of settling in centimeters per minute as a function of the "fraction settled" (the distance settled at the indicated time divided by the distance ultimately settled).

$$\text{Fraction settled} = \frac{Z_0 - Z}{Z_0 - Z_\infty}$$

where Z_0 = initial height.
Z = height at the indicated time.
Z_∞ = ultimate height after infinite time.

The results of sedimentation of slurries of four different concentrations of the same calcium carbonate are plotted in Fig. 100, each from the same

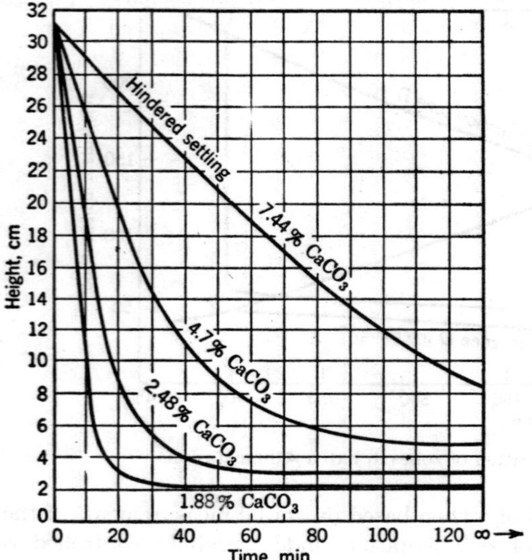

FIG. 100. Height of interface during batch sedimentation.

original height of slurry. The more dense slurries settle at a slower rate, indicating the mutual interference of particles in hindered settling and the deviations to be expected from settling rates estimated from the behavior of individual particles in free settling.

The initial constant rate is clearly evident in each of the curves of Fig. 100. Slurries of very low concentration may settle with a velocity approximately that of free settling for the first few minutes, but a slower rate is usually to be expected.

The transition from the initial constant rate of settling of a slurry of the properties of the feed (original constant rate) to the compression or thickening of the pulp takes place rapidly and is followed by the relatively slow compression period, as indicated in Figs. 99 and 100.

Exercise. Compare the observed constant-rate settling velocities reported in Fig. 100 with maximum settling velocities estimated by equation 14 or equation 17, using the bulk density of the slurry for the density of the fluid, and the bulk viscosity of the slurry for the viscosity of the fluid. Also, compare these velocities with results calculated with the factor F_s from Fig. 71.

The results of sedimentation of identical slurries from two initial heights are given in Table 22. These data indicate that the constant rate of sedimentation at the beginning is the same independent of the height and also that the rates of settling are the same at the same fraction settled.

The elapsed time required to obtain a given fraction settled is proportional to the initial height.

TABLE 22. SEDIMENTATION OF IDENTICAL SLURRIES FROM DIFFERENT HEIGHTS

Data on 2.91 weight % $CaCO_3$ in pure water; particle size about 5 microns.

Time (t), min	Height of Interface, cm Test A	Test B
0	48.33	22.90
2	46.17	20.63
4	43.22	17.86
6	40.31	15.01
8	37.51	12.28
10	34.55	9.63
12	31.72	7.64
14	28.79	6.62
16	26.21	....
21	19.30	4.57
24	16.40	4.05
30	13.55	3.28
39	10.92	2.68
59	7.51	2.41
77	6.21	....
102	5.23	....
∞	4.40	2.29

Exercise. Plot the rate of settling for tests A and B of Table 22 as a function of the fraction settled.

It has been suggested [4] * that equation 17 for laminar flow could be used to compute the rate of sedimentation as follows.

$$v_m = -\frac{dZ}{dt} = \frac{kD^2(\rho_s - \rho_b)}{\mu_b}$$

Rearranging and integrating

$$\int_{Z_0}^{Z} \frac{\mu_b}{\rho_s - \rho_b}\, dZ = kD^2 t \qquad (17b)$$

* The bibliography for this chapter appears on p. 120.

where ρ_b = bulk density of slurry.

μ_b = bulk viscosity of slurry.

k = a factor whose value depends upon concentration of solid and other properties of the systems. In Stokes equation (17), $k = g/18$, a constant.

This equation may be integrated over various limits of time. If integrated over the constant rate period, the value of k so found may be used for computing sedimentation from other heights over the constant rate period as the properties of the suspension remain constant during this interval. The equation may also be used to estimate changes to be expected from varying the particle diameter D, the densities ρ_s and ρ_b, and the bulk viscosity μ_b.

But if the integration is conducted over the time interval corresponding to the whole process the value of k so found will be of no value as it does not correspond to the properties of the suspension at any time and may not be used for any other calculation.

Exercise. Using the data for test B, Table 22, for the sedimentation of $CaCO_3$, construct a curve of height versus time for the same slurry settling from an initial height of 48.33 cm and compare with data reported for test A:

1. Assuming equal settling velocities at equal fractions settled.
2. Integrating equation 17b over constant rate period only to find k, and computing Z as a function of time (t).
3. Integrating equation 17b over the total time interval and using the value of k so determined in equation 17b for computing Z as a function of t (not good practice).

EQUIPMENT

Simple batch settling tanks have been and still are widely used. These operate in the manner described for the small-scale laboratory graduate. The tank is filled, and the slurry is allowed to settle for the desired time. The thickened material may be removed through a valve in the bottom of the tank, or the clarified solution may be withdrawn, either by lowering a swing siphon or by the successive opening of draw-off connections, starting with the uppermost. When the decanted solution begins to show the presence of sludge or the sludge level is exposed, the decantation is stopped.

Continuous sedimentation is conducted in inverted cones, or in cylindrical or rectangular tanks or vessels equipped with slowly revolving rakes for moving the thickened sludge to the central discharge (Fig. 101). In large concrete tanks the mechanism for moving the rakes may be so constructed that the rake arms are automatically raised to lift the rakes off the bottom when subjected to an overload, as might be caused by an interruption to the discharge, and automatically lowered again to the normal operating condition when the overload is passed. The feed is introduced through a feed well at the top center of the cone or cylinder. The slurry settles directly below the feed well, forming a sludge which is withdrawn from the bottom. The clear fluid flows to the periphery and is withdrawn by overflowing a circumferential weir.

A *tray thickener* is an installation of one thickener directly above another, which may be operated independently on the same or different feeds, or in series.

A *filter thickener*, as the name implies, is the combination of a filter and a thickener. The feed is introduced into the tank. The solution is withdrawn through a submerged filtering medium or cloth. The

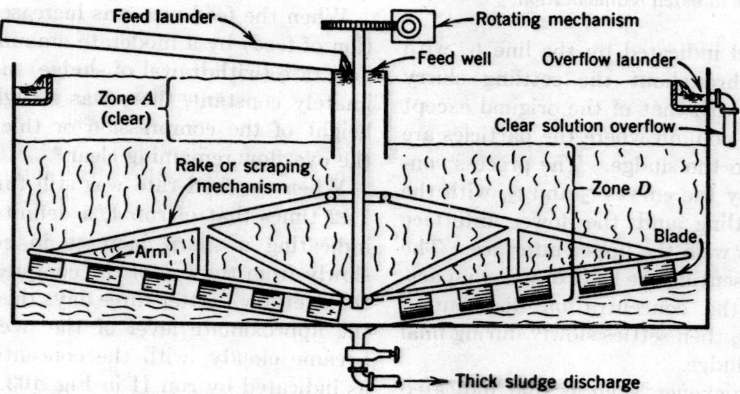

FIG. 101. Sectional diagrammatic drawing of a continuous thickener.

solids that collect on the filtering medium are periodically removed by mechanical scrapers or by low pressure air or water on the reverse side of the filter. They settle to the bottom where they are removed as a sludge.

CONTINUOUS SEDIMENTATION

The operation of continuous sedimentation tanks or *thickeners* depends upon the same characteristics of the slurry as indicated for batch sedimentation. The differences between continuous and batch operations are indicated by the concentration of the solids at different heights in the thickener.

The operation of batch sedimentation starts with a column of slurry having uniform concentration, as indicated by the long dashed line t_0, representing zero time in Fig. 102. Shortly thereafter, the inter-

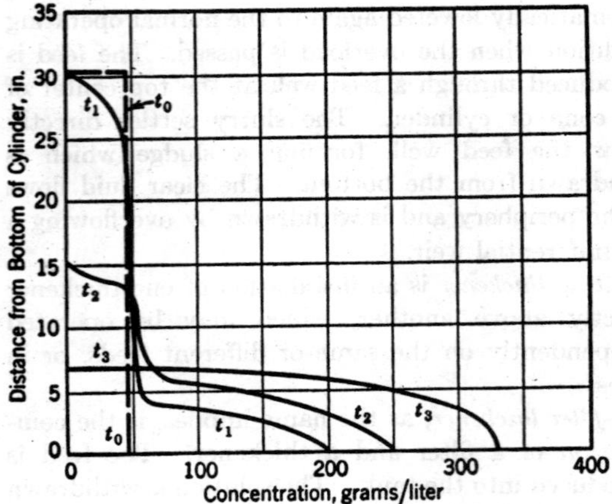

FIG. 102. Concentration as a function of height at various time intervals in batch sedimentation.[2]

face settles to a level indicated by the line t_1, with the concentration throughout the settling slurry practically identical with that of the original except at the bottom of the column where the particles are accumulating to form the sludge. The process continues as indicated by the curves t_2 and t_3, with the upper interface settling and the lower interface rising until it merges with the upper interface. The single interface representing the level of the compacting sludge having the concentration distribution shown by the curve t_3 then settles slowly during final compression of the sludge.

In a continuous thickener, such as that indicated in Fig. 101, the feed enters through the central feed well and is introduced about 1 to 3 ft below the surface of the liquid in a manner to create a minimum of disturbance. The entering fluid flows downward with the suspended solids in the feed well, then radially outward from the center and upward to the overflow. The suspended solids tend to settle down through the liquid which is relatively stagnant, although actually moving slowly radially and upward and rising from the compacting sludge. The sludge is compressed or thickened in the lower part of the cylindrical tank and is moved toward the bottom center by slowly rotating rakes. The action of these rakes in pushing the sludge to the outlet is to compress or thicken the sludge to a greater extent than is accomplished in simple or batch sedimentation.

A major difference between the continuous and batch sedimentations is the complete absence of any zone *B* of the same composition as the feed, unless such conditions are approximated directly below the feed well when the horizontal interface between the clear fluid and sludge is well below the bottom of the feed well. Under these conditions the feed slurry appears to settle in a cylindrical column directly below the feed well. When the feed rate is not excessive and a properly clarified effluent is produced, the upper or clarification zone in the continuous process is a region where the solids are present at such a low concentration that the mechanism approaches that of free settling. Immediately below this clear zone is the zone of sludge compaction.

Figure 103,[2] run I, indicates the concentration of calcium carbonate as a function of height in a continuous thickener which was producing a well-clarified overflow. The high clarification zone provided plenty of distance for the free settling therein to produce a clear solution overflow.

When the feed rate was increased (same composition of feed) by a moderate amount with the underflow rate (withdrawal of sludge) maintained approximately constant, there was a slight increase in the height of the compression or thickening zone with the overflow remaining clear.[2]

When the feed rate was still further increased to 1.42 times that of run I, a definite interface or line indicating a sharp increase in concentration rose slowly from the top of the compression zone toward the overflow. After three days this interface reached the approximate level of the overflow which then became cloudy, with the concentration distributed as indicated by run II in Fig. 103. With the thickener operating in this manner, giving a cloudy overflow containing 4.6 gm of calcium carbonate per liter, from a feed containing 45 gm of calcium carbonate per liter, the approximate constant composition indicated by the almost vertical line of run II was about 70 gm per liter. Under these conditions the concentration-versus-height curve has much the same appearance as that for t_1 in Fig. 102 representing batch sedimentation. However, the practically constant concentration in the continuous sedimentation is not the same as that of the feed.

After continuous running without further change in concentration distribution, the continuous thickener was shut down and the batch rate of sedimentation of the suspension in the thickener was determined. The rate of settling was constant but somewhat less than the rate observed in the batch settling of the feed slurry. This is to be expected as the concentration of solids in the suspension was higher in the thickener than in the feed. On the other hand, the rate of batch sedimentation of the feed slurry was always less than the rate of free settling in the clarification zone when the continuous thickener was operating to produce a clear overflow (run I).

The capacity of continuous thickeners or sedimentation equipment is based on their ability to perform two functions: to clarify the liquid overflow by the elimination of suspended solids, and to thicken the sludge or underflow by the elimination of liquid. The area of the sedimentation equipment controls the time allowed for settling particles out of the liquid for a given rate of feed of liquid and is important in determining the clarification capacity. The depth of the thickener controls the time allowed for thickening the sludge for a given rate of feed of solid and is important in determining the thickening capacity.

The *clarification capacity* of the thickener is determined by the settling rate of the suspended solids. This rate may be estimated from the rate of settling of the upper interface in a batch sedimentation test. By a material balance, the total quantity of fluid in the feed is equal to the sum of the fluid removed as a clear overflow plus the fluid in the thick sludge removed from the bottom. The vertical velocity of the fluid at any height in the thickener is equal to the volume of fluid passing upward at that level divided by the area of the thickener. For successful clarification, the settling velocity of the particles (or of the upper interface in the laboratory batch sedimentation at the pulp density existing at any level in the continuous classifier) must be somewhat greater than the vertical velocity of the fluid at that level. If the settling velocity is less than the upward fluid velocity, the particles will pass out in the overflow and there will be little clarification. If the settling velocity equals the upward fluid velocity, the particles will neither rise nor fall and the concentration of solids in the clarification zone will increase, thereby reducing the settling velocity until

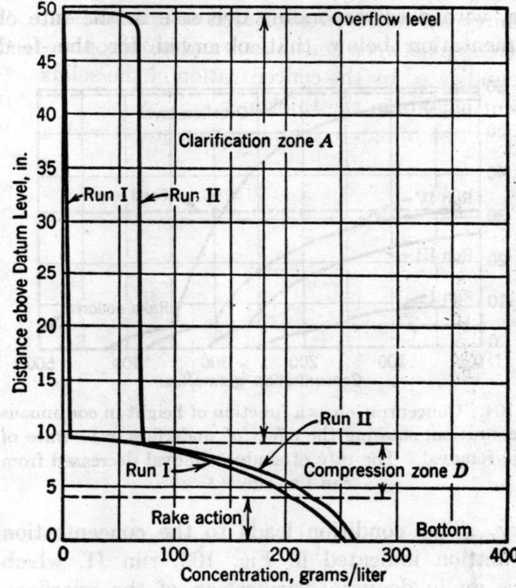

FIG. 103. Concentration as a function of height in continuous sedimentation, showing the effect of feed rate. Run II was made with feed rate equal to 1.42 times that of run I.[2]

the particles are carried out in the overflow as indicated. Therefore the settling velocity of particles must be sufficiently greater than the upward fluid velocity to prevent any increase in concentration in the clarification zone.

Usually the constant rate of settling observed in a batch sedimentation test of the feed slurry may be used as a reasonably conservative value for the settling rate in the less concentrated clarification zone for purposes of design. But under some conditions at a lower level in the thickener where the concentration of solids is so high as to greatly retard the rate of settling, the upward velocity of the fluid may exceed the rate of settling of solids, even when this condition is not encountered in the upper zone, with results as indicated in Fig. 103, run II. In making design calculations it is therefore necessary to consider the rates of settling at different concentrations and the corresponding vertical velocity of the water or fluid to be certain that the area of the thickener is adequate for satisfactory clarification.

When the feed rate to a continuous thickener exceeds the maximum which the thickener can handle and produce a clear overflow, the solid particles are unable to settle down out of the normal clarification zone and therefore build up a higher concentration therein. This causes hindered settling, with a corresponding decrease in the rate of sedimentation below that observed for the feed

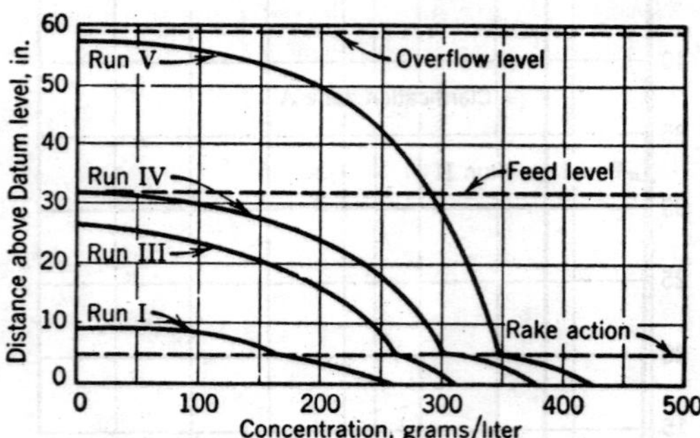

FIG. 104. Concentration as a function of height in continuous sedimentation, showing the effect of underflow rate (rate of sludge removal. The rate of sludge removal decreased from run I to run V.[2]

slurry. This condition leads to the concentration distribution indicated in Fig. 103, run II, which shows an inadequate clarification of the overflow. The feed rate which just fails to initiate this hindered settling is the limiting clarification capacity of the thickener as it is the maximum feed rate at which the suspended solid can reach the compression zone.

Exercise.

Show that the following relationship may be used to estimate the required area of clarifier

$$A = Q\frac{(F - D)}{\rho R} \tag{49}$$

where A = area (sq ft) required to thicken the discharge slurry to consistency of D.

D = parts fluid to one part solids by weight in underflow.

F = parts fluid to one part solids by weight in feed.

Q = mass of solids treated per unit of time (lb/hr).

R = rate of settling of pulp of consistency F (ft/hr).

ρ = density of fluid (lb/cu ft).

It is often desirable to check the estimated cross-sectional area of the thickener by calculations for different concentrations, using as the rate of settling the constant rate of settling of batch sedimentations of slurries of increasing concentrations.

Example. A feed containing 6 lb of water per pound of solids is to be thickened to a concentration corresponding to 1.12 lb of water per pound of solids with the production of a clear overflow. Tests should be made on slurries containing higher concentrations than that of the feed to be certain that the velocity of settling is greater than the upward flow of fluid at all concentrations normally encountered in the thickening of the required feed. In this case the following five concentrations of the solids to be thickened were tested in batch sedimentation with the results indicated in columns 1 and 4.[1]

1	2	3	4	5
Mass Ratio, Fluid to Solids, L/S	Fluid Rising per Pound Solids, lb	Calculated Fluid Rising, cu ft/hr	Observed Rate of Settling, ft/hr	Calculated Minimum Area Required for Clear Overflow, Feeding 1 Ton of Solids per 24 hr, sq ft
6.00	4.88	6.5	2.180	3.0
4.94	3.72	4.96	1.190	4.16
4.00	2.88	3.85	0.893	4.31
3.51	2.39	3.19	0.758	4.22
3.00	1.88	2.51	0.600	4.18

The area of 4.31 sq ft/ton of solids is the correct minimum area to be used for design purposes and not 3.0 sq ft/ton as calculated from a sedimentation test of the feed.

The above data clearly indicate the need of checking the calculated area against all concentrations to be encountered anywhere in the tank, including not only the clarification zone but also the thickening zone.

Although the above tests and calculations indicate that an area of 4.31 sq ft/ton of solids is adequate to produce a clear overflow with an underflow of 1.12 lb of water per pound of solids for feeds containing up to at least 6.00 lb of water per pound of solids, it must not be assumed that this area is satisfactory for all feeds regardless of water content. At high water contents the rate of settling is more nearly independent of water content and may not increase so rapidly with the water to be removed as in the relatively concentrated slurries used in this example.

The *thickening capacity* may be illustrated by considering the effect of varying the underflow rate as indicated in Fig. 104.[2] The overflow from all runs shown was practically clear. The thickness or depth of the compression or thickening zone increased as the underflow rate was decreased, as indicated, and the concentration of the underflow increased with increasing depth of the thickening zone, at least for a constant rate of feed.

The curves indicating concentration as a function of depth in the compression zone in Fig. 104 are substantially vertical displacements of each other, and similar in shape to those for batch sedimentation (compare t_3 of Fig. 102 with I of Fig. 104). The action of the rakes appears to be effective in breaking up a semirigid structure and the accompanying channels in the concentrated sludge. This action extends for about 3 to 4 in. above the top of the rakes and is important in producing a more concentrated underflow.

The concentration of the underflow or sludge from a continuous thickener depends on the depth of the thickening or compression zone and the time that the solids are in this zone. The total height or depth of the thickener depends in large measure on the required depth or time in the compression zone. The *required height of the compression zone* may be estimated from data obtained on a batch sedimentation under the following conditions.

Consider a batch sedimentation starting with a slurry at its *critical concentration*, that is, the slurry has an initial concentration equal to the concentration of the top layer of the compression zone (*D*) during the period of constant rate settling. The sedimentation curve will start at the critical point (normally a point of time between the constant rate settling period and the compression period) and will consist only of a compression curve, with all particles in compression for the same period of time. The time required in this batch experiment for a slurry to pass from this state of critical concentration to the desired underflow concentration may be taken as the retention time for solids in a continuous thickener. This assumes that the concentration of solids at the bottom of the compression zone of a continuous thickener at any time is the same as the average concentration of the compression zone in the batch test described above at a time equal to the retention time of the solids in the continuous thickener. In other words, the concentration at the bottom of the thickener is a function only of the time of thickening.

This retention time can be obtained from a single batch test, irrespective of thickness of sludge, simply by observing the height of the compression zone as a function of time.

The shape of this compression curve indicates a gradual decrease in rate of settling as time increases and may be represented by the equation[3]

$$-\frac{dZ}{dt} = k(Z - Z_\infty) \tag{50}$$

where Z = height of the compression zone at time t.
Z_∞ = height of the compression zone at infinite time.
k = a constant for a particular sedimentation system.

If batch tests are available with the initial slurry at its critical concentration (or critical dilution), then a plot of ln $(Z - Z_\infty)$ versus t should give a straight line with slope equal to $-k$. Integration of equation 50 gives

$$\int \frac{dZ}{Z - Z_\infty} = \ln (Z - Z_\infty)\Big]_{Z_c}^{Z} = -kt\Big]_0^t$$

or

$$\ln (Z - Z_\infty) = -kt + \ln (Z_c - Z_\infty) \tag{51}$$

where Z_c is the height of the compression zone at its critical concentration (in this case $Z_c = Z_0$). Z_∞ may be obtained by experiment or by trial and error to make this plot a straight line.

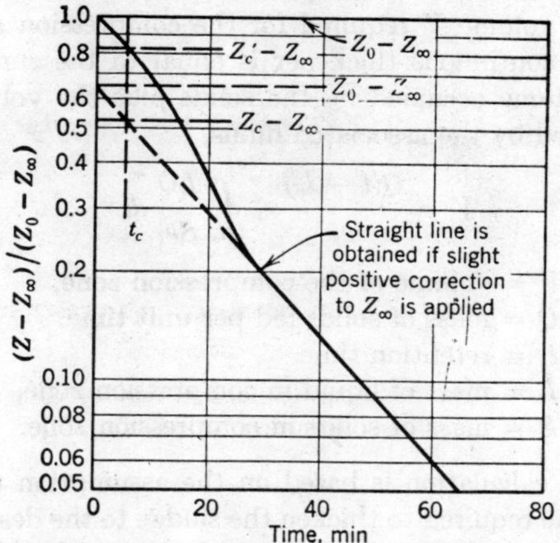

FIG. 105. Plot illustrating the extrapolation of sedimentation curve from data of Table 22 to obtain time for critical concentration.[3]

If the available batch tests are on an initial slurry concentration less than the critical, during the settling period the compression zone will be forming, and at the time corresponding to the critical point (when the two interfaces meet) the average concentration of the compression zone will be greater than the critical concentration since it will be composed of layers that have been in compression for varying lengths of time. A suggested method of obtaining the time required to pass from the critical concentration or critical dilution $(L/S)_c$ to any desired underflow concentration [3] is to extrapolate the compression curve back from the critical point to zero time and then locate the time when the upper interface (between the supernatant liquid and the settling slurry) is at a height Z_c' halfway between the initial slurry height Z_0 and the extrapolated zero-time compression-zone height Z_0' (Fig. 105). This time t_c represents, in effect, the time when all the solids were at their critical dilution and went into compression. Actually, part went into compression sooner and part later. The extrapolation back to the time axis should be made on logarithmic-arithmetic plot as the line is then nearly straight, or with the aid of equation 50. With this value of t_c, the retention time is obtained as $t - t_c$, where t is the time when the solids have reached the desired underflow concentration.

The determination of the necessary volume for the compression zone of the continuous thickener may be obtained from a consideration of the time each layer of solids has been in compression.

The volume V required for the compression zone in the continuous thickener is equal to the sum of the volume occupied by the solids plus the volume occupied by the associated fluids.

$$V = \frac{Q(t - t_c)}{\rho_s} + \int_{t_c}^{t} \frac{LQ}{S\rho_l}\,dt \tag{52}$$

where V = volume of the compression zone.
Q = mass of solids fed per unit time.
$t - t_c$ = retention time.
L = mass of liquid in compression zone.
S = mass of solids in compression zone.

This calculation is based on the assumption that the time required to thicken the sludge to the desired concentration is independent of the thickness of the sludge in the compression zone.

By assuming a constant mean value for L/S so that equation 52 may be integrated with L/S constant, the following *approximate relationship* is obtained.

$$V = Q(t - t_c)\left[\frac{1}{\rho_s} + \frac{1}{\rho_l}\left(\frac{L}{S}\right)_{\text{avg}}\right] \tag{53}$$

where $\left(\frac{L}{S}\right)_{\text{avg}}$ = average mass ratio of fluid to solid in the thickening zone from top to bottom.

Rather than apply equation 53 to the whole sedimentation zone, more reliable results may be obtained by dividing the compression zone in parts and assuming average conditions over each part as suggested by Coe and Clevenger.[1] For thickening of the sludge corresponding to the previous example (p. 116) the observed data from a batch thickening operation were as follows.

Time of Thickening, hr	Consistency Fluid-to-Solids Ratio: Determined	Average over the Indicated Period
0	......	
2	1.70:1	1.70
4	1.59:1	
		1.47
9	1.35:1	
		1.275
14	1.20:1	
		1.16
19	1.12:1	

In order to produce a thickened sludge of 1.12:1 fluid to solids, 19 hr of thickening retention time is indicated. Since an area of 4.31 sq ft is required per ton of solids per 24 hr (previous calculation, p. 116), the solids per square foot in the thickening zone for 1 hr of retention are calculated as 2000/(24 × 4.31) = 19.4 lb. For 19 hr of retention, 19.4 × 19 or 368 lb of solids will be in the thickening zone covering an area of 1 sq ft. From the above data a total of 19 hr of retention is the sum of 5 hr of retention of each of the pulp consistencies averaging 1.16, 1.275, and 1.47, and a 4-hr supply of a pulp averaging 1.7 in consistency. The solids per cubic foot in the above pulps are calculated or determined as 43.2 lb, 37.6 lb, 33.7 lb, and 30 lb, respectively. The depth of each class of pulp is then calculated as $\dfrac{\text{hours} \times 19.4}{\text{pounds solids per cubic foot}}$, or 2.23 ft, 2.57 ft, 2.87 ft, and 2.58 ft, respectively, making a total depth of 10.25 ft.

This method may indicate thick compression zones as it is based on the average concentration of the thickened sludge rather than the concentration at the bottom of the sludge. Also, an increase in concentration of the sludge is brought about by the action of the rakes. These factors probably more than compensate for the turbulence introduced by the rakes. In some cases it may be possible to estimate the desired concentration just above the zone of rake action by making proper allowance for the effect of the rakes.

The total depth of the thickener may be estimated by adding to the estimated thickness of the compression zone the following allowances.

For the pitch of the bottom	1–2 ft
For storage capacity to cover interruptions or irregularities in discharge	1–2 ft
For submergence of feed	1–3 ft

This makes a total depth of thickener about 3 to 7 ft greater than the total compression zone including both the zone of rake action and the zone of thickening by simple settling as estimated.

There is usually an economic balance involved in determining the total depth of the thickener, which involves consideration of the cost of the equipment, interest on capital represented by the material retained in thickener, and desired sludge concentration. If the solids are valuable, filtration is likely to prove more economical.

SEPARATION FROM GASES

Equipment for the separation of suspended solids (or liquids) from gases may be divided into two general classes: (1) those which work "dry," acting on the suspension as received, and (2) those which work "wet," using an additional fluid, usually water, to facilitate the separation.

The most common type of separator is a *filter* which passes the gas and retains the solids on a bed which may be cleared or discarded periodically. A mat of glass fibers, cellulose, or metal mounted in a light frame is used in air-conditioning and home heating systems, the mat being easily replaced with a fresh filter when the old one ceases to operate effectively. The mat of fibers is frequently wet with a nonvolatile oil to assist in retaining the dust particles. In industry a battery of cloth bags is often used, the bags being mounted in groups on independent frames which can be vibrated at intervals, shaking the accumulation of solids into a discharge hopper from which they may be re-used or discarded. Some systems have groups of bags mounted in compartments constructed to permit scavenging air to be blown backwards through the bags for cleaning.

A *dust collector* is essentially a simple enlargement in a pipe line which reduces the velocity enough to permit the solids to settle out in the enlargement. It is similar to the surface velocity classifier (Chapter 8).

A *cyclone separator* (Fig. 106) is a vertical cylinder with the inlet stream introduced tangentially near the top, giving the suspension a spinning motion the cylinder. The centrifugal force acting on the particles tends to throw them radially (equation 44) to the sides of the cylinder as they spiral downward

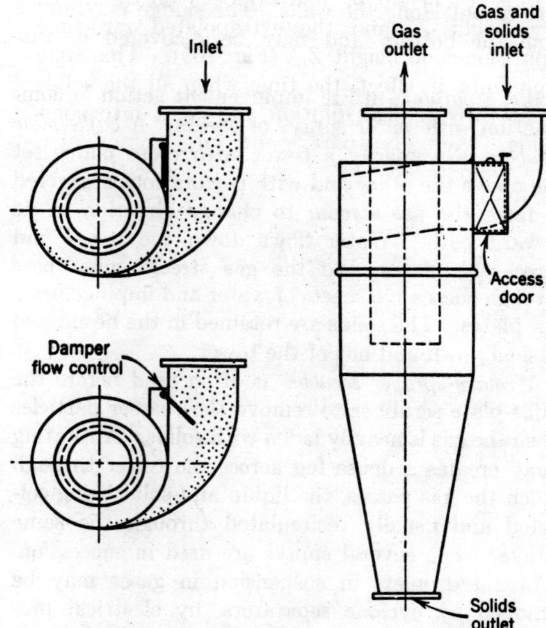

Fig. 106. Diagrammatic drawing of cyclone separator.

to a conical bottom where they are removed. A damper may be used to increase the velocity u_t of the entering stream. The clarified gas leaves the cyclone through a pipe extending down into the center of the cylinder and passes upward and out of the cyclone. A modification in this unit is the "shave-off," a slit in the cylinder placed near the top, one full turn from the inlet, which removes the fine particles immediately without letting them spiral down below the level of the gas exhaust to the cone where they might be re-entrained and carried out in the discharging gas.

A cyclone is quite effective on large particles, removing 99 per cent of the solids coarser than 30 microns. Finer particles are only partially removed, with less than 50 per cent of those smaller than 5 microns being retained in the separator.

A cyclone may be inverted, the suspension entering

near the bottom and the clean gas being removed from the top of the interior through a pipe extending through the bottom. The effectiveness of the inverted cyclone separator when working dry is generally poorer than the normal cyclone described above. This construction is convenient, however, when water is sprayed into the cyclone. The suspension passes upward through the spray, and the solids are trapped in the liquid droplets as well as thrown out along the walls. The sludge is removed from the bottom and may be recovered or discarded.

Wet separators utilize impingement action in combination with water sprays or sheets. A *baffle-plate scrubber* is primarily a tower with cross plates set one above the other and with perforations staggered to force the gas stream to change direction in its upward path. Water flows down the tower and across the plates, and the gas stream must pass through successive sheets of water and impinge upon wet plates. The solids are retained in the liquid and washed down and out of the tower.

A *rotary-sprayer scrubber* is often used before the baffle-plate scrubber to remove the coarser particles when the gas is heavily laden with solids. A rotating spray creates a dense fog across the tower through which the gas passes, the liquid and solid being collected and usually recirculated through the same sprayer when several sprays are used in succession.

Liquid droplets in suspension in gases may be removed by cyclone separators, by electrical precipitators, or by simple impingement baffles. Such devices are frequently built into wet separators as the last stage through which the gas passes, thus removing any liquid picked up in the wet stages.

Electrical precipitation [5] of entrained solid or liquid particles consists in maintaining a high unidirectional difference in potential between two electrodes and passing the gas between these electrodes. One of the electrodes, the discharge electrode, is of small cross section, such as a wire, edge or point, to make a high electrical field at its surface necessary for ionizing the gas. The other electrode, the collecting electrode, has less or no curvature and serves to collect or precipitate most of the separated dispersoids. The ions formed near the discharge electrode are carried through the gas to the collecting electrode at velocities of the order of 100 fps. When the gas between the electrodes carries suspended particles, the ions attach themselves to the particles which then become charged and are attracted to the collecting electrode. The velocity of these particles is much less than that of the ions of the gas.

In the ordinary one-stage method, corona discharge is maintained, giving a high-intensity precipitating field which exerts a pressure on the precipitated material, thus preventing redispersion. It also causes some chemical activity such as production of ozone and nitric acid in air.

The two-stage method uses corona-forming electrodes in the first stage only to form the necessary gas ions for charging the dispersoids. The second stage has nondischarging precipitating electrodes opposing the collecting electrodes between which is maintained a corona-free electric field for precipitating the particles ionized in the first chamber.

The precipitation or collection efficiency of an electrical precipitator is a function of the time that the gas remains in the active field and can be made to approach 100 per cent if desired. But the size and cost of the equipment usually places an economic limit at about 90 to 99 per cent.

For any given dispersoid of uniform size and character in a given precipitator, the precipitation efficiency (eff) is related to the time t (seconds) that the gas remains in the active field of the precipitator by the equation

$$\log (1 - \text{eff}) = t \log K = tEC$$

where K = the so-called precipitation constant, usually 0.05 to 0.50.
E = the voltage.
C = a constant.

BIBLIOGRAPHY

1. Coe, H. S., and G. H. Clevenger, *Trans. Am. Inst. Mining Met. Engrs.*, **55**, 356 (1916).
2. Comings, E. W., *Ind. Eng. Chem.*, **32**, 663 (May 1940).
3. Roberts, E. J., "Thickening—Art or Science?" *Mining Engineering*, **1**, 61 (March 1949).
4. Robinson, C. S., *Ind. Eng. Chem.*, **18**, 869 (1926).
5. Schmidt, W. A., and E. Anderson, *Elect. Eng.*, **57**, 332 (1938).

PROBLEMS

1. Compute the area required of a thickener to handle 20 tons/hr of slurry (Table 22), producing a clear overflow and an underflow or sludge containing 20 per cent by weight of solids. Assume that the constant rate of settling in the batch sedimentation is the rate of settling in the clarification zone of the continuous thickener.

2. Using the data of Table 22 and a density for water of 62.35 lb/cu ft with a specific gravity of 2.71 for $CaCO_3$,

estimate the area and depth of a thickener to treat 4000 lb of solids per hour in a feed containing 33.3 parts of water per 1 part of $CaCO_3$, delivering a clear overflow and an underflow containing 3 parts of water per 1 part of $CaCO_3$.

3. A wet slurry of mix to be burned to make cement is to be thickened to 60 per cent solids in a conventional thickener at the rate of 50 tons/hr of dry solids. A batch sedimentation test made under appropriate laboratory conditions gave the following results.

Time, hr	Graduate Reading, ml volume
0	1017
0.25	925
0.5	815
0.75	700
1	600
1.25	528
1.75	420
3.0	352
4.75	330
6.75	310
12	280
20	251
28	237
∞	220

Graduate height = 35 cm for 1000 ml.
Dry solids in test = 236 grams.
Specific gravity of dry solids = 2.09.

Specify the dimensions of the tank to handle the conditions stipulated, assuming a feed concentration similar to that used in the batch test.

4. Two Dorr thickeners are to be used in preparing grit-free milk of lime from 50 tons of burned limestone per hour. The unslaked lime analyzes as follows.

Material		Mass %	Size
CaO		91.5	100% −200 mesh
SiO_2	Grit	6.0	100% +200 mesh
Al_2O_3		1.5	
Fe_2O_3		1.0	

The unslaked lime is slaked with fresh water and recycle milk of lime to form a milk of lime suspension containing 15 per cent by weight of lime solids (exclusive of grit). The slaking is carried out in the first classifier with the overflow having a maximum size of 200 mesh. The underflow contains all the grit plus 1.5 lb of suspension per pound of grit and passes to the second classifier where it is diluted to milk of lime suspension containing 2 per cent of lime solids. The overflow of this unit is used as recycle milk of lime, while the underflow is discarded, containing 1.5 lb of suspension per pound of grit.

Draw a flow sheet for the process, indicating rates and concentrations of all streams.

(*a*) Calculate the percentage of lime lost.

(*b*) Specify the diameter and depth of the two classifiers.

CHAPTER

11

Transportation of Fluids 1 – Pipes and Fittings

MATERIAL is frequently stored and handled in the fluid state. The fluids in most process engineering problems cannot be handled in open channels but require closed ducts. In ancient times these ducts were hollowed logs, and later they were made of sections of wood or of pottery. The development of iron brought about the manufacture of cast-iron and wrought-iron pipe and permitted some standardization of dimensions of pipes and fittings.

Any structural material now employed in the engineering profession is used for pipe in applications where its peculiar advantages are most valuable. Glass, ceramic, steel, nickel, lead, rubber, brass, copper, concrete, and asbestos pipes are encountered in many processing plants, and wood is still found in many large installations. The methods of joining sections are generally similar for all materials. The principal methods involve threaded, bell-and-spigot, flanged, and welded connections and fittings.

THREADED CONNECTIONS

Threaded pipe is most commonly encountered in industry because practically all small sizes of pipe are joined by this method, whether fabricated of steel, wrought iron, cast iron, brass, or plastic. This system is simple because the outside diameters of the pipe are kept constant with a tolerance of $\frac{1}{64}$-in. oversize and $\frac{1}{32}$-in. undersize, and the inside diameters of fittings are kept within the same limits, regardless of material. Typical standard dimensions of pipe joined by screw threads are given in Table 23. The tolerance for the wall thickness of the different materials varies but is usually 12.5 per cent. Pipe larger than 12 in. is rarely threaded, and the outside diameter corresponds to the nominal pipe size. Standard lengths of pipe are from 16 to 22 ft.

Steel pipe is made by longitudinal shaping of hot steel strips with the butt or lap joints welded together by pressure in the machine.

Steel pipe was originally classed in three thicknesses for different operating pressures, standard, extra-strong (or extra-heavy), and double-extra-strong. These three classes are now obsolete, and thicknesses follow a set formula, expressed as the "schedule number" as established by the American Standards Association. Ten schedule numbers are in current use: 10, 20, 30, 40, 60, 80, 100, 120, 140, and 160, the figures being the approximate value of the expression,

$$1000 \frac{P}{S}$$

where P = internal working pressure (psi).
S = allowable fiber stress (psi) for the particular alloy under the conditions of use.

For example, the schedule number of ordinary steel pipe having an allowable fiber stress of 10,000 psi for use at a working pressure of 350 psi would be $1000 \times (350/10{,}000)$ or 35. This would be the proper schedule for welded joints and steel fittings but not for threaded connections and cast-iron or malleable-iron fittings. In practice, schedule 40 would be used for welded construction and schedule 80 (about twice the computed value) for iron fittings. The higher schedule is required because of weaknesses in the threads and iron fittings.

For all pipe sizes below 10 in., schedule-40 pipe is identical with the former "standard" pipe, and

TABLE 23. DIMENSIONS OF THREADED PIPE

Nominal Pipe Size, in.	Outside Diameter, in.	Low-Carbon Steel Pipe (ASA B36.10) Thickness, in.		Inside Diameter, in.		Transverse Internal Area, sq in.		Underground Water Pipe (AWWA 7A.4), Thickness, in.		Cast-Iron Pipe (ASA A40.5), Thickness, in.	Brass and Copper Pipe (ASTM B42 and B43), Thickness, in.	
		Schedule 40	Schedule 80	Schedule 40	Schedule 80	Schedule 40	Schedule 80				Standard	Extra Strong
⅛	0.405	0.068	0.095	0.269	0.215	0.0569	0.0363	0.068			0.062	0.100
¼	0.540	0.088	0.119	0.364	0.302	0.1041	0.0716	0.088			0.082	0.123
⅜	0.675	0.091	0.126	0.493	0.423	0.1909	0.1405	0.091			0.090	0.127
½	0.840	0.109	0.147	0.622	0.546	0.3039	0.2341	0.109			0.107	0.149
¾	1.050	0.113	0.154	0.824	0.742	0.5333	0.4324	0.113			0.114	0.157
1	1.315	0.133	0.179	1.049	0.957	0.8639	0.7193	0.133			0.126	0.882
1¼	1.660	0.140	0.191	1.380	1.278	1.495	1.283	0.140		0.187	0.146	0.194
1½	1.900	0.145	0.200	1.610	1.500	2.036	1.767	0.145		0.195	0.150	0.203
2	2.375	0.154	0.218	2.067	1.939	3.356	2.953	0.154		0.211	0.156	0.221
2½	2.875	0.203	0.276	2.469	2.323	4.788	4.238	0.203		0.241	0.187	0.280
								Min	Max			
3	3.500	0.216	0.300	3.068	2.900	7.393	6.605	0.125	0.300	0.263	0.219	0.304
3½	4.000	0.226	0.318	3.548	3.364	9.888	8.891	0.125	0.318		0.250	0.321
4	4.500	0.237	0.337	4.026	3.826	12.73	11.50	0.125	0.337	0.294	0.250	0.341
5	5.563	0.258	0.375	5.047	4.813	20.01	18.19	0.156	0.375	0.328	0.250	0.375
6	6.625	0.280	0.432	6.065	5.761	28.89	26.07	0.188	0.432	0.378	0.250	0.437
8	8.625	0.322	0.500	7.981	7.625	50.03	45.66	0.188	0.500	0.438	0.312	0.500
10	10.750	0.365	0.593	10.020	9.564	78.85	71.84	0.188	0.500	0.438	0.365	0.500
12	12.750	0.406	0.687	11.938	11.376	111.93	101.64	0.188	0.500	0.438	0.375	

schedule 80 is identical with the former "extra-strong" pipe. There is no equivalent schedule number for "double-extra-strong" pipe, and schedule-160 pipe is the only other weight in which pipe smaller than 4 in. may be obtained.

Fittings for steel pipe systems are usually of gray cast iron or malleable iron.

For moderately *high pressures* up to about 6000 psi, stainless-steel tubing may be used with threaded forged-steel fittings of heavy walls. For higher pressures, particularly at high temperatures, alloy steels and special compression fittings and needle valves are required, as illustrated in Fig. 120.

Cast-iron pipe is not threaded generally, but it may be in some special applications, particularly on high-pressure water mains. When threaded, the dimensions are almost the same as those for steel pipe, the thickness being different in some cases. Fittings are of gray cast iron or malleable iron.

Cast-iron pipe is made in sand molds of either baked or green sand, or it may be cast centrifugally. Centrifugal casting is becoming more popular, especially by means of metal molds lined with a sprayed refractory material.

Wrought-iron pipe is made of iron made by mechanical puddling, or of a low-carbon steel made by the Byers process in which slag is added to molten iron, followed by refining in a Bessemer converter. It is claimed to have superior resistance to corrosion and is used particularly in some hot-water piping and underground installations. The dimensions are practically identical with those of low-carbon steel pipe. Fittings are usually cast iron.

Brass pipe and *copper pipe* are employed where greater resistance to corrosion is desired. Fabrication is similar to that of steel pipe, and dimensions are similar. Systems of brass or copper pipe usually include brass or copper fittings.

Fig. 107. Threaded pipe fittings. (*Walworth Co.*)

FITTINGS

Sections of threaded pipe are joined together by fittings such as illustrated in Fig. 107. *Couplings* join successive straight lengths of pipe with no change in direction or size. When the size is to be reduced or enlarged, a *reducing coupling* is used. When the direction is to be changed, an *elbow*, either 90-degree or 45-degree, is available; and if both direction and size are changed, a *reducing elbow* is in order. Because of the mechanical difficulty of cutting sharp threads on short pieces of pipe, *nipples* are made at the factory in a series of standard lengths from about four pipe diameters in length to close nipples, whose threads merge from each end of the section.

If more than two branches of piping are to be connected at the same point, *tees* and *crosses* are used. They may be obtained with any reasonable combination of sizes of the openings. Since most piping must be broken at intervals for maintenance and since standard pipe threads are right-hand, thus making it impossible to use right-hand fittings exclusively in connecting pipe from one fixed point to another fixed point, a *union* serves as a connector. The two halves of the union may be tighten the pipe sections independently, and the final connection made by tightening the bonnet of the union. If size reduction is desired at a tapped connection, a *reducing bushing* is the simplest fitting. A simultaneous change in direction and connection to a tapped outlet may be made by a *street elbow* having male threads on one end and female threads on the other. The end of a pipe may be closed with a *cap*, and an opening in a piece of equipment may be closed by a plug or, better, a capped nipple.

Many other special fittings are normally available on demand and can solve almost any piping problem.

VALVES

Flow is controlled by valves. *Gate valves* are shown in Fig. 108. The barrier to flow is a disk- or wedge-shaped dam sliding at right angles to the direction of flow and seating tightly in the valve

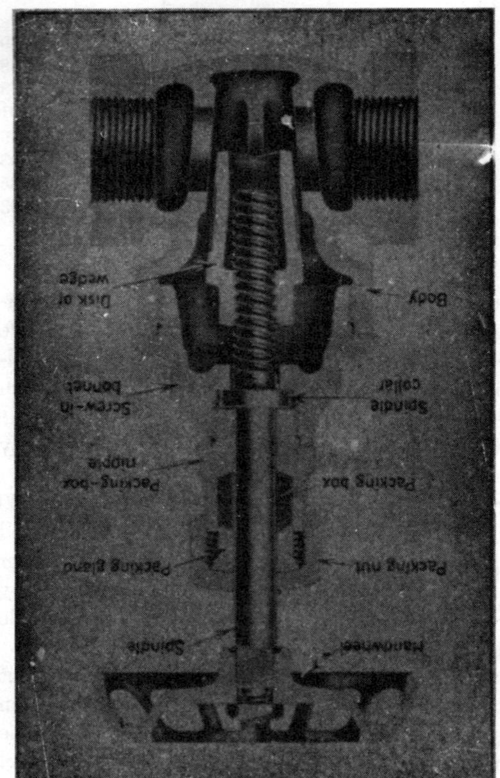

Fig. 108*a*. Sectional view of gate valve with nonrising spindle. (*Jenkins Bros.*)

body. When partially open, this type of valve exhibits a crescent-shaped opening for flow which changes in area extremely rapidly with slight adjustment of the valve handle, thus making this type of valve rather undesirable for partial-flow control although quite suitable for ordinary open-and-shut control. In the larger sizes the disk may swing

FIG. 109. Butterfly valve. (*R-S Products Corp.*)

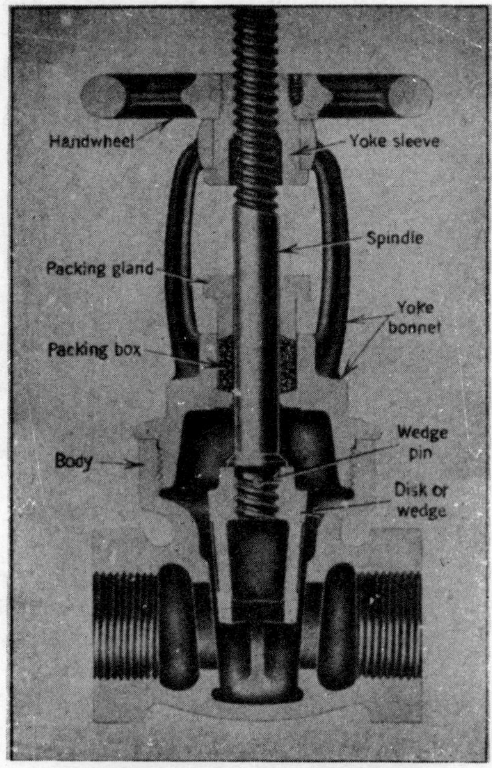

FIG. 108*b*. Sectional view of gate valve with rising spindle and outside screw and yoke. (*Jenkins Bros.*)

rather than slide into place, as indicated in Fig. 109 illustrating the *butterfly* valve.

The *globe valve,* so called because of the bulbous shape of the valve body, is shown in Fig. 110. This valve directs the fluid up or down through a circular opening in the central partition, which may be sealed either by forcing a replaceable composition-fiber disk down upon a flat seat or by inserting a tapered metallic plug into a conical seat, the plug and the seat being of different taper to furnish a line contact for the seal. The plug-type valve is excellent for partial-flow control. This valve may be made with a slender tapering needle seating in a small orifice

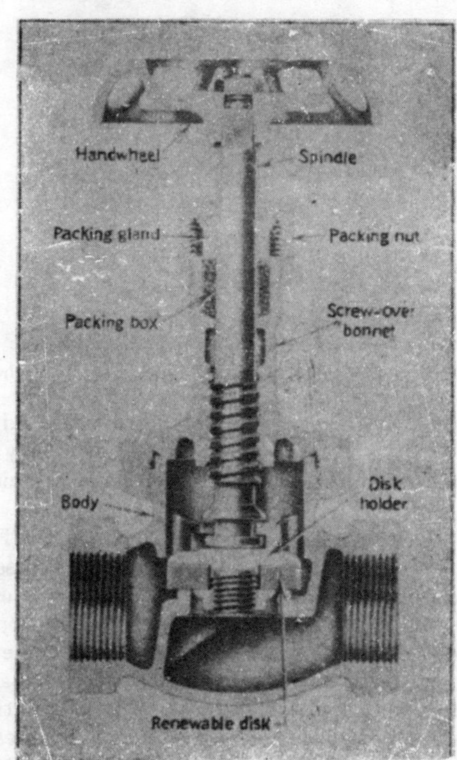

FIG. 110. Sectional view of globe valve with rising spindle. (*Jenkins Bros.*)

drilled in the valve body. Such a valve is termed a *needle* valve (Fig. 111).

Both gate and globe valves vary widely in details of construction. The stem or spindle moves outward (Fig. 108*b* or Fig. 110) or simply rotates without changing its position (Fig. 108*a*) as the valve is opened. The screw threads on the stem or spindle are either inside (Fig. 108*a* or Fig. 110) or outside (Fig. 108*b*) the space under pressure. The seats are replaceable or permanent. The valve is opened or closed by several turns of the handle, or it may be operated in one stroke of a lever handle (quick-opening type), Fig. 112.

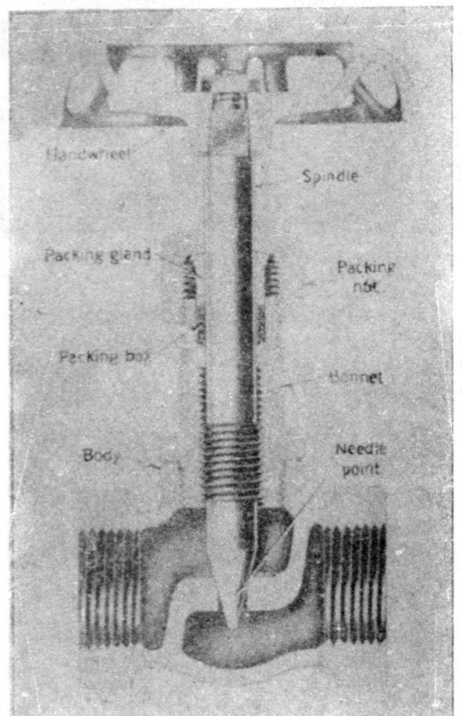

Fig. 111. Needle valve (sectional view). (*Jenkins Bros.*)

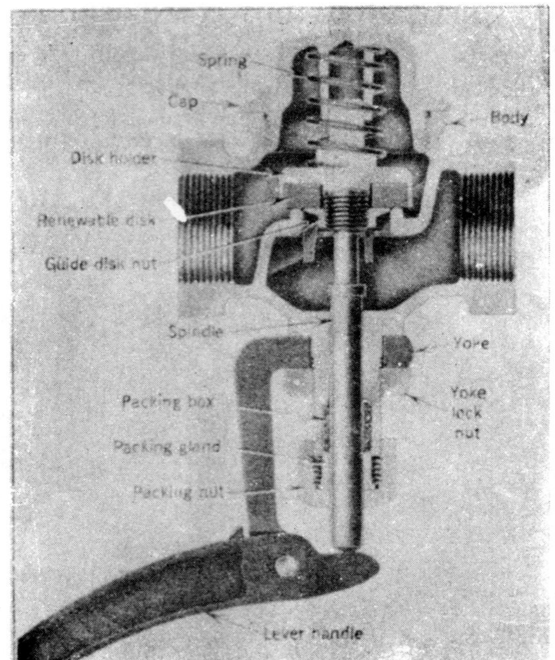

Fig. 112. Quick-opening globe valve (sectional view) (*Jenkins Bros.*)

For simple open-and-shut control, *plug cocks* (Fig. 113) are economical. Inserted into the flow passage is a tapered plug through which an opening is cast or drilled. A 90-degree turn of the plug will open or close the passage. Such valves may be made to accommodate three or four piping connections and direct flow through the different pipes. By application of pressure to the heavy grease lubricant, the grease is forced through the lubricant grooves to the bottom of the plug, thereby lifting the plug free from its seat.

Fig. 113. Cutaway view of Plug cock. (*Walworth Co.*)

BELL-AND-SPIGOT CONNECTIONS

Bell-and-spigot connected pipe handles more fluid than any other connection since it is the most practical type of joint in large pipe sizes when pipes are made of materials other than steel. Some

TABLE 24. DIMENSIONS OF BELL-AND-SPIGOT PIPE

Nominal Size, in.	Cast-Iron Pipe			Concrete and Clay Sewer Pipe (ASTM-C13-C14)		Reinforced-Concrete Pressure Pipe (C) (AWWA Standard)	
	Outside Diameter, in.	Liquids (ASA-A21.2) Thickness, in., Min-Max	Gas (AGA Standard) Thickness, in.	Outside Diameter, in., Min-Max	Thickness, in., Nominal-Min	Inside Diameter, in.	Minimum Thickness, in.
3	3.80	0.37-0.45					
4	4.80	0.40-0.45	0.40	4.88- 5.13	0.50-0.44		
6	6.90	0.43-0.50	0.43	7.06- 7.44	0.63-0.56	6	1.75
8	9.05	0.46-0.57	0.45	9.25- 9.75	0.75-0.69	8	1.75
10	11.10	0.50-0.60	0.49	11.50-12.00	0.88-0.81	10	1.75
12	13.20	0.54-0.65	0.54	13.75-14.31	1.00-0.94	12	1.75
14	15.30	0.54-0.62				14	1.75
16	17.40	0.58-0.67	0.62			16	1.75
18	19.50	0.63-0.72		20.63-21.44	1.50-1.38	18	1.75
20	21.60	0.66-0.82	0.68				
24	25.80	0.74-0.92	0.76	27.50-28.50	2.00-1.88	24	2.50
30	31.74	0.87-0.92	0.85	34.38-35.63	2.50-2.38	30	2.75
36	37.96	0.97-1.03	0.95	40.75-42.25	2.75-2.63	36	3.13
42	44.20	1.07-1.15	1.07			42	3.75
48	50.50	1.18-1.30	1.26			48	4.13
54	56.66	1.30-1.38				54	4.50
60	62.80	1.39-1.45				60	5.00

standard dimensions for various materials are given in Table 24.

Some typical bell-and-spigot joints are shown in Fig. 114. These joints are usually calked with oakum and lead, but the mechanical joint is becoming more popular because of the tighter joint, simplicity of installation, and greater latitude of angular displacement and expansion. These joints may be "locked," with a groove in the spigot which prevents pulling apart of the joint, "roll-on," with a rubber gasket tightened with a bolted ring, or "screwed-gland," with a ring gland drawn up against the gasket when screwed into threads in the bell.

Materials for pipe joined in this manner are usually cast iron, clay, or concrete, although glass, plastic, and cement-asbestos are occasionally employed. Cast-iron pipe is furnished in lengths of 12 to 20 ft, and other materials range in length from 6 to 15 ft.

Fittings of the same material as the pipe are similar in type and function to those for threaded pipe. Typical items are shown in Fig. 115.

Valves are almost always of the gate or butterfly types.

WELDED CONNECTIONS

The modern trend for pipe in sizes above 2 in. is toward more welded connections. No threading, calking, or bolting is needed, and no gaskets are re-

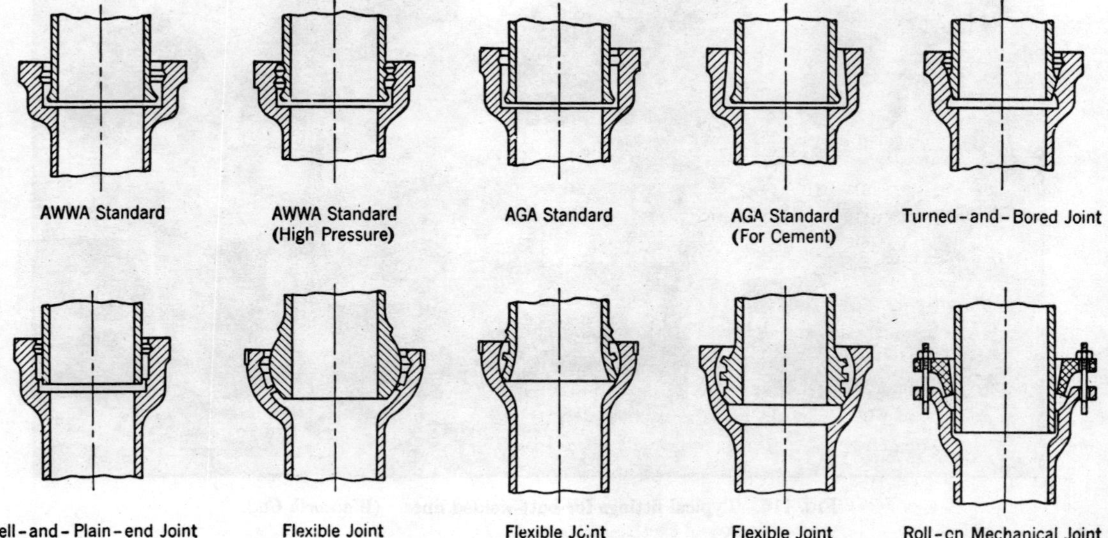

FIG. 114. Sectional drawings of various bell-and-spigot joints.

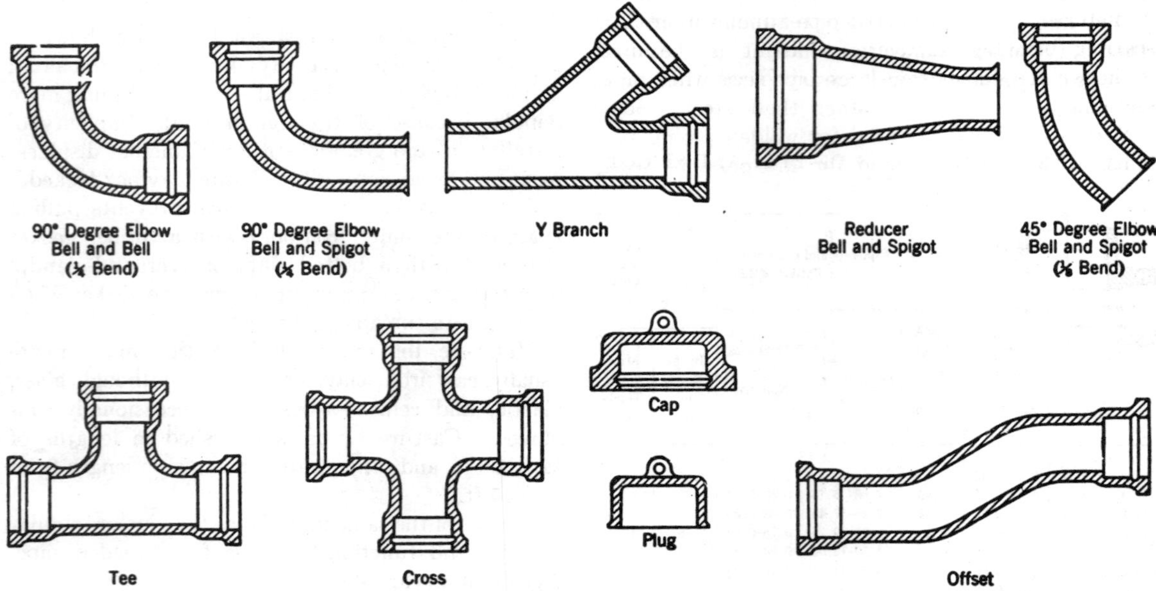

FIG. 115. Sectional drawings of typical bell-and-spigot fittings.

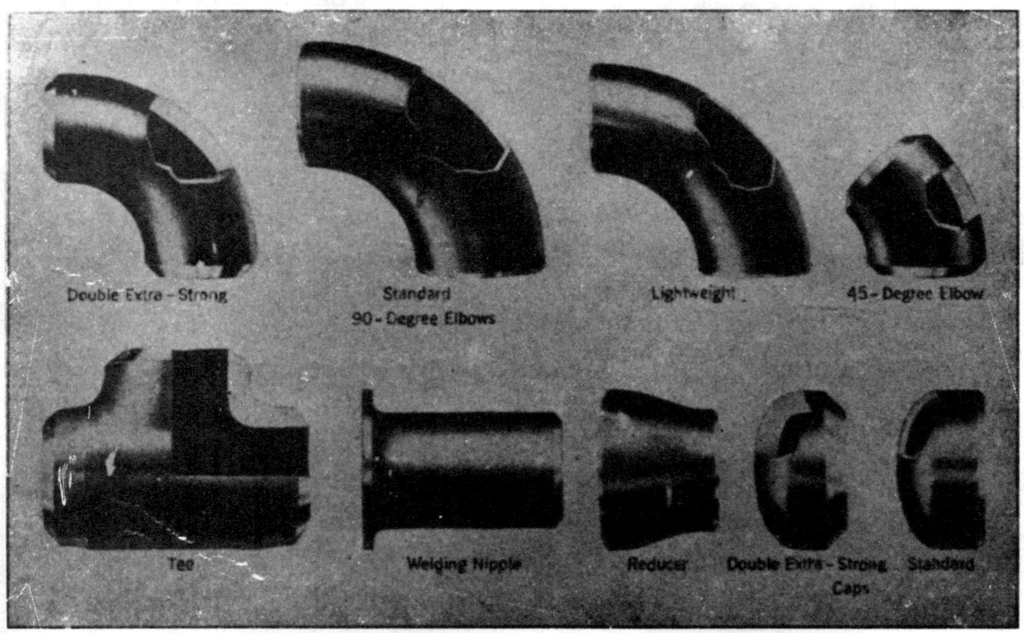

FIG. 116. Typical fittings for butt-welded lines. (*Walworth Co.*)

quired when the system is fused into an unbroken line of material. Pipe ends need no treatment other than scarfing (beveling), and very few fittings are required if the welder shapes the necessary pieces from pipe sections.

Steel is the usual material for welded piping systems, with scheduled pipe for low-pressure work (under 1000 psi) and seamless tubes for higher pressures. Gas welding was universal for some time, but electric-arc welding is increasingly popular.

Fittings and valves are of steel and are of two types, butt-weld and socket-weld. The butt-weld fittings are of the same dimension as the pipe, and the socket-weld fittings have enlarged ends similar to threaded fittings, but the pipe slips into place and is fillet-welded. Typical fittings are shown in Fig. 116.

Brass and copper pipe are joined in socket-weld systems, usually known as "streamlined" piping. The joints are brazed or soldered.

Plastic pipe, if thermoplastic, is easily welded with electric hot plates. The ends are heated to the softening point, then joined and allowed to cool, making a strong fusion joint. Fittings of the simpler types are available for butt-welding.

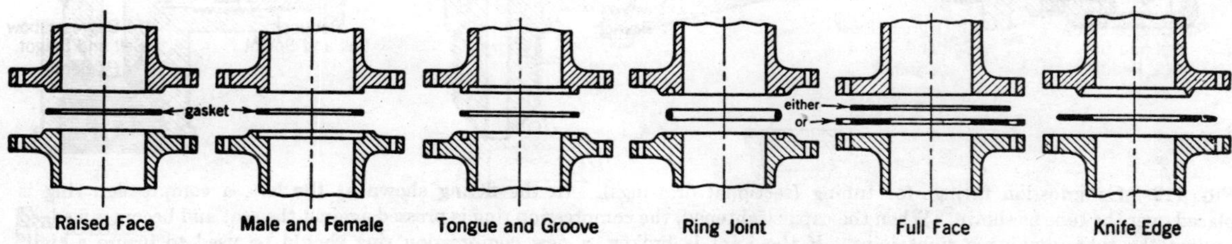

FIG. 117. Typical flanged joints (sectional drawings).

FLANGED CONNECTIONS

Flanged connections are used on larger sizes and higher pressures of piping service where the lines must be disassembled frequently for maintenance or inspection, on larger valve bonnets, and on pipes being connected to equipment. Formerly, steel pipe in sizes over 3 in. was always connected with flanges (threaded onto the pipe), but welded connections have replaced such connections in most new construction and in old as maintenance requires replacement. The various methods of facing the flanges are indicated in Fig. 117. The plain or full-face flange is widely used with either the ring gasket, entirely within the bolt circle, or the full-face gasket, which extends to the outer edge of the flange and is punched for bolt holes.

Flanges may be threaded or welded to the ends of the pipe, in which case the compression face is always on the flange. When threaded or welded flanges are not practical, as in high-silicon cast iron, glass, or sheet metal, the flange is slipped over the pipe and the flared or enlarged ends of the pipe provide the compression face.

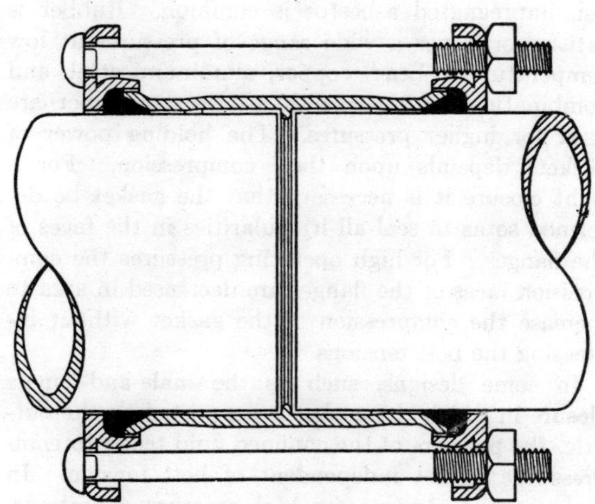

FIG. 118*a*. Dresser coupling (sectional drawing). (*Dresser Industries.*)

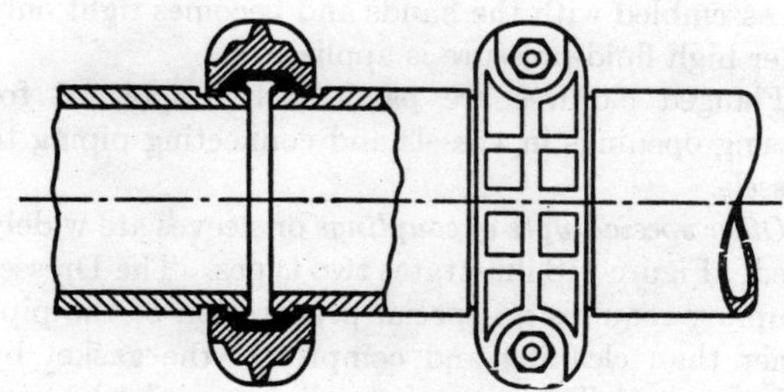

FIG. 118*b*. Victaulic coupling. (*Victaulic Company of America.*)

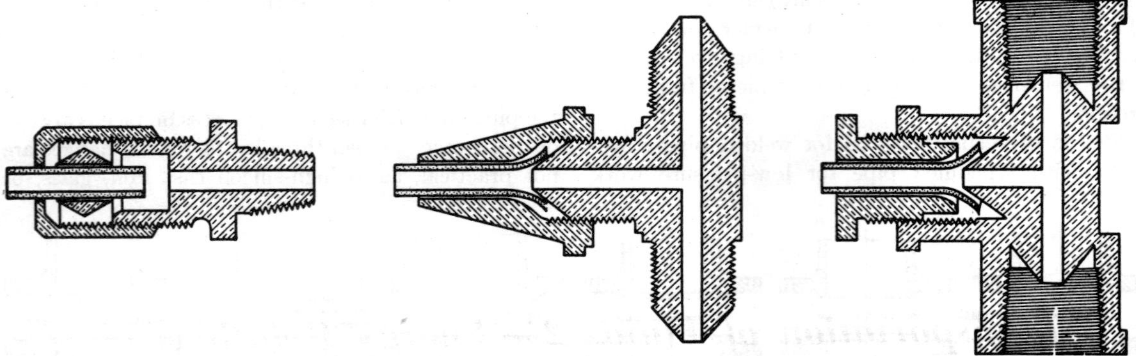

FIG. 119. Compression fittings for tubing (sectional drawings). In the fitting shown at the left, a compression ring is placed over the tube as shown. When the cap is tightened, the compression ring is pressed against the seat and becomes wedged against the tube, making a tight joint. If the joint is broken, a new compression ring should be used to insure a tight joint. In the two fittings shown at the right the tubing must be flanged or spread by a special tool to the form indicated. No compression ring is required.

Gaskets may be made of a wide range of materials from paper to steel. For pressures up to about 500 psi, impregnated asbestos is common. Rubber is satisfactory over a wide range of pressures at low temperatures. Lead, copper, aluminum, steel, and combinations of these with asbestos or paper are used for higher pressures. The holding power of gaskets depends upon their compression. For a tight closure it is necessary that the gasket be deformed so as to seal all irregularities in the faces of the flanges. For high operating pressures the compression faces of the flanges are decreased in area to increase the compression of the gasket without increasing the bolt tensions.

In some designs, such as the male-and-female closure in which the gasket is supported on the outside, the pressure of the confined fluid tends to compress the gasket independent of bolt tension. In some special closures for high-pressure operations, the pressure of the confined fluid alone is depended upon to compress the gasket, and the closure may be assembled with the hands and becomes tight only after high fluid pressure is applied.

Flanged closures are particularly important for closing openings in vessels and connecting piping to vessels.

Other special types of couplings or sleeves are widely used. Figure 118 illustrates two types. The Dresser coupling requires no special preparation of the pipe other than cleaning and compresses the gasket by bolt tension. This type of coupling may also be used as an expansion joint when the sleeve is sufficiently long and adequate clearance is provided between the ends of the pipe. The Victaulic coupling requires that a groove be cut around the pipe and depends upon fluid pressure for compressing the gasket.

Compression fittings are widely used for small-sized tubing at both low and high pressure. These are convenient and efficient, particularly if the connection is to be broken. Figure 119 shows typical fittings of this type.

High-pressure fittings for pressures greater than 6000 psi are illustrated in Fig. 120.

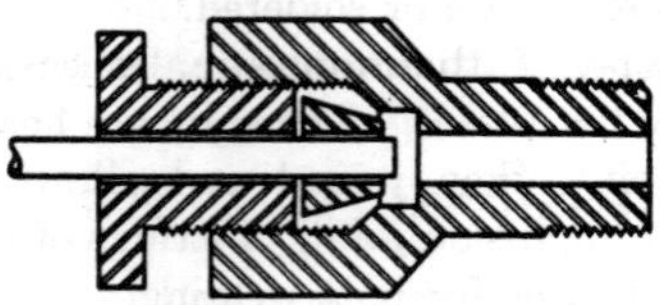

FIG. 120. Special steel fittings and valves for high pressure (sectional drawing). The cone may be threaded on the tubing with a left-hand thread so that it will not be turned off the tubing as the threaded gland is turned in its right-hand threads to tighten the joint.

BIBLIOGRAPHY

1. AMERICAN GAS ASSOCIATION, New York, Standards.
2. AMERICAN SOCIETY FOR TESTING MATERIALS, Philadelphia, Standards.
3. AMERICAN STANDARDS ASSOCIATION, New York, Standards.
4. AMERICAN WATER WORKS ASSOCIATION, New York, Standards.
5. JENKINS BROS., New York, Catalogs.
6. R-S PRODUCTS CORP., Philadelphia, Catalogs.
7. WALWORTH CO., New York, Catalogs.

CHAPTER 12

Transportation of Fluids 2 – Energy Relations

THE energy balance is a most important consideration in making engineering calculations. In making a balance of energy it is essential that all energy factors be expressed in the same units if the calculations are to be correct. If energy absorbed as heat is measured in Btu, and pressure is determined in pounds force per square inch with volume in gallons, and kinetic effects are calculated with mass in pounds mass and velocity in feet per second, the different energy terms must be all converted to the same units before a balance can be made. These units may be either foot-pounds or Btu, for example. The relationship of 1 Btu being equivalent to 778 ft-lb and the application of this conversion factor generally offers no difficulty. But confusion frequently arises when pounds mass and pounds force appear in the same equation.

DIMENSIONS AND UNITS

The relationship between force and mass may be expressed by Newton's law: force equals the time rate of increase of momentum.

$$F' = \frac{d(mv)}{dt} \tag{7}$$

Force has the dimensions (mass)(velocity)/(time), $(m)(L/t)/t$, or (mass)(length)/(time)^2, mL/t^2,

where m = mass in any units.
L = a linear distance in any units.
t = time in any units.

In English units mass is usually expressed in pounds, length in feet, and time in seconds. Then the unit of force is that which gives 1 lb mass an increase in velocity (or acceleration) of 1 fps each second. This unit is 1 poundal. If a mass of 1 lb at mean sea level at 45 degrees latitude is allowed to fall under the influence of gravity, its acceleration will be 32.17 ft per second per second. The force exerted by gravity on 1 lb mass at sea level (32.17 poundals) is called "1 lb force." It is also called the "weight" of the mass of 1 lb at sea level.

The use of the same term *pound* to represent mass and force is frequently confusing. The inclusion of a pound weight is always confusing.

For example, a mass of 100 lb under the influence of gravity at sea level exerts a force (ma) of $100 \times 32.17 = 3217$ poundals or $3217/32.17 = 100$ lb force; it is also said to "weigh" 100 lb. If this same mass is carried to a high elevation where the acceleration due to gravity is 30 ft per second per second, it exerts a force of 3000 poundals or $3000/32.17 = 93.25$ lb force. If the mass of 100 lb is now "weighed" on a spring scale calibrated for use at sea level, its weight (pound force) will be 93.25 lb. But if a beam balance is used with standard "weights" hung on the beam, the 100-lb mass will "weigh" 100 lb, as this type of balance compares forces, which are proportional to mass and equal for equal masses.

Therefore the term *weight* as distinct from mass will not be used in this text because it is indefinite and confusing.

In engineering practice the unit of force is usually 1 lb force, and, if the unit of mass is to be 1 lb mass, a conversion factor must be included to convert pounds force to poundals as follows.

$$F' = g_c F = \frac{d(mv)}{dt} \tag{7}$$

where F' = force in absolute units (as poundals).
F = force in engineering units (as pounds force).
g_c = conversion factor (32.17 poundals per pound in English units).

Usually the mass is constant, and equation 7 may be written

$$g_cF = ma \tag{7b}$$

where a = acceleration or dv/dt.

The choice of mass, length, and time as fundamental dimensions leads to the absolute units for force (poundals) and to the use of the conversion factor g_c as indicated above.

Frequently force, length, and time are chosen as fundamental dimensions. In this system, mass has the dimensions of (force)(time)^2/(length) or Ft^2/L. This unit of mass is called a *slug* when force is expressed in pounds force, length in feet, and time in seconds. The conversion factor g_c is then a multiplying factor to convert slugs mass into pounds mass. Rewriting equation 7*b*

$$F = \frac{m}{g_c} a = m'a \tag{7c}$$

where F = force (lb).
m = mass (lb).
m' = mass (slugs).
m/g_c = mass (slugs).

The acceleration instead of the mass may be divided by the conversion factor g_c. In this case the unit of acceleration is a' when mass and force each have the same units, such as pounds. The numerical value of a' is frequently referred to as the number of gees at sea level where the numerical value of g is equal to g_c.

$$F = m\frac{a}{g_c} = ma' \tag{7d}$$

where F = force (lb).
m = mass (lb).
a' = acceleration (gees).
a/g_c = acceleration (gees).

Any system of fundamental dimensions, mass-length-time or force-length-time, or other system, may be used separately or mixed together and are so found in the technical literature. The conversion factor g_c converts the units corresponding to a system including force as a fundamental dimension to the units corresponding to a system including mass. The dimensions of g_c may be determined from equations 7*b*, 7*c*, or 7*d* and are mL/t^2F.

When heat effects are included, an additional dimension, temperature T, must be added, making four fundamental dimensions necessary to define the quantities ordinarily encountered in engineering operations. But the four dimensions chosen may be almost any combination that is most convenient at the moment, provided they are all independent. For example, as illustrated in Table 25, mass, length, time, and temperature; or force, length, time, and temperature; or energy, force, time, and temperature may be selected as the four fundamental dimensions to express the units and relationships between the various quantities listed.

TABLE 25. DIMENSIONS OF VARIOUS QUANTITIES IN THREE SYSTEMS OF DIMENSIONS

Quantity	System I Mass, m Length, L Time, t Temperature, T	System II Force, F Length, L Time, t Temperature, T	System III Force, F Energy, E Time, t Temperature, T
Temperature (T)	T	T	T
Time (t)	t	t	t
Length (x,y,z,D,L)	L	L	E/F
Mass (m)	m	Ft^2/L	F^2t^2/E
Force (F)	mL/t^2	F	F
Energy (E)	mL^2/t^2	FL	E
Velocity (v)	L/t	L/t	E/Ft
Acceleration (a)	L/t^2	L/t^2	E/Ft^2
Density (ρ)	m/L^3	Ft^2/L^4	F^5t^2/E^4
Pressure (P)	m/Lt^2	F/L^2	F^3/E^2
Surface tension (γ)	m/t^2	F/L	F^2/E
Mass flow rate (W)	m/t	Ft/L	F^2t/E
Mass velocity (G)	m/L^2t	Ft/L^3	F^4t/E^3
Viscosity (μ)	m/Lt	Ft/L^2	F^3t/E^2
"Heat capacity"	mL^2/t^2T	FL/T	E/T
Specific heat (C)	L^2/t^2T	L^2/t^2T	E^2/F^2t^2T
Thermal conductivity (k)	mL/t^3T	F/tT	F/tT
Rate of heat transfer (q)	mL^2/t^3	FL/t	E/t
Coefficient of heat transfer (h)	m/t^3T	F/LtT	F^2/EtT

To convert the dimensions of
System II to system I: multiply by $g_c = mL/t^2F$
System III to system II: multiply by J ($= FL/E$) with the appropriate exponent (from +2 to −4)

The conversion factor g_c has been discussed and applied in the conversion of pounds force (F) (system II) into poundals (mL/t^2) (system I) by equation 7*b*. Pounds mass and poundals force, or grams mass and dynes force, are the absolute units usually derived from the fundamental dimensions of system I.

Pounds force and slugs mass conform to the units usually derived from the fundamental dimensions of system II.

The use of pounds force and pounds mass, as is common in engineering practice, represents the simultaneous use of systems I and II, and the conversion factor g_c must be used either to multiply the units of force or to divide the units of mass. When energy is expressed in foot-pounds, the conversion factor J becomes unity for systems II and III. When force is in pounds, length in feet, and energy in Btu the conversion factor J becomes 778 ft-lb/Btu and must be used either to multiply the energy expressed in Btu or to divide the force expressed in pounds.

If any four fundamental independent dimensions are used, as indicated by any one system of dimensions, no conversion factor is necessary. If any additional dimension is introduced, the result is the equivalent of using an additional system of dimensions for each such additional dimension introduced, and a conversion factor must be included for each such additional dimension beyond the four fundamental dimensions of a single system.

THE FLOW EQUATION

The energy relationships of a fluid or other material flowing through equipment, plant, or piping may be obtained by an energy balance. Energy is carried with the flowing fluid and also is transferred from the fluid to the surroundings, or vice versa.

1. Energy carried with the fluid includes:
   A. The internal energy U, including all energy which is the peculiar property of the fluid, regardless of its relative location or motion.
   B. The energy carried by the fluid because of its condition of flow or position:
      (1) Energy of motion or kinetic energy: $mv^2/2$ in absolute units, or $mv^2/2g_c$ [in engineering practice, $g_c = 32.17$ lb mass ft/lb force (sec)(sec), and the energy of motion is $mv^2/2$ (32.17) in foot-pounds], where m is the mass of material under consideration, and v is the velocity of that material relative to some "stationary" reference.
      (2) Energy of position or potential energy: mgZ in absolute units, or mgZ/g_c ($mgZ/32.17$ in foot-pounds), where Z is the height of the mass m above the reference plane, and g is the acceleration due to gravitational force.
      (3) Energy of pressure PV carried by the material because of its introduction into or exit from flow under pressure, where P is the absolute pressure exerted by the material, and V is the volume of the material.

The necessity of including the term PV may be indicated by the diagrams of Fig. 121, which represent the addition of a quantity of fluid into the flow system in two steps. The first step includes the quantity of fluid by embracing it within the flow system without moving the fluid. It involves the addition of the internal energy U_1 and the energy of motion and position of the added fluid to the energy of the flow system. The second step involves the movement of the quantity of fluid into the flow

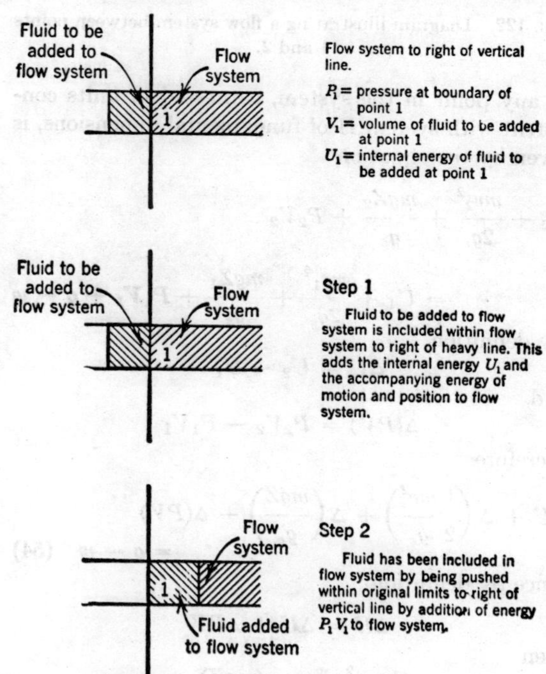

FIG. 121. Diagram illustrating the inclusion of pressure energy in a flow system.

system. This is done by forcing the quantity of added fluid against the pressure of the fluid already in the flow system and the addition of the energy P_1V_1 to the flow system.

2. Energy transferred between a fluid or system in flow and its surroundings is of two kinds:
   A. Heat q absorbed by the flowing material from the surroundings during flow (Fig. 122) between points 1 and 2.
   B. Work w done by the flowing material on the surroundings during flow between points 1 and 2. This is frequently called shaft work w_s to distinguish it from the work done by a stationary batch of material not flowing through a system or not transferred from one state to another.

An energy balance around a flow system, such as between points 1 and 2 in Fig. 122 and the surroundings, assuming no accumulation of material or energy

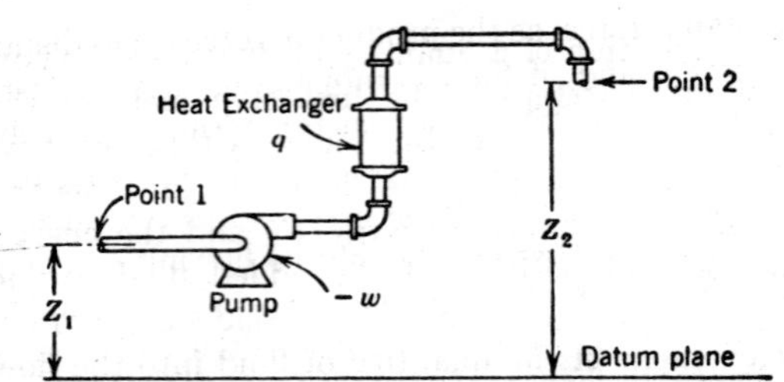

FIG. 122. Diagram illustrating a flow system between points 1 and 2.

at any point in the system, expressed in units consistent with system II of fundamental dimensions, is given by the equation

$$U_2 + \frac{mv_2^2}{2g_c} + \frac{mgZ_2}{g_c} + P_2V_2$$

$$= U_1 + \frac{mv_1^2}{2g_c} + \frac{mgZ_1}{g_c} + P_1V_1 + q - w$$

By definition

$$\Delta U = U_2 - U_1$$

and

$$\Delta(PV) = P_2V_2 - P_1V_1$$

therefore

$$\Delta U + \Delta\left(\frac{1}{2}\frac{mv^2}{g_c}\right) + \Delta\left(\frac{mgZ}{g_c}\right) + \Delta(PV) = q - w \quad (54)$$

Since by definition

$$\Delta H = \Delta U + \Delta PV$$

then

$$\Delta H + \Delta\left(\frac{1}{2}\frac{mv^2}{g_c}\right) + \Delta\left(\frac{mgZ}{g_c}\right) = q - w \quad (54a)$$

The increase in internal energy is the sum of the increases due to all changes considered as taking place in the material in flow, including heat effects, compression effects, surface effects, and chemical effects.

$$\Delta U = \int_1^2 T\,dS + \int_1^2 P(-dV) + \int_1^2 \gamma\,d\sigma + \int_1^2 \mu_A\,dm_A + \int_1^2 \mu_B\,dm_B + \text{etc.} \quad (55)$$

where T = absolute temperature of the material.
S = absolute entropy of the material.

$\int_1^2 T\,dS$ = increase in internal energy due to heat effects between states 1 and 2 or points 1 and 2 in the flow system.

$\int_1^2 P(-dV)$ = increase in internal energy due to compression effects between states or points 1 and 2.

$\int_1^2 \gamma\,d\sigma$ = increase in internal energy due to surface effects between states 1 and 2.

where γ = surface tension of the material.
σ = surface area of the material.

$\int_1^2 \mu_A\,dm_A$ = increase in internal energy due to chemical effects or changes in component or substance A, between states 1 and 2.

The energy term $\Delta(PV)$ is a complete differential

$$\Delta(PV) = \int_1^2 P\,dV + \int_1^2 V\,dP \quad (56)$$

Combining equations 54, 55, and 56 (including surface and chemical effects in the etc. term),

$$\int_1^2 T\,dS + \Delta\left(\frac{1}{2}\frac{mv^2}{g_c}\right) + \Delta\left(\frac{mgZ}{g_c}\right) + \int_1^2 V\,dP + \text{etc.} = q - w \quad (57)$$

In any process the increase in internal energy due to heat effects $\int_1^2 T\,dS$ is equal to the sum of the heat absorbed from the surroundings and all other energy dissipated into heat effects within the system due to irreversibilities such as overcoming friction occurring in the process,

$$\int_1^2 T\,dS = q + (lw) \quad (58)$$

where lw = "lost work," energy that could have done work but was dissipated in irreversibilities within the flowing material.

If equations 57 and 58 are combined and rearranged,

$$\int_1^2 V\,dP + \Delta\left(\frac{mv^2}{2g_c}\right) + \Delta\left(\frac{mgZ}{g_c}\right) + \text{etc.} = -w - (lw) \quad (59)$$

$$\int_1^2 V\,dP + \int_1^2 \frac{mv\,dv}{g_c} + \int_1^2 \frac{mg\,dZ}{g_c} + \text{etc.} = -w - (lw) \quad (59a)$$

These equations (54 through 59*a*) contain no limiting assumptions other than no accumulation of material in the unit and are unrestricted in application to material flowing or transferred from state 1 to state 2. It is mainly a matter of convenience as to which equations are used.

With reference to equation 54*a*, ΔH is a complete differential, and its value depends only upon the initial and final states of the flowing material, being independent of the path followed by the flowing material. This is an important advantage in the use of equation 54*a*. Because there is no term specifically representing friction and other irreversible losses in equation 54*a*, there may be some confusion in handling friction losses with this equation unless it is remembered that the friction losses must occur either within the flowing material, or without in the surroundings.

If the friction losses and other irreversibilities are regarded as occurring within the flowing material, they are included in ΔH in the term $\int_1^2 T\,dS$, which is included in ΔH as part of ΔU, but they are not included in either w or q.

If the friction losses are regarded as occurring outside the flowing material in the surroundings, they are included in w since the energy leaves the flowing material as work which is dissipated into heat effects in the surroundings. In this case frictional losses are not included in ΔH or q. If these heat effects in the surroundings caused by friction losses are absorbed in the flowing material as q, this energy is again in the system and included in ΔH. In this case the friction losses are also included in q and w, canceling out of these two terms because of the difference in sign. Therefore they might better be included only in ΔH for simplicity in calculation.

Equation 59 is applicable whether or not heat is transferred. It is completely independent of q, but it has a term, lw, specifically representing friction losses and irreversibilities, all of which must be included in this term. This eliminates any doubt as to where to include friction losses, but it requires that all friction losses and other irreversibilities be evaluated, or that the exact path followed by the flowing material be known in order to evaluate the incomplete differentials, such as $\int_1^2 V\,dP$, appearing on the other side of the equation.

Friction losses are a form of energy transfer represented by the term lw in equations 59 and 59*a* (included in $\int_1^2 T\,dS$ in equations 55 and 57). This point of view is in harmony with the basic idea of a resistance force, as the product of force times distance is energy, and, if fluid is moving against a resisting force, there is always an energy term, lw, involved which is directly proportional to the resisting force.

In particular cases it is desirable to recognize the peculiar properties of the material flowing as a means of simplifying the calculations. For example, a liquid flowing through a pipe will usually be free of chemical changes, surface effects, etc., and equation 59 may be rewritten as 59*b*.

$$\int_1^2 V\,dP + \Delta\left(\frac{mv^2}{2g_c}\right) + \Delta\left(\frac{mgZ}{g_c}\right) = -w - (lw) \qquad (59b)$$

Writing equation 59*b* for a unit mass of material,

$$\int_1^2 \overline{V}\,dP + \frac{\Delta v^2}{2g_c} + \frac{g}{g_c}\Delta Z = -\overline{w} - \overline{lw} \qquad (59c)$$

If the flow is also approximately isothermal and the fluid is almost incompressible, as are most liquids, the volume of a unit mass may be assumed to be constant and equation 59*c* may be further simplified to

$$\frac{\Delta P}{\rho} + \frac{\Delta(v^2)}{2g_c} + \frac{g}{g_c}\Delta Z = -\overline{w} - \overline{lw} \qquad (60)$$

Equation 60 is limited to a material of approximately constant density and is frequently referred to as Bernoulli's equation when $\overline{w}$ and $\overline{lw}$ are zero.

If equation 60 is divided by g/g_c, and ρ', the specific weight, is substituted for $\rho(g/g_c)$,

$$\frac{\Delta P}{\rho'} + \frac{\Delta v^2}{2g} + \Delta Z = -\overline{w}\left(\frac{g_c}{g}\right) - \overline{lw}\left(\frac{g_c}{g}\right) \qquad (60a)$$

The dimensions of the individual energy terms in equation 60*a* are energy per unit weight of fluid, such as foot-pounds per pound (or feet), or gram-centimeters per gram (or centimeters). In this form (equation 60*a*) the terms have the dimensions of length as of a column of fluid. For this reason these terms are frequently referred to as "head," such as feet or centimeters of the fluid. This common terminology, with pressure head for the term $(\Delta P/\rho')$, velocity head for the term $(\Delta v^2/2g)$, and static head for (ΔZ), must be applied with caution as it is frequently and erroneously applied to the terms of equation 60.

FRICTION LOSSES

As indicated, the use of equation 59 or 60 requires a knowledge of the frictional losses, lw, which cannot be accurately determined except by actual experiment, as was also true in determining the frictional resistance to the motion of solid particles through fluids.

The flow through pipes is a particular example of the relative motion of solids and fluids, and the general equation 10 for the resisting force, $F' = (f'\rho Av^2)/2$, may be applied. The area A, in equation 10, is any area representative of the solid. For a pipe filled with fluid this area is logically taken as the inside surface area of the pipe (πDL).

The frictional force exerted by the pipe against the fluid in the pipe is

$$F' = \frac{f'\rho\pi DLv^2}{2}$$

The energy required to overcome this frictional force in moving the fluid in the pipe a distance equal to δL is the product $F'\,\delta L$. This quantity of energy would push out of the pipe a quantity of fluid represented by the volume $\pi D^2\,\delta L/4$, or the mass of fluid $\rho\pi D^2\,\delta L/4$. Therefore, the energy required to overcome friction (or dissipated as friction losses) per unit mass of fluid discharged from (or entering) the pipe is

$$\overline{lw}_f' = \frac{4F'\,\delta L}{\rho\pi D^2\,\delta L} = \frac{2f'Lv^2}{D} \tag{61}$$

where $\overline{lw}_f'$ is the energy in absolute units dissipated as friction per unit mass of fluid.

$$lw_f' = g_c lw_f$$

Substituting for $\overline{lw}$ in equation 60 with the assumption of no change in velocity, no change in elevation, and no work done by the fluid, the pressure drop due to friction ($-\Delta P_f'$) is obtained in the form of the Fanning equation

$$-\Delta P_f' = \frac{2f'\rho Lv^2}{D} \tag{62}$$

In engineering practice the friction loss is usually written for energy in foot-pounds force per pound mass,

$$\overline{lw}_f = \frac{fLv^2}{2g_cD} = \int_1^2 \frac{fv^2}{2g_cD}\,dL \tag{63}$$

where $f = 4f'$.

The factor f depends only upon the Reynolds number if geometric similarity exists. This requires the same value for the ratio of length to diameter (L/D) for the different systems, and exactly similar surfaces as may be expressed by the ratio ϵ/D; where ϵ represents the height or depth of the elevations or depressions in the surface. Under these conditions, $\phi(Dv\rho/\mu)$ may be substituted for f in equation 63, giving

$$\frac{\overline{lw}_f 2g_c}{v^2} = \frac{(lw)_f 2g_c}{mv^2} = \left(\frac{L}{D}\right)\phi\left(\frac{Dv\rho}{\mu}\right) \tag{64}$$

DIMENSIONAL ANALYSIS

If the friction factor f in equation 63 were not known to depend only upon the Reynolds number in geometrically similar systems, estimation of values for f would appear as difficult as estimation of values for $\overline{lw}$ in equation 59c or equation 60. Where all the significant variables are recognized but the exact relationship between them is not known, because of the difficulty of integrating the differential equations or for other reasons, considerable help in indicating ways of correlating experimental data to obtain an empirical relation between the variables may be obtained by means of dimensional analysis. This is a procedure by which the known variables are arranged in a number of dimensionless combinations or groups. The fundamental equation may be expressed in terms of these groups rather than in terms of the individual variables. The resulting equation will be simpler, as the number of dimensionless groups will be less than the number of variables, generally by the number of dimensions employed.

The essential principles of dimensional analysis are three in number.[1] *

1. Each and every one of the physical quantities or measurements may be expressed as a product of the powers of a very few fundamental dimensions (see Table 25).

2. Equations of physical quantities are homogeneous in the net dimensions of all additive terms. All terms added together must have the same dimensions when expressed in terms of the fundamental dimensions selected.

3. Any general relation between physical quantities can be expressed in a generalized manner independent of any particular units to involve only the

* The bibliography for this chapter appears on p. 146.

dimensionless products of all the physical variables and the necessary dimensional conversion constants. This principle follows from the others and is known as Buckingham's "pi theorem." It is the key to dimensional analysis.

The energy dissipated in overcoming friction is some function of the properties of the flowing fluid and the confining pipe and their relative motion. The significant properties of the pipe may be its length L, its internal diameter D, and the roughness or depth of its surface irregularities ϵ. The properties of the fluid are its mass m, its density ρ, and its viscosity μ. The relative motion between the fluid and pipe is the fluid velocity v. Therefore the energy dissipated in overcoming friction is some function of all these variables or

$$(lw)_f' = \phi_1(L, D, \epsilon, m, \rho, \mu, v) \quad (65)$$

For a given point condition the unknown function, equation 65, may be written in exponential form.

$$(lw)_f' = zD^aL^b\epsilon^cm^d\rho^e\mu^nv^r \quad (66)$$

where the dimensionless coefficient z and all the exponents are of given values only at the point condition. If the condition is changed in any way, as by changing the velocity, z and all the exponents may change. Equation 66 is simply another way of writing equation 65. It is definitely not to be inferred that the true relation between the variables is exponential.

Application of the first principle by substitution of the fundamental dimensions of system I, Table 25, in equation 66 gives

$$\frac{mL^2}{t^2} = zL^aL^bL^cm^d\left(\frac{m}{L^3}\right)^e\left(\frac{m}{Lt}\right)^n\left(\frac{L}{t}\right)^r \quad (67)$$

By application of the second principle, the net dimension of the product on the right side must be the same as on the left, mL^2/t^2, and the summation of the exponents of each fundamental dimension on the right is equal to the exponent of that dimension on the left, as follows.

For m $\quad 1 = d + e + n$

For L $\quad 2 = a + b + c - 3e - n + r$

For t $\quad -2 = -n - r$

Here are three equations and seven variables or unknowns. The energy dissipated as friction is directly proportional to the mass of fluid transported, and d is therefore unity. This leaves six unknowns for three equations. If the three equations are independent, they may then be solved in terms of three arbitrarily chosen variables, such as b, c, and n, giving

$$e = -n \qquad r = 2 - n$$

$$a = 2 - b - c - 3n + n - 2 + n = -b - c - n$$

If these values are substituted for a, e, and r in the exponents of equation 66,

$$(lw)_f' = zD^{-b-c-n}L^b\epsilon^cm\rho^{-n}\mu^nv^{2-n}$$

The groups into which these variables will be assembled in the functional relation are indicated by the exponents. The variables are assembled according to the appearance of each letter in the exponent to assure dimensional homogeneity. For convenience the groups are usually made dimensionless as follows.

All variables having a numerical exponent yield the dimensionless group $(lw)_f'/mv^2$.

All variables having the exponent b yield the dimensionless group L/D.

All variables having the exponent c yield the dimensionless group ϵ/D.

All variables having the exponent n yield the dimensionless group $\mu/Dv\rho$.

These dimensionless groups constitute a function of some description to express the interrelation of variables

$$\left[\frac{(lw)_f'}{mv^2}\right] = \phi_2\left[\left(\frac{L}{D}\right), \left(\frac{\epsilon}{D}\right), \left(\frac{\mu}{Dv\rho}\right)\right] \quad (68)$$

If exponents other than b, c, and n had been chosen for retention in the establishment of dimensional homogeneity, a different set of dimensionless groups would have resulted. Intermultiplication of the dimensionless groups to yield different dimensionless groups is permissible. In this way some other function of different dimensionless groups may be obtained which may be more convenient than the first.

The best choice of exponents for retention, or the manner of intermultiplication to yield the most convenient groups, depends on the use to be made of them. The form of equation 68 is readily adapted to the computation of friction loss for known velocities. But in this form the evaluation of the velocity which will result from known dimensions and friction loss $(lw)_f$ is extremely complex, except by a trial-and-error solution. This may be avoided by a more convenient function established by eliminating velocity from all groups except one, by simply dividing each side of equation 68 by $\sqrt{(lw)_f'/mv^2}$, giving

$$\sqrt{\frac{(lw)_f'}{mv^2}} = \phi_3\left[\left(\frac{L}{D}\right),\left(\frac{\epsilon}{D}\right),\left(\frac{\mu}{Dv\rho\sqrt{(lw)_f'/mv^2}}\right)\right] \tag{69}$$

$$\frac{\sqrt{\overline{lw}_f'}}{v} = \phi_3'\left[\left(\frac{L}{D}\right),\left(\frac{\epsilon}{D}\right),\left(\frac{\mu}{D\rho\sqrt{\overline{lw}_f'}}\right)\right]$$

This is all the information that can be obtained from the application of dimensional analysis to this problem. The nature of the function ϕ_2 or ϕ_3 is not indicated in any way. The analysis simply states that a function exists between these various dimensionless groups, however complicated it may be, provided all the significant physical quantities have been included. This is an important guide to the procedure of determining the function empirically from experimental data. For example, a series of measurements of the friction energy $(lw)_f'$ taken under conditions of constant L/D and constant ϵ/D must yield a single function when $(lw)_f'/mv^2$ is plotted against the Reynolds number $D\rho v/\mu$. All experimental points lie on the same curve, regardless of the velocity, density, viscosity, or mass of the fluid. If the experimental points do not lie on a single curve within the experimental error, one or more significant physical quantities or measurements were omitted in writing the first equation 65.

Equation 68 may be written as follows, with energy given in foot-pounds (or gram-centimeters).

$$\frac{(lw)_f 2g_c}{mv^2} = \phi_4\left[\left(\frac{L}{D}\right),\left(\frac{\epsilon}{D}\right),\left(\frac{D\rho v}{\mu}\right)\right] \tag{70}$$

Equation 70, derived from dimensional analysis, gives fully as much information as equation 64, derived from a theoretical consideration of balancing forces and the principle of dynamic similarity. In either equation, experimental data are required to determine the relation between friction energy and the physical quantities.

In the derivation of equation 64, the friction energy was assumed to vary directly with the length of pipe, and L accordingly appears as the first power. If the same assumption is made in equation 70, it follows that L/D appears as the first power, as this is the only dimensionless group containing the quantity L. This assumption has been found valid for all values of L/D greater than about 50. Therefore

$$f = \frac{(lw)_f D2g_c}{mLv^2} = \frac{\overline{lw}_f D2g_c}{Lv^2} = \phi_5\left[\left(\frac{\epsilon}{D}\right)\left(\frac{D\rho v}{\mu}\right)\right] \tag{71}$$

The accurate estimation of the friction energy from correlation of experimental data has been a goal of engineers for many years. Early work neglected the character of pipe surface, omitting the ratio ϵ/D. The data could not be correlated, particularly at high values of the Reynolds number, until this variable was properly included.

Figure 125 shows graphically the solution to equation 71, obtained by plotting the experimental data. This correlation [4] is in consistent units and may be used with any set of consistent units, English, metric, or otherwise. Similar plots are found in the literature in which other friction factors are plotted as a function of the Reynolds number. Care must be taken to avoid confusion, as the same name and symbol are used for various multiples of f as plotted in Fig. 125. For example, f' of equation 62 is plotted as f in the *Chemical Engineers' Handbook.*[5] * Elsewhere $f/2g_c$ will be found plotted as f.

Figure 125 shows a straight solid line, $f = 64/\text{Re}$, representing the laminar flow region which extends up to a Reynolds number of 2000. In laminar flow the fluid moves only in the direction of flow, as indicated in Fig. 123. The inside surface of the pipe is

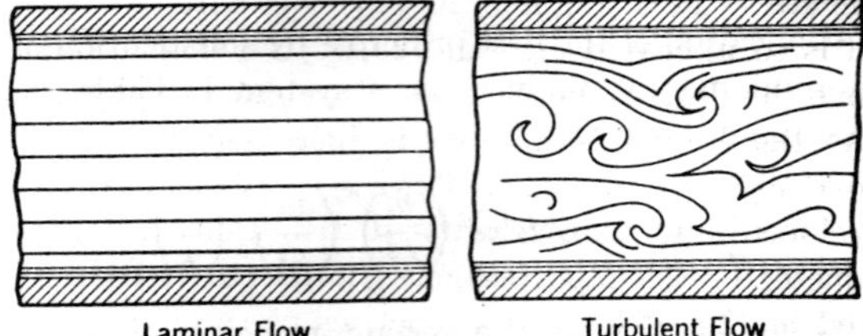

Fig. 123. Laminar and turbulent flow lines.

effectively streamlined by the fluid flowing next to the pipe wall, and the roughness of the pipe has little effect upon the resistance to flow, which is determined largely by the properties of the fluid, particularly the viscosity. The local velocity of the fluid varies across the pipe diameter from zero at the

* In Section 5 of the *Chemical Engineers' Handbook* (first printing) [5] f' of equation 62 is designated by the symbol f. In Section 20 of the same volume, p. 1426, f of equation 71 is designated by the symbol f, and the statement is made that the plot [5] gives values for f, although the plot actually gives values for $f' = f/4$.

wall to the maximum in the center, as indicated in Fig. 124.

As the Reynolds number is increased above 2000, as by increasing the quantity of fluid flowing, the friction factor will follow this straight line into the dotted region only if the flow remains laminar. But near this point, Re = 2000, eddies and turbulence may develop in the core of the stream, which has the greatest velocity. These eddies dissipate more energy and increase the friction losses. Since the eddies are local this increased friction loss is local, tending to decrease the local velocity (decrease Re), and the flow may again become laminar. But the local decrease in velocity causes corresponding local increases in velocity elsewhere, with formation of other eddies. This region of incipient turbulence, from Re = 2000 to about 4000, is therefore unstable and is so indicated on Fig. 125 by the shaded area above the dotted line.

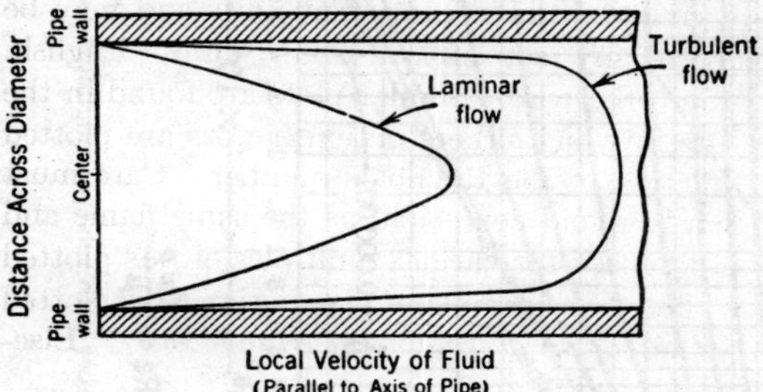

FIG. 124. Relative velocity distribution in laminar and turbulent flow.

As the Reynolds number is increased to values above 4000, the turbulent flowing core becomes well developed and the velocity distribution across a diameter of the pipe becomes similar to that indicated in Fig. 124. The properties of the fluid now have less effect upon the flow conditions than in strictly laminar flow, and the roughness becomes of significance, particularly if it is sufficient to affect the turbulent core in the central part of the stream. Friction losses are stabilized and can be computed. These conditions are represented by the transition zone on Fig. 125.

A small twig or other floating object tossed into a river or small stream running rapidly, as in flood, offers an excellent opportunity to see the laminar flow in the slowly flowing water next to the shore and the turbulent flow in midstream.

As the Reynolds number is increased further, the turbulent flowing core expands at the expense of the laminar flowing annulus until the turbulent flowing core occupies the entire cross section of the pipe right down to the protuberances which make up the roughness. In the completely turbulent region the roughness of the pipe is the most important consideration, and the properties of the fluid, viscosity and density, can be neglected in estimating friction losses.

In Fig. 126 the relative roughness[4] for various pipes has been plotted as a function of pipe diameter for different types of pipe.

The linear velocity of the total fluid flowing through a pipe does not have a definite value. In laminar flow the velocity is linear but varies greatly with distance across the pipe (Fig. 124). In turbulent flow the velocity is neither linear nor constant across the pipe diameter. In making the correlation of Fig. 125, the velocity v was determined by dividing the volume of fluid passing a given point or discharged from the flow system per second by the cross-sectional area A of the pipe. This is frequently referred to as the average velocity and designated by a different symbol. In equations 54 and 59, the velocity v is the actual linear velocity of the fluid. If the average velocity is used for v in the term representing kinetic energy, $mv^2/2g_c$, the result may be in error. In laminar flow the total kinetic energy is more closely represented by mv^2/g_c (without the 2) when using the average velocity for v. But under turbulent conditions the average velocity computed in this manner gives approximately the correct result in the expression $mv^2/2g_c$.

In laminar flow when $f = 64/\text{Re}$, equation 63 may be written

$$\overline{lw_f} = \frac{-\Delta P_f}{\rho} = \frac{32Lv\mu}{g_c D^2 \rho}$$

or

$$-\Delta P_f = \frac{32Lv\mu}{g_c D^2} \qquad (63L)$$

which is known as Poiseuille's equation.[6]

The friction losses computed from Fig. 125 are for straight pipe of circular cross section. In the past, various formulas have been proposed for estimating the friction losses through bends, fittings, etc. But the simplest procedure is to consider each fitting or valve as equivalent to a length of straight pipe[2] as indicated in Fig. 127. This reduces all pipes, valves, and fittings to a common denominator, equivalent length of pipe of the same relative roughness, for purposes of computing friction losses.

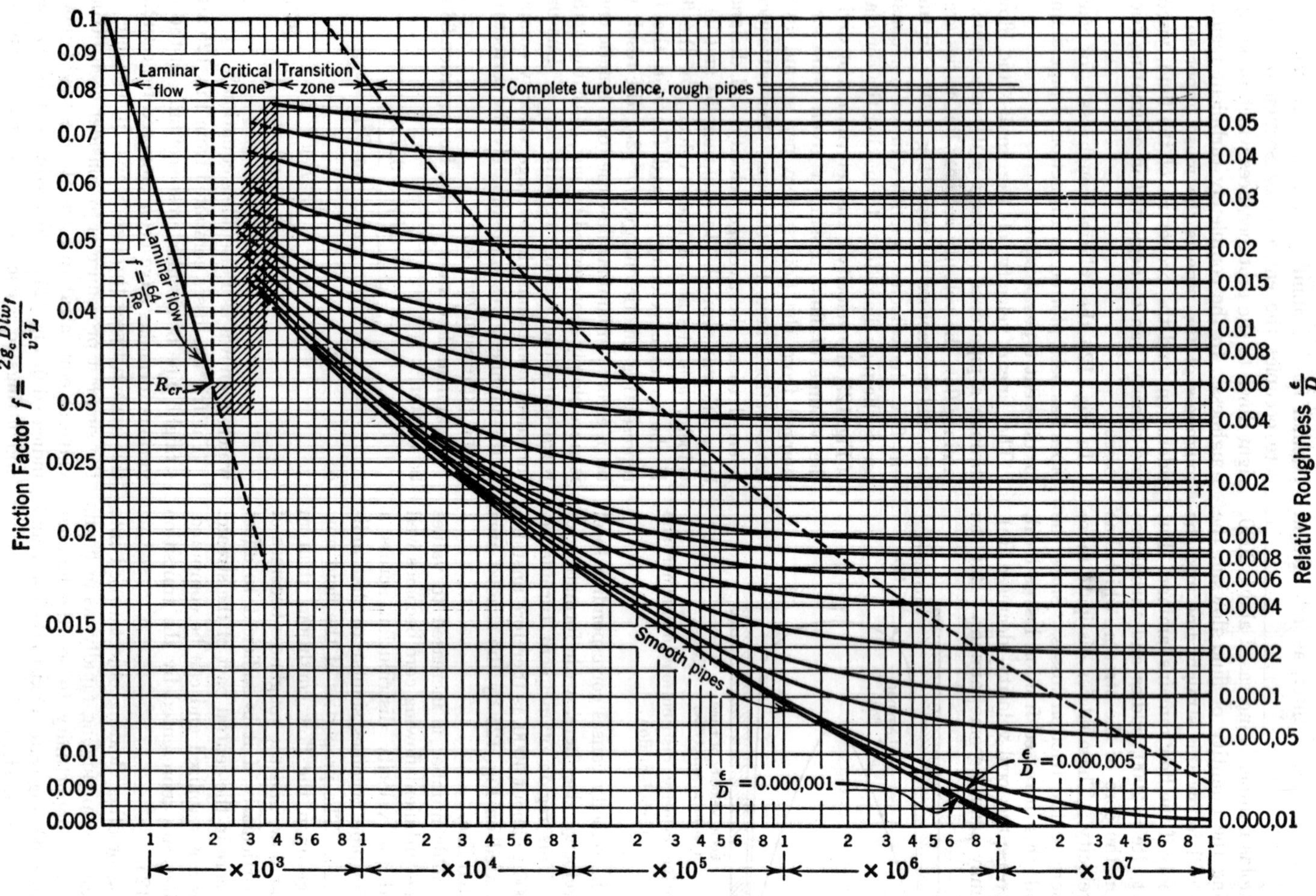

FIG. 125. Friction factor as a function of Reynolds number with relative roughness as a parameter.[4]

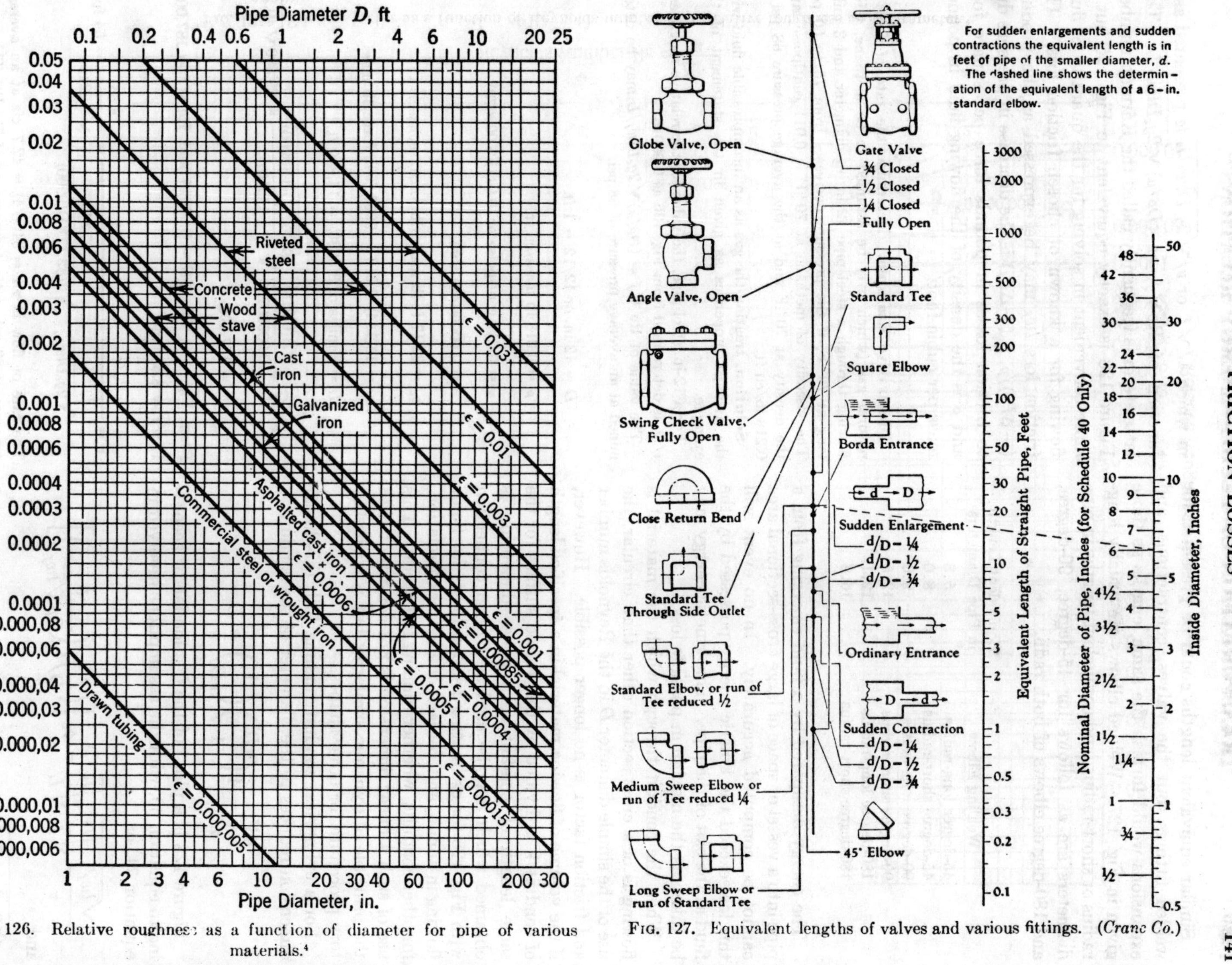

FIG. 126. Relative roughness as a function of diameter for pipe of various materials.[4]

FIG. 127. Equivalent lengths of valves and various fittings. (*Crane Co.*)

Similar equivalent lengths could be presented for welded fittings, but the valves, contractions, and expansions would furnish the same lengths as those given in Fig. 127. Welded elbows are usually long-radius or short-radius, and equivalent lengths in pipe diameters are as follows for 45-degree, 90-degree, and 180-degree elbows of both radii.

Welding Elbow	Equivalent Length in Pipe Diameters
45-degree long-radius	5.8
45-degree short-radius	8.0
90-degree long-radius	9.0
90-degree short-radius	12.5
180-degree long-radius	12.1
180-degree short-radius	16.8

The energy dissipated by a fluid emerging from a pipe into a vessel or space of large cross-sectional area cannot be computed accurately. In no event will this loss exceed the kinetic energy possessed by the fluid at the exit of the pipe. This kinetic energy may be taken as the maximum possible loss.

When the conduit through which the material is flowing is of a cross section other than circular, the use of the simple diameter D in the Reynolds number and friction factor is no longer possible. However, if the volume of flowing material in a specified length of conduit V is divided by the wetted surface for the same length of conduit σ, a linear dimension is obtained which may be used as $D/4$ in connection with Fig. 125 for estimating friction losses for fluids flowing in annuli, and, for lack of a better correlation, for other conduits, whether running full or not. This use of such a linear dimension is not consistent with the principle of dynamic similarity but is a satisfactory procedure for other shapes.

For a cylindrical conduit running full, this linear dimension reduces to the diameter of the conduit.

$$\frac{4V}{\sigma} = 4\left(\frac{\pi D^2}{4}\right)\left(\frac{1}{\pi D}\right) = D$$

Figure 128 is a plot [3] of the friction factor according to equation 69, with the assumptions involved in equation 71, or

$$\frac{\sqrt{\overline{lw}_f'}}{v}\sqrt{\frac{D}{L}}\sqrt{\frac{D}{L}} = \phi_6\left[\left(\frac{\epsilon}{D}\right), \left(\frac{\mu}{D\rho\sqrt{\overline{lw}_f'}}\right)\right]$$

and

$$\frac{\sqrt{2g_c\overline{lw}_f(D/L)}}{v} = \phi_7\left[\left(\frac{\epsilon}{D}\right), \left(\frac{\mu}{D\rho\sqrt{2g_c\overline{lw}_f(D/L)}}\right)\right]$$

in which $1/\sqrt{f}$, or $v/\sqrt{2g_cD\overline{lw}_f/L}$ is presented as a function of $\text{Re}\sqrt{f}$, or $(D\rho/\mu)\sqrt{2g_cD\overline{lw}_f/L}$. This latter term is frequently called the Kármán number. Figure 128 is exactly equivalent to Fig. 125 but is more convenient in solving for the quantity of fluid flowing for a known or chosen friction loss. The friction loss $\overline{lw}_f$ may be expressed as the product $(-\Delta P_f)/\rho$, if $(-\Delta P_f)$ is the decrease in pressure due to friction losses in pounds force per square foot and ρ is the density of the flowing fluid in pounds mass per cubic foot.

Illustrative Example. Determine the rate of flow of natural gas (which may be assumed to be methane, CH_4) flowing through a steel pipe 12 in. in diameter and 3 miles long at 70° F, if the pressure decreases from 75 to 60 psi. The viscosity of methane at 70° F is 0.011 centipoise, and the density at 70° F and at the average pressure 68 psi is 0.218 lb/cu ft.

Solution, treating the gas as an incompressible fluid with the average properties as given in the statement of the example.

For a 12-in. steel pipe, the value for the relative roughness, ϵ/D, is determined from Fig. 126 as 0.00015.

The value of $\text{Re}\sqrt{f} = (D\rho/\mu)\sqrt{2g_cD\overline{lw}_f/L}$ may be determined at an average pressure of 68 psi.

$$D = 12 \text{ in. or } 12/12 = 1 \text{ ft.}$$

$$\rho = 0.218 \text{ lb mass/cu ft.}$$

$$\mu = 0.011 \text{ centipoise or } (0.011)(0.000672)$$

$$= 0.00000739 \text{ Bvu.}$$

$$g_c = 32.17 \text{ ft-lb mass/(lb force)(sec)(sec).}$$

$$\overline{lw}_f = 15(144)/0.218 = 9900 \text{ ft-lb force/lb mass.}$$

$$L = 3 \text{ miles or } 3(5280) = 15{,}840 \text{ ft.}$$

$$\text{Re}\sqrt{f} = \frac{(1)(0.218)}{0.00000739}\sqrt{\frac{2(32.17)(1)(9900)}{15{,}840}} = 29{,}500\sqrt{40.2}$$

$$= 187{,}000.$$

From Fig. 128, for $\epsilon/D = 0.00015$ and $\text{Re}\sqrt{f} = 187{,}000$, $1/\sqrt{f} = 8.55$, and

$$v = \frac{1/\sqrt{f}}{\sqrt{\dfrac{L}{2g_cD\overline{lw}_f}}} = \frac{8.55}{\sqrt{\dfrac{15{,}840}{2(32.17)(1)(9900)}}} = \frac{8.55}{\sqrt{0.0249}} = 54 \text{ fps}$$

The rate of flow is $vA = 54(\pi/4) = 42.7$ cfs at an average pressure of 68 psi and a temperature of 70° F. This corresponds to 236 standard cu ft/sec as measured at 14.7 psia and 60° F.

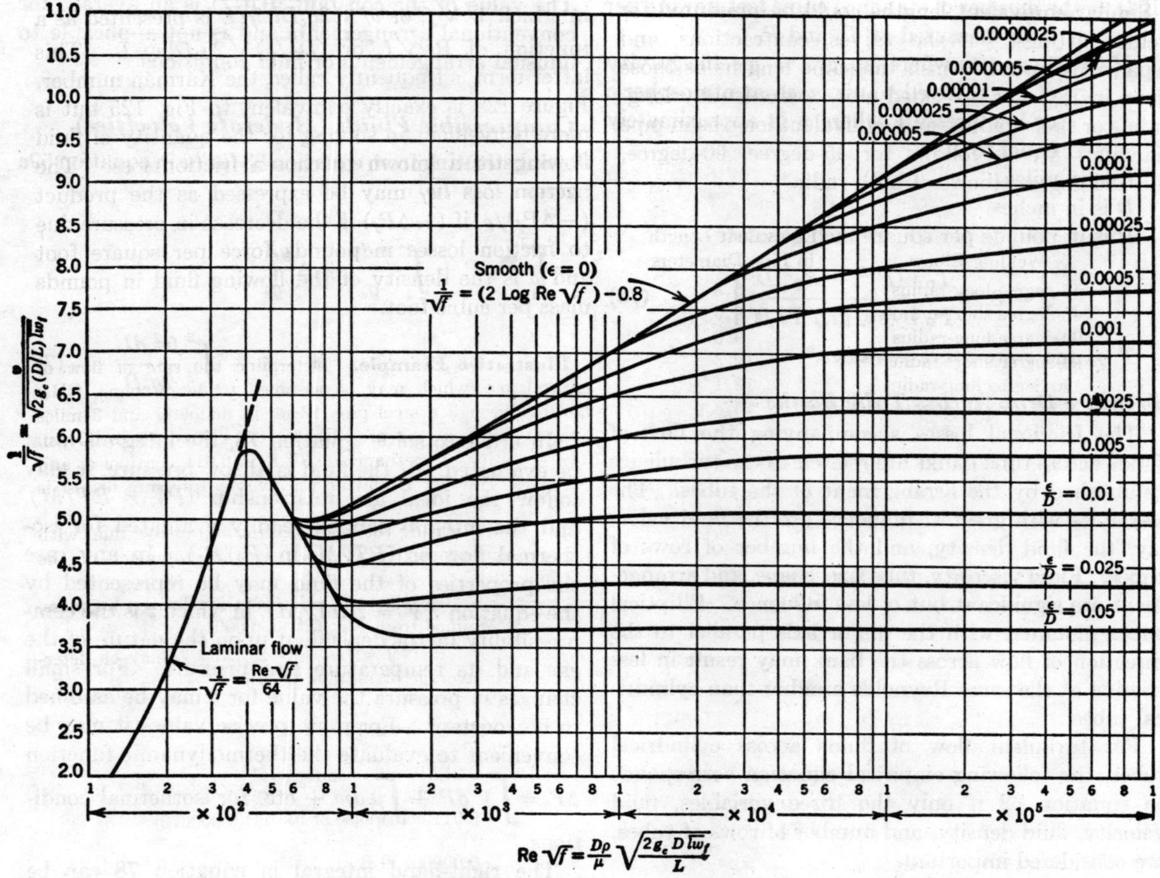

FIG. 128. Friction factor as a function of the Kármán number.

Weymouth's Equation for Gases

Using equations 59*a* and 63 in differential form, assuming horizontal flow, no chemical effects, and no work done, for a unit mass of material flowing

$$\overline{V}\,dP + \frac{v\,dv}{g_c} + \frac{fv^2\,dL}{2g_cD} = 0 \qquad (72)$$

Since

$$v = \frac{\overline{V}W}{A}$$

where W = mass rate of flow,
A = cross-sectional area of flow,

for actual gases

$$\overline{V} = z\,\frac{RT}{MP}$$

where z = compressibility factor to correct for departure from ideal gas behavior (see the Appendix).
R = ideal gas constant.
M = molecular weight.

Substituting from the above equations in equation 72 and multiplying by $2g_c/v^2$ gives

$$\frac{2g_cA^2MP\,dP}{W^2zRT} + \frac{2\,dv}{v} + \frac{f\,dL}{D} = 0 \qquad (73)$$

If $(2\,dv)/v$ is neglected and z, T, and f are assumed to be constant, equation 73 may be integrated

$$P_2{}^2 - P_1{}^2 = \frac{-fLW^2zRT}{Dg_cMA^2} \qquad (74)$$

If Q = the number of standard cubic feet flowing per hour measured at T_0 and P_0

M = 29 (sp. gr.) where (sp. gr.) = specific gravity of gas referred to air, each at atmospheric or low pressure where z is approximately equal to 1

L is in miles

D is in inches

P is in pounds per square inch absolute

$$Q = 3.22 \frac{T_0}{P_0} \left[\frac{(P_1^2 - P_2^2)D^5}{(\text{sp. gr.})\ TLfz} \right]^{1/2} \tag{75}$$

T, f, and z are average values.[9]

Pressure Drop Across Tube Banks

The frictional losses accompanying the flow of fluids across tube banks are related to the turbulence established by the arrangement of the tubes. The variables with greatest influence are the fluid velocity, the fluid density, and the number of rows of tubes. Fluid viscosity, tube size, shape, and arrangement are significant but of less influence. Elliptical tubes mounted with the major axis parallel to the direction of flow across the bank may result in less friction at the same Reynolds number than cylindrical tubes.

For turbulent flow of fluids across cylindrical tubes, the following empirical equation corresponds to equation 62 if only the three variables, fluid velocity, fluid density, and number of rows of tubes, are considered important.

$$(-\Delta P_f) = \phi_8(N, \rho, v_{max}^2, f) \tag{76}$$

where N = number of rows of tubes.

v_{max} = maximum linear velocity or velocity at the minimum cross section.

ρ = fluid density.

f = friction factor dependent upon the Reynolds number of fluid flowing and the arrangement of tubes.

For turbulent flow across the usual tube arrangements, equation 76 has been evaluated [5] as

$$(-\Delta P_f) = \frac{0.047 N G_{max}^{1.8} \mu^{0.2}}{\rho D_e^{0.2}} \tag{77}$$

where D_e = distance between tubes in a transverse row (ft).

G_{max} = mass velocity, [lb mass/(sq ft)(sec)].

μ = fluid viscosity, [lb mass/(ft)(sec)].

ρ = fluid density, (lb mass/cu ft).

P = lb force/sq ft.

The value of the constant (0.047) is an average for conventional arrangements and is not applicable to unusual arrangements or fluid conditions.

Compressible Fluids (Moderate Velocities)

Substituting from equation 63 for $\overline{lw}$ in equation 59c gives

$$\int_1^2 \overline{V}\, dP + \int_1^2 \frac{v\, dv}{g_c} + \Delta \frac{gZ}{g_c} = -\bar{w} - \int_1^2 \frac{fv^2\, dL}{2g_c D} \tag{78}$$

In order to solve equation 78, the integrals must be evaluated. If the fluid is at low pressure, it may follow the ideal gas relationship ($P\overline{V} = RT/M$), and the integral can be readily evaluated for isothermal flow as $(RT/M) \ln (P_2/P_1)$. In any case the properties of the fluid may be represented by the equation $P\overline{V} = z(RT/M)$, in which z is the compressibility factor dependent upon the nature of the gas and its temperature and pressure. For small changes in pressure the value for z may be assumed to be constant. For more precise values it may be convenient to evaluate the thermodynamic function $\Delta F = \int V\, dP + \int \mu\, dm +$ etc. for isothermal conditions.

The right-hand integral in equation 78 can be integrated directly only if fv^2 is known as a function of L. Since this information is not available, a solution of the problem is frequently obtained by stepwise calculations in which equation 78 is applied to successive short sections so that average values of $\overline{V}$ and v may be used for each section without significant error.

In another method of solution the variable v is eliminated by substituting in terms of $\overline{V}$. This is readily done by substituting $G\overline{V}$ for v, where G is the mass velocity and is a constant for a pipe of uniform cross section. Making this substitution in equation 78,

$$\int_1^2 \overline{V}\, dP + \int_1^2 \frac{G^2 \overline{V}\, d\overline{V}}{g_c} + \Delta \frac{gZ}{g_c} = -\bar{w} - \int_1^2 \frac{G^2 \overline{V}^2 f\, dL}{2g_c D} \tag{79}$$

Illustrative Example. Determine the pressure drop accompanying the flow of natural gas (which may be assumed to be methane) at 70° F through a horizontal steel pipe 12 in. in diameter and 3 miles long. The gas enters the pipe at 75 psig, and at a rate of 236 standard cu ft/sec (measured at 14.7 psia and 60° F). The viscosity of methane at 70° F is 0.011 centipoise, and the compressibility factor, $z = P\overline{V}M/RT$, may be assumed constant at 0.995.

Solution. If equation 79 is applied to this problem, the terms including ΔZ and $\bar{w}$ are each equal to zero. Since G is constant, the Reynolds number is constant, the friction factor is constant, and equation 79 may be written as follows, after dividing by $\overline{V}^2$ and letting L equal the length of the pipe.

$$\int_1^2 \frac{dP}{\overline{V}} = -\frac{fLG^2}{2g_cD} - \int_1^2 \frac{G^2}{g_c}\frac{dV}{\overline{V}}$$

Substituting $z(RT/PM)$ for $\overline{V}$, and assuming z to be constant and the last term (velocity effects) to be negligible,

$$\frac{M}{zRT}\int_1^2 P\,dP = -\frac{fLG^2}{2gD_c}$$

$$\left(\frac{M}{zRT}\right)\frac{(P_2^2 - P_1^2)}{2} = -\frac{fLG^2}{2g_cD}$$

$$P_2^2 = P_1^2 - \left(\frac{fLG^2}{g_cD}\right)\left(\frac{zRT}{M}\right)$$

To evaluate the equation:

Molecular weight = 16.

Quantity flowing = 236 standard cu ft/sec.

Since 1 lb-mole = 379 standard cu ft under these conditions,

$$\text{Mass flowing} = \tfrac{236}{379} \times 16 = 10 \text{ lb/sec.}$$

$$\text{Mass velocity} = \frac{10}{\pi/4} = 12.7 \text{ lb/(sec)(sq ft).}$$

$$\mu = 0.011 \text{ centipoise}$$

$$= 0.00000739 \text{ Bvu.}$$

$$\text{Re} = \frac{1 \times 12.7}{0.00000739} = 1.72 \times 10^6.$$

ϵ/D for 12-in. iron pipe = 0.00015 (Fig. 126).

f = 0.0136 (Fig. 125).

$$P_2^2 = [(75 + 14.7)(144)]^2 - \left[\frac{(0.0136)(15{,}840)(12.7)^2}{(32.17)(1)}\right] \times \left[\frac{(0.995)(1544)(530)}{16}\right].$$

$$P_2^2 = (167 \times 10^6) - (54.8 \times 10^6).$$

$$P_2 = \sqrt{112.2 \times 10^6} = 10{,}550 \text{ lb/sq ft absolute or 73.2 psia or 58.5 psig.}$$

The difference in the downstream pressure as computed in this example compared with the stated downstream pressure in the previous example is due to errors introduced into the previous example by treatment of methane as an incompressible fluid.

High Columns of Compressible Fluid (Zero Velocity)

A stationary vertical column of compressible fluid has zero velocity. Equation 78 then becomes

$$\int_1^2 \overline{V}\,dP + \Delta\left(\frac{gZ}{g_c}\right) = 0 \qquad (80)$$

If the column is high, there may be considerable variation in the specific volume of the gas $\overline{V}$ with height Z or pressure P. If actual data are available for the specific volume, a graphical integration of $\overline{V}\,dP$ may be made to obtain the limit P_2 of the integral. If the specific volume is expressed by the gas laws, equation 80 becomes

$$\int_1^2 \frac{zRT\,dP}{PM} = (Z_1 - Z_2)\frac{g}{g_c} \qquad (81)$$

At sea level g is numerically equal to the conversion constant g_c, and, if the compressibility factor z and temperature T may be assumed to be constant along the column, equation 81 may be integrated.

$$\frac{zRT}{M}\int_1^2 \frac{dP}{P} = \frac{zRT}{M}\ln\left(\frac{P_2}{P_1}\right) = Z_1 - Z_2 \qquad (82)$$

Converting equation 82 into the exponential form, solving for P_2, and subtracting P_1 from each side gives the common "atmospheric" formula [7,8] for relating pressure and elevation.

$$P_2 - P_1 = P_1\left(e^{\frac{M(Z_1 - Z_2)}{zRT}} - 1\right) \qquad (83)$$

Equation 83, employing an average compressibility factor and temperature, gives results for pressure gradients in natural-gas wells comparable to the results from equation 80, employing an average value for the specific volume. Integrating equation 80 for constant volume at an average pressure such as $P_a = P_1 + \Delta P/2$

$$\Delta P = P_2 - P_1 = \frac{M(Z_1 - Z_2)}{zRT}\left(P_1 + \frac{\Delta P}{2}\right) \qquad (84)$$

or

$$\Delta P\left[1 - \frac{29\text{ (sp. gr.) }(Z_1 - Z_2)}{(2 \times 1544)zT}\right]$$

$$= \left[\frac{29\text{ (sp. gr.) }(Z_1 - Z_2)}{1544zT}\right]P_1 \qquad (84a)$$

BIBLIOGRAPHY

1. **Bridgman**, P. W., *Dimensional Analysis*, Yale University Press (1946).
2. **Crane Company**, *Tech. Paper 409*, Chicago, Ill. (May 1942).
3. **von Kármán**, T., *Nach. Ges. Wiss. Göttingen, Fachgruppe* I, **5**, 58–76 (1930). Translated in *U. S. Nat. Advisory Comm. Aero. Tech. Memo 611.*
4. **Moody**, L. F., *Trans. Am. Soc. Mech. Engrs.*, **66**, 671–684 (1944).
5. **Perry**, J. H. (editor), *Chemical Engineers' Handbook*, 3rd ed., p. 382, McGraw-Hill Book Co. (1950).
6. **Poiseuille**, J., *Mémoires des Savants Étrangers*, Vol. 9, p. 433 (1846).
7. **Rawlins**, E. L., and M. A. **Schellhardt**, *U. S. Bur. Mines Monograph* 7 (1936).
8. **Rzasa**, M. J., and D. L. **Katz**, *Am. Inst. Mining Met. Engrs. Tech. Pub.* 1814, *Petroleum Technology* (March 1945).
9. **Weymouth**, T. R., *Trans. Am. Soc. Mech. Engrs.*, **34**, 185–231 (1912). (The compressibility factor has been introduced in the derivation in this text.)

PROBLEMS

1. A town derives its water supply from a lake nearby, pumping it up to a standpipe on the highest hill. The intake to the system is 10 ft below the lake surface, the inlet to the pump is 15 ft above the lake surface, and the water level in the tower is kept constant 310 ft above the pump discharge.

The friction loss is 140 ft-lb/lb of water through the 6000 ft of 4-in. pipe which includes the total equivalent length of all piping from lake to water tower.

If the pump capacity is 6000 gph and the water-pump set is 85 per cent efficient, what would be the hourly pumping cost if electric power is 1 cent per kilowatthour?

2. What is the cost at ½ cent per kilowatthour to pump oil with a specific gravity 60/60 of 0.84 and a viscosity of 30 centipoises through a 10-in. line, 30 miles long? The overall efficiency of the pump and motor is 60 per cent, the pipe line is horizontal, and the rate of flow is 800 bbl/hr (42 gal/bbl).

3. A 90 per cent by weight solution of methyl alcohol in water is to be pumped from a storage tank to a process department through a 1¼-in. standard iron pipe 1450 ft long. The line contains 15 standard elbows, 5 gate valves, 6 tees straight through, and 4 tees through a side outlet. The process end of the line must have a pressure of 8 psig and is 22 ft above the storage tank. The temperature of the line can be taken at 40° F as the lowest it will possibly attain, and the flow will be rated at 20 gpm, but it may be overloaded by 50 per cent. Assuming 60 per cent pump efficiency, calculate the output required by the electric motor to be used for this duty.

4. A vertical pipe 100 ft in length is carrying oil downward at the rate of 952 gpm. The pressure at the top is atmospheric and at the bottom is 16 psig. The pipe is 4-in. schedule-40 steel pipe. The oil has a density of 0.85 gram/cc and a viscosity of 8 centipoises.

(*a*) If the oil is to be piped horizontally in the same size pipe from the foot of the vertical section, how far can it be sent without requiring a pump, with the oil discharging at atmospheric pressure?

(*b*) If the pipe line is to be run at an angle down to the same level, how far from the base of the original vertical pipe will be the end of the pipe, with flow at the same rate and discharging at atmospheric pressure?

5. A tank 3 ft ID and 12 ft high filled with water at 68° F is to be emptied through a vertical 1-in. standard pipe, 2 ft long, connected to the tank bottom. How long a time is required for the level to drop from 12 to 2 ft?

6. Water is flowing at a rate of 80 gpm through 50 ft of an annular conduit made of ½-in. and 1¼-in. schedule-40 steel pipe. What is the pressure drop (in pounds per square inch) in the 50 ft? Assume, as average properties, a density of 62.3 lb/cu ft and a viscosity of 0.88 centipoise.

7. A liquid with an effective average density of 0.8 gram/cc and an effective average viscosity of 0.8 centipoise is to be pumped through a straight, smooth, copper heat exchanger of annular cross section at a linear velocity of 10 fps. The outer diameter of the inner wall of the annulus is exactly 1 in.; the inner diameter of the outer wall of the annulus is exactly 1½ in. The length of the exchanger is 20 ft. What is the drop in pressure (in pounds per square inch) from friction within the annulus?

8. Water is retained in a small mountain lake whose surface is 280 ft above the inlet to the turbines in a power house.

Surveyors have laid out the best line location for the piping system to bring the water down to the power house, the route being 3100 ft long.

What is the minimum standard diameter of pipe which may be used in the line so that the natural flow will be 1000 gpm and the pressure at the turbine nozzles will be 100 psig. The normal water temperature is 40° F.

9. One of several pumps at a plant for recovery of bromine from sea water is to have a capacity of 50,000 gpm. The sea water has a specific gravity of 1.03 and a temperature of 55° F. The water is pumped through a pipe line 420 yd long to a tower whose top is 60 ft above sea level.

(*a*) What size pipe would you recommend?

(*b*) What pressure would the pump be required to develop?

(*c*) With an overall efficiency of 70 per cent, what should be the horsepower of the electric motor?

10. A standard 2-in. pipe is conveying an oil of average specific gravity of 0.850 at a rate of 20 gpm. At a certain point in the line the static pressure is 45 psig. Estimate the static pressure at a point in the line which is 200 ft of pipe further along in the direction of flow and which is 50 ft lower in elevation. The oil has a viscosity of 20 centipoises at the temperature of flow.

11. A water main is to be laid from the present pumping station in a small community to a new subdivision. The estimated future population of the subdivision is 1500 persons, each of whom will use an average of 40 gal of water per day, with the maximum rate of 4 gph. The pipe is to be cast iron, bell-and-spigot joint.

The subdivision lies on the crest of a low hill whose summit is 75 ft above the pumping station, and the route of the pipe line is 13,500 ft long to the summit.

The pump delivers the maximum of 100 gpm at a "head" of 145 ft-lb force/lb mass.

If the pressure at the summit must be 25 psig under the maximum flow rate, what size pipe is required?

12. The efficiency of a pipe line is defined as the ratio of the quantity of fluid actually delivered to the quantity which would be delivered by a smooth pipe line of the same length and inside diameter, working with the same inlet and outlet pressures.

(*a*) A level section of a natural gas line transmits 8,500,000 cu ft/hr. The inlet pressure is 831 psig, and the outlet pressure is 636 psig. The section consists of 97 miles of pipe with an outside diameter of 24 in. and a wall thickness of 0.25 in. The barometric pressure is 14 psia, and the temperature of the gas in the line is 40° F. The measured flow rate is expressed in terms of a cubic foot, defined as the quantity of gas which occupies a cubic foot under the pressure of 14.9 psia at a temperature of 60° F. The viscosity of the gas at 40° F and 700 psia is 2.8×10^{-7} (lb force) (sec)/(sq ft). (Note that these units of viscosity are those of system II of Table 25. Viscosity in (lb mass)/(ft) (sec) may be obtained by multiplying by g_c.) The specific gravity referred to air is 0.69 at low pressure. The mass of a cubic foot of the gas at P psia exceeds the mass calculated from the ideal gas law by $0.00018P$. What is the efficiency of this pipe-line section?

(*b*) Two lines, a 20-in. (actual inside diameter, 19.5 in.) and a 22-in. (actual inside diameter, 21.5 in.) are in parallel for a distance of 47 miles. The two together transmit 7 million cu ft/hr, calculated to a base pressure of 16.4 psia and a base temperature of 60° F. The inlet pressure is 550 psig, and the barometer is 14 psia. The specific gravity of the gas at low pressure is 0.65, and its viscosity is 2.5×10^{-7} lb-sec/sq ft at 500 psia and 40° F, 40° F being the flowing temperature of the gas. The weight of a cubic foot of the gas at P psia exceeds the weight calculated from the ideal gas law by $0.00022P$. The 20-in. line has an efficiency of 92 per cent, and the efficiency of the 22-in. line is 88 per cent. The lines are level. What is the outlet pressure?

13. Water at 60° F is pumped from tank T to tanks M and N. The line from tank T to the pump inlet is 3-in. schedule-40 pipe, with an equivalent length of 200 ft. The line from the pump discharge to the tee where the line branches is also 3-in. schedule-40 pipe, with an equivalent length of 100 ft. The line from the tee to tank M is 1½-in. schedule-40 pipe, with an equivalent length of 600 ft. The line from the tee to tank N is 2-in. schedule-40 pipe, with an equivalent length of 600 ft.

The equivalent lengths as given do not include four gate valves, one between tank T and the pump, one between the pump and the tee, and one each at tanks M and N; and two globe valves, A in line from the tee to tank M, and B in line from the tee to tank N. All valves are wide open except either A or B, one of which is throttled to give the rates of flow of 50 gpm to tank M and 90 gpm to tank N. The water level in tank T is 30 ft above the pump inlet, and the levels in tanks M and N are each 8 ft above the pump inlet.

(*a*) What is the horsepower requirement of the pump?

(*b*) Which valve, A or B, is throttled? What is the pressure drop due to friction across the throttling valve in pounds per square inch?

14. Carbon dioxide is forced through a 2-in. schedule-80 steel pipe line with a pressure of 500 psia. The pipe line is 200 ft in length. The reading on the flowmeter at the inlet of the pipe line is 235,000 standard cu ft per hour (60° F, 1 atm).

If the flow is isothermal at 95° F, what pressure will be read on the gage at the discharge end of the line?

PROPERTIES OF CARBON DIOXIDE AT 95° F

Pressure, psia	Density, lb/cu ft	Viscosity, centipoises
14.7	0.111	0.0150
300	2.45	0.0160
400	3.39	0.0165
500	4.45	0.0170
600	5.65	0.0174
700	7.05	0.0177

15. A natural-gas well 7500 ft deep has a pressure at the surface of the ground of 2600 psia. The gas has a specific gravity referred to air of 0.744, the average temperature in the well is 153° F, and the compressibility factor for the gas at the mean conditions is 0.82. Compute the pressure at the bottom of the well, using the stated average value for the compressibility (equation 83) and using an average value for the specific volume.

16. The main system, tank 1, contains water at 50° F and remains at constant level. The water discharges through line A to tee B. Tank 2 contains 40 per cent sucrose solution

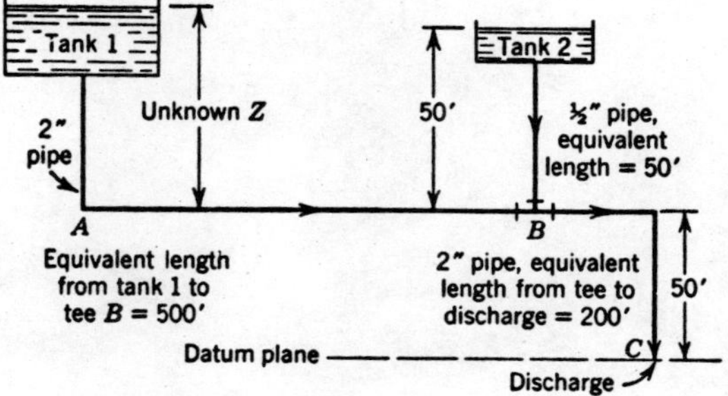

at 60° F and discharges to tee B. At B the water and sucrose mix and then pass as a diluted solution out of line at C. If the velocity of the solution in the pipe at C is 10 fps, what is the concentration of the diluted solution? What is height Z?

VISCOSITY AND DENSITY OF SUCROSE SOLUTIONS AT 50° F

Sucrose, %	Viscosity, centipoises	Density, grams/cc
10	1.8	1.04
20	2.68	1.083
40	9.8	1.178

17. A crude-oil pipe line is carrying 6000 bbl (42 gal each) of crude oil daily through 27 miles of single pipe, 12 in. ID. This capacity is now insufficient to meet the consumers' requirements. Accordingly, a parallel branch of the same size, extending for one-third of the total distance, is to be added. How much has the capacity of the pipe line been increased by the addition of the parallel branch if the line is level throughout, the oil has a specific gravity of 0.91 and a viscosity at flowing temperature of 500 centipoises, and the same pressure is maintained at the inlet to the pipe line?

18. An industrial furnace consumes 10 tons/hr of natural gas which for this purpose may be assumed to be methane. The gas is burned completely with 20 per cent excess air. The pressure of the flue gases leaving the furnace and entering the chimney is 1.0 in. of water below atmospheric pressure. The chimney is specified to be of welded steel plate, 11 ft ID. If the average temperature in the chimney is 500° F, the atmospheric temperature is 70° F, and the pressure is 14.4 psig, how tall a chimney is required?

19. Water is pumped through a 1-in. schedule-40 iron pipe and a strainer to a spray nozzle to be used as part of a humidifier. When spraying into the atmosphere with 90 psig pressure at the pump, the flow rate was found to be 200 gpm. What flow rate is expected when the nozzle discharges into a humidifier operated at 30 psig, if the pressure at pump is maintained at 90 psig and the flow rate of the nozzle is related to the pressure drop across the nozzle as follows?

$-\Delta P$, psi	5	10	15	20	30	40
Gallons per minute	60	100	125	140	175	200

CHAPTER 13

Measurement of Flow of Fluids

A FLOWMETER is a device for measurement of the quantity of fluid flowing per unit time, as in cubic feet per minute or pounds per second, or of the velocity of flow, as in feet per second.

Whatever the construction, a flowmeter is often calibrated by diverting the entire stream of fluid from its usual channel into a receiver arranged to permit accurate measurement of flow by weight or volume during a measured interval of time. A liquid may flow into a "weigh tank" mounted on scales or into a tank calibrated to indicate volume as a function of liquid depth. A gas may flow into a "gas holder," an inverted tank floating in water or oil inside a larger tank. In a sense, all flowmeters are always calibrated in this way, for even a standard calibration meter must be checked by this method.

DISPLACEMENT FLOWMETERS

A displacement meter measures the entire stream of fluid by dividing the stream into segments of known volume. The metering consists of mechanically counting the number of segments per unit of time. The three common types of displacement meters for liquid flows are oscillating-piston, nutating-disk, and multiple-piston meters. For gas flows the two common types are the "wet" or bucket-type meters and the "dry" or diaphragm meters.

Most of the pumps described in Chapter 14 can be operated in reverse as flowmeters, especially the rotary pumps which need no valves. The main shaft is connected to a register, and the moving parts are made of hard rubber or other lightweight material when these designs serve as meters.

Oscillating-Piston Meter

Figure 129 shows the measuring chamber of an oscillating-piston flowmeter, with the piston at four equidistant positions during one revolution. All fluid enters through the inlet port and passes around the annular space between the outer and inner rings to the outlet port. Part of the fluid travels this distance in the space between the piston and the

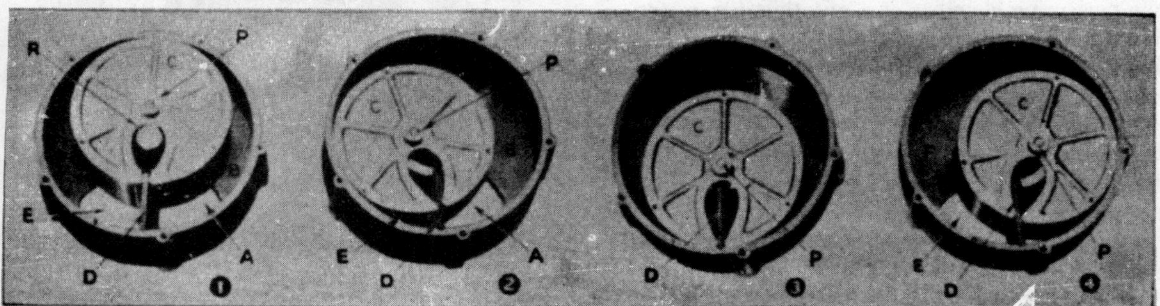

FIG. 129. The operation of the oscillating-piston meter. *A*, inlet port; *B*, annular measuring chamber receiving fluid; *C*, hollow piston; *D*, diaphragm; *E*, outlet port; *F*, annular measuring chamber discharging fluid; *P*, spindle; *R*, roller. The outer ring is the wall of the casing. The inner ring, concentric with the outer ring, is covered by the piston and appears only as short white arcs joining the inner end of the diaphragm in views 2 and 4. (*Neptune Meter Co.*)

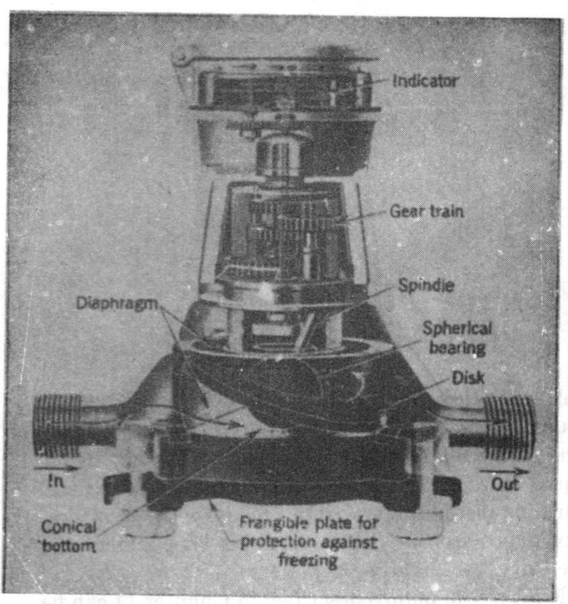

FIG. 130. Cutaway drawing of nutating-disk meter. (*Neptune Meter Co.*)

outer ring, and part moves through the space between the wall of the hollow piston and the inner ring. The piston is guided by the spindle which follows the circular path between the inner ring and the central roller. An extension of the spindle through the cover of the measuring chamber actuates the first of a series of gears which record on a dial the total volume of fluid passed by the meter since it was set at zero.

Nutating-Disk Meter

The measuring chamber of a nutating-disk meter is roughly cylindrical, with a diaphragm extending radially from one side of the chamber wall to the spherical bearing in the center, as in the chamber of the oscillating-piston meter. As seen in Fig. 130, a disk, slotted for the diaphragm, extends from the center of the spherical bearing at right angles to a short spindle extending up to the gear train. The top and bottom of the measuring chamber are conical, pointing in toward the center and forming a line contact with the disk. As the fluid enters the space between the disk and the chamber wall, it forces the disk ahead of it in a nutating, or wobbling, motion. The motion of the disk and spindle is similar to that of a spinning top as the top reaches the end of its action.

Such meters are widely used in water supply linos of moderate capacity.

Multiple-Piston Meter

Figure 131 shows a multiple-piston meter with five pistons and cylinders mounted in a circle and connected to a nutating plate or "wobble-plate." The fluid enters and leaves the top of each cylinder through a rotary valve. The fluid enters the cylinders from the chamber as indicated by the solid arrows and leaves the cylinders through the rotary valve, flowing into a central vertical passage as indicated by the dotted arrows. The rotary valve controls the flow so that the pistons are acting in succession around the circle. One piston, at full withdrawal from the valve, is changing from inlet to outlet while its neighbor on one side is still being pushed down by entering fluid and the neighbor on the other side is being pushed up by the nutating plate on the bottom and is discharging fluid. The connecting rods are of such a length that the plate on the bottom wobbles around its central bearing in an action similar to that of the nutating-disk meter. The rotary valve at the top is driven from the plate and is synchronized with it.

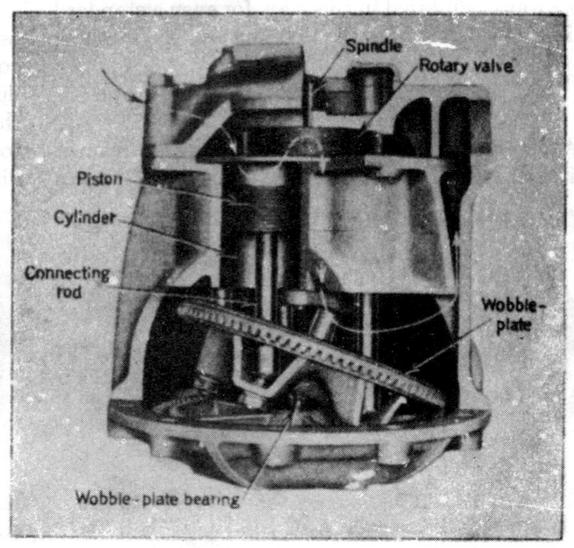

FIG. 131. Cutaway view of wobble-plate piston meter. (*Bowser, Inc.*)

The spindle on the rotary valve drives the first unit in the gear train of the indicating mechanism.

Wet-Test Meter

The bucket-type wet-test meter (Fig. 132) is used principally for testing and laboratory measurement. The main body of the meter contains a cylindrical rotor made as four buckets, each occupying one quadrant of the cylinder. Each bucket receives through a slot in the axis of the rotor and discharges through a narrow slot parallel to the axis on the outer edge of the rotor. The rotor is immersed in water to a level just above the axis. The gas being metered is introduced into the hollow axis and bubbles into the bucket at the right through the inlet slot. The trapped gas lifts the bucket until the discharge slot emerges from the water. At that instant the partition between buckets passes the opening in the hollow axis and shifts the inlet flow to the next bucket on the rotor. The gas washed out of the rotor leaves the case through the opening at the top. The shaft of the rotor drives a gear train indicating

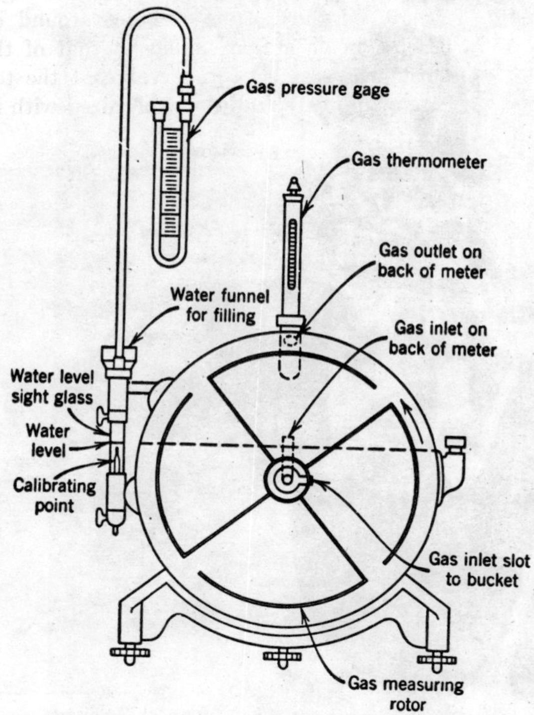

Fig. 132. Sectional view of wet test gas meter.

Fig. 133. Exploded view of diaphragm gas meter. (*Pittsburgh Equitable Meter Division of Rockwell Mfg. Co.*)

the volume of gas passing through the meter at the measured temperature and pressure.

Diaphragm Meter

The working mechanism of a "dry" or diaphragm gas meter is shown in Fig. 133. The two circular diaphragms and diaphragm pans (Fig. 134) are suspended from the valve plate and hang down into the pressure-tight case. The gas enters first the space above the valve plate and is admitted through one side of one of the D-slide valves to either the space between a diaphragm and its pan or between a diaphragm and the case. The diaphragms move in and out in a "breathing" action, being pushed by gas entering on one side and pushing gas out on the other side. The discharged gas leaves through the valve and the discharge port cast into the valve

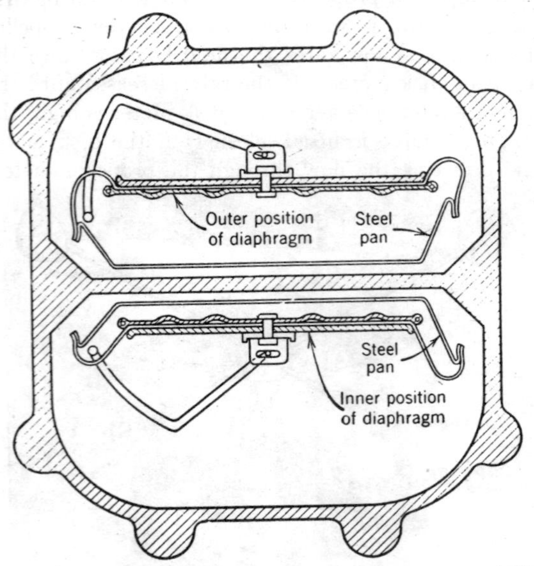

Fig. 134. Transverse view of diaphragm gas meter. (*Pittsburgh Equitable Meter Division of Rockwell Mfg. Co.*)

plate. The motion of the diaphragm, through the flag arms, operates the valve gear and drives the index shaft which actuates the register.

CURRENT FLOWMETERS

A current meter measures the velocity of flow, not the quantity of flow, and is not positive in its action. If a displacement meter should cease operating the total flow would be obstructed, but if a current meter should cease operating the flow would be hindered only slightly. If inserted in a closed conduit the dials of a current meter may be calibrated in terms of quantities.

The three common types of current meters are propeller, cup, and hot-wire meters. The last type is suitable only for measurement of gas flow, but the first two types are used with any fluid.

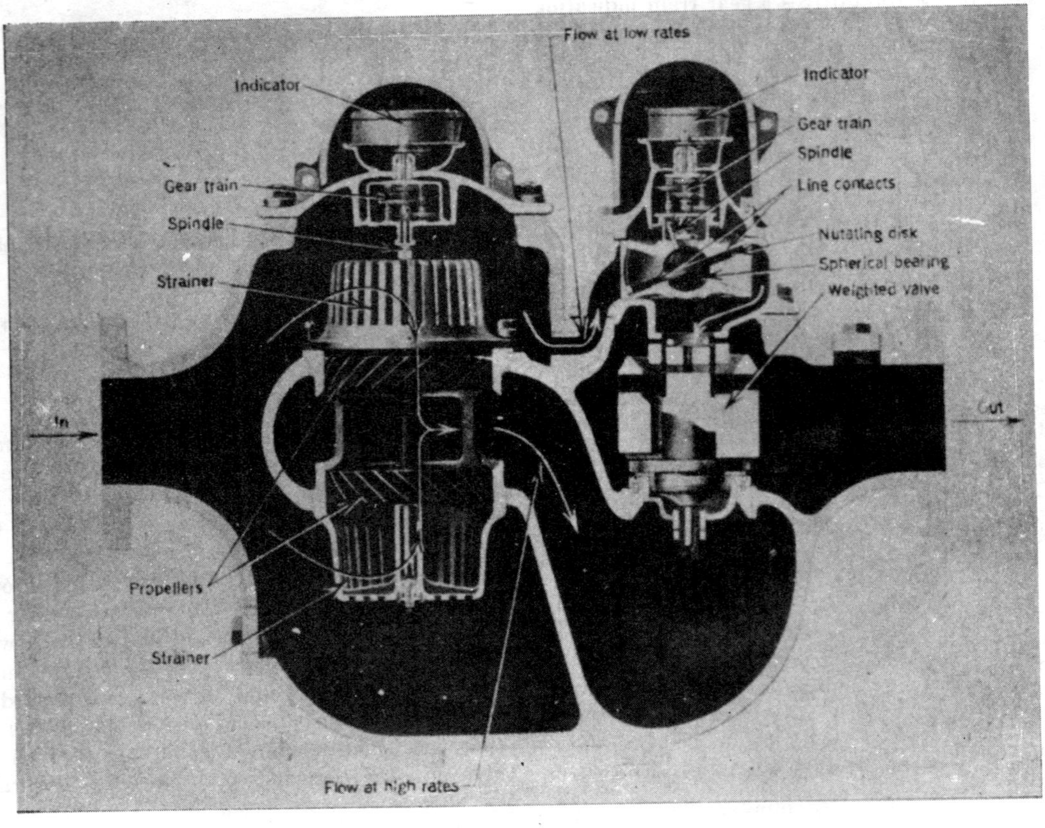

Propeller Meter

Some form of propeller, or screw, placed in a stream of fluid will rotate as the passing fluid acts on the blades. As the density of the fluid decreases, the mass, or inertia, of the propeller and the friction must decrease, since the available energy for operation of the meter is reduced. The propeller is geared to the recording or indicating device included in the meter.

A modified propeller meter, sometimes called a "velocity" or "inferential" meter, is seen in Fig. 135. The entering fluid passes through one of two strainers and comes into contact with one of two propellers. The propellers are of opposite "hand," one threaded to the left and the other to the right, and are attached to the same spindle. The spindle extends through the upper cage into the enclosed gear housing and drives the indicating device.

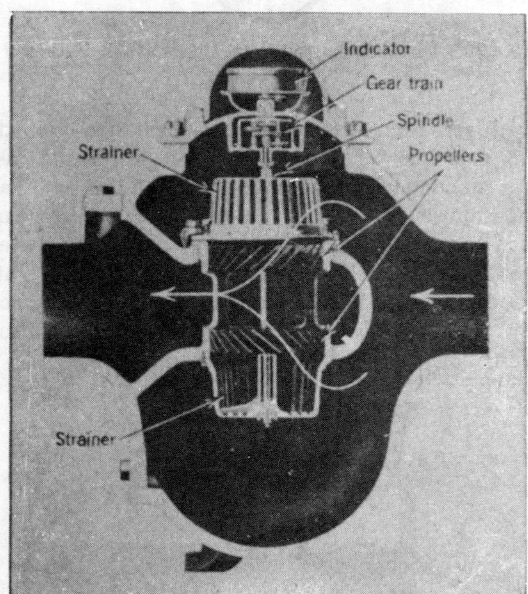

Fig. 135. Cutaway view of propeller meter. (*Neptune Meter Co.*

Compound Meters. A current meter is most accurate at high quantities of flow, and a displacement meter is operable only at low rates of flow. Therefore, metering of flows that vary greatly cannot be done accurately with only one of these types. A compound meter consists of one of each type of meter constructed to permit operation of each at its optimum range. Figure 136 shows a compound meter made up of a propeller meter and a nutating-disk meter. Flow is into the casing of the propeller meter, but at low rates the fluid bypasses into the nutating-disk meter. If the rate increases until the pressure difference across the disk meter is high, the weighted valve is lifted, closing off the disk meter and directing the flow through the propeller meter.

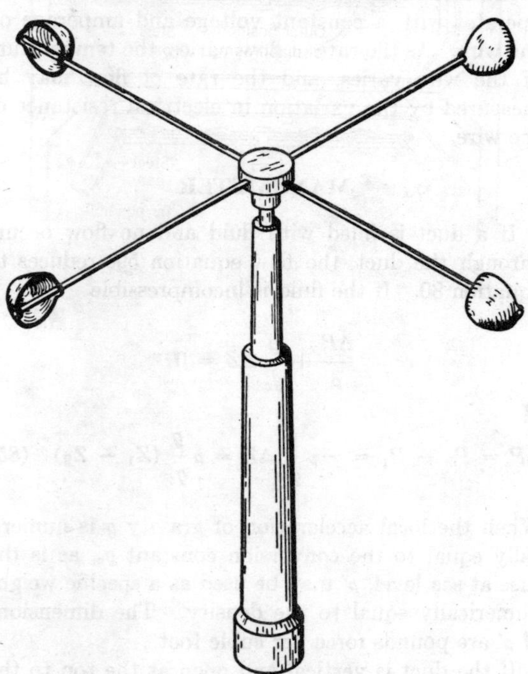

Fig. 137. Diagram of cup-type anemometer.

Cup Meter

The whirling cups of the weatherman's anemometer (Fig. 137) are the commonest example of the cup flowmeter. The fluid striking the concave side of the cup exerts a greater force than the fluid striking the convex side of the opposite cup, resulting in easy rotation of the cup assembly. The central shaft drives the gear train which tells the velocity of flow. If the device is made smaller and more rugged and is enclosed in a housing so that the fluid strikes the cups in a jet, the instrument serves as a liquid meter.

Hot-Wire Anemometer

A hot-wire anemometer relies upon the rate at which the flowing stream absorbs energy supplied by the primary element of the meter. A resistance wire placed in a flowing stream and supplied with electrical energy at a constant voltage gives up heat to the stream. As the rate of flow of fluid increases, heat is absorbed more rapidly by the stream, and, if the temperature of the wire is to be held constant, the electrical current must increase. An ammeter placed in the electrical circuit will vary in reading as the rate of flow changes, and its dial may be modified to read directly the velocity of flow.

Another version of the hot-wire anemometer operates with a constant voltage and amperage on the wire. As the rate of flow varies, the temperature of the wire varies, and the rate of flow may be measured by the variation in electrical resistance of the wire.

MANOMETER

If a duct is filled with fluid and no flow occurs through the duct, the flow equation 59*c* reduces to equation 80. If the fluid is incompressible

$$\frac{\Delta P}{\rho} + \frac{g}{g_c}\Delta Z = 0$$

or

$$\Delta P = P_2 - P_1 = -\rho\frac{g}{g_c}\Delta Z = \rho\frac{g}{g_c}(Z_1 - Z_2) \qquad (85)$$

When the local acceleration of gravity g is numerically equal to the conversion constant g_c, as is the case at sea level, ρ' may be used as a specific weight numerically equal to the density. The dimensions of ρ' are pounds force per cubic foot.

If the duct is vertical and open at the top to the atmosphere, P_1 equals atmospheric pressure and ΔP equals the gage pressure corresponding to P_2, identified by the symbol P_2'.

$$P_2' = \rho'(Z_1 - Z_2) \qquad (86)$$

Equation 86 indicates that the pressure above 1 atm at the bottom of the duct may be measured in terms of the height of liquid above that point.

Consider the vertical duct with the top open to the atmosphere and with the lower end open to a duct carrying fluid in flow. The pressure in the flowing fluid at the junction will sustain a column of water extending above the level of the junction, the force exerted by the column of fluid of height ΔZ on a unit area at the bottom being equal to the pressure P_2' above 1 atm. This height is a difference in vertical elevation and is independent of the length or angle or inclination of the duct, if the duct remains filled with fluid.

The nonflowing fluid need not be identical with the flowing fluid. A nonflowing fluid of high density may reduce the height of the column to a convenient length if the pressure is high. Alternatively, a fluid of lower density may be used to increase the height of the column and increase the accuracy of measurement. Instruments made to measure pressure differences in this way are called manometers.

Manometers always measure differences in pressure by balancing static "legs" or columns of fluid,

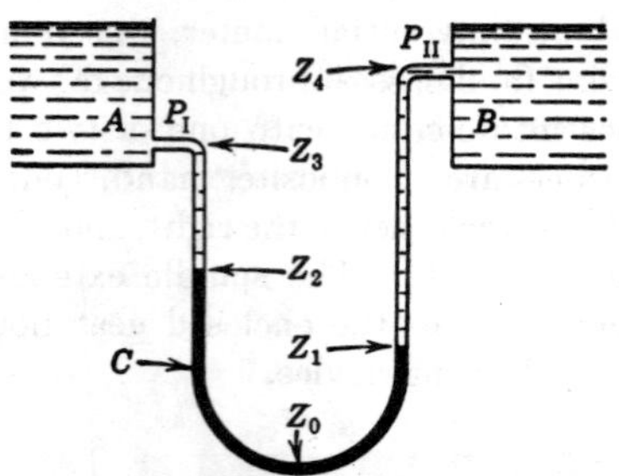

Fig. 138. Diagram illustrating the general application of a manometer.

but a large number of widely varying designs embody this principle. Figure 138 shows the general case of a manometer containing a liquid, C, denser than and immiscible with two other fluids, A and B, being used to measure the difference in pressure between point I in fluid A and point II in fluid B. Points I and II are at different elevations. The application of equation 85 to this system by successive addition of increments of the columns of fluids gives:

$$\begin{aligned}\Delta P &= P_{II} - P_I \\ &= -\rho'_B(Z_4 - Z_2) - \rho'_B(Z_2 - Z_1) \\ &\quad - \rho'_C(Z_1 - Z_0) - \rho'_C(Z_0 - Z_1) \\ &\quad - \rho'_C(Z_1 - Z_2) - \rho'_A(Z_2 - Z_3)\end{aligned}$$

$$\begin{aligned}\Delta P &= -\rho'_B(Z_4 - Z_2) + (\rho'_C - \rho'_B)(Z_2 - Z_1) \\ &\quad - \rho'_A(Z_2 - Z_3) \qquad (87)\end{aligned}$$

where ρ'_A is the specific weight of fluid A.
ρ'_B is the specific weight of fluid B.
ρ'_C is the specific weight of metering fluid C.
P_I is the pressure at the point of connection of one manometer leg to the body of fluid A.
P_{II} is the pressure at the point of connection of the other manometer leg to the body of fluid B.

If the fluids A and B are identical and legs I and II are connected at the same elevation, equation 87 reduces to

$$\Delta P = P_{II} - P_I = (\rho'_C - \rho'_B)(Z_2 - Z_1) \quad (88)$$

An "inverted" manometer, using a metering fluid of lower density than the fluid whose pressure is being measured, may be treated in a similar manner.

If the differential reading of the manometer is too small for satisfactory measurement the instrument may be tilted from the vertical, in which case a small vertical differential will give a much larger differential when measured along the sloping legs.

If one leg of the manometer is made large in cross section in comparison with the other, the level of the metering fluid in the large leg will remain substantially constant and the pressure difference can be read as the level in the small leg alone referred to the constant level in the large leg. This arrangement may be used in either a vertical or an inclined manometer.

Another method of increasing the reading of the manometer for a small pressure difference is indicated in Fig. 139 in which two manometer fluids, B and C, are used. If the density of the fluid C is only slightly

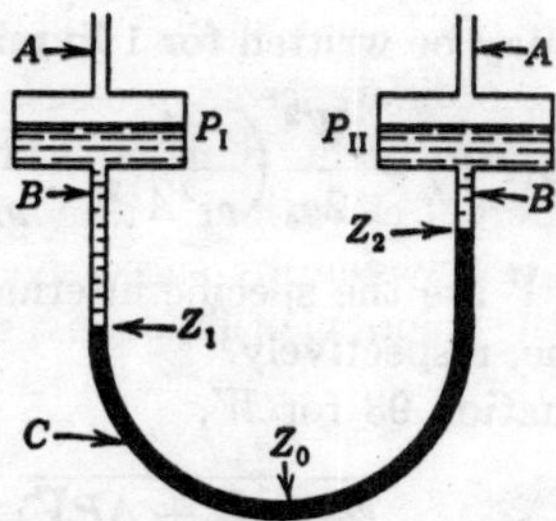

Fig. 139. Diagram illustrating a two-fluid manometer.

greater than that of fluid B, a small difference between the pressures P_I and P_{II}, which causes a small difference between the level of the A–B interface in the two legs, will cause a relatively large difference in the level of the B–C interface in the two legs of the manometer.

Still another method of increasing the sensitivity of the manometer of Fig. 139 is to make the connecting loop of tubing long and flat across the bottom at the position Z_0 and to reduce the quantity of fluid C to a bubble which moves along the horizontal section. The small difference in level of the A–B interfaces in the enlarged sections, corresponding to the pressure difference $\Delta P = P_{II} - P_I$, is magnified by a relatively large horizontal displacement of the bubble. The magnitude of the horizontal displacement depends on the relative cross-sectional areas of the enlarged section and the connecting tube.

PITOT TUBE

If a tube is placed within a flowing stream so that its axis is at right angles to the direction of flow past the open end of the tube, the pressure in the stagnant fluid just within the mouth of the tube will be the same as the pressure in the fluid flowing past the tube. If another tube is placed with its axis parallel to the direction of flow the flowing fluid will tend to enter this second tube with a velocity v_1. If the discharge end of this tube is closed, as by one side of the manometer, the velocity of the fluid in this tube v_2 becomes zero, but the velocity of the fluid in the main stream continues at v_1. Under these conditions application of equation 59c shows that

$$\int \overline{V}\, dP = -\frac{\Delta v^2}{2g_c}$$

Since the velocity just inside the tube, v_2, is zero,

$$\int \overline{V}\, dP = \frac{{v_1}^2}{2g_c} \quad (89)$$

where the subscript 1 refers to a point just outside the tube. If the pressure drop is small, $\overline{V}_2$ may be taken as equal to $\overline{V}_1$, and

$$P_2 - P_1 = \frac{\rho {v_1}^2}{2g_c} \quad (90)$$

If the two tubes are connected to opposite sides of a manometer, a pressure difference will be indicated upon it as a difference in level of the interfaces in the two legs. This pressure difference results from transformation of the kinetic energy of the fluid and by equation 90 can be expressed in units of velocity.

An instrument has been devised which incorporates both these tubes into a single unit, known as a pitot tube, Fig. 140. The two tubes are arranged concentrically and the annular space sealed at the end. The arrangement of two tubes is pointed upstream, so that the flowing stream impinges directly upon the inner tube but cannot flow into the outer tube. Small holes are drilled in the walls of the outer tube a short distance back from the tip, admitting the fluid into that tube but not permitting impingement of the flowing stream. Thus the inner tube transmits to the manometer both the pressure and the pressure equivalent to the kinetic energy of the flowing fluid, whereas the outer tube transmits only the pressure. The manometer thus indicates only the velocity of flow, which permits calculation of the quantity flowing per unit area at the point of measurement.

The pitot tube measures only the local velocity at point 1 in the stream; therefore it can be used to explore the local velocity gradient in a duct carrying a fluid. The "traverse" obtained by measurement of the local velocity at several points across the

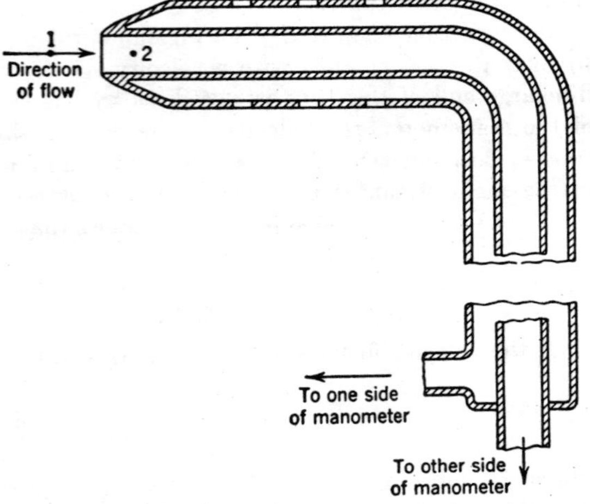

FIG. 140. Sectional diagram of pitot tube.

diameter of the duct furnishes data for calculation of the average velocity of flow based on the entire cross-sectional area of the duct. If the Reynolds number (computed with the average velocity) is above 50,000, the ratio of the average velocity to the local velocity at the center of a circular duct is 0.81, and a single reading from the pitot tube inserted at the center may be used to calculate the average velocity.

In applying equation 90 to actual pitot tubes, a coefficient C is frequently necessary

$$v_1 = C\sqrt{\frac{2g_c(P_2 - P_1)}{\rho}} \qquad (90a)$$

The value of C for a given instrument is determined by calibration.

VENTURI METER

If the fluid to be measured is flowing inside a closed conduit, a constriction in the channel will serve as the primary element of a flowmeter. For example, Fig. 141 shows a constriction with a smooth tapering inlet and a smooth tapering outlet, known as a venturi tube. If this tube is mounted horizontally

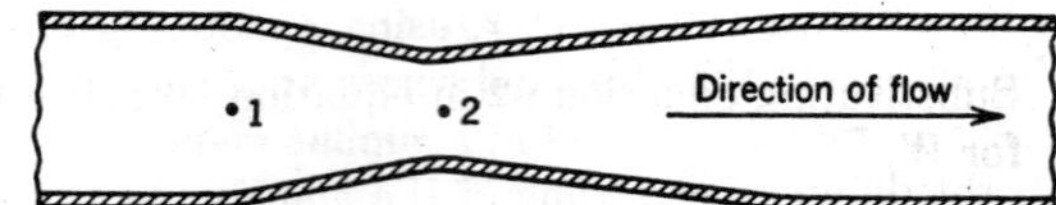

FIG. 141. Sectional diagram of venturi tube.

in a pipe line, there is no difference in elevation, no work is done, and the operation is adiabatic. Equation 54 then reduces to equation 91 when applied between points 1 and 2 as in Fig. 141.

$$\Delta U + \Delta(PV) = -\Delta\left(\frac{mv^2}{2g_c}\right) \qquad (91)$$

Also

$$v = \frac{W}{\rho A} = \frac{G}{\rho} \qquad (92)$$

where W = mass rate of flow per second.
A = cross-sectional area of flow.
G = mass velocity, or mass per second per unit area.

Equation 91 may be written for 1 lb mass:

$$\Delta\bar{U} + \Delta(P\bar{V}) = \frac{W^2}{2g_c}\left(\frac{1}{\rho_1{}^2A_1{}^2} - \frac{1}{\rho_2{}^2A_2{}^2}\right) \qquad (93)$$

where $\bar{U}$ and $\bar{V}$ are the specific internal energy and specific volume, respectively.

Solving equation 93 for W,

$$W = \sqrt{\frac{2g_c(\Delta\bar{U} + \Delta P\bar{V})}{\dfrac{1}{\rho_1{}^2A_1{}^2} - \dfrac{1}{\rho_2{}^2A_2{}^2}}} = \sqrt{\frac{2g_c(\Delta\bar{H})}{\dfrac{1}{\rho_1{}^2A_1{}^2} - \dfrac{1}{\rho_2{}^2A_2{}^2}}} \qquad (94)$$

where $\bar{H}$ is the enthalpy per pound mass.

If the flowing fluid is incompressible or the pressure difference is so small that the density is almost constant, then

$$W = \sqrt{\frac{2g_c\rho^2A_2{}^2(\Delta\bar{H})}{(A_2{}^2/A_1{}^2) - 1}} \qquad (95)$$

This relation is quite useful when the enthalpy of the flowing fluid can be evaluated at points 1 and 2.

When the enthalpies cannot be evaluated, equation 60 may be applied without assuming adiabatic conditions. If $\Delta Z = 0$ and $\bar{w} = 0$,

$$\frac{\Delta P}{\rho} + \frac{\Delta (v)^2}{2g_c} = -\overline{lw} \qquad (96)$$

Substituting equation 92 in equation 96 and solving for W,

$$W = \sqrt{\frac{2g_c\rho(-\Delta P - \rho\overline{lw})}{(1/A_2^2) - (1/A_1^2)}}$$

$$= \sqrt{\frac{2g_c\rho A_2^2(-\Delta P - \rho\overline{lw})}{1 - (A_2^2/A_1^2)}} \qquad (97)$$

The irreversibilities (including friction losses) lw can be expressed as a fraction of the pressure difference ΔP or

$$-\Delta P - \rho\overline{lw} = C^2(-\Delta P) \qquad (98)$$

Equations 97 and 98 combine to

$$W = CA_2\sqrt{\frac{2g_c\rho(-\Delta P)}{1 - (A_2^2/A_1^2)}} \qquad (99)$$

This equation enables the calculation of the quantity of fluid flowing in pounds per second, if the density of the fluid and the cross-sectional area of the tube are known at points 1 and 2 and the pressure difference between points 1 and 2 is measured. Since venturi tubes are difficult and expensive to manufacture, they are made by specialists. Usually the tapering entrance has an interior total angle of 25 degrees and the tapering exit has an interior angle of 7 degrees. For such an instrument the value of the coefficient C is 0.98 if the Reynolds number is greater than 10,000.

FLOW NOZZLE

Because of the size and expense of a venturi tube, a smaller device called a flow nozzle (Fig. 142) is often used as the primary element of a flowmeter. The diverging exit of the venturi is omitted from this device, and the converging entrance is altered to a more rounded form. Many different shapes of the converging entrance of nozzles are used in order to reduce the friction losses, but experiment has shown that the shape of one quadrant of an ellipse (as seen in longitudinal section) is most satisfactory. Equation 99 applies to the flow nozzle, but the value of the coefficient C will vary from 0.70 to 0.98, depending on the shape of the nozzle, its length, and its relative diameter.

FIG. 142. Cross-sectional diagram of a flow nozzle.

ORIFICE

As compared to the venturi tube or the flow nozzle, the orifice (Figs. 143 and 144) is a simple mechanism. A simple flat plate with a central opening made with ordinary tools will suffice if an accurately machined orifice plate is not available.

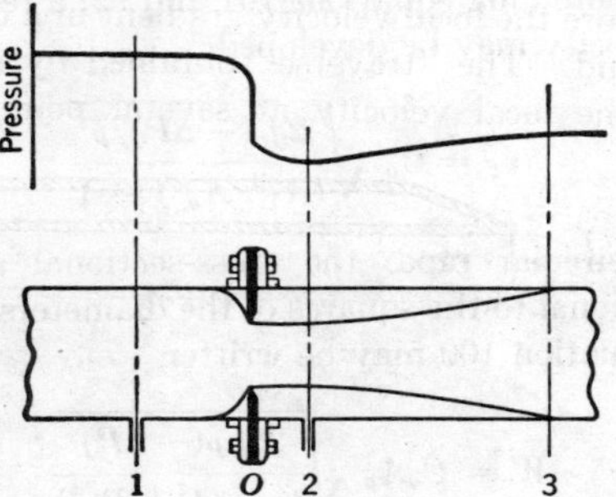

FIG. 143. Cross-sectional diagram of a sharp-edged orifice indicating approximate flow lines and pressure.

The contraction of a stream flowing through an orifice is quite pronounced. The point of minimum cross section is one or two diameters downstream from the orifice plate. This point is known as the "vena contracta" and is indicated in Fig. 143 by the point 2. This point corresponds to the point 2 in the venturi tube and the flow nozzle and is the point at which the pressure is usually measured just downstream from the metering element, thus obtaining the maximum difference in pressure, as shown in Fig. 143.

The cross-sectional area at point 2 is important in flow equations such as equations 95 and 99, but the cross-sectional area of the vena contracta is quite difficult to determine with accuracy. This area may be expressed, however, as a fraction of the area of the orifice opening, which is easily determined. If A_2 in equation 99 is replaced by $C''A_o$, where A_o is the cross-sectional area of the orifice opening, and the coefficients are combined to a new coefficient C_o, then

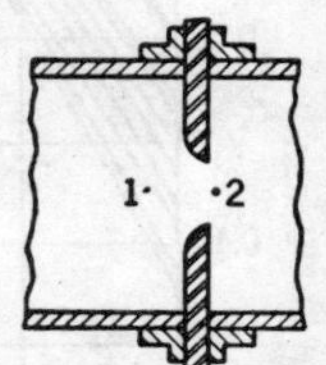

FIG. 144. Cross-sectional diagram of a round-edged orifice.

$$W = C_oA_o\sqrt{\frac{2g_c\rho(-\Delta P)}{1 - (A_o^2/A_1^2)}} \qquad (100)$$

Exercise. Show that

$$C_o = C\frac{A_2}{A_o}\sqrt{\frac{1-(A_o^2/A_1^2)}{1-(A_2^2/A_1^2)}}$$

An alternative equation in terms of the pipe area is

$$W = C_oA_1\sqrt{\frac{2g_c\rho(-\Delta P)}{(A_1^2/A_o^2)-1}} \qquad (101)$$

By combining equations 101 and 92, a relation for the velocity may be developed.

$$v_1 = C_o\sqrt{\frac{2g_c(-\Delta P)/\rho}{(A_1^2/A_o^2)-1}} \qquad (102)$$

For circular pipes the cross-sectional areas are proportional to the squares of the diameters. Therefore, equation 100 may be written

$$W = C_oA_o\sqrt{\frac{2g_c\rho(-\Delta P)}{1-(D_o^4/D_1^4)}} \qquad (103)$$

and equation 102 may be written

$$v_1 = C_o\sqrt{\frac{2g_c(-\Delta P)/\rho}{(D_1^4/D_o^4)-1}} \qquad (104)$$

These equations permit calculation of the quantity or velocity of fluid flowing through an orifice plate in a pipe, provided the dimensions of the system are known, the pressure difference between points 1 and 2 is measured, and the coefficient of the orifice is known.

The friction losses are related to the Reynolds number. Therefore the values of the coefficient C_o may be expressed as a function of the Reynolds number and of the diameter ratio of the orifice and pipe, as shown by the light lines in Fig. 145. For values of the Reynolds number through the orifice greater than 30,000, the coefficient is approximately 0.61 for all ratios of the diameters of the orifice and the pipe, provided the taps, or manometer connec-

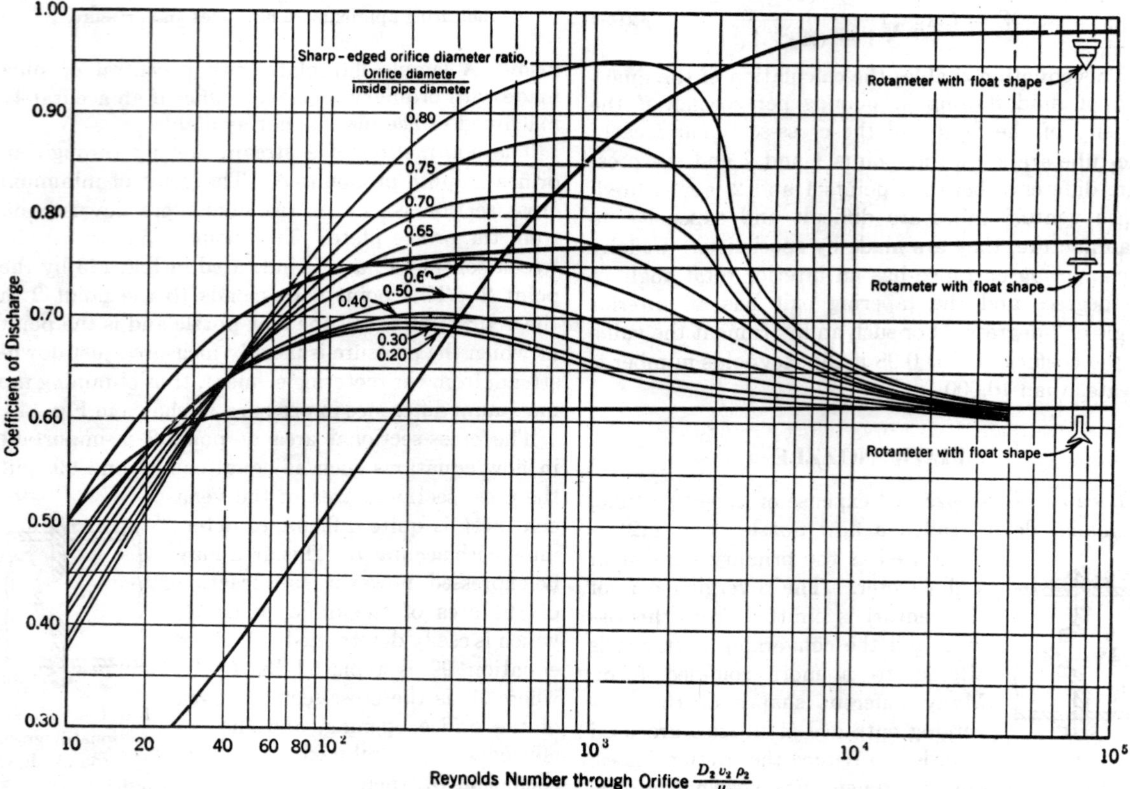

FIG. 145. Variation of discharge coefficient with Reynolds number for sharp-edged orifices[3]* and rotameters.[4]

* The bibliography for this chapter appears on p. 163.

tions, are one-half pipe diameter downstream and upstream from the orifice, and the orifice opening has sharp edges. The Reynolds number through the orifice is usually greater than 30,000. Therefore 0.61 is a reasonably constant value for the coefficient of a sharp-edged orifice.

If the orifice is not sharp-edged but rounded at the upstream face (Fig. 144), the coefficient of discharge C_o has a value varying from 0.70 to 0.88, depending on the degree of curvature and the size of the opening. This increased coefficient results from the approach of the vena contracta to the orifice opening in size.

For any particular orifice in a particular "meter run" (length of pipe containing the orifice), equation 103 may be written

$$W = KA_o\sqrt{\rho(-\Delta P)} \qquad (105)$$

where K is a function of pipe and orifice diameters, of the mechanical setting of the meter run, of the pressure connections, and of the Reynolds number of the fluid.

The value of $\sqrt{1 - (D_o^4/D_1^4)}$ is very close to unity, being 0.96825 if the diameter ratio D_o/D_1 (orifice diameter to pipe diameter) is as large as 0.5. Since the diameter ratio is usually less than 0.5, this factor is frequently omitted in orifice equations such as equations 100 and 103, but not equations 101, 102, or 104.

The location of the orifice in the piping system affects the coefficient of discharge sharply if turbulence is established in a way to modify the general flow lines through the orifice. The orifice should be at least 50 pipe diameters downstream and 10 pipe diameters upstream from any fitting or valve. If this is impossible a bundle of tubes or a straightening vane may be inserted ahead of the orifice, the ratio of length to diameter of the channels in the bundle being 50 or more.

A typical installation of an orifice in a line carrying steam, with a manometer-type indicating instrument connected across it, is shown in Fig. 146. The condensers convert the steam to liquid water and maintain the lines full of water to the same level. Mercury in the manometer carries a float up and down in the float chamber, actuating a lever which protrudes into the mechanism of the indicator. The indicator may be a direct-reading dial or a recorder which makes a continuous inked record of the pressure difference (or rate of flow in any desired units). The lever may control an electrical or compressed-air system which will control the flow automatically.

Orifice meters are widely used for measuring gas flow where it is customary to express the result in

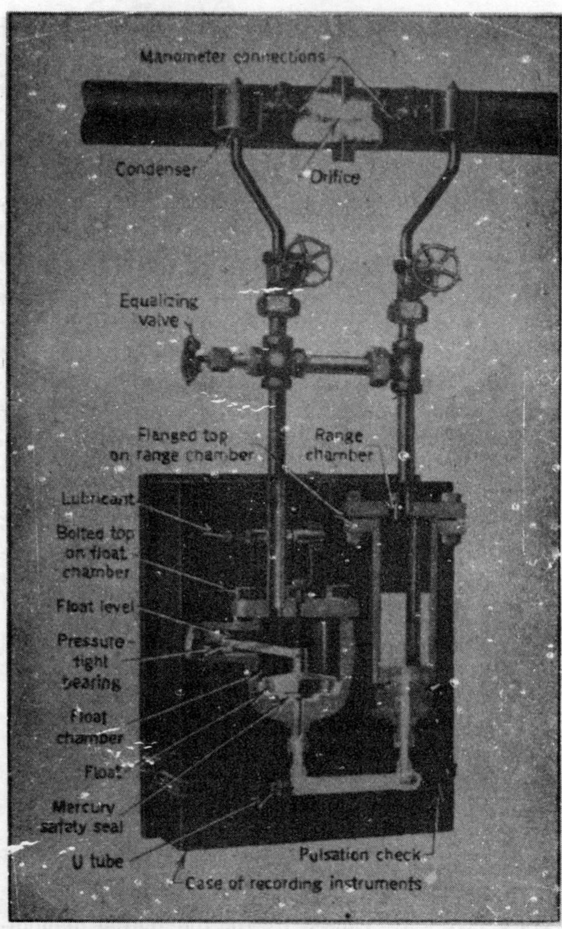

FIG. 146. Phantom illustration of installation of orifice in flow system with manometer used to measure difference in pressure across orifice. (*Brown Instrument Co.*)

standard cubic feet (usually as measured at 60° F and 14.65 psia) per day, or per hour, rather than in pounds mass per second. The pressure drop $(P_1 - P_2)$ is frequently expressed in inches of water or inches of mercury as measured by a manometer. Making these changes in equation 103 and substituting $1/\overline{V} = MP/zRT$ for ρ gives the flow rate in standard cubic feet per hour as measured at T_s in degrees Rankine and P_s in psfa.

Flow rate =

$$C_oA_o\left(\frac{z_sRT_s}{MP_s}\right)(3600)\sqrt{\frac{2g_cMP(h_1 - h_2)_w}{(1 - D_o^4/D_1^4)zRT}\left(\frac{62.4}{12}\right)} \quad (106)$$

where $(h_1 - h_2)_w$ = pressure drop expressed in inches of water.

Assuming z_s is unity and omitting the factor $(1 - D_o^4/D_1^4)$, equation 106 may be simplified to

Standard cubic feet per hour

$$= 2{,}588{,}659C_oA_o\frac{T_s}{P_s}\sqrt{\frac{(h_1 - h_2)_wP_2}{MzT_1}} \quad (107)$$

When the pressure drop through the orifice is not greater than 0.2 of the absolute upstream pressure (P_2/P_1 is not less than 0.8), the error introduced by using P_2 for P in equation 106 does not exceed 1 per cent. Since this error is in the positive direction, it tends to compensate for the simplification incorporated in equatior 107.

Pulsations

Serious errors may be introduced by pulsating flow, as exists in lines connected to gas compressors.

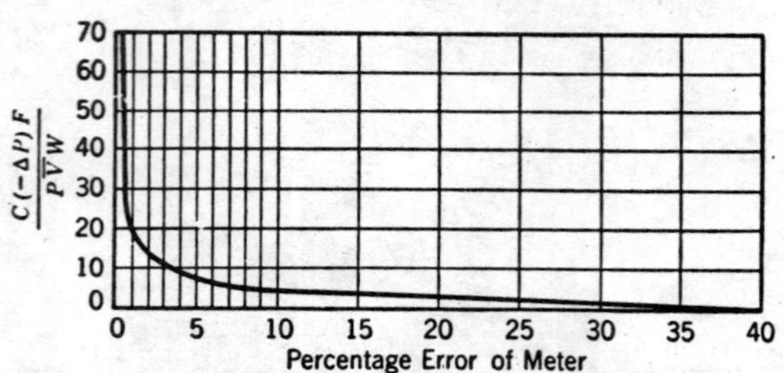

FIG. 147. The approximate percentage error in orifice meter reading due to pulsations.[5]

These errors are caused by the differential manometer, or other recording device, across the orifice, indicating approximately the arithmetic mean differential pressure, whereas the flow varies with the mean square root of the differential pressure. The result is that the recorded differential pressure is too great for the actual flow, causing the meter to read high. This is true even if the manometer taps are throttled so that the indicated differential pressure appears reasonably constant.

These errors may be caused by pulsations either in the form of pressure waves traveling through the fluid with the velocity of sound, or by pulsations in the flow of the fluid itself.

The error due to pressure variation may be reduced by taking pains to have the pressure lines leading to the meter of equal coefficients of discharge for inflow and outflow and the capacity in the lines and meter the same on each side of the mercury or water column.

The error due to velocity variation cannot be eliminated by any meter construction now available. It varies with many factors, and the best plan is to reduce the variation to a minimum as there is no satisfactory manner of making proper corrections.

The approximate error caused by pulsation in the velocity as reported by Hodgson[5] is indicated in Fig. 147 as a function of the product:

$$\frac{C(-\Delta P)F}{P\overline{V}W}$$

where C = the volume capacity in the line (cu ft) between the source of pulsation P and the meter.

$(-\Delta P)$ = the pressure drop (psf) between the entrance to the capacity C and the side of the orifice or venturi meter which is farthest from the source of pulsation.

F = the frequency of pulsation per second.

P = absolute pressure (psf).

$\overline{V}$ = specific volume of fluid (cu ft/lb).

W = mass rate of flow (lb/sec).

As indicated by a study of Fig. 147, the large error due to pulsation in velocity frequently found in gas lines leading to or from compressors can often be reduced to a negligible factor (under 3 per cent) by increasing the capacity of the line between the meter and the source of pulsation and increasing the pressure drop between the source of pulsation and the far side of the meter.

For this reason, small orifices, which have larger pressure drops, are affected less by velocity pulsations than large orifices with smaller pressure drops.

A high frequency F in pulsation also introduces less error than a low frequency. A single compressor giving a low frequency of pulsation will cause a greater error than a battery of compressors giving a high frequency.

In making an installation to reduce errors due to pulsation, the capacity should be inserted between the source of pulsation and the metering point. If the meter itself does not offer sufficient resistance, an additional throttling device may be added.

Friction Losses

As might be expected, the increase in velocity and decrease in pressure at the vena contracta is temporary, except for the friction losses between point 1 and a downstream point, 3, chosen where the main stream of flow fills the pipe completely. This "energy loss" is a function of the ratio of the orifice and pipe diameters and is expressed as a ratio of the permanent pressure loss between points 1 and 3 and the temporary pressure drop between points 1 and 2, as in Fig. 148 for sharp-edged orifices.

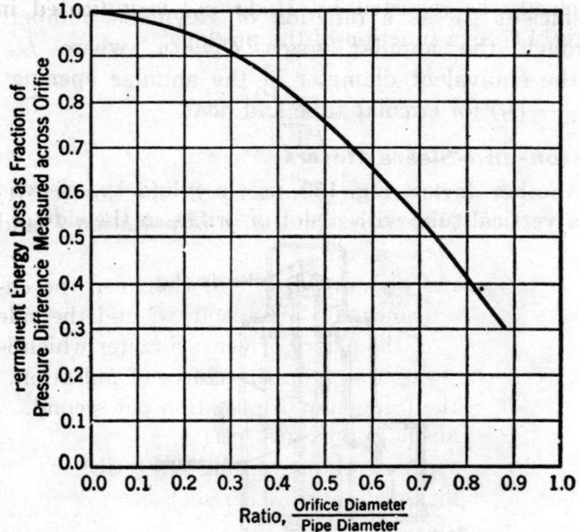

Fig 148. Permanent energy loss in sharp-edged orifice.

The permanent loss in a venturi tube is about one-tenth of the pressure difference between points 1 and 2.

AREA METERS

The venturi tube, flow nozzle, and orifice plate are known as "head" meters. Such meters indicate the quantity of flow by a change in pressure or head, the area of the constriction remaining constant. Inspection of equation 99 shows that the quantity of flow varies as the square root of the pressure difference when the area is constant. When operated over a wide range of flow rates, such an instrument is inaccurate at the low rates of flow. A flowmeter which operates under a constant pressure difference but with a constriction of variable area is known as an "area" meter. The quantity of fluid flowing per unit of time varies directly with the area in such an instrument.

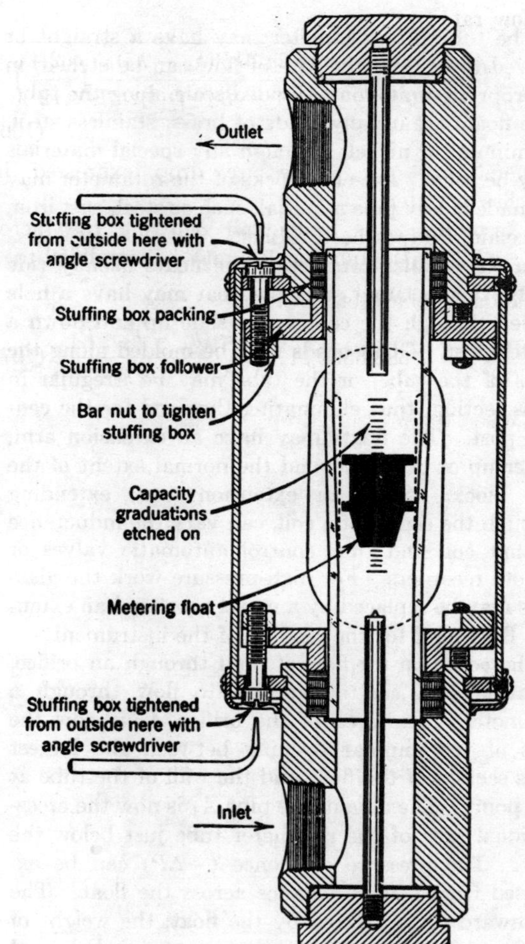

Fig. 149. Sectional drawing showing construction of rotameter. (*Fischer and Porter Co.*)

Rotameters

The rotameter, Fig. 149, consists of a tapered tube, usually glass, rigidly held in place between top and bottom blocks and enclosing a small float. The fluid to be metered enters through the bottom block and flows upward through the tube, leaving through the top block. In its passage upward through the tube, it lifts the float from the bottom stop to some point up in the glass tube. The float serves as a source of constant pressure difference, the amount being varied by varying the mass of the float. The variable area results from the taper in the glass tube which causes the annular space, through which the fluid must flow, to increase in size as the float is raised to a higher point in the tube.

The tube in a rotameter may have a straight or curved taper, and the rate of flow can be etched, in appropriate units, on a divided scale along the tube. The floats are usually made of brass, stainless steel, aluminum, or nickel, although any special materials may be used. The end blocks of the rotameter may be made of any pipe material, such as steel, cast iron, porcelain, lead, or hard rubber.

In the smaller rotameters the floats usually ride freely, but in larger sizes the float may have a hole drilled through the center and slide up and down a center post. Glass beads may be molded along the sides of the tube, or the tube may be irregular in cross section, thus eliminating the need for the central post. The floats may have an extension arm, either up or down, beyond the normal extent of the end block. Such an extension arm, extending through the center of a coil, can vary the inductance of that coil and thus control automatic valves or remote recorders. For high-pressure work the glass tube may be replaced by a metal tube, and an extension float used for the reading of the instrument.

The equation for flow of fluid through an orifice, equation 100, can be applied to flow through a rotameter. The area of the orifice A_o is now the area of the annular opening between the largest cross section of the float and the wall of the tube at any point. The area of the pipe A_1 is now the cross-sectional area of the rotameter tube just below the float. The pressure difference $(-\Delta P)$ can be expressed from a force balance across the float. The downward force exerted by the float, the weight of the float less the buoyant thrust upward, is balanced by the pressure difference across the float times the cross-sectional area of the float:

$$V_f(\rho_f - \rho)\frac{g}{g_c} = A_f(-\Delta P)$$

where V_f = volume of the float.
ρ_f = density of the float.
A_f = maximum cross-sectional area of the float.

Then

$$(-\Delta P) = \frac{V_f(\rho_f - \rho)g}{A_f g_c}$$

and, substituting in equation 100,

$$W = C_R A_o \sqrt{\frac{2g\rho(\rho_f - \rho)V_f}{A_f\left(1 - \frac{A_o^2}{A_1^2}\right)}}$$

Normally the ratio $(A_o/A_1)^2$ is quite small and the term $\left(1 - \frac{A_o^2}{A_1^2}\right)$ approaches a value of 1.0 and is not included in the equation, leaving

$$W = C_R A_o \sqrt{\frac{2g\rho(\rho_f - \rho)V_f}{A_f}}$$

The coefficient of discharge C_R is similar to its counterpart for orifices C_o in that it is sensitive to viscosity and to the flow lines through the constriction. The heavy lines in Fig. 145 show values of the coefficient C_R as a function of Reynolds number through the annular space $D_o v_o \rho/\mu$, where D_o is the equivalent diameter of the annular opening $(D_1 - D_f)$ for circular tube and float.

Piston-and-Sleeve Meters

Another device, Fig. 150, uses a cylindrical sleeve in a vertical tube with a slot or orifice in the side of

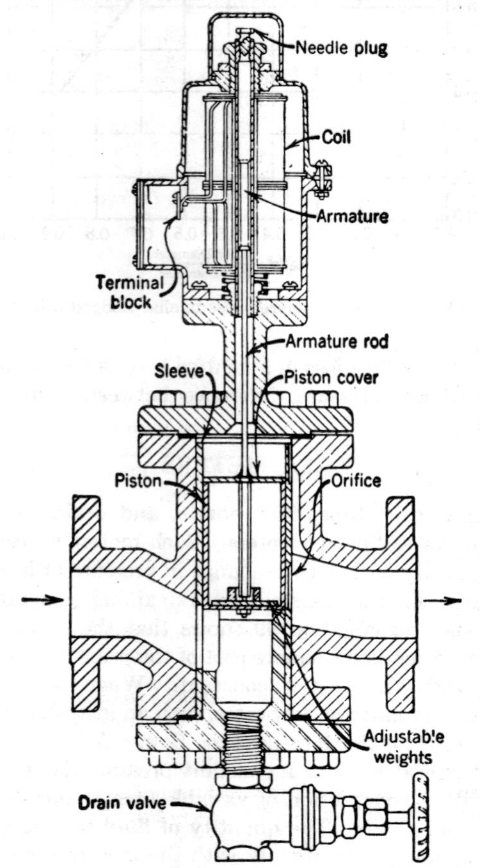

Fig. 150. Sectional drawing showing construction of piston and sleeve meter. (*Brown Instrument Co.*)

the tube. A piston slides in the cylinder with a close fit against the walls. As the fluid enters the bottom of the sleeve, the piston is lifted until enough orifice is uncovered for the fluid to escape. A variation in the rate of flow causes a variation in the level at which the piston will float in the stream, providing a method of measuring the rate of flow by reading the location of the piston. The parts are usually metal, and the piston carries an extension arm which indicates the location of the float by visual observation or by variation in the inductance of a coil surrounding the arm.

BIBLIOGRAPHY

1. AMERICAN SOCIETY OF MECHANICAL ENGINEERS, Special Committee on Fluid Meters, *Fluid Meters, Parts 1, 2, and 3*, New York (1931–1937).
2. BOWSER, INC., Fort Wayne, Ind., Catalogs.
3. BROWN INSTRUMENT CO., Division of Minneapolis-Honeywell Regulator Co., Philadelphia, Pa., Catalogs.
4. FISCHER AND PORTER CO., Hatboro, Pa., Catalogs. Section 98-A (1947).
5. HODGSON, J. L., *Trans. Am. Soc. Mech. Engrs.*, **51**: FSP-51-42, 303–332 (1929).
6. NEPTUNE METER CO., New York, Catalogs.
7. PITTSBURGH EQUITABLE DIVISION, Rockwell Mfg. Co., Pittsburgh, Pa., Catalogs.
8. TUVE, G. L., and R. E. SPRENKLE, *Instruments*, **6**, 201–206 (Nov. 1933).

PROBLEMS

1. A sharp-edged orifice meter is measuring gas flow under such conditions that the relative pressure drop is small (Re = about 10,000). What would be the quantitative effect on the reading of the mercury manometer used to indicate the orifice differential of (*a*) increasing the temperature of the gas from 70° to 100° F, (*b*) an increase in viscosity of 1.5 times, (*c*) doubling the molecular weight. Assume each change can occur independently of any others. The gas volume at standard conditions is to be assumed constant in all cases.

2. Specify the orifice diameter for measuring 100,000 standard cu ft (at 60° F and 1 atm) per hour of propane, flowing through a pipe 12 in. in diameter at a pressure of 10 psig and 80° F. The reading across the orifice should not be more than 5 in. of mercury.

3. Calculate the flow of oil in gallons per minute through a standard sharp-edged orifice ⅝ in. in diameter. The chambers to the orifice are four times the orifice diameter. The oil has a specific gravity of 0.85 and a viscosity of 11 centipoises. The pressure drop across the orifice is measured by a mercury manometer located underneath the orifice meter. The reading on the manometer is 9 in.

4. Oil of specific gravity 0.819 is flowing through a pipe with a constriction. A manometer with taps upstream and downstream of the constriction contains water with lines filled with oil. The manometer has a differential reading on the water of 300 mm. What is the loss in pressure across the constriction in foot-pounds per pound of oil? In pounds per square inch?

5. The manometer in problem 4 was broken and was replaced by another manometer having the upstream leg made of 6-mm glass tubing and the downstream leg made of 8-mm glass tubing. When no oil is flowing, the meniscus is 21 cm above the bottom of the manometer and 40 cm below the top of the manometer. What is the maximum decrease in pressure in pounds per square inch which can be measured? If the connections to the legs are reversed, what is the permissible decrease in pressure?

6. In order to determine the quantity of air flowing through a circular duct whose internal diameter is 20 in., pitot tube readings were taken along diameters at right angles to each other in a cross section of the duct. The data are presented in the table below.

The air was flowing at an absolute pressure of 735 mm of mercury and at a temperature of 250° F. The pitot tube coefficient was 0.98.

(*a*) Calculate the rate of flow of the air in cubic feet per minute, measured at 760 mm and 60° F.

(*b*) Construct a graph showing the variation of the ratio of local velocity to central velocity with the distance from the center of the duct.

Point Location, in. from center of duct	Pitot Tube Reading, in. of water
9.75	0.198
9.00	0.800
8.00	1.40
7.00	1.90
5.00	2.67
3.00	3.27
0.00	3.67
3.00	3.38
5.00	2.81
7.00	2.10
8.00	1.56
9.00	0.891
9.75	0.210
9.75	0.205
9.00	0.850
8.00	1.48
7.00	2.02
5.00	2.73
3.00	3.32
3.00	3.31
5.00	2.75
7.00	1.95
8.00	1.45
9.00	0.872
9.75	0.200

7. Air is flowing through a 12-in. circular duct. A pitot tube is inserted into the duct at the center point of the stream. The temperature of the air is 110° F, and the pressure is 1 atm

absolute. The pitot tube manometer is an oil-water type. The oil, specific gravity 0.835, is above the water, specific gravity 0.998, and contained in relatively large reservoirs compared to the tube. When the manometer reading is 2 in., what is the flow of air in cubic feet per minute at duct conditions?

8. Water is flowing in a horizontal pipe, high up on a mountain ($g = 28$ ft/sec²). An orifice with a mercury manometer is used to measure the flow. The manometer reading is 4.50 in. What is the pressure drop ($-\Delta P$) in pounds per square inch corresponding to this reading?

9. Water at 50° F is flowing through 3-in. schedule-40 pipe at a rate of 150 gpm.

(*a*) If a standard sharp-edged orifice 1¾ in. in diameter is inserted in the line, what would be the reading on a mercury manometer connected across the orifice?

(*b*) If a venturi meter with a throat 1¾ in. in diameter were used instead of the orifice, what would the manometer read?

(*c*) What would be the reading on each instrument if, instead of water, an oil were flowing through the pipe at the same rate, the oil having a density of 0.890 gm/cc and a viscosity of 1.30 centipoises?

10. A 2-in. schedule-40 pipe line carries water at 70° C. A sharp-edged orifice (1.077 in. in diameter) serves as the primary element of a flowmeter. The secondary element is a mercury-water manometer whose two branches have a difference in level of 22 in. at maximum flow.

The manometer is to be inverted and used as an air-water manometer, maintaining the 22-in. differential at maximum flow.

What new orifice diameter is needed?

11. Natural gas consisting of practically pure methane flows through a long straight 10-in. schedule-40 steel pipe in which is inserted a square-edged orifice 2.50 in. in diameter, with pressure taps 5.0 in. from the orifice plate.

Just above the orifice the gas is at 80° F and 5.0 psi. A differential-reading manometer inclined at an angle of 15 degrees with the horizontal, attached across the orifice, reads 6.18 in. of water.

What is the weight rate of flow of gas through this line?

12. A 2-in. schedule-40 steel pipe is carrying hot water at rates of flow varying from 50 to 150 gpm, and a sharp-edged orifice plate with flange taps is being used to meter it. The upstream pressure is 2 psig, and the upstream temperature is 160° F.

At flow rates of 100 gpm and over, however, the manometer across the orifice becomes unreadable because the hot water flashes into vapor at the vena contracta, using the present orifice plate.

What size orifice opening should be used to prevent this?

13. Oil with a density of 55 lb/cu ft and a viscosity of 48.5 lb/hr-ft is flowing through a standard 6-in. schedule-40 pipe at 4 fps.

Calculate:

(*a*) Reynolds number.

(*b*) Power required for a 5-mile pipe line taking 60 per cent overall efficiency.

(*c*) Reading on a mercury manometer measuring the drop across a 2.5-in. diameter sharp-edged orifice.

(*d*) Power requirements for the orifice.

14. A 0.75-in. orifice is installed in a vertical standard 2-in. pipe to measure the upward flow of benzene. With the lead lines supposedly filled with benzene and with mercury in the manometer, the following readings were obtained

Rate, gpm	Manometer reading, in.
30	52.56
15	13.51
5	2.10

After calibration was completed and the meter placed in service it happened that flow was stopped temporarily, and the manometer gave a reading of 0.69 in. as if a small quantity of benzene was flowing. The pressure at the upstream connection of the manometer was 40 psia throughout calibration and the temporary cessation. Bleeding benzene out of the downstream vertical lead line to the manometer disclosed an air pocket, which was entirely displaced and the meter restored to service. What is the present correct calibration?

15. A pump is delivering oil from tank *A* to tank *B* through the piping system shown in the figure, 4-in. pipe from tank *A* to pump and 3-in. pipe from pump to tank *B*.

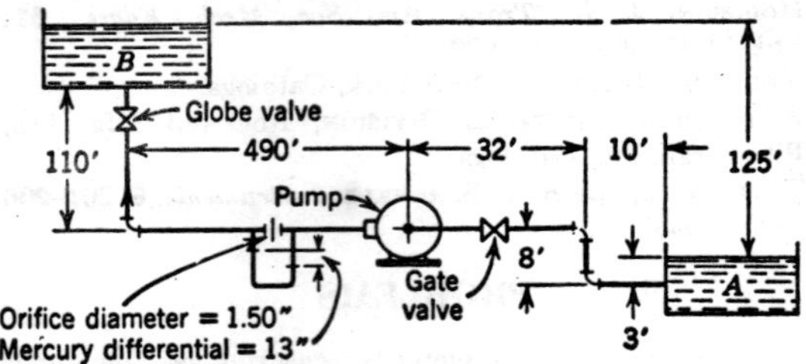

The oil at the pumping temperature has a specific gravity of 0.765 and a viscosity of 1.70 centipoises.

If the pump has a mechanical efficiency of 60 per cent, find the horsepower required to pump the oil.

16. Water at 55° F is flowing through the system shown in the figure. All lengths of pipe are for straight pipe only. Both globe valves are wide open.

Neglecting losses in the tee and assuming constant levels in the tanks, find the flow into tanks *A* and *B*.

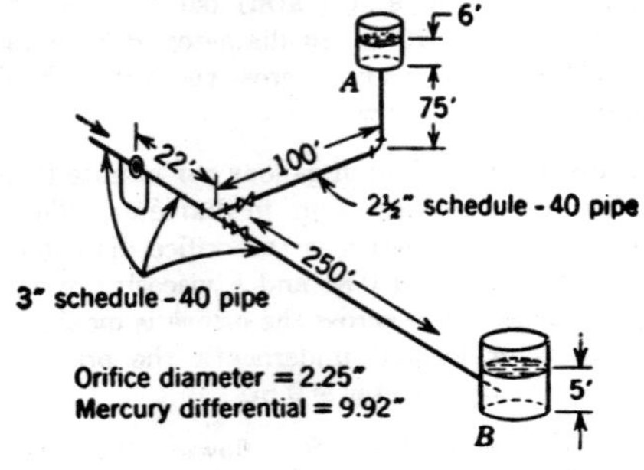

17. Water at 70° F (62.3 lb/cu ft, 0.982 centipoise) is wing in a horizontal 2-in. schedule-40 steel pipe at a rate 200 gpm. Five manometer taps are located along the pipe, 6-in. intervals. An open-end vertical water manometer the downstream tap (No. 5) gives a scale reading of 6 in. hat are the readings in inches and in pounds per square h gage of:

a) A similar manometer at the upstream tap (No. 1)?

b) A simple mercury U-tube manometer connected be- een taps 1 and 4?

c) An inverted U-tube using air as the second fluid con- cted between taps 1 and 5?

d) A manometer using an immiscible fluid with a density 65.2 lb/cu ft connected between taps 2 and 3?

8. The system shown in the figure is carrying water at F from tank A to tank B and through the line leading to C. The line from A to the pump is 3-in. pipe, and the remainder is 2-in. pipe, all schedule 40.

The two valves at the tee are gate valves; all others are globe valves. The elbows and tee are standard.

What is the horsepower to be delivered by the pump?

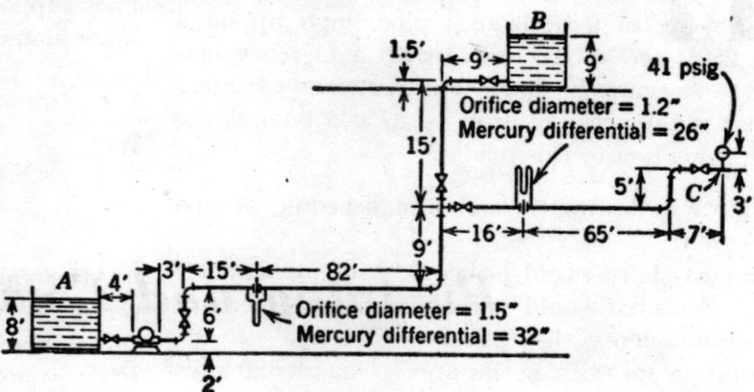

CHAPTER 14

Pumping and Compressing

A MACHINE that does work on a flowing fluid is called a pump, blower, compressor, etc. The quantity of work is represented by the symbol $-w$ in the flow equations.

Machines that remove work from the flowing fluid, w, are known as engines, turbines, water wheels, windmills, etc.

All pumps, blowers, or compressors are rated in terms of four characteristics.

1. *Capacity*, the quantity of fluid discharged per unit time.
2. *Increase in pressure*, frequently reported for pumps as *head*. Head is the energy supplied to the fluid per unit weight and is obtained by dividing the increase in pressure by the fluid specific weight.
3. *Power*, the energy consumed by the machine per unit time.
4. *Efficiency*, the energy supplied to the fluid divided by the energy supplied to the machine.

The effect of most pumping devices is to increase the pressure of the fluid. But some deliver the fluid with an increase in kinetic energy or an increase in elevation.

The majority of all pumps, blowers, and compressors may be classified as reciprocating, rotary, or centrifugal. Reciprocating and rotary pumps do not permit free flow of fluid through the pump except for leakage past close-fitting parts, and are called "positive displacement" pumps.

RECIPROCATING PUMPS AND COMPRESSORS

Reciprocating pumps develop a higher pressure by the direct action of a piston or plunger on the fluid confined in a cylinder, forcing the high-pressure fluid through the discharge valves. The piston or plunger may be actuated directly by a steam-driven piston or by a rotating crankshaft through a crosshead.

Steam-Driven Piston Pump. Figure 151 shows a horizontal steam-driven double-acting pump. The operation of such a pump may be explained in connection with Fig. 152. All valves are shown closed in the drawing. As the liquid piston moves to the right, the high pressure exerted by the liquid in the space ahead of the piston lifts the right-hand discharge valve, allowing the liquid to escape into the discharge chamber. This motion of the piston relieves the pressure on the liquid behind the piston. The pressure of the liquid feed then lifts the left-hand suction valve, and feed liquid enters the cylinder.

As the piston reaches the end of its stroke, it actuates the valve rod, moving the slide valve in the steam chest, reversing the direction of motion of the piston. If only one set of valves is provided, opening into the space on only one side of the piston, the pump would be single-acting and would discharge only when stroking in one direction. The force of steam pushing the steam piston drives the pump. The steam piston may be controlled by the simple slide valve shown in Fig. 152 or by a steam-actuated valve gear as shown in Fig. 153. In either case lost motion is provided in the adjustment of the push rod so that the valve moves only as the piston approaches the end of its stroke and then so as to admit steam ahead of the advancing steam piston to act as a cushion to prevent the piston from hammering the cylinder head. Such pumps use steam at full pressure throughout the stroke and allow no significant expansion of steam in the cylinder.

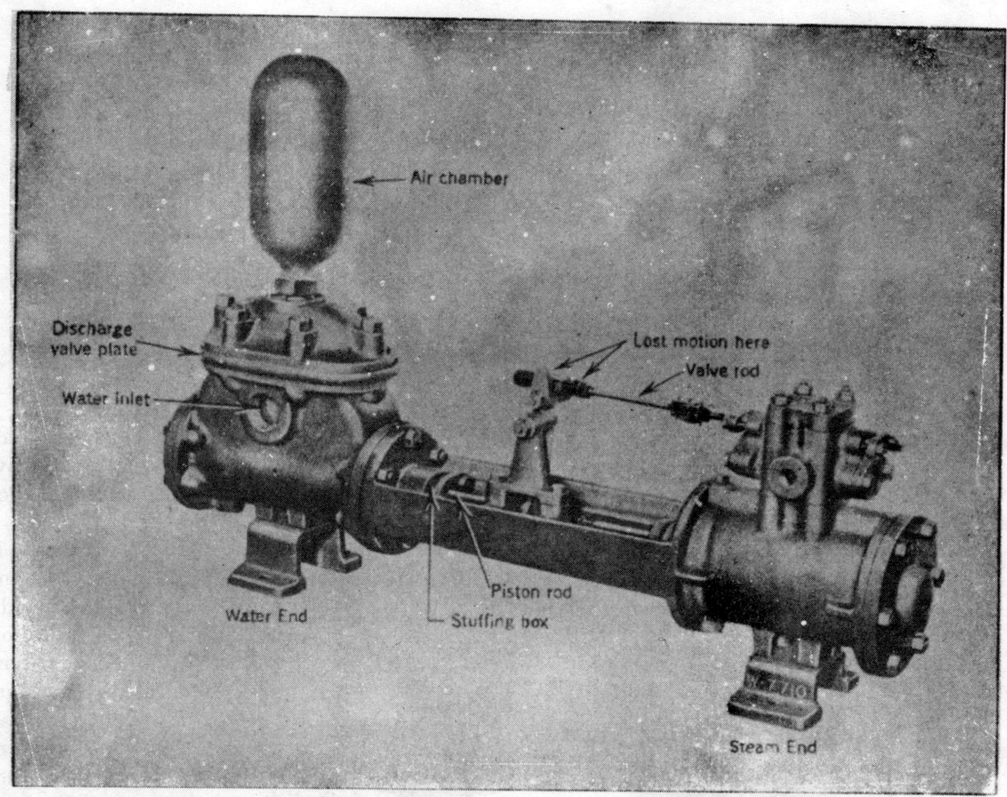

FIG. 151. Horizontal steam-driven single piston pump. (*Worthington Pump and Machinery Corp.*)

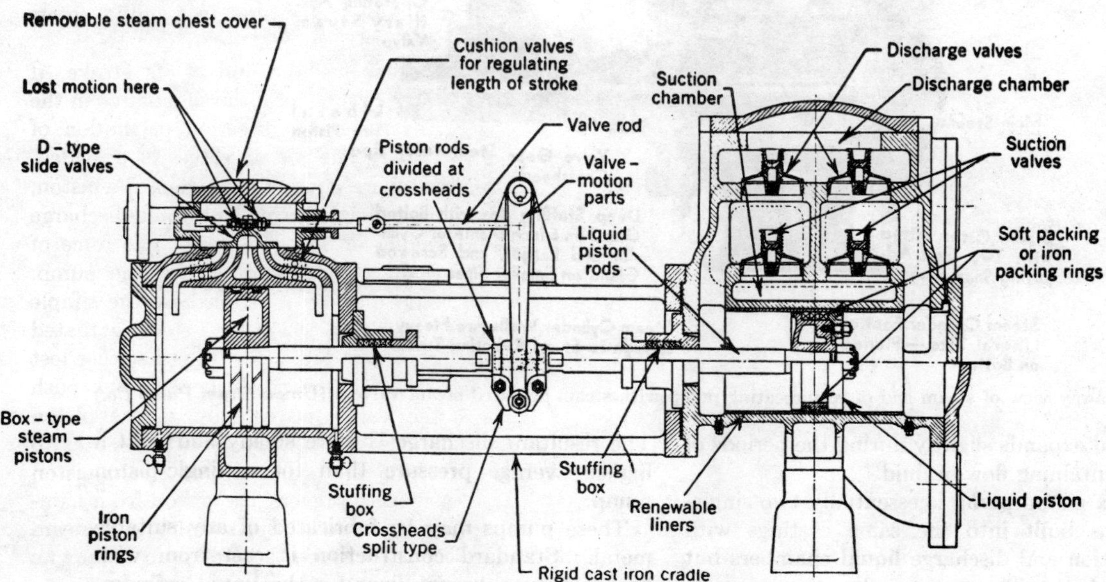

FIG. 152. Sectional view of steam-driven reciprocating pump. (*Worthington Pump and Machinery Corp.*)

The piston comes to a full stop at the end of each stroke. The resulting pulsations in flow and pressure are reduced by the air chamber on the discharge line (Fig. 151). The air trapped in this chamber is compressed during the peak delivery periods of side so timed that at midstroke of one piston steam is admitted ahead of the other piston. This adjustment provides that one piston is in midstroke delivering at its maximum rate when the other piston is at the end of its stroke and not delivering.

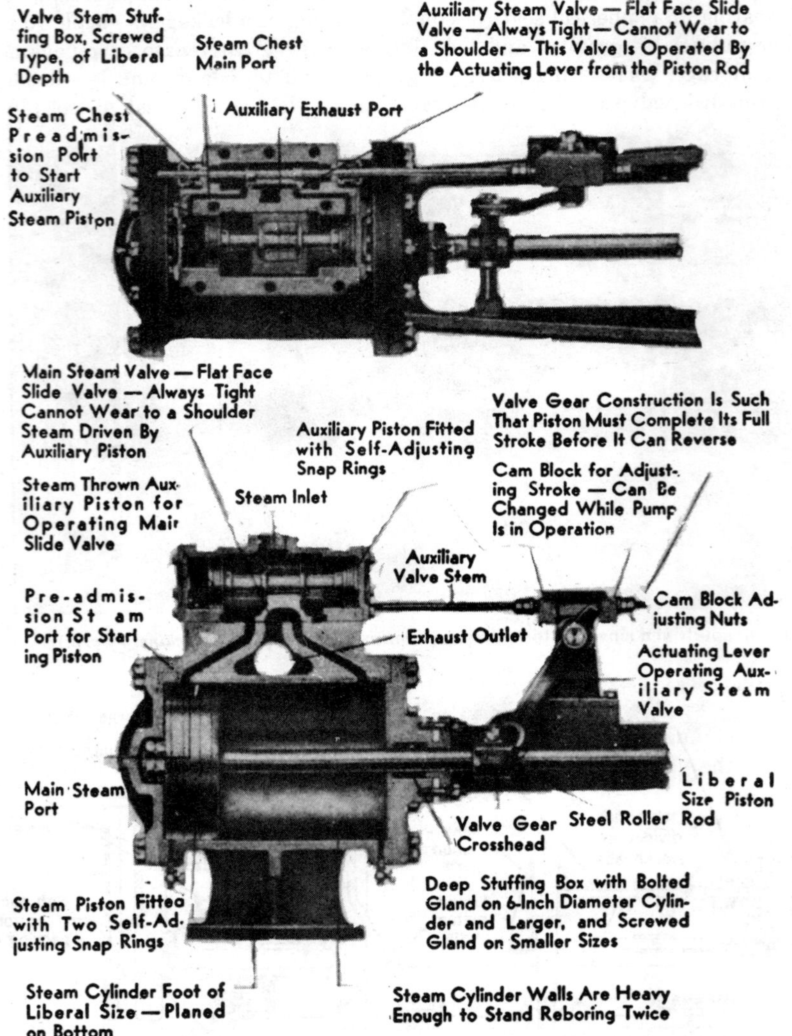

Fig. 153. Cutaway view of steam end of reciprocating pump with steam actuated steam valve. (*Union Steam Pump Co.*)

the pump and expands slightly during the periods of reversal, maintaining flow of fluid.

The duplex piston pump is essentially two single-piston pumps built into the same castings with common suction and discharge liquid chambers but individual valves. The steam valves for one side are actuated by levers on the piston rod of the other The resultant discharge is more steady and is at a higher average pressure than for a single-piston pump.

These pumps may be fabricated of any suitable metal. Standard construction is cast iron, with bronze valves, bronze liner for the liquid cylinder, and steel liquid piston and piston rod. The liner is replaceable if solid particles in the fluid cause scoring.

The fluid valves are generally disks, balls, or wing valves. The disk valve (Fig. 152) is a flat disk of rubber, bronze, or steel, seating on a flat seat, guided by a stem screwed into the valve seat, and held tightly by a spring. The ball valve is a sphere of rubber or metal, seating in a spherical seat (of equal or slightly smaller radius) by gravity, and held in position by a metal cage. The wing valve (Fig. 154) is similar to the disk valve but has guides cast into the disk and extending through the valve seat. It is preferred when the valves do not operate in a vertical position.

The dimensions of the cylinders of reciprocating pumps are usually expressed in the abbreviated form as 12 × 8 × 10, indicating, respectively, the diameter of the steam cylinder, the diameter of the "water" cylinder, and the length of the stroke in inches.

Steam-Driven Plunger Pump. Liquid pistons (Figs. 151 and 152) carry packing to reduce leakage of liquid from the high-pressure to the low-pressure side of the piston. When this packing becomes worn or loose it can be tightened or replaced only by removing the head of the pump, thus putting the pump out of service. In the plunger pump the single-acting plungers work through packing in the heads of the cylinder, using two outside packings instead of one outside and one inside packing. Two opposed plungers (Fig. 154) must be used to give delivery corresponding to a single double-acting piston. The far plunger in Fig. 154 is driven by rods (shown broken) extending around the pump from the crosshead.

Plunger pumps are used for higher delivery pressures than piston pumps of equal size. With outside packing any leakage is visible to the operator. This and the greater ease of repacking results in much better maintenance and less leakage. These factors offset the higher initial cost of plunger pumps.

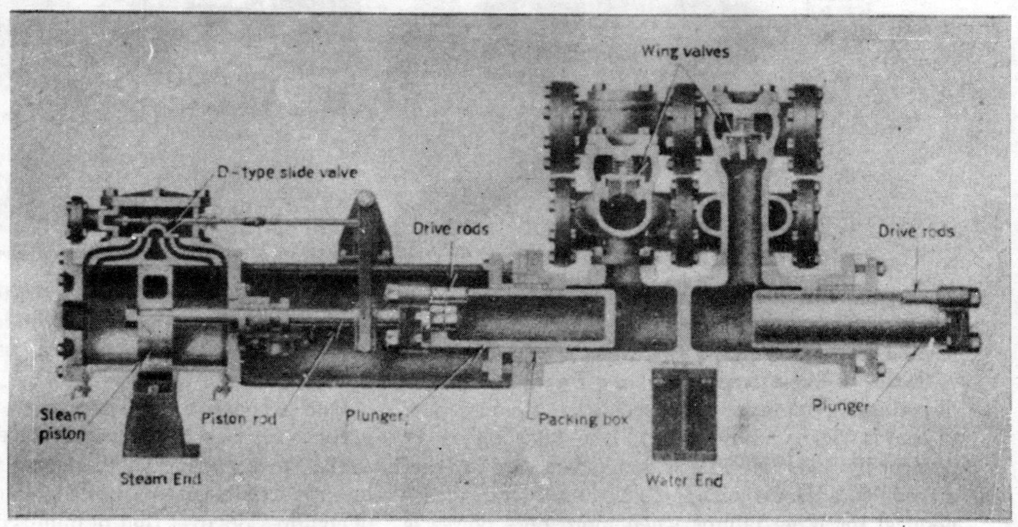

Fig. 154. Longitudinal section of steam-driven duplex plunger pump. *(Worthington Pump and Machinery Corp.)*

Plungers and pumps are frequently constructed of steel, and when special conditions must be met alloys and even porcelain plungers are used.

The proper selection and maintenance of packing is one of the major factors in efficient pump operation. Asbestos rope impregnated with oil and graphite is widely used for packing both steam and fluid stuffing boxes. For higher temperatures and when the fluid acts as a solvent for oil, metallic packing is more satisfactory. This packing is constructed of lead or aluminum foil twisted and formed into split rings. Such packing is available with or without graphite. For special purposes other materials are used, such as leather, rubber, and ropes of various fibers.

The steam pistons are sealed by oil-lubricated piston rings instead of packing. Because the steam vapor is practically free from all suspended solids, the clearances between piston rings and cylinder

FIG. 155. Longitudinal section of crank and flywheel compressor. (*Pennsylvania Pump and Compressor Co.*)

walls can be made very small and sealed with oil. When pumping clean liquid, the liquid pistons can also be sealed by rings.

Power-Driven Reciprocating Pumps. In the direct-acting pumps described, the steam-driven piston is rigidly connected to the fluid piston or plunger by a piston rod. A crankshaft and flywheel driven by a motor or engine is usually more economical for large pumps when more horsepower is required. The flywheel smooths the operation of the pump. The crankshaft permits power to be delivered through a rotating shaft, producing simple harmonic motion of the piston, and through the connecting rod linkage controls the pump stroke. It also permits multiple fluid cylinders which increase the discharge pressure and give less deviation of the flow rate.

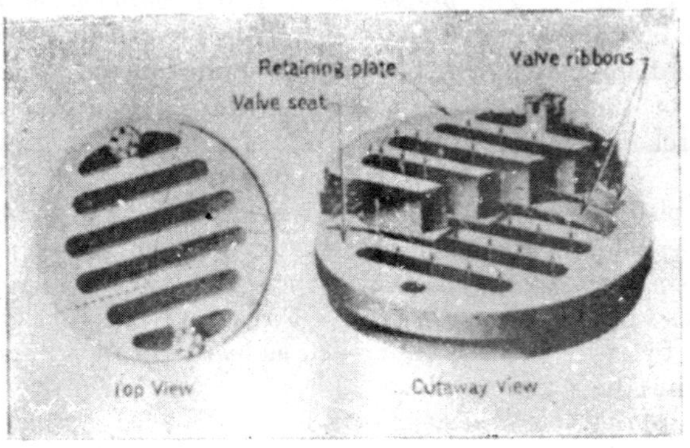

FIG. 156. Cutaway view of ribbon-type compressor valve. (*Worthington Pump and Machinery Corp.*)

Reciprocating Compressors. Compressors for gases, both direct-acting and power-driven, may be similar in construction to pumps for liquids. The lower inertia of gases permits an increase in the speed of compressors over that of pumps of similar physical dimensions. This places the steam-driven direct-acting compressor at a disadvantage and increases the desirability of power drive.

A double-acting crank-and-flywheel compressor is shown in Fig. 155. It can be driven by any engine or motor. The light valves (Fig. 156) of the ribbon type respond very quickly. The passage consists of circular or straight slots. The valves consist of light metal strips in guide grooves. They are closed by gas movement or light springs and opened by gas passing through them.

A double-acting compressor driven by a V-type two-cycle gas engine is shown in Fig. 157. Four pistons are connected to the same connecting-rod bearing on the crankshaft, the two engine pistons, the compressor piston, and the scavenging piston for air supply to the engine, (mounted on the crosshead in the compressor linkage). Such compressors are available with multiple compressor cylinders, discharging at different instants in the cycle, reducing

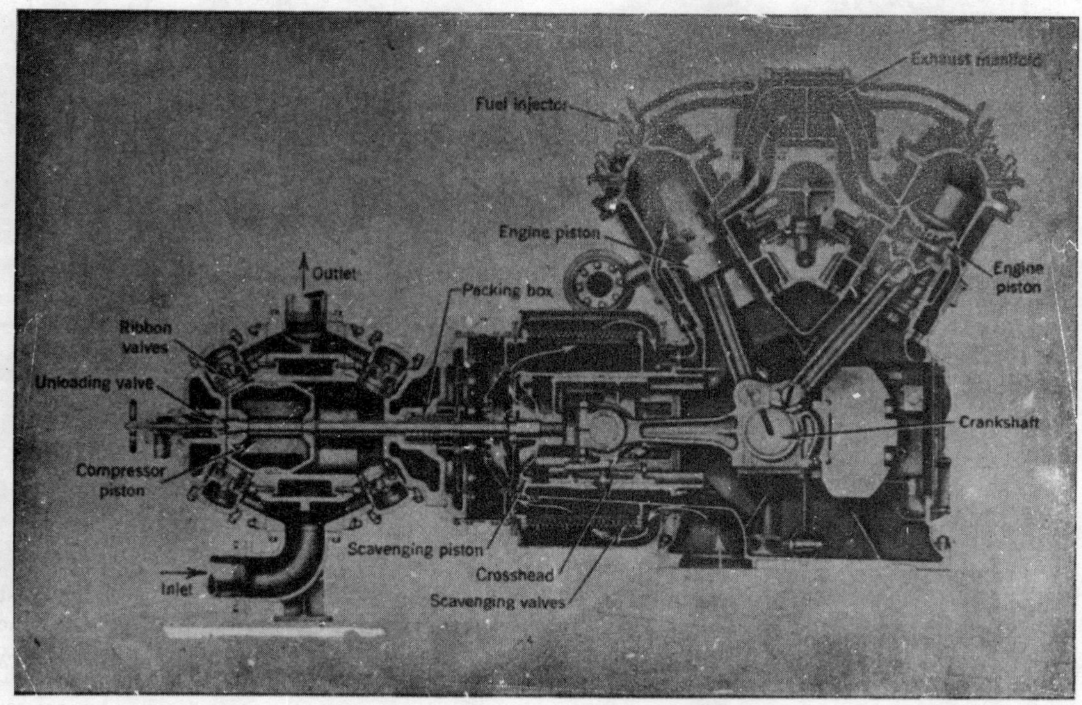

Fig. 157. Cutaway view of compressor direct-driven by V-type internal combustion engine. (*Cooper-Bessemer Corp.*)

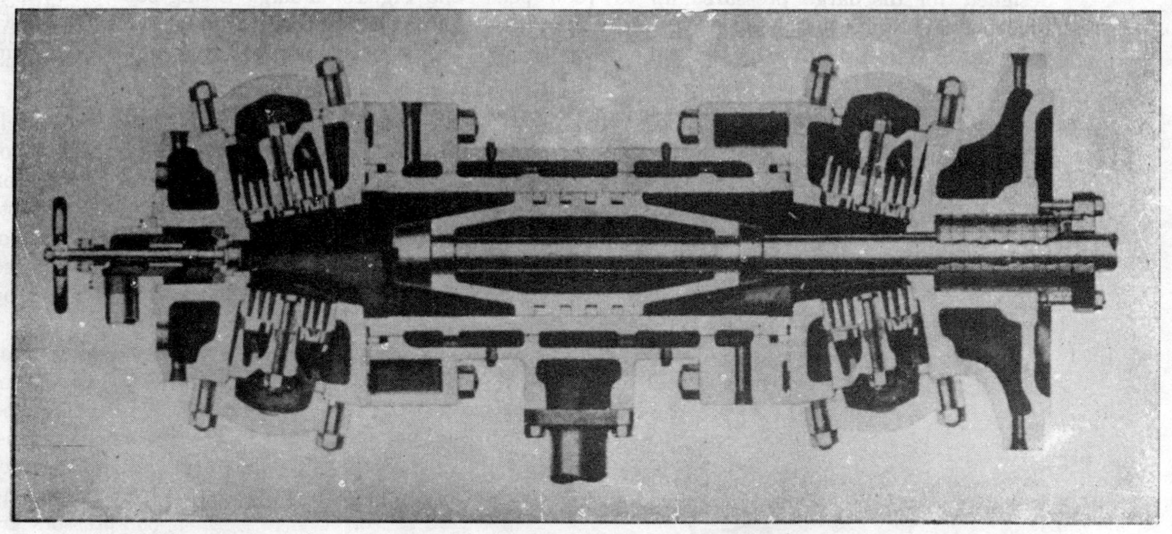

Fig. 158. Cutaway view of compressor cylinder for pressures up to 2000 psi. (*Cooper-Bessemer Corp.*)

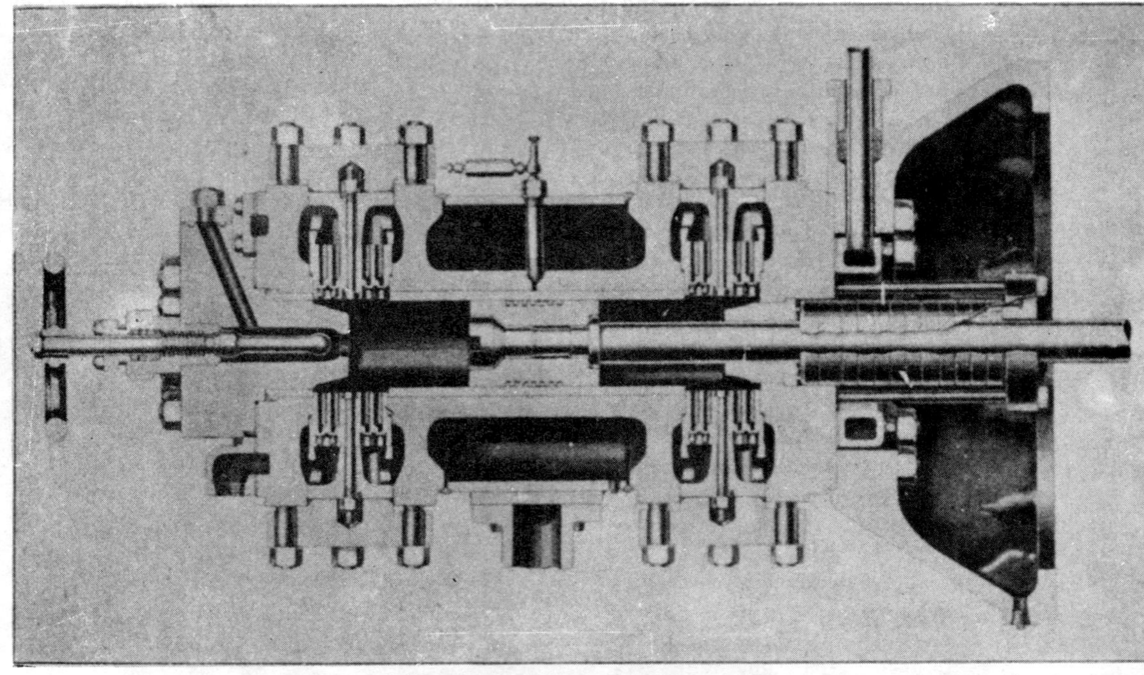

Fig. 159. Cutaway view of compressor cylinder for pressures up to 4500 psi. (*Cooper-Bessemer Corp.*)

the amplitude but increasing the number of pulsations per revolution.

The compressor cylinder shown in Fig. 157 is designed for discharge pressures up to 1000 psi. Figure 158 shows a double-acting cylinder for pressures up to 2000 psi, utilizing the same design principles in a more rugged construction. For st higher pressures, forged-steel construction is use Figure 159 shows a double-acting 4500-psi cylinde and Fig. 160 a single-acting 6000-psi cylinder.

A vacuum pump is identical with a low-pressu compressor (compressing from less than atmosphe

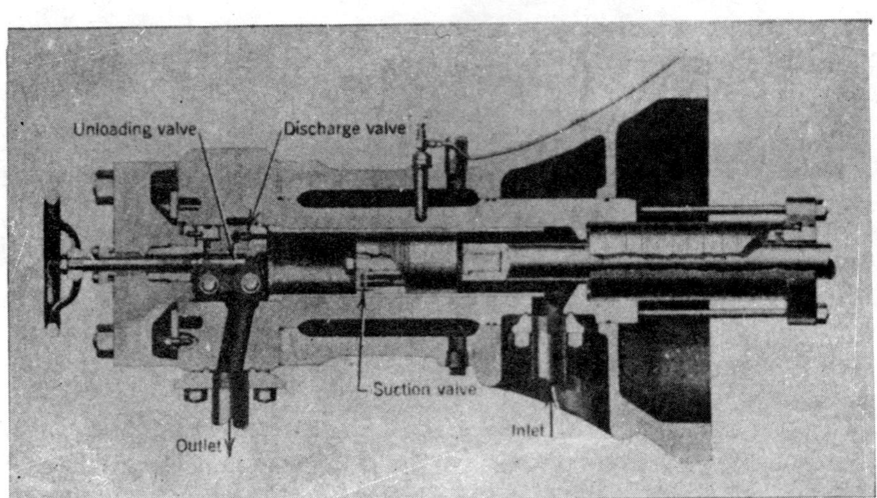

Fig. 160. Cutaway view of compressor cylinder for pressures up to 6000 psi. (*Cooper-Bessemer Corp.*)

to atmospheric pressure), and they may be interchanged in some applications. The vacuum pump, however, is forced to handle larger volumes of gas per unit time when on its designed pressure, and it has a larger ratio of piston diameter to stroke than is usual for compressors working in the pressure range from atmospheric pressure to 500 psi.

Operating Features of Reciprocating Pumps and Compressors

The reciprocating machines deliver a constant volume of fluid against a wide range of pressures, especially if power-driven, and can be built to operate against high pressures. They are self-priming for service on liquids. The valves prevent satisfactory operation on viscous liquids, and the close clearances cause maintenance problems, especially when pumping fluids containing suspended solids.

A relief valve and by-pass lines should be provided to prevent damage to the pump castings upon inadvertent closing of the shut-off valve in the discharge line. Many pumps and compressors have built-in by-pass lines and valve "unloaders," which hold open the suction valves so long as a certain pressure is exceeded in the discharge fluid chamber.

Maintenance problems and the time needed for valve action in pumps limit their speed of operation. The recommended speeds for cold liquids of kinematic viscosity less than 50 centistokes are given in Fig. 161 for single and duplex steam-driven and single, duplex, and triplex power-driven pumps. These speeds should be reduced for liquids of higher viscosity, using 80 per cent of the recommended speeds for liquids with kinematic viscosities of 400 centistokes.

The efficiency of reciprocating pumps and compressors is properly the work done on the fluid divided by the work done on the pump. For a pump driven by an electric motor this definition of efficiency is generally applied to the combined motor and pump, treated as a single unit, in which case the overall efficiency is the work done on the fluid divided by the electrical energy supplied to the motor.

This leads to the attempt to determine the efficiency of steam-driven pumps by dividing the work done on the fluid by the "total energy absorbed," which is sometimes misinterpreted as the energy of the steam supplied or again as the decrease in energy content of the steam as it passes through the steam end. Such calculations may give results varying almost from zero to infinity.

It has been customary to treat some of the individual factors contributing to inefficiency as if they were efficiencies themselves. For example, the steam pressure required for a steam-driven direct-acting pump may be computed as that pressure necessary on the steam piston to counterbalance the discharge pressure of the fluid being pumped. The ratio of

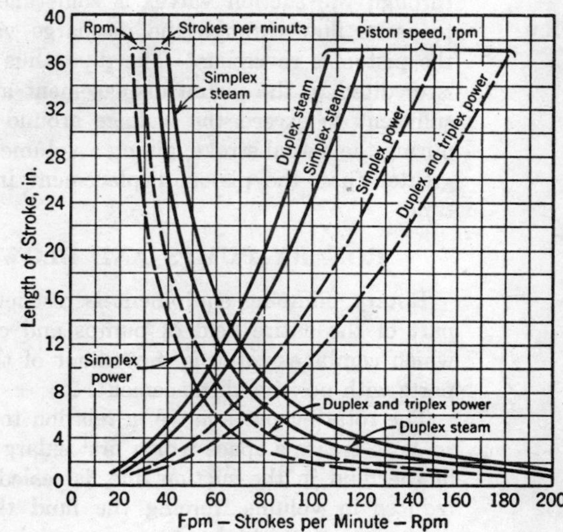

FIG. 161. Recommended speeds for reciprocating pumps for cold liquids with viscosities less than 50 centistokes. (*Hydraulic Institute.*).

this computed pressure to the actual pressure required to operate the pump at the rated delivery and pressure is called the "steam pressure ratio," or sometimes the "steam-end efficiency." This ratio actually indicates the sliding friction of the pistons and rod, the momentum changes to accelerate the pistons and fluids, and the "slip" or leakage of fluid past the piston from the discharge side to the suction side, all in a single term.

Another so-called efficiency is the "water-end" or "volumetric" efficiency, which is supposed to be the actual leakage past the piston. This may be determined by closing the shutoff valve on the discharge of the pump, throttling the steam supply until the discharge pressure is just maintained, and noting the speed of the piston. The volume of leakage per unit of time is the product of the speed and the cross-sectional area of the piston. Another method is to measure the piston speed under operating conditions and compute the volume of fluid which should be delivered if no leakage occurred. The percentage of this volume which is actually delivered is called the "volumetric efficiency." Usually this volumetric efficiency fails to agree with the water-end efficiency computed for the same speed as (piston displacement leakage) ÷ (piston displacement). A volumetric efficiency of greater than 100 per cent (or a negative leakage) may be computed.

The momentum of the fluid being drawn in through the suction valves is sometimes sufficient to cause flow through the discharge valves while the piston is reversing. The fluid thus discharged is credited to the piston displacement and may be sufficient to exceed the leakage around the piston during the actual stroke, giving a volume "pumped" greater than the piston displacement in the same time.

ROTARY PUMPS AND BLOWERS

Rotary pumping equipment is restricted to that part of the entire field of pumps and compressors which combines rotating movement of the working parts with positive displacement.

The rotating parts move in relation to the casing so as to create a space which first enlarges, drawing in the fluid in the suction line, is sealed, and then reduces in volume, forcing the fluid through the discharge port at the necessary higher pressure. Valves are not used in rotary pumps. Small clearances must be maintained for efficient operation.

Rotary pumps and blowers may be divided into five main types according to the character of the rotating parts, as gear, lobe, screw, vane, or cam. But considerable overlapping occurs in specific cases.

Gear Pumps. If two ordinary spur gears are confined by a casing which fits so snugly as to seal effectively the spaces between each pair of adjacent teeth, as shown in Fig. 162, then fluid entering the pump through the suction line is trapped, carried around the outside by the gears, and discharged. One gear is driven by the driver, and the second gear is driven by the first gear. In some pumps very small holes are drilled radially through the teeth of

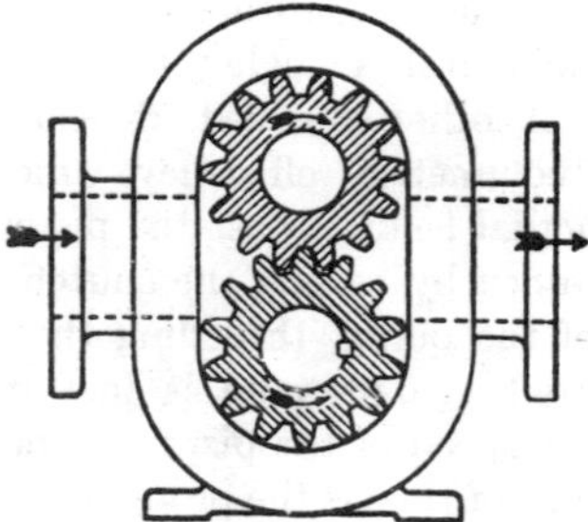

FIG. 162. Cross-sectional drawing of external gear pump. (*Hydraulic Institute.*)

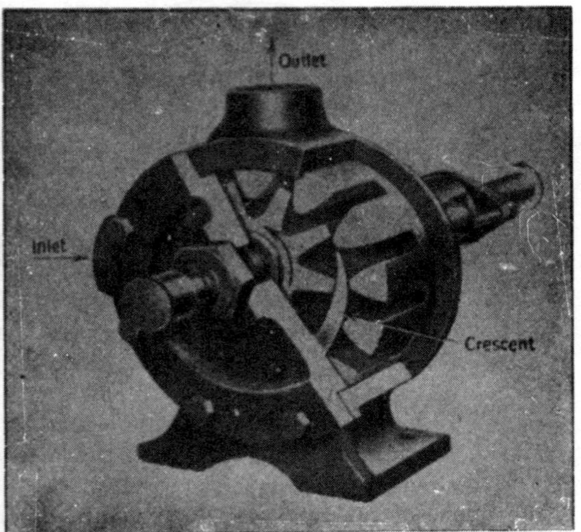

FIG. 163. **Cutaway view of internal gear pump.** (*Fairbanks, Morse and Co.*)

the idler gear, opening into a groove in the idler shaft which discharges back into the suction side of the pump. If fluid is trapped and compressed by the meshing teeth it can then escape back into the suction, preventing strain on the gears and shafts by the pressures created. The groove in the shaft is placed to prevent any by-passing of fluid from the discharge side through the holes to the suction side. Gear pumps may have herringbone teeth, helical teeth, or a combination of these.

Internal gear pumps are of two main kinds. In both kinds a modified spur gear rotates inside a larger gear rotating around an axis parallel to that of the spur gear and displaced sufficiently to mesh snugly at one point on the periphery. Figure 163 shows a common type which is sealed between suction and discharge by a "crescent." Fluid is trapped between the gears, crescent, and casing and swept into the high-pressure discharge port. Either gear may be driven with the other idling.

Figure 164 shows an internal gear pump whose two gears maintain a continuous series of sliding seals with each tooth, forming the pockets entirely between the gears. This type of pump may have any number of teeth in the spur gear, provided one more socket is provided in the ring gear.

Some pumps with the appearance of gear pumps, but with both "gears" driven and using the "teeth"

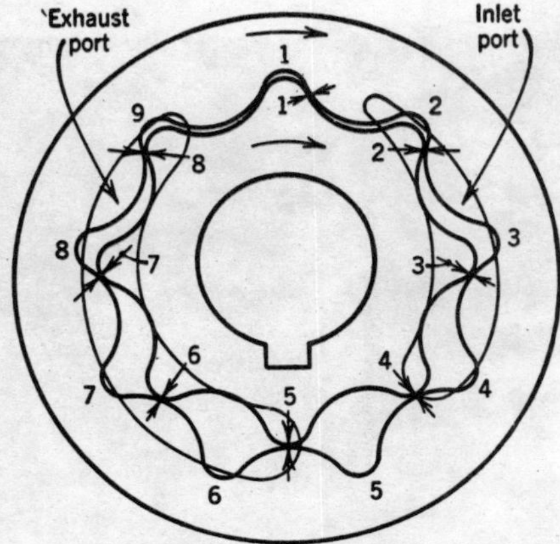

FIG. 164. Diagram of internal gear pump with sliding seal. (*Gerotor Pump Division, May Oil Burner Corp.*)

to form the pump pockets and seal the line of contact between the rotating members, are versions of lobe pumps.

Lobe Pumps. A lobe pump or blower has the general construction of a gear pump with fewer teeth, but the two impellers are driven separately through external gearing. This makes it possible to avoid actual contact of each impeller with the other and with the casing. The clearances are a few thousandths of an inch, sufficient to reduce friction and wear but to maintain minimum leakage from the discharge to the suction side.

FIG. 165. Cutaway view of two-lobe blower. (*Roots-Connersville Blower Corp.*)

A two-lobe blower is shown in Fig. 165. The shape of the lobes of the impeller is not restricted to the particular pattern shown, but no great variety of designs exists.

A three-lobe pump is seen in cross section in Fig. 166. The principle of operation is the same. More

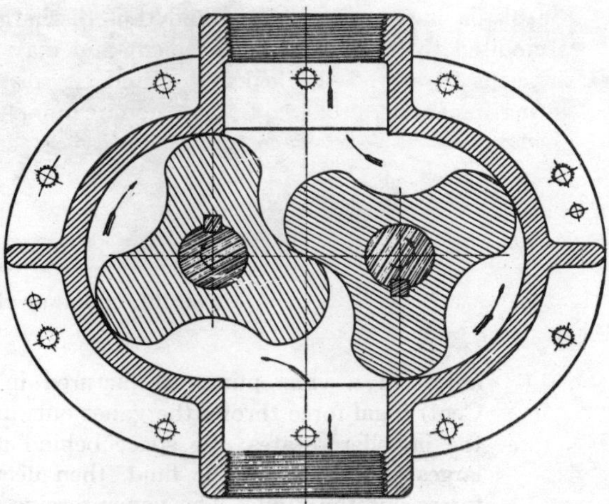

FIG. 166. Sectional diagram of the three-lobe pump. (*Roots-Connersville Blower Corp.*)

lobes may be used in each impeller, and the intricacy of contour may be considerable. The impellers may or may not touch one another and the walls, but if a sliding contact is maintained on the casing the wear may be absorbed in replaceable inserts mounted in the tips of the lobes.

Screw Pumps. One type of screw pump is shown in Fig. 167. The central or power rotor meshes with the two idler rotors, forming fluid-tight closures in the bored housings. Fluid enters at each end of the rotors, is sealed between the rotors and housing, and is forced smoothly to the central discharge. This construction provides almost perfect hydraulic balance. The single stuffing box may be sealed to control leakage. Other types of screw pumps include two screws or one screw, but the principle of operation is similar.

Vane Pumps. A circular disk, fitted with vanes sliding in slots in the disk and maintaining a sliding contact on the elliptical casing, constitutes the mech-

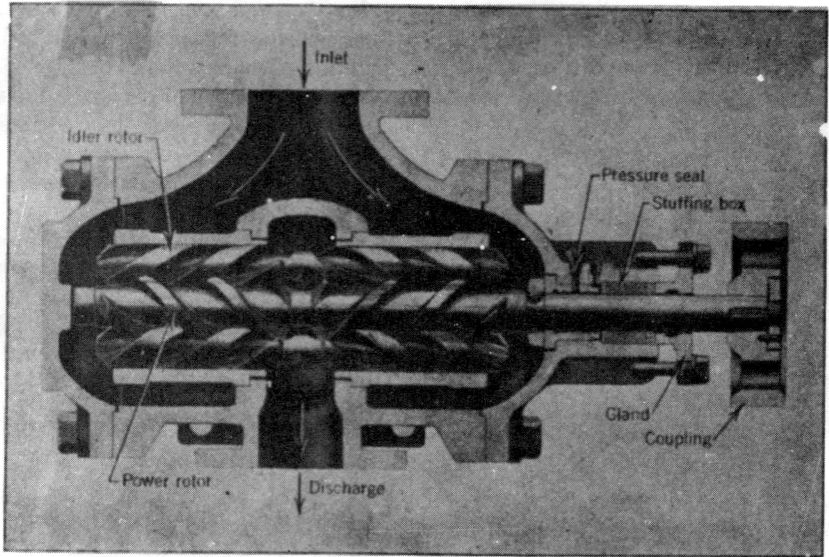

FIG. 167. Cutaway view of screw pump. (*De Laval Steam Turbine Co.*)

anism of a vane pump as pictured in Fig. 168. Centrifugal force throws the vanes outward, and, as the impeller rotates, the space behind a vane enlarges and draws in the fluid, then decreases and forces out the fluid. The vanes receive practically all the wear and are readily replaced. A variation of this idea is the vane shape and attachment of Fig. 169. The vanes swing outward by centrifugal force and seal the trapped fluid. As these vanes wear down, the change in dimension is automatically compensated until the time when the seal is broken by the decrease in size, after which new vanes can be inserted without special tools.

The sliding contact may be avoided by using guided rollers instead of guided vanes. The action is identical to that in Fig. 168, except that the contact with the casing is a rolling action, reducing wear and friction.

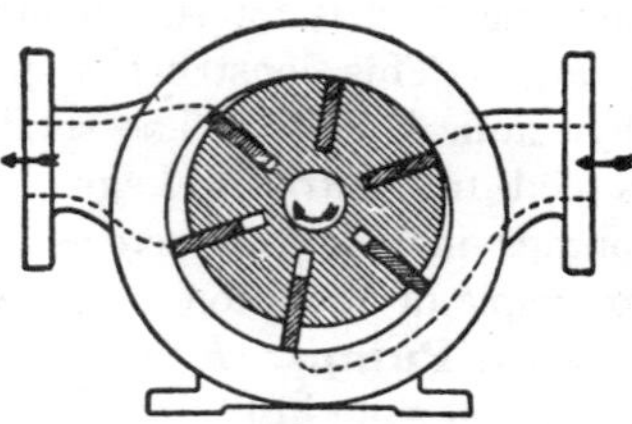

FIG. 168. Diagram of sliding-vane pump. (*Hydraulic Institute.*)

Cam Pumps include a circular rotor mounte eccentrically to sweep a circle whose radius is t sum of the radius of the rotor and the eccentricit Figures 170 and 171 show such a pump, with plunger valve, at the beginning of a stroke and ne the end of a stroke. The circular cam is rigidly fix to the shaft and is inside the rotor ring which is fr to rotate about the cam and which is integral wi the plunger. The plunger slides freely through slide pin to act as the discharge valve. Flow throug the pump is indicated by arrows. The plunger ma be replaced by any form of vane which seals t

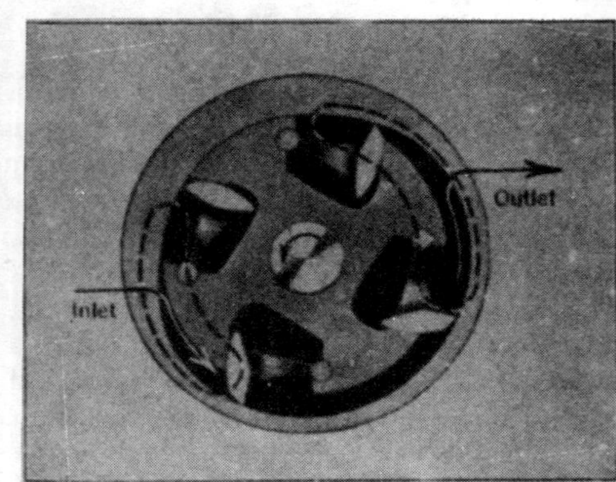

FIG. 169. Cutaway view of end view of vane pump wi bucket-vanes. (*Blackmer Pump Co.*)

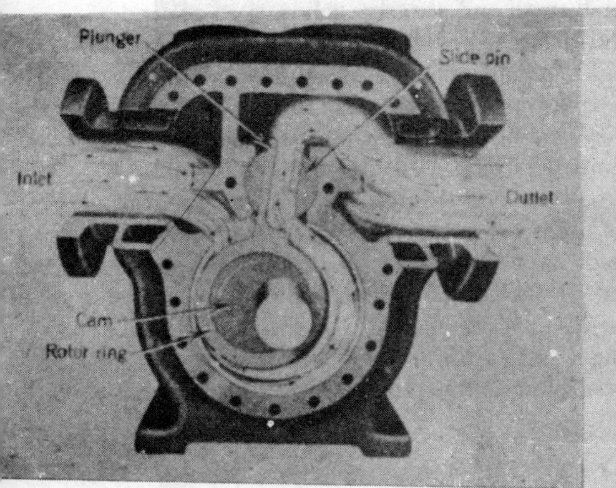

ɢ. 170. Cutaway view of cam pump indicating operation at beginning of stroke. (*Kinney Mfg. Co.*)

Fɪɢ. 171. Cutaway view of cam pump indicating operation approaching end of stroke. (*Kinney Mfg. Co.*)

ıction line from the discharge line. Except for the ·tation of the cam inside the rotor, the contacts ·tween surfaces are almost devoid of friction and ear.

perating Features of Rotary Pumps and Blowers

These pumps and blowers are perfectly adapted · direct connection with the motor. They are sually small enough and balanced well enough not · require a flywheel. The capacity is almost con-ant within the limits of speed of the driving ıechanism for all discharge pressures up to that hich will destroy the pump or stop the driver. otary pumps are made for discharge pressures as igh as 5000 psi, above which the torque drive is ·ss practical than the reciprocating machine.

Rotary machines are capable of operating with ·w absolute suction pressures and require no prim-ıg. The discharge ranges from lightly pulsating in ıe two-lobe blower and some cam pumps, to quite ·eady in the gear pump and vane pump. Contain-ıg no valves, as used in reciprocating machines, all otary pumps are well suited to pumping liquids of igh viscosity.

The extremely close clearances necessary in all otary machines require precise machining and in-rease the cost as well as maintenance problem. 'hese small clearances and frequent rubbing of the ıoving parts preclude pumping of fluids with no ıbricating properties or fluids containing solids in ·uspension. The small size, the high discharge pressures, and the simplicity of construction of rotary pumps give them a wide range of application.

CENTRIFUGAL PUMPS AND COMPRESSORS

Basically, the centrifugal machine is built around an impeller, which is a series of radial vanes of various shapes and curvatures, spinning in a circular casing. Fluid enters at the "eye" or axis of rotation and discharges more or less radially into a peripheral chamber at a higher pressure corresponding to the sum of the centrifugal force of rotation and the kinetic energy given to the fluid by the vanes.

The only moving part in the pump is the impeller. The vanes of the impeller extend from the center of rotation to the periphery, and the shrouds are the disks on each side of the vanes enclosing them. The vanes may be radial, may curve slightly "forward" (in the same direction as that of rotation), or may curve "backward," the usual case.

The common impellers for pumps are shown in Figs. 172–175. Figure 172 shows a high-head close-clearance impeller with narrow passages, six vanes, and a streamlined "eye" or entrance at the axis. This impeller is for use with clear, noncorrosive liquids. Figure 173 shows a similar impeller of more rugged construction and more streamlined passages, for service on corrosive liquids and on liquids containing small solid particles in suspension. Figure 174 shows a two-vane impeller with long sweeping vanes and an open eye, for service on liquids con-

FIG. 172. Cutaway view of high-head close-clearance impeller. (*Nagle Pumps.*)

FIG. 173. Cutaway view of high-bead close-clearance impeller for service with corrosive liquids or suspensions containing small particles. (*Nagle Pumps.*)

FIG. 174. Cutaway view of two-vane impeller with open eye. (*Nagle Pumps.*)

FIG. 175. Semi-open impeller for handling fluids containing pulpy solids (*Nagle Pumps.*)

taining small trash and abrasive solids, such as sand, gravel, and ashes. For sewage service and sump pumps, this type of impeller is often used with wide vanes to increase the passage area still more. Figure 175 shows a semi-open impeller (as contrasted to the enclosed impellers described above) with an abbreviated shroud on one side of the vanes. Such an impeller is used on pulpy solids which tend to clog an enclosed impeller. An open impeller has no shrouds and is better able to handle pulpy solids than the semi-open impeller.

Centrifugal fans, blowers, and compressors * use similar impellers, but the vanes are more often radial

FIG. 176. Open impeller for blower. (*Roots-Connersville Blower Corp.*)

or short straight sections. Curved vanes (usually backward) are used less frequently than for pumps.

The open impeller of Fig. 176 and the semi-open impeller of Fig. 177 have radial vanes. Short straight vanes set at a backward angle are built into the enclosed impeller of Fig. 178. In all these radial impellers the fluid enters at or near the axis of rotation and is thrown radially outward by the vanes.

If the vanes of an open impeller are set at an angle with the axis (Fig. 179) the direction of flow is axial,

* Fans deliver large volumes of gases at low pressures up to 1 psi, blowers (or exhausters) deliver small volumes at pressures up to 35 psi, and compressors deliver smaller volumes at higher heads. Only recently have centrifugal compressors come into extensive use for discharge pressures above 50 psi, but their smaller size and lighter weight give them an advantage over reciprocating compressors for aircraft and locomotives.

FIG. 177. Semi-open impeller for blower. (*Roots-Connersville Blower Corp.*)

FIG. 179. Axial-flow impeller. (*Worthington Pump and Machinery Corp.*)

as in the propeller pump or ordinary electric fan. Axial-flow impellers are excellent for handling large volumes of fluid at low heads, up to 20 ft.

For intermediate conditions of moderately large volumes of fluid at heads of 20 to 60 ft, a mixed-flow impeller (Fig. 180) may be used. The trailing edge of the hub of such an axial-flow impeller is flared out to direct the flow angularly away from the axis, thus adding centrifugal force to the force of direct thrust given the fluid by the propeller blades.

Side-Suction Centrifugal Pumps, are the simplest and most common centrifugal pumps. One version is shown in Fig. 181, a longitudinal section of another version in Fig. 182.

The pump of Fig. 182 is built for belt drive through a pulley keyed on the shaft at the left. The liquid enters through the center of the casing and is discharged tangentially upward from the periphery.

FIG. 178 Enclosed impeller for blower (*Roots-Connersville Blower Corp.*)

FIG. 180. Open mixed-flow impeller. (*Worthington Pump and Machinery Corp.*)

Fig. 181. Side-suction centrifugal pump. (*Aurora Pump Co.*)

The casing of most side-suction centrifugal pumps can be rotated to discharge upward, downward, or horizontally in either direction. The volute, bolted to the frame, is a spiral chamber of increasing cross section in the direction of rotation of the impeller. The volute receives the liquid from the semi-open impeller and converts kinetic energy of the fluid into pressure energy. The pump is built with journal bearings, the "inboard" bearing being grease-lubricated and the "outboard" being oiled with a "slinger ring." The higher-pressure liquid in the volute recirculates behind the shroud of the impeller, causing an end thrust tending to force the impeller off the shaft. A thrust ball bearing is installed to counteract this force. The thrust may be reduced by relief holes through the back shroud, but then the liquid will recirculate through the pumping channel, reducing the efficiency of the machine. The pump pictured has these relief holes but has a side plate which partially seals the discharge side of the pump from the space behind the impeller. Leakage of the liquid from this space through the inboard bearing and out of the pump is reduced and controlled by the stuffing box, aided by the grease lubricant entering from the grease cup.

An enclosed impeller is used for high discharge heads with a single-stage pump to reduce recirculation and to improve efficiency.

The two bearings of the pump in Fig. 181 are both "outboard" (external to the fluid being pumped) and are grease-lubricated ball bearings. The outer bearing is held in place by an adjustable lock nut, whereas the inner bearing is free to float, thus permitting an adjustment of the clearance between the impeller and the casing without dismantling the pump. The illustration shows the pump directly connected through a flexible coupling to an electric motor.

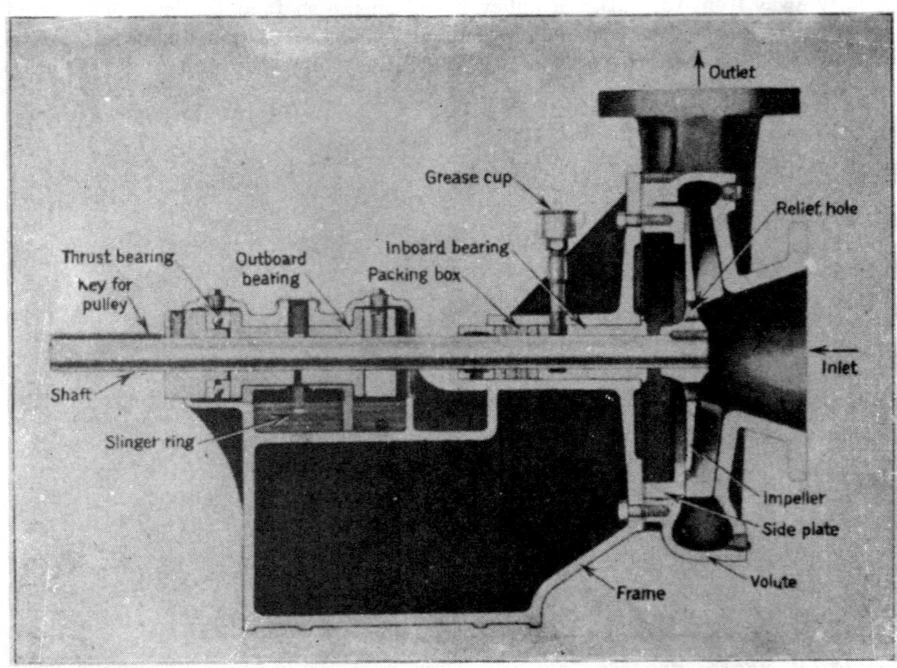

Fig. 182. Sectional view of a side-suction centrifugal pump. (*Fairbanks, Morse and Co.*)

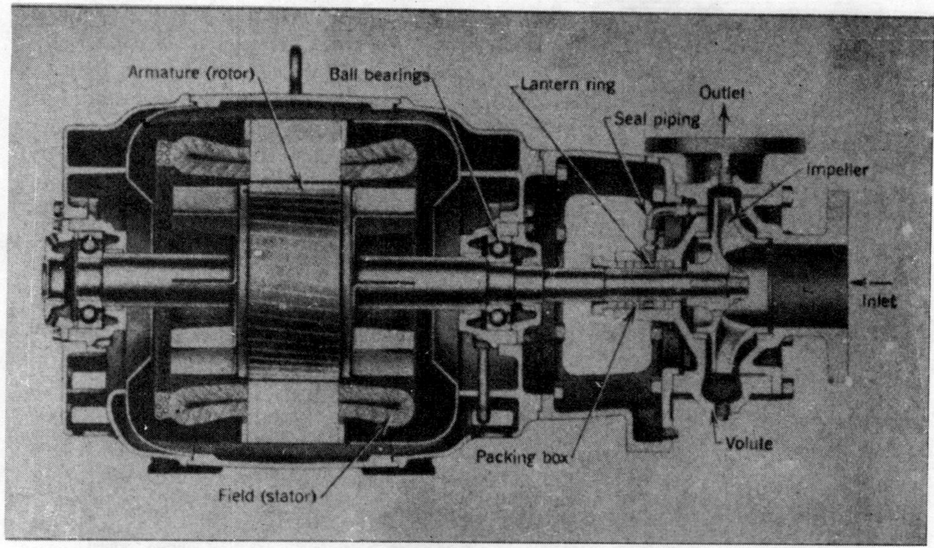

FIG. 183. Sectional view of centrifugal pump built as a unit with motor. (*Fairbanks, Morse and Co.*)

Figure 183 shows a side-suction pump built as a unit with the driver. The rotor of the electric motor and the impeller of the pump are attached to a single shaft riding in ball bearings supported in the motor housing. In smaller pump sizes the motor housing also carries the weight of the pump structure. These combination units are compact, always stay in alignment, and are usually mountable in any position.

A wide-passage impeller in a side-suction pump is shown in section in Fig. 184. The wide openings make this type of impeller useful in pumping storm water, screened sewage, and irrigation water, where large volumes of water containing small solid particles are to be pumped against low heads.

Double-Suction Single-Stage Centrifugal Pumps. A side-suction pump develops a high end thrust when operating at high heads. This thrust may be eliminated by using two impellers in parallel on the same shaft and arranged to oppose one another, or more simply by building the two impellers as one with the liquid entering from each side, as shown in Fig. 185.

The impeller is equivalent to two enclosed impellers back-to-back. The recirculation of high-pressure liquid is hindered by "wearing rings" of soft metal attached to the casing and serving as a seal against the rotating impeller. Leakage around the shaft is reduced by the stuffing boxes containing

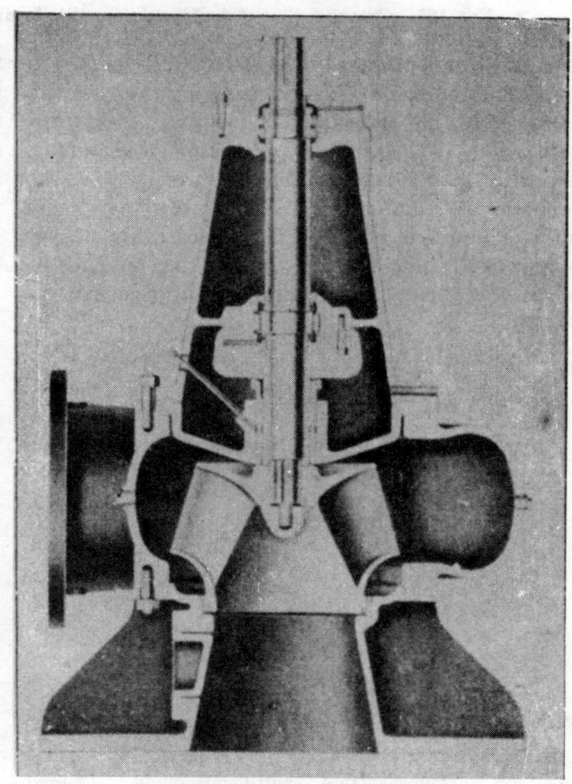

FIG. 184. Sectional view of centrifugal pump with wide-passage impeller. (*Fairbanks, Morse and Co.*)

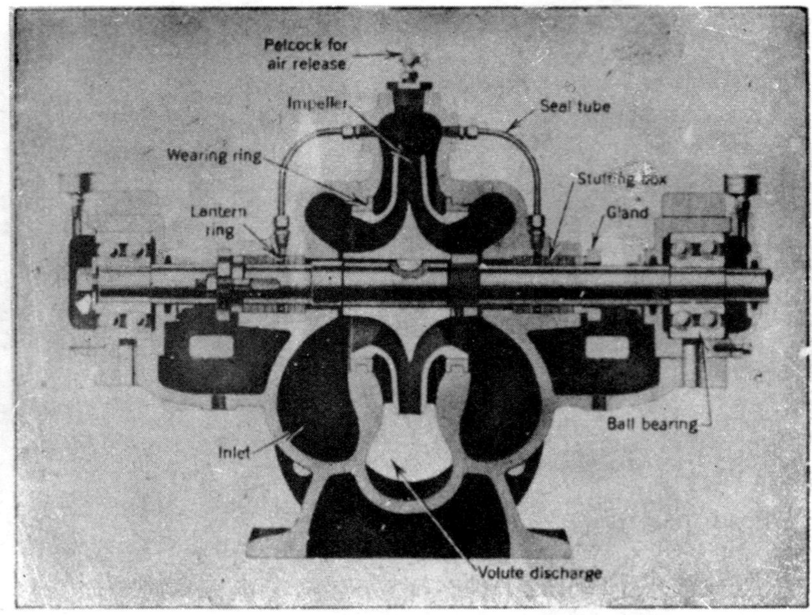

FIG. 185. Cutaway view of double-suction centrifugal pump. (*Pennsylvania Pump and Compressor Co.*)

rings of fibrous packing compressed by a gland. If the suction pressure is less than 1 atm, air would leak in past this packing and "air bind" the pump (see p. 189). This is prevented by sealing the stuffing box with a high-pressure liquid through the seal tube and the "lantern ring." Frequently the high-pressure liquid is obtained from the casing of the pump itself, thus placing the packing under the discharge pressure, but it may be supplied as well from an outside source. Leakage because of loose packing will be visible outside the pump and can be reduced to a trickle by tightening the gland. Some liquid will leak back into the suction of the impeller and be recirculated. This is the price paid for prevention of air leaks and binding. Double-row ball bearings are mounted outboard on this pump equipped for direct drive.

FIG. 186. Double-suction centrifugal pump. (*Aurora Pump Co.*)

Figure 186 shows a pump of this type connected through a flexible coupling to an electric motor.

Multistage Centrifugal Pumps and Blowers. The head delivered by a single impeller (or two impellers in parallel) is restricted by the practical limits to diameter and speed. If high heads are desired, two or more pumps may be connected in series, or two or more impellers mounted on a single shaft may be connected in series.

A four-stage pump is shown in longitudinal section in Fig. 187. The four impellers are mounted to provide balancing thrust, and the casing or piping carries the fluid from the discharge of one stage to the inlet of the next stage.

Multistaging may be applied to blowers and compressors. Figure 188 shows a three-stage compressor.

Deep-Well Centrifugal Pumps. The centrifugal pump is capable of reduction in size to an extent which permits the construction of a multistage unit

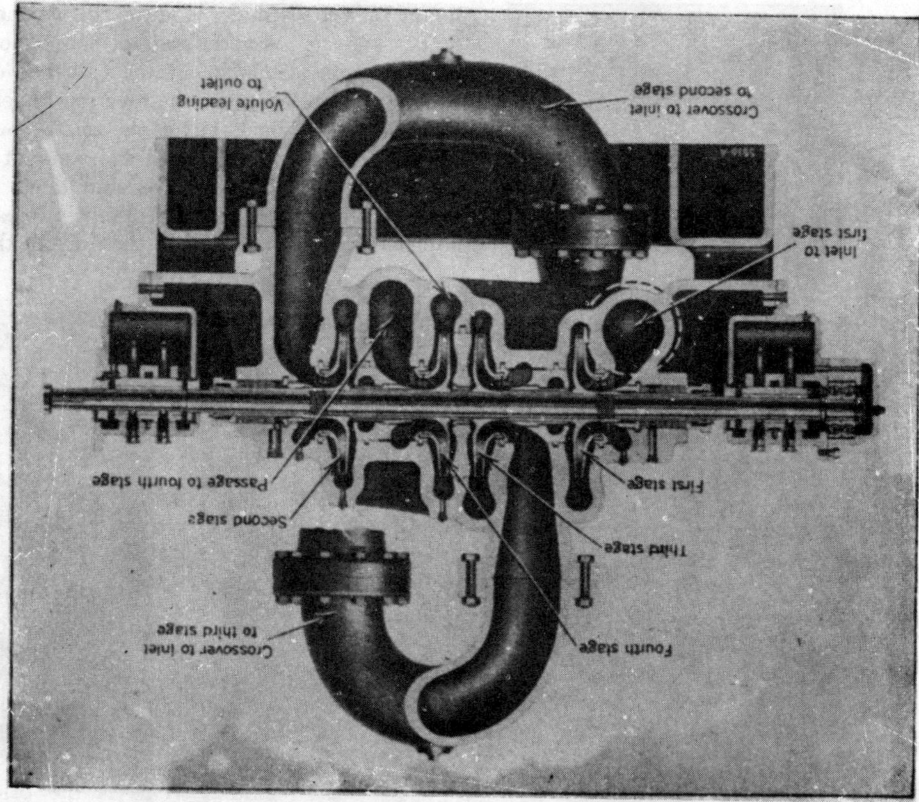

FIG. 187. Cutaway view of four-stage centrifugal pump. (*De Laval Steam Turbine Co.*)

which will fit into well casings as small as 4 in. in diameter. The pump assembly can be lowered down to the water level. The motor may be submerged with the pump or kept at the surface level operating through a long drive shaft extending down to the pump.

A two-stage deep-well pump (Fig. 189) is supported by the discharge piping. The shaft is shown enclosed by tubing in this version. Some pumps use no protective tubing around the shaft and depend upon flowing liquid to lubricate rubber shaft bearings mounted at intervals. In the pump illustrated the oil lubricant is sealed from the water by the labyrinth packing. Figure 190 shows one impeller and bowl, indicating the diffuser channels leading to the bowl above. The replaceable wearing ring for the impeller above is seen at the top of the bowl.

Diffuser Centrifugal Pumps. The guiding channels through which the fluid flows from the discharge of one impeller (Fig. 189 or 190) to the inlet of the next are called diffusers. They are

FIG. 188. Three-stage centrifugal compressor. (*Worthington Pump and Machinery Corp.*)

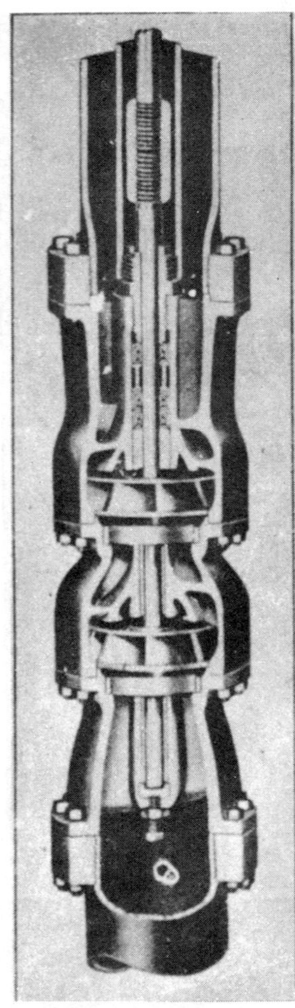

FIG. 189. Longitudinal cutaway view of two-stage deep-well pump. (*Layne and Bowler, Inc.*)

designed to enlarge the cross section of flow gradually and thus convert kinetic energy into pressure head. These diffusers serve the same purpose as the spiral volute of the centrifugal machines previously described. Diffusers are used in pumps only rarely, usually only in deep-well pumps, but they are common in blowers and almost mandatory in compressors.

Self-Priming Centrifugal Pumps. An ordinary centrifugal pump is incapable of priming itself and initiating flow through the system if the pump casing is filled with air or vapor. This handicap can be overcome largely by a modification in casing and volute design, as shown in Figs. 191 and 192. The volute presents a sharp edge to the rotating mass of fluid. The suction and discharge connections are at the top of an oversized housing with two chambers above the impeller (note the reversed impeller, with the eye facing back along the shaft toward the driver), thus keeping the minimum amount of liquid in the pump if the system is shut down and permitted to drain. When the pump is restarted the impeller throws the liquid to the sides of the casing

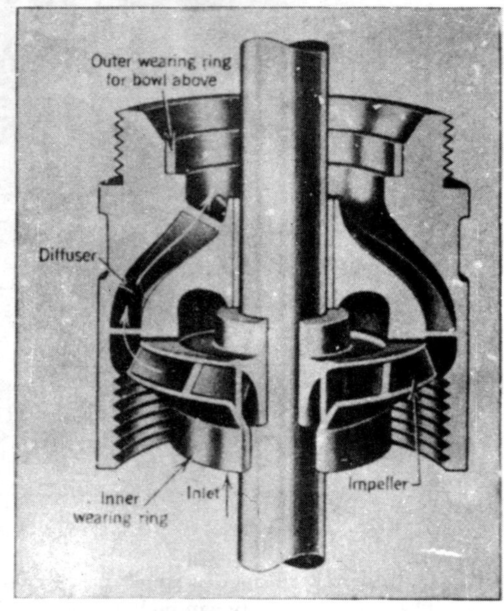

FIG. 190. Cutaway view of one-stage of a deep-well pump. (*Layne and Bowler, Inc.*)

and begins to pump air flowing through the suction pipe and eye (Fig. 193). Just beyond the tips of the vanes the air and liquid are mixed, forming a froth. This froth follows around the impeller to the knife edge, which diverts the frothy mixture away from the impeller into the chamber at the top of the housing. The air then rises and is discharged. In this way the air is exhausted from the suction line, after which time the action is normal. This pump will prime faster if the tips of the impeller vanes are slotted or drilled, as indicated in Fig. 193, to facilitate the flow of air from the eye to the volute.

Operating Features of Centrifugal Machines

The large clear passages in centrifugal machines make them relatively satisfactory for handling fluids

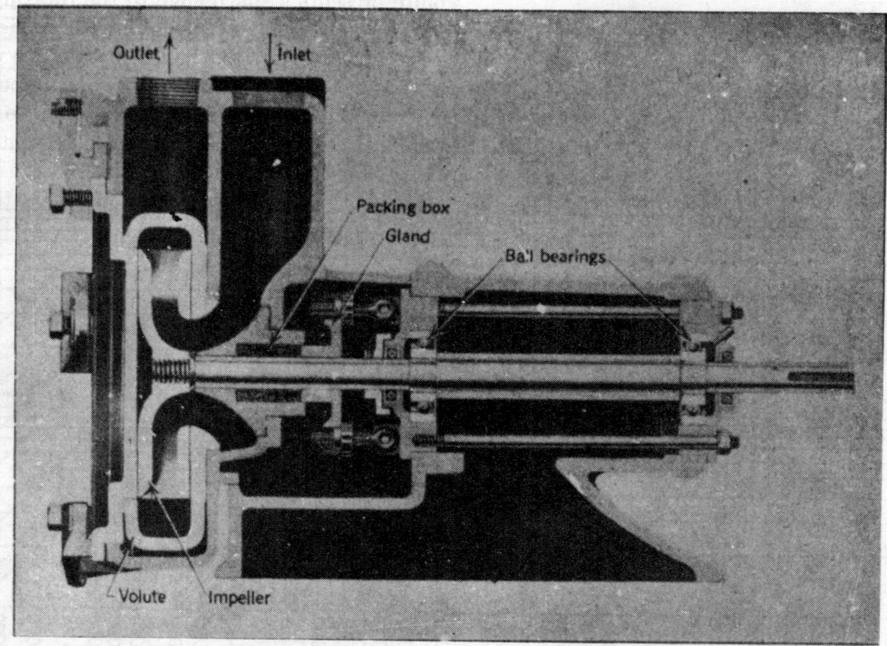

FIG. 191. Sectional view of self-priming pump. (*Nagle Pumps.*)

FIG. 192. Casing and volute of a self-priming pump. (*Nagle Pumps.*)

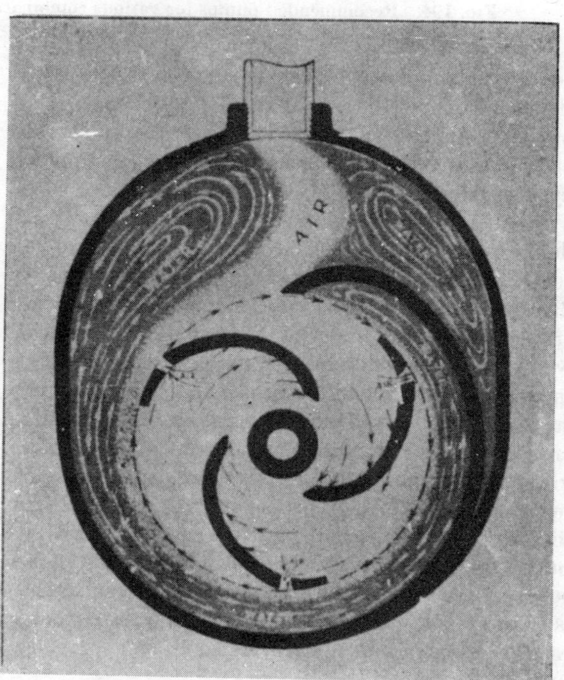

FIG. 193. Drawing illustrating action of a self-priming pump. (*Nagle Pumps.*)

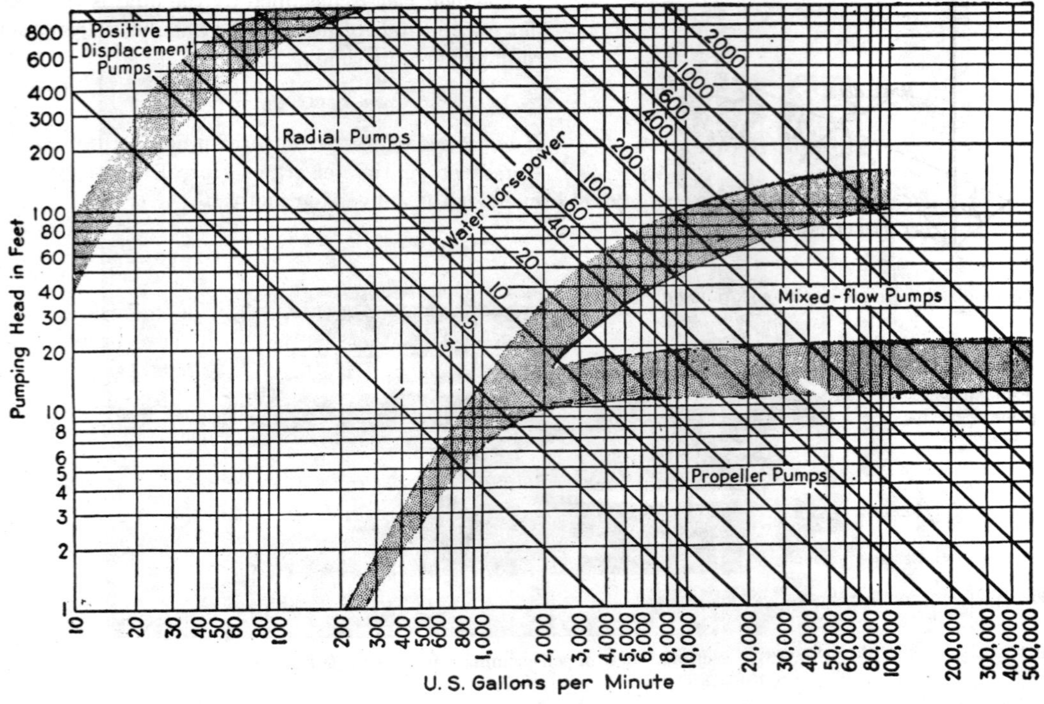

FIG. 194. Recommended pumps for various combinations of head and capacity. (*Fairbanks, Morse and Co.*)

containing solid particles in suspension, and permit throttling and the temporary closing of the valve in the discharge line. The fluid then simply rotates around the casing and absorbs the energy supplied by the driver. The absorption of the total energy by the fluid quickly raises the temperature of the fluid and pump sufficiently to cause distortion of the rotating parts in a short time. The lack of positive displacement restricts the economical use of centrifugals to the conditions indicated on Fig. 194 which represent relatively large volumes of fluid at moderate discharge pressures (heads), for high pressures can be obtained only by excessive multistaging. In general, a centrifugal pump has lower initial and lower maintenance costs and is more easily made of materials that resist corrosion than the positive displacement reciprocating or rotary units, but it has less ability to operate with low absolute pressures on the intake or suction. A centrifugal pump is better able to handle viscous liquids than a reciprocating pump, but less so than a rotary pump.

Virtual Head with Infinite Number of Vanes. If all losses of energy are neglected, the virtual head, or maximum head possible for a given pump, may be derived from the concept of angular momentum. A generalized sketch, Fig. 195, shows an elemental mass of fluid dm moving at a velocity v in the direction and magnitude shown as referred to the center of rotation O. The mass dm is at a distance r from the center of rotation, and its direction of motion forms the angle α with the perpendicular to the radius r.

The linear momentum of the mass dm is

$$(dm)v$$

Its angular momentum is

$$(dm)vr \cos \alpha$$

Its torque dT is

$$dT = dm \frac{d(vr \cos \alpha)}{dt} \tag{108}$$

If an idealized version of a centrifugal pump is considered (Fig. 196), it may be assumed that all the fluid comprising a ring at constant radius r is moving at the same velocity and at the same angle

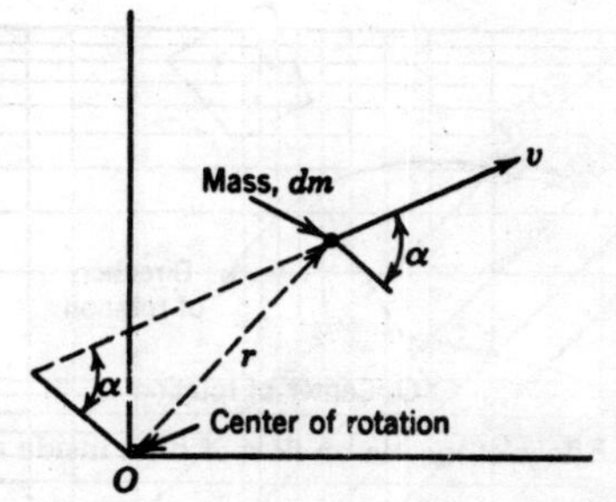

FIG. 195. Generalized sketch of motion in a random direction.

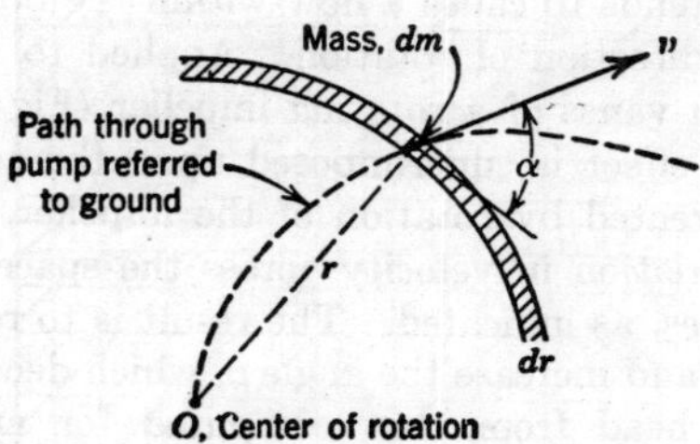

FIG. 196. Sketch of motion through a centrifugal pump.

with the radius. The path followed by an element of fluid as it passes through the pump is shown in relation to a stationary point on the earth's surface. The mass of the ring of fluid is

$$dm = \rho 2\pi rb\, dr \tag{109}$$

where ρ represents the density of the fluid, b represents the width of the ring (perpendicular to this page), and dr is the thickness of the ring along the radius. Combining equations 108 and 109,

$$dT = \rho 2\pi rb \frac{dr}{dt}\, d(vr \cos \alpha) \tag{110}$$

The area of the ring of fluid is $2\pi rb$, and the radial velocity of fluid through the ring is dr/dt; therefore the capacity, or volume of fluid being pumped per unit of time Q is

$$Q = 2\pi rb \frac{dr}{dt} \tag{111}$$

Combining equations 110 and 111, and integrating between points 1 and 2,

$$\int_1^2 dT = T_2 - T_1 = \int_1^2 \rho Q\, d(vr \cos \alpha)$$
$$= \rho Q(v_2 r_2 \cos \alpha_2 - v_1 r_1 \cos \alpha_1) \tag{112}$$

The *total power* consumed by the fluid is the product of the total torque and the angular velocity ω in radians per unit time t.

$$\rho Q\omega(v_2 r_2 \cos \alpha_2 - v_1 r_1 \cos \alpha_1)$$

If energy is expressed in force units, as foot-pounds (force) or centimeter-grams (force), the expression includes the conversion factor g_c.

$$\frac{\rho Q\omega}{g_c}(v_2 r_2 \cos \alpha_2 - v_1 r_1 \cos \alpha_1)$$

Since power is also the product of the mass per unit time and the energy per unit mass, or head, the *virtual head developed* by a centrifugal pump is

$$\frac{\omega}{g_c}(v_2 r_2 \cos \alpha_2 - v_1 r_1 \cos \alpha_1)$$

Figure 197 indicates the vector diagram of the velocity components at the periphery of the impeller of a centrifugal pump. The absolute velocity v of the fluid at any point in the impeller may be expressed in terms of the velocity of the impeller u at that point. The angle between the velocity of the fluid relative to the impeller u' and the velocity of the impeller u is $180 - \beta$, as shown in Fig. 197. Since the angular velocity of the impeller ω is the ratio of the tangential velocity to the radius u/r,

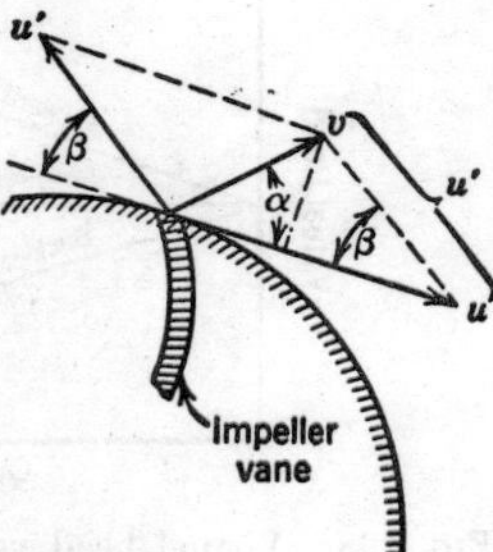

FIG. 197. Vector diagram of velocities. (*After Church.*[3] *)

$$\text{Virtual head} = \frac{(u_2 v_2 \cos \alpha_2 - u_1 v_1 \cos \alpha_1)}{g_c} \tag{113}$$

If point 1 is taken as the eye of the impeller and point 2 is the periphery of the impeller, and if the fluid enters the impeller without any prerotation, then the angle α approaches 90 degrees, its cosine approaches zero, and equation 113 approaches

$$\text{Virtual head} = \frac{u_2 v_2 \cos \alpha_2}{g_c} \tag{114}$$

From Fig. 197, $u = v \cos \alpha + u' \cos \beta$. Therefore equation 114 may be written

$$\text{Virtual head} = \frac{u_2(u_2 - u_2' \cos \beta_2)}{g_c} \tag{115}$$

* The bibliography for this chapter appears on p. 197.

The virtual head is the theoretical maximum head that could be developed under the indicated operating conditions. The quantity of fluid being pumped is more easily determined than the velocity at any point. The quantity of fluid being pumped is

$$Q = 2\pi rbv \sin \alpha = 2\pi rbu' \sin \beta$$

Solving for u' and substituting in equation 115,

$$\text{Virtual head} = \frac{u_2}{g_c}\left[u_2 - \left(\frac{Q \cos \beta_2}{2\pi r_2 b_2 \sin \beta_2}\right)\right]$$

$$= \frac{u_2^2}{g_c} - \frac{u_2 Q}{2\pi r_2 b_2 g_c \tan \beta_2} \qquad (116)$$

If the virtual head is plotted as a function of the quantity of fluid pumped for impellers of the same diameter operating at the same speed, straight lines are obtained as shown in Fig. 198.

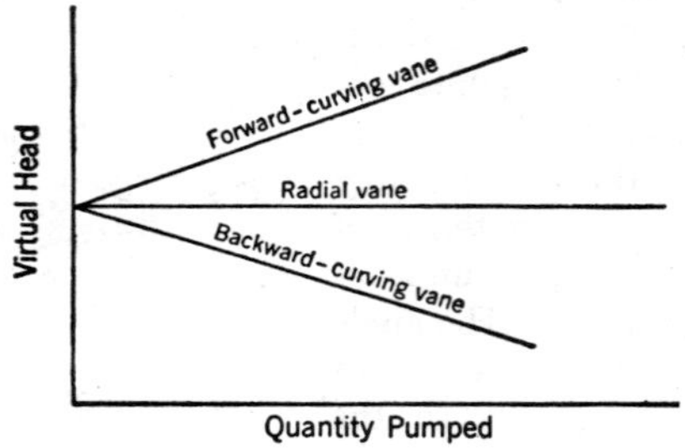

FIG. 198. Virtual head versus capacity for different outlet vane angles. (*After Church.*[3])

The slope of the line is a function of the outlet vane angle β_2. A forward-curving vane, with a negative outlet vane angle β_2, gives a positive slope on Fig. 198 or higher heads for higher throughputs. A backward-curving vane (positive outlet vane angle β_2) gives a negative slope on Fig. 198, and a radial vane gives constant head. At zero flow, $Q = 0$, and

$$\text{Virtual head} = \frac{u_2^2}{g_c} \qquad (117)$$

regardless of vane angle.

Virtual Head with Finite Number of Vanes. The virtual head equations assume no losses and an infinite number of vanes in the impeller to provide perfect guidance to the fluid in its travel through the impeller. Actually, a finite number of vanes must be used, and the fluid is not guided perfectly but tends to circulate in the space between vanes. The fluid within a box rotated about an external axis (Fig. 199) tends to remain in the same position relative

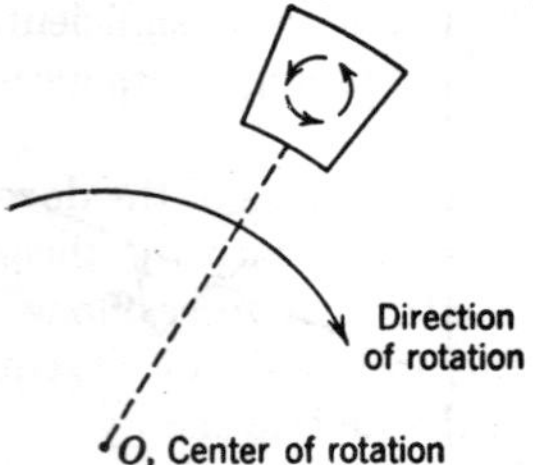

FIG. 199. Effect of inertia on flow of fluid inside rotating box.

to a point on the earth's surface. This effect of inertia tends to cause a flow within the box counter to the direction of rotation. Applied to the space between vanes of a rotating impeller (Fig. 200) the inertial effect is superimposed upon the centrifugal effect created by rotation of the impeller, resulting in a variation in velocity across the space between the vanes, as indicated. The result is to reduce the angle α and increase the angle β, which decreases the virtual head from that computed for an infinite number of vanes.

Actual Developed Heads. An impeller is designed for a particular head and throughput. If the pump or blower is operated at a lower or higher throughput, the inlet and outlet vane angles are not the best for that flow and introduce turbulence which increases as the throughput diverges more and more from the design value. Friction of flow through the impeller itself also reduces the energy convertible into head.

Figure 201 indicates the effect of these factors on the head developed by a pump with backward-curving vanes. This curve is representative of the so-called "rising characteristic" curves, showing an increase in head as the throughput is increased from zero, reaching a maximum and then decreasing in the usual manner. Such a curve is not necessarily the result of forward-curved or radial vanes but can be obtained from backward-curved vanes if the

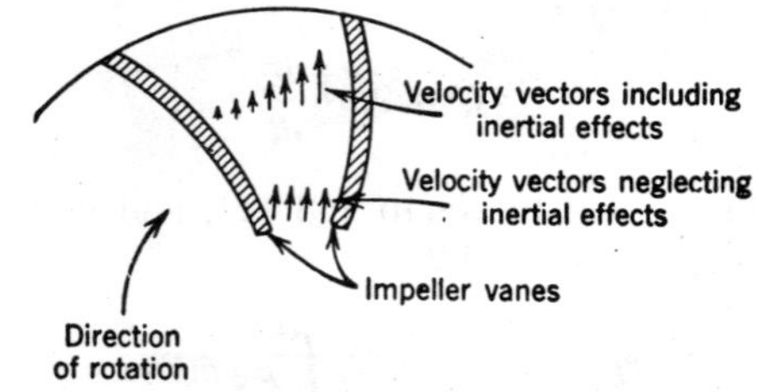

FIG. 200. Effect of inertia on flow through impeller. (*After Church.*[3])

turbulence losses are sufficiently great at low capacities. A "falling characteristic" curve has no maximum.

Experimental values of the developed head at zero capacity agree well with the theoretical head resulting from centrifugal force alone ($u_2^2/2g_c$) which is one-half the virtual head of equation 117. This means that the kinetic energy of the fluid as it leaves the impeller is expended in friction and turbulent losses.

Inspection of equations 114 and 116 show that the head varies with the quantity of fluid pumped, but not with the density. The discharge pressure, however, does vary with the density. In other words, a given centrifugal pump develops the same head at a given capacity, irrespective of the fluid being pumped. If a pump normally handling water is filled with air instead of water, the discharge pressure immediately drops so low that no air can be delivered against the pressure of water filling the discharge piping. The air remains in the impeller because of its low density resulting in cessation of flow through the pump. This is known as air binding. It can be cured in an ordinary pump only by stopping the pump and allowing the air to rise to the top of the casing, from which it is removed through a petcock, as shown in Fig. 185. Flow can be initiated through a centrifugal pump by positive pressure on the suction, by some vacuum priming device, or by a self-priming pump, as described.

Any variation in head accompanying a change in fluid is the result of a change in viscosity. An increase in viscosity of the fluid increases the friction losses shown in Figs. 201 and 202 and thus reduces

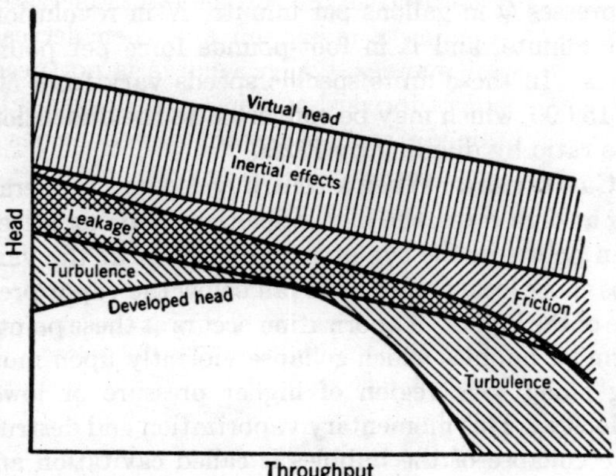

Fig. 201. Factors reducing virtual head to developed head. (*After Church.*[3])

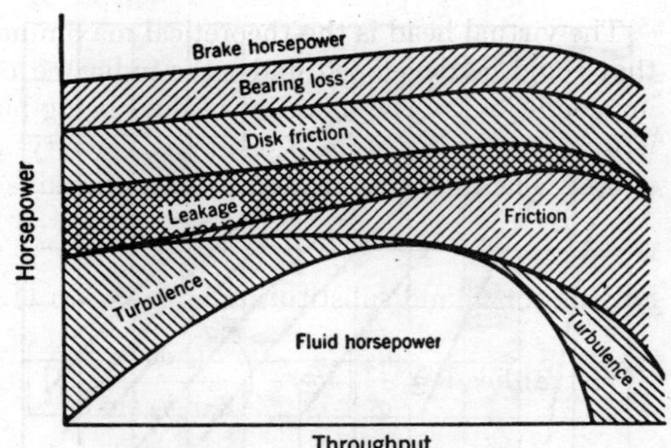

Fig. 202. Factors increasing fluid horsepower to brake horsepower. (*After Church.*[3])

the head and affects the characteristic curves as shown in Fig. 203.

If the pressure in the suction line or within the pumps is less than the vapor pressure of the liquid being pumped, vaporization of the liquid occurs in the pump inlet or impeller. This vapor is persistent and binds the pump exactly as in air binding. This can be corrected only by changing the suction piping to reduce friction losses or eliminate "lift" in the line.

Brake Horsepower. The power required to drive the pump is that required to overcome all the losses and supply the energy added to the fluid. These losses include the friction of flow through the impeller and turbulent losses, the disk friction or energy required just to rotate the impeller in the fluid, the leakage of fluid from the periphery back to the eye of the impeller, and the mechanical friction losses in the bearings, stuffing boxes, and wearing rings as indicated in Fig. 202.

The fluid horsepower is the energy absorbed in the fluid leaving the pump. The brake horsepower is the energy requirement of the pump per unit of time.

Efficiency. The efficiency of a centrifugal machine is the ratio of fluid horsepower to brake horsepower. The effect of operating conditions on the characteristics of a few pumps is indicated in Figs. 203 to 205.

Effect of Speed on Characteristics. Inspection of Fig. 197 shows that an increase in speed of rotation increases tangential velocity u proportionally. The vane angle β and the fluid angle α are unchanged; therefore, all velocities are increased proportional to the increase in speed. The capacity is thus directly proportional to the speed.

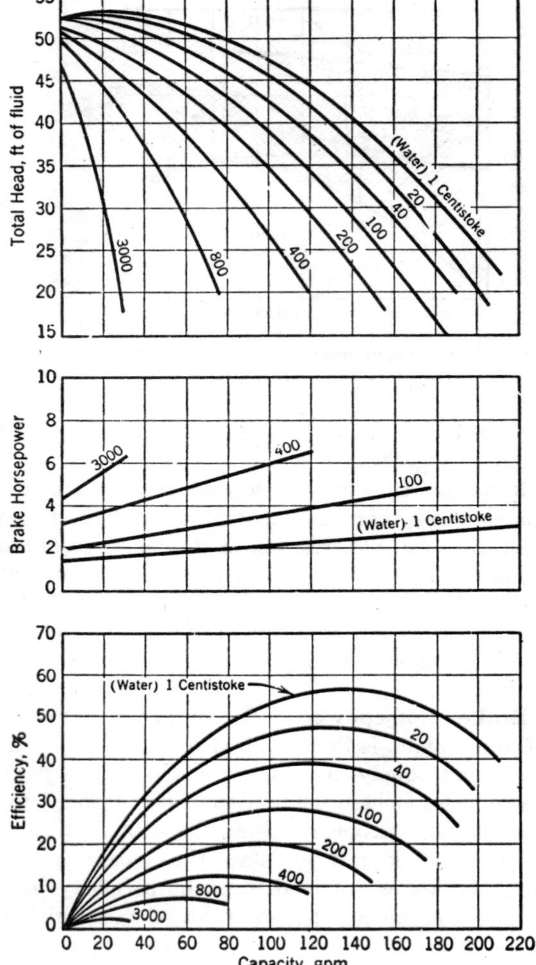

FIG. 203. Characteristic curves of a typical centrifugal pump for fluids of different viscosities. (*Worthington Pump and Machinery Corp.*)

Since the absolute velocity v_2 and the tangential velocity u_2 are increased proportionally, the virtual head and the developed head are increased proportionally to the square of the speed.

The horsepower is proportional to the product of the head and capacity and therefore increases proportionally to the cube of the speed.

Effect of Impeller Diameter on Characteristics. The capacity varies as the area of discharge of the impeller for a given velocity of flow and therefore is directly proportional to the diameter of the impeller.

For the same speed of rotation, the velocity terms u_2 and v_2 are proportional to the radius; therefore, the head of a centrifugal machine is proportional to the square of the diameter.

The horsepower, varying as the product of head and capacity, is proportional to the cube of the diameter.

Specific Speed. For single-stage side-suction impellers, or *one* stage of a multistage pump, the specific speed N_s is a convenient concept

$$N_s = \frac{NQ^{0.5}}{[g_c(-\bar{w})]^{0.75}}$$

where N_s = specific speed.
N = revolutions per second.
Q = volume of fluid per second.
$-\bar{w}$ = "total developed head," or work done by one stage on unit mass of fluid (equation 60 with $\overline{lw} = 0$).

The specific speed is dimensionless if consistent units are used.[17a]

The characteristic curves of a pump represent performance from zero flow to maximum flow ($\bar{w} = 0$), and the specific speed would vary from zero to infinity, respectively. For classifying impellers a single value must be selected. The point of maximum efficiency is usually selected for calculation of the specific speed. The usual range is from 0.03 to 0.87 when so calculated and expressed as the dimensionless ratio given above. The lower values apply to radial-flow centrifugal pumps and the higher specific speeds to axial-flow propeller pumps.

Unfortunately, current practice [3,18a] omits g_c and expresses Q in gallons per minute, N in revolutions per minute, and $\bar{w}$ in foot-pounds force per pound mass. In these units specific speeds vary from 500 to 15,000, which may be converted to the dimensionless ratio by dividing by 17200.

Cavitation. When a centrifugal pump is operating at high rates, the high velocities occurring at certain points in the eye of the impeller or at the vane tips cause local pressures to fall below the vapor pressure of the liquid. Vaporization occurs at these points, forming bubbles which collapse violently upon moving along to a region of higher pressure or lower velocity. This momentary vaporization and destructive collapse of the bubbles is called cavitation and is to be avoided if maximum capacity is to be obtained and damage to the pump prevented. The

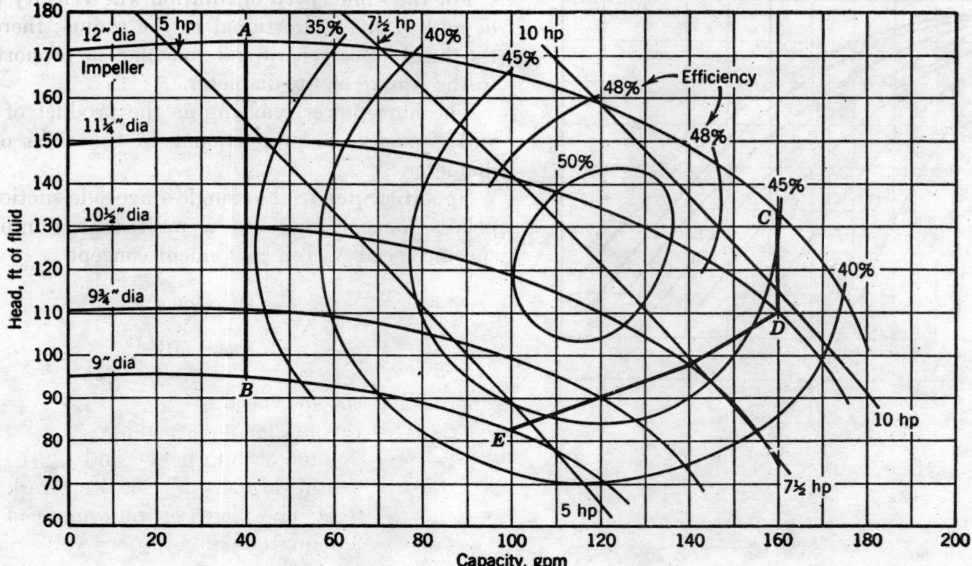

FIG. 204. The effect of impeller diameter on the characteristics of a centrifugal pump with enclosed impellers operating at 1750 rpm. Lines of equal efficiencies and brake horsepowers are shown. (*Aurora Pump Co.*)

shock of bubble collapse causes severe pitting of the impeller and creates considerable noise and vibration. Cavitation may be reduced or eliminated by reducing the pumping rate or by slight alterations in impeller design to give better streamlining.

Cavitation usually does not occur at low flow rates on any given pump, but, when cavitation occurs, all the characteristic curves shown in Fig. 203 "break" suddenly towards zero.

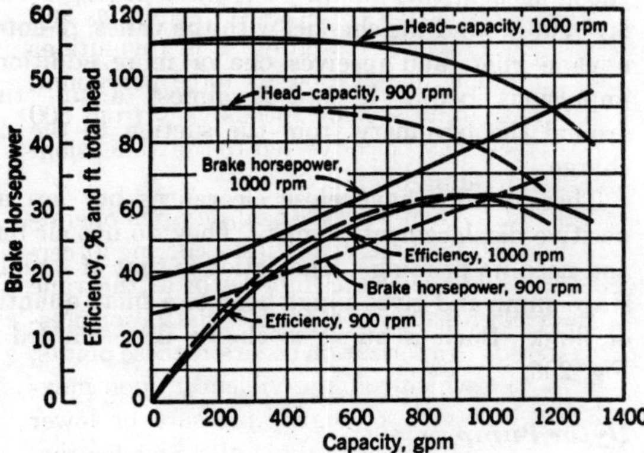

FIG. 205. Head, efficiency, and brake horsepower, plotted as a function of capacity for a centrifugal pump operating at two speeds with the same liquid. (*Worthington Pump and Machinery Corp.*)

Net Positive Suction Head (NPSH). Both vapor binding and cavitation may be eliminated by maintaining a pressure at the pump inlet significantly higher than the vapor pressure of the liquid being pumped. The required margin of pressure is called the net positive suction head. It is a function of the pump design and is usually specified by the manufacturer for pumps to handle liquids such as preheated boiler feed, steam condensate, or volatile liquids. Usually, it is of the order of 2 to 10 ft-lb/lb mass of fluid and increases with increasing throughput.

SPECIAL PUMPS AND BLOWERS

There are many different types of machines for the handling of fluids which do not fall into the three classes discussed above. Some of these, such as the jet ejector, the air lift, and the hydraulic ram, are far from these categories. Some, such as the turbine pump and the Hytor pump, are close relatives.

Turbine Pumps

According to old terminology the diffuser pump was called the turbine because of its similarity to the turbine driving unit, but today the name turbine is restricted to a pump whose impeller is similar to that of Fig. 206. The turbine or "peripheral" pump

FIG. 206. Turbine pump, disassembled. (*Aurora Pump Co.*)

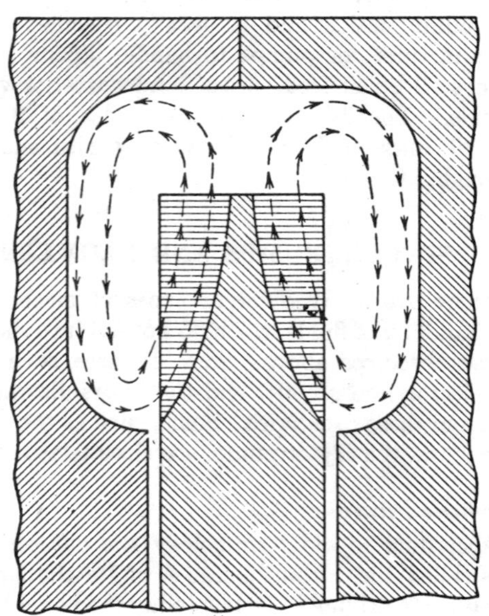

FIG. 207. Drawing of section of fluid channel in turbine pump showing circulation. (*Aurora Pump Co.*)

rotates and acts as a centrifugal with a semi-open impeller, but the "eye" of the impeller is at the point where the flat side of the impeller ceases and the radial vane begins (see Fig. 207). Recirculation is encouraged in this pump. The fluid leaving a vane is swept around the channel with the vanes, re-enters a vane inlet, and receives one or more additional impulsions before it sweeps almost a full turn around the periphery from the suction to the discharge.

These pumps have close clearances but are not positive displacement pumps. They do not air bind and need no priming. The action is that of a multistage pump and gives a high head to a small quantity of fluid. Such a pump is shown disassembled in Fig. 206.

Hytor Pump

A machine whose principle of operation places it in the class of rotary vane pumps but whose construction places it in the class of centrifugal pumps is the Hytor vacuum pump or compressor, shown in section in Fig. 208. The rotor has forward-curved vanes enclosed in shrouds and rotates around a hollow shaft with four ports at the four quadrants, two serving as inlet ports and two as discharge ports, each opposite pair being connected through the shaft with the proper pipe connection. The rotor revolves in an elliptical casing and does not touch the sides at any point. The seal is maintained by liquid which is rotated by the vanes but which follows the shape of the casing as a result of centrifugal action, thus approaching and receding from the shaft twice during each revolution. This motion of the

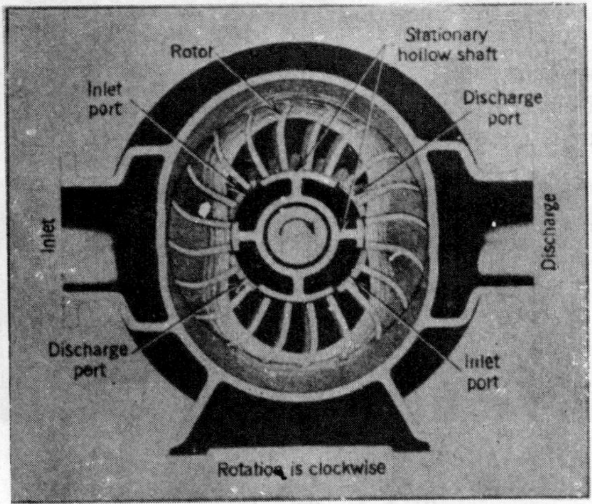

Fig. 208. Sectional view of Hytor vacuum pump indicating operation. (*Nash Engineering Co.*)

liquid acts as a piston in the space between the vanes, sucking in gas and discharging it one-quarter revolution later at a higher pressure.

All seals in this pump are positive, friction of metal against metal is practically eliminated, and wet gas or gas containing slugs of liquid can be handled equally well.

Jet Pumps

The general term, jet pump, includes all machines whose operation is based on the transfer of energy by impact from a fluid jetting at high velocity into a slowly moving or stagnant fluid, giving the mixture of fluids a moderately high velocity which is then reduced carefully so as to give a final pressure greater than the initial pressure of the low-velocity fluid. An *injector* is a jet pump using a condensable gas to entrain a liquid and discharging at a pressure higher than the initial pressure of the motive fluid or the entrained fluid. It is now practically restricted to boiler feed-water injectors. An *ejector* is a jet pump more general in character, using either gas or liquid for either the motive or the entrained fluid and discharging at a pressure intermediate between the motive pressure and the suction pressure. An *exhauster*, *blower*, or *compressor* is an

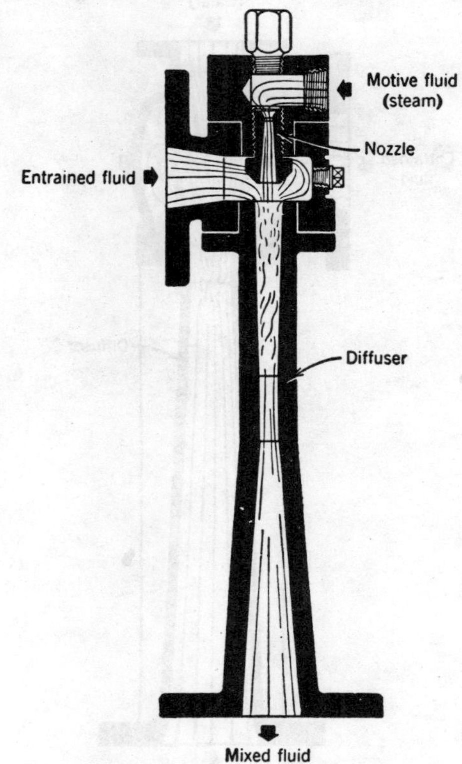

Fig. 209. Longitudinal section of a steam-jet ejector. (*Schutte and Koerting Co.*)

ejector with gases both as motive and entrained fluids; a *siphon* is an ejector with gas as the motive fluid and liquid as the entrained fluid; an *eductor* is an ejector with liquids both as motive and entrained fluids; and a *fume absorber* is an ejector with liquid as the motive fluid and gas as the entrained fluid.

The essential parts of a jet pump are the nozzle and the diffuser. The nozzle is a device for transforming the high pressure of the motive fluid into high velocity; the diffuser is a mixing chamber and a device for converting the residual velocity of the mixture back into pressure. Figure 209 of a steam-jet exhauster and Fig. 210 of a water-jet eductor illustrate the variations in form of these devices.

The nozzle in Fig. 209 is converging-diverging, and the steam passing through it reaches "critical" or "acoustic" velocity at the narrowest part, then increases in velocity still further in the diverging section as the pressure is reduced to a value at the tip approximately equal to that of the slowly moving

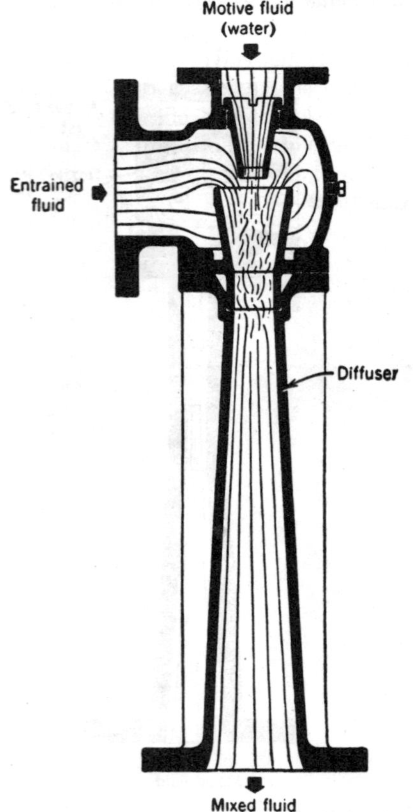

Fig. 210. Longitudinal section of a water-jet ejector. (*Schulte and Koerting Co.*)

fluid into which it jets. The forepart of the diffuser has a slight taper. In this section the two fluids mix, the high-velocity fluid losing momentum to the low-velocity fluid. The mixed fluids enter the diverging section with a velocity not greater than the acoustic velocity. Under these conditions, kinetic energy of the mixed stream is converted to pressure energy and the final discharge of the mixture may be made into a pipe, a tank, or the atmosphere.

The nozzle in Fig. 210 is converging only, giving the water a high velocity by conversion of the high motive pressure and jetting it into the low-velocity liquid. The short forepart of the diffuser serves as a mixing section and is tapered to conform to the increase in average velocity of the mixture passing through it. The diverging section of the diffuser converts the velocity into the pressure at which the mixture is discharged.

Jet pumps have no moving parts and thus practically no maintenance requirements, but they do require a quantity of high-pressure motive fluid, which is usually expensive to supply. They are most useful where need is intermittent and an inexpensive stand-by unit is desired, where corrosion is important, where a combination of heating and pumping is desired, or where a low-pressure exhaust steam can be salvaged and re-used.

The Acid Egg

It is frequently advantageous and possible to eliminate moving mechanical equipment in pumping

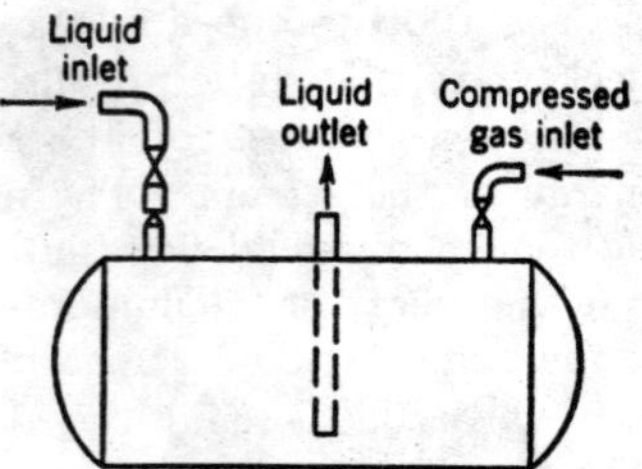

Fig. 211. Sketch of the blow case, or acid egg.

or compressing fluids by the utilization of kinetic, potential, or pressure energy of fluids as a motive force. The acid egg or blow case, Fig. 211, operates by displacement of one fluid by another and includes no moving parts.

The operation of the egg is simple. Liquid which has been introduced through the feed inlet is displaced by a less-dense fluid such as compressed air or inert gas and flows upward through the outlet pipe leading from the bottom of the vessel. If the entrance to the outlet pipe were at the top of the vessel instead of the bottom so that liquid would displace gas, the blow case could function as a gas compressor.

If only one egg is used the operation is intermittent. The operation can be made continuous with respect to the fluid flowing if two or more eggs are used so that at least one is discharging while others are depressuring and filling. Acid eggs are frequently manually operated, although with suitable instrumentation the entire sequence of operation can be made automatic.

The acid egg, as the name implies, is useful if pumps are either unobtainable or too costly because of the corrosiveness of the fluid. Its disadvantages lie in the intermittent action which, in turn, requires manual operation or excessive instrumentation. The egg is inherently inefficient because at the end of the cycle the egg is filled with the motive fluid at its highest pressure; if a compressed gas is used it is usually necessary to vent it to the atmosphere without recovering work. As more corrosion-resistant materials of construction are developed for pumps, the usefulness of the acid egg gradually decreases.

Hydraulic Ram

The hydraulic ram, Fig. 212, uses the kinetic energy of a moving column of fluid, usually water, to raise a portion of the fluid to a higher pressure or elevation. If the fluid flowing in the supply line, or fall pipe, is abruptly stopped by the closure of the escape check valve, the static pressure at the valve suddenly increases because of the conversion of kinetic energy of the stream. The delivery valve opens, and some of the fluid flows into the delivery system. As the energy of the fluid in the fall pipe is absorbed, the static pressure at the base decreases. The escape valve opens, and the delivery valve closes. The flow in the column then increases, and the motive fraction of the fluid flows out, to waste, through the escape valve until its velocity becomes great enough to pick up and seat this valve. The cycle then repeats with a frequency which may be as low as 15 or as high as 200 times per minute.

The efficiency of a properly designed ram may be as high as 90 per cent. The efficiency may be defined:

$$\text{Efficiency (per cent)} = \frac{W'h'}{Wh} \times 100$$

where W' = mass of fluid delivered.

W = mass of motive fluid exhausted through the escape valve.

h = the fall (Fig. 212).

h' = the lift (Fig. 212).

Rams may vary in the details of design, but the essential components are the moving column, the escape and delivery valves, and the air chamber. The efficiency and capacity will depend upon the design of these elements.

The Moving Column. The important factors of this component are the mass of motive fluid and the friction of flow of the motive fluid. These factors are usually expressed as the ratio of the length of the fall pipe to the height of fall. This ratio has an optimum value, for a vertical fall pipe would contain too small a mass of fluid and a relatively flat fall pipe would have too great a friction loss. The optimum value varies with the lift.

The Valves. The variables to be considered in the design of the valves are the weight and the length of the stroke. As a broad generalization it may be stated that the efficiency varies inversely with the length of stroke and the weight of the valve;

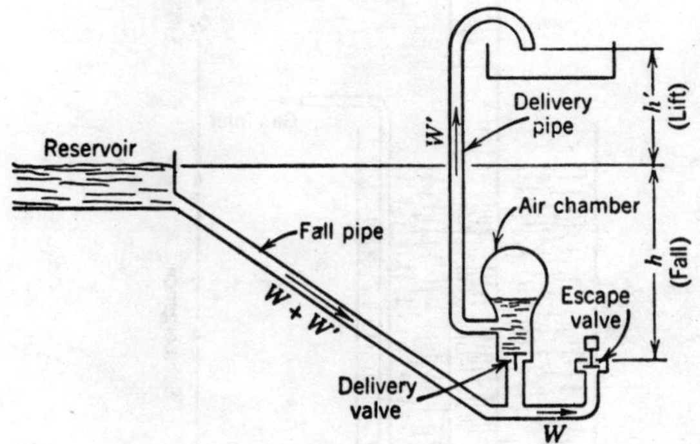

Fig. 212. Sketch indicating principle of operation of the hydraulic ram.

the capacity varies directly with the weight of the valve and the length of the stroke. In addition, decreasing the weight of the valve decreases the length of each cycle.

The Air Chamber. A surge chamber containing an elastic medium such as air is necessary to eliminate intermittent flow in the delivery pipe. In practice, the volume of the air chamber is approximately equal to the volume of the delivery pipe. In order to keep this vessel supplied with air, a small check valve opening inward from the atmosphere is installed just below the delivery valve. At the end of the delivery portion of the cycle a small quantity of air is inspirated and is carried upward at the start of delivery.

The conventional ram may be modified to permit the pumping of one fluid by another. This can be accomplished by replacing the delivery check valve with either a piston or a diaphragm which separates the two fluids. In this manner, for instance, dirty water can be used to pump clean water or some other fluid.

The Air Lift

The air lift is a device for inducing upward flow of a liquid through the introduction of a less dense immiscible fluid. Compressed air is the most common lifting fluid. As shown in Fig. 213 the average density of the material in the submerged column is reduced by the introduction of a gas at the base, and the fluid level will rise above the level of the

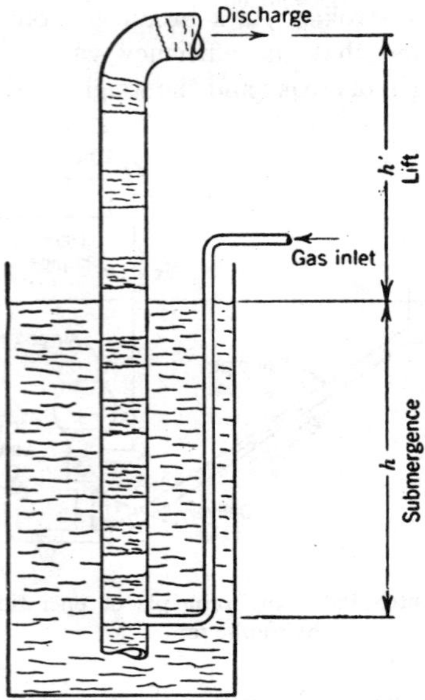

FIG. 213. Sketch indicating principle of operation of the air lift.

submerging liquid. A force balance may be written for the system for the conditions of no net flow of liquid through the column:

$$\rho_l h + \rho_{atm} h' = \left(\frac{V_l \rho_l + V_g (\rho_g)_{avg}}{V_l + V_g} \right) (h + h')$$

$$= \rho_{avg}(h + h')$$

where ρ_l = density of the liquid.

ρ_{atm} = average density of the surrounding atmosphere or fluid.

$(\rho_g)_{avg}$ = average density of the lifting gas (or other lifting fluid) in the column.

ρ_{avg} = the mean density of the contents of the column.

V_l, V_g = the volumes of liquid and gas in the column.

h, h' = the submergence and lift, respectively, as defined by Fig. 213.

This equation states the relation between the height to which the liquid is lifted and the quantity of the lifting fluid required for static equilibrium. As the lifting fluid will always rise because of its lower density, it must be introduced continuously to maintain the liquid level in the column at any desired point.

The quantity of fluid necessary to raise the liquid to any level will be the minimum at the point of no net liquid flow. This minimum quantity which must be continuously charged may be computed if the rate at which the bubbles rise is known.

When quantities of lifting fluid greater than the minimum are used and the liquid is being raised, the flow equation (59) applies to the system. However, there is seldom enough knowledge available to enable a rigorous application of the flow equation. The principal difficulty lies in the determination of the lost work of friction. It is difficult, at best, to estimate accurately the friction loss accompanying the flow of two phases in a pipe. Air-lift systems are further complicated by the existence of a relative velocity, or slip, between the two phases. This factor causes an additional friction loss and an uncertainty as to the velocity to be used in computing the lost work. If the velocities in the column are high enough so that the relative velocity, or slip is small compared to the total velocity, then the lost work may be estimated with sufficient accuracy by conventional methods.

For the air-lift pumping of water an empirical relation has been proposed [19]

$$V_{air} = 0.8 \frac{h'}{C \log \dfrac{h + 34}{34}}$$

where V_{air} is the volume of free air required to lift 1 gal of water, and C is defined by the following list.

h, ft	C
10–60	245
61–200	233
201–500	216
501–650	185
651–750	156

There are many other applications of the same principle as in the tubes of a natural circulatio boiler or evaporator where the lifting fluid is gas generated by vaporizing a portion of the liquid.

For the air lift itself, many different devices, called foot pieces, have been used to inject air into the base of the lift. The common purpose of these various foot pieces is to disperse the air in bubbles as small as possible. As the relative rising velocity, or slip, of a bubble decreases with decreasing bubble diameter, the air consumption will be lowest and the efficiency greatest if the smallest possible bubble is used. The efficiency of a properly designed air lift is at least 75 per cent.

BIBLIOGRAPHY

1. AURORA PUMP CO., Aurora, Ill., Catalogs.
2. BLACKMER PUMP CO., Grand Rapids, Mich., Catalogs.
3. CHURCH, A. H., *Centrifugal Pumps and Blowers*, John Wiley and Sons (1944).
4. COOPER-BESSEMER CORP., Mt. Vernon, Ohio, Catalogs.
5. DAUGHERTY, R. L., *Centrifugal Pumps*, McGraw-Hill Book Co. (1915).
6. DE LAVAL STEAM TURBINE CO., Trenton, N. J., Catalogs.
7. FAIRBANKS, MORSE AND CO., Chicago, Ill., Catalogs.
8. GEROTOR PUMP DIVISION, MAY OIL BURNER CORP., Baltimore, Md., Catalogs.
9. HYDRAULIC INSTITUTE, New York, Standards.
10. INGERSOLL-RAND CO., New York, Catalogs.
11. CARTER, R., I. J. KARASSIK, E. F. WRIGHT, *Pump Questions and Answers*, McGraw-Hill Book Co. (1949).
12. KINNEY MANUFACTURING CO., Boston, Mass., Catalogs.
13. LAYNE AND BOWLER, INC., Memphis, Tenn., Catalogs.
14. NAGLE PUMPS, Chicago Heights, Ill., Catalogs.
15. NASH ENGINEERING CO., S. Norwalk, Conn., Catalogs.
16. PENNSYLVANIA PUMP AND COMPRESSOR CO., Easton, Pa., Catalogs.
17. ROOTS-CONNERSVILLE BLOWER CORP., Connersville, Ind., Catalogs.

17*a*. ROUSE, HUNTER, *Elementary Mechanics of Fluids*, John Wiley and Sons (1946).

18. SCHUTTE AND KOERTING CO., Philadelphia, Pa., Catalogs.

18*a*. STEPANOFF, A. J., *Centrifugal and Axial Flow Pumps*, John Wiley and Sons (1948).

19. SWINDIN, N., *The Modern Theory and Practice of Pumping*, Ernest Benn, Ltd., London (1924).
20. UNION STEAM PUMP CO., Battle Creek, Mich., Catalogs.
21. WORTHINGTON PUMP AND MACHINERY CORP., Harrison, N. J., Catalogs.

PROBLEMS

1. The motor-driven centrifugal pump for which data are given below is to be used for pumping water at 80° F through the piping system shown in the figure which contains 25 ft of 2½-in. schedule-40 straight pipe in the pump suction line and 210 ft of 2-in. schedule-40 pipe in the discharge line. The motor is 220-volt 3-phase alternating current, and the power factor is 0.9. If all the globe valves shown in the sketch are completely open, determine:

(*a*) The flow rate through the piping in gallons per minute.

(*b*) The amperes drawn by the motor driving the pump.

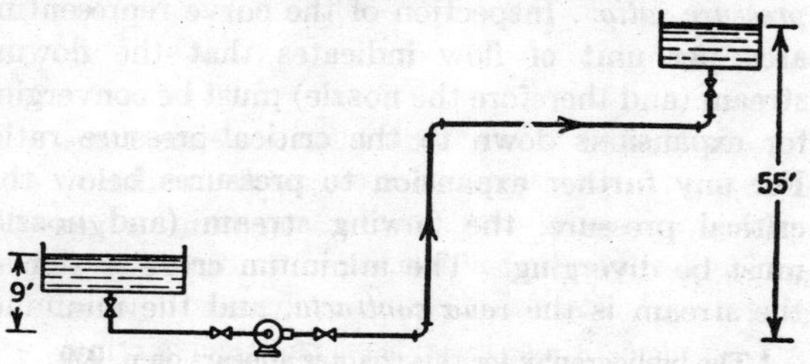

Characteristics of Centrifugal Pump

Capacity, gpm	Developed Head, ft	Efficiency (Pump and Motor)
0	120	0
10	119.5	13
20	117	23.5
30	113	31.6
40	107.5	37.5
50	100.5	42.2
60	93	42.5
70	85	41.7
80	77	39.5

2. A pump is to be selected to replace the old unit in the system below. The pump must pick up 1250 lb of water per minute at 150° F from the condensate hot well and deliver it to the top of the cooling tower at a nozzle pressure of 25 psi. The piping is composed of straight pipe (2½ in. schedule 40), standard elbows, and a gate valve.

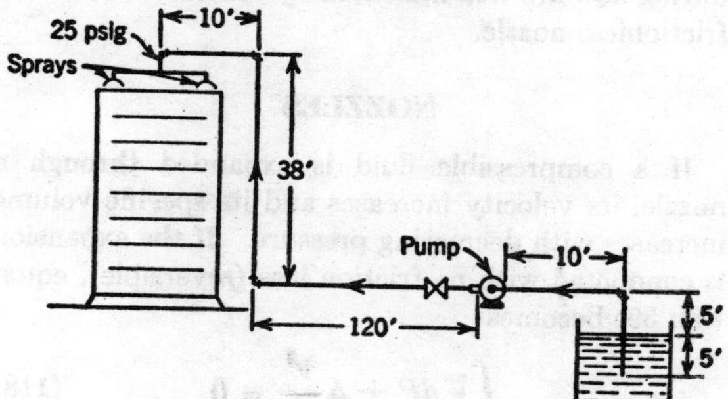

Two used pumps are available, a 12 by 8 by 10 double-acting simplex steam pump with a volumetric efficiency of 80 per cent and a steam pressure ratio of 0.65, and a centrifugal pump with the following characteristic curve:

Gpm	Developed Head, ft of fluid
25	290
50	229
75	186
100	158
125	140
150	127
175	118
200	110

Steam is available at 40 psi.

Would both pumps be suitable for this job? If not, select the more suitable one.

CHAPTER 15

Gas Flow at High Velocity

In the treatment of Chapter 12 the energy absorbed by the increase in velocity of the fluid as it expanded to a lower pressure usually was neglected. For moderate changes in pressure this may not introduce significant error. But, when the pressure drops become large and the gas flows at high velocity, these errors become significant. The importance of these changes in kinetic energy during flow are well indicated by consideration of the frictionless nozzle.

NOZZLES

If a compressible fluid is expanded through a nozzle, its velocity increases and its specific volume increases with decreasing pressure. If the expansion is conducted with no friction loss (reversible), equation 59c becomes

$$\int \bar{V}\, dP + \Delta \frac{v^2}{2g_c} = 0 \qquad (118)$$

and it is possible to convert the drop in pressure into an equivalent increase in velocity of the fluid. This conversion of pressure energy into kinetic energy at relatively low velocities occurs in the venturi meter described in Chapter 13. Throughout any such nozzle operating under steady conditions the mass rate of flow is constant ($W = GA = vA/\bar{V}$), and it is possible to determine the cross-sectional area of the flowing stream as a function of the pressure from the known properties of the fluid (equation 130).

The specific volume $\bar{V}$, velocity v, mass velocity or flow per unit area G, and area per unit mass flowing are plotted for high-velocity flow or large pressure drops as a function of the ratio P/P_0 for a reversible adiabatic expansion[4] * in Fig. 214. At the nozzle inlet, P/P_0 is unity, the velocity is zero, the flow is zero, and the cross-sectional area is infinite. As the pressure decreases along the passage (from right to left in Fig. 214) the flow per unit area reaches a maximum at a value of P/P_0 a little greater than

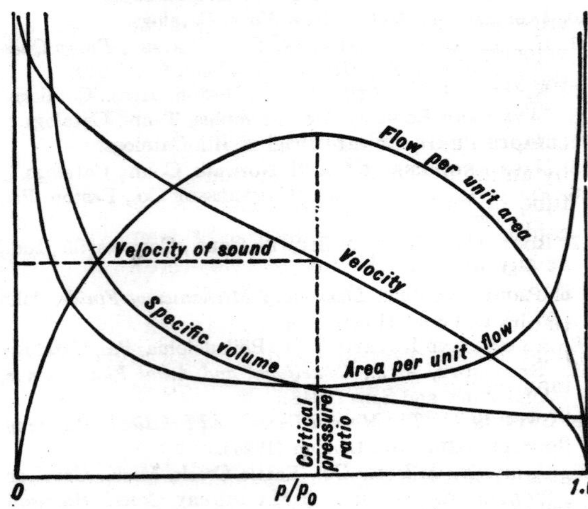

Fig. 214. Characteristics of a reversible adiabatic expansion.[4]

0.5. This pressure ratio corresponding to the maximum flow per unit area, or the minimum cross-sectional area of the stream, is called the *critical-pressure ratio*. Inspection of the curve representing area per unit of flow indicates that the flowing stream (and therefore the nozzle) must be converging for expansions down to the critical-pressure ratio. For any further expansion to pressures below this critical pressure, the flowing stream (and nozzle) must be diverging. The minimum cross section of the stream is the *vena contracta*, and the minimum

* The bibliography for this chapter appears on p. 209.

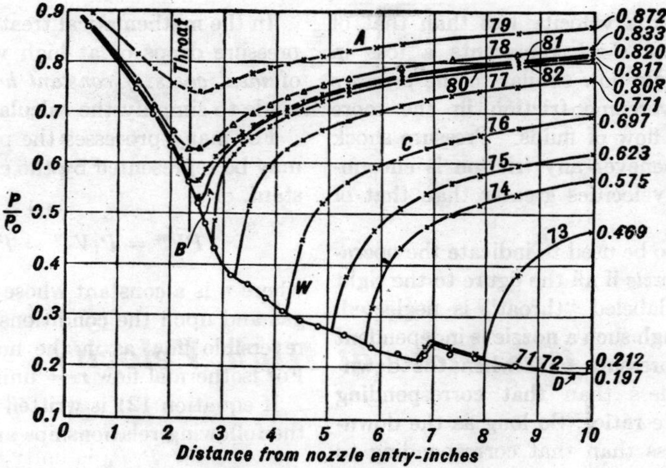

Fig. 215. Pressure distribution in a converging-diverging nozzle expressed in terms of the ratio of pressure to upstream or entrance pressure, P/P_0 for various downstream or exit pressures.[1] The nozzle was designed to operate under the conditions corresponding to runs 71 and 72 represented by curve D. The break in curve D occurring at about 7 in. is caused by sudden condensation of steam to water resulting from supercooling of the steam during expansion and is not a "shock wave."

cross section of the nozzle is called the *throat.* The velocity of the fluid in the throat of the nozzle is the velocity of sound in that fluid.

The velocity of sound is usually derived for a wave comprising an infinitesimal rise in pressure followed by an infinitesimal fall in pressure moving through a tube of uniform cross-sectional area. The fluid is continuous and in a uniform state from wall to wall at any instant. If the observer were to move with the sound wave so that the sound wave would appear to him to be stationary, he could observe a stream of fluid flowing reversibly and adiabatically through an infinitesimal rise and fall in pressure with no change in the cross-sectional area of the stream or in its flow per unit area. Such behavior is found only at a stream velocity corresponding to maximum flow per unit area (a minimum area per unit of flow). Therefore the velocity of sound in the fluid is the velocity of a stream of that fluid at maximum flow per unit area or in the throat of the nozzle. It follows that velocities in a converging nozzle do not exceed the velocity of sound, whereas velocities in, or discharged from, the diverging part of a nozzle may be considerably in excess of the velocity of sound. The smaller the angle used in the divergent nozzle, the less the loss due to lateral expansion, but the greater length of the resulting nozzle leads to increased friction losses. A practical compromise is usually reached in the range between 10 and 30 degrees for the whole angle.

The operation of a *convergent-divergent nozzle* is illustrated in Fig. 215, which shows the pressure as the ratio P/P_0 throughout the length of the nozzle supplied with steam for different downstream pressures.[1] When the pressure ratio is everywhere greater than the critical-pressure ratio and the velocity nowhere exceeds the velocity of sound, the diverging part of the nozzle serves as a diffuser and the nozzle as a whole is in every respect like a venturi meter. This condition is represented by curve A which shows the minimum pressure occurring in the throat of the nozzle. If the exhaust pressure is lowered, the pressure in the throat decreases and the flow through the nozzle increases until the critical-pressure ratio is reached at the throat (curve B). This is the maximum flow rate that can be obtained through the nozzle for the given upstream conditions. No further increase in flow can result from any change in downstream conditions. For all lower exhaust pressures the condition in the throat remains unchanged and the flow remains constant.

When the exhaust pressure is less than that corresponding to curve B but greater than that for which the nozzle is designed (identical expansion ratios of stream and nozzle as represented by curve D), at some section in the diverging passage there will be a standing pressure wave (such as indicated by W on curve C) which is called a *pressure shock.* The fluid enters this wave at low pressure and at a velocity greater than the velocity of sound and leaves it at high pressure at a velocity less than that of sound. This *pressure shock* represents a loss in efficiency for supersonic flow similar to the *pressure drop* required to overcome friction in the more familiar low-velocity flow of fluids. Pressure shock may be expected whenever any friction is encountered with fluids at velocities greater than that of sound.

Figure 215 may also be used to indicate the operation of a *convergent nozzle* if all the figure to the right of the vertical line labeled "throat" is neglected. The rate of flow through such a nozzle is independent of the downstream pressure only when the downstream pressure is less than that corresponding to the critical-pressure ratio. So long as the downstream pressure is less than that corresponding to the critical-pressure ratio, the flow rate is at its maximum and is independent of any change in the downstream pressure. This statement is also true for an orifice, tube, or pipe.

Although the linear velocity may exceed that of sound in a convergent-divergent nozzle, it is impossible to attain a linear velocity in excess of that of sound by expanding a fluid through an orifice, convergent nozzle, tube, or pipe. Any expansion through an orifice or convergent nozzle to a pressure below that corresponding to the critical-pressure ratio takes place irreversibly downstream from the discharge with the energy dissipated in turbulence in the surrounding fluid.

Because the velocity of sound in the fluid is so important in defining the characteristics and limits to flow at high velocities, the ratio of the velocity of a stream to the local velocity of sound in the gas at that point is a useful ratio in treating gases at high velocities. This ratio is called the *Mach number* and is represented by the symbol Ma.

The velocity of an infinitesimal adiabatic pressure wave, or sound wave, in a gaseous medium of pressure P and specific volume $\overline{V}$ is

$$v_c = \sqrt{kg_cP\overline{V}} \tag{119}$$

where $k = C_P/C_V$ (the ratio of the "specific heats").
v_c = acoustic velocity.

The Mach number for an ideal gas is

$$(\text{Ma}) = \frac{v}{\sqrt{kg_cP\overline{V}}} \tag{120}$$

and

$$(\text{Ma})^2 = \frac{v^2}{kg_cP\overline{V}} \tag{120a}$$

In the mathematical treatment of flow under large pressure drops or at high velocity, the assumption of *ideal gases of constant heat capacity* is generally made to simplify the calculations.

For many processes the path followed by the gas may be represented by the expression PV^n = a constant, or

$$PV^n = P_1V_1{}^n = P_2V_2{}^n = P_3V_3{}^n \tag{121}$$

where n is a constant whose value depends upon the gas and upon the conditions of flow. For adiabatic reversible flow as in the nozzle considered, $n = k$. For isothermal flow n = unity.

If equation 121 is written in the logarithmic form the following relationships are apparent

$$\left(\frac{P_2}{P_1}\right) = \left(\frac{V_2}{V_1}\right)^{-n} = \left(\frac{T_2}{T_1}\right)^{\frac{n}{n-1}} \tag{121a}$$

$$\left(\frac{1}{n}\right)\ln\frac{P_2}{P_1} = \left(\frac{1}{n-1}\right)\ln\frac{T_2}{T_1} \tag{121b}$$

When ΔZ, q, and w are each zero, equation 54a may be written in the differential form for unit mass

$$\left(\frac{k}{k-1}\right)\frac{R}{M}\,dT + \frac{v\,dv}{g_c} = 0 \tag{122}$$

since $dH = C_P\,dT$ and $C_P = R[k/(k-1)]$.

Writing equation 120 in logarithmic form, substituting RT/M for $P\overline{V}$, and differentiating,

$$\frac{d(\text{Ma})}{(\text{Ma})} = \frac{dv}{v} - \frac{1}{2}\frac{dT}{T} \tag{123}$$

Frictionless Adiabatic Flow

For reversible adiabatic conditions, equation 121b becomes

$$\frac{1}{k}\,d\ln P = \frac{1}{k-1}\,d\ln T$$

$$\left(\frac{k-1}{k}\right)\frac{dP}{P} = \frac{dT}{T} \tag{124}$$

Combining equations 122, 123, and 124 with equation 120a and $P\overline{V} = RT/M$,

$$\frac{dv}{v} = \frac{-kg_cR\,dT}{(k-1)Mv^2} = \frac{d(\text{Ma})}{(\text{Ma})} + \frac{1}{2}\left(\frac{dT}{T}\right) \tag{125}$$

$$\frac{dT}{T} = \left(\frac{k-1}{k}\right)\frac{dP}{P}$$

$$= -\frac{d(\text{Ma})}{(\text{Ma})}\left[\frac{1}{\frac{1}{(k-1)(\text{Ma})^2} + \frac{1}{2}}\right] \tag{126}$$

Rearranging and integrating from $P = P_0$ to $P = P$, with $v_0 = 0$ or $(\text{Ma})_0 = 0$,

$$\left(\frac{P_0}{P}\right)^{\frac{k-1}{k}} = \frac{k-1}{2}\left((\text{Ma})^2 + \frac{2}{k-1}\right) \tag{127}$$

$$(\text{Ma}) = \sqrt{\frac{2}{k-1}\left[\left(\frac{P_0}{P}\right)^{\frac{k-1}{k}} - 1\right]} \tag{128}$$

Maximum Adiabatic Flow Rate

Equation 127 can be solved for P_0/P.

$$\frac{P_0}{P} = \left[\frac{2 + (k-1)(\text{Ma})^2}{2}\right]^{\frac{k}{k-1}} \tag{129}$$

The cross-sectional area of a flowing stream expanding adiabatically and reversibly through a nozzle may be similarly shown to be given by equation 130

$$\frac{A}{A_0} = \frac{(\text{Ma})_0}{(\text{Ma})}\left[\frac{1 + \frac{k-1}{2}(\text{Ma})^2}{1 + \frac{k-1}{2}(\text{Ma})_0^2}\right]^{\frac{k+1}{2(k-1)}} \tag{130}$$

which may be derived from

$$\frac{A}{A_0} = \frac{\overline{V}v_0}{\overline{V}_0 v}$$

$$\frac{v_0}{v} = \frac{(\text{Ma})_0}{(\text{Ma})}\sqrt{\frac{P_0\overline{V}_0}{P\overline{V}}}$$

$$\frac{V}{V_0} = \left(\frac{P_0}{P}\right)^{\frac{1}{k}}$$

$$\left(\frac{P_0}{P}\right)^{\frac{k-1}{k}} = \frac{(\text{Ma})^2\left(\frac{k-1}{2}\right) + 1}{(\text{Ma})_0^2\left(\frac{k-1}{2}\right) + 1}$$

Inspection of equation 130 or plotting on any arbitrary scale indicates that the area of the stream A is a minimum when Ma is unity or the velocity is equal to the acoustic velocity.

Since $dU = C_V\,dT$; $C_P - C_V = R/M$; and $(R/M)\,dT = d(P\overline{V})$,

$$dU = \frac{d(P\overline{V})}{(k-1)} \tag{131}$$

Substituting in equation 54 in differential form, with q, w, and dZ each equal to zero,

$$\frac{d(P\overline{V})}{(k-1)} + d\frac{v^2}{2g_c} + d(P\overline{V}) = 0$$

Solving for $d(P\overline{V})$,

$$d(P\overline{V}) = -\left(\frac{k-1}{k}\right)d\left(\frac{v^2}{2g_c}\right) \tag{132}$$

Integrating between the limits $v = 0$ and $v = v_1$; and $P\overline{V} = P_0\overline{V}_0$ and $P\overline{V} = P_1\overline{V}_1$,

$$v_1^2 = \frac{2g_ck}{k-1}(P_0\overline{V}_0 - P_1\overline{V}_1)$$

$$= \frac{2g_ckR}{(k-1)M}(T_0 - T_1) \tag{133}$$

$$v_1^2 = \frac{2g_ck}{k-1}(P_0\overline{V}_0)\left[1 - \left(\frac{P_1}{P_0}\right)^{\frac{k-1}{k}}\right] \tag{134}$$

$$G_1^2 = \left(\frac{2g_ck}{k-1}\right)\left(\frac{P_0}{\overline{V}_0}\right)\left(\frac{P_1}{P_0}\right)^{\frac{2}{k}}\left[1 - \left(\frac{P_1}{P_0}\right)^{\frac{k-1}{k}}\right] \tag{135}$$

For maximum mass velocity, $dG_1/dP_1 = 0$. Making the differentiation and changing P_1 to P_c to represent critical flow conditions

$$\frac{P_c}{P_0} = \left(\frac{2}{k+1}\right)^{\frac{k}{k-1}} \tag{136}$$

Substituting equation 136 in equations 128, 134, and 135 gives for the maximum possible discharge

$$v_c^2 = g_ckP_c\overline{V}_c \tag{137}$$

$$(\text{Ma})_c^2 = 1 \tag{138}$$

$$G_{cna}^2 = g_ck\left(\frac{P_0}{\overline{V}_0}\right)\left(\frac{2}{k+1}\right)^{\frac{k+1}{k-1}} \tag{139}$$

where G_{cna} = maximum flow rate in nozzle under adiabatic conditions, which corresponds to a velocity in the throat equal to the acoustic velocity and a Mach number of unity.

The Critical Flow Prover

Since the mass velocity through a nozzle depends only on upstream conditions as indicated by equation 139 when the downstream pressure is less than that corresponding to the critical-pressure ratio, a nozzle operating under critical flow conditions makes a most convenient device for measuring flow. The diverging section is unnecessary. Any carefully streamlined nozzle which may be simply a rounded-entry thick-plate orifice, operating under critical flow conditions may be used for calibrating gas-flow meters or for controlling the rate of flow independent of downstream pressure. The coefficient of discharge may be as high as 0.95 to 0.99. Usually the capacity of the prover is stated in gas volume per unit of time for upstream conditions of the gas, thereby eliminating the difficult determination of the area of the throat of a small nozzle.

Sharp-edged orifices are unsatisfactory as critical flow provers, as the discharge is not entirely independent of the downstream pressure.

Frictionless Isothermal Flow

As has been indicated, $n = 1$ for isothermal flow. Under such conditions the calculations become simple. Lapple [6] has developed a method of calculation for high-velocity flow based on a computed critical mass velocity through a nozzle in isothermal flow. Although such conditions of flow may never exist, such a hypothetical velocity affords a convenient basis for calculation.

If the flow is frictionless and no work is done, equation 59a or 118 becomes

$$\int_{P_0}^{P} \overline{V}\, dP + \int_{v_0}^{v} d\left(\frac{v^2}{2g_c}\right) = 0 \qquad (140)$$

For isothermal flow, $\overline{V} = P_0\overline{V}_0/P$; and for a nozzle, $v_0 = 0$. Integration of equation 140 then gives

$$v^2 = 2g_c P_0 \overline{V}_0 \ln\left(\frac{P_0}{P}\right) \qquad (141)$$

For isothermal conditions $n = 1$, or k may be taken as unity, and the equation corresponding to equation 120a would be

$$(\mathrm{Ma}')^2 = \frac{v^2}{g_c P \overline{V}} \quad \text{for } k = 1 \qquad (120b)$$

Substituting equation 141 in equation 120b,

$$(\mathrm{Ma}')^2 = \ln\left(\frac{P_0}{P}\right)^2 \qquad (142)$$

$$\left(\frac{P}{P_0}\right) = e^{-\frac{(Ma')^2}{2}} \qquad (142a)$$

The mass velocity in the throat of the nozzle may be expressed by substituting in equation 141 G_1 for $v_1/\overline{V}_1$, and $P_0\overline{V}_0/P_1$ for $\overline{V}_1$.

$$G_1^2 = 2g_c\left(\frac{P_0}{\overline{V}_0}\right)\left(\frac{P_1}{P_0}\right)^2 \ln\left(\frac{P_0}{P_1}\right) \qquad (143)$$

If this equation is plotted with G_1 and P_1 as variables for given upstream conditions, P_0 and V_0, the equation indicates that the mass velocity G_1 increases to a maximum as P_1 decreases and then decreases for further reductions in P_1. The condition for maximum discharge is met when $(dG_1/dP_1) = 0$. Differentiation of equation 143 with respect to P_1 and equating $(dG_1/dP_1) = 0$ gives

$$\ln\left(\frac{P_0}{P_c}\right) = \frac{1}{2} \quad \text{or} \quad \left(\frac{P_0}{P_c}\right)^2 = e \qquad (144)$$

where P_c has been substituted for P_1 to represent critical flow conditions.

Substituting equation 144 in equations 141, 142, and 143,

$$v_c^2 = g_c P_0 \overline{V}_0 \qquad (145)$$

$$(\mathrm{Ma}')^2 = 1 \qquad (146)$$

$$G_{cni}^2 = \frac{g_c P_0}{e\overline{V}_0} = P_0^2 \frac{g_c M}{eRT_0} \qquad (147)$$

where G_{cni} = critical mass velocity through a nozzle in isothermal flow.

From equations 145 and 146 it is clear that the maximum flow possible through a nozzle under isothermal conditions corresponds to the fluid attaining an "acoustic velocity" in the nozzle throat equal to the velocity of an infinitesimal isothermal pressure wave through the gas at throat conditions.

Ratio of Flow Rate to Maximum Isothermal Flow

Equation 143 may be divided by equation 147 to give

$$\left(\frac{G_1}{G_{cni}}\right)^2 = e\left(\frac{P_1}{P_0}\right)^2 \ln\left(\frac{P_0}{P_1}\right)^2 \qquad (148)$$

By substituting from equation 142*a*, equation 148 becomes

$$\left(\frac{G_1}{G_{cni}}\right)^2 = e(\text{Ma}')_1{}^2 e^{-(\text{Ma}')_1{}^2} \quad (149)$$

In this way all isothermal flow rates are expressed in terms of a ratio to a calculated maximum discharge through a nozzle under isothermal flow conditions (G_{cni}) which is related to upstream conditions by equation 147. The time of flow through a nozzle is so short that isothermal flow cannot be realized, but the calculated value for G_{cni} makes a convenient rational reference for use in calculating flow.[6]

Frictionless Nonisothermal Flow

For flow under conditions which are not isothermal, the value of n is not unity, and equations corresponding to 148 and 149 may be developed by combining equations similar to 139, 135, and 128 by substituting n for k to give:

$$\left(\frac{G}{G_{cnn}}\right)^2 = \left(\frac{n+1}{n-1}\right)\left(\frac{n+1}{2}\right)^{\frac{2}{n-1}}\left(\frac{P_1}{P_0}\right)^{\frac{2}{n}}\left[1-\left(\frac{P_1}{P_0}\right)^{\frac{n-1}{n}}\right] \quad (150)$$

$$\left(\frac{G}{G_{cnn}}\right)^2 = (\text{Ma}')^2\left[\frac{n+1}{2+(n-1)(\text{Ma}')^2}\right]^{\frac{n+1}{n-1}} \quad (151)$$

The ratio between the maximum flow rates under nonisothermal and isothermal conditions is obtained by combining an equation similar to 139 with n substituted for k, with equation 147.

$$\left(\frac{G_{cnn}}{G_{cni}}\right) = \sqrt{en}\left(\frac{2}{n+1}\right)^{\frac{(n+1)}{2(n-1)}} \quad (152)$$

By substituting k for n in equation 152 the ratio of maximum flow for adiabatic conditions to maximum flow for isothermal conditions is obtained. This ratio is plotted as a function of k in Fig. 216.

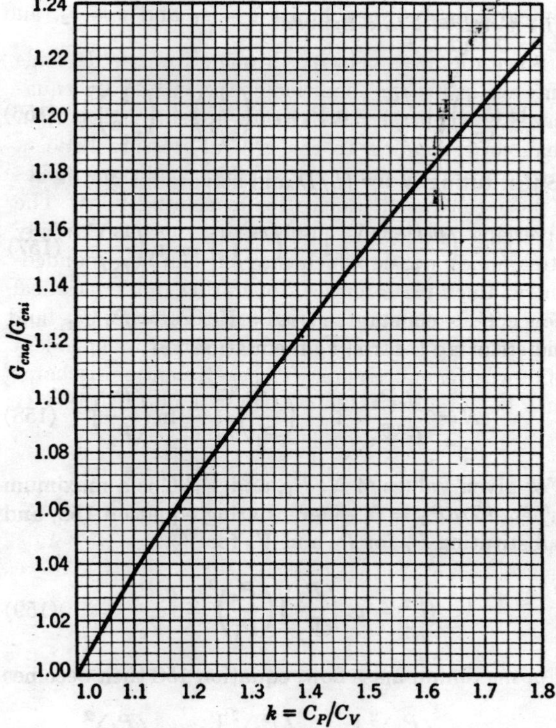

Fig. 216. Ratio of maximum flow for adiabatic conditions to maximum flow for isothermal conditions as a function of k.[6]

PIPES

For flow in a pipe or wherever friction losses must be considered, equation 63 may be substituted in equation 59*a* for unit mass of fluid, giving, when $w = 0$,

$$\overline{V}\,dP + \frac{d(v^2)}{2g_c} + \frac{fv^2}{2g_cD}\,dL = 0 \quad (153)$$

Irreversible Isothermal Flow

For isothermal conditions, $d(P\overline{V}) = 0 = P\,d\overline{V} + \overline{V}\,dP$; therefore,

$$\frac{d(v^2)}{v^2} + \frac{f}{D}\,dL = \left(\frac{2g_c}{v^2}\right)P\,d\overline{V} \quad (154)$$

Since $P = P_1\overline{V}_1/\overline{V}$ for isothermal flow, and for a pipe of constant cross section the mass velocity ($G = v/\overline{V}$) is constant and $d\overline{V}/\overline{V} = dv/v$, equation 154 becomes

$$\frac{d(v^2)}{v^2} + \frac{f}{D}\,dL = 2g_cP_1\overline{V}_1\frac{dv}{v^3} \quad (155)$$

Fig. 216A. Diagram indicating significance of subscripts for flow through pipes.

Integrating between limits $v = v_1$ and $v = v_2$, and $L = 0$ and $L = L$,

$$\ln\left(\frac{v_2}{v_1}\right)^2 + \frac{fL}{D} = g_c P_1 \overline{V}_1 \left[\left(\frac{1}{v_1^2}\right) - \left(\frac{1}{v_2^2}\right)\right] \quad (156)$$

Substituting N for fL/D and solving for N,

$$N = \left(\frac{g_c P_1 \overline{V}_1}{v_1^2}\right)\left[1 - \left(\frac{v_1}{v_2}\right)^2\right] - \ln\left(\frac{v_2}{v_1}\right)^2 \quad (157)$$

Since G is constant, $v_1/v_2 = \overline{V}_1/\overline{V}_2 = P_2/P_1$, and substituting G for $v/\overline{V}$ in equation 157,

$$N = \frac{g_c P_1}{G^2 \overline{V}_1}\left[1 - \left(\frac{P_2}{P_1}\right)^2\right] - \ln\left(\frac{P_1}{P_2}\right)^2 \quad (158)$$

For given values of N, P_1, and $\overline{V}_1$, G is a maximum when $dG/dP_2 = 0$, differentiating equation 158, and substituting P_c for P_2 and $\overline{V}_c$ for $\overline{V}_2$,

$$G_c^2 = g_c\left(\frac{P_1}{\overline{V}_1}\right)\left(\frac{P_c}{P_1}\right)^2 = \frac{g_c P_c}{\overline{V}_c} \quad (159)$$

For maximum mass flow, equation 158 then becomes

$$N_c = \left(\frac{P_1}{P_c}\right)^2\left[1 - \left(\frac{P_c}{P_1}\right)^2\right] - \ln\left(\frac{P_1}{P_c}\right)^2 \quad (160)$$

$$N_c = \left(\frac{P_1}{P_c}\right)^2 - \ln\left(\frac{P_1}{P_c}\right)^2 - 1 \quad (160a)$$

From equation 120a and equation 159

$$(\text{Ma}')_c^2 = \frac{v_c^2}{(g_c P_c \overline{V}_c)} = \frac{(G_c^2 \overline{V}_c)}{(g_c P_c)} = 1 \quad (161)$$

and it follows that the maximum possible mass flow through a pipe including friction effects corresponds to the "acoustic velocity" under the conditions of flow being attained by the gas in the pipe exit.

Irreversible Adiabatic Flow

By a series of operations similar to those followed in deriving equation 159, equation 162 may be derived for adiabatic flow in a pipe,[6]

Stagnation Temperature

In adiabatic flow the temperature is not constant but varies with the pressure or velocity or Mach number, as indicated in equation 126. The flowing temperature would be indicated by a thermometer moving at the same velocity as the fluid whose temperature is being measured. The "stagnation temperature" (or "total temperature") is the temperature attained by the gas upon bringing the flowing gas to rest by a reversible adiabatic process and can be shown to be

$$T_0 = T\left[1 + \left(\frac{k-1}{2}\right)(\text{Ma})^2\right] \quad (163)$$

where T_0 is "stagnation temperature."
T is flowing temperature.

The adiabatic wall temperature T_a is the temperature of the wall that will maintain adiabatic flow of the fluid and is, therefore, equal to the temperature of the stagnant gas adjacent to the wall. If the flowing gas were brought to the stagnant condition reversibly and adiabatically, the adiabatic wall temperature T_a would equal the stagnation temperature T_0. Actually this is not the case. Turbulence introduces irreversibilities evident as friction losses, and heat transfer within the gas tends to equalize temperature differences between the flowing and stagnant gas. As a result, the stagnant gas actually attains a temperature greater than the flowing temperature but less than the stagnant temperature. Therefore, the adiabatic wall temperature is intermediate between the flowing and stagnant temperature.

The recovery factor is the ratio

$$\frac{T_a - T}{T_0 - T} \quad (163a)$$

Heat transfer rates should be calculated using the adiabatic wall temperature T_a as the effective temperature of the gas rather than either the flowing temperature T or the stagnation temperature T_0.

$$N = \left(\frac{1}{k}\right)\left[\frac{2kg_c P_1 + (k-1)G^2\overline{V}_1}{2G^2\overline{V}_1}\right]\left[1 - \left(\frac{\overline{V}_1}{\overline{V}_2}\right)^2\right] - \frac{k+1}{2k}\ln\left(\frac{\overline{V}_2}{\overline{V}_1}\right)^2 \quad (162)$$

$$G_c^2 = \frac{kg_c P_c}{\overline{V}_c} \quad (162a)$$

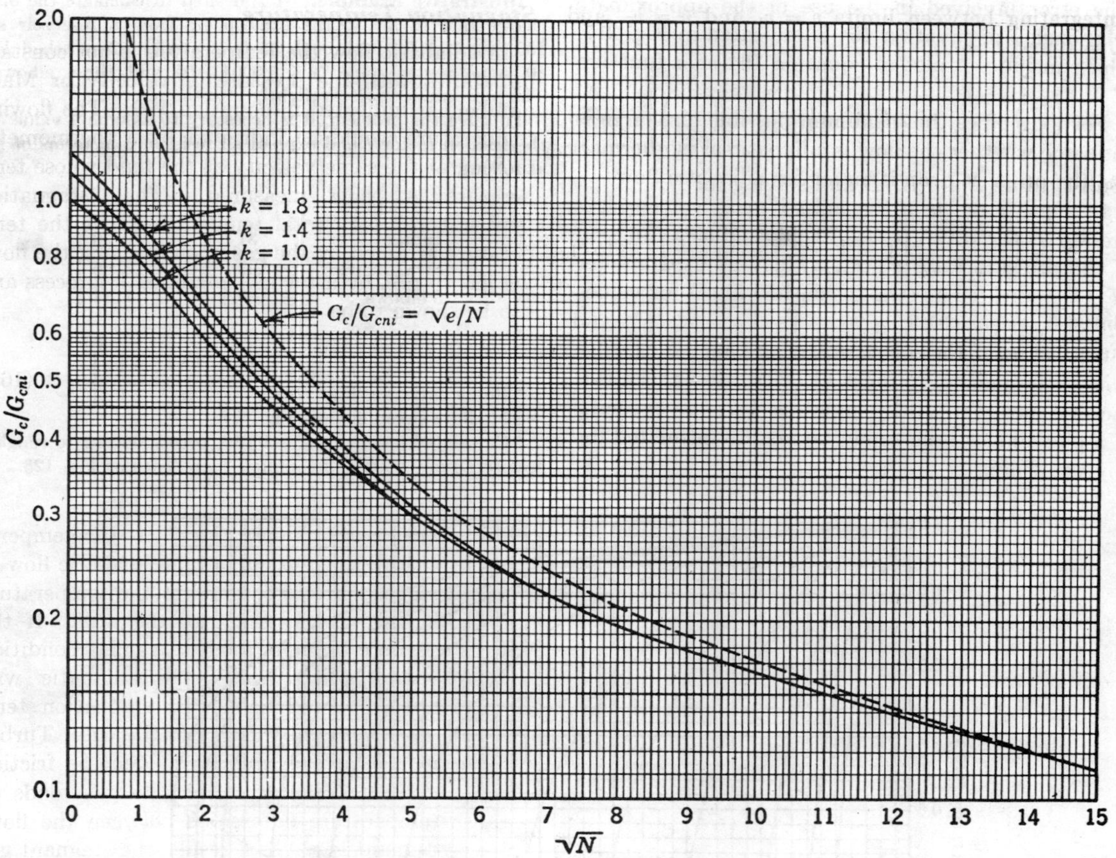

Fig. 217. Ratio of maximum flow to isothermal maximum flow [6] as a function of $\sqrt{N} = \sqrt{fL/D}$ where f = friction factor, L = length of pipe, and D = diameter of pipe.

The distinction between flowing temperature and stagnation temperature is somewhat similar to the distinction between static pressure and impact pressure as discussed in Chapter 14 in connection with the pitot tube.

Methods of Calculating Pipe Capacities [6]

The ratio of maximum flow in a pipe, G_c, to that for isothermal conditions in a nozzle, G_{cni}, is given in Fig. 217 as a function of equivalent pipe length.

Since the isothermal expressions are identical with the corresponding nonisothermal (or adiabatic) expressions for a value of k equal to unity, isothermal flow may be considered as equivalent to adiabatic flow of a hypothetical gas having a value of k equal to unity. For convenience then, all flow conditions are expressed in terms of values for k, N, and P_2/P_c or P_3/P_0, and the mass velocity is expressed as the ratio G/G_{cni}, since G_{cni} may be readily calculated in terms of the entrance conditions in reservoir ahead of the pipe by equation 147.

Figure 217 gives values for the maximum possible discharge (G_c/G_{cni}) as a function of N for various values of k, for given upstream conditions incorporated in G_{cni}. For long pipes the maximum discharge may be estimated by the approximate relationship.

$$\frac{G_c}{G_{cni}} = \sqrt{\frac{e}{N}} \tag{164}$$

and the critical flow pressure may be estimated by the approximation

$$\frac{P_c}{P_0} = \sqrt{\left(\frac{2}{k+1}\right)\left(\frac{1}{kN}\right)} \tag{165}$$

The error involved in the use of the approximate relationship of equation 164 is indicated in Fig. 217, which includes a plot of equation 164 as a function of N.

For long pipes the maximum possible discharge is the same whether the flow is adiabatic or isothermal, but the pressure drop is larger for adiabatic flow.

In Figs. 218*a*, *b*, *c* the actual mass velocity expressed as the ratio (G/G_{cni}) is plotted as a function of the pressure ratio P_3/P_0 with N as a parameter (T_2/T_0 indicated as straight dashed lines) for values of k of 1, 1.4, and 1.8.

By means of these charts and the simple calculation of the maximum mass velocity through an isothermal nozzle (G_{cni}) by equation 147, the capacity of a pipe line or orifice may be readily estimated for any value of k by interpolation.

Illustrative Example. It is desired to calculate the discharge rate of air to the atmosphere from a reservoir at 150 psig and 70° F through 33 ft of straight 2-in. schedule-40 smooth steel pipe (ϵ = 0.00011 ft) and three elbows. The pipe intake projects slightly into the reservoir.

To solve the problem, it is necessary to assume a value of the friction factor f and express all resistance in terms of N as follows:

Resistance	(L/D)	(N)
Intake	31 *	0.50 ‡
Straight pipe	192	3.06 ‡
3 elbows	90 †	1.44 ‡
Total		5.00

* Borda entrance, Fig. 127.
† Medium sweep elbows, Fig. 127.
‡ Calculated, assuming f = 0.016, corresponding to roughness (ϵ/D) of 0.0004 and complete turbulence on Fig. 125.

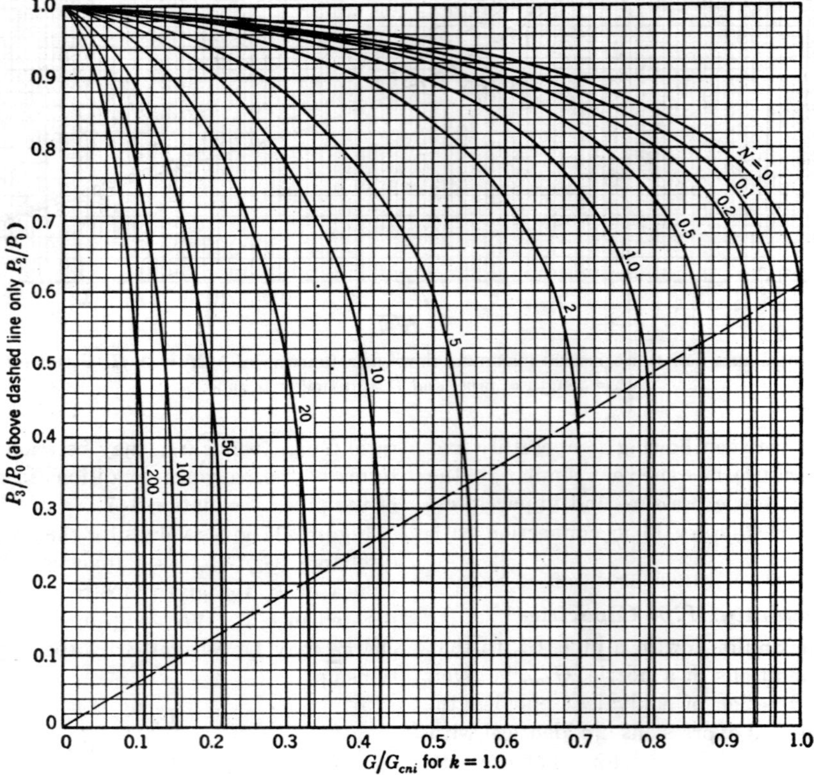

FIG. 218*a*. k = 1.0. Ratio of downstream pressure to upstream pressure as a function of flow ratio G/G_{cni} and resistance factor N.[6]

$$G_{cni} = \sqrt{\frac{g_c P_0}{e \overline{V}_0}} = P_0 \sqrt{\frac{g_c M}{e R T_0}}$$

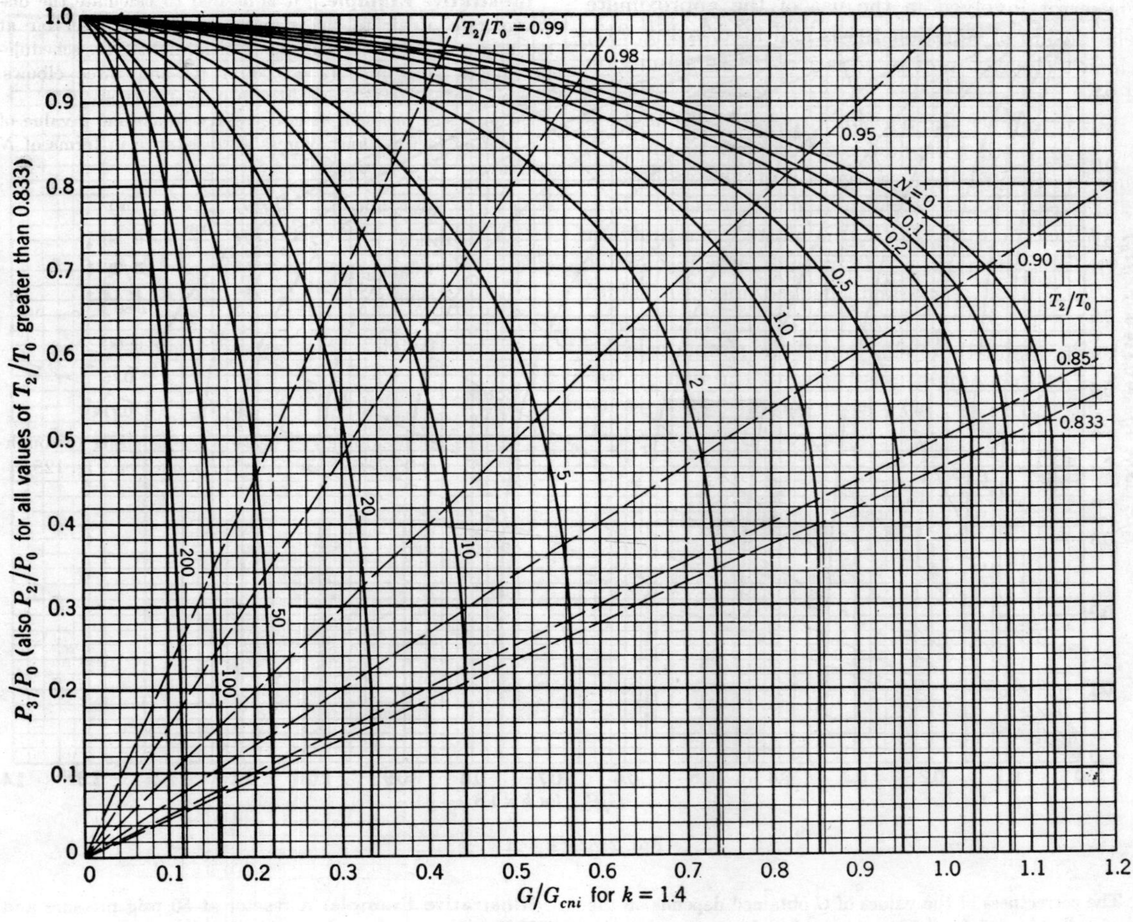

FIG. 218b. $k = 1.4$.

From the conditions of the problem,

$$T_0 = 530° \text{ R}$$

$$P_0 = (150 + 14.7)(144) = 23{,}700 \text{ lb force/sq ft}$$

$$P_3 = (14.7)(144) = 2210 \text{ lb force/sq ft}$$

$$(P_3/P_0) = 0.0893$$

$$M = 29 \text{ (lb mass)/(lb mole)}$$

From the relationship:

$$G_{cni} = P_0\sqrt{\frac{g_c M}{eRT_0}}$$

$$G_{cni} = (23{,}700)\sqrt{\frac{(32.17)(29)}{(2.718)(1546)(530)}}$$

$$= 486 \text{ pounds mass/(sec)(sq ft)}$$

$$D = (2.067/12) = 0.1722 \text{ ft.}$$

$$\text{Pipe cross section} = (0.785)(0.1722)^2 = 0.0233 \text{ sq ft}$$

It is now possible to calculate the discharge by direct use of Fig. 218 as shown by the following tabulation:

	Isothermal	Adiabatic
k	1.0	1.4
(G/G_{cni}), from curve for N = 5.0 and $(P_3/P_0) = 0.0893$ (Fig. 218)	0.545	0.565
G [lb mass/(sec)(sq ft)]	265	275
Discharge rate (lb mass/sec)	6.18	6.41
(T_2/T_0), from dashed line, since (P_3/P_0) is below dashed line at $N = 5.0$	1.0	0.833
T_2 (°Rankine)	530	442
Average gas temperature in pipe (°F)	70	26
μ, at average gas temperature [lb mass/(ft)(sec)]	$1.21(10)^{-5}$	$1.14(10)^{-5}$
Re or (DG/μ)	$3.77(10)^6$	$4.16(10)^6$
(f Fig. 125 for ϵ/D of 0.0004)	0.016	0.016

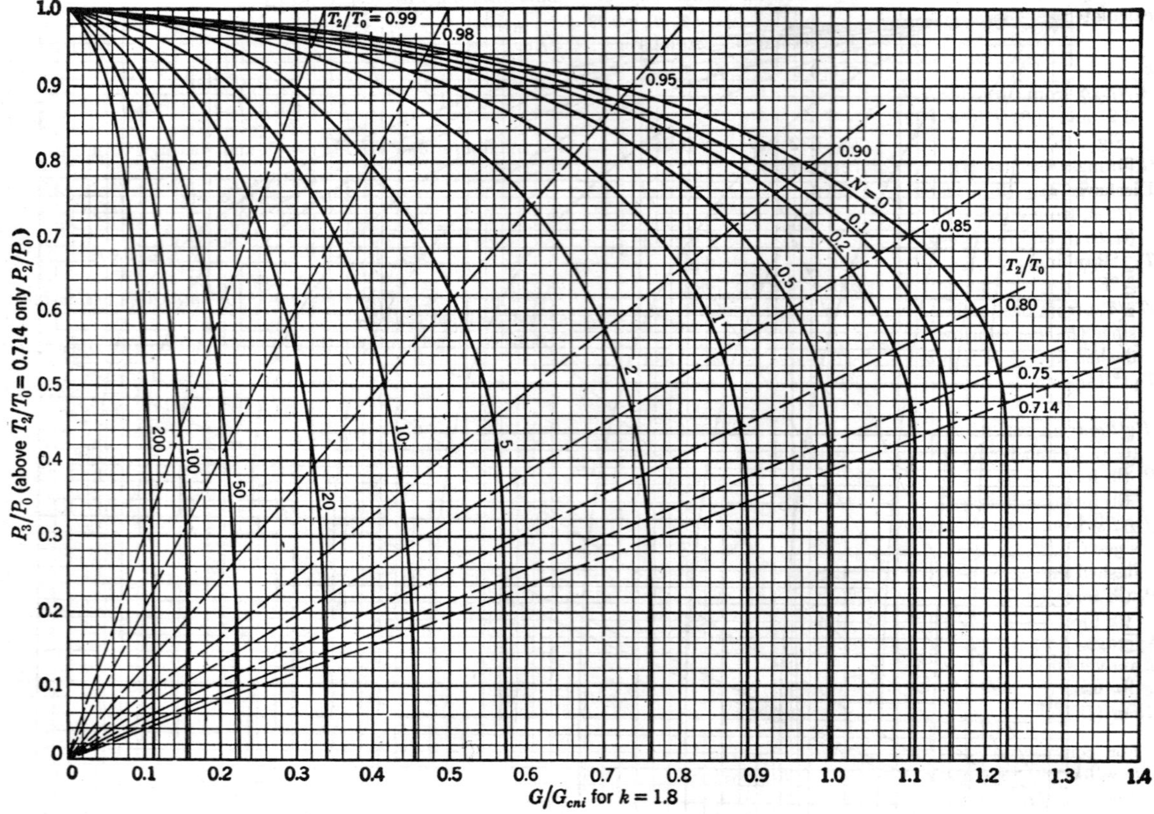

FIG. 218c. $k = 1.8$.

The correctness of the values of G obtained depends on the assumed value of f, which was checked by computing the Reynolds number from the values of G and average temperature obtained from the charts. If the value of f had been incorrectly assumed, a repeat calculation would be required. It should never be necessary to make more than one repeat calculation since the effect of f is not very pronounced. It is seldom necessary to make this check calculation when G/G_{cni} is a maximum, as this condition is usually accompanied by complete turbulence. Although in checking the value of f the average temperature was calculated to obtain the average gas viscosity, sufficient accuracy (better than 2 per cent) will usually result if the viscosity is taken at the upstream temperature.

If the regular friction factor method described in Chapter 12 with all terms defined by the conditions corresponding to the arithmetic average of the upstream and downstream pressures had been employed in the above solution, the calculated discharge would be 356 lb mass/(sec)(sq ft) or 8.30 lb/sec.

Illustrative Example: A reactor at 50 psig pressure and 200° F releases hydrogen chloride gas to a scrubber at substantially atmospheric pressure through a 100-ft straight length of 6-in. ID pipe with a rounded inlet. What is the rate of discharge of hydrogen chloride, and what are the conditions of temperature and pressure within the pipe?

Assuming $f = 0.016$, $N = (0.016)(100)(12)/(6) = 3.2$. From the conditions of the problem

$$T_0 = 660° \text{ R}$$

$$P_0 = (50 + 14.7)(144) = 9330 \text{ lb force/sq ft}$$

$$P_3 = (14.7)(144) = 2120 \text{ lb force/sq ft}$$

$$(P_3/P_0) = 0.227$$

$$M = 36.5 \text{ lb mass/mole}$$

$$G_{cni} = (9330)\sqrt{\frac{(32.17)(36.5)}{(2.718)(1546)(660)}}$$

$$= 192.0 \text{ lb mass/(sec)(sq ft)}$$

$$D = (6.00/12) = 0.500 \text{ ft}$$

$$\text{Pipe cross section} = (0.785)(0.500)^2 = 0.1962 \text{ sq ft}$$

Using Fig. 218 (*a* and *b*), the following terms may be calculated.

	Isothermal	Adiabatic
k	1.0	1.4
(G/G_{cni}), from curve for N = 3.2 and $(P_3/P_0) = 0.227$ (Fig. 218)	0.632	0.662
G [lb mass/(sec)(sq ft)]	121.2	127.2
Discharge rate (lb mass/sec)	23.8	25.0
(T_1/T_0), from curve for (G/G_{cni}) and for $N = 0$	1.0	0.973
T_1 (°Rankine)	660	642
(°F)	200	182
(T_2/T_0) from dashed line, since (P_3/P_0) is below dashed line at $N = 3.2$	1.0	0.833
T_2 (°Rankine)	660	550
(°F)	200	90
(P_1/P_0) from curve for $(G.G_{cni})$ and for $N = 0$	0.916	0.910
P_1 (lb force/sq ft)	8550	8490
(lb force/sq in. gage)	44.6	44.2
(P_2/P_0) from dashed line, since (P_3/P_0) is below dashed line at $N = 3.2$	0.383	0.309
P_2 (lb force/sq ft)	3570	2880
(lb force/sq in. gage)	10.1	5.3
Average of T_1 and T_2 (°F)	200	136
μ, at average gas temperature [lb mass/(ft)(sec)]	$1.15(10)^{-5}$	$1.04(10)^{-5}$
Re or (DG/μ)	$5.28(10)^6$	$6.11(10)^6$

For commercial steel pipe of 6-in. I.D. Fig. 126 gives $\epsilon/D = 0.0003$. From Fig. 125, $f = 0.015$ for this pipe at all values for Re in excess of $5(10)^6$. Therefore the correct value for f should be 0.015 rather than the assumed value of 0.016. The difference, however, is almost negligible as it would decrease the value of N from 3.2 to 3.0 indicating values of

	Isothermal	Adiabatic
(G/G_{cni})	0.64	0.67
G [lb mass/(sec) (sq ft)]	122.9	128.6
Discharge Rate (lb mass/sec)	24.1	25.3

with no significant changes in temperatures or pressures.

ACTUAL GASES

By introducing the compressibility factor, z (see pages 145, 159 and 591) the above equations can be modified to cover actual as well as ideal gases. Substituting zRT for RT; $P_0\bar{V}_0 z = P\bar{V}z_0$ for isothermal conditions and equation 140 when integrated becomes another form of equation 141:

$$v^2 = -2g_c \int_{P_0}^{P} \frac{P_0\bar{V}_0 z}{Pz_0}\, dP = -2g_c P_0 \bar{V}_0 \left(\frac{z_{avg}}{z_0}\right) \ln \frac{P}{P_0} \quad (141)$$

By definition

$$G_1 = \frac{v_1 P_1 z_0}{P_0 \bar{V}_0 z_1}$$

$$G_1^2 = -2g_c \left(\frac{P_1}{P_0}\right)^2 \left(\frac{P_0}{\bar{V}_0}\right) \left(\frac{z_0 z_{avg}}{z_1^2}\right) \ln \left(\frac{P_1}{P_0}\right) \quad (143)$$

differentiating with respect to P_1

$$2G_1 \frac{dG_1}{dP_1} = -2g_c \left(\frac{z_0 z_{avg}}{z_1^2}\right)\left(\frac{P_0}{\bar{V}_0}\right)\left[\left(\frac{P_1}{P_0}\right)^2 \frac{P_0}{P_1} + 2\left(\frac{P_1}{P_0}\right) \ln \left(\frac{P_1}{P_0}\right)\right]$$

For maximum flow $dG_1/dP_1 = 0$, and writing P_c for P_1

$$1 + 2 \ln \left(\frac{P_c}{P_0}\right) = 0, \text{ and}$$

$$\ln \left(\frac{P_0}{P_c}\right) = \tfrac{1}{2} \quad \text{or} \quad \left(\frac{P_0}{P_c}\right)^2 = e \quad (144)$$

This demonstrates that equation 144 is applicable to non-ideal as well as ideal gases.

Substituting from equation 144 in equation 143 gives

$$G_{cni} = g_c \left(\frac{P_0}{\bar{V}_0}\right)\left(\frac{z_0 z_{avg}}{z_c^2}\right)\frac{1}{e} = P_0^2 \frac{g_c M}{eRT}\left(\frac{z_{avg}}{z_c^2}\right) \quad (147)$$

BIBLIOGRAPHY

1. Binney and Woods, *Proc. Inst. Mech. Eng. (London)*, **138**, 200 (1938).
2. Dodge, R. A., and M. J. Thompson, *Fluid Mechanics*, McGraw-Hill Book Co. (1937).
3. Froessel, W., *Forsch. Gebiete Ingenieurw.*, **7**, 75–84 (1936). Translated in *U. S. Nat. Advisory Comm. Aero. Tech. Memo. No.* 844 (1938).
4. Keenan, Joseph H., *Thermodynamics*, John Wiley and Sons (1941).
5. Keenan, Joseph H., *Trans. Am. Soc. Mech. Engrs., J. Applied Mechanics*, **6A**, 11–20 (1939); **A**, 135–6.
6. Lapple, C. E., *Trans. Am. Inst. Chem. Engrs.*, **39**, 385 (June 1943).
7. Lobo, W. E., Leo Friend, and G. T. Skaperdas, *Ind. Eng. Chem.*, **34**, 821 (1942).
8. Stodola-Lowenstein, *Steam and Gas Turbines*, Chapter III, Peter Smith (1945).
9. Thomson, G. W., *Ind. Eng. Chem.*, **34**, 1485 (1942).

PROBLEMS

1. Gas flows through a transmission line 150 miles long. The line consists of two sections, a 90-mile section of 24-in. OD pipe whose wall thickness is 9/32 in., and a 60-mile section of 26-in. OD pipe whose wall thickness is 5/16 in. The inlet pressure is 450 psia, and the outlet pressure is 50 psia. The lines are level, the flowing temperature is 60° F, and the efficiency of the line may be taken as 90 per cent in both sections. The gas has a specific gravity of 0.52, a viscosity of 2.4×10^{-7} lb sec/sq ft, and its deviation from the perfect gas laws is negligible. What is the flow rate in cubic feet per hour measured at 14.7 psi and 60° F?

2. Carbon dioxide at 95° F and 800 psia is contained in a large battery of storage cylinders. A discharge line made of ½-in. steel pipe (schedule 80) is 20 ft long and contains fittings (including the connection between the cylinder and the line) with a resistance to flow equivalent to 30 ft of pipe. What is the maximum rate of discharge (pounds per hour) if the discharge pressure is 14.7 psia? Assume isothermal flow.

CHAPTER 16

Flow of Fluids through Porous Media 1

Single Fluid Phase

FLUIDS flowing through a bed of contiguous particles, such as sand or other porous medium distinct from a conduit, flow through passages between the particles of the bed. The dimensions of these passages depend upon the following variables.

Porosity of the bed.

Diameter of the particles.

Sphericity or shape of the particles.

Orientation or packing arrangement of the particles.

Roughness of the particles.

The actual linear velocity of the fluid through the passages in the porous bed may be expressed in terms of the "superficial" velocity (computed as the rate of flow of fluid through the unobstructed entire cross-sectional area of the bed) and these same variables.

Likewise, the length of the path followed by the fluid may be expressed in terms of the length of the bed and these same variables.

The loss due to friction accompanying the flow of fluids through such a bed of contiguous particles when the particles are in random arrangement may be expressed in the form of an equation similar to equation 71 for cylindrical ducts if the two following conditions are met.

1. The Reynolds number based on the diameter of the duct and the superficial velocity is replaced by a Reynolds number Re′ based on the diameter of the particles and including a factor F_{Re} which is a function of the variables listed above.

2. The friction factor also includes a factor F_f which is another function of these variables.

$$\text{Re} = \frac{D_p F_{\text{Re}} v \rho}{\mu} \tag{166}$$

$$\overline{lw}_f = \frac{v^2 L f F_f}{2 g_c D_p} \tag{167}$$

$$f = \frac{2 g_c D_p \overline{lw}_f}{L v^2 F_f} = \frac{2 g_c D_p (-\Delta P)_f}{F_f L v^2 \rho} \tag{168}$$

where D_p = diameter of the particle. When all particles are of the same size the screen size (D_{avg}) may be used for D_p. For mixed sizes D_p is the mean surface diameter, that is, the diameter of a particle having an area equal to the "average" area of the particles (see p. 21 and equation 169).

v = superficial velocity, linear velocity of the fluid computed on the basis of the total or empty cross-sectional area.

ρ = density of the fluid.

μ = viscosity of the fluid.

L = superficial distance through which the fluid flows, or the thickness of the porous bed in the direction of the flow.

$-\Delta P_f$ = decrease in pressure due to friction losses.

$\overline{lw}_f$ = energy dissipated in friction losses per unit mass of fluid.

F_{Re} = factor included in the Reynolds number Re′ to modify D_p.

F_f = friction-factor factor.

$$D_p = \sqrt{\frac{\sum \frac{m_i}{D_i}}{\sum \frac{m_i}{D_i^3}}} \quad (169)$$

where m_i = mass fraction of a given particle size D_i.

D_i = the diameter of particles in each size fraction taken as the arithmetic average of the screen openings passing and retaining the particles.

The values for the factors F_{Re} and F_f obtained from available experimental data by empirical methods [3] are plotted as a function of the porosity X with parameters of sphericity ψ in Figs. 219 and 220 for random packing of regularly shaped particles.

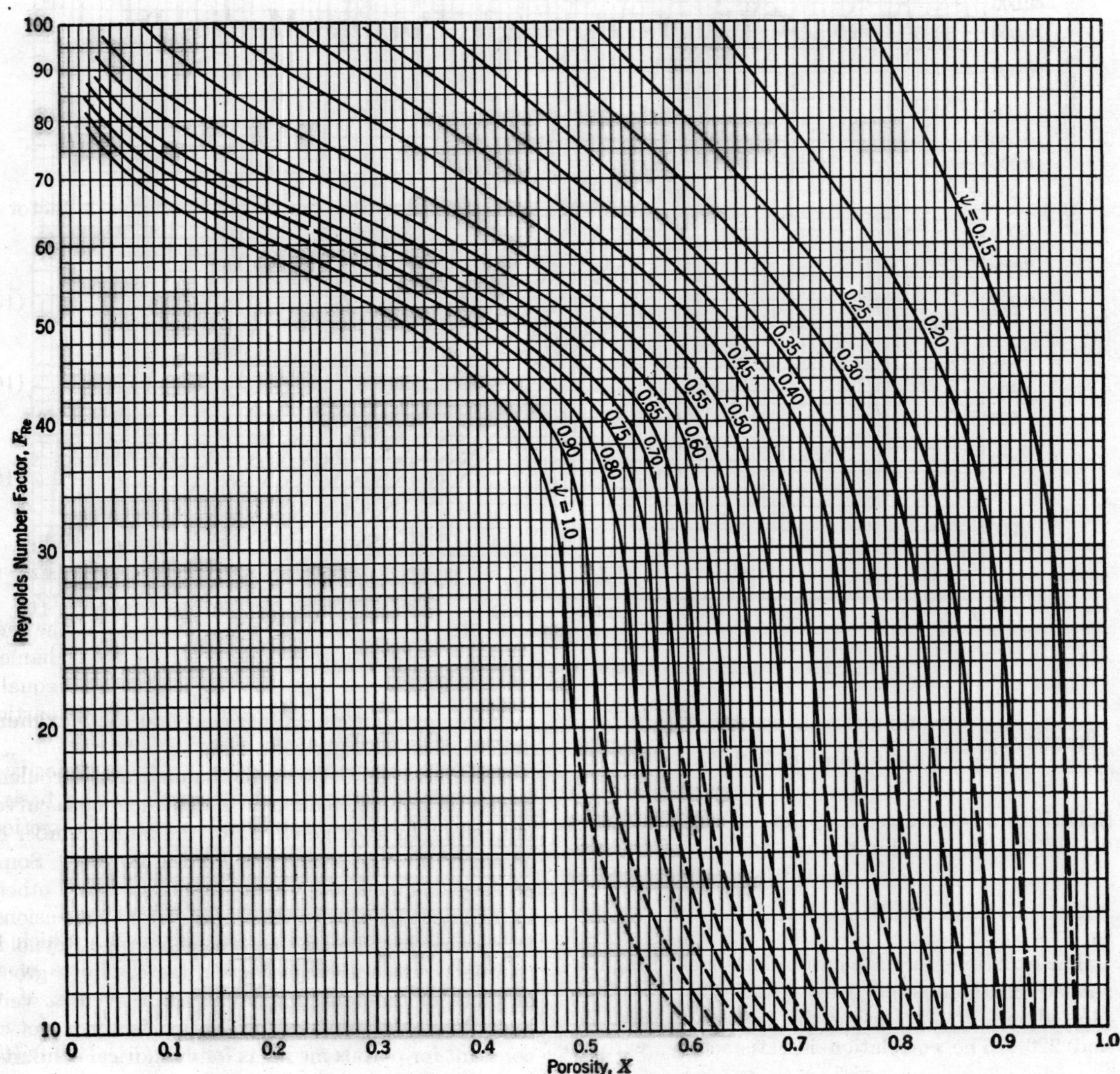

Fig. 219. Reynolds number factor as a function of porosity and sphericity.

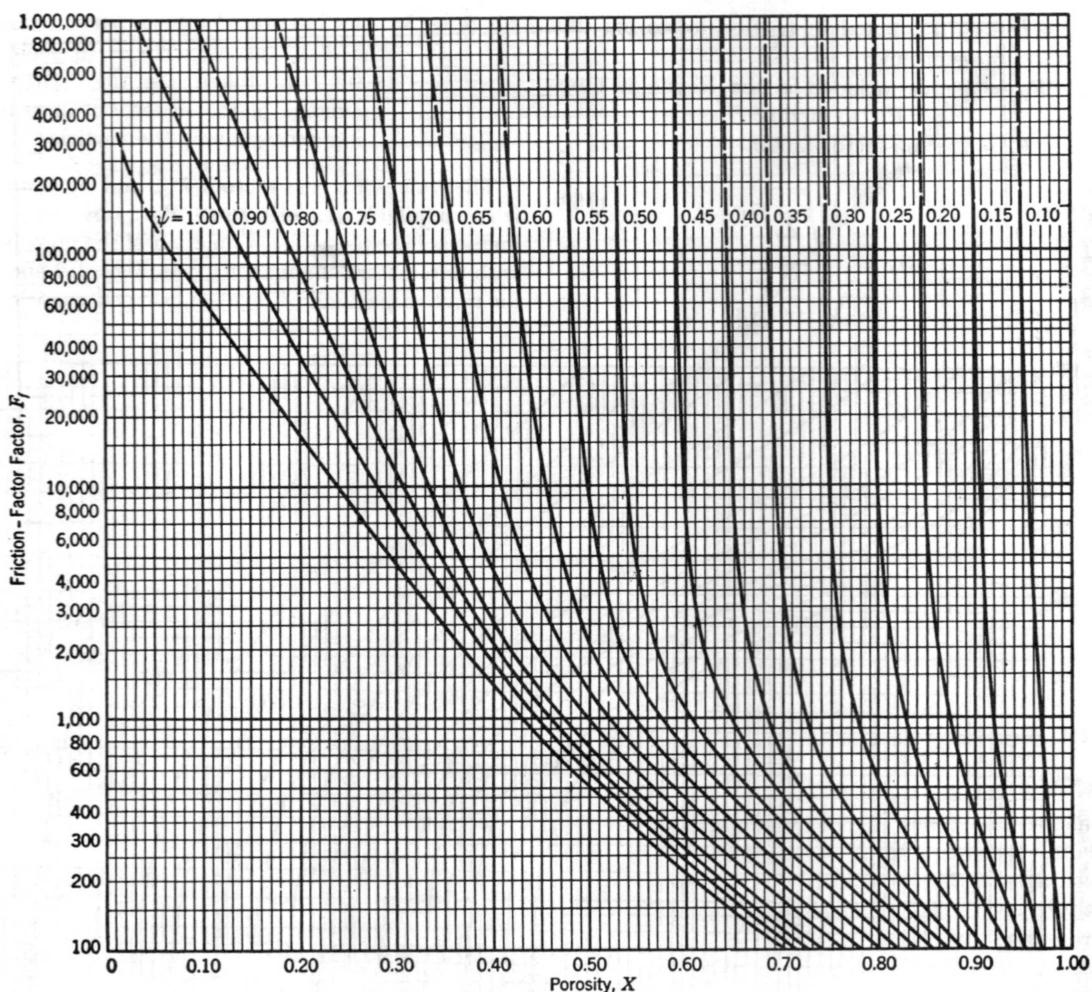

FIG. 220. Friction-factor factor as a function of porosity and sphericity.

$$\text{Porosity } X = \frac{\text{volume of void space}}{\text{total volume of bed}}.$$

Sphericity ψ = surface area of sphere having a volume equal to that of the particle, divided by surface area of particle.

The experimental data on random-packed regular-shaped contiguous particles (Table 26) [2,4,14]* are plotted in Fig. 221, using the factors of Figs. 219 and 220. The correlation is satisfactory, with an average deviation of ±6 per cent and the maximum deviation of ±24 per cent.

The transition between the laminar and turbulent region for porous media is in general a smooth curve, indicating an average effect for the large number of channels through which the fluid is flowing. Some of these channels may be in laminar flow and others in turbulent flow, owing to their different dimensions. It is somewhat similar to a plot of average Reynolds numbers versus average friction factors for a group of pipes of different diameters. But even at very high Reynolds numbers the friction factor is not so constant for porous media as for cylindrical conduits, indicating that the flow through porous media never

* The bibliography for this chapter appears on p. 218.

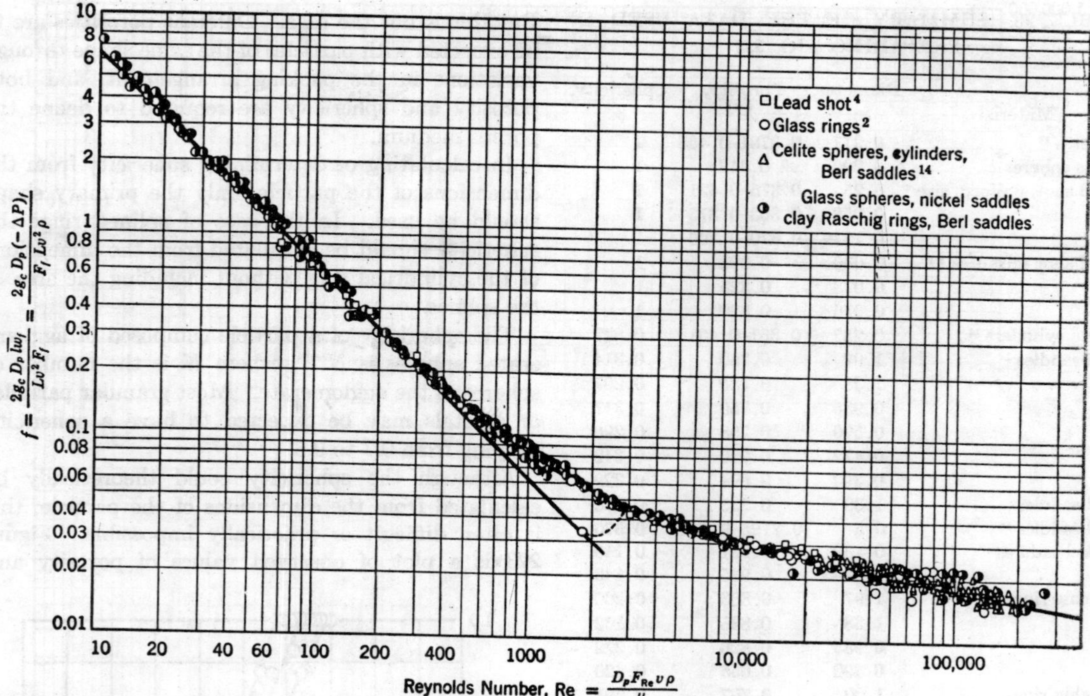

Fig. 221. Experimental data on flow through porous beds of random-packed particles of Table 26 plotted with friction factor (f) as a function of Reynolds number Re.

becomes so completely turbulent as in unobstructed conduits. The existence of very small interstices at the contacts of the particles in the porous medium constitute such small channels that the flow may never become turbulent at these points.

Porosity is the most sensitive variable used in defining a porous medium and must therefore be determined with a high degree of accuracy if the calculations based on Fig. 221 are to be reliable. Particles adjacent to the wall pack more loosely than particles in the central part of the bed and therefore possess a higher porosity. This makes it particularly important that the porosity of a bed be determined in containers having the same cross section as the porous bed or medium for which the computations are desired. This wall effect is included in the curves of Fig. 222, which may be used to estimate the porosity from the ratio of the diameter of a sphere having the same volume as the particle (D_s of Chapter 7) to the diameter of the container or of the bed, D_c.

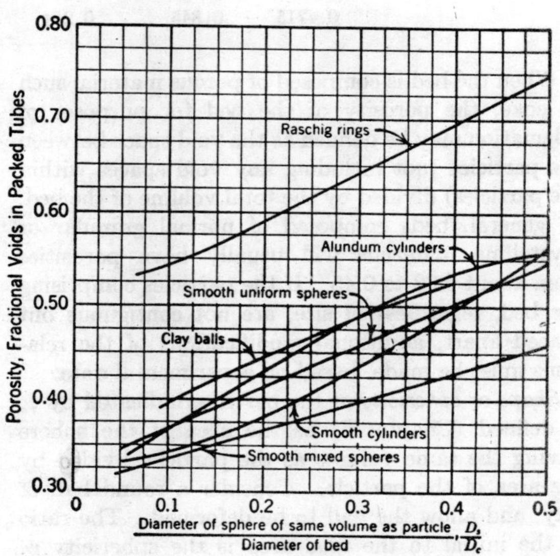

Fig. 222. Porosity as a function of the ratio of diameters D_s/D_c.[13]

TABLE 26. SUMMARY OF THE DATA USED IN PREPARING FIG. 221

Material	Diameter, in.	Porosity, X	Sphericity, ψ
Spheres [14]	0.217	0.3781–0.468	1
Glass spheres	0.209	0.412	1
Lead shot, uniform size [4]	0.25	0.375–0.421	1
	0.058	0.363–0.375	1
	0.121	0.370–0.390	1
Lead shot mixtures [3,4]	0.066	0.303	1
	0.078	0.325	1
	0.076	0.320	1
Celite cylinders [14]	0.267	0.361–0.461	0.877
Berl saddles [1]	1.97	0.780	0.314
	1.38	0.785	0.297
	0.985	0.750	0.317
	0.590	0.758	0.296
	0.472	0.710	0.342
	0.390	0.694	0.329
Berl saddles	1.00	0.725	0.370
Berl saddles [14]	0.5	0.7125–0.761	0.370
Nickel saddles	0.132	0.931	0.140
	0.1295	0.935	0.140
Raschig rings [1]	1.97	0.853	0.260
	1.38	0.835	0.262
	0.985	0.826	0.272
	0.390	0.655	0.420
Raschig rings	1.00	0.707	0.391
Raschig rings [14]	0.385	0.554–0.620	0.531
Glass rings [2]	0.228	0.67	0.411
	0.273	0.72	0.370
	0.3875	0.80	0.294
	0.4715	0.845	0.254

When the bed is composed of porous material such as coke, the porosity of the bed for purposes of calculation may be defined as the void space between the particles (not including any void spaces within the particles) divided by the total volume of the bed. In general, beds composed of normal granular or crystalline materials will usually have porosities from about 0.32 to 0.45. If the particles comprising the bed, regardless of size, are not contiguous but spaced apart, additional modifications of the relations must be made, based on experimental data.

Shape or sphericity of the particle, indicated by ψ, is defined (Chapter 7) as the area of the sphere having the same volume as the particle divided by the area of the particle. Consider a round ball of clay, and allow the ball to be deformed. The ratio of the initial to the final area is the sphericity ψ. The porosity is closely related to the sphericity. The sphericity could be used as the sole determining factor of porosity if the particles of a single size were always oriented in the same spatial arrangement. But this is not the case. Different porosities are to be expected with particles of the same shape through variations in the packing arrangement, and both porosity and sphericity are required to define the porous medium.

In calculating or determining sphericity from the dimensions of the particle, only the primary shape should be used. In the case of splined rings the sphericity should be calculated from the dimensions of the cylindrical ring without including the area of the splines.

The sphericity of a particle composed of agglomerated spheres is $N^{-1/3}$, where N is the number of spheres in the agglomerate. Most granular particles or crystals may be expected to have a sphericity varying from 0.7 to 0.8.

Although the sphericity could theoretically be calculated from the dimensions of the particle, this is often difficult or practically impossible. Figure 223 is a plot of observed values of porosity and

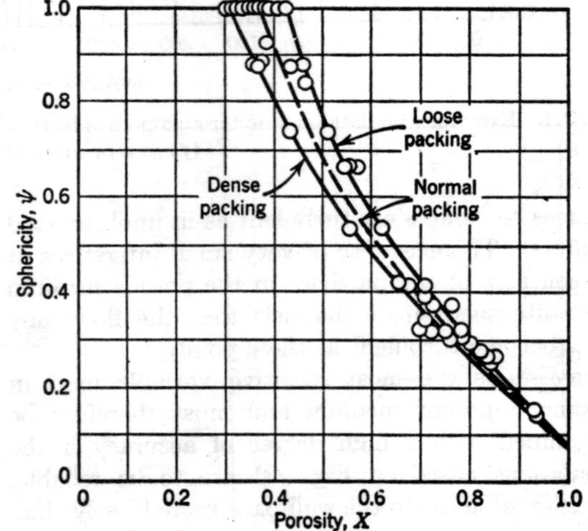

FIG. 223. Sphericity as a function of porosity for random-packed beds of uniform-sized particles.

sphericity for random-packed beds of uniform-sized particles which may be used to estimate particle sphericity provided the bed porosity is known. This relation is recommended for estimating the sphericity of complex particle shapes, or for cases where there is doubt concerning the "effective" particle area and volume. The dashed line should be used for "normal" packing, and the values to the left or right used for dense- or loose-packed beds, respectively. It is not advisable to use values of sphericity

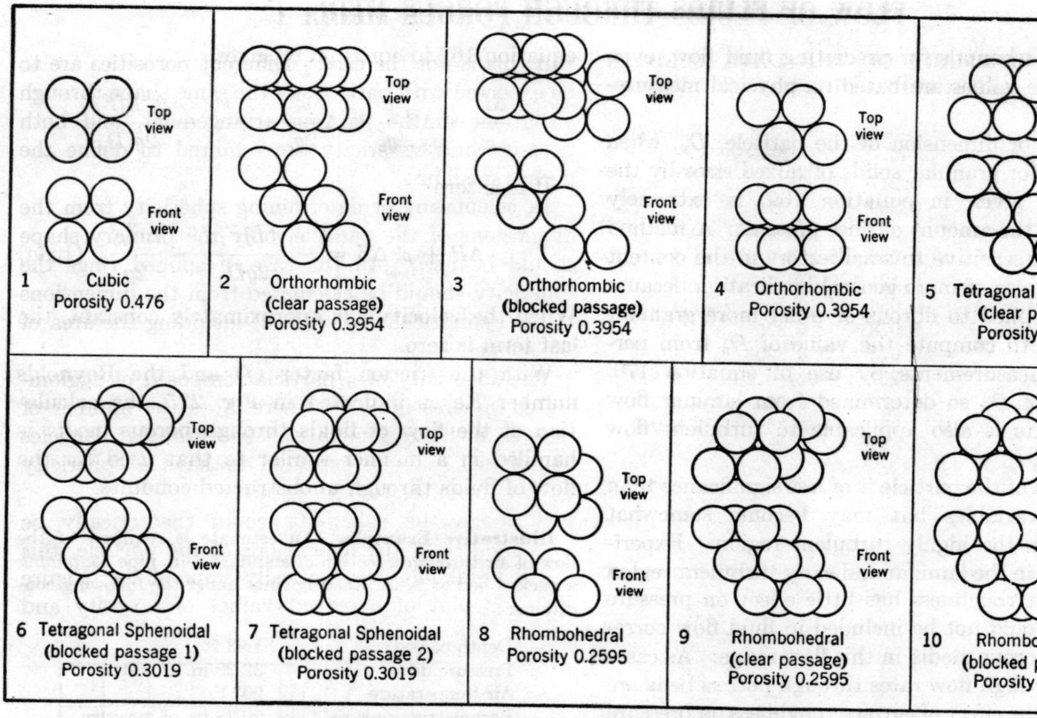

Top view

Front view

1 Cubic
Porosity 0.476

2 Orthorhombic
(clear passage)
Porosity 0.3954

3 Orthorhombic
(blocked passage)
Porosity 0.3954

4 Orthorhombic
Porosity 0.3954

5 Tetragonal Sphenoidal
(clear passage)
Porosity 0.3019

6 Tetragonal Sphenoidal
(blocked passage 1)
Porosity 0.3019

7 Tetragonal Sphenoidal
(blocked passage 2)
Porosity 0.3019

8 Rhombohedral
Porosity 0.2595

9 Rhombohedral
(clear passage)
Porosity 0.2595

10 Rhombohedral
(blocked passage)
Porosity 0.2595

FIG. 224. Systematic arrangement of spheres and their porosities.

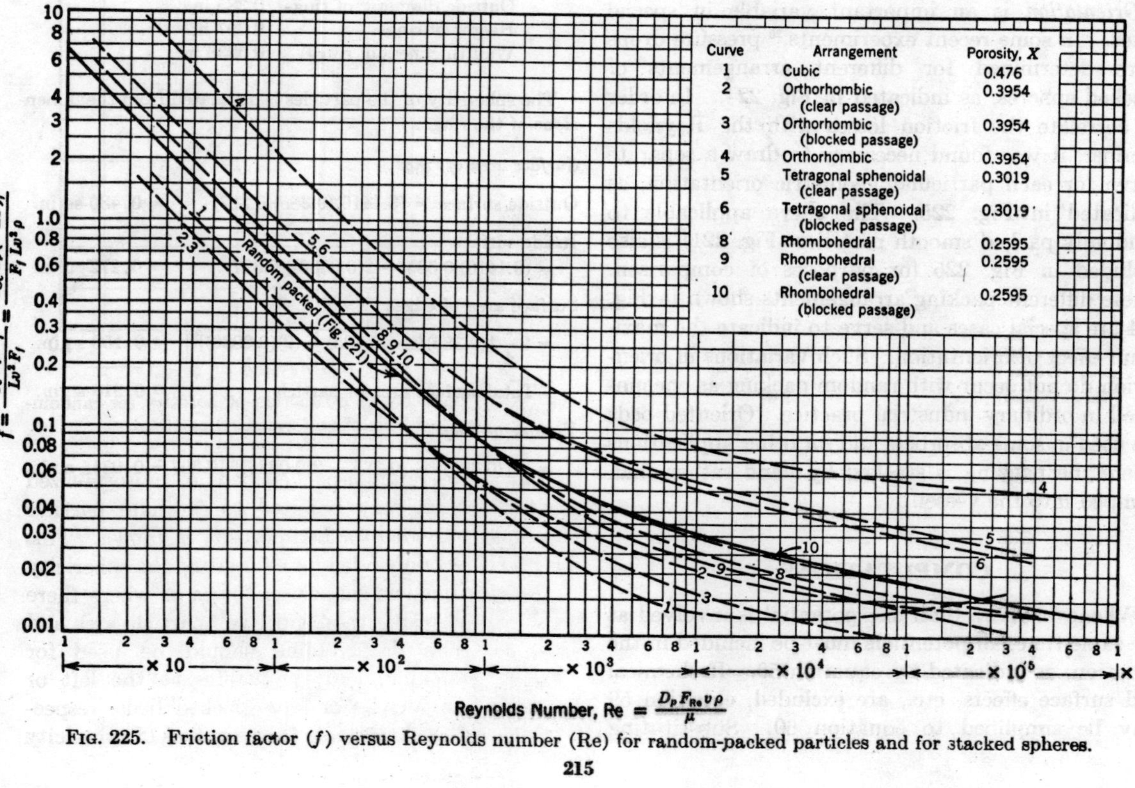

$$f = \frac{2g_c D_p \overline{lw}_f}{Lu^2 F_f} = \frac{2g_c D_p (-\Delta P)_f}{F_f L u^2 \rho}$$

Curve	Arrangement	Porosity, X
1	Cubic	0.476
2	Orthorhombic (clear passage)	0.3954
3	Orthorhombic (blocked passage)	0.3954
4	Orthorhombic	0.3954
5	Tetragonal sphenoidal (clear passage)	0.3019
6	Tetragonal sphenoidal (blocked passage)	0.3019
8	Rhombohedral	0.2595
9	Rhombohedral (clear passage)	0.2595
10	Rhombohedral (blocked passage)	0.2595

Random packed (Fig. 221)

Reynolds Number, $\text{Re} = \frac{D_p F_{Re} v \rho}{\mu}$

FIG. 225. Friction factor (f) versus Reynolds number (Re) for random-packed particles and for stacked spheres.

outside these bounds for predicting fluid flow, even though these values are based on physical measurements.

Diameter, or dimension of the particle, D_p, when determined for granular solids of mixed sizes by the relationship given in equation 169, is extremely sensitive to the amount of fines present. A method which is less sensitive to small errors in the content of fines and also of more general application because it can be applied to fibrous or other more granular materials is to compute the value of D_p from permeability measurements by use of equation 170. The value of D_p so determined from laminar flow measurements is also applicable to turbulent flow conditions.

Roughness of the particle is of less significance than the other variables but may become somewhat important in the highly turbulent region. Experimental data in the laminar and early turbulent region indicate that roughness has little effect on pressure drop and should not be included in fluid flow correlations for porous media in this flow range. As cases of extremely high flow rates through porous beds are infrequent, the effect of surface roughness on pressure drop may, in general, be neglected.

Orientation is an important variable in special cases. In some recent experiments,[20] pressure drops were determined for different arrangements of stacked spheres, as indicated in Fig. 224. In order to correlate the friction losses with the Reynolds number, it was found necessary to draw a separate curve for each particular geometric orientation, as indicated in Fig. 225. The curve applicable to randomly packed smooth particles (Fig. 221) is also included in Fig. 225 for purposes of comparison. These different packing arrangements shown in Fig. 224 are special cases and serve to indicate the maximum effect of orientation. Such variations in orientation do not occur with random packing as encountered in ordinary industrial practice. Oriented beds are used in some absorbers and for other applications where the packing is stacked by hand rather than dumped into the vessel.

COMPUTATIONS

Whenever more than one potential is involved at the same time, all potentials must be included in the equation, as indicated by equation 59. If chemical and surface effects, etc., are excluded, equation 59 may be simplified to equation 60. Substituting equation 167 in equation 60 gives

$$\frac{\Delta P}{\rho} + \Delta Z \frac{g}{g_c} + \Delta \frac{v^2}{2g_c} = -\bar{w} - \frac{v^2 L f F_f}{2g_c D_p}$$

If w is zero

$$\Delta P + \rho' \Delta Z = -\frac{v^2 L f F_f \rho'}{2g D_p} - \frac{\Delta v^2 \rho'}{2g} \qquad (168a)$$

When the velocity v is approximately constant, the last term is zero.

With the friction factor (f) and the Reynolds number Re as indicated in Fig. 225, the calculation of the flow of fluids through porous media is handled in a manner similar to that used for the flow of fluids through unobstructed conduits.

Illustrative Example. An estimate is required of the flow of air through a section of standard 4-in. pipe containing a packed bed of ⅜-in. Raschig rings under the following conditions:

Depth of bed	1.93 ft
Pressure drop	32.29 in. of water
Air temperature	90° F
Barometric pressure	29.38 in. of mercury
Bed porosity	0.5545
Outside diameter of rings	0.385 in.
Height of rings	0.397 in.
Wall thickness of rings	0.0836 in.

The sphericity of the particles is estimated from the dimensions of the rings:

Surface area of rings

Outside surface $= (3.1416)(0.385)(0.397) \qquad = 0.480$ sq in.

Inside surface

$= (3.1416)[0.385 - 2(0.0836)](0.397) \qquad = 0.272$ sq in.

Surface area of edges

$= 2\frac{\pi}{4}\{(0.385)^2 - [0.385 - (2)(0.0836)]^2\} = 0.158$ sq in.

Total surface area of particle $= 0.910$ sq in.

Volume of rings

$= \frac{\pi}{4}\{(0.385)^2 - [0.385 - 2(0.0836)]^2\}0.397 = 0.0315$ cu in.

The area of a sphere having the same volume as the particle is computed by determining the diameter of a sphere having a volume of 0.0315 cu in. and then determining the area of this sphere.

$$V = \frac{\pi D_s^3}{6} = 0.0315 \text{ cu in.}$$

Therefore

$$D_s = 0.392 \text{ in.}$$

$$A = \pi D_s^2 = 0.483 \text{ sq in.}$$

Sphericity (ψ) is obtained by dividing the area of the sphere having the same volume as the particle by the area of the particle, or sphericity = 0.483/0.910 = 0.531.

From Fig. 219 F_{Re} is found to be 50, using $\psi = 0.531$, $X = 0.5545$. From Fig. 220 F_f is found to be 1600.

The pressure drop of 32.39 in. of water is equivalent to 168.5 psf or 2.38 in. of mercury. The arithmetic average density is used because of the relatively small change in pressure. Using a molecular weight of 29 and an average pressure of 30.57 in. of mercury, the average density of the air is computed as an ideal gas.

$$\rho = \frac{MP}{RT} = \frac{29(30.57)(14.7)(144)}{(29.92)(1544)(460 + 90)}$$

$$\rho_{\text{avg}} = 0.074 \text{ lb mass/cu ft}$$

$$\mu = 0.0182 \text{ centipoise}$$

The values of all the factors in the Reynolds number (Re) and friction factor are now known except velocity, which can be found from Fig. 225 as follows.

$$\text{Re} = \frac{D_p F_{\text{Re}} \rho v}{\mu} = \frac{(0.03208)(50)(0.074)v}{(0.0182)(0.000672)} = 9700v$$

$$f = \frac{2g_c D_p \overline{lw}_f}{L F_f v^2} = \frac{(2)(32.2)(0.03208)(168.5)}{(1.93)(0.074)(1600)v^2} = \frac{1.523}{v^2}$$

Therefore

$$f = \frac{(1.523)(9700)^2}{(\text{Re})^2}$$

or

$$\log(f) = \text{Constant} - 2\log(\text{Re})$$

The solution will be obtained by the intersection of a straight line of slope of minus 2 on Fig. 225 with the curve for random-packed particles. At velocity v of 1 fps, Re = 9700 and $f = 1.523$. At v of 10 fps, Re = 97,000 and $f = 0.01523$. Plotting these two points on Fig. 225 and drawing a straight line through them gives the solution for Re and f at the point of intersection corresponding to a Reynolds number (Re) of 89,000. Dividing this by 9700 gives 9.2 fps as the average superficial velocity of the air through the cross-sectional area of the unobstructed 4-in. pipe.

The quantity of air flowing through this packed section of pipe is

$$9.2\,\frac{\text{ft}}{\text{sec}}\,(0.08840\text{ ft}^2)\left(3600\,\frac{\text{sec}}{\text{hr}}\right)\left(0.074\,\frac{\text{lb}}{\text{cu ft}}\right) = 216\,\frac{\text{lb}}{\text{hr}}$$

This value corresponds with an observed rate of 213 lb/hr in an experimental run [14] under the conditions indicated.

EQUATIONS USED FOR LAMINAR FLOW THROUGH POROUS BEDS

In *laminar flow* the same form of equations may be used for the friction losses through porous media as for the flow through conduits; for example,

$$f = \frac{64}{\text{Re}} \quad \text{corresponds to} \quad f = \frac{64}{\text{Re}}$$

$$(-\Delta P)_f = \frac{32 L v \mu F_f}{g_c D_p^2 F_{\text{Re}}} = \rho \overline{lw}_f \quad \text{corresponds to}$$

$$(-\Delta P)_f = \frac{32 L v \mu}{g_c D^2} = \rho \overline{lw}_f \qquad (63L)$$

Solving for v:

$$v = \left(\frac{g_c D_p^2 F_{\text{Re}}}{32 F_f}\right)\left(\frac{\rho \overline{lw}_f}{L\mu}\right) = K\,\frac{\rho \overline{lw}_f}{L\mu} \qquad (170)$$

Equation 170, known as Darcy's equation, corresponds to equation 63L, known as Poiseuille's equation.

For vertical flow in which the energy of position (potential energy) contributes the driving force $\Delta P = 0$ and $\Delta v = 0$, and from equation 60 $\Delta Z g/g_c = -\overline{lw}$. Substituting in equation 170 gives

$$v = \left(\frac{g_c D_p^2 F_{\text{Re}}}{32 F_f}\right)\left[\frac{\rho(-\Delta Z)g}{L\mu g_c}\right] = \frac{K\rho(-\Delta Z)g}{L\mu g_c} \qquad (171)$$

When $L = -\Delta Z$, equation 171 simplifies to

$$v = \frac{K\rho'}{\mu} \qquad (171a)$$

The *permeability* K is defined by equation 172, which is derived from equation 170.

$$K = \frac{g_c D_p^2}{32} \cdot \frac{F_{\text{Re}}}{F_f} \qquad (172)$$

The surface per unit volume of particle s, computed by dividing the surface area of particles by the volume of the particles, is equivalent to a reciprocal diameter and has been used [5, 6, 12] in an equation similar to 170 as follows.

$$v = \frac{g_c s^{-2} X^3}{5(1 - X)^2} \cdot \frac{\rho \overline{lw}_f}{L\mu} \qquad (173)$$

where s = surface area of particles per unit volume of particles.

X = porosity.

Equation 173 has an important advantage in dealing with laminar flow through media such as glass wool, fibers, or packing whose size cannot be expressed in terms of a diameter or screen analysis. For best results the constant, 5, in the denominator may be replaced by other values between 5 and 5.5, depending upon specific conditions.

The *specific resistance* α of the bed is another concept which has been used [16] in laminar flow of fluids through filter beds.

$$v = \frac{\rho \overline{lw}_f A}{\alpha \mu W} = \frac{1}{\alpha} \frac{(-\Delta P)_f}{\mu} \frac{A}{W} \tag{174}$$

where

A = area of bed in plane normal to direction of flow (sq ft).

W = mass of bed (lb).

α = specific resistance of bed $\left[\frac{(\text{sec})^2(\text{lb force})}{(\text{lb mass})^2}\right]$.

The values for α include the variables of porosity, density, and, once determined, simplify the calculations for a given bed.

The relationship between the permeability K of equations 170 to 172, the Kozeny [12] equation (173) using specific surface, and the Ruth [16] equation (174) using specific resistance is

$$K = \frac{vL\mu}{\overline{lw}_f \rho} = \frac{g_c D_p{}^2 F_{Re}}{32 F_f} = \frac{g_c s^{-2} X^3}{5(1-X)^2} = \frac{AL}{\alpha W} \tag{175}$$

BIBLIOGRAPHY

1. Berl, E., *Catalog of Ditt and Frees*, Wiesbaden (1930).
2. Blake, Fr. C., "The Resistance of Packing to Fluid Flow," *Trans. Am. Inst. Chem. Engrs.*, **14**, 415 (1921–22).
3. Brownell, L. E., and D. L. Katz, "Flow of Fluids through Porous Media—Part I," *Chem. Eng. Progress*, **43**, 537 (1947).
4. Burke, S. P., and W. B. Plummer, "Gas Flow through Packed Columns," *Ind. Eng. Chem.*, **20**, 1196 (1928).
5. Carman, P. C., "Fluid Flow through Granular Beds," *Trans. Inst. Chem. Engrs. (London)*, **15**, 150 (1937).
6. Carman, P. C., "The Determination of the Specific Surface of Powders, I," *J. Soc. Chem. Ind.*, **57**, 225 (1938); "II," *ibid.*, **58**, 1 (1939).
7. Chilton, T. H., and A. P. Colburn, "Pressure Drop in Packed Tubes," *Trans. Am. Inst. Chem. Engrs.*, **26**, 178 (1931).
8. Darcy, L., *Les Fontaines Publiques de la Ville de Dijon*, Victor Delmont, Paris (1856).
9. Fair, G. M., and Hatch, L. P., "Fundamental Factors Governing the Steamline Flow of Water through Sand," *J. Am. Water Works Assoc.*, **25**, 1551 (1933).
10. Fancher, G. H., and J. A. Lewis, "Flow of Simple Fluids through Porous Materials," *Ind. Eng. Chem.*, **2,5** 1139 (1933).
11. Furnas, C. C., "Flow of Gases through Beds of Broken Solids," *U. S. Bur. Mines Bull.* 307 (1929).
12. Kozeny, J., "Soil Permeability," *Sitzber. Akad. Wiss. Wien*, **136a**, 271 (1927).
13. Leva, Max, "Pressure Drop through Packed Tubes, I. A General Correlation," *Chem. Eng. Progress*, **43**, 549 (1947).
14. Oman, A. O., and K. M. Watson, "Pressure Drops in Granular Beds," *Natl. Petroleum News, Tech. Sec.*, **36**, 44R795 (1944).
15. Rose, H. E., "An Investigation into the Laws of Flow of Fluids through Beds of Granular Materials," *Proc. Inst. Mech. Engrs.*, **153**, 5, 141 (1945).
16. Ruth, B. F., "Studies in Filtration III; Derivation of General Filtration Equations IV, The Nature of Fluid Flow through Filter Septa and Its Importance in the Filtration Equation," *Ind. Eng. Chem.*, **27**, 708, 806 (1935).
17. Schoenborn, E. M., and W. J. Dougherty, "Pressure Drop and Flooding Velocity in Packed Towers with Viscous Liquids," *Trans. Am. Inst. Chem. Engrs.*, **40**, 51 (1944).
18. Schriever, W., *Trans. Am. Inst. Mining Met. Engrs.*, "Law of Flow of the Passage of a Gas-Free Liquid through a Spherical-Grain Sand," **86**, 329, 333 (1930).
19. Sullivan, R. R., and K. L. Hertel, *J. Applied Phys.*, "The Flow of Air through Porous Media," **11**, 761 (1940).
20. Martin, J. J., *Thesis*, Carnegie Institute of Technology (1948).

1. A cylindrical tower, 2 ft in diameter, packed to a height of 10 ft with activated alumina, is now being used to clean naphtha flowing downward under the following operating conditions.

Temperature	= 80° F
Density	= 49.3 lb mass/cu ft
Viscosity	= 0.916 centipoise
Average diameter (screen size) of alumina	= 0.046 in.
Porosity of bed	= 0.400
Flow rate	= 10 gpm/sq ft, cross section
Pressure difference	= 35 psi

It is desired to convert the tower into an air drier, removing excess moisture from 1000 standard cu ft (measured at 60° F and 14.7 psia) of air per hour flowing upward.

If the drier operates isothermally at 100° F and the air must leave the tower at 25 psig, what must be the inlet pressure?

2. Naphthalene cylinders (0.11 in. in diameter and 0.13 in. long) form a bed 3 in. in diameter and 10 in. long. The weight of 92 cc of the pellets is 58 grams. Compute the pounds of air which will pass through the bed per hour at a flowing temperature of 100° F with a pressure differential of 18 in. of water.

3. A new type of pebble regenerator is used in a process for preheating air from 100° to 250° F. Alternately hot flue gases and then air at 100° F are passed through the heater. The heater is packed with ceramic spheres 1 in. in diameter to a porosity of 40 per cent. The bed is 3 ft in diameter and 8 ft deep.

It has been suggested that the addition of some small spheres 3/32 in. in diameter to fill the interstices of the bed would increase the heat capacity of the bed. Experiments indicate that the 3/32-in. balls would fill one-third of the interstices left between the 1-in. spheres.

Plot a curve of pressure drop in pounds per square foot versus superficial mass velocity (pounds per square foot per hour) of the air for the two beds.

4. A lubricating oil stock is being contacted with a solid adsorbent by gravity percolation. The contacting units are cylindrical beds 6 ft in diameter by 4 ft high. The adsorbent consists of particles 0.02 in. in diameter with a sphericity of 0.72. The porosity of the bed is 34 per cent. During the percolation process, 1 ft of oil stands above the bed with atmospheric pressure above this oil and at the bottom of the adsorbent.

It is suggested that the manufacturing process be changed so that the lubricating oil stock will be diluted with solvent at the time of percolation.

Plot a curve showing the capacity of an adsorbent unit expressed in gallons per hour of lubricating oil versus concentration of solvent in the oil. The properties of the oil solutions at the contacting temperature are as follows:

	Lubricating Stock	95% Lubricating Stock, 5% Solvent	90% Lubricating Stock, 10% Solvent	80% Lubricating Stock, 20% Solvent
Density, grams/cc	0.91	0.90	0.89	0.87
Viscosity, centipoises	6.0	4.6	3.5	2.1

5. A gravity filter consists of a 2-ft bed of 1-in. crushed trap rock covered with a 2-ft layer of 28-mesh sand. Both layers have bed porosities of 40 per cent. The filter is flooded with 2 ft of water above the top of the sand. The pressure at the bottom outlet of the filter is atmospheric.

(*a*) What is the flow of water through the filter, expressed as feet^3/(feet^2) (hour)?

(*b*) If a similar filter bed were enclosed in a pressure tank and operated with a pressure drop of 100 psia, what would be the flow of water?

6. Compute the pressure in the well bore for a natural gas well (6⅝ in. in diameter) at the rate of 34,800,000 cu ft per day of natural gas measured at 60° F and 14.7 psia. The producing sand has a thickness of 1 ft and a porosity of 26 per cent. The sand grains have an average diameter of 0.000791 ft and a sphericity of 0.90. The pressure in the formation at 5000 ft (temperature, 115° F) is constant as 2100 psig at a radial distance of 500 ft from the well bore. The natural gas has a gravity of 0.65 (air = 1), a viscosity at 115° F of 0.0176 centipoise, a pseudo critical temperature of 374° R, and a pseudo critical pressure of 669 psia. The presence of water in the sand may be neglected. (See "A Radial Turbulent Flow Formula," Elenbaas and Katz, *Petroleum Technology*, January 1948.)

CHAPTER 17

Flow of Fluids through Porous Media 2

Two Fluid Phases

In many operations two homogeneous fluid phases may be flowing simultaneously through porous beds. The flow of the two phases may be countercurrent to each other as in packed towers or columns serving as gas absorbers or scrubbers, or the flow of the two phases may be in the same direction as when air is used to displace fluid from a filter cake in the air-flow period of the filter cycle, or in the displacement of oil by gas in the production of oil from permeable sandstone.

When two fluid phases are flowing through a porous medium, one fluid normally wets the solid, flows adjacent to the solid, and prevents contact of the second phase with the solid; the other fluid flows through the void space and is in contact with the first fluid rather than with the solid. This second fluid is then flowing through a porous medium which is different in characteristics from the unwetted porous solid. The fluid wetting the surface of the solid changes the volume of voids available to the gas or nonwetting fluid and also changes the shape of the pore space available to the gas or nonwetting fluid. It is therefore necessary to speak of a wetted porosity and wetted sphericity in considering the flow of the gas or nonwetting fluid in the presence of another fluid which wets the solid surfaces.

Before considering the flow of the nonwetting fluid, it is necessary to establish the flow characteristics of the wetting fluid in order to arrive at values for the wetted porosity and wetted sphericity.

In every porous medium there is a volume of space not occupied by the solid. The volume of these pores is designated by the term *voids*. If the voids are completely filled with a fluid, the porous medium is said to be saturated with that fluid. If the voids are not completely filled by a fluid, the fractional volume of the voids occupied by the fluid is referred to as the *saturation* of that fluid. The saturation is expressed as a percentage or fraction which equals the volume of fluid in the voids divided by the total volume of voids. For simplicity it is the usual practice to express the fluid content of the voids in terms of the saturation of the wetting fluid. For example, if the total void space is filled to the extent of 20 per cent by volume with water and 80 per cent by volume by air, the medium is said to be 20 per cent saturated.

In developing procedures for predicting the flow of two phases through porous media, it is necessary to consider capillary forces in the bed in addition to the degree of saturation.

The calculation of flow for the two phases is based upon the previously described correlation for the flow of a single homogeneous fluid through porous media (Chapter 16) with modifications depending upon which phase is under consideration. The nonwetting fluid is treated as a single phase flowing through a porous bed modified by the presence of the wetting fluid. For the wetting fluid, a correction to the velocity is applied, based upon the saturation.

Figure 226 shows the relative permeability * of

* Relative (sometimes called specific) permeability is the actual quantity of flow of a given phase divided by the quantity of flow of that phase under the same driving force when that phase completely fills the voids, i.e., single-fluid phase flow.

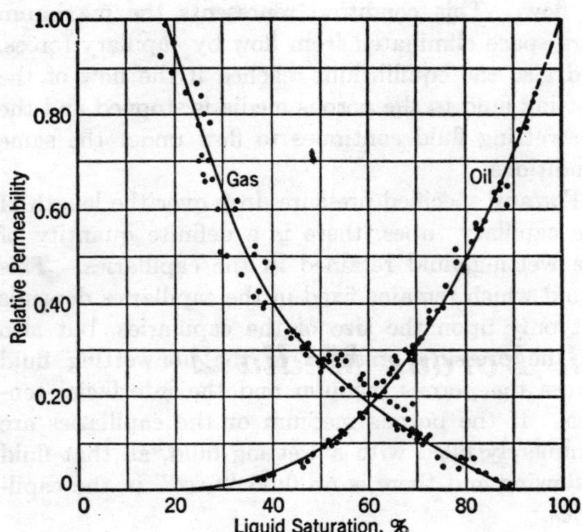

FIG. 226. Permeability of unconsolidated sand to gas and oil as a function of liquid saturation.[12]

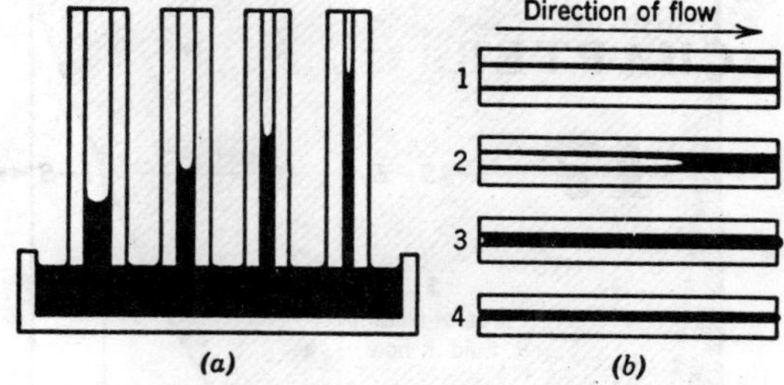

FIG. 227. (a) Capillary tubes in contact with liquid; (b) horizontal capillary tubes.

unconsolidated sand of 70 to 120 mesh to oil and to gas as a function of the liquid saturation.[12] * At saturations less than about 18 per cent, only the gas flows through the porous bed and no oil flows even though the voids are 18 per cent saturated with oil. Similar data for the simultaneous flow of two liquid phases, such as water and oil, are available.[11]

This effect of saturation on the permeability in the simultaneous flow of two fluids through a porous medium can be explained by considering the effect of surface tension of fluid in capillaries. The surface or interface between two fluid phases always tends to contract and generally behaves as if it were in tension. This characteristic is described as the *surface tension* or *interfacial tension*. If four capillary tubes of different internal diameters are placed in a fluid which wets the surface of the capillaries, the fluid rises in the capillaries to different heights, as indicated in Fig. 227*a*. This condition represents a balance of forces. The total surface force tending to pull the liquid upward in the capillary varies directly as the inside perimeter of the capillary, or as the first power of the diameter. The gravitational force tending to pull the liquid downward in the capillaries is the weight of the fluid in the capillary and varies directly as the square of the diameter of the capillary and as the height of the liquid in the capillary.

If the bundle of capillary tubes is placed in a horizontal position, as indicated in Fig. 227*b*, but completely submerged in a flowing stream of the liquid which wets the surfaces of the capillaries, the liquid will flow through each capillary in accord with Poiseuille's equation 63*L* under conditions of laminar flow. The liquid in all capillaries will be flowing under these conditions of 100 per cent saturation.

If the bundle of capillary tubes is raised above the liquid stream and placed in a horizontal position in an air stream, as shown in Fig. 227*b*, with a pressure gradient from one end of the tube to the other, the air will tend to displace the liquid from the capillaries with a force corresponding to the pressure gradient multiplied by the cross-sectional area of the fluid in the capillary. This displacing force is opposed by a force equal to the interfacial tension between the liquid and air multiplied by the internal perimeter of the wetted capillary. Since the cross-sectional area varies as the square of the diameter, and the perimeter as the first power of the diameter, there will be some size of capillary for a given pressure gradient in which these capillary forces balance the pressure gradient, and no fluid will pass through a capillary having a diameter equal to or smaller than this value. In Fig. 227*b*, tube 1 is passing air and the capillary forces in tubes 3 and 4 are sufficient to prevent the flow of air. If the pressure drop is increased over the length of the capillary tubes, air may pass through tube 3 and only tube 4 will be of such small diameter as to prevent air flow.

Further significance of capillary forces is seen by comparison of a porous medium to a bed composed of many capillary tubes of short length, varying diameter, and random distribution. Consider such a bed completely saturated with a wetting fluid such as water and subject to a pressure gradient. Each short capillary conducts flow according to its diam-

* The bibliography for this chapter appears on p. 228.

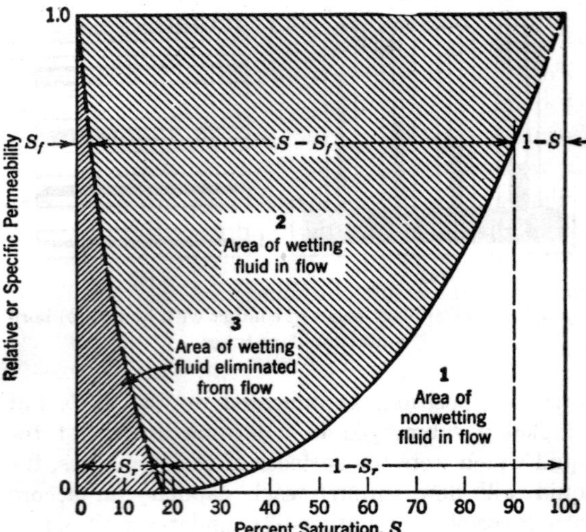

Fig. 228. Graphical representation of the distribution of the two fluids flowing through a porous bed expressed as per cent saturation S and the corresponding relative permeability of the wetting fluid for one particular pressure gradient. The line separating area 1 from area 2 is the oil permeability line of Fig. 226.

eter, pressure gradient, and viscosity of the fluid as stated by Poiseuille's equation (63*L*) for laminar flow. Now replace a portion of the water with a nonwetting fluid such as air so that the saturation is about 80 per cent. In the "unsteady state" representing the change from 100 per cent saturation to 80 per cent saturation, the air flows through the larger interstices, displacing the water as is represented with tube 2 of Fig. 227*b*. When equilibrium is reached, with both fluids being supplied to maintain a saturation of 80 per cent, the air will be flowing in the larger capillaries as represented by tube 1. The water flows by two systems. A portion flows in the annular space as shown for tube 1. At this high saturation many of the small capillaries still are completely flooded and conduct flow in the same manner as if the bed were completely saturated. However, a portion of the short capillaries is exposed to the air stream at one or both ends, producing surface forces which prevent flow as represented with tubes 3 and 4. Thus, water flow decreases for two reasons: (1) in the large interstices the air replaces the water, and (2) a portion of the smaller interstices drops out of flow.

With progressively lower saturations, this phenomenon continues with additional interstices opened to the flow of air and additional interstices eliminated from flow by capillary forces until the liquid ceases to flow. This condition represents the maximum pore space eliminated from flow by capillary forces, and also the equilibrium reached if the flow of the wetting fluid to the porous media is stopped and the nonwetting fluid continues to flow under the same conditions.

For any specified pressure drop over the length of the capillary tubes, there is a definite quantity of the wetting fluid retained in the capillaries. This liquid which remains fixed in the capillaries depends not only upon the size of the capillaries, but also on the pressure gradient of the nonwetting fluid across the porous medium and the interfacial tension. If the porous medium or the capillaries are completely filled with a wetting fluid, all that fluid is flowing and there is no fluid "fixed" in the capillaries.

Figure 228 is presented to clarify the terms for expressing the distribution of the phases for two-plase flow. The voids in a bed are divided into three portions: (1) nonwetting fluid, (2) wetting fluid in flow, and (3) wetting fluid eliminated from flow by capillary forces. Each of these portions of the voids is represented as an area on Fig. 228, which applies specifically to the conditions of liquid saturation indicated by Fig. 226. The oil permeability line of Fig. 226 appears on Fig. 228 as the line separating area 2 from area 1.

The significance of the following definitions is also illustrated on Fig. 228 for the specific case of 90 per cent saturation.

Saturation, S

$$= \frac{\text{Volume of voids filled with wetting fluid}}{\text{Total volume of voids}}$$

Residual saturation, S_r

$$= \frac{\left\{\begin{array}{c}\text{Maximum volume of wetting fluid}\\ \text{eliminated from flow}\end{array}\right\}}{\text{Total volume of voids}}$$

Fixed saturation eliminated from flow, S_f

$$= \frac{\left\{\begin{array}{c}\text{Volume of wetting fluid held stationary}\\ \text{by capillary forces}\end{array}\right\}}{\text{Total volume of voids}}$$

$$= \left(\frac{1 - S}{1 - S_r}\right)(S_r) \tag{176}$$

Effective saturation, S_e

$$= \frac{\text{Voids containing wetting fluid active in flow}}{\text{Voids containing both fluids in flow}}$$

$$= \frac{S - S_f}{S - S_f + 1 - S} = \frac{S - S_r}{1 - 2S_r + SS_r} \quad (177)$$

Equation 176 defines the saturation eliminated from flow S_f in terms of the saturation S and residual saturation S_r and assumes that S_f is proportional to the liquid permeability, i.e.,

$$\frac{S_f}{S_r} = \frac{1 - S}{1 - S_r} \quad (178)$$

This assumes that the liquid removed from flow by capillary forces is proportional to the nonwetting fluid in the bed.

The effective saturation S_e defines the two-phase system with regard to the relative volume of each fluid flowing and is the term used in flow relationships.

RESIDUAL SATURATION

To predict the effective saturation S_e corresponding to a given saturation, it is necessary to know the residual saturation S_r and the total saturation S. The residual saturation may be defined as a function of the permeability of the bed, the interfacial tension of the fluids, and the total force per unit area and length. If flow is horizontal this is the pressure gradient. The result of dimensional analysis gives a combined group,

$$\frac{K(-\Delta P)}{Lg_c\gamma \cos\theta}$$

which is the ratio of forces driving fluids from the bed to the forces retaining fluids in the bed and is called the *capillary number*. Figure 229 is a plot of the residual saturation as a function of the capillary number as calculated from experimental data [3,6,11] for thick beds over 1 ft in thickness (in each case the $\cos\theta$ was taken as unity). Data for low values of residual saturation are for packed towers, whereas data on high residual saturation are primarily from

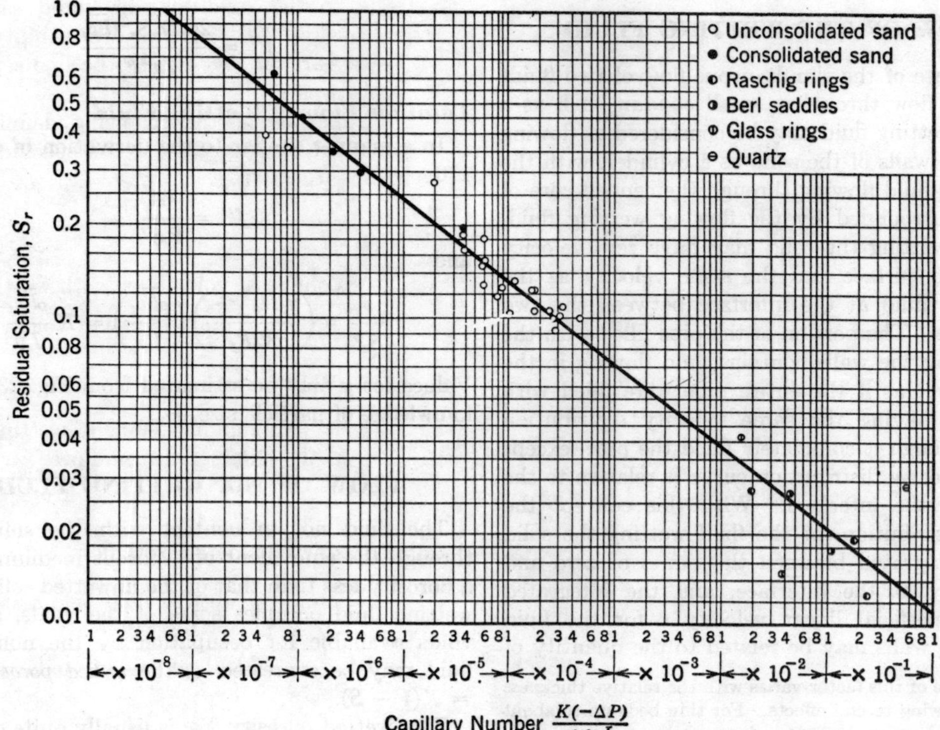

FIG. 229. Residual saturation as a function of capillary number for thick beds.

consolidated sands, thus giving a wide range in the variables involved. Equation 179 is the equation of the line shown in Fig. 229 and is applicable to relatively thick beds where the entrance and exit effects are negligible.

$$S_r = \frac{1}{86.3\,*}\left(\frac{K(-\Delta P)}{g_c L\gamma \cos\theta}\right)^{-0.264}$$

$$= \frac{1}{86.3\,*}\left(\frac{D_p{}^2 F_{\text{Re}}(-\Delta P)}{32L\gamma F_f \cos\theta}\right)^{-0.264} \qquad (179)$$

in which γ = surface tension or interfacial tension (lb/ft).

θ = contact angle measured through the more dense phase.

$\frac{-\Delta P}{L}$ = total force per unit area and length. If the flow is vertical downward and $L = -\Delta Z$, $(\rho' - \Delta P/L)$ is substituted for $(-\Delta P/L)$; if vertical upward and $L = \Delta Z$, $(-\rho' - \Delta P/L)$ is substituted for $(-\Delta P/L)$ (see equations 168*a* and 187).

FLOW OF THE WETTING FLUID

In the case of the simultaneous flow of two fluids in laminar flow through a small conduit such as a pipe, the wetting fluid may be considered as flowing next to the walls of the solid as a cylinder, with the nonwetting fluid flowing through the central core of the pipe surrounded by the flowing wetting fluid. The fluid wetting the solid may have zero velocity at the solid surface and the same velocity as the nonwetting fluid at the interface between the two fluid phases. The portion of the pipe filled with the fluid wetting the walls contains fluid flowing in the same manner as if the entire pipe were filled with that fluid, having the same velocity distribution curve as though it completely filled the pipe, except that its velocity distribution curve is relative to the velocity of the interface. With this concept the velocity distribution for the fluid wetting the solid may be integrated between the limits of zero and the velocity at the interface, and the integrated average superficial linear velocity v for the fluid wetting the walls may be related to the quantity of this fluid flowing per total cross section of pipe in terms of the saturation S as follows.

$$v = \frac{\text{Volume of fluid flowing per second}}{\text{Total cross-sectional area of conduit}}$$

$$= \frac{g_c D_p{}^2 S^2(-\Delta P)}{32L\mu} \qquad (180)$$

Equation 180 is equation 63*L* for laminar flow in a pipe, with v/S^2 substituted for v.

For a porous bed the velocity is proportional to the effective saturation S_e rather than the saturation S. Use of experimental data to determine the exponent applied to the effective saturation indicated that it is variable, having an order of magnitude of 2. Therefore, for the wetting fluid, the modified Reynolds number (Re″) is defined to include the effective saturation as shown by equation 181. With $v/S_e{}^y$ substituted for v in equations 166 to 168, the following modified Reynolds number (Re″) and modified friction factor (f'') result.

$$\text{Re}'' = \frac{D_p F_{\text{Re}} v\rho}{\mu S_e{}^y} \qquad (181)$$

$$f'' = \frac{2g_c D_p S_e{}^{2y}\overline{lw_f}}{Lv^2 F_f} \qquad (182)$$

Laminar Flow of Wetting Fluid

In a manner similar to the derivation of equation 170

$$f'' = \frac{64}{\text{Re}''}$$

and

$$\frac{v}{S_e{}^y} = \left(\frac{g_c D_p{}^2 F_{\text{Re}}}{32F_f}\right)\left(\frac{\rho\overline{lw_f}}{L\mu}\right) = K\frac{\rho\overline{lw_f}}{L\mu} \qquad (183)$$

Values for y may be estimated from Fig. 230 and a knowledge of particle size, D_p.

FLOW OF NONWETTING FLUID

The fluid not in contact with the solid flows through the equivalent of a porous medium having a porosity less than that of the unwetted solid. The wetting fluid occupies some of the voids, and the voids available for occupation by the nonwetting fluid may be expressed as the *wetted porosity*, $X_W = X(1 - S)$.

This wetted porosity X_W is usually quite different in kind from the porosity of the unwetted solid X.

* The value of this factor varies with the relative thickness of the bed, owing to end effects. For thin beds up to about 2 in. in thickness as encountered on rotary vacuum filters, this factor has a value of about 40.

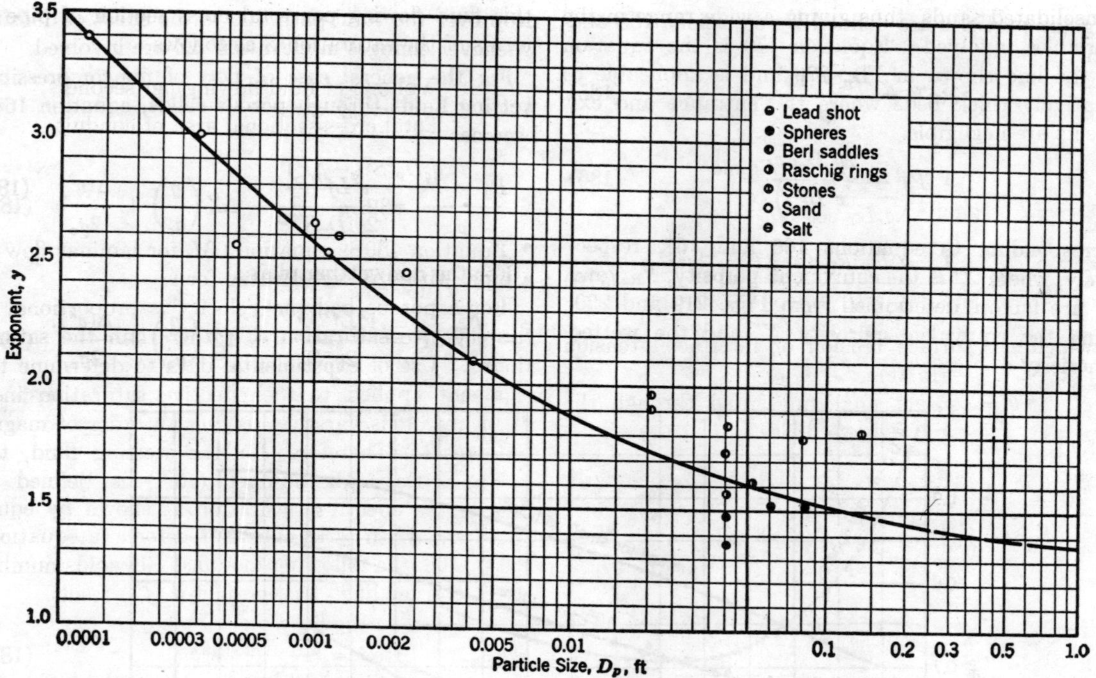

FIG. 230. Exponent y to the effective saturation S_e in equations 181 to 187 as a function of particle size D_p.

The wetting fluid on the surfaces of the solid changes the shape (sphericity) of the particles exposed to the nonwetting fluid and also occupies selectively the small pores and capillaries, leaving only the larger pores open to the nonwetting fluid. The correlation for flow of a single fluid (Fig. 221) is based upon the total porosity of the solid phase including all voids, no matter how small.

In order to use the correlation of Figs. 221 and 225, the porosity must be expressed in terms of a porosity of the same quality as that of the dry medium. The wetted porosity is of a far more effective quality than the porosity of the dry medium because the liquid has selectively filled the smaller, less permeable voids. In order to make this adjustment the wetted porosity must be increased before use in Figs. 221 or 225. The equivalent porosity X' for use in Fig. 225 is obtained by adding the voids eliminated from flow S_fX to the wetted porosity X_W.

$$X' = X(1 - S) + \left(\frac{1 - S}{1 - S_r}\right) S_r X$$

$$= \frac{(1 - S)}{(1 - S_r)} X \tag{184}$$

For example, in Fig. 226 at 100 per cent saturation ($S = 1.0$), the equivalent porosity for the nonwetting fluid is zero. At the residual saturation (about 18 per cent), the equivalent porosity for the nonwetting fluid is the total porosity X as follows from equation 184.

$$X' = X\left(\frac{1 - 0.18}{1 - 0.18}\right) = X$$

At a saturation of 90 per cent the equivalent porosity is 0.122 times the porosity X as follows from equation 184.

$$X' = X\left(\frac{1.0 - 0.90}{1.0 - 0.18}\right) = X\left(\frac{0.10}{0.82}\right) = 0.122X$$

Sphericity of the particles to the nonwetting fluid is altered by the presence of the film of wetting fluid, provided the amount of wetting fluid is greater than the residual saturation. In other words, wetted sphericity is a function of the effective saturation.

Thus, the modified Reynolds number (Re‴) and friction factor (f''') for the flow of the fluid not wetting the particles are defined by equations 185 and 186

$$\text{Re}''' = \frac{D_p v \rho F_{\text{Re}}'}{\mu} \tag{185}$$

$$f''' = \frac{2g_c D_p \overline{lw}_f}{v^2 L F_f'} \tag{186}$$

corresponding to equations 166 and 168, respectively, where X' is the equivalent porosity, F_{Re}' and F_f' are factors determined from Figs. 219 and 220, using the equivalent porosity X' and the wetted sphericity ψ'

drop over the total path due to the effect of density whenever changes in elevation ΔZ are involved.

For the general case of flow of noncompressible wetting fluids through porous solids, equation 168*a* becomes

$$\frac{P_1 - P_2}{\rho} = \frac{v^2 L f'' F_f}{2g_c D_p S_e^{2v}} + \Delta Z\left(\frac{g}{g_c}\right) + \frac{\Delta(v^2)}{2g_c} \tag{187}$$

For the nonwetting phase,

$$-\int \overline{V}\, dP = \frac{v^2 L f''' F_f'}{2g_c D_p} + \Delta Z\left(\frac{g}{g_c}\right) + \frac{\Delta(v^2)}{2g_c} \tag{188}$$

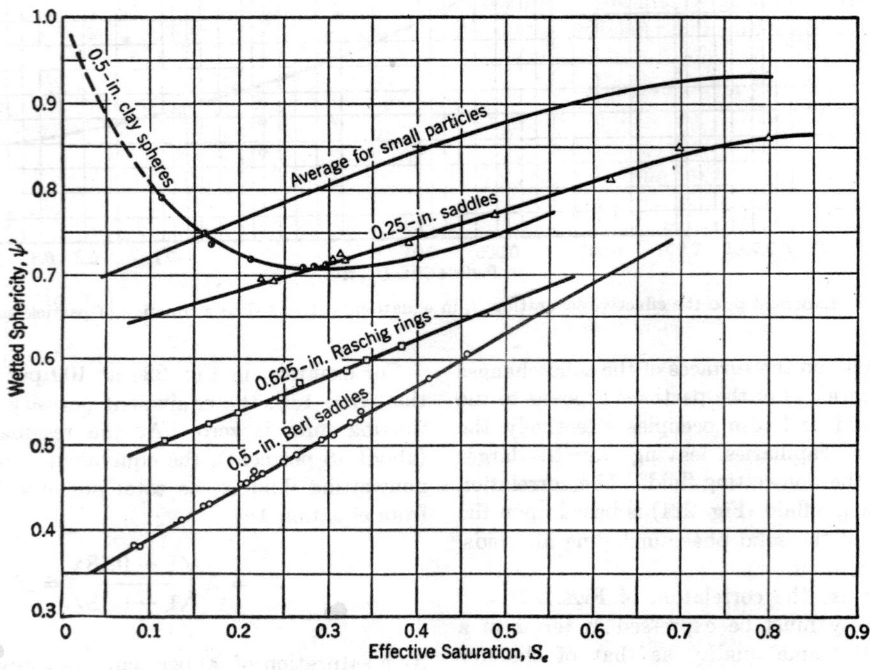

Fig. 231. Wetted sphericity as a function of effective saturation.

Figure 231 shows the wetted sphericity ψ' calculated from experimental data as a function of the effective saturation. For the larger particles the wetted sphericity is reduced with increasing saturation to a greater extent than for smaller particles. In general the larger particles have a lower value of the wetted sphericity ψ' for the same shape and saturation.

Each fluid flows according to its individual properties and the forces acting upon it. The total driving force available to overcome friction may be different for the two fluids even for the same pressure

For noncompressible and nonwetting fluids, equation 188 becomes

$$\frac{P_1 - P_2}{\rho} = \frac{v^2 L f''' F_f'}{2g_c D_p} + \Delta Z\left(\frac{g}{g_c}\right) + \frac{\Delta(v^2)}{2g_c} \tag{189}$$

When applying the flow equations to different streams in the same operation as in this chapter, *the subscript (1) refers to the point upstream and the subscript (2) to the point downstream*; therefore, *in counter-current flow, point (1) for the wetting fluid is not point (1) for the nonwetting fluid.*

Different methods may be used for the simultaneous solution of the two equations, 187 and 188 or

189, to calculate the pressure drop and the velocities of each fluid flowing through the porous medium. The correct solution is obtained when each fluid satisfies its respective flow equation for the operating conditions. For a given solid bed and two fluids of known properties there are four variables: the velocity of the wetting fluid, the velocity of the nonwetting fluid, the effective saturation at steady-state conditions, and the pressure drop across the system. Of these four variables any two may be arbitrarily fixed and the other two determined by trial-and-error calculations.

For the case in which the velocities of the wetting stream and the nonwetting stream are both fixed, it is necessary to compute the effective saturation which gives the same pressure for the two fluids at a selected point in the system such as at the bottom or top of a packed tower, or inlet or outlet of a porous bed.

If the pressure drop is fixed and the velocity of one of the fluids is to be calculated, the procedure is more direct. Such a calculation is demonstrated in the following example problem.

Illustrative Example. Predict the air flow and liquid holdup through a packed bed of 0.500-in. Berl saddles.

Pressure gradient $(-\Delta P/L)$	= 4.0 psf/ft of height
Water rate	= 16,610 lb/hr/sq ft
Temperature	= 21° C
Porosity of bed	= 0.750

The sphericity is estimated from the porosity by Fig. 223, using curve for normal packing, as 0.34.

From Fig. 219 $F_{Re} = 39$

From Fig. 220 $F_f = 660$

From Fig. 230 $y = 1.65$ for a diameter of ½ in. or 0.0417 ft

Applying equation 187 to the liquid (wetting) phase for a height of 1 ft, $L = 1 = -\Delta Z$. Since P_1 refers to the entering point of the liquid

$$\frac{P_1 - P_2}{\rho} = \frac{-4.0}{62.4} = \frac{v^2 L f'' F_f}{2g_c D_p S_e^{2y}} + \Delta Z \frac{g}{g_c} = \frac{v^2 f'' F_f}{2g_c D_p S_e^{2y}} - 1$$

Solving for f''

$$f'' = \frac{2(32.2)(0.0417)(S_e)^{2(1.65)}(62.4 - 4.0)}{\left(\frac{16{,}610}{62.4(3600)}\right)^2 (62.4)(660)} = 0.695(S_e)^{3.3}$$

From equation 181

$$\text{Re}'' = \frac{D_p v \rho F_{Re}}{\mu S_e^y} = \frac{(0.0417)\left[\frac{16{,}610}{62.4(3600)}\right](62.4)39}{[(0.98)(.000672)]S_e^{1.65}} = \frac{11{,}400}{S_e^{1.65}}$$

Since $(S_e^{1.65})$ occurs to the negative first power in Re″ and to the positive second power in f'', the corresponding values of Re″ and f'' will be found on a line with the slope of negative 2. By eliminating $S_e^{1.65}$, $\log f'' = -2 \log \text{Re}'' + \text{constant}$.

Assuming a value of 1.0 for $S_e^{1.65}$, the corresponding value for Re″ is 11,400, and for f'' is 0.695. Therefore the correct solution is found at the intersection of the curve on Fig. 225 with the straight line drawn through the point, Re″ = 11,400, f'' = 0.695, with a slope of negative 2. This intersection is at Re″ = 68,000.

Therefore

$$S_e^{1.65} = \frac{11{,}400}{68{,}000} = 0.168$$

$$S_e = 0.338$$

From equation 179, using $(\rho' + \Delta P/L)$ for the pressure gradient as explained on p. 224,

$$S_r = \frac{1}{86.3}\left(\frac{(0.0417)^2(39)(62.4 - 4)}{(32)(72.8)(6.85 \times 10^{-5})(660)(1)}\right)^{-0.264} = 0.0276$$

From equation 177, using $S_e = 0.338$, $S_r = 0.0276$, S is found to be 0.351, and $S_f = 0.0188$.

The holdup expressed as cubic feet of liquid per cubic foot of voids $= S - S_f = 0.351 - 0.0188 = 0.332$. Expressed as pounds per cubic foot of packing, $(S - S_f)\rho X = (0.332)(62.4)(0.75) = 15.5$ lb of water per cubic foot of packing. An experimental run under these conditions [5] indicated a holdup of 13.3 lb/cu ft of packing.

For the flow of air

By equation 184 $X' = 0.75 \dfrac{1 - 0.351}{1 - 0.0276} = 0.501$

From Fig. 231 $\psi' = 0.535$

From Fig. 219 $F_{Re}' = 54.6$

From Fig. 220 $F_f' = 35{,}000$

From equations 186 and 189, using $(-\rho' - \Delta P/L)$ for the pressure gradient,

$$f''' = \frac{2g_c D_p \overline{lw_f}}{Lv^2 F_f'} = \frac{2(32.2)(0.0417)(-0.076 + 4)}{(1)v^2(0.076)(35{,}000)} = \frac{0.00396}{v^2}$$

From equation 185

$$\text{Re}''' = \frac{D_p v \rho F_{Re}'}{\mu} = \frac{0.0417(0.076)(54.6)v}{(0.0182)(0.000672)} = 14{,}000v$$

Since

$$\log f''' = -2 \log \text{Re}''' + \text{Constant}$$

the solution is obtained as the intersection of the straight line passing through Re‴ = 14,000, f''' = 0.00396, with slope of negative 2 with the curve on Fig. 225. This intersection is at Re‴ = 4350.

Therefore, $v = 4350/14{,}000 = 0.31$ fps.

Experimental results under the same conditions [5] gave a velocity of 0.29 fps.

BIBLIOGRAPHY

1. Boelter, L. M. K., and R. H. Kepner, "Pressure Drop Accompanying Two-Component Flow through Pipes," *Ind. Eng. Chem.*, **31**, 426 (1939).
2. Botset, H. G., "Flow of Gas-Liquid Mixtures through Consolidated Sand," *Trans. Am. Inst. Mining Met. Engrs.*, **136**, 91 (1940).
3. Brownell, L. E., and D. L. Katz, "Flow of Fluids through Porous Media—Part II," *Chem. Eng. Progress*, **43**, 601 (1947).
4. Elgin, J. C., and B. W. Jesser, "Studies of Liquid Holdup in Packed Towers," *Trans. Am. Inst. Chem. Engrs.*, **39**, 277 (1943).
5. Elgin, J. C., and F. B. Weiss, "Liquid Holdup and Flooding in Packed Towers—Part I," *Ind. Eng. Chem.*, **31**, 435 (1939).
6. Leverett, M. C., "Flow of Oil-Water Mixtures through Unconsolidated Sands," *Trans. Am. Inst. Mining Met. Engrs.*, **132**, 149 (1939); *ibid.*, **142**, 152 (1941).
7. Martinelli, R. C., L. M. K. Boelter, T. H. M. Taylor, E. F. Thomsen, and E. H. Morrin, "Isothermal Pressure Drop for Two-Phase Two-Component Flow in a Horizontal Pipe," *Trans. Am. Soc. Mech. Engrs.*, **66**, 139 (1944).
8. Martinelli, R. C., J. A. Putnam, and R. W. Lockart, "Two-Phase, Two-Component Flow in the Viscous Region, *Trans. Am. Inst. Chem. Engrs.*, **42**, 681 (1946).
9. Piret, E. L., C. A. Mann, and T. Wall, Jr., "Pressure Drop and Liquid Holdup in a Packed Tower," *Ind. Eng Chem.*, **32**, 861 (1940).
10. Schoenborn, E. M., and W. J. Dougherty, "Pressure Drop and Flooding Velocity in Packed Tower with Viscous Liquids," *Trans. Am. Inst. Chem. Engrs.*, **40**, 51 (1944).
11. Van Wingen, N., "Influence of Oil Flow on Water Content of Sands," *Oil Weekly* (Oct. 10, 1938); personal letter (Aug. 10, 1945).
12. Wyckoff, R. D., and H. G. Botset, "The Flow of Gas-Liquid Mixtures through Unconsolidated Sands," *Physics*, **7**, 325 (1936).

PROBLEMS

1. A tower filled with 2-in. crushed quartz (porosity 40 per cent) is being drained. The fluid may be considered to have the properties of water. What will be the holdup in the column, expressed as per cent saturation and as volume of fluid per cubic foot of tower packing?

2. Water (60° F) is sprayed onto a bed of 28/32-mesh sand (porosity 40 per cent) at one-half the rate which would flood the bed. What percentage of the voids are filled with water?

3. A horizontal tube of 2-in. inside diameter is packed with sand of the following screen analysis to a porosity of 34.3 per cent. The sand grains are well rounded and have a sphericity of 0.85.

Mesh	24	35	48	65	100	150	200	Pan
Mass % Retained	0.92	0.70	1.80	13.90	57.90	18.68	5.10	1.0

Crude oil saturated with natural gas enters the tube. Pressure taps are located at 5-ft intervals. In one test the pressure readings on two successive taps are 423 and 401 psig. The liquid oil rate between these taps is 1.7 lb/hr. The oil density is 0.78 gram/ml, its viscosity is 2.2 centipoises, and its surface tension is 20 dynes/cm at the flowing temperature of 95° F. The natural gas has a specific gravity of 0.71 air = 1) and a viscosity of 0.012 centipoise.

Compute the average rate of gas flow between the two pressure taps in standard cubic feet per hour and the saturation of the porous bed with oil in this pipe section.

4. An absorber in a brimstone contact plant for sulfuric acid is packed with quartz pebbles to a depth of 20 ft. The pebbles are essentially spherical in shape and have an average diameter of ½ in. The packed quartz has an average porosity of 0.385. Calculate the pressure drop and the liquid holdup as pounds of liquor per cubic foot of packed absorber for an average gas velocity of 160 standard cu ft/sq ft/min and an average liquor rate of 480 lb/hr/sq ft of total cross-sectional area in the absorber. The entering acid is 98 per cent sulfuric acid, and the product is 99 per cent sulfuric acid. Average absorber temperature is 150° F. The gas to the absorber averages 8 per cent sulfur trioxide. The gas has an average viscosity of 0.020 centipoise and the acid a surface tension of 0.005 lbs force per foot and a density of 112 lbs/cu ft.

5. (*a*) Compute the flow of methane upward through a column of sand 4 in. by 4 in. square and 42 in. tall with a pressure drop from 60 psia at the bottom to 51 psia at the top. 0.2 gal of gasoline (56° API, viscosity of 0.56 centipoise, and surface tension of 23 dynes per centimeter) are entering the bottom of the column per hour. Flow is at 80° F. The bed has a porosity of 16.3 per cent. The sand is composed of well-rounded grains of the following screen analysis:

Mesh	Mass Fraction Retained
35	0.17
48	2.44
65	23.23
100	49.28
150	22.45
200	1.77
270	0.44
Over 270	0.22

(*b*) Compute the percentage saturation of the sand bed under conditions of flow.

CHAPTER 18

Filtration

FILTRATION is the operation in which a heterogeneous mixture of a fluid and particles of solid are separated by a filter medium which permits the flow of the fluid but retains the particles of solid. It therefore involves primarily the flow of fluids through porous media.

In all types of filtration the mixture or slurry flows as a result of some driving force, i.e., gravity, pressure (or vacuum), or centrifugal force. In each case the filter medium supports the particles as a porous cake. This cake, supported by the filter medium, retains the solid particles in the slurry which add successive layers to the cake as the filtrate passes through the cake and medium. The several procedures for creating the driving force on the fluid, different methods of cake deposition and removal, and different means for removal of filtrate from the cake subsequent to its formation result in a great variety of filter equipment. In general, filters may be classified according to the nature of the driving force initiating filtration.

GRAVITY FILTERS

Gravity filters are the oldest and simplest type. Gravity-operated sand filters consist of tanks with perforated bottoms filled with porous sand through which the fluid passes in laminar flow, as shown in Fig. 232. They are widely used to process large quantities of fluid containing small quantities of solids, as in water purification.

The tanks may be constructed of wood, steel, or other suitable metal, but for water treatment they are usually of concrete. Ducts beneath the perforated false bottoms lead the filtrate from the bed. The ducts are provided with gates or valves to enable backwashing of the sand bed to remove accumulated solids by reversed flow. The perforated bottom is covered with a foot or more of crushed rock or coarse gravel to retain the sand above. Quartz sand of uniform size is used as the filter medium in water treatment. Graded crushed

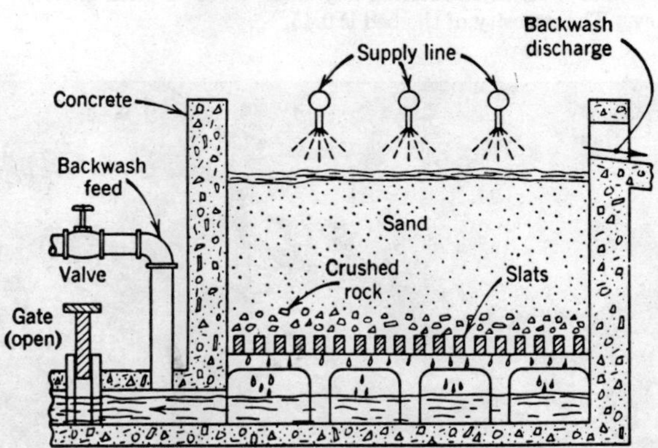

Fig. 232. Sectional diagram of gravity sand filter (not to scale).

coke is used with lead-lined boxes in filtering sulfuric acid, and graded crushed limestone is used for alkaline liquors. Charcoal beds are used to clarify organic liquors both by filtration and adsorption. In all cases it is important that coarse packing be placed at the perforated bottom to support the finer material above. The different sizes of material should be laid in layers so that different sizes of particles are not mixed. The sand used for filtering should be uniform in particle size to provide maximum porosity and maximum filtering rate.

The capacity of these filters may be calculated by the Darcy equation (170) for steady-state flow when the fluid fills the bed and no solids have been deposited on top of the bed. When the bed is just filled (flooded) and the height of the sand bed equals the height of the water column, $-\Delta Z/L$ becomes unity.

The calculation of filter capacity from equation 170 gives the initial capacity when clean. Continuous operation of the filter results in an increased resistance to flow because of the deposition of solids. Cleaning of the bed or removal of these solids by backwashing is required at intervals to maintain or renew the capacity of the filter. The average effective capacity may be even less than 0.5 of the calculated value for clean conditions. For continuous plants, a number of filter beds are used, and a schedule is established for backwashing individual beds in turn.

Illustrative Example. Compute the capacity of a clean sand filter filled to a depth of 1 ft with 35-mesh sand (diameter = 0.001367 ft) when filtering city water at 68° F with gravity flow. The porosity of the bed is 0.45.

The sphericity is estimated from Fig. 223 as 0.75. From Figs. 219 and 220, $F_{\text{Re}} = 48$ and $F_f = 1260$. Solving equation 170 for the superficial velocity of the water (based on empty cross section),

$$v = \frac{g_c D^2 F_{\text{Re}} \rho \overline{lw_f}}{32\mu F_f L}$$

$$= \frac{(32.17)(0.001367)^2(48)(62.3)}{32(1)(0.000672)1260(1)} = 0.00662 \text{ fps}$$

or 2.98 gal/sq ft/min.

The rates of filtration vary from about 2 gal/sq ft/min for water purification to about 5 gal/sq ft/min for some industrial uses. Washing is usually done at a rate of about 15 gal/sq ft/min. For each particular bed there is a definite velocity at which the wash water will begin to fluidize the bed (Chapter 20) and release trapped particles. This is the proper washing velocity which should not be exceeded because of possible loss of sand in the wash water. For high-capacity service the filter bed may be placed in a closed vessel and operated under pressure, as indicated in Fig. 233.

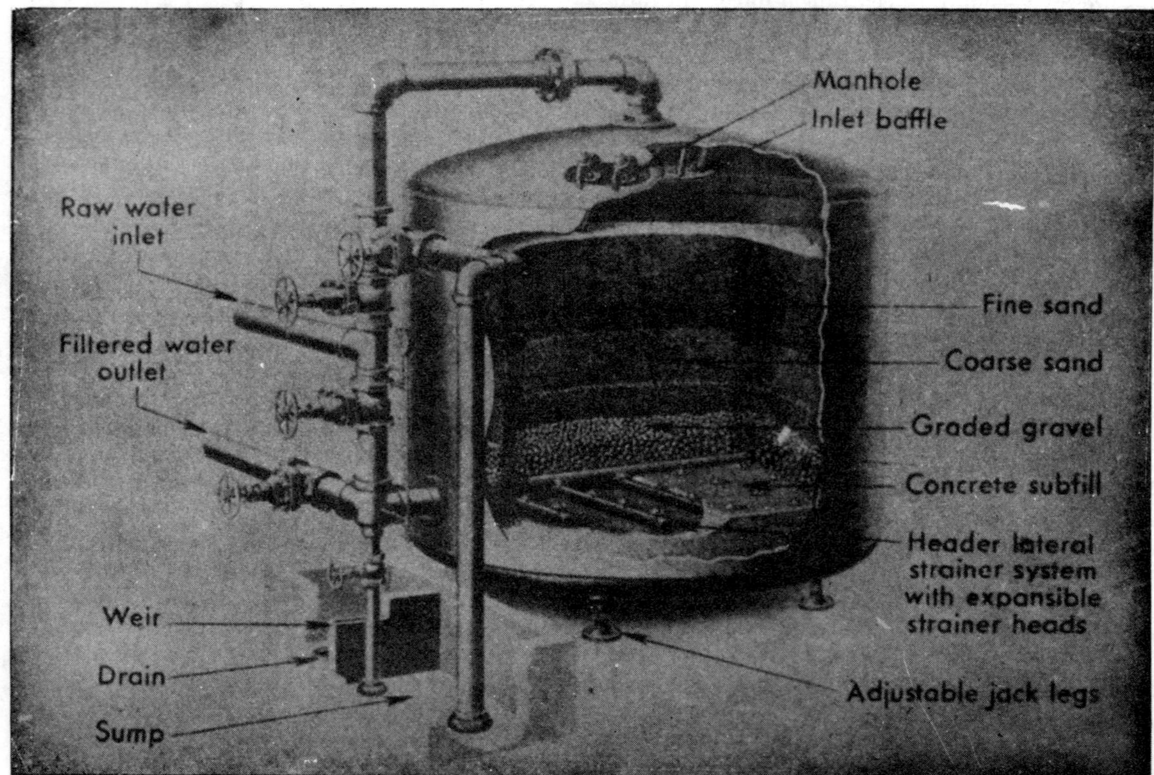

FIG. 233. Vertical pressure sand filter. (*Permutit Co.*)

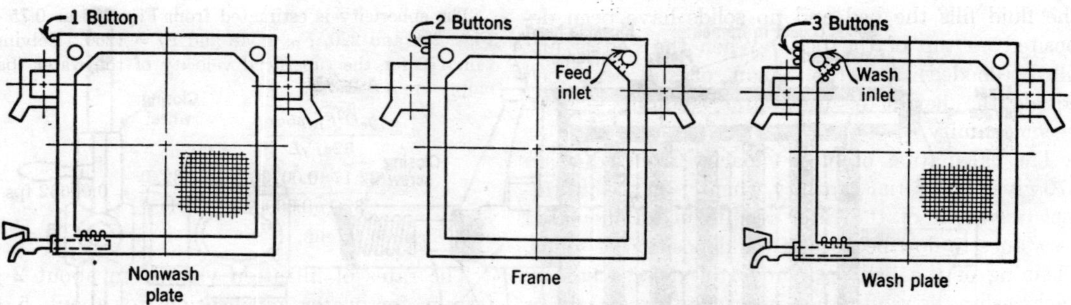

FIG. 234. Plates and frame of an open-delivery filter press.

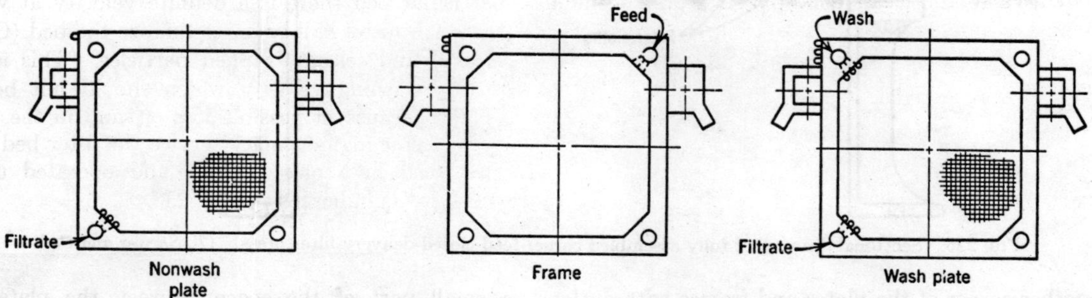

FIG. 235. Plates and frame for closed-delivery washing plate-and-frame filter press.

Bag or "hat" filters consist of cloth bags of felt or flannel with the mixture entering the inside as in a domestic vacuum cleaner. The capacity of such filters is low.

In most industrial operations, the fluid to be filtered carries a high concentration of solids which build up a cake of increasing thickness as the operation proceeds. Such conditions require other types of filters, particularly when the solids are to be recovered. Many varieties of filters are used industrially. Some filters operate batchwise, whereas others operate continuously; some filters require pressure on the slurry entering the press with the filtrate leaving at atmospheric or higher pressure, whereas others allow the slurry to enter under atmospheric pressure and require a vacuum on the filtrate.

PLATE-AND-FRAME FILTERS

The plates and frames of an open-delivery filter press are shown in Fig. 234. The plates and frames are assembled alternately with filter cloths over each side of each plate. The assembly is held together as a unit by mechanical force applied by a screw or hydraulically, as indicated in Fig. 236.

There are many different types of filter presses employing plates and frames. The simplest has a single conduit for introducing the slurry and the wash and a single opening in each plate for removal of the liquid (open delivery), illustrated in section in Figs. 234 and 237. Others have separate conduits for introducing the slurry and wash water. Some also have separate conduits for removing filtrate and wash water (closed delivery), as illustrated in Figs. 235 and 236. The conduits may be at the corners, at the center, or at intermediate locations.

The feed slurry enters through the conduit formed by the holes in the upper right corner of both the plates and frames (Fig. 235). Each *frame* carries an inlet or hole leading from this conduit through which the slurry enters the space between the plates. Pressure on the slurry fed to the press causes the filtrate to pass through the cloths on either side of the plates and run through the space between the cloth and the plate toward the outlet which may be either a spigot, as shown in Figs. 234, 237, and 238, or a second channel formed by holes drilled through

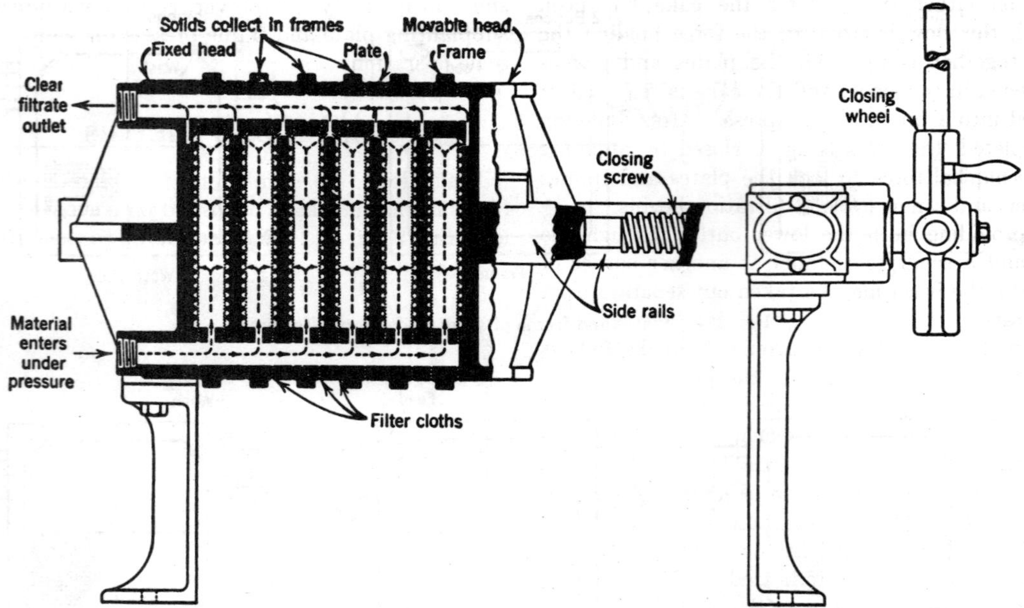

FIG. 236. Sectional drawing of fully assembled corner-feed closed-delivery filter press. (*T. Shriver and Co.*)

another corner of the plates and frames with outlets provided from the *plates* but not the frames (Fig. 236). Whether the outlet is through a conduit (Fig. 236) or through a cock or spigot (Figs. 237 and 238), the plate is so drilled or constructed that the filtrate enters the outlet from both sides of the plate.

The solids in the slurry accumulate on the cloths on opposite sides of the plates. After due time only a small part of the space between the plates is available for the slurry, and the feed is shut off. If the cake is to be washed, clear washing fluid is then passed into the slurry or mixture inlet (Fig. 236 or 237) behind the slurry, enters the cake more or less from the center of the frame, and passes toward the

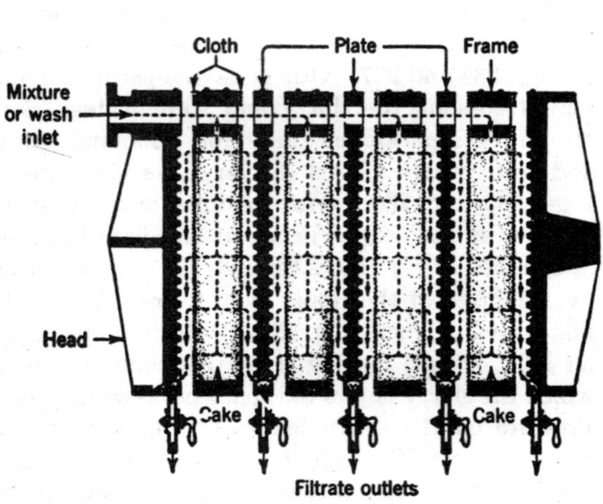

FIG. 237. Exploded sectional drawing of an open-delivery filter press using only one and two button plates. (*T. Shriver and Co.*)

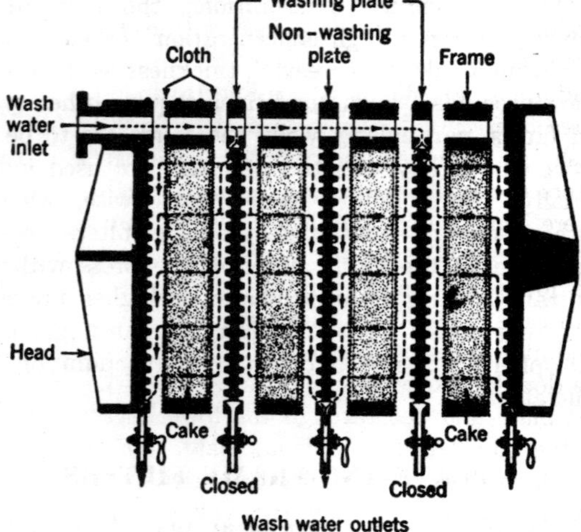

FIG. 238. Exploded sectional drawing of washing-type open-discharge filter press, showing washing operation. (*T. Shriver and Co.*)

plates on either side. After the cake has been washed, this flow is stopped, the force holding the plates together is released, the plates and frames are opened in sequence, and the cake is removed or dumped into a pit below the press. After dumping is completed, the press is again closed by applying the mechanical force to lock the plates and frames together, and a new cycle of filtration begins.

By providing both the lower outlet through the cocks and a separate channel at another corner of the plate, the wash may be taken out separate from the filtrate.

When the wash flowing through the cake follows the same path as the filtrate, the process is called "simple washing." The expression "through washing" or "every-other-plate washing" is applied to the system illustrated in Fig. 238. This requires the use of two different types of plates. The nonwashing plate (one-button) and the wash plate (three-button) are loaded in the press between the frames (two-button), as indicated in Fig. 238. In this process the feed enters the frame as before (upper right corner of frame in Fig. 234) but the wash enters every other plate (the upper left conduit of the three-button wash plate, Fig. 234) and passes through the two cakes in the frames on either side of this plate, leaving through the cock on the nonwash or one-button plate. This method requires that the spigots be closed on those plates (three-button) into which the wash enters.

All these types of plates can be designed to operate on closed delivery by providing a third conduit formed by the holes in the lower right corner of the plates and frames. By providing four conduits (Fig. 235) it is possible to operate with a *closed delivery* with separate outlets for filtrate and wash, as indicated in Fig. 236. The feed slurry enters each frame through the upper right conduit (there being no opening from this conduit to any of the plates). The filtrate leaves each plate through the lower left conduit until the frames are filled with cake. The wash enters through the upper left conduit into every other plate and passes through the double cake within the frames on either side of this plate and out through the lower right conduit on the alternate (one-button) plate. During washing the valves on the filtrate outlet and inlet are closed. During filtering the valves on the wash inlet and outlet are closed.

Plate-and-frame presses are widely used, particularly when the cake is valuable and relatively small in quantity. However, continuous filters are supplanting plate-and-frame presses for many large-scale operations.

BATCH LEAF FILTERS

Leaf filters are similar to the plate-and-frame filters in that a cake is deposited on each side of the leaf (Fig. 239*a*) and the filtrate flows to the outlet in the channels provided by the coarse drainage screen (Fig. 239*b*) in the leaf between the cakes. The leaves are immersed in the slurry. The types of leaf filters shown in Figs. 240, 241, 242, and 243 operate batchwise.

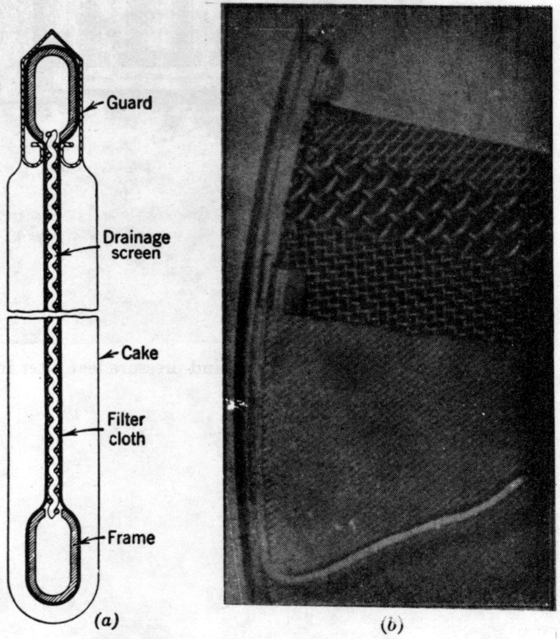

Fig. 239. (*a*) Sectional view of a filter leaf showing construction and approximate location of cake; (*b*) cutaway section of leaf showing details of construction. (*Oliver United Filters.*)

Figure 240 represents a stationary leaf filter (Sweetland type), and Fig. 241 represents a rotating leaf filter (Vallez type). The rotating leaf filter gives a more uniform cake. The incorporation of the screw conveyor permits mechanical discharge of the cake without opening the case.

Figure 242 shows a Kelly filter in the open position. In operation, the filter is closed and the inlet valve is opened, allowing the slurry to enter the shell with the air being displaced through the vent

Fig. 240. A Sweetland pressure leaf filter in open position. (*Oliver United Filters.*)

Fig. 241. Rotary leaf filter. (*Swenson Evaporator Co.*)

FIG. 242. Kelly leaf pressure filter in open position. (*Oliver United Filters.*)

pipe at the top rear of the shell (Fig. 242). This vent may be closed or left open after the shell is filled. If the valve is left open, it acts as a limited overflow valve, returning the excess feed to the feed tank. This may be advisable as it provides better circulation between the filter leaves and tends to keep heavy particles from settling. The cake builds up on both sides of the leaf, as indicated in Fig. 239, and filtration is continued until the desired thickness of cake is formed or until the filtration rate drops off sharply. The feed is then shut off, the drain valve is opened, and the excess slurry is removed. Low-pressure air is introduced into the tank to aid in removing the excess solution. The pressure differential also helps to hold the cake against the filter cloth. After it is emptied, the shell may be filled with wash if desired or blown with air to dry the cake prior to discharge. The excess wash fluid is drained off at the end of the washing period in the same way as the excess slurry, and the cake is then blown with air. The shell is then opened, as indicated in Fig. 242, and the cake is dumped. Air pressure introduced into the filtrate channel, or frequently a water spray, is used to help remove the cake from the leaves.

Leaf filters may be very large. The leaves shown in Fig. 243 suspended on a crane are being transferred to a pit containing the slurry for the recovery of magnesium hydroxide in the process of manufacturing magnesium from sea water. After the cake has been built up to the desired thickness, the crane lifts the leaves and cake out of the pit containing the slurry and immerses the leaves in a second pit containing the wash solution. A vacuum is held on the filtrate or wash side during filtration, transfer, and washing. After the cake has been adequately washed it is similarly transported to a third pit for dumping before the clean leaves are returned to the slurry pit for filtration.

CONTINUOUS ROTARY VACUUM FILTERS

Rotary vacuum filters (Fig. 244) are used where a continuous operation is desirable, particularly for large-scale operations. Figure 245 is a diagrammatic cross section of the filter shown in Fig. 244. The

FIG. 243. Large batch leaf filter press used in producing magnesium from sea water, showing leaves suspended by crane during transfer from filtering point to washing point. (*Dow Chemical Co.*)

filter drum is immersed in the slurry. Vacuum applied to the filter medium causes the cake to deposit on the outer surface of the drum as it passes through the slurry. This part of the cycle is indicated in Fig. 245 as "cake forming." The drum is divided into segments, each of which is connected to the rotating valve through which the vacuum is applied, and the filtrate, wash, and air are removed. These segments are normally about 1 ft in width and in length extend across the entire width of the filter drum.

The position of the twelve segments shown in Fig. 245 is such that segments 1, 2, 3, 4, and 5 are connected to the main filtrate outlet through the rotating valve and the line V_1. B_1 indicates the position of a bridge or blocking shoe which separates the filtrate from the wash. Segments 6, 7, 8, 9, 10, and 11 are connected through the rotating valve to the wash water and air outlet V_2. B_2 indicates the position of the bridge separating washing from air blow, and B_3 indicates the bridge separating air blow from filtering. Air is admitted through the line V_3 to assist in removing the cake.

In the operation as indicated in Fig. 245, the pipe from segment 1 has just passed bridge B_3, and the drum section 1 is completely submerged in the slurry. The vacuum (2 to 26 in. of mercury) causes a thin cake to build up on the outside of the cloth, and the filtrate passes through the piping V_1 to the main filtrate outlet on the valve body.

As the drum rotates in a clockwise manner, the cake becomes progressively thicker, with filtrate still passing out through the main filtrate outlet until station 6 is reached. At this position the cake is fully formed, and the connecting pipe has just passed bridge B_1. The filtrate will now pass into the wash connection and out through the wash outlet of the valve body V_2. The cake is then washed by a series of nozzles as shown mounted on pipes called "wash headers." After being washed, the

Fig. 244. Rotary-drum vacuum filter. (*Oliver United Filters.*)

cake may be rolled in order to increase its density and decrease the moisture content. Wash liquor is removed in positions 10 and 11, primarily by two-phase flow of air and liquor.

At station 12 the segment has passed over bridge B_2 (the blow bridge), and the segment is subjected to a positive air pressure of about 5 psi. This reversal of pressure or "blow" loosens the cake from the filter medium, and the cake is removed by the scraper or "doctor knife."

The cutaway view of a rotary-drum vacuum filter of different design (Fig. 246) may help in clarifying the construction and operation. The drum is divided into segments, each connected through ports in the trunnion to the discharge head. As the drum rotates, the faces of the segments pass successively through the slurry. The vacuum in the segments draws filtrate through the filter medium, depositing the suspended solids on the filter drum as a cake.

As the cake is withdrawn from the slurry by rotation of the drum, it is completely saturated with filtrate and undergoes "dewatering" by the simultaneous flow of air and filtrate. The cake may then

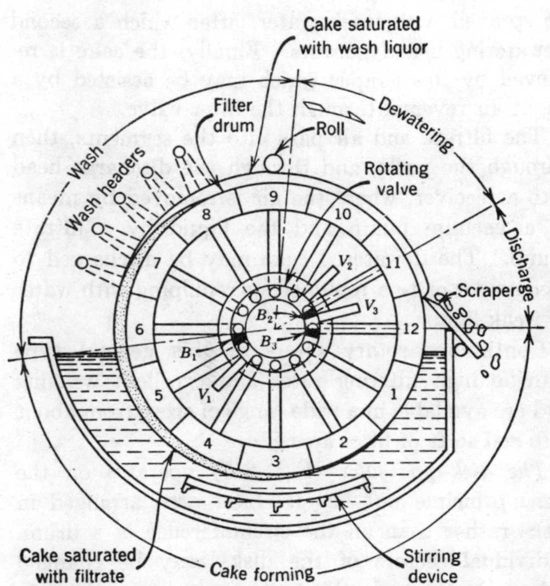

Fig. 245. Diagrammatic cross section of rotary-drum vacuum filter of Fig. 244. (*Oliver United Filters.*)

FIG. 246. Cutaway view of a rotary-drum vacuum filter. (*Swenson Evaporator Co.*)

be sprayed with wash water, after which a second dewatering usually occurs. Finally, the cake is removed by the scraper which may be assisted by a slight air reversal through the filter valve.

The filtrate and air pass into the segments, then through the ports, and through the discharge head into a receiver, where the air is removed by means of a vacuum pump and the liquid by a filtrate pump. The dewatered cake may be discharged to a conveyor or to a repulper for repulping with water or weak liquor.

Continuous rotary filters of this general type provide high filtering rates and excellent washing and are available in a wide range of sizes, from about 3 to 800 sq ft of filter area.

The disk-type filter (Fig. 247) operates on the same principle but has its filter area arranged in disks rather than on the circumference of a drum. Individual sectors of the disks may be changed independently and while other sectors continue in operation. By separating the slurry compartments for the different disks, as by placing one or more division plates in the tank, two or more products may be filtered simultaneously and separately on the same filter, provided the filtrates may be mixed. If the filtrates must be kept separate, only two different products can be handled, as only two valves (one on each end of the trunnion) can be used. Sizes range from about 22 to 2800 sq ft of filter area.

FIG. 247. Rotary-disk filter. (*Oliver United Filters.*)

The internal rotary-drum filter is of the revolving drum type but filters on the inside of the periphery, as indicated in the diagram of Fig. 248. Because of the relatively short arc of travel between the cake forming, washing, and discharge zone, this type of filter is not satisfactory for slow-filtering slurries or where a thoroughly washed cake is required. It is, however, an ideal filter for rapid-settling slurries which do not require a high degree of washing. The difficulty of keeping such slurries in suspension makes the rotary drum vacuum filter less satisfactory for such material. If the slurry has particles of different sizes the cake formed in this filter is properly stratified with the large particles adjacent to the filter medium.

The top-feed filter, illustrated diagrammatically in Fig. 249, is equipped for the use of heated air for drying the solids. The feed is pumped into the hopper wherein suspended solids tend to settle to the bottom while excess liquor overflows and may be returned to the process. The slurry or pulp is carried through nozzles by injecting filtrate. The slurry strikes the spreader apron and flows over feed dams. The baffle spreads the slurry on the surface of the drum, followed by formation of the cake and primary dewatering. The cake is then washed by water from the wash spray nozzles. This is followed by a second dewatering which removes the major part of the liquor before the cake enters the drying section. Heated air is blown through the air duct into the hood surrounding the drum at a positive pressure preventing the inleakage of cold air, then through the cake and out the ports. Drying is followed by removal of the outer crust of the cake by the intermediate scraper. The thinner, partly dried cake continues through the final part of the drying cycle and is completely removed by the final scraper. Spray from cleaning spray nozzles cleans the filter medium, and the cycle is repeated. The dry product drops to the bottom, is picked up by the screw conveyor, and discharged onto a belt conveyor.

Top-feed filters are well adapted for handling of solids which settle readily and filter quickly. They have the advantage of producing a dry product. They are not so flexible nor so simple to operate as a rotary drum filter. A complete installation is indicated in Fig. 250.

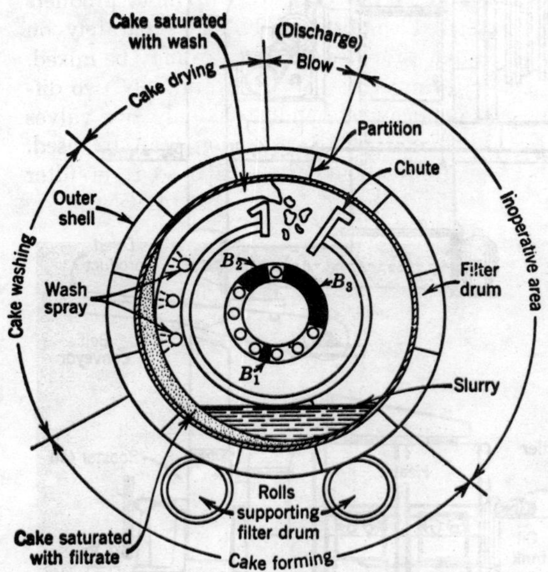

FIG. 248. Diagrammatic cross section of an internal continuous rotary-drum vacuum filter. (*Oliver United Filters.*)

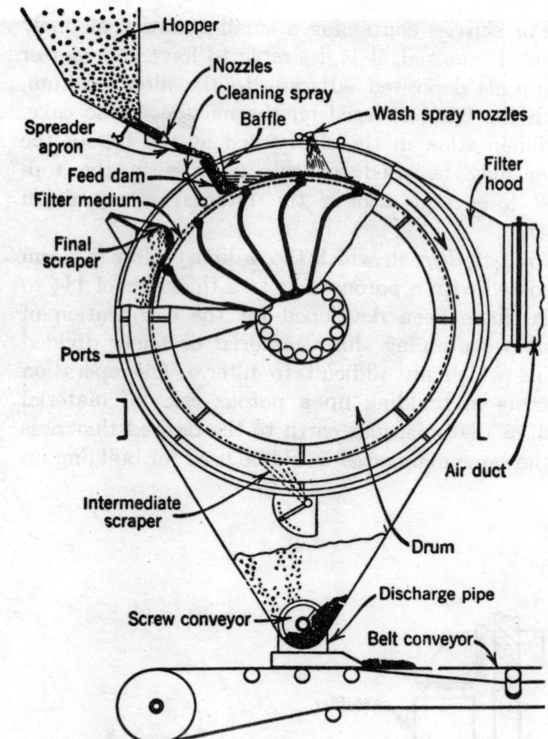

FIG. 249. Diagrammatic section of a top-feed filter. (*Swenson Evaporator Co.*)

For slurries containing a small quantity of finely divided material, it is desirable to have the coarser materials deposited adjacent to the filter medium, with the finer material on the outside of the cake. Sedimentation in the slurry bed at the top of the filter may be used for this purpose in the top-feed filter and also in the internal rotary drum filter.

Precoat filters in which the ordinary filter medium is coated with a porous cake to a thickness of 1½ to 2 in. have been developed for the clarification of liquors containing slimy material or finely divided solids which are difficult to filter. The operation consists in building up a porous cake of material such as diatomaceous earth to the desired thickness in the same manner as would be used for building up any filter cake. The slurry of diatomaceous earth is then removed from the pan, and the liquor to be clarified is admitted. Operation of the filter then causes a cake to build up on the underlying precoat. In this way extremely thin cakes may be built up and removed, thereby maintaining a high rate of filtration. In order to insure complete removal of the thin cake, the scraper or knife is adjusted to advance slowly into the cake or precoat with each rotation of the filter drum. In this way the precoat filter medium is finally removed after a period of hours or days, depending upon the rate of advance of the scraper. The liquor is then removed from the pan, and the precoat is again built up from a slurry of the precoat material. Usually 2 to 3 hr is sufficient allowance for forming the precoat.

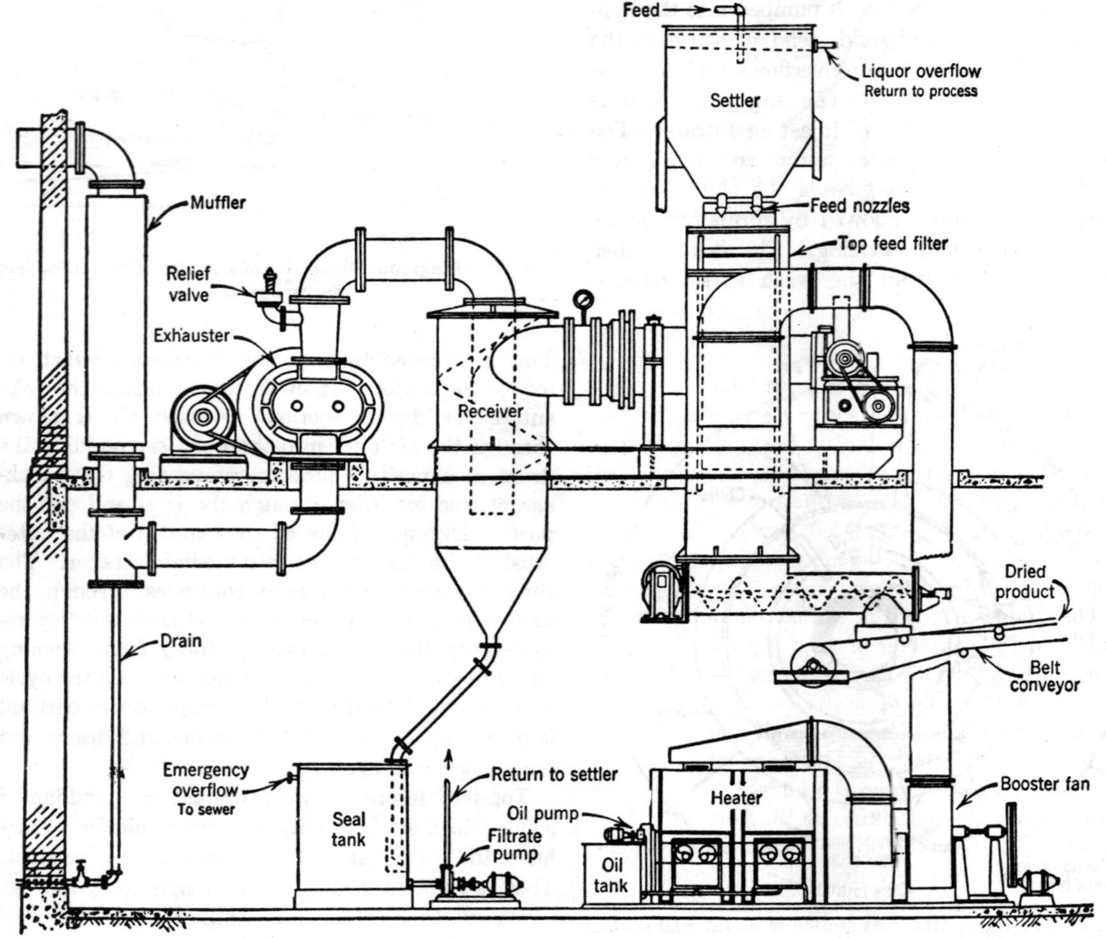

FIG. 250. Flow diagram showing a complete installation of a top-feed filter with accessories. (*Swenson Evaporator Co.*)

The action of the knife on the cake is indicated in Fig. 251.

Rotary filters are used for a wide variety of services. Cakes composed of relatively large crystals may be built up to a thickness of about 4 in. Such cakes normally admit relatively large quantities of air and require a relatively coarse filter medium. In the filtration of finely divided substances such as precipitated calcium carbonate, relatively thin cakes are deposited and the air flowing through the cake is usually negligible in quantity. This means that it is practically impossible to dry these cakes on the filter. Rotary filters are frequently used for removing the solids from the slurry leaving a thickener. In order to maintain high capacity it may be desirable to remove all the cake from the drum at the point of discharge. This results in a turbid filtrate in many cases, but this is not serious if the filtrate is returned to the feed of the thickener for clarification.

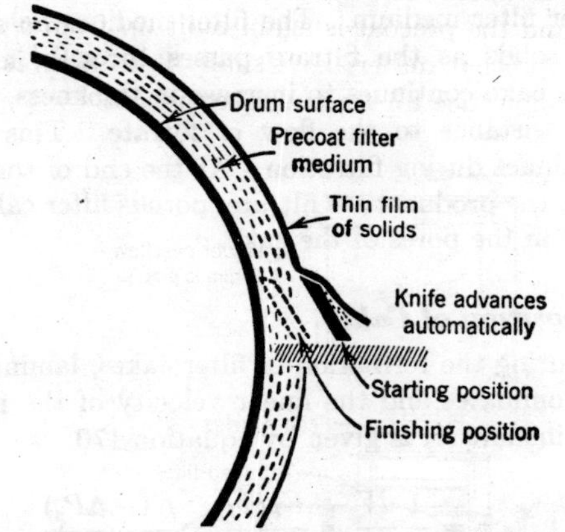

Fig. 251. Drawing showing action of scraper on precoat rotary vacuum filter. (*Oliver United Filters.*)

The sludges handled in commercial filtration vary widely in the character of cake deposits, from granular noncompressible free-filtering materials that filter rapidly to the opposite extreme of slime and colloidal materials that are compressed and tend to plug the filter cloth. The latter are very difficult to separate completely from the filtrate. If the cake is compressible, the resistance to the flow of fluid may increase rapidly as the pressure gradient is increased.

Filter aids are finely divided solids consisting of hard strong particles which in themselves would form a *noncompressible cake.* These materials may be used as described for the precoating filter, or they may be incorporated to a certain percentage with the material of the sludge before it enters the filter press. These filter aids are usually diatomaceous earth or kieselguhr, which consists of the siliceous skeletons of very small marine organisms, diatoms, that were deposited on prehistoric shore lines. They consist of practically pure silica and are very complex in their structure, offering large surfaces for the adsorption of colloids.

OPERATION

The simplest method of operating a *filter press* or a *leaf filter* is to apply the full pressure at the start of the filtration and maintain the pressure constant throughout the run. If the initial pressure is high, the first particles caught in the cloth may be compacted into a tight mass of low permeability which results in a low rate of filtration throughout the rest of the cycle. If the sludge is not homogeneous and contains both crystalline and colloidal particles, the colloidal portion may be forced into the voids between the crystalline solids, greatly decreasing the rate of filtration.

On the other hand, a low initial pressure drop may mean that the first filtrate will not be clear as the cake is formed with a more open structure. However, this permits a more rapid rate of filtration with a cake more readily separated from the cloth, and the initial turbidity of the filtrate disappears with the building up of an adequate cake when the pressure is increased to a constant value which is continued throughout the balance of the run.

Therefore, the common method of operating filter presses is to filter at a constant rate of flow of filtrate during the early part of the cycle. As soon as the cloths are well coated with cake and the filtrate is clear, the pressure is increased to the maximum and the filtration is continued and completed under conditions of constant pressure.

If the material to be filtered is stored in tanks, agitators should be provided so that a uniform suspension is fed to the presses. Reciprocating pumps are not generally adapted to handling slurries and are undesirable as the pulsations tend to make an unduly compact cake. Centrifugal or diaphragm pumps are generally used.

In the operation of *rotary vacuum filters*, the filtrate and the wash run into separate receivers connected to the corresponding ports of the rotary valve. The filtrate and the wash are usually removed through barometric legs and centrifugal pumps. Sometimes pumps of special design are employed for pumping out of a vacuum. Dry vacuum pumps have small clearances and would be ruined if liquid were allowed to enter. In order to protect the vacuum pumps the receivers are provided with float-operated release valves which open and relieve the vacuum if the liquor in the receiver rises above a predetermined level. This destroys the vacuum on the receiver and permits the centrifugal pump to gain its suction. At the same time the operation of the filter ceases until suitable conditions have again been restored.

SELECTION OF FILTERS

Of the various filters, the plate-and-frame filter press is probably the cheapest per unit of filtering surface and requires the least floor space. The cost of labor for opening and dumping such presses is high, particularly in the large sizes. For this reason they are not chosen when a large quantity of worthless solid is to be removed from the filtrate. If the solids have high value and particularly if the quantity to be handled does not justify a continuous automatic filter, the cost of labor per unit value of product is relatively low and the plate-and-frame filter press proves satisfactory. It has a high recovery of solids, and the solid in the form of a cake may be readily handled in a tray or shelf drier which is frequently used for valuable products.

The leaf filter offers the advantages of ease of handling, minimum labor with efficient washing, and discharge of cake without removing any leaves from the filter. The rotary continuous filter offers the additional advantages of continuous and automatic operation for feeding, filtering, washing, and cake discharge.

The rotary continuous filter is widely employed where the precipitate is large in volume and where labor costs must be kept to the minimum. Owing to the method of removal of the cake, gummy or colloidal material is not handled satisfactorily on the continuous rotary filter except by special means, as by the precoating filter. The use of a thickener ahead of the filter greatly decreases the cost of the combined operation if the overflow from the thickener is of sufficient clarity. The operation of the thickener is much cheaper than that of the filter, and the cost of the entire operation is therefore greatly decreased as only a fraction of the material needs to be put through the filter. A continuous filter is frequently used in this connection when the initial cost of such equipment would not be justified if all of the material had to be filtered directly.

FILTER CALCULATIONS

The use of equation 170 for estimating the initial maximum capacity of gravity filters has been indicated. But in all actual filters the resistance to the flow of filtrate varies with time as the precipitate deposits on the filtering sand in sand bed filters, or as the filter cake builds up on the cloth, screen, or other filter medium. The filter medium holds back the solids as the filtrate passes through, and the filter cake continues to increase in thickness, adding its resistance to the flow of filtrate. This action continues during filtration. At the end of the filtration, the products are filtrate, porous filter cake, and fluid in the pores of the cake.

Deposition of Cake

During the formation of filter cakes, laminar flow predominates and the linear velocity of the fluid at any instant (v) is given by equation 170.

$$v = \frac{1}{A}\frac{dV}{dt} = \frac{K\rho \overline{lw}_f}{L\mu} = \frac{K(-\Delta P_c)}{L\mu} \tag{170}$$

where
V = volume of the filtrate.
A = area of the filter medium.
L = thickness of the cake.
t = time.
K = permeability, $g_c D_p{}^2 F_{Re}/32F_f$.
$-\Delta P_c$ = pressure drop through the cake.

In order to obtain an expression relating filtration capacity (expressed as either the quantity of filtrate V, or the cake thickness L) with time of filtering t, it is necessary to obtain a relation between the variables, L and V. This can be done by making a material balance between solids in the slurry filtered and the solids in the cake.

Mass of solids in cake = mass of solids in slurry filtered.

$$(1 - X)LA\rho_s = \frac{(V + XLA)\rho x}{1 - x} \tag{190}$$

where
$$X = \text{porosity of the cake} = \frac{\text{volume of void space}}{\text{total volume of cake}}.$$

XLA = volume of fluid in the cake.
x = mass fraction solids in the feed slurry.
ρ = density of the filtrate.
ρ_s = density of solids in the cake.

Solving equation 190 for V

$$V = \frac{\rho_s(1-x)(1-X) - \rho x X}{\rho x} AL \quad (190a)$$

or

$$L = \frac{V\rho x}{A[\rho_s(1-x)(1-X) - \rho x X]} \quad (190b)$$

Equation 190*b* is a rigorous equation showing the relationship between the filtrate volume V and the cake thickness L. It may be used to eliminate L from equation 170, giving a more convenient relationship between V and t.

$$\frac{dV}{dt} = \frac{KA^2[\rho_s(1-x)(1-X) - \rho x X](-\Delta P_c)}{\mu V \rho x} \quad (191)$$

Equation 191 is an expression for the instantaneous rate of filtration in terms of properties of the slurry, cake, quantity of filtrate, and pressure drop through the cake. For a given slurry, the only variables subject to the control of the operator are pressure drop $(-\Delta P_c)$, filtrate volume V, and time t. Of the remaining terms, the cake porosity X is most likely to vary. Combining many of the terms in equation 191 into a single term, C_V, defined by

$$C_V = \frac{\mu \rho x}{2K[\rho_s(1-x)(1-X) - \rho x X]} \quad (192)$$

a simplified form of equation 191 is obtained.

$$\frac{dV}{dt} = \frac{A^2(-\Delta P_c)}{2C_V V} \quad (191a)$$

Noncompressible Cakes

If the cake porosity remains essentially constant during filtration (as is true with a so-called "noncompressible cake" and may also occur for constant-pressure drop filtration in general) C_V may be considered as a constant, and equation 191*a* is easily integrated. For constant-pressure drop (and constant porosity) this integrates to

$$t = \frac{C_V V^2}{A^2(-\Delta P_c)} \quad (193)$$

Although the differential equation (191*a*) is valid in general, the integrated equation (193) is restricted to constant-pressure drop operation in which the filtration constant C_V remains unchanged.

If it is desired to obtain relationships involving cake thickness instead of filtrate volume, equation 190*a* may be differentiated at constant porosity to obtain

$$dV = \frac{\rho_s(1-x)(1-X) - \rho x X}{\rho x} A\, dL \quad (194)$$

and this equation substituted into equation 170, giving

$$\frac{dL}{dt} = \frac{K\rho x(-\Delta P_c)}{\mu L[\rho_s(1-x)(1-X) - \rho x X]} \quad (195)$$

Defining another filtration constant C_L,

$$C_L = \frac{\mu[\rho_s(1-x)(1-X) - \rho x X]}{2K\rho x} \quad (196)$$

substituting this in equation 195,

$$\frac{dL}{dt} = \frac{(-\Delta P_c)}{2C_L L} \quad (197)$$

and integrating at constant-pressure drop and constant porosity,

$$t = \frac{C_L L^2}{(-\Delta P_c)} \quad (198)$$

In theory, the filtration constants C_V and C_L can be calculated from the properties of the slurry and cake, but the permeability or porosity and particle size of the filter cake are frequently unknown. When operating data with a given slurry have been obtained and when assumption of uniform conditions of cake, filtrate, and slurry can be made, the values of C_V or C_L may be computed from the available operating data and used to estimate other operating conditions.

Illustrative Example. A homogeneous sludge forming a uniform noncompressible cake is filtered through a batch leaf filter at a constant difference in pressure of 40 psi forming a ¾-in. cake in 1 hr with a filtrate volume of 1500 gal. Three minutes are required to drain liquor from the filter. Two minutes are required to fill the filter with water. Washing proceeds exactly as filtration, using 300 gal. Opening, dumping, and closing take 6 min. Assume the filtrate to have the same properties as wash water, and neglect the resistances of the filter cloth and flow lines.

(*a*) How many gallons of filtrate are produced on the average per 24 hr?

According to equation 193,

$$V^2 = \frac{(P_1 - P_2)A^2t}{C_V}$$

$$(1500)^2 = 40\left(\frac{A^2}{C_V}\right)60$$

Therefore

$$\frac{A^2}{C_V} = 938$$

From equation 191*a* the rate of filtration may be found at the time that any known volume of filtrate has been produced. The final rate when $V = 1500$ is desired.

$$\frac{dV}{dt} = \frac{(-\Delta P_c)}{2V}\left(\frac{A^2}{C_V}\right) \tag{191a}$$

$$\frac{dV}{dt} = \frac{40}{2(1500)}(938) = 12.5 \text{ gpm}$$

This final rate of 12.5 gpm equals the rate of washing, since the filtrate has the same properties as wash and the pressure is constant.

$$\text{Time for washing } \frac{300}{12.5} = 24 \text{ min}$$

Total cycle:

Filtering	60 min
Drain	3
Fill	2
Wash	24
Drain	3
Dump	6
Fill	2
Total	100 min

Average gallons of filtrate per 24 hr is

$$\frac{(24)(60)}{(100)}(1500) = 21{,}600 \text{ gal/24 hr}$$

(*b*) How many gallons of filtrate would be produced if a cake of ½-in. thickness were formed, using the same ratio of wash water to filtrate with other conditions the same?

The production of a ½-in. cake would be accompanied by 1000 gal of filtrate. Since $(A^2/C_V) = 938$, equation 193 then becomes

$$(1000)^2 = 40(938)t$$

Therefore

$$t = 26.6 \text{ min}$$

The final filtering, or washing, rate from equation 191*a* as written above is

$$\frac{40}{2(1000)}(938) = 18.76 \text{ gpm}$$

Quantity of wash water = 300(1000/1500) = 200 gal. Time for washing = 200/18.76 = 10.65 min. Cycle time = 26.6 + 3 + 2 + 10.65 + 3 + 6 + 2 = 53.25.

Average gallons of filtrate produced per 24 hr is

$$\frac{(24)(60)}{53.25}(1000) = 27{,}422 \text{ gal/24 hr}$$

Exercise: Determine the optimum cake thickness and maximum daily capacity for the conditions in the above example.

A corresponding treatment [3,8,9]* based on the concept that the rate of filtration is directly proportional to a driving force and inversely dependent upon a resistance may be developed as follows.

A material balance similar to equation 190

$$(1 - X)LA\rho_s = V\rho r \tag{199}$$

where r = mass ratio of dry cake to filtrate.

Solving 199 for L and substituting in equation 170,

$$\frac{dV}{dt} = \frac{A^2K\rho_s(1 - X)}{Vr\rho\mu}(-\Delta P_c) \tag{200}$$

The expression $K\rho_s(1 - X)$ represents the specific conductance of the filter cake and usually is constant during filtration except for changes in porosity of the cake. If the specific resistance, defined as

$$\alpha = \frac{1}{K\rho_s(1 - X)} \tag{201}$$

is substituted in equation 200,

$$\frac{dV}{dt} = \frac{A^2(-\Delta P_c)}{V\alpha r\rho\mu} \tag{202}$$

Combining with equation 191*a*

$$C_V = \frac{\alpha r\rho\mu}{2} \tag{203}$$

In the previous equations the pressure drop $(-\Delta P_c)$ is the pressure on the slurry or upstream side of the cake minus the pressure on the downstream side of the cake. In an operating filter these pressures are usually not available. Instead, the pressures on the slurry in some tank or line considerably upstream from the cake and on the filtrate after it has passed through the cloth or filter medium considerably downstream from the cake are the available data. In the previous equations the resistance is the resistance of the cake alone. In order to use the total pressure drop it is necessary to include the resistances of the filter medium and lines. The resistance contributed by the filter medium, pipe connections, valves, etc., may be expressed as "equivalent cake thickness" L_e, or "equivalent volume of filtrate" V_e, to form a cake of the equivalent thickness L_e.

* The bibliography for this chapter appears on p. 255.

In equation 202 the cake resistance and driving force are $V\alpha r\rho\mu/A^2$ and $(-\Delta P_c)$, respectively. In order to use the total driving force or total pressure drop $(-\Delta P)$, an expression is necessary for the resistance of filter medium, pipe connections, etc. This resistance may be expressed as

$$\frac{V_e\alpha r\rho\mu}{A^2} \quad \text{or} \quad \frac{2C_V V_e}{A^2}$$

The rate equation (202) may then be written as follows:

$$\frac{dV}{dt} = \frac{A^2(-\Delta P)}{\alpha r\rho\mu(V + V_e)} = \frac{A^2(-\Delta P)}{2C_V(V + V_e)} \quad (204)$$

For constant-pressure filtration, equation 204 may be integrated (with $-\Delta P$ constant and V_e constant) to give

$$t = \frac{C_V(V^2 + 2VV_e)}{A^2(-\Delta P)} \quad (205)$$

In a similar manner it may be shown that at constant porosity

$$\frac{dL}{dt} = \frac{K\rho r(-\Delta P)}{\rho_s(1 - X)\mu(L + L_e)} = \frac{(-\Delta P)}{2C_L(L + L_e)} \quad (206)$$

and at constant-pressure drop $(-\Delta P)$ equation 206 may be integrated to

$$t = \frac{C_L(L^2 + 2LL_e)}{(-\Delta P)} \quad (207)$$

Equations 204 and 206 are limited to laminar flow of filtrate through a porous cake. The integrated forms (205 and 207) are also limited to cakes of constant permeability and porosity, formed during a constant-pressure filtration with V_e constant. Since these conditions frequently do not exist, the values for C_L or C_V should be determined experimentally for conditions approximating those for which the equations are to be used. This is most conveniently done by a series of constant-pressure filtrations on the slurry concerned (ΔP constant for each filtration but not for the series).

If equation 204 is rewritten as

$$\frac{dt}{dV} = \frac{2C_V}{A^2(-\Delta P)} V + \frac{2C_V}{A^2(-\Delta P)} V_e \quad (208)$$

and test data are taken in the form of a series of constant-pressure runs, a plot of dt/dV against V, as in Fig. 252, gives a series of lines, each of which will be straight if C_V is constant during that run, which is true for incompressible cakes. These lines are best drawn by differencing the data and plotting $\Delta t/\Delta V$ as the height of the rectangle and ΔV as the base. A smooth line drawn through the top sides of these rectangles so that the areas of the small triangles so formed above the line equal the areas of the triangles below the line gives the best values for dt/dV at any particular value of V. The slope of the line is $2C_V/A^2(-\Delta P)$, and the intercept is $V_e 2C_V/A^2(-\Delta P)$. The corresponding values for

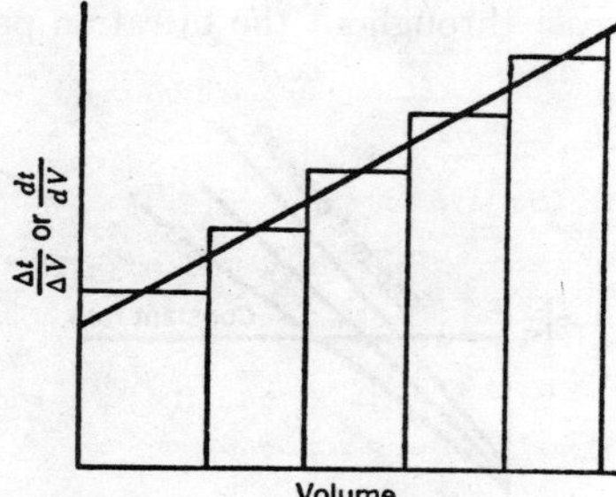

Fig. 252. Plot of derived data for constant-pressure filtrations for evaluating constants.

C_V and V_e may be calculated and used in solving equations such as 204, 205, 206, or 207 for known values of A and ΔP.

A similar procedure may be used for solving equations 206 and 207. Provided the flow through the cake, filter medium, and all connections between the points of pressure measurement is *laminar* and the cake is noncompressible (uniform permeability), the value of V_e is a constant, independent of *pressure* or *rate*.

However, the flow through pipe connections, and probably through the filter medium during the initial high-rate period of a constant-pressure run, may be turbulent, and the initial points on the curves may not indicate the straight-line characteristics of laminar flow. These initial points should be neglected for this reason in evaluating C_V or V_e for use in computing a constant-pressure filtration.

If the test data on an incompressible cake are taken at *constant rate* instead of at constant pressure, V_e is constant for each rate *whether or not the flow is laminar throughout*. Equation 208 may be used in the form

$$(-\Delta P) = \frac{2C_V}{A^2}\frac{dV}{dt} V + \frac{2C_V}{A^2}\frac{dV}{dt} V_e \quad (209)$$

If $(-\Delta P)$ is plotted against V, the straight line has a slope of $\frac{2C_V}{A^2}\frac{dV}{dt}$ and an intercept of $\frac{2C_V}{A^2}\frac{dV}{dt}V_e$. If the area and rate are known, the values for C_V and V_e may be determined.

Since the major part of the resistance to flow between the points of pressure measurement is in the cake itself where the flow is almost always laminar, equations 204 to 208 (as well as 209) with constant values of C_V and V_e give satisfactory results.

The condition of constant-pressure filtration may be said to exist throughout the filtration period of a rotary vacuum filter in the sense that the pressure gage reads a constant value. But actually the pressure drop across the filter medium and cake is zero while the segment is submerged until the valve opens the port to the vacuum and the segment is evacuated. The portion of the cake deposition period during which the pressure varies as described differs with the design of the filter and operating conditions, but it usually varies from about 3 per cent of the total cake deposition period for slow-filtering slurries of fine particles to almost 20 per cent for fast-filtering slurries.

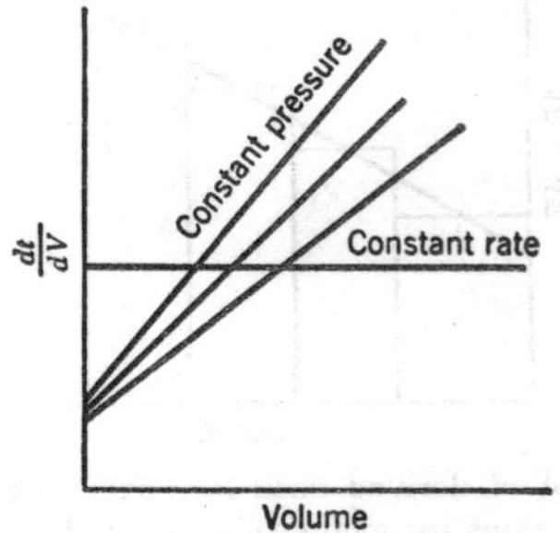

Fig. 253. Reciprocal rates dt/dV for constant-rate and constant-pressure filtration.

In many cases, such as a plate and frame or a leaf press operated with the slurry supplied from a centrifugal pump, the early stages of filtration are conducted at constant rate rather than constant pressure, as the controlling factor is the capacity of the pump to deliver. As the cake becomes thicker and offers more resistance to the flow of fluid, the limiting factor becomes the pressure developed by the pump and the filtration proceeds at constant pressure.

The two types of filtration may be represented by the curves of Fig. 253 in which reciprocal rate dt/dV is plotted as a function of filtrate volume V. Equations 208 or 209 apply equally well to a constant-rate filtration (dt/dV constant) or to a constant-pressure filtration (ΔP constant). The course of the filtration described in the preceding paragraph for a plate-and-frame filter press is represented in Fig. 253 along a horizontal (constant-rate) line at the volume capacity of the pump until the sloping (constant-pressure) line is reached that corresponds to the maximum pressure developed by the pump and then along this constant-pressure line.

The *cake thickness* L is frequently a more convenient variable to use than the volume of filtrate V, as for rotary filters with exposed cake. Equations 206 or 207 may then be used if the cake may be assumed to have constant porosity for a constant-pressure drop. This condition is approximately true even for compressible cakes if the filtration is conducted at constant pressure as on a rotary vacuum filter.

Compressible Cakes

In actual filtrations the combined resistances represented by C_V are seldom independent of the pressure drop $(-\Delta P)$, even when C_V is a constant independent of the thickness of the cake or of filtrate volume, as indicated by the straight lines of Figs. 252 and 253. In other words, practically all cakes are compressible to some degree at least. Large drops in pressure tend to force the solids further into the interstices in the filtering medium, thereby increasing the resistances and the values of C_V and V_e.

If C_V increases with the pressure drop $(-\Delta P)$, there may be a definite value for the pressure drop which gives a maximum rate of filtration. This may be determined by running a number of constant-pressure filtrations and plotting the values of C_V so obtained as $(-\Delta P)/C_V$ against $(-\Delta P)$, as shown in Fig. 254. If C_V is independent of pressure, this results in a straight line passing through the origin with a slope of $1/C_V$. If C_V increases with $(-\Delta P)$ as in compressible cakes, the plot is of a curved line which may have a maximum. The value for $(-\Delta P)$ which corresponds to this maximum is the pressure drop giving the maximum capacity when filtering that slurry.

Compressible cakes are more dense adjacent to the filter medium because of the greater pressure gradient $(-\Delta P/L)$ in the initial stages of cake formation.

For compressible cakes it is necessary to determine values of C_V and V_e for different pressure drops (ΔP) and instantaneous values of either V or L, and to use the proper values corresponding to the known pressure drops in the differential equation 208 which may then be integrated graphically. If mean values for C_V and V_e are determined for a given pressure drop and cake thickness L or the corresponding volume of filtrate V, the integrated equations 205 or 207 may be used with these mean values for these specific conditions *only*.

If the experimental values for C_V (and $C_V V_e$) are plotted as a function of pressure drop for particular

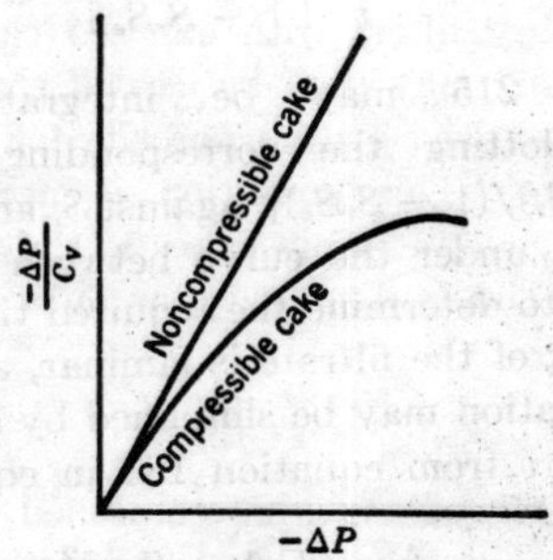

Fig. 254. The effect of pressure drop on capacity of filter for noncompressible and compressible cakes.

values of V, on logarithmic paper a straight line is usually obtained, indicating that C_V may be expressed as an exponential function of the pressure drop $(-\Delta P)$ for substitution in equation 205. Similar procedure is indicated for C_L in equation 207.

If the values for C_V or C_L are exponential functions of the pressure drop for a *given slurry*, the specific resistance α is also an exponential function of (ΔP) for a *given cake*. Under these conditions

$$C_V = (C_V')(-\Delta P)^a \qquad \alpha = \alpha'(-\Delta P)^a$$

$$C_V V_e = c(-\Delta P)^b \qquad \alpha V_e = \beta(-\Delta P)^b$$

where α', β, a, and b are constants for a particular cake, and C_V' and c are constants for a particular slurry. Substituting the appropriate exponential relationships in equation 204 or 208 and integrating between limits of $t = 0$ and $t = t$, and $V = 0$ and $V = V$, for constant ΔP, gives,

$$t = \frac{(C_V')}{A^2}\frac{V^2}{(-\Delta P)^{1-a}} + \frac{2cV}{A^2(-\Delta P)^{1-b}} \qquad (208a)$$

$$t = \frac{r\rho\mu V}{2A^2(-\Delta P)}[(\alpha')(-\Delta P)^a V + 2\beta(-\Delta P)^b] \qquad (204a)$$

These equations may be rewritten as follows.

$$\frac{(-\Delta P)t}{V/A} = (C_V')\left(\frac{V}{A}\right)(-\Delta P)^a + \frac{2c}{A}(-\Delta P)^b \qquad (208b)$$

$$\frac{(-\Delta P)t}{V/A} = \frac{r\rho\mu\alpha'}{2}\left(\frac{V}{A}\right)(-\Delta P)^a + \frac{r\rho\mu\beta}{A}(-\Delta P)^b \qquad (204b)$$

which are equations of straight lines when plotted $\frac{(-\Delta P)t}{(V/A)}$ against V/A for constant-pressure filtration data. The respective slopes are $(C_V')(-\Delta P)^a$ and $\frac{r\rho\mu\alpha'}{2}(-\Delta P)^a$, and the corresponding intercepts are $\frac{2c(-\Delta P)^b}{A}$ and $\frac{r\rho\mu\beta}{A}(-\Delta P)^b$.

The value for the exponent a is obtained by plotting the logarithm of the slopes, $(\log C_V') + a \log (-\Delta P)$, against the logarithm of the pressure drop, $\log(-\Delta P)$, giving a straight line whose slope is a and whose intercept [when $(-\Delta P) = 1$] is $\log C_V'$. Similarly, the value of the exponent b may be obtained by plotting the logarithm of the intercepts, $\log(2c/A) + b \log(-\Delta P)$, against the logarithm of the pressure drop, $\log(-\Delta P)$, giving a straight line whose slope is b and whose intercept is $\log (2c/A)$.

Unfortunately this exponential relationship is only a rough approximation in most cases of compressible cakes, but it may be used to help estimate the effect of changes in operating conditions when experimentally determined coefficients and exponents are available. The values of the exponent a vary from a maximum of about 0.9 for the compressible hydroxide sludges to about 0.01 to 0.15 for calcium carbonate and kieselguhr.

When these equations are used, b is frequently assumed to be equal to a with considerable simplification.

$$t = \frac{(C_V')V(V + 2V_e)}{A^2(-\Delta P)^{1-a}} \qquad (205a)$$

When $a = 0$, equation 205a is identical to equation 205 for incompressible cakes.

Exercise. Derive equation 209a for use in constant-rate filtration.

$$(-\Delta P)^{1-a} = \left[\frac{2C_V'V}{A^2} + \frac{2c}{A^2}\right]\frac{dV}{dt} \qquad (209a)$$

Removal of Filtrate from Cake

After the cake is deposited and built up to the desired thickness, it is usually desirable to remove the filtrate from the voids of the cake. This may be done by washing the cake with another fluid such as water or air.

If the washing fluid is not miscible with the filtrate, as when air is used, the problem becomes one of simultaneous flow of two homogeneous fluids. Under these conditions the quantity of filtrate in the cake may be reduced to the residual saturation S_r as a limit. The quantity of air that must be drawn through the cake to reduce the filtrate content is an important consideration, particularly in vacuum filters as it determines the required capacity of the vacuum pumps. Further reduction of filtrate may be accomplished by some other procedure, such as drying with heated or dried air which removes only the volatile fluid, or washing with a fluid miscible with the filtrate which in time will remove practically all the filtrate. A miscible wash fluid can be used only in limited quantities when the cake is soluble.

The removal of filtrate by another immiscible fluid may be calculated as a function of time, using the relationship for steady-state two-phase flow for instantaneous conditions and integrating as outlined below.

First, select the pressure gradient $(-\Delta P/L)$ to be used. Select values of saturation S of the cake by the filtrate over the range from the initial saturation down to the desired limit. These values are converted to effective saturation S_e by equation 177.

The porosity of the cake X must be determined or estimated. Once the cake is formed the porosity may be assumed to be constant with respect to time during the washing period, provided there is no mechanical compression.

This value for the porosity and the selected values for effective saturation are then used in equations 181 and 182, with the proper values for F_{Re}, F_f, and y determined from the properties of the cake and from Figs. 219 and 220 to determine the corresponding filtrate velocities for the selected pressure gradient.

If V is the volume of filtrate removed from the cake

$$-dS = \frac{dV}{XLA} \tag{210}$$

Solving equation 177 for S

$$S = \frac{S_e - 2S_eS_r + S_r}{1 - S_eS_r} \tag{211}$$

Differentiating equation 211 and substituting in equation 210 as indicated gives

$$-\left[\frac{1 - S_r}{1 - S_rS_e}\right]^2 dS_e = \frac{dV}{XLA} \tag{212}$$

Since

$$v = \frac{dV}{A\,dt} \tag{213}$$

$$dV = vA\,dt = -XLA\left[\frac{1 - S_r}{1 - S_rS_e}\right]^2 dS_e \tag{214}$$

Rearranging,

$$dt = -\frac{XL}{v}\left[\frac{1 - S_r}{1 - S_rS_e}\right]^2 dS_e \tag{215}$$

Equation 215 may be integrated graphically by plotting the corresponding values of $XL/v[(1 - S_r)/(1 - S_rS_e)]^2$ against S_e and determining the area under the curve between the desired limits of S_e, to determine the required time.

If the flow of the filtrate is laminar, as it usually is, the calculation may be simplified by substituting the value of v from equation 183 in equation 215, giving

$$dt = -\frac{32XL^2\mu\left[\dfrac{1 - S_r}{1 - S_rS_e}\right]^2 F_f}{g_cD^2S_e^{\,y}(-\Delta P)F_{\text{Re}}}\,dS_e \tag{216}$$

$$dt = -\frac{XL^2\mu\left[\dfrac{1 - S_r}{1 - S_rS_e}\right]^2}{K(-\Delta P)S_e^{\,y}}\,dS_e \tag{216a}$$

which can be integrated graphically.

The term $[(1 - S_r)/(1 - S_rS_e)]^2$ in equation 216 is unity when the effective saturation S_e equals unity, and it is $(1 - S_r)^2$ when S_e equals zero. As this factor is always close to unity, a good approximation may be made by using the arithmetic average value $\left[\dfrac{(1 - S_r)^2 + 1}{2}\right]$ as a constant value, and the equation becomes

$$t = -\frac{\mu XL^2}{K(-\Delta P)}\left[\frac{(1 - S_r)^2 + 1}{2}\right]\int_1^{S_e}\frac{dS_e}{S_e^{\,y}} \tag{217}$$

Integrating,

$$\frac{t}{C_t} = \left[\frac{(1 - S_r)^2 + 1}{2}\right]\frac{(S_e)^{(1-y)} - 1}{y - 1} \tag{218}$$

where

$$C_t = \frac{\mu XL^2}{K(-\Delta P)} \tag{219}$$

and y = exponent dependent upon particle size, Fig. 230.

The filtration constant C_t has the units of time and incorporates the variables affecting the flow of filtrate. The exponent y varies with particle size from 2.4 for 20 Tyler mesh to 2.8 for 100 Tyler mesh. Using an average value of 2.5 for the exponent y, the time ratio t/C_t is plotted versus saturation S, with parameters of residual saturation S_r as shown in Fig. 255.

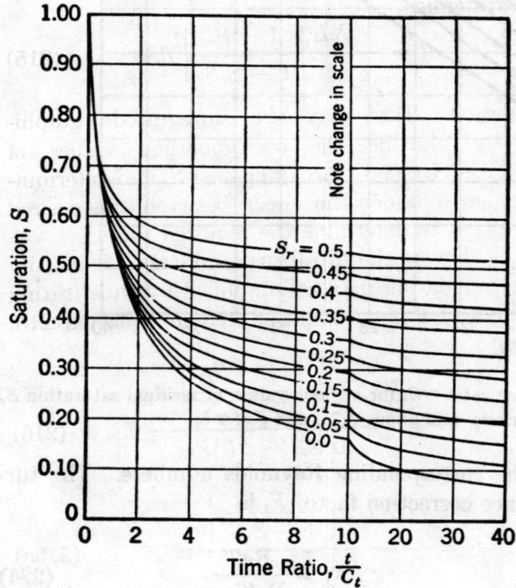

FIG. 255. Graphical solution of equation 218, giving the saturation as a function of the time ratio t/C_t for various values of residual saturation S_r for $y = 2.5$.

Air Flow

As the liquid is removed by fluid flow during the dewatering interval, the rate of flow of air increases rapidly from zero to a maximum, which is virtually the rate obtained with a dry cake.

The rate of flow of air passing through the cake simultaneously with the filtrate may be computed by graphical integration of instantaneous velocities of air flow plotted against time. The instantaneous velocities of air flow are calculated for the selected values of saturation expressed as effective saturation S_e by means of equations 185 and 186 and the relationships of Figs. 219, 220, and 221 or 225. The corresponding values for time are obtained from the previously determined relationship between saturation and time when computing filtrate removal.

The procedure can be simplified by making some approximations and integrations for the general case. As the flow of air is usually in the early transition region between laminar and turbulent flow, relationships may be derived on the basis of laminar flow, with a correction factor for turbulence. Equations 185 and 186 may be combined with $f''' = 64/\text{Re}'''$ for laminar flow of the nonwetting fluid to give

$$v_a' = \left(\frac{g_c D_p^2 F_{\text{Re}}'}{32 F_f'}\right)\left(\frac{-\Delta P}{L\mu_a}\right) \tag{220}$$

where v_a' = superficial air velocity in two-phase flow, cu ft/(sq ft)(sec).

Dividing equation 220 by equation 170,

$$\frac{v_a'}{v_a} = \left(\frac{F_{\text{Re}}'}{F_f'}\right)\left(\frac{F_f}{F_{\text{Re}}}\right) \tag{221}$$

By the solution of equations 221 and 218 for selected values of S and S_r, (v_a'/v_a) may be obtained as a function of t/C_t. By plotting the ratio (v_a'/v_a) against the corresponding ratio (t/C_t) (computed with the same value of saturation S) for the same residual saturation S_r, the total air flow V_a in cubic feet may be determined by graphical integration of these curves, since (V_a) is $\int_0^t v_a'\,dt$, or

$$\int_0^{\frac{t}{C_t}} \left(\frac{v_a'}{v_a}\right) d\left(\frac{t}{C_t}\right) = \int_0^{\frac{t}{C_t}} \frac{v_a'\,dt}{v_a C_t} = \frac{V_a}{C_a} \tag{222}$$

where

$$C_a = v_a C_t = \left(\frac{-\Delta P}{L\mu_a}\right)\left(\frac{\mu_l X L^2}{-\Delta P}\right) = XL\frac{\mu_l}{\mu_a} \tag{223}$$

from equations 170 and 219, canceling K in the numerator with K in the denominator.

In Fig. 256 the accumulative volume ratio V_a/C_a is shown as a function of the ratio t/C_t with parameters of residual saturation S_r. In preparation of these curves, average properties of the cake were used, with porosity about 0.45, sphericity about 0.75, and $y = 2.5$. With Fig. 256 a rapid determination can be made of the accumulative air flow for conditions of laminar flow in two-phase flow.

A correction for turbulence should be made if the modified Reynolds number Re''' for the air flow exceeds a value of 100. In making this correction, $\text{Re}'''_{(l)}$ is first determined, assuming laminar flow. Such a value is on the 45-degree line for laminar flow on Figs. 125, 221, or 225. With reference to

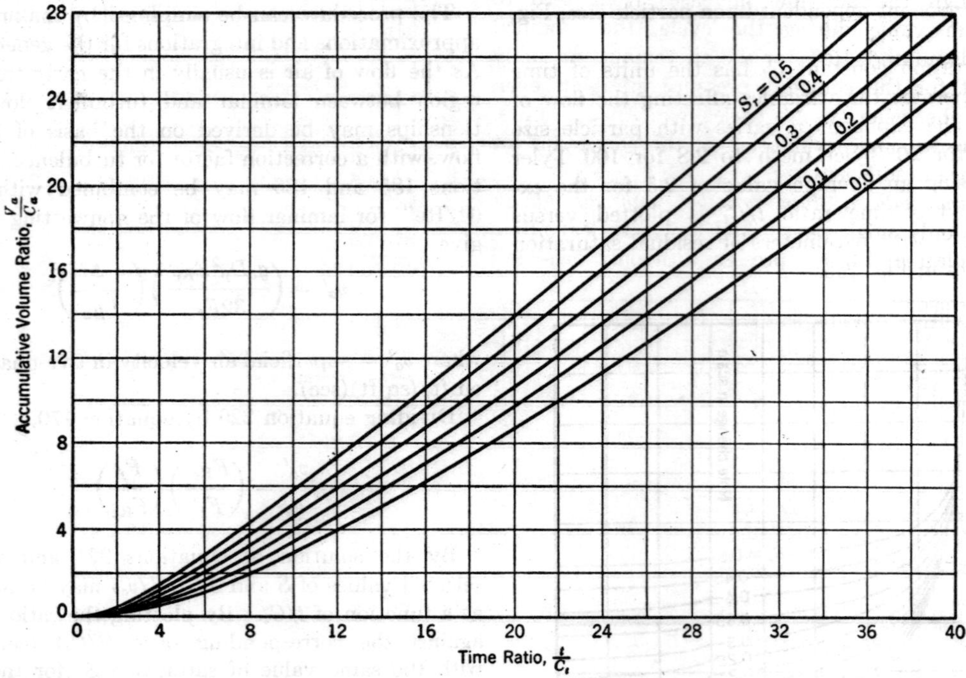

FIG. 256. Accumulative volume ratio V_a/C_a as a function of the time ratio t/C_t for various values of residual saturation S_r for a cake with a porosity X of about 0.45, sphericity ψ of about 0.75, and y of 2.5.

Fig. 257, point 1 represents a Reynolds number above 100. The correct Reynolds number is at point 2, which is the intersection of the friction factor curve and a line with slope of -2 drawn through point 1, because the locus of points in which velocity is the only variable is a line with a slope of -2, as described in Chapter 7 (equation 22).

The ratio of the velocity found by assuming laminar flow to the correct velocity equals the ratio of the corresponding Reynolds numbers. The turbulence correction factor F_t is

$$F_t = \frac{\text{Re}'''}{\text{Re}'''_{(l)}} \tag{224}$$

For convenience, a plot of this correction factor F_t in terms of permeability K and pressure gradient $-\Delta P/L$ is shown in Fig. 258.

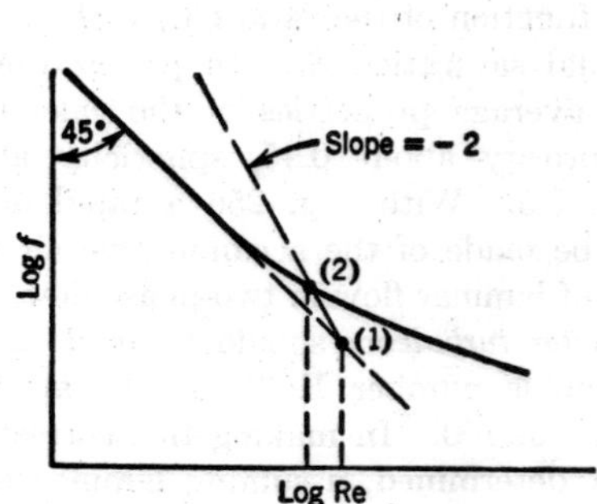

FIG. 257. Diagram illustrating method for correcting for air turbulence.

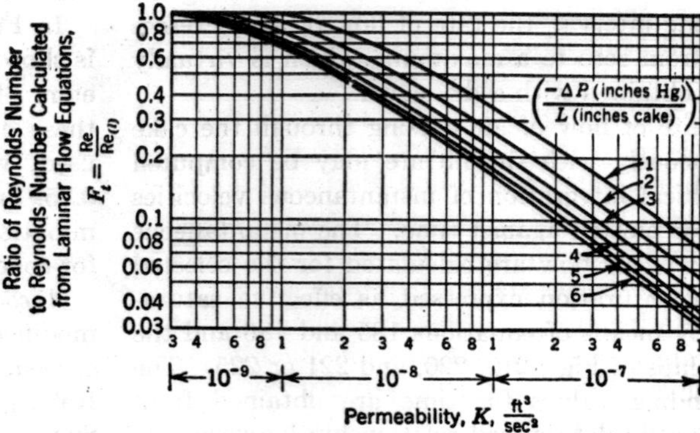

FIG. 258. Correction for turbulence F_t as a function of permeability for various pressure gradients as indicated on the curves.

Although the air flow increases from zero to maximum value during the cycle, the modified Reynolds number Re‴ for the air remains virtually constant for the constant pressure gradient, even though the velocity v_a' varies over wide limits. Therefore, a single evaluation of the correction for turbulence is sufficient.

Washing with a fluid miscible with the filtrate involves only single-phase fluid flow. In leaf filters and in plate-and-frame presses with only one- and

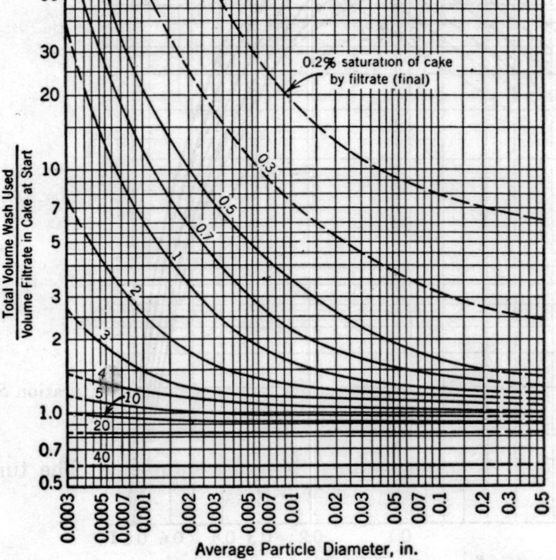

FIG. 259*a*. Filtrate retention by cake (in per cent saturation) after washing with wash of the same viscosity as the filtrate when the cake is flooded and there is no adsorption, expressed as a function of the mean surface diameter (equation 169) of the particles, in inches.

two-button plates, the wash water follows the same path as the filtrate. The properties of the cake at the end of filtration and the properties of the wash fluid control the washing process. When operating in the manner known as through washing, the wash fluid is introduced on every other plate of a plate-and-frame filter press, and the wash fluid passes through a cake twice as thick as that through which the filtrate passed at the end of the filtration. This procedure makes the washing area one-half that of the filtering area. Under these conditions the washing rate is one-fourth that of the final filtering rate for fluids of the same density and viscosity with the same drop in pressure.

The mechanism of washing may be divided into

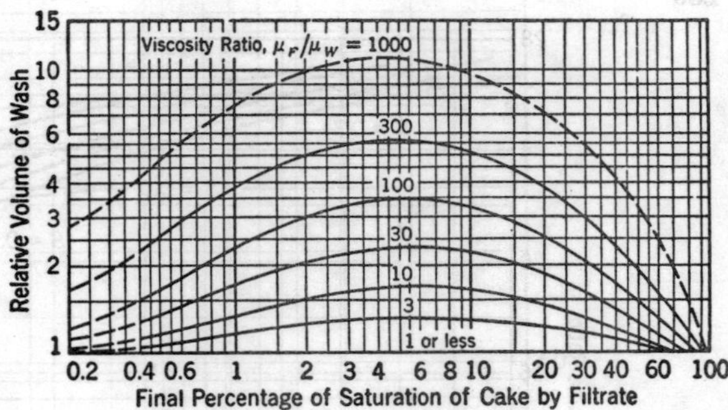

FIG. 259*b*. Relative volume of wash required as affected by viscosity of wash μ_W expressed as a function of per cent final saturation of the cake by filtrate and ratio of viscosity of filtrate μ_F to viscosity of wash, μ_F/μ_W, for flooded cakes with no adsorption.

three stages. During the first stage the filtrate is displaced from the cake without dilution by the wash. During the second stage the concentration of filtrate in the stream leaving the cake decreases continuously. In the final stage the filtrate is slowly leached or washed out of the interstices in the cake, in a manner similar to that described in Chapter 21.

Usually, about 90 per cent of the filtrate may be removed during the first stage if the wash is of approximately the same viscosity as the filtrate. Under these conditions, the volume of wash required to carry the operation through the second stage is equal to about twice the volume of the filtrate originally in the cake. The quantity of filtrate remaining in the cake at the end of the second stage is controlled primarily by the diameter of the particles in the filter cake. Figure 259*a* indicates the filtrate retained in the cake (expressed in terms of filtrate saturation) as a function of the particle size and the volume of wash used; it is applicable only to those cases where

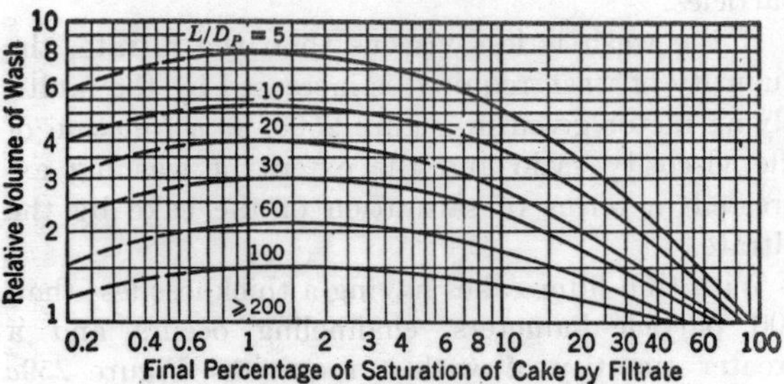

FIG. 259*c*. Relative volume of wash required for thin cakes, expressed as a function of the ratio of cake thickness L to particle size D_p, and final percentage of saturation of cake by filtrate, for flooded cakes with no adsorption.

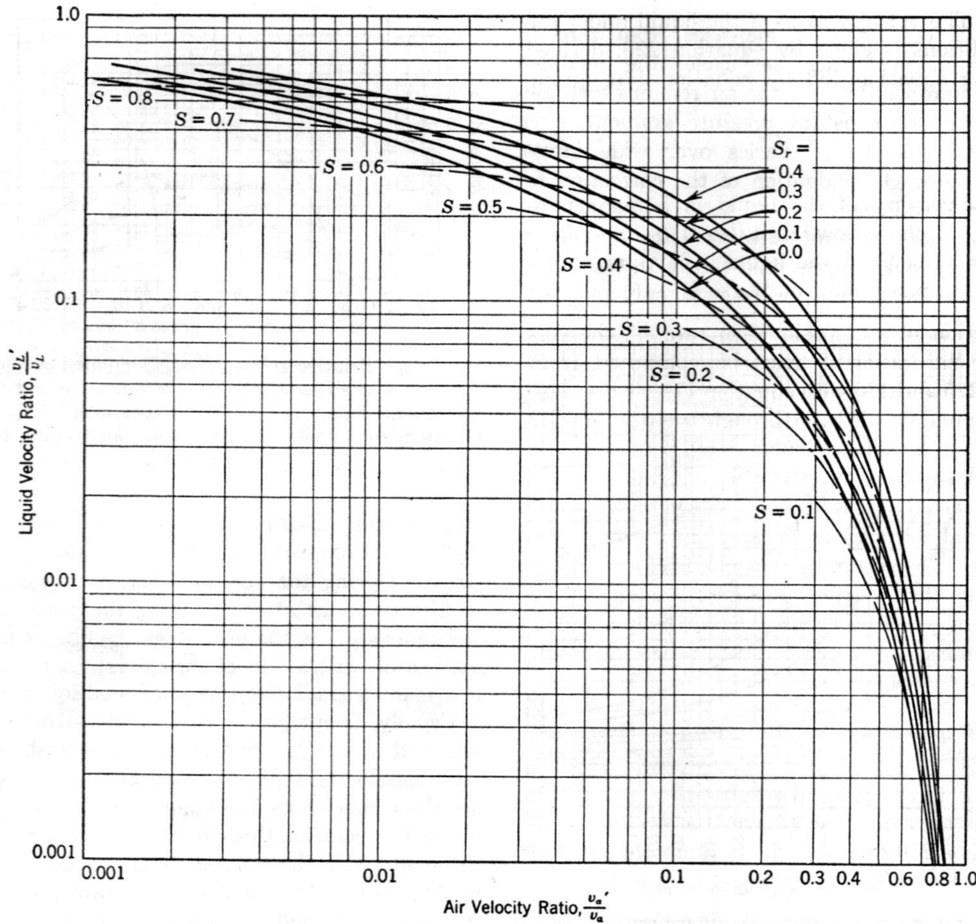

FIG. 259*d*. The ratio of superficial air velocity when liquid is present v_a' to velocity of air when no liquid is present v_a, expressed as a function of the corresponding ratio of superficial liquid velocities, v_l'/v_l for laminar flow of both phases with parameters of residual saturation S_r and saturation S.

there is no adsorption of filtrate on the surface of the particles.

If the wash is less viscous than the filtrate, the quantity of wash required is increased by the multiplying factor given in Figure 259*b*, as a function of the viscosity ratio and the extent of washing expressed in terms of saturation of the cake by the filtrate.

With thin filter cakes having a thickness less than 200 particle-diameters, channeling occurs and a greater quantity of wash is required. Figure 259*c* shows the multiplying factor to be used in estimating the volume of wash required for such thin cakes.

These relationships are based on a flooded cake with the wash following the same path as the filtrate, as may be encountered in plate-and-frame and in leaf filters. When the cake is not flooded because of simultaneous flow of air, the amount of wash required is somewhat reduced.

On a continuous rotary vacuum filter, wash water or weak liquor is sprayed continuously and uniformly onto the cake without flooding the cake and in the presence of air. This creates a condition of steady state *two-phase flow*. The cake attains a saturation dependent upon the rate of supply of the wash water, the properties of the cake and wash, and the pressure gradient. The velocity ratio of the air during washing for laminar flow is given by equation 221. The velocity ratio of the liquid under the same conditions is given by equation 225, derived by dividing equation 183 by equation 170, giving

$$\frac{v_l'}{v_l} = S_e^{\,y} \qquad (225)$$

where v_l' = velocity of the liquid in laminar two-phase flow, cu ft/(sq ft)(sec).

v_l = velocity of the liquid in laminar single-phase flow, cu ft/(sq ft)(sec).

By selection of a suitable wash rate v_l', the corresponding saturation S may be computed from equations 225 and 170 with equation 179, or Fig. 229. The corresponding air rate through the cake during washing may then be calculated by equation 221. The relationship between air velocity ratio (v_a'/v_a) and liquid velocity ratio (v_l'/v_l) is shown on Fig. 259*d* with residual saturation S_r and saturation S as parameters. The accumulative volume of air passing through the cake during washing is equal to v_a' multiplied by washing time. The air flow so determined should be corrected for turbulence by use of Fig. 258, as described.

Blower Requirements for Rotary Vacuum Filters

During a filter cycle, the saturation S and instantaneous air rate v_a' vary over the cycle as indicated in Figs. 260 and 261. The total volume of air passing per square foot of cake can be readily calculated as follows. Determine the total calculated volume per square foot, assuming laminar flow by adding the volume of air for the removal of filtrate during washing and dewatering periods. This total volume is then corrected for turbulence and converted to the pressure at the vacuum pump. Multiplication by the drum area of the filter and division by the cycle time gives the blower capacity in cubic feet per minute.

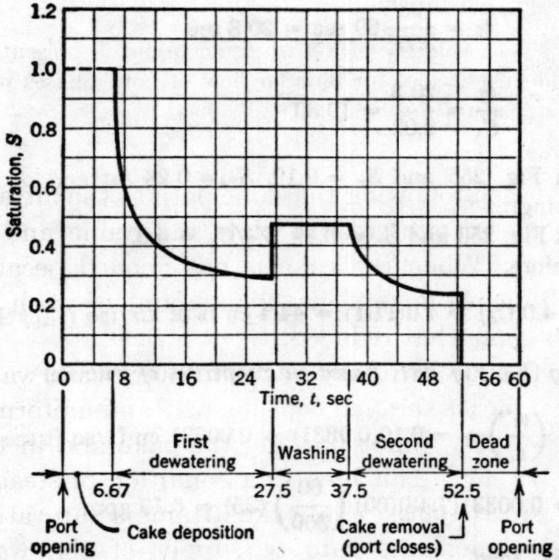

FIG. 260. Typical instantaneous saturation of filter cake as a function of cycle time t for a rotary vacuum filter. This curve is specific only for the conditions given in the illustrative example.

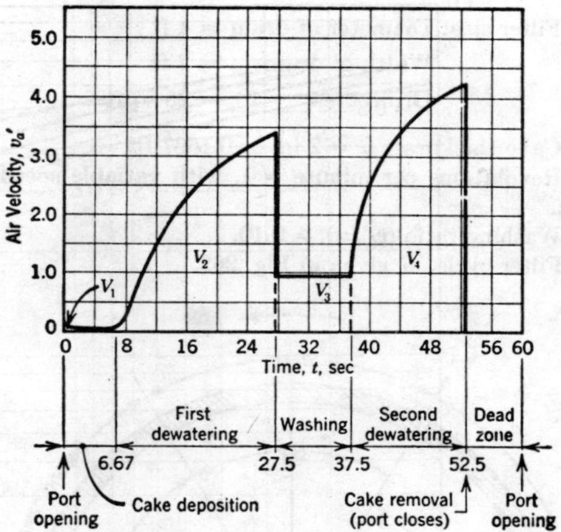

FIG. 261. Typical curve of the instantaneous air velocity v_a' as a function of cycle time t in seconds. The areas V_1, V_2, V_3, and V_4 represent the accumulative volumes of air for the corresponding portions of the cycle. The saturation conditions are represented in Fig. 260, and this curve is specific only for conditions as given in the illustrative example.

Example. Compute the blower requirements for a rotary vacuum filter filtering Glauber's salt from reclaimed rayon spin bath liquor under the following conditions.

Slurry Data

1. Weight fraction of suspended solids, x = 0.20 lb salt per pound of slurry.
2. Density of filtrate, ρ = 82 lb/cu ft.
3. Density of solids in slurry, ρ_s = 91.4 lb/cu ft.
4. Viscosity of filtrate, μ_l = 3.0 centipoises or 0.00202 lb/ft-sec.
5. Viscosity of air, μ_a = 0.018 centipoise or 0.0000121 lb/ft-sec.
6. Surface tension of filtrate, γ = 0.00452 lb force/ft.

Cake Data

7. Porosity, X = 0.4 volume of voids per volume of cake.
8. Sphericity, ψ = 0.75.
9. Average particle size (35 mesh), D = 0.001367 ft.
10. Operating vacuum = 4 in. mercury or 283 psf. If the drop in pressure across filtering medium and connections is neglected, this may be used as the drop in pressure across the cake.

11. Filter size: Diameter of drum = 4 ft
Width of drum = 2 ft
Drum area = 25 sq ft.

12. Cake thickness, L = 2 in. or 0.1667 ft.

13. Revolutions per minute = 1, with variable speed provisions.

14. Washing ratio $(v_l'/v_l) = 0.10$.

15. Filter cycle, as given in Fig. 262.

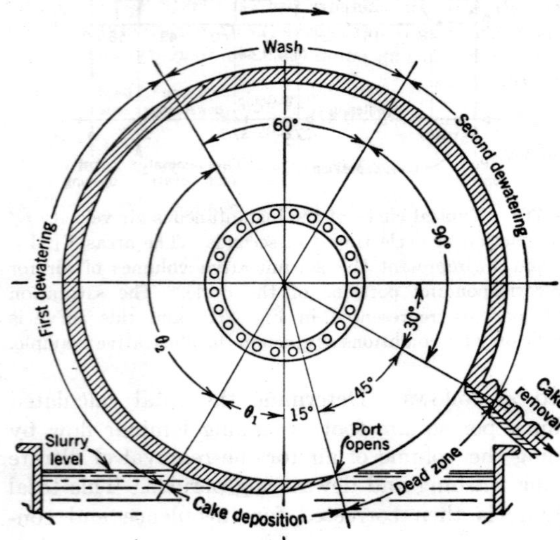

FIG. 262. Filter cycle for illustrative example.

Calculation of Filtration Constants

$$F_{\text{Re}} = 51.8 \text{ from Fig. 219}$$

$$F_f = 2500 \text{ from Fig. 220}$$

In equation 172

$$K = \frac{g_c D_p^2 F_{\text{Re}}}{32 F_f} = \frac{32.2(1.367 \times 10^{-3})^2 \times 51.8}{32 \times 2500}$$

$$= 3.89 \times 10^{-8} \frac{\text{cu ft}}{\text{sec}^2} \left(\frac{\text{lb mass}}{\text{lb force}}\right)$$

Equation 196

$$C_L = \frac{\mu_l}{2K\rho x} [\rho_s(1 - x)(1 - X) - \rho x X]$$

$$= \frac{0.002}{2(3.89 \times 10^{-8})(82)(0.2)}$$

$$\times [91.4(1 - 0.2)(1 - 0.4) - 82(0.2)(0.4)]$$

$$= 58{,}500 \text{ lb force sec/ft}^4$$

Equation 219

$$C_t = \frac{\mu_l X L^2}{K(-\Delta P)} = \frac{0.002(0.4)(0.1667)^2}{3.89 \times 10^{-8}(283)} = 2.02 \text{ sec}$$

Equation 223

$$C_a = XL\frac{\mu_l}{\mu_a} = 0.4(0.1667)\left(\frac{3.0}{0.018}\right) = 11.1 \text{ cu ft/sq ft}$$

Equation 170

$$v_a = \frac{K(-\Delta P)}{L\mu_a} = \frac{3.89 \times 10^{-8}(283)}{0.1667(1.21 \times 10^{-5})} = 5.46 \text{ cu ft/sq ft sec}$$

Equation 170

$$v_l = \frac{K(-\Delta P)}{L\mu_l} = \frac{3.89 \times 10^{-8}(283)}{0.1667(0.002)} = 0.0331 \text{ cu ft/sq ft sec}$$

Equation 179 (footnote)

$$S_r = 0.025 \left[\frac{K(-\Delta P)}{gL\gamma \cos\theta}\right]^{-0.264}$$

$$= 0.025 \left(\frac{3.89 \times 10^{-8}(283)}{32.2(0.1667)(4.52 \times 10^{-3}) \times 1}\right)^{-0.264}$$

$$= 0.19$$

Cake Deposition

Equation 198

$$t_1 = \frac{C_L L^2}{(-\Delta P)} = \frac{58{,}500(0.1667)^2}{283} = 5.74 \text{ sec or 6 sec}$$

$$\theta_1 = \frac{360°(6 \text{ sec})}{60 \text{ sec}} - 15° = 36 - 15 = 21°$$

Use 25 degrees for the angle of submergence θ_1 as providing a somewhat longer submergence than the calculated minimum.

First Dewatering (V_2, Fig. 261)

$$\theta_2 = 150° - \theta_1 = 150 - 25 = 125°$$

$$t_2 = \frac{125°}{360°} 60 \text{ sec} = 20.8 \text{ sec}$$

$$\frac{t_2}{C_t} = \frac{20.8}{2.02} = 10.30$$

From Fig. 255 and $S_r = 0.19$, $S_2 = 0.28$ (at end of first dewatering).

From Fig. 256 and $S_r = 0.19$, $V_2/C_a = 4.0$.

Therefore,

$$V_2 = 4.0(C_a) = 4.0(11.1) = 44.4 \text{ cu ft of air/(sq ft)(cycle)}$$

Washing (V_3, Fig. 261, Based on Properties of Filtrate)

$$v_l' = \left(\frac{v_l'}{v_l}\right) v_l = 0.10(0.0331) = 0.00331 \text{ cu ft/sq ft sec}$$

$$= 0.00331(7.48)(60)\left(\frac{60}{360}\right)(25) = 6.19 \text{ gpm}$$

From Fig. 259*d*

$$S_r = 0.19 \quad \text{and} \quad \frac{v_l'}{v_l} = 0.10$$

$$\frac{v_a'}{v_a} = 0.175, \qquad S_3 = 0.48 \text{ (during washing)}$$

$$v_a' = \left(\frac{v_a}{v_a}\right) v_a = 0.175(5.56) = 0.973 \text{ cu ft/sq ft sec}$$

$$t_3 = \tfrac{60}{360}(60) = 10 \text{ sec}$$

Therefore

$$V_3 = v_a t_3 = 0.973(10) = 9.73 \text{ cu ft/(sq ft)(cycle)}$$

Second Dewatering (V_4, Fig. 261)

$$t_4 = \tfrac{90}{360} (60 \text{ sec}) = 15 \text{ sec}$$

C_t and C_a (corrected for viscosity of wash)

$$C_t' = 2.02 \left(\frac{1.0}{3.0}\right) = 0.673$$

$$C_a' = 11.11 \left(\frac{1.0}{3.0}\right) = 3.70$$

$$\frac{t_4}{C_t'} = \frac{15}{0.673} = 22.3$$

From Fig. 255 and $S_r = 0.19$, $S_4 = 0.245$ (at cake removal).

From Fig. 256 and $S_r = 0.19$, $V_4/C_a' = 12.4$.

$$V_4 = 12.4(3.7) = 45.8 \text{ cu ft/(sq ft)(cycle)}$$

Blower Capacity (Neglecting V_1, Fig. 261)

$$V = V_2 + V_3 + V_4 = 44.4 + 9.7 + 45.8$$

$$= 99.9 \text{ cu ft/(sq ft)(cycle)}$$

From Fig. 258, $F_t = 0.47$, since $K = 3.89 \times 10^{-8}$, $(-\Delta P)/L = 4$ in. mercury per 2-in. cake. Converting from the mean pressure in the cake (27.92 in. mercury) to the pressure at the vacuum pump (25.92 in. mercury) and correcting for turbulence,

$$\text{cfm} = 99.9(0.47) \frac{27.92}{25.92} (25)$$

$$= 1262 \text{ cfm at 4 in. mercury vacuum}$$

Final Moisture Content in Cake (Assuming All Liquid as Water)

$$S\left(\frac{\rho}{\rho_s}\right)\left(\frac{X}{1-X}\right) = 0.245\left(\frac{62.4}{91.4}\right)\left(\frac{0.4}{1-0.4}\right)$$

$$= 0.1115 \text{ lb water/lb Glauber salt}$$

BIBLIOGRAPHY

1. Carman, P. C., "A Study of the Mechanism of Filtration," *J. Soc. Chem. Ind.*, **52**, 280 T (1933); *ibid.*, **53**, 159 T (1934); "The Determination of the Specific Surface of Powders," *ibid.*, **57**, 225 T (1938); *ibid.*, **58** (1939); "Fundamental Principles of Industrial Filtration," *Trans. Inst. Chem. Engrs. (London)*, **16**, 168 (1938).
2. Dickey, G. D., and C. L. Bryden, "Filtration," Reinhold Publishing Corp. (1946).
3. Lewis, W. K., and C. Almy, "Factors Determining the Capacity of a Filter Press," *Ind. Eng. Chem.*, **4**, 528 (1912).
4. Lewis, W. K., "Some Objections to Filtration Formulas," *Chem. Met. Eng.*, **27**, 594 (1922).
5. Pickard, J. A., *Filtration and Filters*, Ernest Benn, London (1929).
6. Ruth, B. F., "Studies in Filtration: III, Derivation of General Filtration Equations; IV, The Nature of Fluid Flow through Filter Septa," *Ind. Eng. Chem.*, **27**, 708, 806 (1935).
7. Ruth, B. F., and L. L. Kempe, "An Extension of the Testing Methods and Equations of Batch Filtration," *Trans. Am. Inst. Chem. Engrs.*, **33**, 34 (1937).
8. Ruth, B. F., G. H. Montillon, and R. E. Montonna, "Studies in Filtration: I, Critical Analysis of Filtration Theory; II, Fundamentals of Constant Pressure Filtration," *Ind. Eng. Chem.*, **25**, 76, 153 (1933).
9. Sperry, D. R., "Principles of Filtrations: I, The Effect of Pressure on Equations When Solids Are Non-Rigid or Deformable," *Chem. Met. Eng.*, **15**, 198 (1916); "Principles of Filtration II," *ibid.*, **17**, 161 (1917); "Constant Pressure Filtration," *Ind. Eng. Chem.*, **20**, 892 (1928); "II, Analysis of Filtration Data," *ibid.*, **36**, 323 (1944).
10. Underwood, A. J. V., "A Critical Review of Published Exponents on Filtration," *Trans. Inst. Chem. Eng.*, **4**, 19 (1926); "Filtration Equations for Compressible Sludges," *J. Soc. Chem. Ind. (London)*, **47**, 325 T (1928); "The Mathematical Theory of Filtration," *Ind. Chemist*, **4**, 463 (1928).

PROBLEMS

1. It is necessary to increase the capacity of a rotary vacuum drum filter. The crystals in the slurry have a density of 100 lb/cu ft and produce a noncompressible cake having a porosity of 45 per cent. The filtrate has a density of 65 lb/cu ft. The following changes have been suggested.

(*a*) Double the vacuum.
(*b*) Double the submergence.
(*c*) Double the speed (rpm).
(*d*) Double the fraction of suspended solids in the slurry.

What will be the percentage change in the quantity of cake produced per unit of time for each of these suggested changes?

2. A disk-type filter handles petroleum residue settlings. A series of tests run at a constant rate of 0.1 gpm yielded the following data:

Pressure differential (psi)	32	67	162
Total volume filtered (gal)	1	2	3

It is proposed that this filter be operated at a constant-pressure differential of 50 psi. If the sludge is homogeneous, calculate the volume of oil filtered in 30 min under such conditions if the change in press resistance with rate is negligible.

Calculate the filtering time if the press were operated at constant rate until the pressure differential became 50 psi, and then at constant-pressure differential until 20 gal of oil were forced through the press.

3. A rotary Oliver filter, 3 ft in diameter and 3 ft 4 in. wide, is operated on a thick slurry. The speed of rotation is 0.254 rpm, and the fraction of the drum periphery submerged is 0.33. The pressure on the vacuum side is 14.5 in. mercury (7.12 psia). Filtration rate is 60 gpm of filtrate. The slurry has a weight fraction of 0.15 lb of solids per pound of slurry. The resultant cake has a porosity of 30 per cent. The specific gravity of the solid is 3.0. The filtrate is water at 70° F. The average resistance of filter cloth and lines may be expressed as $V_e = 1.48$ cu ft. The cake may be considered incompressible.

It is desired to change the filter operation to secure a clearer filtrate, as a small amount of solids is coming through. In order to do this, it is proposed to leave a thin layer of cake 1/16 in. thick on the drum. What would be the expected change in the capacity of the filter if other operating conditions remain the same?

Would it be possible to operate at the same rate under the new condition? If so, at what absolute pressure in the vacuum chest with other conditions constant?

4. A continuous filter dewaxing bright stock solution consists of a horizontal revolving drum 6 ft in diameter and 12 ft long, mounted in a gastight case. The filter cloth covers the peripheral surface of the drum and is in turn covered with wire 1/16 in. in diameter. The cake scraper rests against the wire over the cloth. This arrangement leaves a wax cake 1/16 in. thick on the drum which the scraper cannot remove. The specific gravity of the wax is 0.8. The specific gravity of the bright stock solution is 0.85.

Present operating conditions are:

Filtering rate	= 1.0 gpm/sq ft of filter surface.
Wax concentration	= 0.05 cu ft of wax/cu ft solution.
Rotational speed	= 6 rpm.
Immersion	= 20 per cent of periphery.
Cake porosity	= 20 per cent.

Calculate the filter rate expected, in gpm/sq ft for the following conditions.

(*a*) The level of solution is raised to immerse 40 per cent of the periphery of the drum.

(*b*) The rotation is increased to 10 rpm.

(*c*) A finer wire is used, so that the cake is scraped down to an average of 1/64-in. thickness.

(*d*) New conditions of (*a*), (*b*), and (*c*) are all put into effect together.

5. A leaf filter press with 10 sq ft of filtering area operating at a constant pressure of 40 psig gave the following results:

Filtrate volume (cu ft)	141	215	270	340	400
Time (min)	10	20	30	45	60

The original slurry contained 10 per cent by weight of solid calcium carbonate ($CaCO_3$) and a small amount of dissolved alkalinity in the water.

(*a*) Determine the time required to wash the cake formed at the end of 70 min of filtering at the same pressure, using 100 cu ft of wash water.

(*b*) If the time for dumping and reassembling the press is 1 hr, what is the optimum time in hours for the filtration cycle at the constant pressure of 40 psig. Assume wash water is used in the same proportion to the final filtrate as in (*a*). What is the volume of filtrate obtained per cycle?

6. A certain filter press when tested on a homogeneous slurry *at constant rate* yielded the following data.

Time, min	$P_1 - P_2$, psi	Filtrate, lb
0	..	0
1	6	10
3	8	30
5	10	50
6	11	60
10	15	100
20	25	200
30	35	300

If this press is to be operated at a constant pressure drop of 10 psi and the time for cleaning and washing between cycles is 20 min, compute the optimum cycle time (time for filtration, cleaning, and washing).

7. Calcium carbonate slurry is being filtered in washing plate and frame presses containing 20 frames, 3 ft by 4 ft by 1 in. The calcium carbonate particles are essentially spherical with a diameter of 3.8 microns. The density of the solid is 2.26 grams/cc. The porosity of the filter cake is 32 per cent. The viscosity of the filtrate is 1.65 centipoises, and the density is 66 lb/cu ft at filtration temperature. The concentration of solids in the slurry is 4.8 mass per cent. The filtration pressure is 38 psig.

(*a*) Plot a curve of accumulated filtrate volume versus time.

(*b*) What volume of slurry is required to fill the press with cake?

(*c*) The cake is washed with water at 120° F and a pressure of 51 psig. The volume of wash water equals two times the volume of liquid in the cake at the end of the filter cycle. How long does it take to wash the cake?

8. Ferric hydroxide is to be filtered at constant pressure in a plate and frame press 3 ft square with 1-in. frame thickness. After 6.5 hr of operation, the press is full and 4.5 cu ft of filtrate per square foot of area has collected. It is decided to wash the cake with one-third the volume of filtrate collected. How much time will be required to wash the cake when the same pressure is applied?

9. A leaf-type filter with a filter area of 2 sq ft, handling a compressible sludge at a constant rate of 1 gpm, yields the data given below.

Pressure (psig)	10	20	30	45	60	75	90	110	138	170
Filtrate (gal)	3.5	7	10	15	18.5	22.3	25	27.5	29	30

(a) If this filter is operated on the sludge so that the filtration is a constant-rate operation until the pressure becomes 65 psi, then a constant-pressure operation until 100 gal of filtrate is obtained, then washed at 65 psi with 20 gal of wash liquor (same physical properties as filtrate), then dumped in 30 min, how much time will be required to produce 1000 gal of filtrate? Assume that the resistance of the filter itself is negligible.

(b) A rotary vacuum filter is suggested for the same sludge. The filter is to consist of a cylindrical drum 6 ft in diameter and 6 ft in width, rotating at 6 min per revolution, and immersed in an open tank of the sludge to a depth such that the axis of rotation is 18 in. above the surface of the liquid. The pressure inside the drum is 4.7 psia. Assuming the resistance of the filter itself to be negligible, how much time will be required to produce 1000 gal of filtrate?

10. A leaf filter press operating on a suspension of calcium carbonate yielded 40,000 lb of filtrate in 2 hr when filtering at a constant pressure of 25 psig. The filtrate is water at 70° F. Neglect the resistance of filter cloth and lines. The weight per cent of $CaCO_3$ in suspension is 10 per cent. Compressibility factor a of $CaCO_3 = 0.2$.

(a) At a constant-rate filtration, how long would it take to produce 40,000 lb of filtrate if maximum operation pressure was 25 psig?

(b) What is the 24-hr capacity of suspension for the press, if filtration is carried out at a constant pressure of 45 psig and the cake is washed at the same pressure with 4000 lb of water? Dumping, cleaning, and redressing the press requires 40 min. 40,000 lb of filtrate are obtained per cycle.

11. A leaf filter press is operating at constant pressure, building up a 1-in. noncompressible cake in 5 hr. At the end of 5 hr the press is cleaned and the cycle repeated without washing. Cleaning requires 30 min. If it now be required to wash the cake with an amount of wash water equal to one-half the volume of the filtrate, what will be the maximum capacity of the press obtainable by varying the cake thickness, expressed as a fraction of the present capacity? The nature of the sludge, the pressure, and the time of cleaning will be unchanged.

12. A homogeneous slurry is to be filtered at a *constant rate* of 5 cfm in a Kelly leaf press. Tests made under constant-pressure conditions indicate that the cake has a compressibility coefficient a of 0.5 and that at a pressure of 50 psi the volume of filtrate in cubic feet at any given time in minutes is represented by the equation $V^2 = 15{,}000\ t$. The time of the constant-rate cycle is 300 min. Specify the horsepower of the motor required. What is the theoretical work done by the pump per cycle?

13. A homogeneous compressible sludge was filtered in a leaf filter press of the Kelly type with the following results.

	Run A	Run B
Constant pressure (psig)	50	100
Volume of filtrate (cu ft)	350	380
Time of run (min)	30	25

It is suggested that the press be run on the same sludge at a constant rate of 15 cfm and thus operate at a higher throughput. The present setup has a pump of 4 hp with an overall efficiency of 55 per cent. Compressibility coefficient a may be taken as 0.7 for the filter cake. Is the equipment able to filter at this rate with a total filtration time of $6\frac{2}{3}$ min?

14. Two hundred tons of slurry containing 9 mass per cent solids are to be filtered per hour. The particles of solid are cubical in shape, have an average size of 0.0029 in. on edge, a sphericity of 0.81, and a density of 1.85 grams/cc (115 lb/cu ft). The liquid has a density of 1.08 grams/cc (67.4 lb/cu ft) and a viscosity of 13.4 centipoises. It is estimated that the porosity of the filter cake will be 38 per cent.

Filtration is to be done without washing, and 25 per cent final saturation for the liquid content of the cake is satisfactory.

(a) Specify a filter or filters for the above job, giving area, rotation speed, etc. A vacuum of 24-in. mercury, a cake thickness of 0.5 in., and a submergence of 40 per cent are suggested for first approximations.

(b) What will be the liquid saturation of the cake produced?

(c) What is the capacity requirement (cubic feet per hour at 60° F and 14.7 psia) of the vacuum pump at normal filtering conditions?

15. A series of filters is being fed by centrifugal pump A. By-pass valve B maintains a pressure of 50 psig on the filters. The filter cycle is: 60 min filter, 40 min wash, and 20 min clean and fill. Of the 12 filters in the group, 6 are on stream, one starting every 10 min. The filter presses each deliver 2800 gal of filtrate in the hour.

The slurry is composed of 3.0 volume per cent of solids, filtrate 1.03 grams/cc and solid 2.60 grams/cc.

The orifice has a manometer containing mercury and filtrate. A small amount of filtrate is fed slowly through valves C to keep solids out of the manometer. The orifice coefficient for the slurry is 0.59.

Compute the rate of return of the slurry to tank D through line E at the moment a filter is becoming full and just prior to taking one off stream.

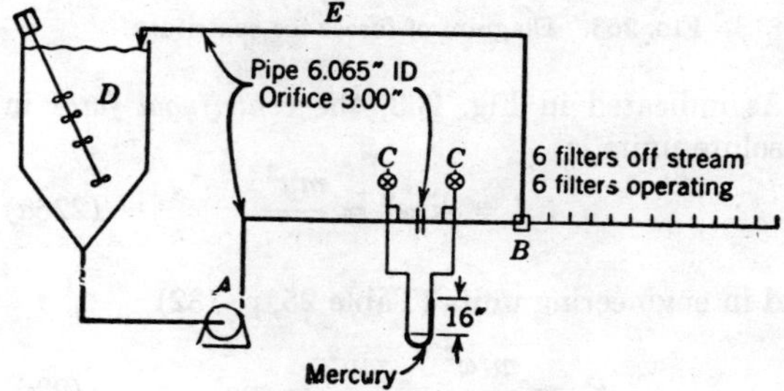

CHAPTER 19

Centrifugation

CENTRIFUGAL force is widely used when a force greater than that of gravity is desired for separation of solids and fluids of different densities, as in settling, or for separations of the nature of filtration. A *centrifugal force* is created by moving a mass in a curved path and is exerted in the direction away from the center of curvature of the path. The *centripetal force* is the force applied to the moving mass in the direction toward the center of curvature which causes the mass to travel in a curved path. If these forces are equal the particle continues to rotate in a circular path around the center.

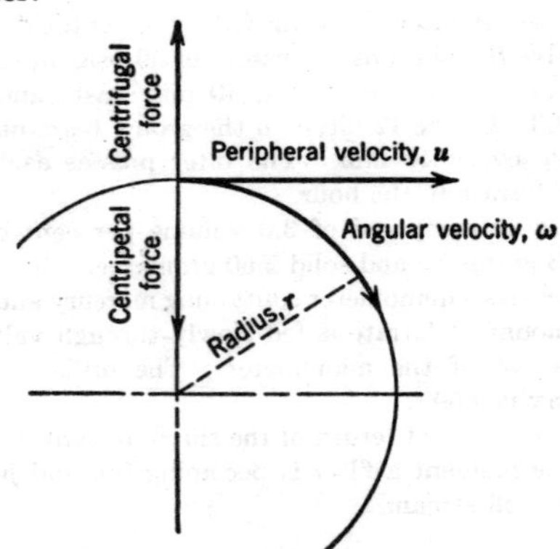

FIG. 263. Diagram of forces for centrifuge.

As indicated in Fig. 263, the *centrifugal force* in absolute units is

$$F_c' = mr\omega^2 = \frac{mu^2}{r} \qquad (226a)$$

and in engineering units (Table 25, p. 132)

$$F_c = \frac{mr\omega^2}{g_c} = \frac{mu^2}{rg_c} = ma' \qquad (226)$$

where F_c' = centrifugal force in absolute units.
F_c = centrifugal force in engineering units.
m = mass.
u = peripheral velocity.
r = radius.
ω = angular velocity in radians.

The terms $(r\omega^2/g_c)$ and (u^2/rg_c) are referred to as the acceleration in gees, designated * as a' in equation 7*d* (Chapter 12). If the speed is expressed in revolutions per minute and the radius r in inches,

$$a' = 2.84 \times 10^{-5}(\text{rpm})^2 r \quad \text{(pounds force per pound mass)} \qquad (226b)$$

The effectiveness of a machine in creating centrifugal force is expressed in terms of a'. By substituting the numerical value † of $\rho a'$ for ρ or for the pressure gradient $-\Delta P/L$ in the equations which have been presented for relative motion between fluids and solids, such equations are made directly applicable to centrifuges.

EQUIPMENT

Probably the earliest industrial equipment utilizing centrifugal force in the processing of materials was the centrifugal filter. Such devices were referred to as centrifugals and consisted of machines of large diameter and relatively slow speed. They were used primarily for separating coarse granular or

* It is common engineering practice to use the term gee also as a dimensionless factor, g/g_s, or $r\omega^2/g_s$, where g_s is the standard acceleration due to gravity and not the conversion factor g_c. The numerical value for the number of gees is the same with either definition since g_s is numerically equal to g_c, but the dimensions are different.

† Actually $\rho a' g_c/g_s$ is substituted for ρ, and $\rho a'$ is substituted for $-\Delta P/L$ for dimensional reasons.

crystalline solids from water or other fluids. The smaller high-speed machines of more recent development are generally referred to as centrifuges.

Batch Centrifugals

The simplest type is the vertical-shaft perforated-basket centrifugal. The basket diameter is usually 30, 40, or 48 in., although models are available having diameters from 12 to 84 in. The shaft is usually direct driven by an electric motor at about 1200 rpm for a 30-in. diameter basket, although greater speeds of about 2000 rpm are frequently used.

Some machines employ centrifugal clutches or hydraulic couplings for more rapid acceleration with a lower starting torque on the motor. Also, V-belt, variable-speed drive, and multispeed motors have been used to increase the flexibility of the batch centrifugal. The motor is generally oversized for quick acceleration. Up to 25 hp is used for a 30-in. diameter basket. A brake is provided for rapid deceleration.

The most common type is a suspended basket with the driving power applied to the vertical shaft above the basket, as indicated in Fig. 264. This overdriven suspended basket type is widely used for separating and rinsing sugar crystals in the refining of sugar and for separating all types of crystalline or granular products. For large-scale production (over 1 ton/8-hr day) automatic centrifugals or vacuum filters are preferred because of the lower labor cost.

Figure 264 is a section of such a machine with a flat-bottom basket indicated for the left side. In operation the basket is rotated at a moderate speed and is fed with a rather thick slurry until sufficient cake is produced on the screen. The liquor or filtrate passes through the cake and filter medium (the screen) as in filtration. The feed is then stopped, and the cake may be spun at high speed to free it of most of the liquor, or it may be washed immediately after the feed is stopped without the period of high-speed operation. After it is washed, the cake may be rotated at high speed to remove most of the wash liquor. The basket is then decelerated, braked, and unloaded through the bottom by raising the flat valve in the bottom.

In the older low-priced machines the cake is loosened by a wooden blade held against the cake while it is rotating at low speed. The use of a hinged blade or plow makes for safer and more rapid operation. The blade may be provided with a rack and pinion to allow vertical motion.

The solids drop through the opening to a conveyor or drier below the centrifugal. The valve in the bottom of the basket is then closed, and the cycle is repeated.

A complete cycle may require from a few minutes to 1 hr, depending upon the extent of washing and dry spinning of the cake.

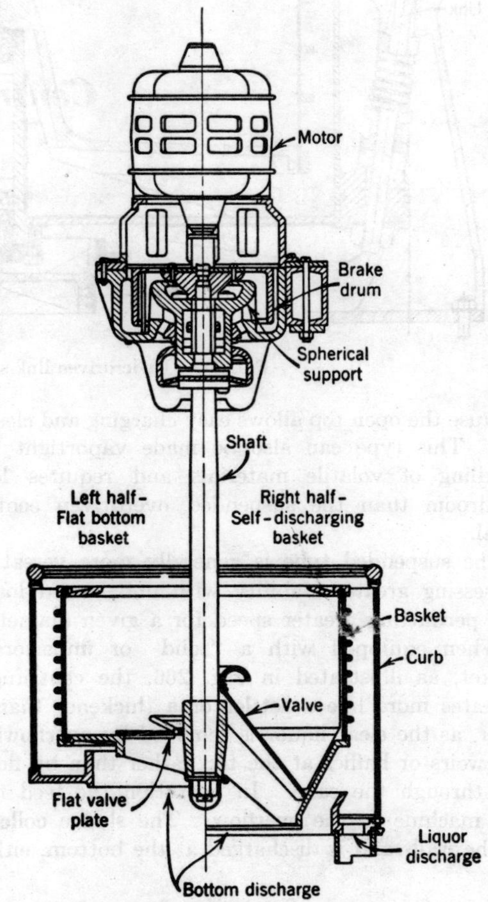

Fig. 264. A composite diagram of a suspended batch basket centrifuge, showing for the left half section a flat-bottom basket, and for the right half section a cone-bottom basket.[1]*

A cone-bottom basket is illustrated for the right side of Fig. 264. This type is preferred for materials which readily leave the basket at low speeds without the necessity of scraping the cake from the screen.

The underdriven link-suspended vertical-shaft batch centrifugal illustrated in Fig. 265 is widely used in laundries and pharmaceutical industries

* The bibliography for this chapter appears on p. 268.

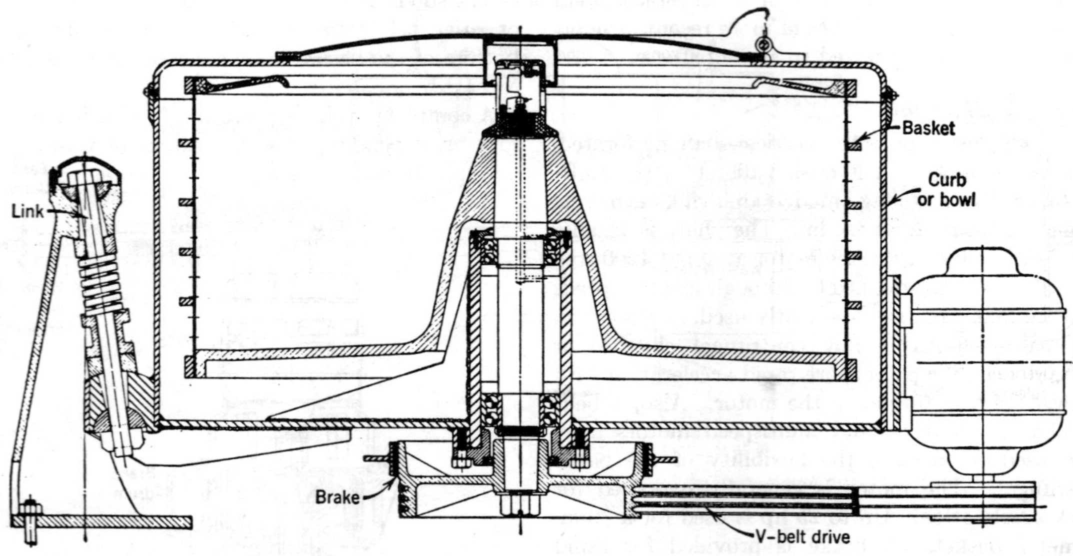

FIG. 265. Underdriven link suspended centrifuge. (*Fletcher Works.*)

because the open top allows easy charging and cleaning. This type can also be made vaportight for handling of volatile materials and requires less headroom than the suspended overdriven centrifugal.

The suspended type is generally more versatile, possessing greater stability with unbalanced loads and permitting greater speed for a given diameter.

When equipped with a "solid" or imperforate basket, as illustrated in Fig. 266, the centrifugal operates more like a settler or a thickener than a filter, as the clear liquor is removed by overflowing the weirs or baffles at the top rather than by flowing through the cake. In operation the feed fills the machine to the overflow. The sludge collects at the walls and is discharged at the bottom, either continuously as shown or at the end of the cycle, while the liquor accumulates until it reaches the flange at the top of the basket where it overflows and is collected in the housing outside the rotating basket. After sufficient cake is formed, a plow may be employed (Fig. 267) to free the solids from the walls of the basket, causing them to drop to the bottom discharge as in the operation of a perforate basket centrifugal. When the quantity of solids is small (up to about 1 per cent of the feed), this

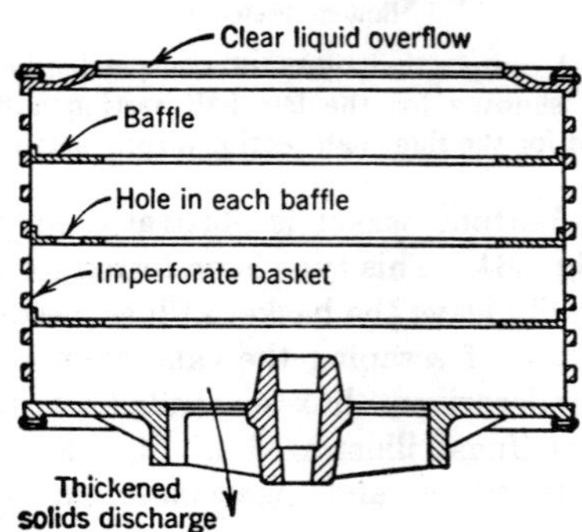

FIG. 266. Solid or imperforate basket.[1] Feed is to center.

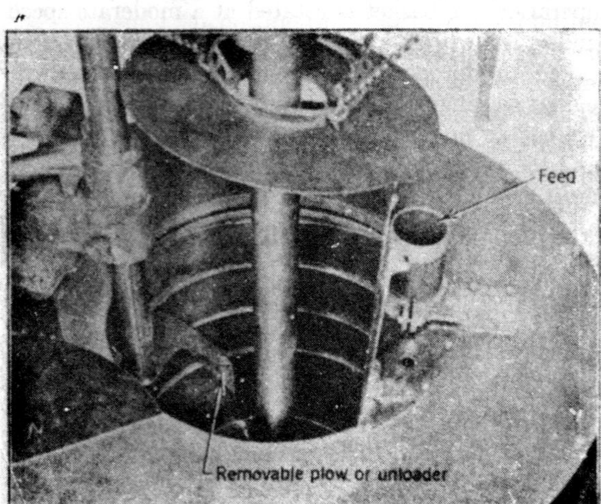

FIG. 267. Solid, or imperforate, basket, showing plow. (*Tolhurst Centrifugals Division, American Machine and Metals, Inc.*)

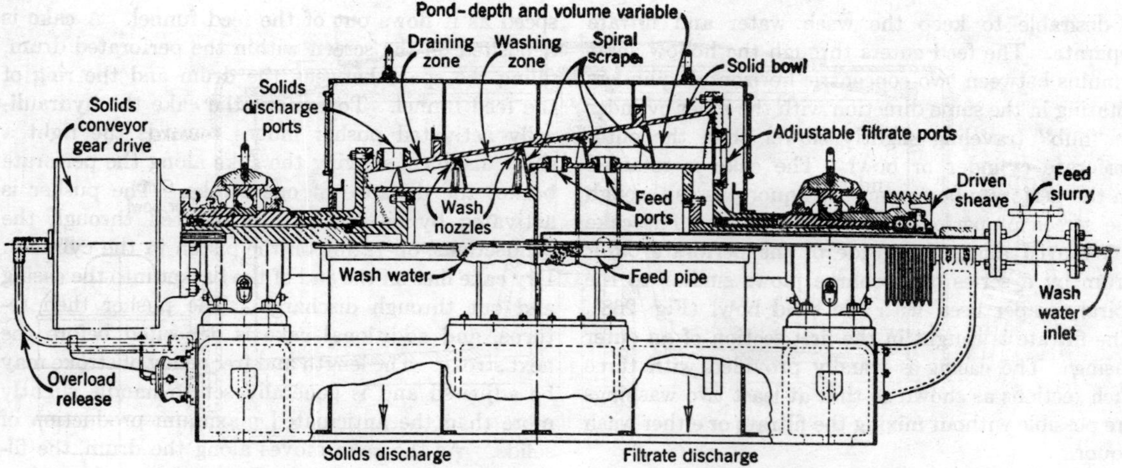

FIG. 268. Continuous horizontal solid-bowl centrifugal with spiral scraper. *(Bird Machine Co.)*

machine can be used satisfactorily for clarification of a fluid with a comparatively long cycle of operation.

The same machine may be used as a classifier. By means of a high rate of feed the fine particles may be caused to overflow with the liquor while the coarser particles are thrown to the wall of the basket and removed through the bottom. In this way the operation may be made continuous for the separation of two liquids, as will be described under continuous centrifuges.

Continuous Centrifugals

The centrifugal filter may be made to operate continuously by the incorporation of a mechanism for the continuous removal of the solids. Two methods are illustrated in Figs. 268 and 269.

The spiral scraper (Fig. 268) is used with both solid and perforate baskets. With the solid horizontal tapered bowl illustrated, the solids thrown against the wall are moved by the spiral scraper out of the bath or pond of liquid up to the draining zone or "beach." The solids are then washed and discharged through the discharge ports.

This machine is actually a continuous settler and drainer. The feed and wash liquor enter through the hollow shaft at the right. The feed passes through holes in the feed pipe and is thrown into the spinning pond. The clear fluid overflows the filtrate ports at the right. These ports can be adjusted to vary the pond depth and volume. The spiral scraper shown only in section is rotated at a slightly different speed from the drum or basket so that there is relative motion between the scraper and drum forcing the solids towards the left and up an incline towards the center until the solids leave the pond. At this point the solids may be washed lightly, drained, and finally thrown out the solids discharge ports at the left. The solids conveyor or spiral scraper is driven from gears at the far left which receive their motion from the hollow main shaft. If the pond depth is increased and speed of operation is sufficient to overflow the fines with the filtrate, this machine can also be used as a classifier with the larger solid particles removed at the solids discharge port.

A perforate horizontal-basket continuous centrifugal (Fig. 269) adapted for handling coarser solids gives more thorough washing, particularly where it

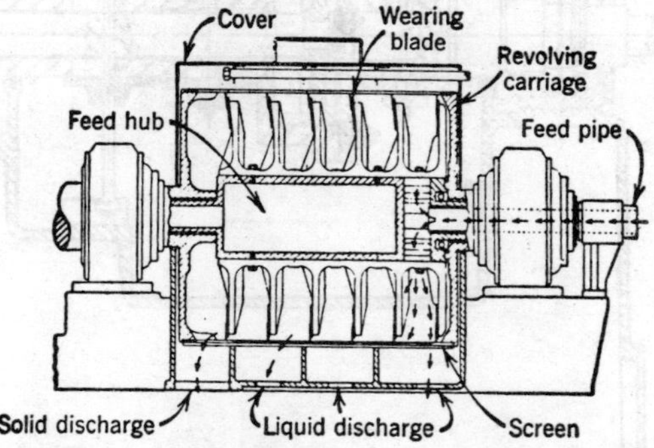

FIG. 269. Perforate horizontai-basket continuous centrifugal with spiral scraper discharge. *(Centrifuge Mechanical Equipment Co., Inc., Division of Condenser Service and Engineering Co., Inc.)*

is desirable to keep the wash water and filtrate separate. The feed enters through the hollow shaft annulus between two concentric horizontal cylinders rotating in the same direction with the inner cylinder or "hub" traveling slightly slower than the outer perforate cylinder or bowl. The cake is retained on the perforate bowl, and the liquor passes through the screen or perforations as in a filter. The cake is removed along the inside of the perforate outer drum by a series of discharge plows similar to the spiral scraper used with the solid bowl (Fig. 268). The filtrate is caught in the first section of an outer casing. The casing is usually provided with three such sections as shown so that at least two washings are possible without mixing the filtrate or either wash liquor.

This type of continuous centrifugal filter is limited to separations of coarse nonfriable crystals, as the scraper tends to break the more friable solids.

A reciprocating shoving mechanism to remove the solids is shown in Fig. 270. The feed enters the feed line at the right and is accelerated gradually to drum speed as it flows out of the feed funnel. A cake is deposited on the screen within the perforated drum, filling the space between the drum and the ring of the feed funnel. To remove the cake the hydraulically activated pusher moves towards the right a short distance, shoving the cake along the perforate basket the distance of one stroke. The pusher is activated by oil pressure, introduced through the connections, operating on the piston in the cylinder. Dry cake flies off the end of the basket into the casing and out through discharge. The pusher then returns, and additional cake is deposited before the next stroke. The length and frequency of stroke may be adjusted and is generally set to handle slightly more than the anticipated maximum production of solids. As the cake moves along the drum, the filtrate is thrown out into the first section of the casing.

The cake may be washed and the wash liquor thrown out in the subsequent sections of the casing. The cake reaches the end of the drum and is thrown off by centrifugal force into the collector housing and discharged through the bottom.

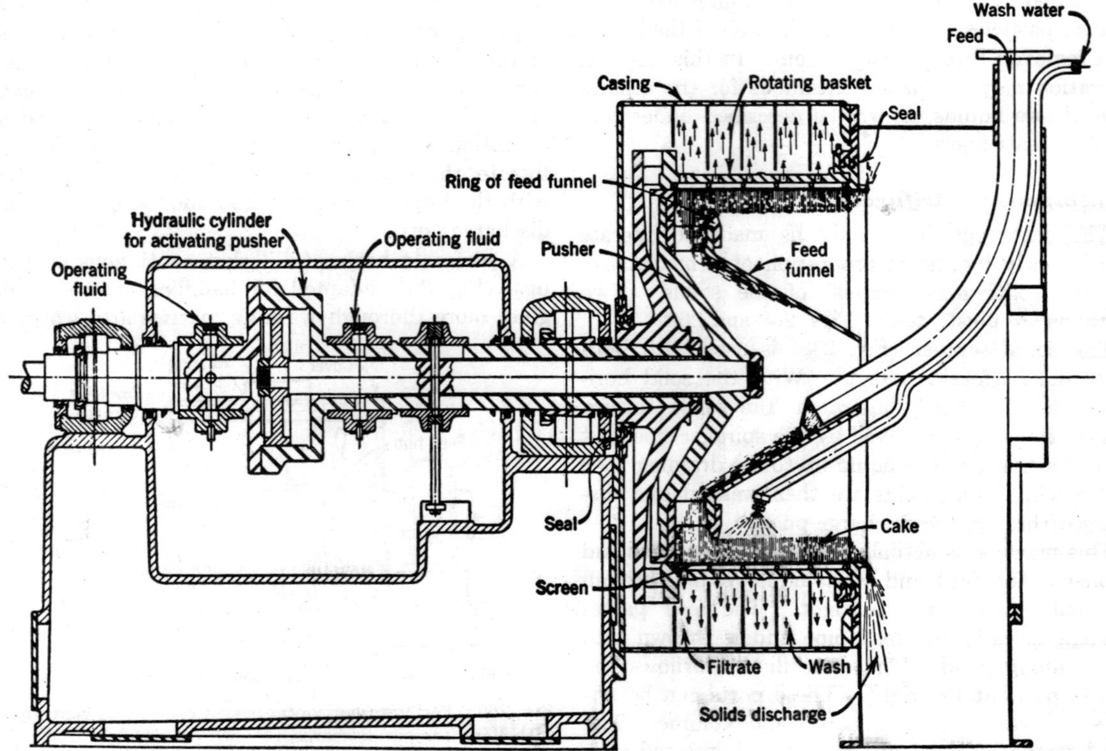

Fig. 270. Sectional drawing of a continuous perforate-basket centrifugal with pusher discharge. (*Baker Perkins, Inc.*)

The thickness of the cake is limited by the annular space between the inside of the perforate basket and the outer circumference of the removable ring of the feed funnel. The diameter of the ring and the setting of the stroke and frequency of the pusher control the capacity of the machine. The low acceleration of the slurry as it is brought up to drum speed and the smooth action of the hydraulic pusher, relatively free from any cutting action on the crystals, make this centrifugal suitable for the more friable crystals. It is made in drum diameters from 24 to 96 in., with capacities from 1 to 20 tons of solids per hour, depending upon operating conditions.

Automatic Batch Centrifugals

In order to save labor, large batch centrifugals are frequently designed for automatic operation on predetermined cycles. The operations are controlled by a timing mechanism and are run continuously with the elimination of the heavy starting torque and the losses of braking inherent in intermittent operation. Figure 271 indicates the construction of a horizontal automatic centrifugal with a perforate bowl. Feed is introduced through the flanged nozzle. After the cake has been built up to the desired thickness, feed is automatically shut off, wash water is introduced through sprays, and, after dewatering, the knife is raised by mechanism, removing the cake through chute. Figure 272 gives an idea of the automatic controls required for such operation.

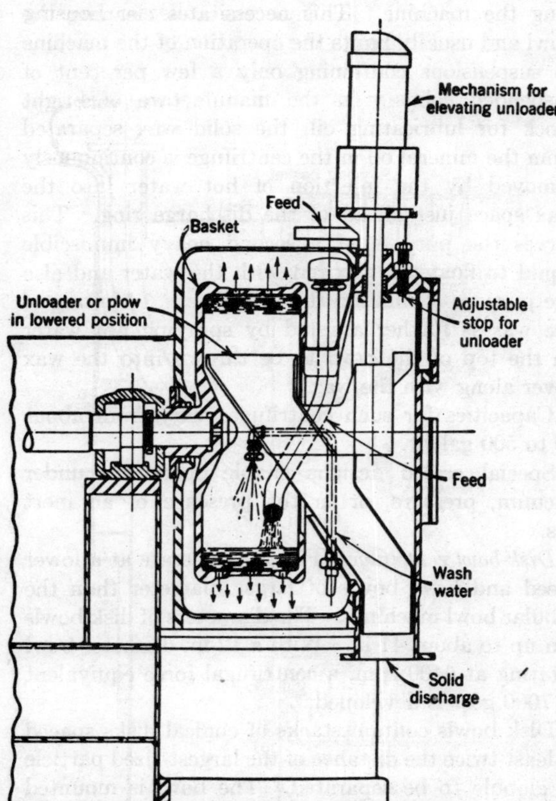

FIG. 271. Sectional drawing of an automatic batch horizontal centrifugal. (*Baker Perkins, Inc.*)

FIG. 272. Automatic horizontal centrifugal showing automatic controls. (*Sharples Corp.*)

These machines are well suited to free-draining crystalline solids not smaller than about 150 mesh. For satisfactory operation the moisture content in the cake should not exceed about 15 to 18 per cent; 2 to 5 per cent is usual. The performance is not strictly equivalent to the ordinary batch basket machine. Loading and unloading at full speed subjects the crystals to vigorous impacts resulting in breakage of fragile crystals. There is always a thin layer of crystals remaining in contact with the screen, which may become rather impervious to drainage after a time.

Centrifuges

The smaller-diameter higher-speed machines used to separate liquids of different densities, to break emulsions, and to remove or classify fine solids in suspension are generally of two types: the tubular bowl illustrated in Fig. 273, and the disk bowl illustrated in Fig. 274. The bottle type, a batch machine, is limited to laboratory and analytical separations. A very high-speed batch machine called the ultracentrifuge is used for difficult separations.

A tubular-bowl centrifuge (Fig. 273) equipped with a bowl 4 in. in diameter and operated at 15,000 rpm develops a centrifugal force per unit mass equivalent to 13,200 gees (a'). The bowls are usually 3 to 6 in. in diameter and up to about 5 ft in length. Some of the smaller laboratory units will develop 100,000 gees. The long bowl, many times the depth of the liquid on the inside wall of the bowl, provides ample time for the desired separation. The bowl is supported vertically with a slender solid flexible shaft and a thrust bearing. It is guided loosely at the bottom by a spring-supported bushing. Drive is usually indirect with a belt from an electric motor, or direct from a steam turbine.

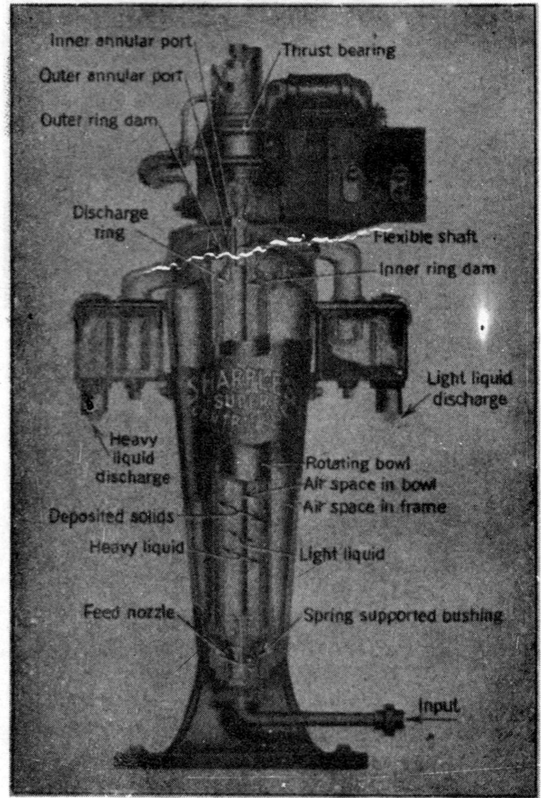

FIG. 273. Cutaway view of a tubular-bowl continuous centrifuge. (*Sharples Corp.*)

The feed enters the bottom of the bowl as a jet from a feed nozzle. It then comes into contact with a set of vertical baffles riding loosely in the bowl, which serve to bring the feed up to bowl speed quickly. When used as a clarifier to remove small amounts of suspended solids from a single liquid, the bowl contains a single discharge port at the top center for the clarified liquid.

When used as a separator (Fig. 273) to separate two liquids of different densities, the bowl contains two discharge ports: an inner annular port for the less dense fluid, and an outer annular port for the more dense fluid. Tubular-bowl centrifuges designed as separators may be employed to remove solids continuously by floating the solids on the heavier liquid, as in the separation of oil and salted-out soap from an oil-soap-water emulsion where the brine is thrown to the wall of the bowl and the soap floats out with it while the oil is discharged through the inner port. Usually the solids to be removed do not float but collect on the wall of the bowl and tend to plug the machine. This necessitates cleaning the bowl and usually limits the operation of the machine to suspensions containing only a few per cent of suspended solids. In the manufacture of bright stock for lubricating oil, the solid wax separated from the mineral oil in the centrifuge is continuously removed by the injection of hot water into the wax space just ahead of the discharge ring. This serves the purpose of a second heavy immiscible liquid to float the wax out with the water and also the purpose of melting the chilled wax. Melting of the wax is further assisted by spraying hot water on the top of the bowl to be thrown into the wax cover along with the wax.

Capacities for such centrifuges vary from about 50 to 500 gal/hr.

Special sealed designs enable operations under vacuum, pressure, or in the presence of an inert gas.

Disk-bowl centrifuges (Fig. 274) operate at a lower speed and have bowls of larger diameter than the tubular bowl machines. The diameters of disk bowls run up to about 41 in. With a 12-in. diameter bowl rotating at 6400 rpm, a centrifugal force equivalent to 7000 gees is developed.

Disk bowls contain stacks of conical disks spaced at least twice the distance of the largest-sized particle or globule to be separated. The bowl is mounted within a cast housing provided with one or more covers, each equipped with a discharge port for a liquid stream. The vertical shaft is usually driven through gears and a centrifugal clutch by a horizontal motor. De Laval originally designed this machine in 1878. It has been widely used in the separation of cream from milk and for many industrial preparations.

The feed enters at the top and flows down the central tube to the bottom of the bowl. During this travel it is accelerated to bowl speed by vertical vanes in the tube. The feed then passes outward through several feed ports into the conical disks and then upward through a series of holes in the conical disks. The location of these holes determines which component is to be recovered with the minimum of contamination by the other. If the less dense component is the more valuable and desired product, as in cream separation, the feed holes will be located in the disks near the center so as to give the greater travel to the more dense component (skim milk), thereby stripping it as completely as possible of the less dense component (butter fat).

As the feed moves upward the centrifugal force throws the heavier liquid or solid outward along the under side of the disks. The less dense liquid then flows toward the center along the upper surfaces of the disks. In this way the disks stratify the liquids into thin layers, creating a shallow settling distance between the disks. The less dense liquid is discharged at the top from the inner annular port. The more dense liquid flows downward and outward, discharging from the outer port. When the centrifuge is used as a clarifier, only the inner discharge port is provided and all the liquid travels inward and upward while the solid particles are thrown outward and collect in the bowl. In order to remove the separated solids continuously, the solids with a small amount of liquid are discharged continuously as a slurry into a discharge cover through radial nozzles in the periphery of the bowl (Fig. 275), or valves which open automatically after a certain amount of solids have collected in the bowl may be used in place of nozzles. In order to control the density of the discharged sludge, some designs permit recirculation of part of the sludge so removed from the bowl.

The *ultracentrifuge* (Fig. 276) is a machine of small diameter, operating at very high speeds. A bowl of 4-in. diameter rotating at 100,000 rpm. develops 500,000 gees. Such machines operate only

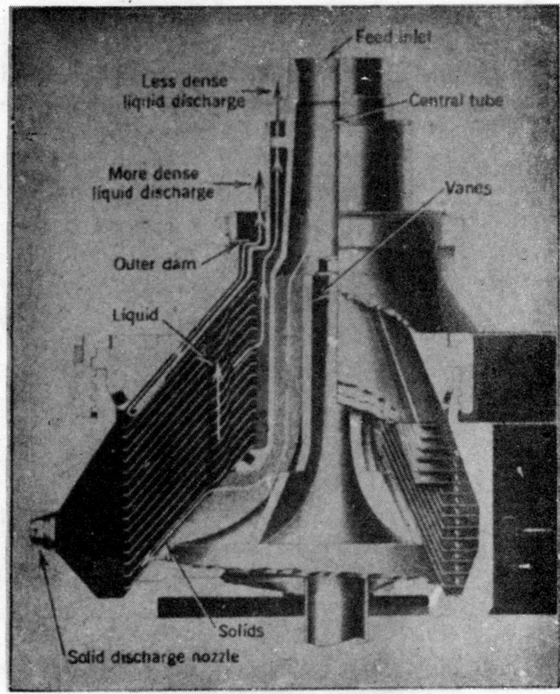

Fig. 275. Cutaway view of disk-bowl centrifuge with nozzle discharge for handling solids. (*De Laval Separator Co.*)

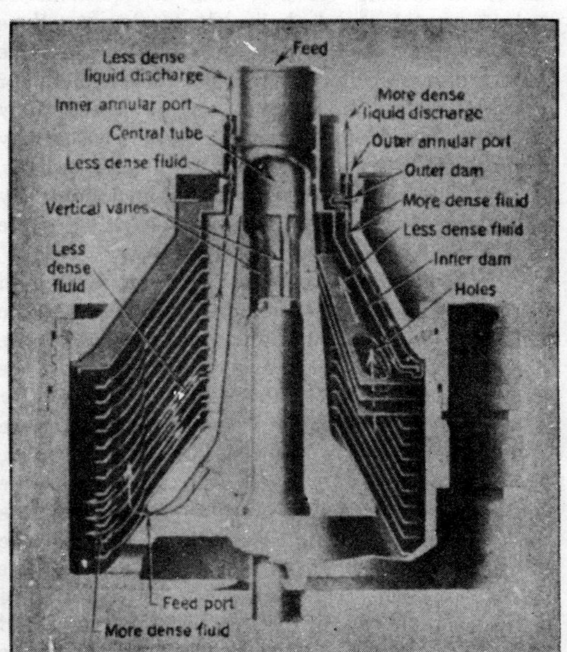

Fig. 274. Cutaway view of disk-bowl continuous centrifuge for separating liquids. (*De Laval Separator Co.*)

on a batch of material and can process only a small quantity in a single run. They are driven either by air or oil turbines, with bearings usually lubricated by a film of compressed air. The rotor may be operated in a vacuum or in an atmosphere of hydrogen at low pressure to minimize heating the rotor by frictional resistance.

FIG. 276. Ultracentrifuge disassembled to show the suspended bowl. (*Sharples Corp.*)

Ultracentrifuges are primarily laboratory machines and are useful in determining particle size or molecular weight by measurement of the rate of separation of suspended particles. They have been used to separate isotopes by means of the slightly different densities (molecular weights) and the tremendous multiplication of this slight difference through centrifugal force.

CALCULATIONS

The capacities and operating conditions of *centrifugal filters* may be computed in the same manner as for pressure or gravity filters by substituting $(\rho a')$ as the driving force for the pressure gradient $(-\Delta P)/L$ in equations such as 170, 179, and 183.

In spinning liquor from a centrifugal filter cake, two fluid phases exist in the pores of the cake, the liquor and air. Neglecting any evaporation of the liquor, the final saturation of the cake after spinning for an indefinite period of time will approach as a low limit the residual saturation as estimated by the application of equation 179 (footnote). When applied to centrifugal filters, the capillary number becomes $K\rho a'/g_c\gamma$ (cos θ), and equation 179 becomes

$$S_r = \frac{1}{40}\left(\frac{K\rho a'}{g_c\gamma \cos\theta}\right)^{-0.264}$$

$$= \frac{1}{40}\left(\frac{D_p{}^2 F_{\text{Re}}\rho a'}{32\gamma F_f \cos\theta}\right)^{-0.264} \qquad (227)$$

In centrifuges the operation is primarily settling under centrifugal force, and the equation giving the maximum settling velocity, such as equations 14 and 17, may be applied directly to the centrifuge by multiplying the densities by a'. Since the value for a' is determined in terms of the conversion factor g_c, these equations are most simply modified by substituting $a'g_c$ for g. Centrifuges are usually applied to the separation of very small solid particles, and the simpler relationships for laminar flow may usually be assumed. But, when the particles are large, turbulent motion is to be expected.

In either case the particle will be removed from the stream of liquid if the settling velocity of the particle is sufficient to bring the particle through the settling distance before it is swept on out with the fluid. The fluid is in laminar flow, and its velocity relative to the rotating bowl varies with the distance from the surface of the rotating bowl or disks. But, if an average linear velocity may be assumed for the fluid relative to the disk or tube, the approximate time that the fluid is in the centrifugal field may be obtained by dividing the length of flow by this average velocity or by dividing the quantity of liquid in the bowl by the rate of flow of liquid through the bowl. The time required for the particle to settle out of the fluid stream must not exceed the time the fluid is in the centrifugal field. The minimum time required for the particle to settle out is equal to the distance the particle moves normal to the flow of fluid divided by the settling velocity in the centrifugal field.

The optimum height of the dam in tubular- or disk-bowl centrifuges may be calculated by balancing the pressures between the two fluids as follows.

First consider a horizontal gravity separator for separating two fluids, as shown diagrammatically in Fig. 277. The mixed liquids are fed in from the upper

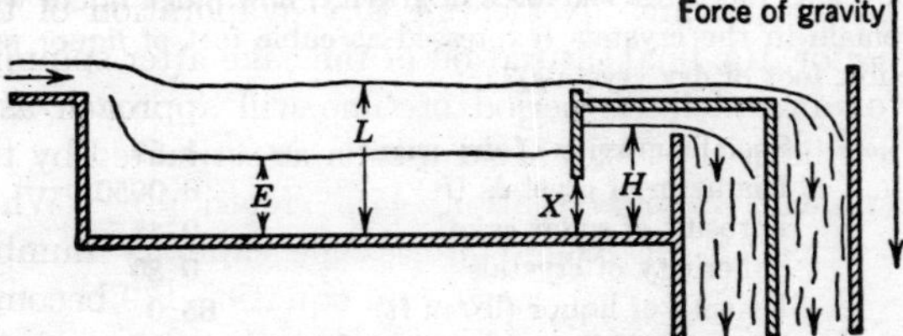

Fig. 277. Diagram illustrating the principle of separation of two flowing fluids by gravity.

left and divided according to their densities into the two phases as indicated. The two levels L and H are determined by the density of the two fluids, and by the level E of the interface between the two fluids. Since atmospheric pressure exists on the free surfaces of the light liquid at elevation L and of the heavy liquid at elevation H,

$$\rho_l(L - E) + \rho_h E = \rho_h H$$

where ρ_l = density of the light liquid.
ρ_h = density of the heavy liquid.
E, H, L, and X = elevations as shown in Fig. 277.

The centrifuge (Fig. 278) may be regarded as the equivalent of the horizontal settler rotated 90 degrees into a vertical position. The same principle of balancing the pressures may be applied with the centrifugal force directed toward the periphery of the bowl instead of gravity acting downward. Since the centrifugal force is not constant but varies with the radius, the pressure exerted by the fluid in a centrifugal field (centrifugal force per unit area) may be expressed in absolute units by dividing the force (equation 226*a*) by the area on which it is acting, or

$$P' = \frac{m}{A}\frac{u^2}{r}$$

$$P' = \int_{r_1}^{r_2} \left[\frac{\rho b(2\pi r)\,dr}{b(2\pi r)}\right]\left[\frac{(2\pi r)^2(N)^2}{r}\right]$$

$$= 2(N)^2\pi^2\rho(r_2{}^2 - r_1{}^2)$$

where P' = pressure in absolute units.
b = linear dimension of the tube in the direction of the axis.
N = revolutions per second.

Balancing pressures at the radius e (see Fig. 278);

$$2(N)^2\pi^2\rho_l(e^2 - l^2) = 2(N)^2\pi^2\rho_h(e^2 - h^2)$$

$$\frac{\rho_h}{\rho_l} = \frac{(e^2 - l^2)}{(e^2 - h^2)} \tag{228}$$

The radius e of the interface between the fluids must be greater than l and less than x if a separation

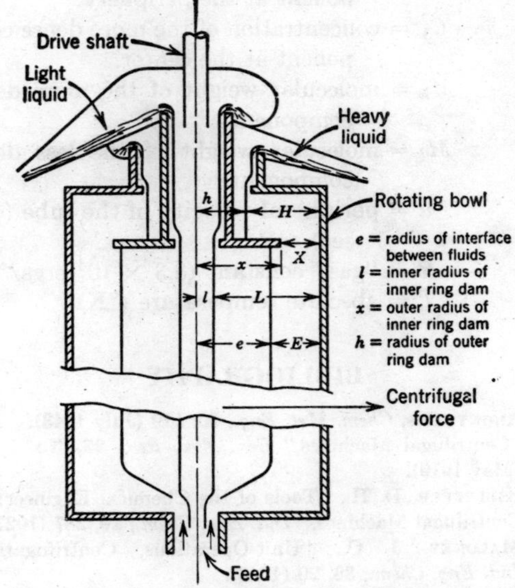

Fig. 278. Diagram illustrating the principle of separation of two flowing fluids by centrifugal force. In operation the radii of the ring dams, h and l, which control the discharge are almost identical with the liquid surfaces indicated in the figure.

is desired. By making e small, approaching l, the heavy liquid will be well stripped of the less dense liquid, thereby insuring complete recovery of the lighter liquid, although it may contain some contamination by heavy liquid, particularly if the emulsion is not completely broken before the liquids approach the inner ring dam. The proper setting of the radius of the outer ring dam h to give the desired position of interface e may be readily calculated from equation 228 from the known densities ρ_h and ρ_l and the known radius l.

If h is made less than l, there will be no discharge of heavy liquid and the centrifuge acts as a clarifier.

In making calculations for the ultracentrifuge, equation 229 [4] may be used to give the ratio of the concentration of the more dense component at periphery to that at the center when equilibrium is attained with a full bowl.

$$\ln \frac{C_2}{C_1} = \frac{(M_h - M_l)u^2}{2RT} \tag{229}$$

or

$$\frac{C_2}{C_1} = e^{\frac{(M_h - M_l)u^2}{2RT}} \tag{229a}$$

in which C_2 = concentration of the more dense component at the periphery.

C_1 = concentration of the more dense component at the center.

M_h = molecular weight of the more dense component.

M_l = molecular weight of the less dense component.

u = peripheral velocity of the tube (cm/sec).

R = "gas" constant (8.3×10^4 ergs/°K).

T = absolute temperature (°K).

BIBLIOGRAPHY

1. ANONYMOUS, *Chem. Met. Eng.*, **50,** 119 (July 1943).
2. "Centrifugal Machines," *Gen. Elec. Rev.*, **22,** No. 5, 416 (May 1919).
3. KILLEFFER, D. H., "Tools of the Chemical Engineer; VI, Centrifugal Machines," *Ind. Eng. Chem.*, **19,** 287 (1927).
4. MALONEY, J. O., "Unit-Operations, Centrifugation," *Ind. Eng. Chem.*, **38,** 29 (1946).
5. SHARPLES, L. P., "Which Centrifugal—and When?" *Ind. Eng. Chem.*, **31,** 1072 (1939).
6. SMITH, JULIAN C., "Selecting Centrifuges for Chemical Processing," *Ind. Eng. Chem.*, **39,** 474 (April 1947).

PROBLEMS

1. What will be the minimum moisture content of a cake of salt crystals if a slurry of saturated brine (60° F) and salt crystals are centrifuged in a basket centrifuge, 30 in. in diameter, rotating at 1800 rpm? (The average diameter of the cake is 28 in.)

2. Wet crystals of sugar are taken from the bottom of an evaporator and charged to a centrifuge. The charge contains 0.4 cu ft of liquor per cubic foot of dry crystals. If the crystals (4-in. thick cake) are centrifuged for 2 min and the centrifuge develops 100 times the force of gravity, how much liquor will remain in the crystals (expressed as cubic feet of liquor per cubic foot of dry crystals)?

Specific gravity of dry sugar	1.59
Diameter of crystals (ft)	0.0050
Porosity of sugar cake	0.45
Sphericity of crystals	0.80
Density of liquor (lb/cu ft)	85.0
Viscosity of liquor (centipoises)	7.0
Surface tension of liquor (lb/ft)	0.0058

3. A tubular-bowl céntrifuge is being used to separate a 50 per cent oil-50 per cent water (by volume) emulsion at a rate of 100 gph. The specific gravity of the oil is 0.945, and that of the water is 0.997. The centrifuge bowl is 3 ft long from entrance to overflow dams and operates at 12,000 rpm. The radii of the inner and outer overflow dams and tube are 1.998, 2.000, and 2.500 in., respectively.

(*a*) What is the smallest oil droplet that can be completely recovered?

(*b*) What would be the effect of decreasing the inner overflow dam radius to 1.995 in.?

4. The tubular-bowl centrifuge of problem 3 is to be used to clarify an oil (density = 58.5 lb/ft^3, and viscosity = 10 centipoises) containing a small amount of fines (density = 98.4 lb/ft^3, and average particle size of 10^{-4} cm). The solids are to be washed out of the bowl with a stream of water. What is the capacity of the machine in gallons of oil per hour?

CHAPTER 20

Fluidization of Solids

FLUIDIZATION of solids by the flow of fluids is an operation intermediate in character between the flow of solids through fluids, as described in Chapter 7, and the flow of fluids through solids, as described in Chapter 10.

When a fluid is passed upward through a bed of closely sized granular solids, a pressure gradient is required to overcome friction. In order to increase the rate of flow, a greater pressure gradient is required. When the pressure drop ($-\Delta P$) approaches the *weight* of the bed over a unit cross-sectional area, the solids begin to move. This motion of the solids is created at superficial velocities far below the terminal free-settling velocities of the solid particles and constitutes the beginning of fluidization. The process is approximately equivalent to the inverse of hindered settling.

PARTICULATE FLUIDIZATION

When the fluid is a liquid such as water and the solid is similar to glass beads in its properties, the motion of the particles at the beginning of fluidization occurs in a minor way throughout the bed. As the velocity of the fluid and the pressure drop are increased, the bed expands and the oscillation of each particle increases in velocity and extent. The mean free path of the particles between collisions with each other increases with increasing fluid velocities. Correspondingly, the porosity of the bed increases. This expansion of the bed continues with increasing fluid velocities until each particle behaves as an individual and is unhindered as a freely settling body by the action of any other solid particle. This entire process is known as particulate fluidization.

A typical variation of pressure drop with superficial velocity is shown in Fig. 279 where the logarithm of the pressure drop is plotted against the logarithm of the fluid velocity. The straight line from A to B represents the variation of the pressure drop through the bed with fluid velocity during the period of fixed-bed operation when no motion of the particles occurs. At the point B the bed becomes

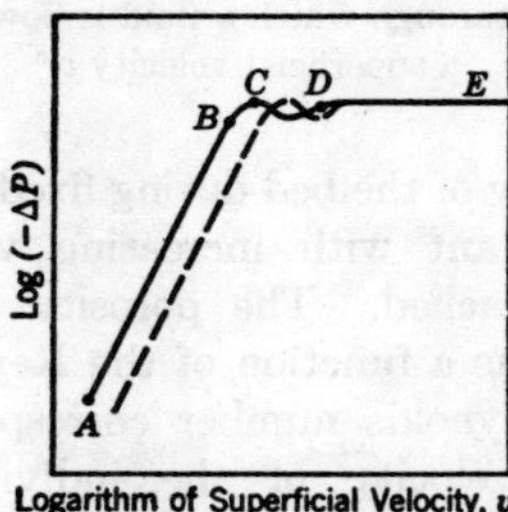

Fig. 279. The effect of superficial velocity on the drop in pressure for a fluid flowing upward through a bed of closely sized particles.[8]

unstable, and a minor movement and readjustment of the particles in the bed begin to take place to offer the maximum cross-sectional area for flow. This change in structure of the bed produces a deviation from the simple relationship between pressure drop and velocity shown in the section A to B. Instability of the bed continues as the velocity is increased, until at point C the loosest arrangement of particles in contact is established. With any further increase in the velocity of flow some of the particles in the bed are no longer in permanent contact with one another and become continuously agitated. This point, C, is known as the point of fluidization.

At this point of fluidization the bed begins to expand with increasing fluid velocities. At point D fluidization is complete and all the particles are in motion. Further increases in fluid velocity beyond the point D are attained by relatively slight increases in pressure drop, merely that required to overcome the increase in friction losses between fluid, suspended solid, and walls of the container. The expansion of the bed during fluidization is indicated in Fig. 280 where the logarithm of the Reynolds number is plotted as a function of the logarithm of the porosity. The Reynolds number in Fig. 280 is based on the dimension of the solid particles and the superficial velocity of the fluid.

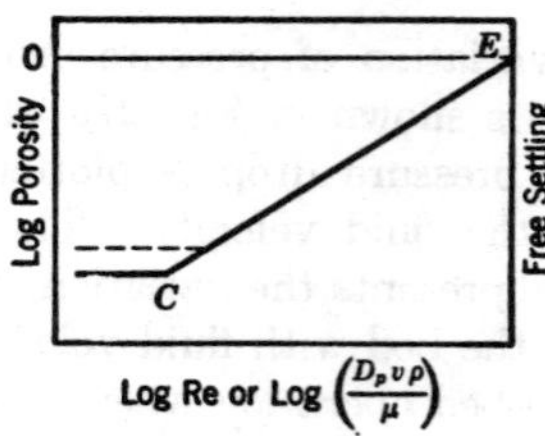

FIG. 280. The effect of Reynolds number on the porosity of a bed of particles through which a fluid is flowing upward with a superficial velocity v.[8]

The porosity of the bed during fixed-bed operation remains constant with increasing velocities until point B is reached. The porosity then increases continuously as a function of the Reynolds number up to the Reynolds number corresponding to the free-settling velocity of the individual particles which may be calculated as described in Chapter 7. At this Reynolds number each particle is moving as an independent individual, the bed has ceased to exist as such, and the porosity may be assumed to be unity. Extensive experimental data [8]* when plotted as in Fig. 280 may be extrapolated as a straight line to the point E where the porosity is unity and the Reynolds number is that corresponding to free settling.

The dotted lines on Figs. 279 and 280 show the relationships for a bed containing the same quantity of the same particles but having a higher initial porosity during the fixed-bed period. In either case fluidization begins at approximately the same value of the pressure drop. Above minimum critical values below which wall and entrance effects may become important, the curve of Fig. 280 is independent of the cross-sectional area or the thickness of bed. This is also true of the curve similar to Fig. 279 obtained by plotting log $[-\Delta P/L_0(1 - X_0)]$ versus log v, where X_0 and L_0 represent respectively the porosity and thickness of the bed at any known condition, and v is the superficial velocity of the fluid.

* The bibliography for this chapter appears on p. 274.

AGGREGATIVE FLUIDIZATION

When the fluid is a gas such as air, the fluidization of material such as closely sized glass beads follows a somewhat different mechanism, although the relationships between the pressure drop and the velocity is similar in all respects to that shown in Fig. 279.

Whereas in particulate fluidization the beginning of fluidization is marked by gentle oscillating motion of some of the particles constituting the bed (B to C, Fig. 279), in aggregative fluidization the fluid literally begins to "bubble" through the solid bed in a manner identical to the action observed in bubbling a gas through a liquid. The bubbles of fluid rise through the bed and break at the surface of the bed, "splashing" a few particles of the solid upwards. As the fluid velocity is increased, the bubbling action becomes more and more violent, "streamers" or "ribbons" of solids being ejected to considerable distances above the bed before returning.

The relationship between the Reynolds number (based on the particles) and the porosity is similar to that shown in Fig. 280, although experimental difficulties have made the procurement of data and Reynolds numbers approaching that of free settling extremely difficult. These experimental difficulties are due primarily to the violent "streaming" action of the bed at high fluid velocities.

CALCULATION OF REQUIRED PRESSURE DROPS

The forces tending to raise the particles are the buoyant force plus the friction force. At the point of fluidization the forces tending to raise the particles are equal to the total weight (force of gravity) of the particles, or,

$$\frac{g}{g_c}(1 - X)(LA)\rho + (-\Delta P_f)A = \frac{g}{g_c}(1 - X)(LA)\rho_s \quad (230)$$

where X = porosity of the bed.
A = cross-sectional area of the bed.
L = thickness of the bed.
ρ_s = density of the solid particles.
ρ = density of the fluid.
$-\Delta P_f$ = pressure drop required for fluidizing.

Solving for $-\Delta P_f$,

$$-\Delta P_f = L(1 - X)(\rho_s - \rho)\left(\frac{g}{g_c}\right) \quad (231)$$

$$\frac{-\Delta P_f}{L(1 - X)} = (\rho_s - \rho)\left(\frac{g}{g_c}\right) \quad (232)$$

The relationship of equation 232 is confirmed by all available data.

Illustrative Example. It is desired to contact in the fluidized state 50 lb of bead catalyst composed of smooth spherical particles 0.174 in. in diameter with 20,000 cu ft/hr of a gas having a density of 0 143 lb/cu ft and a viscosity of 0.011 centipoise at the process conditions. The catalyst density is 1.370 grams/cc. The porosity of the catalyst in a closely packed bed is 0.383.

Estimate the size of the reactor which should be used.

Solution. The reactor required must be large enough to accommodate the bed in its expanded fluidized state. The diameter of the bed may be determined from the mass velocity of the gas under operating conditions. The mass velocity of the gas must be such that the Reynolds number at which the system operates is above the Reynolds number at which fluidization occurs. The operating Reynolds number may be taken arbitrarily as equal to three times the Reynolds number at the point of fluidization. The height of the reactor must be at least equal to the height of the bed in its expanded fluidized state. The height of the bed may be established from a plot such as is shown in Fig. 280.

The steps involved in the solution include

1. Calculation of the Reynolds number at the point of fluidization.
2. Choice of a suitable operating velocity somewhat above that corresponding to the point of fluidization to fix the diameter of the reactor.
3. Construction of the plot of log Reynolds number versus porosity to determine porosity of the bed at operating conditions for estimating the height of the reactor.

The pressure gradient in pounds per square foot per foot of height of bed at the fluidizing point may be calculated by equation 231.

$$\frac{-\Delta P_f}{L} = \frac{g}{g_c}(1 - X)(\rho_s - \rho)$$

$$= (1)(1 - 0.383)[(1.37 \times 62.4) - 0.143] = 52.6$$

The superficial velocity v of the gas at the point of fluidization is obtained by trial and error. Assuming a linear gas velocity of 3.1 fps, the Reynolds number modified for porous beds (ReF_{Re}) is computed as follows, using Fig. 219 to determine the value of F_{Re} as 46.0.

$$\frac{D_p v \rho F_{Re}}{\mu} = \frac{(0.174/12)(3.1)(0.143)46.0}{(0.011)(0.000672)} = 40{,}000$$

The friction factor for gas flowing through a porous bed (f) is then obtained from Fig. 225.

$$f = \frac{2g_c D_p}{v^2 \rho F_f}\left(\frac{-\Delta P_f}{L}\right) = 0.022$$

where D_p = diameter of particle (see equation 168). From Fig. 220, F_f is determined as 1640. Therefore,

$$\frac{-\Delta P_f}{L} = \frac{(0.022)(3.1)^2(0.143)(1600)}{(2)(32.2)(0.174/12)} = 52.6$$

Thus the assumed linear gas velocity of 3 1 fps is correct.

The Reynolds number for a single particle based on superficial velocity of 3.1 fps at the fluidization point is

$$\frac{D_s v \rho}{\mu} = \frac{(0.174/12)(3.1)(0.143)}{(0.011)(0.000672)} = 870$$

Operating Reynolds number chosen arbitrarily as 3×870 = 2610

Operating gas velocity (fps) is therefore 3×3.1 = 9.3

Reactor area (sq ft) is (20,000)/(3600)(9.3) = 0.598

Reactor diameter (ft) is (2) $(0.598/3.14)^{1/2}$ = 0.872 (10.5 in.)

The free-settling Reynolds number is calculated by trial and error as follows.

Assume free-settling velocity (fps) = 32

Reynolds number for free settling

$$\frac{D_s v \rho}{\mu} = \frac{(0.174/12)(32)(0.143)}{(0.011)(0.000672)} = 9000$$

Friction factor for free settling from Fig. 70

$$f_D = \frac{4(\rho_s - \rho)gD_s}{3v^2\rho} = 0.37$$

The free-settling velocity v_m calculated from friction factor

$$v_m = \sqrt{\frac{4(\rho - \rho)gD_s}{3f_D\rho}}$$

$$= \sqrt{\frac{(4)[(1.37 \times 62.4) - 0.143](32.2)(0.174/12)}{(3)(0.37)(0.143)}} = 31.8$$

This is in sufficient agreement with the assumed value of 32. The logarithm of porosity is plotted against the logarithm of Reynolds number, as shown in Fig. 281, by drawing the horizontal line of constant porosity up to the Reynolds number corresponding to the point of fluidization and connecting that point C with the point E representing unit porosity at

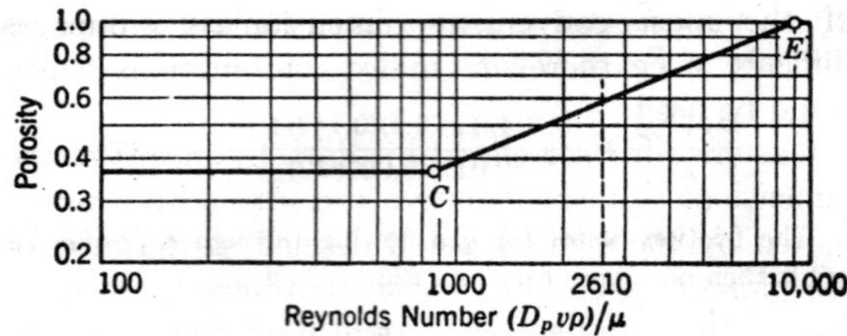

FIG. 281. Solution of illustrative example by plotting logarithm of Reynolds number versus logarithm of porosity.

the Reynolds number corresponding to free settling. From Fig. 281, the porosity at the operating Reynolds number of 2610 is read as 0.58.

Volume of the fluidized bed

$$= \frac{50}{(1.37)(62.4)(1 - 0.58)} = 1.39 \text{ cu ft}$$

Height of the fluidized bed = 1.39/0.598 = **2.33 ft**

The height of the reactor should be sufficient to prevent carry-over of "streamers" of solids ejected above the bed. In the absence of any specific information, the height of the vessel may be fixed at about twice the calculated height of the bed, or 4.66 ft minimum height for a diameter of 0.872 ft or 10.5 in.

Ordinarily the operating conditions are chosen relatively close to those corresponding to point C to insure freedom from carry-over in streamers, etc. A series of calculations similar to the above gives data for different heights of bed, depending upon the diameter of the reactor for the constant gas rate.

In this example two different "diameters" for the same particles were indicated. The pressure gradient through the porous bed is related to D_p, the mean surface particle diameter (equation 169). The free-settling velocity is related to D_s, the diameter of a sphere of the same volume as the particle. In this particular example, both these diameters (D_p and D_s) have the same numerical value since the particles are spheres of a single size.

CRITERIA FOR PARTICULATE AND AGGREGATIVE FLUIDIZATION

It appears that the Froude number (v^2/gD_p) is a convenient criterion for determining whether or not a given system will fluidize in particulate or aggregative fluidization. The value of v in the Froude group is taken at the point of fluidization. Table 27 presents typical data for systems involving air and water as the fluid.

TABLE 27. VALUES OF FROUDE GROUP AT POINT OF FLUIDIZATION IN PARTICULATE AND AGGREGATIVE FLUIDIZATION [8]

System	Particle Diameter, ft	Fluidizing Point	
		Superficial Velocity v, fps	Froude Number, v^2/D_pg
Particulate Fluidization			
Glass beads and water	0.00094	0.0040	0.00052
Sea sand and water	0.0013	0.0050	0.00062
Glass beads and water	0.0017	0.0060	0.00067
Sea sand and water	0.0018	0.0086	0.0013
Sea sand and water	0.0033	0.029	0.0080
Catalyst beads and water	0.011	0.055	0.0088
Catalyst beads and water	0.015	0.068	0.0099
Glass beads and water	0.017	0.14	0.036
Lead shot and water	0.0042	0.13	0.13
Aggregative Fluidization			
Glass beads and air	0.00094	0.18	1.1
Sea sand and air	0.0013	0.30	1.7
Glass beads and air	0.0017	0.33	2.7
Sea sand and air	0.0018	0.59	4.9
Sea sand and air	0.0033	1.1	10
Catalyst beads and air	0.011	2.1	13
Catalyst beads and air	0.015	2.6	14
Glass beads and air	0.017	4.7	40
Lead shot and air	0.0042	3.4	85

From Table 27 it appears that aggregative fluidization will probably occur when the value of the Froude number at the point of fluidization is greater than unity, and particulate fluidization is to be expected when the value of the Froude group at the fluidizing point is less than unity.

In addition to the force of gravity acting downward on the bed and tending to keep the bed compact, and the viscous drag of the fluid acting upward tending to disperse the particles of the bed in aggregative fluidization, there appears to be another force or combination of forces which tends to retain the bed in a compacted although agitated state. These forces may be electrostatic or fluid-dynamic in origin or both. Electrostatic charges have been clearly observed in a number of experiments involving aggregative fluidization. There is some evidence that electrostatic charges may be necessary for fluidization of very small particles. When conditions favor the loss of such charges, as when the fluid is ionized, there is a tendency for the bed to collapse.

TRANSPORTATION OF FLUIDIZED DISPERSED SOLIDS

When the superficial velocity of the fluid exceeds the free-settling velocity of the solid particles, the particles take on a motion relative to the container in the same direction as the motion of the fluid.

For design purposes it is necessary to know the effect on friction losses of such variables as pipe sizes, the amount of fluid and solid flowing, and the properties of the fluid and solid. Relationships developed for the flow of homogeneous fluids are not directly applicable to the flow of suspensions of solids because of the additional variables which are required to describe the solid phase and the effect of gravity.

These variables are:

D_c = diameter of the conduit.
D_s = effective diameter of the particles (Chapter 7).
ρ_s = density of the particles.
Y = mass ratio of solids to fluid in flow.
ψ = sphericity of the particles.
ϵ = roughness of the particles.
g = acceleration of gravity.

The loss due to friction will then be a function of the nine variables as follows (neglecting roughness of the pipe).

$$\overline{lw} = f(D_c, v, \rho, \mu, D_s, \rho_s, Y, \psi, \epsilon, g) \qquad (233)$$

Some simplification might be gained by resorting immediately to dimensional analysis, but the choice of the most convenient dimensionless groups is difficult since there are many combinations which could be used.

Gasterstadt [3] correlated his data by means of the dimensionless term "relative pressure drop," represented by the symbol α and defined as the ratio of the pressure drop obtained in the flow of a suspension to that obtained with pure fluid flowing in the same pipe at the same velocity. Both Gasterstadt [3] and Segler [6] show α to be a linear function of the mass ratio Y for the conveying of wheat in a given pipe with a given air velocity, but they make no generalizations as to the effect of pipe size or air velocity.

Dimensional analysis indicates that α can be expressed as a function of seven dimensionless groups according to the relation

$$\alpha = \phi\left(\frac{D_c v \rho}{\mu}, Y, \frac{D_s}{D_c}, \frac{\rho_s}{\rho}, \psi, \epsilon, \frac{\frac{1}{3}(\rho_s - \rho)\rho g D_s^3}{\mu^2}\right) \qquad (234)$$

Data on the flow of suspensions of sand, wheat, clover seed, and lead shot in air through pipe have been correlated by the following dimensionless equation.

$$\alpha - 1 = B\left(\frac{D_c}{D_s}\right)^2 \left(\frac{\rho Y \mu}{\rho_s D_c v \rho}\right)^k \qquad (235)$$

This equation is dimensionally consistent, and a similar relation may be derived by making a number of assumptions.

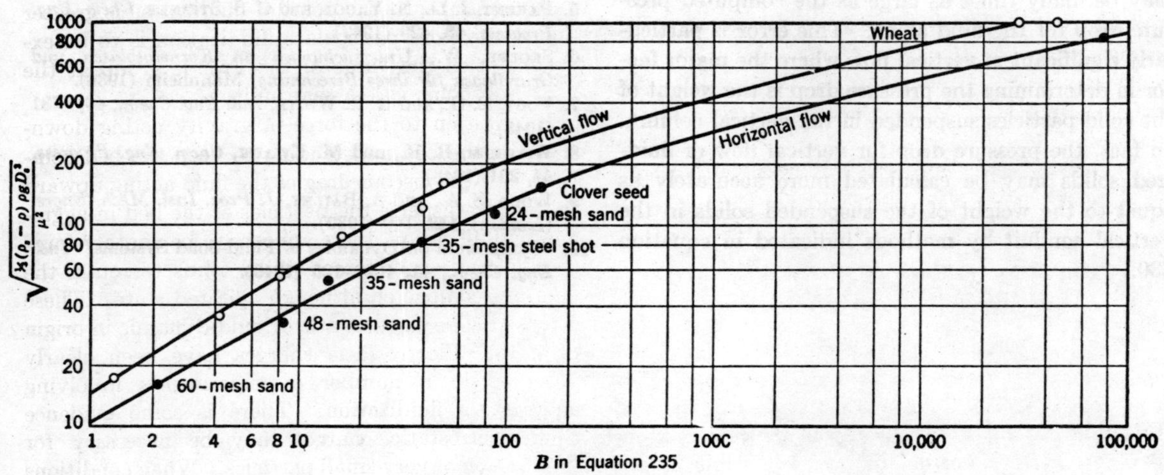

FIG. 282. The relationship between the group $\sqrt{\frac{1/3(\rho_s - \rho)\rho g D_s^3}{\mu^2}}$ and the coefficient B in equation 235

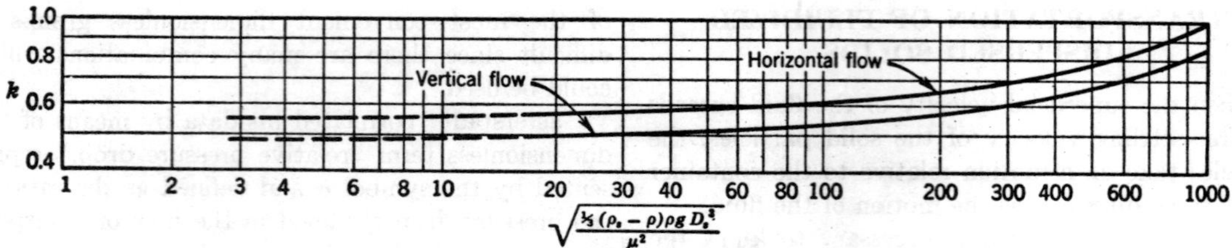

FIG. 283. The relationship between the group $\sqrt{\frac{1/3(\rho_s - \rho)\rho g D_s^3}{\mu^2}}$ and the exponent k in equation 235.[7]

The values of the parameters B and k are expressed as functions of the group $\sqrt{\frac{\frac{1}{3}(\rho_s - \rho)\rho g D_s^3}{\mu^2}}$, which is the product of the Reynolds number based on the solid particles and the square root of the drag coefficient for a spherical particle under free settling conditions (Chapter 7, equation 22) and involves no velocity term.

Figures 282 and 283 show the effect of the group $\left(\sqrt{\frac{\frac{1}{3}(\rho_s - \rho)\rho g D_s^3}{\mu^2}}\right)$ on the values B and k of equation 235 in both horizontal and vertical conveying with air. Data are not yet available by which the effects of particle shape and roughness can be evaluated, and the correlation represented by equation 235 and Figs. 282 and 283 is based on limited data.

Application of the correlations of Figs. 282 and 283 is subject to large error because the value of α may be many times as large as the computed pressure drop for the fluid alone. This error is particularly significant in vertical flow where the major factor in determining the pressure drop is the weight of the solid particles suspended in the vertical column. In fact, the pressure drop for vertical flow of fluidized solids may be calculated more accurately as equal to the weight of the suspended solids in the vertical conduit by methods indicated in equation 230.

In industrial practice the velocity may be limited by damage to the solid particles or by erosion of the conduit which may occur at high velocities. At relatively low velocities unsteady operation or "slugging" may occur in horizontal runs because of the alternate settling and resuspension of the solids. In most industrial applications, solids are transported in a much more dilute or dispersed condition than is represented by the condition of a fluidized bed, and the fluid velocities encountered will usually lie between the two practical limits which represent damage to the particles and conduit on one hand and slugging on the other.

BIBLIOGRAPHY

1. CRAMP, W., *J. Soc. Chem. Ind. (London)*, **44,** 207 (1925).
2. CRAMP, W., and J. F. PRIESTLEY, *Engineer*, **137,** 34 (1924).
3. GASTERSTADT, H., *V.D.I. Forschungsarbeiten*, No. 265 (1924).
4. MURPHREE, E. V., C. L. BROWN, E. J. GOHR, C. E. JAHNIG, H. Z. MARTIN, and C. W. TYSON, *Trans. Am. Inst. Chem. Engrs.*, **41,** 19 (1945).
5. PARENT, J. D., N. YAGOL, and C. S. STEINER, *Chem. Eng. Progress*, **43,** 429 (1947).
6. SEGLER, W., *Untersuchungen an Kornergeblasen und Grundlagen für ihres Berechnung*, Mannheim (1934).
7. VOGT, E. G., and R. R. WHITE, *Ind. Eng. Chem.*, **40,** 1731 (1948).
8. WILHELM, R. H., and M. KWAUK, *Chem. Eng. Progress*, **44,** 201 (1948).
9. WOOD, S. A., and A. BAILEY, *J. Proc. Inst. Mech. Engrs. (London)*, **142,** 149 (1939).
10. "Symposium on Dynamics of Fluid-Solid Systems," *Ind. Eng. Chem.*, **41,** 1099–125 (1949).

PART III

Separation by Mass Transfer: The Ideal Stage Concept

THE movement of one or more components between phases occurs in many operations and is known as mass transfer. The absorption of a volatile component from a vapor phase, the extraction of a component from a solid by solution in a liquid, the separation of volatile components of a liquid by distillation are examples of separation by mass transfer.

The equipment for such separations frequently incorporates a series of stages in which the two phases are brought into intimate contact, to provide an opportunity for the transfer of material between the two phases, and subsequently separated. The effectiveness of the stage in accomplishing the transfer of material depends upon many factors such as the design of the equipment, the physical properties of the phases, and equilibrium relationships.

It is convenient to compare the performance of actual stages with the standard performance of an "ideal stage." This standard performance is such that the two streams leaving an ideal stage are either in equilibrium or of such composition that no further transfer would result from further contact of the phases. The use of this concept of an ideal stage in calculations involves only the fundamental principles of material and energy balances and phase equilibrium relationships, without any consideration of rates of transfer. The effect of the rate of transfer is reflected on the relative performance of the actual stage as compared to that of the ideal stage. This relative performance is usually expressed as a stage efficiency.

In a later section the rate of mass transfer will be considered as applied to these and other operations in which the concept of the ideal stage is not readily applicable.

CHAPTER 21

Solid-Liquid Extraction

EXTRACTION is a term that is used for any operation in which a constituent of a solid or of a liquid is transferred to another liquid (the solvent). The term "solid-liquid extraction" is restricted to those situations in which a solid phase is present and includes those operations frequently referred to as "leaching," "lixiviation," and "washing."

Solid-liquid extraction is important in many industrial processes. Copper is recovered from oxidized copper ores, which are usually low-grade ores containing less than 1.5 weight per cent copper, by extraction with solvents such as dilute sulfuric acid. The extraction by suitable solvents of soybean oil from soybeans, sugar from sugar beets, and fish oils from fish livers are further examples.

Extraction always involves the two steps: (1) contact of the solvent with the solid to be treated so as to transfer the soluble constituent (solute) to the solvent, and (2) separation or washing of the solution from the residual solid.

Liquid always adheres to the solids which must be washed to prevent either the loss of solution if the soluble constituent is the desired material, or the contamination of the solids if the solids are the desired material. This washing of solids is frequently necessary when there is no constituent to be dissolved. For example, the calcium carbonate precipitate formed in the manufacture of caustic soda from soda ash and lime is washed to recover the maximum amount of caustic.

The complete process may also include the separate recovery of the solute and solvent. But this is done by another operation such as evaporation or distillation.

EQUIPMENT

The equipment used for solid-liquid extraction may be classified according to the manner in which the first step is accomplished. The term "solid bed" refers to any operation in which the solid particles are kept in relatively fixed positions with respect to each other while the solvent flows through the bed of solid particles, whether or not the bed of solid material remains stationary with respect to the earth during extraction. The term "dispersed contact" refers to any operation in which the solid particles, suspended in the fluid, are in motion relative to each other and to the solvent during the time of contact. Both types of equipment may be operated either with a batch of solids being treated with one or more batches of solvent or with the solvent flowing successively through the solids being extracted.

Stationary Solid Bed

Open Tanks. Where a fixed bed is used, the simplest type of equipment consists of open tanks made of wood, concrete, or steel with protective linings if the solution used is corrosive. The tanks such as shown in cross section in Fig. 284 may be provided with false bottoms which serve as a support for the bed of solids but allow passage of the solvent or solution. The solid to be treated is placed in the tank in such a way as to prevent size segregation of the solid particles, and the solvent is then introduced at either the top or the bottom of the tank.

In one method of operation the solvent or solution is pumped in until the bed is covered, and the liquid is allowed to remain in contact with the solids for a prescribed length of time which may be until a desired minimum fraction of the solute is dissolved or leached out of the solid phase, or until the concentration of solute in the solvent has attained a desired value. The liquid is then drained from the solids through the false bottom. This cycle may be repeated with successive quantities of fresh solvent, but usually solutions of decreasing solute concentration are used until finally the extracted solids, after drainage, are removed from the tank.

In another method of operation solvent or solution is pumped into the tank continuously and the resulting solution is continuously removed. The concentration of the solute in the effluent from a tank

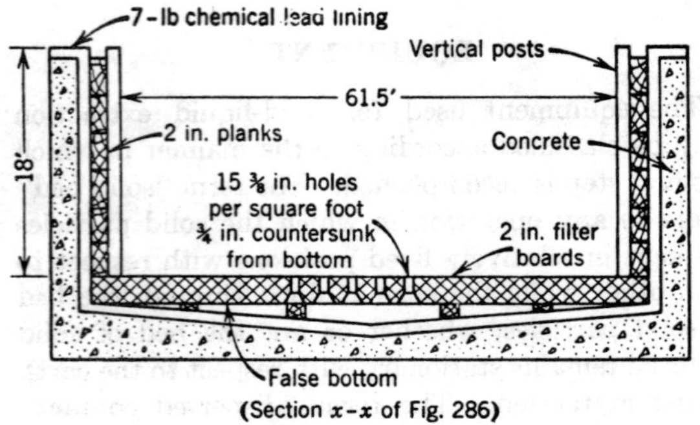

FIG. 284. Diagrammatic cross section (x-x of Fig. 286) of a large leaching tank, indicating construction (not to scale).[1]

becomes less as the extraction proceeds. If a battery of tanks is used, as is customary in continuous operation, the feed to one tank is the effluent from a preceding tank, and the concentrations in both the inlet and outlet solutions change continuously during the process.

An example of this latter method is the extraction of copper from ore as reported by Aldrich and Scott.[1] * The extraction battery consists of thirteen concrete lead-lined tanks, each 175 ft long, 67.5 ft wide, and 18 ft deep, having an approximate capacity of 9000 tons of ore per tank. The tank construction is illustrated in Figs. 284 and 285. The false bottom in each tank consists of 2-in. planks having 15 holes per square foot of surface. Each hole is ⅜ in. in diameter at the upper surface and ¾ in. at the bottom to prevent clogging. A 14-in. lead pipe is "burned" to the lead lining at the end of the tank opposite from the overflow and connected to the discharge of a vertical lead pump which takes suction

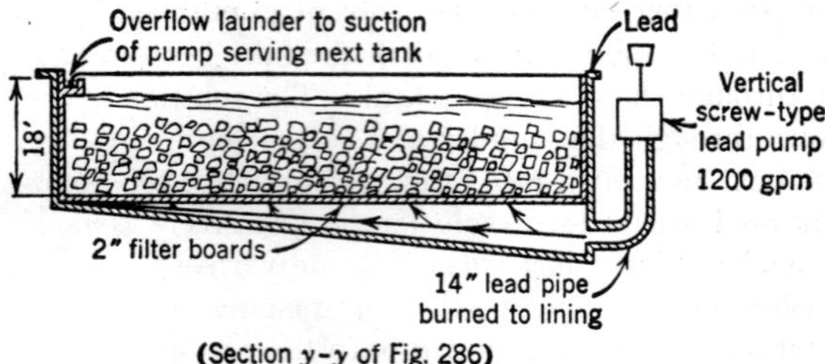

FIG. 285. Diagrammatic lengthwise section (y-y of Fig. 286) of a leaching tank (not to scale).[1]

from the overflow of the next previous tank in battery.

The ore, after crushing and sizing, is delivered to the battery of tanks by a belt conveyor (*A* in Fig. 286) running the length of the battery. The ore is discharged from this belt conveyor to another belt conveyor (*B*) on a movable spreading machine spanning the tanks. An automatically reversing tripper travels on the spreading machine and empties the load of the belt into the tank. At each reversal of the tripper, the spreading machine moves forward 2½ ft until it reaches the side of the tank, when the motion is reversed and the machine travels toward the opposite side.

When the tank is filled with ore to the proper depth, the ore is covered with dilute sulfuric acid (the solvent) pumped in through the bottom of the tank in order to prevent disturbance of the bed. Approximately 175,000 to 200,000 gal of solvent are required to fill one tank. The overflow solution from

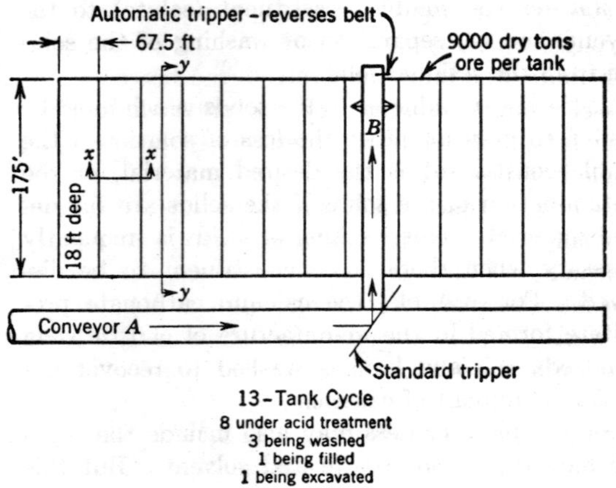

FIG. 286. Diagram of tanks and belt conveyors illustrating operation of an extraction battery leaching copper.[1]

* The bibliography for this chapter appears on p. 294.

the preceding tank or fresh solvent is pumped by a vertical lead pump underneath the false bottom and up through the ore in each tank. The solution overflows through the launder and to the pump serving the next tank at a rate of approximately 1200 gpm. The ore is washed with solutions progressively weaker in acid until the last wash is made with water, ten washes in all being used. In washing, the wash solution or water is pumped onto the ore, circulated by means of the pump, and then drained. The drained solids are removed from the tank by a bridge-type excavator which spans the leaching tanks with the bucket carriage traveling across the long dimension of the tank. This bucket travels to the end of the excavator and discharges its load through a hopper into air-dump standard-gage railroad cars. A complete cycle for one tank takes 13 days. At any one time eight tanks are under acid treatment, three are being washed, one is being filled, and one is being excavated.

Closed Tanks. With volatile solvents, such as ammonia solution in the extraction of copper ores, or hot water, it is necessary to operate at pressures above atmospheric and closed tanks are required. Such vessels are usually constructed of steel and must be provided with adequate openings for charging and discharging the solids.

Digested pulp obtained from the sulfate paper pulp process is washed in the type of closed vessel shown in Fig. 287. The mixture of pulp and hot liquor under pressure in the digester is transferred to the vessel through the feed opening. Steam formed due to the drop in pressure, together with entrained liquor and pulp, escape through the vent at the top. The cone-shaped baffle fastened underneath the pipe through which the pulp is blown into the vessel, distributes the pulp and protects the bottom of the vessel from the impact of the entering material. The false bottom is mounted on I-beams and on angle iron fastened to the circumference of the shell. The empty space underneath the false bottom is partially filled with concrete which supports the I-beams and also decreases the free space beneath the false bottom. The weak liquor and water used to wash the pulp are introduced through a top connection and gradually displace the liquor contained in the pulp, forcing it through the false bottom and through the bottom outlet. The washed pulp is discharged through the discharge opening which is fitted with a heavy cast-steel frame and bolted cover. A water connection is located opposite the discharge opening for injecting a high-velocity stream of water to aid in discharging the pulp.

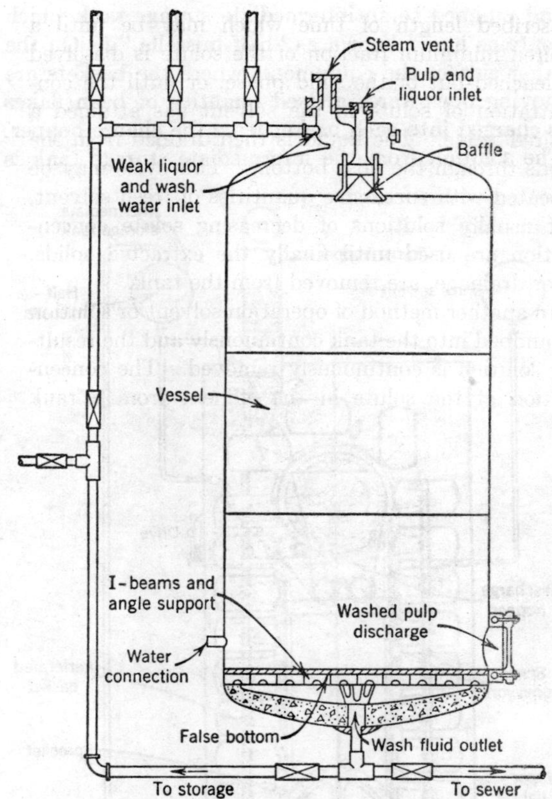

FIG. 287. Diagram of a closed vessel for washing digested pulp from sulfate paper process.[16]

Moving Solid Bed

In some cases it may be advantageous, particularly from the standpoint of handling the solids, to move the bed itself.

The Bollman or *Hansa-Mühle* soybean extractor, Fig. 288, is similar to a bucket elevator, consisting of a series of large perforated baskets suspended on a pair of endless chains which are driven by sprocket wheels. On the side of the equipment where the baskets are moving upward, fresh solvent is sprayed on a basket, near the top, which contains bean flakes almost exhausted of oil. The liquid flows through the bed, through the perforations in the basket, and downward to the next basket. A series of countercurrent multiple-stage contacts is thus obtained until the solution of solvent and oil reaches the bottom of the unit where it is collected in sump *A* and pumped to an intermediate storage tank which contains liquid known as "half miscella." * On the other side of the equipment, where the baskets are moving downward, a fixed quantity of bean flakes is charged into each basket from the charge hopper. The solution from the intermediate storage tank is

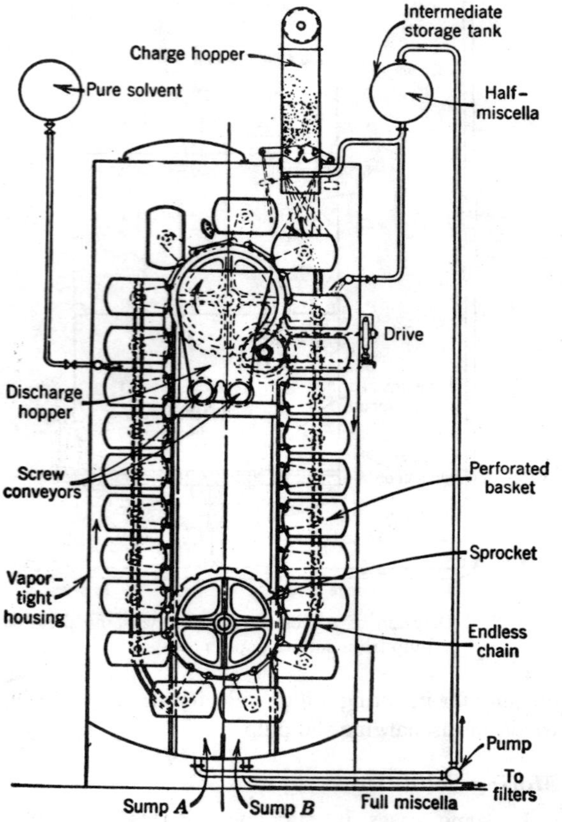

FIG. 288. Diagram of Bollman or Hansa-Mühle continuous moving-bed extractor.[8a]

sprayed onto the top of the freshly charged basket, and both solution and basket move downward. This side of the system corresponds to parallel-flow multiple-contact operation. The strong solution from this side of the system is collected in another sump *B* from which it is pumped to filters and then to storage tanks. When a basket reaches the top of the unit on the ascending side it is automatically inverted and the extracted flakes are dumped into the discharge hopper from which they are taken to driers by the screw conveyors. The entire apparatus is enclosed in a vapor-tight housing to prevent leakage of the solvent vapors.

* Miscella is a term used to designate a mixture of oil and solvent.

Dispersed Contact

A variety of equipment is used when the solid particles are dispersed in the solvent during extraction. Coarse granular materials not fine enough to remain in suspension in the solvent may be handled in *multideck classifiers* composed of units such as illustrated in Fig. 79. Countercurrent flow is used with the solid particles advancing up the inclined deck and liquid flowing down the incline. The drainage which occurs in the upper section of the inclined deck removes the bulk of solution adhering to the solids before they are advanced to the next deck where they come into contact with solution containing less of the soluble constituent.

Where the solids are fine enough to be kept in suspension by agitation, *agitators* (Figs. 88 and 476), *thickeners* (Fig. 101), or a combination of both may be used. If both agitators and thickeners are used, the agitators are used for contacting the solid with the liquid to dissolve the soluble material present in the solids, and the thickeners are used to effect the separation of the solids from the solution.

In some cases the agitation and solution may be accomplished by concurrent flow of solids and solution through a pipe in place of agitation in a tank. For small-scale batch operations, the agitation and separation (sedimentation) may be conducted in the same vessel. Thickeners alone may be used for the separation and washing of a precipitate.

There are many types of equipment in which the solids are moved countercurrent to the liquid by means of screw conveyors, rakes, and other mechanical devices. An interesting development is the soybean extractor,[4,12] shown in Fig. 289, which consists of a vertical cylindrical vessel containing a series of horizontal circular plates equally spaced and fixed to a central shaft which is slowly rotated by a motor. Soybean flakes are fed continuously by the conveyor to the top, and solvent is pumped continuously into the bottom. The flakes introduced at the top are distributed evenly by spreader blades attached to the rotating shaft. Stationary scraper arms fastened to the shell scrape the surface of each plate, which is slotted, so that the soybean flakes are swept through the slots by the stationary scraper arm. The slots are so located that the flakes follow a helical path in moving downward through the unit.

The flakes at the base of the unit are discharged by means of a totally enclosed conveyor which elevates the material a sufficient distance above the solvent level to permit adequate drainage before discharging to the driers. The liquid flows upward through the extractor, countercurrent to the motion of the flakes. A thick layer of fresh flakes is maintained on the top plate, providing a filter bed to retard the fine particles dislodged in the lower parts of the apparatus and carried upward by the liquid. The solution of oil and solvent, together with entrained solid particles, overflows through a screen near the top and leaves the extractor.

An optional design is to fasten the scrapers to the central rotating shaft and make the plates stationary.

In general, the stationary solid bed extractor involves a minimum amount of handling and is often used where large quantities of material have to be treated or where the characteristics of the solid particles are such that continuous movement of the material is undesirable. Mixtures of solid particles which form a bed of low porosity, either originally or during extraction, are usually treated in the dispersed state rather than in a solid bed. As porosity is governed by size distribution rather than the size of the solid particles, porosity may be improved by classification of the solids before extraction. This has been done in the case of metallic ores.[1,5,7]

The Complete Extraction Process

The streams leaving a solid-liquid extraction system usually undergo a series of further operations before the finished product is obtained. Either the solution, the extracted solids, or both may contain the desired material. In addition to the recovery of the desired product or products, recovery of the solvent is usually an important operation.

In the process for the extraction of copper ore which has been described, the desired copper is in the solution while the extracted and washed ore is discarded. The major portion of the copper is recovered by electrolytic deposition from the solution. During the electrolytic deposition a large part of the aqueous sulfuric acid required for the extraction is also formed. The solution from the electrolytic cells, after addition of water and sulfuric acid necessary to make up losses occurring in the process, is then recycled to the extraction system.

In the extraction of soybeans, both the soybean oil and the extracted soybeans (meal) are valuable products. Recovery of the oil, the meal, and the solvent requires additional operations and equipment such as driers, evaporators, stripping columns, water separators, and condensers. Figure 290 is a flowsheet for one process for the extraction of soybeans.

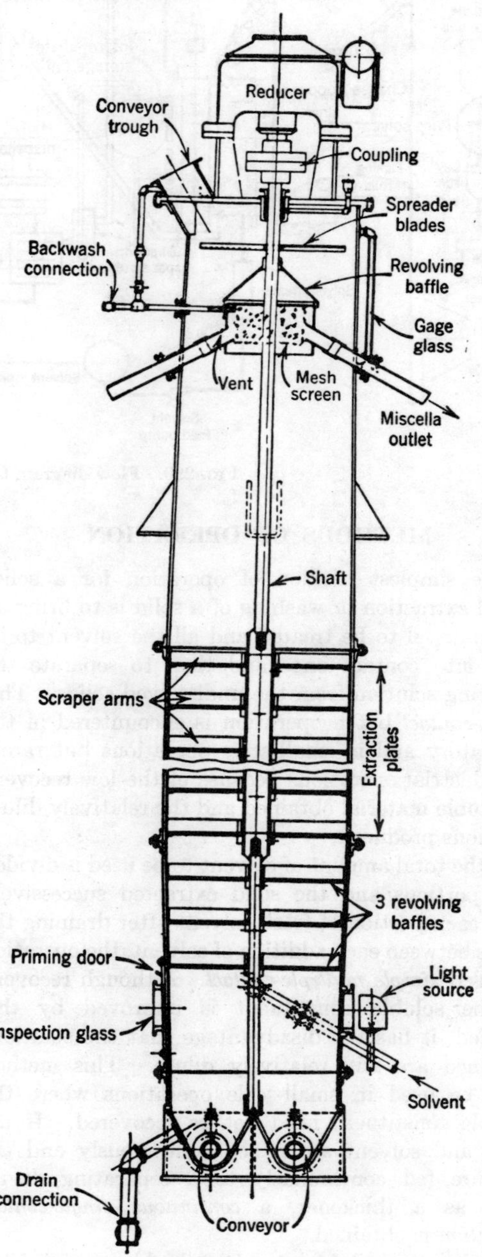

FIG. 289. Diagram of a rotating-plate countercurrent dispersed-contact extractor.[8a]

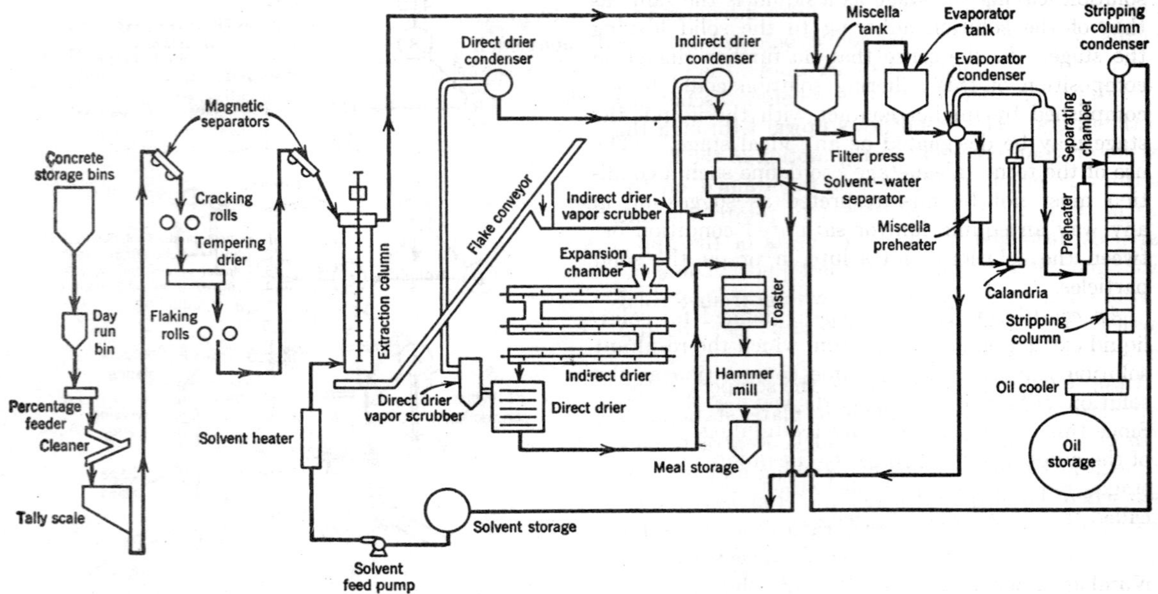

FIG. 290. Flow diagram for solvent extraction of soybeans.

METHODS OF OPERATION

The simplest method of operation for a solid-liquid extraction or washing of a solid is to bring all the material to be treated and all the solvent to be used into contact once and then to separate the resulting solution from the undissolved solids. This *single-contact* batch operation is encountered in the laboratory and in small-scale operations but rarely in industrial operations because of the low recovery of soluble material obtained and the relatively dilute solutions produced.

If the total amount of solvent to be used is divided into portions and the solid extracted successively with each portion of fresh solvent after draining the solids between each addition of solvent, the operation is called *simple multiple-contact*. Although recovery of the soluble constituent is improved by this method, it has the disadvantage that the solutions obtained are still relatively dilute. This method may be used in small-scale operations where the soluble constituent need not be recovered. If the solid and solvent are mixed continuously and the mixture fed continuously to a separating device such as a thickener, a *continuous single-contact operation* is obtained.

High recovery of solute with a highly concentrated product solution can be obtained only by using countercurrent operation with a number of stages. In countercurrent operation the product solution is last in contact with fresh solid feed, and the extracted solids are last in contact with fresh solvent, as illustrated in Fig. 299 which shows a continuous countercurrent battery.

A similar result is obtained by a batch countercurrent operation, illustrated in Fig. 291 which indicates the use of three batch contactors producing the same separation as five stages in a continuous countercurrent battery such as Fig. 299. The six steps indicated in Fig. 291 trace the contacts of the fresh solution added to vessel A in step I through the five countercurrent contacts indicated by asterisks in vessels A, B, C, A, B, after which it must be removed. In general the number of contact units required for a given separation using a continuous countercurrent operation, N_C, is equal to $2N_B - 1$, where N_B is number of contact units available for a batch operation. But the rate of production from the continuous battery of equal-sized units is much greater.

METHOD OF CALCULATION

In most intermittent operations, particularly those in which the solid is agitated with a batch of liquid before separation, the composition of the solution leaving the stage as a liquid is the same as that of the solution adhering to the solid leaving the stage. In the sense that no further change in composition of this adhering solution could be accomplished by further contact with the liquid, the stage may be designated as an "ideal stage." The use of the term "ideal stage" to define such a condition must not be misinterpreted as suggesting in any way an equilibrium or saturated condition between the solutions and solute in or on the solid particles.

An equilibrium or ideal stage is defined for solid-liquid extraction as a stage from which the resultant solution leaving is of the same composition as the solution adhering to the solids leaving the stage. Since this condition is not always fulfilled, the ratio of the number of ideal stages to the number of actual stages required to accomplish the same results is called the *overall stage efficiency*.

$$\text{Number of actual stages} = \frac{\text{Number of ideal stages}}{\text{Overall stage efficiency}}$$

The computation of the number of ideal stages required is an adequate basis for the design of an extraction battery only if the designer also knows, or can estimate, the overall stage efficiency. The stage efficiency is frequently related to the time of contact between the liquid and solid in each stage and therefore tends to decrease with increasing throughput if the equipment is overloaded. Such information can be obtained only from experimental data on the type of equipment to be used when operated with the actual materials involved in the extraction.

In order to simplify the calculation of the number of ideal stages required for a solid-liquid extraction, the following assumptions are usually made.

1. The system is composed of materials which may be treated as three components as follows.

(*a*) Inert, insoluble solids.
(*b*) A single solute which may be liquid or solid.
(*c*) A solvent which dissolves the solute (*b*) but has no effect upon or is saturated with the inert solids (*a*).

2. The solute is not adsorbed by the inert solid.
3. The solute is removed by simple solution in the solvent without what is called a chemical reaction.

These assumptions are not necessary for the calculation. The solid need not be insoluble in the

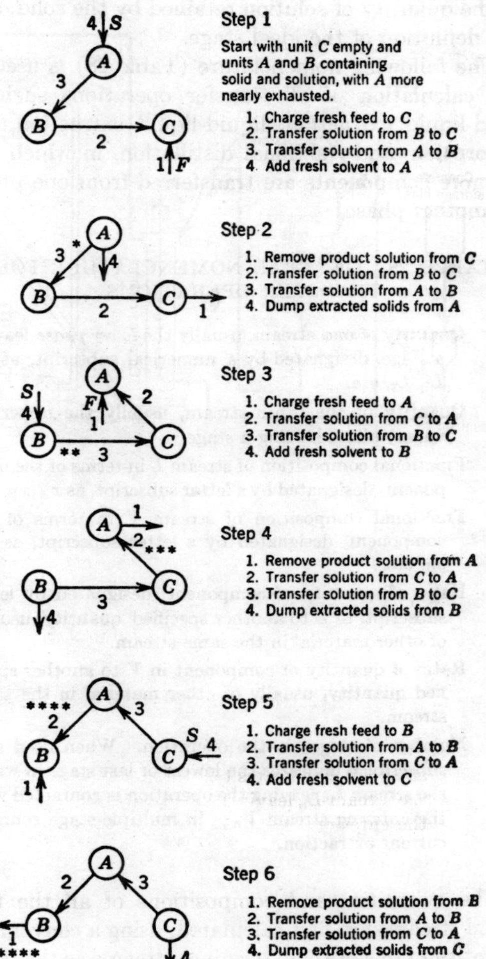

FIG. 291. Diagram illustrating the steps and number of contacts in a batch countercurrent extraction. Each time the solution originally added to vessel *A* in step 1 leaves a countercurrent contact, an asterisk * is added to the stream; * as it leaves *A*, step 2; ** as it leaves *B*, step 3; *** as it leaves *C*, step 4; **** as it leaves *A*, step 5; ***** as it leaves *B*, step 6.

solvent provided that the data are available on the solubility of the inert solid in solutions of varying concentrations. Similarly the solute may be adsorbed on the solid provided that data are available on the amount of solute adsorbed per unit mass of the solid as a function of the concentration of the solution. In the case of electrolytes, adsorption of the solute is usually negligible.

The computation of the number of ideal stages required is based on material balances, a knowledge of the quantity of solution retained by the solid, and the definition of the ideal stage.

The following nomenclature (Table 28) is used in the calculation of all transfer operations such as solid-liquid extraction, liquid-liquid extraction, gas absorption, and fractional distillation, in which one or more components are transferred from one phase to another phase.

TABLE 28. GENERAL NOMENCLATUPE FOR TRANSFER OPERATIONS

L	Quantity of one stream, usually the *lower phase* leaving a stage, designated by a numerical subscript, as L_0, L_1, L_2, etc.
V	Quantity of the other stream, usually the upper (or vapor) phase leaving a stage.
x	Fractional composition of stream L in terms of the component, designated by a letter subscript, as x_A, x_B, etc.
y	Fractional composition of stream V in terms of the component, designated by a letter subscript, as y_A, y_B, etc.
X	Ratio of quantity of component, designated by letter subscript, in L to another specified quantity, usually of other material in the same stream.
Y	Ratio of quantity of component in V to another specified quantity, usually of other material in the same stream.
n	Number of stages in the operation. When used as a subscript n indicates the lowest or last stage in which the stream L_n leaving the operation is contacted with the entering stream V_{n+1} in multiple-stage countercurrent extraction.

The quantities and compositions of all the terminal streams are first calculated, using a convenient quantity of one of these terminal streams as the basis for the calculations. The quantities and compositions of the streams entering or leaving a terminal ideal stage at either end of the extraction system are next calculated. These calculations are repeated for each successive ideal stage from one end of the system until an ideal stage is obtained which corresponds to the desired conditions at the other end of the system. The number of actual stages may be obtained by dividing the number of ideal stages by the overall stage efficiency. These calculations can be made if the quantity of solution retained per unit quantity of inert solids leaving each stage is known. This is not a constant quantity as it depends upon the properties of the solution, particularly its viscosity. It is therefore determined experimentally under conditions similar to those being considered for the commercial operation.

Illustrative Example A (Arithmetic Calculation). It is desired to extract the oil from halibut livers by continuous countercurrent multiple-contact extraction with ethyl ether. The quantity of solution retained by the granulated livers has been determined experimentally as a function of the composition of the solution as follows.

TABLE 29. RETENTION OF OIL BY LIVERS[14]
(FIG. 292)

Pounds of Liver Oil in 1 Lb Solution, y_A	Pounds of Solution Retained by 1 Lb Oil-Free Livers
0.00	0.205
0.10	0.242
0.20	0.286
0.30	0.339
0.40	0.405
0.50	0.489
0.60	0.600
0.65	0.672
0.70	0.765
0.72	0.810

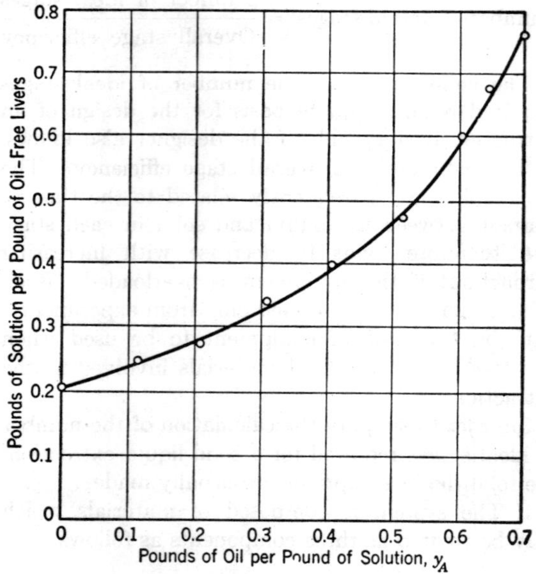

FIG. 292. Retention of solution by livers as a function of solution concentration (Table 29).

The fresh halibut livers contain 25.7 mass per cent oil. If 95 per cent of the oil is to be extracted and the final solution obtained from the operation is to contain 70 mass per cent oil, compute,

1. The pounds of oil-free ether required per 1000-lb charge of fresh livers.
2. The number of ideal stages required.
3. The number of actual stages required if the overall stage efficiency is 70 per cent.

The algebraic calculation on the basis of 1000 lb of fresh livers, L_0, follow.

1. *Calculation of Terminal Conditions*

Oil in fresh livers = $(Lx_A)_0$ = 1000(0.257) = 257 lb
Oil-free livers = 1000 − 257 = 743 lb
Oil remaining in extracted livers = $(Lx_A)_n$ = 0.05 × 257 = 12.85 lb

The livers discharged from the operation will contain 12.85 lb of oil, 743 lb of oil-free livers, and a quantity of ether which is as yet undetermined. The mass ratio of oil to oil-free livers in the discharged livers $(X_A)_n$ will be 12.85/743 = 0.01730 The quantity of ether present is then computed from the experimental data given as follows.

$$\left(\frac{\text{Pounds of oil}}{\text{Pounds of solution}}\right)\left(\frac{\text{Pounds of solution}}{\text{Pound of oil-free livers}}\right) = \frac{\text{Pounds of oil}}{\text{Pound of oil-free livers}} = X_A$$

$$\frac{\text{Pounds of solution}}{\text{Pounds of oil-free livers}} - \frac{\text{Pounds of oil}}{\text{Pounds of oil-free livers}} = \frac{\text{Pounds of ether}}{\text{Pounds of oil-free livers}} = X_S$$

X_A	X_S
0.000	0.205
0.0242	0.218
0.0572	0.229

These values are plotted in Fig. 293 which gives, for $(X_A)_n$ = 0.0173, $(X_S)_n$ = 0.215. Therefore, the *ether present in the discharged livers* is 0.215 (743) = 159.8 lb.

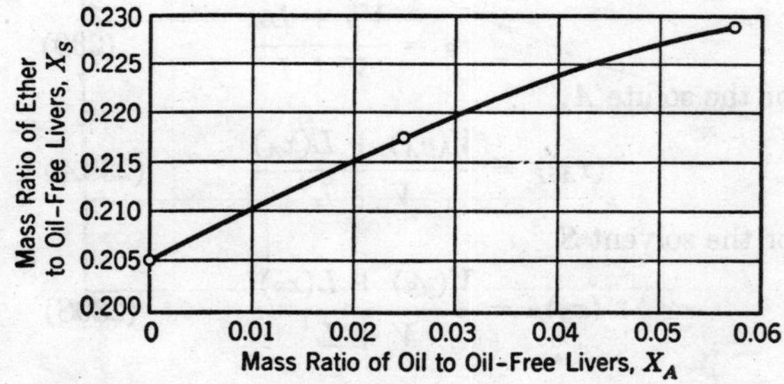

Fig. 293. Retention of solute by livers as a function of retention of solvent by livers.

Total mass of discharged livers, L_n = 743 + 12.85 + 159.8 = 916 lb.

The extract or final solution leaving the operation contains 0.95 mass fraction of the total oil entering in the fresh livers, or 257 × 0.95 = 244.2 lb. The composition of extract was specified as 0.70 mass fraction oil.

Total mass of solution leaving, V_1 = 244.2/0.70 = 349 lb.

OVERALL MATERIAL BALANCE (BASIS, 1000 LB FRESH LIVERS)

	In			Out	
L_0	Fresh livers	1000 lb	L_n	Oil-free liver	743 lb
V_{n+1}	Ether (by difference)	264.65		Oil on liver	12.85
				Ether on liver	159.8
			V_1	Solution	349
		1264.65 lb			1264.65 lb

2. *Material Balances for Stage 1*

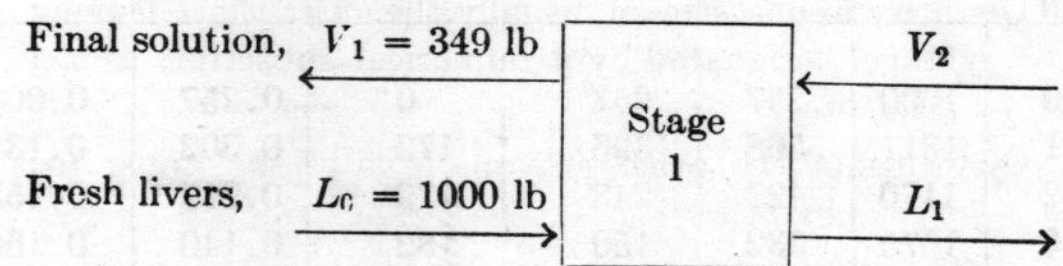

The stream L_1 is a mixture of oil-free livers and solution containing 0.70 mass fraction oil. From the experimental data, plotted in Fig. 292, L_1 contains 0.765 lb of solution per pound of oil-free livers. Therefore,

Solution in L_1 = 0.765 (743) = 568 lb

Oil in L_1 = 0.70 (568) = 396 lb

By difference,

Ether in L_1 = 172 lb

Total mass of L_1 = 568 + 743 = 1311 lb

Since

$$L_0 + V_2 = L_1 + V_1$$

$$V_2 = L_1 + V_1 - L_0 = 1311 + 349 - 1000 = 660 \text{ lb}$$

By a material balance for the oil, around ideal stage 1,

Oil leaving in L_1 and V_1 = 244 + 396 = 640 lb

Oil entering in L_0 = 257 lb

By difference,

Oil entering in V_2 = 383 lb

Composition of V_2 = $\frac{383}{660}$ = 0.580 mass fraction of oil

3. *Material Balances for Ideal Stages 2 to n*

The same procedure is followed in making calculations for ideal stages 2, 3, and so on, as summarized in Table 30. Examination of the values calculated for stream L_6 shows that the discharged livers leaving ideal stage 6 contain 17.1 lb of oil, compared with the required value of 12.85 lb. Stream L_7 contains only 2.4 lb of oil.

If the actual stages were 100 per cent efficient the exact specifications for the discharged livers could not be attained. The designer has the option of readjusting the operating conditions until the calculated composition of the discharged livers agrees with the specified composition, or maintaining the operating conditions and accepting a lower oil content in

TABLE 30. SUMMARY OF CALCULATIONS FOR ILLUSTRATION OF ARITHMETIC CALCULATION

	Underflow Leaving Stage, L_n							Solution Entering Stage, V_{n+1}				
	Quantities, lb				Compositions, Mass Fractions			Quantities, lb			Compositions, Mass Fractions	
n	Total, L_n	Solution	Oil, $(Lx_A)_n$	Ether, $(Lx_S)_n$	Oil, $(x_A)_n$	Ether, $(x_S)_n$	Oil-Free Livers, $(x_C)_n$	Total, V_{n+1}	Oil, $(Vy_A)_{n+1}$	Ether, $(Vy_S)_{n+1}$	Oil, $(y_A)_{n+1}$	Ether, $(y_S)_{n+1}$
0	1000	257	257	0	0.257	0.000	0.743	349	244	105	0.700	0.300
1	1311	568	396	172	0.302	0.131	0.567	660	383	277	0.580	0.420
2	1170	427	248	179	0.212	0.153	0.635	519	235	284	0.453	0.547
3	1075	332	150	182	0.140	0.169	0.691	424	137	287	0.324	0.676
4	1007	264	85.5	178.5	0.085	0.177	0.738	356	72.5	283.5	0.204	0.796
5	955	212	42.4	169.6	0.044	0.178	0.778	304	29.4	274.6	0.097	0.903
6	920	177	17.1	159.9	0.0186	0.174	0.807	269	4.1	264.9	0.0152	0.985
7	896	156	2.4	153.6	0.0026	0.171	0.826	245	Negative	.....	.....	.....

the discharged livers with seven stages. The actual stages are not 100 per cent efficient but are stated to be 70 per cent efficient. Therefore the 6 ideal stages are equivalent to 6 ÷ 0.7 or 8.6 actual stages, and the 7 ideal stages are equivalent to 10 actual stages. Since the number of ideal stages required is slightly greater than 6 and considerably less than 7, it is estimated that 9 actual stages will give the specified result.

GRAPHICAL METHODS

Although any solid-liquid extraction problem can be solved by the repeated use of material balances and the data on the underflow compositions in the above manner, such arithmetical solutions are tedious. Graphical methods not only simplify the calculations and thereby reduce errors but also indicate more clearly the variables involved and their effect on the operation. The same graphical methods are applicable to any operation which involves the transfer of matter (and energy) between two phases. The principles are the same in the arithmetical and graphical solutions. The graphical method is simply a graphical representation of the material balances and the equilibrium data.

If a stream (or quantity of material) designated F is split into the two streams (or quantities) V and L, the quantity of material in stream F, designated F, is equal to the sum of the quantities in streams V and L.

$$F = V + L \tag{236}$$

Conversely, if streams V and L are combined into a single stream F, equation 236 expresses the material balance.

Similarly, for any single component,

$$Fx_F = Vy + Lx \tag{237}$$

Substituting from equation 236 for F in equation 237,

$$(V + L)x_F = Vy + Lx \tag{238}$$

$$x_F = \frac{Vy + Lx}{V + L} \tag{239}$$

For the solute A,

$$(x_A)_F = \frac{V(y_A) + L(x_A)}{V + L} \tag{239A}$$

For the solvent S

$$(x_S)_F = \frac{V(y_S) + L(x_S)}{V + L} \tag{239S}$$

The composition of the streams may be represented on a rectangular (or right-triangular) diagram (Fig. 294). The fraction of the solute in the stream is plotted horizontally along the abscissa and designated x_A or y_A as may be required by the nomenclature given in Table 28. The fraction of solvent is plotted vertically along the ordinate as x_S or y_S, as required. Thus x or y may appear on both the ordinate and abscissa as the nomenclature used is not consistent with the usual geometric plotting of equations involving x and y. This may be confusing until one becomes familiar with the nomenclature. Keep in mind that the subscript controls the coordinate rather than the letter itself. Thus x_A or y_A is always plotted horizontally, and x_S or y_S is always plotted vertically.

Therefore, a single point on the diagram (Fig. 294) indicates the composition of a stream, or the *relative quantities* of the three components of a stream, but not the absolute quantity of the stream. The horizontal distance is the fractional quantity of component A, and the vertical distance is the fractional quantity of component S.

If the slope of a straight line connecting points x and x_F is the same as the slope of the line connecting points x and y, it follows that the three points x_F, x, and y, are on one straight line, and a material balance such as equation 239 can be made by simply drawing a straight line through the points representing the compositions of any two of the three streams.

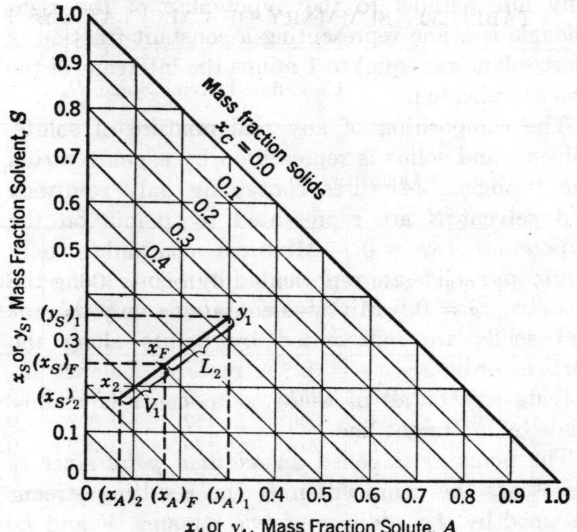

FIG. 294. Representation of material balance on right-triangular diagram.

Keeping in mind that the subscripts A and S determine the coordinate, the slopes of the lines connecting points x_F and x, and connecting points y and x, are

$$\frac{(x_S)_F - (x_S)}{(x_A)_F - (x_A)} \quad \text{and} \quad \frac{(y_S) - (x_S)}{(y_A) - (x_A)}$$

Substituting for $(x_A)_F$ and $(x_S)_F$ from equations 239A and 239S and simplifying,

$$\frac{(x_S)_F - (x_S)}{(x_A)_F - (x_A)} = \frac{V(y_S) + L(x_S) - V(x_S) - L(x_S)}{V(y_A) + L(x_A) - V(x_A) - L(x_A)} = \frac{(y_S) - (x_S)}{(y_A) - (x_A)}$$

Since the slope of the line between points x and x_F is the same as between points x and y, all three points, y, x, and x_F, lie on the same straight line.

The location of the point x_F is always between points y and x if streams V and L are added to form stream F. Similarly, if V is separated from F, leaving by difference the stream L, the resulting product obtained by difference has the composition of the difference point x which always lies on the extension of the line from y through x_F.

The linear distance between these three points representing the composition of the three streams is dependent upon the quantities L and V. This may be demonstrated by simplifying equation 238 as follows.

$$V(x_F - y) = L(x - x_F)$$

$$\frac{V}{L} = \frac{x - x_F}{x_F - y} = \frac{(x_A) - (x_A)_F}{(x_A)_F - (y_A)} = \frac{(x_S) - (x_S)_F}{(x_S)_F - (y_S)} \tag{238a}$$

Exercise. By a similar procedure, eliminating L, and then V, from equation 237 by means of equation 236, show that

$$\frac{V}{F} = \frac{x_F - x}{y - x} \quad \text{and} \quad \frac{L}{F} = \frac{y - x_F}{y - x} \tag{238b}$$

Therefore, the distance between points x and x_F is to the distance between points x_F and y as the quantity V is to the quantity L. In other words, the point x_F is located on a straight line between points y and x so that the ratio of the distance between x_F and the two other points, y and x, is inversely proportional to the quantities of the corresponding streams, V and L.

In dealing with solid-liquid extraction there is present not only the solute A and the solvent S but also the inert solids C. If these three components account for all the material present in any stream or mixture and x is the fraction of the component,

$$x_A + x_S + x_C = 1$$

Rearranging,

$$x_S = -x_A + (1 - x_C) \tag{240}$$

Equation 240 is represented on Fig. 294 as a straight line with a slope of -1 and intercept of $(1 - x_C)$

Any line parallel to the hypotenuse of the right triangle is a line representing a constant fraction of inert solids, x_C, equal to 1 minus the intercept of the line as indicated.

The composition of any real mixture of solute, solvent, and solids is represented by a point within the triangle. Mixtures containing only solute A and solvent S are represented by points on the hypotenuse, $x_C = 0$. Mixtures containing only solute and solids are represented by points along the abscissa, $x_S = 0$. Mixtures containing only solvent and solids are represented by points along the vertical ordinate, $x_A = 0$. A material balance involving any or all of these components is readily made by a straight line.

The point x_F is called an *addition point* since it represents the composition of the resulting stream obtained by the addition of two streams, V and L. Any number of streams may be added together, and the resultant streams may be represented by one addition point by taking pairs of the added streams, obtaining the addition points for these pairs, and then taking pairs of these addition points and combining them until one point representing all the added streams is obtained.

If a mixture, represented by point x_F on the diagram, is to be separated into two streams of fixed compositions, such as represented by points y and x, the process may be considered as the subtraction of one stream from another, such as L from F to give the resultant stream V, or as V from F to give L. When one stream is subtracted from another, the point representing the composition of the resulting stream so obtained is called a *difference point.* The concept of a difference point can be extended to calculations where the subtraction of one mixture from another would result in a stream of imaginary composition, in which case the difference point falls outside the triangle.

An *addition point* always lies between the points representing the streams added together and nearer the original stream of larger quantity. A *difference point* always lies on an extension of the line through the two original points beyond the point representing the larger stream. For example, if the stream (or quantity) V is subtracted from quantity F, the difference point x must lie on the line beyond x_F and nearer x_F than y, since F is larger than V.

The concept of a difference point of negative composition or of composition greater than 1.00 will cause no difficulty if it is recognized that a point on Fig. 294 represents *relative quantities* and that the sign indicates *direction. For example,* consider the two streams V_3 and L_2, flowing in opposite directions to the right and left, respectively, to produce a single stream D, flowing to the right, as indicated in Fig. 295. By a material balance

$$D = V_3 - L_2 = V_2 - L_1 = V_1 - L_0$$

where D is the difference stream between V_3 and L_2 and represents the "net flow" of material to the right at each and every place in the system.

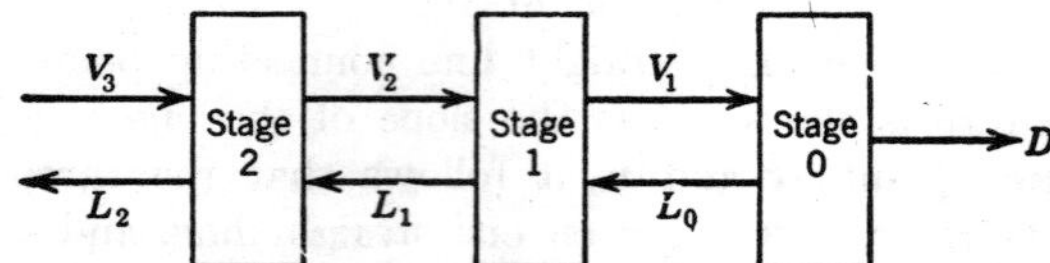

FIG. 295. Diagram representing countercurrent flow of two streams V and L and the significance of the difference stream D as indicating $V - L$.

Similarly, a material balance for any component, such as A,

$$(Dx_A)_D = (Vy_A)_3 - (Lx_A)_2 = (Vy_A)_2 - (Lx_A)_1$$

$$(x_A)_D = \frac{(Vy_A)_3 - (Lx_A)_2}{D}$$

where $(x_A)_D$ is the abscissa for the difference point D.

Since $(x_A)_D$ is the relative quantity of component A moving to the right per unit quantity of D, or the relative net flow of component A to the right, it is clear that the value for $(x_A)_D$ may be any number, positive or negative, depending upon the relative values for $(Vy)_3$ and $(Lx)_2$. Identical equations and statements may be written for any other component.

In a solid-liquid extraction where the solute is originally present as a solid phase, the amount of solute that may be dissolved in a given quantity of solvent is limited to the saturation composition of the solution as determined by the temperature and pressure of operation. The effect of pressure on the solubility is small and may be neglected except for high pressures. At any given temperature, therefore, the right-triangle diagram may be divided into two areas (Fig. 293), one representing all the mixtures in which the solution present is unsaturated, and one representing all the mixtures in which the solution present is saturated. When the solution is saturated the solids present in the mixtures will consist of undissolved solute and inert solids. The point y_{sat}

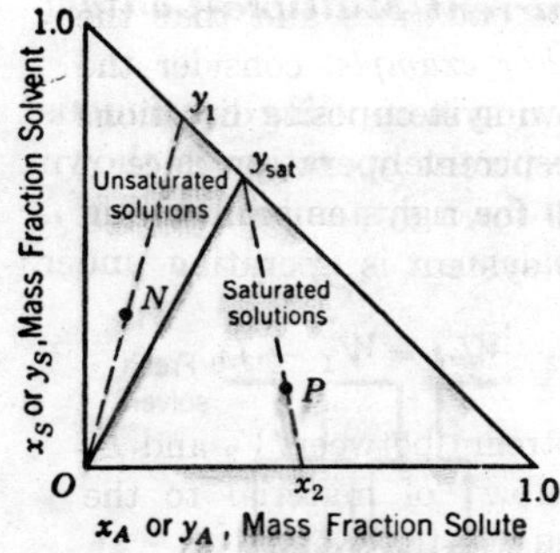

FIG. 296. Isothermal rectangular diagram for solid solute of limited solubility.

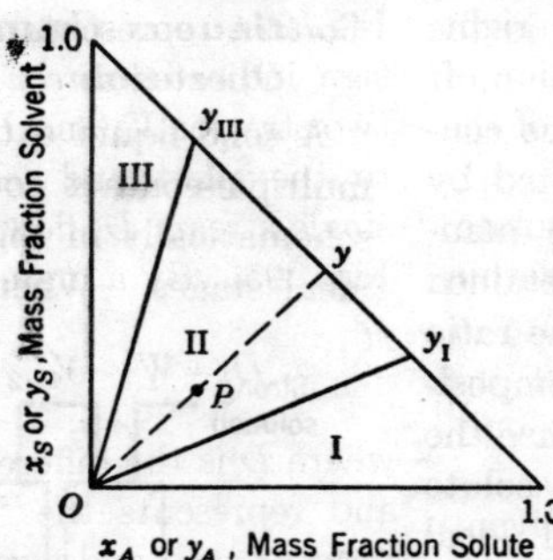

FIG. 297. Isothermal rectangular diagram for a liquid solute that is only partially miscible with the solvent.

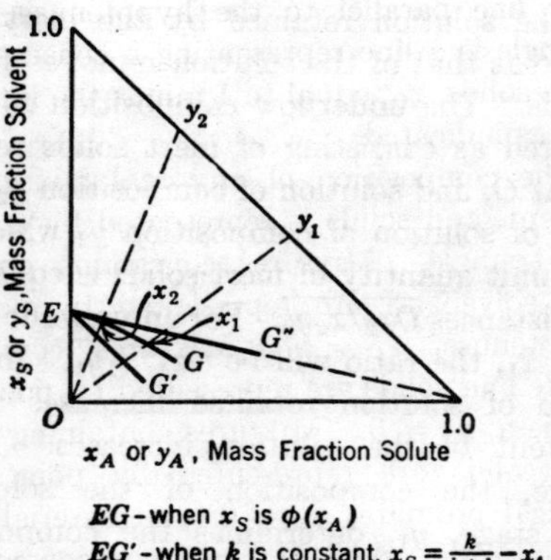

FIG. 298. Loci of underflow compositions.

represents the saturated solution of the solute at the temperature under consideration; the point O represents the inert solids free of solute or solvent. Any mixture of inert solids and saturated solution lies on the line $\overline{Oy_{\text{sat}}}$. Therefore the area to the left of the line, $\overline{Oy_{\text{sat}}}$, labeled "unsaturated" on the diagram, represents mixtures in which the solution is unsaturated since all points in this area are for smaller mass fractions of solute x_A than points to the right of the line $\overline{Oy_{\text{sat}}}$ in the area labeled "saturated."

This is demonstrated by any point such as N representing a mixture in the unsaturated region. If this mixture is separated into a solution free of solids, as represented by a point on the hypotenuse y_1, by removing the inert solids, represented by point O, the difference point y_1 represents a solution containing less solute than the saturated solution represented by point y_{sat}. On the other hand, the area labeled "saturated" on the diagram represents mixtures of saturated solution, undissolved solute, and inert solids. In the absence of supersaturated solutions any mixture in this area, such as P, consists of saturated solution with a solute content represented by point y_{sat}, undissolved solute and inert solids represented by point x_2.

Where the solute is originally present as a liquid, the solute may be miscible with the solvent in all proportions, and, since the mass fraction of solute y_A in the solutions may range from 0 to 1, the entire triangle represents an unsaturated area as the point y_{sat} is then at $x_A = 1$, $x_S = 0$.

If, however, the solute is not completely miscible with the solvent and two different solutions are formed, the triangle may be divided into three areas as shown in Fig. 297. Point y_{I} represents a saturated solution of solvent in the solute. Point y_{III} represents a saturated solution of solute in the solvent. The solution will consist of only one liquid phase in the areas labeled I and III. In the area labeled II, a mixture, such as represented by the point P, will consist of inert solids, represented by point O, and two solutions represented by the point y, indicating a composition between points y_{I} and y_{III}, which requires that point y represent a mixture consisting of two liquid phases, of compositions y_{I} and y_{III}. In this sense y is an addition point for the two immiscible liquid phases of compositions y_{I} and y_{III}.

In most solid-liquid extractions a single unsaturated solution is obtained as a product because of the desire to obtain a high recovery of the solute, or to obtain the inert solids relatively free from the solute.

The quantity of solution retained per unit quantity of the inert solids as determined experimentally under conditions similar to those which are being considered for the full-scale operation may be represented by a curved line such as EG in Fig. 298. This line represents the locus of the compositions of mixtures of inert solids and adherent solution and will be called the locus of underflow, or x, compositions. Point x_2 represents the underflow composition for which the solution retained has the composition represented by point y_2, provided the composition of the solution retained by the inert solids is the same as that of the solution removed from the inert solids. The underflow composition x_2 may be considered as consisting of inert solids represented by point O, and solution of composition y_2. The quantity of solution of composition y_2 which is retained per unit quantity of inert solids is equal to the ratio of distances $\overline{Ox_2}/\overline{x_2y_2}$. For an underflow of composition x_1, the ratio will be $\overline{Ox_1}/\overline{x_1y_1}$. In this case the ratio of solution retained increases as the solute content of the solution increases. For an ideal stage, the composition of the solution leaving the stage, y_1, determines the composition of the underflow x_1, which may be computed by constructing the straight line joining the point y_1 with O and finding its intersection with the curved line $\overline{EG}$ at x_1.

If the quantity of solution (including all of the solute) retained per unit quantity of the inert solids is constant, the locus of underflow compositions x is a straight line, $\overline{EG'}$ parallel to the hypotenuse of the triangle since x_C is constant. The intersection of the x or underflow line with the vertical ordinate $x_A = 0$, point E, represents the limiting case of a mixture containing zero solute and occurs at

$$x_S = \frac{k}{k+1}; \qquad x_A = 0$$

where k equals mass of solution retained per unit mass of inert solids. When k is a constant, the locus of the underflow compositions is a straight line represented by the equation

$$x_S = \frac{k}{(k+1)} - x_A$$

If the quantity of *solvent* (solute free) retained per unit quantity of inert solids is constant, the locus of underflow compositions will also be a straight line $\overline{EG''}$. Since $x_S = Kx_C$, the line passes through the points $[x_S = K/(K+1);\ x_A = 0]$ and $(x_S = 0;\ x_A = 1)$, where K equals mass of solvent retained per unit mass of inert solids. If solute is added to a mixture containing only inert solids and solvent, represented by E in Fig. 298, the resulting mixture by a material balance lies on the straight line joining points E and $(x_A = 1; x_S = 0)$ which line represents a constant ratio of solvent to inert solids and the locus of underflow compositions x for these conditions

Continuous Countercurrent Multiple-Contact Operation

A solid-liquid extraction system using continuous multiple-contact countercurrent operation is shown schematically in Fig. 299 for a system containing n ideal stages. When the system is operating under

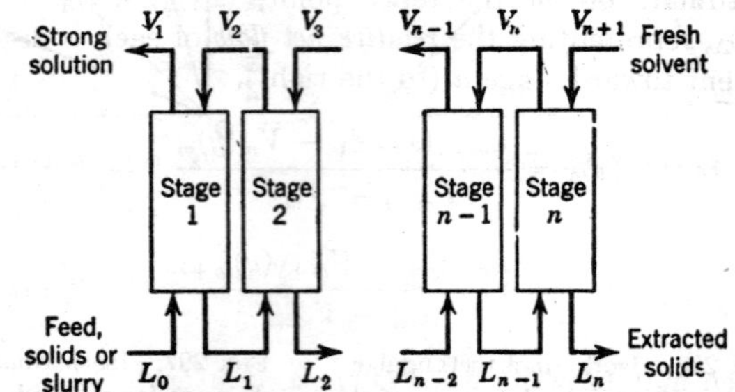

FIG. 299. Schematic diagram indicating the operation of a continuous countercurrent multiple-contact system.

steady-state conditions, the equations for the *material balances over the whole system* are as follows.

$$\text{Total} \qquad L_0 + V_{n+1} = L_n + V_1 \qquad (241)$$

For any component:

$$L_0x_C + V_{n+1}y_{n+1} = L_nx_n + V_1y_1 \qquad (242)$$

$$L_0(x_A)_0 + V_{n+1}(y_A)_{n+1} = L_n(x_A)_n + V_1(y_A)_1 \qquad (242A)$$

The subscripts are used to identify the stage from which a stream originates. The solids or slurry enter the system from an imaginary stage prior to stage one; accordingly the subscript 0 is used. The fresh solvent is introduced from an imaginary stage following stage n, and the subscript $n + 1$ is used.

The extracted solids or slurry leave the operation from the last stage, n. The fresh solvent enters the system on stage n, and the strong solution leaves from stage 1.

A material balance for any stage m in the system gives, with a simple rearrangement to express the differences in the flow of the streams between stages, what might be called the "net flow" toward the nth stage (to the right). For total streams,

$$L_{m-1} - V_m = L_m - V_{m+1} = L_{m+1} - V_{m+2} = \cdots \qquad (243)$$

For any component the net flow toward stage n (to the right)

$$L_{m-1}(x)_{m-1} - V_m(y)_m = L_m(x)_m - V_{m+1}(y)_{m+1} = \cdots \qquad (244)$$

If the difference in mass of any particular component as represented by equation 244 is divided by the corresponding difference in total streams as represented by equation 243, that is, dividing the net flow of a single component by the total net flow, the resulting difference in mass fraction is a coordinate of the difference point, $(x_A)_\Delta$, $(x_S)_\Delta$ or $(x_C)_\Delta$, representing the *relative net flow* of each component toward stage *n* (to the right).

$$(x)_\Delta = \frac{L_{m-1}(x)_{m-1} - V_m(y)_m}{L_{m-1} - V_m}$$

$$= \frac{L_m(x)_m - V_{m+1}(y)_{m+1}}{L_m - V_{m+1}} \qquad (245)$$

which may be written for (x_A), (x_S), or (x_C).

As has been previously indicated the compositions or coordinates of the difference point for the streams on either side of stage *m* are identical. The same difference point represents relative net flow for any stage or for all stages in the operation.

$$(x_A)_\Delta = \frac{L_0(x_A)_0 - V_1(y_A)_1}{L_0 - V_1}$$

$$= \frac{L_1(x_A)_1 - V_2(y_A)_2}{L_1 - V_2}$$

$$= \frac{L_n(x_A)_n - V_{n+1}(y_A)_{n+1}}{L_n - V_{n+1}} \qquad (245A)$$

Similar equations (245*S* and 245*C*) may be written for $(x_S)_\Delta$ or $(x_C)_\Delta$, respectively.

If the quantities and compositions of any two passing streams, as the terminal streams at the same end of the system, are known, the location of the difference point is fixed, since $(x_A)_\Delta$ and $(x_S)_\Delta$ or $(x_C)_\Delta$ are fixed by equations 245*A*, 245*S*, or 245*C*.

In solid-liquid extraction the coordinates of the difference point usually lie outside the triangle corresponding to a "negative" *relative net flow* for one component (negative, because the direction is opposite to the relative net flow of another component, toward stage *n*, or a "positive" relative net flow toward stage 0) and a *relative net flow* greater than unity for another. This is understandable from the definition of the difference point (equation 245) and a realization that the total amount of material corresponding to the difference point, $L_{m-1} - V_m$, or total net flow *toward* stage *n*, may be either positive or negative since the sign indicates the direction.

If the compositions of all terminal streams are fixed by choice or otherwise, the relative quantities of the streams and the number of ideal stages are thereby also fixed and may be determined graphically by the procedure outlined in Fig. 300. The difference point Δ lies on the straight line passing through the points x_0 and y_1 and also on the straight line passing

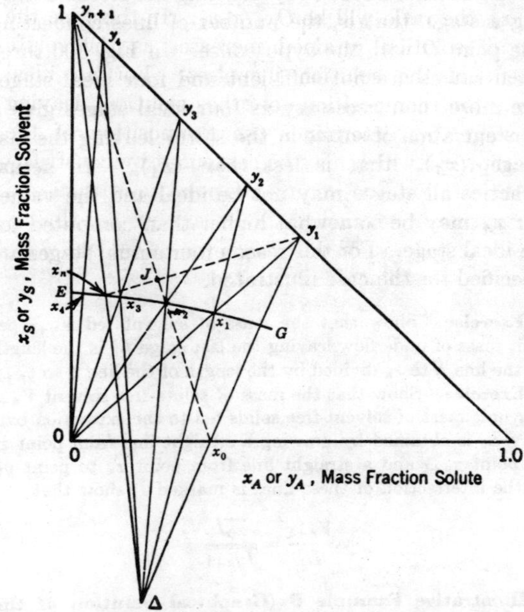

Fig. 300. Graphical solution of countercurrent multiple-contact solid-liquid extraction.

through the points x_n and y_{n+1}, since Δ is the difference point for the two terminal streams at stage 1 and also at stage *n* as shown from equation 245. Therefore, the difference point Δ is the intersection of the straight line passing through the points x_0 and y_1, and the straight line passing through the points x_n and y_{n+1}.

The graphical solution for the number of stages may be started at either end of the system (Fig. 299). Beginning at stage 1, the underflow leaving the ideal stage 1, L_1, has the composition x_1 which is on curve $\overline{EG}$ (Fig. 300) and also on the straight line connecting point y_1 with the point *O*, as has been explained, and therefore is determined as the intersection of these two lines. The composition y_2 of the overflow stream V_2 flowing from ideal stage 2 to stage 1 is on the hypotenuse of the triangle representing zero solids, and also on the straight line connecting the difference point Δ with x_1 or at the intersection indicated. Similarly, x_2 is determined as the intersection of the line $\overline{EG}$ with a straight line drawn through the points y_2 and 0. This procedure is continued as in Fig. 300 until an underflow solute composition x_A is obtained that is equal to or less than the required solute composition $(x_A)_n$.

The number of ideal stages required corresponds to the number of lines representing the streams leaving a stage, that is, the number of lines connecting the point O with the hypotenuse. In Fig. 300 three ideal stages are not sufficient and four ideal stages are more than necessary as four ideal stages give a concentration of solute in the slurry leaving the last stage, $(x_A)_4$, that is less than $(x_A)_n$. In actual practice all stages may not be ideal and the values for x_A may be somewhat higher than computed for an ideal stage. For this reason four actual stages are specified for the case illustrated.

Exercise. Show that the mass of solvent fed V_{n+1} per unit mass of underflow leaving the last stage L_n is the length of the line Δ to x_n divided by the length of the line Δ to y_{n+1}.

Exercise. Show that the mass of solute-free solvent V_{n+1} per unit mass of solvent-free solids fed to the extraction battery L_0 is obtained by drawing a straight line from point x_0 to point y_{n+1} and a straight line from point x_n to point y_1. If the intersection of these lines is marked J, show that

$$\frac{V_{n+1}}{L_0} = \frac{\overline{x_0 J}}{\overline{J y_{n+1}}}$$

Illustrative Example B (Graphical Solution of the Same Problem as Example A). It is desired to extract the oil from halibut livers by continuous countercurrent multiple-contact extraction with ethyl ether. The quantity of solution retained by the granulated livers is given in Table 29 and in the first two columns of Table 31. The halibut livers contain 0.257 mass fraction oil. If 95 per cent of the oil is to be extracted and the strong solution obtained from the system is to contain 0.70 mass fraction oil, determine:

1. The quantity and composition of the discharged solids.
2. The pounds of oil-free ether required per 1000-lb charge of fresh livers.
3. The number of ideal stages required.
4. The number of actual stages required if the overall stage efficiency is 70 per cent.

TABLE 31. CALCULATION OF UNDERFLOW COMPOSITIONS FOR ILLUSTRATIVE EXAMPLE B

Experimental Data		Underflow Compositions				
		Per Lb of Oil-Free Livers			Mass Fraction	
Lb Oil per Lb Solution, y_A	Lb Solution per Lb Oil-Free Livers	Lb of Oil, X_A	Lb of Ether, X_S	Total Mass of Underflow, L	Oil, x_A	Ether, x_S
0.00	0.205	0.0000	0.205	1.205	0.0000	0.1700
0.10	0.242	0.0242	0.218	1.242	0.0195	0.1753
0.20	0.286	0.0572	0.229	1.286	0.0435	0.1781
0.30	0.339	0.1017	0.237	1.339	0.0759	0.1770
0.40	0.405	0.1620	0.243	1.405	0.1152	0.1730
0.50	0.489	0.244	0.245	1.489	0.1642	0.1645
0.60	0.600	0.360	0.240	1.600	0.225	0.1500
0.65	0.672	0.437	0.235	1.672	0.261	0.1405
0.70	0.765	0.536	0.229	1.765	0.303	0.1298
0.72	0.810	0.583	0.227	1.810	0.322	0.1253

Solution by Graphical Method

The locus of the underflow compositions may be either calculated or determined graphically from the experimental data given. Table 31 includes a summary of the necessary calculations. Columns 1 and 2 are the original experimental data. Column 3 is the product of columns 1 and 2. Column 4 is the difference, obtained by subtracting the values in column 3 from those in column 2. The values in column 5 represent the total masses of the underflow streams and are obtained by adding 1.0 to the values in column 2. Columns 6 and 7, representing the underflow compositions, are obtained by dividing the values in columns 3 and 4, respectively, by the values in column 5. The values in columns 6 and 7 are plotted on the triangular diagram thus locating the locus of the underflow compositions.

The graphical method is shown in Fig. 301. For the sake of clarity, only two of the points representing the underflow

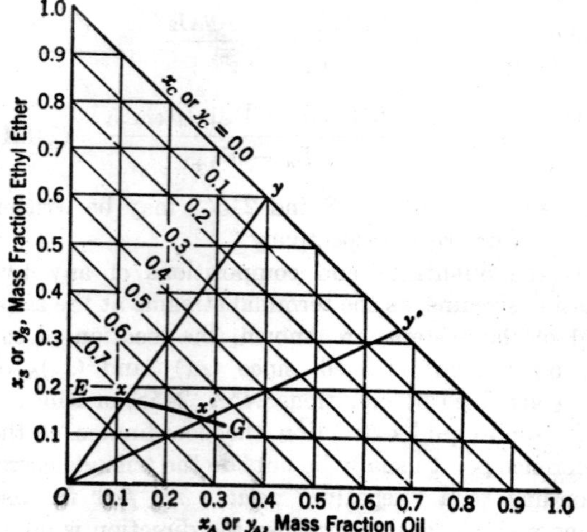

FIG. 301. Graphical determination of locus of underflow compositions for the illustrative example.

compositions are shown. The underflow composition x, corresponding to a solution overflow containing 0.40 mass fraction oil y_A, is obtained by constructing the line $\overline{Oy}$, measuring the total length of this line, and then dividing this length by the total mass of underflow (1 + 0.405), to determine the length $\overline{xy}$ which is measured along the line $\overline{Oy}$ from y and locates point x. The reason for this procedure may be made clearer by the following relationships.

$\overline{Oy}$ corresponds to the total mass of underflow, when the solution retained contains 0.400 mass fraction oil, which is 1.405 lb of underflow L per pound of oil-free livers (Table 31, column 5).

$\overline{xy}$ corresponds to the mass of oil-free livers in the underflow.

$$\frac{\overline{xy}}{\overline{Oy}} = \frac{1}{1.405} \quad \text{or} \quad \overline{xy} = \frac{\overline{Oy}}{1.405}$$

Other underflow compositions may be obtained in the same manner, the construction for point x' also being shown. A smooth curve is then drawn through the points x so obtained, thus locating the locus, EG.

The compositions of only three of the terminal streams are specified, namely, the halibut livers entering the system $(x)_0$, the solvent entering the system, $(y)_{n+1}$, and the strong solution leaving the system $(y)_1$. Since the composition of the fourth terminal stream is fixed by specifying the composition of three terminal streams, the composition of the extracted livers leaving the system $(x)_n$, may be obtained from a material balance. In order to recover 95 mass per cent of the oil in the livers, the total mass of oil left in the extracted livers will be $(1 - 0.95)(0.257) = 0.01285$ lb/lb of fresh liver. Since the oil-free solids are 0.743 lb, the mass ratio of oil to solvent-free liver and oil is

$$\frac{0.01285}{0.01285 + 0.743} = 0.017$$

and may be represented by point P, ($x_A = 0.017$, $x_S = 0$), as shown in Fig. 302. The extracted livers leaving the system contain ether, and the actual composition lies on the straight line joining point P with $x_S = 1.0$, the solvent apex. Since the composition of the extracted livers also lies on the locus of underflow compositions (line $\overline{EG}$), the point x_n is located by the intersection of the line joining P and $x_S = 1.0$ with the underflow composition curve EG. As read from Fig. 302, the composition of the extracted solids is:

Mass fraction ether	= 0.173
Mass fraction oil	0.013
Mass fraction oil-free solids (by difference)	= 0.814

The quantity of discharged solids is therefore

$$\frac{0.743}{0.814} = 0.913 \text{ lb of solids per pound of fresh livers}$$

The quantity of oil-free ether required may be obtained by a material balance. By means of the graphical method, the straight line joining x_0 and y_{n+1}, and the straight line joining x_n and y_1 are constructed. The point J located by the intersection of these lines is the addition point defined by dividing equation 242 by equation 241. The ratio of the mass of ether required to the mass of fresh livers may be obtained by determining the ratio of distances $\overline{x_0J}/\overline{Jy_{n+1}}$ on the diagram. From Fig. 302 this ratio is 0.26. Therefore, 0.260 lb of oil-free ether is required per pound of fresh liver.

The graphical solution for the number of ideal stages, starting from the strong solution end, stage 0, of the system, is shown in Fig. 302. Six equilibrium stages are not sufficient, whereas seven equilibrium stages give a slightly higher recovery of oil than desired. Estimating the equivalent of about 6.3 hypothetical ideal stages, the number of actual stages of 70 per cent overall efficiency required is estimated as 6.3/0.7 = 9 actual stages.

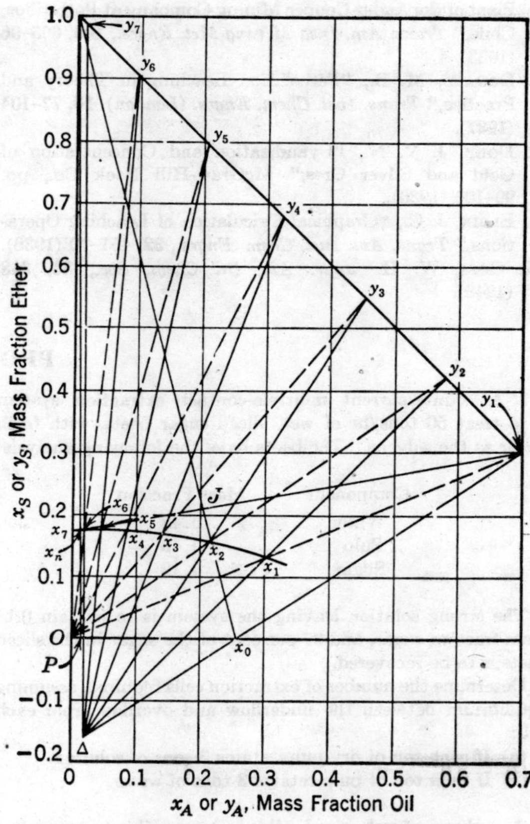

FIG. 302. Graphical solution for the illustrative example.

BIBLIOGRAPHY

1. ALDRICH, H. W., and W. G. SCOTT, "The Inspiration Leaching Plant," *Trans. Am. Inst. Mining Met. Engrs.*, **106**, 650–77 (1933).
2. ARMSTRONG, R. T., and K. KAMMERMEYER, "Countercurrent Leaching—Graphical Determination of Required Number of Units," *Ind. Eng. Chem.*, **34**, 1228–31 (1942).
3. BAKER, E. M., "Calculation Methods for Countercurrent Leaching," *Trans. Am. Inst. Chem. Engrs.*, **32**, 62–72 (1936).
4. BILBE, C. W., "Continuous Solvent Extraction of Vegetable Oils," *Mech. Eng.* **63**, 357–60 (1941).
5. CALLAWAY, L. A., and F. N. KOEPEL, "The Metallurgical Plant of the Andes Copper Mining Company at Potrerillos, Chile," *Trans. Am. Inst. Mining Met. Engrs.*, **106**, 695–96 (1933).
6. DONALD, M B., "Percolation Leaching in Theory and Practice," *Trans. Inst. Chem. Engrs. (London)*, **15**, 77–104 (1937).
7. DORR, J. V. N., "Cyandization and Concentration of Gold and Silver Ores," McGraw-Hill Book Co., pp. 99–106 (1936).
8. ELGIN, J. C., "Graphical Calculation of Leaching Operations," *Trans. Am. Inst. Chem Engrs.*, **32**, 451–70 (1936).

8*a*. GOSS, W. H., *Jour. Am. Oil Chem. Soc.*, **23**, 348 (1946).

9. HAWLEY, L. F., "Discontinuous Extraction Processes," *Ind. Eng. Chem.*, **9**, 866–71 (1917).
10. HAWLEY, L. F., "Numerical Relation between Cells and Treatments in Extraction Processes," *Ind. Eng. Chem.*, **12**, 493–96 (1920).
11. KINNEY, G. F., "Leaching Calculations—A Note on the Graphical Method," *Ind. Eng. Chem.*, **34**, 1102–04 (1942).
12. MARKLEY, K. S., and W. H. GOSS, *Soybean Chemistry and Technology*, Chemical Publishing Co., Inc., Brooklyn, pp. 175–182 (1944).
13. RANDALL, M., and LONGTIN, B., "Separation Processes—Analogy between Absorption, Extraction, Distillation, Heat Exchange, and Other Separation Processes," *Ind. Eng. Chem.*, **31**, 1295–99 (1939).
14. RAVENSCROFT, E. A., "Extraction of Solids with Liquids—Multiple and Countercurrent Extraction," *Ind. Eng. Chem.*, **28**, 851–55 (1936).
15. SANDERS, M. T., "Calculating Washing Efficiency in Countercurrent Decantation," *Chem. Met. Eng.*, **39**, 161–62 (1932).
16. *Manufacture of Pulp and Paper*, McGraw-Hill Book Co. (1937).
17. KENYON, R. L., N. F. KRUSE, and C. P. CLARKE, "Solvent Extraction of Oil from Soybeans," *Ind. Eng. Chem.*, **40**, 186–194 (1948).

PROBLEMS

1. A countercurrent multiple-contact extraction system is to treat 50 tons/hr of wet, sliced sugar beets, with fresh water as the solvent. The beets have the following analysis:

Component	Mass Fraction
Water	0.48
Pulp	0.40
Sugar	0.12

The strong solution leaving the system is to contain 0.15 mass fraction sugar, and 97 per cent of the sugar in the sliced beets is to be recovered.

Determine the number of extraction cells required, assuming equilibrium between the underflow and overflow from each cell:

(*a*) If each ton of dry pulp retains 3 tons of solution.
(*b*) If each ton of pulp retains 3 tons of water.

2. A slurry of pulverized solids in heavy oil is to be washed with a chlorinated solvent in a battery of continuous thickeners. The slurry is initially 60 per cent by volume oil. The underflow from each thickener contains 45 per cent by volume liquid. The specific gravities are: oil, 0.95; solvent, 1.3; solid, 3.0. The oil and solvent are completely miscible and form ideal solutions.

If 98 volume per cent of the oil is to be recovered from the slurry, what is the relation between the number of ideal stages and the volume ratio of solvent to slurry feed?

3. Roasted copper ore, containing the copper as cupric sulfate ($CuSO_4$), is to be extracted in a series of extractors, with water as the solvent. The system is loaded hourly with 10 tons of gangue, 1.2 tons of sulfate, and 0.5 ton of water in the ore. The strong extract solution is to contain 0.07 mass fraction of cupric sulfate, and 98 per cent of the cupric sulfate initially loaded into the system is to be recovered.

If each ton of inert gangue retains 2 tons of water, plus the cupric sulfate dissolved in that water, compute the required number of ideal or equilibrium extraction units.

4. Tung meal containing 55 mass per cent oil is to be extracted at a rate of 4000 lb/hr, using 6000 lb/hr of *n*-hexane, containing 5 mass per cent oil, as the solvent. A countercurrent multiple-stage extraction system which is equivalent to two ideal stages is to be employed. The meal will retain 1 lb of solution per pound of oil-free meal. The overflows will be a mixture of solution and fine meal particles with an estimated ratio of 0.05 lb of solids per pound of solution.

Determine the per cent recovery of the oil obtained under the above conditions. List any assumptions that may be necessary.

5. A slurry consisting of sodium chloride crystals and an aqueous solution of sodium hydroxide and sodium chloride is to be washed with saturated sodium chloride solution to remove the caustic solution from the crystals. The daily production of slurry, of the composition given below, is 200 tons.

Component	Mass %
NaCl (crystals)	40.0
NaCl (dissolved)	8.4
NaOH (dissolved)	9.6
H_2O	42.0

A four-compartment countercurrent washing classifier is to be used for the operation. Twenty-five tons per day of saturated sodium chloride solution (27 weight per cent NaCl) are to be used for the washing. Determine the quantity of sodium hydroxide leaving with the salt crystals from the classifier if the classifier is operated so that the solids discharged from the first three compartments contain 75 weight per cent NaCl crystals (25 weight per cent solution) and the solids discharged from the fourth compartment contain 85 weight per cent NaCl crystals (15 weight per cent solution).

The solubility of sodium chloride in caustic solutions at the temperature of operation is as follows.

Mass % NaOH	Mass % NaCl
0	27.0
2	25.3
4	23.6
6	21.9
8	20.3
10	18.7
12	17.1
14	15.6
16	14.1
18	12.6

6. The causticizing of soda ash follows the reaction:

$$Na_2CO_3 + Ca(OH)_2 = 2NaOH + CaCO_3$$

After most of the liquor containing the NaOH and some residual Na_2CO_3 or $Ca(OH)_2$ is decanted, an appreciable amount of NaOH still remains with the $CaCO_3$ sludge. The $CaCO_3$ is to be recovered and marketed as "precipitated chalk."

The sludge from the precipitation tanks contains 5 per cent $CaCO_3$ on a total mass basis, 0.1 per cent dissolved alkalinity (figured as NaOH), and a balance of water. One thousand tons per day of this sludge is fed continuously to two thickeners in series in which it is washed countercurrently with 200 tons/day of neutral water. The thickened pulp removed from the bottom of each of the thickeners contains 20 per cent solids. A filter takes the pulp from the last thickener and concentrates the solids to 50 per cent, returning the filtrate to the system as wash water.

Calculate the mass per cent alkalinity (as NaOH) remaining in the final $CaCO_3$ after it has been completely dried. Sketch the equipment and show flow paths, rates, and compositions.

7. A countercurrent extraction battery consisting of six Dorr thickeners is to be used to leach 20 tons/hr of material which contains 10 per cent moisture, 30 per cent insoluble gangue, and 60 per cent soluble salt. A saturated solution contains 40 per cent of salt at the leaching temperature. The material to be leached is fed to thickener No. 1. Thickener No. 6 will be fed 19 tons of water per hour. Thickener No. 4 will receive an intermediate feed of 9.5 tons of mother liquor per hour, containing 15.0 mass per cent salt.

Previous experience with this material in Dorr thickeners indicates the following relationships between the solution concentrations and the amount of liquid retained by the solids.

Mass % Salt in Solution	Pounds of Solution Retained per Pound of Dry Solid
0	2.78
5.0	2.37
10.0	2.20
20.0	2.07
30.0	2.03
40.0	2.03

Pure salt crystals carry with them 0.50 lb of solution per pound of dry salt.

Compute:

(*a*) Quantity and concentration of the strong solution produced.

(*b*) The mass per cent recovery as strong solution of all salt fed to the battery.

Recommend any changes which would improve this operation.

8. A continuous countercurrent extraction system is to be used to extract sodium nitrate from a mixture of sodium nitrate and sodium chloride containing 0.520 mass fraction sodium nitrate. The system is to be operated at 100° C. and is to process 50 tons/day of the mixture. The aqueous solution to be used for the extraction contains 0.100 mass fraction sodium nitrate and is saturated with respect to sodium chloride at 100° C. The slurry leaving the system is to contain 0.200 mass fraction sodium nitrate.

If 77.5 tons/day of solution are fed to the system and if the system is operated so that the underflow from each stage contains 1 lb of water per pound of solids, determine:

(*a*) The composition of the solution leaving the system.

(*b*) The number of ideal stages required.

SOLUBILITY DATA AT 100° C

SYSTEM: $NaNO_3$-NaCl-H_2O

Mass %		Solid Phase
$NaNO_3$	NaCl	
63.7	0	$NaNO_3$
60.6	2.8	$NaNO_3$
59.4	4.0	$NaNO_3$
57.5	5.6	$NaNO_3$ and NaCl
56.9	5.8	NaCl
55.7	6.2	NaCl
54.2	6.6	NaCl
50.7	7.7	NaCl
48.7	8.2	NaCl
43.0	10.2	NaCl
33.9	13.4	NaCl
24.0	17.6	NaCl
0	28.6	NaCl

9. A countercurrent extraction battery consisting of five Dorr thickeners gives the following test data when used to leach a soluble salt from an inert solid, with water as a solvent. Solid feed contains 63 per cent insolubles and 37 per cent salt and is supplied at the rate of 10 tons/hr to cell No. 5. Mother liquor containing 8.75 per cent salt is introduced as intermediate feed to cell No. 4 at the rate of 8.97 tons/hr. Water is supplied at the rate of 8.73 tons/hr to cell No. 1. Under these conditions, the extracted solids contain 3.35 per cent salt after drying, and the extract product in the form of a saturated solution containing 27.5 per cent salt is delivered at the rate of 16 tons/hr. The salt content of the overflow streams is as follows:

V_1	3.75 % salt
V_2	9.70
V_3	10.60
V_4	19.30

In view of the relatively high loss of salt in the extracted solids, this operation is considered to be unsatisfactory. Suggest any changes which should be made, and estimate the loss of salt in the extracted solids which you would expect from the corrected unit.

10. The orebody of the Union Minière du Haut-Katanga in the Belgian Congo is composed of malachite ($CuCO_3.Cu(OH)_2$) and gangue. Copper is extracted by crushing the ore to −20 mesh, agitating with a dilute solution of sulfuric acid, followed by multiple-contact countercurrent washing to wash the gangue free of the copper-bearing solution. The rich solution from the washing system is treated for removal of dissolved iron and aluminum and then sent to the electrolytic cells for precipitation of the copper. The dilute acid solution from the cells is recycled to the agitators for treatment of more raw ore.

The countercurrent washing operation uses Dorr thickeners for the recovery of the rich solution. The slurry from the agitators, with the copper in solution as copper sulfate, is fed to the thickeners at the rate of 300 tons per hour.

According to operating records, the underflow from each thickener retains 1.22 tons of solution per ton of gangue, and the streams have the following compositions in mass per cent.

	Feed to Thickener, %	Strong Solution, %	Underflow Leaving System, mass %
$CuSO_4$	6.10	6.69	1.0
Gangue	14.92	.....	99.0 (Gangue + Water)
Water	78.98	93.31	

(*a*) Sketch the flowsheet for the entire process, and include three Dorr thickeners.

(*b*) Determine the number of equivalent equilibrium stages and the quantity of wash water used.

11. A leaching battery, for some reason, operates as shown in the flow diagram below. The solid feed, one-half of which is to stage 1 and one-half to stage 3, contains 50 per cent solute and 50 per cent inert solids.

The solvent is pure water and is fed to stage 2 at a rate equal to the total feed rate. The stages produce slurries which contain 1 lb of solution per pound of solid. The salt forms a saturated solution with water containing 60 per cent salt.

Calculate the composition of all streams shown in the flow diagram.

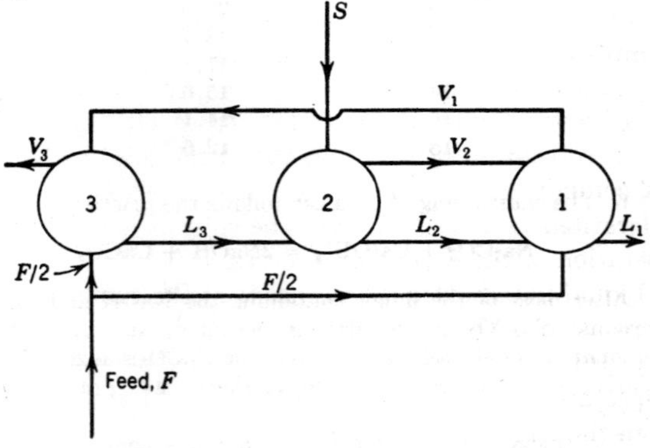

CHAPTER

22

Liquid-Liquid Extraction

LIQUID-LIQUID extraction is the term applied to any operation in which a material dissolved in one liquid phase is transferred to a second liquid phase. The solvent must be insoluble, or soluble to a limited extent only, in the solution to be extracted. If the solvent and the solution were completely miscible, there would be no opportunity for transferring the solute from the original solution to a second liquid phase. The degree of solubility of the solvent in the solution, and of the solution and its components in the solvent, are important considerations in the selection of the solvent and in the operation of the extraction process.

Liquid-liquid extraction consists of the same basic steps as solid-liquid extraction:

1. Intimate mixing or contact of the solvent with the solution to be treated, so as to transfer the solute from the solution to the solvent.

2. Separation of the liquid solution phase from the liquid solvent phase.

The complete extraction process may involve other operations such as the separation and recovery of the solvent from the solute and of the solvent which may be dissolved in the solution. The removal and recovery of the solvent may be more important in determining the successful application of the extraction process than the degree of extraction and separation accomplished in the two basic steps, particularly when special or costly solvents are employed. The separation and recovery of the solvent may be accomplished by various methods such as distillation or simple heating or cooling to diminish the solubility of the solute or of the solvent.

Liquid-liquid extraction is widely used for the separation of the components of a solution, particularly when:

1. The components are relatively nonvolatile.

2. The components have substantially the same volatilities from the mixture.

3. The components are sensitive to the temperatures required for the separation by distillation.

4. The desired less-volatile component is present in the solution only in relatively small amounts. In such a case, the less-volatile component may be extracted into a second solvent, producing a more concentrated solution from which it may be recovered more economically; or similar economies may be accomplished with a solvent less volatile than the desired component, even if the increase in concentration of the desired component is not particularly significant.

Usually the different distribution of the components of the feed mixture between the two layers at equilibrium is depended upon to effect the desired separation. The layer containing the greater concentration of the solvent and the smaller concentration of the feed liquid is referred to as the "extract" layer. The other layer, containing the greater concentration of the feed liquid and the smaller concentration of solvent is referred to as the "raffinate" layer. These terms are arbitrary and are used in different ways in different countries, so the reader of foreign literature should take care not to be confused.

The solvent is said to be selective for the component of the feed which is found in greater ratio to the other components in the *extract* than in either the *raffinate* or the original feed. The solvent is selective for component A if $(x_A/x_B)_F < (y_A/y_B)_1 > (x_A/x_B)_n$.

The residual materials remaining from the extract and raffinate layers after the major part of the solvent is removed are frequently referred to as the extract and the raffinate, respectively.

In operation either layer may be the upper or lower, depending upon their relative densities. Either of the liquid phases may contain dissolved solids. In some operations more than two liquid phases may be formed simultaneously, and solid phases and vapor phases may also be involved. The extraction may involve only the physical process of separation by solution, or chemical reactions between the extracted substances and the solvent or other materials dissolved in the solvent may take place.

Liquid-liquid extraction has been widely used for the removal of the naphthenic and aromatic constituents from the paraffinic constituents of lubricating oil stocks by means of a wide variety of solvents and solvent mixtures.[22] * The recovery of acetic acid from dilute aqueous solutions,[9] the separation of unsaturated constituents from vegetable oils such as soybean oil,[16] the purification of butadiene,[18] and the recovery of penicillin are other examples. A combination of liquid-liquid extraction with fractional distillation known as "extractive distillation" has become of increasing industrial importance in recent years for the separation of close-boiling compounds and azeotropic mixtures.[8]

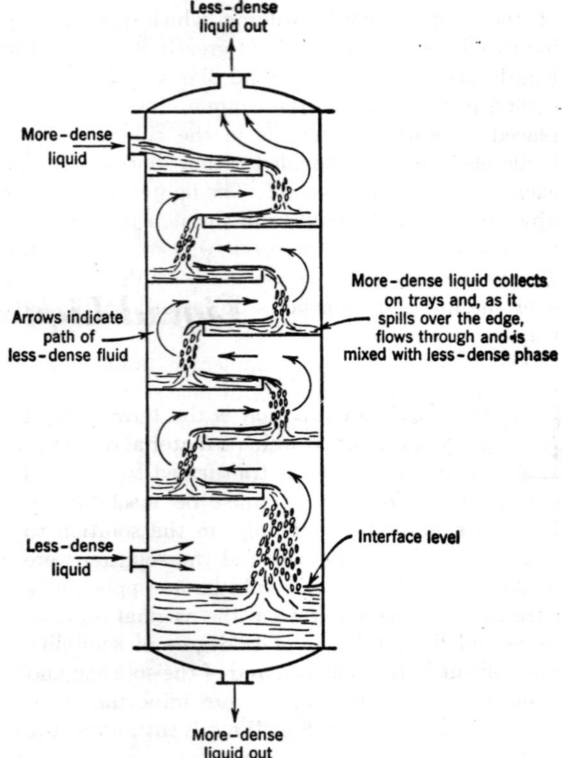

FIG. 303. Diagrammatic cross section of a baffle plate column for liquid-liquid extraction.

EQUIPMENT

All extraction equipment attempts to secure a large contact area between the phases since the rate of transfer of the distributed component is directly proportional to this area. Differentiation between the various types of equipment is based on the method employed for contacting the liquid phases.

Mixing the two phases by subdividing and dispersing one phase forms new surfaces with rapid transfer of the solute across the large contact area. The ease of mixing depends upon the interfacial tension between the two phases, the relative densities of the two phases, and the viscosity of each phase.

Separation of the two phases after they have been brought into contact may be accomplished by gravity or, less frequently, by centrifugal force. The ease of separation of the two phases depends primarily upon the difference in density of the two phases and their viscosities and may be appreciably affected by the presence of impurities which may stabilize emulsions.

* The bibliography for this chapter appears on p. 320.

An alternative procedure for obtaining a large contact area is to have one phase flow past the other but without any attempt to mix the two phases. The equipment used to contact the two phases also acts as the separator, and mixing of the two phases is undesirable. This method depends on the increased length of the path of flow to obtain a large contact area and usually requires less energy consumption than when the two phases are mixed. Such an extended contact area between phases may be obtained by flowing the two liquid phases countercurrently or concurrently in horizontal tubes or ducts with the less dense phase flowing in the top section of the tube, or in vertical tubes or ducts with the more dense phase flowing down the wall of the tube while the less dense phase flows upward through the center of the tube. The term "wetted-wall" tower is often used for the vertical type. In both the horizontal and vertical type of equipment the extended contact area is obtained by having the equipment of sufficient length.

The baffle-plate column (Fig. 303) is a modification of the simple vertical column which provides an increased contact area without greatly increasing the length of the column. Solid baffle plates which extend partially across the column cross section are placed at suitable intervals in the column. Each baffle plate may be provided with a short lip so that each is substantially a tray. The lighter liquid flows upward around the baffle and through the free area between the tray and the inside walls of the column. The heavier liquid flows along the baffle, overflows over the lip, and then flows downward to the next tray.

The capacity of such equipment is limited by the fact that the flow rates of both phases must be low enough to prevent mixing of the phases since both the contacting and the separation of the phases occur in the same piece of equipment. If centrifugal force is utilized to prevent mixing of the phases, higher flow rates may be used. The Podbielniak centrifugal extractor (Fig. 304) is based on this principle.

Subdivision and dispersion of one phase in the second phase may be accomplished by various types of mixers or columns.

Mixers include air agitators, mechanical mixers, and flow mixers and are used to mix the phases in one vessel with the separation accomplished in a separate vessel, except in batch operation when the same vessel may be used first for mixing and then for separation.

Columns may be of the spray, packed or plate type and are usually operated with countercurrent flow of the phases. Separation of the two phases takes place within the column itself and not in a separate vessel. The heavier liquid phase enters at

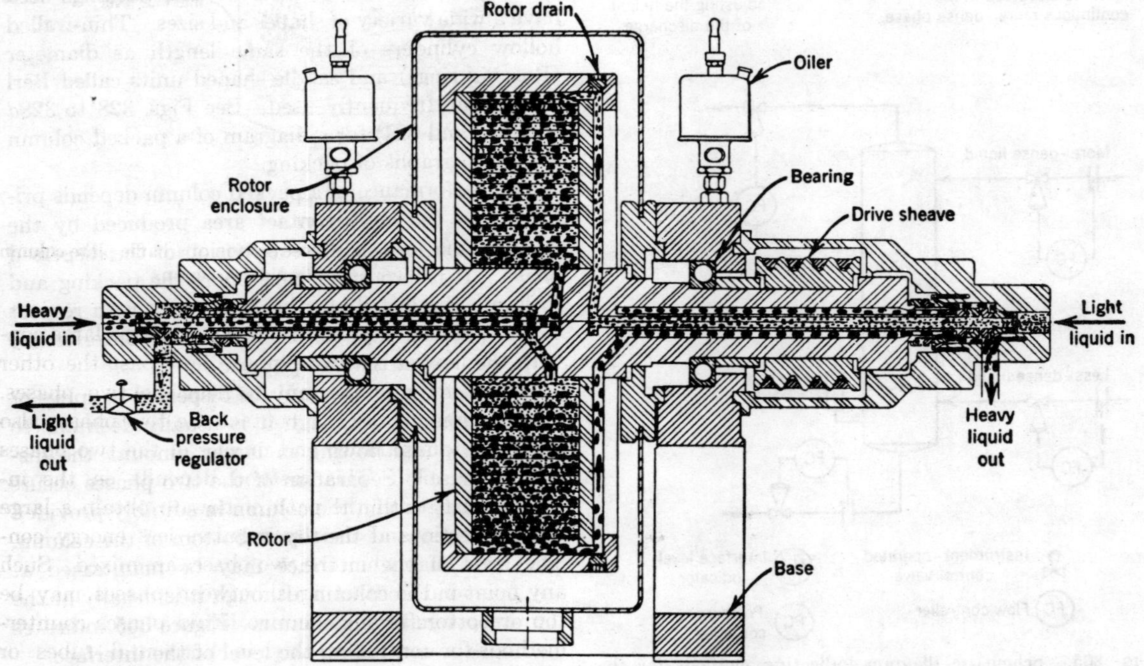

Fig. 304. Diagram of a centrifugal countercurrent extractor. Contacting of two liquid phases is effected continuously and countercurrently in the annular spaces of the rotor passageway while the rotor is revolving at high speeds. The lighter of the two liquid phases is admitted to the outside of the rotor, while the heavier liquid phase enters the center as shown. Because of the density difference between the two liquid phases, the heavier phase will flow to the outside of the rotor, displacing the lighter phase and causing it to flow towards the center. The action of the contacting elements within the rotor is to produce alternately many intimate dispersions of the liquid phases into each other and the immediate separation thereof. Because this process is carried out in a centrifugal force field ranging from 2000 to 5000 times gravity, tendencies to emulsify are greatly reduced. By maintaining a back pressure on the "light liquid out," the interface between the two contained liquids in the rotor can be controlled at any desired position. Capacities are from 25 gph to 1500 gph. (*Podbielniak Inc.*)

the top and leaves at the bottom; the lighter liquid phase flows upward in the opposite direction.

The spray column is the simplest type of equipment and consists of a suitable vessel, usually cylindrical

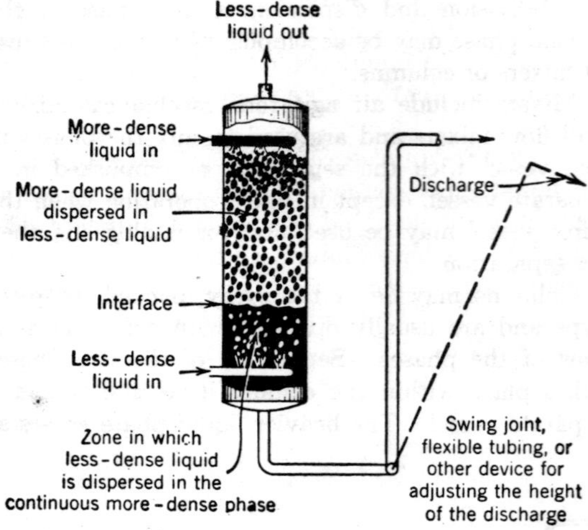

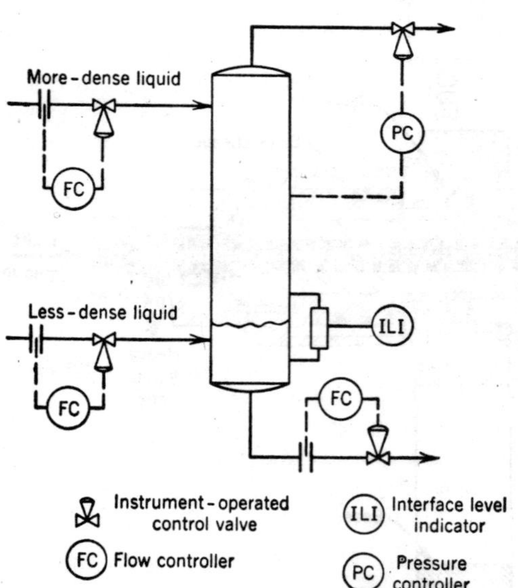

Fig. 305. Schematic diagram indicating methods for the control of the interface level in columns used for liquid-liquid extraction.

and considerably longer in length than in diameter, and equipped with nozzles or distributors for the phase to be dispersed. The continuous phase fills the vessel, and the dispersed phase flows through the continuous phase. The effectiveness of the extraction depends upon the subdivision of the dispersed phase into small droplets. Coalescence of these droplets after they are formed tends to decrease the effectiveness. Coalescence is favored by a high interfacial tension between the liquid phases and by high flow rates of the dispersed phase, that is, when the number of droplets per unit volume is large. Proper design of the spray entrance, so that the entrance is of the gradually expanding diffusion type (funnel type) rather than of the abrupt orifice type, will allow operation at higher rates of flow without coalescence of the droplets.[1] If the column cross section is constricted at the entrance by the introduction of the dispersing nozzle, the countercurrent velocity of the continuous phase between the droplets may tend to promote coalescence instead of keeping the droplets apart.

The packed column differs from the spray column in that the vessel is partially filled with suitable packing, either solid or hollow. The packings used have a wide variety of shapes and sizes. Thin-walled hollow cylinders of the same length as diameter (Raschig rings) and saddle-shaped units called Berl saddles are frequently used. See Figs. 328 to 328*d* (pp. 323 and 324) for a diagram of a packed column and photographs of packing.

The performance of a packed column depends primarily on the large contact area produced by the packing and less on the subdivision of the dispersed phase. It is important to support the packing and to locate the distributor head properly with respect to the packing support, so as to avoid interference with the flow of the two phases.[1]

Either of the phases may be dispersed in spray or packed columns, although it is usually desirable to make the phase flowing in larger amount the dispersed phase.[3] Separation of the two phases occurs continuously within the column in suitably provided settling sections at the top or bottom of the column. The level of the interface may be maintained at any point in the column although it is usually at the top or bottom of the column. Figure 305 indicates methods for controlling the level of the interface.

A plate column consists of a series of plates, spaced at regular intervals along the column, which serve to disperse one of the phases. Circular pipes extending above the bottom of the plates are used for the flow of the continuous phase from plate to plate (Fig. 306). If there is marked difference in the wetting of the column by the two liquids, successful operation is obtained only if the nonwetting phase is the dispersed phase. Separation of the phases occurs in each section between the plates, the interface level being maintained by the length of the downpipes through which the continuous phase flows.

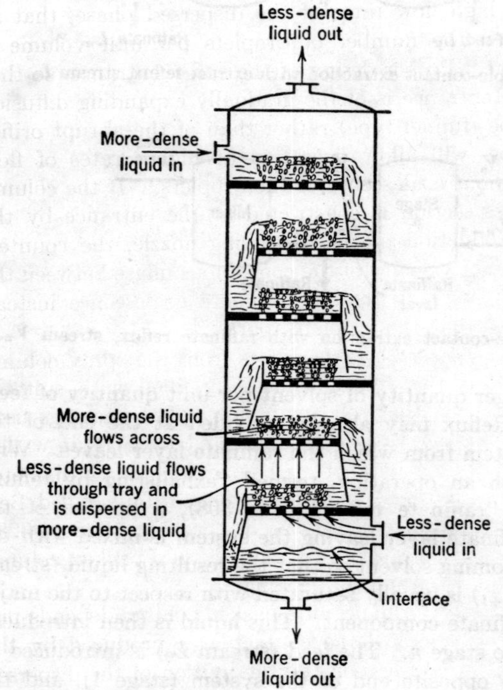

Fig. 306. Perforated plate, or sieve plate, column for liquid-liquid extraction. As shown, liquid flows downward through the pipes or segments and upward through the perforations. Columns are also constructed with the plates inverted, in which case the flow is upward through the pipe or segment and down through the perforations. The arrangement depends upon which phase is to be dispersed and is often determined by the interfacial properties of the fluids and the materials of construction.

The sieve-plate column corresponds, in a sense, to a series of short spray columns mounted one above the other. The dispersed liquid recombines above the plate and is then again subdivided and redispersed in passing through the perforations of the next plate. Since it has been shown that for single drops an appreciable amount (40 to 45 per cent) of the total extraction occurs as the drop is formed and before its release from the nozzle,[17] it is possible that the reported effectiveness of this type of equipment is due to the frequent formation of new surfaces at each of the plates. Decreasing the plate spacing improves the extraction efficiency but decreases the allowable throughput over a range of plate spacings from 3 to 9 in.[19] The recombination and redispersion of the dispersed phase which occurs absorbs energy which can be supplied only when the difference in density of the two phases is sufficient to overcome the increased resistances at the sieve plates, so that the energy consumption for this type may be considerably greater than for a spray column of the same height.

The bubble-cap plate column, Fig. 327, widely used for liquid-vapor contacting has been found unsatisfactory for liquid-liquid extraction as ordinarily designed.[15] The higher interfacial tension between two liquids as compared to that between a liquid and a vapor, and the lower agitation obtained due to higher viscosities when two liquid phases are contacted, probably account for the poor performance of bubble-cap columns when used for liquid-liquid extraction.

The flow rates of the liquid phases in columns used for liquid-liquid extraction are limited by the fact that after certain limiting flows have been reached the column will no longer function satisfactorily. This limiting flow is called the "flooding point" and depends on the flow rates of both the phases. At the flooding point the dispersed phase is backed out of the column and carried along with the continuous phase. In general, the allowable velocity of one phase decreases as the velocity of the other phase is increased. The physical properties of the liquid phases, the type and size of packing (for packed towers), the size of drops (for spray towers), the size and location of perforations (for plate columns), as well as the entrance conditions, all will affect the limiting rates.

Although considerable data on the performance of small-scale equipment are available, little information is available on the performance of large-scale equipment, and the design of such equipment at present is based on empirical procedures and on specific information for the liquid being extracted and the type of equipment being used.

METHODS OF OPERATION

The same methods of operation that were used for solid-liquid extraction may also be applied to liquid-liquid extraction, namely, (1) single contact, (2) simple multiple-contact, and (3) countercurrent multiple-contact. Countercurrent multiple-contact operation may be either batch or continuous. Because of the greater ease in handling liquid phases,

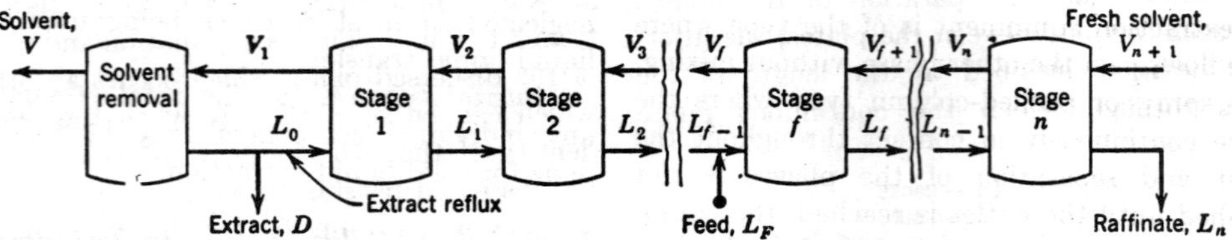

FIG. 307. Schematic diagram illustrating countercurrent multiple-contact extraction with extract reflux, stream L_0.

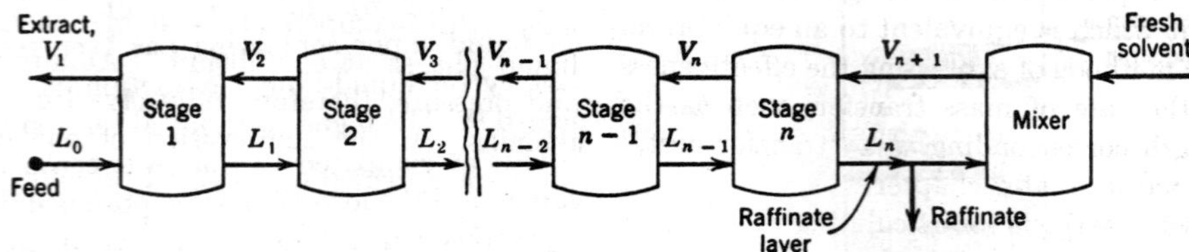

FIG. 308. Schematic diagram illustrating countercurrent multiple-contact extraction with raffinate reflux, stream V_{n+1}.

continuous operation is more commonly used in liquid-liquid extraction than in solid-liquid extraction.

In certain cases, the use of "reflux" with countercurrent multiple-contact operation may be advantageous. When countercurrent multiple-contact extraction without reflux is used, the maximum concentration of solute that can be obtained in the extract layer, even if an infinite number of stages is used, is that corresponding to equilibrium with the incoming feed. If the concentration of the solute in the feed is low, it may be desirable to obtain a higher concentration in the extract layer than corresponds to equilibrium with the feed. This may be accomplished by taking a portion of the extract layer from which some or all of the solvent has been removed and returning this to the system at the same stage from which the extract layer leaves. This type of operation has been termed "enriching by reflux" or "extract reflux[6,21] (Fig. 307). The feed (stream L_F) enters at an intermediate stage of the system. Fresh solvent (stream V_{n+1}) enters at stage n, and the raffinate layer (stream L_n) leaves from this stage. The extract layer (stream V_1) leaves stage 1 and is sent to the solvent removal unit where solvent is removed from the stream. The resulting material is then divided into two portions, the reflux (stream L_0) which is returned to stage 1 and the extract product (stream D). The use of extract reflux allows the extract layer leaving the system to come into contact with material which contains a higher concentration of solute than that of the feed, but accomplishes this by requiring a larger quantity of solvent per unit quantity of feed.

Reflux may also be provided at the end of the system from which the raffinate layer leaves. With such an operation, termed "exhausting by reflux" or "raffinate reflux" (Fig. 308), a portion of the raffinate layer leaving the system is mixed with the incoming solvent so that the resulting liquid (stream V_{n+1}) is usually saturated with respect to the major raffinate component. This liquid is then introduced into stage n. The feed (stream L_0) is introduced at the opposite end of the system (stage 1), and the extract layer leaves this same stage. The use of raffinate reflux has been said to improve the degree of removal of solute from the raffinate.[6,21]

Both extract and raffinate reflux may be employed simultaneously. However, the use of reflux is not generally applicable but is limited to certain types of ternary systems as will be considered under equilibrium relationships.

METHODS OF CALCULATION

As in solid-liquid extraction, the basis of calculation is the ideal stage. Since the two phases leaving an ideal stage are in equilibrium with each other, the "stage" includes that portion of the equipment which contacts and separates the two phases. In the case of liquid-liquid extraction where separate mixing and settling vessels are used, each combination of a mixing vessel and a settling vessel constitutes a stage which may or may not be an equilibrium stage, depending on whether or not equilibrium exists between the two phases leaving the stage.

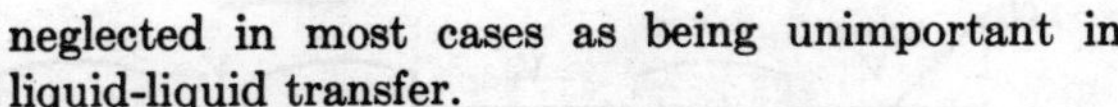

If the extraction equipment is of the type where one phase flows past the other phase without mixing, or of the spray or packed-column type where the phases are continuously in contact throughout the equipment and separation of the phases is not accomplished until the outlet is reached, the equipment may be equivalent to a number of equilibrium stages. In such cases the effectiveness of the equipment may be expressed as the height or length of the equipment which is equivalent to an equilibrium stage. Other methods of expressing the effectiveness in terms of the rate of mass transfer, such as the height or length corresponding to a "transfer unit," will be discussed in a later chapter.

In the present methods for calculating the number of equilibrium stages required for a liquid-liquid extraction, it is assumed that only the two liquid phases are present, and the system consists of three (or more) components, the two (or more) components present in the original solution, and the solvent used for the extraction. The presence of small quantities of impurities may be neglected as far as the equilibrium relationships are concerned. Any single stage operates at a constant temperature.

The calculation of the number of equilibrium stages required for a liquid-liquid extraction is based on the two principles of (1) material balances, and (2) equilibrium relationships.

Energy balances are generally not made because of the lack of data such as heats of solution, heat capacities, and surface energies, and these effects are neglected in most cases as being unimportant in liquid-liquid transfer.

Material balances for the total streams and for any individual component may be made as was done for solid-liquid extraction.

Equilibrium Relationships in Ternary Systems

According to the phase rule there are three degrees of freedom (three independent variables) for a ternary (three-component) system consisting of two liquid phases at equilibrium. At any temperature and pressure, therefore, only one degree of freedom remains. If the composition of one of the liquid phases is specified or fixed, all the independent variables are fixed, and the composition of the second liquid phase is also fixed and is not an independent variable provided that equilibrium conditions exist.

A ternary system composed of components *A*, *B*, and *C* may be regarded as composed of three binary systems—*A* and *B*, *B* and *C*, and *A* and *C*. The most common type of ternary-phase diagram is that for which two of the binary systems are completely miscible liquids and the third binary system is one in which the liquids are only partially miscible. The phase diagram for such a ternary system is shown in Figs. 309 and 310, which represent the data for the diphenylhexane-docosane-furfural system at 45° C and 1 atm.[2] Figure 309 is the equilateral triangle phase diagram customarily used, and Fig. 310 is the right triangle phase diagram for the same system as represented on rectangular coordinates.

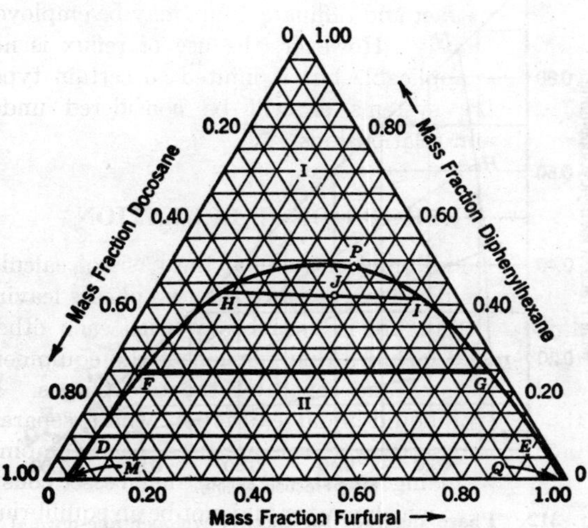

Fig. 309. Equilateral triangular phase diagram for diphenylhexane-docosane-furfural at 45° C and 1 atm.

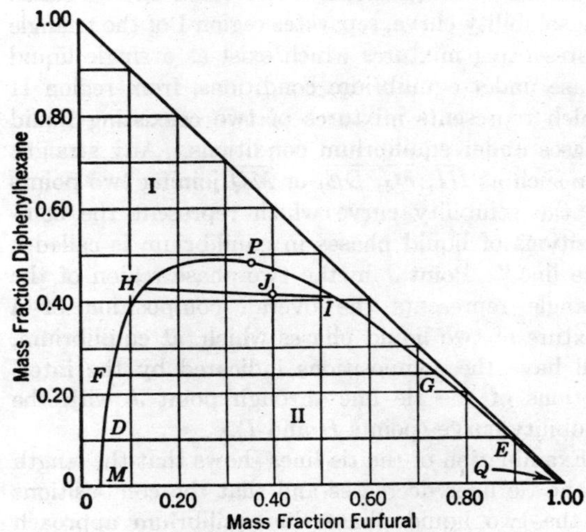

Fig. 310. Phase diagram for diphenylhexane-docosane-furfural at 45° C and 1 atm, plotted on rectangular coordinates.

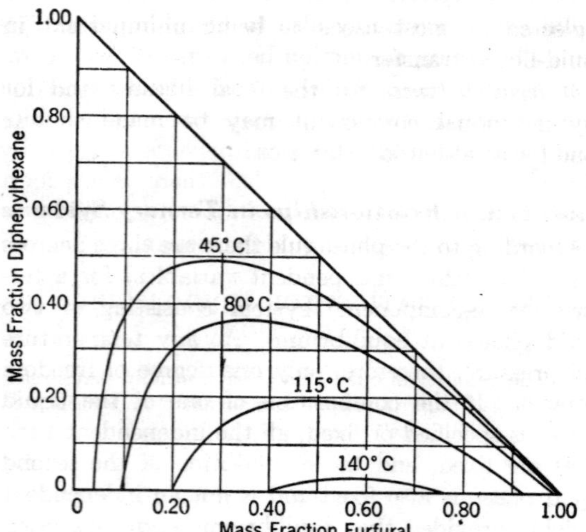

FIG. 311. Phase diagram for diphenylhexane-docosane-furfural, at 45° C, 80° C, and 115° C, at 1 atm.

At 45° C, diphenylhexane and docosane are completely miscible liquids, diphenylhexane and furfural are completely miscible liquids, and docosane and furfural are partially miscible liquids. The diphenylhexane, therefore, may be considered as being distributed between two liquid phases, one phase consisting primarily of docosane, the other phase consisting primarily of furfural. However, since docosane and furfural are partially miscible, all three components are present in each phase.

Any point inside the triangle represents a mixture of all three components. The curve $\overline{MPQ}$, called the solubility curve, separates region I of the triangle representing mixtures which exist as a single liquid phase under equilibrium conditions, from region II which represents mixtures of two coexisting liquid phases under equilibrium conditions. Any straight line such as $\overline{HI}$, $\overline{FG}$, $\overline{DE}$, or $\overline{MQ}$ joining two points on the solubility curve which represent the compositions of liquid phases in equilibrium is called a "tie line." Point J in the two-phase region of the triangle represents the overall composition of a mixture of two liquid phases which, at equilibrium, will have the compositions indicated by the intersections of the tie line through point J with the solubility curve (points H and I).

Examination of the tie lines shows that the length of the tie lines decreases and that the compositions of the two liquid phases in equilibrium approach each other as the diphenylhexane content of the overall mixture increases. The point corresponding to a tie line of zero length, that is, for which the two liquid phases in equilibrium have the same composition, is called the "critical" or "plait" point and is represented by point P.

The influence of pressure on the equilibrium in liquid phases may usually be neglected unless high pressures are considered. The effect of temperature is usually much more important as it may change the type of phase diagram, increase or decrease the area of the two-phase region without affecting the type of diagram, or change the slope of the tie lines. The effect of temperature on the phase diagram for diphenylhexane-docosane-furfural is shown in Fig. 311. For this system, an increase in temperature decreases the area of the two-phase region as indicated. The effect of temperature on the tie lines (not shown), on the other hand, is slight.

Another type of ternary-phase diagram which is not encountered so often is that for two binary systems which have partially miscible liquids and a third binary system in which the liquids are completely miscible. The phase diagram for the system methylcyclopentane-normal hexane-aniline at 25° C and 1 atm [4] is illustrated in Fig. 312. In this case the solubility curve is not a continuous curve but consists of two branches, $\overline{MN}$ and $\overline{QR}$, respectively, which extend from the side of the triangle to the hypotenuse. The two-phase region lies between the two branches, whereas the single-phase regions are

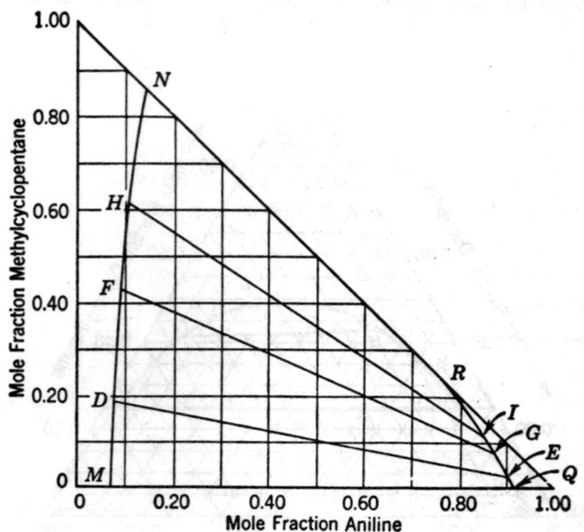

FIG. 312. Phase diagram for methyl-cyclopentane-normal hexane-aniline at 25° C and 1 atm.

outside the two branches. This particular diagram is of special interest because it permits the use of extract reflux. The equilibrium relationships for this system are very sensitive to temperature in the range from 25° to 45° C. If the temperature is increased to 45° C, the phase diagram (Fig. 313) is

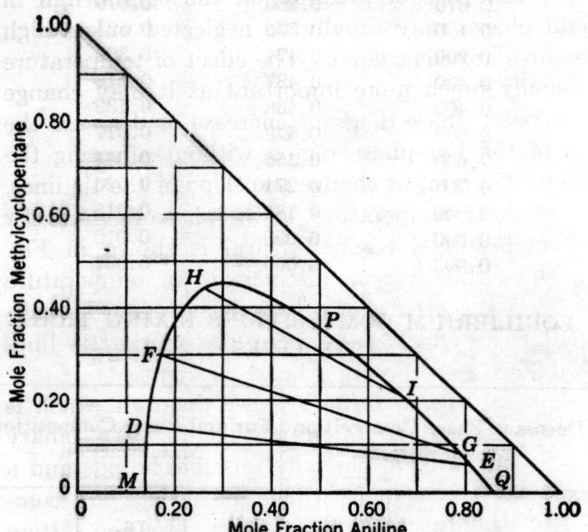

FIG 313. Phase diagram for methylcyclopentane-normal hexane-aniline at 45° C and 1 atm.

similar to the type first discussed, as the methylcyclopentane and aniline are completely miscible at this temperature.

If a separation is to be effected by liquid-liquid extraction, the overall composition of the mixture in any stage must lie in the two-phase region of the phase diagram.

Graphical Methods

Any liquid-liquid extraction problem can be solved by the repeated use of material balances and the equilibrium relationships according to the arithmetical method illustrated for solid-liquid extraction. The graphical method is simply a graphical representation of the material balances and equilibrium data and is similar to that used for solid-liquid extraction. The symbols used for liquid-liquid extraction are similar to those used in Table 28, Chapter 21, and are given below.

The representation of material balances, the concept of addition and difference points, and the general properties of the triangular diagram are the same in both cases. Consequently, all the relationships that have been derived (equations 236 to 245) for solid-liquid extraction apply also to liquid-liquid extraction. The main distinction between the two operations is that the overflow in solid-liquid extraction consists of only two components whereas in liquid-liquid extraction all three components are usually present in the overflow. Also, there is no inert material in liquid-liquid extraction corresponding to the inert solids in solid-liquid extraction.

NOMENCLATURE FOR LIQUID-LIQUID EXTRACTION

- L The quantity of one liquid stream, usually that stream consisting primarily of component C and designated the "raffinate" phase.
- V The quantity of the other liquid stream, usually that stream consisting primarily of component S and designated the "extract" phase.
- x Mass fraction in stream L of the component designated by a letter subscript, as x_A, x_C, or x_S.
- y Mass fraction in stream V of the component designated by a letter subscript, as y_A, y_C, or y_S.

In the diphenylhexane-docosane-furfural system at 45° C (Fig. 310), furfural may be used as the solvent for the extraction of a solution of diphenylhexane and docosane. The furfural will then be designated as component S. The equilibrium data show that the ratio

$$\frac{\text{Mass of diphenylhexane}}{\text{Mass of (diphenylhexane + docosane)}}$$

is greater in the extract (furfural) phase than in the raffinate (docosane) phase. Consequently, the furfural is selective for the diphenylhexane. The diphenylhexane will therefore be designated as component A, and the docosane as component C. The quantity of the extract phase from any stage will be designated by the symbol V, an appropriate subscript being used to indicate the stage from which the stream flows. Similarly, L will represent the quantity of raffinate phase leaving any stage.

Continuous Countercurrent Multiple-Contact Operation

The schematic diagram of a liquid-liquid extraction system using continuous countercurrent multiple-contact operation shown in Fig. 308 may be simplified by omission of the mixer, in which case V_{n+1} represents fresh solvent. When the system is operating under steady-state conditions, the material balances over the whole system are represented by equations identical to those obtained for solid-liquid extraction (equations 241 and 242).

Total material balance:

$$L_0 + V_{n+1} = L_n + V_1 \tag{241}$$

Material balance for any component:

$$L_0x_0 + V_{n+1}y_{n+1} = L_nx_n + V_1y_1 \tag{242}$$

Material balance for component A:

$$L_0(x_A)_0 + V_{n+1}(y_A)_{n+1} = L_n(x_A)_n + V_1(y_A)_1 \tag{242A}$$

Similarly, the equations defining the quantity and composition of the difference point are the same.

$$\Delta = L_0 - V_1 = L_1 - V_2 = \cdots$$
$$= L_n - V_{n+1} \tag{243}$$

$$L_0x_0 - V_1y_1 = L_1x_1 - V_2y_2 = \cdots$$
$$= L_nx_n - V_{n+1}y_{n+1} \tag{244}$$

$$x_\Delta = \frac{L_0x_0 - V_1y_1}{L_0 - V_1} = \frac{L_1x_1 - V_2y_2}{L_1 - V_2} = \cdots$$
$$= \frac{L_nx_n - V_{n+1}y_{n+1}}{L_n - V_{n+1}} \tag{245}$$

The graphical solution, therefore, will follow the same procedure as that described for solid-liquid extraction.

Illustrative Example. Diphenylhexane is to be separated from a mixture containing 0.200 mass fraction diphenylhexane and 0.800 mass fraction docosane by extraction with furfural at 45° C in a continuous countercurrent multiple-contact extraction system. The solvent entering the system contains 0.005 mass fraction diphenylhexane, the balance being furfural.

If the raffinate product leaving the system is to contain 0.010 mass fraction diphenylhexane, determine:

1. The number of equilibrium stages required when the mass ratio of solvent to feed is 1.66.
2. The mass ratio of solvent to feed that must be used if the extraction apparatus is equivalent to three equilibrium stages.
3. The maximum concentration of diphenylhexane that can be obtained in the extract product.

POINTS LOCATING THE SOLUBILITY OR SATURATION CURVE ENCLOSING THE TWO-PHASE REGION IN THE EQUILIBRIUM DIAGRAM[a]

Mass Fraction Furfural	Mass Fraction Diphenylhexane	Mass Fraction Docosane
0.040	0.000	0.960
0.050	0.110	0.840
0.070	0.260	0.670
0.100	0.375	0.525
0.200	0.474	0.326
0.300	0.487	0.213
0.400	0.468	0.132
0.500	0.423	0.077
0.600	0.356	0.044
0.700	0.274	0.026
0.800	0.185	0.015
0.900	0.090	0.010
0.993	0.000	0.007

EQUILIBRIUM COMPOSITIONS FIXING THREE TIE LINES[a]

Docosane Phase Composition Mass Fraction			Furfural Phase Composition Mass Fraction		
Furfural, x_S	Diphenylhexane, x_A	Docosane, x_C	Furfural, y_S	Diphenylhexane, y_A	Docosane, x_C
0.048	0.100	0.852	0.891	0.098	0.011
0.065	0.245	0.690	0.736	0.242	0.022
0.133	0.426	0.439	0.523	0.409	0.068

The composition of the entering feed (stream L_0) and the composition of the entering solvent (stream V_{n+1}) are given and may be located on the triangular diagram as points x_0 and y_{n+1} in Fig. 314, since

For feed stream,

$$(x_A)_0 = 0.200, \qquad (x_C)_0 = 0.800, \qquad (x_S)_0 = 0.00$$

For solvent stream,

$$(y_A)_{n+1} = 0.005, \qquad (y_C)_{n+1} = 0.000, \qquad (y_S)_{n+1} = 0.995$$

The raffinate product leaving the system (stream L_n) is in equilibrium with the extract phase (stream V_n) leaving that equilibrium stage, and the point x_n, representing the composition of the raffinate product, may then be located on the solubility or equilibrium curve since the mass fraction of diphenylhexane is given as 0.010, that is $(x_A)_n = 0.010$. The composition, y_1, of the extract phase leaving the system then remains to be determined.

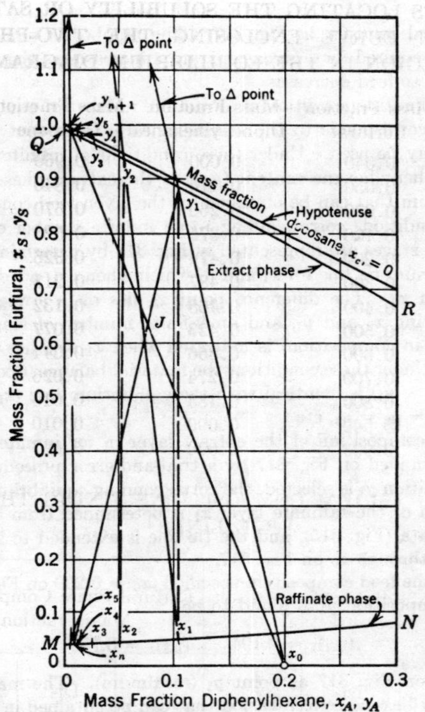

FIG. 314. Graphical determination of number of equilibrium stages required when the solvent-to-feed ratio is 1.66.

Part 1. The addition point x_J (Fig. 314) is located on the line joining points x_0 and y_{n+1} so that

$$\frac{\text{Mass of solvent}}{\text{Mass of feed}} = \frac{\overline{x_0 x_J}}{\overline{x_J y_{n+1}}} = 1.66$$

The point y_1 is then located at the intersection of the straight line through points x_n and x_J with the equilibrium curve. The composition is $(y_A)_1 = 0.1015$, and $(y_S)_1 = 0.882$. The difference point, x_Δ, is located at the intersection of the material balance or operating line through points x_0 and y_1, and that through points x_n and y_{n+1} as indicated in Fig. 314.

The number of equilibrium stages required is determined by making the equilibrium and material balance calculations as described until the desired value for $(x_A)_n = 0.01$ is obtained. From the equilibrium data plotted in Fig. 315, for $(y_A)_1 = 0.1015$, $(x_A)_1$ is found to be 0.104.

The material balance is made by drawing the straight line (Fig. 314) joining points x_1 and x_Δ. The intersection of this line with the equilibrium curve at point y_2 [$(y_A)_2 = 0.051$] gives the composition of the extract stream leaving stage 2. The process of equilibrium calculation to find x_2 and material balance to find y_3 is repeated until a point x is obtained which has a composition (x_A) equal to or lower than $(x_A)_n = 0.010$. Approximately 4.6 such steps corresponding to about 4.6 hypothetical equilibrium stages are required to reach this point as illustrated in Fig. 314. The equilibrium compositions for each stage as so computed are as follows:

Stage Number	Mass Fraction Diphenylhexane	
	Extract Phase, y_A	Raffinate Phase, x_A
1	0.1015	0.104
2	0.051	0.053
3	0.027	0.028
4	0.013	0.0135
5	0.0075	0.0080

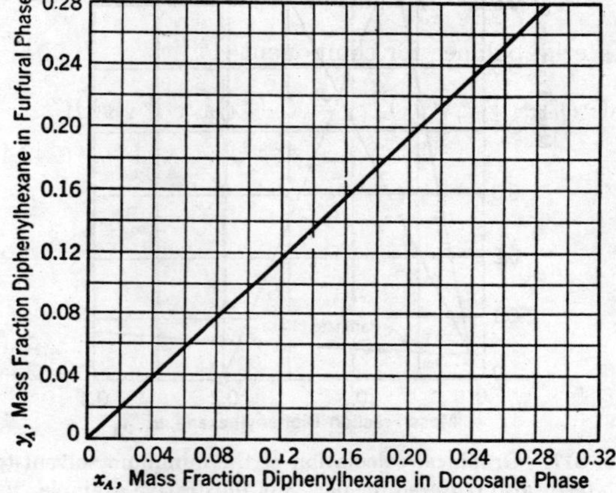

FIG. 315. Equilibrium distribution of diphenylhexane between furfural and docosane at 45° C.

Part 2. When the number of equilibrium stages to be used is specified, a trial-and-error solution is necessary in order to determine the necessary mass ratio of solvent to feed and the composition of the extract product, y_1. The simplest procedure is to assume a value for the composition of the extract product, y_1, and determine the number of equilibrium stages required. This procedure is repeated several times so that the number of equilibrium stages required may be plotted against the solvent-to-feed ratio used, as in Fig. 316. The required solvent-to-feed ratio is then read from this chart for the specific number of equilibrium stages.

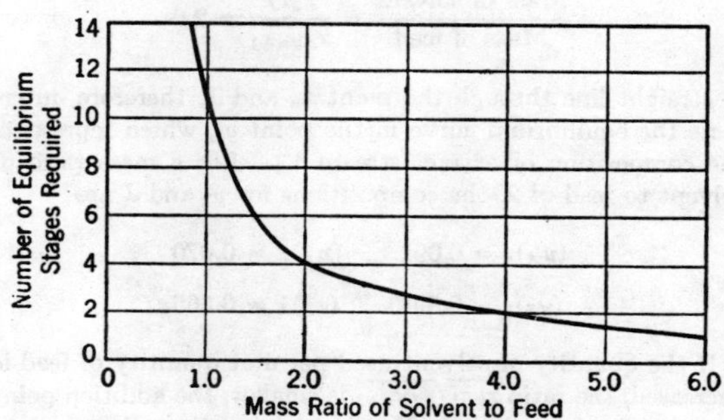

FIG. 316. The number of equilibrium stages required as a function of the solvent-to-feed ratio.

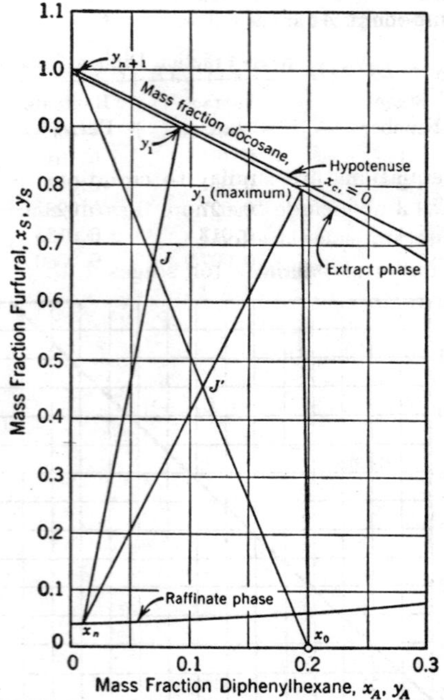

FIG. 317. Graphical calculation of the minimum solvent-to-feed ratio required for part 3 of illustrative example.

From the information obtained from part 1 it is evident that a solvent-to-feed ratio greater than 1.66 must be used if 3 equilibrium stages are desired. This corresponds to a composition, $(y_A)_1$, less than 0.1015 mass fraction diphenylhexane in the extract product. If a value of $(y_A)_1 = 0.070$ is selected, the number of equilibrium stages required is found to be approximately 3, and the mass ratio of solvent to feed is 2.72.

Part 3. Consider the effect of adding various quantities of solvent to the system per unit quantity of feed. As an example, suppose that the mass ratio of solvent to feed is 2.0. If x_J denotes this addition point for streams L_0 and V_{n+1} on the diagram (Fig. 317), it must fulfill the condition

$$\frac{\text{Mass of solvent}}{\text{Mass of feed}} = \frac{\overline{x_0 x_J}}{\overline{x_J y_{n+1}}} = 2.0$$

A straight line through the point x_n and J, therefore, intersects the equilibrium curve in the point y_1, which represents the composition of extract stream V_1. For a mass ratio of solvent to feed of 2, the compositions for y_1 and J are

$$(y_A)_1 = 0.091 \qquad (x_A)_J = 0.070$$

$$(y_S)_1 = 0.894 \qquad (x_S)_J = 0.665$$

If the quantity of solvent used per unit quantity of feed is decreased, the ratio $\overline{x_0 x_J}/\overline{x_J y_{n+1}}$ is smaller; the addition point x_J is located nearer x_0 on the line joining the points x_0 and y_{n+1}; and the composition of the extract layer leaving the equipment y_1 increases in diphenylhexane as indicated by the graphical solution in Fig. 317. The number of equilibrium stages required for a given separation increases as the ratio of solvent to feed decreases.

The limiting condition of an infinite number of equilibrium stages corresponds to the minimum ratio of solvent to feed that may be used. Under these conditions of infinite stages, the diphenylhexane content $(y_A)_1$ of the extract phase is the maximum that can be obtained for the given feed conditions. The conditions corresponding to an infinite number of equilibrium stages are represented in Fig. 317 by determining the equilibrium tie line for x_1 and y_1 which when extended passes through x_0. The difference point x_Δ lies on a straight line connecting x_0 and y_1, and an infinite number of stages (no change in composition) is indicated when x_1 also lies on this line. Under these conditions the material balance (operating) line $(x_0 - y_1)$ is identical with the equilibrium line $(x_1 - y_1)$, and $y_1 = y_2 = y_3$, etc.

The composition of the extract layer y_1 for infinite stages is determined on Fig. 317 by a trial-and-error procedure. A composition y_1 is selected, the corresponding equilibrium composition of the raffinate layer x_1 is determined from equilibrium data (Fig. 315), and the tie line is extended to see if it passes through x_0 on Fig. 317.

For the feed composition specified ($x_0 = 0.200$ on Fig. 317), the composition y_1 is found to be

$$(y_A)_1 = 0.194 \qquad (y_S)_1 = 0.786$$

shown on Fig. 317 as point y_1 (maximum). The maximum diphenylhexane concentration that can be obtained in the extract product is therefore 0.194 mass fraction.

The straight line joining points x_n and y_1 (maximum) intersects the line joining points x_0 and y_{n+1}, at the addition point $x_{J'}$. The minimum mass ratio of solvent to feed, determined from Fig. 317, is

$$\frac{\text{Mass of solvent}}{\text{Mass of feed}} = \frac{\overline{x_0 x_{J'}}}{\overline{x_{J'} y_{n+1}}} = 0.83$$

Exercise. Diphenylhexane is to be separated from a mixture containing 0.200 weight fraction diphenylhexane and 0.800 weight fraction docosane by extraction with furfural at 45° C. A simple multiple-contact extraction system is to be used. The furfural used contains 0.005 weight fraction diphenylhexane. Three equilibrium stages are to be used. The total furfural used is to be 2.7 lb/lb of feed and is to be equally divided among each stage. Determine the quantity and composition of all streams leaving the system based on 1000 lb of feed entering the system. Compare the composition of the raffinate product leaving the system with that obtained for the Illustrative Example, Part 2.

Intermediate Feed. A liquid-liquid extraction system may be operated with more than one feed, using more than one solution from which it is desired to extract a given component. The feed stream containing the highest concentration of the distributed component is introduced into stage 1 and is designated L_0. The feed stream containing the lowest concentration of the distributed component is introduced at a later intermediate stage. The stage into which the intermediate feed is introduced is labeled stage f. The quantity of intermediate feed is designated by L_F.

Under steady-state operating conditions, the material balances may be made graphically as shown in

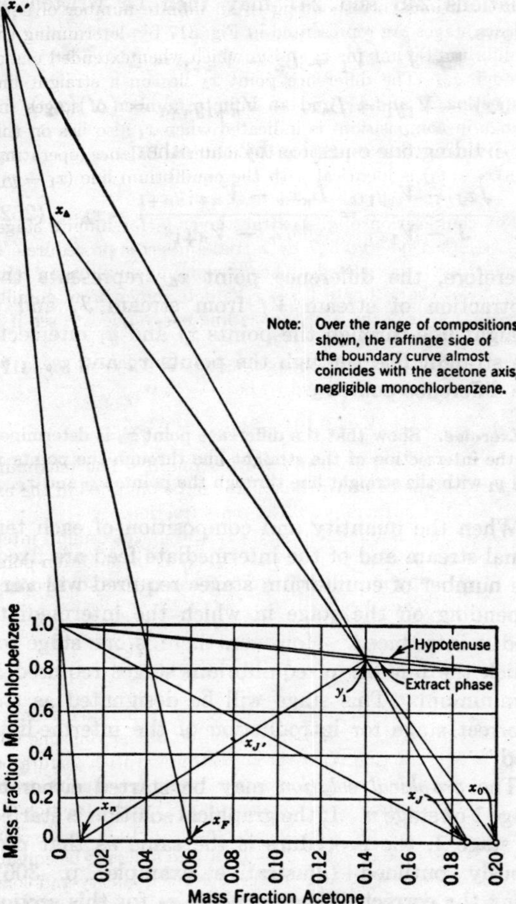

FIG. 318. Graphical material balances to locate difference points for intermediate feed.

Fig. 318, or by the following algebraic equations. In fact, the following equations are simply the algebraic expressions for the points and lines on Fig. 318.

Total material balance:

$$L_0 + L_F + V_{n+1} = L_n + V_1 \quad (246)$$

For any component:

$$L_0x_0 + L_Fx_F + V_{n+1}y_{n+1} = L_nx_n + V_1y_1 \quad (247)$$

For component A:

$$L_0(x_A)_0 + L_F(x_A)_F + V_{n+1}(y_A)_{n+1} = L_n(x_A)_n + V_1(y_A)_1 \quad (247A)$$

These equations are similar to equations 241, 242, and 242A but include one more term because of the additional intermediate feed.

The material balances for stages 1 to $f - 1$ inclusive are:

Total material balance:

$$L_0 + V_f = L_{f-1} + V_1 \quad (248)$$

For any component:

$$L_0x_0 + V_fy_f = L_{f-1}x_{f-1} + V_1y_1 \quad (249)$$

For component A:

$$L_0(x_A)_0 + V_f(y_A)_f = L_{f-1}(x_A)_{f-1} + V_1(y_A)_1 \quad (249A)$$

Equations 248, 249, and 249A may be used to define a difference point Δ.

$$\Delta = L_0 - V_1 = L_1 - V_2 = \cdots = L_{f-1} - V_f \quad (250)$$

$$L_0x_0 - V_1y_1 = L_1x_1 - V_2y_2 = \cdots = L_{f-1}x_{f-1} - V_fy_f \quad (251)$$

$$x_\Delta = \frac{L_0x_0 - V_1y_1}{L_0 - V_1} = \frac{L_1x_1 - V_2y_2}{L_1 - V_2} = \cdots = \frac{L_{f-1}x_{f-1} - V_fy_f}{L_{f-1} - V_f} \quad (252)$$

In the same manner, the material balances for stages $f + 1$ to n are:

Total material balance:

$$L_f + V_{n+1} = L_n + V_{f+1} \quad (253)$$

For any component:

$$L_fx_f + V_{n+1}y_{n+1} = L_nx_n + V_{f+1}y_{f+1} \quad (254)$$

For component A:

$$L_f(x_A)_f + V_{n+1}(y_A)_{n+1} = L_n(x_A)_n + V_{f+1}(y_A)_{f+1} \quad (254A)$$

Equations 253, 254, and 254*A* may be used to define a second difference point Δ'.

$$\Delta' = L_f - V_{f+1} = L_{f+1} - V_{f+2} = \cdots$$
$$= L_n - V_{n+1} \quad (255)$$
$$L_f x_f - V_{f+1} y_{f+1} = L_{f+1} x_{f+1} - V_{f+2} y_{f+2}$$
$$= \cdots = L_n x_n - V_{n+1} y_{n+1} \quad (256)$$
$$x_{\Delta'} = \frac{L_f x_f - V_{f+1} y_{f+1}}{L_f - V_{f+1}}$$
$$= \frac{L_{f+1} x_{f+1} - V_{f+2} y_{f+2}}{L_{f+1} - V_{f+2}} = \cdots$$
$$= \frac{L_n x_n - V_{n+1} y_{n+1}}{L_n - V_{n+1}} \quad (257)$$

The relationship between the two difference points Δ and Δ' may be obtained by rearrangement of equations 246 and 247.

$$L_0 - V_1 + L_F = L_n - V_{n+1} \quad (246)$$
$$L_0 x_0 - V_1 y_1 + L_F x_F = L_n x_n - V_{n+1} y_{n+1} \quad (247)$$

Since

$$\Delta = L_0 - V_1 \quad \text{(equation 250)}$$

and

$$\Delta' = L_n - V_{n+1} \quad \text{(equation 255)},$$

then

$$\Delta + L_F = \Delta' \quad (258)$$

Division of equation 247 by equation 246 gives

$$\frac{L_0 x_0 - V_1 y_1 + L_F x_F}{L_0 - V_1 + L_F} = \frac{L_n x_n - V_{n+1} y_{n+1}}{L_n - V_{n+1}}$$

By substituting from equations 250, 251, 255, and 256 in the above equation

$$\frac{\Delta(x_\Delta) + L_F x_F}{\Delta + L_F} = \frac{(\Delta')(x_{\Delta'})}{\Delta'} = x_{\Delta'} \quad (259)$$

Equations 258 and 259 represent the mathematical conditions for the fact that the points representing the compositions of streams Δ, Δ', and L_F all lie on the same straight line.

Because of the presence of the intermediate feed, two difference points are necessary. One difference point, x_Δ, applies to the stages up to but not including the intermediate feed stage. The other difference point, $x_{\Delta'}$, applies to the stages after the intermediate feed stage. The location of these difference points may be determined from the material balances which have been written. If x_J is the addition point as defined by the following equations:

$$L_0 + L_F = J \quad (260)$$
$$L_0 x_0 + L_F x_F = J x_J \quad (261)$$

equations 246 and 247 may then be written as follows.

$$J - V_1 = L_n - V_{n+1} = \Delta'$$
$$J x_J - V_1 y_1 = L_n x_n - V_{n+1} y_{n+1} = (\Delta')(x_{\Delta'})$$

By dividing one equation by the other

$$\frac{J x_J - V_1 y_1}{J - V_1} = \frac{L_n x_n - V_{n+1} y_{n+1}}{L_n - V_{n+1}} = x_{\Delta'} \quad (262)$$

Therefore, the difference point $x_{\Delta'}$ represents the subtraction of stream V_1 from stream J, and a straight line through the points x_J and y_1 intersects the straight line through the points x_n and y_{n+1} at the difference point $x_{\Delta'}$.

Exercise. Show that the difference point x_Δ is determined by the intersection of the straight line through the points x_0 and y_1 with the straight line through the points $x_{\Delta'}$ and x_F.

When the quantity and composition of each terminal stream and of the intermediate feed are fixed, the number of equilibrium stages required will vary depending on the stage in which the intermediate feed is introduced. However, there is one stage for which the number of equilibrium stages required is a minimum. This stage will be designated as the "correct stage for introduction of the intermediate feed."

The *graphical solution* may be started either at stage 1 or stage n. If the graphical solution is started at stage 1, the procedure is the same as that previously outlined (illustrative example, p. 306), using the correct difference point x_Δ for this section of the system (Fig. 318). This procedure is followed until the stage is reached where the equilibrium tie line for the stage crosses the straight line joining the points x_Δ, x_F, and $x_{\Delta'}$. If the point x representing the composition of the raffinate phase leaving this stage and the difference point x_Δ are used, it is seen that the change in composition effected by the next stage will be less than if the point x and the difference point $x_{\Delta'}$ are used. Consequently, this represents the stage where the intermediate feed should be introduced and the change from difference point x_Δ to the difference point $x_{\Delta'}$ is made at this stage. The same procedure is followed, but using the difference point $x_{\Delta'}$, until a raffinate phase composition is obtained which is equal to or less than the specified value.

Illustrative Example. A countercurrent multiple-stage extraction system at 25° C is to be used for the recovery of acetone from an aqueous solution containing 0.200 mass fraction acetone. Monochlorbenzene has been selected as the solvent. The raffinate product is to contain 0.010 weight fraction acetone.

A saturated water phase containing 0.060 mass fraction acetone is also to be treated in the extraction system at a rate of 0.1 lb/lb of solution containing 0.200 mass fraction acetone.

If the extract product is to contain 0.140 mass fraction acetone, determine:

1. The equilibrium stage in which the saturated water phase should be introduced.
2. The number of equilibrium stages required.
3. The weight of monochlorbenzene required per pound of 0.200 weight fraction acetone solution.

The equilibrium data [13] for the above system are given in Table 32.

TABLE 32. CORRESPONDING EQUILIBRIUM POINTS ON THE SATURATION CURVE ENCLOSING THE TWO-PHASE AREA FOR THE SYSTEM ACETONE-WATER-MONOCHLORBENZENE AT 1 ATM PRESSURE AND 25° C [13]

Compositions in Mass Fraction

Water Layer			Chlorbenzene Layer		
Acetone, x_A	Water, x_C	Monochlorbenzene, x_S	Acetone, y_A	Water, y_C	Monochlorbenzene, y_S
0.00	0.9989	0.0011	0.000	0.0018	0.9982
0.05	0.9482	0.0018	0.0521	0.0032	0.9447
0.10	0.8979	0.0021	0.1079	0.0049	0.8872
0.15	0.8478	0.0024	0.1620	0.0063	0.8317
0.20	0.7969	0.0031	0.2223	0.0079	0.7698
0.25	0.7458	0.0042	0.2901	0.0117	0.6982
0.30	0.6942	0.0058	0.3748	0.0172	0.6080
0.35	0.6422	0.0078	0.4328	0.0233	0.5439
0.40	0.5864	0.0136	0.4944	0.0305	0.4751
0.45	0.5276	0.0224	0.5492	0.0428	0.4080
0.50	0.4628	0.0372	0.5919	0.0724	0.3357
0.55	0.3869	0.0631	0.6179	0.1383	0.2438
0.60	0.2741	0.1259	0.6107	0.2285	0.1508
0.6058	0.2566	0.1376	0.6058	0.2566	0.1376

The first step in the solution of the problem is the location of the difference points, x_Δ and $x_{\Delta'}$. In order to locate the difference point $x_{\Delta'}$, the point x_J as defined by equations 260 and 261 is first located. Since the ratio of intermediate feed (stream L_F) to feed (stream L_0) is 0.1, point x_J (Fig. 313) is located on the straight line joining x_0 and x_F such that

$$\frac{\overline{x_0 x_J}}{\overline{x_J x_F}} = 0.10$$

The difference point $x_{\Delta'}$ is located by constructing the straight line through the points x_J and y_1 and determining its intersection with the straight line through the points x_n and y_{n+1}. The difference point x_Δ is next located by determining the intersection of the straight line through the points x_F and $x_{\Delta'}$ with that through the points x_0 and y_1.

The usual graphical procedure is followed to determine the number of equilibrium stages. Starting from point y_1, the point x_1 is obtained from the equilibrium data. The point x_1 is connected with the difference point x_Δ, and the intersection of this line with the solubility curve determines y_2. This procedure is repeated to locate x_2 and y_3. The point x_3 is found to be on the solubility curve below the point x_F, that is, the tie line joining y_3 and x_3 intersects the line joining points x_F, x_Δ, and $x_{\Delta'}$. Consequently, the point x_3 is joined to the difference point $x_{\Delta'}$ to locate the point y_4.

The complete solution is shown in Fig. 319. The compositions of the streams leaving the equilibrium stages are as follows.

Equilibrium Stage Number	Extract Phase, Mass Fraction Acetone, y_A	Raffinate Phase, Mass Fraction Acetone, x_A
1	0.140	0.130
2	0.0878	0.082
3 *	0.0533	0.0511
4	0.0310	0.0298
5	0.0146	0.0140
6	0.0030	0.0029

* The correct stage for the introduction of the intermediate feed is equilibrium stage 3. The number of equilibrium stages required is slightly greater than 5 and less than 6.

The pounds of monochlorbenzene required per pound of 0.200 mass fraction acetone (stream L_0) may be determined as follows.

Equations 246, 247, and 247A may be used to define another addition point $x_{J'}$. This point lies on the straight line through points x_J and y_{n+1}, and also on a straight line through the points x_n and y_1, and is therefore the intersection of these two lines. By taking the ratio of distances along the line $\overline{x_J, y_{n+1}}$,

$$\frac{J}{V_{n+1}} = \frac{16.7}{18.35} = 0.910$$

Since $J/L_0 = 1.1/1.0$,

$$\left(\frac{V_{n+1}}{L_0}\right) = \left(\frac{V_{n+1}}{J}\right)\left(\frac{J}{L_0}\right) = \left(\frac{1}{0.910}\right)(1.1) = 1.208$$

Exercise. Determine the effect of introducing the intermediate feed at equilibrium stages 2, 4, and 5 on the total number of stages required, using the conditions given in the illustrative example on this page.

Exercise. Determine the number of equilibrium stages required if the 0.200 weight fraction acetone solution and the

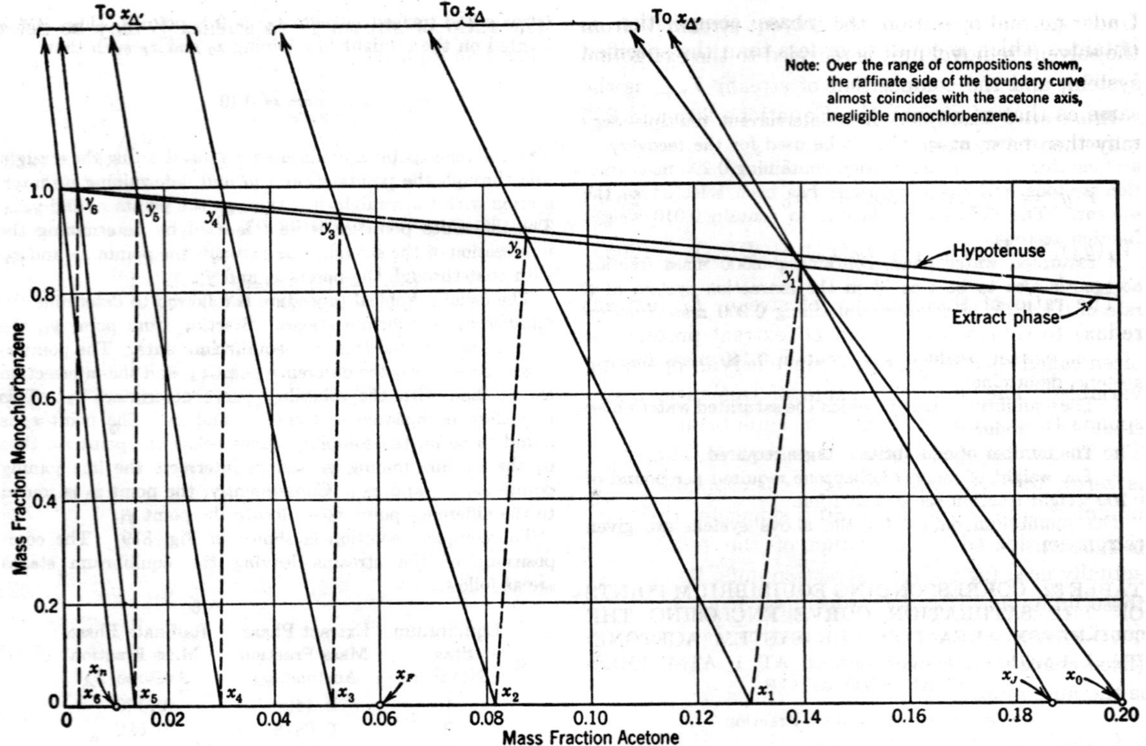

FIG. 319. Complete graphical solution for illustrative example, p. 311.

0.060 weight fraction acetone solution are mixed and the mixture is then extracted in a continuous countercurrent multiple-contact system. All other conditions are to be the same as in the illustrative example, p. 311.

Reflux. The use of extract reflux as described under Methods of Operation (Fig. 307) is usually of no practical importance for a ternary system in which two of the binary systems are completely miscible liquids (Fig. 309), since the overall composition of the mixtures (both phases) in stage 1, to which the reflux L_0 is returned, is very close to the phase boundary curve, making separation of the two phases difficult. This limitation does not occur in a ternary system in which two of the binary systems are partially miscible, as illustrated in Fig. 312.

The solvent removal unit shown in Fig. 307 may include various means of separating the solvent from the extract phase, depending upon the system considered. Regardless of means of separation, the extraction system shown in Fig. 307 (exclusive of the solvent removal) is identical to operating with intermediate feed without reflux (illustrative example, p. 311) and equations 246 through 262 may be used.

When operating with reflux, it will be assumed that:

1. The stream V leaving the solvent removal unit contains only components A and S, with component S being the main constituent.

2. The streams L_0 and D, which have the same composition, are saturated phases so that the point x_D, which represents their composition, lies on the branch MN of the solubility curve (see Fig. 312 or 320).

Under steady-state operating conditions, the material balances written for the solvent removal unit are

Total material balance:

$$V_1 = V + L_0 + D \qquad (263)$$

For any component:

$$V_1 y_1 = Vy + L_0 x_0 + D x_D \qquad (264)$$

For component A:

$$V_1 (y_A)_1 = V(y_A) + L_0 (x_A)_0 + D(x_A)_D \qquad (265)$$

Under normal operation, the solvent stream V from the solvent removal unit is recycled to the extraction system, and the composition of stream V_{n+1} is the same as that of stream V. Equations 264 and 265 may then be written:

$$V_1 y_1 = V y_{n+1} + L_0 x_0 + D x_0 \tag{264a}$$

$$V_1 (y_A)_1 = V (y_A)_{n+1} + L_0 (x_A)_0 + D (x_A)_0 \tag{265a}$$

The ratio of the quantity of stream L_0 (extract reflux) to that of the stream D (extract product) is often called the "reflux ratio" since it is an operating variable. The minimum value of this ratio corresponds to an infinite number of equilibrium stages. The maximum value where the ratio is infinite (D is zero), termed "total reflux," corresponds to the minimum number of equilibrium stages for the separation considered. The value of the reflux ratio actually used for a given extraction must be between these limits and may be determined from economic considerations so that the total cost of the operation (fixed charges on the equipment plus operating cost) is the minimum.

Consider the case of extraction where the following variables have been set:

1. The quantity (L_F) and composition (x_F) of the feed.
2. The composition (x_D) of the extract product; also $x_D = x_0$.
3. The composition (x_n) of the raffinate product.
4. The composition (y_{n+1}) of the solvent; also $y = y_{n+1}$.
5. The reflux ratio (L_0/D).

The composition of stream V_1, y_1 may be determined by constructing the line through the points x_0 and $y = y_{n+1}$ and determining the intersection of this line with the branch QR of the solubility curve (Fig. 320). From equations 263, 264*a*, and 265*a*, the point y_1 lies on the straight line joining x_0 and y, since x_0 represents the composition of both L_0 and D. It also lies on the solubility curve, since it is a stream leaving an equilibrium stage. Hence the intersection represents the composition y_1. The ratio of V_1 to V may be determined as follows.

$$\frac{V_1}{V} = \frac{\overline{x_0 y_{n+1}}}{\overline{x_0 y_1}}$$

The ratio of stream D to stream V may be determined as follows.

$$V_1 = V + L_0 + D$$

$$= V + \left(\frac{L_0}{D} + 1\right) D = \left(\frac{V_1}{V}\right) V \tag{263}$$

Rearranging and dividing by D,

$$\left(\frac{V_1}{V} - 1\right)\frac{V}{D} = \left(\frac{L_0}{D} + 1\right)$$

$$\frac{V}{D} = \frac{\left(\dfrac{L_0}{D} + 1\right)}{\left(\dfrac{V_1}{V}\right) - 1} \tag{266}$$

The difference point x_Δ may be located, using equations 250 and 252, and, by equations 263 and 264,

$$\Delta = L_0 - V_1 = -(V + D) \tag{267}$$

$$x_\Delta = \frac{L_0 x_0 - V_1 y_1}{L_0 - V_1} = \frac{V y_{n+1} + D x_0}{V + D} \tag{268}$$

Exercise. Show from equation 268 that x_Δ always lies within the triangle or on the hypotenuse.

Graphically, x_Δ may be located on the line $x_0 y_{n+1}$, so that

$$\frac{\Delta}{D} = \frac{\overline{x_0 y_{n+1}}}{\overline{y_{n+1} x_\Delta}} = \left(\frac{V}{D} + 1\right)$$

since the direction of measurement (sign) is immaterial when calculating the ratio of distances. Therefore,

$$\overline{y_{n+1} x_\Delta} = \frac{\overline{x_0 y_{n+1}}}{\left(1 + \dfrac{V}{D}\right)} \tag{269}$$

The second difference point $x_{\Delta'}$ may be located in the manner previously described. Figure 320 is the complete graphical solution for the following example.

Illustrative Example. It is proposed to separate a mixture of methylcyclopentane and *n*-hexane containing 0.40 mole fraction methylcyclopentane and 0.60 mole fraction *n*-hexane by extraction followed by fractionation. The extraction is to be conducted at 25° C, with solvent (0.995 mole fraction aniline and 0.005 mole fraction methylcyclopentane) as recovered. The raffinate product is to contain 0.150 mole fraction methylcyclopentane, and the extract product is to contain 0.116 mole fraction aniline and 0.700 mole fraction methylcyclopentane. A continuous countercurrent multiple-contact extraction system with extract reflux is to be used.

1. Determine the number of equilibrium stages required if a reflux ratio (L_0/D) of 10 moles of extract reflux per mole of extract product is used.

2. Determine the minimum reflux ratio $(L_0/D)_{min}$ that may be used.

3. Determine the maximum concentration of methylcyclopentane that may be obtained in the extract product using the same feed composition and the same raffinate product composition but without the use of extract reflux.

The equilibrium data for the above ternary system at 25° C are given in Table 33.

Provided that no chemical reactions occur in the process, so that the number of moles of material in the entering streams is equal to the number of moles of material in the leaving streams, all quantities and compositions may be expressed in terms of moles as well as in terms of mass.

The following values are set by the statement of the problem.

1. Composition of the feed: $(x_A)_F = 0.400$ and $(x_C)_F = 0.600$.

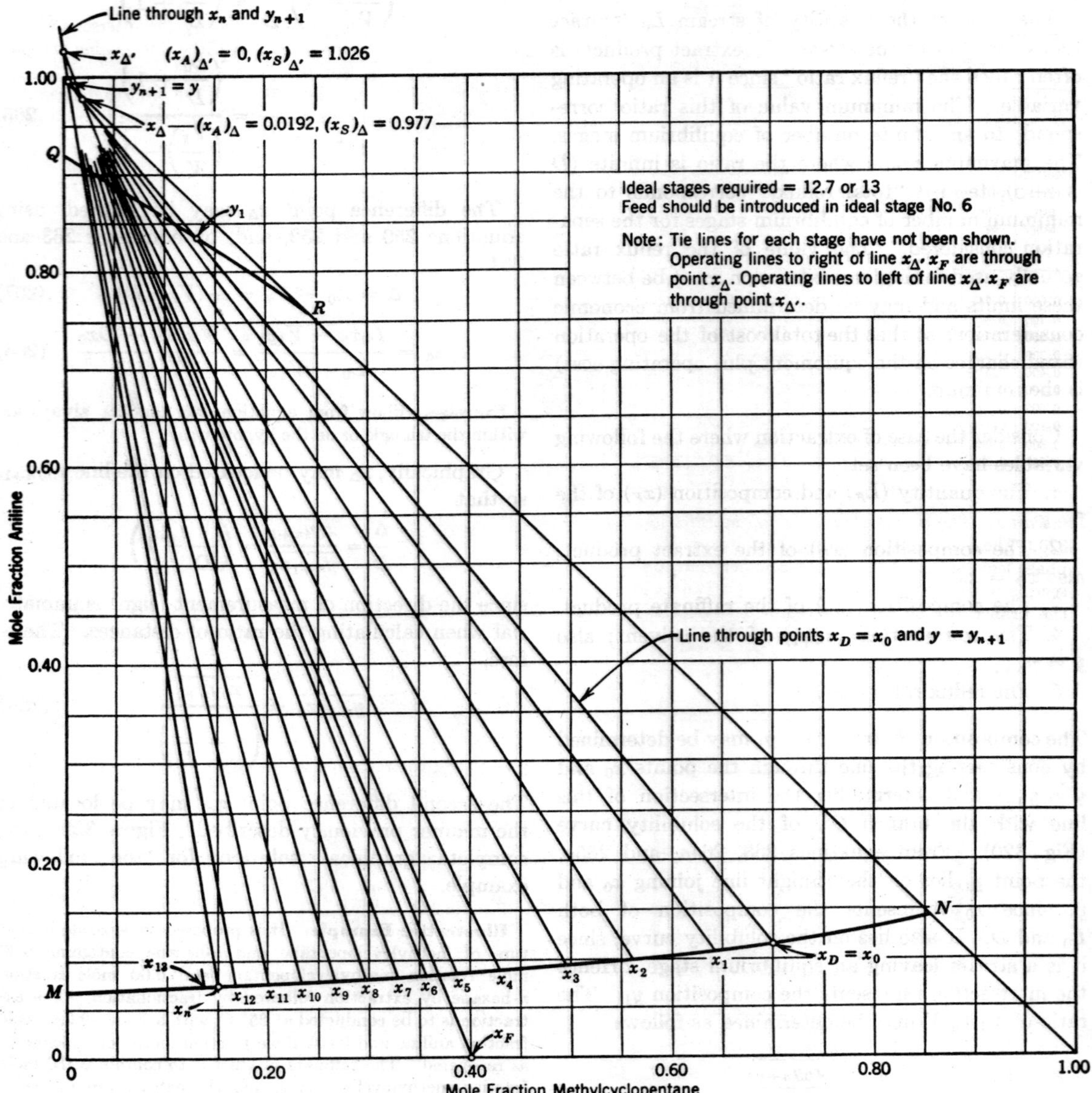

FIG. 320. Complete graphical solution for part 1 of illustrative example.

2. Composition of the extract product: $(x_A)_D = 0.700$, $(x_S)_D = 0.116$, and $x_D = x_0$.
3. Composition of the raffinate product: $(x_A)_n = 0.150$. For an equilibrium stage, x_n must lie on solubility curve so that $(x_C)_n$ is fixed.
4. Composition of the solvent: $(y_A)_{n+1} = 0.005$, and $(y_S)_{n+1} = 0.995$.
5. Reflux ratio $(L_0/D) = 10$.

TABLE 33. SATURATED AND EQUILIBRIUM COMPOSITIONS FOR THE TWO PHASES IN THE SYSTEM METHYLCYCLOPENTANE-N-HEXANE-ANILINE; TEMPERATURE, 25° C [4]

n-Hexane Layer					Aniline Layer				
Mole Fraction			Mole Ratio		Mole Fraction			Mole Ratio	
Methyl-cyclo-pentane, x_A	*n*-Hexane, x_C	Aniline, x_S	X_A	X_S	Methyl-cyclo-pentane, y_A	*n*-Hexane, y_C	Aniline, y_S	Y_A	Y_S
0.000	0.933	0.067	0.000	0.0718	0.000	0.080	0.920	0.000	11.50
0.100	0.829	0.071	0.108	0.0764	0.014	0.076	0.910	0.1556	10.10
0.200	0.725	0.075	0 216	0.0811	0.031	0.069	0.900	0.310	9.00
0.300	0.620	0.080	0.326	0.0870	0.048	0.063	0.889	0.432	8.00
0.400	0.514	0.086	0.438	0.0941	0.068	0.056	0.876	0.548	7.06
0.500	0.406	0.094	0.552	0.1038	0.092	0.048	0.860	0.657	6.15
0.600	0.296	0.104	0.670	0.1161	0.121	0.038	0.841	0.761	5.30
0.700	0.184	0.116	0.792	0.1312	0.157	0.026	0.817	0.858	4.46
0.800	0.068	0.132	0.922	0.1521	0.206	0.011	0.783	0.949	3.60
0.857	0.000	0.143	1.000	0.1669	0.242	0.000	0.758	1.000	3.13

Part 1. The points y_1, x_Δ, and $x_{\Delta'}$ are first located on Fig. 320. The point y_1 representing the composition of the extract phase leaving equilibrium stage 1 is determined by the intersection of the line joining points $x_D = x_0$, and $y = y_{n+1}$, with the branch of the solubility curve representing the extract phase. This intersection corresponds to $(y_A)_1 = 0.132$ mole fraction and $(y_s)_1 = 0.833$ mole fraction. By measurement of distances along the line x_0y_{n+1},

$$\frac{V_1}{V} = \frac{\overline{x_0y_{n+1}}}{\overline{x_0y_1}} = \frac{44.9 \text{ units}}{36.7 \text{ units}} = 1.223$$

$$\frac{V}{D} = \frac{(L_0/D)+1}{(V_1/V)-1} = \frac{11}{0.223} = 49.3 \qquad (266)$$

$$y_{n+1}\Delta = \frac{\overline{x_0y_{n+1}}}{(1+V/D)} = \frac{44.9 \text{ units}}{50.3} = 0.93 \text{ unit} \qquad (269)$$

Similarly, the point $x_{\Delta'}$ is located by the intersection of the line joining points x_F and x_Δ with the line joining the points x_n and y_{n+1} (Fig. 320*a*).

The usual graphical procedure is followed to determine the number of equilibrium stages. The complete solution is indicated in Fig. 320. The approximate compositions of the streams leaving the equilibrium stages are shown in Table 34.

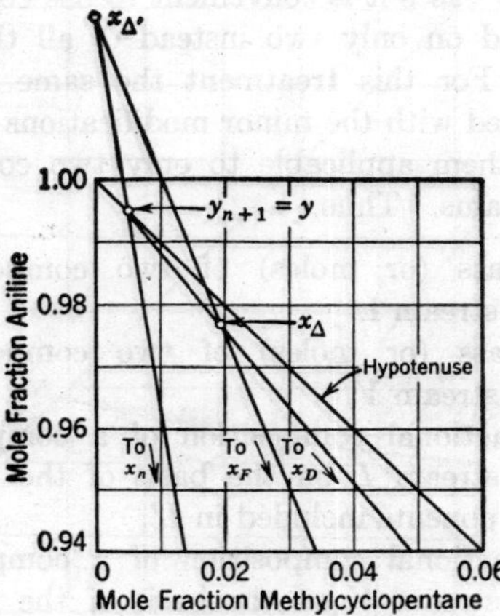

FIG. 320*a*. Enlarged scale showing location of points x_Δ and $x_{\Delta'}$ in Fig. 320.

TABLE 34. COMPOSITION OF STREAMS LEAVING EACH IDEAL STAGE

Ideal Stage Number	Extract Phase		Raffinate Phase	
	y_A	Y_A	x_A	X_A
1	0.132	0.797	0.641	0.710
2	0.109	0.725	0.559	0.620
3	0.090	0.653	0.493	0.543
4	0.076	0.588	0.429	0.471
5	0.064	0.529	0.385	0.422
6*	0.058	0.498	0.356	0.389
7	0.053	0.464	0.327	0.357
8	0.047	0.429	0.295	0.322
9	0.042	0.390	0.264	0.287
10	0.036	0.351	0.232	0.251
11	0.031	0.308	0.200	0.216
12	0.026	0.266	0.167	0.180
13	0.021	0.224	0.141	0.152

x_A = mole fraction methylcyclopentane in raffinate (*n*-hexane) phase.
y_A = mole fraction methylcyclopentane in extract (aniline) phase.
X_A = mole ratio of methylcyclopentane to methylcyclopentane + hexane in raffinate (*n*-hexane) phase.
Y_A = mole ratio of methylcyclopentane to methylcyclopentane + hexane in extract (aniline) phase.

* Ideal stage 6 is the correct stage for the introduction of the feed.

Part 2. Whenever there is an infinite number of stages, there is no change in the composition of a stream as it passes from stage to stage. This condition is indicated by the coincidence of a tie line with an operating line. The required minimum reflux ratio $(L_0/D)_{min}$ is the maximum value of L_0/D computed by means of the coincidence of any tie line with an operating line. The coincidence of an operating line with a tie line does not determine the minimum reflux for the system but simply indicates the required reflux if there is no change in composition from stage to stage at that point in the system.

If the operating line at the feed stage is to coincide with a tie line, it can do so only if the extension of the tie line passes through x_F. This tie line may be found by trial and error to pass through the points

$$(x_A = 0.368), \qquad (x_S = 0.084)$$
$$(y_A = 0.062), \qquad (y_S = 0.880)$$

The extension of this line will also pass through the points x_Δ and $x_{\Delta'}$ for these operating conditions as these points are on the operating line at the feed stage. x_Δ is the intersection

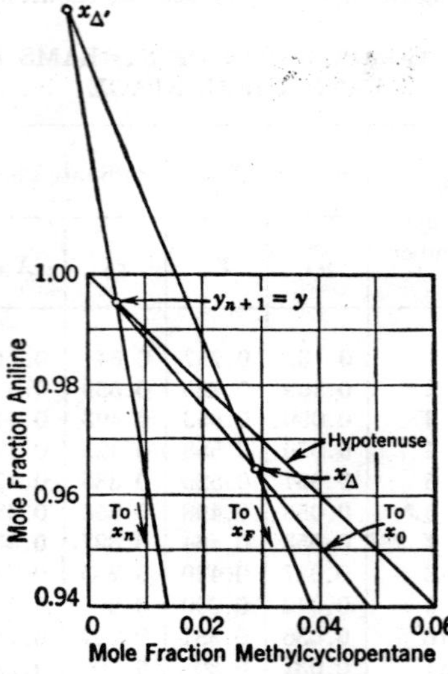

Fig. 321. Graphical solution for reflux ratio corresponding to infinite stages at the feed stage, for illustrative example, part 2.

of this line with the line $\overline{x_0y_{n+1}}$ as indicated in Fig. 321. The relative distances along $\overline{x_0y_{n+1}}$ may be substituted in equation 269, giving

$$\frac{\overline{x_0y_{n+1}}}{\overline{y_{n+1}x_\Delta}} = \frac{44.9}{1.48} = 30.3 = 1 + \frac{V}{D} \tag{269}$$

$$\frac{V}{D} = 29.3 \qquad \frac{V_1}{V} = 1.223 \text{ (from Part 1)}$$

$$\frac{V}{D} = \frac{(L_0/D) + 1}{(V_1/V) - 1} = 29.3 = \frac{(L_0/D) + 1}{1.223 - 1} \tag{266}$$

$$\frac{L_0}{D} = 5.53$$

Therefore, the reflux ratio required when there is no change in composition from stage to stage at the feed stage (infinite stages) is 5.53. With the same procedure at other intermediate positions and at the ends of the system, other reflux ratios are found to correspond to infinite stages in those positions. Since in no case is a reflux ratio greater than the 5.53 indicated, the maximum reflux ratio required for an infinite number of stages anywhere in the extraction system is 5.53, and this is the true "minimum reflux ratio" for the operating conditions considered.

Part 3. If extract reflux is not used continuous countercurrent multiple-contact operation of the extraction system will be followed. The maximum concentration of methylcyclopentane in the extract product will be obtained when the minimum solvent-to-feed ratio is employed. This minimum solvent-to-feed ratio corresponds to a system requiring an infinite number of stages (see part 3, illustrative example, p. 308) and is determined by the coincidence of a tie line with an operating line.

The extract product composition determined by a tie line which, when extended, passes through the point x_F has been determined in part 2 as

$$y_A = 0.062 \qquad y_S = 0.880$$

In the general case it may be possible that a value of y_A lower than 0.062 may be obtained by the coincidence of another tie line with an operating line. In the present case no other tie line leads to values of y_A less than 0.062.

Comparison of the maximum value of $(y_A)_1 = 0.062$ when extract reflux is not used with the value of $(y_A)_1 = 0.132$ which is obtained when extract reflux is used illustrates one of the reasons for the use of extract reflux even though it requires larger amounts of solvent per unit mass of feed. In the present illustration, if extract reflux is not used, the range of solvent to feed ratios which may be used is rather limited. As already determined, the lower limit is set by the requirement of an infinite number of stages. The upper limit is set by the fact that the single-phase region of the equilibrium diagram is approached as the solvent-to-feed ratio is increased.

Mass or Mole Ratio Diagram

In many cases it is convenient to use composition units based on only two instead of all three components. For this treatment the same equations may be used with the minor modifications necessary to make them applicable to only two components of the streams. Thus,

L' = mass (or moles) of two components in stream L.

V' = mass (or moles) of two components in stream V.

X = fractional composition of a component in stream L on the basis of the two components included in L'.

Y = fractional composition of a component in stream V on the basis of the two components included in V'.

For example, a mixture containing 20 lb of A, 30

lb of C, and 50 lb of S may have its composition expressed as follows.

Mass Fraction	Mass Ratio: S Free Basis $X_A + X_C = 1$	Mass Ratio: A Free Basis $X_C + X_S = 1$
$x_A = 0.20$	$X_A = 0.4$	$X_A = 0.25$
$x_C = 0.30$	$X_C = 0.6$	$X_C = 0.375$
$x_S = 0.50$	$X_S = 1.0$	$X_S = 0.625$
1.0		

The mass ratios of the two components may vary from zero to unity, but the third component not included in L' or V' may have mass ratios from zero to infinity. The closed triangle which includes all

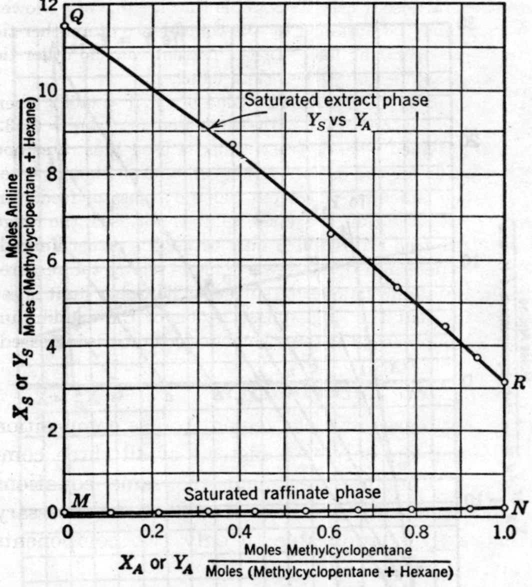

FIG. 322. Phase equilibrium diagram on a mole-ratio basis for the system methylcyclopentane-normal hexane-aniline, at 25° C.

positive values of compositions expressed as mass fractions (x, y) is not adapted to treating values up to infinity. But a rectangle with one side (usually the top) at infinity includes all positive compositions expressed as mass ratios (X, Y). As in the triangular diagram, the compositions of the three streams when plotted as coordinates of mass ratio on the rectangular diagram also lie on a straight line, and the quantities of the two components in the respective streams are inversely proportional to the distances between the corresponding points.

For example the rectangular diagram (Fig. 322) corresponds to the right-triangle diagram (Fig. 312) with the compositions expressed as mole ratios on the solvent-free basis.

$$X_A = \frac{x_A}{x_A + x_C}; \quad X_S = \frac{x_S}{x_A + x_C}; \quad X_C = \frac{x_C}{x_A + x_C}$$

$$Y_A = \frac{y_A}{y_A + y_C}; \quad Y_S = \frac{y_S}{y_A + y_C}; \quad Y_C = \frac{y_C}{y_A + y_C}$$

$$X_A + X_C = 1 \qquad Y_A + Y_C = 1$$

Y_S may have values up to infinity for pure solvent. Under these conditions the material balance equations are most conveniently written in terms of the solvent-free material and not in terms of the total quantity of the streams.

$$L'_0 + V'_{n+1} = L'_n + V'_1 \qquad (270) \text{ or } (241')$$

where L' = moles of components A and C in the stream L.

V' = moles of components A and C in the stream V.

If V_{n+1} is pure component S, the value of V'_{n+1} is zero. Similarly equation 242 becomes

$$L'_0X_0 + V'_{n+1}Y_{n+1} = L'_nX_n + V'_1Y_1 \qquad (271) \text{ or } (242')$$

Illustrative Example. Parts 1 and 2 of the previous illustrative example (p. 313) are to be solved by means of the rectangular mole-ratio diagram.

Part 1. All the known compositions, which are given in mole fractions, are converted to mole ratios.

1. Composition of the feed. Since the feed contains no aniline,

$$(x_A)_F = (X_A)_F; \qquad (x_C)_F = (X_C)_F; \qquad (X_S)_F = 0$$

2. Composition of the extract product:

$$(X_A)_D = \frac{0.700}{0.884} = 0.792; \quad (X_S)_D = \frac{0.116}{0.884} = 0.1312; \quad X_D = X_0$$

3. Composition of the raffinate product:

$$(X_A)_n = \frac{0.150}{0.927} = 0.1618; \qquad (X_S)_n = \frac{0.073}{0.927} = 0.0788$$

4. Composition of the solvent:

$$(Y_A)_{n+1} = \frac{0.005}{0.005} = 1.00; \qquad (Y_S)_{n+1} = \frac{0.995}{0.005} = 199$$

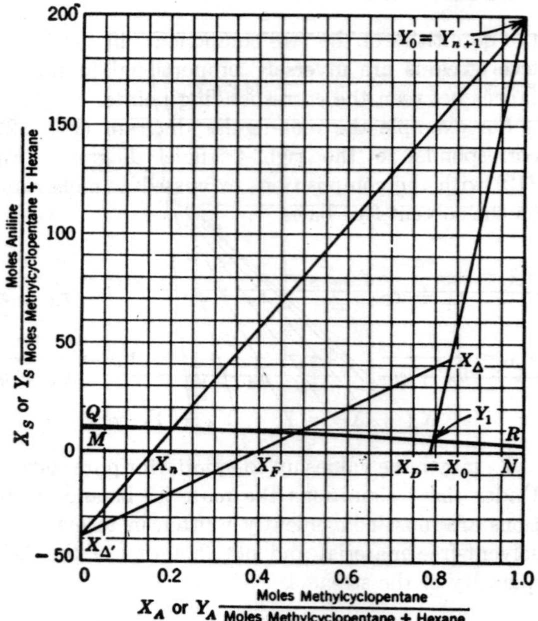

FIG. 323. Overall graphical construction on a mole-ratio diagram for illustrative example, part 1 (p. 315).

The point Y_1 is located by determining the intersection of the line joining the points $X_D = X_0$, and $Y = Y_{n+1}$, with the extract phase branch (QR) of the solubility curve (Fig. 323) at $(Y_A)_1 = 0.797$ and $(Y_S)_1 = 5$.

The difference point X_Δ may be found as follows by similar triangles.

$$\frac{V'_1}{V'} = \frac{\overline{X_DY}}{\overline{X_DY_1}} = \frac{(Y_S) - (X_S)_D}{(Y_S)_1 - (X_S)_D} = \frac{199 - 0.131}{5 - 0.131} = 40.7$$

$$\frac{V'}{D'} = \frac{\left(\frac{L_0'}{D'} + 1\right)}{\left(\frac{V'_1}{V'}\right) - 1} = \frac{11}{(40.7 - 1)} = 0.276 \qquad (266')$$

$$\frac{V'}{D'} = \frac{\overline{X_DX_\Delta}}{\overline{X_\Delta Y}} = \frac{(X_S)_\Delta - (X_S)_D}{(Y_S) - (X_S)_\Delta} = \frac{(X_S)_\Delta - 0.131}{199 - (X_S)_\Delta} = 0.276 \qquad (266'A)$$

Solving, $(X_S)_\Delta = 43.1$.

Since the difference point X_Δ lies on straight line $\overline{X_DY}$, point X_Δ may be located. The value of $(X_A)_\Delta$ is found to be 0.835.

The second difference point $X_{\Delta'}$ may now be located by finding the intersection of lines $\overline{X_\Delta X_F}$ and $\overline{Y_{n+1}X_n}$ at point

$$(X_A)_{\Delta'} = 0 \qquad (X_S)_{\Delta'} = -39$$

The graphical solution for the number of equilibrium stages required is shown in Fig. 325. For the ternary system under consideration, the mole ratio diagram is somewhat more convenient to use than the mole fraction diagram since the difference points X_Δ and $X_{\Delta'}$ are not crowded together (compare with Fig. 320). The compositions of the streams leaving the equilibrium stages correspond with those given in Table 34. The same number of ideal stages (13) are computed as by the triangular diagram of the example on p. 314.

Part 2. The determination of the minimum reflux ratio is illustrated in Fig. 324. For the section to the right of X_F, a series of tie lines (dashed) and their respective intersections with the line $\overline{X_0Y_1}$ are shown. The tie line $\overline{jk}$ corresponds to the tie line which when extended passes through X_F. The coordinates for this tie line are $(X_A) = 0.401$, $(Y_A) = 0.515$. The intersection of this line with the line $\overline{X_0Y_1}$, point l, represents the point X_Δ corresponding to this tie line. The value of $(X_S)_\Delta$ for this tie line is the highest value of $(X_S)_\Delta$ for any tie line in this section and is the limiting case determining the reflux required for an infinite number of stages (so-called minimum reflux).

For the section to the left of X_F, a similar series of tie lines are shown together with their intersections with the line $\overline{X_nY_{n+1}}$. In this case the intersections represent the corresponding $X_{\Delta'}$ points for the respective tie lines. The lowest

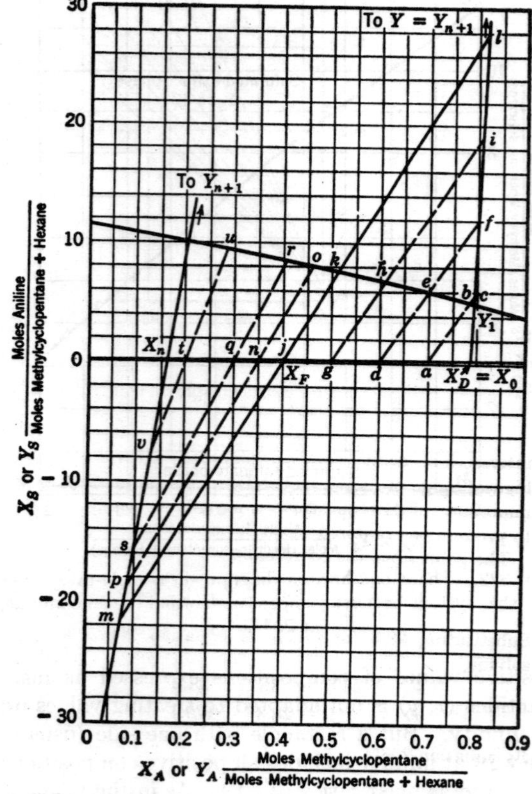

FIG. 324. Graphical solution for minimum reflux for illustrative example, part 2.

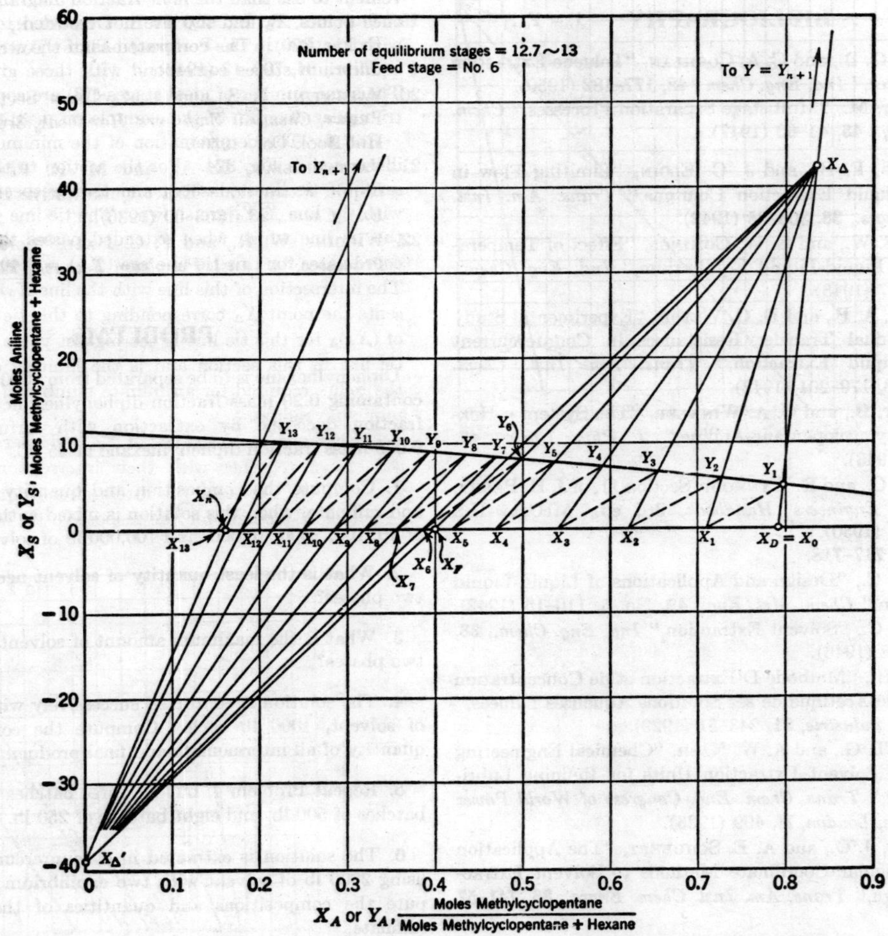

FIG. 325. Graphical solution of illustrative example, part 2, on a mole-ratio basis.

intersection, point m, when connected with X_F, gives the highest value of $(X_S)_\Delta$. Therefore, this tie line indicates the minimum reflux required with an infinite number of stages to produce the required result.

The point l has the coordinates:

$$[(X_S)_\Delta]_{\min} = 28.3, \; [(X_A)_\Delta]_{\min} = 0.820$$

Substituting 28.3 for $(X_S)_\Delta$ in equation 266′A above and solving,

$$\frac{V'}{D'} = \frac{28.3 - 0.131}{199 - 28.3} = \frac{28.2}{170.7} = 0.1658$$

By equation 266′

$$\left(\frac{L'_0}{D'}\right)_{\min} = (39.7)(0.1658) - 1 = 6.58 - 1 = 5.58$$

For a particular extraction, such as has been considered in this illustrative problem, there are two limiting conditions of operation. One extreme requires an infinite number of stages and a minimum reflux ratio. The other extreme is the minimum number of equilibrium stages required which corresponds to total reflux, i.e., no product withdrawn. As the reflux ratio increases, the value for $(X_S)_\Delta$ increases. The maximum value of L'_0 and the minimum value of D' occur when $D' = 0$, $(L'_0/D' = \infty)$. By a material balance on a solvent-free basis corresponding to equation 263,

$$V'_1 = L'_0 + V' + D' \tag{272}$$

and, since $D' = 0$, $\Delta = L'_0 - V'_1 = -V'$, and X_Δ coincides with Y. Figure 325 indicates that this condition requires the minimum number of equilibrium stages.

The conditions of "minimum reflux ratio" and of "total reflux" are the limits of operation. Any practical operation must lie within the limits set by these two extremes.

BIBLIOGRAPHY

a. ARNOLD, G. B., and C. A. COGHLAN, "Toluene Extraction with Water," *Ind. Eng. Chem.*, **42**, 177–182 (1950).

b. BENEDICT, M., "Multistage Separation Processes," *Chem. Eng. Prog.*, **43**, 41–60 (1947).

1. BLANDING, F. H., and J. C. ELGIN, "Limiting Flow in Liquid-Liquid Extraction Columns," *Trans. Am. Inst. Chem. Engrs.*, **38**, 305–35 (1942).
2. BRIGGS, S. W., and E. W. COMINGS, "Effect of Temperature on Liquid-Liquid Equilibrium," *Ind. Eng. Chem.*, **35**, 411–17 (1943).
3. COLBURN, A. P., and D. G. WELSH, "Experimental Study of Individual Transfer Resistances in Countercurrent Liquid-Liquid Extraction," *Trans. Am. Inst. Chem. Engrs.*, **38**, 179–201 (1942).
4. DARWENT, B., and C. A. WINKLER, "The System *n*-Hexane-Methylcyclopentane-Aniline," *J. Phys. Chem.*, **47**, 442–54 (1943).
5. ELGIN, J. C., and R. WYNKOOP, Section 11, of J. H. PERRY, *Chemical Engineers' Handbook*, 3rd ed., McGraw-Hill Book Co. (1950).
6. *Ibid.*, pp. 717–718.
7. ELGIN, J. C., "Design and Applications of Liquid-Liquid Extraction," *Chem. Met. Eng.*, **49**, No. 5, 110–16 (1942).
8. ELGIN, J. C., "Solvent Extraction," *Ind. Eng. Chem.*, **38**, 26–7, 37–8 (1946).
9. GUINOT, H., "Methode D'Extraction et de Concentration de L'Acide Acetique de ses Solutions Aqueuses Diluées," *Chimie & industrie*, **21**, 243–51 (1929).
10. HUNTER, T. G., and A. W. NASH, "Chemical Engineering Design of Solvent Extraction Units for Refining Lubricating Oil," *Trans. Chem. Eng. Congress of World Power Conference, London*, **II**, 409 (1936).
11. MALONEY, J. O., and A. E. SCHUBERT, "The Application of Rectangular Coordinate Methods to Solvent Extraction Design," *Trans. Am. Inst. Chem. Engrs.*, **36**, 741–57 (1940).
12. MILLER, S. A., and C. A. MANN, "Agitation of Two-Phase Systems of Immiscible Liquids," *Trans. Am. Inst. Chem. Engrs.*, **40**, 709–43 (1944).
13. OTHMER, D. F., R. E. WHITE, and E. TRUEGER, "Liquid-Liquid Extraction Data," *Ind. Eng. Chem.*, **33**, 1240–48 (1941).
14. RANDALL, M., and B. LONGTIN, "Separation Processes—Analogy between Absorption, Extraction, Distillation, Heat Exchange, and Other Separation Processes," *Ind. Eng. Chem.*, **31**, 1295–99 (1939).
15. ROGERS, M. W., and E. W. THIELE, "Bubble-Cap Column as a Liquid-Liquid Contact Apparatus," *Ind. Eng. Chem.*, **29**, 529–30 (1937).
16. RUTHRUFF, R. F., and D. F. WILCOCK, "Solvent Extraction of Vegetable Drying Oils," *Trans. Am. Inst. Chem. Engrs.*, **37**, 649–66 (1941).
17. SHERWOOD, T. K., J. E. EVANS, and J. V. A. LONGCOR, "Extraction in Spray and Packed Columns," *Ind. Eng. Chem.*, **31**, 1144–50 (1939).
18. SMITH, A. S., and T. B. BRAUN, "Butadiene Purification by Solvent Extraction," *Ind. Eng. Chem.*, **37**, 1047–52 (1945).
19. TREYBAL, R. E., and F. E. DUMOULIN, "Liquid-Liquid Extraction in a Perforated-Plate Tower," *Ind. Eng. Chem.*, **34**, 709–13 (1942).
20. VALENTINE, K. S., and G. MACLEAN, Section 17, of J. H. PERRY, *Chemical Engineers' Handbook*, 3rd ed., McGraw-Hill Book Co. (1950).
21. VARTERESSIAN, K. A., and M. R. FENSKE, "Liquid-Liquid Extraction—Exact Quantitative Relations," *Ind. Eng. Chem.*, **28**, 1353–60 (1936).
22. WIGGINS, W. R., and F. C. HALL, "Solvent Extraction Processes," *J. Inst. Petroleum Technol.*, **22**, 78–98 (1936).

PROBLEMS

Diphenylhexane is to be separated from 1000 lb of a mixture containing 0.20 mass fraction diphenylhexane and 0.80 mass fraction docosane by extraction with furfural containing 0.005 mass fraction diphenylhexane at 45° C.

1. Compute the composition and quantity of the extract and raffinate, when this solution is mixed with each quantity of 10, 100, 2000, 10,000, and 100,000 lb of solvent.

2. What is the least quantity of solvent necessary to form two phases?

3. What is the maximum amount of solvent that will form two phases?

4. The solution is contacted successively with two batches of solvent, 1000 lb each. Compute the composition and quantity of all intermediate and final products.

5. Repeat Problem 4, but for three batches of 667 lb, four batches of 500 lb, and eight batches of 250 lb.

6. The solution is extracted in a countercurrent extractor using 2000 lb of solvent with two equilibrium stages. Compute the compositions and quantities of the extract and raffinate.

7. Repeat Problem 6, but with three equilibrium stages and five equilibrium stages.

8. Repeat Problem 6, but with an infinite number of equilibrium stages.

9. Compare the raffinate compositions of Problems 1, 4, 5, 6, 7, and 8, by preparing a plot with mass fraction diphenylhexane as ordinate and number of stages as abscissa. There will be one line for multiple-contact extraction and one line for countercurrent extraction.

10. In the extraction of alcohol from an alcohol-benzene solution by water, ten extraction vessels are used in continuous countercurrent extraction of 5 tons/hr of feed containing 0.125 mass fraction alcohol. The extract contains 0.400 mass fraction alcohol, and raffinate is 0.99 mass fraction benzene. Four of the vessels are put out of service by an accident. What is the concentration of alcohol in the raffinate, and what is the increase in loss of alcohol when the six vessels are used, compared with the original ten? (See Appendix for equilibrium data.)

11. Six tons of a 25 per cent solution of acetic acid in water is to be extracted per hour with pure isopropyl ether in a countercurrent multiple-contact system

If less than 5 per cent of the acetic acid is to be lost in the raffinate and the amount of solvent used is to be kept within 25 per cent of the minimum amount permissible, how many ideal stages will be required for the extraction?

12. A countercurrent multiple-stage extraction system is to be used for the recovery of acetic acid from an aqueous solution containing 30 weight per cent acetic acid. Isopropyl ether has been selected as the solvent to be used for the extraction. The raffinate is to contain not more than 1.0 weight per cent acetic acid.

(*a*) What is the maximum acid concentration (mass per cent acetic acid) that can be obtained in the extract from the system? How many ideal stages will be required? What is the required ratio of ether to feed?

(*b*) Determine the quantity of ether required per pound of feed and the composition of the extract if four ideal stages are used.

(*c*) Determine the number of ideal stages necessary, if the extract is to contain 10.5 weight per cent acetic acid.

13. Two sources of aqueous solution of acetic acid are available in a chemical plant, one consisting of 14,000 lb/hr of 0.29 mass fraction acetic acid and one consisting of 3000 lb/hr of 0.12 mass fraction acetic acid.

The acid is to be concentrated by extraction with isopropyl ether, removing the strong extract at 0.08 mass fraction acetic acid, concentrating it in a fractionating column, and returning the ether at 0.004 mass fraction acetic acid for re-use. Raffinate leaves at 0.03 mass fraction acetic acid.

How many ideal stages will be required for the separation? At which stage should each stream be introduced?

14. A lubricating oil stock having a viscosity-gravity constant equal to 0.874 is to be extracted with nitrobenzene in a countercurrent multiple-stage system operating at 10° C. It is proposed to use 100 volumes of nitrobenzene per 100 volumes of stock. Determine the number of stages required if a raffinate product having a viscosity-gravity constant (VGC) of 0.815 is desired. The following data were obtained from experimental single-stage batch extractions of the oil stock with different volumes of solvent.

Solvent Used, Volumes per 100 Volumes of Stock	Raffinate Phase volume % solvent	Raffinate Phase VGC	Extract Phase volume % solvent	Extract Phase VGC
55	29.8	0.866	50.4	0.898
75	23.3	0.849	55.1	0.905
100	20.5	0.840	63.6	0.913
150	16.0	0.831	70.5	0.909
200	14.9	0.826	76.4	0.904
300	15.4	0.818	80.3	0.896
400	16.7	0.814	83.5	0.890
600	21.8	0.809	87.0	0.889

From S. W. Ferris, E. R. Birkheimer, and L. M. Henderson, *Ind. Eng. Chem.*, **23** (1931).

CHAPTER 23

Vapor-Liquid Transfer Operations 1

THE difference in equilibrium composition between liquid and vapor phases may be used to separate individual components or mixtures in much the same manner as the difference in composition between two liquid phases is used in liquid-liquid extraction. Similarly the industrial equipment for vapor-liquid transfer operations has as its main purpose the intimate contact and subsequent separation and handling of vapor and liquid phases. In typical equipment, vapors may bubble through a continuous liquid phase, droplets of liquid may fall through a continuous vapor phase, an extended interface may provide contact between the two phases, or these methods may be used in combination.

Distillation is specifically the process of driving off vapors from liquids (or solids) as by heat in a retort or still, and condensing the vapor products therefrom. It includes evaporation and condensation. Vapor-liquid transfer operations that depend upon distillation for creating the vapor and liquid out of the material being treated are also referred to as distillation.

Absorption refers to an operation in which the significant or desired transfer of material is from the vapor phase to the liquid phase. It usually, but not exclusively, designates an operation in which the liquid is supplied as a separate stream independent of the vapor being treated.

Stripping is the term applied to an operation in which the significant or desired transfer of material is from the liquid to the vapor phase.

These three classifications are arbitrary. They are carried out in the same equipment frequently at the same time and are governed by the same fundamental relationships.

EQUIPMENT

The simplest means of bubbling vapor through liquid is a vapor sparger or perforated pipe laid in the bottom of a tank containing the liquid. The vapor bubbles through the liquid during its passage to the surface where it is liberated and separated and leaves the tank through an overhead vapor line. Liquid may be introduced and withdrawn from the tank continuously if a continuous operation is desired. In general, such equipment is adaptable to small-scale operations where highly efficient contacting is not necessary. Its greatest limitation is the fact that the advantages of countercurrent flow are not obtained, and only one opportunity to approach equilibrium is given. Such spargers are used frequently for heating or stripping the more volatile components from organic liquids, or for absorbing volatile components of the gas in the liquid.

A spray chamber consists of a vessel through which vapors are passed, usually from bottom to top or from side to side, and into which liquid is sprayed through nozzles or spargers installed at the top or side. The liquid after contacting the vapor collects at the bottom of the chamber and is withdrawn. Such chambers are used chiefly for humidification in air conditioning and dust collection applications.

A much more efficient apparatus known as a plate tower permits the use of countercurrent flow of the vapor and liquid phases. A typical arrangement may consist of a vertical shell in which are mounted a large number of equally spaced circular plates as shown in Fig. 326. At one side of each plate a conduit known as a downspout is provided for passing

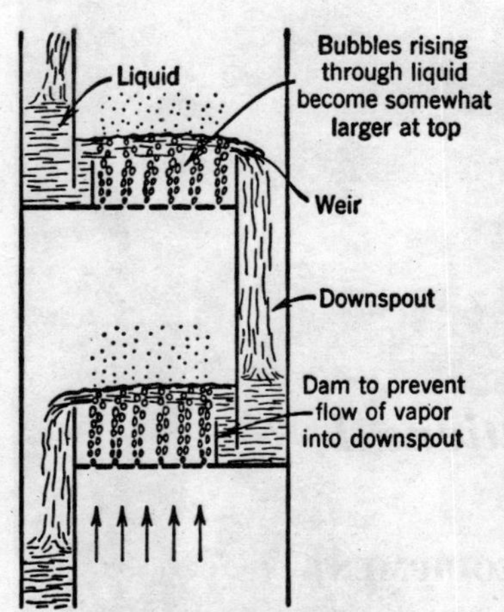

FIG. 326. Diagrammatic representation of a perforated plate tower showing downcomers.

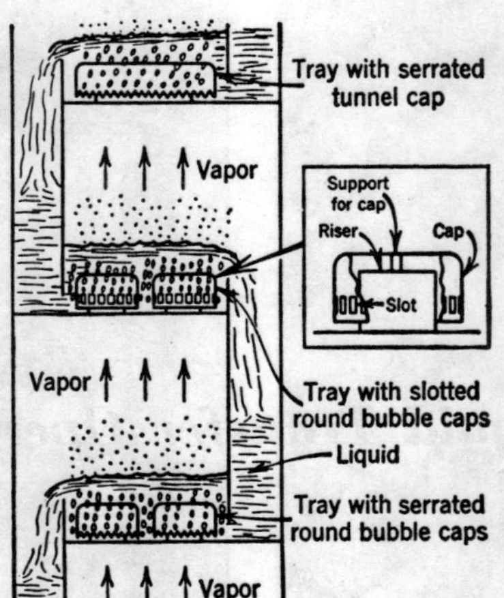

FIG. 327. Diagrammatic representation of a bubble-cap plate tower showing risers, downspouts, and caps, both slotted and serrated.

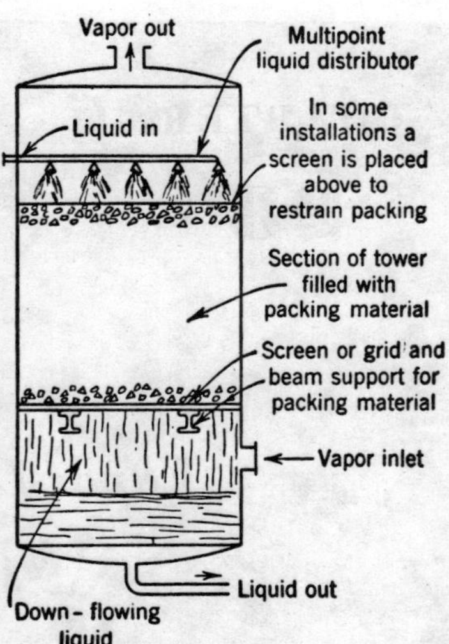

FIG. 328. Diagrammatic representation of packed tower.

liquid from the plate to the plate below. At the opposite side of the plate a similar conduit feeds liquid from the plate above.

A bubble-cap plate is a plate pierced with a number of holes into which are fitted risers or "chimneys" through which vapors from the plate below may pass. Each of the risers is covered by a bell-shaped cap which is fastened to the riser by means of a "spider" or other suitable mounting. The caps are mounted to provide sufficient space between the riser and the cap to allow passage of the vapors. The lower edge of the bubble cap may be serrated, or the "skirt" of the cap may be pierced with a number of slots as shown in Fig. 327.

In operation, vapor rises through the chimney and is diverted downwards by the cap, discharging as small bubbles from the slots or notches at the bottom of the cap beneath the liquid. The liquid level is maintained on the plate by means of a weir so that the liquid surface may be 2 to 2½ in. above the top of the slots of the bubble cap. Liquid is fed to the plate and passes across it and down through the downspout to the plate below, while the vapors pass upward through the plate, mixing intimately with the liquid on the plate because of the dispersion produced by the slots in the bubble caps. The vapors then separate at the liquid surface and pass to the plate above. Thus approximately countercurrent flow of liquid passing downward and vapor passing upward through the column is obtained.

Sieve plates are those in which the bubble-cap assembly is replaced by small holes in the plate, a typical dimension being 3/16 in. in diameter spaced on ½-in. centers. The passage of the vapors through the perforations prevents liquid from draining through the holes. The construction of weirs and downspouts is identical to that used in bubble-cap columns.

The operation of a sieve or perforated plate column in vapor-liquid transfer is similar to that in liquid-liquid transfer as illustrated in Figs. 306 and 326.

The packed tower consists of a vertical shell which is filled with suitable packing material. The liquid flows over the surface of the packing in thin films, thereby presenting a large liquid surface in contact with ascending gases. The packing is supported on a suitable grid. The liquid is introduced at the top of the packing by means of a distributing plate (a perforated plate) and the vapor is introduced beneath the grid which supports the packing. The advantages of multiple-contact and countercurrent flow are obtained in packed towers, although the efficiency of contacting is usually not equal to that obtained in the plate-type towers. Figure 328 shows a typical packed tower. Figure 328*a* is a photograph showing a tower packed with Raschig rings.

FIG. 328*a*. Raschig ring packing in place, as dumped in a packed tower. (*Lapp Insulator Co., Inc.*)

FIG. 328*c*. Cutaway view of spiral partition ring. Various shapes and numbers of turns are available. These rings are stacked manually. (*U. S. Stoneware Co.*)

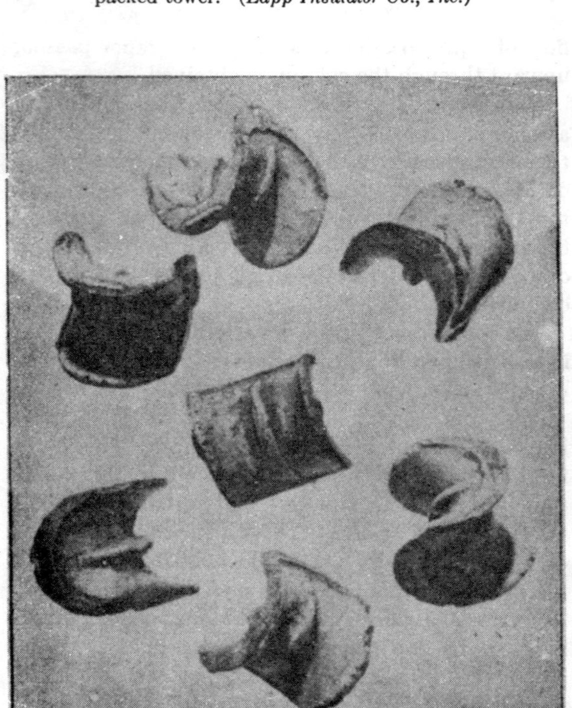

FIG. 328*b*. Various views of Berl saddles used for packing.

FIG. 328*d*. View of stacked partition rings. (*U. S. Stoneware Co.*)

FIG. 328e. Grid packing. This packing is stacked manually. (*U. S. Stoneware Co.*)

Reboil heat may be supplied to the base of a fractionating tower in many ways. Heating coils installed in the bottom section of the column are shown in Fig. 329. The heat exchange surface may become fouled with long, continued use. If the exchanger is arranged so that circulation of the liquid from the bottom of the tower through the exchanger is obtained by thermal convection, fouling

FIG. 329. Bottom section of a 4-ft diameter column, showing heating coils. (*S. Blickman, Inc.*)

may so restrict circulation that the liquid backs up into the column. For this reason external reboilers or heat exchangers are widely used as they are more convenient to clean and maintain in an efficient operating condition. The mixture of vapor and liquid is returned to the base of the column, the liquid recirculating through the exchanger and the vapor passing upwards through the tower. Direct-fired tube stills through which liquid from the bottom of the tower is pumped at high velocities are occasionally used for high-temperature operations.

Condensing steam is widely used as a source of heat, particularly where the material at the base of the column is sensitive to temperature, or the desired temperatures may be attained conveniently at available steam pressure.

For higher temperatures, condensing Dowtherm (diphenyl and diphenyl oxide) vapor, mercury vapor, or high-boiling liquids may be used. The petroleum industry often uses hot oil streams from some other part of the process to heat the bottoms of fractionating columns operating on more-volatile material.

CALCULATIONS BY ENTHALPY-COMPOSITION DIAGRAM

The calculation of the compositions of the vapor and liquid resulting from a contact depends upon the attainment of equilibrium or the use of some factor representing the departure from equilibrium. For the present only the simple case of the equilibrium contact will be considered, and the calculations represent the number of equilibrium stages required to effect the desired separation. The basic concepts for calculating vapor-liquid transfer are the same as those for calculating liquid-liquid transfer. In many cases the energy differences between liquid and vapor are so great that they constitute a major factor and cannot be neglected as was done with liquid-liquid calculations.

The energy term which is of significance in a steady-flow operation is the enthalpy H of each stream. The relative enthalpy of each stream may be readily incorporated with the composition of the stream in an enthalpy-composition diagram, Fig. 330, which shows the enthalpy H per pound mass of liquid, or vapor plotted vertically, and the composition of the liquid or vapor plotted horizontally.

Figure 330 is similar to Fig. 324 or 325 with the enthalpy per unit mass of solution H substituted for the moles of aniline per mole of aniline-free solution X_S or Y_S and the composition of the solution x or y substituted for the composition of aniline-free solution X_A or Y_A. The enthalpy H is independent of the composition of the stream in Fig. 330 in the same sense that the mole ratio of aniline X_S is independent of the composition of the aniline-free solution X_A in Figs. 324 and 325. The enthalpy-composition diagram, Fig. 330, may be used in a manner similar to that demonstrated for Figs. 324 and 325 to compute the number of equilibrium stages and quantities required to accomplish the desired separation.

Enthalpy-Composition Diagrams

The enthalpy-composition diagram shows the relative enthalpy of mixtures as a function of composition along lines of constant temperature. In many cases, the chart is limited to a single pressure, but different pressures may be included if desired. The enthalpy may be expressed in any convenient unit such as Btu's per pound or Btu's per pound mole. Where the enthalpy is plotted in Btu's per pound, the composition is given in mass fraction units, and, where the enthalpy is given in Btu's per pound mole, the composition is given in mole fraction units.

Figure 330 shows the enthalpy-composition diagram for the system of ethanol and water at 1 atm pressure. The reference states of relative enthalpy for this chart are pure water at 32° F, and pure ethanol at 32° F. The line labeled 160 in Fig. 330 shows the enthalpies of various liquid mixtures of ethanol and water at a temperature of 160° F. Similar lines are plotted for other temperatures.

The solid line labeled "saturated liquid" indicates the enthalpies of liquid mixtures of ethanol and water at their saturation temperature corresponding to 1 atm pressure, that is, at their atmospheric bubble point. The solid line labeled "saturated vapor" indicates the enthalpies of vapor mixtures at their saturation temperatures corresponding to 1 atm pressure, that is, at their atmospheric dew point.

A line connecting a liquid composition on the saturated liquid line with the equilibrium vapor composition on the saturated vapor line is an *equilibrium tie line*. There are an infinite number of tie lines which connect all points on the saturated liquid line with the corresponding equilibrium vapor compositions on the saturated vapor line.

The equilibrium compositions which determine the terminal points of a tie line may be represented most accurately by the so-called x-y diagram, which is superimposed on the enthalpy-composition diagram of Fig. 330, as an "equilibrium line." This line shows the relation between the mass per cent of ethanol in the liquid, x, plotted as the abscissa, and the mass per cent of ethanol in the equilibrium vapor, y, plotted as the ordinate to the right.

The construction of an enthalpy-composition chart for a binary mixture is a simple procedure when the following data are available.

1. The heat capacity $C_P = (\partial H/\partial T)_P$ as a function of temperature and composition.
2. The heat absorbed ΔH_S on mixing the two components at some convenient temperature as a function of composition.
3. The initial vaporization (and freezing) temperatures as a function of composition at the desired pressure.
4. The latent heat ΔH of vaporization (and of fusion) between the two phases at known temperatures for different compositions.

Vapor pressures of the solutions and densities are also helpful in extending and checking the data.

For two components, two reference states of different compositions must be chosen and assigned arbitrary values of enthalpy. The arbitrary values are usually taken as zero for the pure components at the same temperature, as, for example, 32° F in constructing Fig. 330.

The values for the base isotherm are obtained from the following equation

$$H = H_A + H_B + \Delta H_S \tag{273}$$

where H_A = enthalpy of m_A mass units of A at temperature T.

H_B = enthalpy of m_B mass units of B at temperature T.

ΔH_S = isothermal heat of solution (heat absorbed) of $m_A + m_B$ at temperature T.

H = enthalpy of $m_A + m_B$ mass units of solution at temperature T.

The enthalpies of the pure components (H_A and H_B) are obtained by integrating heat capacity data from the reference temperature T_0 to T.

$$H_A = m_A \int_{T_0}^{T} (\overline{C_P})_A \, dT = \int_{T_0}^{T} \left(\frac{\partial H_A}{\partial T}\right)_P dT$$

$$H_B = m_B \int_{T_0}^{T} (\overline{C_P})_B \, dT = \int_{T_0}^{T} \left(\frac{\partial H_B}{\partial T}\right)_P dT$$

The heat of solution data may not be in the desired form and must be converted. For example, it is often given in the form of $\Delta H_S/m_A$ as a function of m_B/m_A, where $\Delta H_S/m_A$ is the isothermal heat absorbed per unit of mass of m_A. For use in connection with equation A it is preferable to convert the data to $\Delta H_S/(m_A + m_B)$ as a function of $m_A/(m_A + m_B)$, where $\Delta H_S/(m_A + m_B)$ is the heat of solution per unit mass of solution.

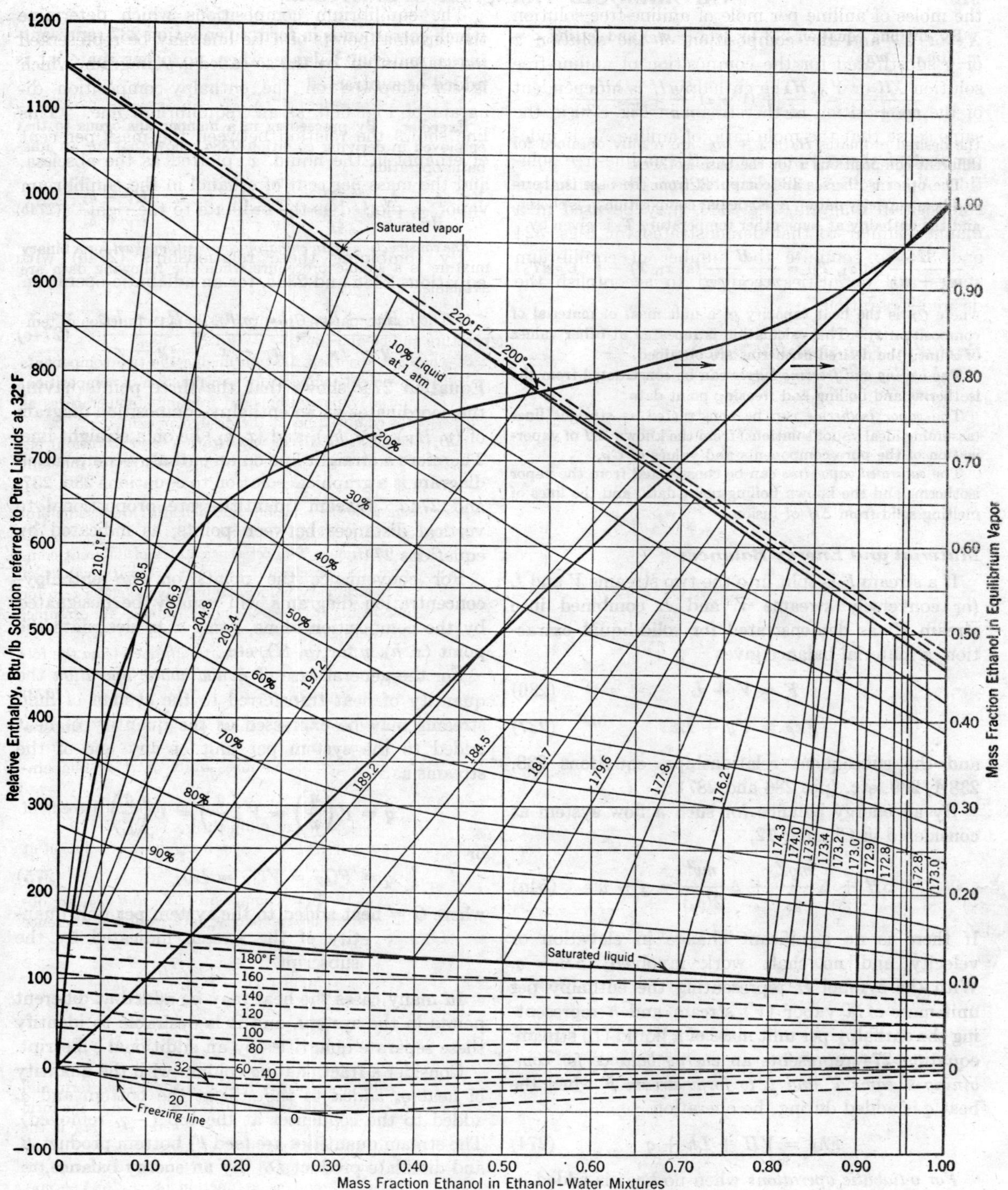

FIG. 330. Enthalpy-composition diagram for the ethanol-water system, showing liquid and vapor phases in equilibrium at 1 atm.

By dividing equation 273 by $(m_A + m_B)$ and letting $x = m_A/(m_A + m_B)$,

$$\frac{H}{m_A + m_B} = x\frac{H_A}{m_A} + (1 - x)\frac{H_B}{m_B} + \frac{\Delta H_S}{m_A + m_B}$$

the desired ordinates $H/(m_A + m_B)$ are readily obtained for different compositions x for the base isotherm.

The other isotherms are computed from the base isotherm and heat capacity data. A particular composition x_1 is chosen, and the enthalpy at some other temperature T_2 is given by

$$\frac{H}{m_A + m_B}\text{ (at } x_1, T_2) = \frac{H}{m_A + m_B}\text{ (at } x_1, T) + \int_T^{T_2} \overline{C_P}\, dT$$

where $\overline{C_P}$ is the heat capacity of a unit mass of material of composition x_1. This calculation is repeated at other values of x until the desired isotherms are obtained.

The *boiling and freezing curves* can be constructed from the isotherms and boiling and freezing point data.

The *vapor isotherms* can be constructed as straight lines (assuming ideal vapor solutions) from the known ΔH of vaporization of the pure components and values of $\overline{C_P}$.

The *saturated vapor line* can be constructed from the vapor isotherms and the known boiling-point data, and the lines of melting solid from ΔH of fusion.

Material and Energy Balances

If a stream F is split into the two streams V and L (or conversely streams V and L combined into stream F) as demonstrated for solid-liquid extraction, a material balance gives

$$F = V + L \tag{236}$$

$$Fx_F = Vy + Lx \tag{237}$$

and the subsequent relationships, equations 238, 238A, 239, etc. (pp. 286 and 287).

By an energy balance for such a flow system as considered in Chapter 12,

$$\Delta H + \Delta\frac{mgZ}{g_c} + \Delta\frac{mv^2}{2(g_c)} = q - w \tag{54a}$$

If there is no significant change in elevation or velocity and no shaft work involved, $\Delta H = q$. With the symbol H representing the enthalpy per unit mass of a vapor (V) stream and h representing the enthalpy per unit mass of a liquid (L) stream, equation 274 represents an *energy balance for combining streams V and L to form stream F* when the heat q is added during the operation.

$$Fh_F = VH + Lh + q \tag{274}$$

For adiabatic operations when no heat is added or removed ($q = 0$), equation 274 may be written

$$Fh_F = VH + Lh \tag{274a}$$

which corresponds in form to equation 237 representing a material balance when no other material is added or subtracted.

Exercise. By proceeding in a manner analogous to that employed in deriving equation 238a, show that for an adiabatic operation

$$\frac{V}{L} = \frac{h_F - h_L}{H_V - h_F};\quad \frac{V}{F} = \frac{h_F - h_L}{H_V - h_L};\quad \frac{L}{F} = \frac{H_V - h_F}{H_V - h_L} \tag{274b}$$

By combining these relationships (274b) with equations 238a and 238b, for an adiabatic operation,

$$\frac{H_V - h_F}{y - x_F} = \frac{h_F - h_L}{x_F - x} = \frac{H_V - h_L}{y - x} \tag{274c}$$

Equation 274c shows that the three points having the coordinates on an enthalpy-composition diagram of (y, H_V), (x_F, h_F), and (x, h_L) lie on a straight line. Therefore a straight line on an enthalpy-composition diagram is a graphical solution to equations 236, 237, and 274a. Stream quantities are proportional to vertical distances between points, as indicated by equations 274b.

For convenience the points on the enthalpy-concentration diagrams will usually be designated by the composition alone, using x to designate the point (x, h), y for (y, H), etc.

For the general case of *nonadiabatic operation* the quantity of heat transferred to the system of fluid streams may be expressed as the quantity of heat added to the system per unit of any one of the streams as:

$$q = F\left(\frac{q}{F}\right) = V\left(\frac{q}{V}\right) = L\left(\frac{q}{L}\right)$$

or

$$q = FQ_F = VQ_V = LQ_L \tag{275}$$

where Q = heat added to the system per unit quantity of the stream indicated by the subscript.

In many cases the heat may be added at different points in the system, and it is desirable to identify these separate quantities by an additional subscript.

Consider a fractionating column with the quantity of heat q_s added to the still at the bottom and q_c added to the condenser at the top ($-q_c$ removed). The stream quantities are feed F, bottom product B, and distillate product D. By an energy balance,

$$Fh_F + q_s + q_c = DH_D + Bh_B$$

$$Fh_F = DH_D - q_c + Bh_B - q_s$$

$$Fh_F = D\left(H_D - \frac{q_c}{D}\right) + B\left(h_B - \frac{q_s}{B}\right)$$

$$Fh_F = D(H_D - Q_{CD}) + B(h_B - Q_{SB})$$

$$F(h_F + Q_{CF}) = DH_D + B(h_B - Q_{SB})$$

$$F(h_F + Q_{SF} + Q_{CF}) = DH_D + Bh_B$$

Substituting equation 275 for q in equation 274 gives

$$F(h_F - Q_F) = VH_V + Lh_L \tag{275a}$$

or

$$Fh_F = VH_V + L(h_L + Q_L) \tag{275b}$$

or

$$Fh_F = V(H_V + Q_V) + Lh_L \tag{275c}$$

These equations are general for the addition of V and L and their subtraction from F; any one may be used. All of them reduce to equation 274*a* for an adiabatic operation.

Exercise. By combining equations 275*a*, *b*, and *c* (instead of equation 274*a*) with equations 236 and 237, show that

$$\frac{V}{L} = \frac{(h_F - Q_F) - h_L}{H_V - (h_F - Q_F)} = \frac{h_F - (h_L + Q_L)}{H_V - h_F} = \frac{h_F - h_L}{(H_V + Q_V) - h_F} \tag{275d}$$

$$\frac{V}{F} = \frac{(h_F - Q_F) - h_L}{H_V - h_L} = \frac{h_F - (h_L + Q_L)}{H_V - (h_L + Q_L)} = \frac{h_F - h_L}{(H_V + Q_V) - h_L} \tag{275e}$$

$$\frac{L}{F} = \frac{H_V - (h_F - Q_F)}{H_V - h_L} = \frac{H_V - h_F}{H_V - (h_L + Q_L)} = \frac{(H_V + Q_V) - h_F}{(H_V + Q_V) - h_L} \tag{275f}$$

Exercise. By combining these relationships (275*d*, *e*, and *f*) with equations 238*a* and 238*b*, show that

$$\frac{h_F - h_L}{x_F - x} = \frac{(H_V + Q_V) - h_F}{y - x_F} = \frac{(H_V + Q_V) - h_L}{y - x} \tag{276a}$$

$$\frac{h_F - (h_L + Q_L)}{x_F - x} = \frac{H_V - h_F}{y - x_F} = \frac{H_V - (h_L + Q_L)}{y - x} \tag{276b}$$

$$\frac{(h_F - Q_F) - h_L}{x_F - x} = \frac{H_V - (h_F - Q_F)}{y - x_F} = \frac{H_V - h_L}{y - x_L} \tag{276c}$$

The equation of the straight line on the enthalpy-composition diagram representing the operation of adding streams V and L to form stream F is either 276*a*, *b*, or *c*, depending upon which stream is the basis for heat transfer. These equations are plotted as indicated in Fig. 331. In order to produce a stream of properties x_F, h_F from streams of properties y, H_V and x, h_L it is necessary to add a quantity of heat, $q = LQ_L = VQ_V = FQ_F$, which may be expressed as Q_L Btu/lb of stream L, or Q_V Btu/lb of stream V, or Q_F Btu/lb of stream F

The point (x_F, h_F) in Fig. 331 represents the properties of the stream formed by the addition of streams V and L when a quantity of heat is *added* equal to $VQ_V = LQ_L = FQ_F$.

The point (y, H_V) represents the properties of the stream formed by the subtraction of stream L from stream F (points x and x_F) when the quantity of heat $VQ_V = FQ_F = LQ_L$ is subtracted. Similarly the point (x, h_L) represents the difference or subtraction of stream V from stream F when the same quantity of heat LQ_L is subtracted.

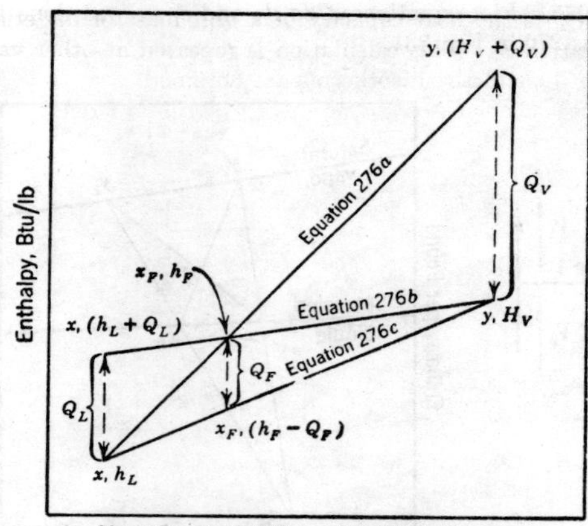

FIG. 331. Energy and material balances on an enthalpy-composition diagram corresponding to equation 276.

The ordinate of a point representing the combination of an entering stream and energy added to a process is obtained by adding the energy per unit mass of the entering stream Q to the enthalpy H of the stream. If the specified stream is leaving (subtracted from) the process, the energy added to the process per unit mass of the stream is subtracted from the enthalpy H of the stream to determine the ordinate of the point representing that stream.

If the stream F is divided into streams V and L which are removed,

$$q = \Delta H = VH_V + Lh_L - Fh_F \tag{277}$$

and the equation of the straight line representing the operation will take one of the following forms:

$$\frac{(h_F + Q_F) - h_L}{x_F - x} = \frac{H_V - (h_F + Q_F)}{y - x_F} = \frac{H_V - h_L}{y - x} \tag{278a}$$

$$\frac{h_F - (h_L - Q_L)}{x_F - x} = \frac{H_V - h_F}{y - x_F} = \frac{H_V - (h_L - Q_L)}{y - x} \quad (278b)$$

$$\frac{h_F - h_L}{x_F - x} = \frac{(H_V - Q_V) - h_F}{y - x_F} = \frac{(H_V - Q_V) - h_L}{y - x} \quad (278c)$$

Exercise. Plot equations 278*a*, *b*, and *c*, as was done for equations 276 in Fig. 331.

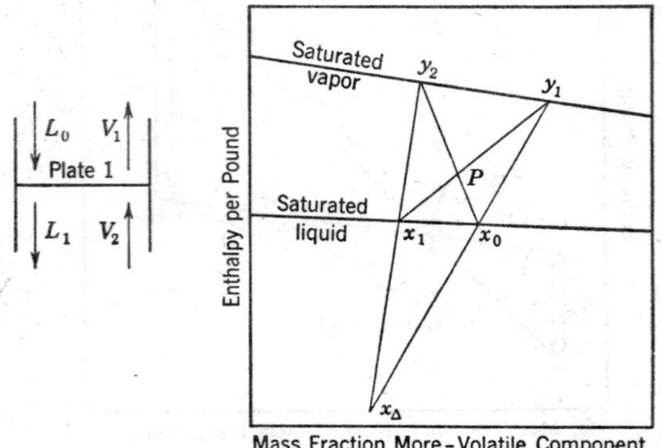

Fig. 332. Graphical representation of the operation of a single ideal stage on an enthalpy-composition diagram.

Calculations for Ideal Stages[45, 46 *]

Figure 332 represents the operation of a single equilibrium stage or plate of a plate fractionating column. The saturated liquid L_0 and the saturated vapor V_2, represented by points x_0 and y_2, respectively, are fed continuously to this stage where they are intimately mixed. The mixed phases separate into the vapor V_1 and the liquid L_1, represented by points y_1 and x_1, respectively, which leave the stage in equilibrium. The point y_2 represents a saturated vapor feed to the stage and therefore lies on the vapor saturation line; the point x_0 represents a saturated liquid and lies on the liquid saturation line.

The point P represents the sum of the liquid and vapor feed to the stage and is located on the straight line y_2x_0 so that the ratio of the distances $\overline{Py_2}/\overline{Px_0}$ equals the ratio of the quantities of the feed streams L_0/V_2. The sum of the liquid and vapor feeds to the stage must equal the sum of the equilibrium liquid and vapor leaving the stage, and therefore the vapor V_1 and the liquid L_1 are represented by the terminal points of the equilibrium tie line which passes through the point P. The fractionation accomplished by the stage may be expressed either as the change in vapor composition produced, $y_1 - y_2$, or as the change in liquid composition, $x_0 - x_1$. Both a liquid and a vapor stream must be fed to the stage in order to produce the fractionation.

The difference between the liquid L_0 and the vapor V_1 passing one another above the stage is equal to the difference between the liquid L_1 and the vapor V_2 passing one another below the stage. This difference is represented by the point x_Δ, which is located by the intersection of the lines x_1y_2 and x_0y_1. The quantity ratios L_1/V_2 and L_0/V_1 are equal to the distance ratios $\overline{x_\Delta y_2}/\overline{x_\Delta x_1}$ and $\overline{x_\Delta y_1}/\overline{x_\Delta x_0}$, respectively.

* The bibliography for this chapter appears on p. 395.

Several Equilibrium Plates or Stages[45, 46]

A system of plates placed one above another, as shown in Fig. 333, constitutes a plate column.

The point x_0 represents a saturated liquid feed to the top plate, and the point y_6 represents a saturated vapor feed under the bottom plate of the column. The sum of the feed streams, L_0 and V_6, is represented by the point P, so located on the straight line x_0y_6 that the ratio of the distances $\overline{Py_6}/\overline{Px_0}$ equals the ratio of the feed quantities L_0/V_6. Since the sum of the feed streams equals the sum of the product streams L_5 and V_1, the point P is also located on the straight line x_5y_1, and the ratio of the distance $\overline{Py_1}/\overline{Px_5}$ equals L_5/V_1.

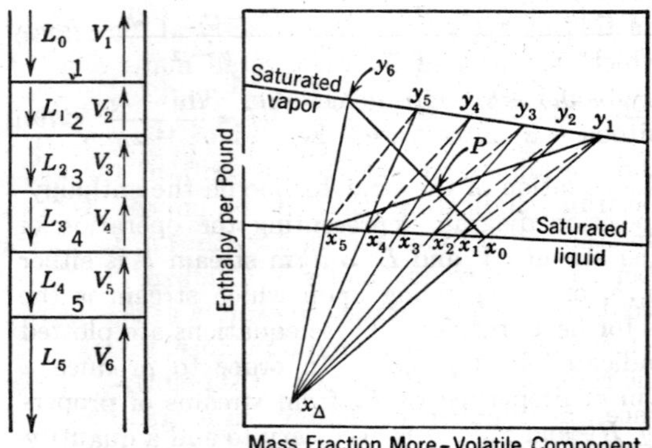

Fig. 333. Graphical representation of the operation of a multiple-stage ideal plate column on an enthalpy-composition diagram.

The difference Δ between the liquid feed L_0 and the vapor product V_1, which pass one another above the top plate, equals the difference between the vapor feed V_6 and the liquid product L_5, which pass one another below the bottom plate. The point x_Δ, which represents this difference, is located at the intersection of the lines $\overline{x_0y_1}$ and $\overline{x_5y_6}$.

The difference between the liquid and vapor passing one another above the top plate and below the bottom plate is also equal to the difference between the liquid and vapor passing one another between any two plates of the column, i.e., between V_2 and L_1, V_3 and L_2, V_4 and L_3, etc. The point x_Δ is therefore the common point of intersection for the straight lines drawn through all the pairs of points representing the liquid and vapor passing one another between any two stages of the column.

The liquid L_1 leaving the top plate is in equilibrium with the vapor V_1 leaving the top plate, and the point x_1 in Fig. 333 is located at the end of the equilibrium tie line from the point y_1 on the vapor saturation line, representing the vapor product. The point y_2, representing the vapor rising from the second plate and passing the liquid L_1, is located at the intersection of the line $\overline{x_\Delta x_1}$ with the vapor saturation line. The liquid L_2 leaving the second plate is in equilibrium with the vapor V_2 leaving the second plate and is represented by the point x_2, located at the end of the tie line from the point y_2. The point y_3 representing the vapor rising from the third plate is located at the intersection of the line $\overline{x_\Delta x_2}$ with the vapor saturation line. The plate-to-plate construction is carried out in this manner until a liquid composition is obtained which is equal to that of the bottom product.

The tie line $\overline{x_5y_5}$ terminates exactly at the point x_5, which was located by the overall material and energy balances around the column. This is entirely fortuitous. In most cases the end of the tie line will not terminate at exactly the bottom product composition, indicating that it is impossible to obtain the separation specified with an integral number of equilibrium plates. Although the tie line could be made to terminate exactly at the point x_5, determined from the material balance, by adjusting the ratio of the feeds slightly, it is pointless to do so since the number of actual plates required is determined by dividing the number of equilibrium plates by a plate efficiency. Therefore, as a general rule, the plate-to-plate construction is continued until a liquid composition is obtained which is less than that of the bottom product composition and the fractional plate estimated by a linear interpolation along the liquid saturation line.

Stripping Columns

A column used to remove the more volatile component from a liquid feed is known as a stripping column. The vapor stream necessary to accomplish the desired fractionation may be generated from the liquid leaving the bottom plate by means of a still, as indicated in Fig. 334. These vapors generated in the still (by the addition of the heat q_S in the still) pass upward to the bottom plate. The desired stripped product B is withdrawn from the still liquid, or in some cases from the still vapor or the liquid from the bottom plate. The feed F is introduced on the top plate, and a distillate product V_1 is withdrawn from the top plate.

The calculation of the number of equilibrium plates required in a stripping column is indicated in Fig. 334, which is the solution to the following example.

Illustrative Example. The ethanol content of a feed containing 50.0 mass per cent ethanol and the balance water is to be stripped in a plate column equipped with a still to produce a bottom product containing 1.0 mass per cent ethanol. Heat is to be supplied to the still at the rate of 440 Btu/lb of feed. The feed, at 70° F, is introduced on the top plate, and the overhead vapors are condensed to give a distillate product. The column is to operate at 1 atm pressure. The bottom product is withdrawn as a liquid from the still.

Compute:

1. The number of equilibrium plates required.
2. The quantity of distillate product per pound of feed.

Solution. The feed is represented by point x_F (Fig. 334); the bottom product by point x_B on the liquid saturation line. The point x_F' represents the feed plus the energy added to the system in the still, the distance $\overline{x_F'x_F}$ being equal to 440 Btu/lb of feed. The point x_F' also represents the sum of the bottom product B and the overall distillate vapor V_1, and therefore the point V_1 is located by extending the line $\overline{x_Bx_F'}$ to the vapor saturation line.

The stripping column shown in Fig. 334 is similar to the continuous process of Fig. 333. In the process of Fig. 333, the vapor stream V_6 was known or specified, whereas in this example the quantity of heat applied to the still for generating vapor is specified. Since the feed stream F entering the column equals the sum of the distillate product V_1 and the bottom product B, minus the energy added in the still ($q_S = BQ_{SB}$), the point $x_{\Delta B}$, having the coordinates x_B and $h_B - Q_{SB}$, lies on the straight line drawn through the points x_F and y_1 to the stated value $x_B = 0.01$.

The point $x_{\Delta B}$ represents the difference between the feed entering the top plate and the distillate leaving the top plate; also the difference between the liquid and vapor passing each other between any two plates of the column or what might

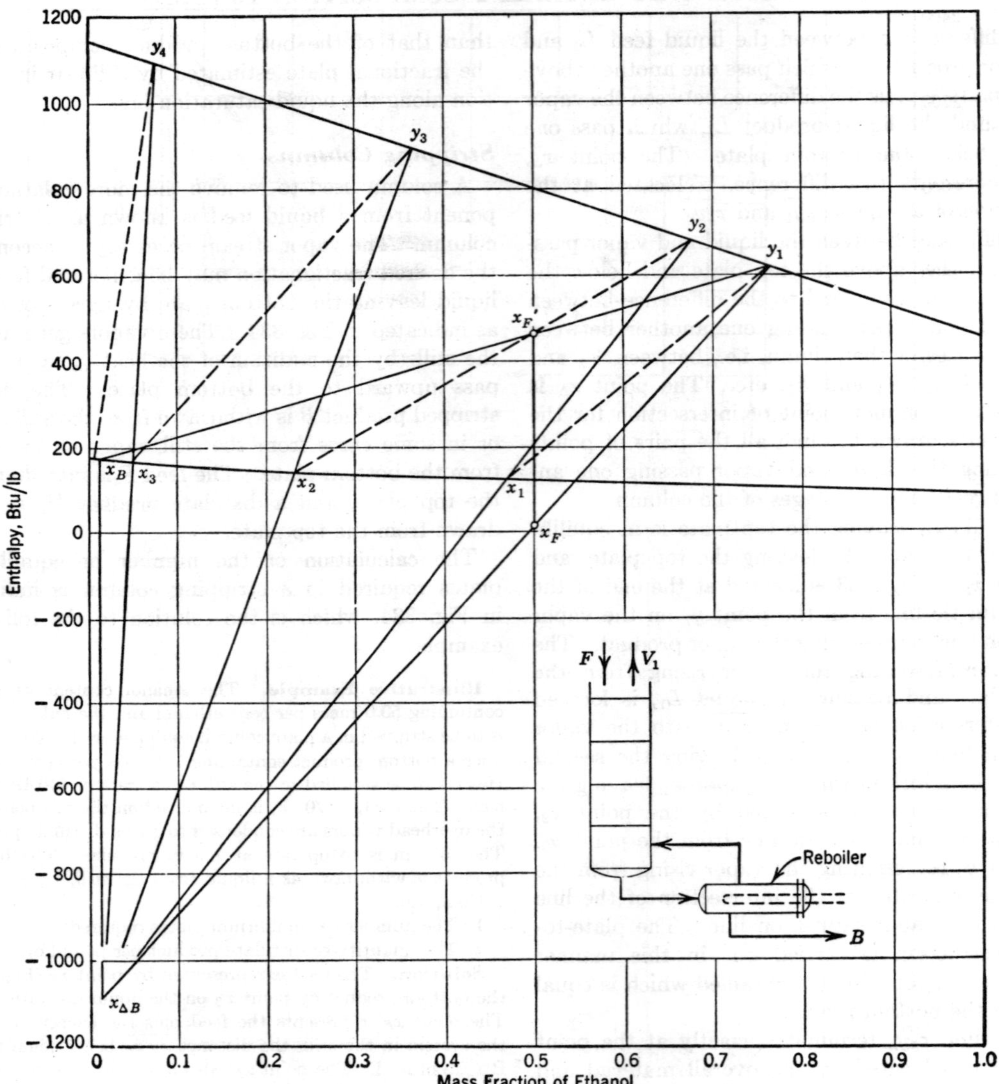

FIG. 334. Graphical representation of the operation of a stripping column on an enthalpy-composition diagram for the ethanol-water system at 1 atm.

be called the "net flow" toward the bottom of the column. The plate-to-plate construction is carried out as described. The number of equilibrium plates required is calculated as 3.8. The vapors leaving the still may usually be assumed to be in equilibrium with the liquid in the still, and therefore the number of equilibrium plates which must be provided in the column equals 3.8 − 1.0 = 2.8. The quantity of distillate product per pound of feed is equal to the ratio of the distances $\overline{x_F x_{\Delta B}}/\overline{x_{\Delta B} y_1} = 0.782$.

The distillate product from a stripping column must, of necessity, have a composition close to that of the feed composition, since the distillate vapors arise from the plate to which the feed is introduced. However the feed plate liquid composition is not equal to the feed composition, as shown in Fig. 334 where the feed composition x_F is 0.500 and the feed plate liquid composition x_1 is 0.476 weight fraction of ethanol.

Rectifying Columns

A column used to remove the less-volatile component from a vapor feed is known as a rectifying column or absorber.

Consider the case where it is desired to increase

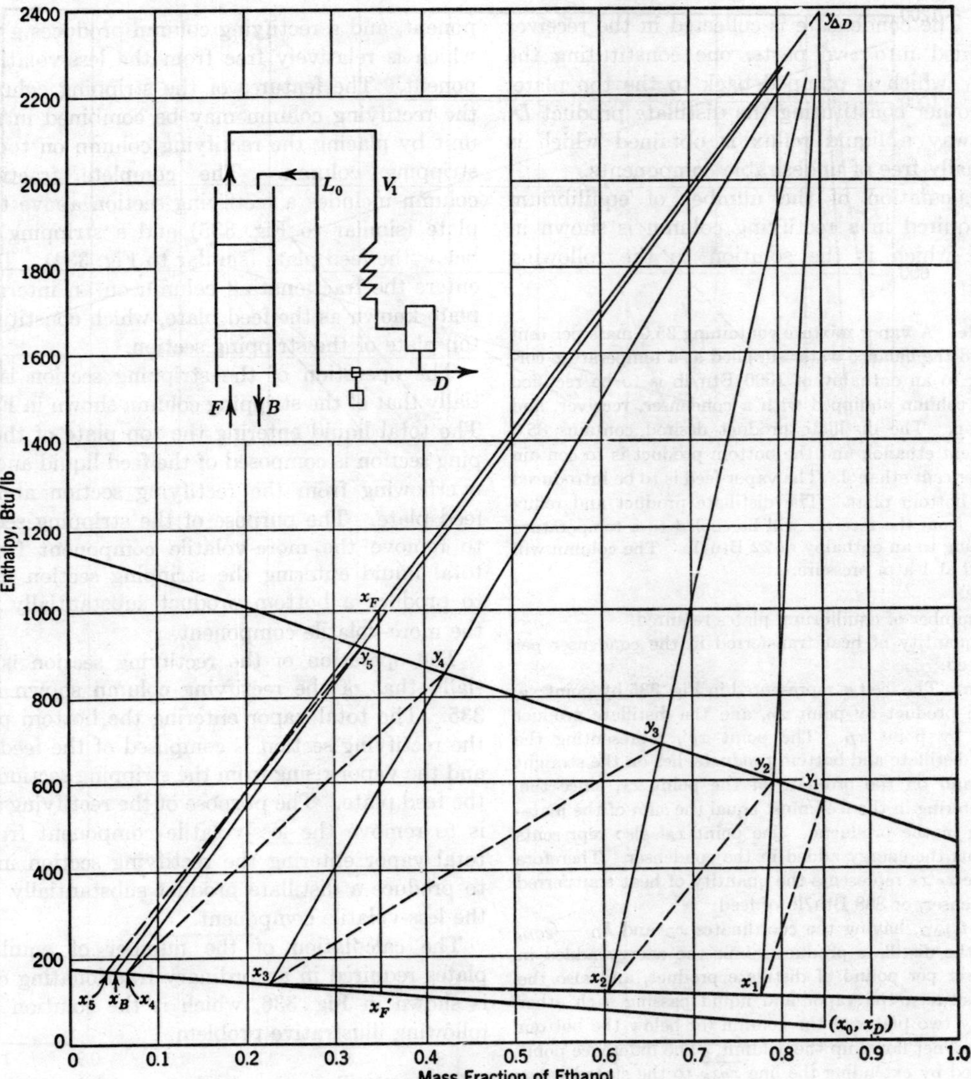

Fig. 335. Graphical representation of the operation of a rectifying column on an enthalpy-composition diagram for the ethanol-water system at 1 atm.

the concentration of ethanol by removing water from a vapor mixture of ethanol and water (such as V_1 of Fig. 334). Since the composition of the distillate vapor product from the column is largely dependent upon the composition of the liquid supplied to the top plate, a low concentration of water or any other undesirable component is required in the liquid L_0 supplied to the top plate of the rectifying column (Fig. 335). If such liquid is available it may be supplied from an independent source. Usually it is impossible to obtain such liquids from independent sources, but they can be obtained readily by condensing and returning a portion of the overhead vapors to the top plate. Liquid so returned is called reflux. A diagrammatic illustration of a rectifying column equipped with a condenser, receiver, and reflux pump to furnish reflux is shown in Fig. 335.

The overhead vapors V_1 from the top plate pass through the condenser where the vapors are condensed. The condensate is collected in the receiver and divided into two parts, one constituting the reflux L_0, which is pumped back to the top plate, and the other constituting the distillate product D. In this way a liquid reflux is obtained which is substantially free of undesirable components.

The calculation of the number of equilibrium plates required in a rectifying column is shown in Fig. 335 which is the solution to the following example.

Example. A vapor mixture containing 35.0 mass per cent ethanol and the balance water supplied at a temperature corresponding to an enthalpy of 1000 Btu/lb is to be rectified in a plate column equipped with a condenser, receiver, and reflux pump. The distillate product desired contains 85.0 mass per cent ethanol, and the bottom product is to contain 5.0 mass per cent ethanol. The vapor feed is to be introduced under the bottom plate. The distillate product and reflux withdrawn from the receiver will be cooled to a temperature corresponding to an enthalpy of 22 Btu/lb. The column will be operated at 1 atm pressure.

Compute:

1. The number of equilibrium plates required.
2. The quantity of heat transferred in the condenser per pound of feed.

Solution. The feed is represented in Fig. 335 by point x_F, the bottom product by point x_B, and the distillate product and reflux by point x_D. The point x_F', representing the sum of the distillate and bottom products, lies on the straight line $\overline{x_Bx_D}$ and on the ordinate of the point x_F, since the material entering in the feed must equal the sum of the material leaving in the products. The point x_F' also represents the feed plus the energy added in the condenser. Therefore the distance $\overline{x_F'x_F}$ represents the quantity of heat transferred in the condenser, or 888 Btu/lb of feed.

The point $y_{\Delta D}$, having the coordinates x_D and $h_D - Q_{CD}$, represents the distillate product minus the energy added in the condenser per pound of distillate product, and also the difference between the vapor and liquid passing each other between any two plates of the column (or below the bottom plate) or the "net flow" up the column. The difference point $y_{\Delta D}$ is located by extending the line $\overline{x_Bx_F}$ to the stated value $x_D = 0.85$. The plate-to-plate construction carried out as described previously indicates the required number of equilibrium plates to be 4.4.

Complete Fractionating Columns

In most commercial operations, a relatively complete separation between two components is desired. This requires a column which is capable of producing from a single feed a distillate product containing only a very small amount of the less-volatile component and a bottom product containing a very small amount of the more-volatile component.

A stripping column produces a bottom product which is relatively free from the more-volatile component, and a rectifying column produces a product which is relatively free from the less-volatile component. The features of the stripping column and the rectifying column may be combined in a single unit by placing the rectifying column on top of the stripping column. The complete fractionating column includes a rectifying section above the feed plate (similar to Fig. 335) and a stripping section below the feed plate (similar to Fig. 334). The feed enters the fractionating column on an intermediate plate known as the feed plate, which constitutes the top plate of the stripping section.

The operation of the stripping section is essentially that of the stripping column shown in Fig. 334. The total liquid entering the top plate of the stripping section is composed of the feed liquid and liquid overflowing from the rectifying section above the feed plate. The purpose of the stripping section is to remove the more-volatile component from the total liquid entering the stripping section in order to produce a bottom product substantially free of the more-volatile component.

The operation of the rectifying section is essentially that of the rectifying column shown in Fig. 335. The total vapor entering the bottom plate of the rectifying section is composed of the feed vapor and the vapor rising from the stripping section below the feed plate. The purpose of the rectifying section is to remove the less-volatile component from the total vapor entering the rectifying section in order to produce a distillate product substantially free of the less-volatile component.

The calculation of the number of equilibrium plates required in an ordinary fractionating column is shown in Fig. 336, which is the solution to the following illustrative problem.

Illustrative Example. A column containing 10 plates and equipped with a still and apparatus for returning reflux is producing a distillate product containing 90 mass per cent ethanol and a bottom product containing 1.0 mass per cent ethanol, from a feed containing 35.0 mass per cent ethanol and the balance water. The column operates at a pressure of 1 atm absolute. The feed is introduced into the downspout of the seventh plate at a temperature of 183.4° F and at the rate of 1000 lb/hr. The distillate product and reflux are withdrawn from the reflux accumulator at 114° F. Reflux is supplied to the top plate of the column at the rate of 1170 lb/hr. It is desired to compute the overall plate efficiency, the rate of heat transfer in the still and condenser, and the rate at which vapor is produced from the still.

Solution. The feed is a mixture of liquid and vapor having an enthalpy of 150 Btu/lb, shown in Fig. 336 by point x_F on the equilibrium tie line for 183.4° F at $x_F = 0.35$ (see Fig.

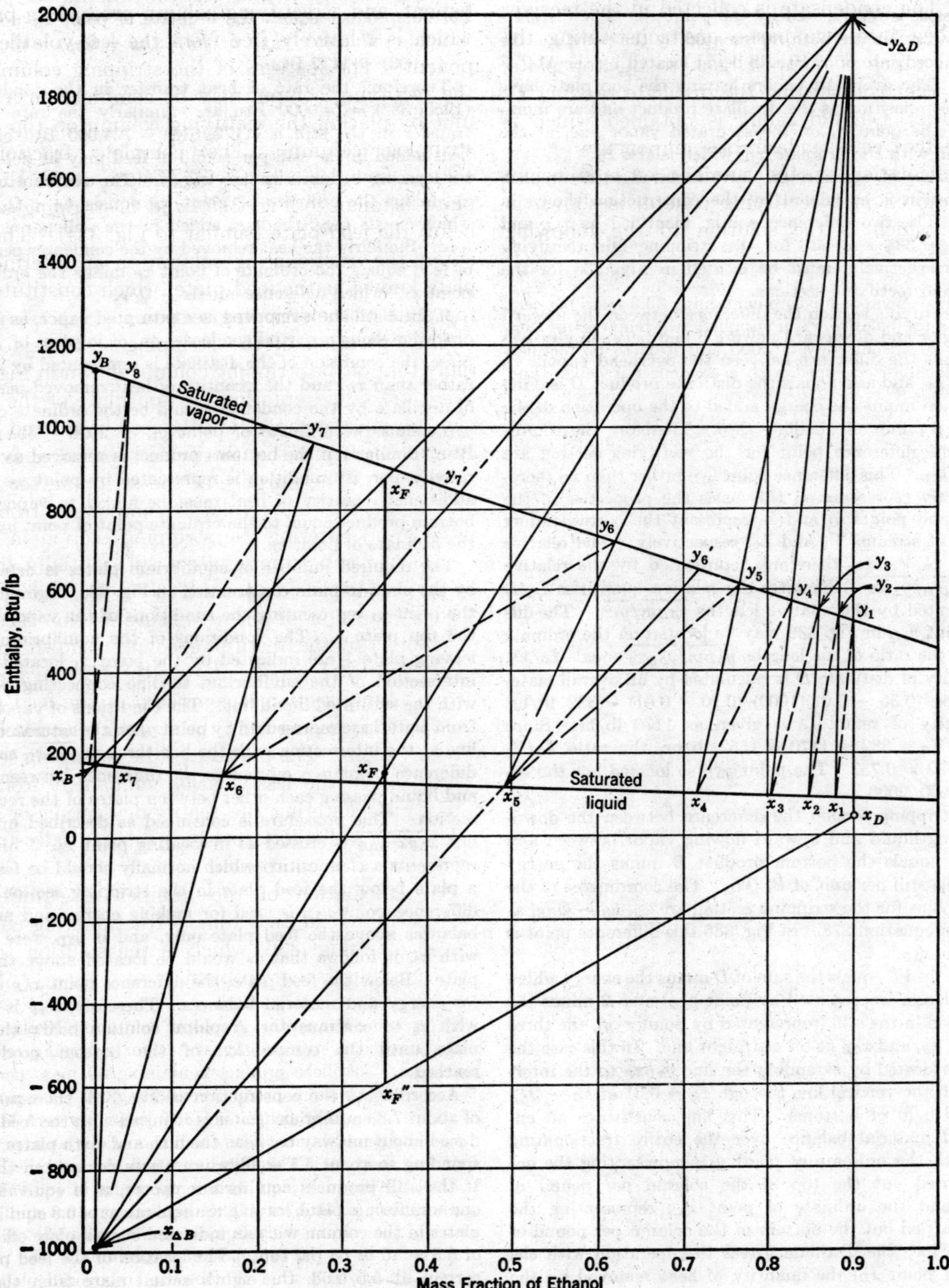

Fig. 336. The application of an enthalpy-composition diagram for the ethanol-water system to the determination of operating conditions for a fractionating column containing stripping and rectifying sections.

330). The bottom product, withdrawn as a saturated liquid from the still, is represented by the point x_B. The distillate product and reflux are both represented by the point x_D, the enthalpy coordinate of 50 Btu/lb being located by the 114° F isotherm. The overhead vapors leaving the top plate have the same composition as the distillate product and are represented by the point y_1 on the saturated vapor line, at the intersection with the ordinate x_D, which is also x_0.

The number of equilibrium plates required to accomplish this separation is calculated by the construction shown in Fig. 336 The two difference points, identified as $x_{\Delta B}$ and $y_{\Delta D}$ in Figs. 334 and 335 for the stripping and rectifying columns, respectively, must be located in Fig. 336 for the stripping and rectifying sections.

In the rectifying section the difference between the upward flowing vapor and downward flowing liquid between any two plates equals the difference between the overhead vapors V_1 and reflux L_0, and also equals the distillate product D leaving the operation minus the energy added to the operation in the condenser per unit of distillate Q_{CD}. Therefore the coordinates of the difference point for the rectifying section are $x_D, h_D - Q_{CD}$. This difference point $y_{\Delta D}$ rather than x_D therefore correctly represents on this basis the properties of the distillate, and points y_1 and x_0 represent the corresponding properties of streams V_1 and L_0, respectively. The relative quantities L_0/V_1 are, therefore, represented by the relative lengths $\overline{y_{\Delta D}y_1}/\overline{y_{\Delta D}x_0}$. Similarly the relative quantities L_0/D are represented by the relative lengths $\overline{y_{\Delta D}y_1}/\overline{y_1x_0}$. The difference point $y_{\Delta D}$ in Fig. 336 may be located on the ordinate x_D so that the ratio of the lengths $\overline{y_{\Delta D}y_1}/\overline{y_{\Delta D}x_0}$ equals L_0/V_1. The quantity of distillate D is calculated by an overall material balance $(0.35 - 0.01)(1000)/(0.90 - 0.01) = 382$ lb/hr. The quantity of reflux L_0 is given as 1170 lb/hr. Since $V_1 = D + L_0 = 382 + 1170 = 1552$ lb/hr, the ratio $L_0/V_1 = 1170/1552 = 0.757$. The point $y_{\Delta D}$ so located has the coordinates 0.90, 2000.

In the stripping section the difference between the downward flowing liquid and upward flowing vapor between any two plates equals the bottom product B minus the energy added in the still per unit of B, Q_{SB}. The coordinates of the difference point for the stripping section are $x_B, h_B - Q_{SB}$, as indicated in equation 278. In Fig. 336 this difference point is identified as $x_{\Delta B}$.

Since the feed F equals the sum of D minus the energy added in the condenser (represented by point $y_{\Delta D}$) and B minus the energy added in the still (represented by point $x_{\Delta B}$), the three points $y_{\Delta D}$, x_F, and $x_{\Delta B}$ lie on a straight line. In this case the point $x_{\Delta B}$ is located by extending the line $\overline{y_{\Delta D}x_F}$ to the intersection with the vertical line through $x_B = 0.01$ at $h_B - Q_{SB} = -993$ Btu/lb of bottoms. This line constitutes an enthalpy and material balance over the entire fractionating column with the ordinate of point $y_{\Delta D}$ representing the enthalpy carried out the top of the column per pound of distillate, and the ordinate of point $x_{\Delta B}$ representing the enthalpy carried out the bottom of the column per pound of bottoms. Since the distillate leaves the operation with the enthalpy of point x_D, the quantity of heat removed by the condenser per pound of distillate equals the ordinate of $y_{\Delta D}$ minus the ordinate of x_D, or $(h_D - Q_{CD}) - h_D = -Q_{CD} = 2000 - 50 = 1950$ Btu. The bottom product leaves the operation as saturated liquid with the enthalpy of point x_B, and the quantity of heat added in the still per pound of bottom product equals the ordinate of point x_B minus the ordinate of point $x_{\Delta B}$, or $h_B - (h_B - Q_{SB}) = Q_{SB} = 179 - (-993) = 1172$ Btu.

Therefore, the rate of heat transfer in the condenser is $(1950)(382) = 745{,}000$ Btu/hr. Similarly the rate of heat transfer in the still is $(1172)(618) = 724{,}000$ Btu/hr. The heat added in the still per pound of feed may be determined by locating x_F' on the line $\overline{x_By_{\Delta D}}$. The ordinate of point x_F' minus the ordinate of point x_F equals $h_F + Q_{SF} - h_F$, which equals Q_{SF}, the heat added in the still per pound of feed. Similarly the heat removed by the condenser per pound of feed equals the ordinate of point x_F minus the ordinate of point x_F'' which difference equals $-Q_{CF}$.

If the distillate is removed as a saturated vapor, as from an operation using a partial condenser on, or in place of, the top plate, the condition of the distillate is represented by point y_1 rather than x_D, and the quantity of heat removed per pound of distillate by the condenser would be the ordinate of point $y_{\Delta D}$ minus the ordinate of point y_1, or $2000 - 520 = 1480$ Btu. Similarly if the bottoms product is removed as a saturated vapor, its condition is represented by point y_B and an additional quantity of heat must be added to vaporize the bottom product equal to the ordinate point of point y_B minus the ordinate of point x_B.

The required number of equilibrium plates is determined by the plate-to-plate construction on Fig. 336, beginning with the point y_1 representing the conditions of the vapor leaving the top plate 1. The conditions of the equilibrium liquid leaving plate 1 are indicated by the point x_1 located at the intersection of the equilibrium tie line connecting point y_1 with the saturated liquid line. The conditions of vapor rising from plate 2 are represented by point y_2 on the saturated vapor line at the intersection with the line through $\overline{y_{\Delta D}x_1}$, since the difference point $y_{\Delta D}$ represents the difference between vapor and liquid passing each other between plates of the rectifying section. This procedure is continued as described until the line $\overline{x_{\Delta B}x_Fy_{\Delta D}}$ is crossed as in locating point x_6. This point represents a composition which normally should be found on a plate below the feed plate in the stripping section. The difference point $y_{\Delta D}$ is used for making energy and material balances above the feed plate only, and if $y_{\Delta D}$ were joined with x_6, it follows that x_6 would be located above the feed plate. Below the feed plate, the difference point $x_{\Delta B}$ is used for energy and material balances. Therefore, $x_{\Delta F}$ is joined with x_6 to continue the graphical solution below the feed plate until the composition of the bottom product is reached.

According to the construction of Fig. 336, the equivalent of about 7.8 equilibrium plates is required with the feed introduced about midway between the fifth and sixth plates, corresponding to about 5.5 equilibrium plates down from the top. If the still produces equilibrium vapors, it is equivalent to one equilibrium plate, leaving the equivalent of 6.8 equilibrium plates in the column with an indicated overall plate efficiency of 6.8 ÷ 10 or 68 per cent. The location of the feed plate is correct at 5.5/0.68, the eighth actual plate from the top. If the actual feed plate location did not correspond to the correct location in terms of equilibrium plates, other equilibrium feed plate locations must be assumed and the corresponding number of equilibrium plates computed. The resulting overall plate efficiency is then used to check the assumed feed plate location.

A tie line through the point x_B intersects the vapor line at the point y_8 which represents the conditions of the vapor rising from the still and passing the liquid from the bottom plate 7 represented by point x_7. Since point $x_{\Delta B}$ represents the difference between the rising vapor y_8 and the liquid x_7, the ratio of the lengths $(\overline{x_{\Delta B}x_7})/(\overline{x_{\Delta B}y_8}) = V_8/L_7$ and, since $L_7 = V_8 + B$, the ratio $(\overline{x_{\Delta B}x_7})/(\overline{y_8x_7}) = V_8/B = 1.24$. Thus $V_8 = (1.24)(618) = 770$ lb/hr.

Feed Plate Location

If the composition and thermal condition of the feed, distillate product, and bottom product are fixed and if the heat transferred in the still or in the condenser is fixed, the energy and material balances around the column and the location of the difference points $y_{\Delta D}$ and $x_{\Delta B}$ are fixed. Under these conditions the number of equilibrium plates required to effect the separation specified is a function of the feed plate location.

In Fig. 336 the feed is introduced between equilibrium plates 5 and 6. This is an optimum location because it requires a smaller number of equilibrium plates than any other location. If the feed is introduced on equilibrium plate 5, the composition of the vapor rising from equilibrium plate 6 is determined in Fig. 336 by the intersection of the line $\overline{x_{\Delta B}x_5}$ with the vapor saturation line at y_6'. The fractionation occurring from plate 5 to plate 6 is then less, and consequently the total number of plates required is greater. Similarly, if the feed is introduced on plate 7 instead of plate 6, the composition of the vapors rising from plate 7 is determined by the intersection of the line $\overline{y_{\Delta D}x_6}$ with the vapor saturation line at y_7'. The fractionation occurring from plate 6 to plate 7 is then less. The optimum feed location is on that plate whose liquid composition is less than or equal to the composition indicated by the intersection of the overall balance line $\overline{x_{\Delta B}x_F y_{\Delta D}}$ with the liquid saturation line.

There are definite limits to the location of the feed plate when a definite separation is specified and the overall energy and material balances around the column are fixed. In such cases the feed must be introduced at least a certain number of plates above the bottom and at least a certain number of plates below the top. If the feed is introduced so far down the column that the equilibrium tie line from the feed plate liquid coincides with the line to the difference point $y_{\Delta D}$, the total number of plates required is infinitely large. A similar case occurs where the feed plate is so far up the column that the equilibrium tie line from the feed plate liquid coincides with the line to the difference point $x_{\Delta B}$. The column will not operate under the conditions fixed if the feed is introduced outside of these limits.

In the discussions of distillation problems, it is usually assumed that the feed is introduced on the optimum plate unless otherwise specified.

Reflux Ratio

The term *reflux ratio* may be applied to the ratio L_0/V_1, as determined, but it is frequently used to describe the ratio of quantity of liquid pumped back to the top plate of a fractionating column to the quantity of distillate product withdrawn (L_0/D). The overhead vapors entering the condenser V_1 are equal to the sum of the reflux liquid L_0 and distillate product minus the energy added in the condenser per unit of distillate Q_{CD}. As explained in the preceding illustrative example the ratio

$$\frac{L_0}{V_1 - L_0} = \frac{L_0}{D} = \frac{\overline{y_{\Delta D}y_1}}{\overline{y_1x_0}} = \frac{(h_D - Q_{CD}) - H_1}{H_1 - h_D} \tag{279}$$

The liquid refluxed to the top of the column is sometimes indicated by the symbol R as well as by L_0. The ratio L_0/V_1 is sometimes called the internal reflux ratio, and the ratio L_0/D the external reflux ratio.

For the column whose operation is represented in Fig. 336, $L_0/D = 3.06$.

If a distillate product of a given composition is to be produced, the effect of increasing the reflux ratio is to increase the value of $h_D - Q_{CD}$. With a fixed thermal condition of the feed, the greater the value of $h_D - Q_{CD}$, the less is the value of $h_B - Q_{SB}$ for a given distillate and bottom product. The greater the distance that the difference points $y_{\Delta D}$ and $x_{\Delta B}$ are removed from the saturation lines, the greater will be the fractionation accomplished by each plate, and consequently the greater the reflux ratio. In general, the greater the reflux ratio, the smaller is the total number of plates required for a given operation.

Total Reflux

When all overhead vapors are condensed and returned to the column as reflux so that no distillate product is withdrawn, the column is operating under total reflux. If no distillate product is withdrawn, no bottom product can be withdrawn and no feed

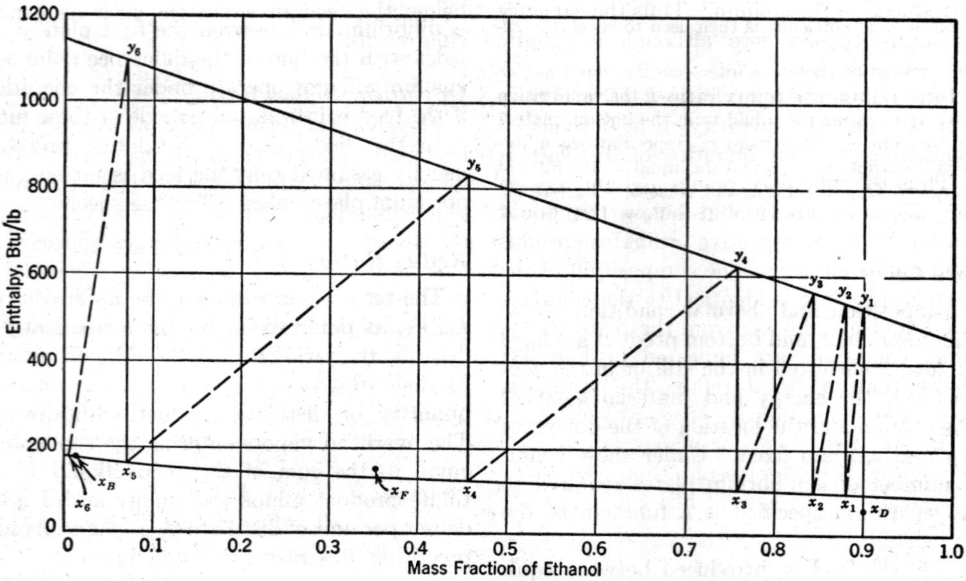

FIG. 337. Graphical representation on an enthalpy-concentration diagram of the operating conditions of a column at total reflux.

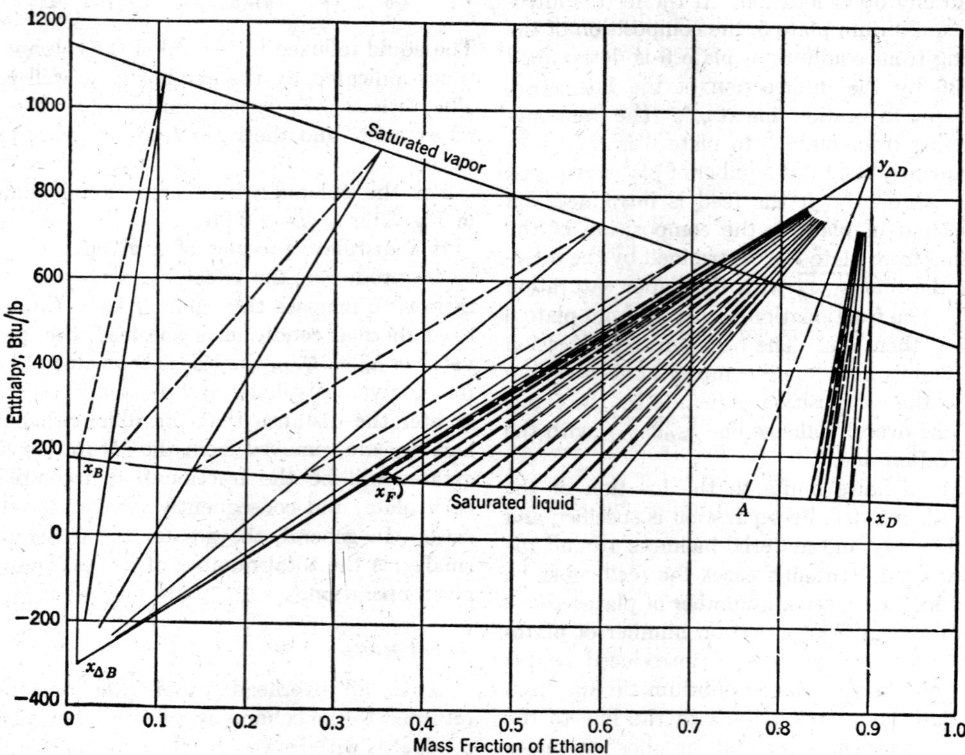

FIG. 338. Enthalpy-composition diagram for the ethanol-water system at 1 atm indicating a column operating at minimum reflux (infinite stages) at point A.

can be introduced to the column. Thus the capacity of the column becomes zero although a definite separation is taking place.

Under total reflux, the reflux ratio L_0/D is infinite (L/V is unity), and the quantity of heat removed in the condenser per unit quantity of distillate is infinite. Thus the difference points $y_{\Delta D}$ and $x_{\Delta B}$ lie at infinite distances above and below the liquid saturation line as their respective ordinates are plus infinity and minus infinity. The composition of the vapor leaving any plate is identical to the composition of the liquid overflowing from the plate above. Total reflux conditions are indicated in Fig. 337 which shows that, at total reflux, the minimum number of plates is required to effect a given separation. The six ideal stages or plates indicated in Fig. 337 accomplish the same separation under total reflux as the eight ideal stages in Fig. 336.

Minimum Reflux

As the reflux ratio (L_0/D) decreases from infinity at total reflux, the number of plates required to produce a given separation increase from the minimum value at total reflux to an infinite number at the minimum reflux ratio.

A zone of an infinite number of plates (where the change in composition from plate to plate is zero) develops wherever the extension of an equilibrium tie line passes through the point $x_{\Delta B}$ or the point $y_{\Delta D}$. This condition frequently arises when the line $x_{\Delta B}x_F y_{\Delta D}$ coincides with the equilibrium tie line that passes through the feed point x_F. In Fig. 338 the reflux ratio is somewhat greater than the reflux ratio which would give infinite plates at the feed plate, but the fractionation per plate near the feed plate has become very small.

A zone of infinite plates in some systems may develop in the rectifying section at reflux ratios higher than that which gives infinite plates at the feed plate. This occurs when one of the equilibrium tie lines in the rectifying section intersects the ordinate x_D at a greater enthalpy than the equilibrium tie line through the feed point (x_F). This is illustrated in Fig. 338 where the line $Ay_{\Delta D}$, the extension of the tie line through A, intersects the line $y_{\Delta D}x_D$ at a higher point than any of the other tie lines in the rectifying section including that through the feed point. Here the zone of infinite plates develops at point A before it develops at the feed plate as reflux ratio is decreased.

Such a zone of infinite plates may also develop below the feed in a system such as benzene and chloroform.

Optimum Reflux Ratio

The choice of the proper reflux ratio should usually be based upon an economic balance. As the minimum reflux ratio is approached, the number of plates required approaches infinity, and therefore the investment and fixed charges approach infinity. On the other hand, the operating costs are at the minimum since the duties of the condenser, still, and reflux pump are at the minimum.

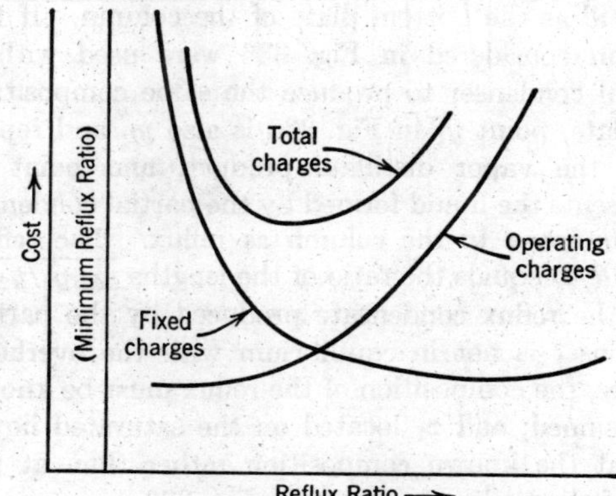

FIG. 339. Cost of operation, including fixed charges for different reflux ratios.[3]

As the reflux ratio is increased from the minimum the column diameter must be increased to maintain constant throughput, but the number of plates is reduced. Accordingly the investment and fixed charges decrease with increasing reflux ratios, go through a minimum, and again rise to infinity at total reflux when a column of infinite diameter is required.

The operating costs similarly rise to infinity at conditions of total reflux, since the duties of the condenser, still, and reflux pump approach infinity for a finite amount of product. Thus, the total charges of the column per unit of feed must pass through the minimum value from infinity to infinity as the reflux ratio is decreased from total reflux to the minimum reflux for any given separation. This is illustrated in Fig. 339 which shows a typical cost analysis for a given separation.

Partial Condensers

A partial condenser at the top of a column corresponds to a still at the bottom. The partial condenser condenses only that part of the upflowing vapor stream required for reflux yielding the distillate in the form of a vapor. The still vaporizes only that part of the downflowing liquid stream required for operation of the stripping section yielding the bottom product in the form of liquid. The coordinates for the difference point $y_{\Delta D}$ are exactly the same whether a partial or total condenser is used.

If the partial condenser produces a liquid in equilibrium with the vapor passing through the condenser, it can be considered as the top plate of the column in a manner equivalent to the treatment of the still as the bottom plate of the column. If the column considered in Fig. 336 were used with a partial condenser to produce the same composition distillate, point y_1 in Fig. 336 is also y_D and represents the vapor distillate product, and point x_1 represents the liquid formed by the partial condenser and returned to the column as reflux. The reflux ratio R/D equals the ratio of the lengths $\overline{y_{\Delta D}y_2}/\overline{y_2x_1}$.

If the reflux condensate produced by the partial condenser is not in equilibrium with the overhead vapors, the composition of the reflux must be known or assumed, and x_1 located on the saturated liquid line at the known composition rather than at the end of the tie line as shown in Fig. 336.

Open Steam

Instead of a still or reboiler at the bottom of the column to generate the vapor stream necessary for operation of the stripping section, the vapor may come from an external source, provided it is relatively free of the component to be stripped from the downflowing liquid. Steam is a very common stripping agent. Open steam eliminates the expense and maintenance of the reboiler which is usually subjected to the most severe corrosive conditions. It is readily available free from the component to be stripped, and introduces no difficulty in the recovery of an immiscible distillate product as a liquid because steam is readily condensed and removed from the vapor phase.

When water is one of the components to be separated, the computations for a column may be readily handled by the construction shown in Fig. 340, which is a solution of the following problem.

Illustrative Example. Compute the number of equilibrium plates required to separate a mixture of 42.2 per cent ethanol in water into a distillate containing 90.0 per cent ethanol and a bottom product containing 10.0 per cent ethanol by the use of open steam having an enthalpy of 1200 Btu/lb fed under the bottom plate at the rate of 0.416 lb of steam per pound of feed. Feed is supplied with an enthalpy of 200 Btu/lb and distillate is to be delivered with an enthalpy of 50 Btu/lb by a total condenser.

Solution. In Fig. 340, conditions of the open steam supplied below bottom plate n are represented by point y_{n+1}, conditions of the feed by point x_F, conditions of the distillate including the heat removed in the condenser by point $y_{\Delta D}$, conditions of the reflux by point x_0 or x_D which represents the distillate excluding the heat removed in the condenser, and the bottoms product by x_B, as stated. The sum of the steam and feed constituting the total feed to the column is represented by point 0 so located on the line $\overline{y_{n+1}x_F}$ that the ratio of the lengths $\overline{0x_F}/\overline{0y_{n+1}}$ equals 0.416 lb of steam per pound of feed. The point 0 also represents the sum of the bottom product represented by point x_B and the distillate minus the energy added in the condenser. The straight line $\overline{x_B0}$ extended to $x_D = 0.90$ intersects the ordinate x_D at an enthalpy equal to $h_D - Q_{CD}$ as represented by point $y_{\Delta D}$. The difference point between the downflowing liquid and the upflowing vapor in the stripping section represents the difference between the steam and the bottom product. It therefore lies on the extension of the line through the points y_{n+1} and x_B. It is also the difference point between the feed and the distillate minus the energy added in the condenser and therefore is located at the intersection of the lines extended through the points $y_{\Delta D}$, x_F and y_{n+1}, x_B at $x_{\Delta B}$, having the coordinates (0.158, −508).

The plate-to-plate construction is made in the manner indicated in Fig. 340 showing the equivalent of 7.7 equilibrium plates required for this separation.

The coordinates of the point $x_{\Delta B}$ are $x_B/(1 - S)$, $(h_B - SH_S)/(1 - S)$ where S is the pounds of steam per pound of bottom product and H_S is enthalpy of steam in Btu per pound.

Comparison of Open and Closed Steam

If the same feed were to be separated into the same distillate, with the same reflux ratio and the same loss of ethanol in the bottoms but with a still or reboiler instead of open steam, the composition of the bottom product would be that of the point $x_{\Delta B} = 0.158$ mass fraction of ethanol instead of $x_B = 0.10$ mass fraction.

The effect of the open steam is that of a diluent. Because of this dilution more fractionation and a greater number of plates are required with open steam (7.7 compared with 7.3) for the case shown in Fig. 340.

Entrainment

The separation of the vapor from the liquid on the plate is never absolutely complete as the vapor rising from a plate will carry with it some liquid as a fine mist or spray. The mass of liquid so carried by a unit mass of vapor is the *entrainment* designated E. Since only equilibrium plates are under consider-

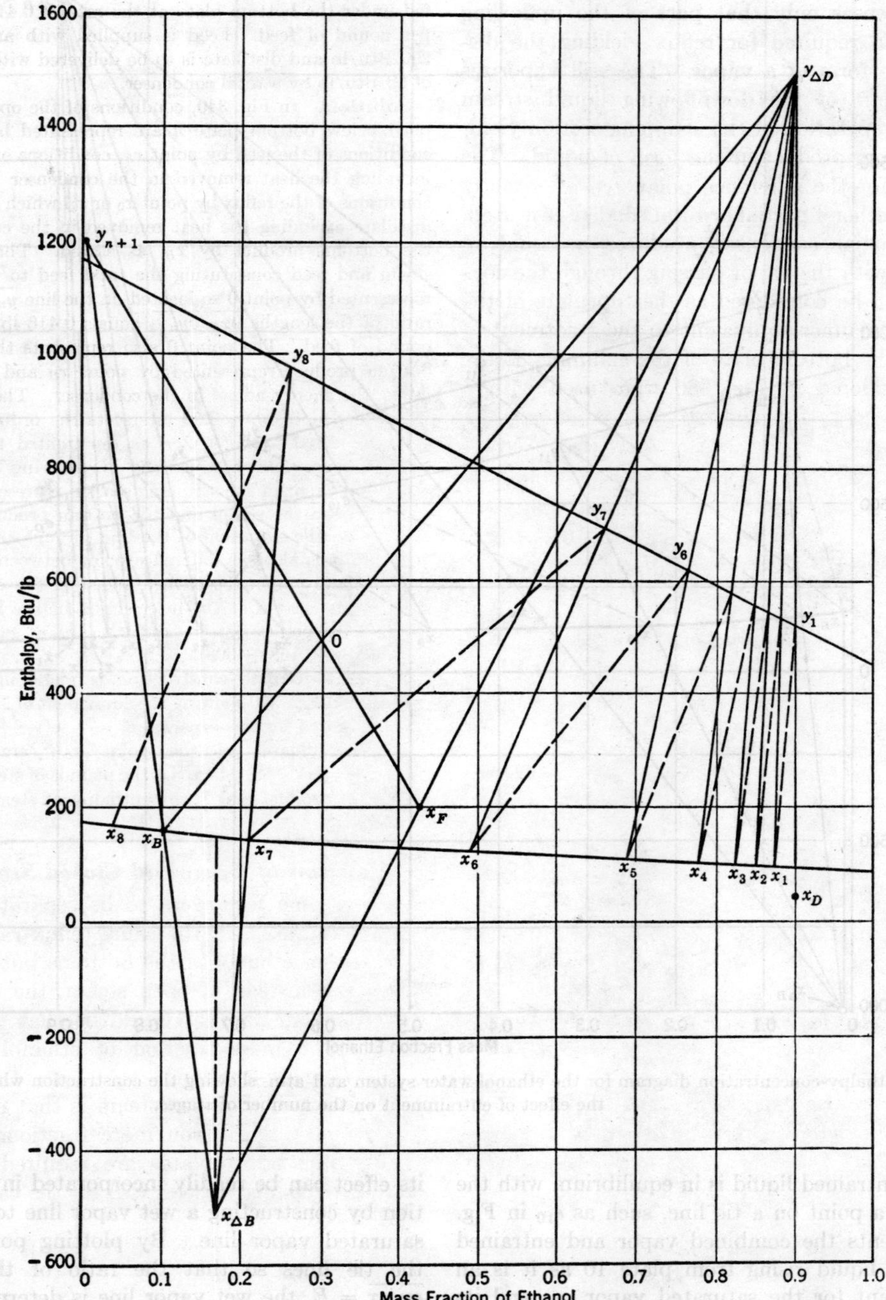

FIG. 340. The operation of fractionating columns using open and closed steam with the same loss of ethanol in the bottoms indicated on an enthalpy-concentration diagram for the ethanol-water system at 1 atm.

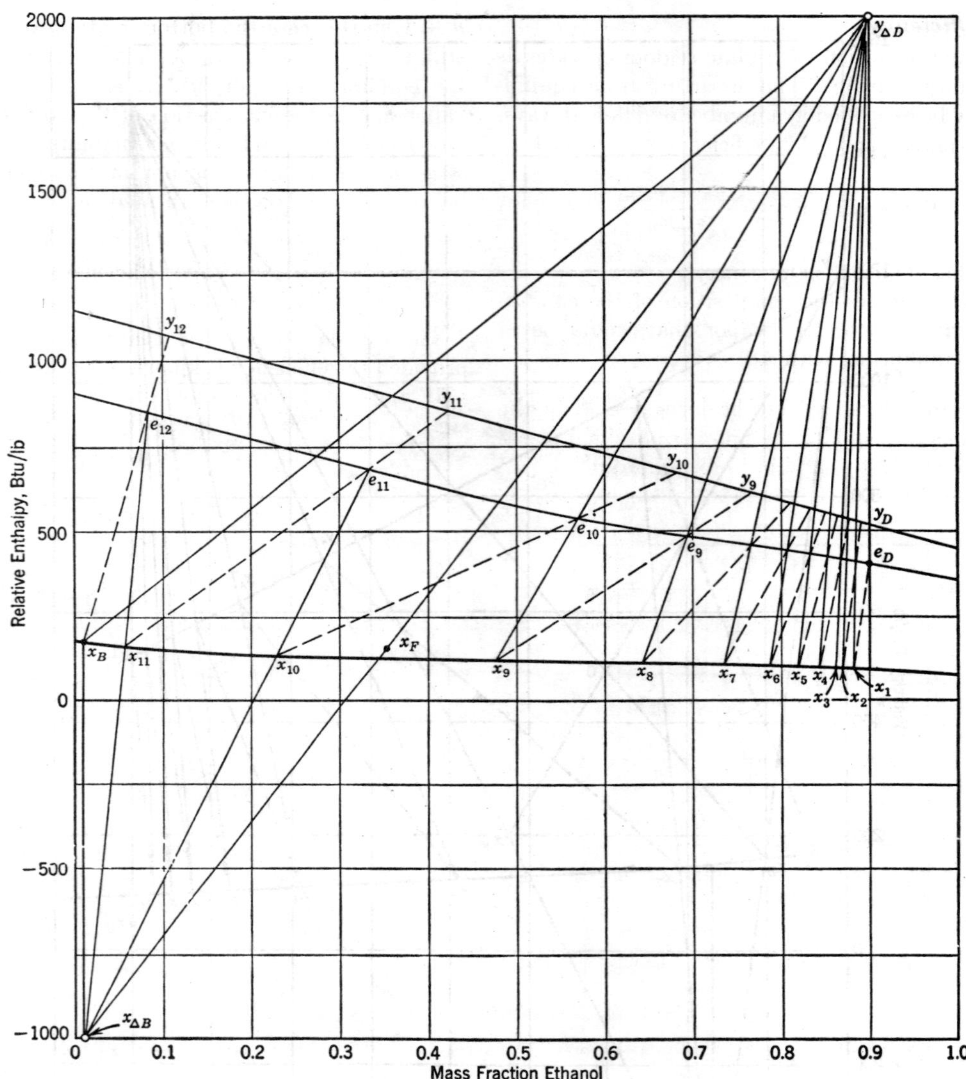

FIG. 341. Enthalpy-concentration diagram for the ethanol-water system at 1 atm, showing the construction when considering the effect of entrainment on the number of stages.

ation, the entrained liquid is in equilibrium with the vapor, and a point on a tie line, such as e_{10} in Fig. 341, represents the combined vapor and entrained equilibrium liquid rising from plate 10 as it is an addition point for the saturated vapor y_{10} and its equilibrium liquid x_{10}. The point e_{10} is located on the equilibrium tie line $\overline{x_{10}y_{10}}$ so that the ratio of the distances $\overline{e_{10}y_{10}}/\overline{e_{10}x_{10}} = E_{10}$.

If the entrainment is known or can be estimated, its effect can be readily incorporated in the calculation by constructing a wet vapor line to replace the saturated vapor line. By plotting points e along the tie lines so that the ratio of the distances $\overline{ey}/\overline{ex} = E$, the wet vapor line is determined as the locus of the points e. The computations are then made in the usual manner except that the wet vapor line is used instead of the saturated vapor line as indicated in Fig. 341.

Plate Efficiency

In actual operation each plate seldom operates as an equilibrium plate. The departure from equilibrium may be expressed in a number of different ways. The Murphree vapor plate efficiency [30] is defined:

$$\text{Vapor efficiency} = \frac{y_n - y_{n+1}}{y_n{}^* - y_{n+1}} \qquad (280)$$

where y has its usual significance.

$y_n{}^*$ indicates the composition of the equilibrium vapor (the vapor that would be in equilibrium with x_n).

This plate efficiency may be incorporated in the calculations to compute the number of actual plates as distinct from the number of equilibrium plates, as indicated in Fig. 342. When dealing with the vapor plate efficiency the calculations are usually started at the bottom. In Fig. 342, y_n represents the vapors rising from the still and are assumed to be in equilibrium with the still liquid or bottoms x_B. In this sense the still is the bottom equilibrium stage, n. The liquid leaving the next higher stage, $n - 1$, in this case the bottom plate next above the still, is represented by x_{n-1}. The point representing the equilibrium vapor, y^*_{n-1}, is on the saturated vapor line at the intersection of the equilibrium tie line from x_{n-1}. The vapor actually rising from stage $n - 1$, however, is not of this composition, but y_{n-1}, located on the saturated vapor line so that

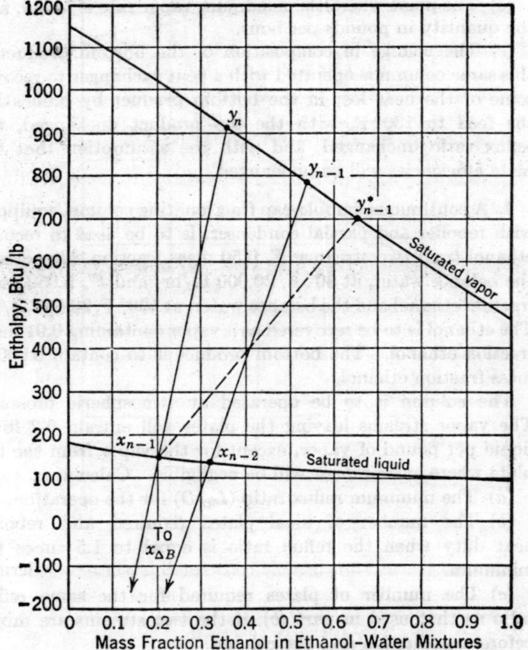

FIG. 342. Enthalpy-concentration diagram, showing construction when considering the effect of Murphree vapor efficiency on the number of stages.

$$\frac{\overline{y_n y_{n-1}}}{\overline{y_n y^*_{n-1}}} = (\text{Vapor efficiency})_{n-1}$$

The calculation may be continued as indicated, provided the efficiency is known or can be estimated.

If the saturated vapor line is approximately straight as in Fig. 342 the total vapor rising from a plate may be regarded as composed of two portions: one portion a vapor in equilibrium with the liquid leaving the plate, and the balance a vapor passing through the plate without any change in composition. The fraction of vapors by-passed in this way would equal the ratio of the distances $\overline{y_n y^*_n}/\overline{y_{n+1} y^*_n}$. Or the Murphree vapor plate efficiency might be expressed

$$\text{Vapor efficiency} = (1 - \phi) \qquad (281)$$

where ϕ = fraction of vapors by-passed.

PROBLEMS

1. A feed at 70° F, 1 atm, containing 50.0 mass per cent ethanol and the balance water, is to be stripped in a plate column to produce a bottom product containing 1.0 mass per cent ethanol. Overhead vapors are withdrawn as a distillate product.

(*a*) What is the minimum heat required per pound of bottom product to effect this separation?

(*b*) What is the composition of the distillate vapors for this operation?

(*c*) What is the minimum ratio of vapors leaving the still pot to bottoms product?

(*d*) What effect will cooling the feed to 0° F have on the answers to questions *a*, *b*, and *c*?

(*e*) With feed at 70° F and V_B/B = 1.5 times the minimum, compute:

(1) Heat required per pound of bottom product.

(2) Composition of distillate vapors.

(3) Number of ideal plates (including still) required.

2. A vapor mixture at 1 atm, containing 35.0 mass per cent ethanol and the balance water (enthalpy = 1000 Btu/lb), is to be rectified in a plate column (with condenser, receiver, and reflux pump) to produce a distillate product at 100° F containing 85.0 mass per cent ethanol.

(*a*) What is the minimum reflux ratio (L_0/D) and minimum condenser duty (Btu per pound of distillate) necessary to achieve this separation?

(b) What is the corresponding composition of the liquid bottom product and quantity per pound of feed?

(c) If the reflux ratio (L_0/D) is increased by 45 per cent over the minimum, compute:

(1) Condenser duty in Btu per pound of distillate.
(2) Number of ideal plates.
(3) Composition of bottom product.

3. A fractionating column is operating at 1 atm to produce a reflux and distillate product at 114° F containing 78 mole per cent ethanol and the balance water, and a bottom product containing 0.4 mole per cent ethanol. The feed contains 17.5 mole per cent ethanol and is introduced at 183.4° F at a rate of 43.5 moles/hr. Reflux at 114° F is returned to the top plate of the column at a rate of 29.4 moles/hr.

(a) How many actual plates (in addition to the still) are required with an overall plate efficiency of 70 per cent? Assume the still is equivalent to one ideal plate.

(b) What are the rates of heat transfer in the still and in the condenser?

(c) At what rate do the vapors leave the still, and what is their composition?

(d) To the downspout of which plate should the feed be introduced? Assume equal plate efficiencies in the stripping and rectifying sections.

4. A feed of 22.2 mole per cent ethanol and the balance water, at 1 atm and an enthalpy of 4860 Btu/mole, is to be separated by distillation in a plate column with open steam at an enthalpy of 1200 Btu/lb and a partial condenser. The bottom product is to contain 3.8 mole per cent ethanol, and the distillate product (saturated vapor) 78 mole per cent ethanol. 10.1 lb of steam are used per mole of feed.

(a) What is the condenser duty per mole of distillate product?

(b) What are the coordinates of the Δ point associated with the bottom of the column? What do these values represent?

(c) How many ideal plates are required (including still and partial condenser)?

(d) How many actual plates are necessary above the still? Assume that the still operates at 30 per cent efficiency, the partial condenser is equivalent to one ideal plate and is separate from the column, and the rest of the tower has a plate efficiency of 75 per cent.

5. A continuous laboratory fractionating unit contains the equivalent of six equilibrium plates and in operation supplies reflux to the top plate at 100° F. It is heated with closed steam. A feed containing 40 weight per cent ethanol and the balance water is divided into two portions, one consisting of 40 parts and the other 60 parts per 100 parts of original feed. The smaller portion is fed to the second plate from the top, and the larger portion to the fifth plate. The temperature of the feed is 130° F. If the column is operated at 1 atm pressure at a reflux ratio (R/D) of 2.0, and the distillate and bottom products are withdrawn at equal rates, what is the composition of the products that will be produced?

6. A bubble-cap column operated at atmospheric pressure has been used for the continuous fractionation of an aqueous solution of ethanol. The column was equipped with a total condenser and a reboiler. The following test data were obtained during an 8-hr period, during which steady-state conditions existed.

Column:
 Actual number of plates, 15
 Plate spacing, 10 in.
Feed introduced on twelfth plate down from top:
 Composition, 0.300 mass fraction ethanol
 Rate, 24,000 lb/hr
 Temperature, 53.6° F
Overhead product: 0.900 mass fraction ethanol
Bottom product: 0.005 mass fraction ethanol
Condenser:
 Water rate, 25,060 gal/hr, as measured by meter in water line
 Water temperature: Inlet, 50° F
 Outlet, 120° F
 Condensate temperature, 80° F

Determine:

(a) The overall plate efficiency of the column during the test.

(b) The point where the maximum vapor rate is found, and the quantity in pounds per hour.

(c) The change in composition of the bottom product if this same column is operated with a heat exchanger to recover some of the heat lost in the bottom product by preheating the feed to 130° F, with the top product unchanged, the reflux ratio unchanged, and with the assumption that the plate efficiencies will be unchanged.

7. A continuous, bubble-cap fractionating column, equipped with reboiler and partial condenser, is to be used to recover ethanol from two streams: F, 0.50 mass fraction ethanol and the balance water, at 80° F, 20,000 lb/hr; and F', 0.050 mass fraction ethanol and the balance water, at 120° F, 20,000 lb/hr. The ethanol is to be recovered as a vapor containing 0.91 mass fraction ethanol. The bottom product is to contain 0.00001 mass fraction ethanol.

The column is to be operated at atmospheric pressure. The vapor streams leaving the plates will entrain 0.2 lb of liquid per pound of vapor, except for the vapor from the top plate where entrainment will be negligible. Calculate:

(a) The minimum reflux ratio (L_0/D) for the operation.

(b) The number of ideal plates required and reboiler heat duty when the reflux ratio is equal to 1.5 times the minimum.

(c) The number of plates required for the same reflux ratio as that used in part (b), if the two streams are mixed before introduction into the column.

8. Ten thousand pounds per hour of a solution at 60° F containing 40 mass per cent ethanol and the balance water are fed to a continuous fractionating column operating at atmospheric pressure. The solution is to be separated into three products containing 90, 66, and 5 per cent ethanol, respectively. The stream containing 66 per cent ethanol is to be withdrawn at a rate of 2000 lb/hr.

The column is to be equipped with a total condenser. Water from a cooling tower is available at 85° F as a cooling medium for the condenser. Tentatively, it is estimated that a reflux ratio (L_0/D) of 2 will be used. Open steam at 310° F is to supply the heat required for the operation.

Specify the following information:

(*a*) Number of ideal plates required for the column.

(*b*) Location of the feed plate.

(*c*) Location of the plate from which the 66 per cent product is to be removed.

9. An aqueous solution of 0.15 mass fraction ammonia is being fed to the top plate of a stripping column at the rate of 21,000 lb/hr. The temperature of the feed is 140° F. Heat is supplied to the reboiler by means of 125 psi steam, the condensate leaving the trap as a saturated liquid. If the column is operated at 80 psia, determine:

(*a*) The maximum concentration of ammonia vapor obtainable.

(*b*) The number of ideal plates required in the column to obtain a top product containing 0.70 mass fraction ammonia and a bottom product containing 0.005 mass fraction ammonia.

(*c*) The steam requirement (pounds per hour) under these conditions.

CHAPTER

24

Vapor-Liquid Transfer Operations 2

Design and Control of Fractionating Columns

THE physical design and layout of fractionating equipment is dependent largely on the judgment of the engineer. The basic information concerning process variables such as temperatures, pressures, energy quantities or reflux ratio, and the number of equilibrium trays required is subject to more or less exact calculation, as has been indicated. The proper design of fractionating equipment includes not only the number of actual plates but, what is even more important, and without specific experience more difficult, the dimensions of the column and its component parts including risers, downspouts, and spacing of plates, or packing and its support.

BUBBLE PLATE COLUMNS

An error in estimating the number of plates is usually not so serious as an error of equivalent magnitude in estimating tower diameter or downspout capacity. Once the tower is erected, these latter items cannot be changed economically, whereas an error in the number of plates can usually be compensated for by an adjustment of operating conditions. A primary consideration in column design is the provision of *sufficient capacity for both the liquid and vapor streams*. Present information [6,16,26] * indicates that the plate efficiency, in most cases, is not sensitive to minor variations in standard tray layouts if ample fluid handling capacity is provided.

The Number of Actual Plates

Although the *overall plate efficiency*, defined as the number of equilibrium stages required for a given separation divided by the number of actual plates required, leaves much to be desired from a theoretical point of view, it has proved to be a convenient and in most cases a reliable method of correlating practical operating results. Other definitions of plate efficiency based on mass transfer theory [30] have made considerable progress.[19,43] But such theoretical methods have not yet reached the point of development where they can be employed to predict actual plate efficiency either conveniently or with confidence.

Since overall efficiencies are based upon performance data, the values reported not only include the effect of the actual plates and materials treated on the approach to equilibrium between the liquid and vapor phases, but also reflect any error or inaccuracy in obtaining the data or in calculating the number of equilibrium stages required. When using overall plate efficiencies reported in the literature, it is important to make proper allowance for such errors and for differences in the methods of calculation, particularly in multicomponent mixtures where approximate methods of calculation are frequently used.[26]

For properly designed and normally operated columns, the overall plate efficiency is largely dependent upon the properties of the material being fractionated. The viscosity appears to be the most important single property affecting the overall plate efficiency. For columns operating on petroleum and similar hydrocarbon materials, the following relationship [16] was found reasonably satisfactory

$$\text{Overall plate efficiency} = 18 - 60 \log \mu \quad (282)$$

where μ = molal average viscosity of feed at the

* The bibliography for this chapter appears on p. 395.

average column temperature in centipoises. The molal average viscosity of the feed is the sum of the products of mole fraction and viscosity for each component of the feed taken at the arithmetic average of the top and bottom temperatures of the column.

Columns of moderate diameters operating on volatile hydrocarbon columns such as natural gasoline show overall plate efficiencies of about 100 per cent. Low-pressure hydrocarbon absorbers operate at about 20 to 45 per cent plate efficiency, high-pressure absorbers (from 100 to 600 psi) at about 45 to 65 per cent plate efficiency, and alcohol-water columns at 70 to 100 per cent overall plate efficiency. Large columns providing multiple contacts of up-flowing vapor with the cross-flowing liquid indicate overall efficiencies up to 50 per cent higher than those given.[11]

Figure 343 is a similar correlation[32] which includes an additional variable, the relative volatility

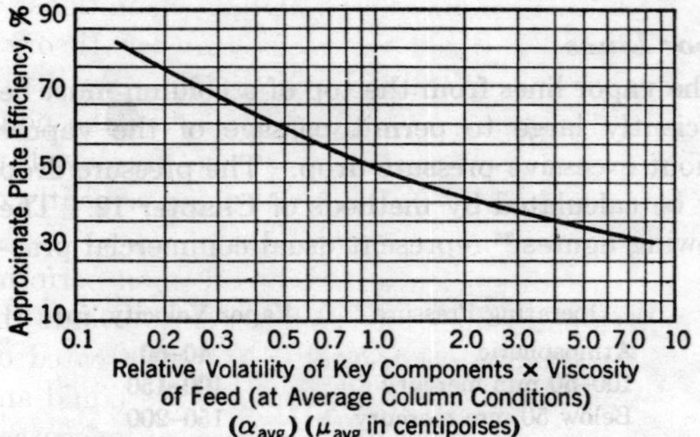

FIG. 343. Approximate overall plate efficiency as a function of the product of relative volatility of the key components multiplied by the viscosity of the feed in centipoises.[32]

of the components being separated, and is applicable to a wide range of materials as well as to commercial sieve plate columns fractionating acetic acid-water mixtures, beer, and alcohol-water mixtures.

The overall relative volatility is obtained by dividing the equilibrium molal ratio y/x of the more-volatile component by the equilibrium ratio y/x of the less-volatile component for which definite limiting concentrations have been specified in the distillate and in the bottoms, at the arithmetic average temperature and pressure of the column. For purposes of correlation this average relative volatility is multiplied by the average viscosity of the feed, and the product used as the abscissa in Fig. 343.

The exact description of what are normal operating conditions is subject to wide variation, and therefore these values must be used with caution according to the judgment and the experience of the engineer.

Column Diameter and Vapor Capacity

The efficiency of the contact between the liquid and vapor on a plate depends largely on the agitation caused by energy supplied by the vapor in passing through the liquid on the plate. At very low vapor velocities, low plate efficiencies are obtained. This may be caused by a decreased liquid level on the plates (a greater proportion of liquid passing down through the drain holes), the channeling of the vapors due to slight differences on liquid head at different points on the plates, and the formation of relatively large bubbles of vapor. Higher vapor velocities produce higher plate efficiencies because of thorough contacting of the liquid and vapor. At very high vapor velocities, entrainment of liquid by the vapor begins and the vapor tends to "cone" the liquid away from the slots of the bubble caps,[33] by-passing the liquid on the plates and producing low plate efficiencies.

Maximum plate efficiencies are obtained over a range of vapor velocities[6] centered approximately at those calculated by the following equation, the form of which was derived by application of equation 14 to the motion of liquid droplets in vapor.[48]

$$G = C\sqrt{\rho_v(\rho_l - \rho_v)} \tag{283}$$

where G is the mass velocity of the vapor [(lb)/(hr)(sq ft) of column cross section].

C is a parameter depending upon the plate spacing and the surface tension of the liquid on the plate.

ρ_l, ρ_v are the densities of the liquid and vapor, respectively, at the column conditions (lb/cu ft).

The values of the parameter C may be determined from Fig. 344 as a function of plate spacing and surface tension.[48] A suitable column diameter may be selected by using the maximum vapor load in the column in pounds per hour. The mass velocity per square foot of column area at which the plates will usually operate efficiently may be computed from equation 283. The quotient of these figures is the total cross-sectional column area which should be used. In making this calculation it is necessary to assume arbitrarily a plate spacing. Subsequent calculations for liquid capacity determine whether or not this assumed plate spacing is satisfactory.

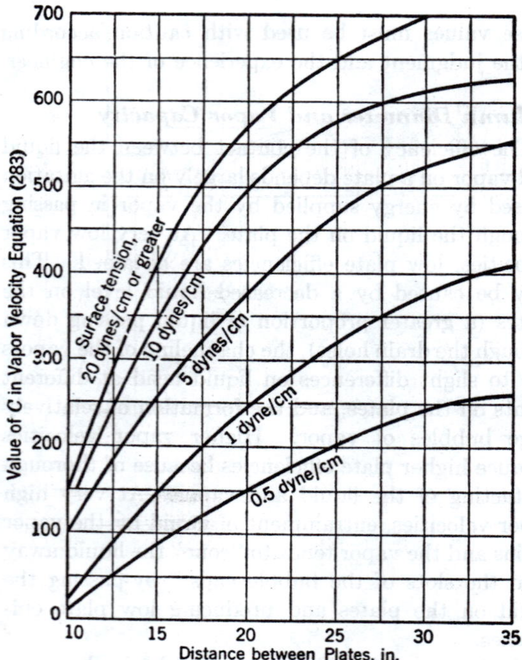

FIG. 344. Value of C in equation 283 for different plate spacings and liquid surface tensions.[48]

Plate efficiency is relatively insensitive to vapor load over a wide range with properly designed bubble-cap plates,[6] and column diameters determined by the above procedure provide a reasonable overload capacity so far as efficiency is concerned, since the vapor velocities so calculated may be exceeded by 30 to 100 per cent without a significant decrease in plate efficiency.

In rare cases the vapor load in different parts of a column will differ so markedly that a column constructed of sections having different cross-sectional area may be justified.

Entrainment

Equation 283 was developed on the concept of a maximum vapor load that would just fail to cause significant entrainment by the suspension of a mist of liquid in the vapor. Where small amounts of entrainment are significant the vapor velocities computed from equation 283 should not be exceeded and the distance from the top of liquid (or froth) on a plate to bottom of plate above should be used as the distance between plates. Entrainment may also be caused by splashing of liquid from one plate into the risers of the plate above. Ordinarily, splashing entrainment is important only at very high vapor velocities and with small plate spacings.

Variations in plate efficiency observed in commercial installations cannot be explained entirely on the basis of entrainment. In one case, entrainment of 0.4 lb of liquid per pound of dry vapor was observed, which theoretically would have reduced the plate efficiency by 25 per cent. Actually, the plate efficiency was reduced two or three times this amount. Apparently, basic changes in the type of contact between the liquid and vapor on the bubble-cap plates such as "coning" occur before entrainment seriously affects plate efficiency, and the most important effect of entrainment is an increase in the liquid loads carried by the downspouts.

Where rigid specifications with respect to color or impurities must be met by distillate products, small amounts of entrainment may be very serious, entirely apart from considerations of plate efficiency or capacity.

Vapor Lines

The vapor lines from the top of a column must be sufficiently large to permit passage of the vapors without excessive pressure drop. The pressure drop may be calculated by methods of Chapter 12. The following figures[34] represent usual commercial practice:

Operating Pressure	Vapor Velocity, fps
Atmospheric	40–60
100–50 mm mercury	100–150
Below 50 mm mercury	150–200

Liquid Capacity

A column operating normally is illustrated diagrammatically in Fig. 345. In order that the vapor from plate 2 may pass through the bubble caps and liquid on plate 1, there must be a pressure differential $P_2 - P_1$. Further, in order that the liquid on plate 1 may pass through the downspout to plate 2, there must be a liquid head in the downspout above the liquid on plate 2 sufficient to balance the pressure differential producing the vapor flow and in addition to produce liquid flow through the downspout.

The first observable effect of exceeding the liquid capacity of a column is an increase in the quantity of liquid on the trays and in the downspouts. When the downspout becomes filled with liquid, any slight increase in the liquid or vapor flow increases the liquid level on the plate. The increase in liquid level on the plate increases the pressure differential across

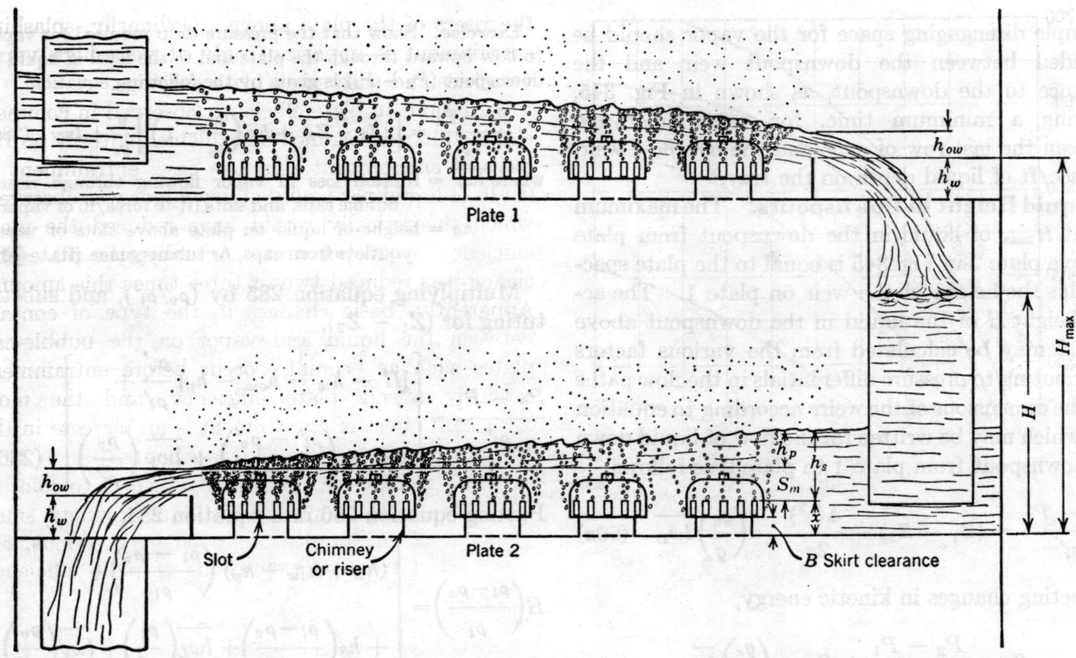

FIG. 345. Diagram illustrating liquid heads in the flow of liquid down a bubble-cap column in normal operation.

the plate which, in turn, causes more liquid to back up through the downspout, producing a further increase in the liquid level on the plate. The result of this accumulative cycle is that the column fills with liquid and ceases to function as a fractionating unit. This phenomenon is called *flooding*.

When the column becomes filled with liquid as a result of flooding, the pressure differentials between the plates disappear to a large extent, while the pressure in the still or reboiler will continue to increase if the temperature of the heating fluid is sufficiently high. Eventually this pressure will become great enough to blow liquid over the top of the column through the vapor lines in more or less violent spasms. This is known as *priming* or *puking*. During such periods vapor generated in the still bubbles through the column of liquid in the column intermittently. While vapor is being generated no liquid flows to the still or reboiler. When vaporization ceases, owing to the increase in pressure, liquid flows to the still. A flooded or partially flooded column cannot fractionate efficiently, and it is essential that the plates of a column be spaced at a sufficient distance so that flooding will not occur under the normal variations in operating conditions.

Where *frothy liquids* are involved this effect is more serious and the flooding occurs more rapidly. When the free surface of the liquid in the downspouts is well below the entrance there is sufficient surface along the walls of the downspout and over the weir for the entrained vapor to disengage from the liquid, and the downspout will operate properly. As the liquid flow increases and the downspout begins to fill with liquid, this surface is decreased, and the point may be reached at which the vapor can no longer disengage from the liquid, thus passing down through the downspout with the liquid. Since the density of this frothy mixture of liquid and vapor is less than that of the liquid alone, a higher head of froth is required to maintain the seal between the plates than is required with liquid alone. Consequently the entire cycle described above occurs more rapidly. Thus it is essential that ample liquid-carrying capacity be provided in the column for materials that tend to froth or foam. A top baffle placed in front of the downspout weirs so that liquid must pass beneath this baffle before passing over the weir has been found helpful in holding the froth out of the downspout,[33] and in preventing the liquid being "blown off" the plate at high vapor rates. This baffle should extend to the level of the slots if it does not seriously restrict liquid flow.

Ample disengaging space for the vapor should be provided between the downspout weir and the entrance to the downspout, as shown in Fig. 345, allowing a minimum time, for vapor liberation between the last row of caps and the overflow weir, of 1 sec/ft of liquid depth on the tray.[26]

Liquid Height in Downspouts. The maximum height H_{max} of liquid in the downspout from plate 1 above plate 2 in Fig. 345 is equal to the plate spacing plus the height of the weir on plate 1. The actual height H of the liquid in the downspout above plate 2 may be calculated from the various factors contributing to pressure differentials in the flow paths and the dimensions of the weirs according to equation 60a, which may be written for the flow of liquid down the downspout from plate 1 to plate 2 as follows.

$$\frac{P_2 - P_1}{\rho_l'} = (Z_1 - Z_2) - \frac{\Delta(v^2)}{2g} - \left(\frac{g_c}{g}\right)\overline{lw_L} \quad (60a)$$

Neglecting changes in kinetic energy,

$$Z_1 = \frac{P_2 - P_1}{\rho_l'} + Z_2 + \left(\frac{g_c}{g}\right)\overline{lw_L}$$

where P_1 = pressure above liquid in downspout (lb force/sq ft).
P_2 = pressure above liquid on plate 2 (lb force/sq ft).
Z_1 = height of free surface of liquid in downspout above plate 2 (ft).
Z_2 = height of liquid on plate 2 surrounding downspout (ft).
$\overline{lw_L}$ = friction losses in liquid flowing down and under downspout (ft-lb force/lb mass of liquid).
ρ_l' = average specific weight of liquid in downspout (lb force/cu ft).

Rewriting, with the following nomenclature,

H = height of liquid in downspout above plate,
h_w = height of weir on plate,
h_{ow} = height of liquid over weir on plate,
h_p = hydraulic gradient (liquid "head" tending to cause horizontal flow of liquid across the plate),

$$Z_2 = h_w + h_{ow} + h_p$$

$$H = \frac{P_2 - P_1}{\rho_l'} + h_w + h_{ow} + h_p + \left(\frac{g_c}{g}\right)\overline{lw_L} \quad (284)$$

Exercise. Show that the pressure drop required for vapor to flow upward through the plate and to the level of liquid in downspout $(P_2 - P_1)$ is given by the following equation.

$$\frac{P_2 - P_1}{\rho_v} = \left[(Z_1 - Z_2) + h_S\left(\frac{\rho_l}{\rho_v} - 1\right)\right]\frac{g}{g_c} + \overline{lw_V} \quad (285)$$

where $\overline{lw_V}$ = friction loss in vapor flowing through risers, bubble caps, and slots (ft-lb force/lb of vapor).
h_S = height of liquid on plate above slots or vapor outlets from caps, or submergence (ft).

Multiplying equation 285 by (ρ_v/ρ_l'), and substituting for $(Z_1 - Z_2)$,

$$\frac{P_2 - P_1}{\rho_l'} = \left[\begin{array}{l}[H - h_w - h_{ow} - h_p]\dfrac{\rho_v'}{\rho_l'} + \\ + h_S\left(\dfrac{\rho_l' - \rho_v'}{\rho_l'}\right) + \overline{lw_V}\left(\dfrac{\rho_v}{\rho_l'}\right)\end{array}\right] \quad (286)$$

Putting equation 286 into equation 284,

$$H\left(\frac{\rho_l - \rho_v}{\rho_l}\right) = \left[\begin{array}{l}(h_w + h_{ow} + h_p)\left(\dfrac{\rho_l - \rho_v}{\rho_l}\right) + \\ + h_S\left(\dfrac{\rho_l - \rho_v}{\rho_l}\right) + \overline{lw_L}\left(\dfrac{\rho_l}{\rho_l'}\right) + \overline{lw_V}\left(\dfrac{\rho_v}{\rho_l'}\right)\end{array}\right]$$

Simplifying by dividing by $(\rho_l - \rho_v)/\rho_l$

$$H = \left[\begin{array}{l}h_S + h_w + h_{ow} + h_p + \\ + \overline{lw_L}\left(\dfrac{\rho_l}{\rho_l' - \rho_v'}\right) + \overline{lw_V}\left(\dfrac{\rho_v}{\rho_l' - \rho_v'}\right)\end{array}\right] \quad (287)$$

Except for high-pressure operations, no significant error is introduced if $(\rho_l' - \rho_v')/\rho_l$ is assumed equal to unity. The downspouts operate as submerged orifices, and the friction losses $\overline{lw_L}$ may be estimated by the orifice equation as

$$\overline{lw_L} = 0.06(v_L)^2 \quad (288)$$

where v_L = maximum velocity of liquid in or under downspout determined at narrowest constriction (minimum cross-sectional area normal to flow) (fps).

Similarly, the value for the friction loss in the flow of vapors through the bubble caps may be approximated by the same formula but with a larger constant as the vapor changes its direction of flow to a greater extent in passing through the bubble caps than does the liquid in passing through the downspout. The value for the constant will depend upon the design of bubble cap. The following is suggested as a fair average.

$$\overline{lw_V} = 0.085(v_V)^2 \quad (289)$$

where v_V = maximum vapor velocity through risers, caps, or slots (fps).

The "head" required for the flow of liquid across the plate, h_p, is relatively small in columns of small diameters and may be approximated or neglected except in the case of large columns which may be calculated by more complex equations [11,15] or estimated in the manner described under Plate Design and Layout, p. 354. The height of the weir h_w is known as it is fixed by the designer. The submergence h_S (see Fig. 345) varies directly with the height over the weir h_{ow} and for the upstream side of the plate may be expressed as

$$h_S = h_{ow} + h_w + h_p - x$$

where x = distance from plate to top of slot.

The height of liquid in the downcomer H can then be estimated if the value for h_{ow} is calculated or estimated. Substituting in equation 287, $H =$

$$2(h_w + h_{ow} + h_p) - x + 0.06(v_L)^2 + 0.085(v_V)^2\left(\frac{\rho_v}{\rho_l}\right) \tag{290}$$

where v_V is the velocity of vapor passing through the upstream row of caps. If v_V is taken as the average maximum velocity through the caps an average value for h_p (such as $h_p/2$ in Fig. 345) is indicated.

The head of liquid over the weir h_{ow} is directly proportional to the quantity of liquid passing over the weir and is inversely proportional to the length of the weir. A given liquid level for a particular liquid load flowing across the plate can be maintained by a long high weir, or by a low short weir. The long high weir gives less variation in liquid level with changes in liquid load and is preferred for that reason.

Fig. 347. A typical bubble-cap plate with circular overflow weirs. (*The Lummus Co.*)

Fig. 346. View of bubble-cap column showing typical installation of round caps with circular downcomers and weirs. (*S. Blickman, Inc.*)

Weirs are usually of two types: (1) circular pipes extending through the plate to the plate below, Figs. 346 and 347, the edge of the pipe serving as a weir, and (2) flat plates extending across the chord of the column circle (chord-type weirs), Figs. 348 and 349. Frequently the downspout is formed by the space between the wall of the column and the extension of the chord-type weir below the plate.

In the equations above the heads have been expressed in feet. In practice the head of fluid over the weir is usually given in inches. The constants in the following equations (through 302) are chosen to give the head over the weir h_{ow} in inches of flowing fluid.

At low flow rates, with ample downspout capacity, the liquid flows over the weir at a low head h_{ow} and passes down the walls of the downspouts as a film. The downspout is largely empty of liquid except for the liquid seal at the bottom of the downspout. In such cases the flow of liquid over the weir may be computed by the Francis weir formula.[4]

$$Q = 0.0067 W h_{ow}^{1.5} \tag{291}$$

where Q = discharge (cfs).
W = length of weir (in.).
h_{ow} = head above edge of weir (in. of flowing liquid).

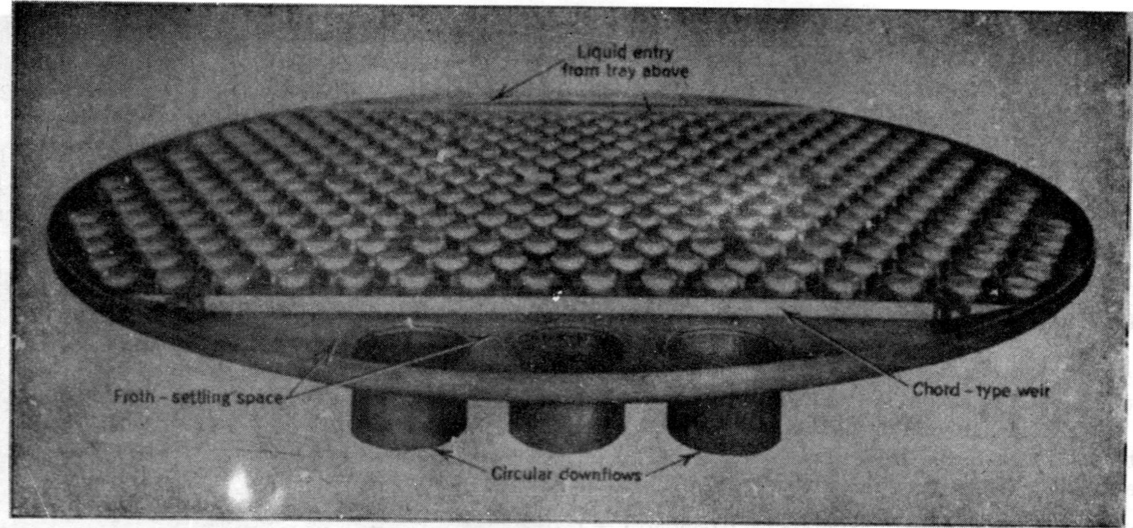

FIG. 348. A large bubble-cap plate with circular caps and downcomers and a chord-type weir. (*E. B. Badger and Sons Co.*)

This equation is applicable to circular weirs provided that the diameter D of the downspout is greater than four times the head of fluid over the weir. If the chord-type weir is relatively close to the wall of the column, the wall exerts a constricting effect on the flow, which makes the calculation a trial-and-error procedure.

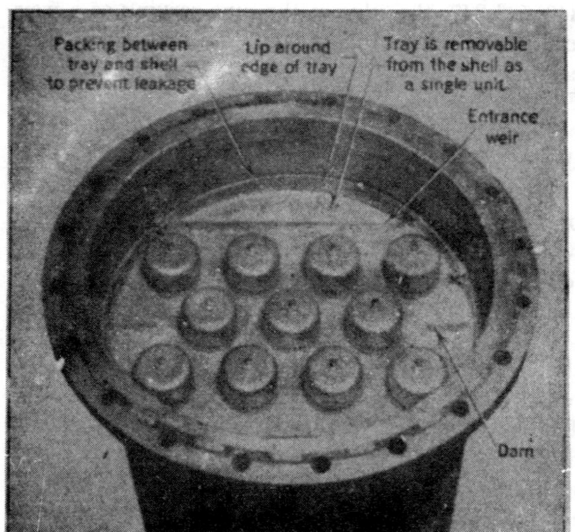

FIG. 349. Chord-type weir in a small bubble-cap column. (*Struthers Wells Corp.*)

Chord Weirs

A convenient method for estimating h_{ow} for chord weirs directly [3] which avoids the trial and error is based on the following form of the weir equation **291.**

$$h_{ow} = 30F\left(\frac{L}{\rho_l W}\right)^{2/3} \tag{292}$$

$$F = \left(\frac{W}{W_1}\right)^{2/3}$$

where h_{ow} = height of liquid over weir (in.).
L = liquid load (lb/sec).
W = length of weir (in.)
W_1 = "effective length" of weir (in.).
ρ_l = liquid density (lb/cu ft).

If the value for F can be obtained by a direct solution, equation 292 can then be solved directly for h_{ow}.

The geometry of these weir equations is indicated in Fig. 350. Application of the Pythagorean theorem to right triangle ABC results [3] in

$$\left(\frac{D}{2} - U\right)^2 + \left(\frac{W}{2}\right)^2 = \left(\frac{D}{2}\right)^2 \tag{293}$$

Similarly, an analogous treatment of triangle AGE gives

$$\left(\frac{D}{2} - U_1\right)^2 + \left(\frac{W_1}{2}\right)^2 = \left(\frac{D}{2}\right)^2 \tag{294}$$

Letting $K = W/D$, equation 293 may be solved for U/D.

$$\frac{U}{D} = \frac{1 - \sqrt{1 - K^2}}{2} \tag{295}$$

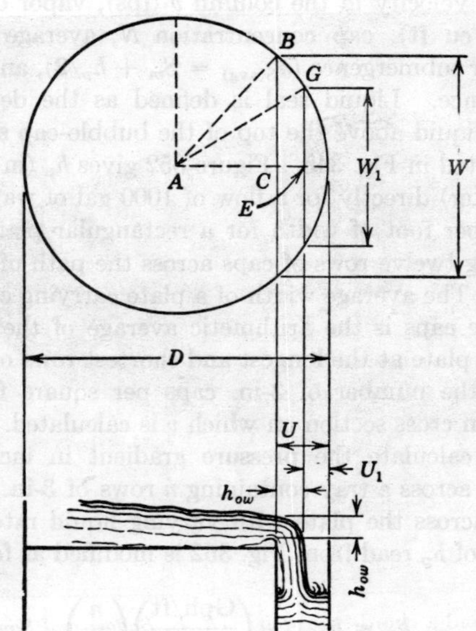

FIG. 350. Diagram illustrating relationship among variables in chord-type weir equations.[3]

And, observing that $U_1 = U - h_{ow}$, equation 294 may be solved for U_1/D.

$$\frac{U_1}{D} = \frac{1 - \sqrt{1 - K^2}}{2} - \frac{h_{ow}}{D} \tag{296}$$

Combining equations 294 and 296,

$$\frac{W_1}{D} = \left[1 - \left(\frac{2h_{ow}}{D} + \sqrt{1 - K^2}\right)^2\right]^{1/2} \tag{297}$$

or

$$\frac{W_1}{W} = \frac{1}{K}\left[1 - \left(\frac{2h_{ow}}{D} + \sqrt{1 - K^2}\right)^2\right]^{1/2} \tag{298}$$

Substituting F for $(W/W_1)^{2/3}$

$$F = K^{2/3}\left[1 - \left(\frac{2h_{ow}}{D} + \sqrt{1 - K^2}\right)^2\right]^{-1/3} \tag{299}$$

Solving equation 299 for h_{ow}/D,

$$\frac{h_{ow}}{D} = \frac{\sqrt{1 - K^2/F^3} - \sqrt{1 - K^2}}{2} \tag{300}$$

Solving equation 292 for $L/W^{2.5}$,

$$\frac{L}{\rho_l W^{2.5}} = \left(\frac{h_{ow}/D}{30KF}\right)^{3/2} \tag{301}$$

And, combining equations 300 and 301,

$$\frac{L}{\rho_l W^{2.5}} = \left[\frac{\sqrt{1 - K^2/F^3} - \sqrt{1 - K^2}}{60KF}\right]^{3/2} \tag{302}$$

Equation 302 is shown graphically in Fig. 351. Since L, W, and K are all known in advance for a given tray design, Fig. 351 may be used to obtain

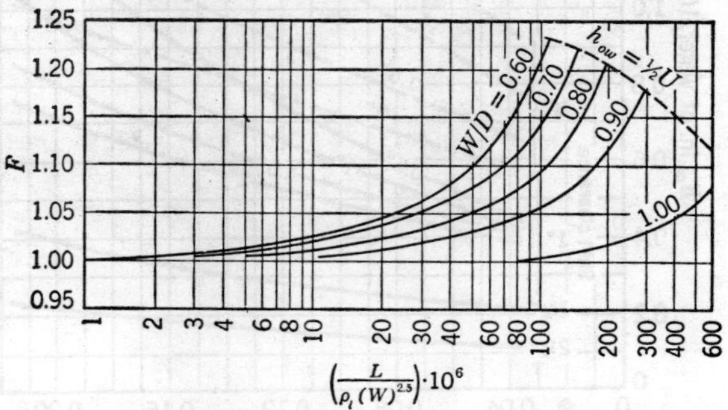

FIG. 351. Correction factor F for constrictions of column wall in a chord-type weir formula, equation 302.[3]

directly the value of F, which may then be substituted in equation 292 to give h_{ow}. The original reference [3] should be consulted for alignment charts for rapid solution.

Plate Spacing

In order to insure adequate liquid capacity, good practice requires that plate spacing be made equal to $2H$ as computed by equation 287. This gives an overload capacity for nonfrothing liquids comparable to the overload capacity for vapor incorporated in the application of equation 283 and Fig. 344 when small quantities of entrainment are not significant.

Plate Design and Layout

One of the important considerations in the design of bubble-cap plates, particularly on large-diameter columns where the liquid cross flow is long and the gradient across the plate is relatively high, is to maintain all bubble caps in operation. If the liquid seal h_S equals or exceeds at any point the drop in pressure of the vapors in passing through the plate, the caps at that point will cease bubbling. No vapor will pass through these caps and the liquid will tend to flow, or *backtrap*, through the slots and down the risers to the plate below. Such a plate is said to be unstable.

In exceptionally large columns the bubble caps may be arranged in cascade or tiers, the plate con-

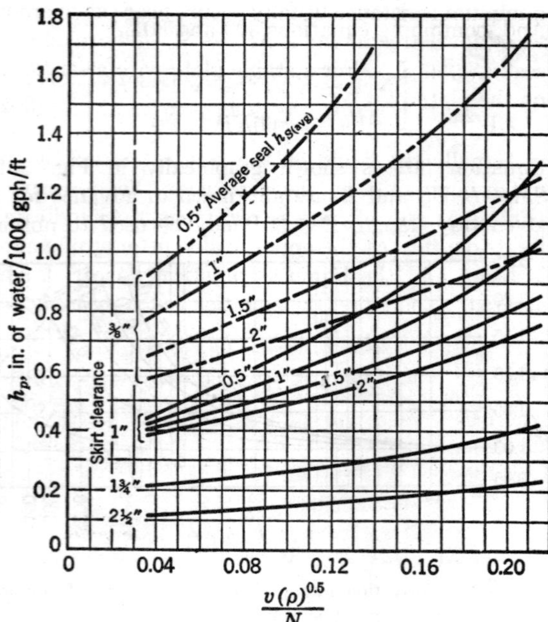

FIG. 352. Correlation of liquid head differentials across bubble caps.[20]

sisting of a series of different levels, the liquid level being maintained in each section by means of an individual overflow weir for each section or tier. In this way the liquid gradient is much less across each tier than it would be across the entire plate if all of one level. Another expedient which may sometimes be employed is the deliberate use of bubble caps which offer a high resistance to the flow of vapors, so that the average pressure drop of the vapors passing through the plate exceeds the maximum expected liquid head above the slots.

The liquid gradient h_p across a plate is not readily computed.[15] It is reduced by increasing the liquid level on the plate and by increasing the skirt clearances of the bubble caps, and it is increased by increases in vapor velocity, in liquid flow rate, or in the number of rows of caps across the path of the liquid flow. It is directly proportional to the cap concentration, i.e., the number of caps per unit column cross section.

These relations determined experimentally for round caps 3 in. in diameter spaced with centers on 4¼-in. equilateral triangles are correlated [20] in Fig. 352. The cap and chimney dimensions are given in Fig. 353. The liquid gradient h_p in inches of water is indicated as a function of linear superficial vapor velocity in the column v (fps), vapor density ρ (lb/cu ft), cap concentration N, average liquid seal or submergence ($h_{S(\text{avg})} = S_m + h_p/2$), and skirt clearance. Liquid seal is defined as the depth of clear liquid above the top of the bubble-cap slots as indicated in Fig. 345. Figure 352 gives h_p (in inches of water) directly for a flow of 1000 gal of water per hour per foot of width for a rectangular plate containing twelve rows of caps across the path of liquid flow. The average width of a plate carrying circular bubble caps is the arithmetic average of the width of the plate at the longest and shortest rows of caps. N is the number of 3-in. caps per square foot of column cross section on which v is calculated.

To calculate the pressure gradient in inches of water across a tray containing n rows of 3-in. round caps across the plate with varying liquid rates, the value of h_p read from Fig. 352 is modified as follows.

$$h_p = h_{p(\text{chart})}\left(\frac{\text{Gph/ft}}{1000}\right)\left(\frac{n}{12}\right)$$

In calculating the gradient, first assume a gradient to estimate the average seal $h_{S(\text{avg})}$. The correct assumed gradient is that which checks the calculated gradient. Usually only one or at the most two trial are necessary.

The maximum liquid flow per row of 3-in. round

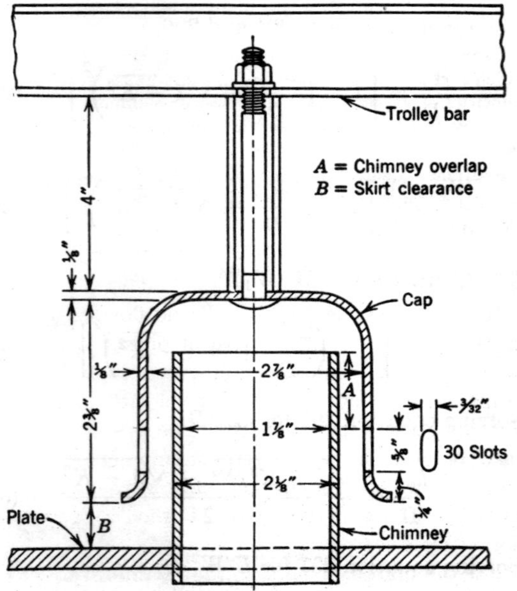

FIG. 353. Cross-sectional detail of the slotted bubble cap and its support as used in obtaining data for Fig. 352.[20]

bubble caps for plate stability or balanced operation with all bubble caps functioning is indicated from experimental data in Fig. 354.

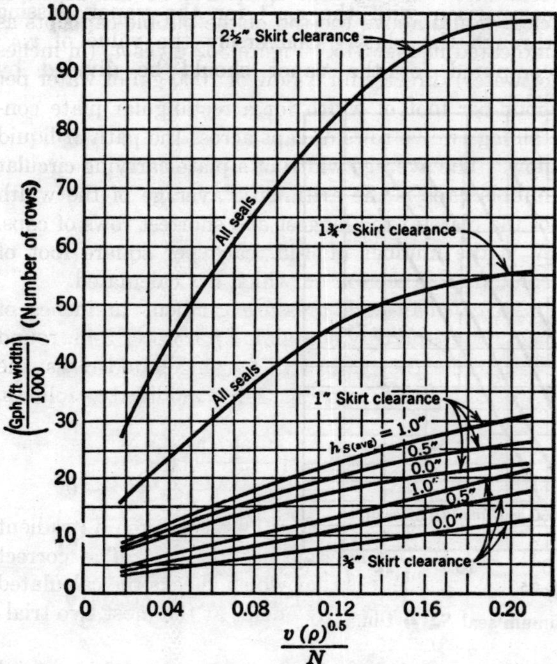

FIG. 354. Maximum liquid flow per row of 3-in. round caps (Fig. 353) for plate stability.[20]

Exercise. Using the dimensions given in Fig. 353, on which Figs. 352 and 354 are based, compute the relative linear velocities of vapor: (1) through riser or chimney, (2) between chimney and bubble cap, and (3) through slots, assuming no vapor flows under the cap skirt.

Exercise. Using dimensions of Fig. 353, derive a conversion factor by which the values plotted along the abscissas of Figs. 352 and 354 ($v\rho^{1/2}/N$) may be multiplied, to give corresponding values of $v_s\rho^{1/2}$ where v_s is linear velocity through slots, assuming no vapor flows under the skirt.

In a later correlation [24] the liquid flow is expressed in gallons per minute per foot of plate width normal to liquid flow. The vapor flow rate is expressed in terms of $v_s\rho^{1/2}$ where v_s is vapor velocity through the slots and ρ is vapor density. The pressure drop in head of clear liquid at zero submergence (the gas pressure drop through the plate with clear liquid level just at top of slots) ΔP_0, is then a direct function of $v_s\rho^{1/2}$.

For the plate to be stable the minimum resistance to flow of vapor through the plate must exceed the maximum hydrostatic pressure or seal, h_S. The minimum resistance usually occurs near the overflow weir (downstream) and is $\Delta P_0 + S_m$. The maximum seal is $h_p + S_m$. For a stable plate with clear liquid on the plate

$$\Delta P_0 + S_m \geqq h_p + S_m$$

Actually the liquid on the tray is aerated by the upflowing vapor at the side near the downspout (Fig. 345) and offers less resistance than indicated by its depth. At the opposite side little or no vapor is rising and the liquid is practically clear or unaerated. Therefore, the equation *for a stable plate* is

$$\Delta P_0 + \alpha S_m \geqq h_p + S_m$$

$$\Delta P_0 \geqq h_p + S_m(1 - \alpha) \qquad (303)$$

where α is the aeration factor $= (\Delta P_2 - \Delta P_1)/(S_2 - S_1)$ or the increase in pressure drop through the plate per unit increase in depth of clear liquid above the top of the cap slots. In practice, α is usually about 0.7, so ΔP_0 must exceed h_p by about $0.3S_m$ for stable operation. Since deep seals are usually employed when handling high liquid rates, $S_m(1 - \alpha)$ may sometimes influence plate stability as much as h_p.[24]

For any given rate of flow of liquid across a given plate there is a definite minimum vapor rate below which the plate becomes unbalanced and some of the bubble caps do not function. It is common practice to design plates with vapor rates corresponding to values of $v_s\rho^{1/2}$ as high as 11, particularly when handling large quantities of liquid. At higher values of $v_s\rho^{1/2}$ (about 12.8 or 13.0), the energy of the vapor streams issuing from the slots is sufficient to "cone" liquid away from the caps.

Figures 355 and 356*a*, *b*, and *c* represent data obtained with caps similar to those shown in Fig. 346 or

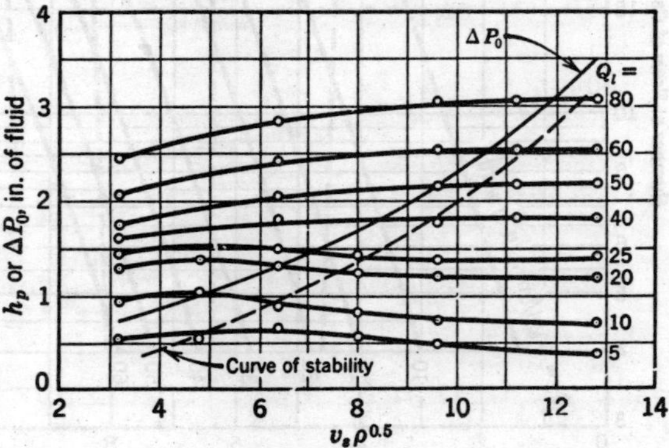

FIG. 355. Plot of h_p and ΔP_0 as function of $v_s\rho^{0.5}$ as used in preparing Fig. 356*a*.[24] Sixteen rows of 3-in. caps, flush, on 4¼-in. equilateral triangular centers at constant minimum seal S_m = 1 in. Q_l = gal/(min)(ft of width).

359, using water as the liquid.[24] The hydraulic gradient h_p in inches of fluid flowing across the plate is almost independent of the density of the liquid, provided the rate of flow of liquid is expressed volumetrically. These figures may, therefore, be used for other liquids by making a proper adjustment in ΔP_0 or in $v_s\rho^{1/2}$ for the different density of the liquid. This adjustment is made by multiplying the value of $v_s\rho^{1/2}$ as read from the charts by $\sqrt{\rho_l/\rho_{H_2O}}$ for comparison with the $v_s\rho^{1/2}$ for the vapor passing through the plate. Conversely the value of $v_s\rho^{1/2}$ computed for the vapor should be divided by

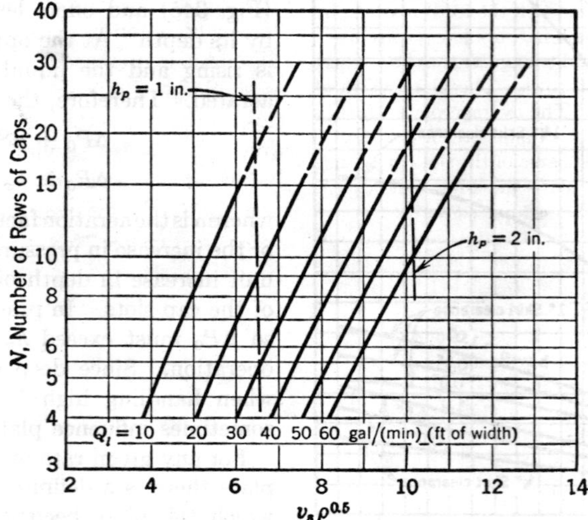

(a) 4¼-in. centers, minimum seal S_m = 1 in.

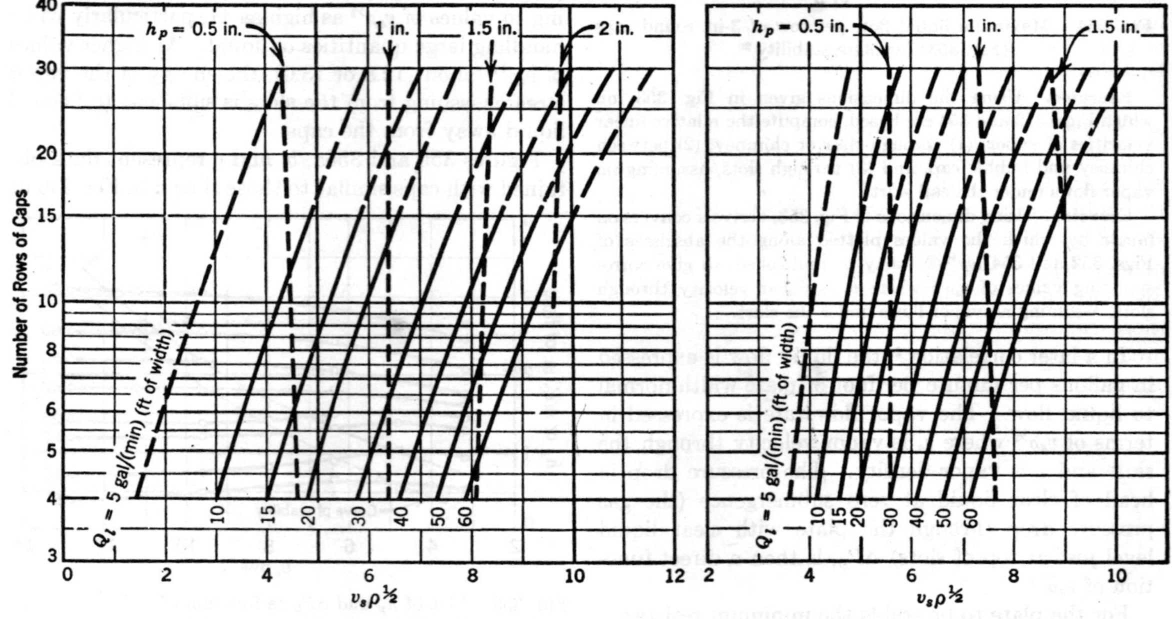

(b) 4¾-in. centers, minimum seal S_m = 1 in.

(c) 4¾-in. centers, minimum seal S_m = 2 in.

Fig. 356. Liquid-handling capacity charts for stable operation with 3-in. bubble caps mounted flush, on equilateral triangular centers.[24, 24a]

$\sqrt{\rho_l/\rho_{H_2O}}$ to obtain the value of $v_s\rho^{1/2}$ to use in the charts.

Illustrative Example.[24] It is desired to design a tower for stripping alcohol out of dilute aqueous solutions. The vapor load or steam consumption requires a column with a diameter of 84 in. The liquid flowing across the plate (the sum of condensed steam and of water entering with the feed) is 230 gpm.

As a first trial, assume that the 3-in. bubble caps are used, mounted flush on 4¾-in. equilateral triangular centers. There are 198 caps per plate in thirteen rows. The plate width at the shortest row of caps is 67 in. The average plate width is $(84 + 67)/(2)(12) = 6.29$ ft, and the liquid rate L is $230/6.29 = 36.6$ gpm/ft of plate width. The area of the slots in a single cap is 3.12 sq in. and for each plate the total slot area is $198 \times 3.12/144 = 4.3$ sq ft. $v_s\rho^{1/2}$ is computed as 7.2.

Figure 356*b* for this cap arrangement and a minimum seal S_m of 1.0 in. shows that at $v_s\rho^{1/2} = 7.2$ and for thirteen rows the liquid-handling capacity is only 27 gpm/ft of width. Therefore the plate will be unstable. According to the chart, the gradient will be slightly more than 1 in. Use of a minimum seal of 2.0 in. would make this plate stable as Fig. 356*c* indicates a liquid capacity of 37 gpm/ft at $S_m = 2$ in.

However, use of 4-in. bubble caps is usual either in large towers (Fig. 360) or where more than moderate liquid rates in smaller towers are necessary. The plate layout, 84 in. in diameter, has seven rows of 4-in. bubble caps, mounted flush on 5¾-in. equilateral triangular centers, using 90 caps per plate. With 6.1 sq in. slot area per cap, the total slot area is $(90)(6.1)/144 = 3.8$ sq ft, or 0.098 of the tower area. At a minimum seal of 2.0 in. and $v_s\rho^{1/2} = 8.1$ this tower should handle satisfactorily a liquid rate of 60 gpm/ft at a gradient of 1 in. Since the average plate width is now 6.75 ft, the liquid-handling capacity of the plate is 405 gpm of water.[24]

The plate layout may vary widely but should be such as to avoid short circuiting of the liquid flow across the plate which reduces the effectiveness of the bubble caps not in the direct line of flow. It should also avoid building up excessive liquid level differentials across the plate which, as has been discussed, may cause the vapor to flow through only those caps having the lower seals. This condition produces spouting and excessive entrainment at the caps with low liquid seals while the caps on the other side of the plate may be entirely inactive. In general, these two difficulties are opposite in their effect on tray layout, since layouts which tend to produce even distribution of liquid flow across the plate may produce excessive liquid gradients.

Figure 357 shows diagrammatically several of the common tray arrangements. The area of the tray which tends to become inactive as the column diameter is increased because of by-passing of liquid is shaded. Figures 346 and 347 show simple cross-flow arrangements of round caps with circular downspouts also serving as weirs. Note the liquid baffles at the ends of the middle row of caps to prevent liquid by-passing the middle row as indicated in Fig. 357*a*. These plates are welded into the column sections. This arrangement is satisfactory for relatively small columns up to 2½ to 3 ft in diameter. For columns in the range of 2 to 6 ft in diameter, it is preferable to include an exit weir, as indicated in Fig. 357*b* and as shown in Fig. 348.

To obtain satisfactory cross-flow distribution in larger columns it is necessary to eliminate a disproportionately large amount of tray space to include a sufficiently long chord or wedge-type weir. This disadvantage is overcome by the use of the arrangements such as indicated in Fig. 357*c*, *d*, and *e*, and illustrated in Figs. 358 and 359, respectively.

The arrangement shown in Figs. 357*c* and 358 is quite satisfactory from the point of view of liquid distribution, but it increases the length of the liquid flow path and tends to produce excessive liquid head gradient across the plate. An important precaution in this arrangement is the removal of some of the caps at the end of the baffle in order to avoid undue constriction of the liquid path at this point, as is illustrated in Fig. 358.

In the radial flow arrangement shown in Figs. 357*d* and 359, the liquid flows to and from the center on alternate plates. In exceptionally large columns, such as in Fig. 360, the arrangement illustrated in

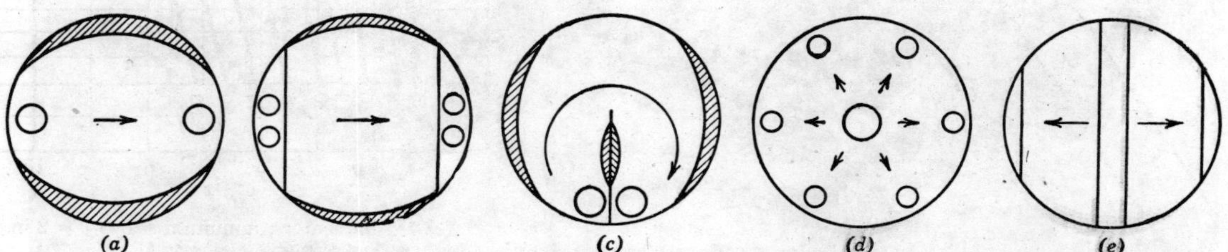

FIG. 357. Diagrammatic illustration of several common types of bubble tray layouts. Shaded areas tend to become inactive as tower diameter is increased because of by-passing of liquid.

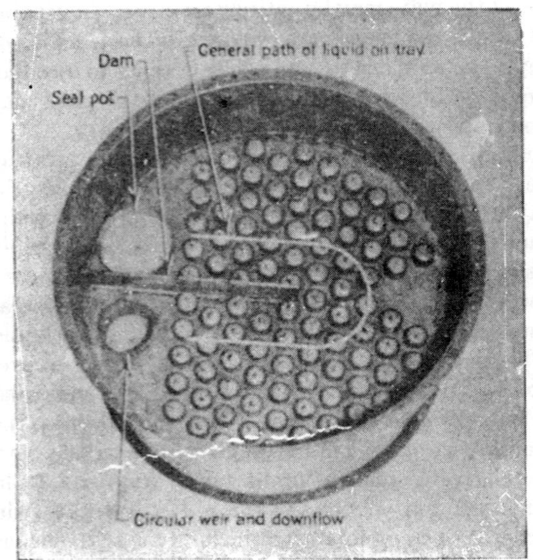

Fig. 358. Bubble-cap tray showing baffle arrangement of liquid distribution. (*The Lummus Co.*)

Fig. 357*e* is effective, particularly when combined with the feature of cascades.

Bubble Caps. A bubble-cap assembly consists of a chimney or riser, the cap which is mounted over the riser, and the means by which the cap is held in place. This is illustrated in Fig. 361 as an "exploded" view of a typical removable cast-metal assembly.

Although a large number of different types of caps are used, the most common type is the round bell-shaped cap, ranging in diameter from 2 to 7 in. The caps are cast iron or sheet steel, as shown in Fig. 362. Where necessary for corrosion resistance, other materials are employed such as copper or nickel. The advantage of cast-iron caps over steel caps in corrosion resistance is not so important as might appear, since the major area of corrosion in a fractionating column is usually at the vapor-liquid interface at the shell. The sheet-metal caps are lighter and may be more closely spaced because of their smaller bulk.

Another type of cap common in the petroleum industry is the so-called tunnel cap which fits over rectangular chimneys from 3 to 6 in. wide and from 12 in. to several feet long. These caps are easy to install and to maintain because of the fewer parts involved. However, better plate efficiency, less entrainment, and more stable performance is usually claimed for the round caps.

The spacing of the caps on a plate should be such as to give ample turbulence to the liquid around the caps but not so close as to produce spouting or jetting between the caps. In general, round caps are spaced on the vertices of equilateral triangles, usually giving a wall-to-wall distance between the caps of

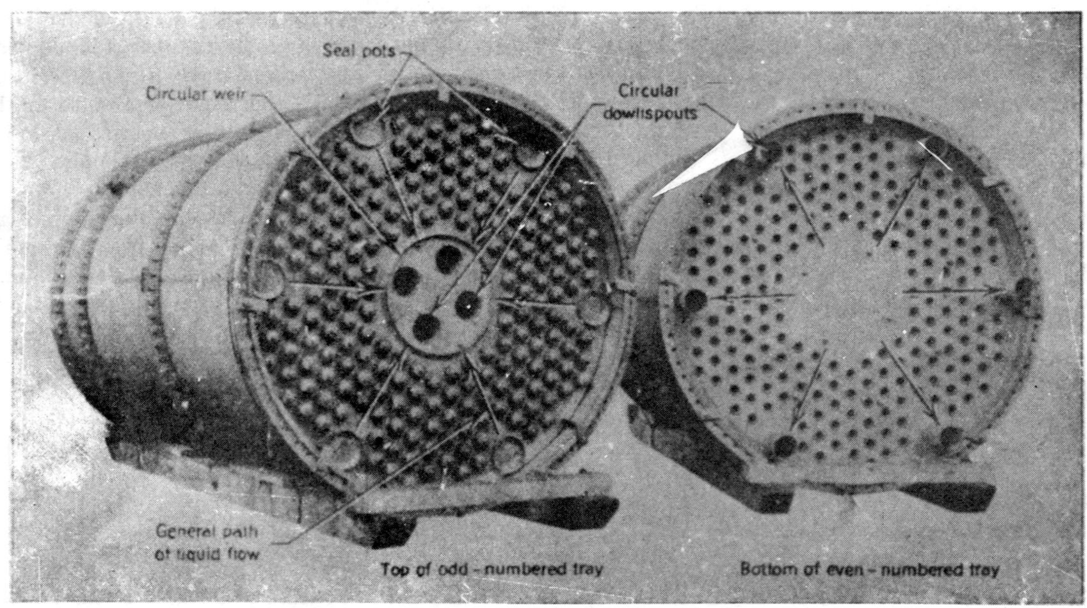

Fig. 359. Radial-flow type of bubble-cap tray. (*Vulcan Copper and Supply Co.*)

Fig. 360. A section of a 33-ft diameter cascade tray. Each tray has over 5000 bubble caps 4½ in. in diameter. (*Fritz W. Glitsch and Sons.*)

1 to 3 in. Particular care should be taken to avoid locating the caps too close to the weirs or downspouts because of spouting of the liquid over the weir into the downspouts, unsteady liquid flow over the weirs, and the possibility that vapors may discharge under the downspouts. It is sometimes desirable to blank off the slots in the sides of the caps nearest the downspouts. In small columns it may also be desirable to place baffle plates from the plate above in front of the exit weir to prevent splashing of liquid over the weir.

In general the design of the cap is such that the area of the chimney is equal to the area of the

Fig. 361. Exploded view of a removable cast-iron bubble-cap assembly. (*E. B. Badger and Sons Co.*)

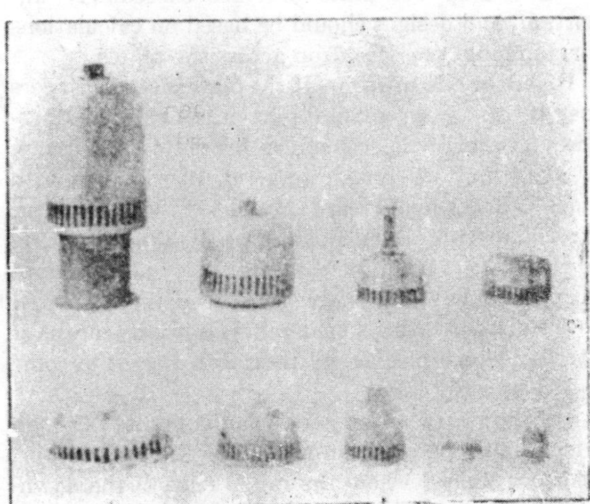

Fig. 362. Typical removable cast-iron bubble caps (upper row) and typical pressed sheet-metal bubble caps (lower row). (*E. B. Badger and Sons Co.*)

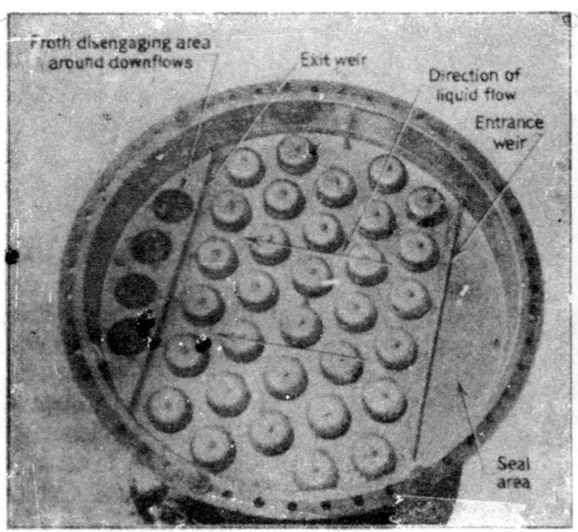

FIG. 363. Cross-flow tray showing inlet and outlet weirs. (*Vulcan Copper and Supply Co.*)

annular space between the cap and the chimney. These areas are also approximately equal to the area of the slots. This relationship provides a path of uniform cross section for the flow of vapors and reduces the pressure drop through the cap to a minimum.

As a rough design figure, the velocity of the vapor passing through the risers and slots should range from 9 to 14 fps for columns operating at atmospheric pressures and above. For vacuum columns, the pressure drop per plate becomes exceedingly important, and designs should be based on calculations for minimum pressure drop across the plates.

Risers or Chimneys. If the plate is cast the risers may be cast as an integral part of the plate. Otherwise they may be welded or brazed to the plate, or they may be rolled, screwed, or clamped into place. Sheet-metal caps are often fastened over the chimneys by tack welding or riveting the skirts to the plate. Where it is desirable for maintenance purposes to have removable caps, they may be bolted to the plate by using a spider, held down by means of "trolley bars," or even by their own weight in some rather unusual cases.

Downspouts may be circular conduits through which liquid passes from a plate to the plate below, or chord shaped where the raised edge of the downspout serves as a weir. In large columns, where liquid is removed from the center of a plate, the downspout may be formed from two parallel plates extending across the diameter of the column as shown in Fig. 357*e*. In all cases the downspout must have an adequate liquid seal on the lower plate to prevent vapor from flowing up the downspout. Usually in moderate-size columns the liquid on the lower plate retained by its overflow weir is considered the best liquid seal if the proper precautions are taken (Fig. 346). The use of a weir to distribute the downflowing liquid (Figs. 348 and 363), necessary in large columns, limits the liquid seal to that liquid behind the weir. A surging column may readily blow a small seal.

In some cases, downspouts are tapered at the bottom to facilitate the maintenance of a liquid seal. In other cases, the minimum cross-sectional area occurs between the bottom of the downspout and the lower plate, as for example with the downspout shown in Fig. 364, so constructed as to prevent vapor from the bubble caps from discharging into the downspout.

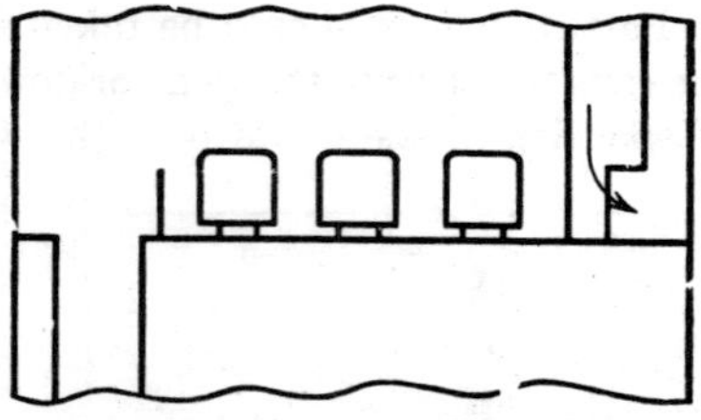

FIG. 364. Cutaway view of downspout construction to prevent vapors from the bubble caps from discharging into the downspout.

As a rough design figure where there are no unusual constrictions, the linear velocity of the liquid in the liquid-filled part of the downspout should be from 2 to 4 in./sec. It is probably good practice to over-design downspouts since they constitute a relatively unimportant cost item but are an important factor in flooding.

The Final Column Design. It has been shown how the diameter of the column may be estimated for a given plate spacing, and how the plate spacing ($2H$) may be estimated for a selected plate layout and column diameter. Many different combinations of these factors (plate layout, plate spacing, and column diameter) will give satisfactory operation. The optimum combination is that corresponding to the minimum cost of column, assuming that the operating conditions have been determined as indicated in Fig. 339.

The most economical design is indicated by plotting the tower cost against plate spacing (2H) estimated for a number of different designs. The following designs were computed for a "de-isobutanizer" of 53 plates handling 4800 gal/hr of 55 per cent isobutane and producing a distillate of 95 per cent isobutane and a bottom product of 95 per cent normal butane.[3]

Plate Spacing, in.	Column Diameter, ft	Column Cost
12.9	15.0	$77,000
13.5	12.0	52,000
14.4	10.0	37,000
15.3	8.7	29,500
20.0	7.5	25,000
25.0	7.0	23,900
30.0	6.75	24,300
35.0	6.6	25,000

As plotted in Fig. 365, a plate spacing of about 26 in. is indicated as the lowest cost column for this job.

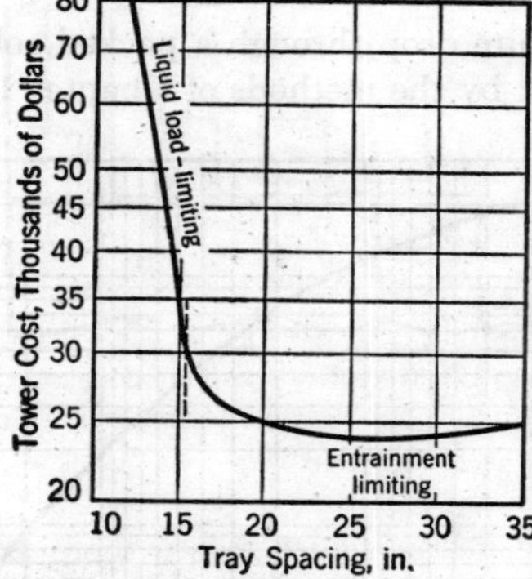

FIG. 365. Effect of tray spacing on original cost of a fractionating column.[3]

Plate Spacing

Other factors than the economics of design enter into plate spacing. Columns which for various reasons are to be housed are frequently designed with close plate spacings of 12 in. or less to save headroom when a large number of plates are required, and it is preferable to minimize height at the expense of cross section. Where columns may be placed outdoors the column height is no longer of prime importance and the plate spacing is frequently set at 24 in. or more, simply to allow the installation of manholes on each plate, an important maintenance feature, since it permits the use of cheaper materials of construction. In general the cost of increased height and smaller cross section as compared to lower heights and wider cross section is more than compensated for by the greater flexibility of the column, since at large spacing overloads can be absorbed without serious consequence, whereas with low plate spacing careful attention must be paid to conditions which might produce flooding.

PACKED COLUMNS

As in the design of bubble plate columns, the engineering design of packed towers requires consideration of the factors influencing the efficiency and capacity. The height of packing required for any particular separation is a function of the number of equilibrium stages necessary and the efficiency of the packing material. The efficiency is, in turn, a function of the size, shape, arrangement, and surface characteristics of the packing material, of the rates of liquid and vapor flowing, and of various physical properties and distribution of the two fluids. The relation between the packing efficiency and these variables is not simple, and this subject is discussed later as a phase of the rate of mass transfer.

The approximate ranges of the height of packing equivalent to an ideal stage in vapor-liquid transfer are given in Table 35.

TABLE 35. HEIGHT OF VARIOUS PACKINGS EQUIVALENT TO IDEAL STAGES

Packings	Height Equivalent to Ideal Stage, ft
Wetted-wall column	0.1 – 5
Rings, saddles, special tiles	
¼ × ¼ in.	0.1 – 0.6
½ in.	0.25– 1
1 in.	0.3 – 1.2
2 in.	0.5 – 2
3 in.	0.5 – 3
Drip-point grid tile	0.5 – 4
Single-turn helices, ⅛ × ¼ in.	0.1 – 0.6
Stedman	0.05– 1
Fiberglas	1 –10

Column Diameter and Capacity

The limiting capacity, or flooding point, of a packed tower is the condition at which the liquid begins to accumulate or back up at any level in the packing. Such a condition may result either at a constant liquid rate by an increase in vapor rate to the point at which the net liquid flow downward begins to decrease, or at a constant vapor rate by an increase in the quantity of liquid flow to the point at which the net liquid flow downward becomes constant or liquid begins to accumulate in the packing. Thus, for any given system, liquid, vapor, and packing, the locus of flooding points can be represented by a single line on a plot of liquid rate versus gas rate (Fig. 366).

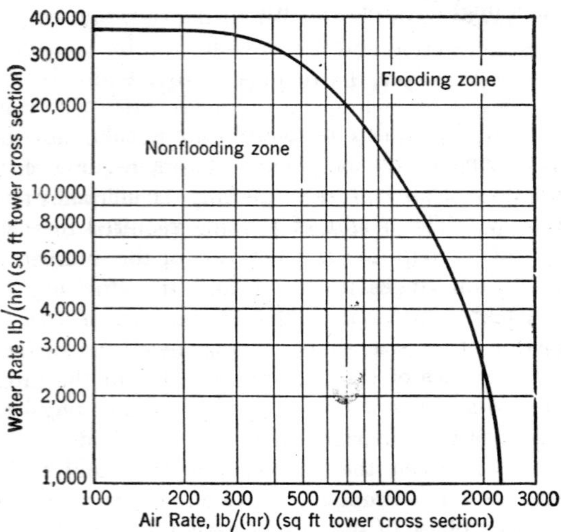

FIG. 366. Locus of liquid and vapor rates at the flooding point of a tower packed with 1-in. Raschig rings ($a/X^3 = 185$) and operating with water and air at 60° F and 1 atm.

For a particular system, any combination of liquid and gas rates which lies to the left of the line in Fig. 366 is physically operable. Any point which lies to the right of the locus represents an impossible condition.

The flooding point for towers filled with dumped packing, i.e., with random orientation of the individual pieces, may be estimated by the correlation in Fig. 367. This relation is based upon a study of the performance of a large number of packed towers [28,36] and generally will allow the prediction of flooding velocities with an accuracy of about ±10 per cent.

In the absence of actual data, the value of the packing characteristic (a/X^3) in Fig. 367 may be taken from Table 36. If the diameter of the packing is greater than one-sixth or one-eighth of the diameter of the tower, a correction for wall effect may be made by a method such as that suggested in Chapter 7, p. 79.

As the liquid and vapor rates frequently vary considerably throughout a column, the diameter of a column will be controlled by the point within the tower at which the severest conditions of liquid and vapor flow exist. Too frequently, this constriction occurs at the packing support. The method of supporting the packing should be so designed that the support has a capacity for passing liquid and vapor at least equal to that of the packing.

TABLE 36. CHARACTERISTICS OF COLUMN PACKING

Packing Material	Value of $\left(\frac{a}{X^3}\right)^*$
¼-in. Raschig rings	2330
½-in. Raschig rings	406
¾-in. Raschig rings	214
1-in. Raschig rings	185
1½-in. Raschig rings	100
2 in. Raschig rings	65
¼-in. Berl saddles	(approx) 4200
½-in. Berl saddles	450
¾-in. Berl saddles	185
1-in. Berl saddles	130
1½-in. Berl saddles	79

* a = surface area of packing in square feet per cubic foot of packed volume.
X = porosity of bulk packing.

The pressure drop through a packed column may be estimated by the methods of Chapter 17.

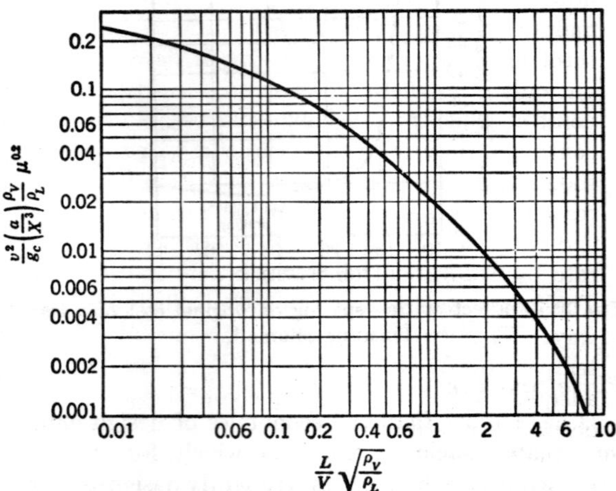

FIG. 367. Correlation of flooding velocities in packed towers,[28] where a = square feet of surface of packing per cubic foot of packed volume; X = fraction of voids in packed volume or the porosity of packing; v = superficial velocity of vapor (feet per second); and μ = viscosity of liquid (centipoises); L and V are mass rates.

INSTRUMENTATION AND CONTROL OF FRACTIONATING COLUMNS

The general procedure is to establish controls so that (1) *each independent variable* is controlled by one and only one primary control, and (2) sufficient flexibility is allowed for variations in feed and specifications of product. It is necessary to control *either directly or indirectly* the rate of flow of all streams (material or energy) to or from the column, and the composition of the desired product. The available impetuses or activators are:

1. The rate of flow of the streams.
2. Temperatures of the streams, or points within the column.
3. Pressures of vapor streams, or points within the column.
4. Liquid levels in flow tanks, separators, reboilers, or bases of towers.

In general, the point of control should be where the volume of material flowing or where the change in temperature is relatively large. When taking overhead a nearly pure distillate the temperature gradient at an intermediate point in the column is steeper than at the top of the column. The control instruments should be selected and located so that the speed of their response is at least equal to the rate of change in feed but not greater than the response of the stream being controlled. For example, if an attempt is made to throttle quickly a long line full of rapidly flowing liquid, the inertia of the liquid tends to maintain a constant velocity even with considerable throttling. Finally the velocity of flow is checked, but only after the motor valve is closed beyond the point at which proper flow could be obtained. The controller then opens the throttle valve and overshoots the mark in the reverse direction for a similar reason, causing surging in the operation.

Steam rates are frequently controlled for a given rate by a flow controller which may be hand reset. The feed stream is then often controlled in the same way. The reflux may then be controlled by the temperature of the product or of the column, insuring adequate reflux for varying conditions to give the desired distillate product.

The steam flow rate is often controlled by the temperature of the bottoms or of the column to insure a satisfactory bottom product. In this case the pressure of the vapors in the lower part of the column may be maintained constant by a pressure controller valve on the overhead vapors actuated by the column pressure.

The operating temperatures are usually controlled directly at one point and indirectly at other points.

Figures 368, 369, 370, and 371 illustrate a few methods of control that are giving satisfactory service. Other combinations and methods may be more satisfactory for other conditions, as these are intended simply to be suggestive.

Figure 368 illustrates instrumentation used in controlling a fractionating column producing a bottom product of constant vapor pressure. Since this vapor pressure is defined by the temperature and pressure in the reboiler (or kettle), the temperature connection and the pressure tap for control are each taken from the reboiler. No attempt is made to control the column-top temperature. The reflux is controlled at a constant rate by the flow-rate controller 2, at a value which is consistent with the liquid capacity of the column. The quantity of bottom product removed from the column is controlled by the liquid-level controller A with the motor valve located at 7, downstream from the heat exchange with the feed. When the bottoms are used to heat the feed as indicated, it is necessary that the streams be maintained as nearly constant in rate as possible in order to maintain the heat input to the feed at a constant value. It is therefore advisable to use a valve positioner in valve No. 7 with a sensitive liquid level controller A. The same precaution should be followed if the bottoms are removed directly from the column and fed into another column. In some cases flow-rate controllers are used to operate the motor valve, such as 7, the flow-rate controllers being reset by the liquid-level controller or the temperature controller.

In all these figures the feed is shown controlled by the feed-rate controller F. This is necessary if the column is to be operated at high capacity with high efficiency, but under some conditions of operation it is not applicable.

If substantially all the overhead is condensed and withdrawn to valve 4 with very little gas released through valve 5, the pressure control will be poor if the arrangement of Fig. 368 is followed. In Fig. 371 the pressure control is applied by the motor valve 9 on the vapor line to the condenser, which insures adequate pressure control even when all the overhead is completely condensed. Pressure controller E may be regarded as optional.

If the column is making a substantially pure distillate, the overhead products composition is fixed and the bottom must be allowed to vary with the feed composition. A constant column pressure can be maintained, and the controls must be arranged so that the bottom temperature may vary to accommodate the variation of the feed. Figure 369 shows one method of accomplishing this result. Tempera-

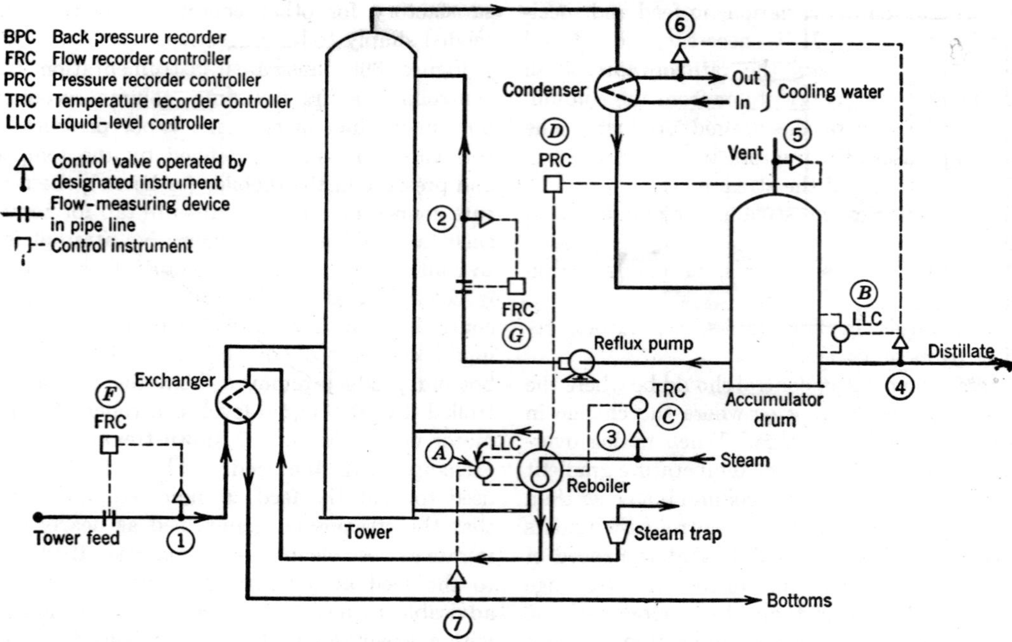

FIG. 368. Control system for a fractionating tower producing bottoms liquid with a constant vapor pressure. The overhead vapors are partially condensed.[11]

ture controller C has its temperature tap located between the feed and the bottom at a point where the temperature gradient is steep. If the liquid distillate withdrawal through valve 4 is very small with respect to the amount of reflux, large variations in pressure may result with the pressure control tap located as shown in Fig. 369. Under these conditions it is better to use the arrangement shown in Fig. 371, controlling the temperature by the large volume reflux stream rather than by the steam supply to the reboiler. It should be clear why the temperature bulb or controller C is placed below the feed when operating on the reboiler (Fig. 369), and above the feed when operating on the reflux (Fig. 371). In either case the pressure-control tap should be located at or close to the location of the thermometer bulb.

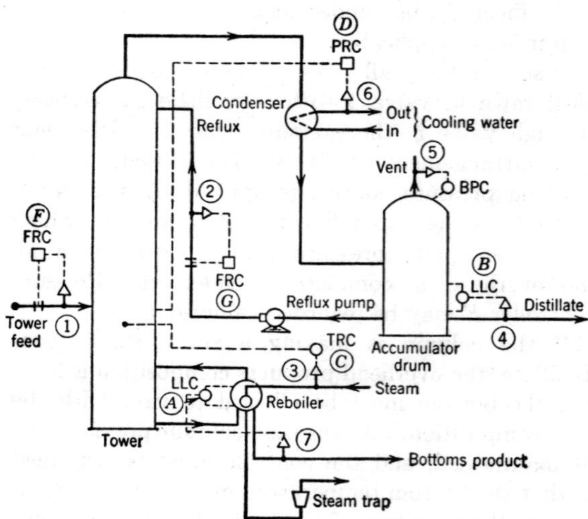

FIG. 369. Control system for a fractionating tower producing a pure overhead product.[11]

Figure 370 shows the temperature bulb located between the top of the column and the feed when operating the motor valve 3 controlling the heat to the reboiler. This type is satisfactory provided there are not so many trays between the point where the temperature bulb is located and the kettle as to cause a lag between the control point and the point of heat application. In any case the lag will be greater than in the arrangement shown in Fig. 369 or 371, which causes this control (Fig. 370) to lack sensitivity compared with the other method.

The pressure control indicated in Fig. 370 depends on flooding the reflux condenser so that only sufficient surface is exposed to vapor condensation to give the desired condensation. This method of pressure control, as well as the method indicated in Fig. 369,

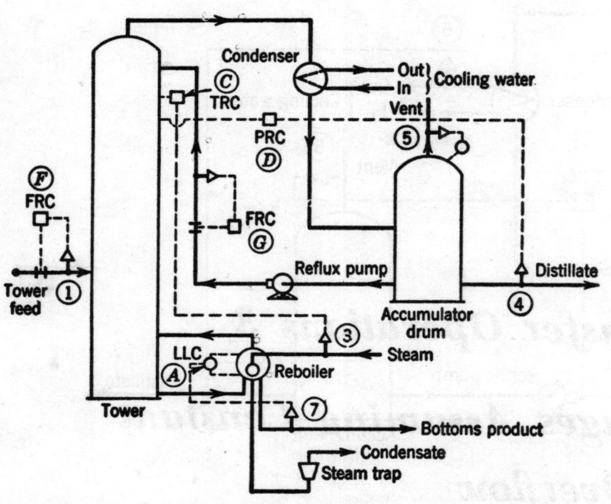

FIG. 370. Control system for a fractionating tower producing a pure overhead product.[11]

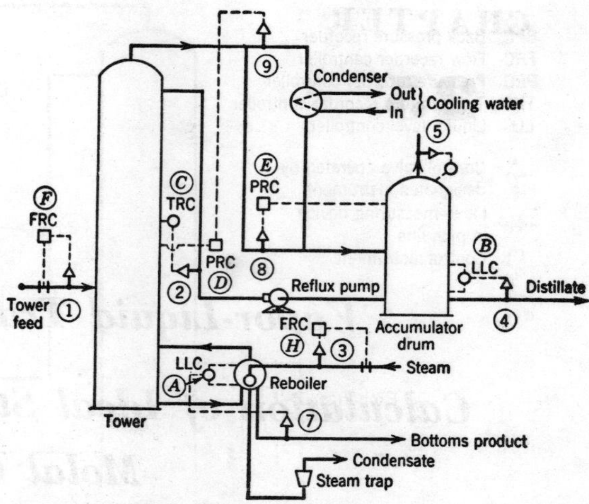

FIG. 371. Control system for a fractionating tower producing a pure overhead product.[11]

is satisfactory and sensitive where the rate of liquid withdrawal through valve 4 equals or exceeds about 15 per cent of the liquid reflux to the top of the column. If there is considerable excess surface in the reflux condenser and water temperatures are low, the reflux going back to the column will be cooled substantially below its bubble point. This results in converting some of the top trays in the column from fractionating trays to heat-transfer trays, with the overall result of reducing the total number of effective plates in the column.

The system illustrated in Fig. 371, in which no attempt is made to control the temperature of the reboiler, causes the column to operate with a constant heat input and substantially constant vapor rate, irrespective of normal changes of feed rate and feed composition. The auxiliary pressure controller *E* operating valve 8 serves to maintain a pressure a few pounds below the column pressure in the reflux accumulator. The temperature in this accumulator is raised above the temperature of the liquid coming from the reflux condenser by the hot vapors from the top of the column. By this means a constant pressure differential is maintained across the reflux valve 9, giving more uniform operation to this piece of equipment.

PROBLEMS

1. Design a bubble-cap column giving all significant specifications including number of plates, plate spacing, column diameter, number, dimensions, arrangement of caps, and downspouts for the operation described in problem 8 at the end of Chapter 23.

2. Specify all significant dimensions for a packed tower and for a bubble-cap tower to operate satisfactorily under the conditions of problem 9*b* at the end of Chapter 23.

CHAPTER 25

Vapor-Liquid Transfer Operations 3

Calculation of Ideal Stages Assuming Constant Molal Overflow

THE use of the enthalpy-concentration diagram as described (Chapter 23) for computing the number of equilibrium stages required is a rigorous method and gives proper emphasis to the energy relationships involved. In engineering practice other methods are more widely used because of their greater convenience without serious loss in accuracy in most cases.

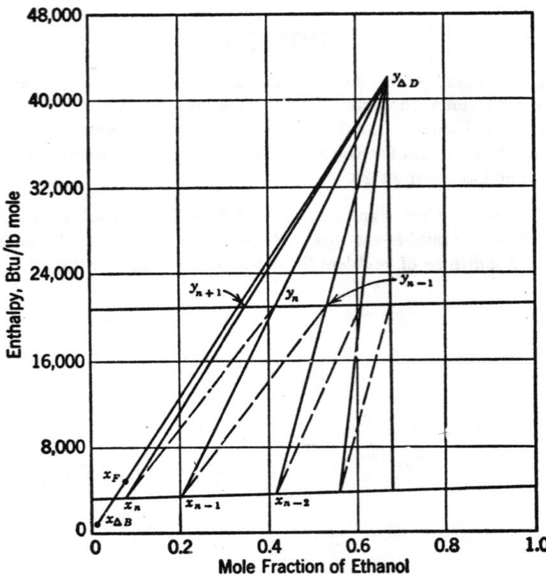

FIG. 372. Diagram illustrating fractionation calculations on an enthalpy-mole-fraction diagram for the ethanol-water system at 1 atm.

If the enthalpy-composition diagram is plotted on the mole basis, such as Btu's per pound mole versus mole fraction, the saturated liquid and vapor lines frequently appear as approximately straight parallel lines. This condition is usual with hydrocarbon solutions at low pressures and is approximated by alcohol-water mixtures as indicated in Fig. 372. The significance of these lines being parallel is that the molal overflow L, expressed in moles, and the vapor load V, expressed in moles, are each constant throughout the rectifying section. Likewise, L', the moles of liquid, and V', the moles of vapor flowing in the stripping section, are constant throughout the stripping section.

These statements may be readily demonstrated by the application of plane geometry to Fig. 372. The ratio of moles of liquid overflowing from plate n, L_n, to the moles of vapor rising from plate $n + 1$, V_{n+1}, equals the ratio of the lengths $y_{\Delta D}y_{n+1}$ to $y_{\Delta D}x_n$ which is the same for all plates. Therefore the ratio of liquid to vapor passing countercurrent between plates L/V is a constant. By a material balance around the top of the column,

$$V_{n+1} = L_n + D \tag{304}$$

Therefore, the ratio L/V may also be written as $L/(L + D)$ which is constant. Since D has a definite constant value for any particular fractionation, L is constant; and similarly V is constant if the saturated liquid and vapor lines are parallel. The same form of proof is applicable to the stripping section based on the material balance around the bottom of the column,

$$V'_{m+1} = L'_m - B \tag{305}$$

The subscript m is used to designate a plate below the feed plate.

It is frequently implied that constant molal overflow is to be expected if the components of the mixture obey Trouton's rule, an empirical approximation which indicates that the molal latent heat of vaporization divided by the absolute temperature of the normal boiling point is a constant. It seems better to regard the condition of constant L/V as a simplifying assumption which gives reasonably reliable results in many cases.

If L/V is constant, energy balances around every stage are unnecessary, and only material balances and equilibrium calculations as used in solid-liquid and liquid-liquid extraction need be used in computing the number of equilibrium contacts required. The energy balance need be made only once around the entire column or some convenient part thereof to determine the moles of external reflux R or the ratio L/V from the known heat supplied to the still or removed from the condenser, or to determine the quantity of heat required in the still from the selected L/V ratio.

GRAPHICAL METHODS

The simplest case is that of two volatile components and constant L/V. If the compositions of the saturated (equilibrium) vapor and liquid at the column pressure are plotted as is done for the alcohol-water system in Fig. 373, the *compositions of the vapor and liquid leaving any equilibrium stage are represented by a point on the equilibrium line.* The material balance or operating line giving the relation between the compositions of the vapor and liquid passing each other at any point (between any two plates) in the rectifying section of a column is represented by a straight line having a slope equal to L/V as follows.[29] *

A material balance for the more-volatile component around the top of the column gives

$$V_{n+1}y_{n+1} = L_n x_n + Dx_D \tag{306}$$

$$y_{n+1} = \frac{L}{V}x_n + \frac{D}{V}x_D \tag{306a}$$

* The bibliography for this chapter appears on p. 395.

Similarly a material balance around the bottom of the column gives

$$V'_{m+1}y_{m+1} = L_m'x_m - Bx_B \tag{307}$$

$$y_{m+1} = \frac{L'}{V'}x_m - \frac{B}{V'}x_B \tag{307a}$$

where B = moles of bottom product per unit of feed F or unit of time.
D = moles of distillate product.
L = moles of liquid flowing down in rectifying section.
L' = moles of liquid flowing down in stripping section.
m = a plate or equilibrium stage in stripping section.
n = a plate or equilibrium stage in rectifying section.
V = moles of vapor flowing up past liquid in rectifying section.
V' = moles of vapor flowing up past liquid in stripping section.
x = mole fraction of a component (the more-volatile component in a two-component system) in the liquid stream.
y = mole fraction of that component in the vapor stream.

These material balance or operating lines have the slopes L/V and L'/V' and the intercepts on the y axes of $(D/V)(x_D)$ and $-(B/V')(x_B)$, respectively, as indicated by equations 306*a* and 307*a*.

The compositions of the vapor and liquid streams passing each other at any point in the column are represented by a point on one of the straight operating lines, equation 306*a* or 307*a*. If the reflux has the same composition x_0 as the vapor rising from the top plate y_1, as is the case when a total condenser is used (Fig. 335), the point representing the compositions of these two streams, V_1 and R, lies on the diagonal $y = x$ (Fig. 373). Since the equilibrium compositions between y and x are given by the equilibrium line, the composition x_1 is obtained by following the horizontal line y_1 to its intersection with the equilibrium line. This value of x_1 is then used in connection with the material balance or operating line to find the value of y_2. The graphical construction as illustrated in Fig. 373 constitutes a series of steps. The compositions of the streams leaving an equilibrium plate are represented by the points on the equilibrium line. The compositions

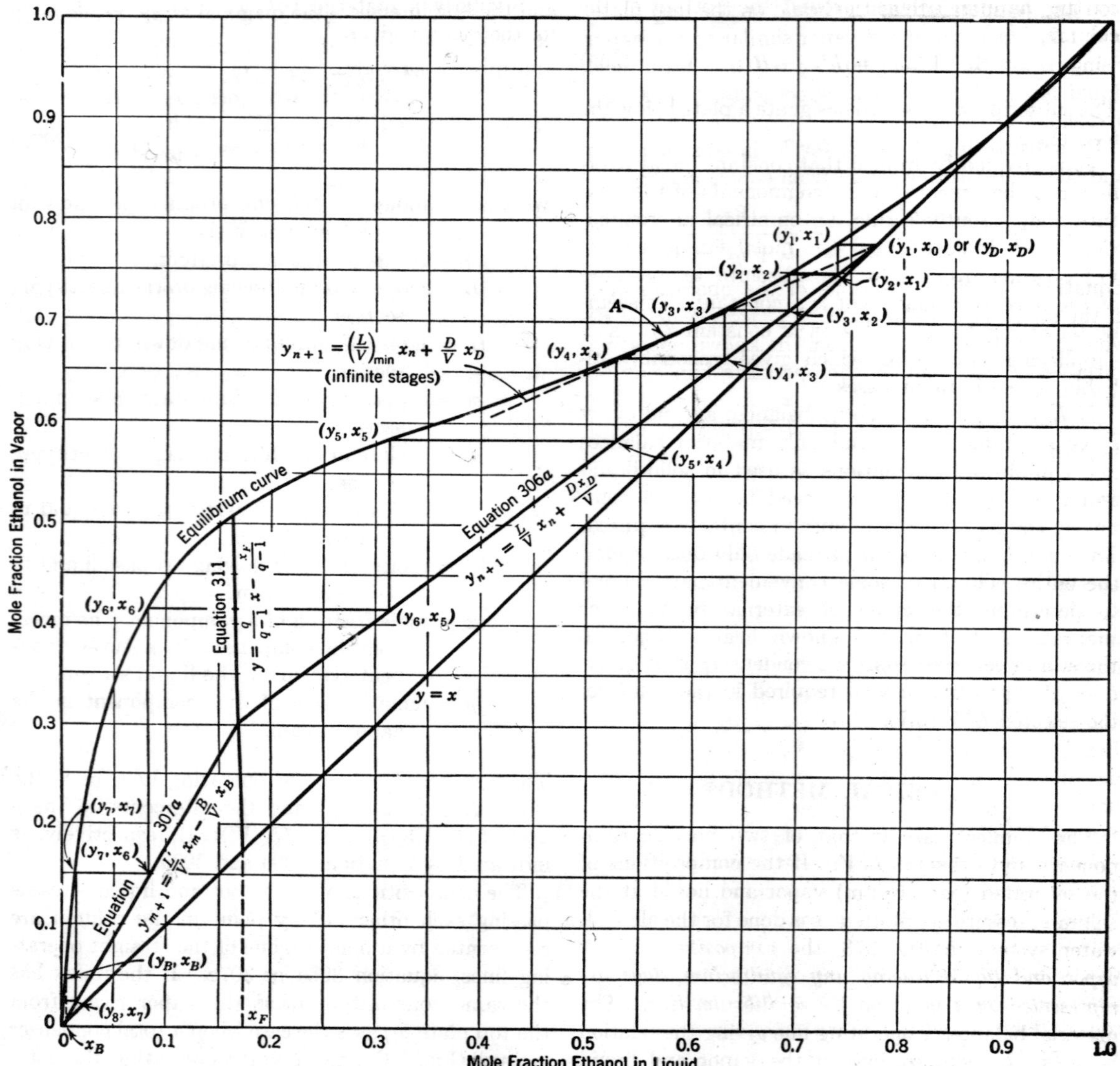

Fig. 373. Fractionation calculations on a vapor-liquid (y versus x) diagram for the ethanol-water system at 1 atm, assuming constant molal overflow.

of the vapor and liquid streams passing each other *below* a plate are represented by the points on the operating line below the corresponding equilibrium point, and the compositions of the streams passing each other *above* a plate are represented by the point on the operating line to the right of the equilibrium point.

The relation between the quantity of liquid and vapor flowing in the rectifying section to that in the stripping section depends upon the condition of the feed. This relation may be expressed as:

$$L' = L + qF \tag{308}$$

or

$$V = V' + (1 - q)F \tag{309}$$

where q = the moles of saturated liquid formed on the feed plate by the introduction of one mole of feed. If the feed is saturated liquid of the same composition and temperature as the liquid on the feed plate, $q = 1.0$; if it is saturated vapor similar to the vapor rising from the feed plate, $q = 0.0$. With cold liquid feed, $q > 1.0$. With superheated vapor feed, $q < 0.0$.

By substituting $(L' - B)$ for V' in equation 307*a* and then substituting for L' from equation 308,

$$y_{m+1} = \frac{L + qF}{L + qF - B} x_m - \frac{Bx_B}{L + qF - B} \quad (310)$$

Equation 310 gives the slope of the operating line in the stripping section as $(L + qF)/(L + qF - B)$ which may be determined from the known operating variables. One point on the operating line may be located by substituting x_B for x_m in equation 307 and solving for y_{m+1}. By substituting from equation 305,

$$V'_{m+1}y_{m+1} = V'_{m+1}x_B$$

and

$$y_{m+1} = x_B$$

and it follows that the point on the operating line representing the compositions of the streams L_B and V_{B+1} lies on the diagonal and has the coordinates x_B, y_B. The operating line may then be drawn through this point with the known slope $(L + qF)/(L + qF - B)$. The vapor from the reboiler is in equilibrium with the bottoms product (Fig. 334) and has the composition represented by the intersection of the line x_B with the equilibrium curve. From this point the stepwise construction may be continued up the stripping column.

A more convenient method when treating a complete fractionating column is to find the locus of the intersection of the operating lines at the feed plate (equation 311). This may be done by eliminating L from equations 306*a* and 310, giving the locus of points common to both operating lines. Substituting from equation 304 for V in equation 306,

$$L(y - x) = D(x_D - y)$$

From equation 310,

$$L(y - x) = qF(x - y) + B(y - x_B)$$

Since $B = F - D$ and $Bx_B = Fx_F - Dx_D$,

$$D(x_D - y) = qF(x - y) + Fy - Dy - Fx_F + Dx_D$$

Simplifying

$$y = \frac{q}{q - 1} x - \frac{x_F}{q - 1} \quad (311)$$

This line giving the locus of the intersections of operating line is frequently called the q line and depends only on q and x_F. The slope is $q/(q - 1)$, and the line intersects the diagonal at y_F, x_F, as may be seen by substituting x_F for x in equation 311 and multiplying through by $(q - 1)$.

The illustrative example on page 334 is recalculated on a mole basis in Fig. 373. The distillate product contains 77.85 mole per cent, the bottom product 0.393 mole per cent, and the feed 17.38 mole per cent of ethanol. From an energy and material balance, the feed of 43.7 moles/hr contains 42.68 moles of liquid. Therefore $q = 42.68/43.7 = 0.9766$, and the slope of the q line, equation 311, is -41.4. From an energy and material balance around the top plate with 29.4 moles/hr of reflux at 114° F, the "internal reflux" L_1 is calculated as 32.9 moles/hr and the vapor rate V_2 as 42.46 moles/hr. The slope of the operating line representing a material balance around the top of the column (equation 306*a*) is thus $L/V = 0.775$. The intersection of this line with the q line (equation 311) determines one point on the operating line giving a material balance around the bottom of the column (equation 307*a*).

The calculation of the number of ideal stages is made by drawing the steps as shown in Fig. 373. The point (y_1, x_0) on the material balance line (equation 306*a*) represents the vapor stream rising from the top plate and the liquid reflux to the top plate. The composition of the liquid x_1 in equilibrium with the vapor y_1 is determined by drawing a horizontal line (y_1) to intersect the equilibrium curve at (y_1, x_1). From point (y_1, x_1) a material balance is made to compute y_2 by drawing a vertical line to intersect the material balance or operating line (equation 306*a*) at (y_2, x_1). This stepwise graphical calculation is continued, using the equilibrium curve and the line of equation 306*a* which is a material balance around the top of the column above the feed, until the q line is crossed. From that point on down the column the material balance is made including the feed, or around the bottom of the column, by using the line of equation 307*a* as indicated in Fig. 373. The completed calculation indicates almost eight equilibrium stages in close agreement with the results of Fig. 336.

At total reflux, $L = V$ and $L' = V'$, so that the operating lines coincide with the diagonal line. As the reflux rate is decreased, the operating lines depart more and more from the diagonal line until at minimum reflux at least one of the operating lines touches the equilibrium line. In the above example as reflux is reduced the operating line becomes tangent to the equilibrium curve at point A, producing a zone of constant composition at that point corresponding to that shown previously (Fig. 338). Where there is no point of inflection in the equilibrium curve, the zone of constant composition develops where the q line intersects the equilibrium line or, in other words, at the feed plate.

This procedure, known as the McCabe-Thiele method,[29] is widely used. It is a simple, rapid method which gives results usually sufficiently accurate for engineering work. In most cases, it is not necessary to calculate the number of equilibrium plates with great precision because of the uncertainty of available data on overall plate efficiency and because additional plates are usually provided as a factor of safety. The method may lead to the unwarranted assumption of constant molal overflow in cases where it may not exist and where the assumption may be critical.

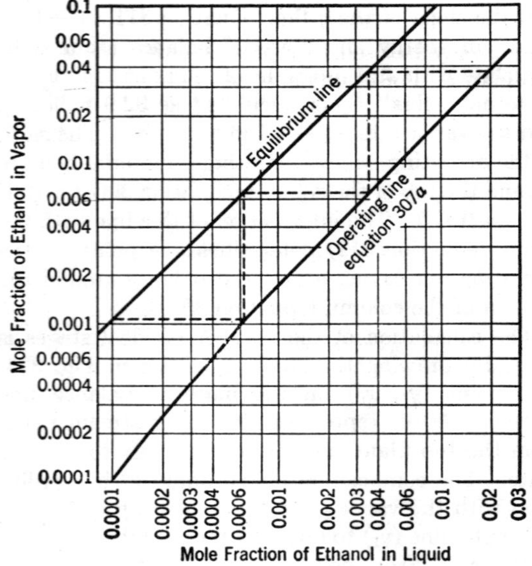

FIG. 374. Fractionation calculations in the dilute region on a logarithmic vapor-liquid (y versus x) diagram for the ethanol-water system at 1 atm.

Dilute Solutions

The McCabe-Thiele method is accurate and particularly advantageous in fractionation calculations involving extremely dilute solutions where the assumption of constant molal overflow approaches actual conditions more closely, as, for example, in reducing the ethanol content of a water solution to a mole fraction of 0.0001.

Calculations in the dilute end of the x–y diagram may be made by expanding the scales, but in general this is not convenient. A more satisfactory procedure is to use a plot of log y versus log x, as illustrated in Fig. 374, which shows the calculation of the number of plates required to reduce the ethanol concentration of liquid in the stripping section of a column from a mole fraction of 0.02 to 0.0001 with reflux ratio and other operating conditions the same as in the illustrative example on pp. 334 and 369.

When the quantity of the volatile component transferred is relatively very small, as in the case of dilute solutions, the operation approximates isothermal conditions. For dilute or ideal solutions at a constant temperature the equilibrium ratio $(y/x)^*$ is a constant for each component. This means that the equilibrium line on an x–y diagram is a straight line passing through the origin with a slope equal to y/x, and on a log x − log y diagram the equilibrium line is a straight line with a slope of 1.0 for isothermal conditions. Thus one point on the equilibrium line ($x = 0.01$, $y = 0.104$ for Fig. 374) suffices to determine the equilibrium line. The position of the operating line is calculated from equation 307*a*.

Three Components

If the temperature is constant, as in liquid-liquid extraction, equilibrium relationships are the same for all stages, and calculations for three components in vapor-liquid transfer may be treated in exactly the same manner as three components in extraction (Chapter 22). Except for dilute solutions or for solutions of components having approximately the same boiling points, the temperature is not the same for all stages in vapor-liquid transfer, and other methods are employed to allow for changing equilibrium conditions from stage to stage.

ANALYTIC EXPRESSIONS

Where the liquid is an ideal solution, the equilibrium between vapor and liquid may be written in the form

$$y_A = K_A x_A$$

where K_A is the equilibrium constant. The "relative volatility" of component A with respect to component B is

$$\frac{(y/x)_A}{(y/x)_B} = \frac{K_A}{K_B}$$

and is frequently designated by a_{AB} or α_{AB}.

Two Components

For a *binary system* where A is the more-volatile component

$$\frac{K_A}{K_B} = a = \left(\frac{y}{x}\right)\left(\frac{1-x}{1-y}\right)$$

The equation of the equilibrium curve, Fig. 373, may be obtained by solving for y

$$y = \frac{ax}{1 + (a - 1)x} \tag{312}$$

The equation of the operating line is given by equation 306*a* or 307*a*.

If the molal overflow is constant, L/V is constant. If the relative volatility K_A/K_B is constant, a is constant. These conditions are approximately true in many cases, particularly when the temperature is approximately constant from plate to plate. The equations 312 and 306*a* may be used in the same manner as the graph of Fig. 373. The starting point is the known or desired distillate overhead, or the desired bottoms product. The solution is obtained by working from one of these products to the other, or, in the case of intermediate feed, from each of these products to the feed.

The use of equations 312 and 306*a* is simplified if the coordinates are transformed so that the points (x_I, y_I) and (x_{II}, y_{II}), representing the intersections of the operating line with the equilibrium line, become (0, 0) and (1, 1) on the new coordinates (Fig. 375).

The points (x_I, y_I) and (x_{II}, y_{II}) lie in the z plane (Fig. 375*a*) and the points (0, 0) and (1, 1) lie in the z' plane (Fig. 375*b*). The transformation from the z plane to the z' plane may be performed by use of the equation

$$z' = \frac{(z - z_I)}{(z_{II} - z_I)} \tag{313}$$

where z may be used to represent either the x or y coordinates. For example,

$$x' = \frac{x - x_I}{x_{II} - x_I} \tag{313a}$$

$$y' = \frac{y - y_I}{y_{II} - y_I} \tag{313b}$$

$$y - y_I = y'(y_{II} - y_I)$$

Substituting for y, y_I, and y_{II} the corresponding values from equation 312,

$$\frac{ax}{1 + (a - 1)x} - \frac{ax_I}{1 + (a - 1)x_I}$$

$$= y'\left[\frac{ax_{II}}{1 + (a - 1)x_{II}} - \frac{ax_I}{1 + (a - 1)x_I}\right]$$

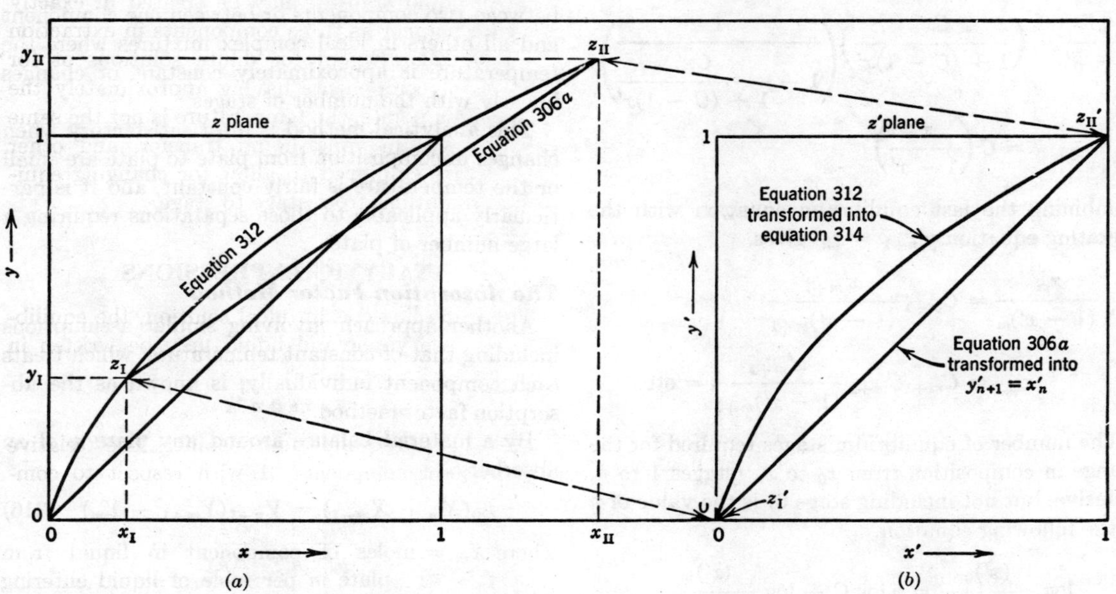

FIG. 375. Transformation of coordinates in connection with equations 312 and 313.

Solving for y' and simplifying,

$$y' = \frac{x - x_{\text{I}}}{x_{\text{II}} - x_{\text{I}}}\left[\frac{1 + (a-1)x_{\text{II}}}{1 + (a-1)x}\right]$$

$$= x'\,\frac{1 + (a-1)x_{\text{II}}}{1 + (a-1)x}$$

Substituting $x'\,(x_{\text{II}} - x_{\text{I}}) + x_{\text{I}}$ for x and simplifying,

$$y' = x'\left[\frac{1 + (a-1)x_{\text{II}}}{1 + (a-1)[(x_{\text{II}} - x_{\text{I}})x' + x_{\text{I}}]}\right]$$

$$y' = \frac{[1 + (a-1)x_{\text{II}}]x'}{1 + (a-1)x_{\text{I}} + (a-1)(x_{\text{II}} - x_{\text{I}})x'}$$

Let [39]

$$C = \frac{1 + (a-1)x_{\text{II}}}{1 + (a-1)x_{\text{I}}}$$

then

$$y' = \frac{Cx'}{1 + (C-1)x'} \tag{314}$$

The operating line (equation 306*a*) is a straight line between points z_{I} and z_{II} and transforms into a straight line between points z_{I}' and z_{II}' which has a slope of unity and an intercept of 0. Therefore equation 306*a* becomes simply

$$y'_{n+1} = x_n'$$

Dividing equation 314 by $1 - y'$ gives

$$\frac{y'}{1-y'} = \left(\frac{Cx'}{1 + (C-1)x'}\right)\left(\frac{1}{1 - \dfrac{Cx'}{1 + (C-1)x'}}\right)$$

$$= C\left(\frac{x'}{1-x'}\right)$$

Combining the last equilibrium equation with the operating equation $y'_{n+1} = x_n'$ gives

$$\frac{x_m'}{(1-x')_m} = C_{m+1}\frac{x'_{m+1}}{(1-x')_{m+1}}$$

$$= C_{m+1}C_{m+2}\frac{x'_{m+2}}{(1-x')_{m+2}} = \text{etc.}$$

The number of equilibrium stages required for the change in composition from x_0 to x_n (stages 1 to n, inclusive, but not including stage 0) is the value of n in the following equation.

$$\log\frac{(x')_0}{(1-x')_0} = n\log C + \log\frac{(x')_n}{(1-x')_n}$$

The number of stages required below stage 0 for a desired separation is

$$n = \frac{\log\dfrac{x_0'(1-x')_n}{x_n'(1-x')_0}}{\log C} \tag{315}$$

The relation between x and x' is given by equation 313*a*. The values for x_{I} and x_{II} are the two roots of the quadratic equation obtained by eliminating y from equations 312 and 306*a*.

Exercise. Show that for the rectifying section the values of x_{I} and x_{II} are the two roots

$$\frac{\left\{-\left[\dfrac{L}{V} - a + (a-1)\dfrac{D}{V}x_D\right] \pm \sqrt{\left[\dfrac{L}{V} - a + (a-1)\dfrac{D}{V}x_D\right]^2 - 4\dfrac{LD}{V^2}x_D(a-1)}\right\}}{2(a-1)\dfrac{L}{V}}$$

A similar procedure may be followed, giving similar equations for the stripping section.

In the special case of total reflux, $D = 0$, $L/V = 1$, and x_{I} and x_{II} become $[(a-1) \mp (a-1)]/2(a-1)$, respectively, or $x_{\text{I}} = 0$ and $x_{\text{II}} = 1$. Also, under these conditions $C = a = K_A/K_B$. This makes equation 315 particularly convenient for determining the minimum number of stages required for separating a binary mixture. But it may be applied with equal ease and reasonable accuracy to the separation between two components or between one component and all others in ideal complex mixtures where the temperature is approximately constant or changes linearly with the number of stages.

This analytical method is most satisfactory when changes in composition from plate to plate are small or the temperature is fairly constant, and it is particularly applicable to those separations requiring a large number of plates.

The Absorption Factor Method

Another approach involving similar assumptions including that of constant temperature, which treats each component individually, is known as the absorption factor method.[7, 8, 23, 27]

By a material balance around any plate m of an absorber,

$$L_0(X_m - X_{m-1}) = V_{n+1}(Y_{m+1} - Y_m) \tag{316}$$

where X_m = moles of component in liquid from plate m per mole of liquid entering absorber (or reflux).

Y_m = moles of component in vapor from plate m per mole of gas entering absorber.
L_0 = moles of liquid entering absorber.
V_{n+1} = moles of vapor entering absorber.

The vapor and liquid leaving an equilibrium plate are in equilibrium, and

$$y_m = K_m x_m$$

where y_m and x_m = mole fractions in equilibrium vapor and liquid, respectively.

$$y_m = \frac{Y_m V_{n+1}}{V_m} \qquad x_m = \frac{X_m L_0}{L_m}$$

$$Y_m = \frac{V_m}{V_{n+1}} \frac{y_m}{x_m} \frac{X_m L_0}{L_m} = K_m \frac{L_0}{L_m} \frac{V_m}{V_{n+1}} X_m \quad (317)$$

By substituting in equation 316 for X_m and X_{m-1} from equation 317

$$Y_m = \frac{Y_{m+1} + \dfrac{L_{m-1}}{K_{m-1}V_{m-1}} Y_{m-1}}{1 + \dfrac{L_m}{K_m V_m}}$$

Defining $A = L/KV$

$$Y_m = \frac{Y_{m+1} + A_{m-1}Y_{m-1}}{1 + A_m}$$

For a one-plate absorber

$$Y_1 = \frac{Y_2 + A_0 Y_0}{1 + A_1} = \frac{Y_2 + \dfrac{L_0}{V_{n+1}} X_0}{1 + A_1} \quad (318)$$

For a two-plate absorber

$$Y_2 = \frac{Y_3 + A_1 Y_1}{1 + A_2} \quad (319)$$

Substituting for Y_1 from equation 318 in equation 319, and rearranging,

$$Y_2 = \frac{(A_1 + 1)Y_3 + A_1 \dfrac{L_0}{V_{n+1}} X_0}{A_1A_2 + A_2 + 1}$$

Similarly, for a three-plate absorber,

$$Y_3 = \frac{(A_1A_2 + A_2 + 1)Y_4 + A_1A_2 \dfrac{L_0}{V_{n+1}} X_0}{A_1A_2A_3 + A_2A_3 + A_3 + 1}$$

And for an absorber of n plates,

$$Y_n = \frac{(A_1A_2 \cdots A_{n-1} + A_2A_3 \cdots A_{n-1} + \cdots + A_{n-1} + 1)Y_{n+1} + A_1A_2 \cdots A_{n-1} \dfrac{L_0}{V_{n+1}} X_0}{A_1A_2 \cdots A_n + A_2A_3 \cdots A_n + \cdots + A_n + 1} \quad (320)$$

If an average effective value may be used for A instead of the individual values A_1, A_2, etc., equation 320 may be simplified to

$$Y_n = \frac{\left(\dfrac{A^n - 1}{A - 1}\right) Y_{n+1} + A^{n-1} \dfrac{L_0}{V_{n+1}} X_0}{\dfrac{A^{n+1} - 1}{A - 1}}$$

$$Y_n = \frac{Y_{n+1}(A^n - 1) + A^{n-1}(A - 1) \dfrac{L_0}{V_{n+1}} X_0}{A^{n+1} - 1} \quad (321)$$

By a material balance around the absorber of n plates,

$$L_0(X_n - X_0) = V_{n+1}(Y_{n+1} - Y_1)$$

Substituting for X_n from equation 317,

$$\frac{L_0}{V_{n+1}} \left[\left(\frac{L_n V_{n+1}}{K_n L_0 V_n}\right) Y_n - X_0\right] = Y_{n+1} - Y_1$$

$$\frac{L_n}{K_n V_n} Y_n - \frac{L_0}{V_{n+1}} X_0 = Y_{n+1} - Y_1$$

Solving for Y_n and substituting A_n for L_n/K_nV_n

$$Y_n = \frac{Y_{n+1} - Y_1 + \dfrac{L_0}{V_{n+1}} X_0}{A_n} \quad (322)$$

Substituting the average effective value A for A_n in equation 322 and equating Y_n from equations 322 and 321,

$$Y_{n+1} - Y_1 + \frac{L_0}{V_{n+1}} X_0 = \frac{Y_{n+1}(A^n - 1)A + A^n(A - 1) \dfrac{L_0}{V_{n+1}} X_0}{A^{n+1} - 1}$$

$$Y_1 = \left[Y_{n+1} \frac{(A^{n+1} - 1) - (A^n - 1)A}{A^{n+1} - 1} + X_0 \frac{L_0}{V_{n+1}} \frac{(A^{n+1} - 1) - A^n(A - 1)}{A^{n+1} - 1} \right]$$

$$Y_1 = Y_{n+1} \left(\frac{A - 1}{A^{n+1} - 1}\right) + X_0 \frac{L_0}{V_{n+1}} \left(\frac{A^n - 1}{A^{n+1} - 1}\right) \quad (323)$$

Equation 323 gives directly the moles of a component (per mole of entering gas) not absorbed, Y_1, as a function of compositions and quantities of the entering streams, the number of equilibrium plates, and effective operating conditions as incorporated in the absorption factor A.

Exercise. By a similar procedure for a stripping column, show that

$$X_n = X_0\left(\frac{S-1}{S^{n+1}-1}\right) + Y_{n+1}\frac{V_{n+1}}{L_0}\left(\frac{S^n-1}{S^{n+1}-1}\right) \tag{324}$$

where S = average effective stripping factor, KV/L.
X_n = moles of a component per mole of entering liquid *not* stripped.

From equation 317, assuming an average effective value for $A = L/KV$,

$$X_0\frac{L_0}{V_{n+1}} = Y_0A$$

Substituting Y_0A for $X_0(L_0/V_{n+1})$ in equation 323 gives

$$Y_1 = Y_{n+1}\left(\frac{A-1}{A^{n+1}-1}\right) + Y_0\left(\frac{A^{n+1}-A}{A^{n+1}-1}\right)$$

Since

$$\frac{A-1}{A^{n+1}-1} = 1 - \frac{A^{n+1}-A}{A^{n+1}-1}$$

$$Y_1 = Y_{n+1} + (Y_0 - Y_{n+1})\left(\frac{A^{n+1}-A}{A^{n+1}-1}\right)$$

$$\frac{Y_{n+1}-Y_1}{Y_{n+1}-Y_0} = \frac{A^{n+1}-A}{A^{n+1}-1} \tag{325}$$

where $Y_{n+1} - Y_1$ = moles of component absorbec per mole of entering gas.
$Y_{n+1} - Y_0$ = number of moles that would be absorbed if it were possible to bring the gas leaving the absorber into equilibrium with the entering liquid.

Similarly, from equation 324

$$\frac{X_0-X_n}{X_0-X_{n+1}} = \frac{S^{n+1}-S}{S^{n+1}-1} \tag{326}$$

Figure 376 provides a convenient graphical solution for equation 325 or 326.

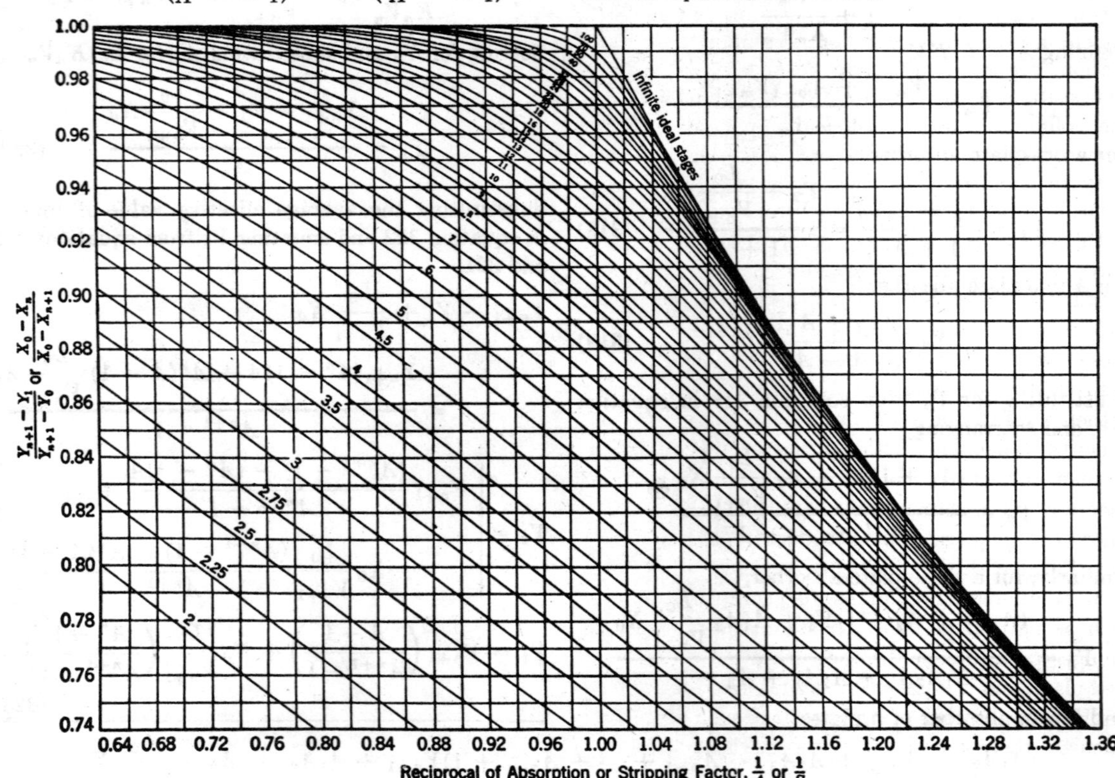

Fig. 376. Solution of absorption and stripping factor, equations 325 and 326.

The absorption factor may be applied to absorption, stripping, or fractionating multiple components. When using the convenient relations based on an average effective value for A, the accuracy of the results depends upon the value chosen for A.[23, 26]

Multicomponent Systems

The absorption factor is well adapted to multiple component systems when the compositions of the streams entering the column are known or can be computed and the use of an average value for the absorption factor A is satisfactory. These conditions are fulfilled in many absorption and stripping operations.[5–8, 23, 27] In this method each component is treated separately. The calculation can be based on the desired absorption of one component to determine operating conditions. The absorption of the other components brought about by these operating conditions may then be computed individually.

The analytic method using equation 315 is also applicable to multiple component systems under certain conditions. The compositions of one component at the extremities of a column, or column section, are represented by x_0 and x_n. Thus the composition of the liquid at the top and bottom of the column in terms of one component must be known. The relative volatility a is then the volatility of the particular component divided by the volatility of the balance of the stream. The balance of the stream is of different composition at different parts of the column, and its volatility is not readily appraised. Equation 315 is accordingly applied to two components of the complex mixture which are present in significant concentrations in the liquid at each end of the column. The relative volatility a is then the ratio of the volatilities of these two components. The mole fraction of the second component is then substituted for $(1 - x)$. This method is generally satisfactory when applied to a column where the temperature is reasonably constant or at least linear with the number of stages.

In fractionating columns for complex mixtures such as natural gasoline, the temperature gradient changes rapidly near the top or bottom or near the feed plate, and it is generally desirable to make plate-to-plate calculations using equations such as 306a to calculate the composition of vapors rising to a plate of known liquid composition. The temperature of the vapors is assumed and the corresponding values for the equilibrium constant K are then used to compute the composition of the liquid on the plate below. The temperature is checked by summing the mole fractions, x's, for all components. This sum should equal 1.00. If not, another temperature is assumed and the corresponding values for K used to compute a new liquid composition from the same vapor composition. In this way it is possible to compute down from the top of the column and up from the bottom for a limited number of plates, or until the presence of the less-volatile components not in the distillate in significant quantities must be considered. It is then necessary to introduce these components into the calculated compositions in order to attain the feed plate composition.[49] A similar procedure is used with equations such as 307a, computing upward from the bottom. Special procedures for simplifying such calculations are available.[5, 26, 42]

Illustrative Example. It is desired to determine the minimum number of plates, and the minimum reflux required to produce a stabilized gasoline having a vapor pressure (Raoult's law) of 28.4 psia at 100° F from the raw gasoline of column 2 of Table 37. There is to be no detectable loss of pentane in the overhead gas, and 97.9 per cent of the normal butane in the raw gasoline is to be recovered in the stabilized gasoline.

It is first necessary to determine the composition of the stabilized gasoline. The vapor pressures of the hydrocarbons at 100° F may be found in the tabulation given below. Let X = moles of isobutane in the stabilized gasoline (bottoms) per mole of feed. The moles of n-butane in the bottoms will be $(0.1935)(0.979) = 0.1895$.

Component	Moles in Bottoms per Mole of Feed	Vapor Pressure, psia at 100° F
$i\text{-}C_4H_{10}$	X	75.8
$n\text{-}C_4H_{10}$	0.1895	51.8
$i\text{-}C_5H_{12}$	0.0732	20.6
$n\text{-}C_5H_{12}$	0.1208	15.4
$C_6H_{14}+$	0.1716	4.84
Total	$0.5551 + X$	

$$\frac{\left\{\begin{matrix}75.8X + (0.1895)(51.8) + (0.0732)(20.6) \\ + (0.1208)(15.4) + (0.1716)(4.84)\end{matrix}\right\}}{(0.5551 + X)} = 28.4$$

From this, X, the number of moles of $i\text{-}C_4H_{10}$ in bottoms per mole of feed, is found to be 0.0368.

In Table 37 the total moles in the finished gasoline per mole of feed is given in the fourth column. After subtracting the moles in the bottoms from the moles in the feed, the

TABLE 37. COMPOSITION OF THE RAW GASOLINE FEED, OF THE DESIRED BOTTOMS PRODUCT (STABILIZED GASOLINE), AND OF THE RESIDUAL OVERHEAD OR DISTILLATE

Component	Saturated Liquid Feed, mole fraction or moles	Bottoms		Overhead Distillate Gas	
		Mole Fraction in Bottoms	Moles of Bottoms per Mole of Feed	Mole Fraction in Distillate Gas	Moles of Distillate Gas per Mole of Feed
(1)	(2)	(3)	(4)	(5)	(6)
C_2H_6	0.0890	......	......	0.2180	0.0890
C_3H_8	0.2714	......	......	0.6650	0.2714
i-C_4H_{10}	0.0805	0.0623	0.0368	0.1073	0.0437
n-C_4H_{10}	0.1935	0.3201	0.1895	0.0097	0.0040
i-C_5H_{12}	0.0732	0.1236	0.0732	......	......
n-C_5H_{12}	0.1208	0.2040	0.1208	......	......
$C_6H_{14}+$	0.1716	0.2900	0.1716	......	......
Total	1.0000	1.0000	0.5919	1.0000	0.4081

Note: The $C_6H_{14}+$ fraction has been assumed to have a molecular weight corresponding to that of normal heptane.

total moles of each component in the overhead distillate is given in column 6. The composition or mole fraction in the overhead distillate is given in column 5.

In this way the material balance for the column can be set up assuming that the high boiling materials desired in the bottoms product are absent from the distillate.

The next step is to determine the operating pressure of the column. If the cooling water temperature is 85° F, the overhead vapors can be cooled to 100° F. If the distillate is to be removed as a vapor as when a partial condenser is employed to form reflux, the minimum operating pressure is computed to be 190 psia as the dew-point pressure of the overhead vapors at 100° F as follows.

TABLE 37A. CALCULATION OF DEW-POINT PRESSURE OF OVERHEAD PRODUCT OF TABLE 37 AT 100° F

Component	Mole Fraction in Distillate Gas	Try 205 psia		Try 190 psia	
		K at 100° F	$y/K = x$	K at 100° F	$y/K = x$
C_2H_6	0.2180	3.0	0.0727	3.22	0.0677
C_3H_8	0.6650	0.95	0.699	1.005	0.660
i-C_4H_{10}	0.1073	0.42	0.257	0.45	0.2415
n-C_4H_{10}	0.0097	0.3	0.0323	0.315	0.0308
Total	1.000		1.0610		1.000

The correct solution for the dew point of a vapor or the bubble point of a liquid is obtained when the summation of the computed mole fractions of all components in the other phase equals unity. Since $y = Kx$, or $x = y/K$, when $\Sigma(y/K) = 1$ the correct solution for a dew point is obtained (see above calculation), and when $\Sigma Kx = 1$ the correct solution for a bubble point is obtained.

The average operating pressure is selected as 215 psia (200 psig) to allow for errors in calculation and also for a reasonable drop in pressure from the bottom to the top of the

TABLE 38. STEPWISE CALCULATIONS FOR TOTAL REFLUX (MINIMUM PLATES) FROM PLATES 1 TO 9

Component	$x_1 = y_2$ Mole Fraction in Vapor from Second Plate = Mole Fraction in Liquid from Top Plate	K at 133 °F and 215 psia	$x_2 = y_3$ Mole Fraction in Liquid on Second Plate or Vapor from Third Plate	K at 150 °F	$x_3 = y_4$ Mole Fraction in Liquid on Third Plate	K at 165 °F	$x_4 = y_5$ Mole Fraction in Liquid on Fourth Plate	K at 176 °F	$x_5 = y_6$ Mole Fraction in Liquid on Fifth Plate	K at 183 °F	$x_6 = y_7$ Mole Fraction in Liquid on Sixth Plate	K at 188 °F	$x_7 = y_8$ Mole Fraction in Liquid on Seventh Plate	K at 192 °F	$x_8 = y_9$ Mole Fraction in Liquid on Eighth Plate	K at 194 °F	$x_9 = y_{10}$ Mole Fraction in Liquid on Ninth Plate
(1)	(2)	(3)	(4)	(5)	(6)	(7)	(8)	(9)	(10)	(11)	(12)	(13)	(14)	(15)	(16)	(17)	(18)
C_2H_6	0.0708	3.6	0.0196	4.0	0.0049	4.35	0.0011	4.6	0.0002	4.78	......		......		......		......
C_3H_8	0.6650	1.26	0.5275	1.45	0.3640	1.625	0.2240	1.76	0.1272	1.84	0.0691	1.91	0.0362	1.97	0.0184	2.0	0.0092
i-C_4H_{10}	0.2343	0.615	0.3820	0.75	0.5080	0.86	0.5900	0.96	0.6150	1.01	0.6070	1.07	0.5680	1.095	0.5190	1.11	0.4660
n-C_4H_{10}	0.0303	0.447	0.0677	0.55	0.1231	0.65	0.1894	0.735	0.2580	0.785	0.3285	0.83	0.3960	0.855	0.4630	0.88	0.5260
Total	1.0004		0.9968		1.0000		1.0045		1.0004		1.0046		1.0002		1.0004		1.0012

TABLE 39. STEPWISE CALCULATIONS FOR TOTAL REFLUX (MINIMUM PLATES) FROM THE BOTTOM UP TO PLATE B−7

Component	x_B Mole Fraction in Bottoms	K at 272° F and 215 psia	$y_B = x_{B-1}$ Mole Fraction in Reboiler Vapor = Mole Fraction in Liquid on Bottom Plate (B − 1)	K at 232° F	$y_{B-1} = x_{B-2}$ Mole Fraction in Liquid on Second Plate	K at 215 °F	$y_{B-2} = x_{B-3}$ Mole Fraction in Liquid on Third Plate	K at 207° F	$y_{B-3} = x_{B-4}$ Mole Fraction in Liquid on Fourth Plate	K at 202° F	x_{B-5} Mole Fraction in Liquid on Fifth Plate	K at 199° F	x_{B-6} Mole Fraction in Liquid on Sixth Plate	K at 197° F	x_{B-7} Mole Fraction in Liquid on Seventh Plate
(1)	(2)	(3)	(4)	(5)	(6)	(7)	(8)	(9)	(10)	(11)	(12)	(13)	(14)	(15)	(16)
$i\text{-}C_4H_{10}$	0.0623	1.94	0.1205	1.5	0.1808	1.32	0.2385	1.25	0.2985	1.2	0.3585	1.175	0.4215	1.15	0.4850
$n\text{-}C_4H_{10}$	0.3201	1.62	0.5195	1.21	0.6280	1.06	0.6650	0.99	0.6590	0.945	0.6230	0.915	0.5700	0.9	0.5130
$i\text{-}C_5H_{12}$	0.1236	1.0	0.1237	0.69	0.0855	0.58	0.0496	0.535	0.0266	0.5	0.0133	0.49	0.0065	0.475	0.0031
$n\text{-}C_5H_{12}$	0.2040	0.83	0.1693	0.565	0.0957	0.47	0.0450	0.425	0.0192	0.4	0.0077	0.386	0.0030	0.38	0.0011
$C_6H_{14}+$	0.2900	0.22	0.0638	0.123	0.0079	0.096	0.0008		......		......		......		......
Total	1.0000		0.9968		0.9979		0.9989		1.0033		1.0025		1.0010		1.0022

column. At this pressure the temperature and composition of the liquid on the top plate are computed as follows.

Component	Mole Fraction in Distillate, y_1	K at 215 psia and 109° F	Mole Fraction in Liquid on Plate 1, $y/K = x$
C_2H_6	0.2180	3.08	0.0708
C_3H_8	0.6650	1.00	0.6650
$i\text{-}C_4H_{10}$	0.1073	0.458	0.2343
$n\text{-}C_4H_{10}$	0.0097	0.32	0.0303
Total	1.000		1.0004

Minimum plates means total reflux or $D = 0$ and $B = 0$. Under these conditions equation 304 gives $V_{n+i} = L_n$ and equations 306 and 307 become $y_{n+1} = x_n$ and $y_{m+1} = x_m$. Using the composition of the liquid on the top plate, x_1, as the composition of the vapor rising from the second plate, y_2, the temperature and composition of the liquid on the second plate may be computed in the manner outlined. This process may be repeated for each successive plate as shown in Table 38 for nine ideal plates.

The sum of the mole fractions is not exactly unity in the calculations of Table 38. Generally this is not necessary so long as these sums average within ±1 per cent of unity.

A similar procedure starting with the liquid bottom product (the stabilized gasoline) of Table 37 and computing upward for seven plates is followed in Table 39.

Comparison of the liquid on plate 9 (Table 38, column 18) with the liquid on plate B−7 (Table 39, column 16) indicates that these compositions are approximately the same in so far as isobutane and normal butane are concerned. The less volatile components are missing in Table 38 because they were assumed to be absent from the overhead distillate. The more volatile components are missing in Table 39 because they were assumed to be absent from the bottoms.

These nondistributed components can be considered in the manner shown in Tables 40 and 41. The isopentane was arbitrarily introduced at plate 6 at a concentration of 0.00026 and n-pentane at a concentration of 0.000048 as indicated in column 2 of Table 40 compared with column 12 of Table 38. The C_6H_{14+} above the feed is present in such small concentration that it may be neglected.

TABLE 40. STEPWISE CALCULATIONS FOR TOTAL REFLUX FROM PLATE 6 TO PLATE 9 AFTER INTRODUCING PENTANES ON PLATE 6

Component	$x_6 = y_7$ Mole Fraction in Liquid on Sixth Plate	K at 188° F	$x_7 = y_8$ Mole Fraction in Liquid on Seventh Plate	K at 192° F	$x_8 = y_9$ Mole Fraction in Liquid on Eighth Plate	K at 195° F	$x_9 = y_{10}$ Mole Fraction in Liquid on Ninth Plate
(1)	(2)	(3)	(4)	(5)	(6)	(7)	(8)
C_3H_8	0.0691	1.91	0.0362	1.97	0.0184	2.0	0.0092
$i\text{-}C_4H_{10}$	0.6070	1.07	0.5680	1.095	0.5190	1.12	0.4630
$n\text{-}C_4H_{10}$	0.3285	0.83	0.3955	0.855	0.4625	0.885	0.5230
$i\text{-}C_5H_{12}$	0.00026	0.425	0.00062	0.445	0.00138	0.46	0.0030
$n\text{-}C_5H_{12}$	0.000048	0.34	0.00014	0.355	0.0004	0.37	0.0011
Total	1.0049		1.00046		1.00168		0.9993

In this way the compositions of plate 9 (Table 40) and plate B−7 (Table 41) are brought into agreement in respect to all components except the two distributed or key components, isobutane and n-butane.

The data resulting from these computations of Tables 38 to 41 are plotted in Fig. 377 which shows the computed

TABLE 41. STEPWISE CALCULATIONS FOR TOTAL REFLUX FROM PLATE B−3 TO PLATE B−7 AFTER INTRODUCING PROPANE ON PLATE B−3

Component	x_{B-3} Mole Fraction in Liquid on Third Plate above Reboiler or Plate B−3	K at 207° F	$y_{B-3}=x_{B-4}$ Mole Fraction in Liquid	K at 201° F	$y_{B-4}=x_{B-5}$ Mole Fraction in Liquid	K at 198° F	$y_{B-5}=x_{B-6}$ Mole Fraction in Liquid	K at 197° F	$y_{B-6}=x_{B-7}$ Mole Fraction in Liquid
(1)	(2)	(3)	(4)	(5)	(6)	(7)	(8)	(9)	(10)
C_3H_8	0.00052	2.18	0.00113	2.09	0.00236	2.04	0.00481	2.02	0.0097
i-C_4H_{10}	0.2385	1.25	0.2985	1.195	0.3575	1.17	0.4180	1.15	0.4810
n-C_4H_{10}	0.6650	0.99	0.6590	0.94	0.6195	0.91	0.5640	0.9	0.5070
i-C_5H_{12}	0.0496	0.535	0.0266	0.495	0.0132	0.48	0.0063	0.475	0.0030
n-C_5H_{12}	0.0450	0.425	0.0192	0.4	0.0077	0.385	0.0030	0.38	0.0011
$C_6H_{14}+$	0.0008		......		....		....		......
Total	0.9994		1.0044		1.0003		0.9961		1.0018

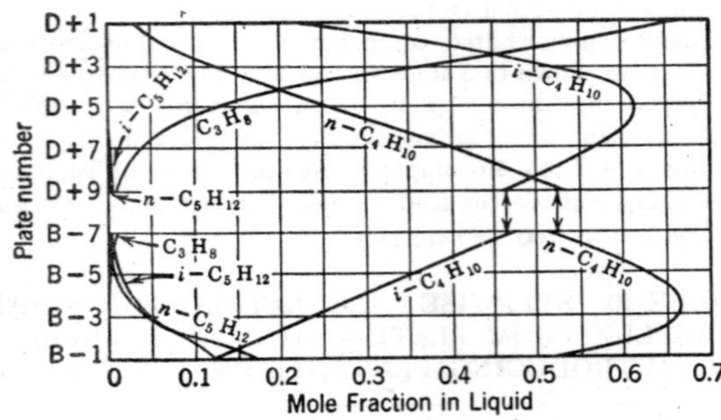

FIG. 377. Composition of liquids on various plates at total reflux for the illustrative example of Tables 38, 39, 40, and 41.[5]

composition of the liquid on each ideal plate when operating at total reflux. It can be seen that the composition of plate 9 (Table 40, column 8) is almost identical with that of plate B−7 as given in Table 41, column 10. By interpolation it would appear that a composition identical to that of ideal plate D+9 would be computed at an ideal plate equivalent to B−6.7. Therefore the minimum number of plates required for the separation of Table 37 would be equivalent to 8 + 6.7, or 14.7 ideal stages including the partial condenser but not the reboiler, or 13.7 ideal plates in addition to the partial condenser and reboiler.

As suggested above, equation 315 may be applied conveniently to two components of a complex mixture to estimate the minimum number of ideal plates. In this case of total reflux equation 315 becomes

$$n = \frac{\log \dfrac{(x_c)_0(x_d)_n}{(x_c)_n(x_d)_0}}{\log \dfrac{K_c}{K_d}}$$

where (x_c) is the mole fraction of isobutane on the plate indicated by the subscript.

(x_d) is the mole fraction of n-butane.

K_c and K_d are average equilibrium constants for isobutane and n-butane, respectively, over the range of temperature in the column.

If this equation is applied over the entire column using composition data taken from Table 37 and equilibrium constants averaged between 109° F and 272° F, the result is

$$n = \frac{\log \dfrac{(0.1073)(0.3201)}{(0.0623)(0.0097)}}{\log \dfrac{0.458 + 1.94}{0.32 + 1.62}} = \frac{\log (56.9)}{\log 1.236} = \frac{1.75511}{0.092018} = 19 \text{ equivalent ideal plates}$$

Since the compositions used were those of the distillate and bottoms, each taken as a liquid, the 19 ideal plates include the partial condenser and the reboiler. Since each of these is calculated as one ideal plate, the result of applying equation 315 over the entire column is 19 − 2 = 17 equivalent ideal plates compared with 13.7 by the rigorous stepwise calculation.

Much better results may be obtained if the values for $a = K_c/K_d$ over the temperature range are averaged rather than using the average of the K's directly. In this case equation 315 becomes

$$n = \frac{\log (56.9)}{\log \left(\dfrac{0.458/0.32 + 1.94/1.62}{2}\right)}$$

$$= \frac{\log (56.9)}{\log (1.314)} = \frac{1.75511}{0.118595}$$

$$n = 14.7 \text{ equivalent ideal plates}$$

Accordingly the number of equivalent ideal stages computed for the column itself is 14.7 − 2 = 12.7, compared with 13.7 computed by the rigorous method.

For best results equation 315 should be used only over that portion of a column where the temperature gradient is linear. Usually this condition exists under conditions of total reflux below the third or fourth plate below the top, and above about the third or fourth plate above the bottom. The plate-to-plate calculations should therefore be made for at least two but not more than three or four plates at each end of the column before applying equation 315. If the stepwise calculations are carried too far errors may be introduced because of neglect of nondistributed components.

For the present problem equation 315 will be applied between plates D+3 and B−2 using data from columns 7 and 8 in Table 38 and columns 8 and 9 in Table 39.

$$n = \frac{\log \dfrac{(0.508)(0.628)}{(0.1808(0.1231)}}{\log \dfrac{0.75/0.55 + 1.32/1.06}{2}} = \frac{\log 14.35}{\log 1.3} = \frac{1.1568}{0.114} = 10.1$$

The 10.1 equivalent ideal plates so computed include plate B−2. Therefore 10.1 + 3 + 1 or 14.1 equivalent ideal plates are computed by this correct use of equation 315 over that part of the column where the temperature gradient is approximately linear. The ratio of the average values of K for each key component is not constant unless the temperature gradient is approximately linear.

Minimum reflux is obtained when the number of plates is infinite. With an infinite number of plates there must be a zone in the column where there is no change in composition from one plate to the next. Equations 306 and 307 may be written for ideal stages

$$V_{n+1}K_{n+1}x_{n+1} = L_n x_n + Dx_D$$

$$V'_{m+1}K_{m+1}x_{m+1} = L_m' x_m - Bx_B$$

In the zones of constant composition $x_{n+1} = x_n$ and

$$x[(D + L)K - L] = Dx_D \quad \text{or} \quad x = \frac{Dx_D}{KD + L(K - 1)} \qquad (327)$$

$$x[(L' - B)K - L'] = -Bx_B \quad \text{or} \quad x = \frac{Bx_B}{KB + L'(1 - K)} \qquad (328)$$

The composition of the zones of constant composition can be computed directly if the following information is available.

1. The composition of the distillate and bottoms.
2. The "minimum reflux" $(L/V)_{\min}$ and $(V'/L')_{\min}$.
3. The temperatures of the zones for evaluating K.

The compositions of the products are given in Table 37 according to the assumption of zero concentration of the undistributed components in one product. This assumption introduces no error in computing minimum reflux.

The minimum reflux is assumed and the assumed value is checked by computation. For computing the minimum reflux for the separation of Table 37, the minimum reflux is assumed to be 0.65 mole of liquid per mole of feed (corresponding to 0.408 mole of distillate or $L/V = 0.615$).

The temperature of the constant composition zone for the assumed minimum reflux is also assumed and then checked by the additional relationship that $\Sigma x = 1$. The correct temperature is indicated when the summation of all the values for x computed by equation 327 or 328 equals unity. As a guide in assuming the temperature, equation 306 indicates that the value of K equals L/V if x_D is zero and x is not zero.

If the mole fraction of n-butane in the distillate were zero, the value of K for n-butane in the constant composition zone would be equal to $(L/V)_m$ or 0.615. Actually the value for K will be somewhat larger than 0.615 because x_D is greater than zero. If K is assumed to have a value of 0.62 a temperature of 160° F is indicated for this zone. Using a temperature of 160° F for the zone of constant composition, the composition of the zone is computed by equation 327.

Component	K at 215 psia 160° F	x from Equation 339	x Revised
C_2H_6	4.25	0.0232	0.0232
C_3H_8	1.59	0.2625	0.2625
i-C_4H_{10}	0.79	0.236	0.2360
n-C_4H_{10}	0.622	0.500	0.4783
Total		1.0217	1.0000

The values for x from equation 327 are high. A slight change in temperature or in K for n-butane makes a relatively large change in mole fraction of n-butane. Therefore, this mole fraction of n-butane is revised as given in the last column above.

The last column gives the composition of the liquid on the plates in the zone of constant composition when the reflux ratio (L/V) is 0.615 and no pentanes are present in the distillate.

TABLE 42. COMPUTATION OF COMPOSITION OF LIQUID IN ZONE OF CONSTANT COMPOSITION BELOW THE FEED

Component	Moles in Bottoms per Mole of Feed Bx_B	K at 215 psia, 228.5° F	KB	$1 - K$	$L'(1 - K)$	(Col. 4 plus Col. 6)	Col. 2 ÷ Col. 7 x
(1)	(2)	(3)	(4)	(5)	(6)	(7)	(8)
i-C_4H_{10}	0.0368	1.425	0.8445	−0.425	−0.701	0.1435	0.2565
n-C_4H_{10}	0.1895	1.155	0.6845	−0.155	−0.256	0.4285	0.4420
i-C_5H_{12}	0.0732	0.68	0.403	0.32	0.528	0.931	0.0786
n-C_5H_{12}	0.1208	0.563	0.334	0.437	0.721	1.055	0.1143
C_6H_{14}+	0.1716	0.118	0.070	0.882	1.46	1.53	0.1120
Total							1.0034

A similar calculation for the stripping section using equation 328 and $L' = 1.65$ (corresponding to $L = 0.65$ and liquid feed $F = 1.00$) and temperatures of about 230° F gives the composition for the zone of constant composition below the feed (Table 42).

The two zones computed above have different compositions and also different ratios for the distributed components (*i*-butane/*n*-butane), and neither zone has the same ratio of distributed components as the feed. These facts do not lead to the conclusion that the assumed minimum reflux ratio ($L/V = 0.615$) is definitely not the correct value. In making the calculations for the zones of constant composition the effects of the less-volatile components in the rectifying section and of the more-volatile components in the stripping section have been correctly neglected. The feed plate composition is then obtained by adding the nondistributed components to the calculated quantities for the zones of constant composition and continuing the calculations plate by plate toward the feed plate as in Tables 43 and 44. The exact quantities added are relatively unimportant in a computation for minimum reflux as the number of plates is not of interest but rather the correct composition at the feed plate for $L/V = 0.615$ with infinite plates.

Figure 378 represents graphically the number of moles of each component in the vapor rising from each plate as computed (1) stepwise down from the top and up from the bottom

TABLE 43. PLATE-TO-PLATE COMPUTATIONS BETWEEN UPPER ZONE OF CONSTANT COMPOSITION (PLATE n) AND FEED PLATE WITH ARBITRARY INTRODUCTION OF PENTANES AND HEXANES FOR THE ASSUMED MINIMUM REFLUX RATIO, $L/V = 0.615$ ($L = 0.65$, $V = 1.0581$)

1	2	3	4	5	6	7	8	9	10	11
Plate		n	$n+1$				$n+2$		$n+3$	
Temperature		160° F	160° F				160° F		160° F	
Component	Dx_D	Liquid Lx_n	Vapor $Lx_n + Dx_D = Vy_{n+1}$	K at 215 psia 160° F	$\frac{Vy}{K}$	Liquid Lx	Vapor Vy	Liquid Lx	Vapor Vy	Liquid Lx
C_2H_6	0.0890	0.0151	0.1041	4.25	0.0245	0.0151	0.1041	0.0151	0.1041	0.0151
C_3H_8	0.2714	0.1708	0.4422	1.59	0.2782	0.1708	0.4422	0.1708	0.4422	0.1708
$i\text{-}C_4H_{10}$	0.0437	0.1533	0.1970	0.79	0.2490	0.1533	0.1970	0.1533	0.1970	0.1533
$n\text{-}C_4H_{10}$	0.0040	0.3105	0.3145	0.622	0.5050	0.3105	0.3145	0.3105	0.3145	0.3105
$i\text{-}C_5H_{12}$	0.000	0.000267	0.000267	0.31	0.000862	0.00053	0.00053	0.00105	0.00105	0.00208
$n\text{-}C_5H_{12}$	0.000	0.0000684	0.0000684	0.246	0.000278	0.000171	0.000171	0.000428	0.000428	0.00107
$C_6H_{14}+$	0.000									
Total		0.6500	1.0581		1.0578	0.6504	1.0585	0.6512	1.0593	0.6528
Correct value for total		0.6500	1.0581		1.0581	0.6500	1.0581	0.6500	1.0581	0.6500

1	12	13	14	15	16	17	18	19	20	21	22
Plate	$n+4$		$n+5$		$n+6$		$n+7$				$n+8$
Temperature	161° F		162° F		164° F		168° F				188° F
Component	Vapor Vy	Liquid Lx	Vapor Vy	Liquid Lx	Vapor Vy	Liquid Lx	Vapor $Lx_{n+6} + Dx_D = Vy_{n+7}$	K at 215 psia 168° F	$\frac{Vy}{K}$	Liquid Lx	Vapor Vy
C_2H_6	0.1041	0.0150	0.1040	0.0149	0.1039	9.0147	0.1037	4.43	0.0234	0.0144	0.1034
C_3H_8	0.4422	0.1702	0.4416	0.1691	0.4405	0.1671	0.4385	1.69	0.2590	0.1593	0.4307
$i\text{-}C_4H_{10}$	0.1970	0.1521	0.1958	0.1507	0.1944	0.1458	0.1895	0.85	0.2230	0.1372	0.1809
$n\text{-}C_4H_{10}$	0.3145	0.3095	0.3135	0.3050	0.3090	0.2934	0.2974	0.675	0.4405	0.2710	0.2750
$i\text{-}C_5H_{12}$	0.00208	0.0041	0.0041	0.00801	0.00801	0.0152	0.0152	0.34	0.0447	0.0275	0.0275
$n\text{-}C_5H_{12}$	0.00107	0.00263	0.00263	0.00647	0.00647	0.0153	0.0153	0.273	0.0560	0.0344	0.0344
$C_6H_{14}+$				0.000032	0.000032	0.00047	0.00047	0.0435	0.0109	0.0067	0.0067
Total	1.0609	0.6535	1.0616	0.6542	1.0623	0.6520	1.0601		1.0575	0.6505	1.0586
Correct value for total	1.0581	0.6500	1.0581	0.6500	1.0581	0.6500	1.0581		1.0581	0.6500	1.0581

using equations 306 and 307, (2) for the two infinite zones, and (3) stepwise from the two infinite zones toward the feed plate in Tables 43 and 44.

The calculations of Table 43 start with the known overhead product Dx_D, column 2 (Table 37, column 6), and the quantities in column 3 obtained by multiplying the mole fractions in the zone of constant composition, x_n, by the moles of liquid, $L = 0.65$, and adding small but reasonable quantities of pentanes. Hexane was added in the liquid from plate $n+5$ in column 15.

The computations of Table 44 start with the known bottom product Bx_B, column 2 (Table 37, column 4), and the quantities of vapor rising from the zone of constant composition, $V'y_{m-1}$, column 3, computed by multiplying x_m (column 8 of Table 42) by K_m (Table 42, column 3) and by 1.0581 (V') and adding a small reasonable quantity of propane. Ethane was added in the vapor from plate $m-10$ in column 23.

Comparison of the quantity of each component in the vapor, Vy_{n+8}, shown in the last column, 22, of Table 43, with that in the vapor, $V'y_{m-14}$, shown in the last column, 33, of Table

TABLE 44. PLATE-TO-PLATE COMPUTATIONS BETWEEN LOWER ZONE OF CONSTANT COMPOSITION (PLATE m) AND FEED PLATE WITH ARBITRARY INTRODUCTION OF ETHANE AND PROPANE FOR THE ASSUMED MINIMUM REFLUX RATIO, $L/V = 0.615$ ($L' = 1.65$, $V' = 1.0581$)

1	2	3	4	5	6	7	8	9	10	11	12	13	14	15	16	17
Plate		$m-1$	$m-2$				$m-3$		$m-4$		$m-5$		$m-6$		$m-7$	
Temperature		228.5° F	228.5° F				228° F		227° F		227° F		227° F		227° F	
Component	Bx_B	Vapor $V'y_{m-1}$	Liquid $V'y_{m-1}$ $+Bx_B$ $=Lx_{m-2}$	K at 215 psia and 228.5° F	$KL'x$	Vapor $V'y$	Liquid $L'x$	Vapor $V'y$	Liquid $L'x$	Vapor $V'y$	Liquid $L'x$	Vapor $V'y$	Liquid $L'x$	Vapor $V'y$	Liquid $L'x$	Vapor $V'y$
C_2H_6	0.000															
C_3H_8	0.000	0.00182	0.00182	2.5	0.00455	0.00292	0.00292	0.00468	0.00468	0.00747	0.00747	0.01193	0.01193	0.0190	0.0190	0.0303
i-C_4H_{10}	0.0368	0.3850	0.4218	1.425	0.6010	0.3845	0.4218	0.3830	0.4198	0.3795	0.4163	0.3780	0.4148	0.3745	0.4113	0.3720
n-C_4H_{10}	0.1895	0.5385	0.7280	1.155	0.8405	0.5380	0.7280	0.5355	0.7250	0.5305	0.7200	0.5280	0.7175	0.5245	0.7140	0.5210
i-C_5H_{12}	0.0732	0.0565	0.1297	0.68	0.0882	0.0565	0.1297	0.0560	0.1292	0.0554	0.1286	0.0551	0.1283	0.0549	0.1281	0.0548
n-C_5H_{12}	0.1208	0.0680	0.1888	0.563	0.1063	0.0680	0.1888	0.0676	0.1884	0.0672	0.1880	0.0670	0.1878	0.0669	0.1877	0.0669
$C_6H_{14}+$	0.1716	0.0140	0.1856	0.118	0.0219	0.0140	0.1856	0.0139	0.1855	0.0138	0.1854	0.0138	0.1854	0.0138	0.1854	0.0138
Total		1.0638	1.6557		1.6625	1.0639	1.6568	1.0607	1.6526	1.0539	1.6458	1.0538	1.6457	1.0536	1.6455	1.0588
Correct value for total		1.0581	1.6500		1.6500	1.0581	1.6500	1.0581	1.6500	1.0581	1.6500	1.0581	1.6500	1.0581	1.6500	1.0581

	18	19	20	21	22	23	24	25	26	27	28	29	30	31	32	33
Plate	$m-8$		$m-9$		$m-10$		$m-11$		$m-12$		$m-13$		$m-14$			
Temperature	225° F		223° F		221° F		217° F		212° F		201° F		188° F			
Component	Liquid $L'x$	Vapor $V'y$	Liquid $L'x$	Vapor $V'y$	Liquid $L'x$	Vapor $V'y$	Liquid $L'x$	Vapor $V'y$	Liquid $L'x$	Vapor $V'y$	Liquid $L'x$	Vapor $V'y$	Liquid $V'y_{m-13}$ $+Bx_B$ $=L'x_{m-14}$	K at 215 psia and 188° F	$KL'x$	Vapor $V'y$
C_2H_6						0.000786	0.000786	0.0028	0.0028	0.00976	0.00976	0.0326	0.0326	4.9	0.1600	0.1025
C_3H_8	0.0303	0.0480	0.0430	0.0748	0.0748	0.1157	0.1157	0.1743	0.1743	0.2545	0.2545	0.3460	0.3460	1.94	0.6700	0.4295
i-C_4H_{10}	0.4088	0.3645	0.4013	0.3525	0.3893	0.3375	0.3743	0.3145	0.3513	0.2835	0.3203	0.2370	0.2738	1.02	0.2780	0.1734
n-C_4H_{10}	0.7105	0.5105	0.7000	0.4955	0.6850	0.4745	0.6640	0.4460	0.6355	0.4070	0.5965	0.3505	0.5400	0.81	0.4365	0.2800
i-C_5H_{12}	0.1280	0.0540	0.1272	0.0525	0.1257	0.0507	0.1239	0.0480	0.1212	0.0446	0.1178	0.0386	0.1118	0.44	0.0492	0.0516
n-C_5H_{12}	0.1877	0.0656	0.1864	0.0636	0.1844	0.0615	0.1823	0.0585	0.1793	0.0540	0.1748	0.0468	0.1676	0.355	0.0595	0.0381
$C_6H_{14}+$	0.1854	0.0132	0.1848	0.0128	0.1844	0.0124	0.1840	0.0117	0.1833	0.0107	0.1823	0.0089	0.1805	0.061	0.0110	0.0070
Total	1.6507	1.0558	1.6477	1.0517	1.6436	1.0531	1.6450	1.0558	1.6477	1.0641	1.6560	1.0604	1.6523		1.6642	1.0671
Correct value for total	1.6500	1.0581	1 6500	1.0581	1.6500	1.0581	1.6500	1.0581	1.6500	1.0581	1.6500	1.0581	1.6500		1.6500	1.0581

44, indicates a satisfactory agreement in the number of moles of all components more volatile than normal butane.

The ethane and propane as computed upward from the bottom of the column (Table 44, column 33) are in agreement with the ethane and propane as determined by computing downward from the top of the column (Table 43, column 22) on the same plate on which the key components are almost in agreement (the iso- and normal butane, of column 33, Table 44, and of column 22, Table 43). This indicates that the arbitrary introduction of propane and ethane on plates $m-1$ and $m-10$, respectively, as indicated in Table 44, is substantially correct and no further adjustment in these compounds is required. Because the pentanes and "hexanes plus" fractions as computed downward from the top of the column (Table 43, column 22) are not in close agreement with the corresponding quantities determined by computing upward (Table 44, column 33) the arbitrary introduction of these components in the computations of Table 43 is not

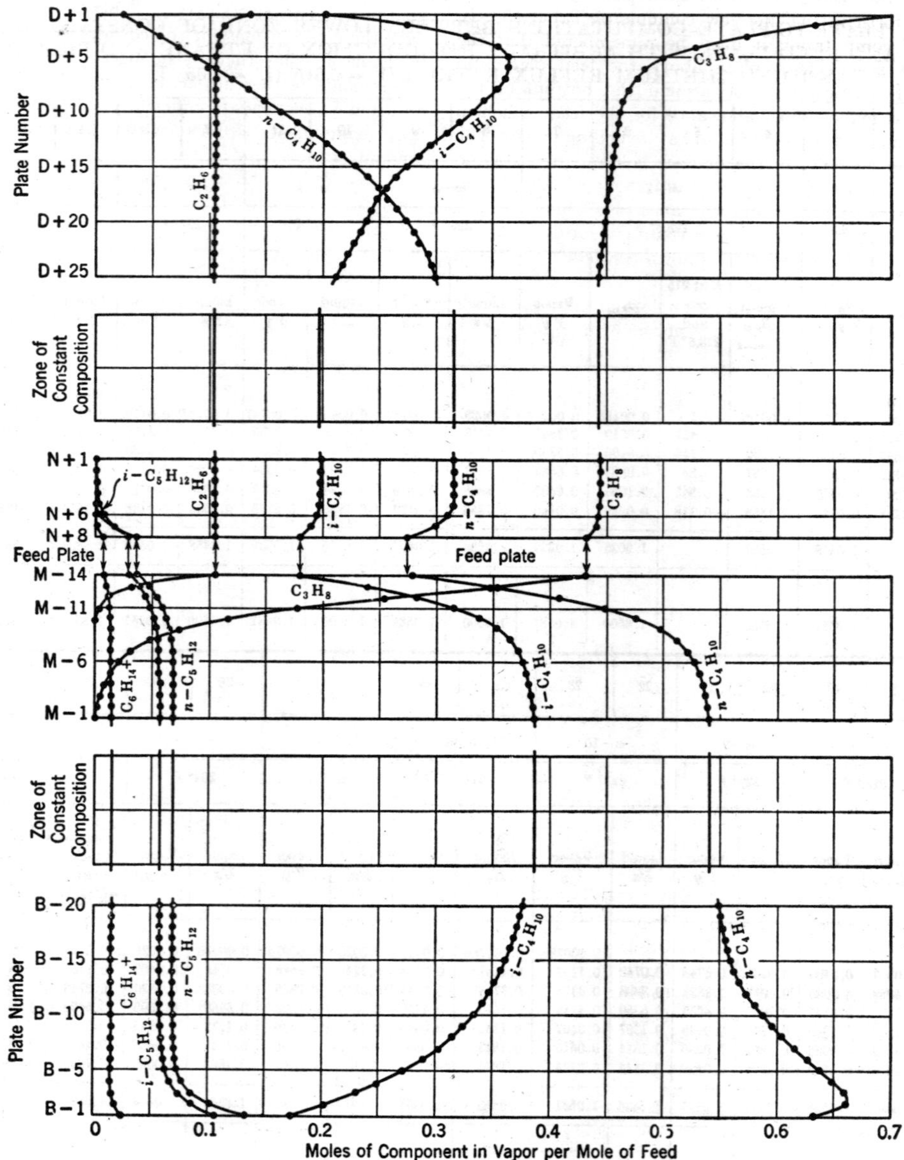

FIG. 378. Composition of vapor rising from plates in a column containing an infinite number of plates expressed in moles of component per mole of feed.

quite correct and some adjustment in the pentanes and "hexanes plus" must be made in the computations for the rectifying section of the column.

The purpose of introducing these components in the computations of Table 43 is to determine what effect their presence might have on the temperature and the fractionation between isobutane and normal butane between the zone of constant composition and the feed plate. The computations of Table 44 may be extended upward to the next plate, which is the plate above the feed plate, designated plate $f-1$, in the manner set forth in the first few columns of Table 45. In making this computation it must be remembered that the plate above $m-14$ is a plate above the feed plate and the material balance, if made around the bottom of the column, must make proper allowance for the feed, which in this case amounts to 1 mole of liquid at its boiling point. The computations set forth in the first six columns of Table 45 have been made in this manner, and the quantity of isopentane, normal pentane, and "hexanes plus" shown in the vapor rising from the plate $f-1$ in column 6 must be correct because they are based upon the known composition of these compounds in the bottoms product.

Because there is no isopentane or normal pentane in the overhead distillate, the quantities of these materials in the liquid overflowing from plate $f-2$ (or, for this computation, plate $n+6$) must be the same as the quantity in the vapor rising from plate $f-1$ as shown in column 6. The total quantity of material in the liquid overflowing from plate $n+6$ is given in column 7, Table 45. This is computed by adding to the ethane, propane, isobutane, and normal butane, as computed for plate $n+6$ in Table 43 (column 17), the quantities of isopentane, normal pentane, and "hexane plus" computed for the vapors rising from plate $f-1$ (column 6, Table 45).

These adjustments or changes in the compositions of the pentanes and hexanes shown in Table 45 are so small compared with the compositions for the corresponding plates computed in Table 43 that there is no necessity in continuing these adjustments farther up the column. The quantity of these adjusted components is sufficiently accurate because the total of all components is 0.656, within 1 per cent of the correct amount, 0.65.

Computations are then continued from this adjusted plate $n+6$ as given in column 7 of Table 45, working down the column in the same manner as set forth in Table 43 to obtain revised quantities for the vapor rising from plate $n+8$ as set forth in column 12 of Table 45.

The temperature of plate $n+7$ in Table 45 is 169° F, compared with 168° F for this plate as computed in Table 43. This difference is due to the different quantities of pentanes

TABLE 45. REVISED PLATE-TO-PLATE COMPUTATIONS ABOVE FEED PLATE FOR THE ASSUMED MINIMUM REFLUX RATIO OF 0.615

1	2	3	4	5	6	7	8	9	10	11	12
Plate	$m-14$	$f-1$				$n+6$ or $f-2$	$n+7$				$n+8$
Temperature	188° F	169° F				164° F	169° F				188° F
Component	(Col. 33, Table 44) Vy_{m-14}	Vy_{m-14} $Dx_D =$ Lx_{f-1}	K at 215 psia and 160° F	KLx	Vapor Vy	Revised Liquid Lx	Vapor Vy_{n+7}	K at 215 psia and 169° F	$\frac{Vy}{K}$	Liquid Lx	Vapor Vy
C_2H_6	0.1025	0.0135	4.45	0.0600	0.0976	0.0147	0.1037	4.45	0.0233	0.0143	0.1033
C_3H_8	0.4295	0.1581	1.7	0.2690	0.4380	0.1671	0.4385	1.7	0.2580	0.1587	0.4301
i-C_4H_{10}	0.1784	0.1347	0.86	0.1159	0.1883	0.1458	0.1895	0.86	0.2200	0.1353	0.1790
n-C_4H_{10}	0.2800	0.2760	0.68	0.1878	0.3055	0.2934	0.2974	0.68	0.4370	0.2690	0.2730
i-C_5H_{12}	0.0316	0.0316	0.346	0.0109	0.0178	0.0178	0.0178	0.346	0.0514	0.0316	0.0316
n-C_5H_{12}	0.0381	0.0381	0.277	0.0106	0.0172	0.0172	0.0172	0.277	0.0620	0.0381	0.0381
$C_6H_{14}+$	0.0070	0.0070	0.044	0.0003	0.0005	0.0005	0.0005	0.044	0.0114	0.0070	0.0070
Total	1.0671	0.6590		0.6545	1.0649	0.6565	1.0646		1.0631	0.6540	1.0621
Correct value for total	1.0581	0.6500		0.6500	1.0581	0.6500	1.0581		1.0581	0.6500	1.0581

and hexanes on this plate, which quantities were not correctly introduced by the arbitrary introduction used in the calculation of Table 43. The discrepancy of 1°, however, is not sufficient to cause any significant difference in the compositions of the isobutane and normal butane.

In making these trial-and-error, plate-to-plate calculations, it is usually best to aim at an agreement between the compounds more volatile than the key components and less volatile than the key components, proceeding in the manner as outlined in Tables 43 and 44. When these components have been brought into approximate agreement, the final adjustment may be made as described in Table 45. If the key components computed down from the top (column 12, Table 45) are also in agreement with the key components computed up from the bottom (column 2, Table 45), the assumed minimum reflux is correct.

This agreement is not perfect as the ratio between *i*-butane and *n*-butane is 0.658 on plate $n+8$ and 0.637 on plate $m-14$. Therefore, the assumed minimum reflux is not quite correct. The normal butane as computed for plate $n+8$ (column 12, Table 45) is about 3 per cent *less* than is required for agreement with plate $m-14$. The higher ratio of *i*-butane to *n*-butane in plate $n+8$ indicates that a slightly greater reflux is required to reduce the *n*-butane in the bottom of the rectifying section with infinite plates when producing the required distillate and bottoms.

To determine the correct minimum reflux, the entire calculation may be repeated, assuming a higher reflux. The results of such calculations for $L = 0.727$ $(L/V = 0.641)$ are given in Table 46.

TABLE 46. RESULTS OF PLATE-TO-PLATE CALCULATIONS, ASSUMING $(L/V)_{\min} = 0.641$

1	2	3	4	5
Temperature	164° F	226.5° F	188° F	187° F
Component	Moles Liquid in Infinite Zone		Feed Plate Vapor Computed	
	Above Feed Lx_n	Below Feed $L'x_m$	Down from Top Vy_{n+10}	Up from Bottom Vy_{m-18}
C_2H_6	0.0153		0.1037	0.1042
C_3H_8	0.1776		0.4356	0.4350
i-C_4H_{10}	0.1568	0.4780	0.1833	0.2170
n-C_4H_{10}	0.3373	0.7440	0.3390	0.3000
i-C_5H_{12}		0.1301	0.0324	0.0324
n-C_5H_{12}		0.1893	0.0390	0.0390
$C_6H_{14}+$		0.1856	0.0072	0.0072
Total	0.7270	1.727	1.1402	1.1348
Correct total	0.7270	1.727	1.1351	1.1351

If the differences between the ratios of the key components in the feed plate vapors are plotted against the assumed minimum reflux ratios (Fig. 379), the correct minimum reflux ratio is indicated as 0.618.

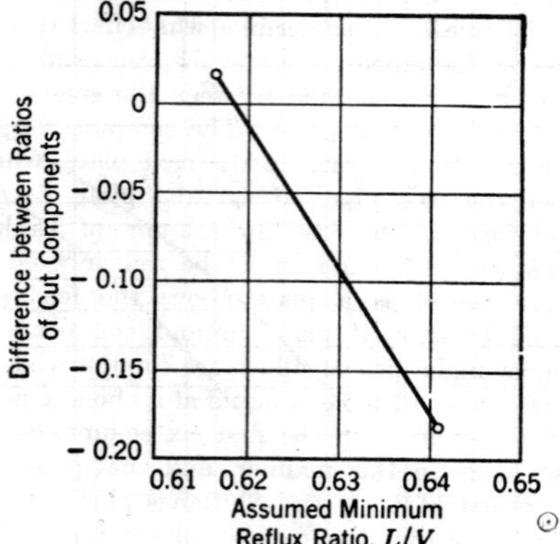

FIG. 379. Graphic interpolation for correct minimum reflux ratio.[5]

For *actual operating conditions* the reflux ratio L/V is less than one and more than the minimum. The number of plates required for a particular separation and reflux ratio may then be computed by a single plate-to-plate calculation starting with the distillate and bottoms with the introduction of the non-distributed components as indicated in the illustrative example.

A reliable short-cut procedure based on empirical relationships which also gives the number of plates required as a function of reflux ratio incorporates the following steps:

1. Compute the minimum number of plates using equation 315 and stepwise calculations as demonstrated in the example above.

2. Assume minimum reflux in the rectifying section and calculate the liquid composition in the zone of constant composition as in the above example or by equation 329.

$$x_n = \frac{x_D}{[(K/K_d)_n - 1](L/D)_{\min} + (K/K_d)_n(x_{dD}/x_{dn})} \tag{329}$$

3. By energy balance calculate the increase in liquid load $(L' - L)$ at feed plate.

4. Calculate the liquid composition in the lower zone of constant composition as in the above example, or by equation 330.

$$x_m = \frac{x_B(K_c/K_d)_m}{(K_c/K_d - K/K_d)_m(L'/B)_{\min} + (K/K_d)_m\left(\frac{x_{cB}}{x_{cm}}\right)} \tag{330}$$

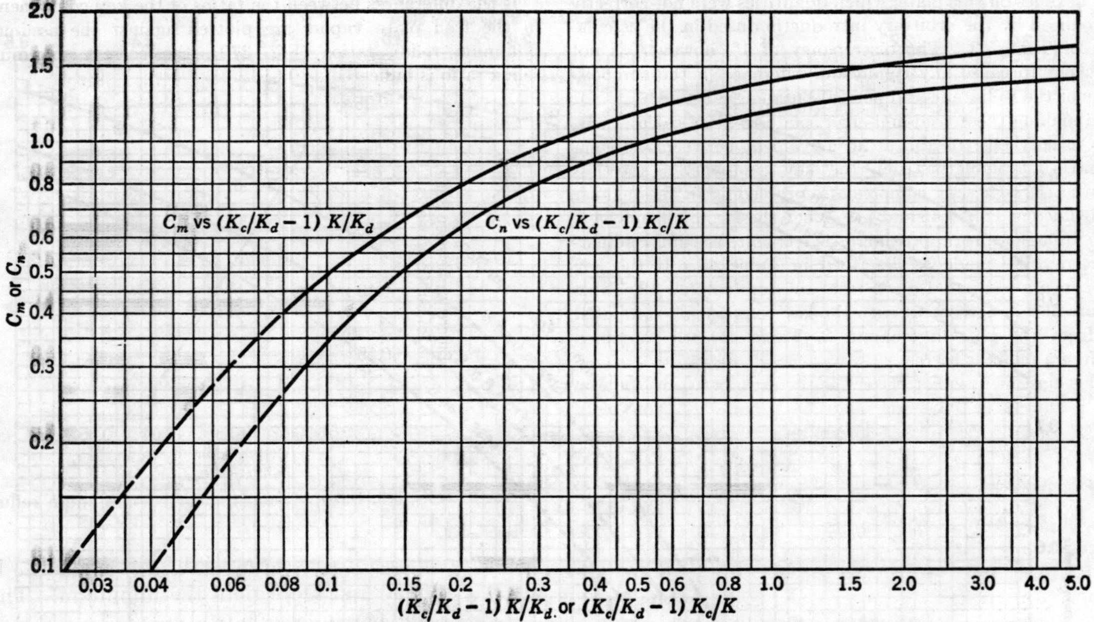

FIG. 380. Chart for determining values of C_n and C_m for equation 331.[14]

5. Check the assumed minimum reflux by equation 331 [14]

$$\frac{(x_c/x_d)_n}{(x_c/x_d)_m} = \left(1 - \Sigma C_m \frac{K_m x_m}{K_d}\right)(1 - \Sigma C_n x_n) \quad (331)$$

where C_m and C_n are obtained from Fig. 380.[14]
subscript c indicates more-volatile key component.
d indicates less-volatile key component.
n indicates upper infinite zone.
m indicates lower infinite zone.
$\Sigma C_m(K_m x_m/K_d)$ is summation of $C_m(K_m x_m/K_d)$ for all components less volatile than the less-volatile key component d.
$\Sigma C_n x_n$ is summation of $C_n x_n$ for all components more volatile than more-volatile key component c.

If the left member of equation 331 is less than the right member, the assumed minimum reflux is too large. The calculation is then repeated with a smaller value for L, and the correct minimum reflux is estimated by graphical interpolation.

6. The required reflux for different numbers of ideal stages, including partial condenser and reboiler if used, is obtained [47] from Fig. 381, using the values for minimum plates and minimum reflux (items 1 and 5).

7. The proper location of the feed may be estimated [26] from equation 332,

$$\log\left(\frac{n}{m}\right) = 0.206 \log\left[\left(\frac{B}{D}\right)\left(\frac{x_d}{x_c}\right)_F\left(\frac{x_{cB}}{x_{dD}}\right)^2\right] \quad (332)$$

where n = number of ideal stages above feed plate.
m = number of ideal stages below feed plate.

With infinite plates the calculated zones of constant composition would come together at the feed plate if the composition of every component in the distillate and in the bottoms were known correctly and each was greater than zero. If one component is absent from the distillate, the upper zone of constant composition is located in the column section above the feed. Similarly, if one component is absent from the bottom product, the lower zone of constant composition is in the stripping section below the feed. The assumption of a number of components completely absent from the distillate, and others absent from the bottoms, can be true only for the case of infinite plates, but it greatly simplifies the calculations and no error is introduced by the procedure outlined as it takes into consideration the effect of these nondistributed components.

By making energy balances around the various plates and the top or bottom of the column, correct values for the liquid and vapor load at any or all points may be computed. The errors introduced by assuming constant reflux may then be reduced

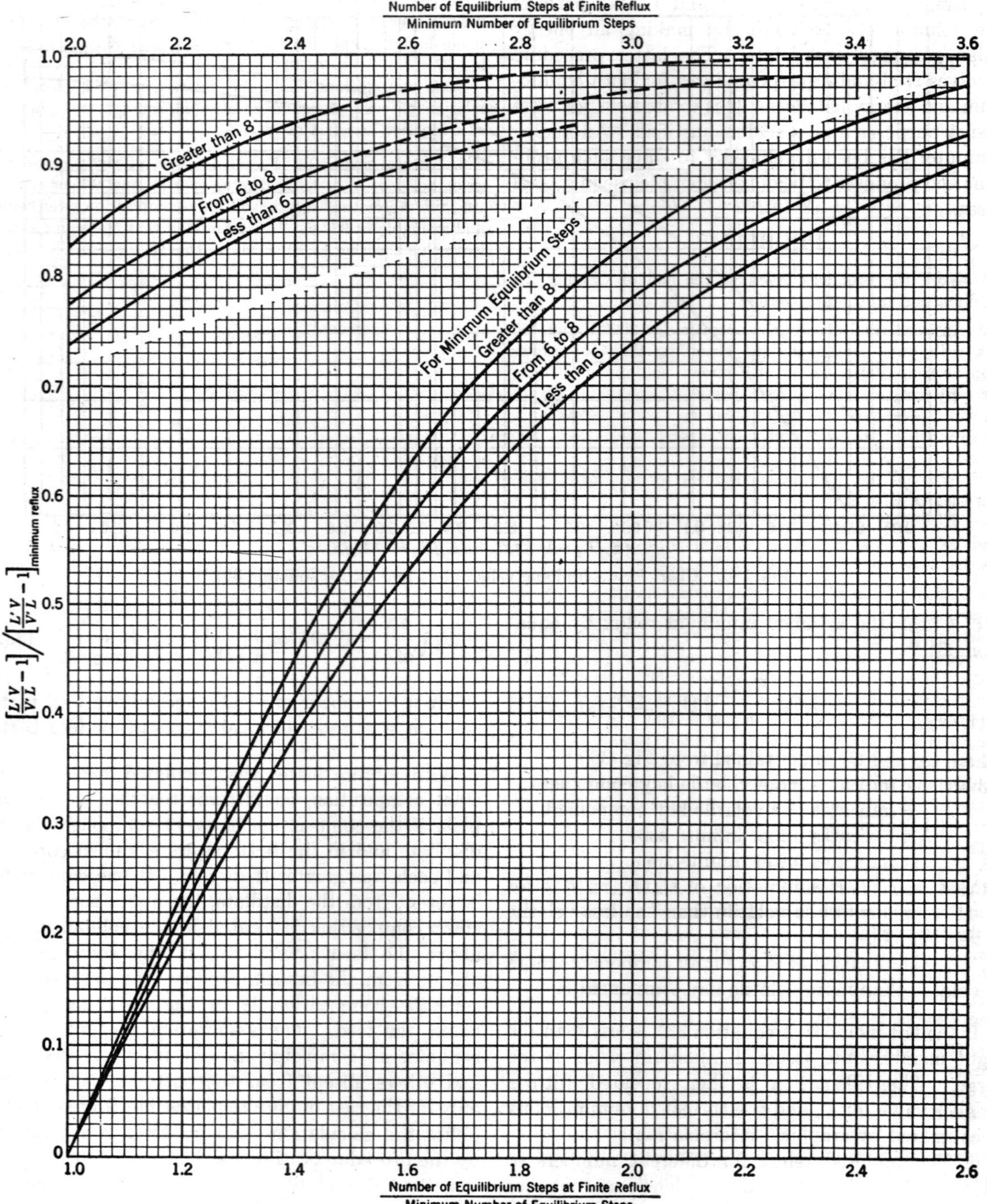

FIG. 381. Relationship between the reflux ratio and the number of ideal stages for a fractionating column. Generally valid if the ratio of stages in the rectifying section to stages in stripping section (n/m) is greater than 0.1 and less than 10 and neither product has practically the same composition as the feed.[47]

by using average values for L and V over parts of the column, or eliminated by making an energy balance around each plate. The latter procedure involves trial-and-error procedures for both equilibrium and energy balances. It is too tedious for most design purposes and is seldom justified. An analog computer has been developed [50] to solve such problems and promises to be of considerable use in special cases.

PROBLEMS

1. Work the problems given at the end of Chapter 23, assuming constant molal overflow.

2. An isothermal absorber is operating at 300° F and 400 psia with pure pentane as the absorbing fluid to recover propane from 100 moles/hr of a stream containing 20 mole per cent methane and 80 mole per cent propane. The off gas contains 80 mole per cent methane on a pentane-free basis.

(*a*) What is the minimum pentane required (moles per hour)?

(*b*) What is the pentane requirement for separation with four theoretical stages?

(*c*) How much pentane is lost in the off gas in part *b*?

(*d*) What is the percentage of propane in the rich (inlet) gas that is recovered in the rich oil?

3. A lithographing concern uses toluene as the vehicle for a transparent coating in a paper-finishing process. The air-toluene mixture obtained from the paper driers is to be treated for recovery of the toluene by being cooled to 90° F and then passed through a "scrubbing tower" in which the toluene is absorbed in an oil which is nonvolatile under these operating conditions. The air-toluene mixture leaving the cooler contains 3 volume per cent toluene. Under present conditions 332,000 standard cu ft (60° F and 760 mm) per hour of the mixture will have to be treated, and 97 per cent recovery of the toluene is desired in a rich oil containing 30 mole per cent of toluene.

The toluene-free absorption oil has a molecular weight of 180, a specific gravity at 90° F (referred to water at 60° F) of 0.865, and a kinematic viscosity of 1.8 centistokes at operating temperature. The absorption oil entering the top of the tower will contain 0 10 mole per cent toluene. The tower is operated isothermally at 90° F. Determine:

(*a*) The number of ideal and of actual plates required.

(*b*) The required column diameter and downspout dimensions if the plate spacing is 18 in.

4. It is proposed to recover 90 per cent of the ethane from a refinery gas stream at 200° F and 200 psia containing 40 mole per cent methane and 60 mole per cent ethane. The absorbing oil has a composition of 98.36 mole per cent pentane, 1.48 mole per cent ethane, and 0.164 mole per cent methane. For each 1000 moles/hr of refinery gas, how much off gas is produced, and lean oil required in a column containing 7 plates of an estimated overall efficiency of 54 per cent? What is the composition of the rich oil?

5. It is proposed to separate by fractionation a mixture containing 50 mole per cent *n*-heptane and 50 mole per cent methylcyclohexane into a distillate product containing 98 mole per cent *n*-heptane and a bottom product containing 2.0 mole per cent *n*-heptane At 1 atm pressure the ratio of the vapor-liquid equilibrium constants $K_{n\text{-heptane}}/K_{\text{methylcyclohexane}}$ is 1.07.

State assumptions made and show in graphical form the relation between the number of equilibrium stages required as a function of reflux ratio.

6. Natural gas at 93° F and 835 psia of the composition given below is to be treated to recover at least 98 per cent of the isobutane with absorption oil of the composition given below and available at 93° F. It is estimated that the lean gas will leave this absorber at 10° F above that of the lean oil.

Component	Mole Fraction in	
	Rich Gas	Lean Oil
CH_4	0.9000	
C_2H_6	0.0518	
C_3H_8	0.0254	
$i\text{-}C_4H_{10}$	0.0047	
$n\text{-}C_4H_{10}$	0.0089	0.00026
$i\text{-}C_5H_{12}$	0.0035	0.00108
$n\text{-}C_5H_{12}$	0.0018	0.00139
C_6H_{14}	0.0024	0.00500
C_7H_{16}	0.0015	0.00925
Oil		0.98302
Total	1.0000	1.00000

Can the desired result be accomplished with 0.18 mole of lean oil per mole of rich gas? If so, how many ideal plates are required?

7. In the alkylation by sulfuric acid catalyst of isobutane with butylene to produce isoöctane, the feed of butylene and isobutane contains some normal butane which builds up in the alkylation and must be removed with the alkylate by fractionation from the isobutane circulated through the alkylation unit. The composition of the feed to the column would be the same as that of the hydrocarbon circulating in the alkylation unit, or about:

$C_2H_6 + C_2H_4$	0.3 mole per cent
$C_3H_8 + C_3H_6$	12.7
$i\text{-}C_4H_{10} + i\text{-}C_4H_8$	51.0
$n\text{-}C_4H_{10} + n\text{-}C_4H_8$	31.8
C_5H_{12}	0.2
Alkylate (equivalent to heptane)	4.0

A fractionating column is available with an allowable working pressure of 160 psia, containing 35 plates not counting the reboiler and partial condenser. Are there enough plates in this present column if they are 100 per cent efficient to fractionate this feed material so that the bottom product will contain substantially all the alkylate and not more than 0.3 per cent of the $i\text{-}C_4H_{10} + i\text{-}C_4H_8$ in the feed and produce an overhead distillate containing not more than 10 mole per cent $n\text{-}C_4H_{10} + n\text{-}C_4H_8$ for return to the alkylation unit?

What operating conditions (temperatures, pressure, reflux ratio, number of stages) are required for this separation?

CHAPTER

26

Vapor-Liquid Transfer Operations 4

Distillation and Condensation

SIMPLE distillation is the operation of producing a vapor from a liquid by boiling the liquid mixture, separating the vapor so formed from the liquid and condensing the vapor without any further transfer of material between vapor and liquid. This operation may be conducted in two ways.

1. *Equilibrium vaporization* (or flash distillation) involves keeping all the vapor and liquid in intimate contact so that the separated vapor is in equilibrium with the residual liquid. By a material balance under these conditions for each component,

$$Fx_F = Vy + Lx \tag{237}$$

The equilibrium relationship may be expressed as

$$\frac{y}{x} = K$$

where K may be an experimentally determined value, or a computed value for the particular temperature and pressure at which the vapor and liquid are separated. For the case of ideal solutions at moderate pressures when y and x are mole fractions and L, V, and F are in moles,

$$yP = xp$$

$$\frac{y}{x} = \frac{p}{P}$$

where P is the total pressure of the distillation.

p is the vapor pressure of the component at the temperature of the distillation.

Substituting y/K for x and solving for y,

$$y = \frac{Fx_F}{V + L/K} = \frac{F}{V}\left(\frac{Kx_F}{K + L/V}\right)$$

$$\Sigma y = 1 = \frac{F}{V}\sum\frac{Kx_F}{K + L/V} \tag{333}$$

Similarly

$$\Sigma x = 1 = \frac{F}{V}\sum\frac{x_F}{K + L/V} \tag{334}$$

The solution involves the assumption of a value for the ratio of vapor to feed V/F and a trial-and-error solution. The example calculation given on p. 389 indicates that the calculated value of this ratio should check the assumed value closely.

It is helpful to remember that, if the calculated V/F differs from the assumed V/F so that a second assumption is necessary, the second assumption should depart from the first assumption in the same direction but to a greater extent than the calculated V/F. A plot of the differences between the computed and assumed values as a function of the assumed value is helpful in indicating the best values.

It is also helpful to make dew-point (Table 37A) and bubble-point calculations to ascertain if the material is in the two-phase region of temperature and pressure before undertaking the calculation for equilibrium vaporization.

2. *Differential distillation* involves the removal of the vapor from contact with the liquid as rapidly as the vapor is formed. In differential vaporization the liquid loses and the vapor or distillate gains a

FLASH VAPORIZATION OF A NATURAL GAS GASOLINE AT 50 PSIA AND 110° F

Component	Mole Fraction in Feed, x_F	K at 110° F and 50 Psia	Kx_F	Assume $V/F = 0.05$: $\left(\frac{L}{V}+K\right)$	Assume $V/F = 0.05$: $\left(\frac{Kx_F}{\frac{L}{V}+K}\right)$	Assume $V/F = 0.06$: $\left(\frac{L}{V}+K\right)$	Assume $V/F = 0.06$: $\left(\frac{Kx_F}{\frac{L}{V}+K}\right)$	Assume $V/F = 0.07$: $\left(\frac{L}{V}+K\right)$	Assume $V/F = 0.07$: $\left(\frac{Kx_F}{\frac{L}{V}+K}\right)$
Ethane (C_2)	0.0079	12.3	0.097	31.3	0.0031	28.0	0.0035	25.6	0.0038
Propane (C_3)	0.1321	3.5	0.462	22.5	0.0205	19.2	0.0240	16.8	0.0275
Isobutane (i-C_4)	0.0849	1.65	0.140	20.65	0.0068	17.35	0.0081	14.95	0.0094
n-Butane (C_4)	0.2690	1.20	0.323	20.2	0.0160	16.9	0.0191	14.5	0.0223
Isopentane (i-C_5)	0.0589	0.50	0.0295	19.5	0.0015	16.2	0.0018	13.8	0.0021
n-pentane (C_5)	0.1321	0.39	0.0512	19.4	0.0027	16.1	0.0032	13.7	0.0037
Hexane plus (C_6+)	0.3160	0.049	0.0155	19.0	0.0008	15.7	0.0010	13.3	0.0012
					0.0514		0.0607		0.0700
$\frac{\text{Computed}}{\text{Assumed}}$			=	$\frac{0.0514}{0.05} = 1.03$		$\frac{0.0607}{0.06} = 1.01$		$\frac{0.07}{0.07} = 1.00$	
Computed − Assumed			=	0.0014		0.0007		0.0000	
Per cent error on calculated value				36%		15%			

differential quantity of material, $dV = -dL$. By a material balance for an instantaneous differential vaporization,

$$-d(Lx) = y\,dV = -y\,dL \tag{335}$$

$$x\,dL + L\,dx = y\,dL$$

$$(x - y)\,dL = -L\,dx$$

$$\int \frac{dL}{L} = \int \frac{dx}{y - x}$$

$$\ln \frac{L_2}{L_1} = \int_{x_1}^{x_2} \frac{1}{y - x}\,dx \tag{336}$$

Equation 336, known as the Rayleigh equation, can be integrated graphically, giving the final composition for a specified quantity of residual liquid if the relationship between y and x is known. For a binary system this relationship is given by the equilibrium diagram such as Fig. 373.

Exercise. Show that, if $y = Kx$,

$$\ln \frac{L_2}{L_1} = \int_{x_1}^{x_2} \frac{1}{K - 1}\,d \ln x \tag{337}$$

Exercise. Show for differential condensation, that is, the formation of liquid from vapor with removal of the liquid from contact with the vapor as rapidly as the liquid is formed, that

$$\ln \frac{V_2}{V_1} = \int_{y_1}^{y_2} \frac{1}{x - y}\,dy \tag{337a}$$

The material balance may be made on the basis of total quantities of each component rather than on the basis of compositions. If $y = Kx$,

$$\frac{y_a}{y_c} = \frac{K_a\,x_a}{K_c\,x_c}$$

For a differential vaporization the instantaneous decrease in moles of any component in the liquid equals the moles present for that instant in the vapor, and, since the ratio between the mole fractions equals the ratio between the number of moles,

$$\frac{-dN_a}{-dN_c} = \frac{K_a}{K_c}\frac{N_a}{N_c} \quad \text{or} \quad \frac{dN_a}{N_a} = \frac{K_a}{K_c}\frac{dN_c}{N_c} \tag{338}$$

$$d \ln N_a = \frac{K_a}{K_c} d \ln N_c \tag{339}$$

where N = number of moles in the liquid.

The ratio K_a/K_c is called the relative volatility of components a and c. It is much less sensitive to changes in temperature than either K_a or K_c. For constant temperature K_a/K_c is constant. If Raoult's law applies, $K_a/K_c = P_a/P_c$. Therefore at constant temperature or when the ratio K_a/K_c may be assumed to have an average constant value,

$$\log \frac{N_{a,2}}{N_{a,1}} = \frac{K_a}{K_c} \log \frac{N_{c,2}}{N_{c,1}} \tag{340}$$

$$\left(\frac{N_{a,2}}{N_{a,1}}\right) = \left(\frac{N_{c,2}}{N_{c,1}}\right)^{\frac{K_a}{K_c}} \tag{341}$$

since

$$\frac{N_{a,2}}{N_{a,2} + N_{b,2} + N_{c,2} + \cdots} = x_{a,2} \tag{342}$$

$N_{a,2}$ may be eliminated from equation 341 which may then be solved directly for $N_{c,2}$ if $x_{a,2}$ is specified.

Exercise. For differential condensation show that

$$\left(\frac{N_{a,2}}{N_{a,1}}\right) = \left(\frac{N_{c,2}}{N_{c,1}}\right)^{\frac{K_c}{K_a}} \tag{341a}$$

when N = number of moles in the vapor.

BATCH FRACTIONATION

Where relatively small quantities of material available at indefinite times or having widely varying compositions are to be separated by distillation, continuous units may be uneconomic or inflexible, leading to the use of batch units. Such units may consist of a still of relatively large volume, a plate column, and the usual auxiliaries for returning reflux and for product storage, as indicated diagrammatically in Fig. 382.

The relationship between the rate of depletion of material in this still, the rate of accumulation of material in product storage, and the rate of transfer of distillate product to the storage tank, neglecting holdup in the system, may be written

$$\frac{-d(Bx_B)}{dt} = \frac{d(Dx_D)}{dt} = x_D' \frac{dD}{dt} \tag{343}$$

where B = quantity of material in the still at time t (moles).
D = quantity of material in product storage at time t (moles).
x_B = mole fraction of a given component in the still at time t.
x_D = mole fraction of a given component in product storage at time t.
x_D' = composition of the distillate product entering product storage at time t.

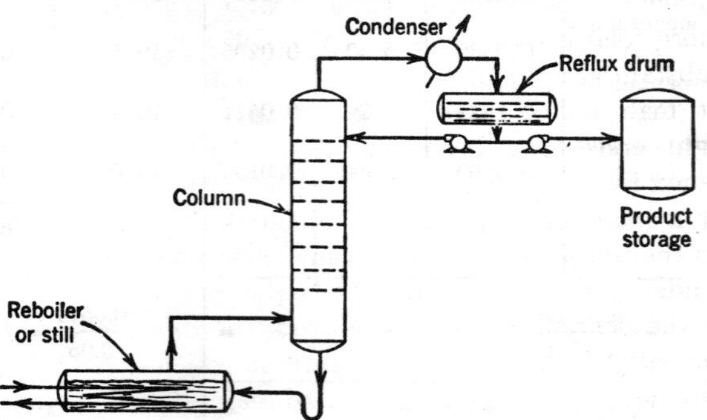

Fig. 382. Flow diagram for a batch distillation system.

Considering the left- and right-hand terms of equation 343, since $-dB/dt = dD/dt$, canceling the time variable dt and rearranging

$$\int_{B_0}^{B} \frac{dB}{B} = \int_{x_{B_0}}^{x_B} \frac{dx_B}{x_D' - x_B} \tag{344}$$

Equation 344 is identical in form with the Rayleigh equation 336 and may be integrated in a similar manner. The simple equilibrium relation between y and x in the denominator of the right-hand side of equation 336 or 337 is replaced by a relationship between x_D' and x_B involving the reflux ratio at which the column is operated and the number of equilibrium stages in the column, such as equation 315 or a graphical relation.

The operation of batch columns may be classified as operation at "constant reflux" or at "constant product composition." As the name implies, in constant reflux operation the distillate product is removed from, and reflux is returned to, the column at a constant rate. As a result the mole fraction of a more-volatile component in the distillate product x_D decreases continuously from a relatively high value at the beginning of the operation. When a column is operated at "constant product composition" the reflux rate is increased continuously from a relatively low value at the beginning of the operation so as to maintain a nearly constant distillate composition.

Most industrial operations, particularly those separating mixtures containing several components, employ both types of operation. For example, in the separation of benzene, toluene, and xylene in the "light oil" obtained from coal, the early stages of the fractionation involve bringing the unit "on stream" by operating at virtually total reflux, followed by the removal of the light ends which are more volatile than benzene. A top plate temperature substantially equal to the bubble point of benzene at high reflux ratios indicates that most of the light ends have been removed. Accordingly the reflux ratio is then reduced, and benzene is removed at a relatively rapid rate. When the temperature on the top plate or, better, some lower plate shows a tendency to increase, thereby indicating that most of the benzene has been removed, the reflux rate is increased by decreasing the rate at which distillate product is removed. In order to maintain the composition of the distillate constant (substantially pure benzene) the operation approaches total reflux (no product withdrawal). As a practical matter, when increasing the reflux rate to maintain the product composition becomes uneconomical, the column is operated at constant reflux, taking overhead a "slop" cut until the temperature on the top plate reaches the boiling point of toluene at which time the entire procedure is repeated.

When a batch unit is operated at constant distillate composition and at maximum capacity, the molal rate at which vapor passes upwards through the column is approximately constant, more and more of this vapor being returned as the reflux rate is increased. Under such circumstances, the time required for a given separation may be determined as follows:

$$B = F\left[\frac{x_D - x_F}{x_D - x_B}\right] \tag{345}$$

where F = moles of material originally charged to the still.

x_F = mole fraction of a given component in the original charge.

Differentiating equation 345 with respect to time t

$$\frac{dB}{dt} = \frac{F(x_D - x_F)\,dx_B}{(x_D - x_B)^2\,dt} \tag{346}$$

Assuming constant molal overflow conditions throughout the length of the column at any instant,

$$\frac{dB}{dt} = L - V = -\left(1 - \frac{L}{V}\right)V \tag{347}$$

when L = moles of liquid per unit time at any section of the column.

V = moles of vapor per unit time at any section of the column.

Substituting the value for dB/dt given by equation 347 in equation 346 and rearranging,

$$\int_0^t dt = \frac{-F(x_D - x_F)}{V}\int_{x_F}^{x_B}\frac{dx_B}{(1 - L/V)(x_D - x_B)^2} \tag{348}$$

If the diameter of a column is known or fixed, the value of V may be estimated and the expression integrated graphically[2] * by relating x_D, x_B, and L/V and the number of stages in the column. The minimum number of stages which can be used is the number given by the separation between x_D and x_B at the end of that separation at total reflux. If this minimum number of plates were used to integrate equation 348, an infinite time would be indicated. Therefore, the number of plates chosen must be greater than the minimum. The larger the number of plates, the less time is required.

The effect of holdup in batch fractionation is at present not fully understood.[12] Although holdup may have an obvious adverse effect on the recovery in product storage because of the presence of desirable material in the column, it may sometime produce a compensating improvement in the fractionating ability of a column.

VACUUM AND STEAM DISTILLATION

Many substances are difficult to purify or separate by distillation at ordinary pressures because such high temperatures would be required to vaporize the material as to decompose the material or make the operation impractical because of excessive temperatures. One solution is to distill in a vacuum or under

* The bibliography for this chapter appears on p. 395.

low pressure, thereby decreasing the required temperature according to the vapor pressure–temperature relationships of the materials being distilled. Another method used alone or in combination with a reduction in pressure is to reduce the partial pressure of the volatile components by introducing an inert vapor which has no significant effect upon the vapor pressure of the liquid components, likewise reducing the temperature. The added vapor should be immiscible with the liquid components being distilled. Steam is widely employed for this purpose as it is readily available; the operation is then called steam distillation. The use of steam is advantageous because it is immiscible with many high-boiling organic compounds, it is readily available at low cost, it may be removed easily from most systems by condensation, and under the proper conditions it can also furnish the necessary heat of vaporization.

Provided the partial pressure of the inert vapor is less than the saturated vapor pressure of the corresponding liquid at the operating temperatures, there will be but two phases present: the liquid components, and the vapor phase including the components and the inert vapor. Under these conditions, when the vapor phase may be regarded as an ideal solution the effect of the inert vapor may be most conveniently taken into consideration by simply recognizing its presence in the vapor phase in computing values for the moles of vapor V, and in computing values for mole fractions of the components y. This assumes that the inert vapor is insoluble in the liquid phase or is soluble to such a limited extent as to have no significant effect on values of L or x, which condition is usually met in such operations.

If the partial pressure of the inert vapor is equal to the saturated vapor pressure of the corresponding liquid at the operating temperature, there will be three phases present: the vapor phase, the liquid components being distilled, and the liquefied inert vapor (water). Under these conditions, the partial pressure of the inert vapor is equal to the saturated vapor pressure of liquid inert, and the mole fraction of inert vapor in vapor phase is equal to $y = p_i/P$, where p_i = saturated vapor pressure of liquid inert, and P = total pressure.

With this value for the mole fraction of inert in the vapor, it is possible to follow the same procedure as outlined in the previous paragraph for computing V and y. The values for L and x are then computed on the basis of the liquid phase carrying the distillable components only, the liquid inert being entirely neglected. But, since the partial pressure of the inert is most easily computed for these conditions, it is usually simpler in the presence of the liquid inert phase to subtract the partial pressure of the inert vapor from the total pressure and then proceed to make the calculations as if the inert vapor were completely absent and the distillation were conducted under a reduced total pressure equal to the partial pressure of the components being distilled, $P - p_i$.

In steam stripping or vacuum distilling operations, the partial pressure of the distilling component in the vapor will in general be low. Therefore, this partial pressure will be sensitive to slight changes in pressure or to changes in the quantity of volatile component in the vapor containing a relatively large quantity of inert vapor. In the vacuum distillation of long-chain fatty acids the absolute pressure at the top of a 30-tray tower is about 5 mm of mercury. Even the moderate pressure drop per tray of 1.5 mm will result in a tenfold increase to a pressure of 50 mm of mercury in the base of the tower. Consequently, the effect of changes in pressure must be considered at each tray in this case. In a graphical solution on an enthalpy-concentration diagram this effect of changing pressure would require different saturated liquid and vapor curves for each tray wherever the changing pressure has an appreciable effect upon the number of equilibrium stages necessary.

The design and operation of a continuous steam-distillation or stripping column may be identical with that of a fractionating tower (Chapter 24). The heat may be supplied by a reboiler or by the condensation of part of the steam supplied, or by the sensible heat of the liquid to be stripped which frequently enters the column superheated with respect to the saturation temperature corresponding to the partial pressure within the tower. If the operating conditions are such that the inert stripping vapor will condense in the tower, provision must be made in the mechanical design of the tower to prevent accumulation of the resulting liquid on the trays. This accumulation will tend to reduce the tray efficiency and may be hazardous as the lower of two liquid layers has a tendency to superheat and "bump" violently.

If water is present beneath a liquid hydrocarbon phase on a tray in a fractionating column, the water may cover the slots in the bubble caps and overflow the weir into the downspouts, preventing proper contact of the vapor with the liquid hydrocarbon and interfering with the flow of distillable liquid. In severe cases, the liquid water may flow down to a lower plate where the temperature is higher and vaporize at that point, only to condense again on the higher plate, in this way short-circuiting those intermediate plates and preventing their functioning by preventing the proper flow of hydrocarbon reflux through that part of the column.

Water may be removed from such plates by passing the liquid overflow through a separator and draining off the lower water layer, allowing only hydrocarbon liquid to flow down to the lower plate. By proper choice of operating conditions, it is usually possible to prevent the condensation of water on the plates within the column.

Batch steam distillation may be conducted as a simple single-stage operation or as a multistage operation such as in a bubble plate column. If volatile material is being distilled from a mixture containing traces of nonvolatile impurities, the equipment may be designed by assuming that steady-state conditions exist. However, if the composition of the material in the still changes appreciably as the batch is distilled the design can be completed only by the integration of the unsteady-state relationships similar to those encountered in differential distillation.

Exercise. For steam distillation of a volatile component from traces of a nonvolatile impurity, show that the following relationships apply.

1. If the inert vapor (steam) does nc . condense,

$$\frac{m_S}{m_A} = \frac{M_S}{M_A}\frac{p_S}{p_A}$$

2. If the inert component is present as a second liquid phase (water),

$$\frac{m_S}{m_A} = \frac{M_S}{M_A}\frac{P - p_A}{p_A}$$

where m = mass of component vaporized or carried in the vapor.
M = molecular weight.
p = partial pressure.
P = total pressure.
S and A as subscripts indicate inert vapor and volatile component, respectively.

AZEOTROPIC AND EXTRACTIVE DISTILLATION

An azeotropic or "constant-boiling" mixture is a liquid mixture whose composition does not change as vapor is generated from the mixture. Since the composition of the liquid does not change as it is converted to vapor, the vapor has the same composition as the liquid, and the boiling point remains constant as vapor is generated. Azeotropes which exist as one liquid phase in equilibrium with vapor may be called homogeneous azeotropes, as distinguished from those existing as two liquid phases in equilibrium with vapor which may be called heterogeneous azeotropes.

Figure 383 is the temperature-composition diagram for the system benzene-ethanol at 1 atm

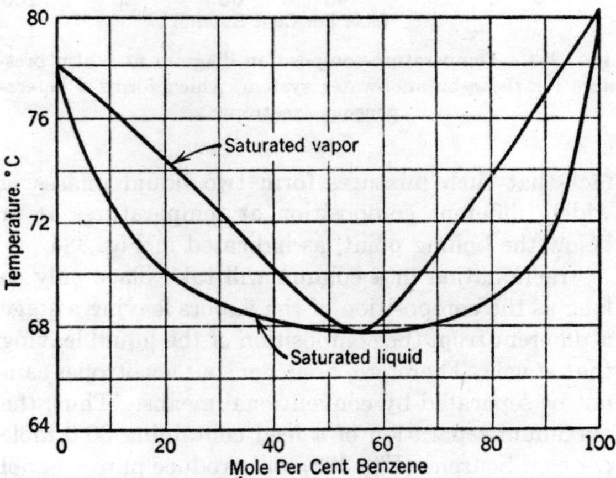

Fig. 383. Temperature-composition diagram at 1 atm pressure for the benzene-ethanol system, which forms a homogeneous azeotrope.

pressure which forms a homogeneous azeotrope at 67.8° C, containing 55.2 mole per cent benzene. Figure 384 shows the temperature-composition diagram for the system n-butanol-water at 1 atm pressure which forms a heterogeneous azeotrope at 92.25° C, consisting of two liquid phases, both of which are in equilibrium with the vapor containing 25.0 mole per cent n-butanol.

Heterogeneous azeotropes may be separated in two conventional columns by taking advantage of the

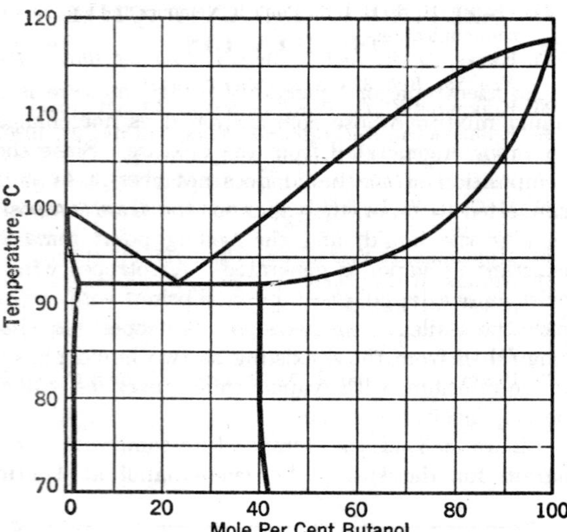

Fig. 384. Temperature-composition diagram at 1 atm pressure for the n-butanol-water system, which forms a heterogeneous azeotrope.

fact that such mixtures form two liquid phases of widely different composition at temperatures at or below the boiling point, as indicated in Fig. 384.

Fractionation in a column will take place only so long as the composition of the vapors leaving a stage is different from the composition of the liquid leaving that stage. Therefore, homogeneous azeotropes cannot be separated by conventional means. Thus, the maximum separation of a feed containing 30.0 mole per cent benzene (Fig. 383) will produce pure ethanol and the azeotrope, whereas the maximum separation of a feed containing 85.0 mole per cent benzene will produce pure benzene and the azeotrope. In both cases, the azeotrope would be the distillate product since it boils at the lower temperature.

The components of a homogeneous azeotrope may often be separated by adding a third component to form a ternary mixture containing no azeotropes or constant-boiling mixtures. If such a third component can be found, one component of the azeotrope may be recovered as the distillate and the other component as the bottom product, either one or both products containing the third component. Where the third component appears in an appreciable concentration in the distillate product, the operation is arbitrarily known as "azeotropic" distillation, whereas, if the third component appears primarily in the bottom product, the operation is arbitrarily termed "extractive" distillation.

Where the third component forms a heterogeneous ternary azeotrope, the separation may usually be accomplished as in the separation of heterogeneous binary azeotropes, since two liquid phases of widely different composition at or below the boiling point are available as feeds to two conventional columns, as indicated in Fig. 385.

The basic principles of azeotropic or extractive distillation are identical to those already explained. It should be remembered that the formation of azeotropes is simply a peculiarity of the particular solutions concerned and offers no particular difficulty in calculation of fractionating columns. In fact, the non-ideal system ethanol-water which forms an azeotrope at about 90 mole per cent ethanol at atmospheric pressure served as the basis of most of the illustrative samples of Chapters 23 and 25.

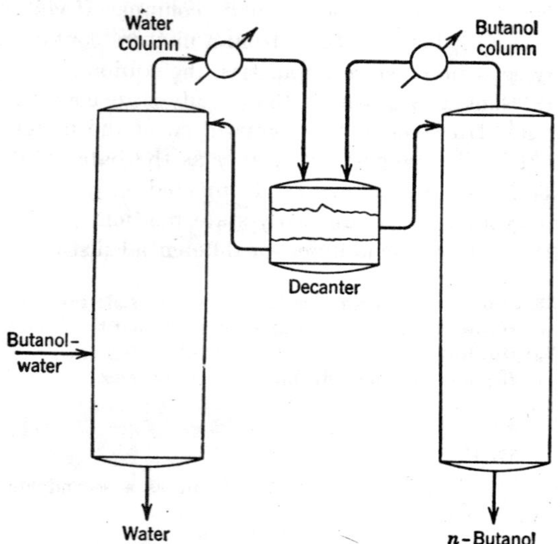

Fig. 385. Flow diagram for the fractionation of the n-butanol-water system, which forms a heterogeneous azeotrope.

1. Ashrof, F. A., T. L. Cubbage, and R. L. Huntington, *Ind. Eng. Chem.*, **26**, 1068 (1934).
2. Bogart, M. J. P., *Trans. Am. Inst. Chem. Engrs.*, **33**, 139 (1937).
3. Bolles, W. L., *Petroleum Refiner*, **25**, 613 (December 1946).
4. Brown, G. G., *Trans. Am. Inst. Chem. Engrs.*, **32**, 321 (1936)
5. Brown, G. G., and D. E. Holcomb, *Petroleum Engr.*, **11**, Nos. 10–13 (1940).
6. Brown, G. G., and F. J. Lockhart, *Trans. Am. Inst. Chem. Engrs.*, **39**, 63 (1943).
7. Brown, G. G., and M. Souders, Jr., *Ind. Eng. Chem.*, **24**, 519 (1932).
8. Brown, G. G., and M. Souders, Jr., *The Science of Petroleum*, Vol. II, p. 1544, Oxford University Press (1938).
9. Carey, J. S., J. Griswold, W. K. Lewis, and W. H. McAdams, *Trans. Am. Inst. Chem. Engrs.*, **30**, 504 (1934).
10 Chillas, R. B., and H. M. Weir, *Trans. Am. Inst. Chem. Engrs.*, **22**, 79 (1929).
11. Cicalese, J. J., J. A. Davies, P. J. Harrington, G. S. Houghland, A. J. Hutchinson, and T. J. Walsh, *Proc. Am. Petroleum Inst.*, **26**, **III**, 180, 1946.
12. Colburn, A. P., and R. F. Stearns, *Trans. Am. Inst. Chem. Engrs.*, **37**, 291 (1941).
13. Colburn, A. P., *Ind. Eng. Chem.*, **28**, 526 (1936).
14. Colburn, A. P., *Trans. Am. Inst. Chem. Engrs.*, **37**, 805 (1941).
15. Davies, J. A., "Bubble Tray Hydraulics," *Ind. Eng. Chem.*, **39**, 774 (June 1947).
16. Drickamer, H. G., and J. R. Bradford, *Trans. Am. Inst. Chem. Engrs.*, **39**, 319 (1943); *Petroleum Refiner*, **22**, 105 (1943).
17. Fenske, M. R., *Ind. Eng. Chem.*, **24**, 482 (1932).
18. Garber, H. J., and F. Lerman, *Trans. Am. Inst. Chem. Engrs.*, **39**, 113 (1943).
19. Geddes, R. L., *Trans. Am. Inst. Chem. Engrs.*, **42**, 79 (1946).
20. Good, A. J., M. H. Hutchinson, and W. C. Rosseau, *Ind. Eng. Chem.*, **34**, 1445 (1942).
21. Harbert, W. D., *Petroleum Processing*, **3**, 610 (August 1947).
22. Holbrook, G. E., and E. M. Baker, *Ind. Eng. Chem.*, **26**, 1063 (1934).
23. Horton, Geo., and W. B. Franklin, *Ind Eng. Chem.*, **32**, 1384 (October 1940).
24. Kemp, H. S., and Cyrus Pyle, "Hydraulic Gradient Across Various Bubble-Cap Plates," *Chem. Eng. Prog.*, **45**, 435 (July 1949).
24*a*. Kemp, H. S., E. I. du Pont de Nemours and Co., private communication.
25. Keyes, D. B., and L. Byman, *Univ. of Illinois Eng. Exp. Sta. Bull.* 328, p. 28 (May 6, 1941).
26. Kirkbride, C. G., *Petroleum Refincr*, **23**, 321 (1944).
27. Kremser, A., *Nat. Petroleum News*, **22** (May 21, 1930).
28. Lobo, W. E., L. Friend, F. Hashmall, and F. Zenz, *Trans. Am. Inst. Chem. Engrs.*, **41**, 693, Fig. 43 (1945).
29. McCabe, W. L., and E. W. Thiele, *Ind. Eng. Chem.*, **17**, 605 (1925).
30. Murphree, E. V., *Ind. Eng. Chem.*, **17**, 747 (1925).
31. Nelson, W. L., *Petroleum Refinery Engineering*, 2nd ed., p. 480, McGraw-Hill Book Co. (1941).
32. O'Connell, H. E., *Trans. Am. Inst. Chem. Engrs.*, **42**, 741 (1946).
33. Peavy, C. C., and E. M. Baker, *Ind. Eng. Chem.*, **29**, 1056 (1937).
34. Perry, J. H., *Chemical Engineers' Handbook*, McGraw-Hill Book Co. (1950).
35. Sherwood, T. K., and F. J. Jenny, *Ind. Eng. Chem.*, **27**, 265 (1935).
36. Sherwood, T. K., G. H. Shipley, and F. A. L. Holloway, *Ind. Eng. Chem.*, **30**, 765 (1938).
37. Siegel, C. L., *Chem. Met. Eng.*, **44**, 493 (1937).
38. Smoker, E. H., *Trans. Am. Inst. Chem. Engrs.*, **34**, 165 (1938).
39. Sorgato, I., "Distillation Equations," *Chimica e industria (Milan)*, **30**, 227 (September 1948).
40. Strang, L. C., *Trans. Inst. Chem. Engrs. (London)*, **12**, 169 (1934).
41. Underwood, A. J. V., *Trans. Inst. Chem. Engrs. (London)*, **10** (112) 1932.
42. Underwood, A. J. V., *Chem. Eng. Progress*, **44**, 603 (August 1948).
43. Walter, J. F., and T. K. Sherwood, *Ind. Eng. Chem.*, **17**, 747 (1925).
44. White, R. R., *Natl. Petroleum News*, **36**, 731 (1944).
45. Ponchon, M., *Tech. Moderne*, **13**, 20 (1921).
46. Savarit, R., *Arts et Metiers*, pp. 65, 142, 178, *et al.* (1922).
47. Martin, H. Z., and G. G. Brown, *Trans. Am. Inst. Chem. Engrs.*, **35**, 679 (October 1939)
48. Souders, M., and G. G. Brown, *Ind. Eng. Chem.*, **26**, 98 (1934).
49. Lewis, W. K., and G. L. Matheson, *Ind. Eng. Chem.*, **24**, 494 (1932).
50. Goelz, G. W., and J. F. Calvert, *Am. Inst. Elect. Engrs.* Technical Paper 50-15 (1950).

PROBLEMS

1. Ten thousand lb/hr of aqueous ammonia solution (0.300 weight fraction ammonia) are flashed from a higher pressure (enthalpy 658 Btu/lb) to 250 psia. The residual liquid is continuously withdrawn, heated, and again flashed to 30 psia.

The system is shown in schematic form. The heat load on the heater is 955,000 Btu/hr.

Determine the quantities and compositions of the two vapor streams and the liquid stream.

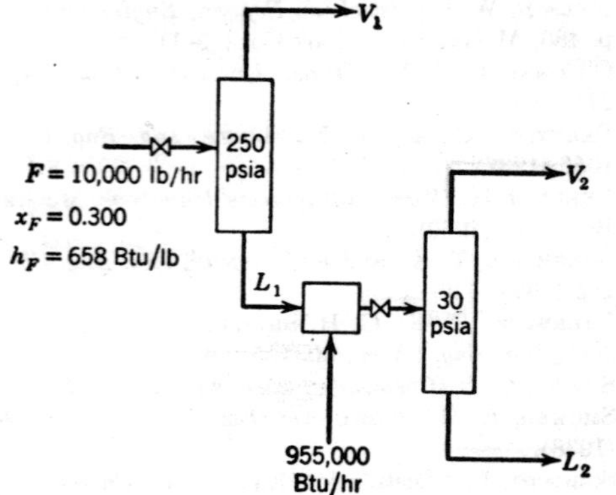

2. A stream of 1000 lb/hr of 30 per cent alcohol by weight at 103° F is fed with 350 lb of saturated steam per hour, supplied at 100 psia to an insulated separator at atmospheric pressure.

What are the quantities and compositions of the resulting equilibrium vapor and liquid?

3. Three streams are fed to a separator at 1 atm pressure as follows:

1090 lb of 20 per cent alcohol at 103° F
233 lb of 15 per cent alcohol at 910° F
128 lb of 5 per cent alcohol, 43 per cent vapor

The vapor from the separator is cooled and fed to separator B, while the liquid is heated and fed to separator C. The liquid from B is also fed to C. The vapors from B and C are combined and form a stream containing 39.5 per cent alcohol. The liquid from C contains 4 per cent alcohol.

Compute the heat duty in Btu per hour, on the cooler and heater.

4. It is necessary to reduce the propane content to 3 per cent in order to market 1000 lb moles of a propane-pentane mixture containing 20 mole per cent propane. Two different methods are being considered: isothermal weathering at 60° F, and efficient fractionation in a fractionating column. For the weathering operation, specify:

(*a*) The original pressure.
(*b*) The final pressure.
(*c*) The per cent recovered as a liquid.
(*d*) The loss of pentane in the vapor.

For the fractionating column, specify:

(*a*) The minimum pressure required if cooling water is available at 60° F.
(*b*) The per cent recovered as a liquid.

Describe very briefly how you would carry out the operations considered.

5. How much and what composition of condensate would be formed if natural gas of the following composition were compressed to 150 psia and 90° F? Composition of gas in mole per cent: CH_4 43, C_2H_6 20, C_3H_8 19, C_4H_{10} 11, C_5H_{12} 5, (C_6H_{14}+) 2.

6. What is the resultant liquid yield obtained from differential weathering of the following high-pressure liquid product at 100° F down to 20 psia? Composition in mole per cent: CH_4 15, C_2H_6 10, C_3H_8 30, i-C_4H_{10} 5, n-C_4H_{10} 10, C_5H_{12} 15, (C_6H_{14}+) 15.

7. What is the resultant liquid yield obtained from the high-pressure liquid of problem 6 if it is flashed to 100° F and 20 psia?

8. A batch fractionating unit contains the equivalent of three equilibrium stages (including the still). A binary mixture containing 50 mole per cent A, the balance B, is charged to the still. The volatility of component A relative to component B is 3.0.

A distillate product is withdrawn as a saturated liquid from the top stage, and reflux is returned to the top stage as a saturated liquid so that the external reflux ratio (L_0/D) is 2.0. A side stream is withdrawn from the second stage at the same rate (moles per unit time) as the overhead distillate product at all times.

Assuming constant molal overflow from the stages and neglecting holdup, what is the composition of the liquid in the side-stream run-down tank when one-half the feed charged to the still has been distilled?

9. A mixture of benzene and toluene containing 0.500 mole fraction benzene and a small amount of nonvolatile impurity is to be purified by batch fractionation at atmospheric pressure.

The equipment to be used consists of a 4-ft diameter column with 20 plates (plate spacing, 18 in.), a still kettle with a capacity of 15,000 gal, and a total condenser. The still kettle is provided with heating coils which are supplied with steam from a 150-psia main.

The fractionation is conducted so that the first cut is taken at a constant distillate composition of 0.995 mole fraction benzene. During this cut, the rate of heat input to the still is maintained constant at the value corresponding to the maximum allowable vapor velocity. The liquid holdup in the column may be neglected.

The relative volatility of benzene with respect to toluene may be considered constant and equal to 2.471.

The overall plate efficiency of the fractionating column is 75 per cent.

The density of 0.500 mole fraction benzene solution at 25° C is 0.868 gram/cc.

If the reflux ratio during the cut is varied continuously until L/V is equal to 0.80 at the end of the cut, determine the per cent recovery of the benzene, where

$$\text{Per cent recovery} = 100 \frac{\text{Quantity of benzene in distillate}}{\text{Quantity of benzene in still charge}}$$

10. A new plant is to produce acetone at an average rate of 13,700 lb/24 hr. As the final step in the process, acetone must be separated from a solution of acetone in acetic acid which contains 0.650 mole fraction acetone.

It is proposed to perform this separation by a batch fractionation at atmospheric pressure conducted so that 95 per cent of the acetone is recovered as top product with a composition of 0.995 mole fraction acetone. An intermediate fraction, or slop cut, of acetone and acetic acid is then taken off as top product at constant reflux ratio until the acetone content of the residue is 0.005 mole fraction. This intermediate fraction is to be recycled to the next batch.

If the plates are spaced 12 in. and of good design, specify the diameter of the column and the number of actual plates required. The effect of liquid holdup may be neglected for the purpose of these calculations. To allow for irregularities in operation, filling and preheating the charge, and pumping out residue, the final design should allow for 50 per cent greater capacity than calculated for ideal operation.

COST DATA ON COLUMNS (ALL COPPER CONSTRUCTION)

Diameter, ft	Cost per Square Foot of Tray Area, dollars
1.5	140
2	125
3	103
4	86
5	73

11. A laboratory fractionating unit consists of a kettle-type still fitted with a closed steam coil and a plate column containing six plates. The overhead vapors are passed through a condenser to a reflux accumulator from which reflux is pumped to the top plate of the column. The distillate product is fed to the still. Previous experience with the unit indicates that with ethanol-water mixtures the overall plate efficiency averages about 67 per cent. The inside diameter of the column is 1.5 ft, and the plate spacing is 12 in. The plates are fitted with wedge-type downspouts so that under normal operating conditions the liquid holdup in the column is about 1 gal per plate. The overhead vapor line is 4-in. standard pipe, 20 ft long, and the liquid holdup in the condenser is estimated at about 1 gal. The total liquid holdup in the distillate and reflux lines is also about 1 gal.

The still is charged with a batch of 42.6 gal (measured at 100° F) of an ethanol-water mixture containing 24.3 weight per cent ethanol. The reflux pump and condenser are operated so that 1 gal of reflux at 100° F is supplied to the top plate per gallon of distillate fed to the still. Steam at 25 psig is supplied to the still coil which contains about 50 sq ft of surface.

(*a*) Estimate the composition of the distillate product when steady conditions have been attained.

(*b*) Estimate the quantity of distillate fed to the still per hour.

12. An absorption unit with a rich oil flash chamber is to be designed for the recovery of ethane from a refinery stream at 100° F and 400 psia containing 300 moles/hr of ethane and 200 moles/hr of methane. The rich oil leaving the absorption column is throttled to a low pressure and sent to a flash chamber to effect a separation between the oil (*n*-octane) and gas (methane and ethane). The oil from the flash chamber is pumped back as lean oil to the top of the absorption tower. Assume that the entire operation is isothermal at 100° F. The off gas shall contain only 20 mole per cent ethane on an oil-free basis. Assume that all methane in the rich oil is flashed and none remains in the lean oil. The flash chamber is to operate at such a pressure that the partial pressure of ethane and octane is 50 psia. The makeup oil is pure *n*-octane. Rich oil contains 1.5 moles octane per mole (methane + ethane).

(*a*) Sketch a neat flowsheet of this process.
(*b*) Make a complete material balance.
(*c*) What is the lean oil recirculation rate?
(*d*) What is the makeup oil rate?
(*e*) How many ideal plates are required in the absorber?
(*f*) What is the total pressure of the flash chamber?

13. Construct a rectangular mole ratio diagram for the system methane, propane and pentane at 300° F and 400 psia and solve problem 2 at end of Chapter 25 by the mole ratio diagram as indicated on page 319. It is best to express compositions on a pentane free basis.

14. Solve problem 4 on page 387 and problem 12 above by a procedure similar to that outlined in problem 13.

CHAPTER

27

Adsorption

THE unit operation of adsorption is concerned with contacting a solid with a fluid mixture under such conditions that some of the fluid is adsorbed on the surface of the solid with a resulting change in composition of the unadsorbed fluid. The adsorption mechanism is relatively complex, and several different types of adsorption are recognized. These types include physical adsorption or condensation of gases on solids at temperatures considerably above the dew point; chemical or activated adsorption, in which definite chemical bonds are produced between the atoms or molecules on the surface of the solid and the adsorbed atoms or molecules; and ion exchange, in which the solid gives up an ion to the fluid for every ion it adsorbs. The ion-exchange reaction is widely used for water softening in which the solid gives up sodium ions to the water in exchange for calcium and magnesium ions. Chemical adsorption is of interest in many catalytic reactions but will not be discussed here. This chapter is primarily concerned with physical adsorption and desorption.

Although physical adsorption can and does occur at all solid-fluid interfaces, it is usually negligible from an engineering point of view unless the fluid is near or below its dew point or the solid is highly porous and filled with fine capillaries. Two of the most important characteristics of the solid adsorbent are its extremely large surface-to-volume ratio, thereby presenting a large area on which the fluid may be adsorbed, and its preferential affinity for certain fluid components. The fluid mixture may be either a liquid or a gas and is usually a single phase.

The commercially important solid adsorbents and some of their uses, in decreasing order of tonnage consumed,[18] * are listed in the table.

Adsorbent	Important Industrial Uses
Fuller's earth	Refining of petroleum fractions, vegetable and animal oils and fats, and waxes
Bauxite	Percolation treatment of petroleum fractions; drying of gases and liquids
Acid-treated clays	Contact filtration and refining of petroleum fractions
Bone char, bone black	Sugar refining; ash removal from solutions
Decolorizing carbons	Refining of sugar, oils, fats, and waxes; decolorizing and purification of water and other liquids
Gas-adsorbent carbon	Solvent recovery; elimination of odors; purification of gases
Alumina	Drying of air, gases, and liquids
Silica gel	Drying and purification of gases; refining of petroleum distillates
Base-exchange silicates	Water treatment
Magnesia	Treatment of gasoline and solvents

Adsorption is used largely as a means of removing one or more components (often present in little more than trace amounts) from a fluid mixture and is in this respect the reverse of solid-liquid extraction. Adsorption is also an analytical tool for separating mixtures which are difficult to separate by distillation, extraction, or crystallization. The separation of close-boiling hydrocarbon mixtures by adsorption on silica gel [11,17] and the separation of mixtures of rare earths by adsorption on resins [21,22] are examples of this. Adsorption and desorption of pure materials is used in gas storage and in refrigeration cycles.[2]

* The bibliography for this chapter appears on p. 411.

In all processes involving adsorption three steps are necessary.

1. Contact of the fluid with solid adsorbent. During this step part of the fluid is preferentially adsorbed on the adsorbent. This adsorbed fluid is termed the adsorbate.

2. Separation of the unadsorbed fluid from the adsorbent-adsorbate.

3. Regeneration of the adsorbent by removing the adsorbate or discarding used adsorbent and replacing it with fresh material.

The principal difference between adsorption and solid-liquid extraction is that extraction involves the transfer of material from the solid phase to the liquid phase whereas in adsorption the fluid phase gives up material to the solid phase. Like solid-liquid extraction, adsorption may be accomplished (1) in a stationary bed of adsorbent as is done in the "percolation" plants, (2) by "dispersed contact" and then separation by sedimentation or filtration as in the "contact" plants, or (3) by a continuous flow of both solid phase and fluid phase, usually countercurrently, as in the Hypersorption process[4,5,13] or in the fluidizing process for separation of gases.[7]

The third step, regeneration of the adsorbent, is performed in a variety of ways, depending on the nature of the adsorbate. Gases and vapors are usually desorbed by either reducing the pressure or raising the temperature of the adsorbent-adsorbate. Steam is often used to perform this desorption, and, in the case of adsorbed water vapor, hot dry gases are sometimes employed. Adsorbed materials from refining oils or petroleum products are often removed by burning. Chemical treatment is also used to regenerate many of the adsorbents.

EQUIPMENT

The *percolation process*[18] uses a fixed bed of adsorbent (Fig. 386*a*). For removing color fron petroleum oil, with fuller's earth or activated bauxite, a cylindrical tank about two or three diameters high contains the solid granular adsorbent. The granular earth may be of various sizes, depending on the viscosity of the oil, and is usually from about 16 mesh to 60 mesh. The oil flow is usually downward by gravity. When the adsorbent is ready for regeneration, the oil is drained and pumped off. Naphtha is then added to recover the oil, and the bed is finally steamed to recover the naphtha. The solid is dumped and sent to kilns for revivication by burning. A typical cycle is as follows.

Charge with oil	6 hr
Shut in, allow air release	2 hr
Running time	100 hr
Draining	2 hr
Soak with naphtha	4 hr
Wash with naphtha	8 hr
Steam	13 hr
Dumping	3 hr

In a typical *contact plant*[18] (dispersed contact), acid-treated reduced crude oil is mixed with adsorbing clay in agitators, and the mixture is heated to 450° F in a pipe still. The hot mixture passes through a vapor separator, is cooled to 300° F, and is filtered. About ½ lb bleaching clay is used per gallon of oil. Further treating is necessary for making lubrication oil.

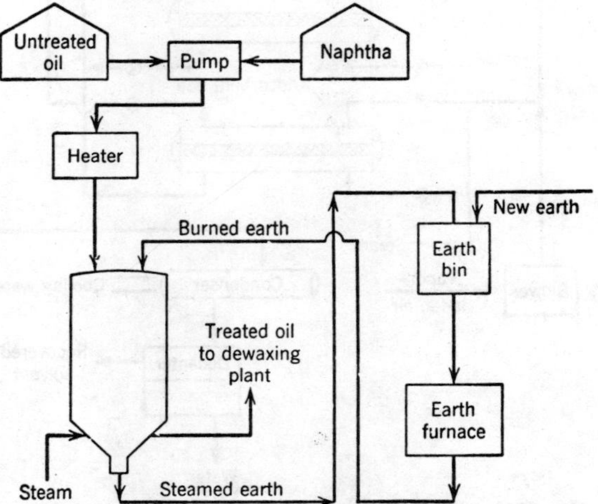

Fig. 386*a*. Flow of a percolation process for treating petroleum oil.[18]

An activated carbon plant for *the recovery of solvent vapors*[3,8] from an air stream consists of at least two adsorbers and condensing and separation equipment (Fig. 386*b*). The cylindrical adsorbers may be either horizontal or vertical, with shallow, horizontal beds of carbon 12 to 30 in. thick. The vapor-laden air is sent down through the beds with the denuded air leaving the bottom of the vessel. After the carbon bed has adsorbed so much solvent vapor that the effluent air contains more than the desired amount of vapor, the vapor-laden air stream is then switched to another adsorber and the first absorber is stripped and reactivated. This reactivation and recovery of solvent is accomplished by passing low-pressure steam up through the bed to desorb the solvent. The steam-solvent mixture is condensed and separated by decantation or distillation.

Dehydration of gases[1] can be carried out in equipment similar to that used for the recovery of solvent vapors. In addition to the two adsorbers, heaters are required to supply the hot gases necessary for regeneration. Since many adsorptive dehydration

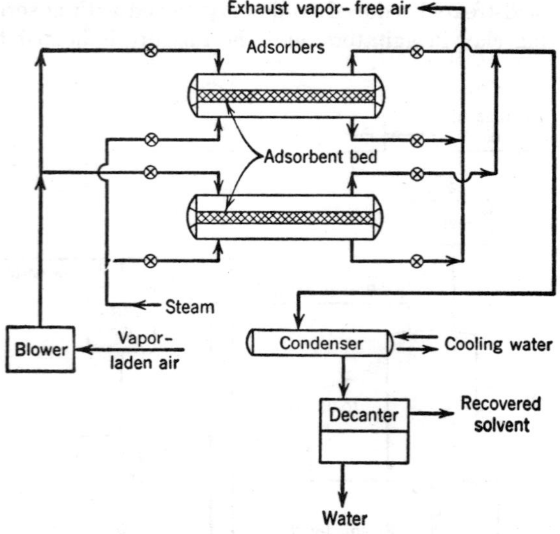

FIG. 386*b*. Flow diagram for the recovery of solvent vapors by activated carbon. (*Carbide and Carbon Chemicals Corp.*)

units operate at high pressures, greater pressure drops through the adsorbers are allowed than for most solvent recovery plants. This permits construction of deeper beds and operation with longer cycles.

Chromatographic Adsorption. When a stream of natural gas containing methane and the higher homologs is passed through an adsorbent bed of active charcoal, all the gas is at first adsorbed. In a few moments the effluent gas appears, composed of pure methane. A few moments later ethane appears with methane in the effluent, later propane appears, and so on, each component appearing later in the order of its adsorption coefficient, i.e., the quantity of adsorbate retained per unit of adsorbent at equilibrium under specified conditions of temperature and fluid phase. This ability of a stationary adsorbent bed to release different compounds in the order of their adsorption coefficients has been recognized for many years and used for analytical purposes under the name of *chromatographic analysis.* It is now being applied as an effective means for the commercial separation[2] of complex materials.

The operation consists essentially in adsorbing the mixture to be separated in a layer at the top of an adsorption column. Pure solvent is then supplied continuously to the top of the column. Each component of the mixture at the top is desorbed into the solvent according to the equilibrium expressed by its adsorption isotherm. As the solvent carrying these solutes moves downward, each compound is again adsorbed by the adsorbent according to the equilibrium conditions. In this way each component moves down the column as a horizontal band at a rate inversely proportional to its adsorption coefficient and may be removed separately from the bottom of the column. As any highly adsorbed component will remain at or near the top of the column, the adsorbent should be mildly active for those compounds it is desired to remove separately from the bottom.

The operation is essentially a countercurrent multiple-stage transfer between a moving solvent phase and a stationary bed of solid adsorbent. In some cases the stationary bed consists of a solvent adsorbed on the solid with the transfer taking place between the stationary solvent phase and a second immiscible solvent flowing through the bed. Because one of the active phases is held stationary a great many contacts or stages are provided in a short travel.

The stationary phase may consist of a strip of filter paper which carries an adsorbed water-rich phase in equilibrium with the immiscible moving phase. The solutes to be separated, identified, or analyzed are then placed at the top of the vertical stationary phase in the form of a small drop or narrow band. The immiscible moving phase free of solute is then supplied at the top of the vertical column. As this second solvent moves down, "countercurrent" relative to the stationary phase, the solutes are carried downward with the moving phase at rates which are proportional to the ratios of solubilities of the solutes in the moving phase to their solubilities in the stationary phase.[19] If the solute is soluble in the stationary phase and insoluble in the moving phase, it remains stationary where it was placed at the top of the vertical column. If the solute is soluble in the moving phase and insoluble in the stationary phase, it is carried downward at

FIG. 387. Two views of the first commercial Hypersorption unit built for the recovery of ethylene. (*Foster-Wheeler Corp.*)

the same rate as the moving phase moves downward.

When used for analytical purposes the operation is stopped when the advancing front of the moving phase approaches the bottom of the stationary phase. The position of the various solutes along the vertical column is determined by observing the positions of centers of color * or other indications of the location of solute. The ratio of distances covered by the solutes to the distance covered by the moving front of the moving phase is characteristic of the solubility ratios of the solutes.

The height equivalent to an equilibrium stage in these chromatographic operations is very small, apparently of the order of 0.002 cm [19] compared with about 1 cm for the best distillation columns in which both fluid phases are in motion relative to the solid phase. The method is applicable to a wide range of materials from simple solutes to complex proteins.[9]

In practical operations considerable variations in the procedure are possible to obtain the desired results. For example, the following steps are reported [2] for an upward flowing operation.

1. The columns are charged with solvent A.

2. Concentrated crude extract in solvent A is charged to the bottom, displacing solvent A to storage, and pumping is continued until the first solute appears in the effluent at the top.

3. Pure solvent A is then charged to the bottom, and pumping is continued until solute 1 is finished or solute 2 appears at the top.

* Hence the name "chromatographic analysis."

4. Pure solvent A is introduced as long as effective separation of compounds is obtained. Then the column is drained.

5. Solvent B is then charged to the bottom, and pumping is continued to separate those compounds not adequately desorbed in solvent A.

6. Solvent C is then charged at the top to desorb and remove the bands of compounds left at the bottom of the column.

The Hypersorption Process.[4, 5, 13] The Hypersorption process also depends upon countercurrent multiple-stage contacts between a fluid phase and an adsorbing solid phase. The operation is continuous and includes a moving bed technique with reflux in a manner quite similar to absorption or fractional distillation with a down-flowing adsorbing carbon bed instead of an absorbing liquid stream. The unit shown in Fig. 387 was built to recover ethylene from a gas stream consisting of hydrogen, methane, about 6 per cent ethylene, and traces of other gases more volatile than ethane. The column is 85 ft high and $4\frac{1}{2}$ ft in diameter. In this operation the activated carbon passes down a column successively through the following sections (see Fig. 388).

1. The tube side of a shell and tube heat exchanger where it is cooled with water on the shell side before entering the adsorbing sections below.

2. An adsorption section, where it first adsorbs from the upflowing gas stream some of the more-volatile components (the less-volatile components are not present in this portion of the column) and then lower down desorbs to the gas stream the most-volatile components while adsorbing some less-volatile components. This adsorption and desorption process is a countercurrent operation with the gas stream flowing upward to a disengaging tray where part of this gas stream which contains only the more-volatile components is removed as a product stream (discharge gas) and the remainder passes up to dry and cool the carbon flowing down through the heat exchanger.

3. A feed distributing plate.

4. A rectifying section where the more-volatile components continue to be desorbed and the carbon adsorbs the less-volatile components. This fractionation operation taking place below the feed is possible because of the reflux produced in the next lower section.

5. A stripping or steaming section, in which desorption occurs. At the top of this section is a disengaging tray, where the stripping steam and less-volatile products are removed in part as a bottoms product and the balance passed up the column as reflux.

6. The tube side of a second shell and tube heat exchanger, where the carbon is heated by a high-temperature fluid, such as Dowtherm, on the shell side to complete the desorption. Steam is introduced to the column below the heat exchanger

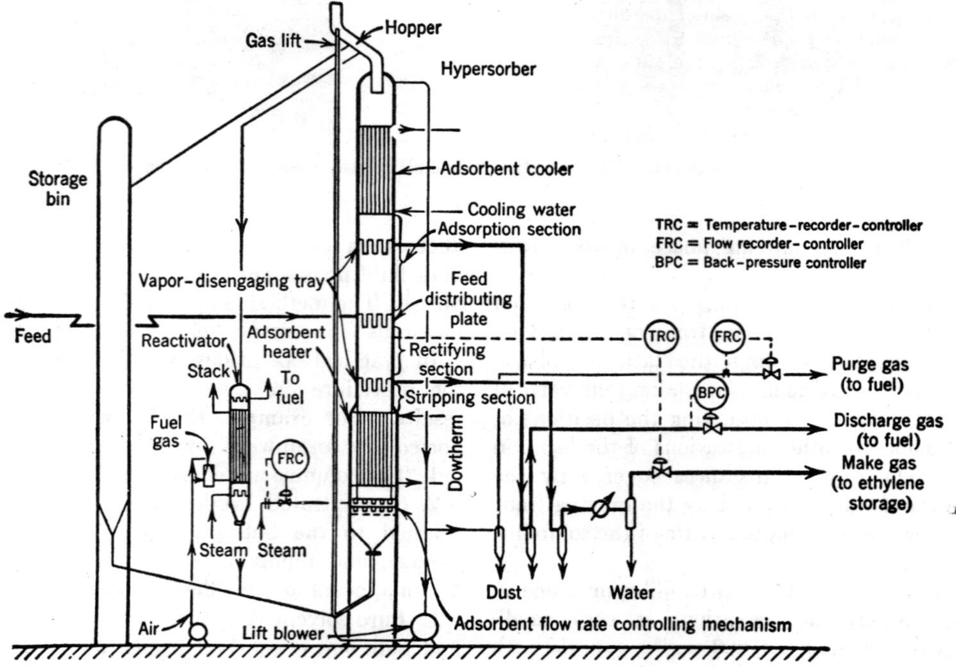

FIG. 388. Flow diagram of a Hypersorption unit built for the recovery of ethylene.[13]

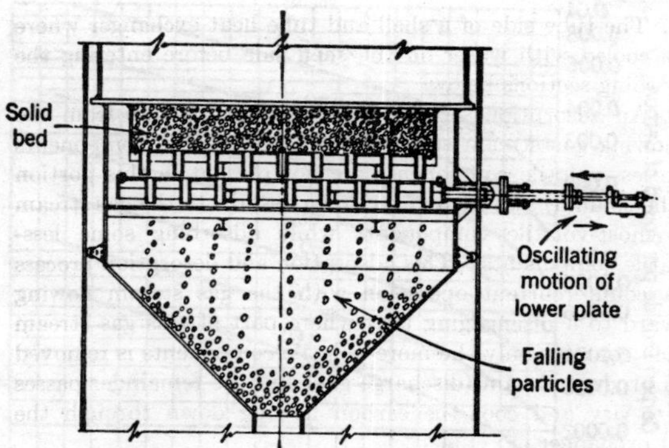

FIG. 389*a*. Carbon flow-rate controlling mechanism in a Hypersorber.[4]

and passes up to the steaming section countercurrent to the carbon. As the carbon passes down through the tubes the gases are completely desorbed by the combination of high temperature and steam.

7. A feed mechanism (see Fig. 389*a*), which controls the carbon circulation rate and maintains even flow over the entire cross section of the column. Then the carbon drops into the sealing leg at the bottom. From the bottom of the column the carbon is transported to the top of the column by a gas lift, and drops into a hopper, located above the water-cooled heat exchanger. The operating carbon-bed level is maintained in a hopper at the top of the unit by addition from the storage bin as required. The lift gas contains carbon fines formed by normal attrition of the carbon in its cycle through the equipment. The fines are separated from the larger particles by elutriation at the top of the tower and removed from the circulation air by a cyclone separator.

Engaging and disengaging tray construction is shown diagrammatically in Fig. 389*b*. This tray has a number of geometrically arranged downspouts which serve to conduct the solids downward in a uniform manner throughout the cross-sectional area of the tray. Disengaging of the gases occurs underneath the tray, and the respective amounts of area inside the tubes and outside the tubes are proportioned in accordance with the flow quantities of gas either disengaged or fed upward in the unit through the downspout section. A number of inlet or outlet points is employed to minimize pressure drop in the gas distribution across large-diameter trays.

The carbon-flow mechanism shown in Fig. 389*a* consists of a circular tray supported at three points and given an oscillating motion. This tray has a number of downspouts which serve as pockets for the conveyance of solids. When in one extreme position of its motion, alternate downspouts of this tray are in position to be filled by the delivery tubes

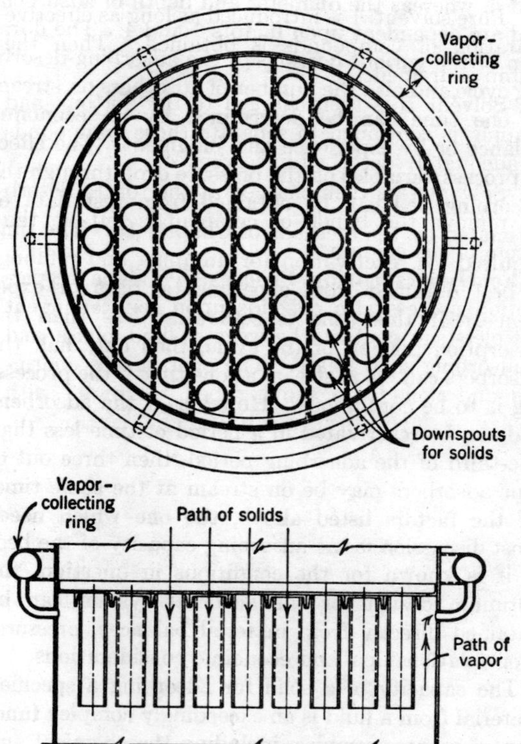

FIG. 389*b*. Typical vapor disengaging tray in a Hypersorber.[4]

in the fixed tray above. These pockets do not discharge because in this position they are sealed by the tray below. As the tray moves to the opposite extreme of its travel, these downspouts pass into position over discharge holes in the tray below and are emptied.

Operating Variables

In the design of adsorbers of the "fixed-bed" type (such as in the "percolation" plants, and adsorbers for vapor recovery and dehydration) the adsorber dimensions can be readily obtained from the following factors.

1. Quantity of fluid handled per unit time.
2. Amount of material to be adsorbed.
3. Adsorbing capacity of the adsorbent.
4. Allowable pressure drop through bed.
5. Duration of adsorption portion of cycle.
6. Time required for reactivation, purging, etc.

Although each of the above items may affect the final design in several ways, the mass or volume of adsorbent is determined primarily by factors 2, 3, and 5, whereas the diameter and depth of adsorbent bed are dependent upon factors 1 and 4. The duration of the various portions of the adsorbing-desorbing cycle and also the number of adsorbers on stream at one time can be determined by an economic balance or by experience and judgment. The effect of process variables on the pressure drop through the adsorbing bed can be obtained from test data or estimated by the methods of Chapter 16. The time required for reactivation (or dumping and refilling) is best obtained from experimental data or experience. If this time is somewhat less than the adsorption portion of the cycle, then only half the adsorbers can be on stream at one time if the processing is to be continuous. However, if the adsorbent bed can be reactivated in a period of time less than one-third of the adsorbing period, then three out of four adsorbers may be on stream at the same time. Of the factors listed above, the one which needs most discussion is the adsorbing capacity of the bed. If it is known for the conditions in question, the optimum design for the adsorber system can be obtained directly from material balances, pressure-drop relationships, and economic considerations.

The capacity of a solid for adsorbing a specified material from a fluid is an exceedingly complex function of many variables, including the chemical and physical characteristics of the solid, the composition of the fluid, the temperature and pressure of the process, and the type and time or rate of contacting. Although many theories have been proposed to explain some of the adsorption phenomena, the simplest and best procedure to obtain information concerning the adsorption capacity of a particular solid is by direct experimental methods on the desired system. The experimental procedure need not always reproduce faithfully the proposed contacting method, but it should make use of samples of the solid and fluid in question and should duplicate the operating temperature and pressure.

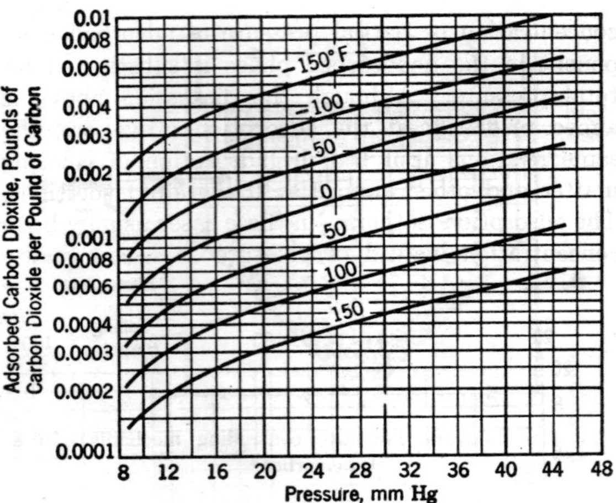

Fig. 391*a*. Adsorption isotherms for carbon dioxide on activated carbon.[16]

If a solid adsorbent is exposed to a pure fluid (usually vapor) under conditions of constant temperature and pressure but for varying periods of time, the rate of adsorption will usually decrease rapidly from an initially high rate,[16] as shown in Fig. 390, where mass of adsorbate per unit mass of adsorbent is plotted against time. The rate of approach to equilibrium varies widely and is greater for smaller size of adsorbent particles, higher temperatures, lower molecular weight adsorbates, and other variables which increase the rate of diffusion. In dynamic adsorption, where the fluid is forced through the adsorbent, equilibrium is reached faster than in static adsorption.

The effects of temperature and pressure on the mass adsorbed at equilibrium are represented in Fig. 391*a*. Although the magnitudes and shapes of these curves vary considerably for different systems, in general the amount adsorbed increases with increased pressure and decreased temperature.[16,18]

If the solid adsorbent is exposed to a vapor mixture in which only one component is adsorbed, the qualitative relations discussed above still apply if the

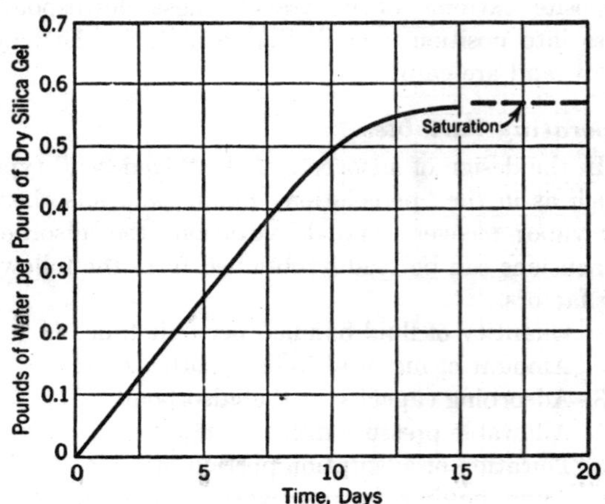

Fig. 390. Isothermal and isobaric adsorption of water on silica gel at 68° F and 1 atm as a function of time.[16]

concentration or partial pressure of adsorbed component in the unadsorbed phase is substituted for total pressure. Although the discussion above is aimed primarily at the adsorption of vapors, the same relations apply to the adsorption of coloring matter and other impurities from liquid solutions. The adsorption isotherms in these cases usually have concentration instead of pressure as a coordinate, and the isobars become lines of constant liquid phase composition.

A convenient way of treating adsorption equilibrium data [12] is indicated in Fig. 391*b*, where the vapor pressure of water adsorbed on silica gel is plotted against the vapor pressure of pure water at the same temperature with lines of constant "saturation" (quantity of adsorbate per unit mass of adsorbent).

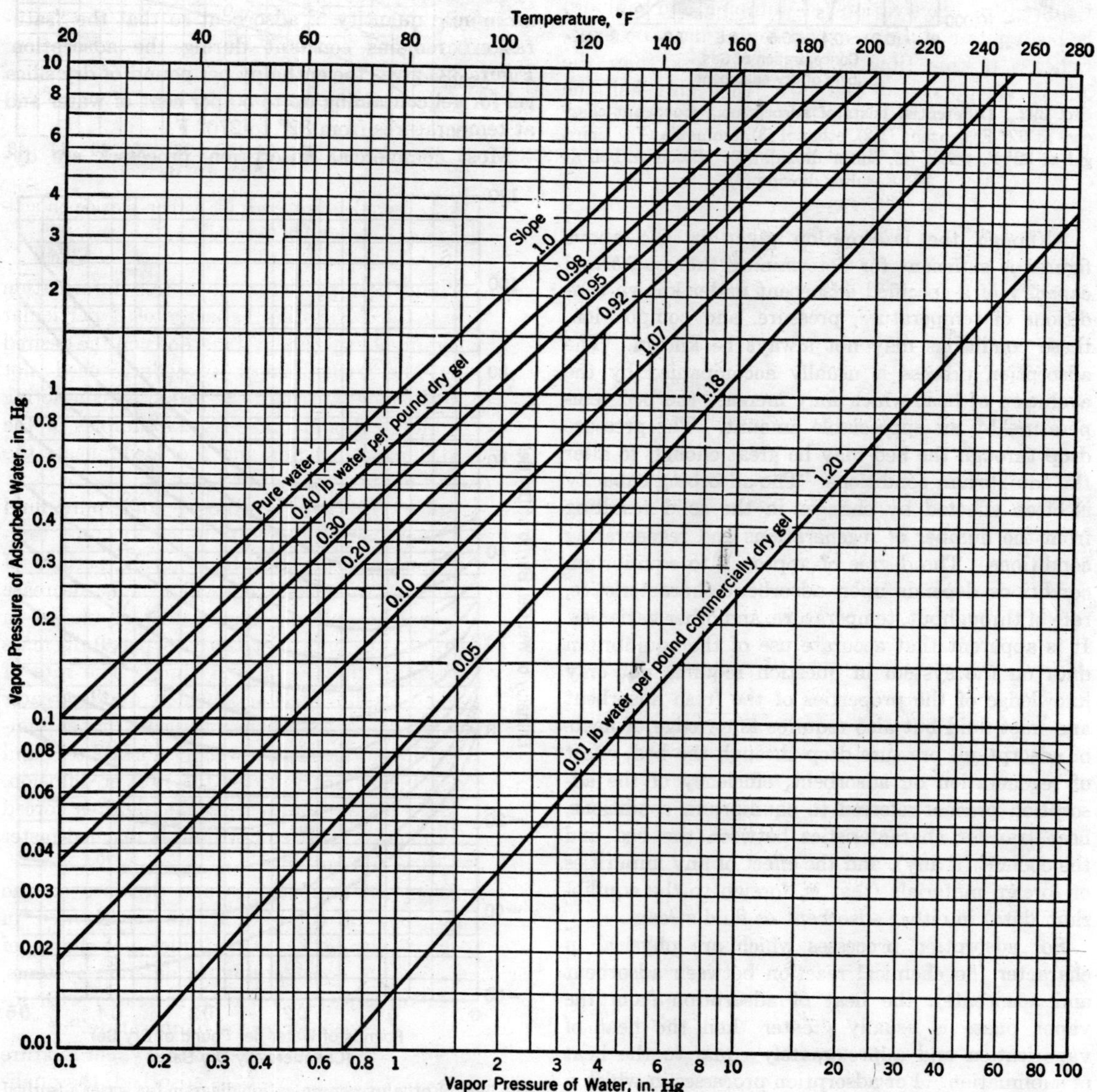

FIG. 391*b*. Vapor pressure of water adsorbed on silica gel as a function of the vapor pressure of pure water at the same temperature.[12]

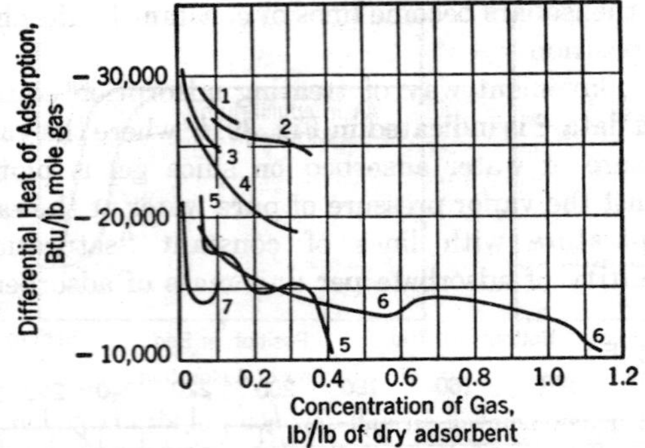

Fig. 392. Differential heats of adsorption. Cocoanut charcoal at 32° F, benzene (1,2), ethanol (3), ammonia (7). Silica gel at 32° F, water (4), sulfur dioxide (5). Blood carbon at 14° F, sulfur dioxide (6).

Although the information given in the above figures is sufficient for determining the *equilibrium* capacity of a specified adsorbent under known conditions of temperature, pressure, and composition, these conditions may not always be known. The adsorption process is usually accompanied by the evolution of heat which may increase the bed temperature by an appreciable amount. The pressure drop through the bed may be great enough to alter the equilibrium conditions. The adsorbing capacity is often affected by changes in the solid resulting from the number of regenerations and regenerating conditions. The degree of approach to equilibrium conditions depends upon adsorbent characteristics, rate of throughput, temperature, and other variables. It is apparent that accurate use of the equilibrium data on the system in question requires not only knowledge of the properties of the fresh adsorbent and inlet fluid but also requires knowledge of heats of adsorption, pressure drop through the bed, effect of regeneration on adsorbent, efficiency of the adsorption process referred to equilibrium conditions, heat transfer characteristics between the bed and the coolant (if any), and the effect of any impurities or foreign materials (that is, foreign to the equilibrium data) in either adsorbent or fluid stream.

For adsorption processes which are physical in character (no chemical reaction between adsorbent and adsorbate), the heat of adsorption from the vapor phase is usually greater than the heat of vaporization and approximately equal to the heat of sublimation. For adsorption processes in which a chemical reaction occurs between the adsorbate and adsorbent, the heat of adsorption is of the same order of magnitude as the heat of reaction.[18] The heat of adsorption may be computed thermodynamically from vapor pressure data by a method similar to the calculation of the heat of vaporization of liquids. For this purpose, vapor pressure-temperature-composition relationships such as those shown by Fig. 391 may be used. Figure 392 gives the differential heat of adsorption at 32° F, that is, the heat liberated when the indicated quantity of gas is adsorbed on an infinite quantity of adsorbent so that the "saturation" remains constant during the adsorption. Figure 393 gives the enthalpy per pound of dry silica gel for gel containing up to 50 per cent of water and at temperatures from 32° to 270° F.

Most commercial adsorption processes are dy-

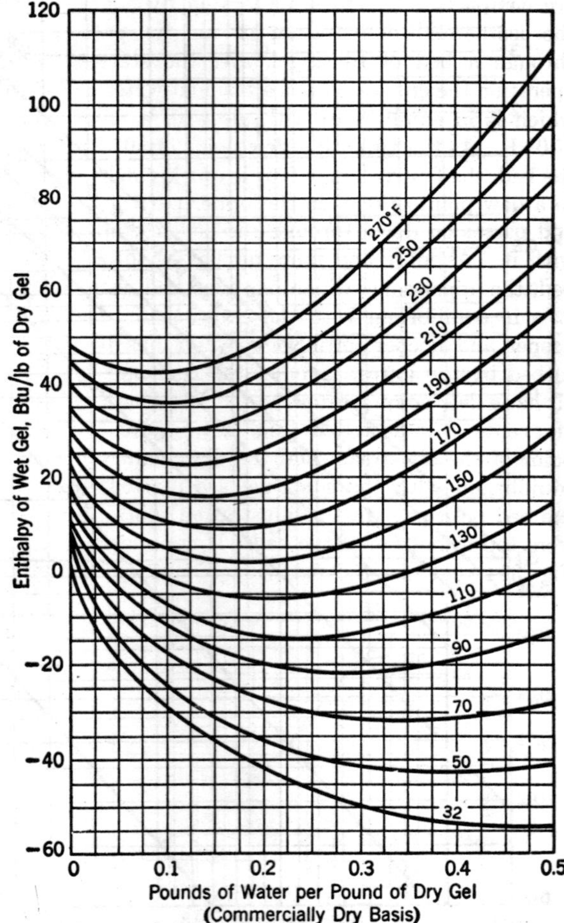

Fig. 393. Enthalpy-concentration diagram for water adsorbed on silica gel. Reference state: commercially dry gel at 32° F and saturated liquid water at 32° F.[12]

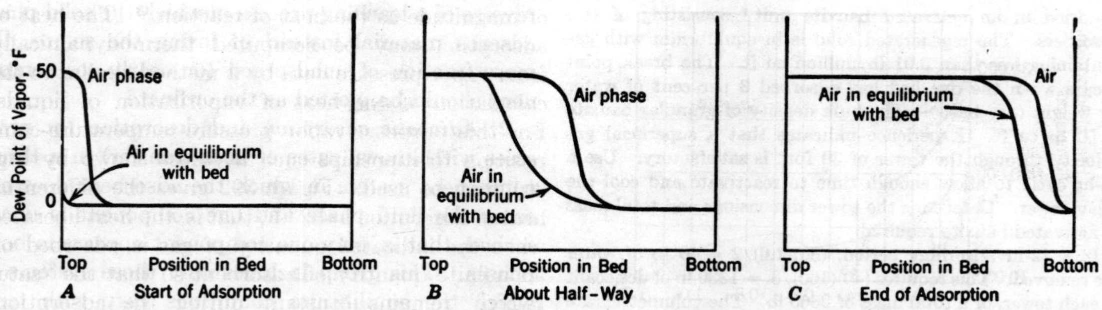

Fig. 394. Conditions in fixed-bed adsorption of water vapor expressed as dew point of air and dew point of air in equilibrium with solid bed.

namic rather than static. Equilibrium conditions are complex and may be illustrated by considering the isothermal dehydration of air by passing it through a bed of silica gel. At the start of the operation the entire bed has a residual water content which is in equilibrium with air with a dew point of 0° F. As the air with a dew point of 50° F enters the bed, the first volume of air is quickly dehydrated to a dew point of 0° F in the upper part of the bed and passes on down through the bed in equilibrium with it. As more and more air passes through the bed, the upper zone, which has adsorbed water until it is in equilibrium with the entering air, becomes deeper and deeper until some water has been adsorbed by the lowest layer of the bed and the exit air has a dew point greater than 0° F. This sequence of events is indicated in Fig. 394, where the dew point of the air stream and the dew point of air in equilibrium with the silica gel are plotted against the bed position. In this illustration the air flow is from top to bottom of the bed. It is apparent from Fig. 394 that, for the particular conditions chosen

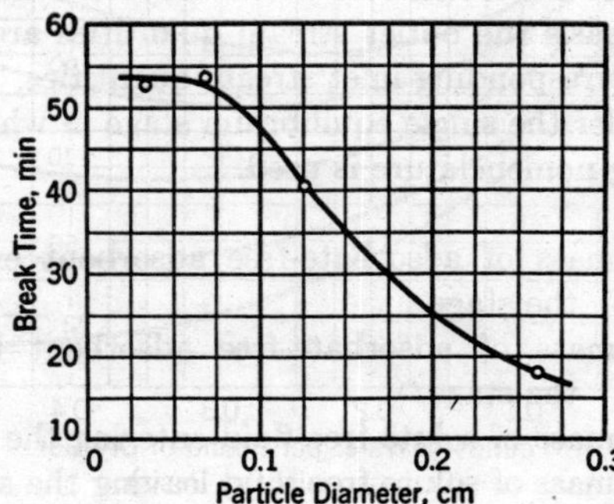

Fig. 395. Effect of particle size on the break time in the adsorption of phosgene on activated carbon in gas-mask canisters.[14]

for the illustration, the exit air has a 0° F dew point until the bed is about 80 per cent saturated with respect to air of 50° F dew point. The bed saturation (actual adsorbate relative to quantity adsorbed if the entire bed is in equilibrium with the incoming fluid) at the "break point" (point of operation when the concentration of adsorbed component in the exit fluid starts to increase) may vary anywhere from under 20 per cent to over 80 per cent, depending on the depth of bed, rate of flow, temperature of operation, kind and size of adsorbent, dew point of entering air, and other variables (see Fig. 395). Figure 394 shows that the adsorbing capacity of a bed of adsorbent (when producing an exit fluid of constant composition) cannot be obtained from equilibrium data alone. A further complication which will not be discussed at this time is the result of adiabatic operation in which the heat of adsorption contributes to a varying temperature which is added to the variables discussed above.

METHOD OF CALCULATION

The variables associated with the design of a *fixed-bed adsorber* are discussed in the preceding paragraphs. If the rise in temperature of the bed and exit fluid during the adsorption portion of the cycle is known or can be neglected, the design calculations can be readily made from feed and product specifications, equilibrium data, and knowledge of the bed saturation at the break point under the desired operation conditions. The interrelation of these variables is so complex that methods of design usually require experimental or operating data.

Illustrative Example. 1,600,000 cu ft/day (standard conditions) of natural gas at 50° F and 225 psia containing 50 lb of water per million cubic feet (standard conditions) are to be dried in an activated bauxite unit[1] consisting of two adsorbers. The regenerated solid is in equilibrium with gas containing less than 0.01 lb/million cu ft. The break point occurs when the dry bed has adsorbed 3 per cent of water by weight (dry basis). The bulk density of granular bauxite is 50 lb/cu ft. Experience indicates that a superficial gas velocity through the tower of 30 fpm is satisfactory. Use a 24-hr cycle to allow enough time to reactivate and cool the other tower. Determine the tower dimensions and total mass of activated bauxite required.

In a 12-hr adsorbing period, (50)(1.6)/2 = 40 lb of water are removed. This requires (40)(100)/3 = 1330 lb of desiccant in each tower, or a total mass of 2660 lb. The volumetric rate of flow is

$$\frac{(1{,}600{,}000)}{(24)(60)} \times \frac{14.7}{225} \times \frac{510}{492} = 70 \text{ cfm}$$

With an allowable velocity of 30 fpm the required area is 70/30 = 2.33 sq ft. This corresponds to a diameter of 20.7 in. Use a tower diameter of 24 in. (radius = 1 ft).

The height of the towers can be calculated from the mass of bauxite, its bulk density, and the tower diameter.

$$\frac{1330}{(50)(3.14)} = 8.5 \text{ ft of bauxite}$$

Make the towers 10 ft high to allow for free space above and below the bed.

Ideal or Equilibrium Stages

The calculation of adsorption equipment of the contact type or of the countercurrent type exemplified by the Hypersorption process may be facilitated by means of the concept of the equilibrium stage as used previously. For convenience the operations are discussed in two groups.

1. *Isothermal adsorption* from, and desorption to, a binary fluid phase in which three components are involved, all of which may be in the condensed phase. This group also includes all processes in which either the heat of adsorption is negligible or the effect of temperature on equilibrium compositions is negligible, and it may include adsorption from a multicomponent fluid phase if the components can be combined into a pseudo-binary system.

For these operations, the methods developed for solid-liquid extraction and liquid-liquid extraction can be used directly. The equilibrium data may be plotted on a right-triangle diagram with the mass fraction of adsorbed component as the abscissa and the mass fraction of either adsorbent or unadsorbed component as the ordinate. If the adsorption operation is considered as an extraction process in which the adsorbent is the solvent, then the mass fraction of adsorbent should be the ordinate. On the other hand, if the adsorption process is considered as the reverse of a leaching process in which the solid phase adsorbs material instead of losing the solute, the mass fraction of unadsorbed (or nearly so) component should be plotted as the ordinate.

2. *Adiabatic adsorption* and desorption (or processes with known rates of heat transfer) in a three-component system in which the adsorbent does not enter the fluid phase and one component is unadsorbed, that is, only one component appears in both phases. This group includes only those systems in which the equilibrium conditions vary with temperature. The straightforward solution of a process consisting of a single equilibrium stage can be obtained graphically on a modified enthalpy-composition diagram.

Since the adsorbent is assumed to be entirely in the solid phase and the third component is completely unadsorbed, each phase contains only two components. The usual enthalpy-composition diagram can be constructed for each phase from the necessary thermal data (Fig. 392 and steam tables). If the diagram for the solid phase is plotted with enthalpy per unit mass of adsorbent (adsorbate-free basis) as the ordinate and mass of adsorbate per unit mass of adsorbent as abscissa, the lower half of Fig. 396 is the result. In a similar manner, if the diagram for the fluid phase is plotted on a solute-free basis with enthalpy per unit mass of the unadsorbed component and mass ratio of solute to unadsorbed component, the upper half of Fig. 396 results. From the available equilibrium data, Fig. 391*b*, the necessary tie lines (several for each isotherm) are plotted on this diagram. In order to separate the two diagrams for convenience in use, the reference temperature for the unadsorbed component may be arbitrarily chosen at a low value. The coordinates are on an adsorbate-free or solute-free basis since on this basis the outlet stream quantities are equal to the corresponding inlet stream quantities.

Consider the single equilibrium stage in which the following nomenclature is used.

L_0 = mass of adsorbate-free adsorbent entering the stage.
L_1 = mass of adsorbate-free adsorbent leaving the stage.
V_2 = mass of solute-free fluid entering the stage.
V_1 = mass of solute-free fluid leaving the stage in equilibrium with stream L_1.
X = mass ratio of adsorbate to adsorbent in an L stream.

Y = mass ratio of solute to solute-free fluid in a V stream.

h = enthalpy of an L stream, per unit mass of adsorbate-free adsorbent.

H = enthalpy of a V stream, per unit mass of solute-free fluid.

J = an addition of two streams, such as $L_0 + V_2$.

Writing a material balance on an adsorbate-solute-free basis,

$$L_0 + V_2 = L_1 + V_1 = J \tag{349}$$

Since the adsorbent does not appear in the fluid phase, and the third component is unadsorbed,

$$L_0 = L_1 \quad \text{and} \quad V_2 = V_1$$

Writing a material balance on the adsorbed component,

$$L_0X_0 + V_2Y_2 = L_1X_1 + V_1Y_1 = JX_J \tag{350}$$

Writing an enthalpy balance,

$$L_0h_0 + V_2H_2 = L_1h_1 + V_1H_1 = Jh_J \tag{351}$$

By means of the algebraic procedures employed on similar equations (274–278) for vapor-liquid transfer operations, equations 349, 350, and 351 may be rearranged to yield relations showing that a straight line passing through the points (X_0, h_0), (X_J, h_J), and (Y_2, H_2) and a straight line through (X_1, h_1), (X_J, h_J), and (Y_1, H_1) represent solutions to the equations. Furthermore, these relations indicate that the tie line (the line through X_1 and Y_1) is divided by the point (X_J, h_J) into two segments which are in the ratio L_1/V_1.

If the temperature, composition, and quantity of both L_0 and V_2 are known, the temperature and composition of L_1 and V_1 are obtained by locating the one tie line which passes through the point (X_J, h_J) and is divided in two parts in the ratio L_1/V_1 which equals L_0/V_2. This tie line can be located directly by plotting on the enthalpy-composition diagram several isotherms, each representing an equilibrium mixture of the two phases in the known ratio $L_1/V_1 = L_0/V_2$. These isotherms are easily plotted by dividing the known tie lines in the specified ratio L_0/V_2. Three isotherms for the case $L_0/V_2 = 1.0$ are plotted in Fig. 396. The equilibrium temperature of the point (X_J, h_J) is then obtained by interpolation among these isotherms, and the desired tie line is immediately obtained. This method of solution may be extended to more than one equilibrium stage, but then it requires a trial-and-error procedure.

Allowance for heat transfer may be made by relocating the J point according to the equation,

$$L_0X_0 + V_2Y_2 = J(h_J - Q_J) \tag{352}$$

where Q_J is the heat added to the stage per unit mass of (adsorbate-free adsorbent plus solute-free fluid).

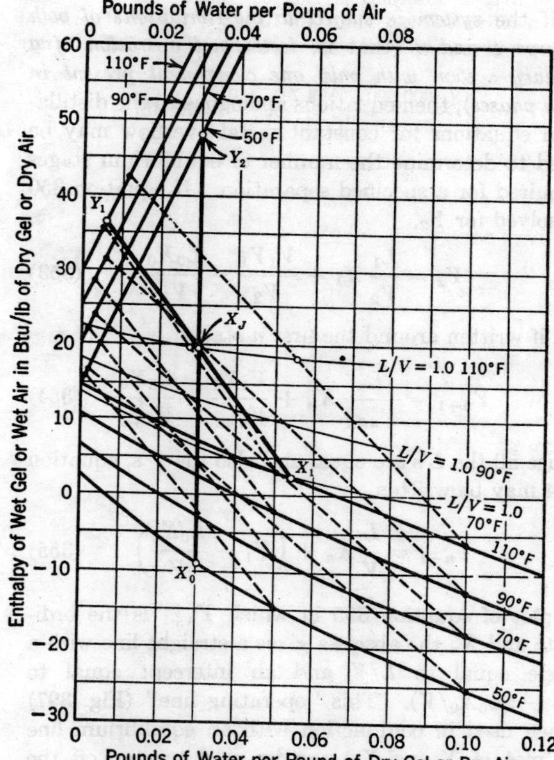

FIG. 396. Graphical solution of illustrative example below by material and energy balances, using an enthalpy-mass ratio diagram for the water-silica gel and water vapor-air systems.

Illustrative Example. Water vapor is to be removed from 20 lb (dry basis) per minute of air by adsorption on 20 lb (dry basis) per minute of silica gel in a continuous process. The effective height of the tower is sufficient to provide for one equilibrium stage. In other words, the silica gel passing down from the feed plate to the stripper is assumed to be in equilibrium (both thermal and chemical) with the air leaving the column at the upper disengaging plate. If the entering air at 50° F contains 0.03 lb of water vapor per pound of dry air, and the entering silica gel at 50° F contains 0.03 lb of water per pound of dry gel, what are the temperature and composition of the gel passing down from the feed plate and the air leaving the column?

Solution. The graphical solution to this problem (Fig. 396) is obtained by connecting the two points representing the feed stream and entering adsorbent with a straight line. The addition point J is located so that this line is divided into two segments in the ratio $L_0/V_2 = (20)/(20) = 1.00$. The temperature of the leaving streams is obtained by interpolation between the isotherms for $L/V = 1.0$. This temperature is found to be 110° F. Also by interpolation the tie line corresponding to this temperature and passing through the addition point J indicates $X_1 = 0.054$ and $Y_1 = 0.006$.

If the *system is subject to the limitations of both groups 1 and 2* (that is, *isothermal operation of a ternary system with only one component present in both phases*), then equations analogous to the distillation equations for constant molal overflow may be used to determine the number of equilibrium stages required for a specified separation. If equation 350 is solved for Y_2,

$$Y_2 = \frac{L_1}{V_2} X_1 + \frac{V_1 Y_1}{V_2} - \frac{L_0 X_0}{V_2} \tag{353}$$

or, if written around the first n stages,

$$Y_{n+1} = \frac{L_n}{V_{n+1}} X_n + \frac{V_1 Y_1}{V_{n+1}} - \frac{L_0 X_0}{V_{n+1}} \tag{354}$$

Since all the L's are equal and also the V's, equation 354 may be written

$$Y_{n+1} = \frac{L}{V} X_n + \left(Y_1 - \frac{L_0 X_0}{V}\right) \tag{355}$$

A plot of equation 355 in which Y_{n+1} is the ordinate and X_n the abscissa gives a straight line with a slope equal to L/V and an intercept equal to $Y_1 - (L_0X_0/V)$. This "operating line" (Fig. 397) when used in conjunction with an equilibrium line (a plot of Y_n vs X_n) can be used to step off the number of equilibrium stages in a manner similar to the McCabe-Thiele method for distillation calculations (Chapter 25). This procedure is not necessarily limited to isothermal operation, but in any case the temperature variation and corresponding equilibrium data must be known.

Illustrative Example. Two hundred pounds per minute of dry air at 70° F, carrying 2 lb of water vapor per minute is to be dehumidified with silica gel to 0.001 lb of water vapor per pound of dry air. This operation is to be conducted isothermally and countercurrently with 10 lb/min of dry silica gel. What is the water content of the silica gel leaving the bottom stage, and how many ideal stages are required?

Solution. By material balance the water content of the exit gel stream is $(2/10) - 0.001(200/10) = 0.18$ lb of water/lb of dry gel. In Fig. 397, the operating line is drawn through the points (0.01, 0.18) and (0.001, 0.0) which correspond respectively to (Y_{n+1}, X_n) and (Y_1, X_0). By a stepwise procedure the required number of equilibrium stages is found to be between 3 or 4.

These methods as applied to adsorption (solid-vapor) operations are identical in principle to those for solid-liquid, liquid-liquid, and vapor-liquid transfer operation. In fact the last method desciibed is used for vapor-liquid absorbers. All these transfer operations are computed in the same manner by ideal stages, the different operations exemplifying different applications of the same basic principles.

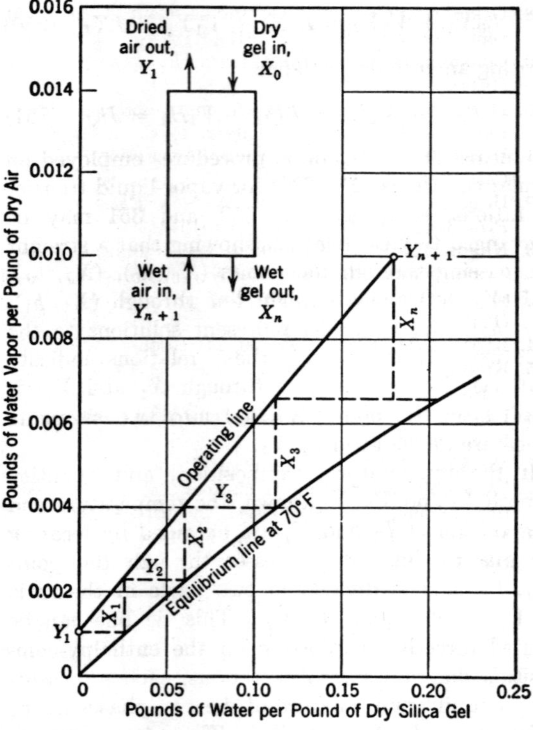

Fig. 397. Graphical solution of illustrative example on mole ratio plot, assuming isothermal operation with only one component soluble in two phases.

BIBLIOGRAPHY

1. Amero, R. C., J. W. Moore, and R. G. Capell, "Design and Use of Adsorptive Drying Units," *Chem. Eng. Prog.*, **43**, 349–70 (1947).
2. Anonymous, "Process Engineering—Chromatography," *Chem. Eng.*, **55**, No. 5, 117–8 (1948).
3. Benson, R. E., and P. H. Courouleau, "European Practice for Solvent Recovery in Printing Industry," *Chem. Eng. Progress*, **44**, 459–468 (1948).
4. Berg, Clyde, "Hypersorption Process for Separation of Light Gases," *Trans. Am. Inst. Chem. Eng.*, **42**, 665–80 (1946).
5. Berg, Clyde, *Hypersorption*, presented at ASEE Summer School, University of Wisconsin (Sept. 2, 1948).
6. Boyd, G. E., *et al.*, "The Exchange Adsorption of Ions from Aqueous Solutions by Organic Zeolites," *J. Am. Chem. Soc.*, **69**, 2818–29, 2836–38, 2849–59, 2800–12 (1947).
7. Campbell, D. L., H. Z. Martin, E. V. Murphree, and C. W. Tyson, "Separation of Gases," U. S. Patent 2,446,076 (July 27, 1948).
8. Carbide and Carbon Chemicals Corporation, "Solvent Recovery," Booklet (1947).
9. Consden, Gordon, and Martin, *Biochem. J.*, **38**, 224 (1944).
10. Deitz, V. R., *Bibliography of Solid Adsorbents*, National Bureau of Standards, Washington, D C. (1944).
11. Hirschler, A. E., and S. Amon, "Adsorption," *Ind. Eng. Chem.*, **39**, 1585–96 (1947).
12. Hougen, O. A., and K. M. Watson, *Chemical Process Principles*, John Wiley and Sons (1943).
13. Kehde, H., R. G. Fairfield, J. C. Fránk, and L. W. Zahnstecher, "Ethylene Recovery Commercial Hypersorption Operation," *Chem. Eng. Progress*, **44**, 575–582 (1948).
14. Klotz, I. M., *Chem. Revs.*, **39**, 241 (1946).
15. Kunin, R., "Ion Exchange," *Ind. Eng. Chem.*, **40**, 41–5 (1948).
16. Ledoux, E., *Vapor Adsorption*, Chemical Publishing Co., Brooklyn (1945).
17. Mair, B. J., A. L. Gaboriault, and F. D. Rossini, "Assembly and Testing of 52-Foot Laboratory Adsorption Column," *Ind. Eng. Chem.*, **39**, 1072–81 (1947).
18. Mantell, C. L., *Adsorption*, McGraw-Hill Book Co. (1945).
19. Martin, A. J. P., and R. L. M. Synge, *Biochem. J.*, **35**, 1358 (1941).
20. Ray, A. B., "Adsorption," *Ind. Eng. Chem.*, **39**, 12–13, 32–5 (1947).
21. Spedding, F. H., *et al.*, "The Separation of the Rare Earths by Ion Exchange," *J. Am. Chem. Soc.*, **69**, 2777–81, 2786–92, 2812–18 (1947).
22. Tompkins, E. R., *et al.*, "Ion Exchange as a Separations Method," *J. Am. Chem. Soc.*, **69**, 2769-2777, 2792–2800, 2859–2865, 2866–2874 (1947).
23. Tryhorn and Wyatt, *Trans. Faraday Soc.*, **22**, 135 (1926).
24. Zechmeister and Cholnoky, *Die Chromatographische Adsorptionsmethode*, Vienna (1936).

PROBLEM

By the graphical method developed for solid-liquid extraction problems, calculate the number of countercurrent ideal stages and mass of fresh char required for the isothermal removal of 99 per cent of the colored material from 100 lb of sugar syrup per hour by agitation with finely divided color-free activated vegetable char. The colored syrup contains 0.001 mass fraction of color. The spent char is to contain 0.0004 mass fraction of color. After agitation the slurry is allowed to settle and the syrup drawn off before the char is contacted with a second batch of more highly colored syrup. Between steps syrup adheres to the char to the extent of 0.2 lb of syrup per pound of char. The char, after settling from the last agitation (contacted with the raw syrup), is filtered and washed to recover the adhering syrup.

EQUILIBRIUM DATA (MASS FRACTIONS)

Solid Phase		Liquid Phase	
Color	Syrup	Color	Syrup
0.0001	0.0005	0.00002	0.99998
0.0002	0.0005	0.00008	0.99992
0.0003	0.0004	0.00020	0.99980
0.0004	0.0004	0.00050	0.99950
0.0005	0.0003	0.00100	0.99900

PART IV

Energy and Mass Transfer Rates

THE time rate with which energy or mass is transferred from one stream to another determines the size of the equipment required to accomplish a specified result in a specified time. The rate of transfer is therefore a most important consideration in engineering operations.

In general, the rate of transfer is regarded as equal to a driving force or difference in potential divided by a resistance, or, as usually expressed, the rate of transfer equals the product of a "transfer coefficient," the area of contact, and the drop in potential.

The major problem in calculating the rate of transfer is usually the evaluation of the transfer coefficient. This is usually based on an empirical correlation of transfer coefficients calculated from an extensive series of tests in which the rate of transfer has been measured.

PART IV

Energy and Mass Transfer Rates

THE rate with which energy or mass is transferred from one stream to another determines the size of the equipment required to accomplish a specified result in a specified time. The rate of transfer is therefore a most important consideration in engineering operations.

In general, the rate of transfer is regarded as equal to a driving force or difference in potential divided by a resistance, or, as usually expressed, the rate of transfer equals the product of a transfer coefficient, the area of contact, and the drop in potential.

The major problem in calculating the rate of transfer is usually the evaluation of the transfer coefficient. This is usually based on an empirical correlation of transfer coefficients calculated from an extensive series of tests in which the rate of transfer has been measured.

CHAPTER 28

Heat Transfer 1

THE transfer of energy as heat occurs in practically every engineering process. Each branch of engineering has its special problems of heat transmission, and no one group of engineers has been solely responsible for the many developments made in this field. The fundamental concepts and theory of heat transfer developed herein are applicable to all engineering problems.

Heat is itself a symbol created by the human mind, as are all forms of energy. It is the modern counterpart of the medieval philosophers' caloric. The concept of heat is dependent upon its transfer from a body at a high temperature to another at a lower temperature, and accordingly heat may be defined as the energy which is so transferred by means of the difference in temperature. When two bodies are in contact with each other, heat may be transferred directly from one to the other by conduction. When two bodies are not in direct contact, heat may still be transferred from one to the other by convection or by radiation.

These three modes of heat transmission may be defined as follows.

Conduction is the transfer of heat from one part of a body to another part of the same body, or between two bodies in physical contact, without significant displacement of the particles of the bodies.

Convection is the transfer of heat from one point to another within a fluid, or between a fluid and a solid or another fluid, by the movement or mixing of the fluids involved. If the motion of the fluid is due entirely to differences in density resulting from differences in temperature, the operation is called natural convection. If the motion of the fluid is produced by mechanical means, the operation is referred to as forced convection.

Radiation is the transfer of heat by the absorption of radiant energy. Electromagnetic waves emanate from all bodies in all directions at all temperatures. The most important property of electromagnetic waves is that they convey energy. When the waves impinge on a body, part are reflected, part are transmitted, and part are absorbed. The absorbed waves may be converted into high-grade forms of energy as in photochemical changes, but they are more commonly converted into heat. Light is also due to similar radiant energy but of shorter wavelength or higher frequency. Electromagnetic waves emitted by the sun travel through space in straight lines to furnish both heat and light to the earth.

In most cases the actual transfer of energy as heat is accomplished by more than one of these modes of heat transfer. In all cases the total rate of heat transfer may be expressed in terms of a driving force which is a decrease in temperature, and a resistance.

$$\frac{dQ}{dt} = \frac{\text{Driving force}}{\text{Resistance}}$$

$$= \frac{T_1 - T_2}{1/UA} = UA(T_1 - T_2) \qquad (356)$$

where $\frac{dQ}{dt}$ = rate of heat transfer (as in Btu/hr) in the direction from point 1 to point 2.

t = time (as in hr).

T_1 = temperature at point 1, as in °F.

T_2 = temperature at point 2, as in °F.

U = average overall coefficient of heat transfer [as in Btu/(hr)(°F)(sq ft)].

A = area of surface through which heat is transferred, or cross-sectional area of material through which heat is being conducted (as in sq ft).

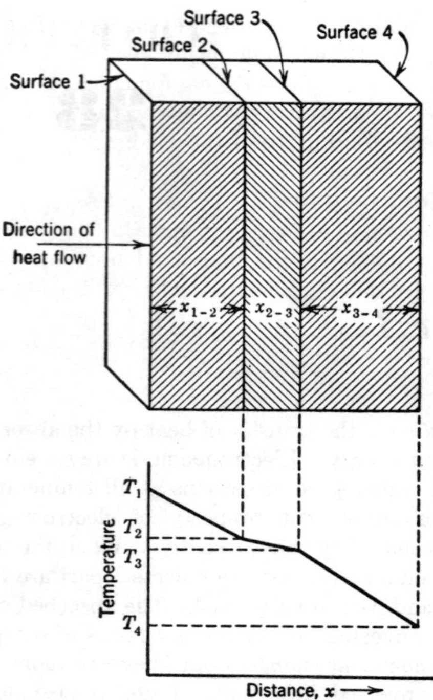

FIG. 398*a*. Diagram representing heat transfer by conduction through solids in series.

The product of the overall coefficient U times the area A is the reciprocal of the resistance $(1/UA)$. The overall coefficient U is a conductivity corresponding to the reciprocal of the resistivity and is a convenient method of coordinating and expressing heat transfer rates.

The usual problem in heat transfer gives a quantity of heat to be transferred per unit time under given conditions of temperature. The coefficient of heat transfer is calculated or estimated from experience and used in equation 356 to compute the heat transfer area.

The flow of any energy such as heat or electricity may be considered as controlled by a driving force and a resistance. The driving force for electrical energy is the difference in potential called voltage, and for heat energy the driving force is the difference in temperature. The rate of flow of electricity through several resistances in series is obtained by dividing the voltage by the sum of the resistances.

$$I = \frac{E}{R} \tag{357}$$

where I = rate of electrical flow.
E = driving force, voltage.
R = resistance.

Likewise, the rate of heat flow for a series of resistances, Fig. 398*a* is expressed by the equation:

$$\frac{dQ}{dt} = \frac{T_1 - T_4}{R_{1-2} + R_{2-3} + R_{3-4}} = \frac{T_1 - T_4}{\dfrac{1}{K_{1-2}} + \dfrac{1}{K_{2-3}} + \dfrac{1}{K_{3-4}}} \tag{358}$$

where R = resistance. Not to be confused with resistivity.
K = conductance. Not to be confused with conductivity.

Subscripts indicate the steps in the heat transfer as indicated in Fig. 398*a*.

The proper overall coefficient U is determined by the expression

$$\frac{1}{UA} = \frac{1}{K_{1-2}} + \frac{1}{K_{2-3}} + \frac{1}{K_{3-4}} \tag{359}$$

or the total resistance is equal to the sum of the individual resistances for series flow. The resistance to heat transfer from fluid A at 1 (Fig. 398*b*) to the

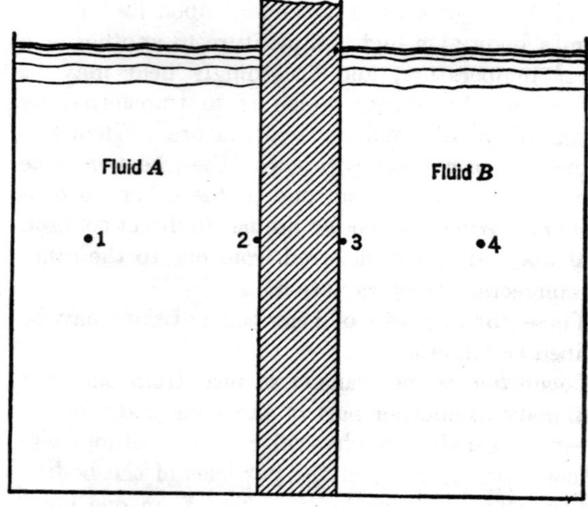

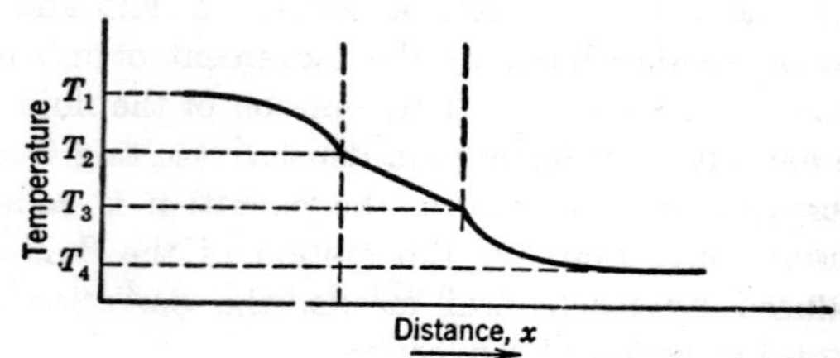

FIG. 398*b*. Diagram representing heat transfer between fluids which are separated by a solid.

metal surface at 2 involves convection heat transfer. The transfer of heat through the solid is by conduction, and the transfer from 3 to 4 is similar to that for fluid A. Coefficients for heat transfer from a fluid to a solid surface are sometimes called film coefficients because the major resistance to heat transfer is in the fluids adjacent to the metal, frequently called a "fluid film." In series heat transfer, Fig. 398(a and b), the total driving force, temperature drop, is divided by the sum of the resistances to obtain the rate.

For flow in parallel the total rate of flow equals the sum of the individual rates. Walls 2 and 3 in Fig. 398c, at different temperatures and facing each other, transfer heat by radiation and by air circulating because of density differences accompanying temperature changes. The total rate of flow of heat equals the sum of the rates by radiation and by natural convection. When encountered, parallel flow is easily computed as the sum of the transfers by the different mechanisms.

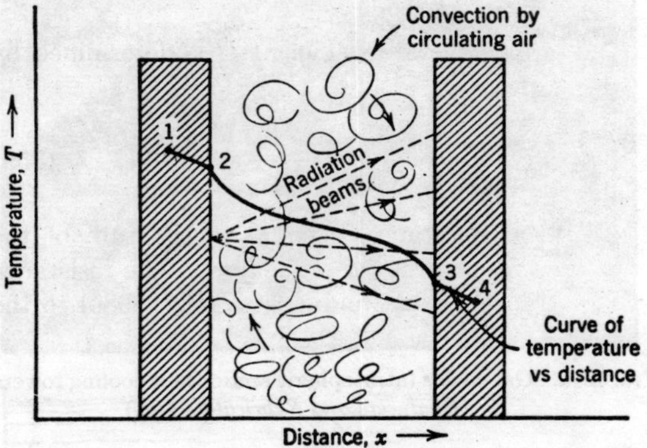

FIG. 398c. Diagram illustrating heat transfer by parallel mechanisms of radiation and convection from surface 2 to surface 3.

HEAT EXCHANGE EQUIPMENT

There is an almost endless variety of heat transfer devices varying from the electric heater in the home to giant boilers in utility power plants. A limited number of heat transfer devices likely to be encountered by the engineer have been selected for description. The major portion of the units will transfer heat from one fluid to another fluid, with the heat passing through a solid interface such as a tube wall. The size, shape, and material employed to separate the two fluids is of primary importance. The second problem in a heat exchanger is to confine one or both of the two fluids involved in the transfer process. In some cases the equipment is constructed to control the nature of the flow and the fluid velocity adjacent to the surface of the solid.

One Fluid Confined

A simple type of heat exchanger in which only one fluid is confined is the household hot-water "radiator." In this equipment the water usually circulates between the furnace and the radiator by natural convection (thermosiphon). The air in the room circulates by natural convection to heat the room. Natural circulation or convection is produced solely by the differences in density resulting from heating and cooling of the fluids.

FIG. 399. Air heater constructed of finned tubes. (*Unifin Tube Co.*)

Another common type of heat exchanger is the air heater. Heat from condensing steam inside the tubes is transferred to air flowing outside the tubes. Figure 399 shows such a unit employing finned tubes to increase the transfer surface in contact with the air. Steam enters the top of the unit, and the condensate drips leave the bottom of the unit. A fan usually blows the air across the tube bank to increase the rate of heat transfer between the air and the metal surface. In this case the confinement of the air is limited to the casing around the tubes which assists in promoting the flow of the air through the tube "bundle" or "coil." In recent years these "blast" heaters have found extensive use in industrial space heating.

Figure 400 presents overall heat transfer coefficient U in Btu/(hr)(°F)(sq ft of outside surface) including the total extended surface of the fins, measured on the finned tube unit shown in Fig. 399 and on a similar unit having one row of tubes.[23] * The

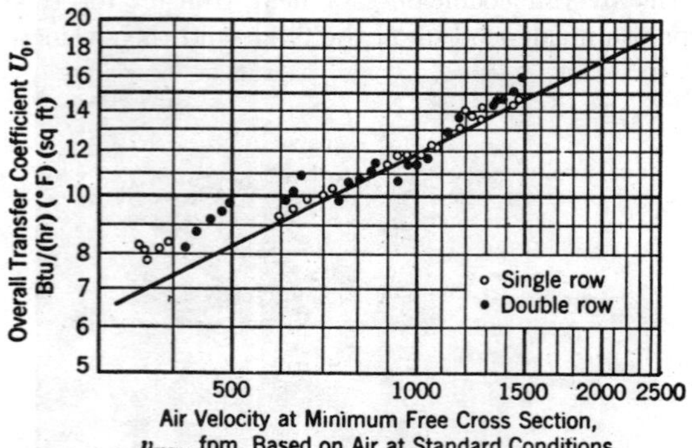

Fig. 400. Overall coefficients of heat transfer for heating air in finned tube banks, such as illustrated in Fig. 399, as a function of the maximum air velocity (the velocity of the air at the minimum free cross section).

coefficient is a function of the air velocity at the minimum cross section expressed as feet per minute of air measured at a standard air temperature of 70° F. With condensing steam inside the tubes, the metal temperature approaches the steam temperature and the overall coefficients are only 1 to 3 per cent lower than the convection coefficients from the finned surfaces to air.

Another type of unit in which the fluid on one side of the heat transfer surface is not confined is the cooling unit used in *spray towers*. Figure 401*a* shows a cooler for gases at high pressure assembled from sections such as shown in Fig. 401*b*. Gases enter the header at the top, pass in parallel flow through the upper tube bundle, through the headers at the back down to the middle bundle, and finally through the lower bundle. The unit is placed in the lower part of a water-cooling tower. Water is sprayed over the tubes to cool them. Evaporation of a portion of the water reduces the temperature rise which would otherwise occur for the cooling water. The plugs in the header may be removed to clean the inside of the tubes or to replace one in case failure occurs.

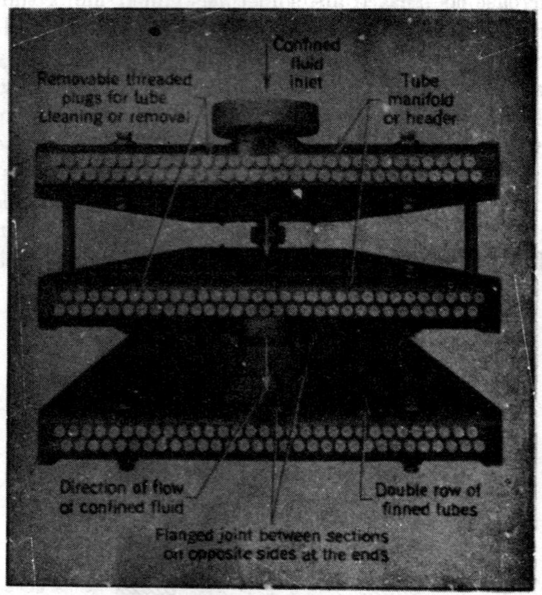

Fig. 401*a*. Gas cooler (atmospheric section) for cooling towers. (*Engineers and Fabricators Inc.*)

Such units (Fig. 401) are also used for condensing vapors, often called evaporative condensers, as well as for cooling oils or gases. The coefficients of heat transfer depend upon the fluid inside the tubes and upon the cleanliness of the surfaces. Overall coefficients of 250 Btu/(hr)(°F)(sq ft) are

Fig. 401*b*. Sections of the cooler, shown in Fig. 401*a*, before assembling. (*Engineers and Fabricators Inc.*)

attainable for clean surfaces with high inside coefficients, but overall coefficients for design purposes are usually taken to be in the range from 25 to 125 Btu/(hr)(°F)(sq ft).

* The bibliography for this chapter appears on p. 472.

Many varied applications are made of pipe coils inserted into tanks or vessels to cool or heat the contents. Steam or cooling water is used in the coil, and the fluids in the tank or vessel may be agitated to increase the heat transfer rate on the outside of the pipe. Pipe coils are easily installed or replaced and are well adapted to small-scale, temporary, or experimental installations.

Both Fluids Confined

The simplest heat transfer device which confines both fluids is *the double-pipe heat exchanger*. One fluid flows inside of a pipe or a tube while a second fluid flows either co-, or counter- currently in the annulus between a larger tube and the outside of the inner tube carrying the first fluid. Figure 402*a* shows a double-pipe heat exchanger constructed of glass for handling corrosive fluids. Figure 402*b* shows details of the stuffing box of an exchanger in which the inner tube is glass and the outer tube is ordinary pipe. The length of tube in such installations is limited, and a bank of tubes is used to obtain the desired area. Special connections are required to conduct the fluids from tube to tube.

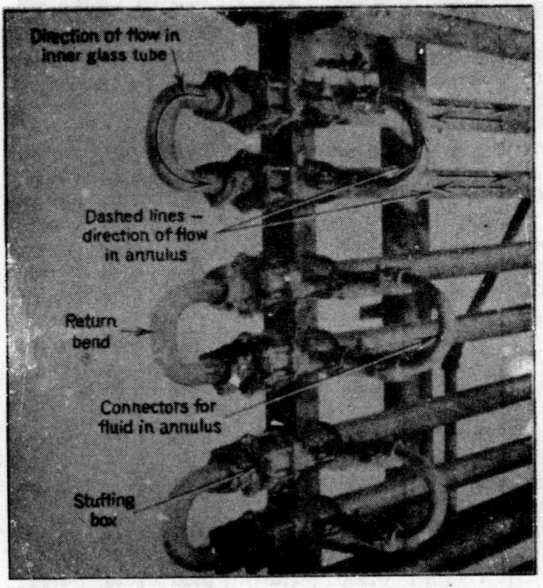

FIG. 402*a*. Double-pipe heat exchanger. The inner pipe is fabricated of Pyrex glass for corrosive service. (*Corning Glass Works.*)

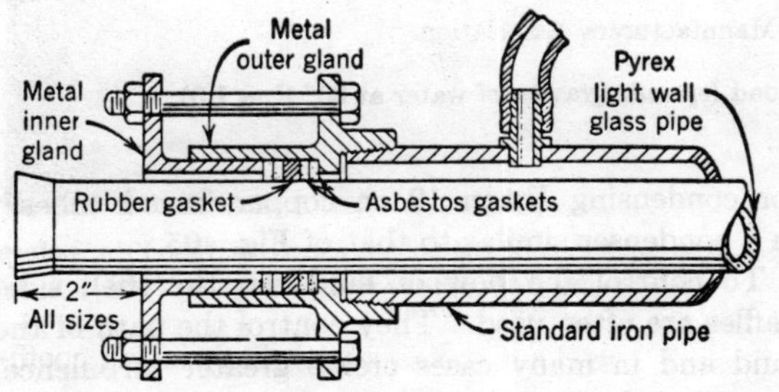

FIG. 402*b*. Details of the stuffing box assembly of the double-pipe heat exchanger shown in Fig. 402*a*. (*Corning Glass Works.*)

The overall coefficients of heat transfer for this type of unit depend upon the fluids and their velocities inside the tubes as well as in the annulus. A coefficient of 90 to 100 Btu/(hr)(°F)(sq ft) would be normal for cooling a chlorine solution inside tubes with water in the jacket, while a coefficient of 3 to 5 would be expected when cooling a dry gas inside the glass tube. The inner tubes when made of metal may carry fins to increase the surface and hence the transfer from one fluid to the other in case the coefficients are low such as for a viscous or low-density fluid passing through the annulus. A condensing vapor may replace either of the fluids in a double-pipe heat exchanger when the equipment is constructed to allow proper drainage of condensate.

Shell and tube exchangers are widely used. Figure 403 is a single-pass shell and tube unit with baffles to control the flow of the fluid outside the tubes. The tube sheets for this exchanger are nonferrous castings in which the holes for inserting the tubes have been drilled and reamed before assembly. Often steel tube sheets are drilled and reamed and then welded on the ends of steel shells. The seal between the tubes and the tube sheet is usually made by a rolling operation in which the tube is expanded by rollers inside the tube to fit tightly in the reamed hole in the tube sheet, as indicated in Fig. 404. Concentric grooves may be made in the tube sheet to assist in the seal. Tubes may also be welded, brazed, or soldered into the tube sheet. Figure 403 illustrates a single-pass unit on the tube side as well as on the shell side since the fluid inside the tubes passes through tubes only once and the fluid outside the tubes passes along and across the tubes only once. Table 47 lists the dimensions of standard exchanger and condenser tubes.

Figure 405 shows a shell and tube condenser with the vapors outside the tubes and water passing through the tubes in six passes. Water enters the

TABLE 47. CHARACTERISTICS OF TUBING *

OD Tubing, in.	Thickness: Birmingham Wire Gage	Thickness: In.	Internal Transverse Area, sq in.	Square Feet External Surface per Foot Length	Square Feet Internal Surface per Foot Length	Weight per Foot of Length, Brass, lb	Weight per Foot of Length, Copper, lb	Weight per Foot of Length, Steel, lb	ID Tubing, in.	Constant, C †	$\frac{OD}{ID}$	Metal Area (Transverse Metal Area), sq in.
3/8	16	0.065	0.0471	0.0982	0.0641	0.233	0.245	0.218	0.2450	74	1.530	0.063
3/8	20	0.035	0.0731	0.0982	0.0799	0.149	0.156	0.139	0.3050	114	1.230	0.040
9/16	21	0.032	0.1952	0.1473	0.1304	0.196	0.206	0.183	0.4985	305	1.128	0.053
5/8	12	0.109	0.130	0.1636	0.107	0.649	0.683	0.605	0.407	202	1.53	0.177
5/8	16	0.065	0.1924	0.1636	0.1295	0.420	0.44	0.39	0.495	300	1.26	0.114
5/8	20	0.035	0.2419	0.1636	0.146	0.238	0.25	0.22	0.555	377	1.13	0.0649
3/4	12	0.109	0.223	0.1963	0.14	0.807	0.845	0.75	0.532	348	1.41	0.219
3/4	16	0.065	0.302	0.1963	0.163	0.514	0.54	0.48	0.620	471	1.21	0.140
3/4	20	0.035	0.3632	0.1963	0.179	0.289	0.306	0.267	0.680	567	1.10	0.0786
7/8	10	0.134	0.2890	0.2297	0.158	1.134	1.24	1.06	0.607	451	1.44	0.312
7/8	14	0.083	0.394	0.2297	0.186	0.759	0.798	0.702	0.709	615	1.23	0.207
7/8	18	0.049	0.474	0.2297	0.2034	0.467	0.49	0.432	0.777	740	1.12	0.127
1	10	0.134	0.421	0.2618	0.192	1.34	1.41	1.24	0.732	657	1.37	0.364
1	14	0.083	0.5463	0.2618	0.2183	0.88	0.923	0.813	0.834	852	1.20	0.239
1	18	0.049	0.639	0.2618	0.236	0.54	0.568	0.50	0.902	997	1.11	0.146
1¼	10	0.134	0.757	0.3272	0.258	1.73	1.83	1.597	0.982	1180	1.27	0.470
1¼	14	0.083	0.923	0.3272	0.284	1.12	1.173	1.04	1.084	1440	1.15	0.304
1¼	20	0.035	1.092	0.3272	0.308	0.498	0.521	0.454	1.180	1704	1.06	0.135
1½	10	0.134	1.192	0.3927	0.322	2.12	2.23	1.98	1.232	1860	1.22	0.575
1½	12	0.109	1.291	0.3927	0.335	1.76	1.85	1.64	1.282	2020	1.17	0.476
1½	16	0.065	1.474	0.3927	0.359	1.08	1.13	0.996	1.37	2300	1.10	0.293
2	11	0.12	2.433	0.5236	0.460	2.62	2.76	2.45	1.76	3780	1.14	0.709
2	13	0.095	2.573	0.5236	0.474	2.09	2.20	1.933	1.81	4015	1.10	0.569
2½	9	0.148	3.80	0.654	0.575	4.09	4.30	3.82	2.20	5930	1.14	1.109

* Condensed from a table published by the Tubular Exchanger Manufacturers' Association.

† Liquid velocity $= \dfrac{\text{Pounds per tube per hour}}{C \times \text{specific gravity of liquid}}$ in feet per second (specific gravity of water at 60° F = 1.0).

cast header in the inlet at the bottom and passes through the tubes as indicated. The two chambers shown in section above the inlet convey the water to the next row of tubes. The water finally leaves the outlet at the top, making a total of six passes. Figure 406 shows the effect of velocity on the overall coefficients of heat transfer based on the outside area for condensing Freon 12 on copper finned tubes [24] in a condenser similar to that of Fig. 405.

To control the flow of fluids on the shell side, baffles are often used. They control the path of the fluid and in many cases create greater turbulence, increasing both the heat transfer and the pressure drop. Figures 407, 408, and 409 indicate three

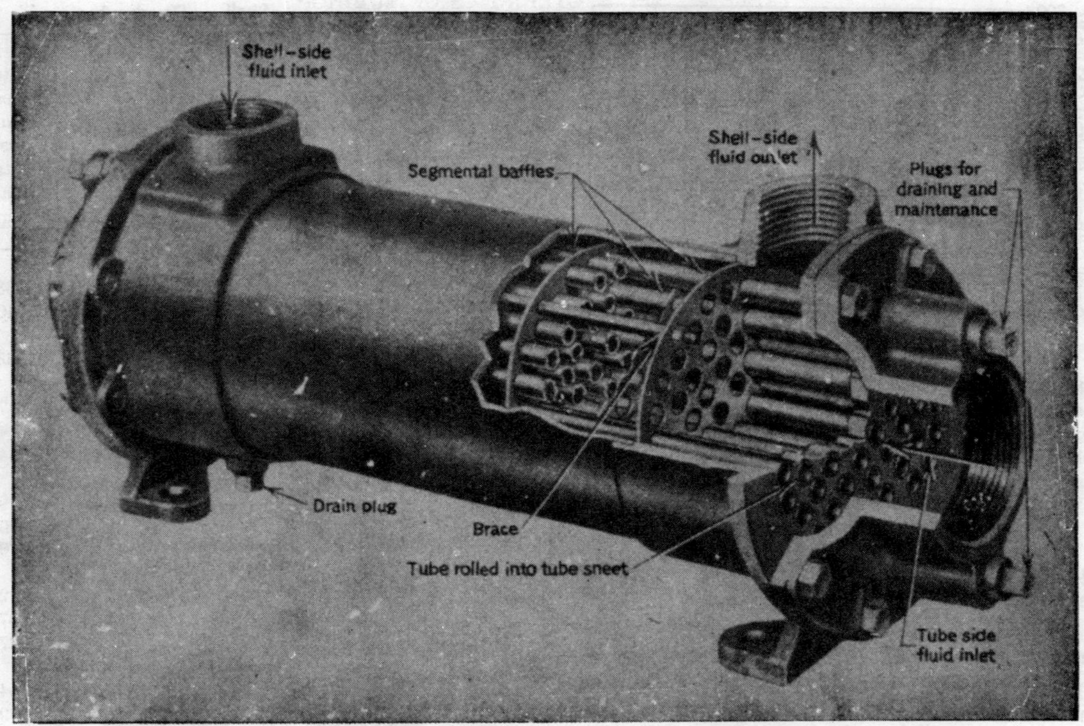

FIG. 403. Cutaway view of a single-pass shell and tube exchanger. (*Ross Heater and Mfg. Co.*)

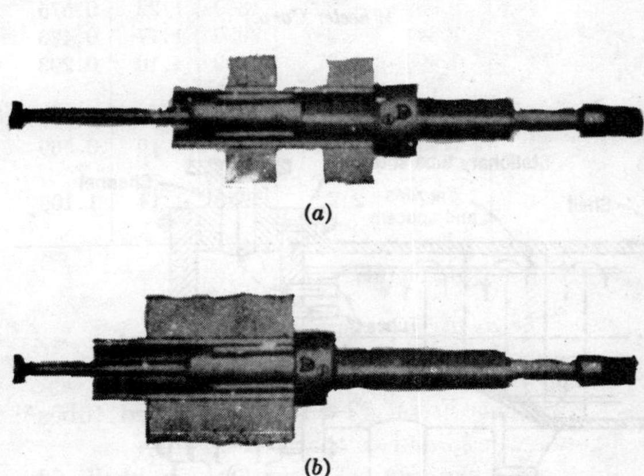

FIG. 404(*a*) Tube rolled into inner tube sheet by an adjustable tube expander; (*b*) same tool used for a single thick tube sheet. (*Gustave Wiedecke Co.*)

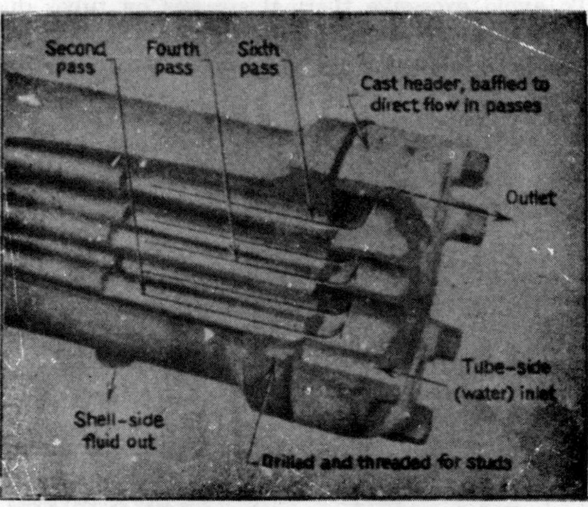

FIG. 405. Cutaway view of a multipass shell and tube condenser. (*Wolverine Tube Co.*)

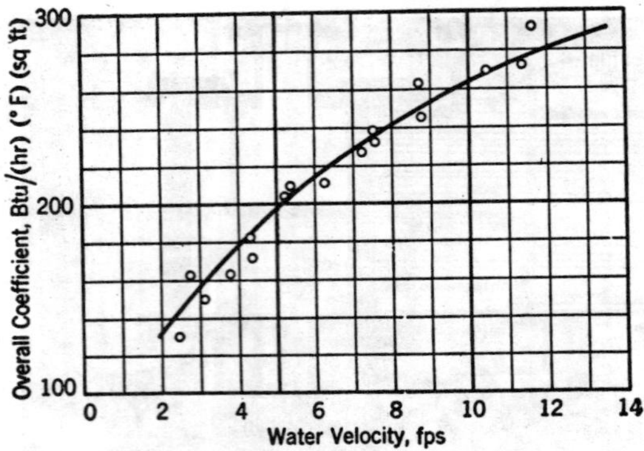

FIG. 406. Overall heat transfer coefficients based on outside area of tubes for condensing Freon 12 in a finned tube exchanger as a function of water velocity inside the tubes. The exchanger has tubes with 16 fins per inch. The total heat transfer area is 65.6 sq ft. The fin diameter is 0.750 in.; fin height is 0.063 in.

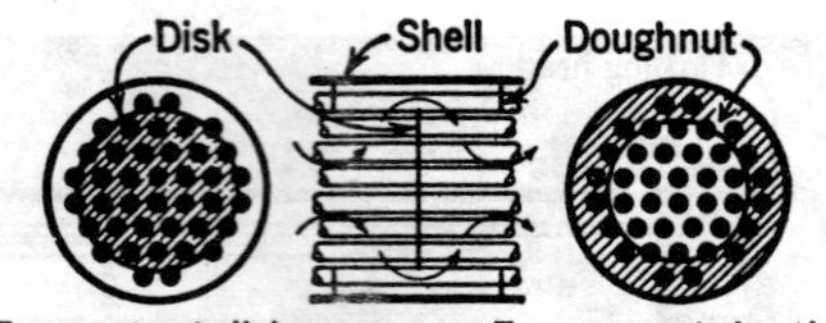

FIG. 407. Disk-and-doughnut transverse shell side baffle. (*Foster-Wheeler Corp.*)

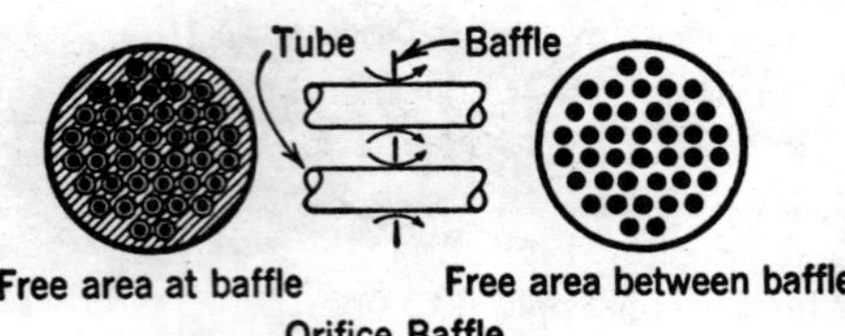

FIG. 408. Orifice transverse shell side baffle. (*Foster-Wheeler Corp.*)

types of baffles. The exchanger of Fig. 403 has segmental baffles.

There are limitations on the permissible temperature range of the fluid for shell and tube exchangers having tube sheets fixed at both ends of the shell. This limitation may be serious when the tubes are of a material having a different coefficient of expansion from the shell or when the tubes are of the same material as the shell but at a different temperature. To avoid excessive thermal stresses on tubes due to differences in temperature, one tube sheet frequently is made movable. Figure 410 is a section

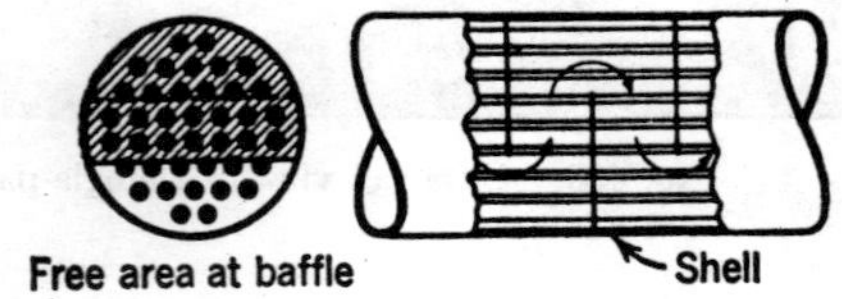

FIG. 409. Segmental transverse shell side baffle. (*Foster-Wheeler Corp.*)

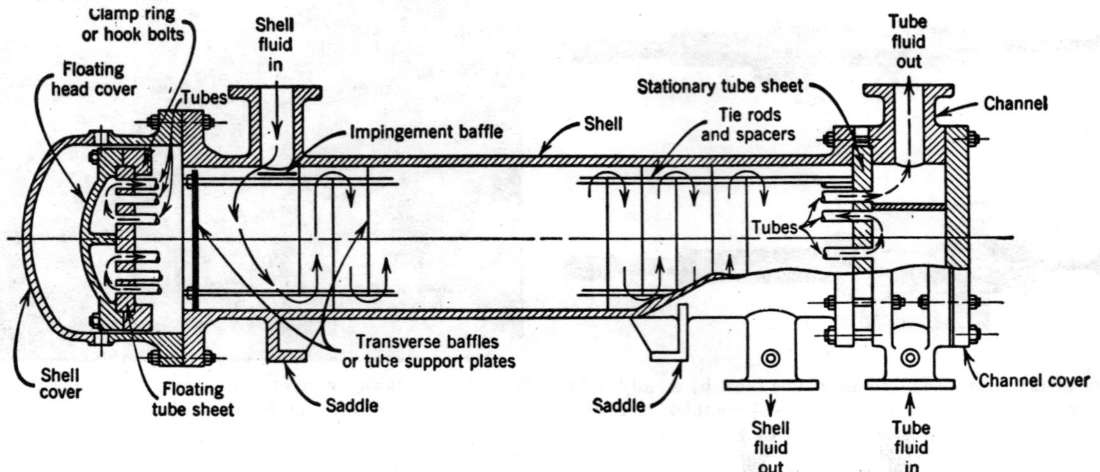

FIG. 410. Cross-sectional drawing of a typical four-pass tube side, single-pass shell side, floating head heat exchanger. (*Tubular Exchanger Manufacturers Association.*)

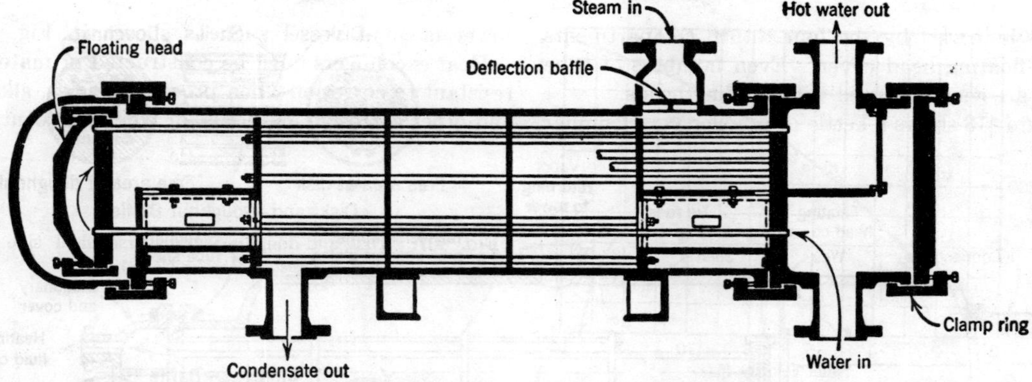

Fig. 411. Cross-sectional drawing of a feed water heater for high-pressure service. (*American Locomotive Co.*)

of a typical 4-pass tube, 1-pass shell, floating head heat exchanger. Figure 411 is a section of a feed water heater constructed to operate at high pressures (2350 psi) and temperatures (900° F). The floating head is enclosed within the shell and permits the tubes to expand or contract to a different degree from that of the shell without causing undue stress. If the change in temperature of the fluid is very large the floating head may be split into two or more sections. Figure 412 shows the component parts of a feed water heater with a single floating head, as in Fig. 411. This unit has eight passes on the water

Fig. 412. Component parts of a feed water heater. (*American Locomotive Co.*) 1, shell; 2, assembled tube bundle; 3, clamp ring for floating head; 4, channel with integral tube sheet; 5, clamp ring for channel cover; 6, shell cover; 7, floating head tube sheet; 8, floating head cover; 9, channel cover.

side as indicated by the four return compartments in the floating head cover. Even numbers of tube passes are generally used with floating heads.

Figure 413 shows a kettle or reboiler vessel similar to a floating head exchanger with an enlarged shell to accommodate liquid and vapor. Steam condensing inside the tubes, or occasionally hot oil, supplies the heat.

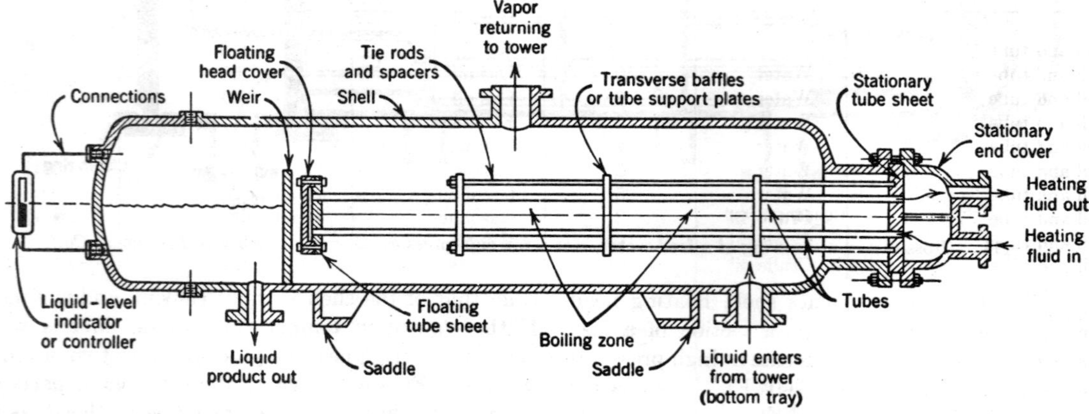

Fig. 413. Sectional diagram of a kettle reboiler. (*Tubular Exchanger Manufacturers Association.*)

Hairpin heaters are essentially 2-pass tube heaters with a single tube sheet. The tubes are in the shape of U. Such heaters are often employed to boil liquids by insertion in the bottom section of a fractionating tower or in a vessel such as shown in Fig. 413.

Fig. 414. Sectional cascade cooler fabricated of Karbate for corrosion resistance. (*National Carbon Co.*)

Heat exchangers must be constructed of materials resistant to corrosion when processing acids, alkalies and other corrosive materials, or whenever contamination of the product such as food must be held to the lowest possible level. There are few heat transfer media which have high thermal conductivity other than metals and alloys. Bimetallic tubes are available and exchanger shells may be clad or lined with corrosion-resisting materials. A graphitic base material, Karbate, has been developed as a material of construction for exchangers. It is resistant to most acids, alkalies, and solvents and has a thermal conductivity as high as 75 Btu/(hr)(°F/ft)(sq ft). A cascade cooler of Karbate is shown in Fig. 414.

Approximate Coefficients

Approximate overall coefficients of heat transfer for estimating the performance or required size of equipment are given in Table 48. Special conditions may give higher or lower values. The coefficients given in Table 48 are for operating equipment except when marked "clean" and may not represent new, clean conditions.

THEORY AND FORMULATION

Equations 356 and 358 give the instantaneous rate of heat transfer. If the temperatures do not change with time, the instantaneous rate remains constant with time. Such a *steady-state condition* exists when water is boiling in a kettle on an electric stove. The heat is transferred largely by radiation and conduction from the electric heater to the kettle, by con-

TABLE 48. TYPICAL OVERALL COEFFICIENTS OF HEAT TRANSFER

Type of Exchanger	Inside or Tube: Fluid	Inside or Tube: Velocity, fps	Outside or Shell: Fluid	Outside or Shell: Velocity, fps	Overall Coefficient, Btu/(hr)(°F)(sq ft)
Shell and tube	Brine	1–3	Water	1–5	50–400
Shell and tube	Water	2	Gas oil	3.0	50–70
Shell and tube	Water	2	Gas oil	6.0	120
Shell and tube	Water	2	Kerosene	0.5	40
Shell and tube	Water	2	Lubricating oil	0.2	15
Shell and tube	Water	3	Kerosene	Condensing	50
Shell and tube	Water	5	Gasoline	Condensing	90
Shell and tube	Crude oil	2	Gasoline	Condensing	20–30
Shell and tube	Crude oil	10	Wax distillate	2.5	65
Shell and tube	Crude oil	10	Gas oil	6.0	80–90
Shell and tube	Water	3	Butane	2.0	90
Shell and tube	Water	4–6	Steam	Condensing	400–800
Shell and tube	Water	3–5	Ammonia	Condensing	150–300
Shell and tube	Water	4	Freon 12	Condensing	150–175
Double pipe	Water	3–8	Brine	3–8	150–300
Coil in vessel	Water	1–3	Water	Natural convection	20
Coil in box	Gasoline	Condensing	Water	Natural convection	15–25
Coil in box	Kerosene	Condensing	Water	Natural convection	10–20
Coil in box	Gas oil	Condensing	Water	Natural convection	8–20
Horizontal tube **reboiler,**					
Commercial	Steam	Condensing	Butane	Boiling	100–150
Laboratory, **clean**	Steam	Condensing	Butane	Boiling	1000–1500
Radiator	Steam	Condensing	Air	Natural convection	1–4
Tube bank	Steam	Condensing	Air	10	9
Tube bank	Steam	Superheat	Flue gas		2–6
Basket evaporator	Brine	Boiling	Steam	Condensing	150–225
Vertical tube evaporator (clean)	Water	Boiling	Steam	Condensing	400–1000

duction through the bottom of the kettle, and by convection from the kettle through the boiling water. The temperature of the boiling water is constant, and the temperature of the electric heater is constant. Under such steady-state conditions the rate of heat transfer remains constant with time, and a measurement of the total heat transferred over a measured time interval gives the rate of heat transfer for the conditions existing during the measurements.

Many industrial heat transfer operations are in continuous processes and represent steady-state conditions.

The *unsteady state* is more difficult to treat mathematically than the steady state. When a kettle of cold water is placed over the hot burner, there is a change in the temperature of the water and kettle with time (an unsteady state). Due to the changing temperature the rate of heat transfer changes with time, and the total quantity of heat transferred over any time interval may be computed only by integrating equations 356 and 358 over the appropriate time interval. The quenching of hot steel is an important problem in unsteady-state heat transfer. Such problems are satisfactorily handled by graphical integration in simple cases and by empirical relationships for the more complex problems of unsteady-state conduction.[31] The simpler methods of steady-state conditions may be used to approximate some unsteady-state conditions by calculating a series of instantaneous values for transfer rates and properly averaging the results. In many cases this is the only practical method of handling unfamiliar problems in unsteady-state heat transfer.

The bulk of the heat transfer data has been obtained for steady-state conditions, and the correlations of data are primarily for these conditions. Therefore the general relationships are presented for steady-state continuous operation.

Conduction

As has been indicated, the transfer of heat through a solid is usually one step in a more complicated path of heat transfer. The basic concept of equation 356 when applied to conduction gives the following equation known as Fourier's law.

$$\frac{dQ}{dt} = -kA\frac{dT}{dx} \tag{360}$$

where k = thermal conductivity of the solid in Btu/(hr)(sq ft)(°F/ft) and must not be confused with conductance, $K = kA/\Delta x$, which may be expressed in Btu/(hr)(°F).

A = area at right angles to the direction of heat flow (as in sq ft).

x = distance (as in ft).

Thermal conductivity k in the equation for heat conduction varies widely for different materials, as indicated in the Appendix. For porous materials, such as are used for insulation, the material is in reality a discontinuous phase of fluid maintained stagnant by the supporting porous solids, and k represents a combined effect of fluid and solid heat transmission. Gaseous fluids such as air are low in thermal conductivity, although they may transfer heat readily by convection; but the minute film of air in the interstices of a porous insulating material exhibits a minimum of convection currents. Therefore, the heat transfer through such insulating material takes on the characteristics of heat conduction through the fluid phase.

Since the density of the insulating material will reflect the proportion of the solid to fluid, the conductivity of porous material varies directly with the density and approaches the low value for the conductivity of the fluid as the solid path for heat flow becomes smaller, so long as the gas spaces are small enough to minimize convection.

The conductivity of water is 0.34 Btu/(°F/ft)(hr)(sq ft) at 60° F and is 25 times that of air. Accordingly, the conductivity of wet porous solids is greater than that of dry porous solids. This is quite noticeable in the loss of insulating power of woolen clothing on damp days. The data on thermal conductivities tabulated in the Appendix indicate the change in thermal conductivity of plywood with moisture and the effect of density of cork board on its conductivity.[39]

Figure 398*a* represents an element of a solid of uniform cross section from surface 1 to surface 2 under the uniform temperature gradient dT/dx and length $x_2 - x_1$. The instantaneous rate of heat transfer dQ/dt depends upon the temperature gradient dT/dx and the nature of the solid.

Steady-State Conduction

Under steady-state conditions the rate of heat transfer is constant, and the right-hand side of equation 360 remains constant with time. If the symbol q is used to represent the constant value for dQ/dt, equation 360 may be written in the more convenient form:

$$q\int\frac{dx}{A} = -\int k\,dT \tag{361}$$

For significant variations of area A with distance x and of thermal conductivity k with temperature T, it is necessary to integrate equation 361. For those cases in which the area is constant, such as Fig. 398, and the thermal conductivity is constant independent of temperature, equation 361 becomes

$$\frac{q}{A}(x_2 - x_1) = -k(T_2 - T_1) \tag{362}$$

or

$$q = \frac{kA(T_1 - T_2)}{(x_2 - x_1)} = \frac{T_1 - T_2}{\Delta x/kA} \tag{362a}$$

Often there is a variation of area with distance in many solids conducting heat. Rather than integrating for each type of solid, it is more convenient to use an average area (A_m) for the area A in the form shown in equation 362*a*.

If the conductivity k varies with the temperature T, the temperature gradient would not be constant and the integral of equation 361 may be evaluated graphically or algebraically. If the conductivity is a linear function of temperature T, as when represented by an equation of the form $k = a + bT$, or if k varies slightly with temperature over the temperature range involved, an average or mean value k_m may be used and the integrated form of the equation written as follows.

$$q = \frac{k_mA_m(T_1 - T_2)}{x_2 - x_1} = \frac{k_mA_m(-\Delta T)}{\Delta x} = \frac{-\Delta T}{\Delta x/k_mA_m} \tag{362b}$$

It is common practice to omit the minus sign in equation 362*b* using the expression ΔT to represent the drop in temperature, $T_1 - T_2$, although this is mathematically improper. The mathematically correct statement will be used in this textbook, and ΔT will represent $T_2 - T_1$. Also the symbol k is frequently used to indicate the average conductivity k_m, and the symbol L is frequently used for the distance $x_2 - x_1$.

The resistance to heat transfer by conduction under these steady-state conditions is represented by $\Delta x/k_m A_m$ or L/kA_m.

Illustrative Example. A furnace wall is to be composed of three layers: 6 in. of firebrick (with conductivity $k = 0.95$), insulating brick ($k = 0.14$), and common brick ($k = 0.8$). It is estimated that the surface of the firebrick, corresponding to T_1 in Fig. 398*a*, is 1800° F. The maximum temperature of the insulating brick, corresponding to T_2, is 1720° F and the temperature corresponding to T_3 under these conditions is 280° F. Assume negligible resistance to the flow of heat across the interfaces between the bricks.

1. Compute the thickness of the layer of insulating brick.
2. If the common brick are 9 in. thick, compute their outside temperature.

Solution. For each layer of brick (equation 362*b*)

$$\frac{q}{A} = \frac{k_m(T_1 - T_2)}{\Delta x}$$

Equating the heat transfer per unit area for firebrick and insulating brick,

$$\frac{0.95 \times 12(1800 - 1720)}{6} = \frac{0.14 \times 12(1720 - 280)}{\Delta x \text{ (insulating brick)}}$$

$$\Delta x \text{ (insulating brick)} = \frac{0.14 \times 12(1720 - 280) \times 6}{0.95 \times 12(1800 - 1720)} = 16 \text{ in.}$$

The same procedure could be used for computing the outside temperature of the common brick, or one can solve for q/A from the data on the fireclay brick and use this value to compute T_2 for the common brick. For the firebrick,

$$\frac{q}{A} = \frac{0.95 \times 12(1800 - 1720)}{6} = 152 \text{ Btu/(hr)(sq ft)}$$

For the common brick,

$$\frac{q}{A} = 152 = \frac{0.8 \times 12}{9} (280 - T_2)$$

$T_2 = 138°$ F, outside temperature of common brick

A common shape of solid encountered in heat transfer is a cylinder such as a thick pipe insulation, represented in Fig. 415. If the heat is flowing radially in such a solid, the area A varies with the distance x.

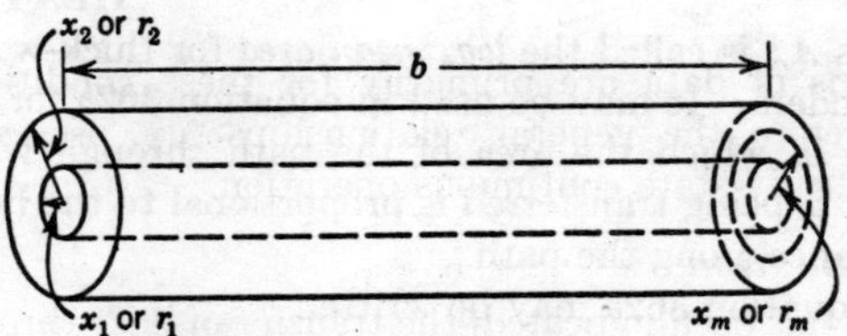

Fig. 415. Diagram indicating dimensions of significance in heat transfer through a thick-walled cylinder by radial conduction.

For steady-state heat transfer by radial conduction through a cylinder, $2\pi rb$ corresponds to A of equation 360, and

$$q = \frac{dQ}{dt} = -k(2\pi rb)\frac{dT}{dr} \qquad (363)$$

where b = length of the cylinder.
r = radial distance.

Rearranging,

$$\frac{dr}{r} = \frac{-k}{q} 2\pi b\, dT \qquad (363)$$

Integrating with k constant at the value k_m,

$$\ln\left(\frac{r_2}{r_1}\right) = \frac{k_m}{q} 2\pi b(T_1 - T_2) \qquad (364)$$

$$q = \frac{2\pi b k_m(T_1 - T_2)}{\ln (r_2/r_1)} \qquad (364a)$$

Equation 364*a* may be written

$$q = \frac{2\pi b k_m(T_1 - T_2)(r_2 - r_1)}{(r_2 - r_1) \ln (r_2/r_1)} \qquad (364b)$$

Substituting $2\pi r_m b$ for A_m in equation 362*b* gives

$$q = \frac{2\pi r_m b k_m(T_1 - T_2)}{r_2 - r_1} \qquad (362c)$$

Equating these expressions (364*b* and 362*c*) for q gives *the* $\log_e$ *mean radius*

$$r_m = \frac{r_2 - r_1}{\ln (r_2/r_1)}$$

This equation may be written

$$(2\pi b)r_m = \frac{(2\pi b)r_2 - (2\pi b)r_1}{\ln (2\pi b r_2/2\pi b r_1)}$$

Since $A = 2\pi br$ for any cylinder

$$A_m = \frac{A_2 - A_1}{\ln (A_2/A_1)}$$

This A_m is called the *log$_e$ mean area* for thick-walled cylinders. It may be used in equation 362*b* for any case in which the area of the path through which heat is being transferred is proportional to the linear distance along the path.

Equation 362*b* may be written

$$q = \frac{2\pi b r_m}{\Delta x}[k_m(T_1 - T_2)] \qquad (365)$$

This procedure for finding the proper average value for a linear dimension such as r_m for use in equation 365, may be generalized by mathematical procedures. The following common types of integrals may be encountered in which an alternate solution to the integration may be made by multiplying an average value of the dependent variable by a Δ of the independent variable. Thus

$$\int_{x_1}^{x_2} x\,dx = \frac{x_2^2 - x_1^2}{2} \qquad (366)$$

However, the same numerical value can be obtained by use of a proper value for x_m in the following expression.

$$\int_{x_1}^{x_2} x\,dx = x_m\,\Delta x \qquad (367)$$

Combining and solving for x_m, it is found that the proper value for x_m is $(x_2 + x_1)/2$ for this case. $(x_2 + x_1)/2$ is called the arithmetic mean.

Similarly, for other integrals the following mean values of the dependent variable may be used as indicated in Table 49.

The log$_e$ mean radius must not be confused with that radius at which the temperature is the arithmetic mean between the inner and outer wall of the cylinder. This latter radius may be determined as follows:

Referring to Fig. 415, the heat transfer from inner surface (1) to the cylinder at radius r equals the heat transfer from r to surface 2, or $q_{1-r} = q_{r-2}$.

$$q_{1-r} = \frac{(T_1 - T_r)k2\pi b(r - r_1)}{(r - r_1)\ln(r/r_1)}$$

$$q_{r-r_2} = \frac{(T_r - T_2)k2\pi b(r_2 - r)}{(r_2 - r)\ln(r_2/r)}$$

$q_{1-r} = q_{r-r_2}$ for steady-state series transfer of heat, and $T_1 - T_r = T_r - T_2$ by the statement of the problem.

$$\frac{k2\pi b(r - r_1)}{(r - r_1)\ln(r/r_1)} = \frac{k2\pi b(r_2 - r)}{(r_2 - r)\ln(r_2/r)}$$

$$\ln\frac{r}{r_1} = \ln\frac{r_2}{r}$$

$$r^2 = r_1 r_2$$

$$r = \sqrt{r_1 r_2}$$

Exercise. Plot the error caused by using the arithmetic mean radius $(r_1 + r_2)/2$ as r_m in equation 365 as a function of r_2/r_1 for a cylinder.

Exercise. Show that for use in equation 362*h* $A_m = \sqrt{A_1 A_2}$ for radial heat flow through a spherical shell whose inside surface is A_1 and outside surface is A_2.

TABLE 49. MEAN VALUES OF DEPENDENT VARIABLES FOR DETERMINING VALUES OF INTEGRALS

1. $\int_{x_1}^{x_2} x\,dx = \frac{x_2^2 - x_1^2}{2} = x_m(x_2 - x_1)$; therefore $x_m = \frac{x_2 + x_1}{2} \equiv$ Arithmetic mean

2. $\int_{x_1}^{x_2} y\,dx = y_m(x_2 - x_1)$ if $y = ax + b$; therefore $y_m = \frac{y_2 + y_1}{2} \equiv$ Arithmetic mean

3. $\int_{x_1}^{x_2} \frac{dx}{x} = \ln\frac{x_2}{x_1} = \frac{x_2 - x_1}{x_m}$; therefore $x_m = \frac{x_2 - x_1}{\ln(x_2/x_1)} \equiv$ Logarithmic mean

4. $\int_{x_1}^{x_2} \frac{dx}{y} = \frac{x_2 - x_1}{y_m}$ if $y = ax + b$; therefore $y_m = \frac{y_2 - y_1}{\ln(y_2/y_1)} \equiv$ Logarithmic mean

5. $\int_{x_1}^{x_2} \frac{dx}{x^2} = \frac{x_2 - x_1}{x_2 x_1} = \frac{x_2 - x_1}{x_m^2}$; therefore $x_m = \sqrt{x_2 \cdot x_1} \equiv$ Geometric mean

6. $\int_{x_1}^{x_2} \frac{dx}{y^2} = \frac{x_2 - x_1}{y_m^2}$ if $y = ax + b$; therefore $y_m = \sqrt{y_2 \cdot y_1} \equiv$ Geometric mean

7. $\int_{x_1}^{x_2} \frac{dx}{xy} = \frac{x_2 - x_1}{(xy)_m}$ if $y = ax + b$; therefore $(xy)_m = \frac{y_1 x_2 - y_2 x_1}{\ln(y_1 x_2 / y_2 x_1)} \equiv$ Mixed mean

The following empirical equations[26,35] applicable to rectangular thick-walled bodies, such as furnaces, whose shortest inside dimension does not exceed twice the wall thickness were derived from the determination of the electrical resistance of solutions contained in a vessel having the same relative dimensions as the rectangular solid.

When the length E of each of twelve inside edges is between 0.2 and 2.0 times the wall thickness (Δx),

$$A_m = A_1 + 0.54(\Delta x)\Sigma E + 1.2(\Delta x)^2 \quad (368)$$

where A_1 = inside surface area.

When one inside dimension is less than $0.2(\Delta x)$, the corresponding four lengths are omitted in calculating ΣE in the following equation.

$$A_m = A_1 + 0.465(\Delta x)\Sigma E + 0.35(\Delta x)^2 \quad (369)$$

When two inside dimensions are less than $0.2(\Delta x)$, only the longest inside edge E is used in the equation.

$$A_m = \frac{6.4\,E(\Delta x)}{\ln (A_2/A_1)} \quad (370)$$

When each interior edge is less than $0.2(\Delta x)$,

$$A_m = 0.79\sqrt{A_1A_2} \quad (371)$$

Conduction through a Series of Solids

For steady heat flow through a series of solids as illustrated in Fig. 398*a*, the quantity of heat flowing from 1 to 2 is the same as that flowing from 2 to 3 and from 3 to 4, provided no heat is transferred to the surroundings except through the surfaces A_1 and A_4.

Writing equation 362 for each solid

$$q = \frac{k_{1-2}A_{1-2}(T_1 - T_2)}{x_2 - x_1} = \frac{k_{2-3}A_{2-3}(T_2 - T_3)}{x_3 - x_2} = \frac{k_{3-4}A_{3-4}(T_3 - T_4)}{x_4 - x_3} \quad (362)$$

Solving for the temperature drop $-\Delta T$ across each solid, and substituting for temperature drop in

$$T_1 - T_4 = (T_1 - T_2) + (T_2 - T_3) + (T_3 - T_4) \quad (372)$$

$$T_1 - T_4 = \frac{q}{\dfrac{k_{1-2}A_{1-2}}{x_2 - x_1}} + \frac{q}{\dfrac{k_{2-3}A_{2-3}}{x_3 - x_2}} + \frac{q}{\dfrac{k_{3-4}A_{3-4}}{x_4 - x_3}} \quad (373)$$

$$q = \frac{T_1 - T_4}{\dfrac{x_2 - x_1}{k_{1-2}A_{1-2}} + \dfrac{x_3 - x_2}{k_{2-3}A_{2-3}} + \dfrac{x_4 - x_3}{k_{3-4}A_{3-4}}} \quad (373a)$$

Equation 373 corresponds to equation 358 for series flow where the total resistance equals the sum of the individual resistances. The denominator may be expressed as the reciprocal of the product of an overall coefficient times area,

$$\frac{1}{UA} = \frac{x_2 - x_1}{k_{1-2}A_{1-2}} + \frac{x_3 - x_2}{k_{2-3}A_{2-3}} + \frac{x_4 - x_3}{k_{3-4}A_{3-4}} \quad (374)$$

When using the overall coefficient U, it is necessary to base the calculations on a single area A.

When all areas are equal, no difficulty arises, but when the areas are different, as in radial flow through pipes or insulation, it is customary, though not necessary, to express the overall coefficient on the basis of that area applicable to the major resistance.

Exercise. Show that, for series flow, the temperature drop ($-\Delta T$) across any portion of the flow system is proportional to the resistance thereof.

Unsteady-State Conduction. In heating and cooling solids when a steady-state process is put on or off stream and in all processes in which the steady-state period is relatively short, the temperature at a given point varies with time, and the process is said to be under unsteady-state operation. The introduction of time as an additional variable makes the calculations for the unsteady state somewhat more complicated than for the steady state.

In many cases it is practical to subdivide the unsteady-state period into a finite number of relatively short periods in each of which the methods of steady-state heat transfer calculations are employed, using the average values of the properties over each period. In this manner the heat transfer for the unsteady-state period is approximated by simple summation.

In the more rigorous method, the general differential equation for unsteady-state heat transfer by conduction is derived from the basic Fourier equation, 360.

Considering a cubical element of volume of sides dx, dy, and dz, the rate at which heat enters along the x, y, and z axes is given by the expression

$$\frac{dQ_{\text{in}}}{dt} = -\left(k_x\,dy\,dz\frac{\partial T}{\partial x} + k_y\,dz\,dx\frac{\partial T}{\partial y} + k_z\,dx\,dy\frac{\partial T}{\partial z}\right) \quad (375)$$

whereas the rate at which heat leaves along the x, y, and z axes is given by the expression

$$\frac{dQ_{\text{out}}}{dt} = -\left[k_x\frac{\partial T}{\partial x} + \frac{\partial}{\partial x}\left(k_x\frac{\partial T}{\partial x}\right)dx\right]dy\,dz - \left[k_y\frac{\partial T}{\partial y} + \frac{\partial}{\partial y}\left(k_y\frac{\partial T}{\partial y}\right)dy\right]dz\,dx - \left[k_z\frac{\partial T}{\partial z} + \frac{\partial}{\partial z}\left(k_z\frac{\partial T}{\partial z}\right)dz\right]dx\,dy \tag{376}$$

Therefore the net rate at which heat is accumulated in the element of volume is

$$\frac{dQ_{\text{net}}}{dt} = \frac{dQ_{\text{in}}}{dt} - \frac{dQ_{\text{out}}}{dt} = \left[\frac{\partial}{\partial x}\left(k_x\frac{\partial T}{\partial x}\right) + \frac{\partial}{\partial y}\left(k_y\frac{\partial T}{\partial y}\right) + \frac{\partial}{\partial z}\left(k_z\frac{\partial T}{\partial z}\right)\right]dx\,dy\,dz \tag{377}$$

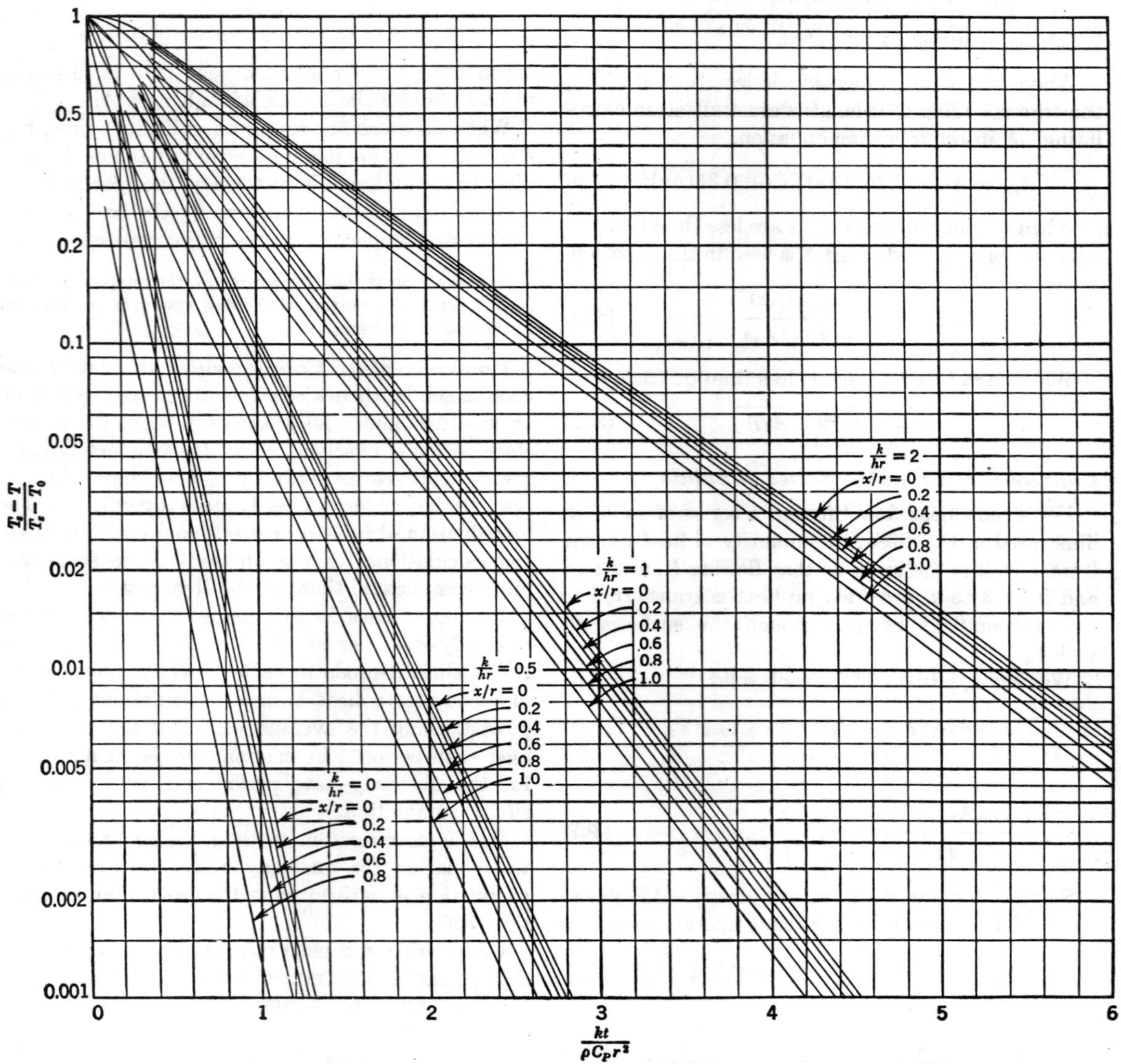

FIG. 416. The relationships between the dimensionless groups indicated resulting from the integration of equation 380 when applied to the heating and cooling of a solid cylinder having an infinite ratio of length to diameter.[13]

The rate of heat accumulation equals the rate of increase in enthalpy and is related to the rate of rise of temperature of the element $dx\,dy\,dz$ as follows.

$$\frac{dQ}{dt} = \frac{dH}{dt} = \left(\rho \overline{C_P} \frac{\partial T}{\partial t}\right) dx\,dy\,dz \tag{378}$$

It therefore follows that the general differential equation for unsteady-state transfer by conduction is

$$\frac{1}{\rho \overline{C_P}}\left[\frac{\partial}{\partial x}\left(k_x \frac{\partial T}{\partial x}\right) + \frac{\partial}{\partial y}\left(k_y \frac{\partial T}{\partial y}\right) + \frac{\partial}{\partial z}\left(k_z \frac{\partial T}{\partial z}\right)\right] = \frac{\partial T}{\partial t} \tag{379}$$

If the material is homogeneous and isotropic and if the variation of k with temperature within the element of volume is neglected, the above equation reduces to

$$\frac{k}{\rho \overline{C_P}}\left[\left(\frac{\partial^2 T}{\partial x^2}\right) + \left(\frac{\partial^2 T}{\partial y^2}\right) + \left(\frac{\partial^2 T}{\partial z^2}\right)\right] = \frac{\partial T}{\partial t} \tag{380}$$

In the case of an infinite slab, in which the ratio of surface to thickness is large, the flow of heat is unidirectional through the thickness x and the equation reduces to

$$\frac{k}{\rho \overline{C_P}}\left(\frac{\partial^2 T}{\partial x^2}\right) = \frac{\partial T}{\partial t} \tag{381}$$

The above equations are all of second order, and their integrated forms depend upon the boundary conditions which introduce considerations of heat transfer by convection across the boundaries.

Fortunately for the practicing engineer, the above equations have been integrated by various writers for various boundary conditions, and the results of their labors are available in graphical form in most of the standard references.[31,38] The following dimensionless groups are used as variables for Fig. 416 which is applicable to unsteady-state heat transfer in cylinders.[13]

$$\left(\frac{T_s - T}{T_s - T_0}\right) \quad \left(\frac{kt}{\rho C_P r^2}\right)$$

$$\left(\frac{k}{hr}\right) \quad \left(\frac{x}{r}\right)$$

where T_s = temperature of the surroundings.
T_0 = the initial uniform temperature of the body.
T = the temperature at a given point in the body at time t.
k = thermal conductivity of the body.
ρ = density of the body.
C_P = specific heat of the body.
h = coefficient of heat transfer between the surroundings and the surface of the body.
x = distance in the direction of heat conduction from the midpoint of the body to the point under consideration.
r = radius of the cylinder.
t = time.

Such charts are available for the heating and cooling of a solid sphere, of a solid cylinder having infinite ratio of length to diameter, of a solid slab having a large face area relative to that of the edges, and of a solid of infinite thickness neglecting edge effects.

Illustrative Example. Determine the time required to quench a billet of aluminum, 8 in. in diameter, from an initial uniform temperature of 1000° F to a final mid-axis temperature of 300° F, if during quenching the temperature of the quenching medium remains essentially constant at 250° F while the surface coefficient of heat transfer has a value of 840 Btu/(hr)(°F)(sq ft).

Specific gravity = 2.70; specific heat capacity = 0.186 Btu/(lb)(°F); thermal conductivity = 140 Btu/(hr)(°F/ft)(sq ft); end effects may be neglected.

Solution.

$$\left(\frac{x}{r}\right) = 0 \qquad \left(\frac{k}{hr}\right) = \frac{140}{(840)(\frac{1}{3})} = 0.50$$

$$\left(\frac{T_s - T}{T_s - T_0}\right) = \left(\frac{T - T_s}{T_0 - T_s}\right) = \frac{300 - 250}{1000 - 250} = 0.067$$

From Fig. 416

$$\frac{kt}{\rho C_P r^2} = 1.20 = \frac{(140)t}{(2.70)(62.4)(0.186)(\frac{1}{3})}$$

Therefore t = 0.0298 hr = 107.2 sec = 1 min, 47 sec.

Other methods[11,13,17,31,33,34,37] for the solution of the complex problems in unsteady-state heat transfer have been developed. The most satisfactory methods for reliable results for all different shapes are probably the algebraic and graphical methods of Dusinberre[11] and Hawkins and Agnew.[17]

Convection

Although the mechanism of heat transfer through a fluid, or from a solid to a fluid, by convection is quite different from heat transfer by conduction, the same form of equation has been found most convenient. The rate of heat transfer by convection for steady-state transfer between a solid surface and a fluid is usually expressed as

$$q = hA(T_1 - T_2) \tag{382}$$

where h = the convection heat transfer coefficient between the solid and the fluid as in Btu/(hr)(°F)(sq ft).

This h is equivalent to $k/\Delta x$ in equation 362*a*.

For heat transfer through a series of resistances which may include one or more convection coefficients and conduction through solids, it is convenient to use the overall coefficient U which has the same dimensions as h. For such conditions the following equation (383) may be written as equivalent to equation 374.

$$\frac{1}{UA} = \frac{1}{h_1A_1} + \frac{x_3 - x_2}{k_mA_m} + \frac{1}{h_2A_2} \tag{383}$$

An important case of this kind is indicated diagrammatically in Fig. 417, representing a condition

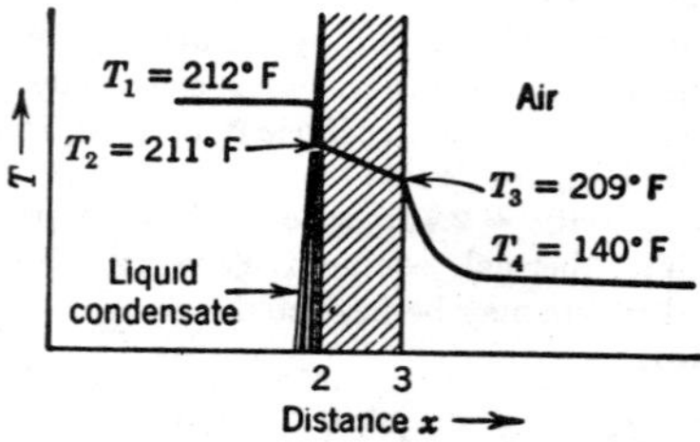

FIG. 417. Diagrammatic representation of the conditions accompanying the transfer of heat through a condensing film by conduction and convection.

for the transfer of heat from condensing steam through a metal wall to air. The steam is at constant pressure and therefore is at a constant temperature such as 212° F. As the steam condenses it forms a liquid condensate film on the metal surface. The heat must then be transferred from the steam through the liquid film, the metal wall, and then from the outside surface of the metal wall to the surrounding air. In this case the overall resistance is the sum of the resistances through the liquid film, the metal wall, and for the transfer into the air.

The resistance in the condensate film depends upon its thickness and characteristics such as its density, viscosity, thermal conductivity, and degree of turbulence. The transfer coefficient from the outside of the metal wall to the air is also dependent upon similar physical characteristics of the air. For this reason convection coefficients are frequently referred to as film coefficients even though no physically separable film exists. The air immediately adjacent to the metal is in laminar flow, and there is no sharp transition between this laminar flowing air and the air which may be in turbulent motion away from the surface. Increased agitation or turbulence in the condensate film on the steam side or in the air on the air side increases the heat transfer coefficient or decreases the resistance to heat transfer.

The values for the convection coefficient, and also for the thermal conductivity of various materials, are determined from experimental measurements. Convection coefficients are most satisfactorily correlated in terms of the physical properties and operating conditions which are recognized as influencing the transfer. By means of dimensional analysis, these factors are collected into dimensionless groups. The influence of each group is evaluated empirically from correlations of experimental measurements in a manner similar to that employed in evaluating friction factors in fluid flow. If proper account is taken of all factors, the evaluation will be applicable to any condition but will require empirical evaluation of constants; application to conditions which introduce factors not included in the original dimensional analysis vitiates the results.

Calculation of Heat Transfer Coefficient. The simplest method of treating calculations is by means of the overall coefficient as indicated in equation 356. In the case of a vapor such as steam condensing outside tubes through which cooling water is flowing, the temperature of the condensing steam depends upon the pressure on the steam side (actually the partial pressure of the steam if other vapors are present) and is usually constant throughout the test as T_1. The cooling water enters at a lower temperature and is heated to a higher exit temperature in its passage through the condenser. In this case the arithmetic average temperature of the cooling water may be used as T_2 with reliable results if the increase in temperature of the water is relatively small compared with the difference in temperature between the water and steam.

The quantity of heat transferred per unit of time is obtained from measurements of the quantity of water flowing through the condenser and the accompanying increase in temperature of this water. The overall coefficient U may then be computed using equation 356 and this measured rate of heat transfer, the difference between the steam temperature and the average temperature of the water $(T_1 - T_2)$, and either the inside area (water side) or the outside area (steam side) of the heat transfer surface. It is clear that a different value for the overall coefficient will be obtained, depending upon the area chosen.

In order to compute convection coefficients from experimental data or to compute overall coefficients from predicted convection coefficients, it is necessary to understand the relationship between the overall coefficient and the individual coefficients for convection or conduction.

Figure 418 represents a general case of heat transfer with a cylindrical pipe as the conducting solid separating the two fluids A and B. With reference to section C, the overall coefficient represents the rate of heat transfer from point 1 to point 4, and the individual coefficients refer to the transfer from 1 to 2, 2 to 3, 3 to 4. The first and last of these are convection coefficients, and the intermediate is for conduction. Since this is steady-state heat transfer,

$$q_{1-2} = q_{2-3} = q_{3-4}$$

and the same procedure may be used as in the case of conduction through a series of solids represented in Fig. 398a. The following equation may be written for the conditions of Fig. 418 in the same manner that equation 373 was applied to the conditions of Fig. 398a.

$$q = \frac{T_1 - T_4}{\dfrac{1}{h_{1-2}A_2} + \dfrac{x_3 - x_2}{k_{2-3}A_{m,\,2-3}} + \dfrac{1}{h_{3-4}A_3}} \qquad (384)$$

It will be noticed that there are three areas, A_2, A_m (the mean area through which the heat is transferred by conduction), and A_3. The overall coefficient must be expressed in terms of one area only. If it is desired to obtain the overall coefficient based on the inside area of the metal tube A_2, each side of equation 384 may be divided by A_2 and by the overall temperature drop $T_1 - T_4$, giving the overall coefficient based on area A_2, as written in the equation,

$$\frac{q}{A_2(T_1 - T_4)} = U_2 = \frac{1}{\dfrac{1}{h_{1-2}} + \dfrac{(x_3 - x_2)A_2}{k_m A_m} + \dfrac{A_2}{h_{3-4}A_3}} \qquad (385)$$

Since the surface areas of cylinders such as pipe vary directly as the diameter, the ratio of diameters may be substituted for the ratio of the areas in equation 385 as follows.

$$U_2 = \frac{1}{\dfrac{1}{h_{1-2}} + \dfrac{(x_3 - x_2)D_2}{k_m D_m} + \dfrac{D_2}{h_{3-4}D_3}} \qquad (386)$$

By reference to equation 386 it is clear that some additional information is required in order to compute the two convection coefficients, h_{1-2} and h_{3-4}.

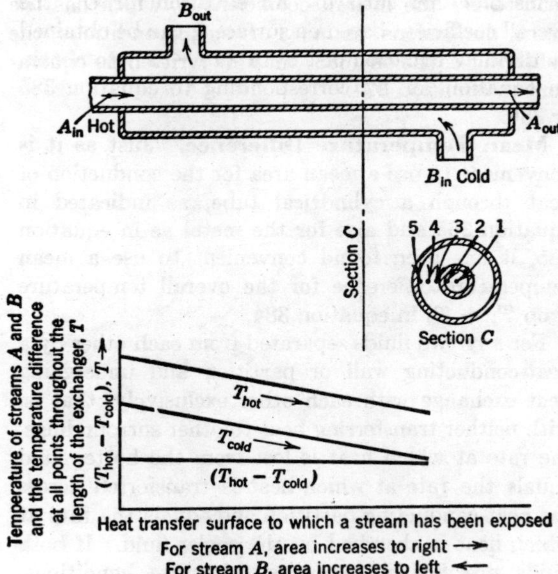

FIG. 418. Diagram indicating the relative temperatures and temperature differences in a countercurrent heat exchanger.

If the overall coefficient has been determined, the diameters are known, and the conductivity of the metal tubes is available, there are still the two unknowns, h_{1-2} and h_{3-4}, in the single equation 386. If the temperature of the center of the metal tube is measured by suitable means, this value may be taken as the average temperature of the metal wall, and T_2 and T_3 may be determined from the temperature drop through the tube computed by means of equation 365. Then equation 382 may be applied to compute the two convection coefficients.

The measurement of the tube temperature is a reasonably difficult determination in itself. The average tube temperature may be computed by measuring the electrical resistance of the tube, provided the resistance has been determined as a function of temperature. Attempts to measure the tube temperature by imbedding thermocouples in the tube so as to present a minimum of disturbance at the point of measurement have met with limited success.

In practice, equation 385 or 386 may be used to calculate overall coefficients from predicted convection coefficients. It must be kept in mind that the area for a condensing film or for a convection coefficient from fluids to a solid is always the area of the surface of the solid adjacent to the fluid. In the case of the overall coefficient it may be expressed in terms of either surface. An equation for U_3, the overall coefficient based on surface 3, can be obtained by dividing equation 384 by $A_3(T_1 - T_4)$ to obtain an equation for U_3 corresponding to equation 385 or 386.

Mean Temperature Difference. Just as it is convenient to use a mean area for the conduction of heat through a cylindrical tube, as indicated in equation 365 and also for the metal as in equation 385, it has been found convenient to use a mean temperature difference for the overall temperature drop $T_1 - T_4$ in equation 384.

For any two fluids separated from each other by a heat-conducting wall or partition and undergoing heat exchange with each other exclusively, that is, with neither transferring heat to other surroundings, the rate at which heat is lost from the hotter fluid equals the rate at which heat is transferred across the heat-conducting partition and equals the rate at which heat is absorbed by the cooler fluid. If both fluids are flowing under steady-state conditions, regardless of whether their relative motion be parallel or countercurrent, the relationship may be expressed mathematically for a differential element of heat transfer area.

$$-(C_P)_h W_h \, dT_h = U(T_h - T_c)\, dA \qquad (387)$$

$$= (C_P)_c W_c \, dT_c$$

where C_P = specific heat capacity [Btu/(lb)(°F)].
W = steady-state rate of mass flow (lb/hr).
dT = differential increase in temperature (°F).
U = overall coefficient of heat transfer (based on dA) [Btu/(hr)(°F)(sq ft)].
dA = differential element of heat transfer area (sq ft).

Subscripts h and c refer to hotter and cooler streams, respectively.

Rearranging and integrating equation 387 between the inlet and outlet temperatures,

$$\int_{T_{c1}}^{T_{c2}} \frac{W_c(C_P)_c\, dT_c}{U(T_h - T_c)} = \int_0^A dA = A = -\int_{T_{h1}}^{T_{h2}} \frac{W_h(C_P)_h\, dT_h}{U(T_h - T_c)} \qquad (388)$$

When $W_c(C_P)_c$ and $W_h(C_P)_h$ are each constant, T_h is a linear function of T_c.

$$dT_h = -\frac{W_c(C_P)_c}{W_h(C_P)_h}\, dT_c \qquad (389)$$

or

$$T_h = aT_c + b \qquad (390)$$

$$T_h - T_c = (a - 1)T_c + b \qquad (391)$$

If U is also constant

$$\frac{W_c(C_P)_c}{U}\int \frac{dT_c}{T_h - T_c} = A \qquad (392)$$

Since $T_h - T_c$ is a straight-line function of T_c, the integral in the equation is of the same form as the fourth integral in Table 49, and the proper mean temperature difference is the logarithmic mean temperature difference,

$$(T_h - T_c)_m = \frac{(T_h - T_c)_2 - (T_h - T_c)_1}{\ln \dfrac{(T_h - T_c)_2}{(T_h - T_c)_1}} \qquad (393)$$

where the subscripts 1 and 2 refer to the two ends of the heat transfer surface, or

$$q = \int W_c(C_P)_c\, dT_c = UA(T_h - T_c)_m \qquad (394)$$

If U is not constant but may be expressed as a linear function [6a] of T_c or of $(T_h - T_c)$,

$$U = \alpha T + \beta \qquad (395)$$

Equation 388 becomes

$$W_c(C_P)_c \int_{T_{c_1}}^{T_{c_2}} \frac{dT_c}{U(T_h - T_c)} = A \qquad (396)$$

Substituting for U from equation 395 and for $(T_h - T_c)$ from equation 391

$$W_c(C_P)_c \int \frac{dT_c}{\alpha(a-1)T_c^2 + [\alpha b + \beta(a-1)]T_c + \beta b} = A \qquad (397)$$

Equation 397 may be integrated and simplified.

$$W_c(C_P)_c\left[\frac{1}{\alpha b-\beta(a-1)}\ln\frac{2\alpha(a-1)T_c+2\beta(a-1)}{2\alpha(a-1)T_c+2\alpha b}\right]_{T_{c_1}}^{T_{c_2}}=A$$

$$W_c(C_P)_c\left[\frac{T_{c_2}-T_{c_1}}{U_2(T_h-T_c)_1-U_1(T_h-T_c)_2}\ln\frac{U_2(T_h-T_c)_1}{U_1(T_h-T_c)_2}\right]=A \quad (398)$$

Equation 396 may be written

$$W_c(C_P)_c\frac{T_{c_2}-T_{c_1}}{[U(T_h-T_c)]_{\text{mean}}}=A \quad (396a)$$

Solving for $[U(T_h - T_c)]_{\text{mean}}$ by equations 398 and 396*a*

$$[U(T_h-T_c)]_m=\frac{U_2(T_h-T_c)_1-U_1(T_h-T_c)_2}{\ln\dfrac{U_2(T_h-T_c)_1}{U_1(T_h-T_c)_2}} \quad (399)$$

or

$$q=W_c(C_P)_c\,\Delta T_c=A[U(T_h-T_c)]_m \quad (400)$$

Since the same relationships result where T_h is taken as the independent variable, the following equation which does not distinguish between the hot or cold fluid is the usual form.

$$q=A\frac{U_1\,\Delta T_2-U_2\,\Delta T_1}{\ln\dfrac{U_1\,\Delta T_2}{U_2\,\Delta T_1}} \quad (401)$$

It is important to keep in mind that each product in the above equation contains the overall transfer coefficient at one end and the temperature difference at the other end of the exchanger. If $U_1 = U_2$, equation 401 reduces to equation 394.

The correct solution is always given by rigorous integration of equation 388. Use of the logarithmic mean temperature difference is simply one convenient means of obtaining a practical solution to equation 388. In certain cases, such as when the ratio between the temperature differences is less than 2, the arithmetic mean temperature difference also gives a satisfactory solution. Use of the arithmetic mean temperature difference is strongly recommended whenever the ratio of the temperature differences approaches 1.

In commercial heat exchangers, the flow is usually neither co- nor counter-current throughout, and no simple mean temperature difference represents the conditions within the exchanger. Careful analysis of the conditions within the exchanger [4] has provided an empirical procedure using the logarithmic mean temperature difference and a multiplying factor Y as indicated in equation 402.

$$q = UAY(T_B - T_A)_m \quad (402)$$

Values for the factor Y for two typical heat exchangers are given in Fig. 419.

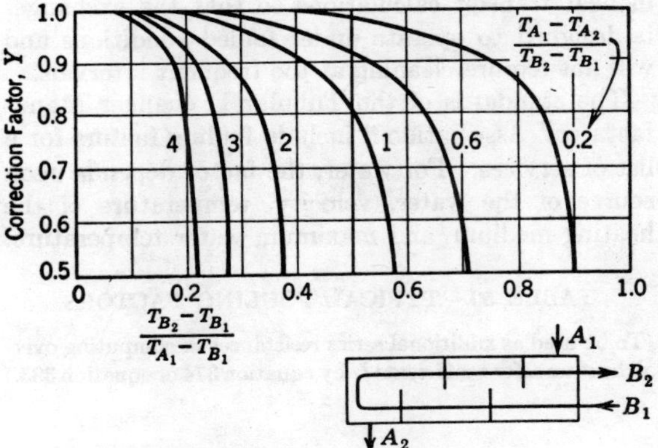

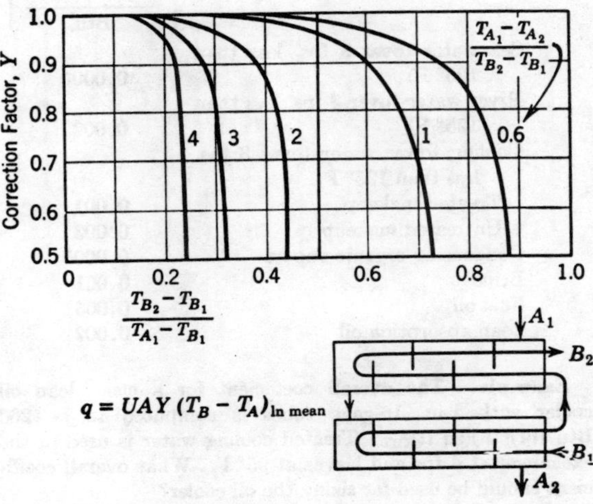

FIG. 419. Values for Y factors in equation 402, or for determining the effective mean temperature difference in shell and tube exchangers using $\log_e$ mean temperature difference.[4]

FOULING FACTORS

Heat transfer rates through dirty equipment are less than through clean equipment. A scale deposit on any heat transfer surface adds another resistance to the transfer of heat. If the thickness and thermal conductivity of the scale are known, the resistance of the scale can be computed in a manner similar to that used for a pipe wall. In service, heat exchangers become dirty or "fouled" and must be cleaned by partially dismantling and mechanically removing the deposits or by circulating solvents through the unit. Generally, the fouling of a heat transfer surface is described by a fouling factor for a given service, which factor is added as an additional resistance to heat transfer that should be included in heat transfer calculations so that the exchanger is designed to operate under fouled conditions and will not require cleaning at too frequent intervals.

The standards of the Tubular Exchanger Manufacturers' Association [44] include fouling factors for a list of services. For water, the factor depends upon source of the water, velocity, temperature of the heating medium, and maximum water temperature.

TABLE 50. TYPICAL FOULING FACTORS

(To be used as additional series resistances in computing over-all heat transfer coefficients U by equation 374 or equation 383.)

	Fouling Factor, R, $\left[\frac{(\text{hr})(°\text{F})(\text{sq ft})}{\text{Btu}}\right]$
Sea water (over 3 fps, less than 125° F)	0.0005
River water (over 3 fps, less than 125° F)	0.002
Cooling tower water (over 3 fps, less than 125° F)	
Treated makeup	0.001
Untreated makeup	0.003
Condensing organic vapors	0.0005
Brine	0.001
Fuel oil	0.005
Lean absorption oil	0.002

Example. The overall coefficient for a clean lean oil cooler with 1-in. 16-gage tubes is computed to be 260 Btu/(hr)(°F)(sq ft)$_{OD}$. Treated cooling water is used in the exchanger at 4 fps and leaves at 95° F. What overall coefficient should be used for sizing the oil cooler?

Solution.

$$\frac{1}{U_{\text{fouled}}} = \frac{1}{260} + R_{(\text{oil})} + R_{(\text{water})}\frac{1.00}{0.87}$$

$$\frac{1}{U_{\text{fouled}}} = 0.00385 + 0.002 + 0.001 \times \frac{1.00}{0.87}$$

$$= 0.007\,\frac{(\text{hr})(°\text{F})(\text{sq ft})_{\text{OD}}}{\text{Btu}}$$

$$U_{\text{fouled}} = 143\,\frac{\text{Btu}}{(\text{hr})(°\text{F})(\text{sq ft outside surface})}$$

PROBLEMS

1. A furnace wall is constructed of firebrick, 6 in. thick. The temperature of the inside of the wall is 1300° F, and the temperature of the outside of the wall is 175° F. If the mean thermal conductivity of the brick under these conditions is 0.17 Btu/(hr)(°F/ft)(sq ft), what is the rate of heat loss through 10 sq ft of wall surface?

1*a*. If the coefficient of the heat transfer between the outside of the wall and the air is 2.2 Btu/(hr)(°F)(sq ft), the inside temperature is 1300° F, and the air temperature is 70° F, what is the rate of heat loss through 10 sq ft of wall surface? What is the outside wall temperature under these conditions?

1*b*. If the furnace wall were constructed of an outside layer of firebrick 2 in. thick and an inside layer of chrome brick 4 in. thick [thermal conductivity = 0.80 Btu/(hr)(°F/ft)(sq ft)], what would be the answers to Problem 1*a*? Also what is the temperature between the two layers of brick?

2. A furnace wall is constructed of firebrick, insulating brick, and common brick, each 4 in. thick. The temperature on the inside surface is 1600° F and on the outside is 95° F. The resistance of the joints may be neglected. Conductivity k of firebrick = 0.7, of insulating brick 0.046, and of common brick 0.4, in Btu ft/(hr)(°F)(sq ft).

(*a*) If the furnace has 450 sq ft of wall of this type, how many Btu's are lost per 24-hr day by conduction through the walls?

(*b*) If a thermocouple were placed in the center of the insulation brick, what temperature should it indicate?

3. Heat is being transferred under steady conditions through a shape of fire clay whose thermal conductivity varies with temperature according to the following table.

Temperature, °F	Thermal Conductivity, k, Btu/(hr)(sq ft)(°F/ft)
392	0.58
1112	0.85
1832	0.95
2552	1.02

The shape is 2 ft thick. The large end face of the shape (7 in. by 5 in.) is maintained at a temperature of 2200° F, and the small end face of the shape (4 in. by 3 in.) is maintained at a temperature of 250° F. Calculate the rate of heat transfer through the shape if all the other faces of the shape are perfectly insulated.

4. Compute the total length of 1-in. OD 18-gage plain copper tubes required to condense 20,000 lb/hr of Freon 12 at its saturation temperature of 90° F. The cooling water enters the tubes at 60° F with a velocity of 4 fps and leaves at 64° F. The latent heat of condensation of Freon 12 is 59 Btu/lb; the overall coefficient of heat transfer may be taken from Fig. 406.

5. In a petroleum refinery propane is recovered from a stream containing 0.30 mole fraction propane, the balance being n-butane, by distillation at 250 psig. The feed, 1800 gph, is supplied at 60° F. The distillate contains 95 per cent of the propane in the feed and is 99 per cent propane. A shell and tube heat exchanger is available for preheating the feed by exchange with the bottoms from the still. The exchanger has thirty 1-in. OD tubes 8 ft long, single pass on shell side and two pass on tube side. An overall heat transfer coefficient of 100 Btu/(hr)(sq ft)(°F) had been predicted with the cold stream flowing through the tubes. Assuming the heat losses to be negligible, what is the expected temperature of the preheated feed?

6. A shell and tube exchanger is designed for vaporizing n-butane at 300 psig, using condensing Dowtherm at atmospheric pressure (500° F) as a source of heat. Reasonable estimates of the condensing and boiling coefficients are 130 and 150 Btu/(hr)(sq ft)(°F), respectively. The exchanger is to be insulated with 2 in. of 85 per cent magnesia. The ambient air temperature is taken as 0° F. The tentative design of the exchanger calls for thirty ¾-in. OD 14-gage mild steel tubes, 8 ft long, welded into tube sheets 1 in. thick. The shell is to be fabricated from 8-in. schedule-40 pipe. Considering normal on-stream operation and start-up conditions when the butane may be vaporizing at 100° F, determine whether fixed or floating tube sheets are required.

7. Tallow is stored in vertical tanks 10 ft in diameter and 30 ft high which are insulated with 2 in. of 85 per cent magnesia. To prevent freezing of the tallow (melting point 112° F) a heating coil of 2-in. schedule-40 pipe is to be installed in each tank. Steam at 5 psig is to be used for heating. The tanks are exposed to temperatures as low as 0° F. Compute the required length of pipe for each tank, using the following heat transfer coefficients.

Steam condensing inside pipe	800 Btu/(hr)(°F)(sq ft)
Pipe wall to molten tallow	40
Outer surface of tank to surroundings	2

CHAPTER 29

Heat Transfer 2

Transfer Coefficients between Fluids and Tubes

THE most common type of heat transfer encountered by the engineer involving convection coefficients is the transfer of heat between a fluid and the metal walls of the tube containing the fluid. The mechanism of heat transfer from the fluids to the metal surface depends largely upon whether the fluids are in turbulent flow or in "streamline" or viscous flow. For this reason certain factors have a more marked effect upon the rate of heat transfer in one type of flow than in the other. The more common type of flow is turbulent motion, and this is discussed first.

FLUIDS INSIDE TUBES

This section is concerned with heat transfer by the combined mechanism of conduction and convection between clean, smooth metal tubes and fluids flowing in turbulent flow inside the tubes.

Most of the experimental data reported in the literature are presented as individual coefficients h and correlated into relations expressing the coefficient as a function of the flow characteristics and the physical properties of the stream.

Since fluid friction and convection heat transfer both involve the exchange of energy between a flowing stream and a surface—in one case, kinetic energy, and in the other, heat energy—both are related to the flow properties of the stream. It will be recalled from the study of fluid friction (Chapter 12) that the flow properties are the diameter and the velocity of the stream, the density and viscosity of the fluid flowing, and the length of the conduit. All these factors influence the turbulence of the stream. Heat transfer also depends upon the thermal conductivity and specific heat of the fluid. Therefore, for convection heat transfer,

$$\frac{q}{A \cdot \Delta T} = h = \Phi(D, v, \rho, \mu, L, k, C_P) \qquad (403)$$

If h is expressed as a function of seven physical variables, as in equation 403, the fundamental dimensions selected must all be independent, for there are no dimensional constants included. On the other hand, if the systems I and II, Table 25, are to be used for the analysis, then

$$h = \Phi_1(D, v, \rho, \mu, L, k, C_P, g_c) \qquad (404)$$

It makes no difference whether equation 403 or 404 is selected, except that the final result may have the dimensional constant if a mixed system is chosen. The dimensional constant may drop out if each of the dimensions which it relates appears more than once; in such cases no error is introduced by omitting it from the analysis, but this omission is discouraged. Equation 404 will be used for this analysis, with density in pounds mass per unit volume and other units in accordance with system II, Table 25. For a given point condition, the unknown function yielding h may be written in exponential form.

$$h = zD^a v^b \rho^c \mu^d L^e k^i C_P{}^j g_c{}^n \qquad (404a)$$

The dimensionless coefficient z and all the exponents are of given values only at the point condition; if any one condition, such as velocity, is changed, then

z and all the exponents may change. This exponential equation simply states that there is a relation between h and the variables and dimensional constant listed on the right.

Equation 404a sheds no light on the form of the function. Since it must be dimensionally homogeneous, if all pertinent variables have been included, it will be useful in establishing the relative size of the various exponents necessary to preserve that dimensional homogeneity. The sum of the exponents on terms involving any one dimension must be the same on both sides of the equation. One may substitute the fundamental dimensions into equation 404a, obtaining

$$\frac{F}{LtT} = z(L)^a\left(\frac{L}{t}\right)^b\left(\frac{m}{L^3}\right)^c\left(\frac{Ft}{L^2}\right)^d(L)^e\left(\frac{F}{tT}\right)^i\left(\frac{L^2}{t^2T}\right)^j\left(\frac{mL}{t^2F}\right)^n \quad (405)$$

Since the net dimensions on each side must be the same, the sum of the exponents involved on any one dimension must be the same on both sides of the equation, or,

Considering F: $1 = d + i - n$

Considering L:

$-1 = a + b - 3c - 2d + e + 2j + n$

Considering t: $-1 = -b + d - i - 2j - 2n$

Considering T: $-1 = -i - j$

Considering m: $0 = c + n$

Here are five equations, involving eight unknowns. If the five equations are all independent, as they are in this case, it will be possible to solve for a maximum of five unknowns in terms of the remaining three. If the exponents b, e, and j are chosen as the independent ones, all other exponents may be determined as functions of these, giving

$$a = b - e - 1$$

$$c = b$$

$$d = j - b$$

$$i = 1 - j$$

$$n = -b$$

Substituting in equation 404a,

$$n = zD^{(b-e-1)}v^b\rho^b\mu^{(j-b)}L^e k^{(1-j)}C_P{}^j g_c{}^{-b} \quad (406)$$

Dimensional homogeneity is preserved if the variables are collected into groups having like exponents. These groups are as follows.

Variables having numerical exponents: $\dfrac{hD}{k}$

Variables having b in exponents: $\dfrac{Dv\rho}{\mu g_c}$

Variables having e in exponents: $\dfrac{L}{D}$

Variables having j in exponents: $\dfrac{\mu C_P}{k}$

Each of these groups is dimensionless; hence the general relationship (equation 404) can now be written

$$\frac{hD}{k} = \Phi_2\left[\frac{Dv\rho}{\mu g_c}, \frac{L}{D}, \frac{\mu C_P}{k}\right] \quad (407)$$

Equation 407 gives all the information that can be obtained from dimensional analysis of this problem. The nature of the function Φ_2 is in no way indicated. Although equation 407 may not appear at first to be particularly useful, it is a guide to the procedure to be followed in determining the function from experimental data in the absence of a solution from theoretical considerations, and the number of variables involved has been reduced from a total of eight to four.

In the case of heat transfer by convection, equation 407 shows, for example, that, if a series of experiments is run varying only in the velocity and the measured values of h, the relation of the dimensionless groups hD/k and $Dv\rho/\mu g_c$ is determined for the particular values of the other dimensionless groups. By making another series of runs at varying velocities with a different set of values of the other dimensionless groups, the relation is determined for the new conditions. If the form of the relation of hD/k to each of the dimensionless groups in turn is independent of the values of the remaining groups, equation 407 may be simplified considerably to

$$\frac{hD}{k} = \Phi_3\left(\frac{Dv\rho}{\mu g_c}\right)\Phi_4\left(\frac{\mu C_P}{k}\right)\Phi_5\left(\frac{L}{D}\right) \quad (408)$$

Use of a mixed system of units involving both force and mass results in retention of the conversion factor g_c in the equation. If both density and viscosity are defined in terms of either mass pounds or force pounds, the conversion factor is not retained.

Most engineering literature uses density in pounds mass per cubic foot, and viscosity in pounds mass per foot-second (or hour), hence avoiding the necessity for using g_c. This convention is followed here, and the Reynolds number is written as $Dv\rho/\mu$. Conversely, if an energy unit E, as in system III, Table 25, is used for one or two of the variables, h, k, and C_P, while foot-pounds (force) are used for the others, the conversion factor J will appear in that group containing both units. Chemical engineering literature uses almost exclusively an energy unit for all three of these quantities.

It is frequently convenient, and is permissible, to use different units for a variable in different groups. Many engineers work in terms of velocities in feet per *second*; if this is used in the group $Dv\rho/\mu$, then the time unit in viscosity must be in *seconds*. This decision does not preclude use of thermal conductivity on an *hourly* basis, but then the time unit in the viscosity term of the group $\mu C_P/k$ must be in *hours*. In fact, metric units may be used within one dimensionless group, and English in the remainder without error. The important consideration is that each group must in itself be dimensionless.

If any other set of independent exponents than b, e, and j is chosen, a different set of dimensionless groups will result. These are convertible one to another by intermultiplication, and, in accordance with the theory of functions, any such set of groups may be used, depending on the requirements. Many of these groups are known by the names of investigators in the fields of fluid friction and heat transfer as follows.

Group	Name	Symbol
$\frac{Dv\rho}{\mu}$ or $\frac{DG}{\mu}$	Reynolds number	Re
$\frac{h}{v\rho C_P}$ or $\frac{h}{C_P G}$	Stanton number	St
$\frac{hD}{k} = (\text{St})(\text{Re})(\text{Pr})$	Nusselt number	Nu
$\frac{\mu C_P}{k}$	Prandtl number	Pr
$\frac{Dv\rho C_P}{k} = (\text{Re})(\text{Pr})$	Peclet number	Pe
$\frac{WC_P}{kL}$	Graetz number	Gz

If it is assumed that the function, equation 408, can be expressed as a simple exponential function, using consistent units for density and viscosity and eliminating g_c, it may be written in the form

$$\frac{hD}{k} = z\left(\frac{Dv\rho}{\mu}\right)^a \left(\frac{\mu C_P}{k}\right)^b \left(\frac{L}{D}\right)^c \tag{409}$$

The soundness of this assumption can be demonstrated only by its ability to correlate experimental data. The minimum of experimental data is required if three series of experiments are made: one determining the relation of the Nusselt number (or Stanton number) with the Reynolds number, while the other groups are maintained constant; another determining the relation of the Nusselt number with the Prandtl number, with the other groups constant; and a third determining the relation to L/D.

The form presented in equation 409 is most useful in studying the influence of changes in velocity on the heat transfer coefficient in a given system, since each of these variables appears only once. In determining the constants z, a, b, and c from experimental data, the normal procedure involves equating the heat transferred to the gain in enthalpy of the fluid. With the proper mean temperature difference $(-\Delta T_m)$ substituted for $T_1 - T_2$ in equation 382,

$$q = hA(-\Delta T_m) = GA_F C_P(T_{\text{out}} - T_{\text{in}}) \tag{410}$$

where $G = v\rho$, the mass velocity.

A_F = cross-sectional area for the flowing material.

Rearranging,

$$\frac{h}{C_P G} = \frac{A_F(T_{\text{out}} - T_{\text{in}})}{A(-\Delta T_m)} \tag{410}$$

Thus the Stanton number h/C_PG is evaluated directly from experimental data as indicated in equation 410. The Stanton number, St, can be derived by intermultiplication of the groups: $\text{Nu} \times \text{Re}^{-1} \times \text{Pr}^{-1}$, and equation 409 becomes

$$\frac{h}{C_P G} = z\left(\frac{DG}{\mu}\right)^{(a-1)} \left(\frac{\mu C_P}{k}\right)^{(b-1)} \left(\frac{L}{D}\right)^c \tag{411}$$

The values for the physical properties of fluids vary with temperature as shown in the appendix, and temperature is not constant either across or along the tube. The most convenient temperature for evaluating the physical properties is the arithmetic average of the inlet and outlet bulk temperatures of the fluid. The bulk temperature is the temperature that would be indicated if the flowing stream at any point were thoroughly mixed.

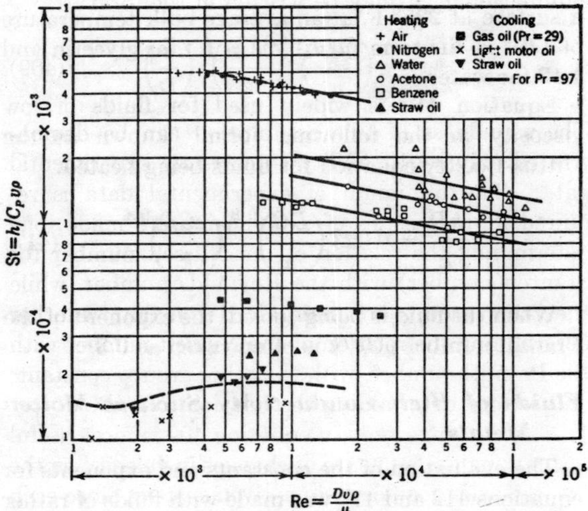

FIG. 420. Plot of experimental data on fluids being heated or cooled in tubes in which the Stanton number ($h/\rho v C_P$) is plotted as a function of the Reynolds number ($Dv\rho/\mu$) for selected constant values of the Prandtl number ($\mu C_P/k$) and L/D.

The soundness of the assumption of an exponential relation and the value of the exponent of the Reynolds number in equation 411 are then determined by plotting the Stanton number as a function of the Reynolds number for selected constant values of the two other groups ($\mu C_P/k$) and (L/D) on logarithmic cross-sectional paper as indicated in Fig. 420 for fluids being heated. Since the data are represented by straight lines, all having the same slope (-0.2), the use of equation 411 is justified and the value of the exponent ($a - 1$) is -0.2. This is made clear by writing the equation in the logarithmic form,

$$\log \frac{h}{C_P G} = \log z + (a - 1) \log \left(\frac{DG}{\mu}\right) + (b - 1) \log \left(\frac{\mu C_P}{k}\right) + c \log \left(\frac{L}{D}\right) \tag{411}$$

The fact that a single line on Fig. 421 represents all the data with Reynolds number greater than 6000 for various values of L/D indicates that the last group (L/D) has no significance for fully developed turbulent flow (Re > 6000).

The exponent ($b - 1$) of the Prandtl number $\mu C_P/k$ is then determined in a similar manner by plotting (Fig. 421) [$\log (h/C_P G) + 0.2 \log (DG/\mu)$] as a function of $\log (\mu C_P/k)$ for the same data used in Fig. 420. These data are best represented by a line of a slope of -0.6 and intercept of 0.023 at

$$\log \left(\frac{\mu C_P}{k}\right) = 0 \quad \text{or} \quad \frac{\mu C_P}{k} = 1$$

Therefore, these experimental data for fluids being heated may be represented by the following equation.

$$\frac{h}{C_P G} = 0.023 \left(\frac{DG}{\mu}\right)^{-0.2} \left(\frac{\mu C_P}{k}\right)^{-0.6} \tag{412}$$

or more conveniently

$$\frac{h}{C_P G} = 0.023 \left(\frac{\mu}{DG}\right)^{0.2} \left(\frac{k}{\mu C_P}\right)^{0.6} \tag{412a}$$

If the fluid is being cooled instead of heated, a similar procedure leads to the following correlation.

$$\frac{h}{C_P G} = 0.023 \left(\frac{\mu}{DG}\right)^{0.2} \left(\frac{k}{\mu C_P}\right)^{0.7} \tag{413}$$

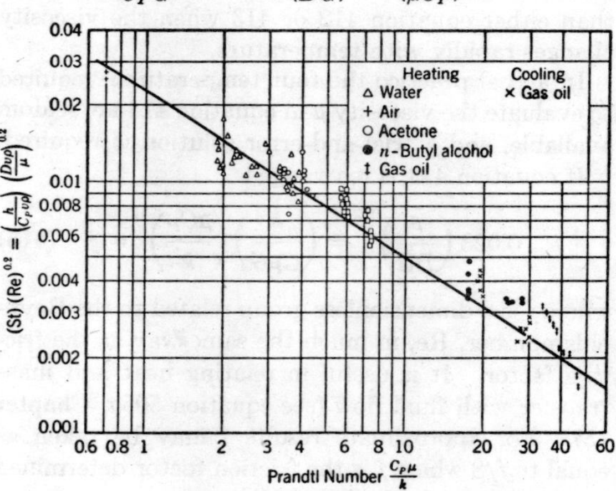

FIG. 421. Plot of data on fluids being heated or cooled inside tubes in which $(\text{St})(\text{Re})^{0.2}$ is plotted as a function of the Prandtl number ($\mu C_P/k$).

The lines of Figs. 420 and 421 representing equation 412 reproduce about 90 per cent of the data within a deviation of about ± 10 per cent.

Temperature traverses in turbulent streams being heated or cooled indicate the existence of:

1. A laminar or viscous-flowing layer of fluid adjacent to the surface.

2. A transition zone adjacent to the viscous layer in which some turbulence exists.

3. The core of the stream in which turbulence is fully developed.

Since the major part of the resistance to heat transfer is in the viscous layer, it would appear better to evaluate the physical properties at a temperature more representative of the viscous layer than the bulk temperature. Colburn [6] * suggested that this temperature be chosen as the mean of the average bulk temperature and the average surface temperature. The use of physical properties, particularly the viscosity, evaluated at this temperature leads to the following correlation.

$$\frac{h}{C_P G} = 0.023 \left(\frac{\bar{\mu}}{DG}\right)^{0.2} \left(\frac{k}{\bar{\mu} C_P}\right)^{2/3} \qquad (414)$$

where $\bar{\mu}$ = viscosity determined at the described mean temperature.

Equation 414 appears to apply equally well to fluids being heated or cooled and is generally better than either equation 412 or 413 when the viscosity changes rapidly with temperature.

In actual practice the four temperatures required to evaluate the viscosity $\bar{\mu}$ in equation 414 are seldom available, and a trial-and-error solution is required.

If equation 414 is rearranged

$$0.023 \left(\frac{\bar{\mu}}{DG}\right)^{0.2} = \left(\frac{h}{C_P G}\right)\left(\frac{\bar{\mu} C_P}{k}\right)^{2/3} = j \qquad (415)$$

where j is a dimensionless group related to the Reynolds number, Re, in much the same way as the friction factor. It is useful in relating heat and mass transfer with fluid flow (see equation 505*a*, Chapter 35). For approximate results j may be taken as equal to $f/8$ where f is the friction factor determined for the flow of fluids (Fig. 125).

A more convenient method [42] for including the effect of radial variations in viscosity due to temperature gradients, which is reliable when Re is greater than 8000 where the effect of the ratio of length to diameter becomes negligible, is

$$\frac{h}{C_P G} = 0.027 \left(\frac{\mu}{DG}\right)^{0.2} \left(\frac{k}{\mu C_P}\right)^{2/3} \left(\frac{\mu}{\mu_s}\right)^{0.14} \qquad (416)$$

where μ = viscosity at average bulk temperature of fluid.

μ_s = viscosity at average temperature of the heating or cooling surface.

The importance of the viscosity variation with radial temperature gradient for fluids having a rapid change in viscosity with temperature is emphasized by comparing the $(\mu/\mu_s)^{0.14}$ term for water and glycerin. When each fluid is heated by contact with a surface at 225° F, at an average bulk temperature of 140° F, the term $(\mu/\mu_s)^{0.14}$ is 1.4 for glycerin and 1.12 for water.

Equation 412 is widely used for fluids of low viscosity in the following form [10] known as the Dittus-Boelter equation for fluids being heated.

$$\frac{hD}{k} = 0.023 \left(\frac{DG}{\mu}\right)^{0.8} \left(\frac{\mu C_P}{k}\right)^{0.4} \qquad (417)$$

When the fluid is being cooled, the exponent of the Prandtl number $\mu C_P/k$ is often taken as 0.3.

Fluids of High Conductivity Such as Molten Metals

The evaluation of the constants and exponents for equations 412 and 417 was made with fluids of rather low conductivity, with a Prandtl number in the range of about 1 to 5. When molten metals such as mercury or sodium, of high thermal conductivity and low Prandtl number (Pr < 0.1), are transferring heat, the thermal conductivity is more significant than the degree of turbulence in affecting heat transfer.[30] The influence of the turbulence on the coefficient for low-conductivity fluids (high Pr), as indicated by the 0.8 power of the Reynolds number, does not apply to molten metals.

By making necessary assumptions and using the velocity profile of the fluid, a relationship has been derived between the dimensionless groups, Nu, Re, and Pr, which includes fluids of low Pr. For substances whose Prandtl numbers are less than 0.1, the following equation is recommended.[28]

$$\text{Nu} = 7 + 0.025(\text{Pe})^{0.8} \qquad (417a)$$

Laminar Flow

At low Reynolds numbers, when the flow is laminar or viscous, the ratio D/L becomes significant and must be included in the correlation. The following equations were obtained by the methods described above when extended to include experimental data at low Reynolds numbers.

$$\frac{h}{C_P G} = 1.86 \left(\frac{\mu}{DG}\right)^{2/3} \left(\frac{k}{\mu C_P}\right)^{2/3} \left(\frac{\mu}{\mu_s}\right)^{0.14} \left(\frac{D}{L}\right)^{1/3} \qquad (418)$$

or

$$\frac{hD}{k} = 1.86 \left(\frac{DG}{\mu}\right)^{1/3} \left(\frac{\mu C_P}{k}\right)^{1/3} \left(\frac{D}{L}\right)^{1/3} \left(\frac{\mu}{\mu_s}\right)^{0.14} \qquad (418a)$$

* The bibliography for this chapter appears on p. 472.

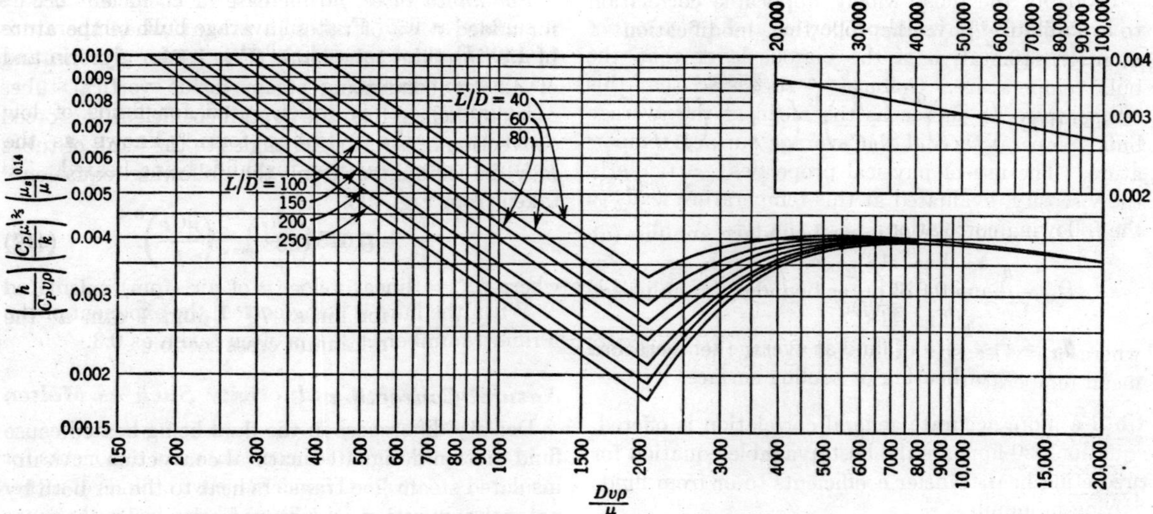

Fig. 422. Correlation of experimental results on heating and cooling single-phase fluids flowing inside tubes.[42]

When the flow is neither clearly viscous nor turbulent, particularly in short tubes where disturbances due to entry may be significant, it is impractical to develop formal expressions for the coefficients, and graphical representations of the experimental results must be relied upon, as given in Fig. 422.

Simplified Equation for Isothermal Conditions

In the case of a particular fluid, the properties, viscosity, specific heat, and thermal conductivity may be combined and expressed as a single function of temperature over the limited range of temperature usually encountered for convection transfer inside tubes.

Rewriting equation 417 for completely turbulent flow,

$$h = 0.023\left(\frac{G^{0.8}}{D^{0.2}}\right)\left(\frac{C_P{}^{0.4}k^{0.6}}{\mu^{0.4}}\right) \qquad (417b)$$

The last term may be expressed as a function of temperature, and for any particular temperature of a particular fluid the last term is a constant which may be combined with the factor 0.023 in the constant a in the equation

$$h = a\frac{G^{0.8}}{D^{0.2}} \qquad (419)$$

Exercise. Compute the value of a in equation 419 for benzene at 100° F.

Exercise. Plot the value of a for water as a function of temperature for the range 60° to 200° F.

FLUIDS OUTSIDE TUBES

The relationships developed for convection coefficients for fluids flowing inside tubes have proved fairly reliable, and numerous efforts have been made to apply these relationships to the estimation of convection coefficients for fluids flowing outside tubes. The first question that arises in such use of these equations is the value to be used for D. Usually an equivalent diameter D_e is substituted for D in equations 409 to 417, but additional modifications of the equations are required to obtain reasonably satisfactory results when applied to fluids outside tubes.

Forced Convection

In predicting heat transfer to or from a fluid flowing in the annular space between concentric pipes, the principal problem hinges around the selection of a linear dimension to characterize the system. The original correlations were based on the equivalent diameter applicable to fluid flow, which becomes, in terms of Fig. 418, $x_5 - x_3$. It has been proposed that the equivalent diameter would be more accurately expressed if related to the ratio between the cross-sectional area and the heated perimeter, as $(x_5{}^2 - x_3{}^2)/x_3$. Neither of these "equivalent diameters" appears to be generally applicable with dependable results. Predictions of heat transfer coefficients based on equation 412, 416, or 417, with the equivalent diameters indicated, will in most instances be below the measured values.

Probably the most widely applicable correlation now available[9,22] is the following modification of equation 416.

$$\frac{h}{C_P G} = 0.029 \left(\frac{\mu}{D_i G}\right)^{0.2} \left(\frac{k}{\mu C_P}\right)^{2/3} \left(\frac{\mu}{\mu_s}\right)^{0.14} \left(\frac{D_o}{D_i}\right)^{0.15} \quad (420)$$

where D_i = diameter of inner boundary annulus (at x_3, Fig. 418).
D_o = diameter of outer boundary annulus (at x_5, Fig. 418).
μ_s = viscosity of fluid at average temperature of heating or cooling surface.

Until a more accurate general correlation is offered, equation 420 appears the best available equation for predicting heat transfer coefficients to or from fluids flowing in annuli.

Convection coefficients for heat transfer to or from a *fluid flowing at right angles to banks of plain tubes* may be estimated by a further modification of these equations. For fluids in forced convection perpendicular to *a single row of tubes spaced on two diameters*, the following equation[14,31] derived from data on air and gases may be used.

$$\frac{hD}{k} = 0.21 \left(\frac{DG_{max}}{\mu}\right)^{0.6} \left(\frac{\mu C_P}{k}\right)^{0.333} \quad (421)$$

where D = outside diameter of tube.
G_{max} = maximum mass velocity, or mass velocity at minimum cross section.

All properties of fluid are taken at the mean temperature between fluid and wall, $\frac{1}{2}(T_{fluid} + T_{wall})$. The constants and exponents of equation 421 depend upon the arrangement of the tubes.[14]

For staggered plain tubes spaced on equilateral triangles of two diameters, the constant (0.21) in equation 421 has the values given below:

Rows of Tubes	Constant in Equation 421
1	0.21
2	0.23
3	0.27
5	0.30
10 or more	0.33

This increase in coefficient is due to the turbulence at the succeeding rows of tubes caused in the fluid passing the earlier row of tubes.

For finned tubes, no increase in coefficient occurs for added rows of tubes in a bank when the air is blown into the tube bank (Fig. 399). For heating of air by condensing steam inside vertical tubes carrying helical fins, the overall coefficient for forced convection of air at right angles to the bank of tubes may be estimated from the following equation[23] taken from Fig. 400.

$$U = 0.308 v_{max}^{0.53} \quad (422)$$

where v_{max} = linear velocity of air (fpm) calculated for air at 70° F and 1 atm at the minimum cross section.

Natural Convection

Density differences in the fluid being heated cause fluid motion designated natural convection. An uninsulated steam line transfers heat to the air both by natural convection and by radiation. The transfer of heat to air by convection from cylinders or plates is given by the following equations.[31]

For cylinders,

$$h = 0.27 \left(\frac{T_s - T_{air}}{D}\right)^{0.25} \quad (423)$$

For vertical plates over 3 ft high,

$$h = 0.3(T_s - T_{air})^{0.25} \quad (424)$$

where T_s = temperature of surface of cylinder or plate (°F).
D = outside diameter of cylinder (ft).

Equation 423 is to be used cautiously. It may be unsound at diameters larger than 1 ft. If the two equations were consistent equation 423 would become identical with equation 424 at D = infinity.

PROBLEMS

1. A method has been suggested for making lead shot involving heating the lead to well above its melting point and forcing it through a multiple nozzle under a pressure of 250 psig. Give the process design for conveying continuously 1 ton of lead per hour from the melting furnace to the nozzle. Suggest an operating temperature for the metal leaving the furnace.

Temperature, °F	Viscosity, centipoises	Approximate Density, grams/cc	Specific Heat, Btu/(lb)(°F)
620	.....	11.2	.....
825	2.12	11.0	0.041
1023	1.70	10.8	.....

The melting point of lead is 620° F.

2. A new type of insulation is suggested for cabins of airplanes. It is constructed of three sheets of bright aluminum, each 0.04 in. thick between which are two air gaps, each 0.10 in. thick.

Compute the rate of heat transfer through the insulation in Btu/(hr)(sq ft) for a vertical section of this covering with an outside metal temperature of −40° F and an inside air and surroundings temperature of 60° F. The air gaps may be assumed to carry heat by conduction rather than convection.

3. A shell-and-tube exchanger of the dimensions given below is to be used to heat 60,000 lb/hr of an aqueous solution (whose physical properties may be considered as substantially equal to those of water) before the solution is sent to a reaction system. Saturated steam at 5 psig is available as the heating medium and will be condensed on the outside of the tubes. From previous experience, the steam-side coefficient may be assumed constant and equal to 2000 Btu/(hr)(°F)(sq ft).

Determine the temperature of the solution leaving the exchanger if the inlet temperature is 60° F.

Dimensions of Exchanger

Number of tubes	52 horizontal copper
Tube length	6 ft
OD of tubes	1 in.
Thickness of tubes	16 Birmingham wire gage
Number of passes (tube side)	4

4. An oil stream enters a 12-ft heated section of 1-in. steel pipe, schedule 30, at a rate of 674 gph and at a temperature of 100° F. The 12-ft section of pipe is in a furnace whose temperature is 1800° F. Estimate the exit temperature of the oil in degrees Fahrenheit.

The composite coefficient for radiation and convection heat transfer between the furnace and the outside surface of the pipe = 14.0 Btu/(hr)(°F)(sq ft). The convection coefficient of heat transfer between the pipe and the oil is

$$h_c = 0.2(350 + 0.01T)v^{0.8}$$

where h_c = convection coefficient, pipe wall to the oil [Btu/(hr)(°F)(sq ft)].
T = average oil temperature (°F).
v = average linear velocity of oil (fps).

The specific gravity of the oil is $0.7 - 0.0001T$, where T = average oil temperature (°F). The specific heat of the oil is 0.8 Btu/(lb)(°F).

5. In an oil-processing plant, cottonseed oil is heated in a shell-and-tube exchanger by the condensation of a mixture of diphenyl oxide and diphenyl (Dowtherm). The oil flows inside the tubes, and a recent inspection of the shell indicates that the shell-side operating pressure must not exceed 1 atm, absolute.

To what temperature will the oil be heated under the following conditions?

Exchanger

Effective length of a tube	7 ft 9 in.
Total number of tubes	36
Number of tubes per pass	6
Tube material	18-8 Cr-Ni alloy steel
Size of tubes	¾-in. OD, 0.065-in. wall

Dowtherm

Condensing temperature at 0 psig	500° F
Heat transfer coefficient based on OD	$150 \dfrac{\text{Btu}}{(\text{hr})(°\text{F})(\text{sq ft})}$

Cottonseed Oil

Flow rate (lb/hr)	7500
Inlet temperature (°F)	250
Density gm./ml.	$0.855 - 0.0004T$ (°F)
Thermal conductivity, [Btu/(hr)(°F/ft)(sq ft)]	$0.078 - 0.000024T$ (°F)
Specific heat [Btu/(lb)(°F)]	$0.462 + 0.00045T$ (°F)
Viscosity, centistokes	38.8 at 100° F
	8.3 at 210° F
	4.8 at 260° F
	3.0 at 320° F
	2.0 at 400° F

6. An experimental double-pipe heat exchanger consists of a horizontal copper pipe (ID = 0.593 in., OD = 0.750 in., length = 9.03 ft) inside and concentric with a 4-in. schedule-40 steel pipe. The outside of the 4-in. pipe is well insulated with 85 per cent magnesia. The temperature of the outer surface of the copper pipe is measured by means of fine copper constantan thermocouples attached to the surface. Condensing steam is used on the outside surface of the copper pipe, while the fluid under investigation flows through the copper pipe The results for two runs in which water was flowing through the copper pipe are given below.

Run 1

Inlet water temperature	39.5° F
Outlet water temperature	82.5° F
Average water velocity through pipe	15.30 fps
Temperature of condensing steam	244.3° F
Heat given up by steam	299,000 Btu/hr
Mean temperature of outer surface of copper pipe	151.7° F

Run 2

Inlet water temperature	164.1° F
Outlet water temperature	197.2° F
Average water velocity through pipe	3.21 fps
Temperature of condensing steam	222.4° F
Heat given up by steam	47,700 Btu/hr
Mean temperature of outer surface of copper pipe	211.1° F

(*a*) Calculate the heat transfer coefficient between the water and the inside of the copper pipe for each run.

(*b*) Compare these experimentally determined coefficients with values obtained from equation 417.

7. An experimental double-pipe heat exchanger was constructed by using an inner pipe of Karbate No. 2 (1½-in. OD, 1-in. ID) 60 in. long (effective heating length = 48 in.) inside a 2-in. schedule-80 steel pipe. The data reported below were obtained during a series of runs in which water was used on both sides, and the fluid flowing through the inner pipe was cooled.

EXPERIMENTAL DATA

Run No.	Average Water Temperature, °C *		Temperature Change, °C		Flow Rates, lb/hr		Heat Transfer, Btu/hr based on annulus flow	Log Mean Temperature Difference, °C
	Inner Pipe	Annulus	Inner Pipe	Annulus	Inner Pipe	Annulus		
5	40	26	−0.60	1.13	24,900	14,300	29,000	13.7
9	50	28	−1.10	1.80	26,200	14,400	46,700	21.4
43	63	36	−5.50	0.95	3,070	14,900	25,500	22.4
44	62	30	−8.35	0.84	1,420	14,700	22,200	32.0
45	68	32	−3.90	2.33	9,070	14,500	60,800	36.1

* Arithmetic average of inlet and outlet bulk temperatures.

Thermal conductivity of Karbate No. 2 = 50 Btu/(hr)(°F/ft)(sq ft).

Determine the value of the individual heat transfer coefficients for the annulus side for the flow rates considered, and compare these values with those calculated from the empirical correlations equations 417 and 420.

8. A double-pipe heat exchanger made up of 1½-in. and 2½-in. schedule-40 steel pipe has an effective heating surface of 25.8 sq ft, based on the outside surface of the inside pipe. This exchanger has been used for preheating water, and tests indicate that there is a scale deposit on the heating surface. The heat transfer coefficient for this deposit, based on the inside surface of the 1½-in. pipe, is 505 Btu/(hr)(°F)(sq ft).

It is now proposed to use this exchanger to preheat benzene from an initial temperature of 68° F by means of hot water which will enter the exchanger at 190° F. The benzene will flow through the annular space at a rate of 11,000 lb/hr; the hot water flow rate will be 12,500 lb/hr.

Determine the temperature of the benzene leaving the exchanger if countercurrent flow is to be used.

9 A series of heat transfer tests in which liquids were heated by passage through an electrically heated tube has been completed. The liquids were water, commercial butanol, and AN-E-2 ethylene glycol (composition in weight per cent: glycol = 94.5, water = 3.0, triethanolamine phosphate = 2.5).

The data obtained are given in the table below, in which the temperature reported for each run is the arithmetic mean of the inlet and outlet bulk temperatures of the liquid. The physical properties of the liquids used for calculating the various dimensionless quantities were taken at this temperature.

Evaluate the constant and the exponents, if the data are to be correlated by the relationship

$$\frac{h}{C_P G} = z\left(\frac{DG}{\mu}\right)^a\left(\frac{\mu C_P}{k}\right)^b$$

AN-E-2 ETHYLENE GLYCOL

Run No.	Average Temperature, °F	$\frac{\mu C_P}{k}$	$\frac{DG}{\mu}$	$\frac{h}{C_P G}$
476	199.5	25.2	14,900	0.00052
477	200.5	25.1	7,500	0.00057
478	197.9	25.7	21,900	0.00046
479	197.7	25.8	26,900	0.00045
480	198.7	25.5	14,600	0.00050
503	150.5	41.3	31,500	0.00033
505	150.5	41.3	25,000	0.00036
506	150.5	41.3	21,400	0.00034
508	151.0	41.1	14,400	0.00038
512	149.0	42.1	11,000	0.00037
514	151.5	40.9	6,200	0.00041

COMMERCIAL BUTANOL

Run No.	Average Temperature, °F	$\frac{\mu C_P}{k}$	$\frac{DG}{\mu}$	$\frac{h}{C_P G}$
523	151.0	18.0	17,600	0.00056
524	152.3	17.8	26,300	0.00053
525	154.0	17.5	38,100	0.00049
526	153.5	17.6	44,200	0.00047
527	153.0	17.7	50,200	0.00046
528	152.0	17.8	58,500	0.00043

WATER

Run No.	Average Temperature, °F	$\frac{\mu C_P}{k}$	$\frac{DG}{\mu}$	$\frac{h}{C_P G}$
119	152.0	2.7	38,700	0.00168
120	149.5	2.7	57,600	0.00148
121	150.5	2.7	78,500	0.00133
122	151.0	2.7	98,600	0.00126
124	149.8	2.7	137,500	0.00110
134	149.5	2.7	22,600	0.00189
348	149.8	2.7	154,800	0.00101
495	150.5	2.7	299,600	0.00081
497	151.0	2.7	232,900	0.00089
498	149.5	2.7	201,800	0.00091
501	151.0	2.7	121,300	0.00115

10. Dry air is fed at the rate of 350 lb/hr at 70° F and 0 psig to a compressor which delivers the air at 100 psig and is passed through the tubes of a single-pass shell and tube exchanger. Water enters the shell at 85° F and leaves at 90° F. The tubes have a 1-in. OD and 0.065-in. wall, spaced on equilateral triangles 1½ in. from center to center and 8 ft long. There are 60 tubes in all. The shell is equipped with ten equally spaced segmental-type baffles.

What is the temperature of the air leaving the cooler?

11. Propane at 1500 psia is being heated by passing through a horizontal ¾-in. OD copper tube, 12 gage. The propane enters at 7 fps and 100° F and leaves the tube at 175° F. Saturated steam at 2.5 psig is condensing on the outside of the tube (h = 1200 Btu/(hr)(sq ft)(°F)).

PROPANE DATA

Temperature, °F	Assumed k, (Btu)(ft) / (hr)(°F)(sq ft)	C_P, Btu / (lb)(°F)	Density, grams / cc	Viscosity, centipoises
100	0.074	0.67	0.506	0.09
125	.....	0.735	0.481	0.08
150	.....	0.83	0.460	0.072
175	0.069	1.00	0.438	0.065

Compute the length of the tube:

(*a*) By using mean properties and a log mean temperature difference.

(*b*) By integrating the instantaneous coefficients.

12. The expansion program of the Ajax Refining Company requires the installation of 1200 ft of 2-in. schedule-80 steam piping. In addition, 18 valves, 20 ells, and 26 tees are necessary. Two brands of insulation are being considered. Given the data below:

(*a*) Which insulation do you recommend and in what thickness?

(*b*) What is its installed cost?

(*c*) What quantity of steam is condensed for the specified insulation (in pounds per hour)?

Steam pressure: 400 psig (saturated).

Steam cost: $0.80 per 1000 lb.

Coefficient of heat transfer for condensing steam: 900 Btu/(hr)(°F)(sq ft).

Average temperature of air: 40° F (for design purposes).

Installation cost of insulation: $0.10 per lineal foot, $1.75 per valve or fitting.

Maintenance: 4 per cent of initial cost per year.

Amortization period: 10 years.

COEFFICIENT OF HEAT TRANSFER FROM SURFACE OF INSULATION TO SURROUNDING AIR, BTU/(HR)(SQ FT)(°F)

Diameter of Insulation, in.	Temperature Difference between Surface and Air				
	50° F	100° F	150° F	200° F	250° F
1	2.3	2.5	2.75	3.0	3.3
3	2.0	2.2	2.45	2.7	3.0
5	1.95	2.15	2.35	2.6	2.9
10	1.85	2.05	2.3	2.5	2.8

PRICE OF INSULATION

Wall Thickness, in.	Type A			Type B		
	Ells or Tees	Valves	Pipe	Ells or Tees	Valves	Pipe
1			$0.80/ft	$1.30 ea.	$2.50 ea.	$0.36/ft
1½	$ 2.50 ea.	$ 5.40 ea.	1.60	1.80	3.75	0.64
2	4.90	10.30	2.40	2.30	5.50	1.00
2½				3.20	6.05	1.25
3	7.20	12.20	3.80	4.00	7.10	1.65
4	12.60	16.80	4.20			

Thermal Conductivity, k, Btu/(hr)(sq ft)(°F/ft). T = °F
Type A = $0.03 + 0.00006T$; Type B = $0.05 + 0.00015T$

13. A single-pass shell and tube exchanger contains 20 copper tubes, 16 ft long, 1-in. OD, 0.065-in. wall, and is used to heat 12,000 lb/hr of chlorobenzene from 90° F to 200° F by saturated steam at 25 psig which is condensed in the shell

The exit temperature of the chlorobenzene must be maintained at 200° F. This is done by controlling the pressure, and hence the temperature, at which the steam is condensed on the outside of the tubes. The control is an automatic temperature controller actuated by the hot chlorobenzene and operating on a throttle valve in the steam supply.

Assume that the resistance to heat transfer of the tube walls and the condensing steam, and the viscosity correction $(\mu/\mu_s)^{0.14}$ are negligible.

Because of reduced demand, the plant capacity is to be reduced to 3000 lb/hr of chlorobenzene.

(*a*) What are your recommendations concerning this heat exchanger?

(*b*) For your recommended conditions, what is the operating steam pressure in the shell? Give the supporting arguments.

14. Air at a pressure of 100 psig is heated in a single-pass exchanger consisting of 1-in OD 16-gage copper tubes, 4 ft long. The air flows inside the tubes, entering at a velocity of 20 ft/sec, and is heated by steam condensing at 5 psig outside the tubes.

(*a*) What is the exit air temperature?

(*b*) To increase the heat exchange the tubes are filled with ⅛-in. diameter steel shot to a porosity of 0.30. Using the analogy between heat transfer and friction (equation 415) to predict heat transfer rates, what exit air temperature might be expected if the original rate of air flow is maintained?

CHAPTER

30

Heat Transfer 3

Condensing Vapors and Boiling Liquids

CONDENSATION of a saturated vapor is accomplished by bringing the vapor into contact with a surface whose temperature is below the dew point of the vapor. Condensates of organic substances commonly wet metallic surfaces, forming a film of liquid on the cold surface which is called "filmwise condensation." Water condensate usually wets metallic surfaces, but under some conditions may not and in such cases "dropwise condensation" occurs.

FILMWISE CONDENSATION

The thickness of the condensate layer in filmwise condensation depends on the configuration of the surface, the rate of condensation, and the rate at which the liquid flows from the surface. The nature of the condensate film on vertical and horizontal surfaces is shown diagrammatically in Fig. 423.

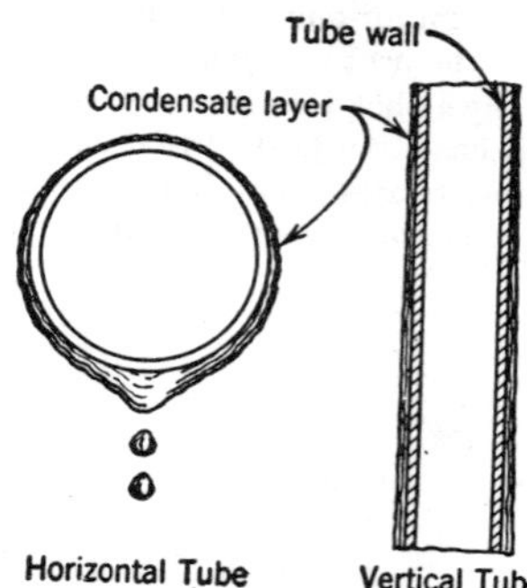

FIG. 423. Diagram representing filmwise condensation on horizontal or vertical tubes.

Saturated Vapors

The following equation for the coefficient of heat transfer for a pure saturated vapor condensing on a cold surface was derived [36] * by assuming that the vapor condenses in the form of a continuous film flowing in laminar flow down the vertical cooling surface due to effects of gravity, and that the temperature difference between the vapor and the cold surface is constant. The effects on the condensate film thickness caused by the vapor velocity or by turbulent flow of the liquid on long vertical tubes were neglected. A reduction in the film thickness would increase the condensing coefficient. *For vertical surfaces,*

$$h = 0.943 \left[\frac{k_f^3 \rho_f^2 g(\Delta H)}{L\mu_f(T_{sv} - T_s)} \right]^{1/4} \tag{425}$$

where T_{sv} = temperature of saturated vapor (° F).
T_s = temperature of surface (° F).
ΔH = latent heat of condensation (Btu/lb).
L = length of tube or vertical surface (ft).
g = acceleration due to gravity [ft/(hr)(hr)].

At standard conditions $g = 4.18 \times 10^8$ ft/(hr)(hr).

Based on the assumption that $1/\mu$ is linear in T and that there is a linear temperature gradient through the condensate film, it has been stated [31] that the mean temperature to be used in evaluating the physical properties, k_f, ρ_f, and μ_f, is

$$T_f = [T_{sv} - \tfrac{3}{4}(T_{sv} - T_s)]$$

* The bibliography for this chapter appears on p. 472.

The values predicted by equation 425 are lower than those obtained in practice. Consequently, the following modified forms of equation 425 are recommended for practical use.

$$h = 1.13 \left[\frac{k_f^3 \rho_f^2 g(\Delta H)}{L\mu_f(T_{sv} - T_s)} \right]^{1/4} \quad (426)$$

$$h = 1.18 \left[\frac{k_f^3 \rho_f^2 g \pi D}{\mu_f W} \right]^{1/3} \quad (427)$$

where D = outside diameter of tube (ft).
W = pounds of condensate per hour.
Properties of condensate taken at the mean condensate temperature.

Equations 426 and 427 are equivalent; whichever form is more convenient may be used. If the length of the vertical surface L is known, equation 426 is used; if the quantity of vapor condensed per hour is known, equation 427 is used.

For horizontal tubes, the following equation may be derived under the same assumptions that were made in obtaining equation 425.

$$h = 0.725 \left[\frac{k_f^3 \rho_f^2 g(\Delta H)}{ND\mu_f(T_{sv} - T_s)} \right]^{1/4} \quad (428)$$

where N = the number of tubes in the vertical row of tubes.

The variation in condensate thickness from the top to the bottom of the tube causes the temperature of the tube surface to vary considerably.[1] An average temperature both circumferentially and lengthwise is normally used.

Fig. 424. Diagram indicating accumulation of condensate on horizontal tubes placed in a vertical row.

The decrease in the average condensing coefficient for several tubes in a vertical row follows directly from the increase in the quantity of liquid flowing over the tube. The thickness of the condensate film increases with successive tubes until essentially only liquid condensate flows over the tubes, as indicated in Fig. 424.

Cylindrical condensers with horizontal tubes (Fig. 405) normally have a larger number of tubes in the vertical row at the center than in the rows near the sides. The average coefficient may be computed from equation 428 for the entire condenser by calculating the average value of N to be used as follows.

$$N_{avg}^{1/4} = \frac{N_1 + N_2 + N_3 + \text{etc.}}{N_1^{3/4} + N_2^{3/4} + N_3^{3/4} + \text{etc.}} \quad (429)$$

where N with a subscript 1, 2, 3, etc., is the number of tubes in the row indicated by the subscript.

Exercise. Compute the condensing coefficient for the bottom tube in a row of 3 tubes when the condensation coefficient for the top tube is 350, assuming laminar flow of the condensate film.

Illustrative Example. From the following data,[15] compute the overall coefficient, the condensing film coefficient for hexane, and the heat transfer coefficient for water inside the condenser tube.

Position No.	1	2	3	4	5	6	7
Average Temp. °F.	57.20	62.48	104.5	105.6	106.4	107.4	155.0

The single copper tube condenser (1.049-in. ID, 1.315-in. OD) was used to condense hexane at atmospheric pressure. The thermocouples in the center of the wall of the copper tube (Fig. 425) and the thermometers gave the indicated average readings during a run with water flowing through the tube at 22.83 lb/min. The tube was turned to take wall temperatures at eight evenly spaced angles, as indicated in Fig. 426 for No. 3. The trough collected the hexane from the tube at the rate of 0.833 lb/min. The effective length of the tube is 20⅝ inches.

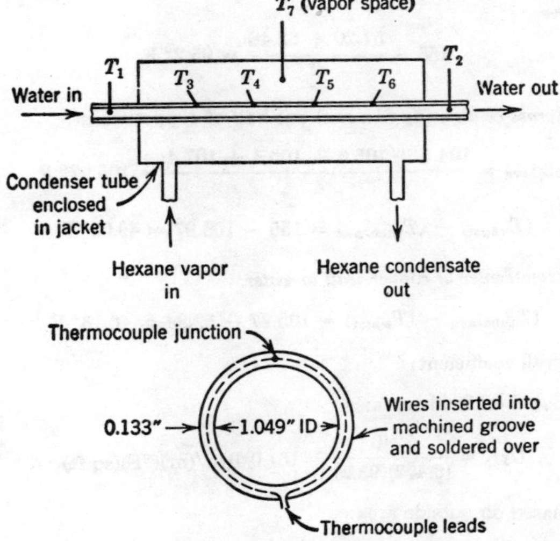

Fig. 425. Diagrammatic representation of thermocouples used in determining condensing coefficients in the example.

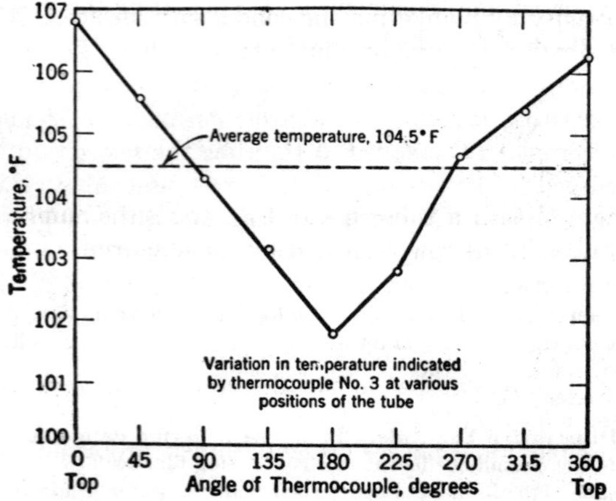

FIG. 426. Temperatures indicated by thermocouple 3 at various angles from the vertical. (See Fig. 425.)

Solution.

Quantity of heat transferred:

$$q = \int_{T_1}^{T_2} WC_P \, dT = 60 \times 22.83 \times 10(62.48 - 57.20)$$

$$= 7240 \text{ Btu/hr}$$

Area:

$$\text{Inside tube} = \frac{(20.625)\pi(1.049)}{(12)(12)} = 0.472 \text{ sq ft}$$

$$\text{Outside tube} = \frac{(20.625)\pi(1.315)}{(12)(12)} = 0.593 \text{ sq ft}$$

Temperature difference:

Overall:

$$155 - \frac{57.20 + 62.48}{2} = 95.2° \text{ F}$$

Across condensing film and one-half of copper wall:

$$(T_{\text{tube}})_{\text{avg}} = \frac{104.5 + 105.6 + 106.4 + 107.4}{4} = 105.97° \text{ F}$$

$$(T_{\text{vapor}}) - (T_{\text{tube}})_{\text{avg}} = 155 - 105.97 = 49.03° \text{ F}$$

From center of copper wall to water:

$$(T_{\text{tube}})_{\text{avg}} - (T_{\text{water}}) = 105.97 - 59.84 = 46.13° \text{ F}$$

Overall coefficient:

Based on inside area:

$$U_{\text{ID}} = \frac{7240}{(0.472)(95.2)} = 161.0 \text{ Btu/(hr)(°F)(sq ft)}$$

Based on outside area:

$$U_{\text{OD}} = \frac{7240}{(0.593)(95.2)} = 128.2 \text{ Btu/(hr)(°F)(sq ft)}$$

Condensing coefficient including resistance of *one-half* the metal based on outside area, U'_{OD}:

$$U'_{\text{OD}} = \frac{q}{A_{\text{OD}}(T_{\text{vapor}} - T_{\text{tube}})} = \frac{7240}{(0.593)(49.03)}$$

$$= 248 \text{ Btu/(hr)(°F)(sq ft)}$$

$$U'_{\text{OD}} = \frac{1}{\frac{1}{h_{\text{OD}}} + \frac{L}{k}\frac{D_{\text{OD}}}{D_{\text{avg}}}} = \frac{1}{\frac{1}{h_{\text{OD}}} + \frac{0.066}{(12)(220)}\frac{(1.315)(2)}{1.315 + 1.1825}}$$

$$= 248 \text{ Btu/(hr)(°F)(sq ft)}$$

Condensing film coefficient h_{OD}:

$$\frac{1}{h_{\text{OD}}} = 0.00403 - 0.0000263 = 0.004004$$

$$h_{\text{OD}} = \frac{1}{0.004004} = 250 \text{ Btu/(hr)(°F)(sq ft)}$$

The resistance of the metal is so small as compared with that of the condensing film that it could have been neglected. The heat transfer coefficient for copper tube to water, including the resistances of one-half of the tube, and of transferring heat from the tube to the water:

$$U'_{\text{ID}} = \frac{q}{A_{\text{ID}}(T_{\text{tube}} - T_{\text{water}})} = \frac{7240}{(0.472)(46.13)}$$

$$= 332 \text{ Btu/(hr)(°F)(ft)}^2$$

$$U'_{\text{ID}} = \frac{1}{\frac{1}{h_{\text{ID}}} + \left(\frac{L}{k}\right)\frac{D_{\text{ID}}}{D_{\text{avg}}}} = \frac{1}{\frac{1}{h_{\text{ID}}} + \frac{0.066}{(12)(220)}\left(\frac{1.049}{1.115}\right)}$$

$$= 332 \text{ Btu/(hr)(°F)(sq ft)}$$

$$\frac{1}{h_{\text{ID}}} = \frac{1}{332} - \frac{0.066 \times 1.049}{12 \times 220 \times 1.115} = 0.00301 - 0.000023$$

$$h_{\text{ID}} = 334 \text{ Btu/(hr)(°F)(sq ft)}$$

Another equally satisfactory procedure would be to calculate the temperatures of the inside and outside surfaces of the copper tube from the known temperature at the center of the tube, the thermal conductivity and thickness, and the heat flux, q/A.

An alternate method for measuring condensing film coefficients which does not involve measurement of tube wall temperatures is that of taking a series of measurements at increasing water velocity at constant condensing coefficient. A plot of the overall resistance against the reciprocal of the water velocity to the 0.8 power gives a straight line which may be extrapolated to infinite water velocity to give the resistance of the tube wall plus the condensing film from which the condensing film coefficient may be computed as described by Wilson.[45]

Turbulent Flow in Condensate Layer

Although the assumption that the liquid condensate layer flows on the surface in laminar flow probably applies to the usual horizontal tube, the accumulation of condensate on a vertical tube may develop turbulent flow.[25]

When the dimensionless ratio $4W/\mu_f \pi D$ evaluated at the lowest point on the condensing surface is less than 2100, the flow is laminar and equations 426 and 427 may be used. If the dimensionless ratio is greater than 2100, the flow of the condensate layer is turbulent at least at the bottom of the tube, and the following equation may be used.

$$h = 0.0077 \left(\frac{4W}{\mu_f \pi D}\right)^{0.4} \left(\frac{k_f^3 \rho_f^2 g}{\mu_f^2}\right)^{1/3} \quad (430)$$

where W = pounds of condensate flowing per unit time,
D = outside diameter of completely wet tube, or
πD = wetted perimeter.

None of these equations, 425 to 430, considers turbulence or changes in film thickness as affected by velocity of approaching vapors. For high vapor velocities the coefficients may be considerably greater than those predicted by these equations.

Superheated Vapors

Condensation of superheated vapors involves the additional step of reducing the temperature of the vapor to the saturation temperature before condensation takes place. During the reduction in superheat, down to the temperature at which the tubes are wet with condensate, the vapor behaves in a manner predicted for fluids flowing outside or inside of tubes. In this case the convection coefficient is relatively low [10 to 40 Btu/(hr)(°F)(sq ft)]. For the condensing portion of the heat transfer, the temperature difference is the saturation temperature minus the surface temperature. When the degree of superheat is small as compared to the latent heat, the coefficient for saturated vapors may be used for superheated vapors, provided the temperature difference between the saturation temperature and the surface temperature is used.[24,32]

Mixtures

Gases which are noncondensable at the condenser temperature can cause severe reduction in condenser capacity by blanketing the condensing surface.[7] Air in steam is a common illustration of this. For condensers operating on a closed system such as a refrigerating system, the noncondensable gas should be completely removed at the time the refrigerant is charged to the system. Continuous removal of air is also required if there is any leak into the system to prevent blanketing of the condensing surface by accumulated air, even though the concentration of air is very low. For steam condensers operating at subatmospheric pressure, a vacuum pump is required for removing the noncondensables continuously. Condensation of mixed vapors which result in immiscible liquid phases presents similar problems.[7]

DROPWISE CONDENSATION

Dropwise condensation occurs when the condensate does not wet the surface but forms droplets of liquid which roll off the surface.

Actually the liquid forms a definite contact angle with the solid surface, as indicated on Figs. 427 and 428. If the contact angle (the angle measured through the liquid) becomes much less than 50 degrees, the droplets spread unevenly and areas will be covered with a continuous film.

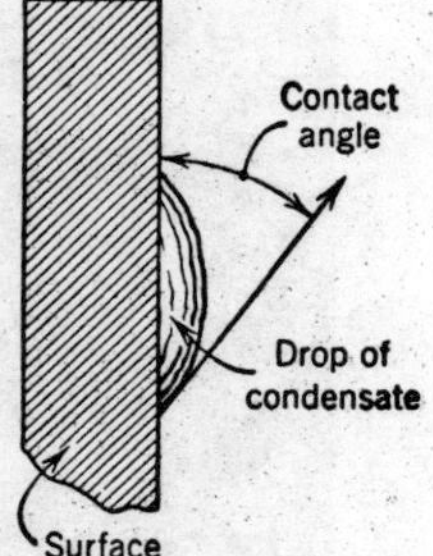

Fig. 427. Diagram indicating contact angle in dropwise condensation.

Normally, organic substances form continuous films, and water condenses filmwise except on surfaces covered by a fatty acid or other similar material not wet by water. Accordingly the knowledge of dropwise condensation is limited to experiments with steam on metal surfaces containing some adsorbed organic substance such as a fatty acid to control the contact angle between the liquid condensate and the solid.

A study of dropwise condensation of steam on vertical and inclined surfaces (Fig. 429) under dynamic conditions[12] indicated the following mechanism. Vapor condenses on the surface in small drops which are generally of uniform size and shape. The small drops grow both by condensation on their surfaces and by coalescence until one or more drops reach the maximum size possible for adhering to the surface. These large drops then roll and grow rapidly by coalescing with other drops in the path. In this way the surface is cleared of condensate periodically by droplets sweeping down the surface, exposing a bare strip of metal. The cycle is then repeated, starting with the condensation of vapor on the surface in small drops.

During the growing process, the surface area covered by droplets remains substantially constant (about 45 per cent for the systems studied).[12] High rates of heat transfer occur when small droplets roll to clear the surface at more frequent intervals. Figure 428 shows the dropwise condensation of steam on polished chromium covered with stearic acid. The light spots on the droplets were used to measure contact angles. The stearic acid is washed off the surface by the condensate in a few hours, and, unless it is replenished, filmwise condensation occurs.

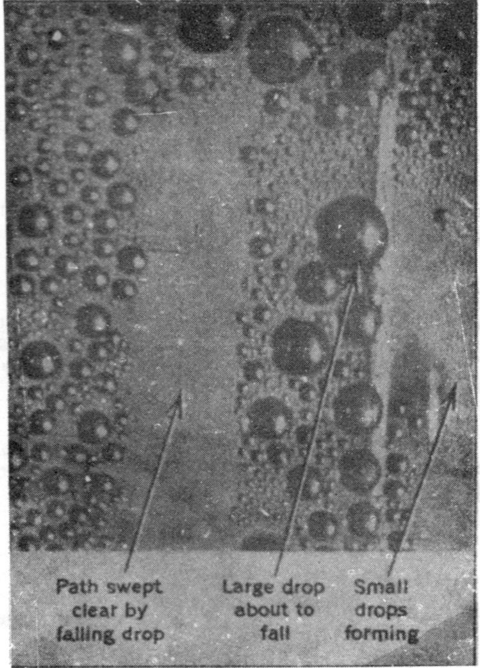

FIG. 428. Photograph showing dropwise condensation of steam on polished chromium which has been covered with stearic acid.

Figure 429 presents successive frames of moving pictures of the dropwise condensation process. The rolling of droplets off the surface to clear the surface below can be seen in the successive photographs from top to bottom. Figure 429*b*, taken near the top of the surface, indicates that irregular-sized droplets may occur in this area which has not been cleared by rolling droplets. Figure 429*a* is typical of the center portion of a surface, all of which is cleared at frequent intervals.

The rates of heat transfer are very high for dropwise condensation, with coefficients in the range from about 10,000 to 70,000 Btu/(hr)(°F)(sq ft). Data typical for dropwise condensation[40] of steam are plotted in Fig. 430. That part of the surface not covered by drops is free of the resistance offered by the condensate liquid. The heat transfer through the drops is a conduction process through segments of spheres of changing size. The prime variable which influences the size of the droplets through which heat must be conducted is the difference between the advancing and the receding contact angle

FIG. 429. Successive frames from motion pictures taken of dropwise condensation of steam on a vertical, polished surface.[12]

since it controls the size attained by droplets before they begin to roll.

Reproducible results with dropwise condensation are difficult to obtain because the contact angles cannot be controlled. In practice, a copper tube with the normal amount of oil on its surface that comes from handling in a shop will often start condensing in a dropwise manner. If steam is not oil-free, portions

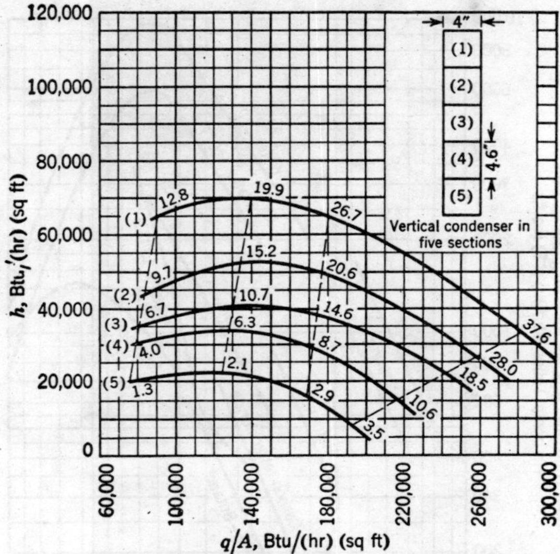

The five different vertical sections of a vertical copper plate condenser exhibit different coefficients as indicated. The numbers in parentheses refer to the corresponding section of the condenser. The other numbers on the curves indicate the steam velocities in feet per second at the intersection with the dashed line. Benzyl mercaptan was used as a promoter of dropwise condensation for the experiments reported. All sections showed a maximum coefficient at about 120,000 to 150,000 Btu/(hr) (sq ft) heat flux. At these high rates the droplets rolling off the surface become in effect a stream of condensate much like a film. The condensate soon removes the promoter from the surface under conditions of high condensation rates, thereby further increasing the tendency for film condensation.

FIG. 430. Heat transfer coefficients for dropwise condensation of steam on a vertical copper plate condenser.[40]

of the tube surface may maintain a dropwise condition while the remainder of the tube is covered with a film of condensate.

BOILING COEFFICIENTS

Liquid in contact with a solid at a higher temperature increases in temperature until the boiling point at the existing pressure is reached. At some temperature above the boiling point, bubbles of vapor form with absorption of heat. Numerous bubbles rise and agitate the liquid, creating a condition favorable for heat transfer, as indicated in Figs. 431, 433*a*, and 433*b*. The ease with which the nuclei of vapor form regulates the degree of superheat in the liquid required to initiate boiling and is a function of the interfacial forces between the vapor and liquid and the solid. Bumping in a flask is a result of superheating. This occurs because no nucleus of vapor is formed until the temperature is considerably above the equilibrium boiling point. Finally the superheated liquid flashes, and bumping occurs.

In general, the coefficient of heat transfer between the solid and the liquid increases with increased temperature difference between the solid and the main

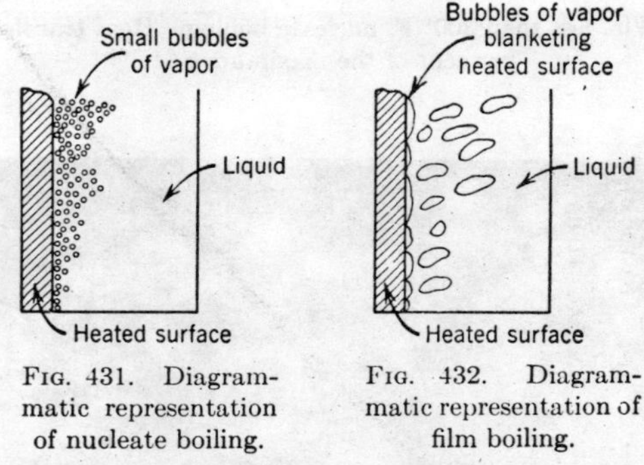

FIG. 431. Diagrammatic representation of nucleate boiling.

FIG. 432. Diagrammatic representation of film boiling.

body of the liquid. Figure 434 is a plot of boiling coefficients as a function of the temperature difference between the solid and the liquid. As the temperature difference increases the heat flux increases and the quantity of vapor bubbles increases, causing the bubbles to increase in size by coalescence. This action in itself tends to decrease the proportion of solid surface which is in contact with the liquid. Since the liquid absorbs heat much more readily than the vapor, owing to its higher conductivity, the reduction in liquid contact tends to reduce the transfer rate and to overcome the tendency of increased agitation by bubbles to increase the transfer rate. These two phenomena of increased agitation and vapor blanketing cause the heat transfer coefficients to exhibit a maximum when plotted against the temperature drop or heat flux (Fig. 434).

The change in interfacial properties between the solid and liquid with increased temperature difference also contributes to vapor blanketing. The liquid has less tendency to wet the solid so that a bubble of vapor covers more surface and is not pinched off by the liquid as quickly as at low temperature differences. Boiling in the presence of the vapor blanket over the solid, Figs. 432 and 433*c*, at

(*a*) Wire less than 300° F, nucleate boiling. Heat transfer 11 per cent of the maximum q/A.

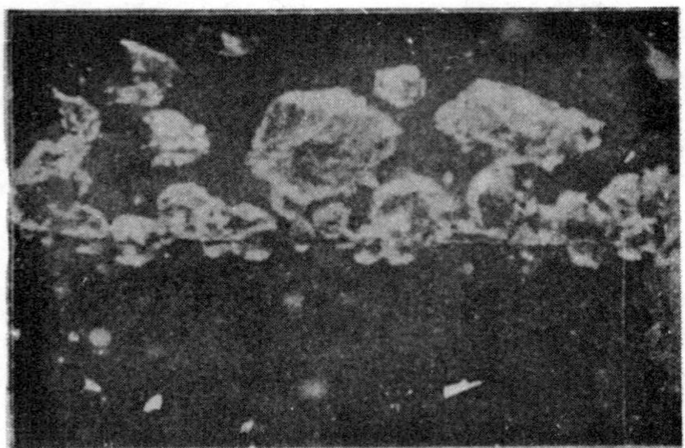

(*b*) Wire less than 300° F, nucleate boiling. Heat transfer 80 per cent of the maximum q/A.

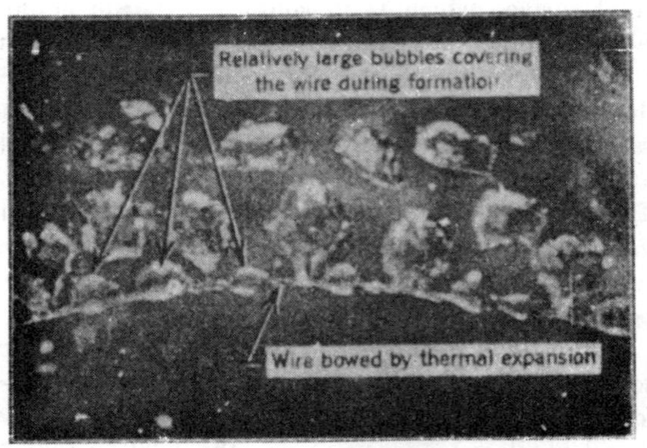

(*c*) Wire red-hot, film boiling. Heat transfer 72 per cent of the maximum q/A.

Fig. 433. Boiling on a heated wire immersed in water.[31a] (*Courtesy Prof. W. H. McAdams, Mass. Inst. Tech.*)

high heat flux has been termed "film boiling" and is characterized by decreasing coefficients with increased temperature difference.

Many forms of equipment are used for transfer of heat by boiling. A bundle of horizontal tubes submerged in boiling liquid is used in a reboiler for a fractionating column or in a horizontal tube evaporator. Liquids may be boiling inside vertical tubes,

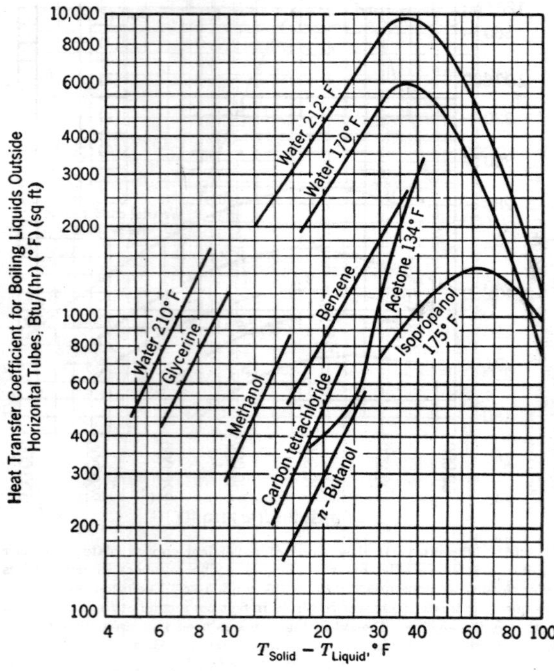

Fig. 434. Coefficients of heat transfer to boiling liquid as a function of the temperature drop between the solid surface and the boiling liquid.[2, 8, 31]

in which case the flow of two phases inside the tube becomes a factor in the mechanism of boiling. Fluids may be circulated inside either horizontal or vertical tubes such as in a petroleum pipe still or in a forced circulation evaporator. The influence of the interfacial tensions for different fluids in different equipment and the effects of agitation either by the bubbles or by fluid circulation have made it impossible to formulate a general relationship for predicting boiling coefficients.

Since the boiling coefficient varies with the temperature difference between the surface and the liquid, temperature changes around the circumference of a horizontal tube would vary the local coefficients for boiling. The temperature around the circumference of a horizontal tube submerged in a boiling liquid and containing condensing steam on the inside depends upon the local condensing coefficient which in turn is a function of the quantity of condensate which has accumulated in the lower portion of the tube.[15]

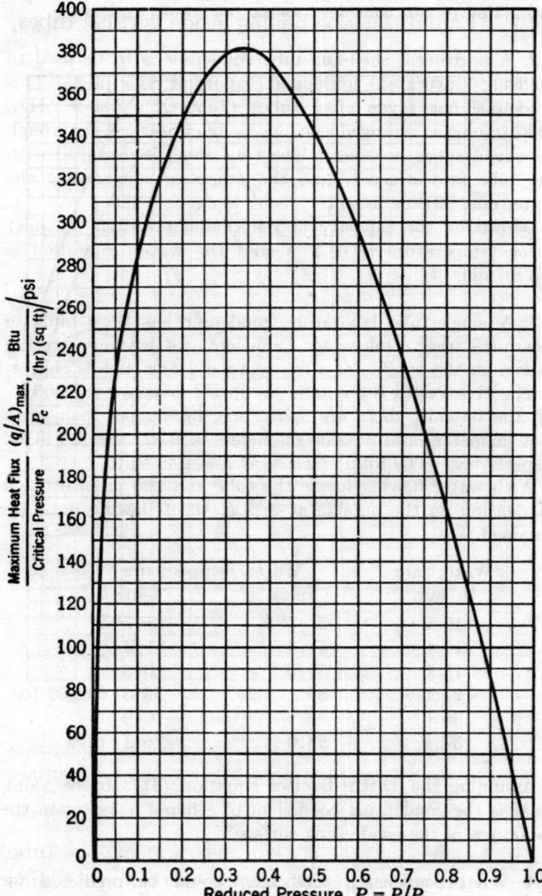

FIG. 435. The ratio of the maximum heat flux (q/A) to the critical pressure as a function of the reduced pressure.[3]

Initial adsorption of gases on a surface or contamination of the surface by foreign materials may give abnormal behavior in short experiments. The addition of wetting agents to lower the vapor-liquid interfacial tension may increase the boiling coefficient. High temperature levels which reduce viscosity and surface tension give higher boiling coefficients when nucleate boiling takes place. Since higher temperatures are required to boil fluids at higher pressures, boiling coefficients increase with increased pressure for a given substance. A heat flux of 2,100,000 Btu/(hr)(sq ft) was obtained between a platinum wire and water under 1200 psi.[32]

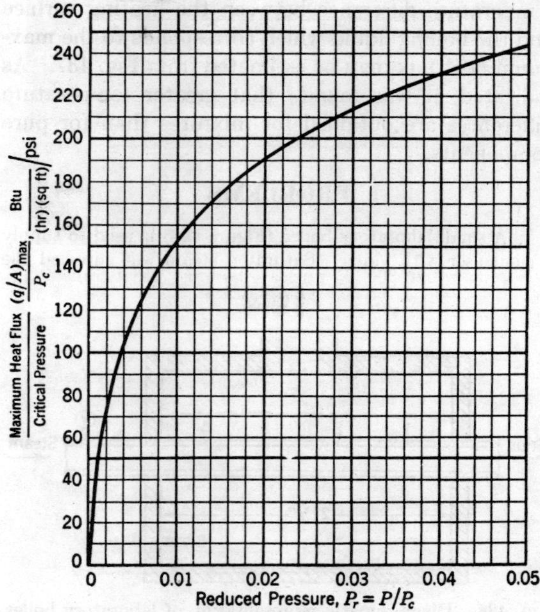

FIG. 436. The ratio of the maximum heat flux to the critical pressure as a function of the reduced pressure, in the low-pressure region.[3]

Boiling as a commercial operation normally is conducted with temperature differences across the boiling film of less than 40° or 50° F and hence is nucleate boiling. The relatively high boiling coefficients mean a low resistance for the boiling film and the temperature difference across the boiling film may be a small portion of the overall temperature difference.

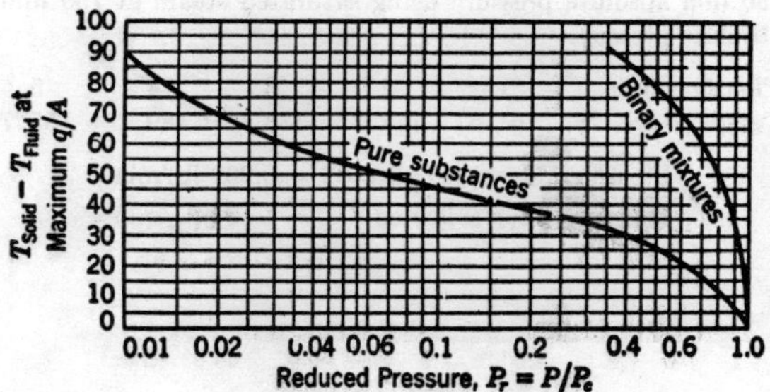

FIG. 437. The temperature difference at the maximum heat flux as a function of the reduced pressure.[3]

The maximum heat flux at which nucleate boiling may still be obtained may be estimated from the graphical correlation[3] of Figs. 435 and 436 which may be used for either pure liquids or mixtures. The temperature difference between the heating surface and the boiling liquid which corresponds to the maximum heat flux may be estimated from Fig. 437. As indicated, it was found[3] that greater temperature differences are obtained for mixtures than for pure components.

PROBLEMS

1. A small laboratory boiler (Figure 438) is used to supply a condenser with vapor. Saturated steam was supplied the

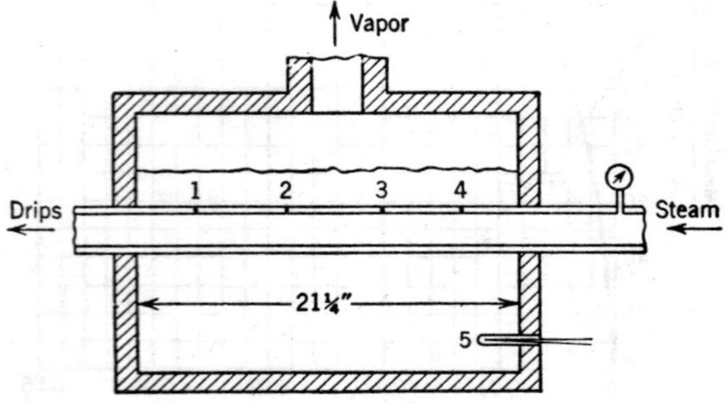

Fig. 438. Diagrammatic representation of laboratory boiler. Numbers indicate locations of thermocouples.

copper tube (1.315-in. OD, 0.951-in. ID), and drips were removed from the other end at the same pressure as the steam inlet. Four thermocouples were inserted in the center of the copper tube wall, and provisions were made so that the tube could be rotated to obtain average temperatures around the tube as well as along the tube.

The following data were taken when boiling hexane at 750 mm absolute pressure using saturated steam at 750 mm absolute pressure.

Thermocouple	1	2	3	4	5
Temperature, °F	161.6	172.3	183.5	189.1	155.7

Steam condensate collected = 0.299 lb/min
Hexane vapor collected = 1.400 lb/min

(*a*) Compute the overall coefficient of heat transfer between steam and hexane.

(*b*) Compute the boiling coefficient for hexane, neglecting the resistance of the copper.

(*c*) Compute the condensing coefficient for steam, neglecting the resistance of the copper.

(*d*) Compute the true boiling coefficient for hexane, assuming that the thermocouples are in the center of the copper wall and giving proper consideration to the copper resistance.

2. A horizontal shell-and-tube condenser is to be used to condense saturated ammonia vapor at 145 psig. This condenser has seven steel tubes (2-in. OD, 1.81-in. ID), 12.82 ft long, so arranged that the cooling water passes through the seven tubes in series. The tube sheet is arranged with one tube at the center and the other six surrounding the center one, all equidistant.

Determine the capacity of the condenser when the inlet water temperature is 68.2° F and the water rate used is 375 lb/min.

3. A single-tube laboratory condenser has been built to determine the condensing coefficients of various organic vapors on the outside of a finned tube. The tube is copper, its effective length is 3 ft, and the inside diameter is 0.555 in. On the outside, there are spiral fins, fifteen per inch, with root diameters and outside diameters of 0.621 and 0.746 in. respectively. The total external area is 1.45 sq ft.

With water flowing inside the tube and 190 proof ethanol condensing on the outside at 3 psig, the following data are obtained.

Water rate lb/min	Water Temperature Inlet	Outlet
10.9	51.3° F	81.7° F
11.0	51.6	81.3
15.6	54.5	80.6
24.6	56.1	78.5
36.6	58.1	76.9
37.4	58.0	76.4

Assuming the Dittus-Boelter equation (417) to be valid, what is the condensing coefficient of ethanol based upon the total area of the condensing surface?

4. What condensing coefficient would be predicted for ethanol condensing upon a plain, horizontal copper tube, ¾ in. OD, under the conditions of problem 3?

CHAPTER

31

Heat Transfer 4

Radiation

ENERGY is transferred by radiation in the form of electromagnetic waves which travel in straight lines at the speed of light. The sun emits radiant energy which travels through space and the atmosphere, reaching the earth where it is absorbed. The air is relatively transparent to these electromagnetic waves and absorbs only a small amount of the radiant energy. Other gases such as carbon dioxide and water vapor do absorb radiant energy in significant quantities. All substances which absorb radiation also are capable of emitting energy as electromagnetic waves.

The geometric patterns involved in radiation are the same as those evident to the eye for beams of light. The concept of shade as related to light is also directly applicable to the radiation of heat, as when one individual stands behind another in a cold room beside an open fire or hot stove. A body can receive heat by radiation only if it can "see" the source of the thermal radiation, either directly or by reflection.

The distribution of wavelengths of the emission from a surface depends upon the temperature of the surface (Fig. 439). At about 1000° F the distribution is such that the concentration of the dull red waves is high enough to be perceived by the eye.

At higher temperatures such as 2500° F, objects become "white" hot because the distribution of the intensity of the waves in the visible region is such as to be seen as white light. Although visible light is emitted, the greater proportion of radiant energy is in the nonvisible infrared region.

Waves of the frequency range classified as thermal radiation (1- to 20-micron wavelength) are not absorbed by diatomic gases having symmetrical molecules such as oxygen and nitrogen at temperatures of ordinary industrial interest. Substances transparent to thermal radiation are called diathermanous. Water vapor and carbon dioxide absorb energy of characteristic frequencies, as indicated in Fig. 439. Substances which absorb radiation also have the property of emitting radiant energy in the same frequency bands.

The radiant energy impinging upon a body is in part absorbed by the body, is in part reflected, and may in part pass through the body. The fraction of the incident energy which is absorbed cannot exceed unity. Those bodies which absorb all the incident energy are called *black bodies.* Although no real surface behaves as a black body, it is a useful reference for comparison. The best example of a black body is a small hole in a large empty container. All the radiant energy entering the hole is absorbed by the inner walls of the container, either on the first contact or by subsequent contacts after continued reflection within the empty space.

If one or more bodies are placed within an enclosed space in a container and thermal equilibrium exists between the body and the container, any area of surface is receiving and emitting energy at the same rate. If one of the surfaces absorbs all energy incident on it (a black body), then the rate at which it emits must be the maximum possible at that temperature; if this were not true, thermal equilibrium would be attained between two bodies at different temperatures. The *black body* must accordingly radiate, or emit, the maximum amount of radiant

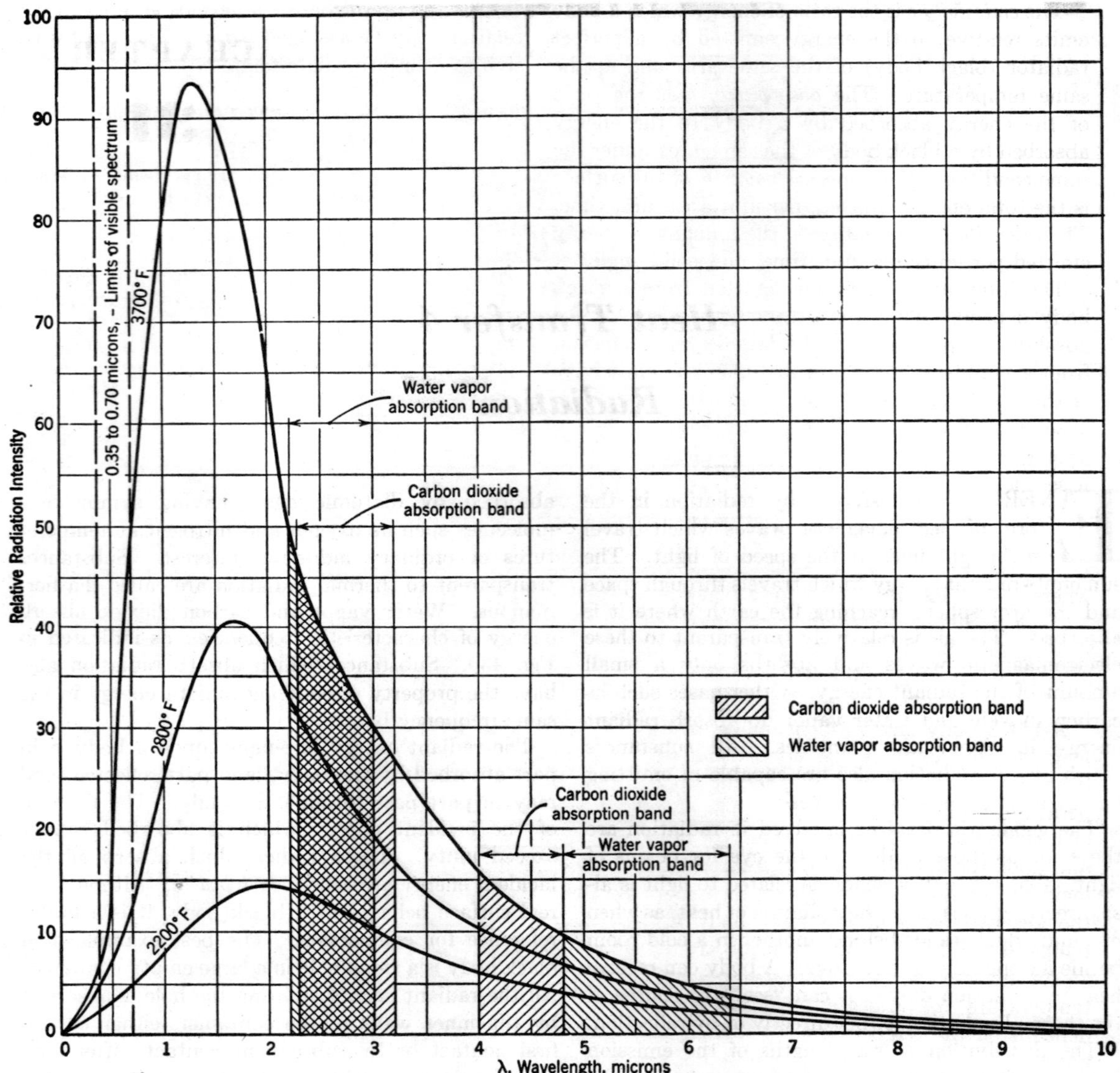

FIG. 439. Relative intensity of radiation from black bodies as a function of wavelength and temperature.

energy per unit of surface for a given temperature. It is also a perfect absorber. If the body is a non-black body, absorbing only a fraction of the radiant energy reaching the surface, it must also emit that same fraction of the amount of energy which a perfect radiator (black body) of the same size emits under the same condition. The principle, that *at thermal equilibrium the emissivity of a body equals the absorptivity*, is known as Kirchhoff's law.

If I represents the radiant energy incident on unit surface of any body in the enclosed space, an energy balance at thermal equilibrium gives

$$Ia_1 = E_1$$

For a black body $a = 1$ and the emissive power of a black body $E_B = I$. Therefore

$$e_1 = \frac{E_1}{E_B} = \frac{Ia_1}{I} = a_1$$

The *emissivity* e is the ratio of energy which a body emits relative to the energy emitted by a perfect radiator (black body) of the same area and at the same temperature. The *absorptivity* a is the ratio of the energy absorbed by a body to the energy absorbed by a black body of the same area under the same condition. The *emissive power* E of the surface is the total energy emitted, per unit area, unit time. The *intensity* i of radiation is the amount of energy emitted per unit area, unit time, unit solid angle.

The intensity of radiation emitted from a black body depends solely upon the fourth power of the absolute temperature.* The intensity of radiation from a nonblack body depends upon the emissivity of that body as well as upon the fourth power of its absolute temperature. The energy emitted from a unit area of a body to the whole hemisphere which it sees is

$$E = \frac{q}{A} = e\sigma T^4 \tag{431}$$

where e = emissivity (unity for a black body).
σ = Stefan-Boltzmann constant [1730×10^{-12} in Btu/(sq ft)(hr)(degree Rankine)^4].
T = absolute temperature of the body.

A black body emits a continuous series of wavelengths with the maximum intensity at wavelengths from about 1 to 5 microns, depending upon the temperature as indicated in Fig. 439. The absorptivity a of a surface depends upon the nature of the surface and upon the distribution of the wavelengths in the incident radiation. Although the emissivity e equals the absorptivity a of the same body at the same temperature, the absorptivity at a body temperature T_1 for incident radiation from a source at a different temperature T_2 indicated by the symbol $a_{1\leftarrow 2}$ depends on the temperature T_2 as well as the temperature T_1. If the monochromatic absorptivity of surface 1 (absorption of radiation of a single wavelength) varies considerably with wavelength and much less with the temperature of surface 1, total absorptivity $a_{1\leftarrow 2}$ will vary more with T_2 than with T_1. Since this condition generally exists, the value of $a_{1\leftarrow 2}$ may be taken as equal to the emissivity of the body (1) at the temperature of the incident radiation T_2. Emissivities for a number of surfaces over specified temperature ranges are listed in Table 51.

*This relationship, known as the Stefan-Boltzmann law, may be derived from thermodynamic relationships. (See reference 10*a*, p. 472.)

In the absence of better information, a straight-line relation may be assumed for interpolation for emissivities at intermediate temperatures.

TABLE 51. NORMAL TOTAL EMISSIVITIES OF SOLID SURFACES[31, 38]*

Surface	Temperature, °F	Emissivity,
Metals		
Aluminum		
Highly polished plate	440–1070	0.039–0.057
Oxidized at 1110° F	390–1110	0.11 –0.19
Brass		
Highly polished (73–27)	476– 674	0.028–0.031
Polished	100– 600	0.096–0.096
Dull plate	120– 660	0.22
Copper		
Polished	242	0.023
Plate heated at 1110° F	390–1110	0.57 –0.57
Iron, polished	800–1880	0.144–0.377
Cast iron		
Polished	392	0.21
Turned on lathe	1630–1810	0.60 –0.70
Oxidized at 1100° F	390–1110	0.64 –0.78
Steel oxidized at 1100° F	390–1110	0.79 –0.79
Rough ingot iron	1700–2040	0.87 –0.95
Steel plate, rough	100– 700	0.94 –0.97
Molten steel	2910–3270	0.28 –0.28
Mercury	32– 212	0.09 –0.12
Nickel polished plate	74	0.045
Miscellaneous Building Materials		
Asbestos board	74	0.96
Brick, red rough	70	0.93
Silica, glazed	2012	0.85
Lampblack, 0.003 in. or thicker	100– 700	0.945
Enamel, white fused on iron	66	0.897
Glass, smooth	72	0.937
Marble, polished	72	0.931
Paints		
Black, shiny lacquer	76	0.875
White lacquer	100– 200	0.80 –0.95
Oil paints, 16 colors	212	0.92 –0.96
10% Al, 22% lacquer body	212	0.52
26% Al, 27% lacquer body	212	0.30
Paper	66	0.924
Plaster, rough lime	66	0.944
Refractories		
Poor radiators	1110–1830	0.65 –0.75
Good radiators	1110–1830	0.80 –0.90
Water	32– 212	0.95 –0.963

Since bodies which are emitting radiant energy are also absorbing radiant energy, the net quantity of energy transferred by radiation equals the energy emitted minus the energy absorbed. For a body at

* The bibliography for this chapter appears on p. 472.

temperature T_1 enclosed by surroundings at the temperature T_2, the rate of heat transfer from 1 to 2 is *

$$q_{1,2} = A_1\sigma[e_1 T_1^4 - a_{1\leftarrow 2}T_2^4] \quad (432)$$

or

$$\frac{q_{1,2}}{A_1} = 1730\left[e_1\left(\frac{T_1}{1000}\right)^4 - a_{1\leftarrow 2}\left(\frac{T_2}{1000}\right)^4\right] \quad \text{Btu/(hr)(sq ft)} \quad (433)$$

It is convenient for slide-rule manipulation to place arbitrarily the 10^{-12} of the Stefan-Boltzmann constant in the denominator of the absolute temperature as in equation 433 if T is greater than 1000; if T is less than 1000, σ may be written 0.173×10^{-8}, with the temperature term written $(T/100)^4$. This equation is useful for computing the radiation between an enclosed object and its surroundings when the object is small compared to its surroundings.

If the body absorbing the radiation absorbs all wavelengths with the same absorptivity, the body is called a *gray body* and its total absorptivity a is independent of the energy distribution in the incident radiation. In such a case the emissivity e_1 may be used as the absorptivity $a_{1\leftarrow 2}$, since $a_{1\leftarrow 2} = a_{1,1} = e_1$. The assumption of *gray body conditions* simplifies the calculations, as equation 433 then becomes

$$q_{1,2} = 0.173Ae_1\left[\left(\frac{T_1}{100}\right)^4 - \left(\frac{T_2}{100}\right)^4\right] \quad (434)$$

Illustrative Example. A furnace with a vertical door 3 ft by 4 ft is located in the middle of a large room. The door is made of cast iron and is oxidized. The outer surface of the door is at 340° F. The walls of the room and its contents which the door "sees" are at 95° F. How much heat is being transferred from the door by radiation and convection?

Solution. Equation 433 may be used directly, substituting the values for the absolute temperatures of 800° Rankine for T_1 and 555° Rankine for T_2. Some judgment is required in estimating the value for the emissivity e_1 and the absortivity $a_{1\leftarrow 2}$. By referring to Table 51 for cast iron oxidized at 1100° F and assuming a linear relationship for the emissivity as a function of temperature, the emissivity from the oxidized cast iron door is estimated as 0.63 at 340° F. Assuming the absorptivity of the iron surface at 340° F to be dependent primarily upon the character of the radiation incident to it, the absorptivity $a_{1\leftarrow 2}$ for the heat-absorbing surface at the temperature of the incident radiation (95° F) is taken as equal to the emissivity of the iron surface at 95° F or 0.58.

$$q_{1,2} = 1730A\left[e_1\left(\frac{T_1}{1000}\right)^4 - a_{1\leftarrow 2}\left(\frac{T_2}{1000}\right)^4\right] \quad (433)$$

$$= 1730 \times 12\left[0.63\left(\frac{460+340}{1000}\right)^4 - 0.58\left(\frac{460+95}{1000}\right)^4\right]$$

$$= 20{,}800(0.63 \times 0.408 - 0.58 \times 0.0910)$$

$$= 4250 \text{ Btu/hr}$$

or, if the door is assumed to be "gray,"

$$q = 20{,}800 \times 0.63(0.408 - 0.091) \quad (434)$$

$$= 4160 \text{ Btu/hr}$$

In addition to the heat transferred by radiation, in this case with the furnace surrounded by air at 95° F, there will be an additional 3480 Btu transferred by natural convection. This is computed by equation 382, using the value for the convection coefficient as computed from equation 424 as follows.

$$q_c = hA(T_1 - T_2) \quad (382)$$

$$h = 0.3(340 - 95)^{0.25} = 1.185 \quad (424)$$

$$q_c = 1.185 \times 12(340 - 95) = 3480 \text{ Btu/hr}$$

This would make a total heat transfer from the vertical cast iron door of 7730 Btu/hr by radiation and convection.

The overall results for solids are about the same whether they are assumed to be gray bodies or not. The uncertainty due to difficulties in describing the surface is probably less than that involved in estimating emissivities from tabulated values. For these reasons solids are often assumed to be gray bodies because of the greater ease in computation.

A flat surface, such as the furnace door, emits radiation throughout the hemisphere which the surface "sees." The greatest intensity of this radiation is in a direction normal to the surface, and the intensity in the direction of a small angle with the surface may be quite different. The total normal emissivities given in Table 51 include the total spectrum for waves normal to the surface but may be used for most cases as the average hemispherical radiation. For most nonmetallic or unpolished metal surfaces the total hemispherical emissivity is approximately equal to the normal value given in Table 51.

The radiation considered in the illustrative example is a special case in which the body, or surface, is completely surrounded by a black body. Under such conditions the entire hemisphere which the surface "sees" is substantially uniform, and the emissivities given in Table 51 may be used directly. The same condition exists between two infinite parallel planes.

* The subscripts indicate the direction of energy transfer from the first to the second, as from 1 to 2 in equations 432 and 433, in all cases except that of absorptivity, a.

When radiation between two solids or surfaces is being considered, each of which has limited dimensions, the entire hemisphere seen by the solids does not represent uniform radiating conditions, and the interchange of energy by radiation between surfaces depends to a large extent upon the geometry of the system.

Illustrative Example. Indicate the computation of the interchange of radiant energy between surfaces dA_1 and dA_2 in Fig. 440.

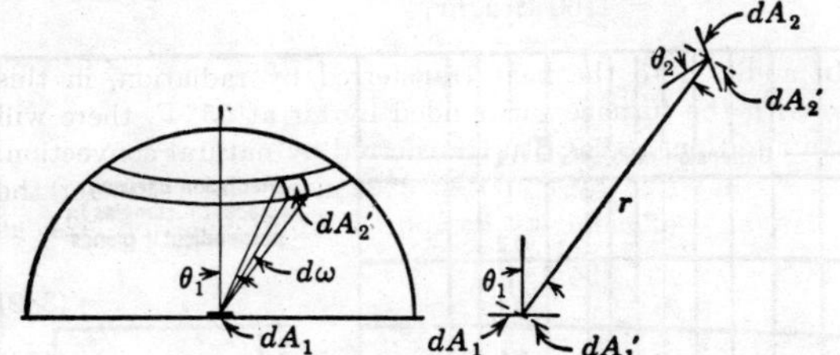

Fig. 440. Geometrical factor in radiation; derivation of cosine law by integration through a solid angle.

Solution. The portion of the energy emitted by dA_1 which is intercepted by dA_2 is proportional to the projected areas dA_1' and dA_2' of the two surfaces normal to radiant beam and inversely proportional to the square of the distance r between the surfaces. If θ_1 and θ_2 are the respective angles between the normals to the surfaces and the radiant beam, and i indicates the intensity of radiation, the energy intercepted by dA_2 is

$$\frac{i\,dA_1'\,dA_2'}{r^2} = \frac{i\,dA_1 \cos\theta_1\,dA_2 \cos\theta_2}{r^2} \tag{435}$$

Equation 435 is frequently called the cosine law. The proportionality constant i may be evaluated by expressing equation 435 in terms of the solid angle $d\omega$ of Fig. 440 as follows

$$i\,dA_1 \cos\theta_1 \left(\frac{dA_2'}{r^2}\right) = i\,dA_1 \cos\theta_1\,d\omega$$

Since a solid angle is defined as the area subtended upon the surface of a hemisphere of radius r divided by the square of the radius.

The total energy emitted by one side of the surface dA_1 will be absorbed by a surrounding hemisphere of radius r and is therefore

$$E_1\,dA_1 = i\int_0^{2\pi} dA_1 \cos\theta_1\,d\omega = i\pi\,dA_1 \tag{435a}$$

The emissive power E is defined from equation 431

$$E_1 = \sigma e_1 T_1^4 = i\pi \frac{dA_1}{dA_1}$$

Solving for i,

$$i = \frac{E_1}{\pi} = \frac{\sigma e_1 T_1^4}{\pi}$$

The amount of radiant energy emitted by dA_2 which is intercepted by dA_1 may be evaluated in a manner similar to equation 435. Substituting the above expression for i and combining the net interchange of energy from dA_1 to dA_2 is

$$q_{1,2} = \frac{\sigma e_1 a_{2\leftarrow 1} T_1^4 dA_1 \cos\theta_1 dA_2 \cos\theta_2 - \sigma e_2 a_{1\leftarrow 2} T_2^4 dA_1 \cos\theta_1 dA_2 \cos\theta_2}{\pi r^2}$$

$$q_{1,2} = \sigma(dA_1 \cos\theta_1 dA_2 \cos\theta_2)(e_1 a_{2\leftarrow 1} T_1^4 - e_2 a_{1\leftarrow 2} T_2^4)/\pi r^2 \tag{436}$$

Exercise. Show that for two black rectangular planes intersecting at right angles, as a wall and a ceiling, with a common length x and widths y and z, the net radiant energy transfer is given by evaluating the integral (see Fig. 441).

$$q_{1,2} = (i_1 - i_2)\int_0^x\int_0^x\int_0^y\int_0^z \frac{yz\,dz\,dy\,dx\,dx}{(x^2+y^2+z^2)^3} \tag{437}$$

Formal integration of such expressions is usually too tedious for engineering application. But, when necessary, an approximate value for the interchange of energy may be obtained as follows:

1. Subdivide each surface into portions.
2. Compute the net interchange of energy per unit area of surface 1 between one of these portions on surface 1 and each portion on surface 2.
3. The summation of these interchanges is the total interchange per unit area of that portion of surface 1.
4. Repeat steps 2 and 3 for a number of representative portions of surface 1.
5. Average the summations and multiply this average value by the total area of surface 1.

GEOMETRIC FACTORS

For finite black bodies, the net exchange can be evaluated by including in the basic interchange equation 432 a geometry factor which is the ratio of interchange per unit area of one surface to the interchange per unit area that would take place between two infinite parallel black planes. The symbol $F_{1,2}$ is used to define the fraction of the radiation leaving surface A_1 in all directions that is intercepted by surface A_2.

The basis for the derivation of the geometrical factors is indicated in Fig. 440 and the illustrative example and exercise above. Consider the lines 1 and 2 as representing areas of planes normal to the paper. The radiation per unit time $q_{1\rightarrow 2}$ from area A_1 intercepted by the area A_2 is proportional to the apparent area A_1' as viewed from A_2. Also, the proportion of the beam emitted from A_1 that is intercepted by A_2 is proportional to the apparent area A_2' of area A_2 taken normal to the beam. These apparent areas are frequently called projected areas as they are the projection of the area on a plane normal to the beam. Furthermore, the radiation received at area A_2 varies inversely as the square of the distance separating A_1 and A_2, as indicated in Fig. 440. Values of the geometrical factor F have been calculated for various surface arrangements, assuming that the emissivity e is constant and independent of temperature.[18,31] This assumption is exact for black surfaces and reasonably good for most nonmetallic or tarnished, unpolished metal surfaces which are approximately gray bodies. The choice of either area and its corresponding geometric factor is arbitrary.

Also, when all bodies or spaces receiving radiant energy from the body or surface A_1 are considered, it follows that

$$F_{1,1} + F_{1,2} + F_{1,3} + F_{1,4} + \text{etc.} = 1 \quad (439)$$

And, when A_1 can see no part of itself, $F_{1,1} = 0$.

Figure 441 gives the numerical value for the geometric factor F for radiation between adjacent rec-

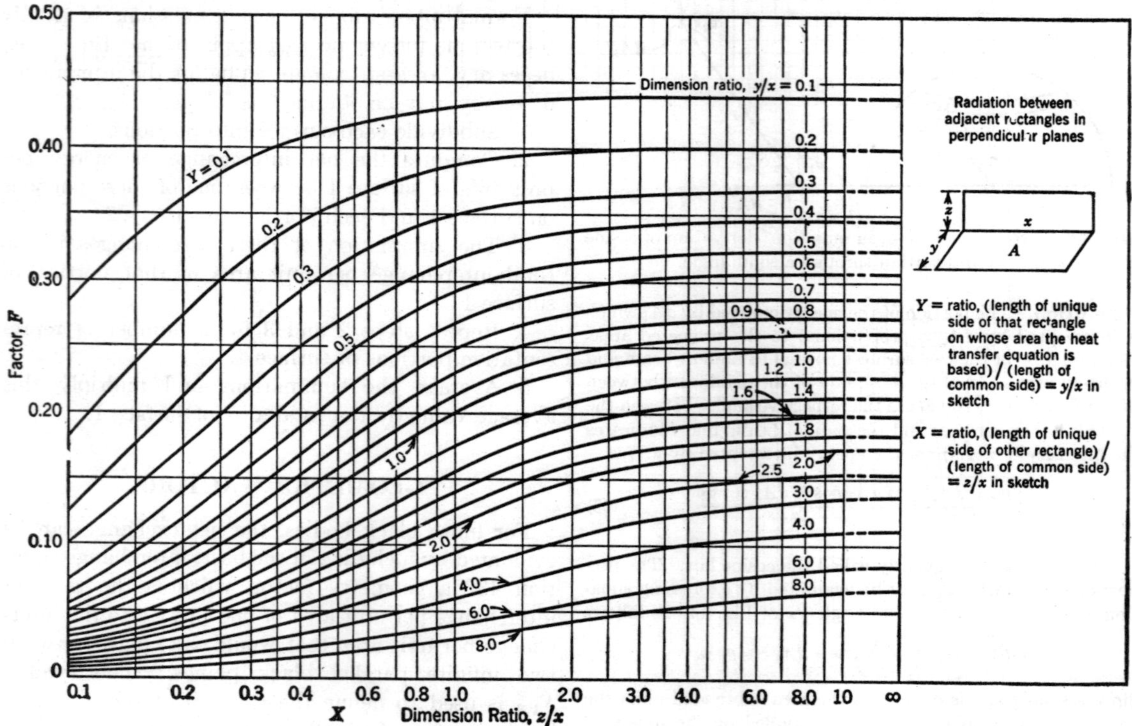

FIG. 441. Geometric shape factor for radiation between adjacent rectangles in perpendicular planes.[31]

Some of these values are plotted in Figs. 441 and 442 as indicated. For problems involving different areas of surfaces 1 and 2, the values of $F_{1,2}$ will differ from $F_{2,1}$. This is evident from a consideration of black bodies.

$$q_{1,2} = \sigma F_{1,2} A_1(T_1^4 - T_2^4)$$
$$= \sigma F_{2,1} A_2(T_1^4 - T_2^4) \quad (438)$$

The product $A_1F_{1,2}$ equals the product $A_2F_{2,1}$, as otherwise there would be a net transfer of energy when the two surfaces are at the same temperature.

tangles in perpendicular planes, obtained by application of equation 436.

If the radiating surfaces are directly opposed parallel planes not connected by refractory or other radiating or reflecting surfaces, the value for geometric factor F is given by the lower four curves in Fig. 442 as indicated.

If there are refractory surfaces between the parallel radiating planes, but with no net radiant heat flux to or from the refractory surfaces, the unknown refractory surface temperatures may be eliminated by energy balances, giving equation 440 for the energy transferred by radiation from A_1 to A_2 by the combined mechanism of direct radiation plus reradiation from the refractory surfaces.

$$q = \sigma\overline{F_{1,2}}A_1(T_1^4 - T_2^4) = \sigma\overline{F_{2,1}}A_2(T_1^4 - T_2^4) \quad (440)$$

where $\overline{F}$ is the geometrical factor, including the effects of refractory walls between black parallel planes.

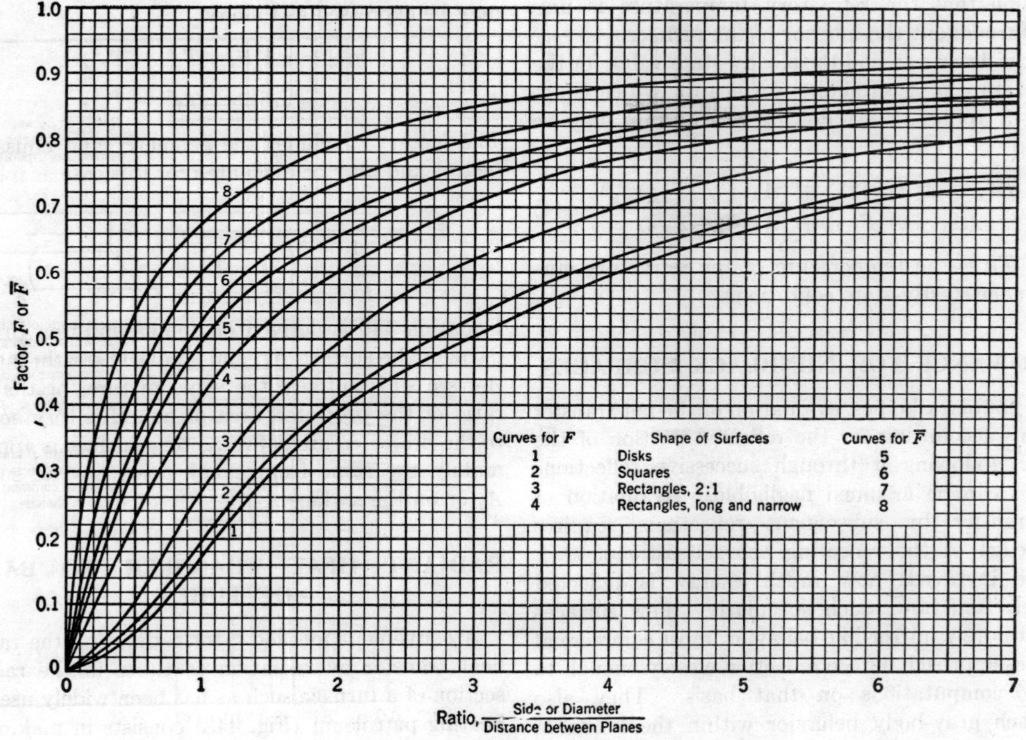

FIG. 442. Geometric factors F and $\overline{F}$ for radiation between parallel planes.[31]

The effect of the refractory walls in increasing the rate of heat transfer between the parallel planes is indicated by comparing the four upper lines in Fig. 442 with the four lower lines. For example, consider two square parallel planes with the distance between them equal to one side of the square. The direct radiation between these two planes is given by curve number 2 in Fig. 442 as 20 per cent of the radiation that would be transferred between unit areas in two infinite parallel planes. If these two parallel planes are connected by nonconducting but reradiating walls, about 53 per cent as much energy would be transferred per unit area between the two parallel surfaces as would be transferred between unit areas in two infinite planes. The addition of the refractory walls has more than doubled the rate of heat transfer.

If an enclosure such as the interior of the furnace may be divided into several radiating surfaces such as A_1, A_2, etc., and the rest of the enclosure may be regarded as reradiating refractory surface A_R at a uniform temperature T_R, the factor $\overline{F_{1,2}}$ may be expressed in terms of the direct geometric factors F as follows.

$$\overline{F_{1,2}} = F_{1,2} + \frac{F_{1,R}F_{R,2}}{1 - F_{RR}} \quad (441)$$

If there are only two heat transfer surfaces involved, equation 441 may be reduced to a more convenient form (equation 442) by substituting from equations $A_1F_{1,R} = A_RF_{R,1}$, $A_2F_{2,R} = A_RF_{R,2}$, and $F_{R,1} + F_{R,2} + F_{R,R} = 1$, as obtained from equations 438 and 439.

$$\overline{F_{1,2}} = F_{1,2} + \frac{1}{\dfrac{1}{F_{1,R}} + \dfrac{A_1}{A_2}\dfrac{1}{F_{2,R}}} \quad (442)$$

If neither A_1 nor A_2 can see itself, that is, if each surface is free of negative curvature, $F_{1,1}$ and $F_{2,2}$ are each equal to zero, and equation 443 applies.

$$F_{1,R} = 1 - F_{1,2} \quad \text{and} \quad F_{2,R} = 1 - F_{2,1}$$

$$\overline{F_{1,2}} = \frac{A_2 - A_1 F_{1,2}^2}{A_1 + A_2 - 2A_1 F_{1,2}} \tag{443}$$

The use of equation 443 requires the evaluation of only one geometric factor. It is based on the assumption that the refractory temperature is uniform throughout the interior of the furnace.

If it is desired to find the steady-state value for the temperature of a refractory surface for which equation 443 applies, the following relationship may be derived.

$$T_R^4 = \frac{(A_1 - A_1 F_{1,2})T_1^4 + (A_2 - A_1 F_{1,2})T_2^4}{(A_1 - A_1 F_{1,2}) + (A_2 - A_1 F_{1,2})} \tag{444}$$

where T_R = the temperature of the refractory surface under steady-state conditions.

ALLOWANCE FOR NONBLACK SURFACES

For nonblack bodies, evaluation of the net transfer requires accounting for the reflected portion of the energy, following it through successive reflections until it can be assumed negligible. Evaluation of the transfer by subsequent reflections requires knowledge of the reflecting characteristic of each surface involved: how much energy is reflected diffusely and how much specularly. Few surfaces exhibit purely mirror-like behavior; most commercial surfaces approach diffuse reflection closely enough to justify computations on that basis. They also approach gray-body behavior within the limits of accuracy of usual practical requirements.

Gray Surfaces

If it may be assumed that all surfaces are gray, a simple and adequate allowance may be made for the emissivity and absorptivity of the nonrefractory surfaces. The emissivity of the refractory surfaces is not a factor, as it is immaterial whether the refractory surface maintains its thermal equilibrium by complete absorption and black-body reradiation or by complete reflection with no absorption.

The rigorous calculation of a factor to express the net transfer between actual surfaces of known geometry and emissivity requires the tracing of typical rays of energy from typical portions of surfaces through multiple reflections and summing up appropriate terms of these infinite series to represent the total flux in each zone. This extremely tedious procedure can be simplified in some cases, for instance, in an enclosure containing only a heat source and a heat sink, both gray, the remainder being diffuse-reflecting refractories.

The transfer of heat due to the combined mechanisms of direct radiation, reradiation from refractory surfaces, and multiple reflection within the enclosure may be expressed as follows.

$$q_{1,2} = \sigma A_1 \mathfrak{F}_{1,2}(T_1^4 - T_2^4)$$
$$= \sigma A_2 \mathfrak{F}_{2,1}(T_1^4 - T_2^4) \tag{445}$$

The factor $\mathfrak{F}$ includes the geometry and emissivity factors and may be evaluated for this case as follows.

$$\mathfrak{F}_{1,2} = \frac{1}{\dfrac{1}{\overline{F_{1,2}}} + \left(\dfrac{1}{e_1} - 1\right) + \dfrac{A_1}{A_2}\left(\dfrac{1}{e_2} - 1\right)} \tag{446}$$

Equation 446 is derived on the assumption that all parts of the surface A_1, and similarly for the surface A_2, see substantially the same picture, or that the ratio of the projected area of nonrefractory surface to the projected area of refractory surface is approximately the same from every point on the surface A_1 or on the surface A_2.

RADIANT HEAT TRANSFER TO BANKS OF TUBES

A convenient method [29] for estimating the rate of heat transfer by radiation to tubes in the radiant section of a furnace such as has been widely used for heating petroleum (Fig. 443) consists in making an

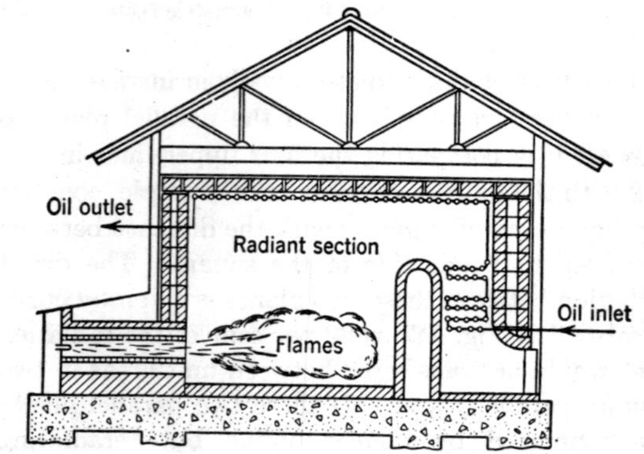

FIG. 443. Diagrammatic section of a typical "radiant heat" furnace for heating petroleum.

energy balance in connection with the rate of heat transfer as suggested in Fig. 445. If A_1 represents an infinite plane, and A_2 represents one or two rows of parallel tubes in parallel planes constituting the equivalent of an infinite plane parallel to A_1, and the only other surface is a refractory surface behind the tubes, $\overline{F_{1,2}}$ is given by Fig. 444. The value for A_2 is the area of the continuous plane in which the tubes are located and not necessarily the projected area of the tubes themselves. For two rows of tubes spaced on centers two diameters apart the projected area of the tubes equals the area of the plane A_2. The rate of heat transfer to a square foot of the plane in which the tubes are located (A_2) is plotted as a function of the furnace temperature for the proper tube wall temperature, incorporating the combined geometrical factors of Figs. 442 and 444, as well as the emissivity of the oxidized surface of the tubes. This curve constitutes a graphical solution of equation 438 if the furnace temperature and tube temperature are known.

An energy balance is then made on this plot, Fig. 445, for the fuel, air, and combustion products supplied to the furnace per hour per square foot of plane surface A_2. If the quantity of fuel supplied to the furnace per hour per square foot of plane surface A_2 and the air-fuel ratio are known, the energy in the form of heat developed by the combustion of this fuel, plus the "sensible heat" introduced as preheat in the air and fuel above the specified datum such as 60° or 0° F, are totaled for 1 sq ft of plane surface A_2 and plotted as a point M on the temperature ordinate representing this datum. This point represents the quantity of heat liberated in the furnace per hour per square foot of plane surface A_2 above the datum temperature.

If all this energy remains in the products of combustion, that is, if no heat is transferred from the burning gases in the radiant section of the furnace, the gases would leave the radiant section of the furnace at a temperature equal to the computed flame temperature. This computed temperature is plotted on the ordinate of zero heat transfer (point N) and represents the furnace temperature if all the heat liberated by the gases in the furnace remains in the products of combustion corresponding to zero heat transfer rate in the radiant section.

Intermediate points are then plotted by computing the temperature of the products of combustion for different quantities of heat transferred to the plane

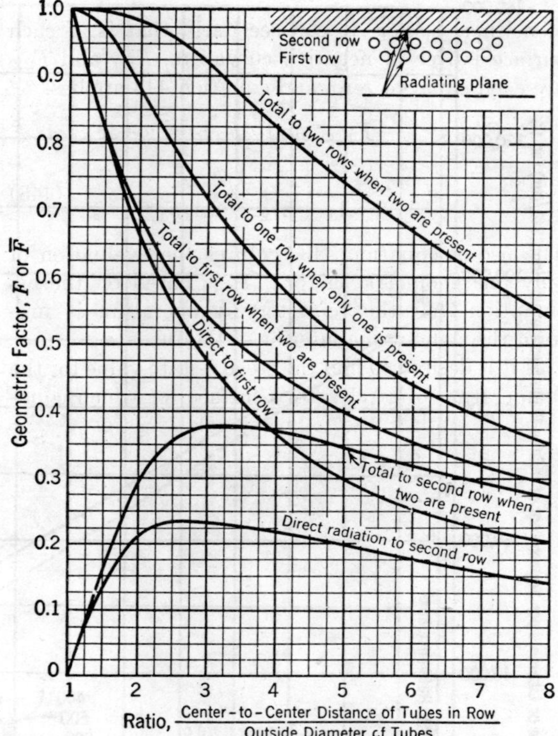

FIG. 444. Geometric factor F or $\overline{F}$ for use in equations such as 440 or 446 for radiation to rows of tubes, based upon infinite parallel planes.[18]

surface A_2 during combustion. These temperatures are plotted as a function of the quantities of heat transferred per hour per square foot of A_2 as indicated in Fig. 445. The intersection of the locus of these points with the heat transfer rate curve indicates the necessary energy balance, assuming that the temperature of the furnace is substantially uniform throughout the radiant section and equal to the temperature of the products of combustion when leaving the radiant section. This assumption closely approximates actual conditions in the type of furnace shown in Fig. 443 where the flame is usually adjusted to extend a good part of the distance across the bottom of the furnace when burning oil, or becomes an extremely short flame when burning gas with preheated air. The differences between these two types of flames can be adjusted by using a higher emissivity for the long radiant flame.

The same chart, Fig. 445, may also be used to determine the quantity of heat to be liberated in

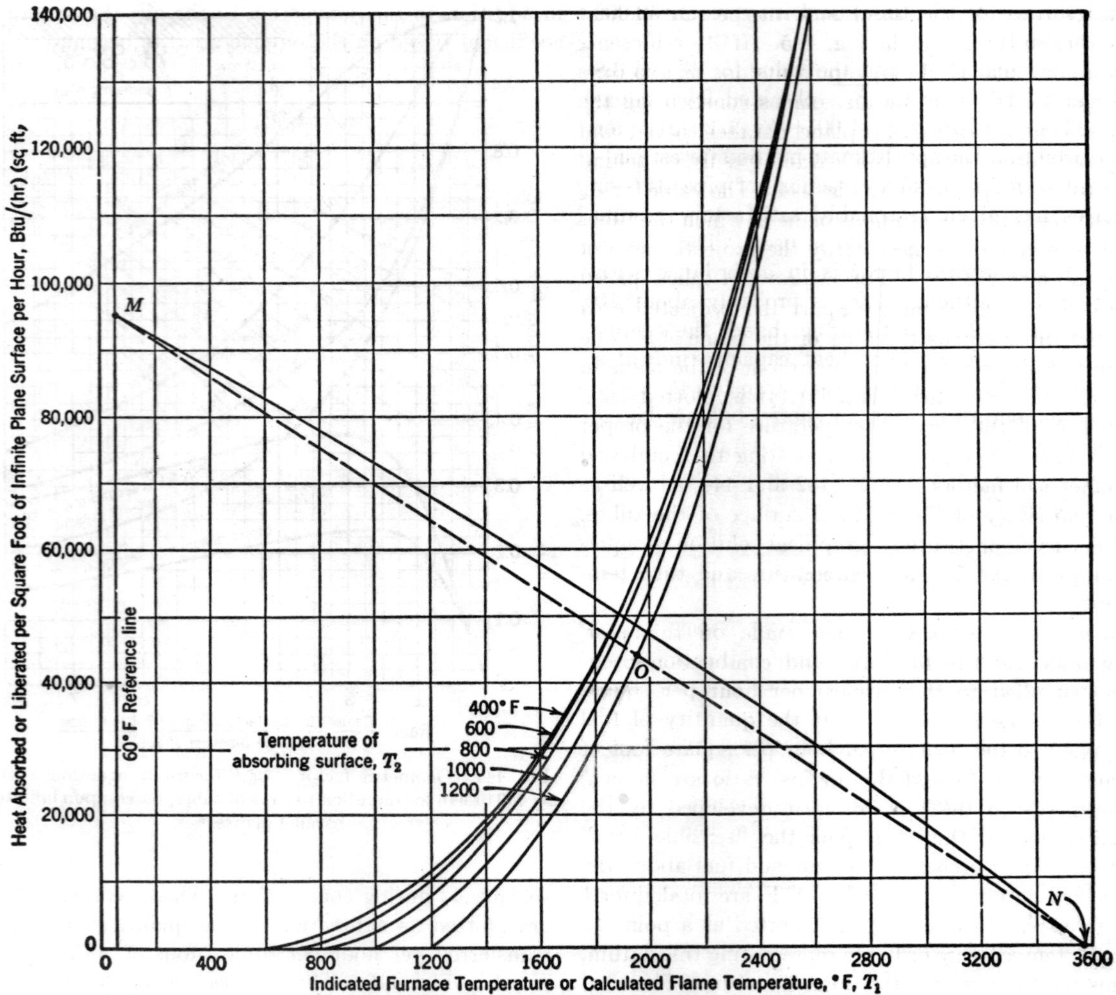

FIG. 445. Graphical solution for an energy balance and rate of heat transfer by radiation in a furnace.[29]

the furnace for a desired heat transfer rate. In this case the computed flame temperature and the intersection of the curve is known. The solution is obtained by extending the locus of the temperature of the products of combustion to the ordinate of the datum temperature. If it is assumed that the heat capacity of the products of combustion is constant, this is readily done by drawing a straight line through the computed flame temperature for zero heat absorption (point N) and through the desired value for the heat absorption per square foot of surface A_2 (point O) and reading the intersection (point M) of the extension of this straight line with the ordinate corresponding to the datum temperature used in making the energy balance. This general method of approach has been further developed for design of furnaces.[27]

With a little study of Fig. 445, it can be seen that the greater the area of the heat-absorbing plane A_2 for the same fuel consumption per hour, the more mild are the heating conditions. When radiant furnaces were first designed, this relationship was not generally appreciated and very small radiant heat-absorbing surfaces were used because the early furnace builders were "afraid" of the high heat transfer rates. In modern furnaces heat transfer rates in the radiant sections of oil-heating furnaces are usually 10,000 to 20,000 Btu/(hr)(sq ft of external surface of the tube) and may reach 50,000 Btu/(hr)(sq ft).

If A_1 is equal to A_2 and the value for $F_{1,2}$ is 0.8 or greater, the value for $\overline{F_{1,2}}$ obtained from equation 443 is approximately $(1.00 + F_{1,2})/2$. It is for this reason that the approximate method for estimating heat transfer in radiant sections of pipe stills or furnaces has given reasonably satisfactory results when a shape factor of 1.00 is used. The use of the high geometrical factor is in accordance with equation 443, although 1.00 is probably about 10 per cent too high. On the other hand, the energy balance assuming constant heat capacity indicates a somewhat lower rate of heat transfer (a lower furnace temperature) than an energy balance based on more reliable specific heats. Therefore the use of a shape factor of 1.00 and a straight-line energy balance gives reasonably satisfactory results.

RADIANT HEAT TRANSFER COEFFICIENTS

In many cases the transfer of heat includes convection and radiation. An example of this kind was indicated in the case of heat transfer from the cast iron furnace door to the surrounding space. Computations are usually best made in the manner indicated by determining the heat transferred by radiation and the heat transferred by convection and adding the two quantities to obtain the total heat transferred.

In many cases, however, it is convenient to compute the rate of heat transfer by radiation according to the same equation 382 as is used for convection transfer. This may be done by combining equations 382 and 434 to solve for the radiant heat transfer coefficient h_r.

$$h_r = \frac{q_r}{A_1(T_1 - T_2)}$$

$$= \frac{0.173e_1[(0.01T_1)^4 - (0.01T_2)^4]}{T_1 - T_2} \quad (447)$$

Figure 446 is the solution of equation 447 for $e = 1$. It gives h_r in Btu/(hr)(°F)(sq ft) as a function of T_1 and T_2 for black-body conditions, $e = 1$. As indicated in equation 447, the value for h_r given in Fig. 446 may be modified by multiplying by the value of e, F, $\overline{F}$, or $\mathfrak{F}$.

For the heat transfer indicated in Fig. 446a, considering radiation from surface 3 to the surroundings in addition to the convection to the air, the overall coefficient based on the outside area A_3 becomes

$$U_3 = \frac{1}{\dfrac{A_3}{A_2 h_{1,2}} + \dfrac{(x_3 - x_2)}{k}\dfrac{A_3}{A_{2,3}} + \dfrac{1}{h_c + h_r}} \quad (448)$$

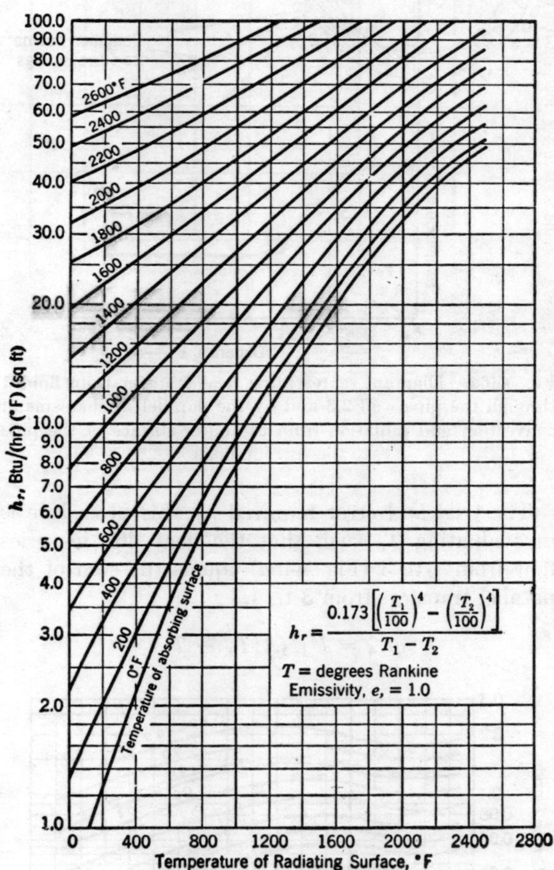

FIG. 446. Coefficient of heat transfer by black body radiation ($e = 1$).

Application of this concept that temperature differences are proportional to resistances must recognize that the resistance to the flow of heat from surface 3 involves both radiation and convection in parallel or that $R_{3,4} = 1/(h_{c,3,4} + h_{r,3,4})$.

For such cases involving uncertain surface temperatures, it is often convenient to equate the heat flux from fluid 1 to surface 3 (Fig. 446a) to the sum of the radiation and convection from surface 3 to the fluid 4, for assumed value of T_3, that is,

$$q_{1,3} = q_{3,4,\text{ conv.}} + q_{3,4,\text{ rad.}} = U_3A_3(T_1 - T_4) \quad (449)$$

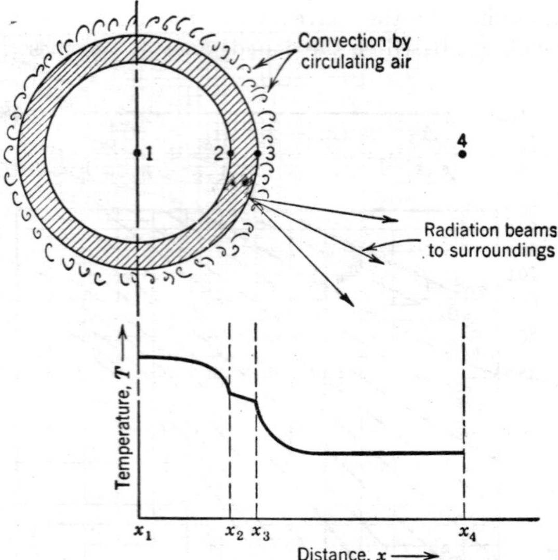

FIG. 446a. Diagram representing heat transfer from fluid 1 through the pipe wall 2,3 and by the parallel mechanisms of convection and radiation from the outer surface of the pipe to the surroundings.

The trial and error required by this procedure is in evaluating T_3 such that the heat flux in series flow from 1 to 3 (Fig. 446a) equals the sum of the parallel transfers from 3 to 4.

$$q_{1-4} = U_{1-4}A(T_1 - T_4)$$

RADIATION FROM NONLUMINOUS GASES

In dealing with solids and refractory surfaces, the assumption of gray-body conditions, that is, constant emissivity and absorptivity independent of the wavelength of the radiation, does not introduce appreciable error. But in the treatment of radiation from nonluminous gases, it is necessary to recognize the dependence of the emissivity or absorptivity on wavelength. If black-body radiation passes through a mass of gas, absorption occurs in certain regions or wavelengths of the infrared spectrum. Similarly if the gas is heated it radiates energy of these same wavelengths. This radiation and absorption originate in quantum energy changes in the energy of rotation and interatomic vibration within the molecules. Industrial furnace gases such as carbon monoxide, carbon dioxide, water vapor, and sulfur dioxide, also ammonia, hydrogen chloride, the hydrocarbons, and alcohols, possess emission bands of sufficient magnitude to be important. Symmetrical diatomic molecules such as hydrogen, oxygen, and nitrogen do not show emission or absorption bands of sufficient importance at temperatures encountered in industrial operations.

The intensity of radiation from a gas mass is a function of the number of molecules contributing to it, and of the number of intervening molecules which can absorb some of the emitted energy before it leaves the gas. Consider the radiation from a hemispherical gas mass to an element of surface located

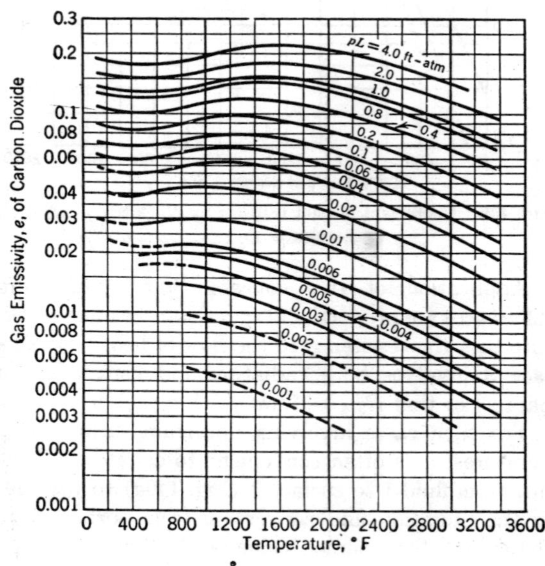

FIG. 447. Emissivity of carbon dioxide.[20]

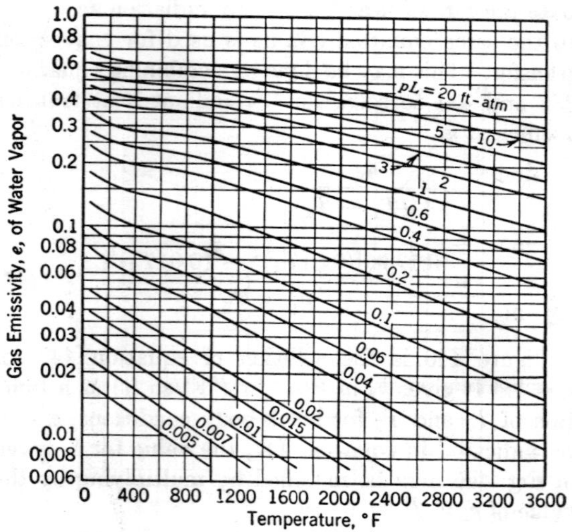

FIG. 448. Emissivity of water vapor.[20]

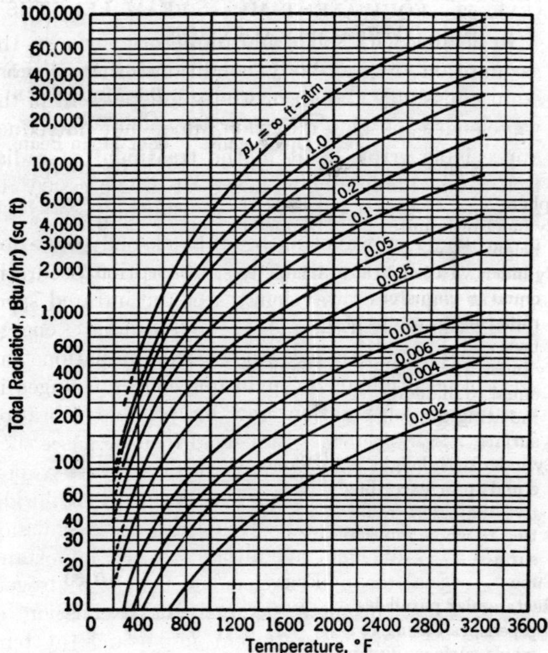

FIG. 449. Total radiation q/A due to sulfur dioxide.[31] $q/A = e\ (0.173)\ (T/100)^4$, where e = emissivity of SO_2.

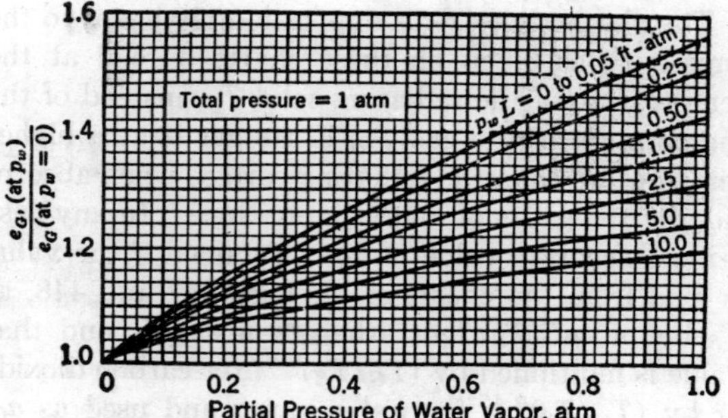

FIG. 450. Correction factor to the emissivity of water.[20]

at the center of the base of the hemisphere. If the radius of the hemisphere is L, according to Beer's law the intensity of the radiation from a gaseous component on the surface is proportional to the product of L and p, the partial pressure. The distance L is called the beam length. For shapes other than hemispheres an effective beam length may be determined.

Experimental measurements indicate that Beer's law is sufficiently accurate for carbon dioxide but quite inaccurate for water vapor; hence its general applicability must be questioned. This contribution of layers beneath the surface is in sharp contrast with solids and most liquids, for which radiation is strictly a surface phenomenon.

The emissivity of the gas also depends upon the temperature of the gas T_G and the total pressure. Figures 447, 448, and 449 give the emissivity of carbon dioxide,[21] water vapor,[20] and sulfur dioxide[31] as functions of the temperature and of the product pL, for a total pressure of 1 atm. The failure of Beer's law for water vapor requires that a correction also be made for partial pressure of water vapor in rich gases, as indicated in Fig. 450.

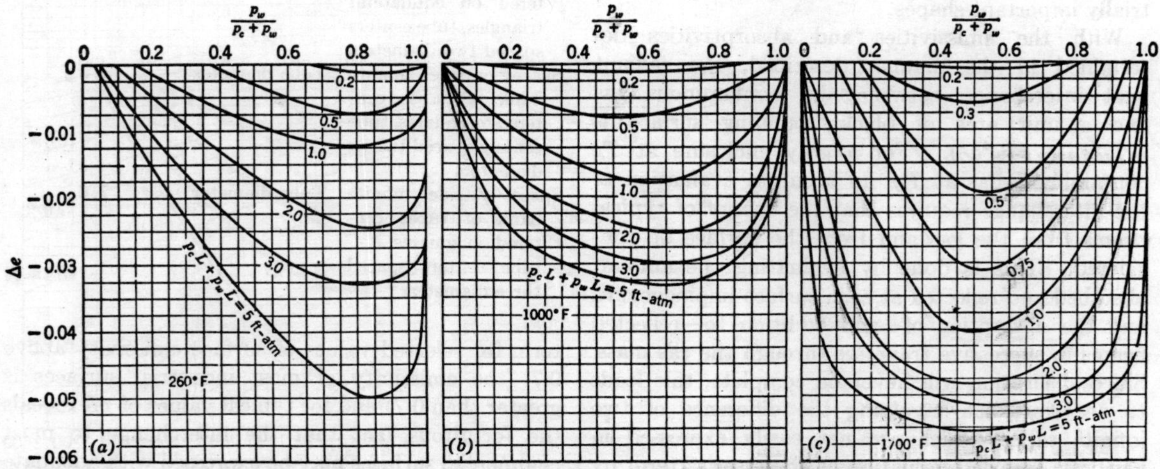

FIG. 451. Corrections to added emissivities of carbon dioxide and water vapor:[21] (a) at 260° F; (b) at 1000° F; (c) at 1700° F and higher.

The absorptivity a_G is approximately equal to the emissivity as given on these charts if read at the temperature of the radiating solid T_S, instead of the temperature of the gas T_G, if the gas is at a higher temperature than the radiating surface when absorption by the gas is of minor importance. In any case a more accurate value of a_G is obtained if the value is read from the charts of e_G, Figs. 447 and 448, at T_S but at $pL(T_S/T_G)$ rather than at pL, and that value is multiplied by $(T_G/T_S)^{0.65}$ for carbon dioxide or by $(T_G/T_S)^{0.45}$ for water vapor and used as a_G.

When carbon dioxide and water vapor are present together, the total emissivity of the mixture is somewhat less than the sum of the emissivities of the two gases.[21] This is caused by the gases having common wavelengths of absorption, as indicated in Fig. 439, which makes each gas somewhat opaque to the other. The correction to be applied to the sum of the emissivities is given in Fig. 451.

The charts of Figs. 447, 448, and 449 are based on the radiation emitted by a hemisphere of gas to an element of surface at the center of the base, in which case the length of the beam L is the same in all directions. In industrial operations the shape is usually quite different from that of the hemisphere, but an equivalent mean beam length L may be determined from the dimensions and shape of the radiating system, as indicated in Table 52, and used for determining the emissivities in the charts of Figs. 447, 448, and 449. The values given in Table 52 were obtained by integrating the basic differential equations within the proper limits for various industrially important shapes.

With the emissivities and absorptivities determined in the manner described, the radiant heat interchange between the nonluminous gas and a unit area of black bounding surface is $(\sigma T_G^4 e_G - \sigma T_S^4 a_G)$. For a gray enclosure at T_S containing a gas at T_G, an accurate evaluation of the interchange requires that the energy of typical beams from the gas and from the surface be considered. This is done by evaluating the amount absorbed and reflected at each surface impingement, and the absorption of—and addition to—reflected beams in successive traverses through the gas mass. Since absorption will never be complete, this leads to an expression involving the difference of two infinite series, which are not easily expressed in algebraic form. Evaluation of these series term by term for selected values of surface emissivity above 0.7 (the emissivity of most industrial surfaces is greater than 0.7) and for typical values of pL reveals the fortuitous fact that the interchange to most commercial surfaces may be expressed with adequate

TABLE 52. EQUIVALENT MEAN BEAM LENGTHS FOR GAS RADIATION

Shape	Characterizing Dimensions, D	Factor by Which D is Multiplied to Obtain Equivalent Mean Beam Length L
Sphere	Diameter	0.60
Cylinder of infinite height	Diameter	0.9
Cylinder with height equal to diameter radiating to center of base	Diameter	0.77
Cylinder with height equal to diameter radiating to whole surface	Diameter	0.6
Cylinder with height equal to or exceeding two diameters radiating to whole internal surface	Diameter	1.0
Cube	Edge	0.60
Rectangular parallelepipeds, radiating to whole surface, dimension ratios as follows:		
1:1:1 to 1:1:3	$\sqrt[3]{\text{Volume}}$	0.66
1:2:1 to 1:2:4	$\sqrt[3]{\text{Volume}}$	0.66
1:1:4 to 1:1:∞	Shortest edge	1
1:2:5 to 1:2:8	Shortest edge	1.3
1:3:3 to 1:∞:∞	Shortest edge	1.8
Space between infinite parallel planes	Distance between planes	1.8
Space outside infinite bank of tubes centered on equilateral triangles, tube centers spaced two diameters	Tube diameter	2.8
Space outside infinite bank of tubes centered on squares, tube centers spaced two diameters	Tube diameter	3.5
Space outside infinite bank of tubes centered on squares, tube centers spaced three diameters	Tube diameter	7.6

accuracy if surface emissivity is taken as half-way from its actual value to 1, and only the first impingement considered, that is,

$$\frac{q_{G,S}}{A} = 1730\left(\frac{e_S + 1}{2}\right)\left[e_G\left(\frac{T_G}{1000}\right)^4 - a_G\left(\frac{T_S}{1000}\right)^4\right] \tag{450}$$

The total heat transfer to the ultimate receiving surface through intermediate heat receivers and reradiators such as refractory surfaces and the interchange between gases and the surfaces becomes a complicated problem involving energy balances, radiation, and convection coefficients. However, a satisfactory approximation can be made by the method of Fig. 445. The effective area is assumed to be that of the heat-absorbing surface which may be modified by various shape factors, depending upon the ratio of refractory to nonrefractory surfaces within the radiant system. The effective temperatures are assumed to be those of the gas and the ultimate receiving surface. The approximate allowance for the ratio of refractory surface to heat-absorbing surface can be made by adding to the heat-absorbing surface a certain fraction of the refractory surface. The value of this fraction of the refractory surface to be added to the area of the heat-absorbing surface varies from zero, when the ratio of refractory surface to ultimate receiving surface is very high, to unity, when this ratio is very low and the value of the emissivity is also low. When the areas are of the same order of magnitude, approximately 0.7 times the refractory surface may be added to the heat-absorbing surface. In this manner equation 445 or 450 may be solved using only the two temperatures, those of the receiving surface and of the gas.

RADIATION FROM LUMINOUS FLAMES

If the hot gas contains soot particles formed in the flame by the combustion or decomposition of fuel, these particles radiate energy and the flame is said to be luminous. If the total emissivity of the flame as computed by Figs. 452 and 453 is added as an additional multiplying factor to equation 434 or 450, using $(e_S + 1)/2$ for the emissivity of the solid surface, satisfactory results may be obtained.

The absorptivity and emissivity of luminous flames decrease with an increase in wavelength, and the total emissivity is less than the emissivity in the visible spectrum.[19] This makes visual estimation of luminous flame emissivity misleading. By means of

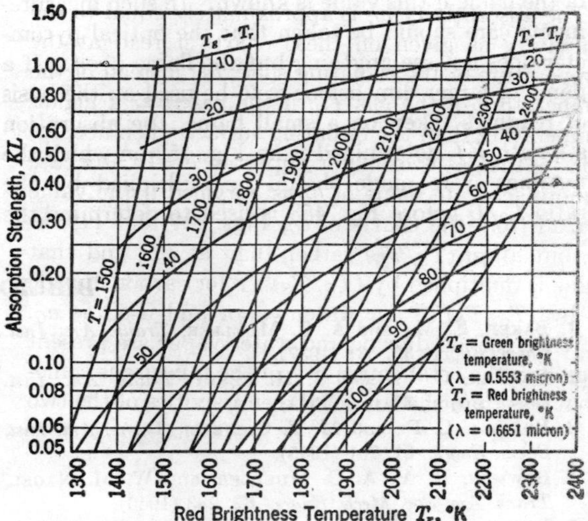

FIG. 452. Absorption strength KL of luminous flames.[19]

an optical pyrometer containing color screens of different wavelengths, red and green, the two apparent temperatures, the red brightness temperature T_r and the green brightness temperature T_g, permit a calculation of the true flame emissivity as indicated in Fig. 452. The absorption strength KL is indicated as the product of K, which measures the soot concentration of the flame, and L, which is the effective thickness of the flame through which the pyrometer is sighted. An optical pyrometer with only the red screen available may still be used in connection with Fig. 452 and the true temperature

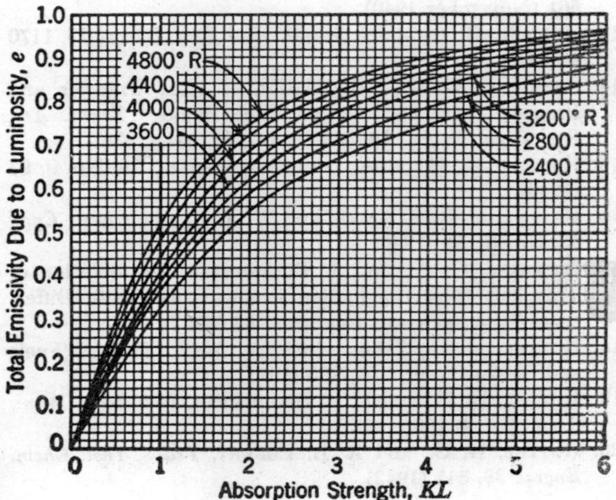

FIG. 453. Emissivity of luminous flames.[19]

of the flame if this value is known. In such measurements care should be taken that the optical pyrometer does not see anything but the flame itself. If a flame of larger dimensions is to be used on the basis of readings taken on a small flame, the absorption strength KL determined from Fig. 452 on which the readings were made should be multiplied by the ratio L_2/L before Fig. 453 is used to determine the total emissivity of the luminous flame, where L_2 represents the effective thickness of the flame whose emissivity it is desired to estimate, and L equals the effective thickness of the flame on which measurements were taken.

With large combustion chambers having highly luminous flames, the emissivity approaches unity.

BIBLIOGRAPHY

1. Baker, E. M., and A. C. Mueller, *Trans. Am. Inst. Chem. Engrs.*, **33**, 531 (1937).
2. Bonilla, C. F., and C. H. Perry, *Trans. Am. Inst. Chem. Engrs.*, **37**, 685 (1941).
3. Bonilla, C. F., and M. T. Cichelli, *Trans. Am. Inst. Chem. Engrs.*, **51**, 761 (1945).
4. Bowman, R. A., A. C. Mueller, and W. M. Nagel, *Trans. Am. Soc. Mech. Engrs.*, **62**, 283 (1940).
5. Breidenbock, E. P., and H. E. O'Connell, *Trans. Am. Inst. Chem. Engrs.*, **42**, 761 (1946).
6. Colburn, A. P., *Trans. Am. Inst. Chem. Engrs.*, **29**, 174 (1933).

6a. Colburn, A. P., *Ind. Eng. Chem.*, **25**, 873 (1933).

7. Colburn, A. P., and O. A. Hougen, *Ind. Eng. Chem.*, **26**, 1178 (1934).
8. Cryder, D. S., and A. C. Finalborgo, *Trans. Am. Inst. Chem. Engrs.*, **33**, 346 (1937).
9. Davis, E. S., *Trans. Am. Soc. Mech. Engrs.*, **65**, 755 (1943).
10. Dittus, F. W., and L. M. K. Boelter, *Univ. Calif. Pub. Eng.*, **2**, 443 (1930).

10a. Dodge, B. F., *Chemical Engineering Thermodynamics*, p. 358, McGraw-Hill Book Co. (1944).

11. Dusinberre, G. M., "Numerical Solution of Heat Conduction Problems," *Trans. Am. Soc. Mech. Engrs.*, **67**, 703 (1945).
12. Fatica, N., and D. L. Katz, *Chem. Eng. Progress*, **45**, 661 (November 1949).
13. Gurney, H. P., and J. Lurie, *Ind. Eng. Chem.*, **15**, 1170 (1923).
14. Grimison, E. D., "Correlation of Heat Transfer and Pressure Drop Data across Tube Banks," *Trans. Am. Soc. Mech. Engrs.*, **59**, 583 (1937); **60**, 381 (1938).
15. Hanson, G. H., *et al.*, *Petroleum Refiner*, **25**, No. 9, 93 (1946).
16. Happel, J., R. S. Aries, and W. J. Barns, *Chem. Eng.*, **53**, No. 10, 99 (1946).
17. Hawkins, G. A., and J. T. Agnew, "The Solution of Transient Heat Conduction Problems by Finite Differences," *Purdue Univ. Eng. Bull.* 98 (1947).
18. Hottel, H. C., *Trans. Am. Soc. Mech. Engrs.*, "Fuels and Steam Power," **53**, 265 (1931).
19. Hottel, H. C., and F. P. Broughton, *Ind. Eng. Chem., Anal. Ed.*, **4**, 166 (1932).
20. Hottel, H. C., and R. B. Egbert, *Trans. Inst. Chem. Engrs.*, **38**, 541 (1942).
21. Hottel, H. C., and H. G. Mangelsdorf, *Trans. Am Inst. Chem. Engrs.*, **31**, 517 (1935).
22. Jacob, M., *Trans. Am. Inst. Chem. Engrs.*, **42**, 1015 (1946).
23. Katz, D. L., K. O. Beatty, Jr., and A. S. Foust, *Trans. Am. Soc. Mech. Engrs.*, **67**, 665 (1945).
24. Katz, D. L., R. E. Hope, S. Datsko, and D. B. Robinson, *Refrig. Eng.*, **53**, 315 (1947).
25. Kirkbride, C. G., *Trans. Am. Inst. Chem. Engrs.*, **30**, 170 (1933–34).
26. Langmuir, I., E. Q. Adams, and G. S. Meikle, *Trans. Am. Electrochem. Soc.*, **24**, 53 (1913).
27. Lobo, W. E., and J. E. Evans, *Trans. Am. Inst. Chem. Engrs.*, **35**, 743 (1939).
28. Lyon, R. N., *PhD. Thesis*, University of Michigan (1949).
29. Maker, F. L., *Notes on Design of Furnaces*, Private Papers.
30. Martinelli, R. C., *Trans. Am. Soc. Mech. Engrs.*, **69**, 947 (1947).
31. McAdams, W. H., *Heat Transmission*, 2nd ed., McGraw-Hill Book Co. (1942).

31a. McAdams, W. H., "Heat Transfer," *Chem. Eng. Progress*, **46**, 121 (March 1950).

32. McAdams, W. H., *et al.*, *Chem. Eng. Progress*, **44**, 639 (1948).
33. Moore, A. D., *Ind. Eng. Chem.*, **28**, 704 (1936).
34. Nessi, A., *Bull. soc. encour. ind. natl.*, **131**, 289 (1932).
35. Nukiyama, S., and H. J. Yosikata, *Soc. Mech. Eng. (Japan)*, **33**, No. 1 (1930).
36. Nusselt, W., *Z. Ver. deut. Ing.*, **60**, 541 (1916).
37. Paschkis, V., and H. D. Baker, *Trans. Am. Soc. Mech. Engrs.*, **64**, 105 (1942).
38. Perry, J. H., *Chemical Engineers' Handbook*, 3rd ed., p. 485, McGraw-Hill Book Co. (1950).
39. Rowley, F. B., R. C. Jordan, and R. M. Lander, *Trans. Am. Soc. Refrig. Eng.*, **53**, 35 (1947).
40. Shea, F. L., and N. W. Krase, *Trans. Am. Inst. Chem. Engrs.*, **36**, 463 (1940).
41. Short, B. E., *Univ. Texas Pub.* 3819, May 1938.
42. Sieder, E. N., and G. E. Tate, *Ind. Eng. Chem.*, **28**, 1429 (1936).
43. Sinnott, M. J., and C. A. Siebert, *Ind. Eng. Chem.*, **40**, 1039 (1948).
44. Tubular Exchanger Manufacturers Association, Standards (1941).
45. Wilson, E., *Trans. Am. Soc. Mech. Engrs.*, **37**; 47 (1915).

1. An uninsulated 3-in. steel schedule-40 pipe carrying saturated steam at 60 psia is suspended from the ceiling of a large room. The walls of the room and objects in the room are at 70° F.

(*a*) Compute the heat lost per hour by radiation and convection for a 10-ft length of black pipe.

(*b*) If galvanized pipe were substituted for the black pipe, how much heat would be lost by radiation?

e and *a* for black iron surface = 0.88 at 60° F, 0.92 at 400° F.

e and *a* for galvanized iron = 0.23 at 60° F, 0.30 at 400° F.

2. In a special test, a vertical lime kiln 6 ft in diameter has an empty section at the top from which the combustion gas leaves at 620° F. The gas analysis is 30% CO_2, 2% H_2O to 68% N_2. The brick wall of the kiln is at 240° F. Brick emissivity = 0.78 at 240° F.

How much heat is being transferred from the gases by radiation per foot of kiln height?

3. A horizontal hot-oil line with 6-in. OD and 5.5-in. ID is carrying oil at 750° F. The pipe is lagged with a layer of magnesia 1 in. thick; emissivity $e = 0.90$. The surrounding air and objects are at 90° F. How much heat is being lost per hour per foot of pipe if the heat transfer coefficient from oil to inside pipe wall is 120 Btu/(hr)(°F)(sq ft)?

4. It is known that polished or reflecting surfaces emit and absorb little radiant heat as compared to dark surfaces. If 1-in. iron pipe (oxidized) were carrying steam at 20 psig through an enclosed room at 50° F and heat units were valued at 0.0002 cent/Btu, how much could you afford to pay per foot of pipe for a polished nickel surface which would last 7 years?

How much could you pay for a specific aluminum paint which would last 4 years?

5. If the convection coefficient of heat transfer of a nickel wire of 0.035-in. diameter at 1700° F to air is 6 Btu/(hr)(°F)(sq ft), how much heat is lost by 1 ft of wire inside an oxidized copper container, the air and container being at 160° F?

6. A 3-ft laboratory fractionating column of 0.176-in. ID and 0.276-in. OD made of Pyrex glass [$k = 0.668$ Btu/(hr)(°F)(ft)] is separating a small impurity from methane at 1 atm pressure with the methane boiling throughout the column at −258° F. Around the column is a 2-in. ID jacket, constructed of glass ⅛ in. thick, and evacuated to eliminate conduction. In some cases the outside surface of the column and the inside surface of the jacket are silvered to reduce the heat input to the fractionating column. What is the heat input (Btu per hour) to the boiling methane in the 3-ft length of column with and without the two surfaces coated with silver?

State any assumptions made. The boiling coefficient for methane may be taken as 1000 Btu/(hr)(°F)(sq ft). Air and surroundings are at 70° F.

7. A furnace is placed in a laboratory in such a position that the face of the door is parallel to the wall and removed from the wall by a distance of 4 ft. The door of the furnace is 2 ft square, is maintained at 500° F, and has an emissivity of 0.9. The room and air are at 90° F. The wall is covered with a white lacquer.

What temperature (average) will the area (2 ft by 2 ft) of wall opposite the furnace door attain?

8. A billet of steel at 100° F is placed in a furnace with refractory walls maintained at 1800° F Plot the radiation coefficient for transfer of radiant heat to the billet as a function of the temperature of the surface of the billet.

9. The gas from a lime kiln contains 30 volume per cent carbon dioxide and a negligible amount of water vapor. This gas flows through a horizontal 12-in. schedule-40 steel pipe, which serves as a gas cooler, at a rate of 0.6 lb/(sq ft)(sec). At a given section of this cooler, the gas temperature is 1000° F and the pressure is 14.7 psia. The temperature of the air on the outside of the pipe and surroundings is 80° F. Determine the temperature of the gas at a section of the cooler 100 ft downstream from the first section.

10. The return line to the ammonia compressor in a refrigeration system is a 4-in. schedule-80 steel pipe covered with 1.70 in. of molded cork insulation. At a given point in this line, the ammonia vapor is at a temperature of −10° F and a pressure of 20 psia. The temperature of the air and the surroundings is 80° F.

Calculate the heat transferred from the surroundings to the ammonia vapor per foot length of the line for a velocity of ammonia vapor of 6000 fpm at the stated point in the line.

11. Two parallel disks, 6 and 12 in. in diameter on a common axis and 8 in. apart, are maintained at temperatures of 600° and 1000° F, respectively. If each disk may be considered to be a black body, what is the net interchange of radiant energy between the two?

CHAPTER 32

Evaporation

THE operation of evaporation is usually defined as the concentration of solutions by evaporation in vapor-heated equipment. The evolution of the present-day evaporator from the direct-fired salt-boiling pan of the Middle Ages represents the contribution of many men over many generations toward improvement of the efficiency of the operation. The salt boiler of the Middle Ages charged his brine to a simple rectangular tank under which a fire was built of whatever fuel was available. The salt was expensive, and the purity control negligible.

The principle of evaporation is not of necessity restricted to this class of processes, since the evaporation of a solvent is important in processes which range all the way from humidification where small amounts of water in the absence of a dissolved solute are evaporated into a gas, through the range of commercial solutions which must be concentrated, to the evaporation of water from a product predominantly solid. Those operations at the extremes of this range are usually treated as other unit operations, and evaporation is here restricted to concentration of typical solutions.

Evaporators depend entirely upon heat transfer for their operation. Evaporation equipment has been developed largely as an art, and the heat transfer behavior is difficult to predict because of the many factors involved. Heat and material balance equations are important and useful in computing the size and behavior of evaporators. The effect of dissolved solids on the vapor-liquid equilibria must be recognized, especially for multiple-effect operation. Scaling and salting are often expected when evaporating solutions saturated with respect to one or more constituents.

Scaling is the term normally used in reference to deposition of a solid whose solubility decreases with increasing temperature on the heating surface from the hotter liquor adjacent to the heating surface, and in an amount directly proportional to the amount of evaporation. This phenomenon results in the overall resistance $(1/U)$ increasing with time. It has been shown [8] * that the square of the overall resistance $(1/U^2)$ increases linearly with time. Removal may require shut-down and mechanical removal.

Salting is less clearly understood, in that it is a rapid build-up of a normally soluble material on the heating surface in the zone of vaporization. It is aggravated by small fluctuations in the operating conditions and by any condition that will encourage crystal nucleation, rather than the deposition of materials on previously formed crystals. Removal can usually be effected by boiling with the solution rendered unsaturated by adding an excess of solvent. Fouling by heat-sensitive organics is frequently a serious problem.

HORIZONTAL-TUBE EVAPORATOR

The simplest application of vapor heat to the boiling of a solution resulted in the development of the horizontal-tube evaporator, which has been widely used for more than 50 years. The characteristics of this type of evaporator are less favorable than those of newer developments, and few are being built at present.

In principle, the horizontal-tube evaporator is a closed vessel in the bottom of which horizontal tubes

* The bibliography for this chapter appears on p. 491.

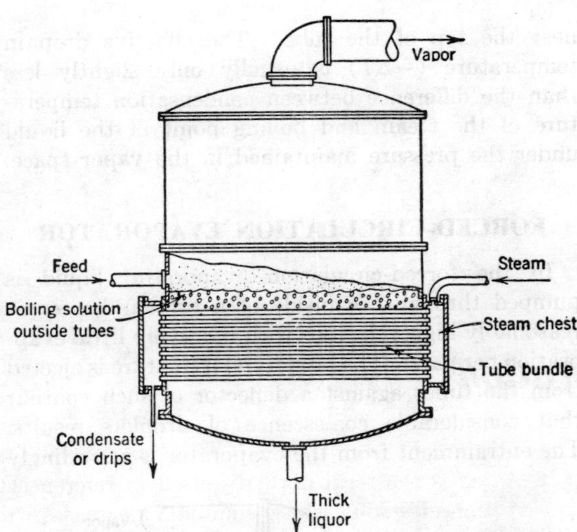

Fig. 454. Cross-sectional diagram of horizontal-tube evaporator. (*Swenson Evaporator Co.*)

are mounted (Fig. 454). The body of the evaporator was first built of cast-iron plates bolted together. Then cylindrical sections of castings were used, and present-day practice predominantly employs welded steel construction. The tubes are usually mounted into the tube sheet with gaskets, in contrast with the usual practice of rolling tubes into the tube sheets of heat interchange equipment. The heating medium is usually steam which is introduced into a steam chest connecting with the inside of the horizontal tubes. The condensate is normally removed at the opposite end of the tube bank. The liquid level is carried somewhat above the top row of tubes in the body of the evaporator, and no effort is made to improve circulation.

The absence of enforced circulation in such an evaporator results in low coefficients, particularly for viscous liquids. The horizontal evaporator is not generally adaptable for foamy liquids, since there is no means of breaking the foam formed by vaporization. Since the liquid is outside the tubes, this type of construction is not adaptable for a liquid which deposits scale on the heating surface because of the extreme inconvenience of removing such deposits. But the horizontal-tube evaporator is economical in first costs for a small machine handling a liquid which is subject to neither foaming nor deposition of solid, and is of relatively low viscosity.

VERTICAL-TUBE EVAPORATOR

The standard vertical and basket evaporator are so similar in their characteristics that they may be discussed together. The vertical-tube evaporator is normally a cylindrical evaporator with a steam chest at the bottom submerged in the liquor. Tubes through the steam chest provide the heating surface, with the liquid inside them. Liquid circulation is upward through the tubes, due to the pumping action of water vapor formed in the tubes, or thermal expansion of the liquid if boiling in the tubes is suppressed by high liquid-level operation. The liquid return to the space below the steam chest may be annular, or it may be a central cylindrical section or several cylindrical downtakes in extremely large installations. If the steam chest is a closed cham with annular liquid return space, the evaporator is classified as a *basket* evaporator (Fig. 455). If the steam chest is annular in configuration with central downtake, the evaporator is classified as *standard vertical* (Fig. 456).

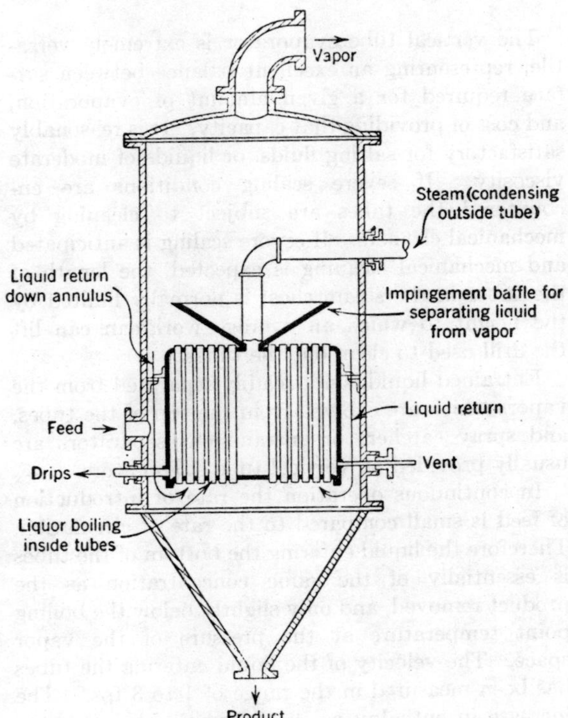

Fig. 455. Cross-sectional diagram of basket-type evaporator. (*Swenson Evaporator Co.*)

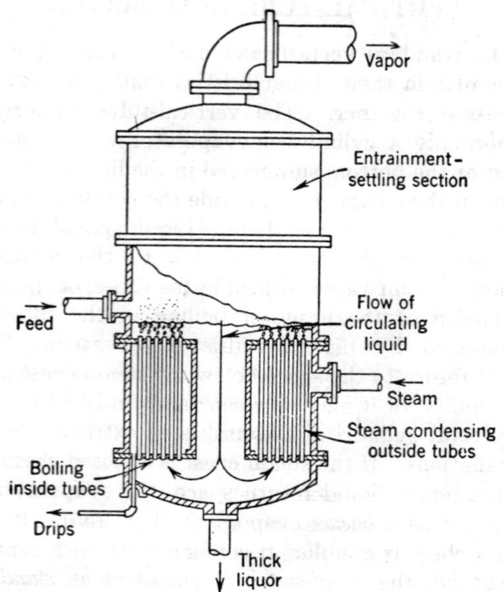

Fig. 456. Cross-sectional diagram of standard vertical-tube evaporator with natural circulation. (*Swenson Evaporator Co.*)

The vertical tube evaporator is extremely versatile, representing an excellent balance between surface required for a given amount of evaporation, and cost of providing that capacity. It is reasonably satisfactory for salting fluids, or liquids of moderate viscosity. If severe scaling conditions are encountered the tubes are subject to cleaning by mechanical cleaners. If severe scaling is anticipated and mechanical cleaning is expected, the length of the tubes in the steam chest is normally limited by the height to which an average workman can lift the drill used to clean out the tubes.

Entrained liquid is not readily separated from the vapors which are evolved from the top of the tubes, and spray catchers or entrainment separators are usually provided on vertical-tube evaporators.

In continuous operation the rate of introduction of feed is small compared to the rate of circulation. Therefore the liquid entering the bottom of the tubes is essentially of the same concentration as the product removed, and only slightly below the boiling point temperature at the pressure of the vapor space. The velocity of the liquid entering the tubes has been measured in the range of 1 to 3 fps.[6] The increase in enthalpy as the liquid rises in the tube is sufficient to raise it to its boiling point and usually to generate some vapor under the existing pressure near the top of the tube. The effective drop in temperature ($-\Delta T$) is usually only slightly less than the difference between condensation temperature of the steam and boiling point of the liquid under the pressure maintained in the vapor space.

FORCED-CIRCULATION EVAPORATOR

In the forced-circulation evaporator, liquid is pumped through the inside of a tubular heater at reasonably high velocities with relatively little evaporation per pass. The liquid-vapor mixture is ejected from the tubes against a deflector of such contour that considerable coalescence of droplets results. The entrainment from the evaporator is accordingly

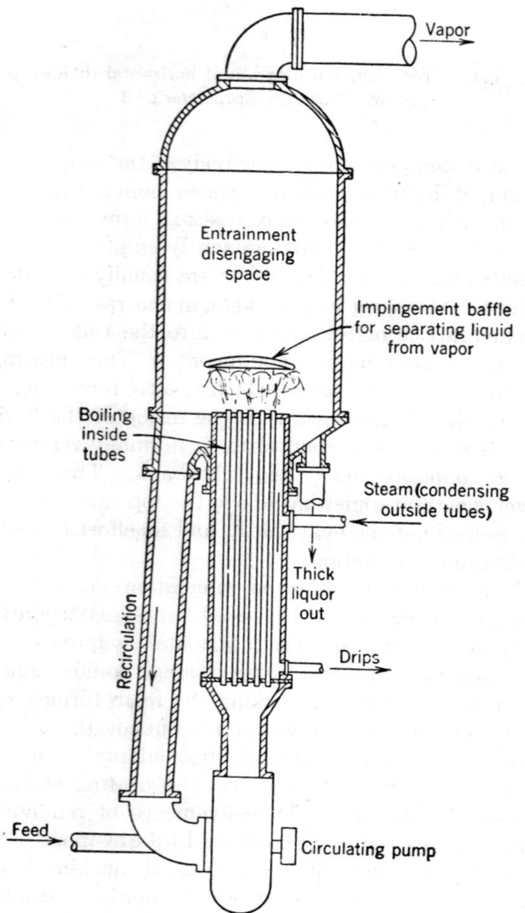

Fig. 457. Cross-sectional diagram of vertical-tube evaporator forced circulation. (*Swenson Evaporator Co.*)

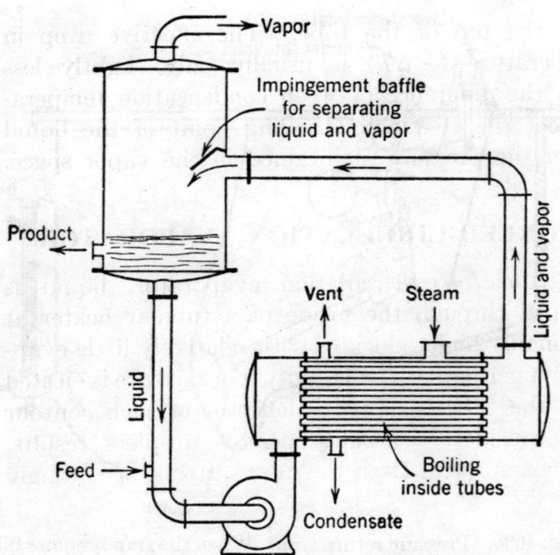

FIG. 458. Cross-sectional diagram of forced-circulation evaporator with an external horizontal heater. (*Swenson Evaporator Co.*)

controlled by the shape of the deflector. The unevaporated liquid is allowed to recirculate, normally through an external piping arrangement to the pump which forces the liquid through the heating element, as shown in Fig. 457. Present practice tends toward building the heating element as a separate tubular heater as shown in Fig. 458, rather than the integral vertical construction of Fig. 457. The external heating unit has considerable advantage in allowing greater ease of cleaning or replacement of tubes, particularly if it is horizontal, but it requires somewhat more complicated piping. This is more than compensated by the improved average coefficient of heat transfer through a cycle of operation since it is possible to install the heating element sufficiently far below the liquid level in the vapor head to avoid boiling on the heating surface, greatly decreasing the rate of deposition of solids.

The selection of a forced-circulation evaporator for an installation depends upon a favorable balance between the cost of energy to circulate the liquid versus the improved coefficient which is secured by the greater velocities possible. One efficient practice is the use of a steam turbine to drive the circulation pump, with the exhaust from the turbine providing the heat source to the evaporator system. The complication of operating a steam turbine compared with electric drive eliminates this setup for small installations.

The circulation pump is most frequently a centrifugal pump; however, for extremely viscous liquids a positive action pump is frequently installed. For a colloid which might be damaged in structure by the rapid agitation in a centrifugal pump, it is possible to substitute a slower-moving type of pump which will not have a deleterious effect upon the colloidal structure of the liquid being evaporated.

LONG-TUBE VERTICAL EVAPORATOR

The long-tube vertical evaporator (LTV or Kestner evaporator) uses natural circulation of the liquid through tubes which are surrounded by steam in the steam chest. The tubes are usually in the range of 12 to 20 ft long and discharge against a deflector in the vapor head, shown in Fig. 459. The

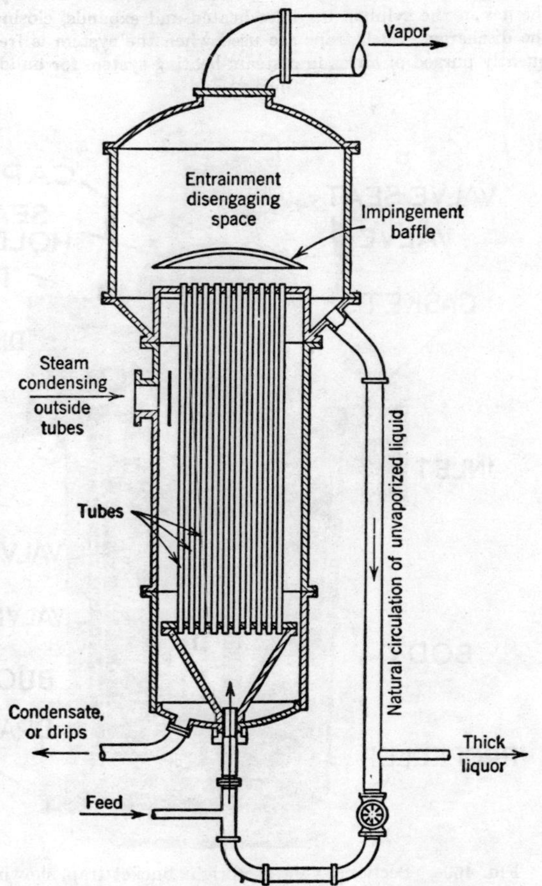

FIG. 459. Cross-sectional diagram of long-tube vertical evaporator. (*Swenson Evaporator Co.*)

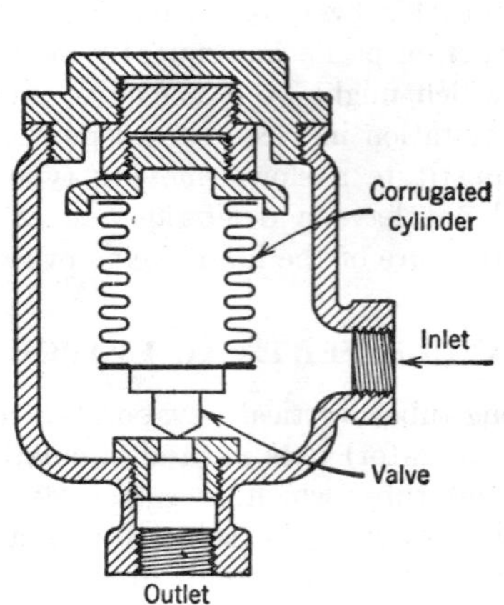

Fig. 460*a*. Thermostatic trap. When steam enters the trap, the gas in the sylphon becomes heated and expands, closing the discharge. Such traps are used when the system is frequently purged of air as in a steam-heating system for buildings.

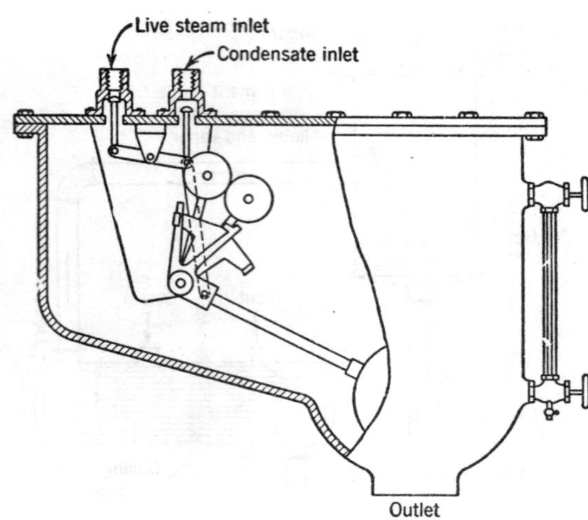

Fig. 460*b*. Pressure return trap. When the trap becomes filled with liquid, the float rises, closing the feed to the trap and opening the steam pressure line, thereby increasing the pressure in the trap sufficiently to return the liquid through the discharge line of the trap.

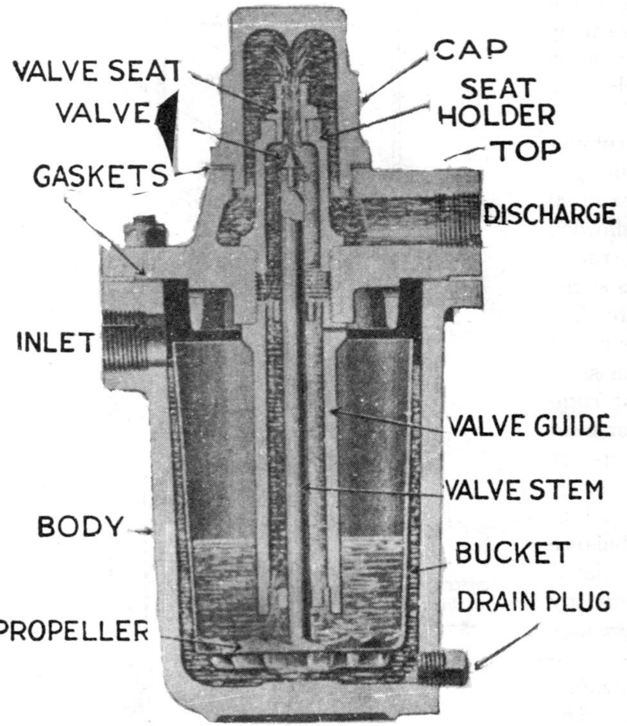

Fig. 460*e*. Sectional view of upright bucket trap, showing bucket in lower position with the discharge valve open and discharging liquid. As the bucket empties, it becomes buoyant and floats on the liquid in the trap, closing the valve. The propeller on the outside bottom of bucket causes the bucket and valve to rotate with each vertical movement of bucket. (*H. O. Trerice Co.*)

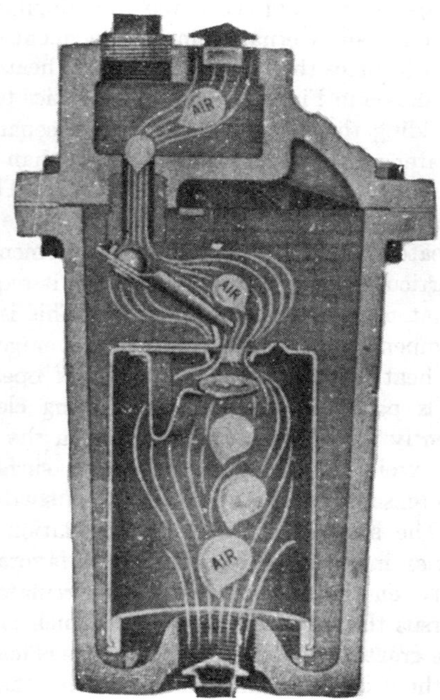

Fig. 460*d*. Sectional view of inverted-bucket blast trap for handling large quantities of air. Air and water flow into trap from bottom until steam enters and causes the bimetallic strip to bend upward, closing the large vent in the top of the bucket. The trap then operates in a manner similar to that shown in Fig. 460*c*. (*Armstrong Machine Works.*)

(1)

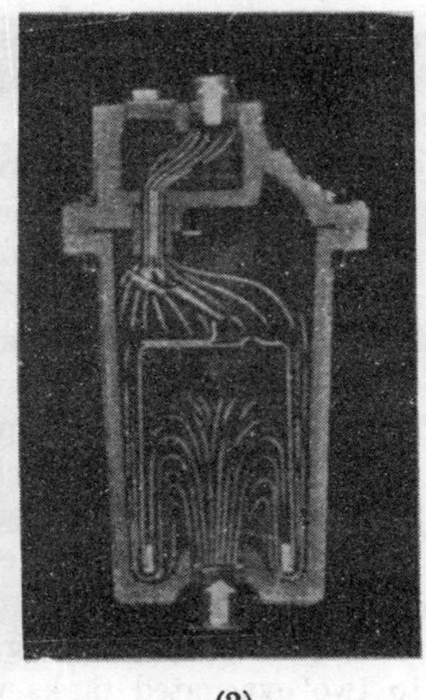

(2)

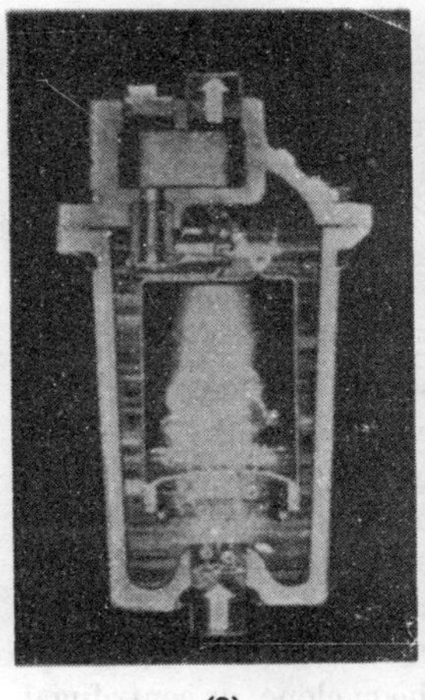

(3)

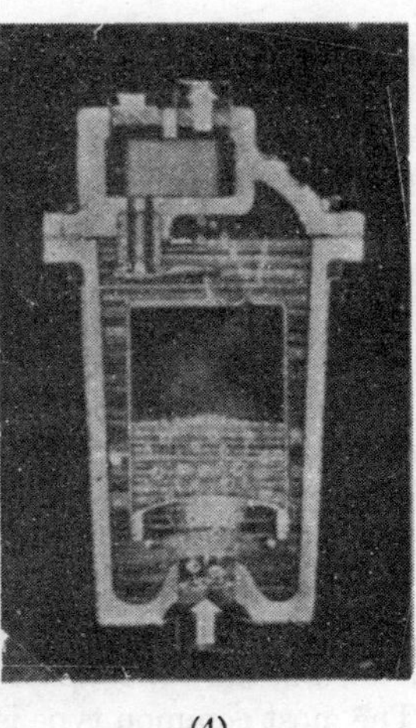

(4)

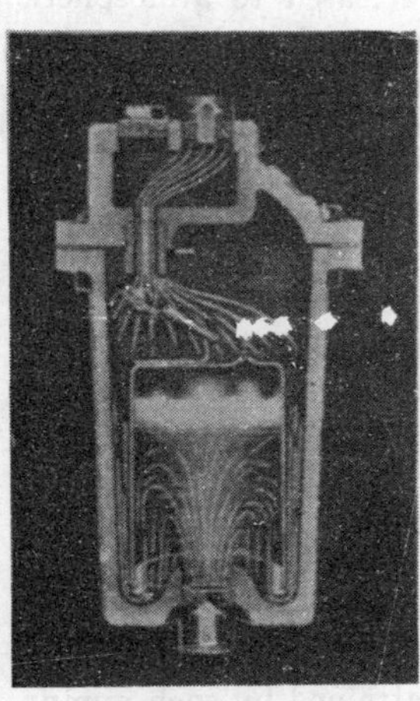

(5)

Fig. 460c. Inverted-bucket steam trap. Inlet at the bottom, outlet at the top. (*Armstrong Machine Works.*)

(1) Steam off. Bucket down, outlet valve open.

(2) Steam on. Water fills body of trap. Excess water escapes through orifice at top.

(3) When steam reaches trap, it fills bucket, displacing water. The bucket floats and closes outlet orifice.

(4) As steam condenses, water enters trap and fills bucket. Air escapes through small vent in top of bucket. Bucket loses buoyancy.

(5) When weight of bucket times leverage equals pressure on valve, trap opens and discharges through outlet at top.

long-tube vertical evaporator is in general not recommended for salting or scaling liquids nor for liquids with extremely high viscosity. It can be adapted excellently to the concentration of foamy liquids. This is believed due to the breakup of the foam when the liquid-vapor mixture is ejected from the tubes at high velocity against a properly shaped deflector.

The fraction of the liquid evaporated per pass is in general considerably higher in an LTV evaporator than in a forced-circulation evaporator.[4] Operation may be with recirculation, as shown, or it may be one pass, with thick liquor removed from the head. The coefficients are usually not so high, but the expense of installing and operating the circulation pump is avoided. Special-purpose evaporators or vacuum pans may have the heating surface in coils or in extremely long units.

MATERIALS OF CONSTRUCTION

Evaporators are built of mild steel unless corrosion conditions dictate a more resistant material. Welded construction is almost universal. For severe conditions or for products subject to damage by slight corrosion, they may be built of any material: lead for rayon spin baths, nickel for strong caustic, Karbate for certain acids, or stainless steel for food products.

EVAPORATOR AUXILIARIES

The feed pumps, condensate withdrawal pumps, and liquor withdrawal pumps, as necessary, for evaporator operations in general, and the circulating pump for forced-circulation evaporators, are standard equipment of a design appropriate to the temperatures of operation encountered, and frequently to the necessity of pumping from a space at subatmospheric pressure. Temperature, pressure, and liquid-level controllers are standard equipment. For removal of condensate from spaces above atmospheric pressure, conventional thermal or bucket-type traps are used (Fig. 460*a*, *c*, *e*); for removal from spaces below atmospheric pressure, a pump or pressure return trap is necessary. One common type of pump trap, using high-pressure steam as the activating fluid, is shown in Fig. 460*b*.

Where the solute is valuable or where it is necessary to re-use condensate in as high a state of purity as possible, an entrainment catcher should be installed in the vapor space or line from an evaporator. The most common type is the cyclone, or centrifugal separator, but baffles or screens are also used. The vapor formed in the evaporator must be condensed, for which a jet or a surface condenser may be used. Removal of cooling water plus condensate from a jet condenser is usually accomplished by a "barometric leg" to avoid the expense of pumping. The condenser is mounted high enough above the discharge point that the water column established in the discharge pipe more than compensates the difference in pressure between the condensing space and the atmosphere. About 34 ft is the usual requirement. A seal pot at the bottom should have a capacity equal to that of the tail pipe; otherwise evacuation and flow of cooling water must be skillfully balanced to establish the water column.

The surface condenser is of conventional tube and shell design, with water and vapor nozzles properly proportioned for the most economical handling of the streams. Since pressure drop on the vapor side is frequently important, vapor lanes through the tube bundle are usual to facilitate distribution of vapor and promote uniform effectiveness of all surfaces. Venting of noncondensables may be from any well-cooled point in the condenser. In large units, or those operating with small temperature differences between cooling-water temperature and saturation temperature, special features are included to assure close approach to cooling-water inlet temperature by the condensate.

Removal of condensate from surface condensers and steam chests below atmospheric pressure may be accomplished by pumping, or by flowing alternately to two evacuated tanks which can be periodically isolated from the system, released to atmospheric pressure for draining, re-evacuated, and returned to service.

When a jet condenser is used, condensate and cooling water are mixed and removed together. The jet condenser is simply a chamber for intimate contacting of vapors with cooling water (Fig. 461). Sprays, trays, and counterflow are used in various designs to improve the contact. Inherently, they are smaller than the surface condenser and much cheaper. A modification of the jet condenser permits elimination of the vacuum pump and the barometric leg in some instances. It is possible to inject a part of the cooling water with sufficient directed velocity into a properly shaped discharge throat to function as a jet pump for removal of the noncondensables. Pressures below 2 psia are attained by such pumps.

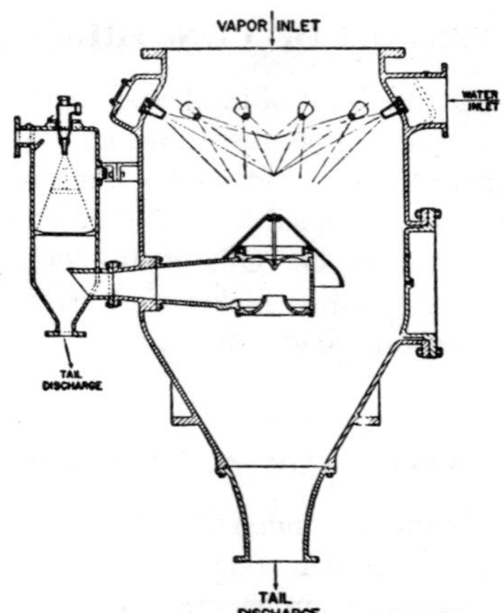

Fig. 461. Two-stage barometric condenser. (*Schutte and Koerting Co.*)

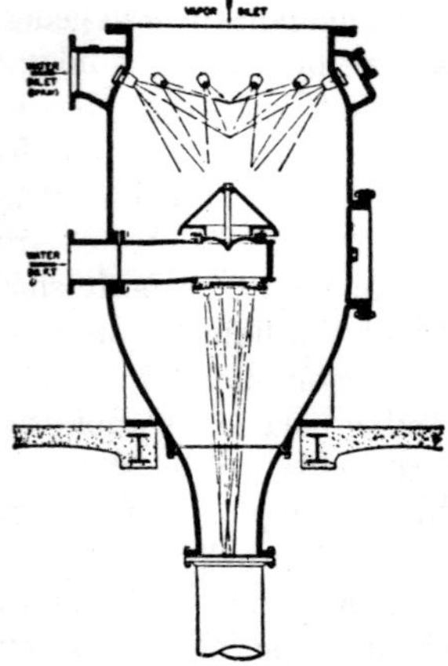

Fig. 462. Barometric jet condenser. (*Schutte and Koerting Co.*)

Figure 462 illustrates one design in which a part of the cooling water is sprayed into the vapor stream for maximum effectiveness in condensation. The remainder is injected into jets directed into the discharge throat to complete the condensation and to eject the noncondensables. These condensers are normally recommended for installation with a barometric leg for evaporator service because of immunity from flooding. Designs are also available in which the jet pump action is sufficient also to eject the cooling water to atmospheric pressure without a barometric leg.

When solids separate out during evaporation, they must be removed if present in sufficient quantity to interfere with free circulation of the liquor. Various means have been used for this purpose, all of which have been discarded in favor of auxiliary settling chambers through which a stream of the slurry is circulated, with solid removed as a thickened slurry for filtration, and clarified liquor returned to the system. The amount of the circulated stream, the reduction in solid content, the dimensions and arrangement of the settlers, the consistency of the thickened slurry, and the extent to which it is washed during the filtration are variables which must be fitted to the individual requirements.

EVAPORATOR OPERATION

The largest economies which can be effected in the operation of evaporators result from the re-use of the energy of vaporization as a heat source for further evaporation at a lower temperature, as indicated diagrammatically in Fig. 463. This repetitive use of the heat supplied is usually referred to as multiple-effect operation, each re-use being accomplished in an additional body, or effect.

Multiple Effect

Although it is theoretically possible to re-use the energy as many times as desired, practical considerations limit the number of effects which may be used. The original energy supply, assuming steam heat is employed, is at a fixed temperature. The minimum attainable temperature in the system is usually limited by the temperature of cooling water available. This overall available temperature drop can be partitioned across as many steps in series as desired. The decrease in temperature drop across each additional effect decreases the heat flux through the heating surface and proportionately reduces the evaporation per unit of heating surface.

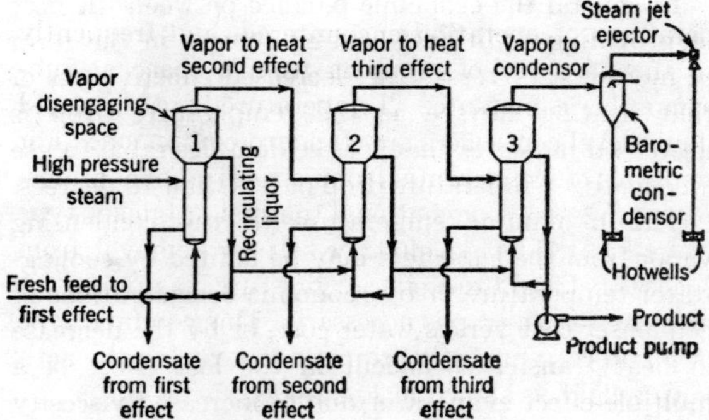

Fig. 463. Flow diagram of a forward-feed triple-effect evaporator.

For solutions which exhibit an "elevation in the boiling point" above that of pure water, there is a loss in potential of the energy in each effect between the temperature at which the solvent is evolved from the boiling solution and the temperature at which the solvent will condense in the steam chest of the next effect where it is essentially pure. For solutions of strong electrolytes this boiling-point elevation is appreciable, and if repeated in more than three or four effects it will absorb a large fraction of the overall available temperature drop. Figure 464 indicates graphically the principle of lost energy potential because of boiling-point elevation for a three-effect evaporator with steam at 5 psig as heat source and with condenser temperature of 125° F saturation. The boiling-point elevations used are those for saturated solutions of sodium chloride.

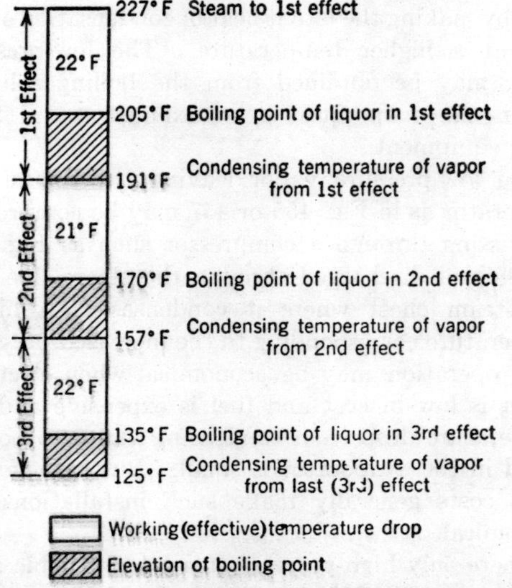

Fig. 464. Boiling-point elevations in a triple-effect evaporator.

In general the economic balance between the cost of providing additional heating surface in the form of more effects versus the decreased effectiveness of that heating surface as the temperature drop is decreased indicates that it is seldom desirable to use an effective temperature drop of less than 10 degrees.

The minimum temperature of the condensing vapor from the last effect may be limited by cooling-water temperature or by economic considerations of condenser cost versus water cost, or by the decrease in heat transfer coefficient in the last effect of a multiple-effect evaporator due to increased viscosity of the liquor at the boiling point of the liquor therein.

Steam is valuable in industry principally according to its availability or temperature level. Although it may appear that large quantities of heat are being wasted when low-pressure vapor is sent to a condenser, this is not necessarily true. Work can be obtained from heat only when there is a sink as well as a source of heat. An evaporator receiving steam under pressure and discharging vapor or steam at lower pressure to a condenser is essentially a heat engine with the work evident in the desired separation of the components of the feed. It is possible to supply this work mechanically by vapor recompression rather than by a temperature drop in heat.

Vapor Recompression

The energy potential of low-pressure vapor may be increased by mechanical or steam-jet compression, thereby making the latent heat of condensation available at a higher temperature. The low-pressure vapor may be obtained from the boiling solution within the evaporator or as exhaust steam from other equipment.

The low-pressure vapor leaving the top of the evaporator as in Fig. 455 or 457 may be compressed by passing through a compressor such as Fig. 165 or 188 and discharged through the steam line into the steam chest where it condenses at a higher temperature corresponding to the increased pressure. Such operation may be economical when electrical power is low in cost and fuel is expensive and the temperature drop from condensing vapor to boiling liquid in the evaporator is small. But high investment costs generally make such installations uneconomical.

Where only high-pressure steam is available a jet compressor (Fig. 209) may be used with some saving in high-pressure steam compared to the simpler scheme of throttling the high-pressure steam directly to the steam chest. But the low efficiency of steam-jet compressors when operating with a large difference in pressure between the high- and low-pressure steam limit the ecnomy of this arrangement.

In any case the use of vapor compression requires a complete economic balance to determine its feasibility in any given application.

Feeding

The feed to a multiple-effect evaporator is most conveniently fed to the effects in series in the same order as the steam flows. This is called *forward feed.* Under these conditions it is not necessary to pump the liquid between effects, and the only pump required to operate from a space below atmospheric pressure is the discharge pump of the last effect. If the feed is cold, this results in the absorption of considerable heat in the first effect, which heat is not available in vapor for re-use in the second and subsequent effects, except insofar as flash vaporization occurs when the liquid boiling in the first effect is transferred to the second effect which will be at a lower pressure. Some flash vaporization occurs in each effect after the first when feeding forward. If the feed is at or above the boiling point of the hottest effect, it may be most economical to feed into that effect to utilize the flash vapor in heating subsequent effects.

The feed is called *backward* when it is introduced into the effect at the lowest temperature and pressure and pumped successively into the effects at higher temperatures and pressures. This requires more equipment and more care in operation, but with a cold feed or a liquor which increases in viscosity during evaporation it gives considerably improved overall economy.

It is possible and sometimes most practicable in evaporation of saturated liquids to feed into each effect of a multiple-effect evaporator in parallel, but the economics seldom indicate that this is the cheapest method in so far as heat and investment costs are concerned. This is called *parallel* feed. In solutions having considerable change in viscosity with temperature over the concentration range involved, it is frequently most economical to feed forward with the fresh feed entering the second or third effect, thence returned for evaporation of the thickest liquor at the high temperature. This type of feeding is called *mixed* and represents a compromise for a viscous thick liquor between the operational advantage of forward feed and the greater economy of backward feed.

Selection and Operation

No generalization can be made which will serve as a reliable guide in the selection of type of evaporator, number of effects, method of feeding, or types of equipment most satisfactory for different solutions under the variety of economic limitations which will be imposed in various plants. Accordingly, evaporator design has remained a specialty in which each installation is tailor-made to fit the requirements of the problem at hand.

Hot condensate from an early steam chest may be "flashed" to the pressure of later steam chests to yield additional vapor for heating if the amount of such heat recovery justifies the cost of additional equipment and the extra complication of operation. Vapor may be withdrawn between any effects for process heating if economical, or extra steam from other sources may be added if available. In most plants, steam condensate is recovered for boiler feed; danger of contamination usually precludes use of vapor condensate except for process water.

The heat transfer on the steam side is retarded by the presence of noncondensable gas or air which must be thoroughly removed for satisfactory operation. The size and location of vent lines for removal of noncondensables from steam spaces below atmospheric pressure with maximum effectiveness and minimum loss of steam is a matter of considerable importance. These may lead directly to the condenser or to any subsequent effect at a lower pressure. Control of the amount of venting may be indicated by the temperature of the vent line at a distance from the space being vented.

HEAT TRANSFER COEFFICIENTS

As the solution flows upward through the tubes of a vertical evaporator its pressure continually decreases, its temperature rises until boiling starts, when the temperature may begin to fall, accompanying the decreasing pressure. For computing overall heat transfer coefficients by equation 356 under these conditions, an appropriate temperature drop must be used. The temperature of the condensing steam is normally constant and easily determined from its pressure. The problem is to select a suitable mean temperature for the solution as it passes through the tubes. A simple and common practice is to use the boiling point of the solution in the evaporator at the pressure of the vapor space. The difference between the temperature of the condensing steam and the temperature of the boiling solution at the pressure of the vapor space is called [1] the net temperature drop or the "temperature drop corrected for elevation in boiling point" and is usually indicated whenever temperature drop is mentioned in connection with an evaporator, as in Fig. 465.

Some confusion exists in early literature on evaporators owing to the use of a "coefficient not corrected for boiling-point elevation," i.e., calculated on the basis of a temperature drop equal to the difference between steam temperature and saturation temperature of pure water at the pressure existing in the vapor space. Further confusion exists because some investigators made the fallacious assumption that boiling occurs through the full length of the tubes and attempted to evaluate an average boiling point under the varying hydrostatic head, usually without regard for the acceleration head due to vaporization, leading to a "coefficient corrected for hydrostatic head." There appears to be no justification for perpetuating either of these practices.

The overall heat transfer coefficients in vertical-tube evaporators follow the general pattern indicated in Fig. 465, where coefficients [1] are presented for a basket evaporator having a 30-in. ID shell and 24 2-in. tubes 48 in. in length in the 26-in. steam

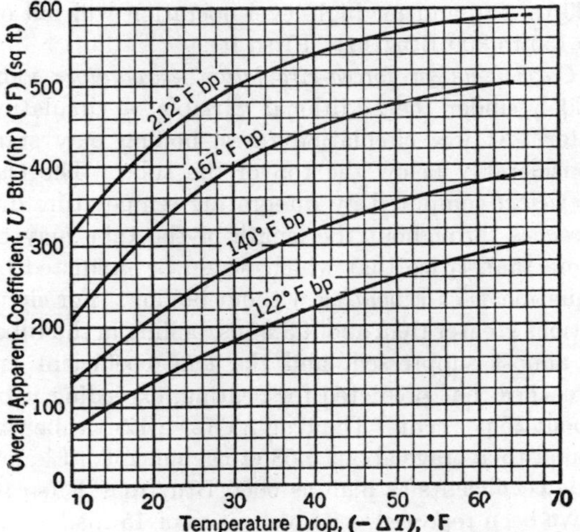

Fig. 465. Typical overall coefficients of heat transfer in vertical-tube evaporators expressed as a function of the temperature drop ($-\Delta T$) and the boiling point (bp) of the liquor.[1]

basket. The coefficient increases with lower liquid levels down to a level that results in insufficient liquid being carried up by the evolved vapor to wet the entire heating surface. At low-level operation, however, the rate of salting or scaling is greatly increased, and, for a liquid which deposits solids on the heating surface, the operating cycle between cleanouts may be unreasonably short. At the other extreme, with the liquid level considerably above the tube sheet, negligible evaporation takes place in the tubes because of the hydrostatic head of the liquid above the top tube sheet, and the rate of scaling or salting is greatly reduced. The choice of the appropriate liquid level with balance between benefit to coefficient and decrease in the length of the operating cycle must be decided for each installation. For the majority of installations, the optimum liquid level is near the top tube sheet, though sometimes as much as 2 ft above. Predictions of the rate of decrease of coefficient with scaling is difficult on the basis of the few tests which are available.

Horizontal-tube and submerged-coil evaporators give coefficients from about 200 Btu/(hr)(°F)(sq ft) at moderate temperature drops and boiling points in the vicinity of 160° F to about 400 Btu/(hr)(°F)(sq ft) for atmospheric boiling point and overall temperature drops of 70° to 80° F.

Standard vertical and basket-type evaporators give coefficients of the same order, or slightly higher for operation with liquid level at the top tube sheet. For low-level operation, these coefficients will reach 300 to 500, and for high-level operation will fall off to 150 to 300 Btu/(hr)(°F)(sq ft).

Coefficients in forced-circulation evaporators vary widely under the additional variable of circulation rate.[2] At low circulation rates, boiling may start considerably below the top of the tube. The *true coefficient* computed by integrating temperature differences throughout the length of the tube may be more than twice that which would be predicted by equation 417 for heating without boiling. For circulations greater than about 3.5 fps, boiling in the tubes is almost suppressed, and the true coefficient approaches that predicted for heating, exceeding it by about 25 per cent. The true ΔT for a 12-ft tube was found to average about 0.63 of the net ΔT. Liquor-side coefficients as high as 6000 Btu/(hr)(°F)(sq ft) have been reported, with velocities of 15 fps.

In designing evaporators the true or integrated temperature difference is not available and the net ΔT is used with overall coefficients.

Overall coefficients in *long-tube vertical evaporators* range from 200 to 600 Btu/(hr)(°F)(sq ft), based on the net temperature drop. These are less than can be obtained in a forced-circulation evaporator at high circulation rates—say 10 fps and higher. The typical feed rates for LTV evaporators range from 0.1 to 0.5 fps entering velocity, and the fraction evaporated per pass is much higher than for any other common type.

The values of heat transfer coefficients listed here are intentionally conservative and do not represent maximum values obtainable with scrupulously clean heating surface, or with promotors used to secure dropwise condensation of steam.

It is well established that heat transfer coefficients in most types of evaporators increase with increasing temperature drops, increasing circulation, decreasing viscosity, and decreasing surface tension. The relations are not clear in the case of the long-tube natural-circulation evaporator because of the uncertainty as to what fraction of the tube represents boiling and what represents nonboiling length; all factors which increase the boiling length seem to increase the coefficients, as expected.

CALCULATIONS

If the overall heat transfer coefficient for a given solution in a given type of single-effect evaporator is available, determination of the size required to accomplish the necessary evaporation with the specified steam temperature and condenser temperature is a straightforward computation.

For multiple-effect evaporation, the temperature and pressure at intermediate heating surfaces in the system cannot be fixed in advance and will hunt their own level appropriate to the equipment being used. The determination of heating surface needed in each effect can be accomplished by a series of heat balances around each effect and an overall heat and material balance. These equations can be solved after assumptions are made as to the temperature and amount of evaporation in each effect. The classical method of determination of required area has been to solve these heat balance equations simultaneously; the area required to accomplish the assumed evaporation in each effect is then evaluated, and conditions readjusted and recomputed to attain the desirable condition of equal heating surface in all effects. This last condition is not necessary, but for multiple-effect evaporation the installation cost is considerably lessened if the effects can all be built from the same designs.

Vapor-Liquid Equilibria

Boiling in an evaporator normally creates a vapor which is in equilibrium with the liquid. The temperature of the vapor is the same as that of the liquid, and the vapor is saturated relative to the liquid phase. Figure 466 is an approximate phase equilibrium diagram at a constant pressure of 1 atm for water-sodium hydroxide solutions. Most vapors in equilibrium with liquids containing dissolved solids normally concentrated in commercial evaporators are essentially pure water. Since the temperature of the vapor is above the boiling point of pure water at the existing pressure, the vapor is superheated relative to pure water.

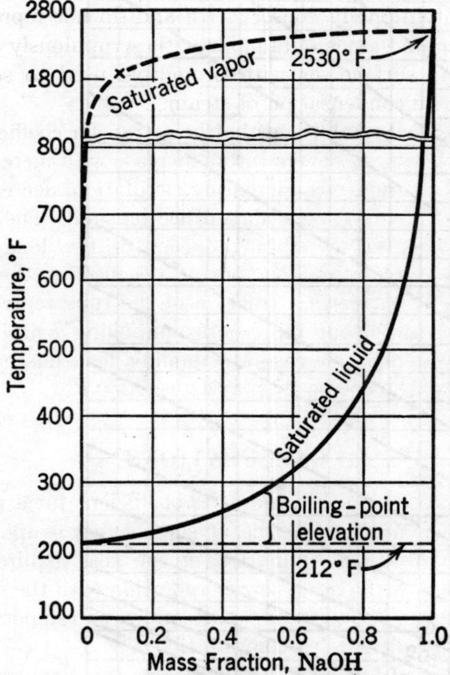

FIG. 466. Approximate phase equilibrium diagram for the water-sodium hydroxide system at 1 atm.

The boiling point of water containing dissolved solids may be given by its *boiling-point elevation* above that of pure water. This term is convenient in evaporation since it gives the temperature decrease which will take place between evolution of the vapor from a liquid and condensation to pure water at constant pressure. Boiling-point elevations are significant for such dissolved solids as caustic, sodium chloride, etc., but they may be negligible when the solids consist of collodial suspensions or organic substances.

The relationship between the boiling point of saturated liquids and pure water is best shown by a Dühring plot. Figure 467 shows the boiling point of caustic solutions plotted against the boiling point of pure water at the same pressure.[10] Individual lines are required for each concentration of sodium hydroxide.

Heat Required

The heat required to vaporize a pound of water from a solution is not necessarily the same as the latent heat of the pure water, since the vaporization process includes a concentration of the solution. For substances like caustic which have appreciable heat of dilution, the enthalpy requirements are best determined from enthalpy-concentration charts (see p. 326, Chapter 23). Figure 468 is an enthalpy-concentration chart for water-sodium hydroxide solutions based on data for heats of dilution at constant temperature and specific heat at constant composition.[7] The datum for water is the same as the steam tables so that they may be used for pure water. The datum for NaOH on this chart is for NaOH in an infinitely dilute solution, and pure caustic has an enthalpy at 68° F of 455 Btu/lb above this datum.

Example. Compute the heat required to vaporize 1 lb of water in an evaporator producing caustic solution containing 40 per cent NaOH by weight, at a pressure of 3.716 psia, when feeding 20 per cent NaOH solution to the evaporator at the boiling temperature in the evaporator. What is the boiling-point elevation?

Solution. From properties of water, the boiling point of water = 150° F.

Boiling point 40 per cent NaOH = 197° F from Fig. 467
Boiling point elevation = 47° F
Enthalpy of steam at 197° F and 3.716 psia is 1148 Btu/lb

Enthalpy in	Enthalpy out
$q + 2 \times 142$	$174 + 1148$

$\Delta H = q = 174 + 1148 - 284 = 1038$ Btu to evaporate 1 lb of water and concentrate 20 per cent NaOH to 40 per cent NaOH. From steam tables, the latent heat at 3.716 psia = 1008 Btu, which is only a fair approximation of the heat requirement.

If the vapor from this evaporator were sent to the heating element of another effect, without any pressure drop, the condensing temperature would be 150° F.

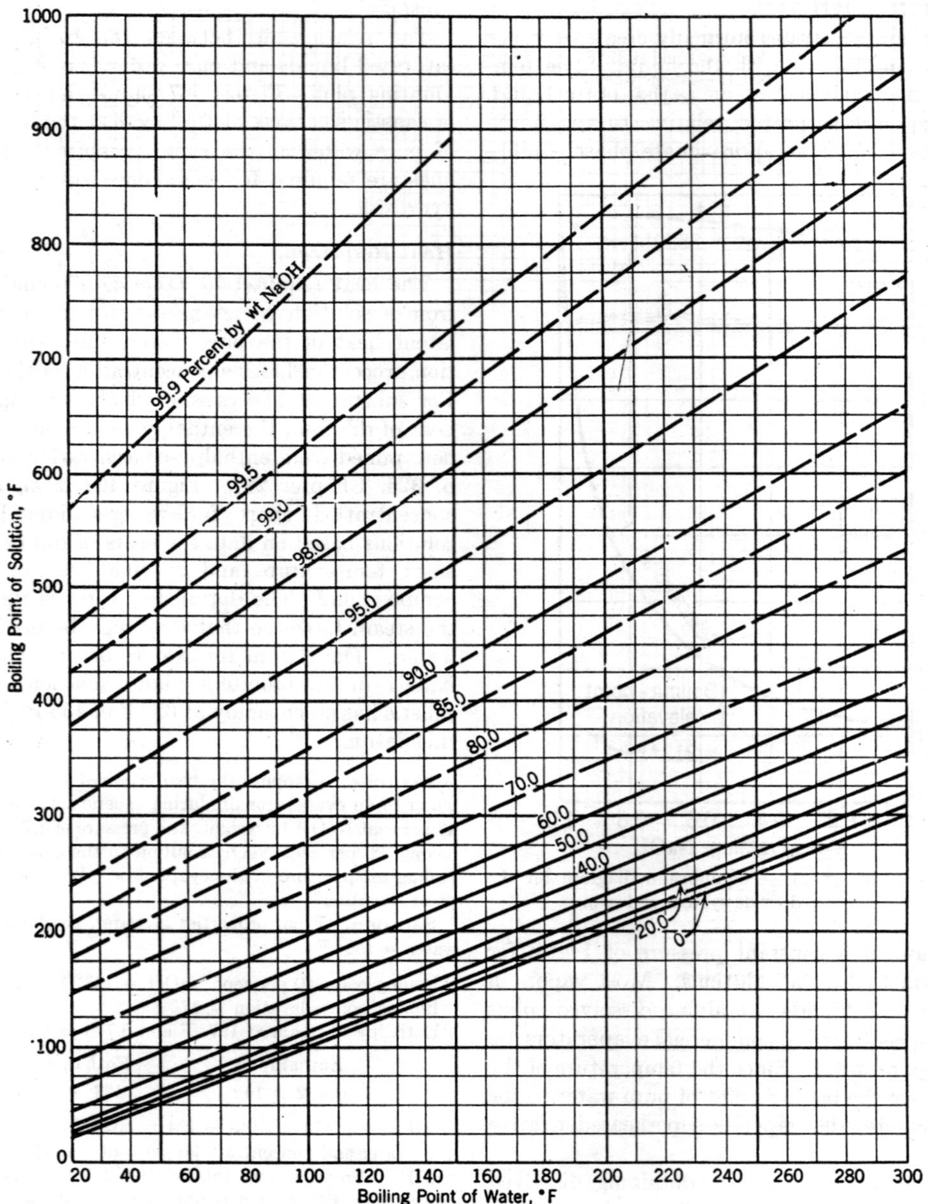

FIG. 467. Dühring lines for aqueous solutions of sodium hydroxide.

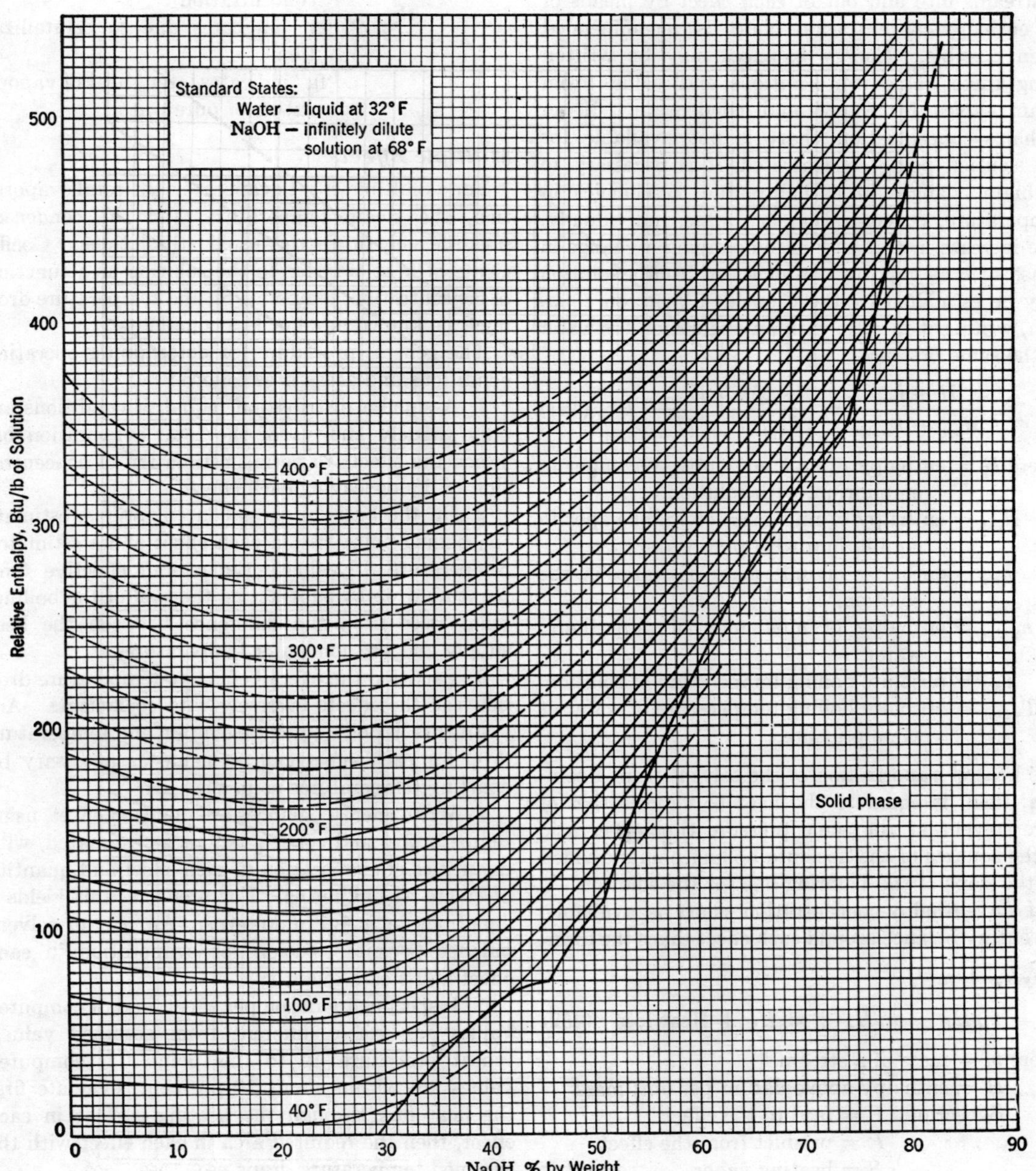

FIG. 468. Enthalpy-concentration diagram for aqueous solutions of sodium hydroxide. Reference state liquid water at 32° F under its own vapor pressure.

This method of evaluating the enthalpy content of streams into and out of each effect by means of an enthalpy-concentration chart is recommended when possible. Energy balances may be written using these values for solutions and values from steam tables for vapor and condensate. If an enthalpy-concentration chart is not available for the solution being concentrated, it is necessary to evaluate changes in enthalpy due to changes in temperature, concentration, or phase, individually. The increase in enthalpy accompanying a change in phase such as vaporization of water from a solution may be estimated by the Clausius-Clapeyron equation, writing it in ratio with an evaluation for water at the same pressure.

$$\frac{H - h_{\text{sol}}}{H_{\text{sat}} - h_{\text{water}}} = \frac{(T_{\text{sol}})^2}{(T_{\text{water}})^2}\left(\frac{dT_{\text{water}}}{dT_{\text{sol}}}\right) \tag{451}$$

where H = enthalpy of unit mass of water vapor in equilibrium with solution.
H_{sat} = enthalpy of unit mass of water vapor in equilibrium with pure water.
h_{sol} = enthalpy of unit mass of water in the solution.
h_{water} = enthalpy of unit mass of pure liquid water.
T_{sol} = absolute boiling temperature of solution.
T_{water} = absolute boiling temperature of water at same pressure.

The heat of vaporization of the solvent from a solution, then, is equal to the heat of vaporization of pure solvent at the same pressure times the ratio of the squares of the absolute boiling points divided by the slope of the Dühring line for the solution.

If an enthalpy-concentration chart is available, the energy balance around each effect takes the form:

$$m_FH_F + m_SH_S = m_PH_P + m_VH_V + m_SH_C + \text{Heat loss} \tag{452}$$

where m = mass of a stream.
H = enthalpy above datum per unit mass.
Subscripts: F = feed to the effect.
P = product from the effect.
S = heating vapor.
V = evolved vapor.
C = condensate.

When an enthalpy-concentration chart is not available, the energy balance takes the form:

$$m_S(H_S - H_C) = m_PC_P(T_P - T_F) + m_VC_P(T_V - T_F) + m_V \cdot \Delta H_{\text{fg}} + m_P\,\Delta H_{\text{dil}} + m_{\text{cr}}\,\Delta H_{\text{cr}} + \text{Heat loss} \tag{453}$$

where subscripts "dil" indicates due to dilution (or concentration).
"cr" indicates due to crystallization.
"fg" indicates latent heat of vaporization of pure water.

Multiple Effects

For a multiple-effect evaporator, the total evaporation, the steam temperature, and the condenser pressure constitute the usual stipulations. Coefficients of heat transfer to the solution as a function of concentration, temperature, and temperature drop must be known more or less completely.

The general procedure for determining operating conditions and areas is as follows:

1. Assume a distribution (equal, if conditions are only vaguely known) of the total evaporation between the effects to permit estimation of concentration of liquor leaving each effect.

2. On the basis of these concentrations, estimate the boiling-point rise in each effect at its estimated temperature. Evaluate the overall effective temperature drop by subtracting the sum of the boiling-point rises from the difference between the heat supply and the condenser temperatures.

3. Partition the overall effective temperature drop between the effects inversely with coefficients. Adjustments in boiling-point elevation and temperature drop for each effect can be made as necessary by charting temperatures as in Fig. 464.

4. Write energy balances around each effect, using temperatures and concentrations as estimated, with quantities of evaporation in each effect and quantity of steam as unknowns. For n effects, this yields n equations in $n + 1$ unknowns. An overall solvent balance permits solution for evaporation in each effect and steam quantity.

5. Inspect evaporation per effect, and, if computed values are *widely* different from assumed values, repeat procedure up to this point. If computed values are within about 10 per cent, evaluate first the heat flux through the heating surface in each effect, then the required area in each effect with the assumed temperature drops.

6. If the divergence of areas from the desired area in each effect—normally they should be equal—is of the order of 2 per cent or so, the average area may be taken and minor adjustment made in the tem-

perature drops to compensate. If the necessary adjustment in temperature drops is large enough so that error is introduced in enthalpy values, repeat the whole calculation, using the calculated conditions instead of the previously assumed values, until calculated values check assumptions within the desired accuracy.

This procedure of successive approximation using simultaneous equations is highly recommended to the student, since it emphasizes the inevitable seeking of the most efficient operating level of temperatures and concentrations in the evaporator. Any attempt to force other operating conditions will result in inefficiencies.

The experienced designer frequently uses a simpler method.[5] Temperatures and approximate concentrations are estimated, and a series of energy balances are calculated around each effect in series. With relative areas decided upon in advance, these energy balances and concurrent heat transfer calculations for each effect in turn give sufficient information to permit specification of areas, temperatures, and concentrations with more than adequate accuracy.

The following procedure emphasizes one characteristic which is of importance. By far the largest quantity in each energy balance is the latent heat term, which represents the heat available to the next effect. If this were the only heat effect, the flux would be identical in all effects. The minor heat effects—sensible heat of liquor, heats of dilution, crystallization, etc.—which are rendered unavailable to the next effect may be significant. They represent a decrease in heat flux through successive effects in almost all cases except the sensible heat of a hot feed, flowing forward through the evaporator. The area and concentration calculations can be executed[9] on the basis of a series of terms of the form (1 − fraction of heat rendered unavailable) for each effect. This method of evaporator calculation emphasizes the decrease in flux in successive effects. It is more time consuming than simultaneous solution of heat balances when enthalpy-concentration charts are used.

In general, multiple-effect evaporators are built with all bodies identical to save in the cost of preparing plans and fabrication and for the general advantages of uniformity of equipment. Where a large fraction of the feed is evaporated or where its characteristics change radically, differences between effects may be indicated. Under some conditions, design to equal areas may not represent minimum total cost.[3]

Illustrative Example. Sodium hydroxide solution, at 100° F, is to be concentrated from 10 per cent to 50 per cent by weight in a triple-effect forced-circulation evaporator, feeding 2, 3, 1 (Fig. 469). Steam at 100 psig will be the heat source. The condenser will be maintained at a 26-in. vacuum referred to a 30-in. barometer. Thirty-six tons per day of NaOH will be processed. Circulation rates will be maintained sufficient to give coefficients of 600, 500, and 400 Btu/(hr)(°F)(sq ft) in effects 1, 2, and 3, respectively. Evaluate temperatures and concentration in each effect. If all effects are of the same size, determine the heating surface in each. What is the heat load on the condenser?

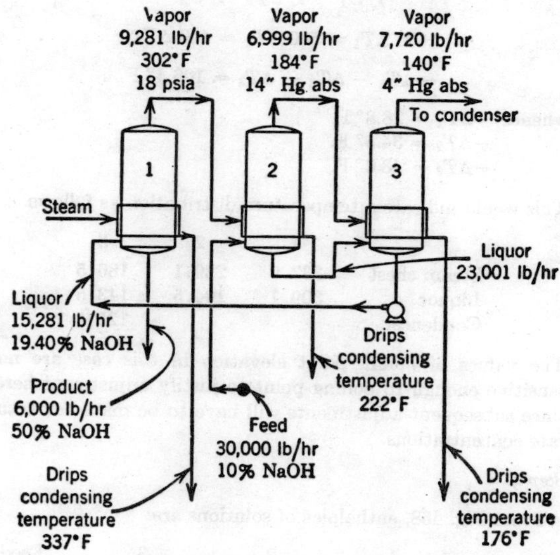

FIG. 469. Flow diagram for the triple-effect evaporator of the illustrative example, showing operating conditions.

Solution. (Based on 1 hr. Radiation is assumed negligible in the first approximation.)

Step 1

$$\text{Total evaporation: } \frac{36 \times 2000}{24 \times 0.1} - \frac{36 \times 2000}{24 \times 0.5} = 24{,}000 \text{ lb/hr}$$

Assume evaporations, E:

$E_1 = E_2 = E_3 = 8000$ lb/hr.
F_2 = feed to 2 = 30,000 lb/hr.
F_3 = feed to 3 = 22,000 lb/hr.
F_1 = feed to 1 = 14,000 lb/hr.
P = product = 6000 lb/hr.

C_2 = concentration in 2 = $\frac{3000}{22{,}000}$ or 13.63 per cent.

C_3 = concentration in 3 = $\frac{3000}{14{,}000}$ or 21.43 per cent.

C_1 = concentration in 1 = 50 per cent.

Step 2

Steam temperature = 337.9° F
Condenser temperature = 125.5° F
Total overall $(-\Delta T)$ = 212.4° F

From Fig. 467:

Boiling point rise of 50 per cent NaOH at 300° boiling point = 80° F.

Boiling point rise of 13.63 per cent NaOH at 190° boiling point = 8° F.

Boiling point rise of 21.43 per cent NaOH at 125.5° boiling point of water = 18° F.

Overall effective $(-\Delta T) = 212.4 - (80 + 8 + 18) = 106.4°$ F.

Step 3

$$\frac{\Delta T_1}{1/U_1} = \frac{\Delta T_2}{1/U_2} = \frac{\Delta T_3}{1/U_3}$$

$$600\,\Delta T_1 = 500\,\Delta T_2 = 400\,\Delta T_3$$

$$-\Delta T_1 - \Delta T_2 - \Delta T_3 = 106.4° \text{ F}$$

whence $-\Delta T_1 = 28.8°$ F.
$-\Delta T_2 = 34.5°$ F.
$-\Delta T_3 = 43.0°$ F.

This would indicate a temperature distribution as follows.

	1	2	3
Steam chest	337.9	229.1	186.5
Liquor	309.1	194.5	143.5
Condenser			125.5

The values of boiling-point elevation in this case are not sensitive enough to boiling point to justify adjustment here, since subsequent adjustments will have to be made for accurate concentrations.

Step 4

From Fig. 468, enthalpies of solutions are:

	1	2	3	Feed
	50% at 310° F	13.6% at 195° F	21.4% at 144° F	10% at 100° F
H Btu/lb	310	145	96	60

From steam tables:

	Steam	From 1 to 2	From 2 to 3	To condenser
H_{vap}	1189	1195	1146	1124
h_{liq}	308.8	199	153	

Heat balances may now be written for each effect:

Around 1:

$$(30{,}000 - E_2 - E_3)96 + 1189S = (6000)(310) + 1195E_1 + 308.8S$$

Around 2:

$$(30{,}000)60 + 1195E_1 = (30{,}000 - E_2)145 + 1146E_2 + 199E_1$$

Around 3:

$$(30{,}000 - E_2)145 + 1146E_2 = (30{,}000 - E_2 - E_3)96 + 1124E_3 + 153E_2$$

Solvent balance:

$$E_1 + E_2 + E_3 = 24{,}000$$

whence $E_1 = 9439$ lb/hr.
$E_2 = 6845$ lb/hr.
$E_3 = 7715$ lb/hr.
$S = 13{,}244$ lb/hr.

Check for equality of areas:

$$A_1 = \frac{(13{,}244)(880)}{(600)(28.8)} = 688 \text{ sq ft}$$

$$A_2 = \frac{(9439)(996)}{(500)(34.5)} = 543 \text{ sq ft}$$

$$A_3 = \frac{(6845)(993)}{(400)(43.0)} = 395 \text{ sq ft}$$

Step 5

Inspection reveals that the evaporation is far from equal, because of the large sensible heat requirement in the first and second effects. A preliminary test of areas indicates that the first estimate of $-\Delta T$ in effect 1 was too low and in effect 3 was too high. A second estimate may be made, using concentrations indicated by the computed evaporations, and ΔT's adjusted as indicated by deviation of areas from an average value (this is an approximation only, since the flux through the effects is not equal).

$$-\Delta T_1 = 29 \times \tfrac{688}{550} = 36.4, \text{ say } 37.9$$

$$-\Delta T_2 = 34.6 \times \tfrac{543}{550} = 34.3, \text{ say } 37.0$$

$$-\Delta T_3 = 43.0 \times \tfrac{395}{550} = 31.0, \text{ say } 35$$

Fitting these $-\Delta T$'s with boiling-point elevation values for the calculated concentrations yields temperatures:

	1	2	3
Steam chest	337.9	220	175
Liquor	300	183	140
Condenser			125.5

From Fig. 468, enthalpies of solutions are:

	1	2	3	Feed
	50% at 300° F	12.95% at 183° F	19.40% at 140° F	10% at 100° F
H Btu/lb	302	135	92.5	60

From steam tables, enthalpies are:

	Steam	From 1 to 2	From 2 to 3	To condenser
H_{vap}	1189	1191	1140	1124
H_{liq}	308.8	188	142	

It is possible to make a rough estimate of radiation losses for inclusion in the final heat balance if desirable. The heaters will be not over 600 sq ft heating surface. If this is to be as 1-in. tubes, 10 ft long, about 270 tubes will be required. These will require a shell about 30 in. in diameter, having an external surface about 100 sq ft/effect. If insulated with 2 in. of 85 per cent magnesia, the heat loss per square foot will average 25 Btu/(hr)(sq ft). This indicates a radiation loss of 7500 Btu/(hr)(effect). Even if a factor of safety of 5 is used, the radiation loss becomes only ½ per cent of the heat through each effect, and its omission from the heat balance equations is almost always justifiable. A value of 25,000 Btu/effect will be included primarily to show the method of handling.

Enthalpy balances around each effect:

Around 1:

$$(30{,}000 - E_2 - E_3)92.5 + 1189S$$
$$= (6000)(302) + 1191E_1 + 308.8S + 25{,}000$$

Around 2:

$$(30{,}000)(60) + 1191E_1$$
$$= (30{,}000 - E_2)135 + 1140E_2 + 188E_1 + 25{,}000$$

Around 3:

$$(30{,}000 - E_2)135 + 1140E_2$$
$$= (30{,}000 - E_2 - E_3)92.5 + 1124E_3 + 142E_2 + 25{,}000$$

Solvent balance:

$$E_1 + E_2 + E_3 = 24{,}000$$

whence $E_1 = 9281$ lb/hr.
$E_2 = 6999$ lb/hr.
$E_3 = 7720$ lb/hr.
$S = 12{,}076$ lb/hr

and

$$A_1 = \frac{(12{,}076)(880.2)}{(600)(37.9)} = \frac{10{,}629{,}295}{22{,}740} = 467 \text{ sq ft}$$

$$A_2 = \frac{(9281)(1003)}{(500)(37.0)} = \frac{9{,}308{,}843}{18{,}500} = 503 \text{ sq ft}$$

$$A_3 = \frac{(6999)(998)}{(400)(35)} = \frac{6{,}985{,}002}{14{,}000} = 498.9 \text{ sq ft}$$

Condenser duty, assuming condensate leaves at 100° F:

$$-q_c = 7720(1124 - 78) = 8{,}075{,}120 \text{ Btu/hr}$$

The maximum deviation of calculated evaporation from that used in estimating concentrations is slightly over 2 per cent in effect 2. It is seldom justifiable to attempt to refine such values further, but, if necessary, the calculation can be repeated to the desired precision.

Step 6

The stipulation of equal areas in each effect is not satisfied. This is handled by trial-and-error redistribution of the total effective temperature drop between the effects, assuming that boiling-point elevation and heat flux values remain unchanged. The accuracy of data almost never justifies refinement closer than 1°. A few trials result in $-\Delta T$'s of 36, 38, 35, in that order, whence,

$$A_1 = 467 \times \frac{37.9}{36} = 492$$

$$A_2 = 503 \times \frac{37}{38} = 490$$

$$A_3 = 499 \times \frac{35}{36} = 485$$

Design area, without factor of safety = 490 sq ft.

These changes of 1 and 2 degrees obviously do not justify reevaluating enthalpies and solving for new evaporations.

Summarized values are:

	1	2	3	Condenser
Feed, lb/hr	15,281	30,000	23,001	
Thick liquor, lb/hr	6,000	23,001	15,281	
Thick liquor, per cent NaOH	50%	12.95%	19.40%	
Evaporation, lb/hr	9,281	6,999	7,720	
Temperature of steam spaces, °F	337.9	222	176	125.5
Temperature of liquor, °F	302	184	140	
Heat flux (Btu/hr) $\times 10^{-6}$	10.63	9.31	6.98	8.07
Area, sq ft	490	490	490	

The assumption that solvent evolved at the boiling point of a solution retains the same temperature to the condensing space in the next effect is questioned by many engineers. The superheat credited to the condensing vapor in this calculation represents only a few per cent of its available heat, and disregard of it represents a small factor of safety. Obviously, the temperature drop $(-\Delta T)$ in the next effect must be evaluated from condensation temperature. The condensate is assumed to leave the steam chest at its saturation temperature, but it is likely to lose additional sensible heat by subcooling to the extent of one-fourth to one-half of the temperature drop.

BIBLIOGRAPHY

1. BADGER, W. L., and W. L. MCCABE, *Elements of Chemical Engineering*, 2nd ed., pp. 186–195, McGraw-Hill Book Co. (1936).
2. BOARTS, R. M., W. L. BADGER, and S. J. MEISENBERG, *Trans. Amer. Inst. Chem. Eng.*, **33**, 363 (1937).
3. BONILLA, C. F., *Trans. Amer. Inst. Chem. Eng.*, **41**, 529 (1945).
4. BROOKS, C. H., and W. L. BADGER, *Trans. Amer. Inst. Chem. Eng.*, **33**, 392 (1937).
5. CALDWELL and KOHLINS, *Trans. Amer. Inst. Chem. Eng.*, **42**, 495 (1946).
6. FOUST, A. S., W. L. BADGER, and E. M. BAKER, *Trans. Amer. Inst. Chem. Eng.*, **35**, 45 (1939).
7. MCCABE, W. L., *Trans. Amer. Inst. Chem. Eng.*, **31**, 129 (1935).
8. MCCABE, W. L., and C. S. ROBINSON, *Ind. Eng. Chem.*, **16**, 478 (1924).
9. RAY and CARNAHAN, *Trans. Amer. Inst. Chem. Eng.*, **41**, 253 (1945).
10. VON ANTROPOFF and SOMMER, *Z. physik. Chem.*. **123**, 161 (1926).

PROBLEMS

1. An aqueous solution (at 60° F) containing 10 weight per cent $MgSO_4$ is to be concentrated to 37.5 weight per cent $MgSO_4$ in a double-effect evaporator using backward feed. No crystallization is to take place in the evaporator. The thick liquor from the first effect is to be at 170° F. This thick liquor from the evaporator is then cooled to 80° F in a crystallizer with negligible evaporation. The slurry from the crystallizer is centrifuged, the liquid removed being returned to the evaporator (first effect). Twelve tons per day of solids from the centrifuge are to be produced. The composition of these solids is:

$MgSO_4 \cdot 7H_2O$	96 weight %
H_2O	4 weight %

The boiling-point elevation of the $MgSO_4$ solution may be neglected. The pressure on the vapor space of the second effect is 2 psia and each effect has an overall coefficient of 250 Btu/(hr)(°F)(sq ft). Determine the heat transfer surface required in each effect.

Enthalpies of the solution are given in J. H. Perry, *Chemical Engineers' Handbook*, 3rd ed., p. 1052, Fig. 147, McGraw-Hill Book Co. (1950).

2. A double-effect evaporator is to be constructed to deliver 20,000 lb/hr of 50 per cent by weight NaOH. The feed is 30 per cent NaOH supplied at 90° F preheated to 140° F by heat exchange with product leaving the first effect. Backward feed is used with a vacuum of 24 in. of mercury on the second effect with a barometer of 29.4 in. Saturated steam is available at 30 psig. The overall heat transfer coefficients based on actual temperatures at the liquid-vapor interfaces are 280 Btu/(hr)(°F)(sq ft) for the first effect, 310 Btu/(hr)(°F)(sq ft) for the second effect.

(*a*) What heat transfer surface do you recommend for the evaporators?

(*b*) What are the temperatures of the caustic streams leaving the evaporators and the heat exchanger?

3. A triple-effect forced-circulation evaporator is to be designed to concentrate 90,000 lb/hr of 10 per cent by weight NaOH, at 120° F to 50 per cent by weight. Backward feed is to be used, and saturated steam at 125 psig is to be used in the steam chest of the first effect. The pressure on the last effect is to be 3 in. of mercury, absolute.

The overall coefficients of heat transfer, corrected for boiling-point elevation, are assumed to be equal at 500 Btu/(hr)(°F)(sq ft) for all effects when backward feed is used. The condensate leaving the steam chest of each effect may be assumed to be saturated at the corresponding pressure in the steam chest. The heat loss from each effect is estimated to be 4 per cent of the total heat input to that effect (based on vapor to the effect).

Determine:

(*a*) The heating surface required for each effect, assuming that equal areas are used.

(*b*) The pounds of water evaporated per pound of steam used in the first effect.

4. As a plant engineer the following data are available to you.

Feed 90,000 lb/hr of 10 per cent NaOH solution to the first effect of a triple-effect evaporator (forward feed). Steam at 50 psia is supplied to the first effect. Solution leaving the third effect is 50 per cent NaOH. Each effect has a heat transfer surface of 1340 sq ft, and operating conditions are as follows with a barometer of 14.47 psia.

Effect	First	Second	Third
Condensate, lb/hr	36,200	23,500	23,350
Operating pressure, psig	+13.33	−0.35	−13.2

The 50 per cent caustic is concentrated to 75 per cent in a single-effect evaporator having 210 sq ft of heating surface operating at a vacuum of 13.2 psi (pressure of −13.2 psig).

Neglect radiation losses, assume all condensates leave at condensing temperature, and compute

(*a*) Temperature of feed to the first effect.

(*b*) Overall coefficients of heat transfer for each of the three effects and for the single effect.

5. A triple-effect forced-circulation evaporator is to concentrate NaOH solution from 9.5 per cent NaOH to 50 per cent NaOH, producing 200 tons of 50 per cent solution per 24 hr.

Steam to the first effect: 35 psig, dry and saturated.
Vacuum on the third effect: 28 in. referred to a 30-in. barometer.
Feed solution: 100° F. No appreciable impurities; no salt or scale on evaporation.

Coefficients:

	First Effect	Second Effect	Third Effect
Forward feed, (Btu/(hr)(°F)(sq ft)	1100	1000	700
Backward feed, (Btu/(hr)(°F)(sq ft)	800	1000	1000

Costs:
Steam, 30 cents per 1000 lb.
Water, 3 cents per 1000 gal.
Evaporators, $40.00 per square foot, complete with all accessories.
Fixed charges, 15 per cent per year.

Water for condensers: Enters at 70° F; leaves at 96° F.
Plant operates 350 days per year.

What is the difference in cost, per ton of actual NaOH, between forward and backward feed?

CHAPTER

33

Crystallization

SOLID crystalline salts are handled more conveniently and economically than solutions when the material must be moved outside of the plant of origin. A crystalline product also has in general more "sales appeal" than a solution. Although crystallization is ordinarily thought of as the separation of a solid crystalline phase from a liquid phase by cooling, evaporation, or both, the same principles apply to crystal formation by precipitation caused by the addition of a third substance, which may react to form the precipitate or simply decrease the solubility of the precipitated material.

Crystallization is important in the preparation of a pure product, since a crystal usually separates out as a substance of definite composition, from a solution of varying composition. The impurities in the mother liquor are carried in the crystalline product only to the extent that they adhere to the surface or are occluded within the crystals which may have grown together during or after the crystallization operation.

The problems concerning crystallization which are most frequently encountered by the engineer are:

1. Yield of a given product.
2. Purity of the product.
3. Energy requirements for cooling, evaporation, etc.
4. Shape of the individual crystals.
5. Size of the crystals.
6. Uniformity or distribution of the size of the crystals.
7. Rate of production of the desired crystals.

RATE OF CRYSTALLIZATION

The last of these items is discussed first in view of its importance in the analysis of crystallizer operation. The rate of crystallization involves two distinct actions: first, the rate of formation of new crystals, or nucleation, either in a clear solution or in a solution containing solids; and second, the rate of precipitation on crystals already present, usually called crystal growth.

The deposition of a solid from a solution onto a crystal can take place only if there is a state of unbalance with a driving force or decrease in chemical potential (or concentration) between the bulk of the solution and the crystal interface. This means that the solution must be supersaturated with respect to crystals of the size on which deposition is to occur before the crystals can grow by deposition from the solution. Since the solubility of crystals increases as the size decreases, it is often possible so to control the concentration that the large crystals grow or become larger by deposition from the solution while the small crystals do not grow but may actually dissolve.

Miers [2] * postulated the existence of a supersolubility curve, characteristic of each solid, and of approximately the same shape as the normal solubility curve but corresponding to a higher concentration of solid dissolved in a solution, at which new crystals would form from a clear solution. The theory postulates that the precipitation of solid on a crystal from a solution (crystal growth) cannot occur until the concentration exceeds that of the normal solubility or saturation curve, and that the formation of new crystals does not occur unless the concentration is equal to that indicated by the supersolubility curve. This would be very convenient, if true, as it would be necessary only to maintain the concentration less than that indicated by the supersolubility curve and greater than that indicated by the normal

* The bibliography for this chapter appears on p. 501.

solubility or saturation curve to insure the growth of existing crystals without the formation of new crystals. Attempts to establish such definite supersolubility curves have not been successful, and the existence of this supersolubility limit is doubtful.

A satisfactory explanation of the experimental facts may be based on the theory that the formation and continued existence of crystal nuclei depend upon two conditions: first, the probability of the requisite number of atoms or molecules coming together in such proximity and ordered arrangement as to establish the structure of the solid phase even though this aggregation does not yet exist as a separate solid phase; second, the solution being supersaturated with respect to crystals of the size of this aggregation.

The exact mechanism of arrangement of molecules from random distribution in a solution into the regular crystal lattice is not clearly understood, and the energy effects involved in such an orientation are elusive. In discussing the growth of a crystal from the first orientation of a sufficient number of molecules to constitute the lattice unit up to what might be called a crystal even of submicroscopic size, the formation of the first crystal lattice unit is considered as a probability function. If the concentration maintained in the solution exceeds the solubility of crystals of the size of this first crystal lattice unit, it will remain and grow to visible or practical size. If the concentration is less than the solubility of crystals of this size, such crystals cannot grow but will dissolve. As the crystal grows from this submicroscopic size, its solubility in the solution decreases. If the solution in which this crystal is being grown is maintained at such concentration as to be slightly supersaturated with respect to that crystal, a concentration difference is established between the bulk of the solution and the surface of the crystal, resulting in transfer of dissolved material to the surface.

Since the solubility of small crystals is greater than that of larger crystals, material may be simultaneously depositing on larger crystals and dissolving from smaller crystals when both are exposed to the same solution. But in practice it is difficult, even in special laboratory operations, to maintain conditions so completely uniform while growing crystals that at no spot in the solution is supersaturation allowed to reach that point at which new nuclei will form and persist. In ordinary operations it is usually impossible to avoid locally the greater degree of supersaturation which will maintain and grow the crystal lattice units which are constantly forming.

This explanation suggests that a normal crystal nucleus is an extremely small crystal with a solubility greater than that of ordinarily recognized crystal sizes and that the growth or deposition of solid on these submicroscopic crystals may depend upon any of a number of factors such as irregularities of distribution of solute in the solution, impact of existing crystals, or breakage of a growing crystal, as well as upon a high degree of supersaturation throughout the solution. If a high supersaturation is maintained during a crystallization operation, the probability that each crystal which forms will persist is higher, and accordingly more of the original crystal units which are formed will have a chance to grow to such size as to be recognized as crystals.

In order to control the number of crystals onto which a given amount of material is deposited, it is common practice to inject into a solution, immediately prior to incipient crystallization, small crystals known as *seeds* on which the salt will be deposited more easily since the solubility with respect to crystals of that size is less than with respect to the submicroscopic crystals. It is impossible to avoid completely the formation of new crystals for most substances, and there is inevitably some breaking of crystals in an agitated solution. The operation of a crystallizer is accordingly planned to retain the solution supersaturated with respect to the seeds which may be introduced and grown but not supersaturated with respect to crystals of the size of the aggregations which will inevitably form under these conditions.

YIELD OF A GIVEN OPERATION

Computation of the yield of a crystallization operation involves a material balance. The yield can usually be predicted from the solubility of the solid phase being precipitated by assuming equilibrium or saturated conditions. In some cases crystallization occurs very slowly, and equilibrium will not be attained in a reasonable length of time. The commercial separation of potassium chloride from the more slowly crystallizing borax present in certain brines depends on the rapid crystallization of potassium chloride before significant quantities of borax precipitate. In some cases no precipitation occurs unless seed crystals are present. If solubilities are used in the form of an equilibrium diagram the prediction of the equilibrium yield can be made by applying the usual "lever arm ratio" principle (as described in connection with phase diagrams for liquid-liquid and vapor-liquid transfer operations) between the phases which are known to be present at the specified final temperature.

The solubilities[3] commonly reported in the literature are expressed in terms of mass of anhydrous salt per 100 or 1000 masses of pure solvent. The prediction of a yield of a nonhydrated salt from a solution is simple since the amount of solvent present during crystallization is constant and the quantity remaining at the terminal temperature is known.

If the material crystallizes with water of crystallization or if evaporation occurs, the total amount of solvent changes during the process, but the amount of solvent present in excess of that required to hydrate all salt present does remain constant or decreases by a known amount of evaporation. On the basis of this solvent whose quantity is known, a material balance can be written relating the amount of hydrate per pound of excess water in the original solution to the final solution. Solubilities are conveniently expressed in terms of hydrate per unit of excess water. This value may be subtracted from the amount of hydrate in the original solution expressed on the same basis to evaluate the yield.

Exercise. Derive the following formula for yield, assuming equilibrium conditions.

$$C = R \times \frac{100A_o - X(S_o + \Delta S)}{100 - X(R - 1)} \tag{454}$$

where C = mass of crystals in the final magma.
R = the ratio, molecular weight of the hydrated solute over molecular weight of the anhydrous solute.
X = solubility (parts by weight of the anhydrous solute per 100 parts by weight of the total solvent) of the material at the final temperature.
A_o = mass of anhydrous solute in the original batch.
S_o = total mass of solvent in the batch at the beginning of the process.
$-\Delta S$ = evaporation during the process in pounds mass of solvent.

PURITY OF PRODUCT

Even though a double salt may be deposited, the crystal which is formed during the crystallization operation is pure, except in the freezing of a melt which solidifies as a solid solution in which an atom or molecule of one substance is substituted for the other in an identical crystal lattice. Impurities in crystallization operations are introduced only from the mother liquor which is not removed from the solid product. The extent of occlusion and the completeness of washing are important factors in the determination of purity of a crystalline product. The agglomeration of crystals into large grains which can occur during the growth of the crystals makes washing more difficult and accordingly makes for a lower purity of product. Agitation decreases the tendency to agglomerate.

ENERGY EFFECTS IN THE PROCESS

The usual crystallization operation is carried out either by cooling a solution or by evaporative concentration thereof, or by both simultaneously. The cooling of a solution may be accomplished by removal of sensible heat and heat evolved during crystallization of the product to cooling water or air. If the solution is crystallized by evaporation, the required heat may be supplied by the sensible heat of the solution as in vacuum crystallizers, or it may be supplied from an external source as in salt evaporators.

The values for "heats of solution" ($-\Delta H$) of most salts are available in standard tables as heats of solution into infinitely dilute solutions. Although the use of these values without correction for heat of dilution as representing the heat of crystallization from a saturated solution may lead to errors, fortunately the heat of dilution is usually small compared to the heat of solution, and most industrial crystallizations can be evaluated as far as energy effects are concerned by the use of heats of solution as recorded in the standard tables without serious error in the overall energy balance. The assumption that the heat of vaporization of the solvent from the solution is numerically equal to the heat of vaporization of the pure solvent at the same temperature is convenient and introduces no significant error for those solutions having a slight elevation of the boiling point.

The most satisfactory method of evaluating heat effects during a crystallization operation is the use of the enthalpy-concentration chart for the solution and various solid phases which are possible for this system. Unfortunately, complete enthalpy-concentration charts are available for relatively few substances. When the complete chart is available such as in Fig. 470, it is necessary only to read the enthalpy of the original solution at the initial temperature and the enthalpy of the final magma including crystals and liquor to evaluate the difference ΔH as the energy effect per unit mass. If evaporation has occurred during the process, the enthalpy of the water vapor formed in the process is obtained from steam tables. The net increase in enthalpy ΔH during the operation is easily evaluated by subtracting the enthalpy of a given amount of feed from that of the corresponding amount of final magma plus vapor, if any. Since $w = 0$, $\Delta H = q$, the heat absorbed.

The relative size of the faces as the crystal grows is independent of the crystallographic structure. Occasionally the crystal habit (the relative sizes of the faces of the crystal) must be modified to satisfy a particular consumer demand, as for medicinal magnesium sulfate, or Epsom salt. This may force the use of a particular type of equipment for making crystals of the shape demanded for sales appeal. The crystal habit may also be modified by the addition of an extraneous substance or impurity. Crystals as formed are essentially perfect in shape except under conditions of vigorous agitation when there may be mechanical breakage, and in certain types of crystallizer equipment a rounded structure results rather than the typical sharp crystal.

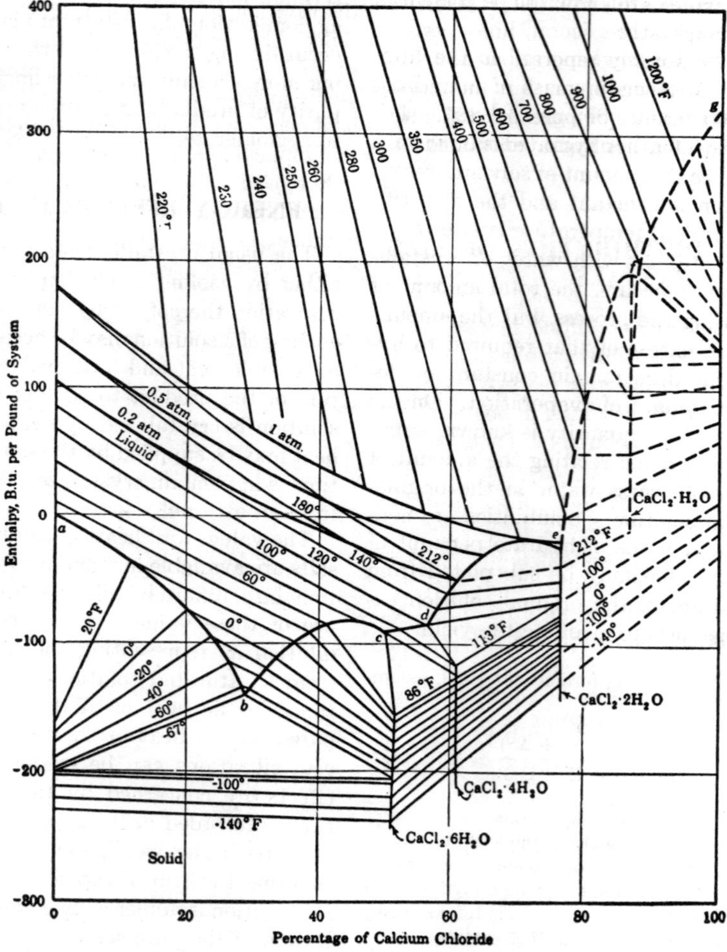

FIG. 470. Enthalpy-concentration diagram for the calcium chloride-water system. (*Hougen and Watson, "Chemical Process Principles," John Wiley and Sons, 1948, p. 281, Fig. 54.*)

AVERAGE SIZE OF THE CRYSTALS

The average size to which crystals are grown may be dictated by the use to which they are to be put, such as the granulation of sugar or the preparation of salt for different uses, as, for example, table, ice cream, butter, or pretzels. In general, fine crystals have less opportunity for agglomeration and accordingly occlude less mother liquor; but fine crystals have greater surface and offer greater difficulty of complete removal of mother liquor entrained on the crystals.

UNIFORMITY OR DISTRIBUTION OF SIZE OF CRYSTALS

The caking tendency of crystalline materials in storage is predominantly a result of the formation and evaporation of a solution with recrystallization on the surface of the crystals, resulting in bonding together of crystals at points of contact. The number of points of contact is significantly less for uniform crystals. Accordingly, hygroscopic salts, or salts carrying hygroscopic impurities, will have considerably less tendency to cake in storage if the crystal size is uniform. When salts are to be dissolved, it is advantageous if the size is uniform in order that the solution of all crystals will be completed at approximately the same time. A uniformly crystallized product usually offers considerably better "sales appeal" and conveys the impression of care in the manufacture of the product.

Solution of crystalline materials is likely to occur if the humidity of the air with which they are in contact exceeds a critical value corresponding to the vapor pressure of a saturated solution which may be formed on the surface of the crystal. If the humidity fluctuates above and below this critical value the saturated solution formed on the surface of the crystal is evaporated, the recrystallized solid bonds the crystals together, and caking becomes troublesome.

If it were possible to carry out crystal growth on seed crystals without the formation of additional nuclei, it would be possible to predict exactly the size distribution of the product from the size distribution of the seeds, since the rate of growth of a linear dimension of the crystal is constant for all crystals in the batch.[1] This computation could be made by plotting the diameter of the particles versus the number of crystals in the original seed batch (as obtained from a screen analysis) and increasing the diameters by the amount of growth to obtain directly the size distribution for the product. This procedure is not convenient, as the size distribution of crystals is not readily computed on the basis of numbers of crystals. The usual practice is to express the size distribution in a screen analysis by mass fractions. Since an increase by a constant amount in a chosen linear dimension results in a greater fractional increase in mass for a small crystal than for a large one, it is necessary to integrate the change in mass over the applicable size distribution.

For producing crystalline material of a given size range there is a definite upper and lower limit to the number of such crystals which may be grown by the deposition of a given amount of solid. If the quantity of seeds added to a specific quantity of solution exceeds the upper limit, all crystals in the product may be smaller than desired. If the quantity of seeds added is less than the lower limit, all crystals in the product may be larger than desired. Some quantity of seeds between these limits will yield the maximum quantity of product of the desired size range.

For transfer of solute to a crystal freely suspended in the slightly supersaturated mother liquor, there must be a driving force or decrease in potential (or concentration) between the bulk of the liquor and that wetting the surface of the crystal which is approximately saturated with respect to crystals of that size. As we do not know how much of the resistance to deposition is in the liquid and how much is in the interface, the resistance is handled as an overall unit.

The rate of deposition is equal to a driving force divided by a resistance or, as usually stated, a function of the concentration difference times a conductance which includes shape and surface factors and differences in rate of growth of different faces.

$$\frac{dw}{dt} = k[f(x - x_i)] \tag{455}$$

where w = mass of crystal.
t = time.
k = a constant.
x = concentration of solute in the body of the liquid.
x_i = concentration of solute in the interface.

For crystals of widely different sizes in contact with the same solution, the driving force for transfer and deposition, $f(x - x_i)$, will vary because of the increasing solubility for crystals of decreasing size. It is possible, but not probable, that where the driving force is such as to result in deposition on large crystals, small crystals will dissolve, that is, $(x - x_i)$ may have a different sign for different crystals.

In most cases the linear dimension of the crystal D as may be determined by a screen analysis is a more convenient value to use than the mass of the crystal.

$$\frac{dD}{dt} = K[f(x - x_i)] \tag{456}$$

where K is a conductance including k and the relation between the dimension D and the mass w of the crystal.

If the crystals are all of the same shape the relationship between the mass of a crystal and its dimension D is the same for all sizes. If the seed crystals are of substantially uniform size (infinitesimal range of sizes) and $(x - x_i)$ is constant,

$$\frac{dw_s}{D_s^3} = \frac{dw_p}{(D_s + \Delta D)^3} \tag{457}$$

where w = the mass of crystal.
subscript s refers to the seeds.
subscript p refers to the product.

Summation over the mass of seeds yields

$$w_p = \int_0^{w_s} \left(1 + \frac{\Delta D}{D_s}\right)^3 dw_s \tag{458}$$

There is always the possibility of dusty material adhering to the seed crystals, of breakage of crystals, of dust-like crystals floating in the air, and of submicroscopic crystals growing to recognizable size to the extent that fine material is almost always included in the product. The virtual impossibility of attaining the same number of product crystals as seeds vitiates the accuracy of this prediction, but the approximate prediction is highly useful.

The graphical integration of equation 458 can be accomplished directly for a known value of ΔD_{avg}. For a known ratio of crop, or yield, to seeds, it is necessary to fit by trial and error an appropriate value of ΔD_{avg} to determine the growth. A plot is made of w_p versus w_s from which the mass of product corresponding to any given mass of seeds may be read. In order to determine the screen analysis of the product, it is most convenient to evaluate w_s from zero to a mass corresponding to an even screen size opening decreased by the value of ΔD_{avg}.

Illustrative Example. A batch of seed crystals has the size distribution shown in the following table. If in a crystallizer the seeds grow so that ΔD is 0.009 cm, compute the quantity of product per pound of seeds.

Screen Analysis of Seed Crystals

Mesh	Mass Fraction of Seeds: Differential	Cumulative
−14 +20	0.028	0.028
−20 +28	0.176	0.204
−28 +35	0.293	0.497
−35 +48	0.336	0.833
−48 +65	0.128	0.961
−65	0.039	1.000
	1.000	

Solution. To determine the amount of product, the integral in equation 458 must be evaluated. Since the relation between D_s and w_s is given in a table rather than by an equation, the least difficult solution will be that of graphical integration of equation 458.

To obtain the relation between D_s and w_s, the cumulative size analysis is plotted in Fig. 471*a*. On the basis of 1 lb of seed crystals, the value of D_s at any value of w_s may be obtained directly from Fig. 471*a*. The value of $[1 + (\Delta D/D_s)]^3$

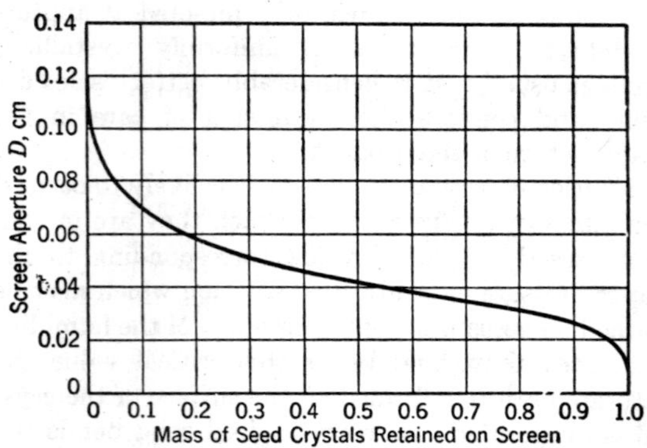

FIG. 471*a*. Cumulative size analysis of seed crystals for the illustrative example.

may then be computed as a function of w_s from Fig. 471*a* and plotted, $[1 + (0.009/D_s)]^3$ as the ordinate versus w_s, in Fig. 471*b*. The integral is evaluated by determining the area under the curve between the limits of $w_s = 0$ and $w_s = 1$. This is best done by dividing the area into rectangles (Fig 471*b*) such that the areas of the rectangles equal the areas under the curve as described on p. 80, and summing the areas as given in Table 53.

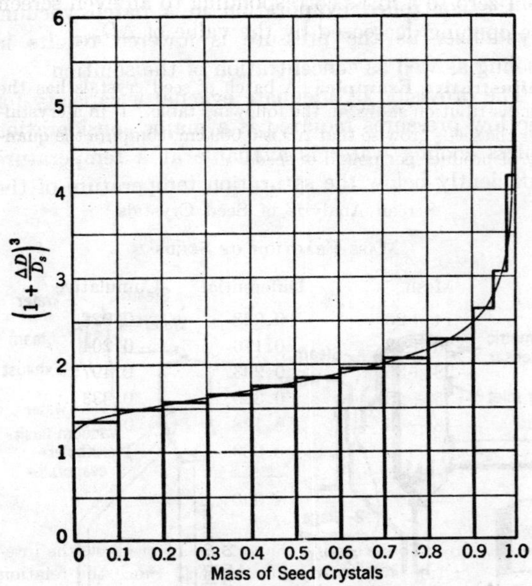

FIG. 471*b*. Graphical integration for equation 458 for the illustrative example.

TABLE 53. INTEGRATION OF EQUATION 458 BY SUMMING THE AREAS OF THE TWELVE RECTANGLES IN FIG. 471*b*

Δw_s	$\left(1+\frac{\Delta D}{D_s}\right)^3$	Area
(0.1 –0.0) ×	1.36	= 0.136
(0.2 –0.1) ×	1.49	= 0.149
(0.3 –0.2) ×	1.58	= 0.158
(0.4 –0.3) ×	1.66	= 0.166
(0.5 –0.4) ×	1.75	= 0.175
(0.6 –0.5) ×	1.84	= 0.184
(0.7 –0.6) ×	1.94	= 0.194
(0.8 –0.7) ×	2.07	= 0.207
(0.9 –0.8) ×	2.28	= 0.228
(0.95–0.9) ×	2.62	= 0.131
(0.98–0.95) ×	3.04	= 0.091
(1.0 –0.98) ×	4.15	= 0.083
		1.902

Therefore

$$\int_0^1 \left(1+\frac{\Delta D}{D_s}\right)^3 dw_s = 1.902$$

and the yield of product is 1.90 lb/lb of seeds.

EQUIPMENT

The simplest type of equipment for crystallization is a tank in which natural cooling is allowed to lower the temperature of the solution, with whatever simultaneous evaporation may occur to the atmosphere. This type of crystallizer is inefficient in regard to quantity produced per unit of floor space or per unit of time, since the cooling rate is inevitably slow. It offers no control of size range of crystals except that it favors formation of large crystals and moreover is subject to occlusion of mother liquor as crystals tend to grow together under the stagnant conditions which obtain. The provision of agitation during the cooling process results in improved crystal structure at a moderate increase in operating costs. In the crystallization of small quantities of material, where size range of product is not of particular consequence, the simple batch crystallizer may represent the most economical process as it is most economical in first cost.

The Swenson-Walker crystallizer (Fig. 472) operates by cooling of the solution. The semicircular cross section of the bottom of the trough provides the heat interchange surface between the annular cooling water jacket and the crystallizing solution which is inside the trough. The flights on the central shaft rotate slowly and serve the dual purpose of keeping crystallization from blanking off the heat transfer surface and of continually agitating the magma to secure more uniform crystallization conditions. The crystallizer is normally built in 10-ft

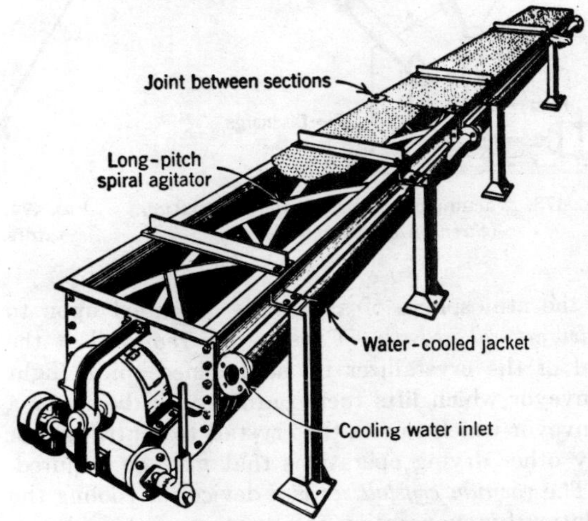

FIG. 472. Swenson-Walker continuous crystallizer. (*Swenson Evaporator Co.*)

lengths, of which up to four sections may be operated in tandem from a single drive. If more than four units are required, it is customary to provide drives for groups not to exceed four units, with flow of the magma from one section usually cascading down into the head of the next section. The shock of cascading usually increases nucleation and should be avoided if possible. Seed crystals are sometimes added, or spontaneous nucleation or the inevitable seeding by microscopic crystals of the salt floating in the atmosphere of a plant is depended upon to start crystal growth. Crystals are removed at the end of the crystallizer by an inclined spiral flight conveyor which lifts them onto a drain board or a conveyor which carries the crystals to centrifuges or any other drying operations that may be required.

The *vacuum crystallizer* is a device for cooling the solution by evaporation of a portion of the solvent. It may be batch or continuous in operation. In principle, the vacuum crystallizer is little more than a vessel which may be evacuated to extremely low pressure, ordinarily by steam jet ejectors, into which the feed may be introduced. The spontaneous flashing of feed in a continuous vacuum crystallizer or the controlled evaporation in a batch vacuum crystallizer as the pressure is lowered results in cooling as well as concentration of the solution.

The provision of adequate evacuator capacity for the low pressures required is a major consideration unless cooling water is available at a temperature sufficiently below the saturation temperature of the vapor so that direct condensation of the solvent vapor is possible. Under such conditions a vacuum pump or a one- or two-stage steam ejector is adequate to maintain the reduced pressure by simply removing the noncondensable gases. Usually the operating pressure required to attain the desired crystallization is so low that the evolved vapor cannot be condensed directly by the cooling water available. In this case a steam jet booster, as shown in Fig. 473, is used to compress the vapor and noncondensables before condensation. The removal of noncondensables saturated at the temperature of condensation is accomplished by steam jets as described.

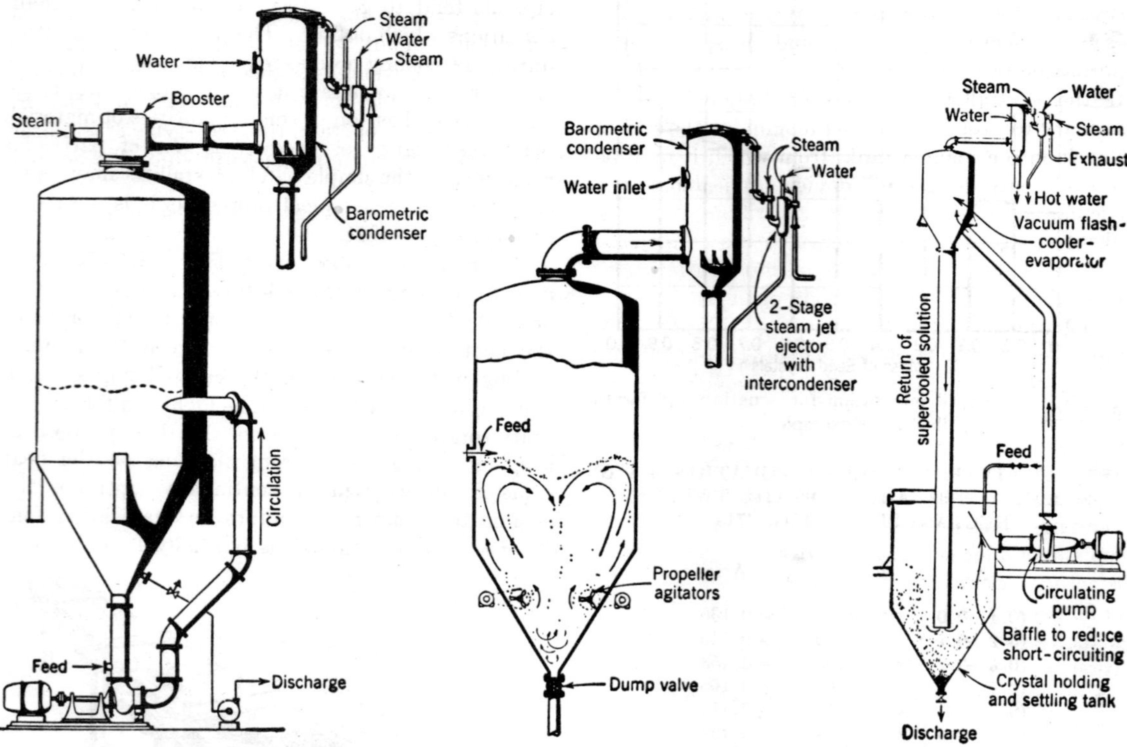

FIG. 473. Vacuum crystallizer with recirculation. (*Swenson Evaporator Co.*)

FIG. 474. Vacuum crystallizer with agitators. (*Swenson Evaporator Co.*)

FIG. 475. Oslo vacuum crystallizer. (*Swenson Evaporator Co.*)

The design of the crystallizer itself must be such that stratification of the feed cannot occur. At the low temperatures involved, the hydrostatic head of a few inches of the solution results in raising the boiling point to such an extent that liquor well above the saturation temperature at the pressure of the vapor space may pass through the crystallizer. Agitation necessary to avoid this is usually accomplished by the joint effect of propellers and of spontaneous vaporization of feed (Fig. 474).

If the crystallizer is operated batchwise, it is customary to drop the finished magma at the end of the cycle into a magma tank, from which it may be pumped to any dewatering device applicable to the particular installation, that is, centrifuge, filter, or thickener. If the crystallizer is operated continuously, it is usual practice to pump continuously some of the magma from the body of the crystallizer to the dewatering filter or centrifuge. The type of pump is usually any standard centrifugal pump with adequate clearance for the crystals being handled. The limitations on slurry density are usually imposed by the viscosity of the magma which will permit adequate circulation in the crystallizer body, rather than the characteristics of the removal pump. Depending primarily upon the relative density of the crystals and solution, the critical slurry density for satisfactory operation of the crystallizer may be from 30 per cent solids by weight for heavy crystals to as high as 55 per cent for crystals approaching the solution in density. The mother liquor separated in the filter or centrifugal can be returned to process or discarded as desired.

The types of equipment illustrated in Figs. 472, 473, and 474 have no provision for the removal of large crystals of the desired size with retention in the equipment of small crystals for further growth. The combination of classifying action with crystallization operations is incorporated in several crystallizers such as that shown in Fig. 475. Classifying crystallizers are available in which the crystal formation results from either cooling of the solution or evaporation of solvent therefrom. In either case the liquid being circulated within the crystallizer is carried into the region of supersaturation in the absence of crystals. The circulation then returns this slightly supersaturated solution to a bed of crystals on which salt is deposited. This bed is supported in such fashion that crystals are retained in the crystallizer until the size is such that they will drop out of the circulating liquid stream and then be removed as product. The classifying action retains the small crystals in the circulating liquor stream in the crystal bed until their size has reached the desired range.

The *classifying crystallizer* is subject to operation under vacuum for low temperature if desired or at moderate vacuum for intermediate temperature if that is necessary for phase stability or for rate of deposition. Thus, if it is desired to produce $MgSO_4 \cdot 6H_2O$, which is stable in contact with saturated solution only between about 120° and 155° F, it is possible to operate the Oslo-type crystallizer within this particular range to produce the desired phase.

Multiple-effect evaporators are also operated for the production of a crystalline product. The operation of the evaporator seldom allows much flexibility as to the size and shape of crystals produced. Some classifying action does occur in that fine crystals are usually carried in the circulating mother liquor until they attain such size as will result in their dropping out of the circulating stream into the salt-removal equipment. Where careful control of the crystal size is necessary, or where the desired range of crystal size is different from that which will result from crystallization in the evaporator, it is common practice to transfer the hot concentrated mother liquor from the evaporator directly to a crystallizer in which the desired crystal size can be attained.

BIBLIOGRAPHY

1. McCabe, W. L., "Crystal Growth in Aqueous Solutions," *Ind. Eng. Chem.*, **21**, 30, 112 (1929).
2. Miers, Sir H. A., "The Growth of Crystals in Supersaturated Solutions," *J. Inst. Metals*, **37**, 1, 331 (1927).
3. Perry, J. H., *Chemical Engineers' Handbook*, 3rd ed., p. 196, McGraw-Hill Book Co. (1950).

PROBLEMS

1. The solubility of sodium sulfate is 40 parts Na_2SO_4 per 100 parts of water at 30° C, and 13.5 parts at 15° C. The latent heat of crystallization (liberated when crystals form) is 18,000 gram-calories per gram mole Na_2SO_4. Glauber's salt ($Na_2SO_4 \cdot 10H_2O$) is to be made in a Swenson-Walker crystallizer by cooling a solution, saturated at 30° C, to 15° C. Cooling water enters at 10° C and leaves at 20° C. The overall heat transfer coefficient in the crystallizer is 25 Btu/(hr)(°F)(sq ft) and each foot of crystallizer has 3 sq ft of cooling surface. How many 10-ft units of crystallizer will be required to produce 1 ton/hr of Glauber's salt?

2. One ton of sodium thiosulfate ($Na_2S_2O_3 \cdot 5H_2O$) is to be crystallized per hour by cooling a solution containing 56.5

per cent $Na_2S_2O_3$ to 30° C in a Swenson-Walker crystallizer. Evaporation is negligible. The product is to be sized closely to approximately 14 mesh. Seed crystals closely sized to 20 mesh are introduced with the solution as it enters the crystallizer.

How many tons of seed crystals and how many tons of solution are required per hour?

3. A continuous adiabatic vacuum crystallizer is to be used for the production of $MgSO_4 \cdot 7H_2O$ crystals from 20,000 lb/hr of solution containing 0.300 weight fraction $MgSO_4$. This solution enters the crystallizer at 160° F.

The crystallizer is to be operated so that the mixture of mother liquor and crystals leaving the crystallizer contains 6000 lb/hr of $MgSO_4 \cdot 7H_2O$ crystals. The estimated boiling-point elevation of the solution in the crystallizer is 10° F.

Solubility data are found in reference 3.

How many pounds of water are vaporized per hour?

4. Compute the screen analysis expected of the product obtained from the crystallizer as operated in the illustrative example, p. 498.

5. A solution of potassium chloride contains 30 per cent potassium chloride (KCl) by weight and insufficient impurities to alter its solubility significantly from that in pure water. It is cooled from its original temperature (over 200° F) in a tank and fed to a Swenson-Walker crystallizer just above the temperature at which it is saturated. It is seeded with ½ lb of KCl crystals per 100 lb of solution.

The seeds have the following screen analysis:

Mesh	Mass Per Cent
On 48	0
48–65	8
65–100	57
100–150	23
Pan	12
	100

Predict the screen analysis of the crop if the magma is cooled to 70° F.

6. Sodium nitrate is to be purified by preparing a solution saturated at 212° F, filtering, and crystallizing to a final temperature of 85° F. The dried product will be sized by screening. The material retained on a 20-mesh screen or passed by a 28-mesh screen will be reprocessed. Seeds will have the following screen analysis:

Mesh Size	Mass Per Cent
On 35	2
35–42	5
42–48	22
48–60	37
60–65	15
65–80	8
80–100	5
Through 100	6

Specify the weight of seeds per 1000 lb of hot filtered solution for the maximum yield of the 20–28 mesh screen size.

7. Potassium chloride may be produced from sylvinite (a natural mixture of 42.7 per cent sylvite, KCl, and 56.6 per cent halite, NaCl) by the following process. The ore is crushed and screened to −4 mesh and sent to measuring bins. The crushed ore from the measuring bins, together with recycle mother liquor, is sent to the dissolver where the KCl is dissolved at 110° C and the remainder of the ore is sent to waste. The liquor at 110° C and saturated with both NaCl and KCl is cooled in vacuum coolers and crystallizers to 27° C. Water is added to the crystallizers (in amount just equal to the water evaporated) to prevent NaCl from crystallizing. The suspension of KCl in a saturated solution formed in the crystallizers is allowed to settle, decanting the mother liquor which is heated and recycled to the dissolver. The thickened crystal mass is filtered, washed, dried, crushed, screened, and conveyed to the warehouse. The filtrate and wash water are added to the decanted mother liquor and recycled.

(*a*) Sketch a flowsheet of the operation.

(*b*) Plot an accurate phase diagram (temperature versus mass ratio of KCl to KCl + H_2O) for the system KCl–NaCl–H_2O for a constant NaCl to H_2O ratio of 0.274.

(*c*) Calculate, per ton of dried KCl, the quantities and compositions (mass fraction) of all important streams, and indicate these on the flowsheet.

(*d*) Indicate and justify all necessary assumptions.

(*e*) Indicate on the phase diagram the path of the crystallization process, points representing stream compositions, and line lengths representing quantities.

(*f*) Discuss the formation or lack of formation of NaCl crystals during this process.

SOLUBILITY OF NaCl AND KCl IN H_2O

Temperature, °C	Mass % NaCl	Mass % KCl	Solid Phases
0	19.6	8.75	KCl
0	22.4	7.2	KCl + NaCl
20	19.05	11.35	KCl
20	21.0	10.2	KCl + NaCl
40	18.5	14.0	KCl
40	19.6	13.4	KCl + NaCl
60	17.95	16.55	KCl
60	18.6	16.0	KCl + NaCl
80	17.4	19.1	KCl
80	17.6	19.0	KCl + NaCl
100	16.8	21.75	KCl
100	16.9	21.7	KCl + NaCl
110	16.6	22.8	KCl + NaCl

CHAPTER 34

Agitation

AGITATION as used in the process industries is the production of irregular disturbances or turbulent motion within a fluid by means of mechanical devices acting on that fluid. Agitation finds widespread application in industry as a means of promoting certain operations such as extraction, mixing, absorption, heat transfer, and chemical reaction. Although agitation might be discussed in connection with each of these operations in which it is employed, it is so common to all of them that it may be regarded as a unit operation itself.

OBJECTIVES AND REQUIREMENTS

Many different requirements are made of agitation systems, depending upon the desired result. There are four general types of operations which may be distinguished by their objectives, and each type may require different agitation equipment.

Mass Transfer in Heterogeneous Systems

This category includes chemical reactions, solution of solids, extraction, absorption, and adsorption. The requirements of the agitator are twofold: It must disperse or suspend the discontinuous phase throughout the continuous phase; and it must produce turbulence of high intensity around the suspended droplets, bubbles, or particles to promote mass transfer between the phases. This demands an agitator producing velocities of flow sufficiently high to prevent the settling out of particles or the stratifying of phases. There must be no dead spaces in the agitation system which would allow one of the phases to concentrate there. The intensity of turbulence should be uniform throughout the whole tank if mass transfer is to be promoted at all points of interfacial contact. Small impellers operating at high speeds produce high stream velocities but concentrate the turbulent zone near the impeller; on the other hand, large impellers operating at low speeds produce uniform turbulence throughout the tank but may develop velocities too low for good suspension or dispersion of phases. Consequently medium-size impellers operating at medium speeds are recommended for this type of operation.

Mixing or Blending of Two Liquids

In continuous flow two liquids may be mixed simply by passage through a centrifugal pump where intimate contact results. In batch mixing or blending in a large tank, the agitator must pump large streams of the liquids to all points in the system. After the masses of liquids have been pumped around and mixed on a bulk scale, the final localized mixing depends upon the intensity of turbulence at all points or upon molecular diffusion. The most important part in this operation is the large-scale flow to all points in the system, for without this any high intensity turbulence at any point will accomplish nothing. Therefore, for mixing or blending two or more fluids large impellers operating at low speeds are generally superior to small impellers operating at high speeds.

Physical Change or Emulsification

Some operations, such as the emulsification of two immiscible liquids, require an extremely high rate of shear or intensity of turbulence at some point in the system. Usually the liquids will be fairly well mixed before arriving at this point of high shear. If the operation is conducted in a batch in a large tank, a gradual turnover of all the tank contents is required to bring all portions into the highly turbulent zone. Small impellers operating at high speeds with small clearances between the impeller and fixed surroundings are especially suited to produce the high shearing stresses required for this operation.

Heat Transfer and Uniformity of Temperature

Heat transfer is often a necessary accessory to certain mass transfer operations, especially those involving chemical reactions. Consequently, reaction tanks are equipped with heating (or cooling) coils or jacketed walls, and adequate velocities past these heat exchange surfaces are necessary to promote heat transfer. In such cases, the agitator impeller should be located near the coils or jacketed walls. Moreover, the impeller must produce large-volume streams of flow so that all the contents of the tank will be brought in the neighborhood of the heat transfer surfaces if uniform temperatures are to prevail throughout the system. Large impellers operating at low speeds will usually be suitable for this operation. In a few instances it may be necessary to prevent localized overheating of a certain area of heat transfer surface by allowing the impeller to scrape such a surface or by employing a small high-speed impeller to develop high turbulence in the immediate vicinity of the surface.

In considering any of the above requirements, one must bear in mind that the zone of high turbulence surrounding the impeller is contained in a smaller volume at higher viscosities. Therefore, large impellers are required in viscous fluids if flow to all points in the system is desired.

TYPES OF AGITATION EQUIPMENT

The mechanical devices which have been employed for the production of agitation are extremely varied, ranging all the way from the familiar egg beater to the high-speed colloid mill. The following are the more important types of agitators found in use today.

Rotating Impellers

Impellers of many different shapes and sizes (Fig. 476) revolve on shafts inserted in tanks which may be cylindrical, conical, hemispherical, or rectangular. Marine propellers, paddles, and curved-blade turbines are good examples of this type of equipment. More than one impeller may be mounted on a revolving shaft, and more than one shaft may be used in a given tank. Sometimes two adjacent impellers rotate in opposite directions, forming a beater, and sometimes the impellers actually touch the walls of the tank, giving a positive scraping action which is desirable when thick layers of material tend to stick to the wall. The tanks are often constructed with baffles, draft tubes, deflector rings, or even heating and cooling coils, all of which serve to deflect or guide the fluid after it leaves the impeller. In small closed glass installations the impeller may be made of iron and driven by a rotating magnetic field with no shaft to connect the impeller with the outside. This eliminates any packing problems in systems which must be airtight.

Circulation Pump Systems

Tanks may be connected to pumps which remove the fluid at one point and return it to the tank at some other point, thus setting up circulation in the tank. Centrifugal pumps by themselves are good agitators in continuous-flow systems. If two liquids are fed to the suction side of the pump, they will come from the discharge side in a well-mixed condition, for the high-speed centrifugal impellers create considerable turbulence in the small casings.

Reciprocating Paddles

Paddles or blades may be moved back and forth through rectangular tanks.

Revolving Tanks or Pans

A whole pan of material may be revolved while the blades or baffles remain stationary. Usage is generally confined to mixing of very pasty materials.

Air Lifts and Air Agitators

Air is passed upwards through a large-diameter tube, open at both ends and submerged in the liquid, setting up a circulation as a result of the decreased density of the air-liquid mixture in the tube. Air or other gas may be admitted at the bottom of a tank through small holes or from the end of pipe, causing irregular disturbances as the gas bubbles rise to the surface of the liquid.

Colloid Mill, Homogenizer, and Mixing Jet

In the *colloid mill* fluid is fed to a minute clearance space between a high-speed rotor and its casing. Extremely high shearing forces are set up, and very intimate mixing results. The clearances are of the

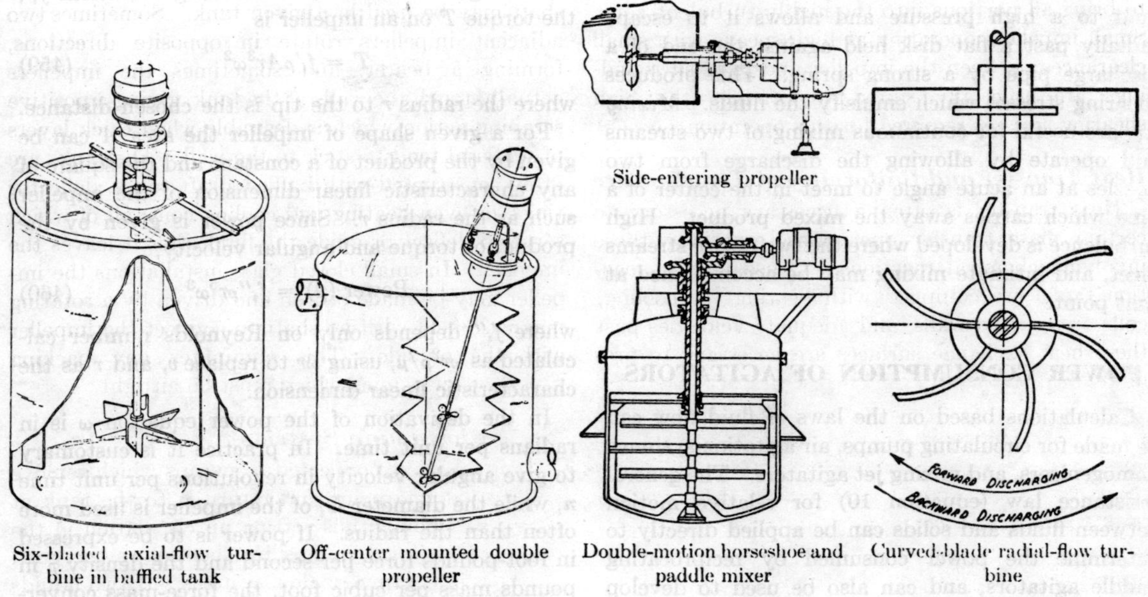

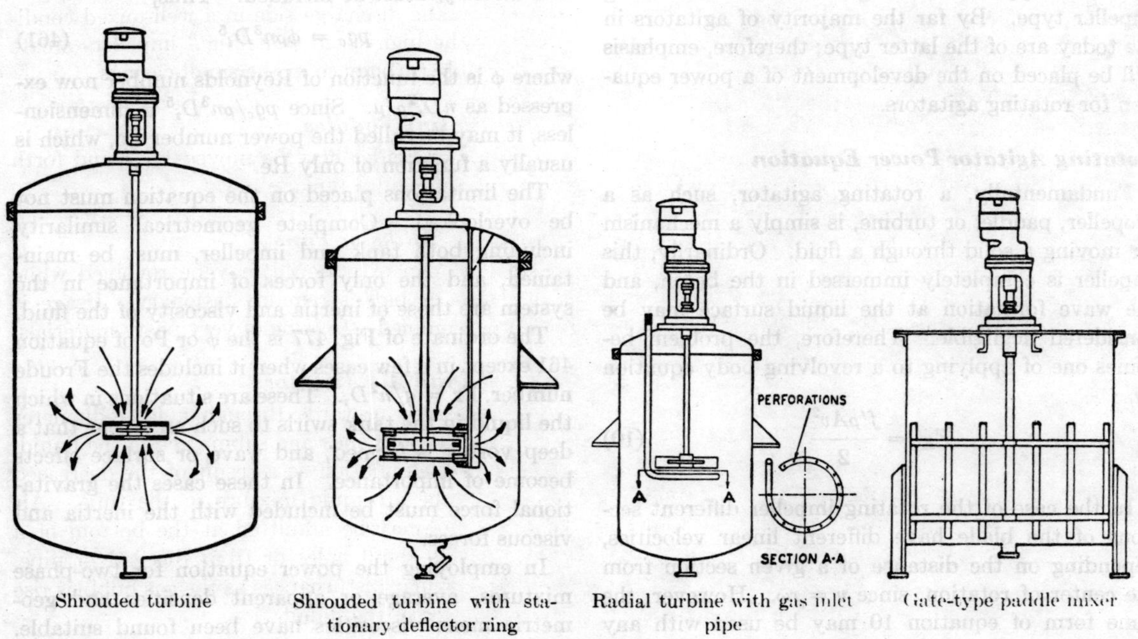

FIG. 476. Representative rotating-impeller agitators. (*Mixing Equipment Co.* and *International Engineering Inc*)

order of 0.001 in. The *homogenizer* compresses a fluid to a high pressure and allows it to escape radially past a flat disk held against the end of a discharge pipe by a strong spring. This produces shearing stresses which emulsify the fluids. *Mixing jets* are useful for continuous mixing of two streams and operate by allowing the discharge from two nozzles at an acute angle to meet in the center of a pipe which carries away the mixed product. High turbulence is developed where the two nozzle streams meet, and intimate mixing may be accomplished at that point.

POWER CONSUMPTION OF AGITATORS

Calculations based on the laws of fluid flow can be made for circulating pumps, air agitation systems, homogenizers, and mixing jet agitators. The general resistance law (equation 10) for relative motion between fluids and solids can be applied directly to determine the power consumed by reciprocating paddle agitators, and can also be used to develop equations applicable to agitators of the rotating impeller type. By far the majority of agitators in use today are of the latter type; therefore, emphasis will be placed on the development of a power equation for rotating agitators.

Rotating Agitator Power Equation

Fundamentally, a rotating agitator, such as a propeller, paddle, or turbine, is simply a mechanism for moving a solid through a fluid. Ordinarily, this impeller is completely immersed in the batch, and the wave formation at the liquid surface may be considered negligible. Therefore, the problem becomes one of applying to a revolving body equation 10,

$$F_R' = \frac{f'\rho A v^2}{2} \tag{10}$$

In the case of the rotating impeller different sections of the blade have different linear velocities, depending on the distance of a given section from the center of rotation, since $v = r\omega$. However, the same form of equation 10 may be used with any convenient relative velocity between the fluid and the solid if the appropriate drag (friction factor) coefficient is used. The torque on the impeller is the product of an average force and moment arm or radius to the action line of the force. Throwing all corrections for variable distances or moment arms to given sections into the drag coefficient f_i', the torque T on an impeller is

$$T = f_i'\rho A r^3 \omega^2 \tag{459}$$

where the radius r to the tip is the chosen distance.

For a given shape of impeller the area A can be given by the product of a constant and the square of any characteristic linear dimension of the impeller such as the radius r. Since power is given by the product of torque and angular velocity,

$$\text{Power } (p) = f_i''\rho r^5 \omega^3 \tag{460}$$

where f_i'' depends only on Reynolds number calculated as $\omega r^2 \rho/\mu$, using ωr to replace v, and r as the characteristic linear dimension.

In the derivation of the power equation ω is in radians per unit time. In practice it is customary to give angular velocity in revolutions per unit time n, while the diameter D_i of the impeller is used more often than the radius. If power is to be expressed in foot-pounds force per second and the density ρ in pounds mass per cubic foot, the force-mass conversion factor g_c must be included. Thus,

$$pg_c = \phi\rho n^3 D_i^5 \tag{461}$$

where ϕ is the function of Reynolds number now expressed as $nD_i^2\rho/\mu$. Since $pg_c/\rho n^3 D_i^5$ is dimensionless, it may be called the power number Po, which is usually a function of only Re.

The limitations placed on the equation must not be overlooked. Complete geometrical similarity including both tank and impeller, must be maintained, and the only forces of importance in the system are those of inertia and viscosity of the fluid.

The ordinate of Fig. 477 is the ϕ or Po of equation 461 except in a few cases when it includes the Froude number, $\text{Fr} = g/n^2 D_i$. These are situations in which the liquid in the tank swirls to such an extent that a deep vortex is formed, and wave or surface effects become of importance. In these cases the gravitational force must be included with the inertia and viscous forces.

In employing the power equation for two-phase mixtures, average or apparent densities and geometric mean viscosities have been found suitable. The geometric mean viscosity is calculated as $(\mu_1^x)(\mu_2^y)$, where μ_1 and μ_2 are the viscosities of the individual phase liquids and x and y are their respective volume fractions in the two-phase mixture.

Systems Not Geometrically Similar. If the system in use is not geometrically similar to one of

Type of Impeller	$\frac{D_t}{D_i}$	$\frac{Z_l}{D_i}$	$\frac{Z_i}{D_i}$	Baffles No.	Baffles w/D_i	No.	Ref.
Turbine with 6 flat blades. $0.25D_i$ → $0.2D_i$	3	2.7–3.9	0.75–1.3	4	0.17	1	7
Same as No. 1.	3	2.7–3.9	0.75–1.3	4	0.10	2	7
Same as No. 1.	3	2.7–3.9	0.75–1.3	4	0.04	4	7
Same as No. 1. $a = 1$, $b = 40$.	3	2.7–3.9	0.75–1.3	0		14°	7
Turbine with 6 curved blades. Blade sizes same as No. 1.	3	2.7–3.9	0.75–1.3	4	0.10	3	7
Turbine with 6 arrowhead blades. Blade sizes same as No. 1.	3	2.7–3.9	0.75–1.3	4	0.10	5	7
Radial turbine with deflector ring.				0		7	9
Shrouded turbine with 6 blades, 20-blade deflector ring.	2.4	0.74	0.9	0		11	6
Similar to No. 11, but not identical.	3	2.7–3.9	0.75–1.3	0		12	7
Same as No. 12, but no deflector ring.	3	2.7–3.9	0.75–1.3	4	0.10	13	7
Axial turbine with 8 blades at 45° angle. See No. 17.	3	2.7–3.9	0.75–1.3	4	0.10	9	7
Axial turbine with 4 blades at 60° angle. $0.25D_i$	3	3	0.50	0		17	2
Axial turbine with 4 blades at 45° angle. See No. 17.	5.2	5.2	0.87	0		19	2
Same as No. 19.	2.4–3.0	2.4–3.0	0.4–0.5	0		22	2
Disk with 16 vanes $0.1D_i$ $0.35D_i$	2.5	2.5	0.75	4	0.25	6	1

Type of Impeller	$\frac{D_t}{D_i}$	$\frac{Z_l}{D_i}$	$\frac{Z_i}{D_i}$	Baffles No.	Baffles w/D_i	No.	Ref.
Paddle with 2 blades. $0.25D_i$	4.35	4.3	0.29	3	0.11	8	3
Paddle with 4 blades. See No. 8.	3	3	0.5	0		16	2
Paddle with 2 blades. See No. 8.	3	3.2	0.33	0		20	4
Paddle with 2 blades. See No. 8.	3	2.7–3.9	0.75–1.3	4	0.10	10	7
Paddle with 2 blades. See No. 8. Blade width = $0.13D_i$.	1.1	0.5	0.19	0		29	10
Paddle with 2 blades. See No. 8. Blade width = $0.17D_i$.	1.1	0.4	0.10	0		29	10
Marine propeller with 3 blades. Pitch = $2D_i$.	3	2.7–3.9	0.75–1.3	4	0.10	15	7
Same as No. 15. $a = 1.7$, $b = 18$.	3.3	2.7–3.9	0.75–1.3	0		21°	7
Same as No. 15, but pitch = $1.33D_i$.	16			3	0.06	18	5
Same as No. 15, but pitch = $1.09D_i$.	9.6			3	0.06	23	5
Same as No. 15, but pitch = $1.05D_i$. $a = 2.3$, $b = 18$.	2.7	2.7–3.9	0.75–1.3	0		27°	7
Same as No. 15, but pitch = $1.04D_i$, $a = 0$, $b = 18$.	4.5	2.7–3.9	0.75–1.3	0		25°	7
Same as No. 15, but pitch = D_i.	3	2.7–3.9	0.75–1.3	4	0.10	24	7
Same as No. 15, but pitch = D_i. $a = 2.1$, $b = 18$.	3	2.7–3.9	0.75–1.3	0		26°	7
Same as No. 15, but pitch = D_i.	3.8	3.5	1.0	0		28	8

D_i = diameter of impeller, D_t = diameter of tank, n = revolutions per second, w = width of baffle, Z_i = elevation of impeller above tank bottom, Z_l = height of liquid in tank.

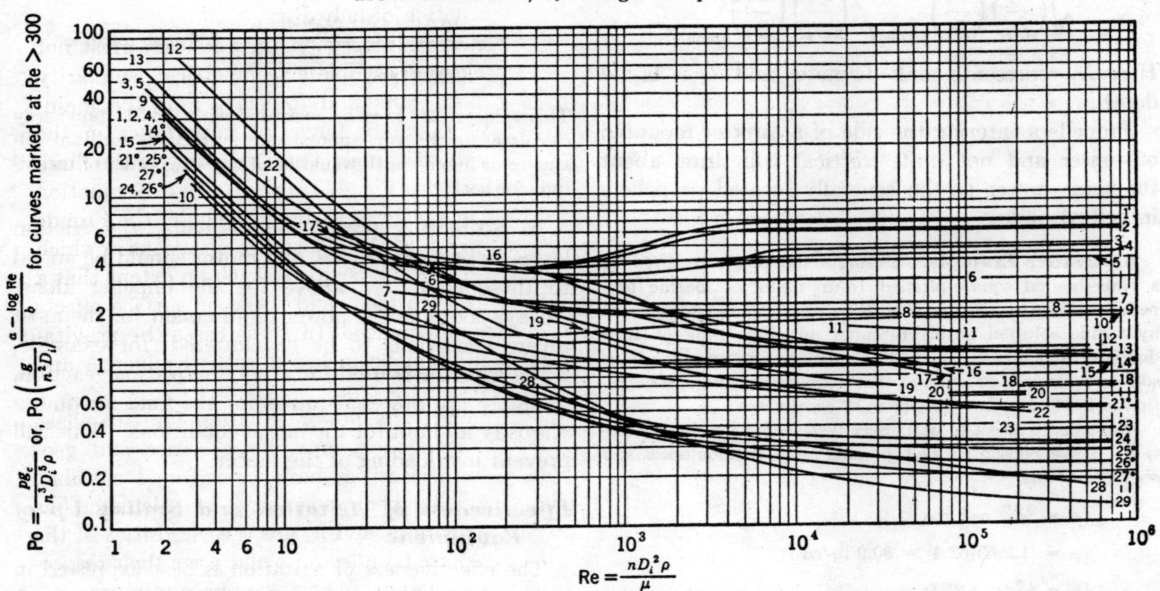

Fig. 477. Power consumption of various agitators expressed in terms of Po as a function of Reynolds number, Re. For curves marked with ° surface effects become important and the Froude number $Fr = g/n^2D_i$ is included as indicated for Re > 300.

those given in Fig. 477, the graphs may still be used to make the necessary power calculations in the majority of installations if the impeller alone is similar to an impeller for which a curve is given, provided some estimate is made of the effect of any changed conditions around the impeller. The three most important changes to be considered are baffling, liquid depth, and tank diameter.

If an impeller is to operate in a tank with several baffles at the wall and data are given in the graph for that same impeller operating in a baffled tank, the graph will probably give the correct power consumption even if the liquid depth and tank diameter to be used are different from those given. If the impeller is to be used in a well-baffled tank and the data of Fig. 477 are for an unbaffled tank, the power so predicted will be only one-half to one-fourth of the power actually required by the baffled agitator. The reverse is also true; the graph will predict two to four times as much power consumption if the data of the graph are for a baffled agitator and the intended use is in an unbaffled tank. Increasing the tank diameter or the liquid depth in cases where no baffles are present usually increases the power consumption. Corrections for such changes in design may be made approximately by multiplying the predicted power consumption from the graph by

$$\sqrt{\left(\frac{D_t}{D_i}\right)\left(\frac{Z_l}{D_i}\right)_{\text{desired}} \bigg/ \left(\frac{D_t}{D_i}\right)\left(\frac{Z_l}{D_i}\right)_{\text{graph}}}$$

Here D_t represents tank diameter and Z_l is liquid depth.

Propellers entering the side of a tank or mounted off-center and not quite vertical will draw about the same power as the centrally located propellers in baffled tanks.

Illustrative Example. Estimate the power required by a three-bladed square-pitched 15-in. diameter marine propeller rotating at 300 rpm in a 30 per cent (by weight) sodium hydroxide solution at 65° F in an unbaffled tank of 10-ft diameter and 8-ft liquid depth. The specific gravity of the sodium hydroxide solution at this temperature and concentration is 1.297, and its viscosity is 13 centipoises.

Any consistent set of units may be selected for the calculation. Those chosen are feet, pounds force, pounds mass, and seconds. Thus,

$$n = \tfrac{300}{60} = 5 \text{ rps}$$

$$\rho = (1.297)(62.4) = 80.9 \text{ lb/cu ft}$$

$$D_i = \tfrac{15}{12} = 1.25 \text{ ft}$$

$$\mu = (13)(0.000672) = 0.00874 \text{ lb mass/ft-sec}$$

$$\text{Re} = \frac{nD_i^2\rho}{\mu} = \frac{(5)(1.25)^2(80.9)}{0.00874} = 72{,}300$$

$$g_c = 32.2 (\text{lb mass})(\text{ft})/(\text{lb force})(\text{sec})^2$$

This impeller is similar to the one used indicated for curve No. 28 of Fig. 477. At a Reynolds number of 72,300 the value of ϕ is read as 0.22 and the power consumption from the graph is

$$p = \frac{\phi\rho n^3 D_i^5}{g_c} = \frac{(0.22)(80.9)(5)^3(1.25)^5}{32.2} = 211 \text{ ft-lb force/sec}$$

This result will probably be on the low side because the tank is not baffled and the diameter and liquid depth are greater than those corresponding to the graph. Therefore, the approximation factors are used as follows to obtain the predicted power consumption.

$$p = 211\sqrt{\left(\frac{10}{1.25}\right)\left(\frac{8}{1.25}\right) \bigg/ (3.8)(3.5)}$$

$$= 414 \text{ ft-lb force/sec}$$

$$= \tfrac{414}{550} = 0.75 \text{ hp}$$

The result obtained directly from curve 26°, Fig. 477, which allows for wave action with an impeller operating in a tank only three times the diameter of the impeller, is obtained by use of the Froude number as well as the Reynolds number, as follows.

$$\text{Fr} = \frac{g}{n^2 D_i} = \frac{32.2}{(5)^2(1.25)} = 1.03$$

From the graph, $\phi = 0.245$, and, since

$$\phi = \left(\frac{pg_c}{n^3 D_i^5 \rho}\right)\left(\frac{g}{n^2 D_i}\right)^{\frac{a - \log \text{Re}}{b}}$$

from Fig. 477, $a = 2.1$, $b = 18$,

$$p = \frac{(0.245)(5)^3(1.25)^5(80.9)}{(32.2)(1.03)^{\frac{2.1 - \log 72.300}{18}}} = 236 \text{ ft-lb force/sec}$$

This value of 236 is to be compared with the 211 previously calculated, since any approximate corrections for tank diameter or liquid depth would be of the same order of magnitude for both.

In estimating power requirements, the friction losses in bearings, drive, and motor should be added to those computed above for the impeller alone. Always compute the power requirement for the most severe conditions to be encountered; for example, if the agitator is used for a polymerization reaction, calculate the power required for the final maximum viscosity and not for the initial conditions. This will prevent overloading of the motor.

Effectiveness of Agitation and Scaling Up of Equipment

The effectiveness of agitation is best expressed in terms of its effect upon the desired operation, such as promotion of heat, or mass transfer, solution of solids, or blending of two liquids. Various tests have been reported in the literature, but the results are more or less specific for the particular process and equipment employed. Consequently no completely general correlation has been developed.

Ordinary industrial practice assumes that the effectiveness of similar agitation equipment will remain more or less the same if the same power per unit volume of fluid is supplied the impeller. Thus, the design of plant-size equipment requires a laboratory experiment on the agitation problem in question in equipment which is geometrically similar to that of the commercial installation. Knowing that the small agitator will do the required job, the large one may be specified with the same geometrical proportions and the same power per unit volume of fluid as was used in the small one.

At best the power per unit volume theory is only a good approximation. It does not take into account the actual scale and intensity of turbulence or the maximum shearing rate developed in the liquid, which are the real factors governing the operation in question. For example, in emulsification a definite minimum shearing rate must be maintained to emulsify two immiscible liquids. This shearing rate might be represented by the tip speed of the impeller. Consider the construction of a piece of equipment twice the size of an experimental unit in which successful emulsification was accomplished. At high Reynolds numbers the required power is proportional to $n^3D_i^5$, and in geometrically similar systems the volume of the tank is proportional to D_i^3. Therefore, power per unit volume is proportional to $n^3D_i^5/D_i^3$ or just $n^3D_i^2$. The tip speed of the impeller is proportional to nD_i. The same tip speed of the impeller is preserved if its diameter D_i is doubled in the new equipment and the speed of rotation n is halved. This condition calls for one-half the power per unit volume, $(\frac{1}{2})^3(2)^2 = \frac{1}{2}$, used in the experimental unit. Experiments on the blending of two miscible liquids also indicate a greater efficiency of blending in large-scale than in small-scale equipment. However, in the absence of any known controlling factor, the power per unit volume theory is recommended as the best means of scaling up agitation equipment.

Although in only a few cases is it possible to determine beforehand how much power should be supplied a given system to accomplish a desired result, a rough approximation may be made from the following empirical relationship.

$$\text{Horsepower per cubic foot} = \frac{(\text{Specific gravity})(10 + \sqrt[4]{\mu})}{C}$$

where μ is in centipoises.

C varies from about 330 for high-level agitation to 3300 for low-level agitation.

BIBLIOGRAPHY

1. Cooper, C. M., G. A. Fernstrom, and S. A. Miller, *Ind. Eng. Chem.*, **36**, 504 (1944).
2. Hixson, A. W., and S. J. Baum, *Ind. Eng. Chem.*, **34**, 194 (1942).
3. Mack, D. E., and A. E. Kroll, *Chem. Eng. Progress*, **44**, 189 (1948).
4. Miller, S. A., and C. A. Mann, *Trans. Am. Inst. Chem. Engrs.*, **40**, 709 (1944).
5. Miller, F. D., and J. H. Rushton, *Ind. Eng. Chem.*, **36**, 499 (1944).
6. Olney, R. B., and G. J. Carlson, *Chem. Eng. Progress*, **43**, 473 (1947).
7. Rushton, J. H., E. Costich, and H. J. Everett, Presented at Annual Meeting of the American Institute of Chemical Engineers, Detroit, 1947.
8. Stoops, C. E., and C. L. Lovell, *Ind. Eng. Chem.*, **35**, 845 (1943).
9. Valentine, K. S., and G. MacLean, in J. H. Perry, *Chemical Engineers' Handbook*, 3rd ed., Section 17, McGraw-Hill Book Co. (1950).
10. White, A. M., E. Brenner, G. A. Phillips, and M. S. Morrison, *Trans. Am. Inst. Chem. Engrs.*, **30**, 570 (1934).

CHAPTER

35

Mass Transfer 1

THE movement of one or more components within or between phases occurs in many unit operations and is known as mass transfer. Absorption, crystallization, extraction, distillation, stripping, humidification, and drying are all examples of mass transfer operations. On the other hand, the flow of a fluid through a conduit is not usually considered as a mass transfer operation, since here matter is not transferred between phases unless a change of phase occurs, nor is one component of a phase in motion relative to other components of the phase.

Several of the unit operations involving mass transfer have already been considered from the point of view of the ideal or equilibrium stage. A more fundamental approach to the mass transfer operations is through the rate of transfer of the components between phases under the conditions encountered in a given unit of equipment and the correlation of the factors influencing these rates. This method becomes particularly useful when applied to equipment such as the packed tower whose performance bears only a slight resemblance to that of the equilibrium stage. On the other hand the use of the ideal stage is convenient and satisfactory in many applications where stage efficiencies are known or where the equipment approaches the operation of the equilibrium stage, such as a bubble-cap plate in a fractionating tower.

THE RATE EQUATION

Whenever there is a concentration gradient of a component within a phase, there is a potential available tending to transfer the component in the direction of decreasing concentration. Under steady conditions of transfer at constant temperature, the energy relations of the component being transferred within the phase and across the concentration gradient are expressed by equation 59 which may be written as follows.

$$\Delta F_T + \Delta m \frac{Zg}{g_c} + \frac{\Delta m v^2}{2g_c} = -w - (lw) \quad (462)$$

where $\Delta F_T = \int V\,dP + \int \mu_B\,dm_B + \text{etc.}$, frequently called the isothermal increase in free energy.

In most cases where a component is being transported in a phase, ΔmZ and Δmv^2 are negligible and w is zero, so that equation 462 may be written

$$-\Delta F_T = RT \ln \frac{f_1}{f_2} = (lw) \quad (463)$$

where f_1, f_2 = fugacity of the transferring component at points 1 and 2.
R = universal gas constant.
T = absolute temperature.

Equation 463 indicates that the energy $(-\Delta F)_T$ required to overcome the resistance to transfer (lw) from point 1 to point 2 is dependent upon a difference in fugacities.

The decrease in fugacity ($f_1 - f_2$, or $-\Delta f$) represents a driving force tending to transfer the component from point 1 to point 2 similar to the pressure drop ($-\Delta P$) in transporting fluids.

Mass transfer involves chemical energy, whereas fluid transportation involves pressure energy. The rate of transfer equals the driving force divided by the resistance, or the driving force multiplied by the conductance, or, in general,

$$\text{Rate} = (\text{Drop in potential})(\text{Conductance}) \qquad (464)$$

$$\text{Conductance} = (\text{Area})(\text{Transfer coefficient})$$

The following equation for the rate of mass transfer is equivalent to equation 382 for the rate of heat transfer.

$$N = kA(f_1 - f_2) = kA(-\Delta f) \qquad (465)$$

where N = mass (or moles) of component transferred per unit time.

k = transfer coefficient, mass (or moles) per unit time, unit drop in fugacity, and unit cross-sectional area.

$(-\Delta f)$ = decrease in fugacity of the transferring constituent.

The mass transfer coefficient is written k_g or k_l when applying equation 465 to mass transfer in a gas or in a liquid phase, respectively. When equation 465 is applied to mass transfer through two or more phases in series, the coefficient is called an overall mass transfer coefficient and is written K, which corresponds to the overall heat transfer coefficient U.

Mass transfer through two phases is illustrated in Fig. 478, which shows diagrammatically the potential of the transferring component as a function of distance.

Under steady conditions, the rate of transfer is the same in each phase and may be written

$$N = (K)(A)(f_g - f_l) = k_gA(f_g - f_i)$$
$$= k_lA(f_i - f_l) \qquad (466)$$

where k_g = coefficient of mass transfer in gas phase [moles/(hr)(sq ft)(atm)].

k_l = coefficient of mass transfer in liquid phase [moles/(hr)(sq ft)(atm)].

In writing equation 466 it is assumed that equilibrium conditions exist at the interface between the two phases so that the fugacity of the transferring component f_i is the same in both phases at the interface. This is equivalent to postulating that the interface itself offers no resistance to the transfer process. It is difficult and probably impossible at present to verify this assumption by direct experimental measurement. The successful correlation of mass transfer data based upon this assumption is highly presumptive evidence of its validity in simple mass transfer uncomplicated by chemical reactions and selective adsorption.

In most applications to fluid phases there is turbulent relative motion of the phases. In such cases, the fugacity of the transferring component in each

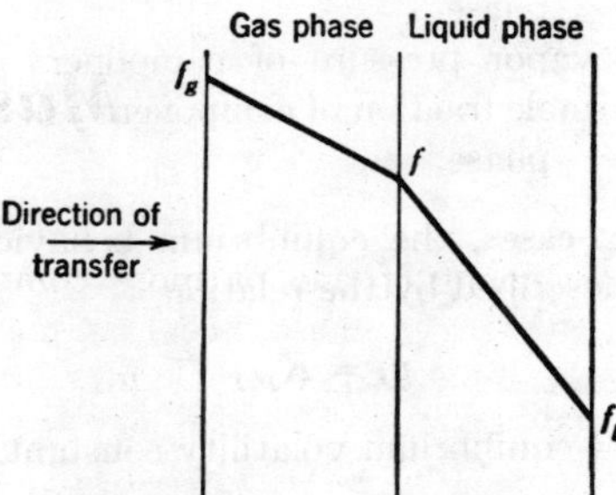

Fig. 478. Diagrammatic representation of driving forces in mass transfer.

phase, f_g or f_l, is determined from the mixed average composition of the phase.

If the fugacity of the transferring component at the interface f_i is eliminated in equation 466, the relation between the overall transfer coefficient and the phase coefficients is similar to that expressed by equation 385 or 386 for heat transfer, or

$$K = \frac{1}{(1/k_g) + (1/k_l)} \qquad (467)$$

THE DRIVING FORCE, FUGACITY, AND CONCENTRATION

The *fugacity* of a component in a phase may be calculated rigorously by known thermodynamic methods. In many of the applications of mass transfer involving contacting of vapor and liquid phases, the gases are at relatively low temperatures and pressures, and, *where the gas phases behave as ideal gases, the fugacity of a component in the phase is equal to its partial pressure.* Since the fugacity of a component in the gas phase is equal to the fugacity of the component in a liquid phase in equilibrium with the gas phase, the fugacity of components of a liquid phase may be calculated from equilibrium data. Thus, *when the gas phase behaves as an ideal gas and the liquid phase follows Raoult's law,* for any component

$$f_j = Py_j = p_j^* x_j \tag{468}$$

where f_j = fugacity of component j in a liquid solution (atm).
P = total pressure in the equilibrium gas phase (atm).
y_j = mole fraction of component j in the gas phase.
p_j^* = vapor pressure of component j (atm).
x_j = mole fraction of component j in the liquid phase.

In many cases, the equilibrium behavior of the system is described by the relation

$$y_j = K_j x_j \tag{469}$$

where K_j = equilibrium volatility constant.

For ideal gases

$$f_j = Py_j = PK_j x_j \tag{470}$$

Another form of this relation, known as Henry's law, is expressed as

$$f_j = Py_j = H_j C_j \tag{471}$$

where H_j = Henry's law constant.
C_j = concentration of component j (lb moles /cu ft of solution).

In operations involving the transfer of material between two liquid phases or between solids and fluids, equilibrium relationships must be known or estimated.

Wetted-Wall Column as an Adiabatic Humidifier

One of the simplest systems of mass transfer is the vaporization of various liquids from a thin film of liquid flowing down the wall of a cylindrical tube countercurrent to a stream of gas flowing through the center of the tube, as illustrated diagrammatically in Fig. 479. The transfer section is usually made of glass about 2 in. in diameter and 4 ft long. Liquid is introduced at the top of the circular section which serves as a weir and distributes the liquid as an even film over the walls. The liquid is removed through the seal at the bottom from the slightly flared lower end. Calming sections are provided for the gas stream at each end, as shown, to minimize turbulence.

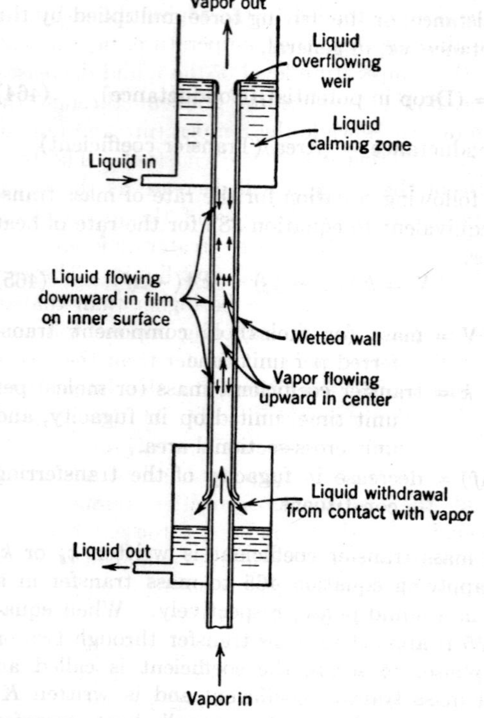

Fig. 479. Cross-sectional diagram of a typical wetted-wall column.

When a wetted-wall column is insulated from its surroundings so that its operation is adiabatic and the liquid is recirculated from the bottom of the column through the reservoir to the top of the column, the operation is described as "adiabatic humidification." Under these circumstances, the relationship between the gas composition and the temperatures of the gas and liquid may be calculated from thermodynamic properties and mass and energy balances.

Consider the wetted-wall column as an adiabatic humidifier with provision for temperature control of the liquid in the reservoir and the addition of "make-up" liquid to the reservoir at a controlled temperature. Assume that the gas and liquid throughout the system are initially at the same temperature. As mass is transferred from the liquid by vaporization, a drop in temperature is required to supply the latent heat of vaporization. As soon as the liquid temperature falls below the gas temperature, heat is transferred from the gas to the liquid. In this way the gas is cooled and humidified.

If the liquid supplied to the top of the column is now maintained at the temperature of the exit liquid, the temperature level of the liquid decreases, and the liquid temperature gradient through the column decreases while the temperature and humidity of the entering gas is held constant. The temperature of the exit gas will decrease as the liquid temperature decreases because of the greater heat transfer rates obtained with greater differences in temperature between the gas and liquid. The exit gas temperature will always be higher than that of the entering liquid. This cooling process will continue until the rate of heat transfer from the gas to the liquid is just equivalent to the latent heat required to vaporize the liquid. At this state of dynamic equilibrium, the temperature gradient in the liquid will have disappeared, all of the liquid being at the "adiabatic saturation temperature" corresponding to the properties of the entering gas.

If the unit contained an infinite extent of contacting surface, the exit wet gas would be in equilibrium with the entering liquid and therefore would be saturated with vapor at the adiabatic saturation temperature. If the reference states are liquid at the adiabatic saturation temperature T_{as} and dry gas at the adiabatic saturation temperature, an enthalpy balance gives, for a tower of infinite height,

$$G_g'\mathcal{H}_{as}\lambda_{T_{as}} - [G_g'\mathcal{H}_1\lambda_{T_{as}} + G_g's_1(T_1 - T_{as})] = 0 \quad (472)$$

where G_g' = mass velocity of the dry gas [lb/(hr)(sq ft)].
$\mathcal{H}_1$ = humidity of the entering gas (lb vapor/lb dry gas).
$\mathcal{H}_{as}$ = humidity of saturated exit gas at T_{as} (lb vapor/lb dry gas).
$\lambda_{T_{as}}$ = latent heat of vaporization at the temperature T_{as} (Btu/lb).
T_{as} = adiabatic saturation temperature.
T_1 = entering gas temperature.
s_1 = specific humid heat capacity of the entering gas (Btu)/(lb dry gas)(°F).

Equation 472 may be rearranged to give equation 473.

$$T_{as} = T_1 - \frac{(\mathcal{H}_{as} - \mathcal{H}_1)\lambda_{T_{as}}}{s_1} \quad (473)$$

Equation 473 demonstrates that for a given entering temperature and composition (T_1, $\mathcal{H}_1$) there is a unique adiabatic saturation temperature.

An equation similar to equation 473 can be written for the gas at any point of a unit of any size but with an entering gas of temperature (T_1, $\mathcal{H}_1$) producing the same adiabatic saturation temperature T_{as}, so that

$$T = \frac{(\mathcal{H}_{as} - \mathcal{H})\lambda_{T_{as}}}{s} + T_{as} \quad (474)$$

Equation 474 represents the relationship between the temperature and humidity of a gas passing through an adiabatic humidifier. Equation 473 may be used to calculate the adiabatic saturation temperature and humidity required in equation 474 from the entering gas conditions. These relationships are represented on a humidity chart as adiabatic cooling curves. For example, the adiabatic saturation temperature and mixing ratio (humidity) corresponding to a temperature of 144° F and a humidity of 0.02 are 60° F and 0.07, respectively, as found by following the adiabatic cooling curve of Fig. 498 from 144° F and 0.02 humidity to the saturation line.

Because the temperature of the liquid is constant in the wetted-wall column when operated as an adiabatic humidifier, there is no fugacity gradient in the liquid phase, the transfer process operates only through the gas phase, and the device is well suited to a study of mass transfer in the gas phase.

The transfer coefficient k as defined by equation 465 may be calculated from experimental data by integrating equation 475 over the measured conditions of the transfer section of the wetted-wall column.

For a differential element of interfacial area dA, measuring A from the bottom of the transfer section,

$$V'\,dY = k_g(-\Delta f)\,dA \quad (475)$$

where k_g = mass transfer coefficient [moles/(hr)(sq ft)(atm)].
V' = gas rate of flow of solute-free gas (moles/hr).
A = interfacial area (sq ft).
Y = gas composition (moles solute/mole solute free gas).
$-\Delta f = f_i - f_g$ = fugacity of the transferring component at the interface minus its fugacity in the gas phase.

Since

$$dY = \frac{dy}{(1 - y)^2} \tag{476}$$

and

$$V = \frac{V'}{(1 - y)} \tag{477}$$

where y = gas composition (mole fraction solute),
V = gas rate of flow (moles of total gas per hour),

substituting equations 476 and 477 in equation 475 and rearranging

$$\int_{y_1}^{y_2} \frac{dy}{(1 - y)(f_i - f_g)} = \int_0^A \frac{k_g}{V} dA \tag{478}$$

Anticipating the results of the wetted-wall experiments somewhat, it has been found that the gas phase mass transfer coefficient is approximately proportional to the total mass rate of gas flow so that the quotient k_g/V may be considered as a constant in the integration of the right-hand side of equation 478. This procedure is further justified in most cases since the variation in V is small through the transfer section. In the integration of the left-hand side of equation 478, the fugacity of the transferring component at the interface f_i is a constant and may be evaluated at low pressures as the vapor pressure of the liquid at the adiabatic saturation temperature. Thus, the integrated form of equation 478 is

$$\frac{1}{y_i - 1} \ln \left(\frac{1 - y_1}{1 - y_2}\right)\left(\frac{y_i - y_2}{y_i - y_1}\right) = \frac{k_g PA}{V} \tag{479}$$

In many cases, the gas mixture contains only a small amount of solute, the values of y become negligible compared to 1.0, and equation 479 may be simplified to

$$\ln \left(\frac{y_i - y_1}{y_i - y_2}\right) = \frac{k_g PA}{V} \tag{480}$$

or

$$k_g = \frac{V}{A}\left(\frac{y_2 - y_1}{(-\Delta f)_{lm}}\right) \tag{481}$$

where $(-\Delta f)_{lm}$ = log mean value of $(f_i - f_g)$ at the top and bottom of the transfer section.

MASS TRANSFER BY MOLECULAR DIFFUSION

The vast majority of transfer operations occur, at least in part, within or between fluids in motion. The rate at which mass may be transferred in a fluid is determined by the physical properties of the fluid, its state of motion, and the driving potential. As has been shown previously, the rate at which heat may be transferred in a fluid is determined by the physical properties of the fluid, its state of motion (as may be described by the Reynolds number), and the temperature difference producing the heat transfer.

When the fluid in which mass transfer is occurring is stagnant, or is in laminar motion normal to the direction of mass transfer, the mechanism of transfer is that of *molecular diffusion* which has been defined[43] * as "the spontaneous intermingling of miscible fluids placed in mutual contact, accomplished without the aid of mechanical stirring." The analog of molecular diffusion in heat transfer is thermal conduction. Similarly the shear stresses in fluids in laminar motion, which can be thought of as the transfer of momentum in the fluids, exist through the action of viscosity.

The relations which describe the process of molecular diffusion are derived from the kinetic theory of gases. According to this theory,[36, 43] in an ideal gas the resistance to the diffusion of a gas A through a gas B is proportional to (1) the number of molecules of gas A per unit volume, (2) the number of molecules of gas B per unit volume, (3) the mean velocity of gas A relative to gas B, and (4) the length of the diffusing path. The potential gradient producing diffusion is proportional to the resistance to diffusion, and, since in an ideal gas the fugacity is equal to the partial pressure, this relationship may be stated mathematically as follows.

$$-dp_A = \alpha_{AB}\left(\frac{\rho_A}{M_A}\right)\left(\frac{\rho_B}{M_B}\right)(v_A - v_B)\, dx \tag{482}$$

where p_A = partial pressure of the diffusing gas A (atm).
α_{AB} = proportionality constant.
ρ_A, ρ_B = partial densities of gas A and B (lb/cu ft).
M_A, M_B = molecular weights of gas A and B.
v_A, v_B = mean velocities of gas A and B (ft/hr).
x = length of the diffusion path (ft).

* The bibliography for this chapter appears on p. 575.

The Diffusion of Gas A through Gas B

When one gas A is diffusing through a second stagnant gas B, the term v_B in equation 482 becomes zero. Substituting $\rho_B/M_B = p_B/RT$ and $N_A/A = v_A\rho_A/M_A$ (by definition) in equation 482 and rearranging

$$\frac{N_A}{A} = \left(-\frac{RT}{\alpha_{AB}}\right)\frac{dp_A}{p_B\,dx} \qquad (483)$$

The diffusivity D_G of the system of gas A and gas B is defined as $D_G = R^2T^2/\alpha_{AB}P$, so that equation 483 becomes

$$\frac{N_A}{A} = -\frac{D_GP}{RT}\frac{dp_A}{p_B\,dx} \qquad (484)$$

In the general case of unsteady-state diffusion, in an element of volume $A\,dx$, the rate of accumulation of gas A in time dt is

$$\frac{\partial}{\partial t}\left(\frac{\rho_A A\,dx}{M_A}\right) = \frac{\partial}{\partial t}\left(\frac{p_A A\,dx}{RT}\right)$$

The rate at which gas A diffuses into the element is N_A, and the rate at which gas A diffuses out of the element is $[N_A + (\partial N_A/\partial x)\,dx]$. A material balance around the element then becomes

$$\frac{\partial}{\partial t}\left(\frac{p_A A\,dx}{RT}\right) = N_A - \left[N_A + \frac{\partial N_A}{\partial x}dx\right]$$

$$= -\frac{\partial N_A}{\partial x}dx \qquad (485)$$

or substituting from equation 484 in equation 485

$$-\frac{\partial N_A}{\partial x} = \frac{A}{RT}\left(\frac{\partial p_A}{\partial t}\right) = +A\frac{D_GP}{RT}\frac{\partial}{\partial x}\left(\frac{\partial p_A}{p_B\,\partial x}\right) \qquad (486)$$

and

$$\frac{\partial p_A}{\partial t} = D_GP\frac{\partial}{\partial x}\left(\frac{\partial p_A}{p_B\,\partial x}\right) \qquad (487)$$

Under steady-state conditions, the rate of diffusion N_A in equation 484 is a constant, and, since $p_A + p_B = P$, the integrated equation becomes

$$\frac{N_A}{A} = \frac{D_GP}{RTx}\ln\left(\frac{p_{B2}}{p_{B1}}\right) = \frac{D_GP}{RTx}\frac{p_{A1} - p_{A2}}{(p_B)_{lm}} \qquad (488)$$

where $(p_B)_{lm} = \log_e$ mean value of p_{B1} and p_{B2}.

Counter Diffusion of Gas A and Gas B

In the case of equimolar countercurrent diffusion of gases A and B

$$\frac{N_A}{A} = -\frac{N_B}{A} = \frac{v_A\rho_A}{M_A} = -\frac{v_B\rho_B}{M_B} \qquad (489)$$

Under these circumstances equation 482 becomes

$$\frac{N_A}{A} = -\frac{D_G}{RT}\left(\frac{dp_A}{dx}\right) \qquad (490)$$

Substituting from equation 490 in equation 485 describing the unsteady-state material balance,

$$\frac{\partial p_A}{\partial t} = D_G\frac{\partial^2 p_A}{\partial x^2} \qquad (491)$$

Under steady conditions where N_A is constant, equation 490 may be integrated to give

$$\frac{N_A}{A} = \frac{D_G}{RTx}[(p_A)_1 - (p_A)_2] \qquad (492)$$

Diffusivity

Values for the diffusivity D_G in the diffusion equations are preferably obtained from experimental values when available. The most satisfactory generalized correlation [1,22] of the diffusivities of gas systems appears to be

$$D_G = 0.0166\left[\frac{T^{3/2}}{P(V_A^{1/3} + V_B^{1/3})^2}\right]\sqrt{\frac{1}{M_A} + \frac{1}{M_B}} \qquad (493)$$

where D_G = diffusivity or diffusion coefficient [sq ft/hr or moles/(hr)(sq ft)(moles per cu ft per ft)].
T = absolute temperature (°K).
P = pressure (atm).
V_A, V_B = molecular volumes of gases A and B.
M_A, M_B = molecular weights of gases A and B.

TABLE 54. ATOMIC VOLUMES

Bromine	27.0
Sulfur	25.6
Oxygen	
(O=, as in aldehydes, ketones, and doubly linked oxygen)	7.4
(—O—, as in methyl esters)	9.1
(—O—, as in higher esters and ethers)	11.0
(H—O—, as in acids)	12.0
Carbon	14.8
Chlorine	24.6
Hydrogen	3.7
Nitrogen	15.6
In primary amines	10.5
In secondary amines	12.0

For benzene ring formation deduct 15; for naphthalene deduct 30.

For the hydrogen molecule use $V = 14.3$; and for air use $V = 29.9$.

The values of the molecular volumes which should be used with equation 493 are obtained from the atomic volumes shown in Table 54, which, by Kopp's law of additive volumes with the appropriate rules for special cases, gives molecular volumes from the atomic volumes.

Equations 482 to 492 are generally satisfactory when applied to systems which closely approximate ideal gases.

Molecular diffusion is usually of minor importance in mass transfer in gases, as the gas is usually in a highly turbulent state. The main interest in the diffusion equations as applied to gases occurs in academic and theoretical work where the experimental conditions are carefully selected so as to eliminate turbulence.

Diffusion in Liquids

Cases of mass transfer by diffusion occur somewhat more frequently where liquids are involved. However, a liquid is far from being an ideal gas, and no satisfactory description of the liquid state, such as the kinetic theory of gases, is available. In spite of this fact and for the lack of any better method, diffusion equations similar to those derived for ideal gases are often applied to diffusion in the liquid state. In such applications it is customary to substitute for the partial pressure potential a concentration potential expressed as moles per unit volume so that equations 487, 488, 491, and 492 become, respectively,

$$\frac{\partial C_A}{\partial t} = D_L C \frac{\partial}{\partial x} \frac{\partial C_A}{C_B \, \partial x} \tag{494}$$

$$\frac{N_A}{A} = \frac{D_L C}{x} \frac{[(C_A)_1 - (C_A)_2]}{(C_B)_{lm}} \tag{495}$$

$$\frac{\partial C_A}{\partial t} = D_L \frac{\partial^2 C_A}{\partial x^2} \tag{496}$$

$$\frac{N_A}{A} = \frac{D_L}{x} [(C_A)_1 - (C_A)_2] \tag{497}$$

where D_L = liquid diffusivity.

C_A, C_B = moles per unit volume for A and B, respectively.

$(C_B)_{lm}$ = $\log_e$ mean value of C_{B1} and C_{B2}.

C = total number of moles per unit volume.

The validity of these equations for liquids is rather doubtful in view of the assumptions that the kinetic theory describes the liquid state and that the total moles per unit volume of liquid are independent of composition. The difference between equations 494 and 496 and equations 495 and 497 is often neglected, and the simpler form of equations 496 and 497 are often used for all cases. Equation 496 is often called Fick's law.

No generalized correlation of liquid phase diffusivities such as equation 493 has been found, but certain empirical relationships are available.[54] This probably reflects the inadequacy of the kinetic theory when applied to liquids. It is therefore necessary to use experimental data for liquid phase diffusivities. Since these diffusivities assume a diffusion equation in their calculation from experimental data, the use of the values in diffusion equations tends to cancel any errors in the application of kinetic theory to liquids, provided the conditions of the application approximate the conditions under which the diffusivities were measured. In effect, the treatment of diffusion in liquids is, therefore, essentially empirical.

Table 55 presents a number of typical values for liquid phase diffusivities.[40] The range of variation of these values is not large so that in the absence of data a reasonable guess can often be made.

TABLE 55. DIFFUSION COEFFICIENTS FOR DILUTE SOLUTIONS OF GASES IN WATER AT 20° C [40]

Solute Gas	D_L, sq ft/hr
CO_2	0.000069
Cl_2	0.000062
H_2	0.000202
H_2S	0.000063
N_2	0.000074
NH_3	0.000071
N_2O	0.000068
O_2	0.000081

Diffusion in Solids

The diffusion equations have also been applied with dubious success to the transfer of mass within solids. Where the operation involves a truly homogeneous phase such as the solid diffusion that occurs in metals at high temperatures reasonable results may be obtained. Where the operations involve the transfer of fluid phase through the interstices and pores of a solid phase such as encountered in the drying of solids the application of diffusion equations is open to considerable question because of the undoubted influence of other forces such as capillarity.[32]

Diffusion in Fluids in Conduits

When one component A diffuses in more than one direction through a second nontransferring component B and is present in such low concentrations that the partial pressure p_B is approximately equal to the total pressure P and C_B is approximately equal to C, equations 487 and 494 may be written

$$\frac{\partial p_A}{\partial t} = D_G\left[\frac{\partial^2 p_A}{\partial x^2} + \frac{\partial^2 p_A}{\partial y^2} + \frac{\partial^2 p_A}{\partial z^2}\right] \qquad (487a)$$

$$\frac{\partial C_A}{\partial t} = D_L\left[\frac{\partial^2 C_A}{\partial x^2} + \frac{\partial^2 C_A}{\partial y^2} + \frac{\partial^2 C_A}{\partial z^2}\right] \qquad (494a)$$

These equations are exact for equal molar counter diffusion.

For radial diffusion in a cylinder where the concentration of the transferring component A is symmetrical about the axis

$$\frac{\partial p_A}{\partial t} = D_G\left[\frac{\partial^2 p_A}{\partial r^2} + \left(\frac{1}{r}\right)\frac{\partial p_A}{\partial r} + \frac{\partial^2 p_A}{\partial z^2}\right] \qquad (487b)$$

where z = distance along the axis of the cylinder.
r = radial distance from the axis of the cylinder.

For steady motion in which the nontransferring component B moves only in the z direction with a velocity v, a material balance around a differential length of conduit gives

$$\frac{\partial (vp_A)}{\partial z} = D_G\left[\frac{\partial^2 p_A}{\partial r^2} + \frac{1}{r}\frac{\partial p_A}{\partial r} + \frac{\partial^2 p_A}{\partial z^2}\right] \qquad (498)$$

If v is known as a function of r as in the case of laminar flow when the parabolic velocity distribution of Figs. 126 and 482 may be assumed, if D_G is constant, if p_A is uniform along the cylindrical fluid interface and if diffusion in the direction of flow is negligible $\left(\frac{\partial^2 p_A}{\partial z^2} = 0\right)$ equation 498 may be integrated over the length of the conduit or wetted wall section to give equation 499.[17]

$$\frac{p_2 - p_1}{p_i - p_1} = \left\{\begin{array}{l} 1 - 0.81904e^{-14.6272\frac{\pi}{4}\frac{\rho D_G L}{W}} \\ - 0.0976e^{-89.22\frac{\pi}{4}\frac{\rho D_G L}{W}} \\ - 0.0135e^{-227.8\frac{\pi}{4}\frac{\rho D_G L}{W}} - \cdots \end{array}\right\} \qquad (499)$$

where p_1 = partial pressure of component A in the entering gas (atm).
p_2 = partial pressure of component A in the exit gas (atm).
p_i = partial pressure of component A at the interface (atm).
W = mass rate of gas flow.
L = length of wetted-wall section.

Typical data from wetted-wall column experiments are plotted in Fig. 480 which also shows the curve given by equation 499.[43] A better agreement is obtained [16] by the relation

$$\frac{p_2 - p_1}{p_i - p_1} = 1 - 4\sum^{\infty}\frac{1}{\alpha_n{}^2}e^{\left(\frac{-\alpha_n{}^2\pi\rho D_G L}{W}\right)} \qquad (500)$$

where α_n = nth root of the Bessel function J_0. Equation 500 was derived by assuming a uniform velocity across the diameter instead of a parabolic velocity distribution.

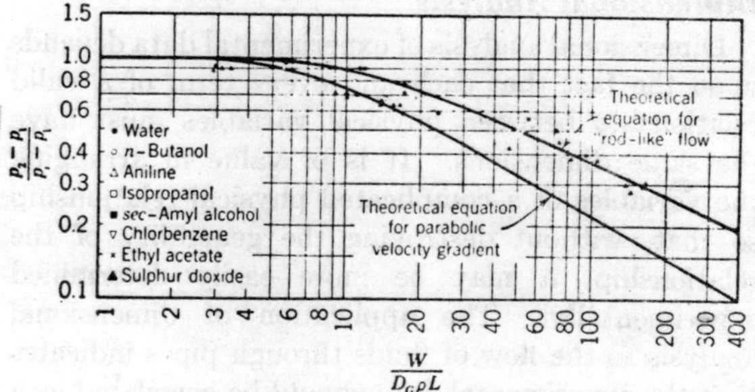

Fig. 480. Experimental data from wetted-wall column.[43]

MASS TRANSFER IN TURBULENT FLOW

When the fluid in which mass or heat transfer is occurring is in turbulent motion, the secondary motions of the fluid in the form of vortices physically transport the mass or heat through the fluid, this action being superimposed on the laminar mechanism of diffusion or conduction. As the turbulence of the fluid is increased the turbulent mechanism may become the predominant factor. The turbulent mechanism is called "eddy diffusion," or, in the case of heat transfer, "eddy conduction," although it is clear that the transfer is accomplished by convection and that *diffusion* or *conduction* is not involved.

If it were possible to describe completely the motion of turbulence in mathematical terms it would probably be possible to derive relations similar to the diffusion relations, giving the rate of mass transfer by the turbulent mechanism. Such a description of turbulence is at present impossible, although considerable progress along these lines is being made in the field of fluid mechanics. As a result the treatment of mass transfer in fluids undergoing turbulent motion is almost entirely empirical. Although our present knowledge of turbulent fluids is inadequate to calculate rates of transfer, certain theoretical ideas are very useful in relating the associated processes of heat, mass, and momentum transfer in fluids. In many cases, if the pressure drop generated in a fluid is known, it is possible to estimate the mass or heat transfer rates which might be expected. Similarly, if the heat or mass transfer rates in a given system are known, it may be possible to estimate the pressure drop in the fluid phase.

Under conditions of turbulent flow in the absence of a justifiable theoretical mechanism of transfer, empirical methods are used in much the same manner as for friction losses and heat transfer.

Dimensional Analysis

Dimensional analysis of experimental data depends upon the fact that each and every term of a valid relationship between physical variables must have the same dimensions. It is of value in arranging the variables of a complicated physical relationship so that, without destroying the generality of the relationship, it may be more easily determined experimentally. The application of dimensional analysis to the flow of fluids through pipes indicates that the experimental data should be correlated as a function of the Reynolds number (see equations 65 to 70). Dimensional analysis applied to the correlation of heat transfer coefficients indicates the use of various dimensionless groups.

Similarly, for the correlation of mass transfer coefficient in wetted-wall columns, the experimental data indicate the following functional relationship.

$$k' = \phi_1(G, D, \mu, D_G, \rho) \qquad (501)$$

where k' = mass transfer coefficient in concentration units [moles/(hr)(sq ft)(moles/cu ft)].

G = mass velocity [(lb mass)/(hr)(sq ft)].

Expressing the variables in terms of mass, length, and time,

$$Lt^{-1} = z(mL^{-2}t^{-1})^a(L)^b(mL^{-1}t^{-1})^c(L^2t^{-1})^d(mL^{-3})^e$$

Σ of exponents of m $\quad 0 = a + c + e$

Σ of exponents of t $\quad -1 = -a - c - d$

Σ of exponents of L $\quad 1 = -2a + b - c + 2d - 3e$

Retaining the exponents a and e,

$$k' = z(G)^a(D)^{a-1}(\mu)^{-a-e}(\rho)^e(D_G)^{1+e}$$

and

$$\frac{k'D}{D_G} = \phi\left[\left(\frac{DG}{\mu}\right)\left(\frac{\mu}{\rho D_G}\right)\right] = \phi[(\text{Re})(\text{Sc})] \qquad (502)$$

The dimensionless ratio $\mu/\rho D_G$ is called the Schmidt number. When experimental data from wetted-wall columns [24] are plotted as a function of the dimensionless ratios, indicated by the dimensionless equations, the excellent agreement from widely varying types of systems indicates that a power series form of the dimensionless function is apparently correct.

The assumption made in the integration of the rate equation over the transfer section of the wetted-wall column to the effect that the quotient of the coefficient and the molal mass velocity was almost constant is also justified by the fact that k_g is a function of $G^{0.8}$, all other variables being held constant, as indicated in equation 503.

The similarity is apparent between equation 503 which correlates mass transfer through the gas phase flowing in turbulent flow through a wetted-wall column and the Dittus-Boelter equation 417 correlating heat transfer coefficients for fluids flowing through pipe in turbulent flow. In place of the Prandtl number ($\mu C_P/k_q$) in the heat transfer equation there is the Schmidt number ($\mu/\rho D_G$) in the mass transfer equation. Similarly instead of the Nusselt number in the heat transfer equation (hD/k_q) there is the group ($k'D/D_G$). The Reynolds number is common to both correlations.

$$\frac{k'D}{D_G} = 0.023\left(\frac{DG}{\mu}\right)^{0.8}\left(\frac{\mu}{\rho D_G}\right)^{0.44} \qquad (503)$$

$$\frac{hD}{k_q} = 0.023\left(\frac{DG}{\mu}\right)^{0.8}\left(\frac{\mu C_P}{k_q}\right)^{0.3\text{ to }0.4} \qquad (417)$$

Both equations [7-9] may be arranged as shown to give the same function of the Reynolds number equal to the product of a pair of dimensionless groups.

$$j = \frac{k'}{v}\left(\frac{\mu}{\rho D_G}\right)^{0.56} = 0.023\left(\frac{DG}{\mu}\right)^{-0.20} = \frac{1}{8}f \qquad (504)$$

$$j = \frac{h}{v\rho C_P}\left(\frac{\mu C_P}{k_q}\right)^{0.6\text{ to }0.7} = 0.023\left(\frac{DG}{\mu}\right)^{-0.20} = \frac{1}{8}f \qquad (505)$$

This function of the Reynolds number is called the j factor when applied to heat or mass transfer. The identity of the j factor for mass transfer and the j factor for heat transfer has been verified experimentally, and certain theoretical considerations to be discussed later support this idea. The j factor is fundamentally a friction factor. For flow in pipe it is equal to one-eighth of the friction factor f used in equation 63 and plotted in Fig. 125, or one-half of the friction factor f' used in equation 62. In more complex geometry such as fluids flowing through beds of packed solids, the identity of the j factors for heat and mass transfer and friction has not been established with certainty. This is due probably to the inability to characterize properly the state of motion of the fluid in such cases by a simple term such as the Reynolds number.

Thus, heat, mass, and momentum transfer in fluids are intimately related and are accomplished by similar mechanisms. Furthermore the j factor idea can be used to predict any two of these three items from a third, provided that the transfer processes are operating through the same paths.

Since $k' = RTk_g$, and $RT = PM_{avg}/\rho$ the expression k'/v, which appears in equation 504, equals $k_gM_{avg}P/v\rho$ or $k_gM_{avg}P/G$, and with the exponent ⅔ instead of 0.56, equation 504 becomes

$$j = \left(\frac{k_g M_{avg} P}{G}\right)\left(\frac{\mu}{\rho D_G}\right)^{2/3} \tag{504a}$$

where M_{avg} = mean molecular weight of the gas.

In much of the literature the log$_e$ mean partial pressure of the non-transferring component replaces P in equation 504a because mass transfer equations were originally written as the analog of equation 488 for molecular diffusion. The mechanism of transfer is rarely pure diffusion, and the available experimental data are confined to dilute gases where the value for the log$_e$ mean partial pressure is approximately equal to the total pressure within the limits of accuracy of the data.

Similarly, for heat transfer, equation 505 becomes

$$j = \left(\frac{h}{C_P G}\right)\left(\frac{\mu C_P}{k_q}\right)^{2/3} \tag{505a}$$

These equations correlated data on absorption and evaporation in wetted-wall columns, absorption on cylinders placed at right angles to a gas stream, and vaporization from a plane surface.

The mass transfer coefficient k' is given in equation 504 as proportional to $(D_G)^{0.56}$. If the transfer occurred entirely by molecular diffusion, k' would be directly proportional to D_G. If the transfer occurred entirely by turbulent action, k' would be independent of D_G (D_G would appear with the exponent of zero). The exponent of 0.56 on the diffusivity is an indication of the dual mechanism of mass transfer. Gilliland [23] correlated mass transfer data by the expression

$$-\frac{\Delta p_A}{N_A} = \left[903 + \frac{34.7}{D_G}\right]\frac{1}{(\mathrm{Re})^{0.8}}$$

which may be rearranged to give

$$k_g = (\mathrm{Re})^{0.8}\left[\frac{1}{903 + 34.7/D_G}\right] \tag{506}$$

The term $\dfrac{1}{903 + (34.7/D_G)}$ has the form of two resistances to mass transfer operating in series, one a resistance to molecular diffusion varying inversely with D_G, and the other a turbulent resistance which is not a function of diffusivity.

In momentum transfer (friction losses) at high values of Re the friction becomes almost independent of viscosity (see Fig. 125). Similarly, conditions may be encountered in heat transfer and in mass transfer where the rate may be almost independent of thermal conductivity or molecular diffusivity.

ANALOGY BETWEEN MOMENTUM, HEAT, AND MASS TRANSFER

The analogy between momentum, heat, and mass transfer may be approached in a theoretical manner by considering the mechanisms of molecular and turbulent transfer processes.

Momentum Transfer in Laminar Motion

The existence of shear stresses within a fluid is reflection of the fact that momentum is being transferred within the fluid. The pressure drop resulting from the flow of fluid through pipes produces shear stresses distributed throughout the fluid. A section of a circular conduit through which fluid is flowing is shown in Fig. 481.

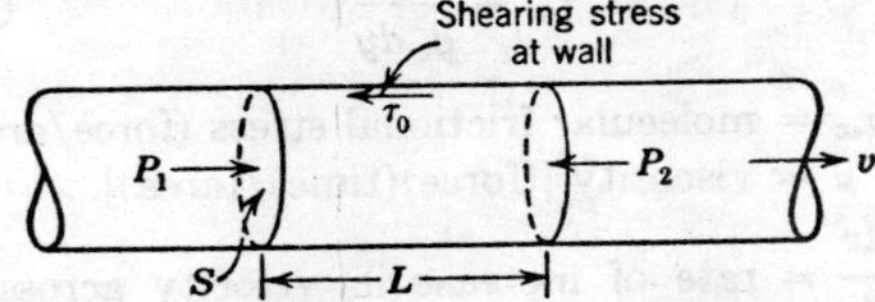

Fig. 481. Diagram illustrating shear stresses in fluid flowing in a cylinder.[5]

Under steady-flow conditions, the difference between the forces acting on the fluid $(P_1 - P_2)S$ must be in equilibrium with frictional forces acting at the interface between the wall and the fluid so that

$$(P_1 - P_2)S = \psi L \bar{\tau}_0 \tag{507}$$

where ψ = length of the wetted perimeter (ft).
L = length of section (ft).
$\bar{\tau}_0$ = average friction stress, force per unit area.
S = cross-sectional area.

Defining the hydraulic radius r_h as S/ψ

$$\bar{\tau}_0 = \frac{-\Delta P}{L} r_h \tag{508}$$

In circular conduits, the stress τ_0 will be the same at all points of the wall so that $\tau_0 = \bar{\tau}_0$. In conduits of other shapes the shearing stress τ_0 will vary from point to point on the interface between the wall and the fluid, and $\bar{\tau}_0$ is an average value.

It is clear that a similar reasoning would apply to an inner cylinder of fluid with radius r, so that the internal frictional stress between the layers of fluid at radius r would be

$$\tau = -\left(\frac{\Delta P}{L}\right)\frac{r}{2} \tag{509}$$

since for a circular conduit $r_h = r/2$. Since r_h in equation 508 for a circular conduit is $r_0/2$, dividing equation 509 by equation 508,

$$\tau = \tau_0 \frac{r}{r_0} = \tau_0 \left(1 - \frac{y}{r_0}\right) \tag{510}$$

where y is the distance from the wall $(r_0 - r)$.

Equation 510 is known as the law of linear stress distribution. It is a perfectly general relationship which indirectly governs the velocity distributions produced under various conditions of flow.

The mechanism of laminar resistance is comparatively simple, the tangential stress being expressed by the law of viscous friction.

$$\tau_m = \frac{\mu}{g_c}\frac{dv}{dy} \tag{511}$$

where τ_m = molecular frictional stress (force/area).
μ = viscosity [(force)(time)/(area)].
$\frac{dv}{dy}$ = rate of increase in velocity across the flow.

Thus, whenever motion ceases, the viscous forces disappear. If equation 511 is applied to flow in circular pipe, the Poiseuille equation (63L) giving the pressure drop required to produce laminar flow is obtained.[5]

$$\frac{-dP}{dL} = 32\frac{\mu v}{g_c D^2} \qquad (63L) \tag{512}$$

Velocity Distribution

The forms of equation 512 and similar expressions for turbulent flow such as equation 62 indicate the basic differences in the laws of resistance in laminar and turbulent flow. The difference is also apparent in the velocity distribution. Laminar motion leads to a parabola as illustrated in Fig. 126. Typical relative velocity distribution curves for laminar and turbulent flow are shown in Fig. 482.

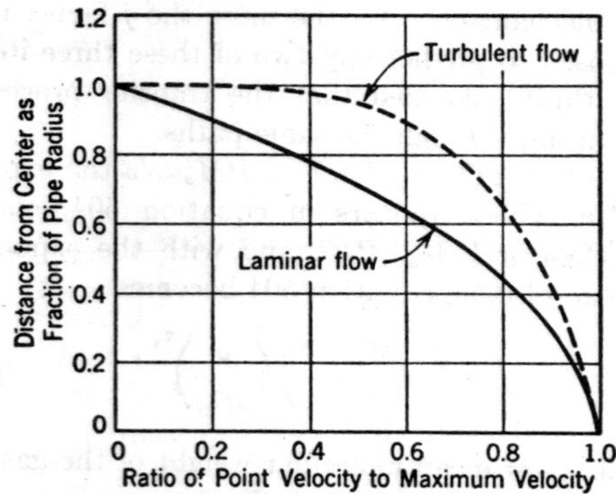

Fig. 482. Velocity distribution of fluids flowing in cylindrical ducts.

Local velocities throughout the central portions are far more uniform in turbulent than in laminar flow. The average velocity in turbulent flow is about 0.8 to 0.85 times the maximum velocity as compared to 0.5 times the maximum velocity in laminar flow.

Distribution of Molecular and Turbulent Shear Stress

When the velocity distribution within a flowing fluid is known it is possible to determine the proportion between shear stresses generated by turbulent action by means of the law of linear stress distribution, equation 510, and the viscosity of the fluid, equation 511. As illustrated in Fig. 483, the total shear stresses have been calculated from the measured pressure drop and the law of linear stress distribution, equation 509. The amount of molecular shear stress τ_m has been calculated from a measured velocity distribution curve such as that of Fig. 482 by means of equation 511. The amount of shear stress generated by turbulent action is then determined by subtracting the molecular shear stress from the total shear stress.

Under laminar flow conditions, the molecular shear stress is equal to the total shear stress, and the combination of the law of linear stress distribution and the definition of viscosity is what leads to the parabolic velocity distribution in laminar flow.

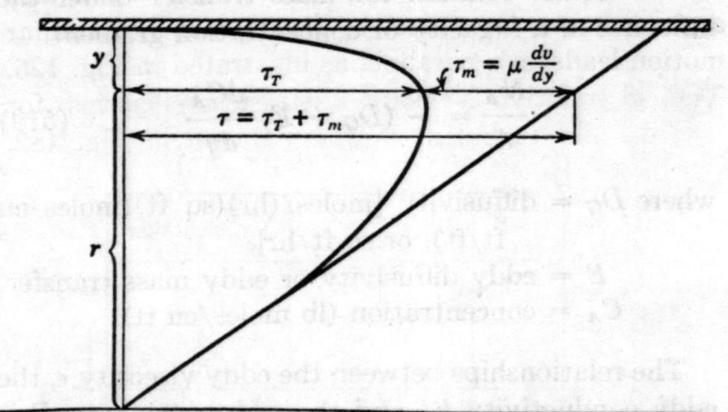

FIG. 483. Diagram illustrating distribution between molecular and turbulent shear.

Under conditions of turbulent flow, the total shear stress at the wall (velocity = zero) is due to molecular action. Toward the center of the pipe, the proportional part of the total shear that is due to molecular action steadily decreases to an exceedingly small value at the center. Furthermore, the more turbulent the flow, i.e., the higher the Reynolds number, the less the proportion of molecular shear stress to total shear stress at any point away from the wall.

Mechanism of Turbulent Shear Stresses[5]

When water flows at 60° F through 12-in. commercial pipe, 100 ft long, at an average velocity of 10 fps, the loss in head is approximately 3.9 ft. If by some means turbulence could be suppressed and the water was made to flow in laminar motion at 60° F, the loss in head would be approximately 0.0117 ft. The ratio of the resistance heads and therefore the ratio of the average shearing stresses in the two cases is 3.9/0.011 = 335. Thus, turbulence causes tangential stresses to be over 300 times as large as those produced by purely viscous resistance. These large tangential forces in turbulent flow are produced by a process of momentum exchange between two layers of fluid.

In Fig. 484, two separate bodies a and b are shown moving on parallel tracks at the respective speeds v_a and v_b, or b is moving with a velocity $v_b - v_a = \Delta v$ relative to a. Imagine body a frictionlessly mounted and attached to an immobile block through a dynamometer D. Let body b be provided with a row of machine guns set at right angles to the trajectory, which fire bullets into a at a regular rate of w lb/sec. Under these circumstances the force

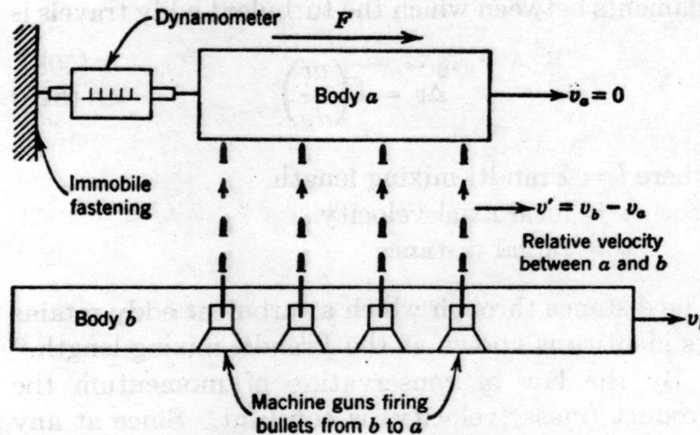

FIG. 484. Diagrammatic analogy illustrating mechanism of turbulent shear.[5]

registered on the dynamometer in the direction of the motion of b relative to a will be

$$F = \frac{w}{g_c}(v_b - v_a) \tag{513}$$

Thus a stream of bullets fired at the rate of 200 lb/min from a body passing at a relative speed of 30 mph will produce a force equal to

$$F = \frac{200}{60 \times 32.2} \times 44 = 4.57 \text{ lb}$$

If both bodies were in motion, firing from the faster body would accelerate the slower one whereas a stream of bullets coming from the slower body would retard the faster one.

This reasoning may be applied to two adjacent filaments a and b in a moving fluid. Let the adjacent surface area under consideration be A. Suppose further that in addition to a relative difference of velocity in the axial direction Δv there is a cross-current motion, with a velocity u. This cross-current motion will cause an exchange of particles from b to a, and, if this cross-current velocity were distributed uniformly over the surface A, then $(u)(A)$ would be the volume and ρuA the mass transported from b to a per unit of time. Thus the force generated by the exchange would be $\rho Au\,\Delta v$, and the stress between the filaments is

$$\tau = \rho u\,\Delta v \tag{514}$$

Prandtl Mixing Length

If l is defined as the cross-current distance through which a turbulent eddy retains its identity, the relative axial velocity difference Δv of the adjacent filaments between which the turbulent eddy travels is

$$\Delta v = l\left(\frac{dv}{dy}\right) \tag{515}$$

where l = Prandtl mixing length.
v = local axial velocity.
y = radial distance.

The distance through which a turbulent eddy retains its identity is known as the Prandtl mixing length.[41]

By the law of conservation of momentum the product (mass)(velocity) is constant. Since at any axial section in the fluid there is constant mass, any cross-current velocity u is numerically equal to Δv, and both are equal to $l\,dv/dy$ by equation 515. Therefore

$$\tau = \rho(\Delta v)u = \rho l^2\left(\frac{dv}{dy}\right)^2 = \epsilon\frac{dv}{dy} \tag{516}$$

where $\epsilon = \rho l^2(dv/dy)$ is called eddy viscosity, mechanical friction coefficient, or turbulence factor.

The total stress in a fluid in motion is

$$\tau = \tau_0\left(1 - \frac{y}{r}\right) = \mu\frac{dv}{dy} + \rho l^2\left(\frac{dv}{dy}\right)^2$$

$$= (\mu + \epsilon)\frac{dv}{dy} \tag{517}$$

The shear forces in a fluid are equal to the rate at which momentum is transferred in a cross-current direction through a fluid. It has been shown that a dual mechanism of molecular and turbulent action operates to transfer the momentum. When heat or mass is transferred through a fluid in motion, it is to be expected that the same dual mechanism of molecular and turbulent action will operate in the transfer process.

Thus, corresponding to the equation for the transfer of momentum at any point in a fluid there will be a similar relation for heat transfer under the influence of a temperature gradient.

$$\frac{q}{A} = -(k_q + k_e)\frac{dT}{dy} \tag{518}$$

where k_q = thermal conductivity.
k_e = eddy conductivity or eddy heat transfer factor.

and a similar relation for mass transfer under the influence of a fugacity or concentration gradient.

$$\frac{N_A}{A} = -(D_G + E)\frac{dC_A}{dy} \tag{519}$$

where D_G = diffusivity [moles/(hr)(sq ft)(moles/cu ft/ft), or sq ft/hr].
E = eddy diffusivity or eddy mass transfer.
C_A = concentration (lb moles/cu ft).

The relationships between the eddy viscosity ϵ, the eddy conductivity k_e, and the eddy diffusivity E is expressed by the following equations.

$$k_e = \beta\epsilon C_P \tag{520}^*$$

* Equations 520 and 521 may be derived in the following manner.

The rate of heat transfer by turbulent action expressed in terms of the eddy conductivity is

$$\frac{q_t}{A} = -k_e\frac{dT}{dy} \tag{a}$$

Considering the heat transferred by the motion of an eddy through its mixing length l, the increase in temperature is $l(dT/dy)$, the quantity of material transferred across a unit area is $u\rho$, so that

$$\frac{q_t}{A} = -\left(l\frac{dT}{dy}\right)(u\rho)C_P \tag{b}$$

Since $u = \Delta v$, substituting equation 515 in equation 516 gives

$$u = l\frac{dv}{dy} \tag{c}$$

Substituting equation c in b, and equating b to a,

$$k_e = l^2\rho C_P\frac{dv}{dy}$$

$$\epsilon = l^2\rho\frac{dv}{dy} \tag{516}$$

Therefore,

$$k_e = \beta\epsilon C_P \tag{520}$$

The proportionality constant β is introduced to account for inadequacies in the theory. Similarly,

$$E = \frac{\alpha\epsilon}{\rho} \tag{521}$$

$$E = \frac{\alpha\epsilon}{\rho} \qquad (521)$$

where α and β are proportionality constants.

ϵ = eddy viscosity.
C_P = heat capacity.
ρ = density.

In general (for Newtonian fluids) it has been found by experiment that the viscosity, thermal conductivity, and diffusivity are physical properties of the particular fluid not affected by the motion of the fluid or the geometrical structure of the equipment containing the fluid. On the other hand, the eddy viscosity, eddy conductivity, and eddy diffusivity are primarily functions of the state of motion of the fluid as described by the Reynolds number and are not functions of the physical properties of the system except in so far as they may affect the Reynolds number.

From a study of transport of water vapor across a duct through which carbon dioxide, hydrogen, and air were flowing over a wide range of Reynolds numbers, it was found [45, 48, 49] that the proportionality constant in equation 521 was 1.6. The relationship between eddy viscosity and eddy diffusivity calculated from these measurements is shown as a function of the Reynolds number is Fig. 485.

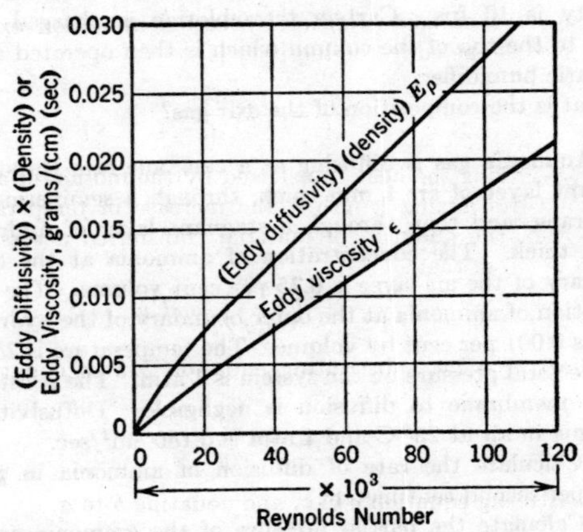

FIG. 485. Comparison of eddy viscosity and eddy diffusivity.[48]

The eddy viscosity $\epsilon = \rho l^2(dv/dy)$ can be determined as a function of Reynolds number for a given fluid in a given system provided measurements of pressure drop and velocity distribution data are available (see equation 517). If the proportionality constants α and β are known, the rates of heat or mass transfer through the fluid can also be determined. This has been done for heat transfer between a fluid in a pipe and the pipe [39, 51] with good agreement between measured and calculated heat transfer rates; and for the mass transfer from a gas to the wetted walls of a cylindrical column (wetted-wall column) with similar results.[45]

Physical Significance of Dimensionless Groups

Since a number of dimensionless groups continually reappear in the correlation of momentum, heat, and mass transfer operations, it is helpful to place a physical interpretation on the various common groups.[50, 51]

Reynolds Number (DG/μ)

The shear stress generated by turbulent motion of a fluid may be written

$$\tau_t = \rho u(\Delta v) = \rho(\Delta v)l\frac{dv}{dy} \qquad (516)$$

whereas the shear stress generated by molecular action is

$$\tau_m = \mu\frac{dv}{dy} \qquad (511)$$

Dividing equation 516 by 511,

$$\frac{\tau_t}{\tau_m} = \frac{\rho(\Delta v)l(dv/dy)}{\mu(dv/dy)} = \frac{\rho(\Delta v)l}{\mu} \qquad (522)$$

The right-hand side of equation 522 is equal to the product of a cross-current mass velocity and the Prandtl mixing length (the distance through which the cross-current mass velocity exists) divided by viscosity, and is therefore a Reynolds number. The Reynolds number is therefore a measure of the ratio of the rate of momentum transfer by turbulent action to the rate of momentum transfer by molecular action, describing the state of motion of the fluid.

Nusselt Number (hD/k_q)

The rate of heat transfer by molecular action may be written as

$$\frac{q_m}{A} = -k_q\frac{dT}{dy} \qquad (523)$$

and the total rate of heat transfer as

$$\frac{q}{A} = -h\,dT \qquad (524)$$

Dividing equation 524 by 523,

$$\frac{q}{q_m} = \frac{h\,dT}{k_q(dT/dy)} = \frac{h\,dy}{k_q} \qquad (525)$$

The right-hand term is a Nusselt number hD/k_q, since y is a linear distance or a diameter. Thus the Nusselt number is a measure of the ratio of the total rate of heat transfer to the rate of heat transfer by molecular action.

Similarly the dimensionless group $(k'D/D_G)$ is a measure of the ratio of the total rate of mass transfer to the rate of mass transfer by molecular diffusion.

Prandtl Number $(\mu C_P/k_q)$

If the equation 511 for shear stress by molecular action is divided by equation 523 for the rate of heat transfer by molecular action,

$$\frac{\tau_m}{q_m} = \frac{\mu}{k_q}\left(\frac{dv/dy}{dT/dy}\right) = \frac{\mu C_P}{k_q}\left(\frac{dv/dy}{dT/dy}\right) \tag{526}$$

There is a constant relation between the gradient dv/dy and dT/dy, so that the Prandtl number

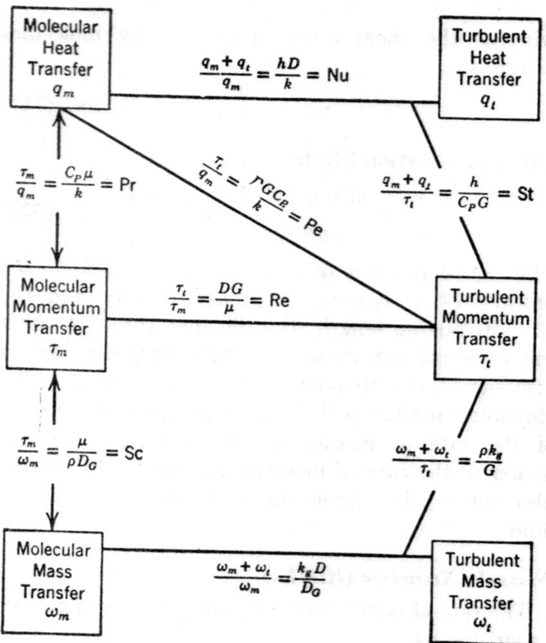

FIG. 486. Diagram illustrating the relation between dimensionless groups and various transfer operations.[50]

$(\mu C_P/k_q)$ is a measure of the ratio of the rate of molecular transfer of momentum and heat transfer (or heat transfer by conduction).

Similar relationships for other common dimensionless groups are shown in Fig. 486.

PROBLEMS

1. A wetted-wall column has an internal diameter of 2 in and is to be supplied with water at the top and air at the bottom. The air velocity is to be fixed at 8.3 fps and will enter at a temperature of 20° C. The wet bulb temperature of the air will be 9.7° C. The entering water temperature will be 10.2° C.

Estimate the heat and mass transfer coefficients from the friction factor expected, and compare the coefficients with those predicted from correlations for heat transfer and correlations for mass transfer.

2. Ethyl acetate is to be evaporated into a stream of dry air in a wetted-wall column which is 2.67 cm in inside diameter and 117 cm long. The wetted-wall column is to be operated adiabatically with recirculation of the liquid. The operating pressure is to be 800 mm of mercury absolute. Countercurrent flow of gas and liquid is to be used.

If dry air enters the bottom of the column at 21.1° C and a rate of 357 grams/min, determine the temperature and composition of the gas phase leaving the column.

3. Pure carbon dioxide at 100° F and 1 atm absolute pressure is fed to the bottom of a wetted-wall column 3 ft long and 2 in. inside diameter at a rate such that the entering velocity is 10 fps. Carbon tetrachloride is charged as a liquid to the top of the column which is then operated as an adiabatic humidifier.

What is the composition of the exit gas?

4. Ammonia gas is diffusing at a constant rate through a stagnant layer of air 1 mm thick, through a semipermeable membrane, and then through a stagnant layer of hydrogen 1 mm thick. The concentration of ammonia at the outer boundary of the air layer is 0.75 per cent volume. The concentration of ammonia at the outer boundary of the hydrogen layer is 0.001 per cent by volume. The temperature is 25° C, and the total pressure on the system is 1 atm. The resistance of the membrane to diffusion is negligible. Diffusivity of ammonia in air at 25° C and 1 atm is 0.186 cm^2/sec.

(*a*) Calculate the rate of diffusion of ammonia in gram moles per second centimeter^2.

(*b*) Calculate the partial pressure of the ammonia at the membrane.

CHAPTER

36

Mass Transfer 2

Coefficients in Packed Towers

IN ORDER to increase the interfacial area for mass transfer and the intimacy of contact of phases between which mass transfer is to be effected, packed towers are often used. Consider the absorption of a solute from a gas phase by a nonvolatile solvent, as illustrated in Fig. 487.

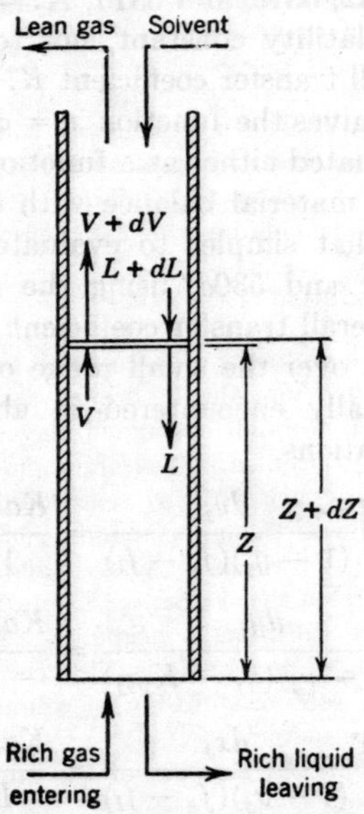

FIG. 487. Diagrammatic representation of a packed tower.

The rate of absorption or mass transfer which is accomplished in the differential section of height dZ may be written from a material balance as

$$dV = dL$$

$$d(Vy) = d(Lx)$$

where V and L = rate of flow of gas and liquid, respectively (moles/hr).

y and x = mole fraction of solute in gas and liquid streams, respectively.

If the gas carrying the solute is considered to be insoluble in the liquid phase, and if the solvent is considered to be nonvolatile, the material balance for the solute is more conveniently expressed as

$$V'\,dY = L'\,dX \tag{527}$$

where V' and L' = rate of flow of solute free gas and liquid, respectively (moles/hr).

Y and X = composition of gas and liquid phases, respectively (moles solute per mole of solute-free gas or liquid).

The rate of absorption may also be expressed by equation 465 written for the gas phase, the liquid phase, or for both phases.

$$\begin{aligned} \text{Rate of absorption} &= k_g aS\,dZ(f_g - f_i) \\ &= k_l aS\,dZ(f_i - f_l) \\ &= KaS\,dZ(f_g - f_l) \end{aligned} \tag{528}$$

where k_g, k_l = mass transfer coefficients for gas and liquid phases, respectively [moles/(hr)(sq ft)(atm)].

K = overall mass transfer coefficient for gas and liquid phase.

a = interfacial area between gas and liquid per unit empty tower volume (sq ft/cu ft).

S = empty tower cross section (sq ft).
Z = height in tower (ft).
f_g, f_i, f_l = fugacity of the solute (transferring component) in the gas, at the interface, and in the liquid, respectively.

Equating the rate of transfer based on the material balance, equation 527, to the rate of absorption given by equation 528, and assuming V' and L' to be constant,

$$\begin{aligned} -V'\,dY &= -L'\,dX = k_g aS\,dZ(f_g - f_i) \\ &= k_l aS\,dZ(f_i - f_l) \\ &= KaS\,dZ(f_g - f_l) \end{aligned} \tag{529}$$

In these relations the square feet of interfacial area per cubic foot of tower volume a is difficult to determine experimentally since it is made up of the interfacial area between drops of liquid falling through the gas phase, between bubbles of gas rising through liquid, and between the gas and the surface of the liquid on the packing. The surface area of the packing has little relation to the interfacial area between the phases. Therefore a is usually combined with and evaluated with k as a product ka.

It is convenient to group the variables related primarily to a single phase so that the following equations result.

$$-\int \frac{dY}{f_g - f_i} = \int \frac{k_g aS}{V'}\,dZ$$

or, by equations 476 and 477,

$$-\int \frac{dy}{(1-y)(f_g - f_i)} = \int \frac{k_g aS}{V}\,dZ \tag{530a}$$

$$-\int \frac{dX}{f_i - f_l} = \int \frac{k_l aS}{L'}\,dZ$$

or

$$-\int \frac{dx}{(1-x)(f_i - f_l)} = \int \frac{k_l aS}{L}\,dZ \tag{530b}$$

$$-\int \frac{dY}{f_g - f_l} = \int \frac{KaS}{V'}\,dZ$$

or

$$-\int \frac{dy}{(1-y)(f_g - f_l)} = \int \frac{KaS}{V}\,dZ \tag{530c}$$

$$-\int \frac{dX}{f_g - f_l} = \int \frac{KaS}{L'}\,dZ$$

or

$$-\int \frac{dx}{(1-x)(f_g - f_l)} = \int \frac{KaS}{L}\,dZ \tag{530d}$$

In integrating equations 530a through 530d, it is assumed that the mass transfer coefficients k_g and k_l are approximately proportional to V and L, respectively, as in the case of the wetted-wall column, so that for the gas and liquid phase, respectively,

$$-\int \frac{dy}{(1-y)(f_g - f_i)} = \frac{k_g aSZ}{V} \tag{531a}$$

$$-\int \frac{dx}{(1-x)(f_i - f_l)} = \frac{k_l aSZ}{L} \tag{531b}$$

In general, the integration of the left-hand side of equations 531a and 531b must be carried out graphically, and the fugacity of the solute at the interface f_i must be determined as a function of composition. If the mass transfer coefficients for each phase are known, this may be accomplished by equating

$$k_g aS\,dZ(f_g - f_i) = k_l aS\,dZ(f_i - f_l) \tag{528}$$

and, solving for f_i and substituting equation 470,

$$f_i = \frac{k_l a f_l + k_g a f_g}{k_l a + k_g a} = \frac{(k_l a)(PK_j x_j) + k_g a P y_j}{k_l a + k_g a} \tag{532}$$

in equations 532, 531c, and 531d, $K_j = (y_j/x_j)^*$, the equilibrium volatility constant, not to be confused with the overall transfer coefficient K. Since a material balance gives the function $x = \phi(y)$, the term f_i may be evaluated either as a function of x or y by combining the material balance with equation 532.

It is somewhat simpler to evaluate the integral equations 530c and 530d, using the overall coefficient. The overall transfer coefficient K is assumed to be constant over the small range of variation in L and V usually encountered in absorption and stripping operations.

$$-\int \frac{dy_j}{(1-y_j)(f_g - f_l)} = \frac{KaSZ}{V}$$

or

$$-\int \frac{dy_j}{(1-y_j)(y_j - K_j x_j)} = \frac{KaPSZ}{V} \tag{531c}$$

$$-\int \frac{dx_j}{(1-x_j)(f_g - f_l)} = \frac{KaSZ}{L}$$

or

$$-\int \frac{dx_j}{(1-x_j)(y_j - K_j x_j)} = \frac{KaPSZ}{L} \tag{531d}$$

Again the relation between y and x is obtained by a material balance, and the integration is completed graphically.

A graphical interpretation of the relation between the driving potentials across each phase and the overall driving potential may be shown by plotting the fugacity of the solute in the gas phase as a function of the fugacity of the solute in the liquid phase, as in Fig. 488.

The 45-degree line on the plot represents the interface conditions of equilibrium. The point A represents a point on an operating line relating the composition of the gas phase expressed as the solute fugacity to the composition of the liquid phase passing the same point, also expressed as the solute fugacity. This operating line may be calculated from the material balance around one end of the column.

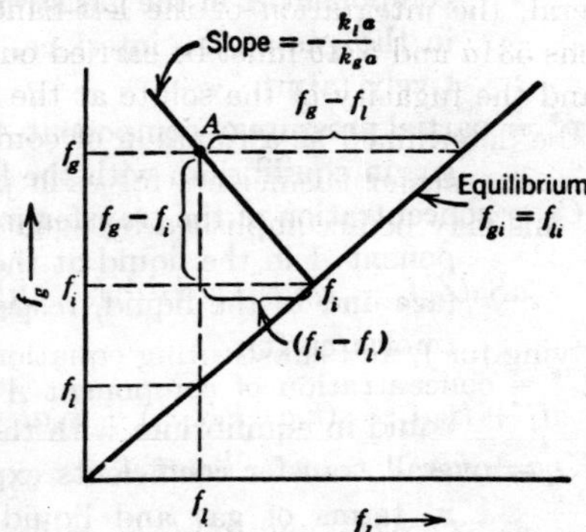

Fig. 488. Diagram illustrating fugacity potentials in mass transfer.

Rearranging equation 528,

$$\frac{-k_l a}{k_g a} = \frac{f_g - f_i}{f_l - f_i} \tag{528a}$$

shows that a straight line drawn from the point f_g, f_l with a slope $(-k_l a/k_g a)$ will intersect the equilibrium line at the interface fugacity f_i.

EXPERIMENTAL MASS TRANSFER COEFFICIENTS

Since both $k_g a$ and $k_l a$ are required for the integration of the equations relating to one phase only, equations 531*a* and 531*b*, it is not possible to calculate $k_g a$ or $k_l a$ directly from one experimental run in an absorption tower. The average overall mass transfer coefficient K, however, may be evaluated from one experimental run by the application either of equation 530*c* or 530*d*. The individual phase coefficients $k_g a$ and $k_l a$ may be determined from a series of properly selected data giving values of Ka as described below.

The magnitude of $k_g a$ is primarily a function of the properties of the phase as described by the Schmidt number $(\mu/\rho D_G)$, and the state of motion of the phase as described by Reynolds number. Liquid flowing over the packing complicates the definition of Reynolds number and undoubtedly has some, at present unknown, effect on the magnitude of the interfacial area per unit volume a. The product $k_g a$ is affected only slightly by liquid rate of flow and should be primarily dependent upon the Reynolds number of the gas phase with a given system. Similarly the value of $k_l a$ is primarily a function of the Reynolds number of the liquid phase and is only slightly affected by the gas rate of flow, provided flooding and loading conditions are avoided.

Thus, if overall transfer coefficients are computed from a series of data in which the gas rate is varied and the liquid rate is held constant, equation 533 indicates that the overall transfer coefficient Ka varies as a function of the Reynolds number of the gas.

$$\frac{1}{Ka} = \frac{1}{k_g a} + \frac{1}{k_l a} = \frac{1}{k_g a} + \text{Constant}$$
$$= \phi(\text{Re})_{\text{gas}} + \text{Constant} \tag{533}$$

Although it is difficult to define the Reynolds number and its function in equation 533, theoretical considerations previously discussed in connection with equation 503 indicate that $k_g a = \alpha G_g{}^n$ for a given system, so that $1/Ka$ should be a straight-line function of $1/G_g{}^n$, provided the correct value of the exponent n is found. Usually a value of 0.8 for n will produce such a straight-line relationship. A typical plot [27, 28] * for the absorption of SO_2 in water from air with 3-in. spiral tile packing is shown in Fig. 489. An extrapolation of the straight line to the intercept ($G_g = \infty$ and $k_g a = \infty$) gives the value of $1/k_l a$ corresponding to the average gas velocity of the data on which the straight line was based. Similarly, when $1/Ka$ is plotted as a function of $1/G_l{}^m$ at constant gas velocity, the intercept will give the value of $1/k_g a$ corresponding to the average liquid rate of the data.

Controlling Resistances

The equation

$$\frac{1}{Ka} = \frac{1}{k_g a} + \frac{1}{k_l a} \tag{467}$$

indicates that, where $k_g a$ is large compared to $k_l a$,

* The bibliography for this chapter appears on p. 575.

the term $1/k_g a$ is negligible compared to the term $1/k_l a$, and therefore the overall transfer coefficient becomes approximately equal to the liquid phase coefficient. In such cases, it is said that the liquid phase constitutes the controlling resistance of the operation. Similarly, when $k_l a$ is large compared to $k_g a$, Ka is approximately equal to $k_g a$ and the gas phase is said to be the controlling resistance.

When the liquid phase constitutes the controlling resistance, a large fugacity potential $(-\Delta f)$ across the liquid phase is required to transfer a given mass

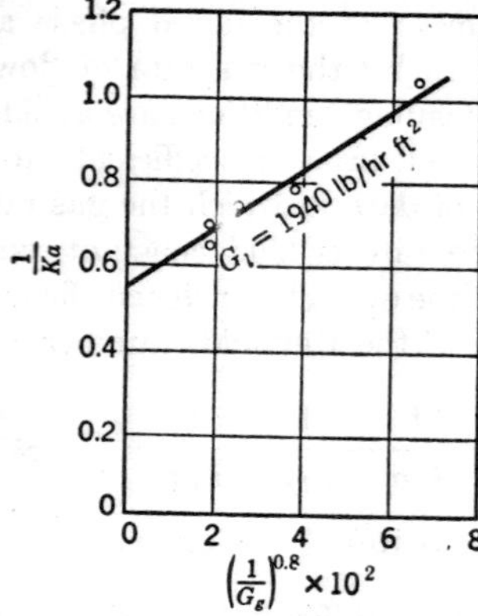

FIG. 489. Absorption coefficients of sulfur dioxide from air by water,[27, 28] plotted as a function of gas rate for extrapolation to obtain the liquid transfer coefficient $k_l a$.

compared to a relatively small potential $(-\Delta f)$ across the gas phase for the transfer of the same mass, and the potential $(-\Delta f)$ for the liquid phase is substantially equal to the total $(-\Delta f)$ across both phases.

For a gas highly insoluble in the liquid, a given fugacity potential across the liquid phase is equivalent to a relatively small concentration gradient in the liquid phase, and the more insoluble the gas the smaller will be the concentration gradient equivalent to a given fugacity gradient. Similarly, with a highly soluble gas, a given fugacity gradient is equivalent to a rather large concentration gradient, and, the more soluble the gas, the larger the concentration gradient in the liquid phase.

In early work on mass transfer,[36, 43] it was assumed that the diffusion equations applied to the transfer processes in form, whether or not the transfer processes were actually accomplished by the mechanism of diffusion. As a result it became customary to express the driving potential across liquid phases in terms of concentration units rather than fugacity. Thus, the transfer equations corresponding to equation 529 were written

$$\begin{aligned} -V'\,dY = -L'\,dX &= k_g aS\,dZ\,(p - p_i) \\ &= k_l' aS\,dZ\,(C_i - C_l) \\ &= K_G aS\,dZ\,(p - p^*) \\ &= K_L aS\,dZ\,(C^* - C_l) \end{aligned} \tag{534}$$

where p, p_i = partial pressure of the transferring component A in the gas stream and in the gas at the interface, respectively (atm).

p^* = partial pressure of component A in the gas in equilibrium with the liquid.

C_i, C_l = concentration of the transferring component A in the liquid at the interface and in the liquid, respectively (moles/cu ft).

C^* = concentration of component A in the liquid in equilibrium with the gas.

K_G, K_L = overall transfer coefficients expressed in terms of gas and liquid phase driving potentials respectively [moles/(hr)(sq ft)(atm), and moles/(hr)(sq ft)(moles/cu ft)].

The overall coefficient could then be based either on liquid phase units of potential, i.e., moles per cubic foot, or upon gas phase units of potential in atmospheres. The resulting overall coefficients were then known as overall gas phase coefficients or overall liquid phase coefficients and were described by the following relations.

$$K_G = \frac{1}{(1/k_g) + (H/k_l')} \tag{535}$$

$$K_L = \frac{1}{(1/Hk_g) + (1/k_l')} \tag{536}$$

where $H = (p/C)^*$, Henry's law constant.

Where highly insoluble gases are involved, the value of H in these equations is very high so that a superficial examination of equation 536 indicates that the term $(1/Hk_g)$ will be negligible as compared to the term $(1/k_l')$, and the overall liquid phase coefficient will be equal to the liquid phase coefficient. If this were true, the liquid phase resistance would be controlling and the gas phase resistance could be neglected. This idea is reinforced by the fact that at a given transfer rate the concentration potential across the gas phase is exceedingly small with insoluble gases and small potential values indicate low resistance.

This entire argument, however, is predicated on the assumption that the high value of H makes the term $(1/Hk_g)$ negligible as compared to the term $(1/k_l')$. But this is true only if the gas and liquid phase coefficients are of the same order of magnitude.

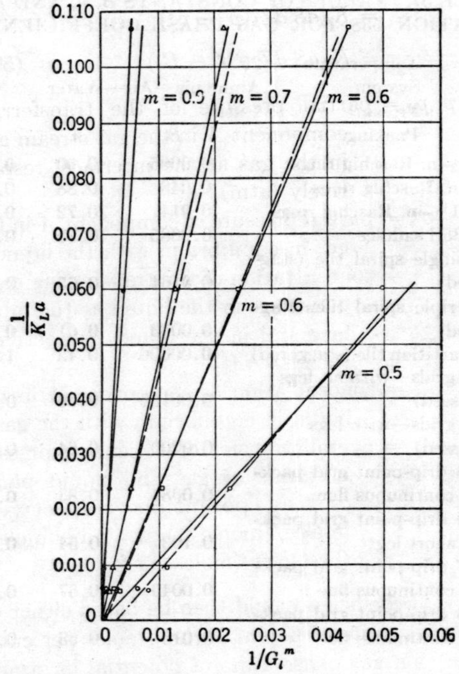

Fig. 490. Coefficients for stripping oxygen from water [47] plotted against the liquid rate for extrapolation to evaluate the gas transfer coefficient k'_ga. Solid lines are drawn through the origin. Dashed lines are drawn to represent the data.

However, it is impossible to determine the values for the individual phase coefficients from a single experimental run on a packed column, and an extrapolation similar to that of Fig. 489 is far from satisfactory in obtaining such information because it involves selecting a value for an exponent of liquid or gas rate which will make a plot of $1/K_Ga$ versus $1/G_g^n$ or $1/K_La$ versus $1/G_l^m$ a straight line.

Mass transfer data have an unfortunate tendency to scatter, and the amount of data taken under suitable conditions is rather meager, so that a considerable range of values of the exponents of liquid or gas rate will give apparently equally satisfactory straight lines. This is indicated in Fig. 490 by the family of straight lines, each line representing the same data plotted for different values of the exponent m. The range of exponents m on the liquid rate G_l which gives reasonably straight lines from the data is from 0.5 to 0.9. The corresponding values of the intercept $1/Hk_ga$ for the dashed lines vary from 0.0056 to -0.00212.

The negative values of the intercept can have no physical validity. If the liquid phase resistance were actually "controlling," the value of the intercept would be zero. This may or may not be the case, but unfortunately this is in no way proved by the data which would also correlate on the basis of a considerable gas phase resistance.

Correlation of Liquid Phase Coefficients

Data [47] are available on the absorption and desorption of carbon dioxide, oxygen, and hydrogen with air in water at gas mass velocities of from 30 to 1300 lb/(hr)(sq ft) and liquid velocities from 200 to 32,000 lb/(hr)(sq ft) at temperatures from 5° to 40° C, using a large number of packings. These data have been correlated on the assumption that the gas phase resistance for these insoluble gases is negligible and that the liquid phase is completely controlling. This is tantamount to making an extrapolation similar to that of Fig. 489, as shown in Fig. 490, drawing the straight line through the origin and thus evaluating the exponent of the liquid rate. On this basis, which *may or may not be true*, a general correlation of the results with insoluble gases has been presented in the form of the following equation.

$$\frac{k_l'a}{D_L} = \alpha\left(\frac{G_l}{\mu}\right)^{1-n}\left(\frac{\mu}{\rho D_L}\right)^{1-s} \qquad (537)$$

where $k_l'a$ = liquid phase mass transfer coefficient [lb moles/(hr)(cu ft)(lb mole/(cu ft)].

G_l = mass velocity of liquid, LM_{avg}/S[lb/(hr)(sq ft)].

s is usually constant at 0.5

Equation 537 is a modification of the functional equation 502 indicated by dimensional analysis in which the linear dimension D has been omitted in the correlation because of the difficulty in defining an effective diameter for the flow of liquid through the packed column.

The values of the constants, α, n, and s, are listed in Table 56 for the various packings investigated. Since the equation is not dimensionless, the constant α has dimensions, and the values listed in Table 56 are valid only when units of pounds mass, hours, and feet are employed. The effect of temperature on the liquid phase coefficients is determined through its effect on the physical properties in equation 537. The liquid phase diffusivity D_L for liquid systems at temperatures at which experimental data are not available may be estimated by use of the Stokes-Einstein relation [12, 54] that the liquid phase diffusivity is directly proportional to the absolute temperature and inversely proportional to liquid viscosity.

TABLE 56. VALUES OF α AND n IN EQUATION 537 FOR DIFFERENT PACKINGS [38] (WITH s CONSTANT AT 0.5)

Packing	α	n
2 x 2-in. Raschig rings	80	0.22
1.5 x 1.5-in. Raschig rings	90	0.22
1 x 1-in. Raschig rings	120	0.24
0.5 x 0.5-in. Raschig rings	280	0.35
⅜ x ⅜-in. Raschig rings	550	0.46
1.5-in. Berl saddles	160	0.28
1-in. Berl saddles	190	0.31
0.5-in. Berl saddles	150	0.28
3 x 3-in. single spiral tile (staggered)	29	0.15
3 x 3-in. triple spiral tile (staggered)	86	0.28
3 x 3-in. partition spiral tile (staggered)	16	0.09
#6295 drip-point grid packing, continuous flue	138	0.31
#6897 drip-point grid packing, continuous flue	61	0.25
#6146 drip-point grid packing, continuous flue	65	0.23

Equation 537 is used for the prediction of liquid phase coefficients for systems on which no experimental data are available. The experimental data on which it is based covered a tenfold range of variation in Schmidt number and a wide range of (G_l/μ), so that the relation should be capable of extrapolation.

Correlation of Gas Phase Coefficients

A correlation of gas phase coefficients similar to equation 537 for liquid phase coefficients has not yet been prepared. Gas phase information is presented in the literature in a variety of forms.

Data on the absorption of ammonia from air in water over a wide variety of packing have been correlated [37, 38] by estimating the liquid phase coefficients $k_l'a$ by equation 537 and calculating k_ga from values of K_Ga by equation 535.

The values of k_ga were then correlated as a function of the gas and liquid rates according to the relation

$$k_ga = BG_g{}^nG_l{}^m \tag{538}$$

where G_g and G_l are gas and liquid mass velocities, respectively [lb/(hr)(sq ft)].

B, n, and m are empirical constants specific to the system of ammonia, air, and water and functions of the packing used as given in Table 57.

TABLE 57. VALUES OF CONSTANTS B, n, AND m IN EQUATION 538 FOR GAS PHASE COEFFICIENTS [38]

Temperature 25° C
System Ammonia—Air—Water
Pressure 1 atm

Packing	B	n	m
½ x ½-in. Raschig rings	0.0065	0.90	0.39 [15]
1 x 1-in. Raschig rings	0.048	0.88	0.09
1½ x 1½-in. Raschig rings	0.014	0.72	0.38 [15]
1-in. Berl saddles	0.0085	0.75	0.40
3-in. single spiral tile (staggered)	0.0164	0.65	0.29
3-in. triple spiral tile (staggered)	0.0083	0.61	0.44
3-in. partition tile (staggered)	0.00006	0.42	1.06
Wood grids—without legs (crossed)	0.00158	0.66	0.61
Wood grids—with legs (crossed)	0.0209	0.64	0.31
#6295 drip-point grid packing, continuous flue	0.0084	0.83	0.27
#6295 drip-point grid packing, short leg	0.193	0.54	0.11
#6897 drip-point grid packing, continuous flue	0.0049	0.57	0.51
#6146 drip-point grid packing, continuous flue	0.010	0.63	0.39

Gas phase coefficients for systems where no experimental data are available may be estimated from the data on the absorption of ammonia from air into water at 25° C, by means of the j factor concept. According to this concept, at a given gas phase Reynolds number, packing, and rate of liquid flow,

$$k_ga = (k_ga)_{NH_3}\left[\frac{(\mu/\rho D_G)_{NH_3}}{(\mu/\rho D_G)}\right]^{2/3} \tag{539}$$

There are a few data on the vaporization of water, methanol, benzene, and toluene in a tower 3.6 in. ID, packed with 5 in. of ⅝-in. rings [47] which indicate that k_ga is proportional to $D_G{}^{0.17}$ rather than $D_G{}^{0.67}$, as indicated by the j factor expressions. This discrepancy casts doubt as to the general applicability of the exponent ⅔ on the Schmidt number at least for packed towers. Possibly the significance of the diffusivity is less under the turbulent conditions in packed towers than in wetted-wall columns where k_ga has been established as proportional to $D_G{}^{0.56}$.

THE TRANSFER UNIT

An equivalent method of dealing with mass transfer problems based upon the concept of a transfer unit has proved convenient.

In the integral form of the rate equation as applied to an absorption or stripping tower,

$$-\int \frac{dy}{(1-y)(f_g - f_i)} = \frac{k_g a S}{V} Z \qquad (531a)$$

the left-hand side of the equation is a function only of composition and is a measure of the separation effected. Assuming ideal gases, the number of gas phase transfer units n_g is defined [8,13,14] as

$$n_g = -\int \frac{dy}{(1-y)(y-y_i)} = \frac{k_g a P S}{V} Z \qquad (540)$$

The height equivalent to one gas phase transfer unit, $(HTU)_g$ is

$$(HTU)_g = \frac{Z}{n_g} = \frac{V/S}{k_g a P} = \frac{G_g}{k_g a P M_{avg}} \qquad (541)$$

Similarly, the number of liquid phase transfer units is defined as

$$n_l = -\int \frac{dx}{(1-x)(x_i - x)} \qquad (542)$$

By equation 529,

$$\begin{aligned} -L'\,dX &= -\frac{L\,dx}{(1-x)} = k_l a(f_i - f_l) S\,dZ \\ &= k_l a P(y_i - y^*) S\,dZ \\ &= k_l' a(C_i - C_l) S\,dZ \\ &= k_l' a C(x_i - x) S\,dZ \qquad (543) \end{aligned}$$

where $k_l a$ = transfer coefficient in fugacity units [moles/(hr)(cu ft)(atm)].

$k_l' a$ = transfer coefficient in concentration units [moles/(hr)(cu ft)(moles/cu ft)].

C = average molal density (total moles/cu ft solution).

y^* = mole fraction of solute in gas in equilibrium with liquid containing mole fraction x of solute.

From equation 529

$$n_l = -\int \frac{dx}{(1-x)(x_i - x)} = \frac{k_l' a C}{L/S} Z \qquad (544)$$

so that

$$(HTU)_l = \frac{Z}{n_l} = \frac{L/S}{k_l' a C} \qquad (545)$$

Also, since

$$\frac{k_l a P(y_i - y^*)}{(x_i - x)} = k_l' a C \qquad (546)$$

$$(HTU)_l = \frac{L/S}{k_l a P m} \qquad (547)$$

where m = the average slope of the equilibrium curve $y = \phi(x)$ between x_i and x or approximately the slope at x_i or x, since $x_i - x$ is usually small.

Similarly, the number of overall transfer units based on the gas phase concentration may be defined

$$n_G = -\int \frac{dy}{(1-y)(y-y^*)} = \frac{K_G a P M_{avg}}{G_g} Z \qquad (548)$$

and

$$(HTU)_G = \frac{Z}{n_G} = \frac{G_g}{K_G a P M_{avg}} \qquad (549)$$

where K_G = overall gas phase transfer coefficient [moles/(hr)(sq ft)(atm)].

The number of overall transfer units based on the liquid phase concentration may be defined as

$$n_L = -\int \frac{dx}{(1-x)(x^* - x)} = \frac{K_L a C}{L/S} Z \qquad (550)$$

and

$$(HTU)_L = \frac{L/S}{K_L a C} \qquad (551)$$

where x^* = mole fraction of solute in liquid in equilibrium with gas containing mole fraction y of solute.

K_L = overall liquid phase transfer coefficient [moles/(hr)(sq ft)(moles/(cu ft)].

Exercise. Show that

$$HTU_G = HTU_g + \frac{mV}{L} HTU_l \qquad (552)$$

$$HTU_L = HTU_l + \frac{L}{mV} HTU_g \qquad (553)$$

$$HTU_G = \frac{mV}{L} HTU_L \qquad (554)$$

In equations 552, 553, and 554, the values of m are not quite identical. In equation 552, m is the average slope of the equilibrium curve $y = \phi(x)$ between x_i and x. In equation 553, m is the average slope between x^* and x_i. In equation 554, m is the average slope between x^* and x. Ordinarily, the variation in the slope of the equilibrium curve between x^*

and x is small so that m may be considered to be constant over the range.

The height equivalent to a transfer unit is a convenient and simple means for expressing the performance of various packings, is practically independent of flow rates, and varies only slightly from system to system. Its single linear dimension makes it easy to estimate. A series of experimental runs made with a given system in a tower may be correlated by plotting HTU_G as a function of mV/L, (Fig. 491). The slope of the resulting line is HTU_l, and its intercept is HTU_g. Similarly, HTU_L might be plotted as a function of Lm/V to give a line having a slope of HTU_g with an intercept of HTU_l.

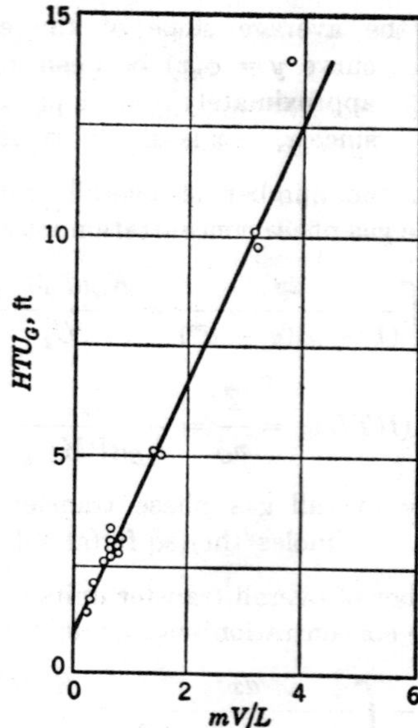

FIG. 491. The height of a transfer unit (HTU_G) versus mV/L for absorption of sulfur dioxide in water on 3-in. spiral rings, stacked staggered.[13, 40]

Where the gas phase behaves as an ideal gas, and the equilibrium relations can be expressed as $y = mx$ and the concentration of the solute in the gas and liquid phases is small, the number of overall gas phase transfer units may be expressed approximately by the relation

$$n_G = -\int_{y_1}^{y_2} \frac{dy}{y - y^*} = \int_{y_2}^{y_1} \frac{dy}{y - mx} \quad (555)$$

and a material balance by

$$x = \frac{V}{L} y + x_2 - \frac{V}{L} y_2 \quad (556)$$

Substituting the value of x given by equation 556 in equation 555, the integration may be performed analytically to give the relation

$$n_G = \frac{1}{1 - (mV/L)} \ln \left(\frac{y_1 - mx_1}{y_2 - mx_2} \right) = \frac{y_1 - y_2}{(-\Delta y)_{lm}} \quad (557)$$

where

$$(-\Delta y)_{lm} = \frac{(y_1 - y_1^*) - (y_2 - y_2^*)}{\ln \left(\dfrac{y_1 - y_1^*}{y_2 - y_2^*} \right)}$$

Many commercial absorption problems involve the recovery of a solute from a gas by absorption, followed by the recovery of the solute from the solvent by a distillation operation. If the recovery of the solute from the gas is fixed, the values y and y_2 are fixed. The concentration of solute in the absorbent is also known, fixing x_2, which is usually negligibly small. Under such circumstances the value of the term mV/L is approximately equal to y_1^*/y_1 or the fractional approach of the exit liquid

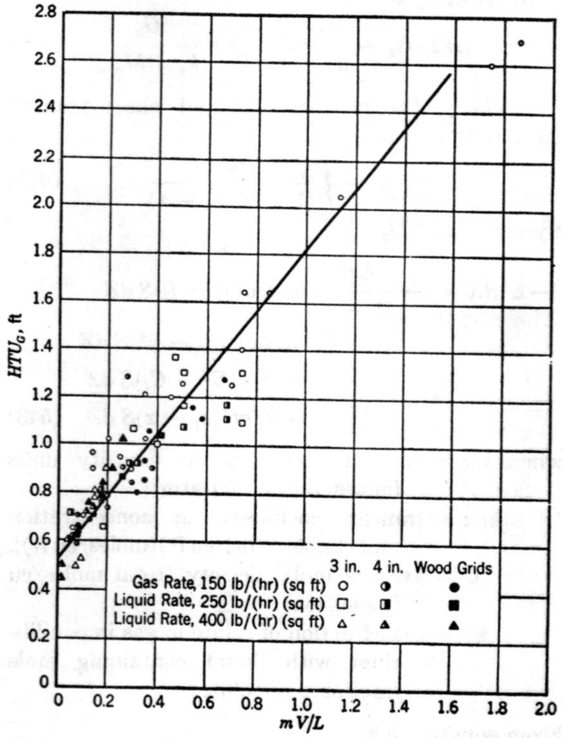

FIG. 492. The height of a transfer unit (HTU_G) in feet versus mV/L for absorption of ammonia in water in 3-in. spiral rings, 4-in. partition rings, and ½-in. wood grids.[13, 40]

to equilibrium with the inlet gas. Thus, for a given system the greater the value of mV/L or of V/L, the greater the concentration of the solute in the exit liquid and the lower the cost of the subsequent recovery operation. On the other hand the greater the concentration of solute in the exit liquid, the

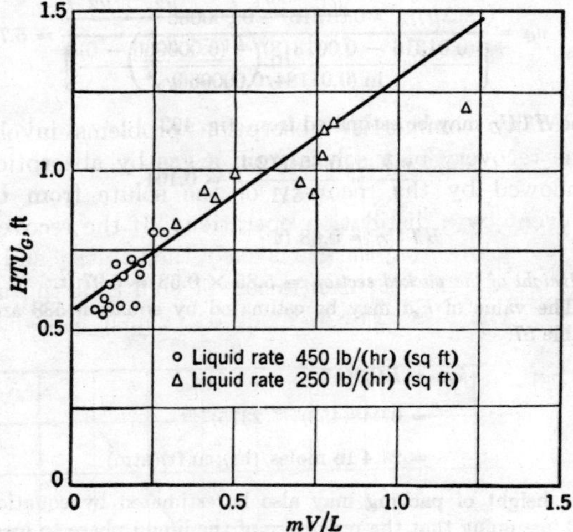

FIG. 493. The height of a transfer unit (HTU_G) in feet versus mV/L for absorption of ammonia in water on 1¼-in. to 1¾-in. broken quartz.[13, 40]

lower the value of $(-\Delta y)_{lm}$ for a given system. There is, therefore, an optimum value of mV/L for minimum total cost of recovering solute.

In general, with valuable solutes, the optimum value of mV/L will range from 0.5 to 0.8 in absorption operations and from 1.5 to 2.0 for stripping operations.

Some typical values of HTU_G for the absorption of sulfur dioxide and ammonia in water are shown in Figs. 491, 492, and 493 as a function mV/L, as indicated by equation 552. A summary of the values of HTU_g and HTU_l, as determined from the slopes and intercepts of the lines on these figures is presented in Table 58.[13] The variation in the values from one type of packing to another and even from one system to another is very small. It is this fact which makes the HTU concept convenient in estimating values for systems and packing on which no experimental data are available.

TABLE 58. APPROXIMATE VALUES OF HTU_g AND HTU_l OBTAINED FROM PLOTS[13] OF HTU_G VERSUS mV/L

Packing	Solute	G_g*	G_l*	HTU_g	HTU_l
3-in. spiral tile, stacked staggered	SO_2	100	2500	0.70	2.85
3-in. spiral tile, stacked staggered	NH_3	200	350	0.58	1.27
4-in. partition rings, stacked staggered	NH_3	200	350	0.55	1.22
0.5-in. x 3-in. wood grids, 0.5-in. space	NH_3	200	350	0.55	1.16
1.5-in. quartz, dumped	NH_3	200	350	0.53	0.77
Vermorel sprays	NH_3	200	350	1.0	4.0

* Rough average values.

Illustrative Example. It is desired to recover 99.5 per cent of the ammonia NH_3 from a stream of air which is saturated with water. The partial pressure of the NH_3 is 10 mm of mercury. The gas is supplied at 72° F and in a quantity which contains 2100 lb dry air per hour. Use a packed column supplied with water at 72° F. Compute:

1. The minimum water flow required.
2. The dimensions of the column if 1-in. ceramic Raschig rings are used and it is operated at ½ of the flooding gas velocity with ten times the minimum water flow.
3. For comparison, the height of a wetted-wall column (no packing) of the same diameter with the same water rate as in 2.

Solution. Assumptions:

(a) No transfer of water between phases.
(b) Atmospheric pressure of 760 mm mercury.
(c) The molal equilibrium ratio y^*/x is constant.
(d) The heat effect of absorption of ammonia has a negligible effect on fluid properties.
(e) The thickness of the falling film in the wetted-wall column may be considered negligible.

1. The minimum water rate is obtained when NH_3 in exit liquid is in equilibrium with NH_3 in entering gas. From Fig. 551 (Appendix) the ratio y^*/x is estimated to be 0.85 as follows: At 72° F the partial pressure of NH_3 in equilibrium with a solution containing 2 lb NH_3 per 100 lb H_2O is 13.5 mm, calculated by finding the mass fraction of NH_3 in the vapor in equilibrium with NH_3 in aqueous solution containing 0.02 mass fraction at 72° F under the equilibrium pressure of 0.7 psia, and converting to partial pressure of NH_3 as follows:

$$y^* = \frac{13.5}{760} = 0.01777$$

$$x = \frac{\frac{2}{17}}{\frac{2}{17} + \frac{100}{18}} = 0.0208$$

$$\frac{y^*}{x} = \frac{0.01777}{0.0208} = 0.854$$

For a partial pressure of 10 mm

$$y_1 = \frac{10}{760} = 0.01316 \quad \textbf{and} \quad x_1 = \frac{0.01316}{0.854} = 0.0154$$

For 99.5 per cent recovery of NH_3 a material balance gives a water rate of 61.1 moles or 1100 lb H_2O per hour as follows: At these low concentrations the moles NH_3 per mole dry air

Y may be assumed to be equal to the mole fraction y of NH_3. 2100 lb dry air = 72 moles.

$$V'(Y_1 - Y_2) = L'(X_1 - X_2)$$

$$L' = \frac{V'(Y_1 - Y_2)}{X_1 - X_2} = \frac{72(0.01316)(0.995)}{0.01542 - 0}$$

$$L' = 61.1 \text{ lb moles or } 1100 \text{ lb } H_2O \text{ per hour.}$$

2. The dimensions of the column for a water rate of 11,000 lb/hr and a gas rate of 2160 lb ($NH_3 + H_2O$ total about 60 lb in gas). The flooding rate may be estimated from Fig. 367.

$$\text{Density of gas } \rho_V = 0.075 \text{ lb/cu ft}$$

$$\text{Density of liquid } \rho_L = 62.3 \text{ lb/cu ft}$$

$$\frac{G_g}{G_l}\sqrt{\frac{\rho_V}{\rho_L}} = \frac{11{,}000}{2160}\sqrt{\frac{0.075}{62.3}} = 0.177$$

From Fig. 367 for flooding, and Table 36 for (a/X^3):

$$\frac{v^2}{g_c}\left(\frac{a}{X^3}\right)\frac{\rho_V}{\rho_L}\mu^{0.2} = 0.078$$

$$\frac{v^2}{32.2}(185)\left(\frac{0.075}{62.3}\right)(0.96)^{0.2} = 0.078$$

$$v = 3.36 \text{ ft/sec at flooding}$$

At the flooding point the superficial mass velocities are

Gas $\quad G_g = (3.36)(3600)(0.075) = 906 \text{ lb/(hr)(sq ft)}$

Liquid $\quad G_l = 906\left(\frac{11{,}000}{2100}\right) = 4750 \text{ lb/(hr)(sq ft)}$

These may be compared with values on Fig. 366.

At a design gas velocity of one-half the flooding velocity:

Design mass velocity $G_g = (\frac{906}{2}) = 453$ lb/(hr)(sq ft)

Cross-sectional area $S = (\frac{2160}{453}) = 4.77$ sq ft

Tower diameter $D = 2.46$ ft

Therefore, use a column 30 in. in diameter as a standard size.

The height of the column depends upon the required number of transfer units. Equation 548,

$$-\int \frac{dy}{(1-y)(y-y^*)} = n_G = \frac{Z}{HTU_G}$$

may be integrated graphically by evaluating $\dfrac{1}{(1-y)(y-y^*)}$ for various values of y as follows: By a material balance

$$x = \frac{72.4}{611}(y - 0.000066)$$

y	x	y^*	$1-y$	$y-y^*$	$\frac{1}{(1-y)(y-y^*)}$
0.01316	0.001542	0.001318	0.987	0.01184	85.5
0.010066	0.00118	0.001016	0.990	0.00905	111.7
0.005066	0.00059	0.000495	0.995	0.004571	220
0.001066	0.000118	0.0001016	0.999	0.000964	1,040
0.000566	0.000059	0.0000495	1.000	0.0005165	1,936
0.000166	0.0000118	0.00001016	1.000	0.0001558	6,420
0.000066	0	0	1.000	0.000066	15,100

Integration of equation 548 for the above values between the limits of $y_1 = 0.01316$ and $y_2 = 0.000066$ gives the number of transfer units $n_G = 5.85$.

An approximation giving satisfactory results when y is small is based on the assumption that $(1 - y) = 1.00$; then by equation 557

$$n_G = \frac{y_1 - y_2}{(-\Delta y)_{\ln}}$$

$$n_G = \frac{0.01316 - 0.000066}{\left[\dfrac{(0.01316 - 0.001318) - (0.000066 - 0)}{\ln(0.01184/0.000066)}\right]} = 5.77$$

The HTU_G may be estimated from Fig. 492:

$$\frac{mV}{L} = \frac{0.854(72.4)}{611} = 0.101$$

$$HTU_G = 0.68 \text{ ft}$$

Height of the packed section = 5.85 × 0.68 = 3.97 ft.

The value of k_ga may be estimated by equation 538 and Table 57.

$$k_ga = BG_g^nG_l^m$$

$$= 0.048(453)^{0.88}(2375)^{0.09}$$

$$= 21.4 \text{ lb moles/(hr)(cu ft)(atm)}$$

The height of packing may also be estimated by equation 540, assuming that the resistance of the liquid phase to mass transfer is negligible.

$$n_G = \frac{k_gaPSZ}{V} = \frac{K_GaPSZ}{V}$$

$$Z = \frac{(\frac{2100}{29})5.85}{(21.4)(1)(4.90)} = 4.05 \text{ ft of packing}$$

The total height of the column will be greater than that of the computed packed height. A typical design might be as follows, based on $n_G = 5.85$ and 4 ft of packing.

Computed height of packing	4.0 ft
Additional packing to insure liquid distribution	1.7 ft
Height above packing for entrainment separation	2.5 ft
Height below packing for vapor introduction and distribution and packing support	2.5 ft
Five minutes liquid retention for control	3.0 ft
Total height of column excluding heads	13.7 ft

3. The height of a wetted-wall column by equation 478,

$$-\int_{y_1}^{y_2} \frac{dy}{(1-y)(y-y^*)} = \int_0^A \frac{k_gP\,dA}{V}$$

Upon integration and substitution this becomes

$$\frac{k_gP\pi DZ}{V} = n_G$$

The value of k_g may be estimated by equation 503.

$$\frac{k'D}{D_G} = \frac{RTk_gD}{D_G} = 0.023\left(\frac{DG}{\mu}\right)^{0.8}\left(\frac{\mu}{\rho D_G}\right)^{0.44}$$

The diffusivity D_G may be estimated by equation 493 and Table 54.

$$D_G = 0.0166\left[\frac{T^{3/2}}{P(V_A^{1/3} + V_B^{1/3})^2}\right]\sqrt{\frac{1}{M_A}+\frac{1}{M_B}}$$

$$= 0.0166\left[\frac{295^{3/2}}{1(29.9^{1/3} + 26.7^{1/3})^2}\right]\sqrt{\frac{1}{17}+\frac{1}{29}}$$

= 0.708 sq ft/hr (An experimental value [40] of 0.915 sq ft/hr at 25° C and 1 atm is equivalent to 0.90 sq ft/hr at 72° F.)

Viscosity μ of air = 0.0178 centipoise

$$k_g = \frac{(0.023)(0.708)}{(0.7302)(532)(2.5)}\left(\frac{2.5 \times 453}{0.0178 \times 2.42}\right)^{0.8}\left(\frac{0.0178 \times 2.42}{0.075 \times 0.708}\right)^{0.44}$$

k_g = 0.055 lb moles/(hr)(sq ft)(atm)

The height of the column if no packing is used:

$$Z = \frac{(5.85)(\frac{2100}{29})}{(0.055)(1)(2.5)(\pi)} = 980 \text{ ft}$$

DISTILLATION

The rate equation may be applied to a differential section of a packed tower in which distillation is occurring in the same manner as in absorption or stripping. The equations, which are analogous to equation 529, take the following form.

$$-d(Vy) = -d(Lx) = k_g aS\,dZ\,(f_g - f_i)$$
$$= k_l aS\,dZ\,(f_i - f_l) = KaS\,dZ\,(f_g - f_l) \quad (558)$$

When the molal rate or flow of liquid and of vapor through the column is the same at all sections, the column equation 558 leads to the following integral forms, analogous to equations 530*a* to 530*d*.

$$\int \frac{dy}{f_g - f_i} = \int \frac{k_g aS\,dZ}{V} \quad (559a)$$

$$-\int \frac{dx}{f_i - f_l} = \int \frac{k_l aS}{L}\,dZ \quad (559b)$$

$$-\int \frac{dy}{f_g - f_l} = \int \frac{KaS}{V}\,dZ \quad (559c)$$

$$-\int \frac{dx}{f_g - f_l} = \int \frac{KaS}{L}\,dZ \quad (559d)$$

The integration of equation 559 is complicated by the wide variation in the mass rate of flow of liquid and vapor through the column section. This is due to the wide variation in the molecular weight of the material as the composition changes, so that, although the mole rate of flow may be constant, the mass rate of flow may vary widely. As has been pointed out previously, the transfer coefficients are approximately proportional to mass rates of flow. There is the further complication that the physical properties of the fluids may change markedly, owing to the relatively large temperature changes which frequently occur. Such variations in physical properties may also be reflected in a relatively large variation in the mass transfer coefficients.

Under these circumstances, for lack of a better procedure, it is customary to determine average values of the transfer coefficient by a constant, K_{avg} for K.

The left-hand side of equations 559*a* and 559*b* may be integrated graphically by calculating the interface conditions as described previously. Again, it is usually more convenient to use equations 559*c* and 559*d*, the left-hand sides of which may be integrated readily by relating y, x, f_g, and f_l through a material balance and the equilibrium relationships.

These difficulties in obtaining precise integration of the rate equations apparently have somewhat discouraged research directed towards the evaluation of transfer coefficients for distillation.

The HTU concept may also be applied to distillation to give the following relationships.

$$n_g = \frac{Z}{(HTU)_g} = -\int \frac{dy}{y - y_i} = \frac{Z}{V/k_g aPS} \quad (560a)$$

$$n_l = \frac{Z}{(HTU)_l} = -\int \frac{dx}{x_i - x} = \frac{Z}{L/k_l aPSm_{avg}}$$
$$= \frac{Z}{L/k_l' aCS} \quad (560b)$$

$$n_G = \frac{Z}{(HTU)_G} = -\int \frac{dy}{y - y^*} = \frac{Z}{V/KaPS} \quad (560c)$$

$$n_L = \frac{Z}{(HTU)_L} = -\int \frac{dx}{x^* - x} = \frac{Z}{L/KaPSm_{avg}}$$
$$= \frac{Z}{L/K_L aCS} \quad (560d)$$

also

$$(HTU)_G = (HTU)_g + \frac{mV}{L}(HTU)_l \quad (561)$$

$$(HTU)_L = (HTU)_l + \frac{L}{mV}(HTU)_g \quad (562)$$

$$(HTU)_G = \frac{mV}{L}(HTU)_L \quad (563)$$

It should be emphasized that in distillation the value of m may vary widely in a section of the column.

An investigation[23] of the distillation of the ethanol-water system over packings of ⅜-in., 1-in., and 2-in. Raschig rings and ½-in. and 1-in. Berl saddles in a column section operated as a rectifier indicated the data were correlated by the equations

$$(HTU)_G = \frac{145\overline{G_g}}{\left[\frac{1}{m}\left(\frac{\overline{G_l}}{\mu}\right)\right]^{1.21}} \tag{564}$$

or

$$K_Ga = 0.0069\left[\frac{1}{m}\left(\frac{\overline{G_l}}{\mu}\right)\right]^{1.21} \tag{565}$$

where $m = \left[\frac{(HTU)_G}{(HTU)_L}\right]\left(\frac{\overline{G_l}}{\overline{G_g}}\right)$

K_Ga = overall gas phase transfer coefficient [lb moles/(hr)(cu ft)(atm)].

$\overline{G_l}$ = arithmetic mean or average liquid mass velocity [lb/(hr)(sq ft)].

$\overline{G_g}$ = arithmetic mean gas mass velocity [lb/(hr)(sq ft)].

μ = liquid viscosity [lb/(hr)(ft)].

These results show little effect of the size or type of packing and a large effect of the rate of liquid flow on the transfer coefficient. This implies that a major resistance to transfer is in the liquid phase. Concentration is also an important factor through its effect on m, but this was not brought out in this work which was limited to rectification over a concentration range of from 6.1 to 51.6 mole per cent in the vapor phase.

Additional investigations[18] showed a successful empirical correlation of the results over ½-in. rings by plotting $(HTU)_G$ versus m as defined by equation 566 for systems having equilibrium curves typical of relatively ideal systems, as in Fig. 494.

$$m_{\text{avg}} = \frac{\int_{y_1}^{y_2}\left(\frac{dy}{dx}\right)dy}{y_2 - y_1} \tag{566}$$

However, this method of correlation was not successful when applied to systems having irregular equilibrium curves such as ethanol-water.

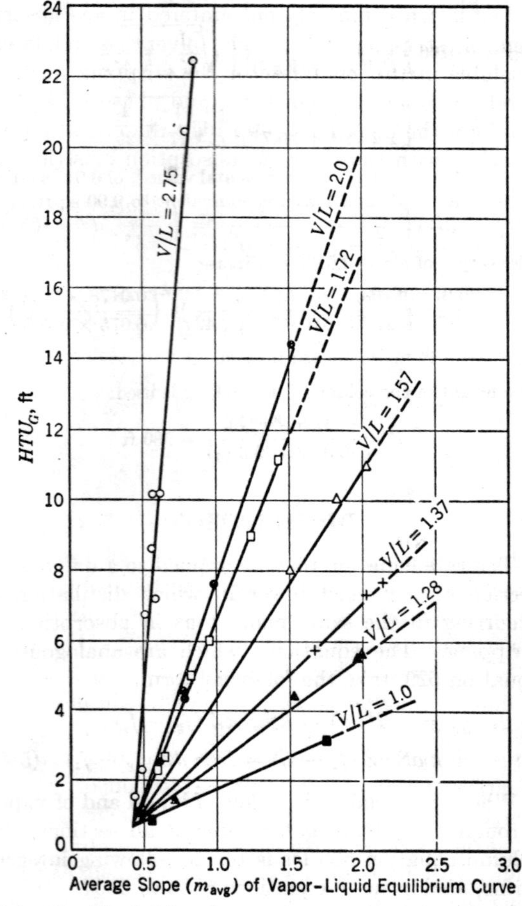

Fig. 494. The height of a transfer unit (HTU_G) versus slope of equilibrium diagram ($y - x$) for various reflux ratios (V/L) in the methanol-water system.[18]

LIQUID-LIQUID EXTRACTION

The rate equations may be written for a differential section of a packed tower or similar contacting equipment in which extraction is occurring in the same manner as for distillation or absorption. The equations are identical to equation 558, provided the symbols V, y, f_g, etc., refer to one liquid phase and the symbols L, x, f_l, etc., refer to the other liquid phase.

Where the two solvents C and S are only slightly miscible and retain their immiscibility regardless of solute content over the operating range, the ratio of the quantity of solvent C in the raffinate phase to the quantity of solvent S in the extract phase passing one another at any point in the unit will be substantially constant. This is strictly analogous to the situation commonly encountered in gas absorption or stripping when the solvent is relatively insoluble in the gas phase or the stripping medium is relatively insoluble in the liquid phase. Thus, in this case the equations describing the operation are identical with those for gas absorption or stripping.

$$-\int \frac{dy}{(1-y)(f_g - f_i)} = \int \frac{k_g aS}{V} dZ \qquad (531a)$$

$$-\int \frac{dx}{(1-x)(f_i - f_l)} = \int \frac{k_l aS}{L} dZ \qquad (531b)$$

$$-\int \frac{dy}{(1-y)(f_g - f_l)} = \int \frac{KaS}{V} dZ \qquad (531c)$$

$$-\int \frac{dx}{(1-x)(f_g - f_l)} = \int \frac{KaS}{L} dZ \qquad (531d)$$

In the general case, the liquid-liquid extraction system is of the type in which the two solvents are appreciably miscible and to a varying degree throughout the range of operation, as has been described previously. In this case, the molal rate of flow of each phase and the molal rate of flow of a component of each phase vary over a wide range. Thus, the previous methods of arranging the rate equations for integration for distillation or absorption or stripping are no longer satisfactory. The equations analogous to equations 531 are written below, without assuming V and L to be constant and using the subscript v to represent the V phase.

$$-\int \frac{d(Vy)}{(f_v - f_i)} = \int k_v aS\, dZ \qquad (567a)$$

$$-\int \frac{d(Lx)}{f_i - f_l} = \int k_l aS\, dZ \qquad (567b)$$

$$-\int \frac{d(Vy)}{f_v - f_l} = \int KaS\, dZ \qquad (567c)$$

$$-\int \frac{d(Lx)}{f_v - f_l} = \int KaS\, dZ \qquad (567d)$$

The left-hand sides of equations 567a to 567d may be integrated by methods previously described and by relating the quantity of the phases to their compositions, provided the fugacities of the components of the phases may be evaluated.

It is usually necessary, because of the lack of thermodynamic data for most liquid-liquid systems of interest, to assume that the fugacities are proportional to the molal concentrations and to write equations 567a to 567d as follows.

$$-\int \frac{d(Vy)}{\overline{C} - \overline{C}_i} = \int k_v' aS\, dZ \qquad (568a)$$

$$-\int \frac{d(Lx)}{C_i - C} = \int k_l' aS\, dZ \qquad (568b)$$

$$-\int \frac{d(Vy)}{\overline{C} - \overline{C}^*} = \int K_V aS\, dZ \qquad (568c)$$

$$-\int \frac{d(Lx)}{C^* - C} = \int K_L aS\, dZ \qquad (568d)$$

where the transfer coefficients are all based on liquid phase concentrations and have the units, lb moles/(hr)(sq ft)(moles/cu ft). The bar over the concentration term $\overline{C}$ designates a concentration in one phase V as distinct from the unbarred C, designating a concentration in the other phase L.

$\overline{C}^*$ = concentration of the solute in the phase V which would be in equilibrium with the concentration C in phase L.

C^* = concentration of the solute in the phase L which would be in equilibrium with the concentration $\overline{C}$ in phase V.

However, the right-hand side of equations 568a to 568d is no longer suitable for integration because wide variation in the mass rates of flow of the phases may produce wide variation in the transfer coefficients. This situation might be improved by writing equations 567a to 567d as

$$-\int \frac{d(Vy)}{V(f_v - f_i)} = \int \frac{k_v aS}{V} dZ \qquad (569a)$$

$$-\int \frac{d(Lx)}{L(f_i - f_l)} = \int \frac{k_l aS}{L} dZ \qquad (569b)$$

$$-\int \frac{d(Vy)}{V(f_v - f_l)} = \int \frac{KaS}{V} dZ \qquad (569c)$$

$$-\int \frac{d(Lx)}{L(f_v - f_l)} = \int \frac{KaS}{L} dZ \qquad (569d)$$

MASS TRANSFER IN SYSTEMS OF FLUIDS AND GRANULAR SOLIDS

Data on the rate of mass transfer in fluid phases flowing through beds of granular solids are of fundamental importance in such operations as the adsorption of liquids or gases by solids, ion exchange reactions, the transfer of heat and mass in packed columns, the drying or leaching of solids by circulating fluids through solids beds, and the interpretation of kinetic data for chemical reactions between fluids and solids, or for chemical reactions in fluids which are catalyzed by solids.

Fixed Beds

The experimental basis of present correlations of the rate of mass transfer in fluid phases flowing through fixed beds of granular solids is data on the drying of Celite catalyst particles by air,[21,53] the vaporization of naphthalene particles by air, carbon dioxide, and hydrogen,[33,42] and the leaching of organic solvents from Celite catalyst particles by water.[30] The correlation is shown in Fig. 495 where the j factor is plotted as a function of Reynolds number based on particle size as a parameter. The j factor is defined by equation 504a, discussed previously.

In calculating the j factor from the experimental data, it is assumed that all the surface of the particles was effective in the transfer process.

The Reynolds number is based on the mean surface diameter (p. 20) or the diameter of a sphere having the same surface area as the average surface area per particle.

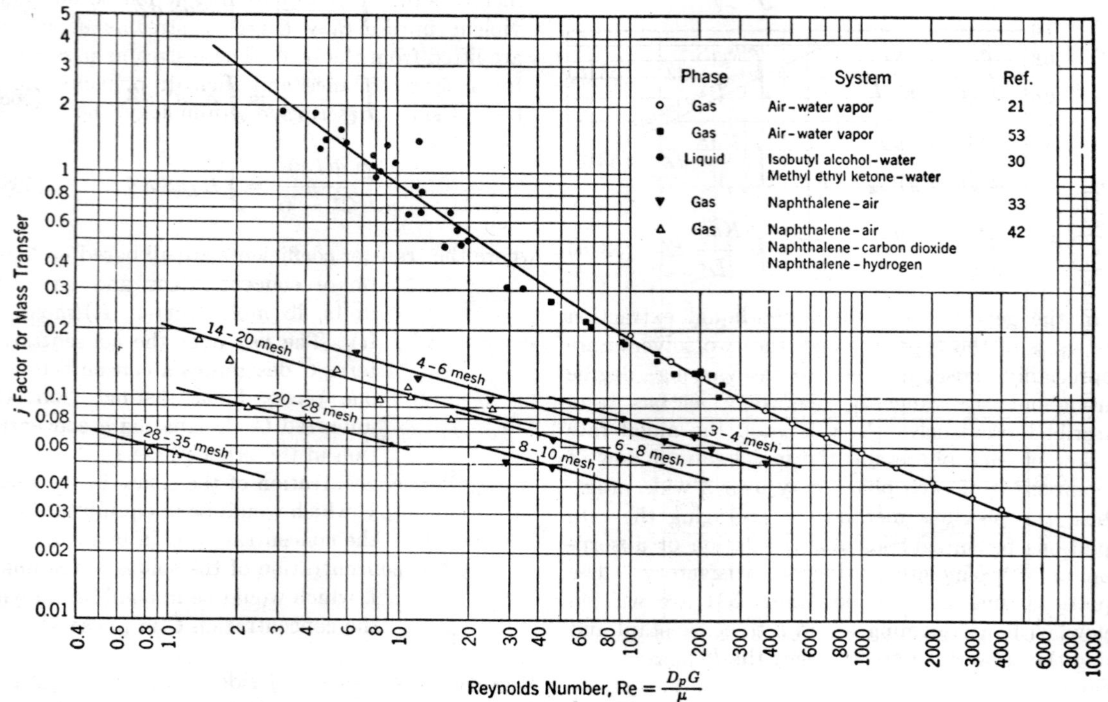

FIG. 495. j factor for mass transfer as a function of Reynolds number and particle size.[13,42,53]

The upper curve of Fig. 495 is based on data for the drying of Celite particles with air and the leaching of Celite particles with water. All the particles in these experiments were greater than 0.09 in. in diameter (D_p), so that above this value there is no effect of particle diameter in the correlation other than that included in the Reynolds number. On the other hand, the experiments on the vaporization of naphthalene in air, carbon dioxide, and hydrogen show an effect of particle size. The particles in these experiments were all small, as indicated in Fig. 495. The inconsistencies between the data of Hurt [33] and those of Resnick and White [42] shown in Fig. 495 on the vaporization of small particles of naphthalene are due, probably, to the fact that the incompletely reported properties of one bed[33] were assumed to be the same as those of the other beds.[42]

The similarity between Fig. 495 and a plot of friction factor versus Reynolds number for flow in pipes of varying roughness (Fig. 125) is interesting. Figure 125 shows a series of lines having the relative

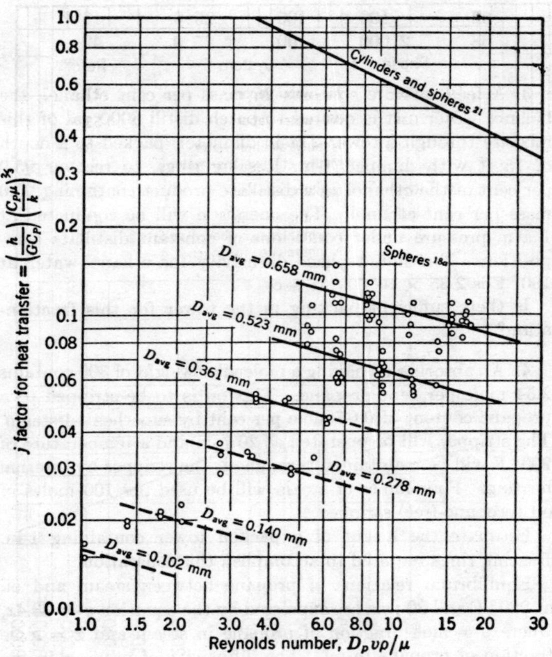

FIG. 495a. *j* factor for heat transfer as a function of Reynolds number and particle size.[18a]

roughness, ϵ/D as a parameter, the lines corresponding to large values of roughness being closer together than those corresponding to small values. Similarly Fig. 495 shows a series of lines having particle size as a parameter, the lines corresponding to large particles being closer together than those corresponding to small particles. Figure 495a is a similar plot of heat transfer data[18a] which show the same characteristics. The dimensionless parameter for fixed beds corresponding to the relative roughness in pipes has not yet been determined.

Fluidized Beds

Mass transfer in fluidized beds has been investigated by vaporizing naphthalene with air, carbon dioxide, and hydrogen. Visual observation of fluidized beds of granular naphthalene indicates two types of fluidization. These two types of aggregative fluidization are characteristic of the behavior of closely sized fractions of particles. With low gas velocities there is no observable motion of the bed. As the velocity increases, the bed begins to expand and bubbles of gas pass through the bed which resembles a gently boiling liquid. As the velocity increases further, the bubbling action becomes more violent until finally slugs or streams of material are lifted from the bed. These streams pass upward for a short distance and then drop back onto the bed. Using a descriptive terminology, the two types

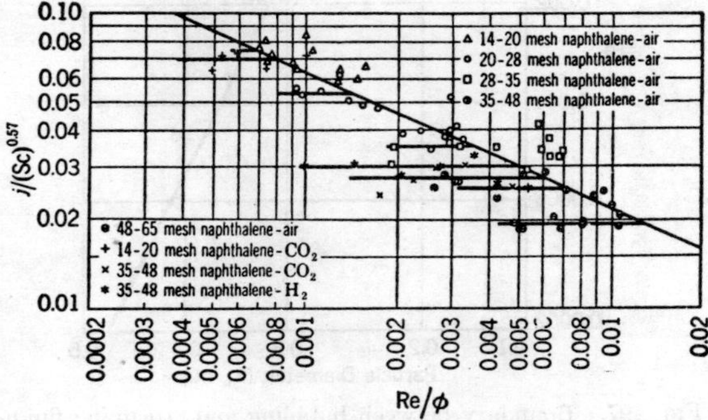

FIG. 496. Correlation of data on mass transfer in streaming fluidization[42] by plotting $j/(Sc)^{0.57}$ versus Re/ϕ.

of aggregative fluidization are termed "bubbling" and "streaming" fluidization. When the material being fluidized is not closely sized the bubbling type of fluidization persists at much higher gas velocities than when the material is closely sized. The wider the size range of the fraction, the higher the velocities to which the so-called bubbling region persists.

The mass transfer data[42] in fluidized beds in the streaming region are correlated as shown in Fig. 496 which plots $j/(Sc)^{0.57}$ as a function of Re/ϕ. The *j* factor is defined by equation 504*a*, and again it is assumed that all the area of the particles is active in the mass transfer process. The Reynolds number is modified as in the fixed-bed correlation. The group ϕ is defined

$$\phi = \frac{D_p{}^3 \rho_g g(\rho_s - \rho_g)}{\mu^2} \tag{570}$$

The curve shown in Fig. 496 correlated the data for closely sized fractions ranging from 14 to 48 Tyler screen mesh.

The mass transfer data for fluidized beds in the bubbling region are as yet too meager to permit development of a generalized correlation. The short horizontal lines in Fig. 496 present the data available at the present time. An approximate correlation of the velocity at which the bubbling region transforms to the streaming region for closely sized fractions of naphthalene is shown in Fig. 497. In this figure, the area above and to the right of the

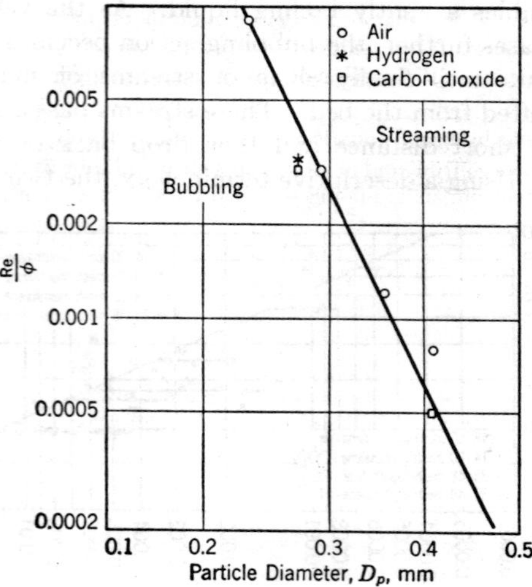

FIG. 497. Boundary between bubbling and streaming fluidization as indicated by Re/ϕ and particle diameter D_p.[42]

curve represents conditions of streaming fluidization, whereas the area below and to the left of the curve represents conditions of bubbling fluidization as long as fluidization exists.

PROBLEMS

1. A tower having an inside diameter of 1 ft is packed to a depth of 2.0 ft with 1-in. Raschig rings. Gas consisting of 3.0 per cent by volume acetone and 97.0 per cent air (dry basis) and saturated with water is introduced under the bottom of the packing at a rate of 300 lb of dry gas per hour and square foot of column cross section, at a temperature of 80° F and a pressure of 760 mm mercury. Pure water is introduced at the top of the packing at the rate of 400 lb/(hr)(sq ft) at a temperature of 80° F.

The partial pressure p of acetone in millimeters of mercury in equilibrium with dilute acetone-water mixtures at 80° F is $p = 266x$, where x is mole fraction of acetone. Diffusivity of acetone in water at 80° F is 4.88×10^{-5} sq ft/hr.

Estimate the fraction of acetone which is recovered in the liquid phase.

2. The table gives data [45] on the desorption of oxygen from water into air. Estimate the resistance of the gas phase to mass transfer relative to the total resistance to transfer at a liquid rate of 10,000 lb/(hr)(sq ft).

Run No.	G_l, lb/(hr)(sq ft)	G_g, lb/(hr)(sq ft)	Temperature, °C	K_La, lb moles/(hr)(cu ft)(lb moles/cu ft)
13.2 in. of dumped 0.5 in Berl saddles				
58	16,000	100	25.0	154
61	27,500	100	25.0	148
64	8,000	100	25.0	104
67	500	100	25.0	16.7
68	2,000	100	25.0	42

3. A feed mixture contains 25 mass per cent ethanol, the balance water. It is desired to batch distill 8000 gal of this mixture through a tower 2 ft in diameter, packed to a depth of 15 ft with dumped 1-in. Raschig rings, to recover 95.0 per cent of the ethanol as a distillate product containing 90.0 mass per cent ethanol. The operation will be conducted at 1 atm pressure under conditions of constant distillate composition. The liquid phase diffusivity for ethanol-water at 180° F is 2.35×10^{-8} sq ft/sec.

Is there sufficient packing in the tower for this fractionation?

4. An absorber oil having a molecular weight of 300 contains 2.54 mole per cent propane. This oil is to be stripped to a propane content of 0.05 mole per cent by superheated steam. The stripper will be operated at 20 psia and a temperature of 280° F will be maintained throughout the stripper by internal heating. Four moles of steam will be used per 100 moles of oil (propane-free) stripped.

Estimate the height of a packed tower containing 1-in. Raschig rings required to accomplish this separation.

Equilibrium relations of propane between steam and oil at 280° F and 20 psia are expressed by the equation $y = 33.4x$, where y = mole fraction of propane in steam and x = mole fraction of propane in oil. The diffusivity of propane in the absorber oil at 86° F is 2.0×10^{-8} sq ft/sec, the oil having a viscosity of 1.0 centipoises. The viscosity of the oil at 280° F is 0.29 centipoises.

5. Ammonia is produced at a rate of 7240 grams/liter of catalyst per hour at 300 atm and 380° C, in a 1:3 molal mixture of N_2 and H_2 entering a bed of a granular iron catalyst. The gas flows through the catalyst at a rate of 600 lb/(sq ft)(hr). The catalyst is in the form of cylindrical pellets ¼ in. by ¼ in., and the void in the bed is 38 per cent.

Estimate the temperature and partial pressure of nitrogen at the surface of the catalyst at the entrance of the bed.

6. A liquid mixture containing 35.0 mole per cent ethanol and 65.0 mole per cent benzene is to be extracted with pure water to produce a raffinate product (benzene-rich phase) containing 1.0 mole per cent ethanol, and an extract product containing 35.0 mole per cent ethanol and 59 mole per cent water. The extraction operation is to be conducted isothermally at 25° C in a column packed with 1.0-in. Raschig rings, the benzene being the dispersed phase. The mass velocity

of the feed (35 per cent EtOH, 65 per cent C_6H_6) will be 500 lb/(hr)(sq ft) of column cross section.

The pertinent data on the physical properties and equilibria of the saturated solutions are given in the Appendix. The viscosities of the saturated phases at 25° C may be assumed constant at 0.9 and 0.75 centipoise for the water-rich and benzene-rich phases, respectively. The diffusivity of ethanol at 25° C in the water-rich and benzene-rich phases may be assumed constant as 3.9×10^{-5} and 5.8×10^{-5} sq ft/hr, respectively. The fugacity of ethanol in the feed is 46.0 mm of mercury.

Estimate the height of packing required.

7. Dry air at 100° F is to be preheated to 2180° F in a "pebble heater." The pebble heater consists essentially of a cylindrical shaft having an inside diameter of 19 in. Cold air enters at the bottom. Hot refractory pebbles at 2230° F enter at the top and move downward countercurrent to the air. The air rate is 1800 lb/hr. The pebbles are ½-in. diameter spheres with 86.5 sq ft of surface per cubic foot of packed volume. The average specific heat is 0.27 Btu/(lb)(°F). The average specific heat of air is 0.267 Btu/(lb)(°F). $\mu = 0.095$ lb/(ft)(hr), and $\mu C_P/k = 0.65$. Determine the effective height of shaft required if heat losses may be considered to be negligible, and pebble rate is 2150 lb/hr.

8. A waste gas from an amination process contains 500 lb/hr of ammonia and 50 lb/hr of hydrogen. The gas is saturated with water vapor at 1 psig and 140° F. An absorbing stripping operation using water in columns packed with 1-in. Raschig rings has been proposed to recover the ammonia for re-use in the process. The loss of ammonia should not exceed 3 lb/hr, and the ammonia should be recycled to the amination process as a gas at 50 psia and should not contain more than 0.1 mole fraction of water. Cooling water is available at 80° F.

Prepare a flow diagram specifying all operating conditions, i.e., temperatures, pressures, flow rates, compositions, energy quantities, and dimensions of absorber and stripper.

9. Five hundred pounds per hour of pure propane are to be condensed from and at 70 psia in a shell and tube exchanger. The refrigerant is ammonia supplied by an absorption refrigeration process. A drop in pressure of 2 psi is permissible in either or both propane and ammonia streams.

(*a*) What condenser area is required if the tubes are to be ¾-in. OD and 8 ft in length?

(*b*) Recommend a flow diagram for the absorption refrigeration unit.

(*c*) What are the operating conditions and sizes of major items of equipment in the refrigeration unit?

CHAPTER

37

Simultaneous Heat and Mass Transfer 1

Psychrometry

PSYCHROMETRY refers mainly to the systematized knowledge of the properties of air and water, but its broader meaning includes the principles which govern the properties and behavior of mixtures of the so-called "fixed" gases with condensable vapors.

The system of air and water is important, not only with regard to air conditioning for physiological comfort, but also with regard to water cooling and drying. There are frequent opportunities to apply psychrometric principles to other systems in connection with problems of solvent recovery and the removal of organic vapors. Examples of this are in the dry-cleaning industry where recovery of the Stoddard's solvent, carbon tetrachloride, and other valuable materials is essential for economical operation. Again there are cases where it is desired to recover organic vapors but at the same time to prevent air from coming in contact with these vapors because of the explosion and fire hazard. In such cases the carrier gas may be nitrogen, or a flue gas, instead of air. All these operations are concerned with the vapor-carrying capacity of the carrier gas and with the engineering variables which affect this capacity.

DEFINITIONS

The following definitions are consistent and avoid the present confusion in psychrometric terminology.

The *mixing ratio* or mass ratio Y expresses the mass of one component carried by unit mass of the vapor-free gas.

The *humidity ratio* $\mathcal{H}$ expresses the mass of water vapor carried by unit mass of dry air. The same symbol, $\mathcal{H}$, may be used to express the mass ratio for other systems by special designation as "ethanol humidity."

The term *humidity* is normally restricted to water vapor in air.

The humidity of any ideal gas is related directly to the partial pressure of the vapor being carried, if

p = partial pressure of the vapor.
P = total pressure of the gas-vapor mixture.
M_v = molecular weight of the vapor.
M_g = molecular weight of the gas.

$$\mathcal{H} = \left(\frac{p}{P - p}\right)\left(\frac{M_v}{M_g}\right) \qquad (571)$$

The *saturated humidity* $\mathcal{H}_s$ of a gas-vapor mixture expresses the pounds of vapor carried by 1 lb of vapor-free gas when the vapor is in equilibrium with liquid (condensed vapor) or solid (as ice) at the temperature and pressure of the gas.

$$\mathcal{H}_s = \left(\frac{p_s}{P - p_s}\right)\left(\frac{M_v}{M_g}\right) \qquad (572)$$

where p_s = saturated partial pressure or vapor pressure of the condensed vapor.

The *relative saturation* $\mathcal{S}$ is the ratio, usually expressed as a percentage, of the pounds of vapor carried by 1 lb of dry carrier gas at any given conditions to the pounds of vapor carried by 1 lb of dry gas when in equilibrium with water or other liquid under the same conditions of temperature and pressure.

$$\mathcal{S} = \frac{\mathcal{H}}{\mathcal{H}_s} = \frac{\left(\frac{p}{P - p}\right)\left(\frac{M_v}{M_g}\right)}{\left(\frac{p_s}{P - p_s}\right)\left(\frac{M_v}{M_g}\right)} = \frac{\frac{p}{P - p}}{\frac{p_s}{P - p_s}} \quad (573)$$

The *relative partial pressure* is the ratio of the partial pressure of the vapor to the vapor pressure of the liquid at the same temperature.

$$\text{Relative partial pressure} = \frac{p}{p_s} \quad (574)$$

In the past the relative partial pressure and the relative saturation have both been referred to as relative humidity, and care must be taken to avoid confusion with relative saturation.

The relationship between relative partial pressure and relative saturation is indicated as follows:

$$\text{Relative partial pressure} = \mathcal{S}\left(\frac{P - p}{P - p_s}\right) \quad (575)$$

The *dry volume* V_g is the volume in cubic feet occupied by 1 lb of the carrier gas. For ideal gases,

$$V_g = \frac{(359)T}{PM_g(492)} = \frac{0.73T}{PM_g} \quad (576)$$

where T = temperature [°R,(°F + 460)].
P = pressure (atm).

The *humid volume* $V_{\mathcal{H}}$ is the volume of moist gas per unit mass of dry gas. It is the volume occupied by 1 lb of the carrier gas and the vapor carried by it. For ideal gases,

$$V_{\mathcal{H}} = V_g + \mathcal{H}\left(\frac{0.73T}{PM_v}\right) \quad (577)$$

$$= \frac{0.73T}{P}\left(\frac{1}{M_g} + \frac{\mathcal{H}}{M_v}\right)$$

The *saturated volume* V_s is the volume occupied by 1 lb of the carrier gas and the vapor which saturates it.

$$V_s = \frac{0.73T}{P}\left(\frac{1}{M_g} + \frac{\mathcal{H}_s}{M_v}\right) \quad (578)$$

The *humid heat* s of a gas-vapor mixture is the energy required to raise the temperature of 1 lb of the carrier gas and its accompanying vapor, 1° F at constant pressure. Thus, if

C_g = heat capacity of the carrier gas at the stated pressure (Btu/lb °F),
C_v = heat capacity of the vapor at the stated pressure (Btu/lb °F),

$$s = C_g + \mathcal{H}C_v \quad (579)$$

since the specific heats of the gas and vapor are additive.

The *enthalpy of the humid gas mixture* $\mathcal{H}$ expressed as Btu per pound of dry carrier gas is defined as follows, where the reference state for the vapor is the pure liquid at the reference temperature.

$$H = C_g(T - T_r) + \mathcal{H}[C_v(T - T_r) + \lambda_r] \quad (580)$$

where T = temperature,
T_r = reference temperature,
λ_r = latent heat of vaporization of the vapor at temperature T_r,

or

$$H = s(T - T_r) + \mathcal{H}\lambda_r$$

HUMIDITY CHART

The most convenient method of handling psychrometric relationships makes use of the so-called humidity chart, which is simply a plot of the properties of the mixtures under consideration at a specified total pressure, usually one atmosphere. Figure 498 shows such a plot for the system air and ethyl alcohol, and Fig. 499 is a humidity chart for air and water. As indicated, the following properties are usually shown.

1. The humidity $\mathcal{H}$ as pounds of vapor per pound of dry gas is plotted against temperature with lines of constant relative saturation. This relationship is shown by the family of curves beginning in the lower left-hand corner of the plot and sloping upwards to the right, each curve being labeled according to its relative saturation (equation 573).

2. The specific volume of dry and saturated carrier gas expressed as cubic feet per pound of dry carrier gas is plotted as a function of temperature in degrees Fahrenheit. Only the two curves for the dry volume (equation 576) and the saturated volume (equation 578) are usually shown. Specific volumes of gas-vapor mixtures which are unsaturated may be evaluated by a linear interpolation between these lines. These curves are labeled dry volume (or specific volume) and saturated volume, respectively.

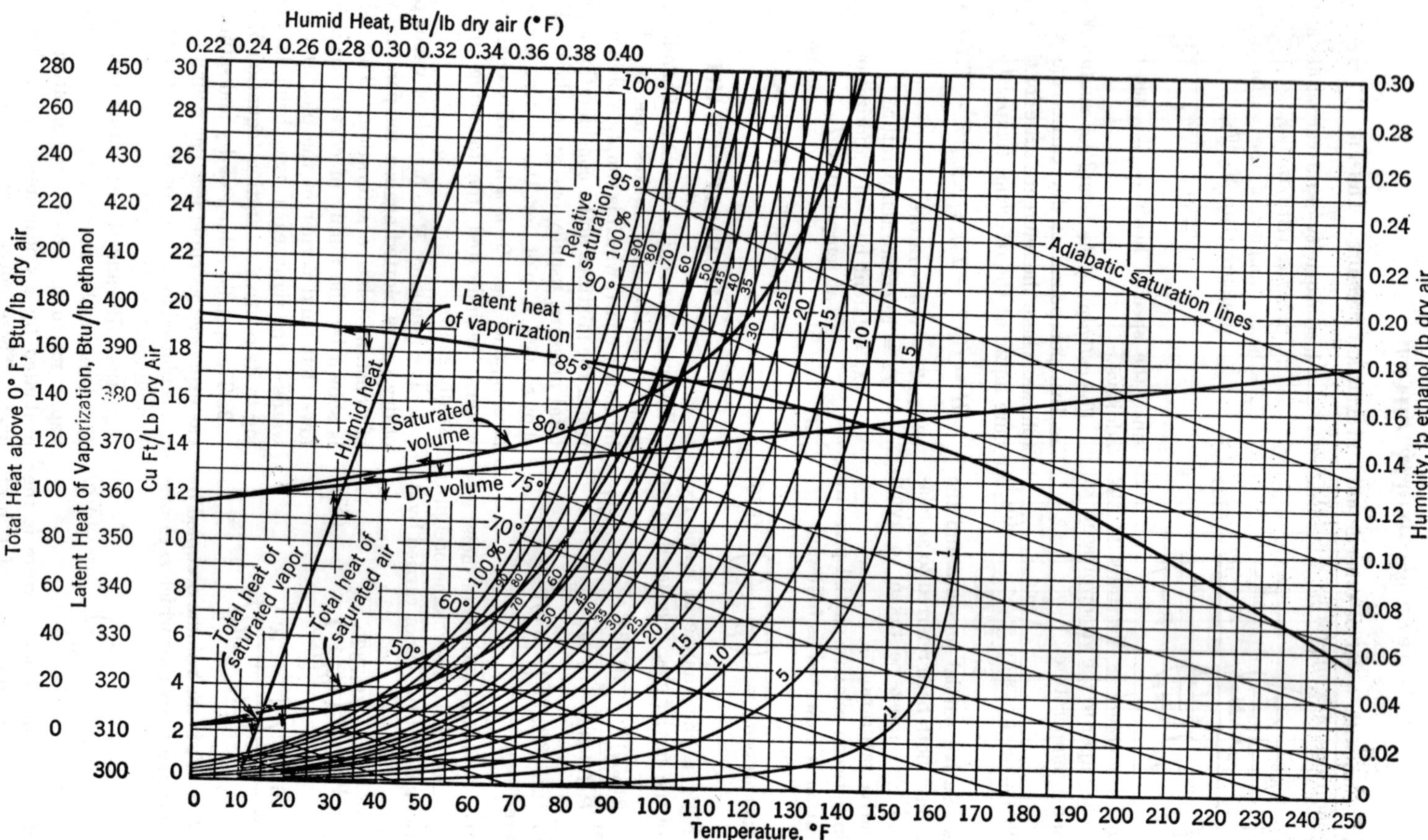

FIG. 498. Humidity chart for the air-ethanol system.[6]

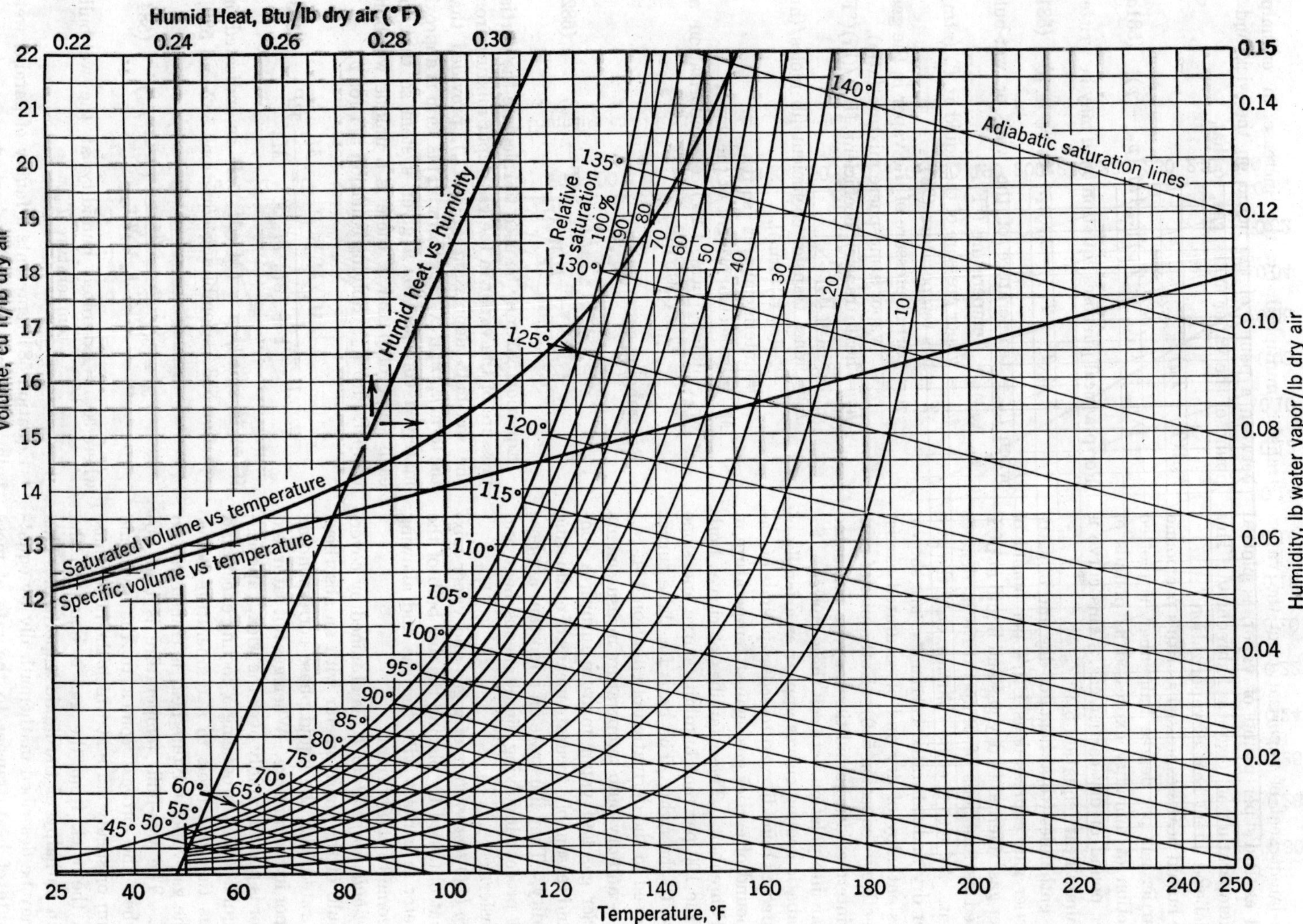

Fig. 499. Humidity chart for the air-water system[25]

3. The latent heat of vaporization of the vapor, expressed as Btu's per pound of vapor, is plotted against temperature in degrees Fahrenheit. This curve is labeled latent heat of vaporization.

4. The humid heat s expressed as Btu's per pound of dry carrier gas per degree Fahrenheit is plotted as a function of humidity expressed as pounds of vapor per pound of dry carrier gas. This curve is labeled humid heat (equation 579).

5. The enthalpies (total heat) of dry and of saturated carrier gas, expressed as Btu's per pound of dry carrier gas, referred to dry gas and liquid at 0° F are plotted as functions of temperature in degrees Fahrenheit. As with specific volumes, only the two curves for dry and saturated carrier gas are shown, labeled as enthalpy of dry and of saturated carrier gas, respectively (equation 580). Enthalpies of mixtures of intermediate humidities may be evaluated by a linear interpolation between these curves.

6. Adiabatic cooling curves appear as a family of lines plotted as humidity expressed as pounds of vapor per pound of carrier gas against temperature in degrees Fahrenheit. These families of curves both begin at the 100 per cent humidity curve and slope downward to the right. The basis for these curves was discussed in the derivation of equation 474.

The wet- and dry-bulb temperature measurements of a carrier gas and vapor mixture furnish a convenient and reasonably accurate means of evaluating the humidity. They illustrate an important application of a process involving simultaneous heat and mass transfer.

The *dry-bulb temperature* of a gas and vapor mixture is that obtained by placing the dry bulb of the thermometer in contact with the gas and allowing them to come to equilibrium.

The *wet-bulb temperature* T_w is obtained by covering the bulb of the thermometer with an absorbent wick saturated with the liquid phase corresponding to the vapor in the mixture. When gas not saturated with vapor is passed rapidly over the wick, vaporization of liquid occurs, producing a cooling effect, and as soon as the wick cools to a temperature below that of the gas-vapor mixture heat is transferred to the wick. The wet-bulb temperature is that obtained when the process comes to the dynamic equilibrium or steady state at which the heat transferred to the wick from the gas is equal to the increase in enthalpy of the liquid vaporized. This process may be described mathematically by equating the rate of heat transfer to the rate of mass transfer multiplied by the increase in enthalpy between a pound of the liquid on the wick and a pound of the vapor in the gas, so that

$$h_g A(T - T_w) = [\lambda_{T_w} + C_v(T - T_w)][k_g A M_v(p_w - p)] \quad (581a)$$

For practical purposes equation 581*a* may be written

$$h_g A(T - T_w) = \lambda_{T_w} k_g A M_v(p_w - p) \quad (581)$$

when T_w = temperature of the wick or wet-bulb temperature (°F).
p_w = vapor pressure of the liquid at T_w (atm).
T = gas temperature (°F).
p = partial pressure of the vapor in the gas vapor atmosphere mixture (atm).
h_g = heat transfer coefficient [Btu/(hr)(°F)(sq ft)].
k_g = mass transfer coefficient [lb moles/(hr)(sq ft)(atm)].
A = interfacial area (sq ft).
M_v = molecular weight of the vapor.
λ_{T_w} = latent heat (ΔH) of vaporization at T_w (Btu/lb).

Rearranging equation 581

$$p = p_w - \frac{h_g}{k_g M_v \lambda_{T_w}}(T - T_w) \quad (582)$$

Equation 582 may be used to calculate the partial pressure of the vapor in the gas-vapor mixture from the wet-bulb depression $(T - T_w)$, provided that the term $h_g/k_g M_v \lambda_{T_w}$ is known. This term is approximately a constant for a given system so that for each value of p and T there is a unique wet-bulb temperature T_w. Solving equation 571 for p,

$$p = \frac{(M_g/M_v)\mathcal{H}P}{1 + (M_g/M_v)\mathcal{H}} \approx \frac{M_g}{M_v}\mathcal{H}P \quad (583)$$

If $\mathcal{H}$ is small, $p \approx (M_g/M_v)\mathcal{H}P$ as indicated in equation 583. Combining equations 582 and 583,

$$\mathcal{H} = \mathcal{H}_w - \frac{h_g}{k_g M_g P \lambda_{T_w}}(T - T_w) \quad (584)$$

where $\mathcal{H}_w$ = saturated humidity at the wet-bulb temperature T_w.

Equation 584 represents a family of parallel wet-bulb curves on the humidity chart which may be used to determine humidity from wet- and dry-bulb temperature measurements in the same way that the adiabatic cooling lines are used. Knowing the value of T_w and consequently $\mathcal{H}_w$ on the saturation line, the wet-bulb line is followed downward and to the right to the intersection of the ordinate corresponding to T. This locates the humidity $\mathcal{H}$.

In the specific case of air and water at temperatures up to about 150° F, the wet-bulb curves are practically coincident with the adiabatic cooling curves because the term h_g/k_gM_gP is practically equal to the term s in the adiabatic cooling equation 474. Under these circumstances, the adiabatic cooling curves may be used as wet-bulb lines.

The wet-bulb temperature of the mixture represented by the point A in Fig. 500 is T_{w1} as given by

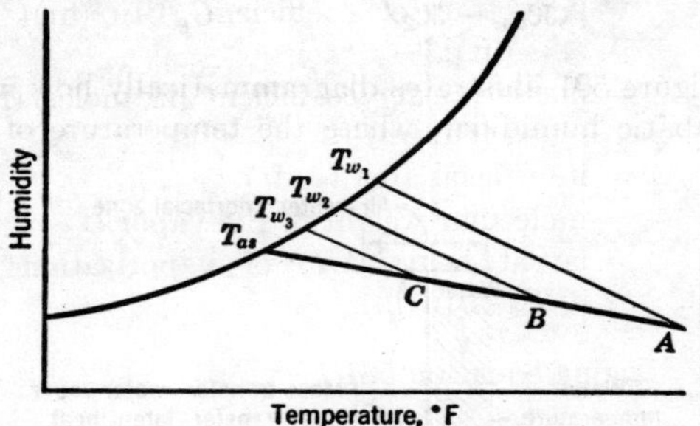

Fig. 500. Diagram showing construction of adiabatic lines on humidity chart.

equation 584. If the mixture represented by the point A is fed to an adiabatic humidifier, its properties will follow the path AT_{as} as given by equation 474. When the mixture has reached the condition represented by the point B, its wet-bulb temperature will have decreased along the saturation curve to T_{w2}, and at point C the wet-bulb temperature will have decreased to T_{w3}. Since the wet-bulb temperature and the adiabatic saturation temperature are both on the saturation curve, when the original mixture becomes saturated, its wet-bulb temperature becomes equal to the adiabatic saturation temperature.

There are relatively few experimental data on wet-bulb temperature for systems other than air and water. These data are summarized in Table 59.

The values of h_g/k_gM_gP for systems where no experimental data are available can be estimated from the j factor analogy between heat and mass transfer. According to equations 504a and 505a where heat and mass transfer occur across the same path,

$$j = \frac{k_gM_{avg}P}{G_g}\left(\frac{\mu}{\rho D_G}\right)^{2/3} = \frac{h_g}{C_PG_g}\left(\frac{\mu C_P}{k_q}\right)^{2/3} \qquad (585)$$

In the equation for the j factor the average molecular weight of the flowing stream M_{avg} is used. If the humidity $\mathcal{H}$ is small, the average molecular weight of the flowing stream is approximately equal to M_g, and equation 585 may be simplified as

$$\frac{h_g}{k_gM_gP} = C_P\left(\frac{\mu/\rho D_G}{C_P\mu/k_q}\right)^{2/3} \qquad (586)$$

Values calculated from equation 586 are shown in the last column of Table 59.[43] *

TABLE 59. SUMMARY[43] OF VALUES OF h_g/k_gM_gP CALCULATED FROM WET-BULB DETERMINATIONS IN AIR *

Vapor	Mark[35]	Sherwood and Comings[46]	Awbery and Griffiths[3]	Hilpert[29]	Arnold[1] *	Calculated from j Factor Analogy
Benzene	0.41	0.40	....	0.49	....	0.44
Brombenzol	....	....	....	0.46	....	0.47
Carbon tetrachloride	0.44	0.50	....	0.50	....	0.49
Chlorobenzene	0.44	....	....	0.51	....	0.48
Ethyl acetate	0.42	....	....	....	....	0.46
Ethylene bromide	....	....	....	0.53	....	0.47
Ethylene tetrachloride	0.50	....	....	....	....	0.51
Ethyl propionate	....	....	....	....	0.50	0.46
Methyl alcohol	....	....	....	....	0.35	0.31
n-Propyl acetate	....	0.52	....	....	....	0.49
Propyl alcohol	....	....	....	....	0.43	0.41
Toluene	0.44	0.50	....	0.46	0.50	0.47
Water	0.26	0.29	0.36	....	0.27	0.21

* The values tabulated have not been corrected for radiation from the surroundings, since in most cases the conditions of the tests were not sufficiently well described to make this correction with any accuracy. Mark's data[35] were obtained with high air velocities, and the correction is negligible; the radiation correction would reduce Arnold's experimental values[1] by 12 to 16 per cent.

THE INTERACTION OF AIR AND WATER

The evaporation of water into air for the purpose of increasing the air humidity is known as humidification. Closely allied to this is the evaporation of water into air for the purpose of cooling the water. Dehumidification consists of condensing water from air to decrease the air humidity. All these processes are of considerable industrial importance and involve the contacting of air and water accompanied by heat and mass transfer.

* The bibliography for this chapter appears on p. 575.

Consider a packed tower to which air at a temperature T_{g1} and humidity $\mathcal{H}_1$ is fed countercurrent to water entering at temperature T_{l2}. The air leaves the contactor at a temperature T_{g2} and humidity $\mathcal{H}_2$, and the water at a temperature T_{l1}.

A material balance over a differential element of height dZ gives

$$dG_l = G_g' \, d\mathcal{H} \qquad (587)$$

where G_l = mass velocity of water [lb/(hr)(sq ft)].
G_g' = mass velocity of dry air [lb/(hr)(sq ft)].

An enthalpy balance, assuming no heat transfer between the tower and its surroundings, gives

$$G_l C_l \, dT_l = G_g' s \, dT_g + G_g' \lambda \, d\mathcal{H} \qquad (588)$$

The rate of heat transfer between the air stream and the interface between the air and water is given by

$$-G_g' s \, dT_g = h_g a \, dZ \, (T_g - T_i) \qquad (589)$$

and the rate of heat transfer between the liquid stream and the interface is given by

$$C_l G_l \, dT_l = h_l a \, dZ \, (T_l - T_i) \qquad (590)$$

Finally, the rate of mass transfer between the air stream and the interface is given by

$$-G_g' \, d\mathcal{H} = k_g a M_v \, dZ \, (f_g - f_i) \qquad (591)$$

or, since the fugacity of the water in the air is practically proportional to the humidity for the system air and water

$$-G_g' \, d\mathcal{H} = k_{\mathcal{H}} a M_v \, dZ \, (\mathcal{H} - \mathcal{H}_i) \qquad (591a)$$

where $k_{\mathcal{H}}$ = mass transfer coefficient [(lb moles)/(hr)(sq ft)(lb water/lb dry air)].

Equations 587 and 591a are of general applicability to all processes where air and water are brought into countercurrent contact, the equipment being insulated from the surroundings, and operated under steady conditions.

Adiabatic Humidification

The operation of the adiabatic humidifier, where make-up water enters the system at the adiabatic saturation temperature, has been discussed in connection with the wetted-wall column. Under these conditions, the temperature of the water in the system is constant at the adiabatic saturation temperature, and the air temperature and humidity follow the adiabatic saturation curve whose equation is

$$T = \frac{(\mathcal{H}_{as} - \mathcal{H})}{s} \lambda_{T_{as}} + T_{as} \qquad (474)$$

For an adiabatic humidifier $dT_l = 0$, and equation 588 becomes $-dT_g/d\mathcal{H} = \lambda/s$, and, since $T_l = T_i = T_{as}$, equation 590 becomes indeterminate. Equations 587, 589, and 591a may be integrated separately to give

$$G_{l2} - G_{l1} = G_g'(\mathcal{H}_2 - \mathcal{H}_1) \qquad (592)$$

$$\ln \left(\frac{T_{g1} - T_{as}}{T_{g2} - T_{as}} \right) = \frac{h_g a (Z_2 - Z_1)}{G_g' s_{\text{avg}}} \qquad (593)$$

$$\ln \left(\frac{\mathcal{H}_{as} - \mathcal{H}_1}{\mathcal{H}_{as} - \mathcal{H}_2} \right) = \frac{k_{\mathcal{H}} a M_v (Z_2 - Z_1)}{G_g'} \qquad (594)$$

Figure 501 illustrates diagrammatically how in an adiabatic humidifier, where the temperature of the water remains constant, all the heat required to vaporize the water is supplied from the sensible heat of the air.

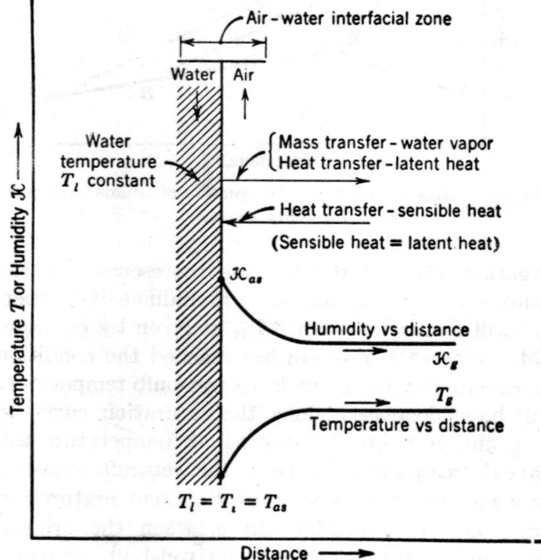

FIG. 501. Diagram representing conditions in an adiabatic humidifier.[4] The term adiabatic humidifier is applied to a humidifier in which the temperature of the circulating water is constant.

Where the make-up water enters an adiabatic humidification system at a temperature T_r different from the adiabatic saturation temperature, an energy balance around a section of the contactor gives equation 595 instead of equation 474 for infinite contacting surface.

$$T = \frac{(\mathcal{H}_{as} - \mathcal{H})[\lambda_{T_r} + C_v(T_{as} - T_r)]}{s} + T_{as} \quad (595)$$

Equation 595 indicates that the temperature T_{as} is no longer a function only of the entering gas temperature and humidity but is also dependent upon the temperature of the make-up liquid T_r. Thus, a family of "saturation" curves with T_r as a parameter results for a given entering air temperature and humidity instead of the single adiabatic saturation curve.

Usually T_r is not greatly different from T_{as}, so that the term $C_v(T_{as} - T_r)$ is negligible compared to the term λ_{T_r}. This means that the family of "saturation" curves with T_r as a parameter is ordinarily very closely spaced around the adiabatic saturation curve. Therefore, for practical purposes, the adiabatic saturation curve and equations 592, 593, and 594 for the adiabatic humidifier will give results well within the range of engineering accuracy for these cases. Where the temperature T_r is greatly different from T_{as}, it is necessary to use the methods described in the next section for design purposes.

Dehumidification

The decrease in both the temperature and humidity of air by contacting it with cold water is illustrated in Fig. 503.

In a dehumidifier both sensible and latent heat are transferred from the air to the water, resulting in an increase in the water temperature from the top to the bottom. The general relationships are illustrated diagrammatically in Fig. 502.

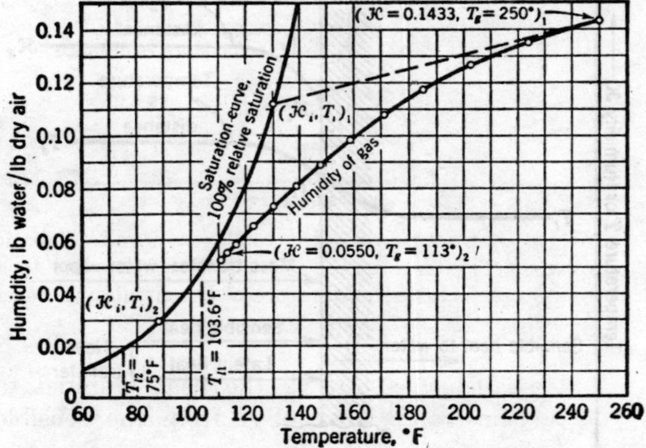

FIG. 502. Diagram of humidity chart indicating dehumidification by contact with cold water as computed in the illustrative example.

In the design of a dehumidifier, all the general equations (587 to 591a) are applicable simultaneously, and a trial-and-error numerical integration of the equations is necessary. This calculation is illustrated in the problem of estimating the height of the dehumidifier required to change the humidity of air at temperature T_{g1} from $\mathcal{H}_1$ to $\mathcal{H}_2$, with G_{l2} lb/(hr)(sq ft) of water at a temperature T_{l2}. The air rate is G_g' lb per hour of dry air per square foot of column cross section, and all the rate coefficients $h_g a$, $h_l a$, and $k_g a$ are known. The calculation may be carried as follows.

Step 1. The mass velocity of the liquid leaving is calculated from a material balance.

Step 2. An energy balance is made to determine the relationship between the exit temperatures T_{l1} and T_{g2}.

Step 3. The exit gas temperature T_{g2} is assumed, and the exit liquid temperature T_{l1} is calculated.

Step 4. By combining equations 588, 589, 592, and 591a,

$$h_l a(T_i - T_l) = h_g a(T_g - T_i) + k_{\mathcal{H}} a M_v \lambda(\mathcal{H} - \mathcal{H}_i) \quad (596)$$

which is used to compute the interface conditions T_i, $\mathcal{H}_i$ at any point in the column where values of T_g, T_l, and $\mathcal{H}$ are known or assumed. These values are initially available only at the top and bottom. Starting at the bottom the interface conditions are calculated.

Step 5. A second point on the curve representing humidity $\mathcal{H}$ as a function of gas temperature T_g may be obtained by assuming average values for the following derivatives over small increments.

$$\frac{dT_g}{dZ} = \frac{-h_g a(T_g - T_i)}{G_g' s} \quad (589)$$

$$\frac{dT_l}{dZ} = \frac{h_l a(T_l - T_i)}{G_l C_l} \quad (590)$$

$$\frac{d\mathcal{H}}{dZ} = \frac{k_{\mathcal{H}} M_v a(\mathcal{H}_i - \mathcal{H})}{G_g'} \quad (591a)$$

(The values of the above derivatives as computed for the bottom conditions may be used as a guide for a first approximation for the average values in computing values for T_g and T_l at $Z_0 + \Delta Z$.)

Step 6. Values for T_i and $\mathcal{H}_i$ are then computed at $Z_0 + \Delta Z$ by equation 596.

This process is repeated over successive increments until the air humidity $\mathcal{H}_1$ equals that specified. If, at the same time, the calculated entering air temperature T_{g1} agrees with that specified, the original assumed exit air temperature T_{g2} was correct. Otherwise, the entire process must be repeated.

Illustrative Example. A mixture of air and water vapor at 250° F, containing 0.1433 lb of water vapor per pound of dry air, is to be dehumidified by countercurrent contact with water in a packed tower. The tower is to be packed with 1-in. Raschig rings and is to be operated at atmospheric pressure. The temperature of the water entering the tower is 75° F. The air leaving the tower is to contain 0.0550 lb of water per pound of dry air.

The mass velocity of the water at the top of the tower is to be 2000 lb/(hr)(sq ft), and that of the air-water vapor mixture at the bottom of the tower is to be 500 lb/(hr)(sq ft). For these rates of flow the average values of the heat transfer coefficients are estimated [34a] to be

$$h_g a = 290 \text{ Btu/(hr)(cu ft)(°F)}$$

$$h_l a = 2620 \text{ Btu/(hr)(cu ft)(°F)}$$

Determine the temperature of the air leaving the tower and the height of packed section required.

Solution. The mass velocity of dry air is

$$G_g' = \frac{G_{g1}}{1 + \mathcal{H}_1} = \frac{500}{1.1433} = 437 \frac{\text{lb dry air}}{\text{(hr)(sq ft)}}$$

Step 1. The mass velocity of the liquid leaving the tower is calculated from an overall material balance (equation 587).

$$G_g'(\mathcal{H}_1 - \mathcal{H}_2) = G_{l1} - G_{l2}$$

$$G_{l1} = 437(0.1433 - 0.0550) + 2000 = 2036 \text{ lb/(hr)(sq ft)}$$

Step 2. The humid heat (equation 579) for air-water vapor mixtures is

$$s = 0.24 + 0.45\mathcal{H}$$

At bottom, $s_1 = 0.304$ At top, $s_2 = 0.265$

Average $s = 0.285$

$\lambda_0 = 1094$ Btu/lb (obtained by extrapolation of the latent heat of vaporization of liquid water to 0° F).

The overall energy balance using reference states of liquid water at 0° F, air at 0° F, is

$$G_g'(s_1 T_{g1} + \lambda_0 \mathcal{H}_1) + G_{l2} C_l T_{l2} = G_g'(s_2 T_{g2} + \lambda_0 \mathcal{H}_2) + G_{l1} C_l T_{l1}$$

$$437[(0.304)(250) + (1094)(0.1433)] + 2000(1)(75)$$

$$= 437[(0.265)(T_{g2}) + (1094)(0.0550)] + 2036(1)T_{l1}$$

Therefore,

$$T_{l1} = 110.0 - 0.0570 T_{g2}$$

Step 3. The first trial is based on an assumed temperature for the outlet gas T_{g2} of 111° F. Then $T_{l1} = 103.7°$ F.

Step 4. For the particular system of water and air only, the numerical value of the humid heat,

$$s = h_g / k\mathcal{H} M_v \tag{597}$$

Substituting equation 597 in equation 596,

$$2620(T_i - T_l) = 290(T_g - T_i) + \frac{290}{s}(\mathcal{H} - \mathcal{H}_i)(1094)$$

Using an average value of 0.285 for s, this equation reduces to

$$10.04T_i + 3840\mathcal{H}_i = 9.04T_l + T_g + 3840\mathcal{H}$$

For the conditions at the bottom of the tower, using the saturation curve of Fig. 499 or Fig. 502 for values of T_i and $\mathcal{H}_i$, by trial and error $T_i = 130.1°$ F and $\mathcal{H}_i = 0.1118$.

Step 5.

$$\frac{dT_g}{dZ} = -\frac{290(T_g - T_i)}{437(0.285)} = -2.35(T_g - T_i) \tag{589}$$

$$= -282 \quad \text{at bottom of tower}$$

$$\frac{dT_l}{dZ} = \frac{2620(T_l - T_i)}{2018(1)} = -1.299(T_i - T_l) \tag{590}$$

$$= -34.3 \quad \text{at bottom of tower}$$

$$\frac{d\mathcal{H}}{dZ} = \frac{290(\mathcal{H}_i - \mathcal{H})}{0.285(437)} = -2.35(\mathcal{H} - \mathcal{H}_i) \tag{591a}$$

$$= -0.0740 \text{ at bottom}$$

Step 6. Table 60 gives the average values for the derivatives used over different increments in tower height, together with the results obtained from the calculations. The desired

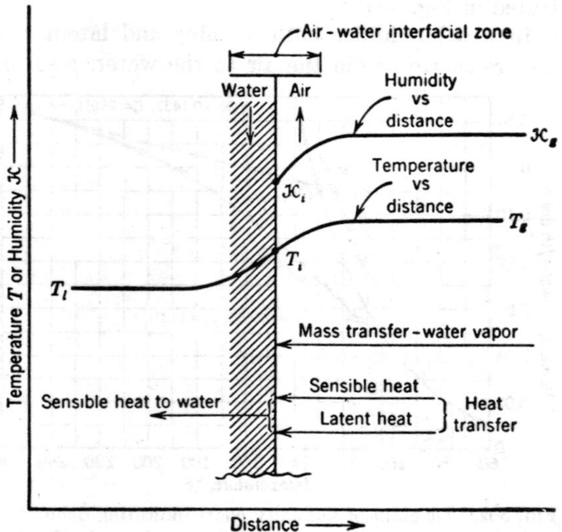

Fig. 503. Diagram illustrating conditions in a dehumidifier.

TABLE 60. SUMMARY OF CALCULATIONS FOR ILLUSTRATIVE EXAMPLE

First Trial: Outlet Gas Temperature, T_{g2}, Assumed to be 111° F, T_{l1} Calculated from Energy Balance, 103.7° F

Z ft	T_g °F	$\mathcal{H}$	T_l °F	T_i °F	$\mathcal{H}_i$	$\frac{dT_g}{dZ}$	Avg $\frac{dT_g}{dZ}$	$\frac{dT_l}{dZ}$	Avg $\frac{dT_l}{dZ}$	$\frac{d\mathcal{H}}{dZ}$	Avg $\frac{d\mathcal{H}}{dZ}$
0	250.0	0.1433	103.7	130.1	0.1118	−282		−34.3		−0.0740	
							−256		−34.0		−0.0794
0.1	224.4	0.1354	100.3	126.3	0.0993	−230		−33.7		−0.0848	
							−210		−33.3		−0.0882
0.2	203.4	0.1266	97.0	122.3	0.0876	−190		−32.9		−0.0915	
							−175		−32.3		−0.0931
0.3	185.9	0.1173	93.8	118.2	0.0770	−159		−31.7		−0.0946	
							−147		−31.0		−0.0948
0.4	171.3	0.1078	90.7	114.0	0.0674	−135		−30.3		−0.0950	
							−125		−29.5		−0.0938
0.5	158.8	0.0984	87.7	109.7	0.0590	−115		−28.6		−0.0926	
							−107		−27.7		−0.0907
0.6	148.1	0.0893	84.9	105.5	0.0515	−100		−26.8		−0.0888	
							−93.5		−25.9		−0.0858
0.7	138.7	0.0807	82.3	101.5	0.0454	−87.0		−25.0		−0.0829	
							−82.5		−23.8		−0.0801
0.8	130 5	0.0727	79.9	97.4	0.0398	−78.0		−22.7		−0.0774	
							−73.4		−21.9		−0.0739
0.9	123.2	0.0653	77.7	93.9	0.0354	−68.8		−21.1		−0.0703	
							−65.1		−20.0		−0.0671
1.0	116.7	0.0586	75.7	90.2	0.0314	−62.3		−18.8		−0.0638	
							−59.1		−17.9		−0.0606
1.1	110.8	0.0526	73.9	87.0	0.0282	−55.9		−17.0		−0.0573	

humidity ($\mathcal{H}$ = 0.0550) is reached when the outlet temperature T_{g2} is 113.3° F, as compared to the assumed value of 111° F.

The calculations for the second trial, based on an assumed value of T_{g2} = 113° F (T_{l1} = 103.6° F), are summarized in Table 61. Figure 502 shows the saturation curve ($\mathcal{H}_i$ versus T_i) and the curve of $\mathcal{H}$ versus T_g for this trial. In this case, the value of T_{g2} corresponding to the desired humidity is 113.4° F. This may be considered as a satisfactory check of the assumed value. The required height of packed section is between 1.0 and 1.1 ft.

Comparison of the calculations summarized in Tables 60 and 61 shows that only minor variations occur between the two, indicating that the results are not sensitive to small variations between the assumed and calculated values of the outlet gas temperature.

A similar graphical procedure which is more convenient has been described.[36a] Unfortunately the transfer rates required for these calculations are not generally available, and most commercial units are sized on the basis of previous operating experience as indicated in the following discussion of cooling towers.

TABLE 61. SUMMARY OF CALCULATIONS FOR ILLUSTRATIVE EXAMPLE

Second Trial: Outlet Gas Temperature, T_{g2}, Assumed to be 113° F, T_{l1} Calculated from Energy Balance, 103.6° F

Z ft	T_g °F	$\mathcal{H}$	T_l °F	T_i °F	$\mathcal{H}_i$	$\frac{dT_g}{dZ}$	Avg $\frac{dT_g}{dZ}$	$\frac{dT_l}{dZ}$	Avg $\frac{dT_l}{dZ}$	$\frac{d\mathcal{H}}{dZ}$	Avg $\frac{d\mathcal{H}}{dZ}$
0	250.0	0.1433	103.6	130.0	0.1116	−282		−34.3		−0.0745	
							−256		−33.8		−0.0805
0.1	224.4	0.1352	100.2	126.0	0.0984	−231		−33.3		−0.0864	
							−211		−33.0		−0.0892
0.2	203.3	0.1263	96.9	122.1	0.0871	−191		−32.7		−0.0921	
							−175		−32.1		−0.0938
0.3	185.7	0.1169	93.7	117.9	0.0763	−159		−31.4		−0.0955	
							−147		−30.7		−0.0952
0.4	171.0	0.1074	90.6	113.8	0.0670	−134		−30.0		−0.0950	
							−125		−29.2		−0.0940
0.5	158.7	0.0980	87.7	109.6	0.0587	−115		−28.4		−0.0924	
							−107.5		−27.4		−0.0904
0.6	148.0	0.0890	85.0	105.4	0.0514	−100		−26.4		−0.0884	
							−93.5		−25.5		−0.0856
0.7	138.6	0.0804	82.4	101.4	0.0452	−87.0		−24.6		−0.0827	
							−82.0		−23.6		−0.0795
0.8	130.4	0.0724	80.1	97.5	0.0399	−77.0		−22.7		−0.0763	
							−73.0		−21.6		−0.0732
0.9	123.1	0.0651	77.9	93.8	0.0353	−69.0		−20.6		−0.0700	
							−65.5		−19.6		−0.0667
1.0	116.6	0.0584	75.9	90.2	0.0314	−62.0		−18.5		−0.0634	
							−59.0		−17.6		−0.0601
1.1	110.7	0.0524	74.1	87.0	0.0282	−56.0		−16.7		−0.0568	

COOLING TOWERS

In cooling towers a stream of hot water is cooled by exchange of sensible and latent heat and water vapor with a stream of relatively cool, dry air. The basic relationships developed for dehumidifiers also apply to cooling towers although the transfer is in the opposite direction, the tower being a humidifier rather than a dehumidifier of air.

The operation of a cooling tower is indicated diagrammatically on the humidity chart in Fig. 504. The hot water at T_{l2} is introduced at the top of the tower and leaves the bottom at T_{l1}. The air flows countercurrent to the water, entering the bottom at conditions indicated by point 1 and leaving the top

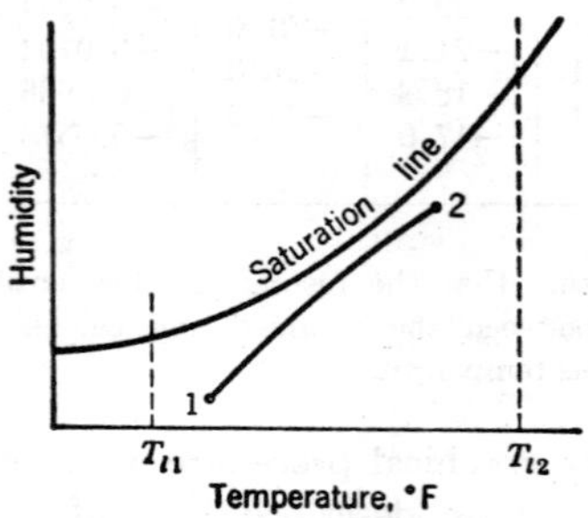

Fig. 504. Diagram of humidity chart illustrating operation of a cooling tower.[52]

at conditions corresponding to point 2, following the path shown.

As in a humidifier, the temperature of the interface tends to approach the adiabatic saturation or wet-bulb temperature of the air. At the top of the tower (Fig. 505) heat is being transferred from the hot water to the air, the temperature of the water is higher than that of the interface, and the interface temperature is usually higher than that of the air. This sensible heat removed from the water appears as sensible and latent heat of the air-water mixture. At the bottom of the tower the temperature of the water and of the interface may both be lower than that of the air as shown in Fig. 506, with sensible heat being transferred both from the liquid and air to the interface where it is absorbed as latent heat in vaporizing water. Thus, water may be cooled by air at a higher temperature provided a humidity gradient which produces evaporation is maintained.

The general equations (587 to 591a) may be used to design cooling towers in the manner described for dehumidifiers. This procedure suffers from the disadvantages mentioned previously, lack of data and laborious trial-and-error calculation. It is customary in practice to estimate the size of cooling towers

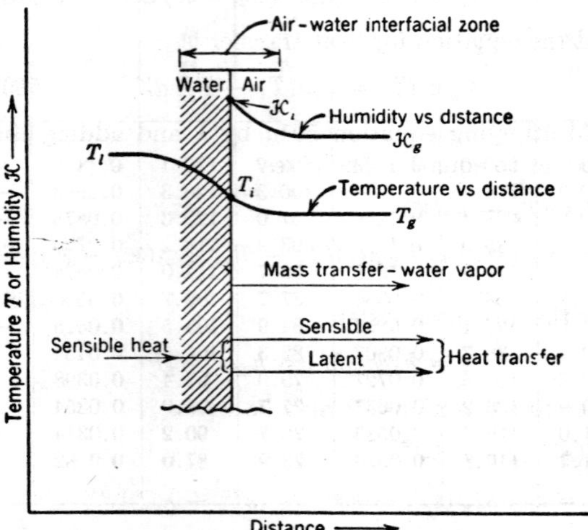

Fig. 505. Diagram representing conditions at the top of a cooling tower.[4]

according to empirical correlations of operating experiences.

An approximate method of handling cooling tower calculations which has shown some promise involves

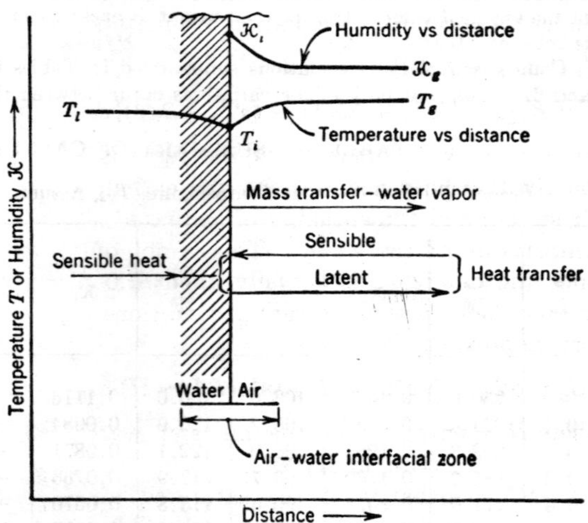

Fig. 506. Diagram representing conditions at the bottom of a cooling tower.[4]

the water temperatures and the enthalpy of the air passing through the cooling tower.[44] This relationship may be derived as follows.

The equation for mass transfer may be written

$$G_g'\,d\mathcal{H} = k_{\mathcal{H}}aM_v(\mathcal{H}_i - \mathcal{H})\,dZ \qquad (591a)$$

and the equation for heat transfer is

$$G_g's\,dT_g = h_g a(T_i - T_g)\,dZ \qquad (589)$$

Multiplying equation 591*a* by λ and adding the product to equation 589 gives

$$G_g'(s\,dT_g + \lambda\,d\mathcal{H}) = k_{\mathcal{H}}aM_v[(h_g/k_{\mathcal{H}}M_v)(T_i - T_g) + \lambda(\mathcal{H}_i - \mathcal{H})]\,dZ \qquad (598)$$

For the air-water system

$$s = h_g/k_{\mathcal{H}}M_v = h_g/k_gM_gP$$

Therefore equation 598 may be written

$$G_g'(s\,dT_g + \lambda\,d\mathcal{H}) = k_{\mathcal{H}}aM_v[(sT_i + \lambda\mathcal{H}_i) - (sT_g + \lambda\mathcal{H})]\,dZ \qquad (599)$$

If the interface temperature T_i is assumed to be equal to the temperature of the water T_l and if the humidity at the interface $\mathcal{H}_i$ is assumed to be equal to that of air in equilibrium with water, or $\mathcal{H}^*$, the overall transfer coefficient $K_{\mathcal{H}}$ is equal to $k_{\mathcal{H}}$. These ssumptions are equivalent to assuming that the liquid phase offers no significant resistance to heat or mass transfer relative to that in the gas phase. Under these conditions equation 599 may be written

$$G_g'(s\,dT_g + \lambda\,d\mathcal{H}) = K_{\mathcal{H}}aM_v[(sT_l + \lambda\mathcal{H}^*) - (sT_g + \lambda\mathcal{H})]\,dZ \qquad (600)$$

The quantity in brackets on the right is approximately the difference between the enthalpy of the air stream and the enthalpy of saturated air at the temperature of the water. This is not quite true, since the difference is actually about 7.0 per cent greater under actual operating conditions. Therefore, approximately

$$G_g'(s\,dT_g + \lambda\,d\mathcal{H}) = K_{\mathcal{H}}aM_v(H^* - H)\,dZ \qquad (601)$$

where H = enthalpy of the air and water vapor (Btu/lb dry air).

H^* = enthalpy of the saturated air at the water temperature (Btu/lb dry air).

and when equation 601 is combined with the enthalpy balance

$$C_l\,G_l\,dT_l = G_g's\,dT_g + G_g'\lambda\,d\mathcal{H} \qquad (588)$$

equation 602 results.

$$C_l\,G_l\,dT_l = K_{\mathcal{H}}aM_v(H^* - H)\,dZ \qquad (602)$$

The application of equation 602 is illustrated in Fig. 507 in which the upper curved line represents the enthalpy curve for saturated air plotted as a function of temperature. The lower curve is an operating line derived from the enthalpy balance

$$C_l\,G_l(T_l - T_{l1}) = G_g'(H - H_1) \qquad (603)$$

The vertical difference between the operating line and the saturation curve represents the potential producing cooling in the tower.

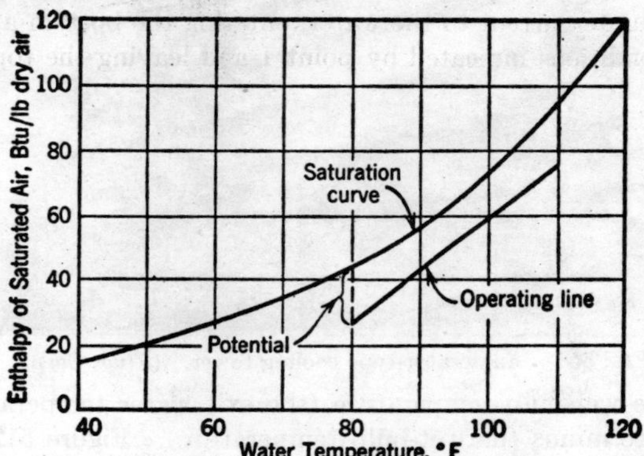

Fig. 507. Diagram illustrating the application of equation 604 to the design of a cooling tower.[44]

Although this method is approximate, it is simple and helpful in understanding the operation of towers. One consequence of these relations is the fact that, since enthalpy is dependent primarily on the wet-bulb temperature, the outlet water temperature is found to be independent of the dry-bulb temperature. This method presents the possibility of designing a tower on the basis of transfer coefficients rather than on empirical performance charts and may permit a comparison between the performance of cooling towers and similar equipment when used in absorption or stripping.

Natural Draft

Some idea of the construction and operation of natural-draft cooling towers may be gathered from Figs. 508, 509, and 510.

An empirical correlation [19,34] for estimating sizes and capacities of conventional louvre-type natural circulation towers is illustrated by the use of Figs. 511 to 516. Figure 511 shows the temperature drop produced in the hot water as a function of the tower capacity expressed as gallons of water per square foot of deck area, in a 35-ft high 12-deck tower with a 3-mph wind and an air wet-bulb temperature of 70° F, and also as a function of the approach to

Fig. 508. An aerator-type cooling tower. (*Fluor Corp.*)

the wet-bulb temperature (the exit water temperature minus the wet-bulb temperature). Figure 512 presents a correction factor by which the capacity

Fig. 509. Spray heads in operation in a cooling tower. (*Fluor Corp.*)

(quantity of water fed to the tower in gallons of water per unit deck area) should be multiplied when the wet-bulb temperature is other than 70° F. Figure 513 presents a similar correction factor to capacity when the tower height is other than 35 ft. Figure 514 shows the relation between the water lost due to windage (not evaporation) as a function of wind velocity. Figure 515 shows the relative tower size as a function of wind velocity, the size at a 3-mph wind velocity being taken arbitrarily as one.

The size of a tower required to handle a given heat load varies greatly with the performance required, and care should be used in specifying the wet-bulb temperature, the final approach of the tower

Fig. 510. Up-spray head in operation in cooling tower. (*Fluor Corp.*)

water to the wet-bulb temperature, the wind velocity, and the quantity and temperature range of the tower water being circulated.

If plant operation requires the maintenance of a definite minimum temperature the highest prevailing wet-bulb temperature, usually occurring during the summer, must be used in rating the tower or provision must be made for adding cold water when the tower cannot maintain the minimum temperature. Then, when the wet-bulb temperature drops and colder tower water is obtained, the quantity of cooling water being circulated may be decreased. Final temperature of tower water should be well above the wet-bulb temperature to avoid unnecessarily large towers. Figure 516 illustrates the above point by showing the relation between tower size

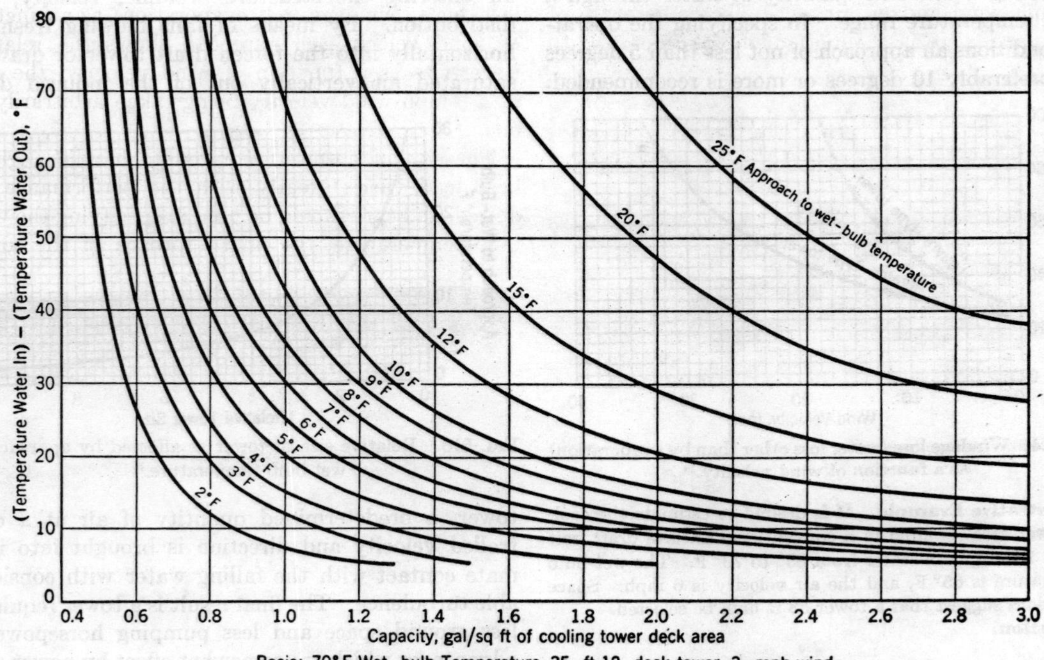

FIG. 511. Capacity of cooling towers as controlled by temperature drop and approach to wet-bulb temperature at 70° F wet-bulb temperature and 3-mph wind velocity, for a 35-ft 12-deck tower.[19]

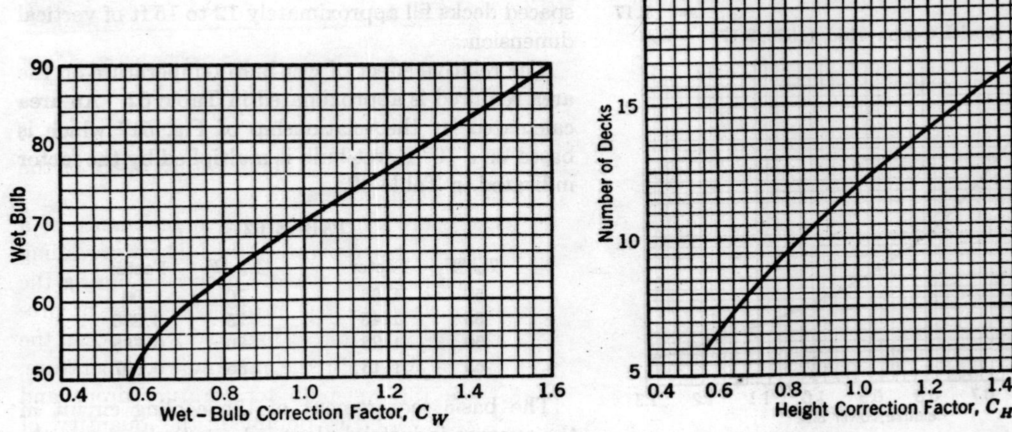

FIG. 512. Correction factor for use with Fig. 511 for wet-bulb temperature other than 70° F.[19]

FIG. 513. Correction factor for use with Fig. 511 for tower height other than 35 ft.[19]

and approach to the wet-bulb temperature while circulating a constant quantity of water through a given temperature range. In specifying the operating conditions an approach of not less than 5 degrees and preferably 10 degrees or more is recommended.

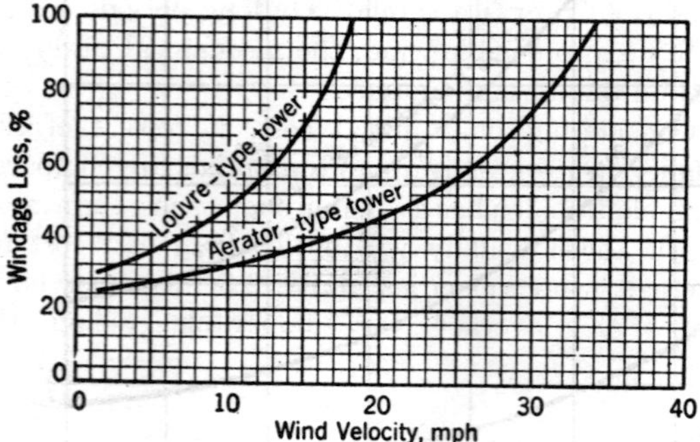

FIG. 514. Windage loss (water loss other than by evaporation) as a function of wind velocity.[19]

Illustrative Example. It is desired to estimate the cooling tower area required in a conventional natural draft unit to cool 1500 gpm of water from 85° to 70° F. The wet-bulb temperature is 65° F, and the air velocity is 6 mph. Space limitations suggest that a tower 53 ft high be selected.

Solution.

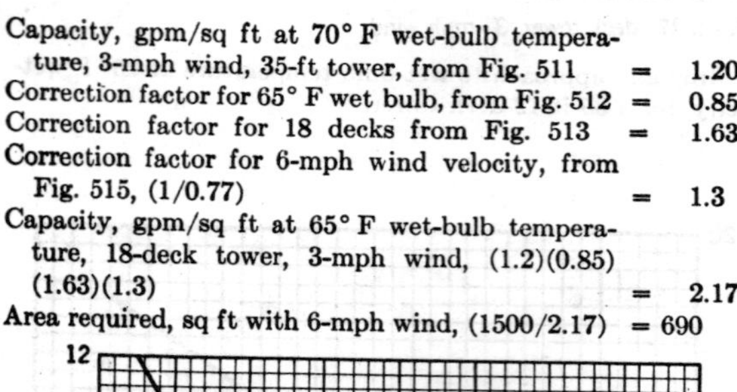

Capacity, gpm/sq ft at 70° F wet-bulb temperature, 3-mph wind, 35-ft tower, from Fig. 511	=	1.20
Correction factor for 65° F wet bulb, from Fig. 512	=	0.85
Correction factor for 18 decks from Fig. 513	=	1.63
Correction factor for 6-mph wind velocity, from Fig. 515, (1/0.77)	=	1.3
Capacity, gpm/sq ft at 65° F wet-bulb temperature, 18-deck tower, 3-mph wind, (1.2)(0.85)(1.63)(1.3)	=	2.17
Area required, sq ft with 6-mph wind, (1500/2.17)	=	690

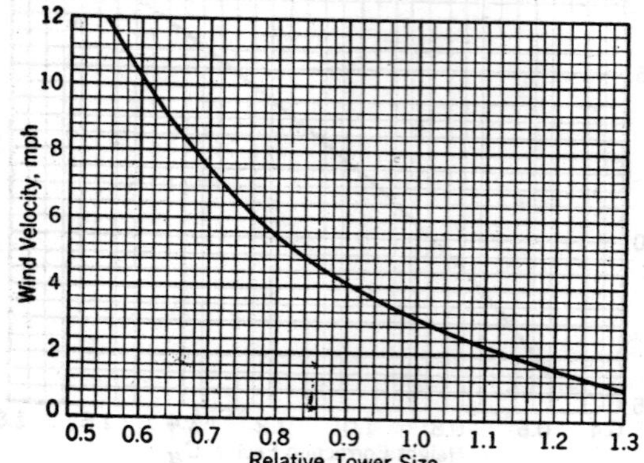

FIG. 515. Variation in size of cooling tower with wind velocity relative to unity for a wind velocity of 3 mph.[19]

Mechanical Draft

The mechanical draft towers, both forced and induced draft types, bring into the cooling tower design three important elements of control of the air entering the structure—volume, velocity, and distribution. By means of fans blowing fresh air horizontally into the forced draft tower or drawing saturated air vertically out of the induced draft

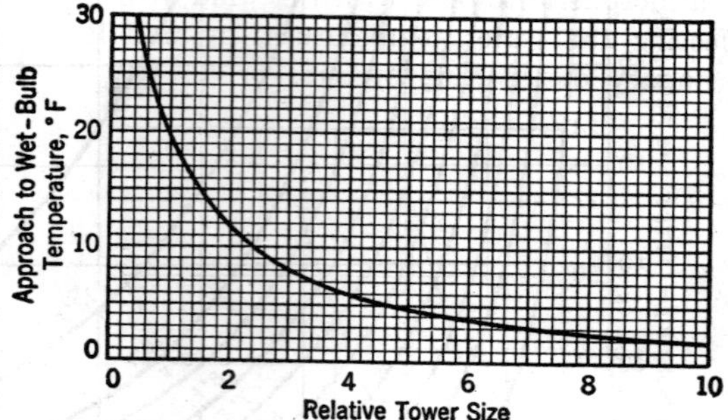

FIG. 516. Relative size of tower as affected by approach to wet-bulb temperature.[19]

tower, a predetermined quantity of air at a controlled velocity and direction is brought into intimate contact with the falling water with considerable turbulence. The final result is a tower requiring less ground space and less pumping horsepower—advantages which are somewhat offset by power consumed by fans and by the inherent mechanical problems of maintaining equipment in saturated vapor.

Design data for mechanical draft towers are presented in Fig. 517 which shows the nominal area required per gallon of circulating water for an average tower 20 to 26 ft high, in which closely spaced decks fill approximately 12 to 15 ft of vertical dimension.

The relative effect of wet-bulb temperature on the area required is approximated in Table 62. An area calculated by the relationship of Fig. 517 which is based on a 70° F wet-bulb is multiplied by the factor indicated in Table 62.

TABLE 62

T_w, °F	Factor	T_w, °F	Factor
50	1.66	70	1.00
55	1.45	75	0.886
60	1.28	80	0.785
65	1.13	85	0.697

The basic requirement of the cooling circuit in the average industrial plant is heat dissipation without critical regard for temperature. In some plants, constant temperature of water is essential. Temperature fluctuations will occur in both types of towers with fluctuations in wet-bulb temperature. In mechanical draft towers the change in off-tower water temperatures can be held to a minimum by adjusting the pitch of the fan blades, reducing the speed of the fans, or by by-passing some of the hot water directly into the basin by means of thermostatically controlled, mechanically actuated values. In atmospheric towers, with the wind velocity also affecting the water temperature, by-pass lines to secondary distributing systems or directly to the

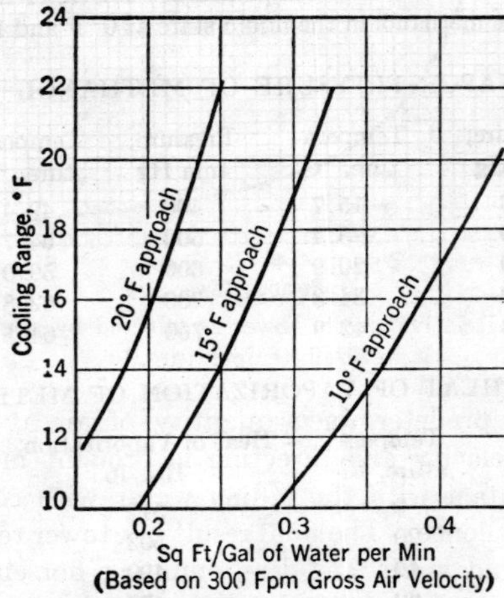

FIG. 517. Design relationship for a mechanical draft tower.[34]

basin can be installed to effect the same degree of temperature control attainable with the mechanical draft towers.

Although wind definitely affects mechanical tower performance, the effect is seldom more than a 10 per cent decrease in capacity due solely to reduced air flow through the tower. However, another element of design enters the picture when the wind causes the hot, saturated vapors discharging from the tower to recirculate back into the air intake of the tower. This means the "entering air" is at a higher wet-bulb temperature and higher relative saturation. The water temperature rises in approximate proportion to the increase in wet-bulb temperature.

One large user of both induced and forced draft towers determined by actual field test that it was true operating economy to build the discharge stacks of their towers up to a point 50 ft above the ground, or 25 ft above the top of a 25-ft tower. By thus discharging the hot saturated vapors the moisture and heat were so diffused into surrounding fresh air that the effect of recycling was negligible.

If open space near the plant site is available, for normal thermal ranges an atmospheric tower will do the same cooling job for a lower initial cost and less operating and maintenance cost than a mechanical draft tower. If power costs are exceptionally low and a maintenance crew is available, the mechanical draft tower might be the economical selection. If unusually high (80° F or more) wet-bulb conditions prevail for relatively long periods of time with no appreciable wind, the mechanical tower has certain distinct advantages.

Considering only initial cost, if towers of like quality in workmanship and materials are compared for a plant which has available open space, the atmospheric tower usually will be lower in first cost for average wind velocities above 2.5 miles per hour. Below this average, for comparable cooling, the mechanical tower will represent a lesser investment.

In operating cost, if both fan horsepower and pump horsepower are evaluated, the atmospheric-type unit will show a saving in almost all cases. In locations in which the operating limits of the atmospheric unit are approached—in high wet-bulb areas where high relative saturations prevail during periods of unusually low wind velocity—the mechanical draft towers show a distinct advantage to the operator.

COOLING PONDS

The cooling pond or reservoir is used for removing a relatively small amount of heat from water over a small temperature range. The pond area required may be estimated from a relationship giving the rate of evaporation of water into still air. One such relation[34] is

$$w = 240 + 3.7\left(\frac{T_1 + T_2}{2}\right)(\bar{p}^* - p) \qquad (604)$$

where w = water evaporated [grains/(hr)(sq ft)].

$\frac{T_1 + T_2}{2}$ = arithmetic average of entering and exit water temperature (°F).

$\bar{p}^*$ = vapor pressure of water at $(T_1 + T_2)/2$ (in. of mercury).

p = partial pressure of water vapor in air (in. of mercury).

The evaporation required for the dissipation of a given heat load is calculated from the latent heat of water. Equation 604 is based on zero wind velocity. Any wind will increase the capacity of a given pond so that equation 604 gives conservative results.

SPRAY PONDS

If a system of spray nozzles is installed above the surface of a cooling pond, the cooling capacity is

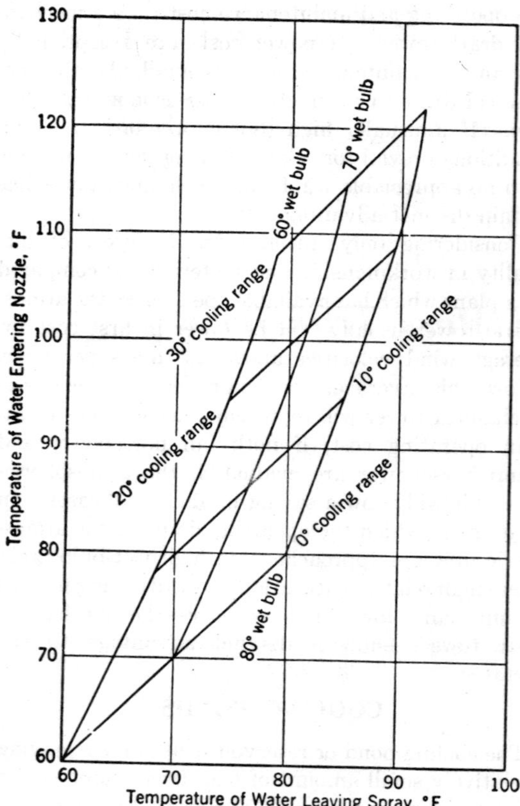

FIG. 518. Operating conditions for a particular spray pond with wind velocity of 5 mph using a 1-in. orifice with a nozzle pressure of 7 psig.[34]

greatly increased, at the expense of pumping costs. A typical cooling curve for spray ponds is shown in Fig. 518 in which a particular type of nozzle at one particular pressure is rated at a wind velocity of 5 mph.

PROBLEMS

1. Determinations of humidity in rotary driers used for drying water-soluble salts are difficult to make because of the effect of the salts on the wet-bulb measurements. It is proposed to overcome this difficulty by making the wet-bulb measurements with water saturated with the salt to be handled in the drier. Derive the corresponding form of the wet-bulb equation for such a case. Indicate all work clearly and define all symbols used.

2. Construct the following parts of a methanol humidity chart for air and methanol for a barometric pressure of 760 mm.

(*a*) The relative saturation lines for 100 and 10 per cent.

(*b*) The saturated volume versus dry-bulb temperature curve.

(*c*) Humid heat as a function of humidity.

(*d*) The adiabatic cooling line for an adiabatic saturation temperature of 60° F.

(*e*) The humidity as a function of the dry-bulb temperature for a wet-bulb temperature of 60° F.

(*f*) The enthalpy of the dry gas as a function of the dry-bulb temperature; the enthalpy of the saturated gas as a function of the dry-bulb temperature. The reference state will be pure methanol in the liquid state at 0° F and 1 atm.

VAPOR PRESSURE OF METHANOL

Pressure, mm Hg	Temperature, °C	Pressure, mm Hg	Temperature, °C
10	−15.7	400	49.4
50	+8.4	500	54.7
100	20.9	600	59.0
200	34.3	700	62.8
300	42.9	760	64.5

LATENT HEAT OF VAPORIZATION OF METHANOL

Temperature, °C	Heat of Vaporization, Btu/lb
0	512
20	503
40	490
60	477
64.5	473
80	456

Specific heat of methanol vapor = 0.39 Btu/(°F)(lb). For methanol vapor,

$$\frac{h_g}{k_{\mathcal{H}} M_v} = 0.35$$

where h_g = heat transfer coefficient, Btu/(hr)(°F)(sq ft).
$k_{\mathcal{H}} M_v$ = mass transfer coefficient, lb/(hr)(sq ft), (lb methanol/lb dry air).

3. It is desired to remove methanol from naphthalene crystals by passing them, on a belt, through a chamber in which warm air is blown countercurrent to the crystals. The air, on leaving the drying chamber, passes through a condenser and then is recirculated back into the drier through a bank of steam coils which bring it back to its original temperature. The air is to enter the drying chamber at 110° F and with a methanol saturation of 10 per cent. It is to leave the drier at a temperature 5° F above its adiabatic saturation temperature. The crystals enter at 70° F and may be assumed to remain at the adiabatic saturation temperature throughout the process and leave methanol-free.

(*a*) To what temperature must the air be cooled in the condenser?

(*b*) If 1000 lb/hr of wet crystals, containing 10 weight per cent methanol on the dry basis, enter the drier, specify the volume of air (under conditions at which it enters the drier) recirculated per minute, the heat removed per hour in the condenser, and the heat added per hour in the steam coils

CHAPTER

38

Simultaneous Heat and Mass Transfer 2

Drying

DRYING is usually distinguished from other operations, such as evaporation, as the removal of relatively small amounts of water from relatively large amounts of solids into a gas or air stream. The removal of water from solids is similar to the evaporation of water in humidifiers and cooling towers in that it involves both the transfer of heat and material. Not only is the transfer of heat and material through the drying medium involved but also the transfer of water through solid material.

EQUIPMENT

Although many materials may be economically dried by proper exposure to the atmosphere, modern industry usually requires more control of the drying medium and its circulation.

Tray or Shelf Driers

Wet solids such as granular material are frequently dried in tray or shelf driers (Fig. 519). The wet solids are placed on trays or shelves within an enclosure or cabinet, and the drying medium is circulated either across (Fig. 520) or through the trays, depending on the nature of the material (Fig. 521) until the desired dryness has been reached. The driers may be equipped with dampers and heaters to control the humidity and temperature of the operation.

This general type of drier is frequently designed as a continuous unit in which the trays are mounted upon trucks which are moved continuously through the drying tunnel. The conveyor-drier is another variation in which the solids are transported through the drying medium on a belt conveyor. Frequently, as in the drying of paper and textiles, the belt itself becomes the material being dried.

FIG. 519. Tray drier. (*National Drying Machinery Co.*)

Rotary Driers

Granular solids and some slurries may be dried inside rotating cylinders set at a small angle with the horizontal (Fig. 522). The material to be dried is fed at the upper end, and the dried material is discharged from the lower end. The rotation of the cylinder which may be equipped with vanes lifts the solid material and drops it through the drying medium to a position each time lower or nearer the

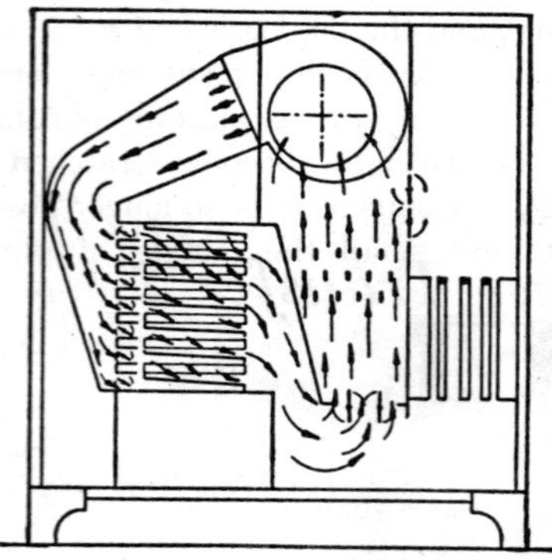

Fig. 520. Diagram illustrating air drying in a tray drier with air circulating over or across the trays and through the heater. (*National Drying Machinery Co.*)

discharge end. The drying medium is usually introduced at the lower end and leaves at the upper end, flowing countercurrent to the solid. The time required for a particle to pass through a rotary drier depends upon a number of factors, including the angle of the axis with the horizontal, the lift given

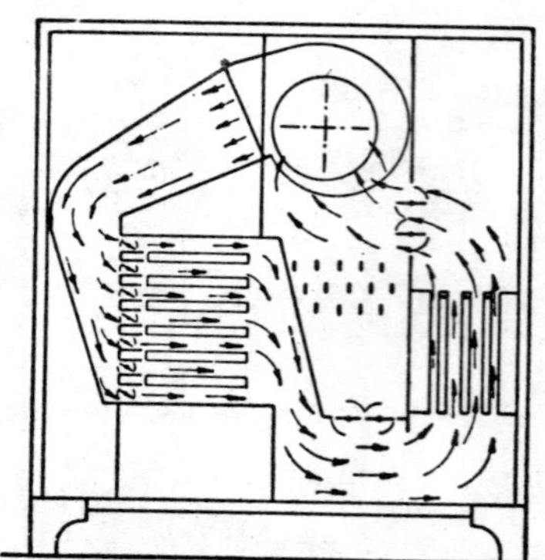

Fig. 521. Diagram illustrating a tray drier with drying air circulating through the trays and by-passing the heater. (*National Drying Machinery Co.*)

the particle by the rotation, the size and density of the particle, and the velocity of the drying medium.

Spray Driers

Solutions and suspensions which can be pumped may be sprayed into a stream of the drying medium. The large specific surface area of the small particles allows the solvent to evaporate quickly, leaving a dry product in the form of a powder. The essential components of a spray drier are the (1) means of

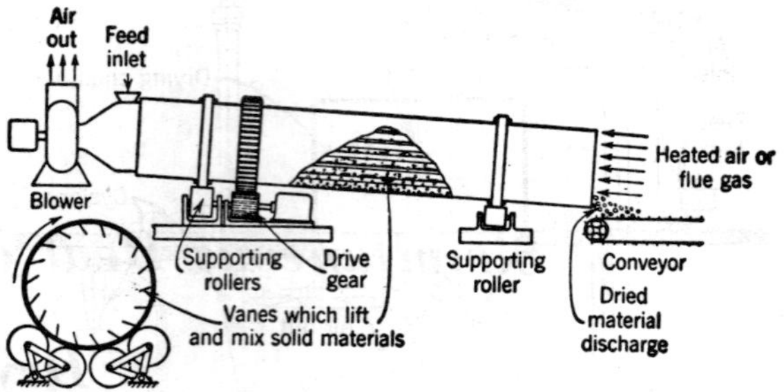

Fig. 522. A direct-heat, countercurrent rotary drier.

fluid dispersion, (2) contact of spray and drying medium, and (3) recovery of the dried product. Figure 523 shows one arrangement of parts and the flow of material in a spray drier used for drying a solution or suspension. The direct air heater is replaced by an indirect steam heater if the product

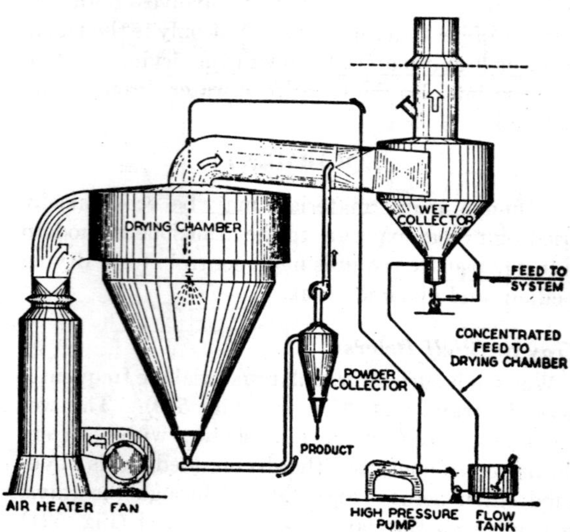

Fig. 523. Typical spray drier using direct heat and pressure spray nozzle. (*Swenson Evaporator Co.*)

must be free of contamination. Air filters may also be used. Different arrangements and flow are indicated in Figs. 524 and 525. In some cases the dried product and exhaust air are both removed from the bottom of the drying chamber, and the product is separated in a separate cyclone.

The fluid may be dispersed or sprayed by pressure nozzles or by centrifugal disk or bowl nozzles. In *pressure nozzles* the dispersion is obtained by forcing

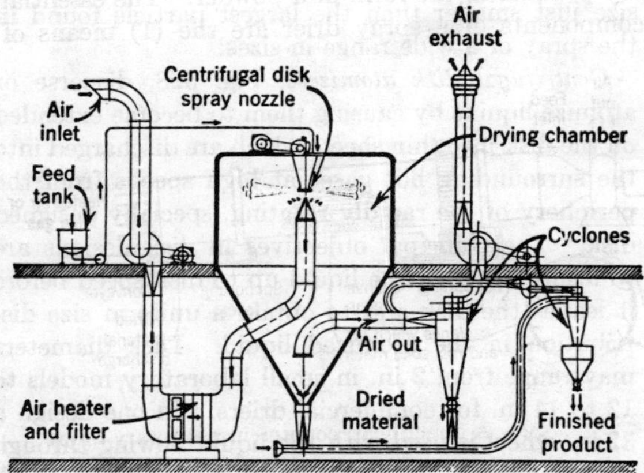

Fig. 524. Spray drier installation with centrifugal disk sprayer and steam-heated air. (*Niro Co.*)

the fluid through a very small orifice at a high velocity. There are two general types of pressure nozzles, the single-fluid type exemplified in a crude form by

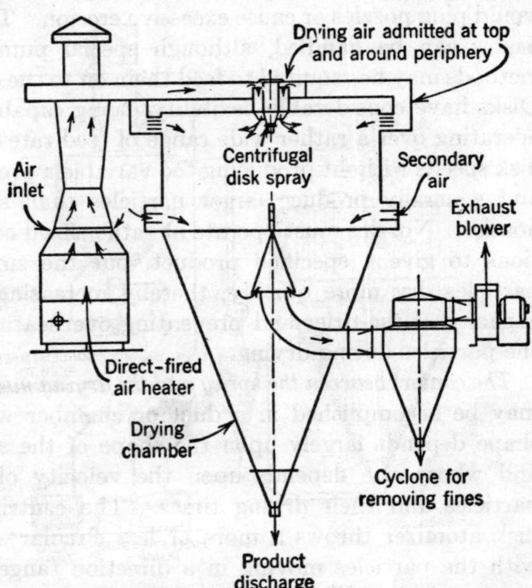

Fig. 525. Spray drier with direct heat and centrifugal disk sprayer with drying air admitted at top around periphery. (*Bowen Engineering Co.*)

the ordinary garden hose, and the two-fluid nozzle which is illustrated by a paint sprayer or a perfume atomizer.

In the *single-fluid nozzle*, Figs. 526 and 527, the spray is formed by the break-up of the high-velocity fluid stream upon encountering the surrounding vapor. Single-fluid nozzles will not, therefore, produce a spray in a vacuum. Pressure may range from 50 to 10,000 psi, depending on the degree of atomization, the capacity, and the fluid properties. Maintenance of pressure nozzles is important since erosion will occur with even the hardest material of construction, and, once the orifice has become scratched and nonuniform, good atomization is no longer possible. As the orifices are small the danger of plugging by

Fig. 526. Single-fluid pressure spray nozzle. (*Schutte and Koerting Co.*)

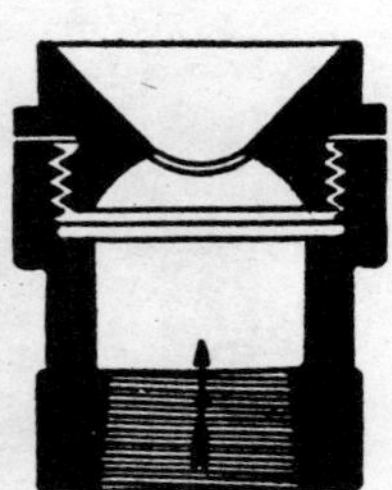

Fig. 527. Single-fluid pressure spray nozzle. (*Schutte and Koerting Co.*)

foreign matter is always present. Spray characteristics of pressure nozzles depend on the pressure and nozzle orifice size. The pressure drop affects not only the spray characteristics but the capacity as well. If it is desired to reduce the amount of liquid sprayed by reducing the pressure drop, then the spray may become coarser than desired. To correct this a smaller orifice would be inserted. The use of multiple nozzles tends to overcome this inflexible characteristic of pressure atomization, although several nozzles in a drier complicate the chamber design and the air flow pattern and will cause collision of particles, resulting in nonuniformity of spray and particle size.

In a *two-fluid atomizing nozzle* the liquid is delivered to the nozzle and dispersed by an impinging stream of gas. The two-fluid nozzles operate at relatively low pressures, the liquid being introduced at pressures up to about 60 psig while the atomizing fluid is under only slightly greater pressures. The atomizing fluid is usually steam or air. Atomizers of this type are classified as either outside mixers or inside mixers. An example of outside mixing is the

FIG. 528. Centrifugal disk sprayer. (*Bowen Engineering Co.*)

impingement of a horizontal jet of compressed air on a stream of liquid issuing slowly from a vertical tube. The perfume atomizer is an inside mixer. The use of low pressures is in itself an advantage, and the large orifices make for low maintenance cost compared with the single-fluid type.

Pressure nozzles produce a relatively wide range of particle sizes. This characteristic may be unsatisfactory from the standpoint of sales appeal and may lead to losses through excessive dustiness or fines. Since coarse particles dry more slowly than fine particles, the throughput of a drier is controlled by the time required to dry the largest particles. If the coarsest particle is satisfactorily dried, all smaller drops will have remained in the drier for an excessively long period. The capacity of the drier would be greater if the spray particles were of a uniform size just smaller than the largest particle found in the spray of a wide range in sizes.

Centrifugal disk atomizers, Fig. 528, disperse or atomize liquids by causing them to become extended on the disk into thin sheets which are discharged into the surrounding hot gases at high speeds from the periphery of the rapidly rotating, specially designed disk. The principal objectives in disk designs are to insure bringing the liquid up to disk speed before it leaves the disk and to obtain a uniform size distribution in the atomized liquid. Disk diameters may range from 2 in. in small laboratory models to 12 to 14 in. for commercial driers. In one design a 32-in. wheel is used with the liquid flowing through the spokes of the wheel to the periphery. Disk speeds range from 3000 to 50,000 rpm, depending upon the diameter. Large driers usually use disk speeds from 6000 to 12,000 rpm, corresponding to about 7000 to 12,000 fpm peripheral velocity. Centrifugal disk atomization is particularly advantageous for atomizing suspensions and pastes which would plug nozzles or cause excessive erosion. Thick pastes can be handled, although special pumping methods may be required to feed them on to the disk. Disks have considerable flexibility, being capable of operating over a rather wide range of feed rates and disk speeds without producing too variable a product and generally produce larger particles than spray nozzles. Nozzles must operate at rather fixed conditions to give a specified product, but the smaller particles dry more quickly, thereby increasing the capacity of the drier and preventing overheating of the particles during drying.

The contact between the spray and the drying medium may be accomplished in a duct or chamber whose shape depends largely upon the shape of the spray and whose size depends upon the velocity of the particles and their drying times. The centrifugal disk atomizer throws a more or less circular spray with the particles moving in a direction tangential to the circumference of the disk. Such a spray requires a circular chamber of sufficient diameter and depth to permit the particles to become dry before reaching the circumference, or bottom, of the chamber. Pressure nozzles may be designed to give sprays of different shapes and therefore permit a wider range in shapes and sizes of the drying chamber. In some cases relatively small cylinders can be used.

The gases entering a spray drier may flow down around the atomizer (Fig. 525) or be introduced beneath the atomizer (Fig. 524) when the latter is a centrifugal disk, combined, in some designs, with tangential flow into the chamber from side inlets (Fig. 525). Countercurrent flow is difficult to use with the disk type of atomizer because of the flat or horizontal particle trajectories it produces. In the case of nozzles, the direction of flow may be co- or counter-current to the spray, or a combination of both, because of the more nearly vertical particle trajectories.

The drying medium may be the product of combustion from oil-, gas-, or coal-fired furnaces, or waste gases from plant boiler houses, or it may be air or other gas heated by indirect steam heaters or indirect coal-fired heaters. Inlet gas temperatures may be as high as 1400° F.

Some designs may provide cooling air to enter around the chamber, wet dust collectors to recover excessively fine dust, and air sweepers or mechanical rakes to remove dry product from within the chamber. These last are essential for continuous removal of dry product from a flat-bottom drier. Wherever sufficient headroom is available and particularly when the product is sensitive to heat and should be removed from the drier as soon as it is dry, the cone bottom drier is preferred. In some cases the dried product may all be carried along with the exit gases and recovered in a separate device.

Spray drying is useful if a uniform powdery product is desired, as the spraying in the process breaks up the feed into small particles which then dry individually. Materials which are heat sensitive are often spray dried because the residence time in the drying chamber is short. Moreover, the rate of transfer is sufficiently high so that the incoming hot air is cooled quickly and the zone of high temperature is relatively small. Spray drying usually results in a finely divided product; consequently the process may accomplish both drying and pulverizing in a single operation and it often eliminates the need for a grinding operation that might necessarily follow another method of drying. The uniformity and the good appearance of the product is often of considerable importance in the selection of a spray drier.

Spray driers have been used for drying many chemicals, resins, pigments, soaps, miscellaneous by-products, and wastes. Because of the short exposure of the material to high temperatures, spray driers are common in the food industries for materials such as milk, eggs, soups, and starches.

This type of equipment is also readily adaptable to processes in which mass transfer is absent or negligible and only heat transfer is of any consequence, as in the solidification or crystallization of liquids such as sulfur or molten metals.

The design of a spray drier must include consideration of the rates of drying of the particles, the trajectory of the particles, and the pattern of flow of the drying medium in the drying chamber. The time required for drying will determine the necessary residence time of a particle in a drier. The gas flow patterns and the particle trajectories will, in turn, define the residence time. The residence time for drying may be computed by the methods outlined in this chapter if the physical properties and particle size of the spray are known. The particle trajectories and residence time may be computed by the methods described in Chapter 7 if the gas properties and the particle sizes are known.

Unfortunately, little is known of the particle size distribution of sprays, and either pilot-plant experience or empirical methods are usually relied upon to design spray-drying equipment.

The first calculation to make in designing a spray drier is an estimate of the ratio of hot air to feed. This may be done approximately by making an energy balance on the basis of the weight of water to be evaporated from 1 lb of feed and the known or desired temperatures of the streams in and out of the drier. The desired moisture content of the product should then be checked by assuming the vapor pressure of water in the exit product equal to the partial pressure of water in the exit gas, and appropriate revisions made for increasing (or decreasing) the air-feed ratio to obtain the desired product. The energy balance is then repeated on the basis of the revised exit conditions to check or determine the required temperature of inlet air, making any allowance for heat loss from the drier.

The capacity or size of the drier depends upon the path of the particles being dried and the time required to dry the largest particles. Usually no definite information is available about either of these important variables, and it is necessary to estimate the time required to dry the wet particles in a manner which is little more than a guess, unless the results of pilot-plant tests on the material to be dried are available. As a crude basis for a preliminary guess of the "residence time," the following expressions are offered as a rough guide.

For disk sprays,

$$t_d = 50\sqrt{\frac{L}{S}}$$

For pressure spray nozzles,

$$t_p = 17\left(\frac{L}{S}\right)$$

where t_d or t_p = residence time of the feed (or drying gas) in the drier (sec).

L/S = mass ratio of water to solids in feed.

These relationships were indicated from an inspection of about forty tests on commercial driers from 10 to 18 ft in diameter and were based on the assumption that the residence time of the particles in the drier was equal to the residence time of the gases at their average temperatures. This is equivalent to assuming that the particles are distributed radially through the gas and flow axially with the gas down or up) through the drier.

The required residence time as estimated above may be assumed to vary directly with the diameter of the largest particles produced by the spray. The values for t_d or t_p estimated in this way should be multiplied by 2 or 3 for those materials which form large drops or particles, and it may be necessary to multiply t_d by 5 or more in extreme cases.

It is always necessary to provide sufficient distance in the line of the particle trajectory to enable the particles to become dry before striking any surface.

Drum or Roller Driers

All or part of the heat required for drying may be supplied directly to the material being dried. The drum drier, Fig. 529, is one example of such a device which is widely used for drying a variety of fluids and suspensions. The fluid is spread in a thin film upon the outer surface of a heated, rotating drum. As the drum rotates the fluid becomes dried by evaporation of the solvent and is removed from the drum as the drum moves past a knife or scraper. As the drying is accomplished in thin films which remain in the heated zone for relatively short periods, this type of drier is suitable for handling heat-sensitive products.

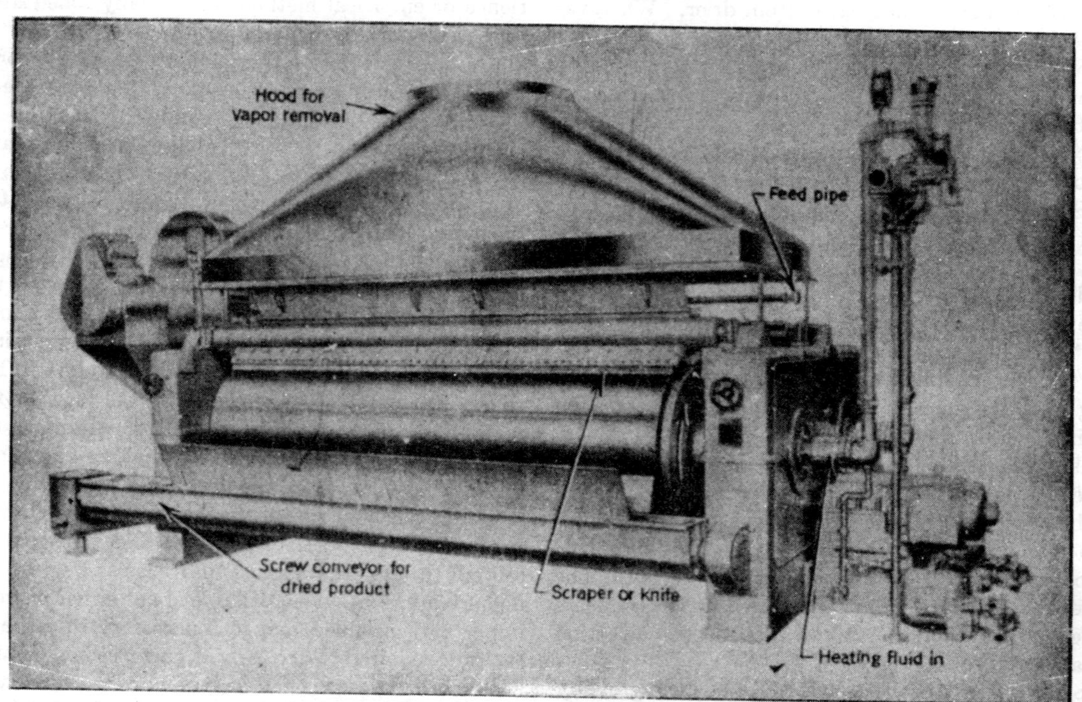

FIG. 529. A double drum drier, atmospheric pressure. (*Buflovak Equipment, Division of Blaw Knox Co.*)

FIG. 530. Twin drum dip-feed drier. The solution is fed into the shallow pan below the drums. The liquid level is adjusted and controlled so that drums pick up the desired quantity of slurry. The knives are located to provide the maximum drying area. (*Buflovak Equipment, Division of Blaw Knox Co.*)

FIG. 531. Twin drum center-feed drier. The solution is fed directly between the drums and is retained by end boards. Center feed is recommended when a product of high moisture content and coarse granular structure is desired. (*Buflovak Equipment, Division of Blaw Knox Co.*)

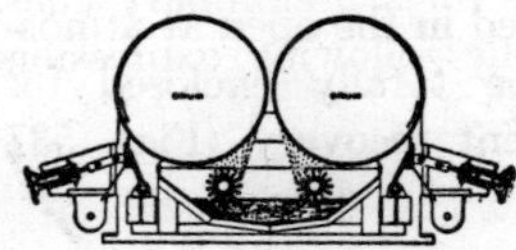

FIG. 532. Twin drum splash-feed drier. By impinging the wet material on the hot drums, the tendency of some slurrys to be repelled by heated surfaces is overcome, and capacities may be increased. (*Buflovak Equipment, Division of Blaw Knox Co.*)

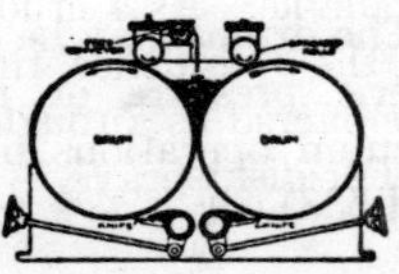

FIG. 533. Twin drum center-feed drier with spreader rolls. (*Buflovak Equipment, Division of Blaw Knox Co.*)

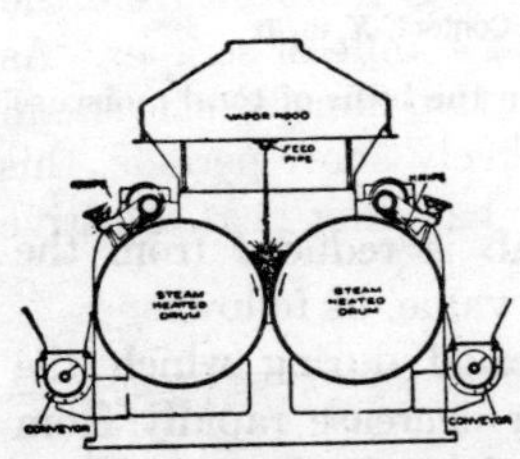

FIG. 534. Double drum drier with perforated pipe feed. (*Buflovak Equipment, Division of Blaw Knox Co.*)

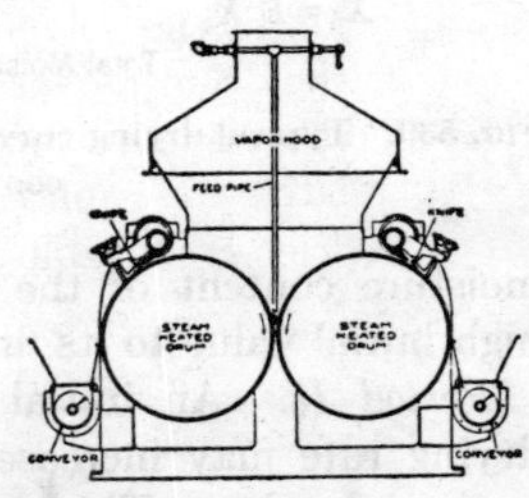

FIG. 535. Double drum drier with pendulum feed. (*Buflovak Equipment, Division of Blaw Knox Co.*)

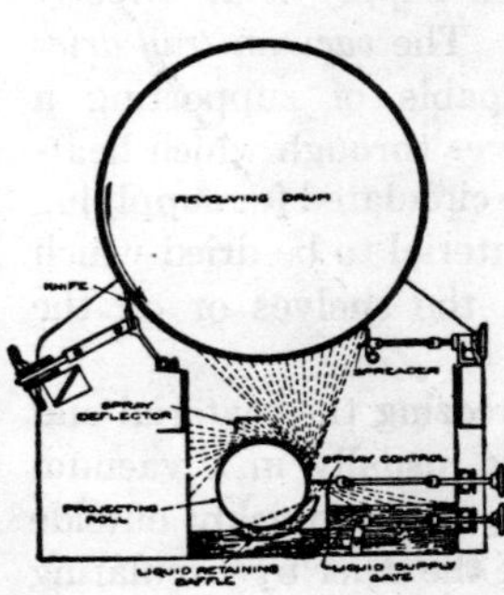

FIG. 536. Single drum drier with spray film feed. (*Buflovak Equipment, Division of Blaw Knox Co.*)

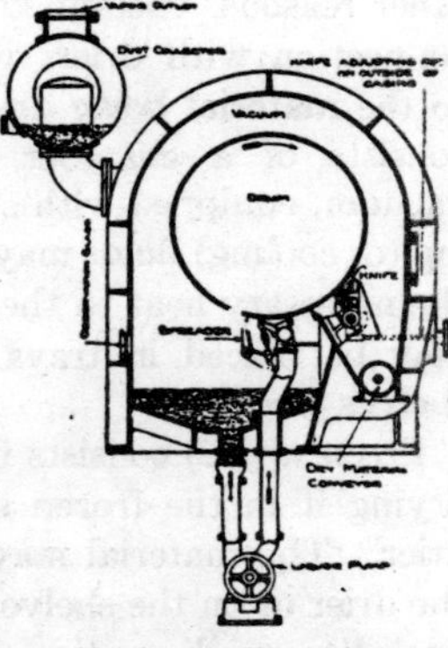

FIG. 537. Vacuum drum drier with pan feed. (*Buflovak Equipment, Division of Blaw Knox Co.*)

Because of drying in a thin film, products are removed as flakes, which in many cases enhances the "sales appeal" of the material. The dryness of the product may be easily controlled by the temperature and the speed of the drums. As illustrated by Figs. 529 to 538, there are many variations of drum driers which have been developed for varying services. For example, dip feeding (Fig. 530) is not satisfactory for suspensions of solids which are apt to settle; trough or overflow feeding (Figs. 533 and 534) to the top of a drier is better in this case. For certain materials which do not adhere well to the heated surface, a more forceful method of feeding, either by splashing (Fig. 532) or spraying (Fig. 536) is used. If a coarse, granular product with a high moisture content (up to 25 per cent) is desired, spreading rolls may be used in conjunction with center feeding (Fig. 533).

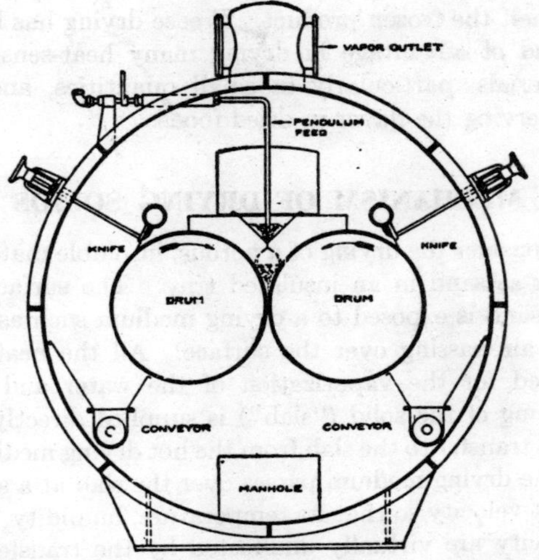

FIG. 538. Vacuum double drum drier with pendulum feed. (*Buflovak Equipment, Division of Blaw Knox Co.*)

The drums may be operated in the open at atmospheric pressure or may be totally enclosed for vacuum operations or solvent recovery (Figs. 537 and 538).

Vacuum Driers

As has been indicated the drum drier may be operated under a vacuum to decrease the temperature required to vaporize the solvent from the solution on the drums. In the case of air circulating this lowering of the required temperature is attained by circulating more air (or decreasing the feed) to lower the partial pressure of the solvent. In other cases where the solvent must be recovered, or for other reasons, vacuum drying may be desirable in connection with driers which supply heat directly to the material being dried. The *vacuum tray drier* consists of a chamber, capable of supporting a vacuum, equipped with shelves through which heating (or cooling) fluids may be circulated for supplying the necessary heat to the material to be dried which may be placed in trays on the shelves or on the shelves directly.

Freeze drying consists in freezing the material and drying it in the frozen state, usually in a vacuum drier. The material may be frozen quickly outside the drier or on the shelves of the drier by circulating a cooling medium through the shelves and jacket of the drying chamber. When frozen, vacuum is applied and a heat-supplying medium is circulated to supply the required latent heat of vaporization of the evaporating solvent but not sufficiently warm to melt the frozen product. Freeze drying has been found of advantage in drying many heat-sensitive materials, particularly in small quantities, and in preserving the flavor in dried foods.

MECHANISM OF DRYING SOLIDS

Consider the drying of a porous, insoluble material such as sand in an insulated tray. The surface of the sand is exposed to a drying medium such as hot dry air passing over the surface. All the heat required for the vaporization of the water and the heating of the solid ("slab") is supplied directly by heat transfer to the slab from the hot drying medium. If the drying medium passes over the slab at a sufficient velocity so that its temperature, humidity, and velocity are virtually unaffected by the transfer of water vapor from the slab to the drying medium, data are obtained as plotted in Fig. 539 which indicates the water content of the slab (expressed as pounds of water per pound of water-free solid) as a function of the drying rate (expressed as pounds of water per hour per pound of water-free solid). This is a typical drying rate curve. Such a curve may be divided into several distinct periods as the

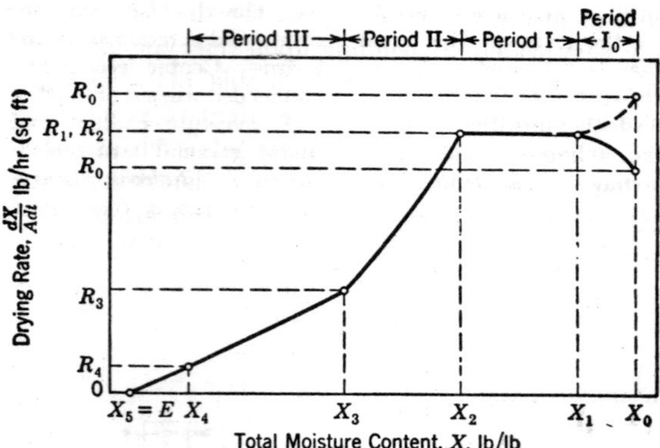

Fig. 539. Typical drying curve on the basis of total moisture content.

moisture content of the slab is reduced from the high initial value to its final value, as follows.

Period I_0. An initial period during which the drying rate may increase or decrease rapidly from an initial value. This period is of relatively short duration and in some experiments may be unobservable.

Period I. An early stage of drying during which the drying rate remains at a constant value.

Period II. A period of drying during which the drying rate decreases, more or less linearly with the continued decrease in the water content.

Period III. A stage of drying immediately following Period II but not always clearly distinguishable from it, during which the drying rate decreases more or less linearly with decreasing water content, but in general at a different rate than during Period II.

On prolonged drying under constant conditions of the drying medium, the rate at which the slab is dried becomes zero. The limit to which the water content can be reduced by drying for an infinite length of time is known as the *equilibrium moisture content*.

The *equilibrium moisture content* X^* is the limit to which a given material can be dried by means of a drying medium of given temperature, humidity, etc. It is determined by the nature of the solid material, the temperature of the drying medium, and the partial pressure of water vapor in the drying medium.

The relationship even for a given solid material is exceedingly complex and in general must be determined experimentally. Typical equilibrium moisture relationships for various humidities are shown in Fig. 540. The equilibrium moisture contents at any given humidity of the drying medium may vary widely, depending upon the type of material. A nonporous highly insoluble material may have an equilibrium moisture content of practically zero, whereas some organic materials such as soap, leather and wood have equilibrium moisture contents that may vary over exceedingly wide ranges, depending upon the temperature and humidity of the drying medium. The water that makes up the equilibrium moisture content of a given solid may be adsorbed on the solid, it may be held by capillary forces in the pores of the solid, or it may be in chemical combination with the solid.

The *free moisture content* F is the difference between the total moisture content X and the equilibrium moisture content X^*, expressed as pounds of water per pound of dry solid.

$$F = X - X^* \tag{605}$$

For a given total moisture content X, the free moisture content F will be a function of the same variables as the equilibrium moisture content X^*. The value of free moisture content is meaningless unless the conditions of temperature and partial pressure of water vapor in the drying medium are stated and the value of the equilibrium moisture content is known.

Usually, the value of the equilibrium moisture content is small compared to the total moisture content, so that the variation in free moisture content with the temperature and humidity of the drying medium is also small. The drying rate data are frequently plotted on the basis of a free moisture content rather than a total moisture content because the drying rate of a given material at a constant temperature and humidity of the drying medium becomes zero at a zero free moisture content.

Period I_0 is essentially a period of unsteady operation during which the drying conditions within the slab are adjusting themselves to the steady state represented by Period I.

When the slab is initially cold, i.e., below the adiabatic saturation temperature of the drying medium, the mechanism may be explained as follows.

Heat is transferred to the relatively cold wet surface of the slab from the relatively hot drying medium. This heat raises the temperature of the water in the slab to the temperature at which evaporation takes place and supplies the latent heat of vaporization. This heat also serves to increase the

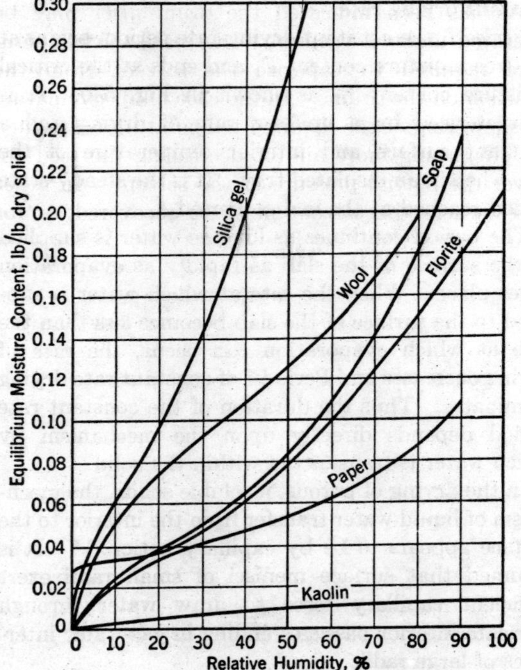

FIG. 540. Equilibrium moisture content of various materials.

surface temperature and average temperature of the slab. During this process, evaporation of the water from the surface of the slab into the drying medium takes place under the influence of the driving potential of the difference in the vapor pressure of the water on the surface of the slab and the partial pressure of the water vapor in the drying medium.

As this process continues, the increase in the surface temperature causes a corresponding increase in the evaporation rate and a decrease in the rate at which heat is transferred to the slab. After a relatively short time, the rate of heat transfer becomes equal to the heat required for evaporation, and the rate of evaporation reaches a constant value at approximately the wet-bulb temperature.

When the slab is initially hotter than the wet-bulb temperature of the drying medium, the mechanism of adjustment is similar except that the latent heat requirement is greater than that transferred to the slab and the drying rate decreases to the steady condition.

Under special conditions where the initial temperature of the slab is at the adiabatic saturation temperature of the drying medium, no initial varying rate period is observed, the drying starting with a constant drying rate.

Period I, the constant drying-rate period, begins at the free moisture content F_1 and ends at the critical moisture content F_2 as shown in Fig. 539. It is characterized by a uniform rate of drying and a constant surface and interior temperature of the slab when in an insulated tray. It is the steady-state period reached at the end of Period I_0.

The period continues as long as water is supplied to the surface of the slab as rapidly as evaporation takes place. When the rate at which water is supplied to the surface of the slab becomes less than the rate at which evaporation can occur, the rate of drying decreases and Period I of constant rate drying terminates. Thus the duration of the constant rate period depends directly upon the mechanism by which water is transmitted within the solid.

In the drying of porous, insoluble solids, the mechanism of liquid water transfer from the interior to the surface appears to be by capillary action.[32] * It is assumed that surface menisci of small radii exert sufficient capillary force to draw water through intricate interior passages ending in gas-water interfaces of large radii.

The water drawn to the surface is necessarily replaced by air which enters the solid through the larger passages connected with the larger openings at the surface. Because of the complicated interconnecting passages beneath the surface, it is possible for the necessary air to enter through a relatively few surface openings and thus for the moisture concentration near the surface to remain relatively high. . . . The water will continue to rise to the surface through any system of interconnecting passages until all of the various menisci at the lower ends of the water column have the same radius of curvature as the small menisci at the surface from which evaporation is taking place. When this stage is reached, a small amount of evaporation from the surface menisci may result in a retreat of these surface menisci into passages of smaller cross section, and the increased capillary tension is sufficient to draw additional water to the surface. It is possible, in fact, for the increased tension caused by this retreat of the surface menisci to draw some of the menisci at the lower ends of the water column through the narrow constrictions into larger cavities and thus reduce the tension necessary to cause movement toward the surface. The menisci in the passages at the surface can then rise to the former position, and the process can continue.

As the drying process proceeds, a time will be reached when the menisci at the lower ends of the water column in any system of interconnecting passages, are, in general, about the same size as the smallest cross section of the surface openings, and water will no longer be drawn to the surface through these passages.[46]

With the above mechanism of transfer of water through the solid, the critical moisture content is a function not only of the material being dried but also its thickness and the rate at which it is dried.

During the constant rate period the wet surface of the solid behaves as a free water surface in that the water on the surface exerts a pressure equal to the vapor pressure of water at the surface temperature p^*.

Thus, for a drying medium of a given temperature T_g, partial pressure of water vapor p, and velocity, the drying rate may be expressed

$$-\frac{dX}{A\,dt} = k_g M_v(p^* - p) \tag{606}$$

Similarly, the rate of heat transfer may be written

$$\frac{q}{A} = (h_c + h_r)(T_g - T_i) \tag{607}$$

where h_c and h_r = heat transfer coefficients by convection and radiation, respectively.

T_i = temperature of solid surface.

Since the temperature of the surface is constant during the constant rate period, assuming that the enthalpy increase of the water vaporized is equal to the latent heat,

$$k_g M_v(p^* - p) = \frac{(h_c + h_r)}{\lambda_{T_i}}(T_g - T_i) \tag{608}$$

and

$$p = p^* - \frac{h_c + h_r}{k_g M_v \lambda_{T_i}}(T_g - T_i) \tag{609}$$

It is apparent that equation 609 is identical to the wet-bulb temperature equation 582. It may be concluded, therefore, that when material is drying during the constant rate period in insulated trays, the surface temperature is the wet-bulb temperature of the drying medium.

Period II, the first falling rate period, starts at the free moisture content F_2 and ends at F_3 (Fig. 539). It is characterized by a decreasing rate of evaporation which results from the spot-wise recession of the evaporation surface into the first layers of the stock, with the consequent exposure of small

* The bibliography for this chapter appears on p. 575.

radii of curvature. The momentarily undiminished heat supply causes an increase in the temperature at the receded zone of evaporation. This temperature approaches a changing equilibrium value determined primarily by the radii of curvature of the menisci slightly below the surface. As drying proceeds, the fractional surface area accounted for by the receded water menisci increases to unity and the fraction of "wetted surface" decreases to zero. At this time, all evaporation becomes subsurface and Period II terminates.

The second critical moisture content F_3, like the first F_2, is not only a "property" of the stock but is dependent also upon drying rate and stock thickness.

Period III, the second falling rate period, starts at F_3 when capillary flow to the surface has ceased [30,44] and continues, under prolonged time, to $F = 0$, when the stock is at its equilibrium moisture content X^*. This period is characterized throughout by subsurface evaporation from a continuously receding plane, and by the necessity for the heat for evaporation to penetrate increasing thickness of partially dried stock. The surface of the stock approaches, but does not reach, the temperature of the drying gases T_g. The temperature at the receding plane of evaporation approaches a changing equilibrium value determined primarily by the radii of the exposed menisci.

The *overall drying process* may include all periods or only parts of one or more, depending upon the initial and final moisture contents. Although the rates of heat transfer, rates of mass transfer, and the heat and material balances may be set up in mathematical form for all periods, the mathematical equations are of little practical utility because virtually no data on many of the important properties and rate coefficients are available and the solution of the mathematical relationships would be too complicated and tedious for practical use.

CALCULATIONS

As our understanding and knowledge of mass and energy transfer improves, better design of drying equipment is possible. But in all cases the drying properties of the material must be determined, or known.

Tray Driers

In designing tray driers and operations to be conducted in tray driers, it is usually desired to determine the time required and the tray area. At the present time, except in rare cases, it is necessary to have a reasonably complete series of drying data relating the time required to dry the material in layers of known thickness under certain drying conditions. The details of such calculations are illustrated by the following examples.

Illustrative Example (Insulated Trays). It is desired to determine the total time required to dry sand containing 35.0 per cent water (dry basis) to a final moisture content of 0.377 per cent water (dry basis) by passing air at 60° C, with a relative saturation of 10 per cent over the surface of the sand, at a velocity of 10 meters/sec. The dry sand has a density of 1.38 grams/cc. The sand is to be dried in layers 6 cm thick in insulated trays, and it will be assumed that the sand is initially at the wet-bulb temperature of the air. Experiments in the laboratory have resulted in the data shown in Figs. 541 and 542, and the fact that the rate of

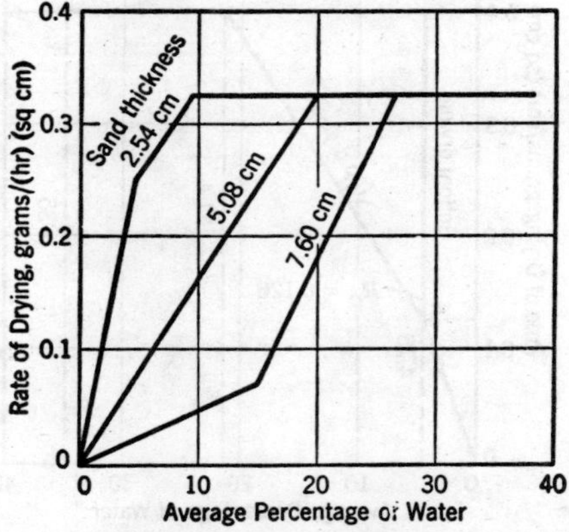

Fig. 541. Drying rates of sand as determined in layers of different thickness, drying from the top surface only.[32]

drying during the constant rate period is correlated by equation 610. The equilibrium moisture content of sand is negligible. The drying rate is

$$R = -\frac{dX}{A\,dt} = 0.00433v^{0.8}(p^* - p) \qquad (610)$$

where R = the drying rate [grams/(hr)(sq cm)].
v = air velocity (meters/sec).
p^* = partial pressure of water at the drying surface (mm Hg).
p = partial pressure of water in the air (mm Hg).
A = drying surface area per unit mass of dry solid (sq cm/gram).

Solution. The data of Fig. 541 are interesting in that the three important stages of drying are clearly evident, and the critical moisture contents separating the periods are markedly dependent upon the thickness of the layer of sand being dried.

During the constant rate period the surface of the sand remains wet even though the moisture content on a dry-weight basis is diminishing. The first critical point occurs when the surface water film breaks and exposes a gradually reduced surface to evaporation with a correspondingly reduced rate of evaporation. The structure of the water film during this period is sometimes referred to as the funicular state. Another break in the drying curve may occur when continuity of the water film is no longer maintained in the water wedged between the grains, the so-called pendular state, and the surface of water evaporating begins to retreat progressively below the surface of the sand.

When the sand is drying from the bottom surface only, giving the data shown in Fig. 542, the rate of drying remains constant until the pendular state has been reached throughout the entire layer. The pendular state starts first at the top surface, the effect of gravitational forces maintaining the bottom surface wet until the pendular state has been reached

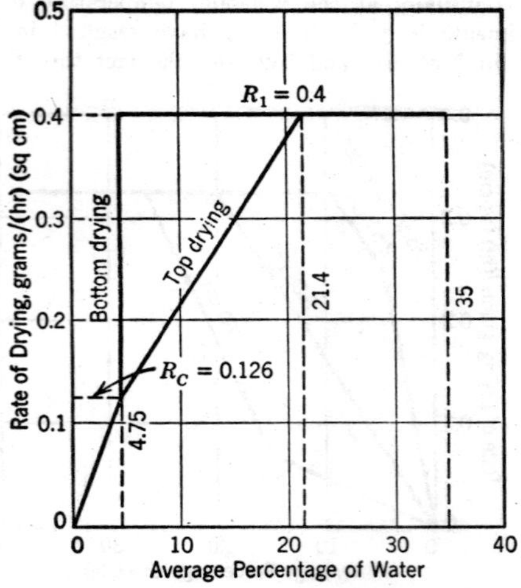

FIG. 542. Drying rates for sand in layer 6 cm thick, drying from the bottom only.[32]

throughout the sand above. For this reason less time is required when drying from the bottom than when drying from the top, other conditions remaining the same.

Constant rate period. At 60° C, and 10 per cent humidity, the wet-bulb temperature of air is 30.6° C, and the dew point is 20.6° C. Therefore, during the constant rate period,

Partial pressure of water at the surface

$$= \text{Vapor pressure of water at } 30.6^\circ \text{ C}$$

$$= 32.6 \text{ mm Hg}$$

Partial pressure of water in the air

$$= \text{Vapor pressure of water at } 20.6^\circ \text{ C}$$

$$= 18.0 \text{ mm Hg}$$

$$\text{Rate of drying} = (0.00433)(10^{0.8})(32.6 - 18.0)$$

$$= 0.40 \text{ gram/(hr)(sq cm)}$$

$$\text{Dry sand per unit of drying surface} = (1.38)(6.0)$$

$$= 8.28 \text{ grams/sq cm of area}$$

Water removed during the constant rate period

$$= (8.25)(0.35 - 0.214)$$

$$= 1.13 \text{ grams/sq cm}$$

$$\text{Time required during the constant rate period} = \frac{1.13}{0.40}$$

$$= 2.82 \text{ hr}$$

Second drying period. Since the rate of drying is a function of the moisture content, the rate equation for a differential increment of time may be obtained from equation 611.

$$\int_0^t dt = -\frac{1}{A}\int_{X_1}^{X_2} \frac{dX}{R} \tag{611}$$

For the second period R is a linear function of X. Therefore, as indicated in Table 49, p. 428, the logarithmic mean is the correct mean value of R to use in evaluating equation 611 as follows.

$$t = \left(\frac{1}{A}\right)\frac{X_1 - X_2}{R_{lm}} \tag{612}$$

Time required for the second period,

$$t = (8.28)\frac{0.214 - 0.0475}{\dfrac{0.40 - 0.126}{2.303 \log \dfrac{0.40}{0.126}}} = \frac{(8.28)(0.214 - 0.0475)}{0.238} = 5.80 \text{ hr}$$

Third drying period. Since R is a linear function of X during the third period and since the equilibrium moisture content, X^*, is zero, the rate at a moisture content of 0.377 per cent (dry basis) will be (see Fig. 542)

$$R = \left(\frac{0.126}{0.0475}\right) 0.00377 = 0.01$$

Therefore, from equation 612 the time required for the third period,

$$t = (8.28)\frac{0.0475 - 0.00377}{\dfrac{0.126 - 0.01}{2.303 \log \dfrac{0.126}{0.01}}} = 6.06 \text{ hr}$$

The total time required is $2.82 + 5.80 + 6.06 = 14.68$ hr.

If the sand were dried from the bottom only, under the same conditions, the time for the constant rate period would be

$$t = \frac{(8.28)(0.35 - 0.0475)}{0.40} = 6.28 \text{ hr}$$

The time required for the final period would be the same as that in the previous calculation, or 6.06 hr. Thus the total time required would be $6.28 + 6.06 = 12.34$ hr.

If the sand were dried from both the top and the bottom simultaneously,

Time for constant rate period

$$= \frac{8.28(0.35 - 0.214)}{0.4 + 0.4} = 1.41 \text{ hr}$$

$$\text{Time for second period} = \frac{8.28(0.214 - 0.0475)}{0.4 + 0.238} = 2.16 \text{ hr}$$

$$\text{Time for third period} = \frac{8.28(0.0475 - 0.0038)}{0.046 + 0.046} = 3.93 \text{ hr}$$

$$\text{Total time} = 7.50 \text{ hr}$$

Illustrative Example (Uninsulated Trays). Although laboratory experiments designed to establish drying rate curves are usually carried out with trays that are insulated against heat transfer on all nondrying surfaces, in commercial practice uninsulated trays are used.

It is desired to determine the time needed to dry a granular material 1 in. thick from 35.0 per cent to 10.0 per cent moisture (dry basis) in a tray drier. The stock is to be dried by blowing air horizontally over the top surface, with the air at average conditions of 140° F, 10 per cent relative saturation, and a velocity of 1960 fpm. The stock is to be placed in large galvanized iron trays with uninsulated bottoms exposed to the air stream and is initially preheated to the temperature of the constant rate period.

In the laboratory it has been determined that the critical moisture content of this material for top drying of a layer 1 in. thick in insulated trays is 9.2 per cent (dry basis). The density of the dry material is 86.5 lb/cu ft, and the thermal conductivity of the wet material is estimated to be 2.0 Btu/(hr)(°F/ft)(sq ft). The rate of drying during the constant rate period was correlated by equation 610, which for engineering units becomes

$$R = -\left(\frac{1}{A}\right)\frac{dX}{dt} = 2.68v^{0.8}(p^* - p) \qquad (613)$$

where R = rate [lb/(hr)(sq ft)]
v = air velocity (fps).
p^* = vapor pressure or equilibrium partial pressure of water at the surface (atm).
p = partial pressure of water in the air (atm).

Solution. Where the tray is uninsulated, heat flows into the solid from two opposite sides, whereas evaporation takes place from the surface only. It is necessary to calculate the heat transferred to the evaporating surface. The coefficient of heat transfer between the air and the drying surface may be estimated from the results of the laboratory tests in the insulated trays.

In the laboratory tests, where the tray is insulated, the surface of the stock is at the wet-bulb temperature of the air. The resulting temperatures and pressure are as follows:

Dry-bulb	140° F
Wet-bulb	87° F
Dew point	69° F
Vapor pressure p^*	= 0.0429 atm
Partial pressure p	= 0.0237 atm
$(p^* - p)$	= 0.0192 atm

According to equation 613

$$R = (2.68)(\tfrac{1960}{60})^{0.8}(0.0192) = 0.84 \text{ lb/(hr)(sq ft)}$$

The total water evaporated per square foot of drying surface = $(86.5)(\frac{1}{12})(0.35 - 0.10) = 1.80$ lb, and the corresponding time of drying in the laboratory would be

$$t = \frac{1.80}{0.84} = 2.14 \text{ hr}$$

The heat of vaporization including superheat is 1068 Btu/lb at 87° F. The corresponding heat transfer coefficient from air to the wetted surface is therefore

$$h_c = \frac{(0.84)(1068)}{140 - 87} = 16.9 \text{ Btu/(hr)(°F)(sq ft)}$$

With heat flowing into the bottom of the tray, the temperature of the sand rises above the wet-bulb temperature of the air. The total rate of heat transfer to the evaporating surface directly from the air and from the air through the tray to the surface is

$$\frac{q}{A} = h_c(T_g - T_i) + \frac{T_g - T_i}{\dfrac{1}{h_c} + \dfrac{x}{k_q}} \qquad (614)$$

where h_c is the heat transfer coefficient by convection.
x is the thickness of the drying material through which the heat flows.

In this solution it is assumed that the same value of the heat transfer coefficient may be applied to both the top and bottom surfaces, and the heat transfer through the sides of the tray is neglected. Substituting the numerical values in equation

$$\frac{q}{A} = 16.9(140 - T_i) + \left(\frac{140 - T_i}{\dfrac{1}{16.9} + \dfrac{1}{(12)(2.0)}}\right) \text{Btu/(hr)(sq ft)}$$

where T_i = temperature at surface (°F).

The rate of drying from equation 613 is

$$R = (2.68)(\tfrac{1960}{60})^{0.8}(p^* - 0.0237) \text{ lb/(hr)(sq ft)}$$

Multiplying this equation by the latent heat of evaporation, 1068 Btu/lb, and equating the product to the preceding equation because $q/A = R\lambda$,

$$p^* = \frac{4850 - 26.8T_i}{46{,}500} \qquad (615)$$

Equation 615 gives the partial pressure of water at the evaporating surface as a function of temperature. The surface temperature must also satisfy the relationship between the vapor pressure of water and its temperature. Therefore solving equation 615 simultaneously with vapor pressure data for water, $T_i = 92.2°$ F and $p^* = 0.0510$ atm.

The corresponding rate of drying during the constant rate period is

$$R = (43.5)(0.0510 - 0.0237) = 1.19 \text{ lb/(hr)(sq ft)}$$

and, since the drying specified lies entirely within the constant rate drying region, the time required for drying in an uninsulated tray is

$$t = \frac{1.80}{1.19} = 1.51 \text{ hr}$$

Drum Drier

When the required heat is supplied to the material independently of the circulating air, as in the drum drier, the drying rate may be calculated primarily from consideration of the heat transfer rates.

Illustrative Example (Material Heated Directly). In the drying of paper on a 4-ft steel drum, steam is supplied to the drum at 200° F, and air is passed over the surface of the paper on the drum at 100° F and 40.0 per cent relative saturation at a velocity of 300 fpm. The paper is thin so that constant rate drying conditions may be assumed to apply. For the removal of the free moisture during the constant rate period the following relation holds.

$$R = 2.45v^{0.8}(p^* - p) \tag{616}$$

when R = rate of drying [lb water/(hr)(sq ft)].
v = air velocity (fps).
p^* = vapor pressure of water at the surface temperature (atm).
p = partial pressure of water in the air (atm).

Estimate the rate of drying of the paper.

Solution. Resistance to the flow of heat is offered by the condensate formed on the inside of the metal drum, by the metal wall, by the paper, and by the outside surface. Heat flows to the surface of the paper by three resistances in series and into the air from the surface by three parallel paths, namely, by radiation, by convection, and by evaporation.

The rate of heat flow is given by the equation

$$\frac{q}{A} = \frac{200 - 100}{\dfrac{1}{h_a} + \dfrac{1}{h_m} + \dfrac{1}{h_p} + \dfrac{1}{h_c + h_r + h_e}} \tag{617}$$

where h_a (condensate) = 800 Btu/(hr)(°F)(sq ft)
h_m (metal) = 1000 Btu/(hr)(°F)(sq ft)
h_p (paper) = 800 Btu/(hr)(°F)(sq ft)
h_c (convection) = 1.1 Btu/(hr)(°F)(sq ft)
h_r (radiation) = 1.5 Btu/(hr)(°F)(sq ft)
h_e (evaporation) = $\dfrac{2.45v^{0.8}(p^* - p)\lambda}{T_i - T_g}$

It is now assumed that the surface temperature (T_i) is 183° F. Then

$$p^* = 0.547 \text{ atm}$$

$$p = 0.033 \text{ atm}$$

$$p^* - p = 0.514 \text{ atm}$$

$$\lambda = 988 \text{ Btu/lb}$$

$$v = 5.0 \text{ fps};\ v^{0.8} = 3.62$$

$$h_e = \frac{(2.45)(3.62)(0.514)(988)}{83} = 54.2$$

$$\frac{q}{A} = \frac{200 - 100}{\dfrac{1}{800} + \dfrac{1}{1000} + \dfrac{1}{800} + \dfrac{1}{1.1 + 1.5 + 54.2}} = \frac{100}{0.0211}$$

$$= 4740 \text{ Btu/(hr)(sq ft)}$$

Then

$$(T_i - 100) = (200 - 100)\frac{\dfrac{1}{1.1 + 1.5 + 54.2}}{0.0211} = 83$$

$$T_i = 183° \text{ F}$$

Therefore the assumed value was correct. The rate of drying is (2.45)(3.62)(0.514) = 4.55 lb/(sq ft)(hr).

Continuous Driers

In the design of continuous driers, it is usually desired to determine the length of the drier required to reduce the moisture content of a given quantity of wet material per hour.

Illustrative Example (Continuous Drier). A counter-current progressive drier is desired for drying 10,000 skins of chrome leather per 24 hr. The inside vertical cross section of the drier is 10.7 ft wide by 7 ft high. The skins are hung crosswise, three abreast, in rows averaging 8.6 skins per foot of length. Propeller fans are placed above the stock at 10-ft intervals to circulate the air downward through the vertical sheets, and to by-pass it through side channels on each side of the drier to the adjoining 10-ft section. The cross section of the air stream is thus 107 sq ft as it passes down through the leather. This spiral flow of air is necessary to insure a uniform circulation without channeling or temperature stratification. Air is taken from the outside at 80° F, 70.0 per cent relative saturation, and heated to 140° F before it enters the drier. Steam coils in the drier maintain the air at a temperature of 140° F. The average weight of dry leather per skin is 1.162 lb. The initial moisture content of the stock is 1.562 lb per pound of dry leather. The average thickness of the skin is 1.03 mm. It is desired to dry this leather to a final moisture content of 9.0 per cent.

In drying this material, evaporation is the controlling factor and drying takes place from both sides of the sheet. The rate of drying is correlated by the following equation.

$$-\frac{dX}{dt} = \left(\frac{0.00239}{x}\right)\left(\frac{G}{60}\right)^{0.6}(F)(\mathcal{H}^* - \mathcal{H}) \tag{618}$$

where F = free moisture content (lb water/lb dry solid).
x = thickness (ft).
G = air mass velocity [lb/(hr)(sq ft)].
$\mathcal{H}$ = air humidity (lb water/lb dry air).
$\mathcal{H}^*$ = air humidity when in equilibrium with the water on the skins (approximately the saturated humidity at the wet-bulb temperature T_w).
$\dfrac{-dX}{dt}$ = rate of drying per pound of dry solid [lb water/(min) (lb)].

The equilibrium moisture content varies with both temperature and relative saturation. Values of the equilibrium moisture content are presented in the following table expressed as pounds of moisture per pound of dry stock.

TABLE 63. EQUILIBRIUM MOISTURE CONTENT, LB WATER/LB DRY STOCK

Relative Saturation, %	80° F	100° F	120° F	140° F
10	0.071	0.060	0.048	0.040
20	0.103	0.090	0.080	0.070
30	0.128	0.115	0.105	0.095
40	0.152	0.140	0.129	0.118
50	0.173	0.162	0.153	0.142
60	0.198	0.190	0.181	0.170
70	0.232	0.222	0.212	0.200
80	0.282	0.270	0.255	0.245
90	0.352	0.338	0.323	0.305

Estimate the length of the progressive drier required when the mass velocity of the air flowing through the drier of the above dimensions is 4.0 lb/(min)(sq ft).

Solution. In calculations dealing with continuous driers, it is convenient to consider a differential section of the drier dL. Equating the rate of drying as given by a moisture balance around the differential section of the drier to the rate given by drying equation 611 gives

$$-W\,dX = RA_L\,dL = -\left(\frac{dX}{dt}\right)\left(\frac{A_L}{A}\right)dL \qquad (619)$$

where W = rate of flow of dry material (lb/hr).
X = moisture content (lb water/lb dry solid).
L = drier length (ft).
R = rate of drying per unit exposed area [lb water/(hr)(sq ft)].
A_L = sq ft of drying surface per foot of drier length.
A = sq ft of drying surface per lb of dry solid.

Therefore, substituting the value of $-dX/dt$ as given by equation 618 in equation 619,

$$\int_0^L dL = -\frac{W}{\left(\frac{0.00239}{x}\right)\left(\frac{G}{60}\right)^{0.6}\left(\frac{A_L}{A}\right)}\int_{X_{\text{initial}}}^{X_{\text{final}}}\frac{dX}{(F)(\mathcal{H}^* - \mathcal{H})} \qquad (620)$$

In order to perform the integration indicated in this equation it is necessary to determine F, $\mathcal{H}^*$, and $\mathcal{H}$ as a function of X. The relation between $\mathcal{H}$ and X is given by the material balance around any section of the drier from one end, so that

$$X_1 - X = \frac{GS}{W}(\mathcal{H} - \mathcal{H}_1)$$

where S = cross section of the path of air flow (sq ft).
1 = the condition at one end of the drier.

The value of $\mathcal{H}^*$ is readily determined from the humidity $\mathcal{H}$ and temperature T_g of the air and a humidity chart. The free moisture content F is equal to the total moisture content X less the equilibrium moisture content X^*, and the latter is given as a function of the air temperature and humidity in Table 63. These relationships may be tabulated as in Table 64 and integrated graphically by plotting $1/F(\mathcal{H}^* - \mathcal{H})$ as a function of X and determining the area under the curve between the limits of X.

TABLE 64

Total Moisture Content X, lb water/lb solid	Air Humidity $\mathcal{H}$, lb water/lb dry air	$\mathcal{H}^*$, lb water/lb dry air	Equilibrium Moisture Content X^*, lb water/lb solid	Free Moisture Content ($F = X - X^*$), lb water/lb solid	$\frac{1}{F(\mathcal{H}^* - \mathcal{H})}$
0.099	0.0152	0.0278	0.040	0.059	1341
0.300	0.0190	0.0310	0.049	0.251	332
0.500	0.0231	0.0350	0.057	0.443	190
0.700	0.0265	0.0372	0.063	0.637	147
0.900	0.0303	0.0410	0.070	0.830	112
1.100	0.0341	0.0440	0.077	1.023	98.5
1.300	0.0378	0.0480	0.083	1.217	77
1.562	0.0428	0.0519	0.093	1.469	69

The area under the curve equals 300.

$$\text{The drier length equals } \frac{\frac{10{,}000}{(24)}(1.162)(300)}{\frac{(0.00203)(12 \times 25.4)}{1.03}(8.6 \times 1.162)(4)^{0.6}} = 150 \text{ ft.}$$

If no heating coils had been incorporated into the drier and the operation had been adiabatic, all heat required would have been supplied by the air and the humidity and temperature of the drying air would have followed an adiabatic cooling curve on the humidity chart instead of remaining at 140° F with increasing humidity as in the example problem.

ESTIMATING DRYING RATES

The engineer is sometimes required to estimate the time needed for drying where available laboratory data are inadequate. He must then estimate the drying rate curves for conditions other than those applying to the data available.

As has been indicated in the example problems, the drying rate during the constant rate period, where the surface of the wet stock exhibits the same equilibrium partial pressure characteristics as a free water surface in so far as the partial pressure of water is concerned, is proportional to the mass transfer coefficient and is equal to the rate of mass transfer during the constant rate period. The mass transfer coefficient is in turn proportional to the air velocity raised to the power 0.8. It is possible, therefore, to estimate the rate of drying from correlations of the rate of mass transfer to flat surfaces from fluids passing over them.

The rate of heat transfer from a fluid to a flat surface can also be estimated, and the rate of heat transfer divided by the latent heat of vaporization is equal to the rate of drying during the constant rate period. One difficulty in these estimates lies in the fact that in general the experimental correlations of the heat or mass transfer coefficients between fluids and flat surfaces are based for the most part on relatively smooth surfaces. In most drying operations the drying surface is far from smooth and the actual drying rate is generally higher than such estimates.

If a drying rate curve is available for one set of air conditions, the rate during the constant rate period at another set of air conditions may be estimated from the fact that the drying rate is proportional to the 0.8 power of air velocity and directly proportional to the partial pressure driving force for mass transfer.

The most important datum is a knowledge of the first critical moisture content ending period I.

Although in general the value of the critical moisture content is a function of many variables, when necessary in the absence of complete data it may be assumed to be independent of drying rate and other variables. When only the first critical moisture content is known, the drying rate curve during the falling rate period may be assumed to be a straight line connecting the origin and the critical point terminating the constant rate period when plotted on a free moisture content basis. On the basis of these latter assumptions, the drying rate at any given moisture content becomes proportional to the partial pressure driving force and the 0.8 power of air velocity, regardless of whether or not the drying is being conducted in the constant or falling rate periods.

These assumptions are undesirable and should be avoided where possible. The assumption that the critical moisture content is independent of drying variables and is a property only of the material being dried is most in error for insoluble porous materials such as sand, and is least in error for materials which hold a large proportion of their water by chemical means.

PROBLEMS

1. Design a co-current adiabatic tunnel drier to handle 2000 lb/hr of wet sand from a free moisture content of 1.0 to 0.1 lb of water per pound of dry solid. The equilibrium moisture content of the material is negligible. The air available for drying has a dry-bulb temperature of 200° F and a wet-bulb temperature of 94° F. The air leaves the drier at 100° F. The stock enters and leaves the drier at 94° F. A sample of this sand when dried with the above hot air under constant drying conditions had a critical moisture content of 0.5 lb of water per pound of dry solid, and the rate above the critical moisture content was 1.0 lb of water per square foot per hour. Below the critical moisture content the rate fell to zero at zero free moisture content along a straight line. The effective drying area in the continuous drier is 0.3 sq ft/lb of dry solid. Neglect shrinkage.

Compute the length of the drier if it holds 20 lb of dry material per foot of length.

2. A countercurrent continuous drier is to be designed to remove water from 8.15 lb/min of dry material. Dry air enters the system at 428 lb/min and a temperature of 140° F and is maintained at that temperature by steam coils within the drier. The entering air humidity is 0.0155. Twenty-five per cent of the dry air passing through the drier is recirculated to the inlet.

The entering material has a total moisture content of 1.562 lb of moisture per pound of dry solid and is to be dried to a total moisture content of 0.099. The equilibrium moisture content is constant at 0.05 lb of water per pound of dry solid.

Laboratory data for this material were correlated on the basis of the following equation:

$$\frac{\partial F}{\partial t} = 1.625F \, \Delta \mathcal{H}$$

where F = free moisture content.
t = time (min).
$\Delta \mathcal{H} = \mathcal{H}^* - \mathcal{H}$.
$\mathcal{H}^*$ = humidity of air when in equilibrium with water on the solid (approximately saturated humidity at the wet-bulb temperature T_w).
$\mathcal{H}$ = humidity of air.

If the drier holds 10 lb of dry material per foot of length, how long should the drier be?

3. Compute the dew-point pressure of a mixture of air-carbon tetrachloride-water at 38.7° C. The carbon tetrachloride mixing ratio X is 2.00, and the water humidity is 0.05. The vapor pressures of carbon tetrachloride and water at 38.7° C are 200 and 51.6 mm of mercury, respectively.

BIBLIOGRAPHY

1. ARNOLD, J. H., *Ind. Eng. Chem.*, **22**, 1091 (1930).
2. ARNOLD, J. H., *Trans. Am. Inst. Chem. Engrs.*, **40**, 361 (1944).
3. AWBERY, J. H., and E. GRIFFITHS, *Proc. Phys. Soc. (London)*, **44**, Pt. 2, 132 (1932).
4. BADGER, W. L., and W. L. MCCABE, *Elements of Chemical Engineering*, McGraw-Hill Book Co., (1936).
5. BAKHMETEFF, B. A., *The Mechanics of Turbulent Flow*, Princeton University Press (1936).
6. BARTA, E. J., and H. G. GARBER, *Chem. & Met. Eng.*, **47**, 292 (1940).
7. CHILTON, T. H., and A. P. COLBURN, *Ind. Eng. Chem.*, **26**, 1183 (1934).
8. CHILTON, T. H., and A. P. COLBURN, *Ind. Eng. Chem.*, **27**, 255 (1935).
9. COLBURN, A. P., *Ind. Eng. Chem.*, **22**, 967 (1930).
10. COLBURN, A. P., *Trans. Am. Inst. Chem. Engrs.*, **28**, 105 (1932).
11. COLBURN, A. P., *Trans. Am. Inst. Chem. Engrs.*, **29**, 174 (1933).
12. GLASSTONE, S., K. J. LAIDLER, and H. EYRING, *The Theory of Rate Processes*, p. 517, McGraw-Hill Book Co. (1941).
13. COLBURN, A. P., *Trans. Am. Inst. Chem. Engrs.*, **35**, 211 (1939).
14. COLBURN, A. P., *Ind. Eng. Chem.*, **33**, 459 (1941).
15. DWYER, O. E., and B. F. DODGE, *Ind. Eng. Chem.*, **33**, 485 (1941).
16. DREW, T. B., J. J. HOGAN, and W. H. MCADAMS, *Ind. Eng. Chem.*, **23**, 936 (1931).
17. DREW, T. B., *Trans. Am. Inst. Chem. Engrs.*, **26**, 26 (1931).
18. DUNCAN, D. W., J. H. KOFFOLT, and J. R. WITHROW, *Trans. Am. Inst. Chem. Engrs.*, **38**, 259 (1942).

18*a*. EICHHORN, J., Ph.D. dissertation, University of Michigan (1950).

19. FLUOR CORP. LTD., *Bulletin T* 337 (1939).
20. FURNAS, C. C., and M. L. TAYLOR, *Trans. Am. Inst. Chem. Engrs.*, **36**, 135 (1940).
21. GAMSON, B. W., G. THODOS, and O. A. HOUGEN, *Trans. Am. Inst. Chem. Engrs.*, **39**, 1 (1943).
22. GILLILAND, E. R., *Ind. Eng. Chem.*, **26**, 681 (1934).
23. GILLILAND, E. R., *Ind. Eng. Chem.*, **30**, 506 (1938).
24. GILLILAND, E. R., and T. K. SHERWOOD, *Ind. Eng. Chem.*, **26**, 516 (1934).
25. GROSVENOR, W. M., *Trans. Am. Inst. Chem. Engrs.*, **1**, 184 (1908).
26. JOHNSTONE, H. F., and A. D. SINGH, *Ind. Eng. Chem.*, **29**, 286 (1937).
27. HASLAM, R. T., R. L. HERSHEY, and R. H. KEAN, *Ind. Eng. Chem.*, **16**, 1224 (1924).
28. HASLAM, R. T., W. P. RYAN, and H. C. WEBER, *Trans. Am. Inst. Chem. Engrs.*, **15**, Pt. I, 177 (1923).
29. HILPERT, R., *Forschung, Forschungsheft*, **355**, 21 (1932).
30. HOBSON, M., and G. THODOS, *Chem. Eng. Progress*, **45**, 517 (August 1949).
31. HIGBIE, R., *Trans. Am. Inst. Chem. Engrs.*, **31**, 36 (1935).
32. HOUGEN, O. A., "Teaching of Chemical Engineering," *Proc. Chem. Eng. Div., Soc. Prom. Eng. Ed.*, 2nd Summer School in *Chem. Eng.*, June 1939, chapter on "Drying."
33. HURT, D. M., *Ind. Eng. Chem.*, **35**, 522 (1943).
34. KELLY, R. C., paper published by the Fluor Corp., presented before the California Natural Gasoline Association, Dec. 3, 1942.

34*a*. MCADAMS, W. H., J. B. POHLENZ, and R. C. ST. JOHN, *Chem. Eng. Progress*, **45**, 241 (April 1949).

35. MARK, J. C., data presented by SHERWOOD, *Trans. Am. Inst. Chem. Engrs.*, **28**, 107 (1932).
36. MAXWELL, J. C., Scientific Papers, **2**, p. 625, Cambridge University Press, Cambridge, England (1890). Reprinted from *Encyclopedia Britannica*, article on "Diffusion," 9th ed. (1877).

36*a*. MICKLEY, H. S., *Chem. Eng. Progress*, **45**, 739 (December 1949).

37. MOLSTAD, M. C., R. G. ABBEY, A. R. THOMPSON, and J. F. MCKINNEY, *Trans. Am. Inst. Chem. Engrs.*, **38**, 387 (1942).
38. MOLSTAD, M. C., J. F. MCKINNEY, and R. G. ABBEY, *Trans. Am. Inst. Chem. Engrs.*, **39**, 605 (1943).
39. MURPHREE, E. V., *Ind. Eng. Chem.*, **24**, 726 (1932).
40. PERRY, J. H., *Chemical Engineers' Handbook*, 3rd ed., pp. 539–696, McGraw-Hill Book Co. (1950).
41. PRANDTL, L., *Z. Physik*, **11**, 1072 (1910); **29**, 487 (1928).
42. RESNICK, W., and R. R. WHITE, *Chem. Eng. Progress*, **45**, 377 (1949).
43. SHERWOOD, T. K., *Absorption and Extraction*, 1st ed., McGraw-Hill Book Co. (1937).
44. SHERWOOD, T. K., *Ind. Eng. Chem.*, **33**, 424 (1941).
45. SHERWOOD, T. K., *Trans. Am. Inst. Chem. Engrs.*, **36**, 817 (1940).
46. SHERWOOD, T. K., and E. W. COMINGS, *Trans. Am. Inst. Chem. Engrs.*, **28**, 88 (1932).
47. SHERWOOD, T. K., and F. A. L. HOLLOWAY, *Trans. Am. Inst. Chem. Eng.*, **36**, 21, 39 (1940).
48. SHERWOOD, T. K., and B. B. WOERTZ, *Trans. Am. Inst. Chem. Eng.*, **35**, 517 (1939).
49. TOWLE, W. L., and T. K. SHERWOOD, *Ind. Eng. Chem.*, **31**, 457 (1939).
50. VAN DREIST, E. R., *J. Applied Mechanics*, **13**, 34 (1946).
51. VON KÁRMÁN, THEODORE, *Trans. Am. Soc. Mech. Engrs.*, **61**, 705 (1939).
52. WALKER, W. H., W. K. LEWIS, W. H. MCADAMS, and E. R. GILLILAND, *Principles of Chemical Engineering*, 3rd ed., McGraw-Hill Book Co. (1937).
53. WILKE, C. R., and O. A. HOUGEN, *Trans. Am. Inst. Chem. Engrs.*, **41**, 445 (1945).
54. WILKE, C. R., "Estimation of Liquid Diffusion Coefficients," *Chem. Eng. Progress*, **45**, 218 (March 1949).
55. WILKE, C. R., "Diffusional Properties of Multicomponent Gases," *Chem. Eng. Progress*, **46**, 95 (Feb. 1950).

Nomenclature

Except where noted otherwise, the following nomenclature is used throughout the text.

A	Area; cross-sectional area of pipe, or that area through which transfer takes place.
$A = L/KV$	Absorption factor.
A'	Projected area.
A_m	Mean or average area.
A_o	Cross-sectional area of orifice opening.
A_n	Angle of nip.
a	Area per unit volume; a constant; relative volatility.
a	Acceleration (L/t^2); absorptivity in radiation.
a'	Acceleration in "gees."
B	Quantity of bottom product as from fractionating column; a constant.
b	Width of an element of fluid; linear dimension in direction of axis; a constant.
C	A constant; concentration; volume capacity in piping between source of pulsation and meter.
C^*	Concentration of component in liquid in equilibrium with vapor.
C, C', C''	Dimensionless coefficients for primary metering elements.
C_o	Dimensionless coefficient of discharge for orifice.
C_L, C_V	Filtration constants for noncompressible cakes and constant pressure filtration; concentration in liquid and vapor phases, respectively.
C_V'	Filtration constant for compressible cakes.
$C_P = (dH/dT)_P$	Heat capacity; also specific heat $\bar{C}_P$ at constant pressure.
$C_V = (dU/dT)_V$	Heat capacity; also specific heat $\bar{C}_V$ at constant volume.
c	Coefficient of performance of air lift; concentration.
$c = C_V'V_e$	A constant in filtration of compressible cakes.
D	Diameter or dimension; quantity of distillate or extract product.
D_{avg}	Average screen size of particles.
D_c	Diameter of vessel, container, or conduit.
D_e	Distance between tubes in a bank.
D_f	Diameter of feed particles.
D_G	Molecular diffusivity in gas phase.
D_i	Particular diameter or average size of particles.
D_L	Molecular diffusivity in liquid phase.
D_o	Diameter of orifice.
D_p	Diameter or dimension of particle; distance between rolls.
D_r	Diameter of rolls in roll crushing.
D_s	Diameter of sphere of same volume
d	Differential operator.
E	Electrical potential volts; evaporation (lb/hr); energy; emissive power or energy emitted per unit of area in radiation; eddy diffusivity.
e	Base of natural logarithms, 2.718, dimensionless; emissivity in radiation.
F	Force (dimensions, System II, Table 25); quantity of feed material; shape factor in radiation; frequency of pulsation; $H - TS$; free moisture content in drying.
$\bar{F}$	Shape factor in radiation including effect of connecting refractory walls.
F'	Force (dimensions, System I, Table 25).
F_c	Centrifugal force.
F_f	Friction-factor factor for flow of fluids through porous beds.
F_f'	Friction factor for nonwetting fluid based on wetted sphericity ψ' (two-phase flow).
F_N	Normal force.
F_R	Resultant force.
F_T	Tangential force.
F_{Re}	Reynolds number factor for flow of fluids through porous beds.
F_{Re}	Reynolds number factor for nonwetting fluid in two-phase flow based on effective porosity X'.
F_s	Factor for hindered settling.
$\mathfrak{F}$	Factor for radiation including shape and emissivity and effect of refractories.
f	Friction factor for flow of fluids in conduits or pipe; feed stage (the stage which receives the feed); fugacity of a component.
f'	Friction factor for general relative motion of solids and fluids.
f''	Modified friction factor for wetting fluid.
f'''	Modified friction factor for nonwetting fluid.
f_D	Friction factor, or drag coefficient, for relative motion of solid particles and fluids.
G	Mass velocity [lb mass/(sec)(sq ft)].
G'	Mass velocity of solute-free stream [lb mass/(sec)(sq ft)].
g	Acceleration due to force of gravity.
g_c	Conversion factor between dimensions of Systems I and II, Table 25.
H	Enthalpy ($U + PV$); $\bar{H}$, specific enthalpy, particularly for vapor.
H	Henry's law constant $(Py/C)^*$.
$\mathcal{H}$	Humidity (lb vapor per lb dry gas).
$\mathcal{H}_s$	Saturated humidity.
$\mathcal{H}_{as}$	Saturated humidity at adiabatic saturation temperature.
$\mathcal{H}^*$	Humidity of saturated gas at a liquid temperature.

$\mathcal{H}_w$	Saturated humidity at the wet-bulb temperature T_w.
h	"Fall" of hydraulic ram; or "submergence" of air lift; coefficient of heat transfer; enthalpy of liquid.
h_p	Liquid gradient across plate in fractionating column.
h_s	Submergence, height of liquid above top of slots on bubble cap tray.
h_t	Coefficient of heat transfer by turbulent motion exclusively.
h'	"Lift" of hydraulic ram or of air lift.
h_c, h_r	Coefficient of heat transfer by "convection" and by "radiation," respectively.
I	Current of electricity.
i	Intensity of radiation.
J	Conversion factor between dimensions of Systems II and III, Table 25; addition of quantities of streams as $L_0 + L_F$.
j	j factor, approximately $f/8$ or $f'/2$ or (0.023 $\mathrm{Re}^{-0.2}$).
K	Conductance; permeability; equilibrium constant or ratio, as y/x; overall mass transfer coefficient.
k, k_q	Thermal conductivity.
k	Ratio of heat capacities C_P/C_V, dimensionless factor or constant; mass of solution retained per unit mass of wet solids; coefficient of mass transfer.
k_e	Eddy conductivity.
k_g	Coefficient of mass transfer in gas phase.
$k_{\mathcal{H}}$	Coefficient of mass transfer in humidity units.
k_l	Coefficient of mass transfer in liquid phase.
k'	Coefficient of mass transfer in concentration units [moles/(hr)(sq ft)(moles/cu ft)].
L	Length or linear dimension; thickness of bed or cake; quantity of liquid or lower stream.
L'	Quantity of solute-free (or inert) liquid, quantity of liquid or lower stream below feed.
L_e	Equivalent length or thickness.
lw	Lost work (energy that might have done work); all irreversibilities.
M	Molecular weight.
Ma	Mach number v/v_c.
m	Mass; any stage.
m'	Mass (dimensions of System II, Table 25).
m_i, m_j	Mass or mass fraction of material of particular properties.
N	Number; revolutions per minute; mass (or moles) transferred per unit time; resistance factor, dimensionless, fL/D.
Nu	Nusselt number, hD/k.
n	Ratio of specific surfaces (of particles to spheres); exponent for V in equation for gas expansion ($PV^n = C$); number of stages; lowest or last stage; number of transfer units.
P	Pressure (lb force/sq ft); absolute pressure.
P	Product, in screening, in evaporation.
P'	Gage pressure.
P_0	Reference pressure for gas measurement.
Pe	Peclet number, $Dv\rho C_P/k$ = (Re)(Pr).
Po	Power number, $pg_c/\rho n^3 D_i^5$
Pr	Prandtl number, $\mu C_P/k$.
$-\Delta P$	Pressure drop; $(-\Delta P_f)$ due to friction, or required for fluidizing.
$(-\Delta P)_c$	Pressure drop through filter cake.
p	Partial pressure; power.
p_s	Vapor pressure of pure component at the temperature of the gas.
p^*	Partial pressure of a component in a gas in equilibrium with a liquid. Vapor pressure of a liquid.
p_w	Partial pressure of vapor in a saturated gas at the wet-bulb temperature.
Q	Quantity of material; quantity of heat; volume of fluid pumped per unit time; number of standard cubic feet of gas measured at T_0 and P_0 per unit time.
q	Heat absorbed; heat transferred per unit time, dQ/dt.
q_c	Heat absorbed in condenser.
R	Universal gas constant [energy/(mass)(degree temperature), as, 1544 (ft-lb)(lb mole)(°F)]; resistance, in ohms, or due to fouling of heat transfer surfaces; reject in screening; rate of drying; reflux ratio L_0/D.
Re	Reynolds number, $Dv\rho/\mu$.
Re	Reynolds number, $D_p F_{\mathrm{Re}} v\rho/\mu$.
Re″	Reynolds number for wetting fluid, $D_p F_{\mathrm{Re}} v\rho/\mu S_e^y$.
Re‴	Reynolds number for nonwetting fluid, $D_p F_{\mathrm{Re}}' v\rho/\mu$.
r	Radius, or radial distance; distance from center of rotation.
r_h	Hydraulic radius, cross-sectional area ÷ wetted perimeter, S/ψ.
S	Entropy; saturation volume of voids filled with wetting fluid ÷ total volume of voids; area; cross-sectional area; quantity of solids; quantity of solvent.
S	Stripping factor (KV/L).
S	Relative saturation.
S_e	Effective saturation.
S_m	Minimum seal or submergence, or height of liquid above slots on bubble-cap tray.
S_r	Residual saturation.
S_f	Fixed saturation.
Sc	Schmidt number, $\mu/\rho D_G$.
St	Stanton number, $h/C_P v\rho$.
s	Distance; surface per unit volume of particles; humid heat (increase in enthalpy of wet gas per lb dry gas per °F).
T	Temperature, particularly absolute; tons per hour; torque.
T_i	Temperature of interface, or of solid surface.

T_0 Reference temperature for gas measurement.
T_{as} Adiabatic saturation temperature.
T_w Wet-bulb temperature.
t Time.
U Internal energy (not including energy of position, motion, or electrical); average overall heat transfer coefficient.
u Velocity of fluid derived from particle; peripheral velocity; tangential velocity of impeller.
u' Velocity of fluid relative to impeller.
V Volume; quantity of vapor or upper stream.
V_e Equivalent volume of filtrate.
V' Quantity of inert, insoluble, or solute-free gas; quantity of vapor or upper stream below feed.
v Velocity; average superficial velocity; velocity of fluid relative to pump casing.
v_c Acoustic or critical velocity.
v_h Horizontal component of velocity of particle.
v_m Maximum velocity of freely falling particle.
v_r Radial velocity.
v_s Velocity of vapors through slots in bubble cap.
v_t Tangential velocity of particle.
v_v Vertical component of velocity.
v_c' Velocity (superficial) of air in two-phase flow.
v_L' Velocity of liquid in two-phase flow.
W Mass flow rate; mass rate of condensate in condensing vapors; mass of fluid exhausted by hydraulic ram.
W' Mass of fluid pumped by hydraulic ram.
w Work done by system, or material flowing, on surroundings (also w_s); mass of fluid displaced by solid.
X Volume fraction of slurry occupied by fluid.
X Porosity, volume of void space ÷ total volume of bed; ratio of quantity of component in stream L to quantity of other specified component.
X Mass ratio of one component to another, as pounds of water per pound of dry solid.
X^* Equilibrium mass ratio, as pounds of water per pound of dry solid in equilibrium with a particular medium.
X' Equivalent porosity of wetted bed.
x Distance; mass (or mole) fraction, particularly in stream L.
Y Volume fraction of slurry occupied by solid; ratio of quantity of component in stream V to quantity of other specified component (as moles of solute per mole of solute-free gas); correction factor for multiple-pass heat exchangers (equation 402).
y Distance; distance from wall; exponent for S_e in two-phase flow; mass (or mole) fraction, particularly in stream V.
Z Height, vertical distance above reference plane.
z Distance; dimensionless coefficient in dimensional analysis; compressibility factor for real fluids ($PV = zNRT$).

Greek Letters

α (alpha) Angle between velocity of fluid (v) and tangential velocity of impeller (u) or fluid angle; angle between direction of motion and horizontal; specific resistance of filter cake; constant.
β (beta) Angle between negative velocity of fluid relative to impeller ($-u'$) and tangential velocity of impeller (u) or vane angle; constant.
γ (gamma) Surface tension (force per unit of length).
Δ (delta) Operator signifying the finite increase in value from initial to later or final state; the difference in value between two streams, as $L_0 - V$.
δ (delta) Operator signifying a small increment or value.
ϵ (epsilon) Eddy viscosity; roughness of surface as height or depth of irregularities.
η (eta) Proportionality constant in eddy friction.
θ (theta) Angle; angle from the normal; contact angle (measured in more dense phase).
λ (lambda) Latent heat of vaporization (ΔH).
μ (mu) Coefficient of viscosity.
μ_b Bulk viscosity, as viscosity of suspension or emulsion.
π (pi) Geometric ratio of circumference to diameter of circle, 3.1416.
ρ (rho) Density (mass per unit volume).
ρ_{avg} Mean or average density of contents of column of air lift.
ρ_b Bulk density.
ρ_h Density of heavy liquid.
ρ_l Density of light liquid.
ρ_s Density of solid.
σ (sigma) Surface area; Stefan-Boltzmann constant in radiation.
τ (tau) Frictional stress (force per unit area); average frictional stress.
ϕ (phi) Operator signifying "function of" (different operators distinguished by different numerical subscripts); fraction of vapors bypassed in column; a term representing a complex group.
ψ (psi) Length of wetted perimeter; sphericity, surface of sphere of equal volume ÷ surface of particle.
ψ' Wetted sphericity.
ω (omega) Angular velocity; solid angle.

Subscripts

A, B, C, *etc.*	Designating the component, stream, or phase of material to which the symbol is restricted.
f	Designating irreversibility due to friction; fluidization; feed stage.
g	Designating gas or gas phase.
i	Designating interface.
l	Designating liquid phase, or phase L.
m	Designating any stage; by molecular action; mean or maximum value.
t	Designating by turbulent action.
v	Designating vapor, or phase V.
0, 1, 2, 3, etc.	Designating points or stages in the operation.

In gas flow:

a	Adiabatic flow.
c	Conditions at end of pipe corresponding to conditions of maximum mass; discharge or "critical flow."
i	Isothermal flow.
n	For nozzle alone ($L = 0$, $N = 0$).
0	In reservoir ahead of pipe or nozzle.
1	In throat of nozzle, or outlet to pipe.
2	At any point distant L downstream from throat of nozzle inlet or exit of pipe.
3	In reservoir into which pipe discharges.
—	Bar over a symbol indicates per unit mass, or an average value.

Appendix

LIQUID-LIQUID EQUILIBRIA

Acetic Acid-Water-Isopropyl Ether System

Observed Data at 20° C and 1 Atm

Water Layer, mass %			Ether Layer, mass %		
Acid	Ether	Water	Acid	Ether	Water
0.0	1.0	99.0	0.0	99.4	0.6
1.0	1.4	97.6	0.33	98.95	0.72
2.0	1.6	96.4	0.63	98.60	0.77
5.0	1.9	93.1	1.4	97.6	1.0
10.0	2.15	87.85	3.6	95.0	1.4
20.0	2.80	77.20	7.0	90.8	2.2
30.0	3.5	66.5	11.5	83.9	4.6
40.0	5.8	54.2	25.9	65.4	9.7
44.0	9.6	46.4	30.1	59.1	10.8
45.0	11.6	43.4	31.9	56.3	11.8
46.5	16.6	36.9	36.20	48.7	15.1
44.8	31.3	23.9	44.8	31.3	23.9

Benzene-Ethanol-Water System

Observed Data at 25° C and 1 Atm *

Light Layer Mass Fraction		Heavy Layer Mass Fraction	
Benzene	Ethanol	Benzene	Ethanol
0.9935	0.0060	0.0013	0.1120
0.9800	0.0185	0.0028	0.2200
0.9630	0.0350	0.0053	0.3090
0.9525	0.0445	0.0102	0.3600
0.9520	0.0450	0.0110	0.3640
0.9200	0.0740	0.0335	0.4350
0.9180	0.0755	0.0370	0.4410
0.9075	0.0845	0.0480	0.4590
0.8440	0.1385	0.1370	0.5220
0.8000	0.1745	0.2045	0.5215
0.7060	0.2475	0.3320	0.4800
0.6905	0.2585	0.3490	0.4720
0.6401	0.2930	0.4095	0.4400
0.520	0.374	0.520	0.374

* K. A. Varteressian and M. R. Fenske, *Ind. Eng. Chem.*, **28**, 928 (1936).

n-Heptane-Methylcyclohexane-Aniline System

Observed Liquid-Liquid Equilibrium Compositions at 25° C and 1 Atm *

Hydrocarbon-Rich Layer Mass Fraction		Aniline-Rich Layer Mass Fraction	
Methyl-cyclo-hexane	*n*-Heptane	Methyl-cyclo-hexane	*n*-Heptane
0	0.926	0	0.062
0.092	0.831	0.008	0.060
0.186	0.734	0.027	0.053
0.220	0.698	0.030	0.051
0.338	0.576	0.046	0.045
0.409	0.504	0.060	0.040
0.460	0.450	0.074	0.036
0.597	0.307	0.092	0.028
0.672	0.228	0.113	0.021
0.716	0.182	0.127	0.016
0.736	0.160	0.131	0.014
0.833	0.054	0.156	0.06
0.881	0	0.169	0

* K. A. Varteressian and M. R. Fenske, *Ind. Eng. Chem.*, **29**, 270 (1937).

VAPOR-LIQUID EQUILIBRIA

n-HEPTANE-METHYLCYCLOHEXANE SYSTEM

Observed Data at 1 Atm *

Mole Fraction n-Heptane in Liquid	Saturation Temperature, °C	Mole Fraction n-Heptane	
		in Liquid	in Vapor
0	100.80	0.0310	0.0350
0.0787	100.55	0.0580	0.0620
0.1638	100.35	0.0950	0.1030
0.2486	100.15	0.1330	0.1430
0.4126	99.70	0.1800	0.1920
0.5186	99.20	0.2160	0.2290
0.6056	99.00	0.2175	0.2890
0.6993	98.85	0.3170	0.3330
0.7942	98.60	0.3630	0.3810
0.9338	98.50	0.4010	0.4200
1	98.40	0.4560	0.4750
		0.5010	0.5210
		0.5590	0.5780
		0.5990	0.6180
		0.6470	0.6660
		0.7090	0.7280
		0.7560	0.7710
		0.7960	0.8100
		0.8430	0.8535
		0.8790	0.8900
		0.9060	0.9130
		0.9310	0.9400
		0.9540	0.9625
		0.9800	0.9860

* E. C. Bromiley and D. Quiggle, *Ind. Eng. Chem.*, **25**, 1136 (1933).

ETHANOL-WATER SYSTEM AT 1 ATM *

Reference states for enthalpies are pure liquids at 32° F under their own vapor pressures

Saturation Temperature, °F	Ethanol Concentration		Enthalpy, Btu/lb of Mixture †	
	Mass Fraction in Liquid	Mass Fraction in Vapor	Liquid	Vapor
212	0	0	180.1	1150
210.1	0.010	0.103		
208.5	0.020	0.192		
206.9	0.030	0.263		
204.8	0.040	0.325		
203.4	0.050	0.377	169.3	1115
197.2	0.100	0.527	159.8	1082
189.2	0.200	0.656	144.3	1012.5
184.5	0.300	0.713	135.0	943
181.7	0.400	0.746	128.2	873
179.6	0.500	0.771	122.9	804
177.8	0.600	0.794	117.5	734
176.2	0.700	0.822	111.1	664
174.3	0.800	0.858	103.8	596
174.0	0.820	0.868		
173.7	0.840	0.877		
173.4	0.860	0.888		
173.2	0.880	0.900		
173.0	0.900	0.912	96.6	526
172.9	0.920	0.926		
172.8	0.940	0.942		
172.7	0.960	0.959		
172.8	0.980	0.978		
173.0	1	1	89.0	457.5

* Cornell and Montonna, *Ind. Eng. Chem.*, **25**, 1331 (1933); Noyes and Warfel, *J. Am. Chem. Soc.*, **23**, 463 (1901).

† Composition for liquid and vapor is the same as given in the second column under mass fraction in liquid.

ACETONE-ACETIC ACID SYSTEM

Observed Data at 1 Atm *

Temperature, °C	Mole % Acetone	
	Liquid	Vapor
118.1	0	0
110.0	5	16.2
103.8	10	30.6
93.1	20	55.7
85.8	30	72.5
79.7	40	84.0
74.6	50	91.2
70.2	60	94.7
66.1	70	96.9
62.6	80	98.4
59.2	90	99.3
56.1	100	100

* D. F. Othmer, *Ind. Eng. Chem.*, **35**, 617 (1943).

DENSITIES OF ETHANOL-WATER MIXTURES, GRAMS/CC

Mass % Ethanol	Saturated Liquid	Liquid at 100° F
0	0.952	0.992
20	0.905	0.959
40	0.859	0.920
60	0.813	0.873
80	0.768	0.827
100	0.721	0.772

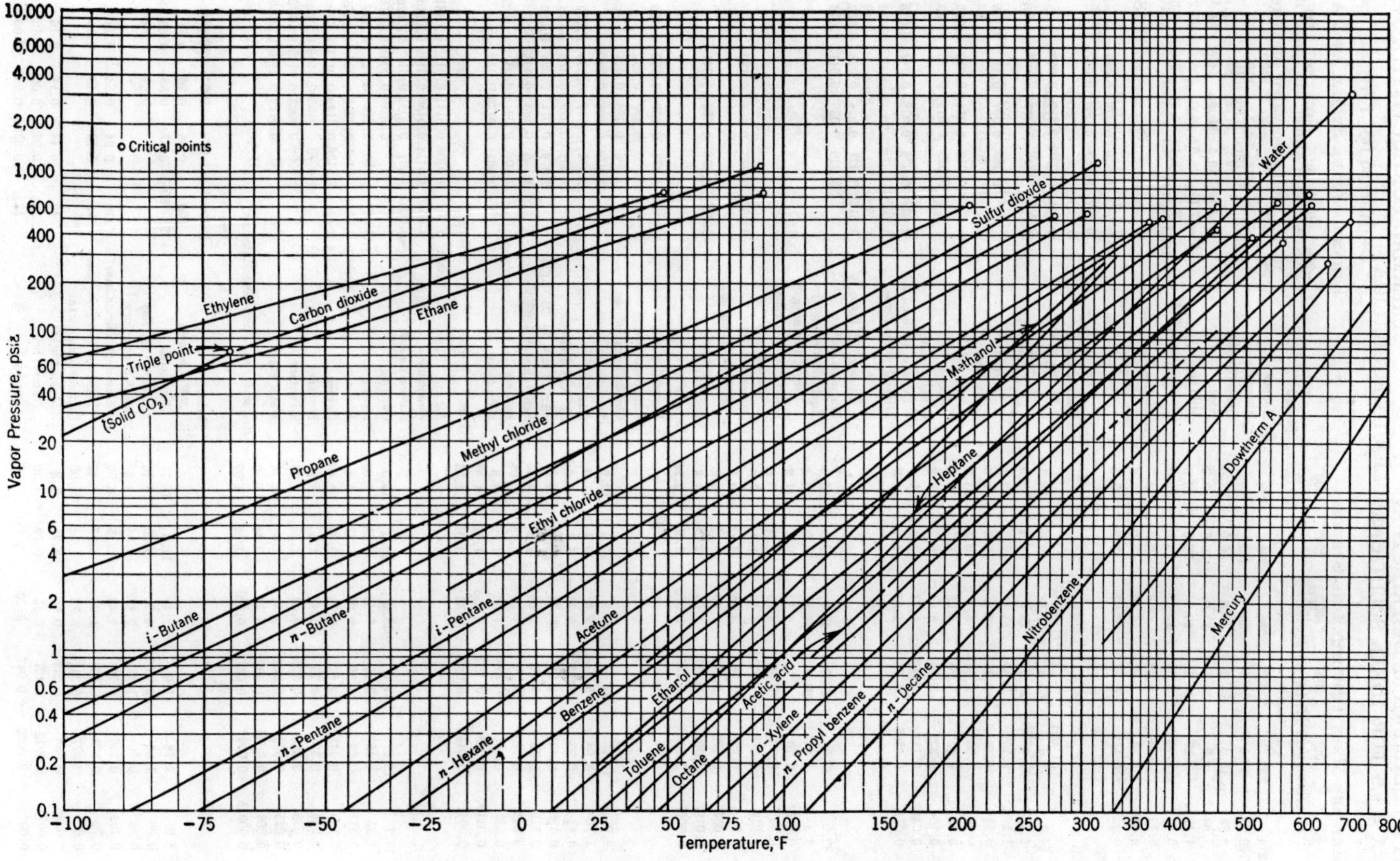

FIG. 543. Vapor pressure as a function of temperature.

VOLATILITY EQUILIBRIUM DISTRIBUTION RATIOS, $K = y/x$ FOR IDEAL SOLUTIONS

Pressure, psia	Temperature, °F 40°	100°	200°	300°	400°	500°
Pressure = 14.7 psia						
CH_4	214	252	276	286	291	296
C_2H_6	22.5	38.5	69.0	94.0	110	124
C_3H_8	4.95	10.0	25.0	41.0	56.0	71.0
i-C_4H_{10}	1.83	4.6	13.7	25.0	38.5	52.0
n-C_4H_{10}	1.19	3.27	10.4	20.5	32.5	46.0
i-C_5H_{12}	0.41	1.40	5.60	12.2	21.8	32.0
n-C_5H_{12}	0.30	1.02	4.26	7.15	14.2	28.3
n-C_6H_{14}	0.091	0.34	1.89	6.00	12.0	19.6
n-C_7H_{16}	0.028	0.112	0.82	3.23	7.5	13.4
Pressure = 50 psia						
CH_4	64.3	76.4	83.8	87.0	89.0	89.5
C_2H_6	6.3	11.4	21.0	28.5	34.4	39.0
C_3H_8	1.5	3.15	7.85	13.1	17.8	22.3
i-C_4H_{10}	0.59	1.45	4.25	8.1	12.0	16.0
n-C_4H_{10}	0.40	1.03	3.24	6.7	10.4	13.9
i-C_5H_{12}	0.14	0.43	1.70	4.05	6.75	9.8
n-C_5H_{12}	0.104	0.322	1.39	3.44	5.85	8.8
n-C_6H_{14}	0.034	0.113	0.605	1.93	3.8	6.0
n-C_7H_{16}	0.011	0.0395	0.27	1.04	2.46	4.3
Pressure = 100 psia						
CH_4	32.0	37.8	41.8	43.9	45.0	45.5
C_2H_6	3.4	5.8	10.6	14.8	18.0	20.4
C_3H_8	0.795	1.70	4.17	6.87	9.4	11.8
i-C_4H_{10}	0.31	0.76	2.30	4.35	6.50	8.6
n-C_4H_{10}	0.217	0.545	1.77	3.5	5.55	7.4
i-C_5H_{12}	0.08	0.235	0.94	2.24	3.74	5.4
n-C_5H_{12}	0.058	0.172	0.745	1.89	3.23	4.8
n-C_6H_{14}	0.0197	0.062	0.327	1.05	2.08	3.3
n-C_7H_{16}	0.0067	0.022	0.143	0.575	1.38	2.25
Pressure = 200 psia						
CH_4	15.9	19.1	21.4	22.1	22.5	22.9
C_2H_6	1.76	3.04	5.5	7.6	9.4	11.0
C_3H_8	0.43	0.95	2.17	3.71	5.2	6.6
i-C_4H_{10}	0.180	0.425	1.25	2.40	3.60	4.8
n-C_4H_{10}	0.119	0.30	0.97	2.03	3.15	4.25
i-C_5H_{12}	0.045	0.128	0.52	1.28	2.15	3.05
n-C_5H_{12}	0.033	0.096	0.414	1.08	1.89	2.8
n-C_6H_{14}	0.0118	0.035	0.180	0.595	1.22	1.92
n-C_7H_{16}	0.0041	0.013	0.080	0.33	0.84	1.35
Pressure = 300 psia						
CH_4	10.7	13.0	14.6	15.0	15.2	15.5
C_2H_6	1.25	2.15	3.75	5.40	6.60	7.60
C_3H_8	0.317	0.69	1.55	2.67	3.77	4.8
i-C_4H_{10}	0.130	0.315	0.81	1.75	2.64	3.60
n-C_4H_{10}	0.086	0.22	0.72	1.48	2.3	3.2
i-C_5H_{12}	0.035	0.095	0.375	0.93	1.60	2.34
n-C_5H_{12}	0.025	0.072	0.305	0.80	1.4	2.12
n-C_6H_{14}	0.009	0.0265	0.132	0.44	0.94	1.50
n-C_7H_{16}	0.00315	0.0097	0.057	0.243	0.66	1.09
Pressure = 400 psia						
CH_4	8.0	9.78	11.0	11.4	11.6	11.7
C_2H_6	0.96	1.66	2.93	4.20	5.20	6.10
C_3H_8	0.26	0.56	1.23	2.17	3.04	3.9
i-C_4H_{10}	0.104	0.250	0.74	1.43	2.15	3.00
n-C_4H_{10}	0.70	0.178	0.585	1.20	1.88	2.65
i-C_5H_{12}	0.029	0.077	0.31	0.78	1.32	2.00
n-C_5H_{12}	0.0205	0.058	0.243	0.66	1.2	1.8
n-C_6H_{14}	0.0074	0.0213	1.107	0.36	0.79	1.30
n-C_7H_{16}	0.00265	0.0080	0.047	0.20	0.55	0.9

THERMAL CONDUCTIVITIES, k, Btu/(hr)(sq ft)(°F/ft)

	Temperature, °F −200°	−100°	0°	100°	200°	300°	400°	500°
Metals								
Copper	245	228	224	220	218	215	212	210
Aluminum		112	114	117	119	122	128	130
Brass (70-30)		54	56	57.2	59	61	63	65
Nickel		37.6	36.5	35.4	34.5	34.0	33.0	32.0
Mild steel		27.0	26.6	26.2	26.0	25.7	25.5	25.2
Stainless steel (304)		7.7	8.1	8.55	9.0	9.5	10.0	10.4
Insulating Materials								
Fir plywood								
17% moisture		0.062	0.069	0.076	0.083			
0% moisture		0.047	0.053	0.060	0.067			
Cork board								
12.2 lb/cu ft		0.021	0.0235	0.026	0.0285			
6.5 lb/cu ft		0.018	0.020	0.022	0.024			
Asbestos fiber, 36 lb/cu ft			0.082	0.097	0.11	0.115	0.12	0.125
Glass fiber								
9 lb/cu ft			0.010	0.011	0.020	0.023	0.031	0.043
3 lb/cu ft			0.013	0.015	0.025	0.035	0.047	0.065
Cellular Glass								
Density, 10 lb/cu ft		0.019	0.028	0.038	0.042	0.046	0.050	0.055
Mineral wool (Zero-lite)								
Density, 15 lb/cu ft		0.016	0.021	0.024	0.028	0.032	0.035	0.038
Magnesia (85%)			0.036	0.039	0.041	0.043	0.046	0.049
Equivalent conductivity of 1-in. air gap								
Between materials of ordinary emissivity		0.052	0.078	0.104	0.130			
With aluminum on warm side		0.037	0.038	0.039	0.041			
Liquids								
Water (liquid)			0.298	0.350	0.402	0.470		
Ammonia (liquid)			0.290	0.290	0.290			
Carbon tetrachloride			0.098	0.086	0.074			
Acetone			0.108	0.103	0.0975			
Gasoline			0.091	0.089	0.085	0.082	0.080	
37° API distillate			0.082	0.079	0.076	0.075	0.072	
SAE 10 lube oil (21° API)			0.074	0.071	0.068	0.066	0.064	
Ethyl alcohol (100%)			0.109	0.103	0.0983			
Benzene			0.10	0.09	0.08			
n-Pentane			0.083	0.0775	0.072			
Gases								
Air	0.006	0.0089	0.0108	0.0141	0.0162	0.0181	0.0197	0.0211
Ammonia		0.0088	0.0118	0.0150	0.0185	0.0215	0.025	0.029
Carbon dioxide		0.0057	0.007	0.0085	0.01	0.0126	0.0135	0.0156
Chlorine				0.0049	0.0061	0.0075		
Hydrogen		0.072	0.095	0.113	0.127	0.14	0.153	0.168
Methane	0.0080	0.0125	0.0165	0.0205	0.025	0.0295	0.0347	0.0398
Oxygen	0.0074	0.011	0.0126	0.0149	0.0171	0.019	0.0206	0.0221
Propane			0.0076	0.0105	0.0146	0.0190	0.024	0.029
n-Pentane			0.0066	0.0093	0.0125	0.0160	0.0198	0.024
Water vapor				0.0115	0.0140	0.0165	0.0192	0.0225

Refractory and Insulating Materials

Brick	Density, lb/cu ft	0°	500°	1000°	1500°	2000°
30% Chromia	200	0.56	0.70	0.81	0.89	0.96
Fire brick	125–130	0.47	0.55	0.66	0.76	0.86
Missouri fire clay			0.64	0.82	0.91	0.97
Castable refractory						
Aluminum silicate base	110	0.24	0.28	0.31	0.34	0.38
Lightweight silicate base	75	0.13	0.13	0.15	0.15	0.16
Diatomaceous earth						
Molded brick	40	0.13	0.14	0.16	0.18	0.19
Natural brick	28	0.07	0.08	0.10	0.11	0.12
Powder	17	0.035	0.05	0.06	0.07	0.08

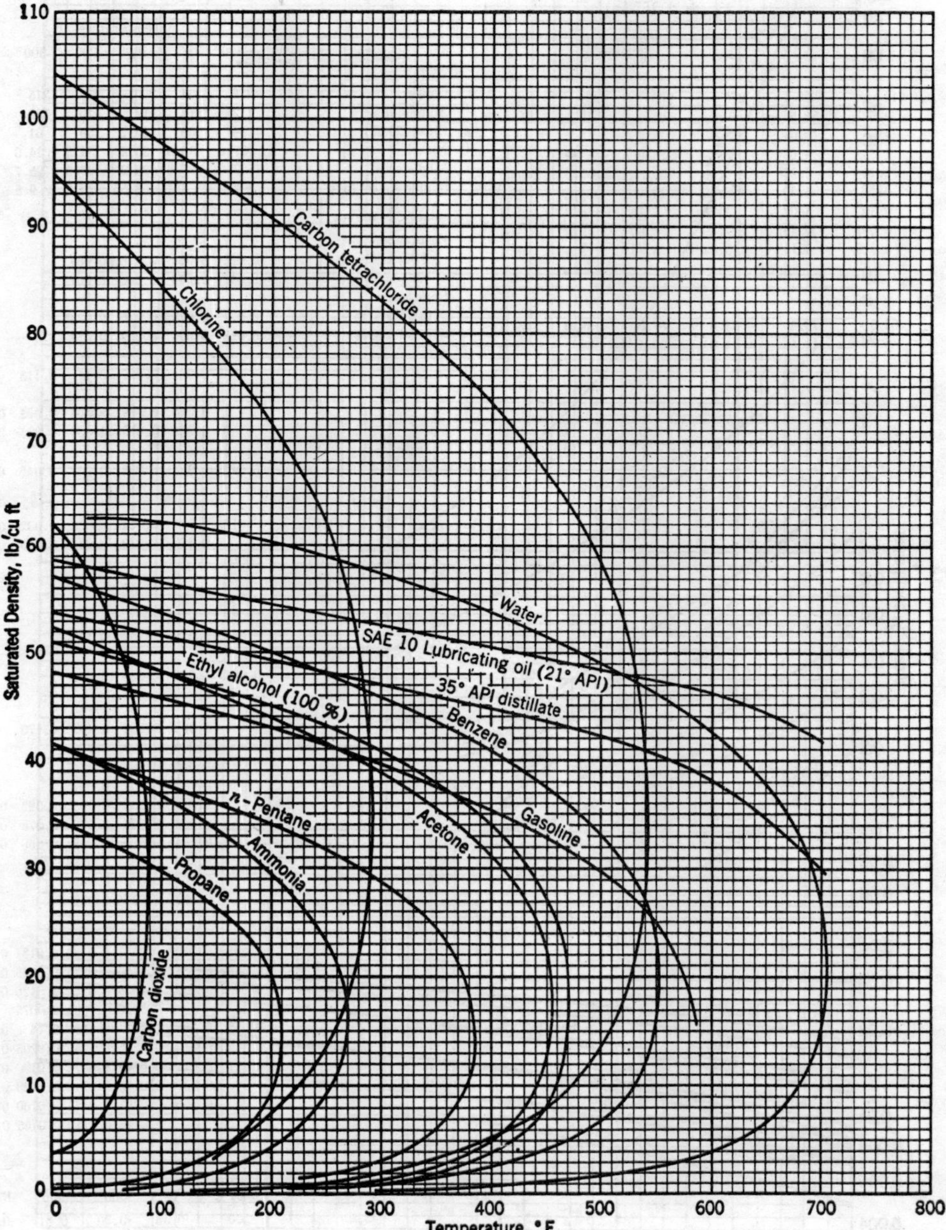

FIG. 544. Saturated densities of various liquids and gases as a function of temperature.

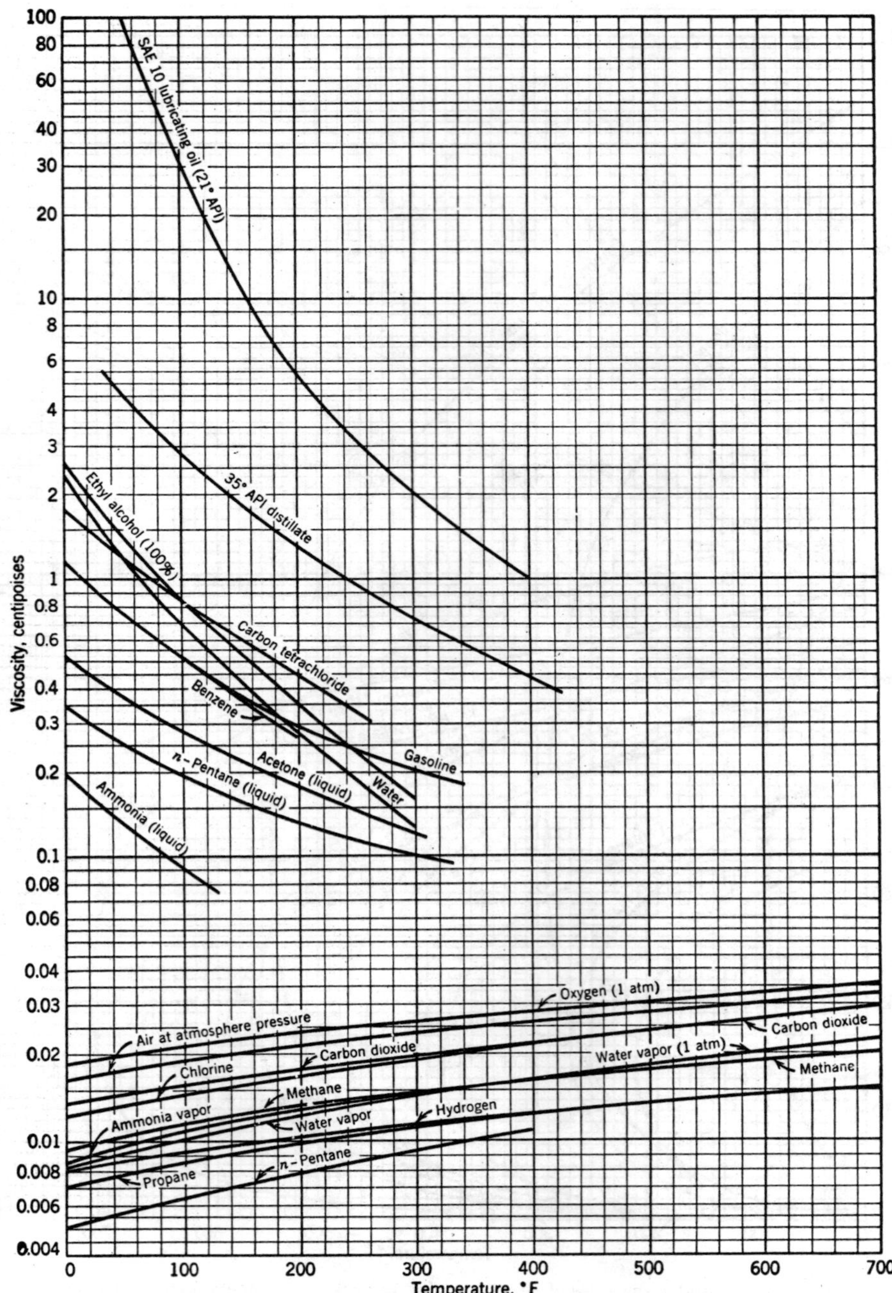

FIG. 545. Viscosities of various liquids and gases as a function of temperature at 1 atm.

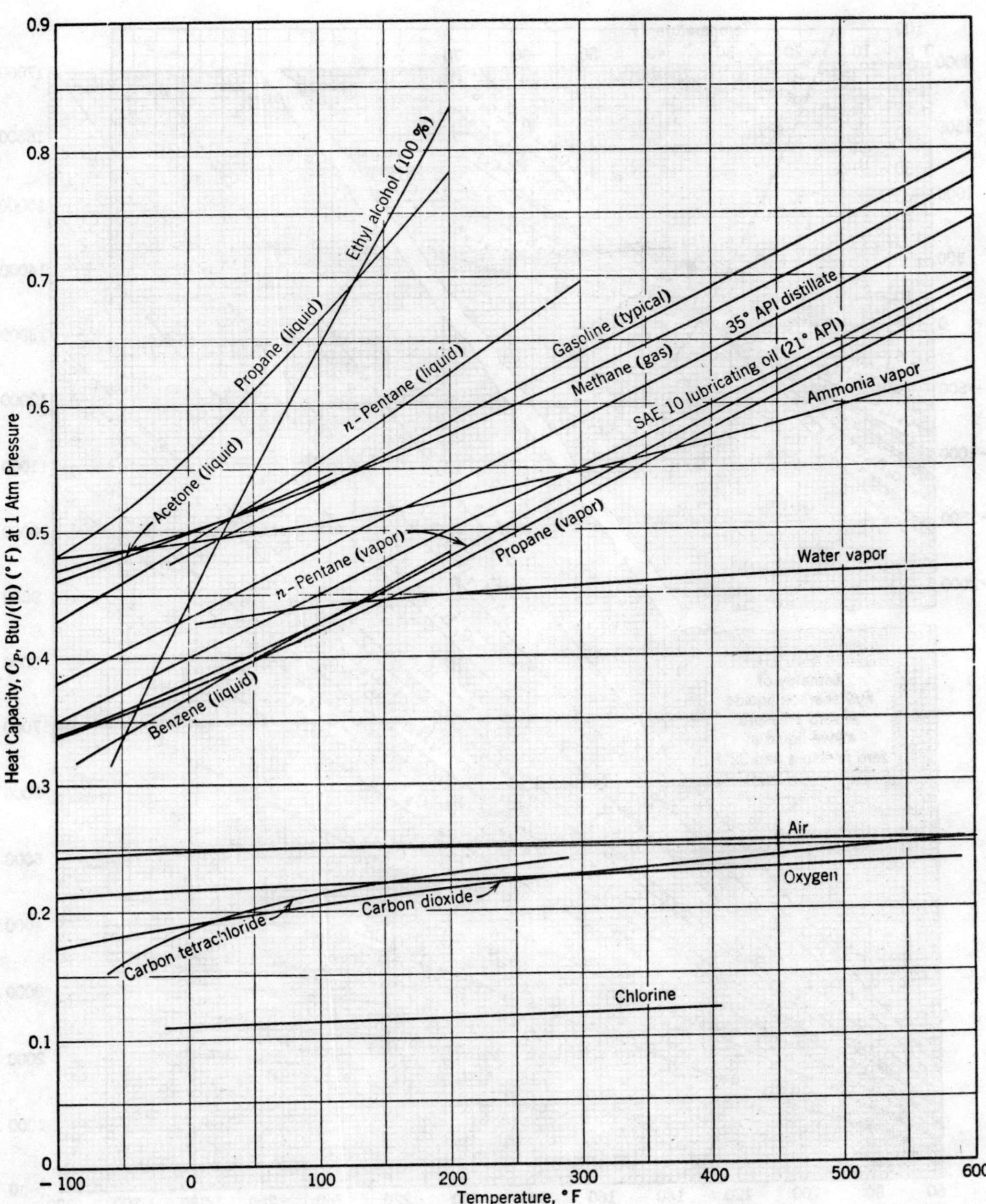

FIG. 546. Specific heats of various liquids and gases at 1 atm pressure as a function of temperature.

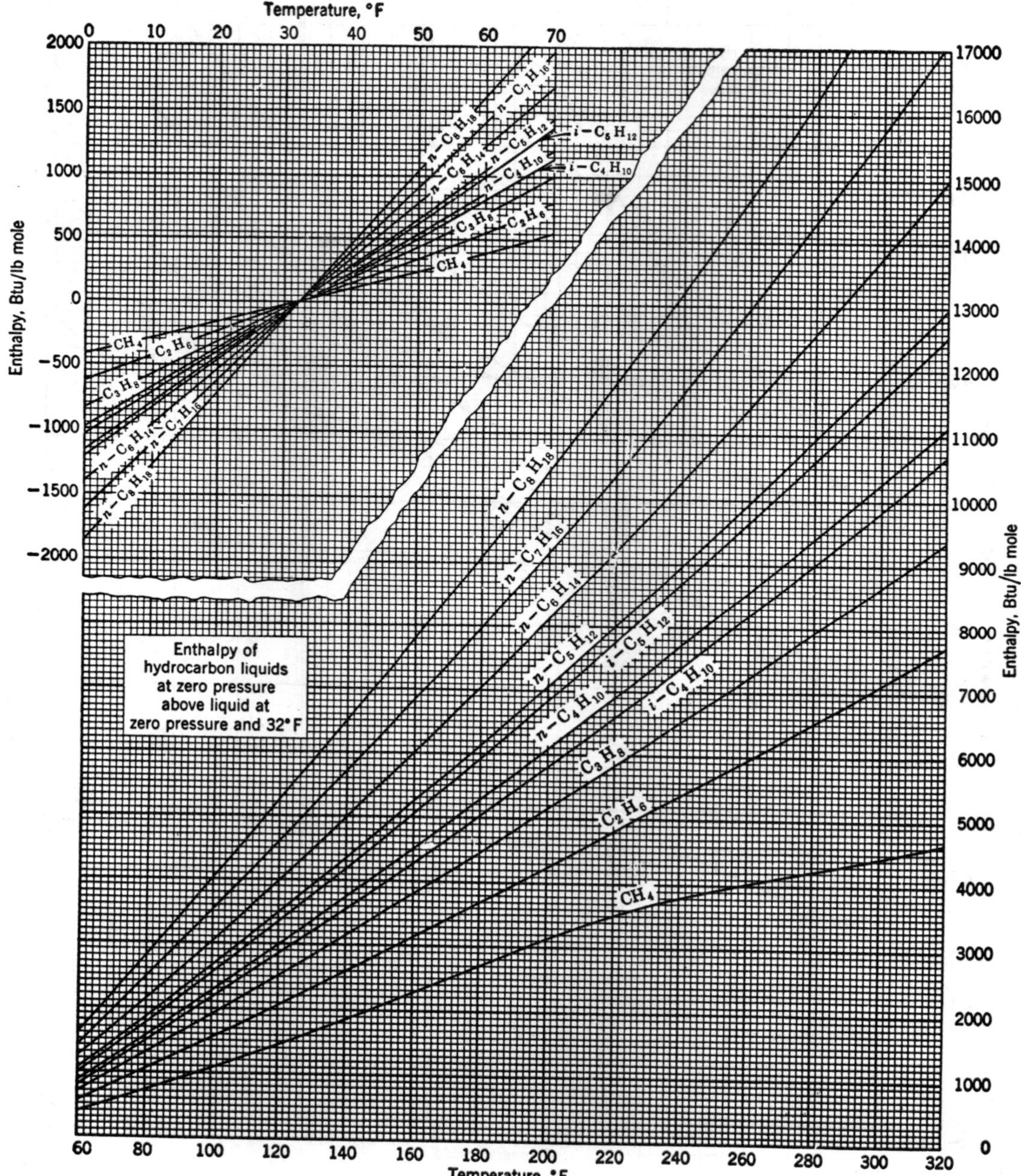

FIG. 547. Enthalpy of paraffin hydrocarbon liquids as a function of temperature in Btu per pound mole.

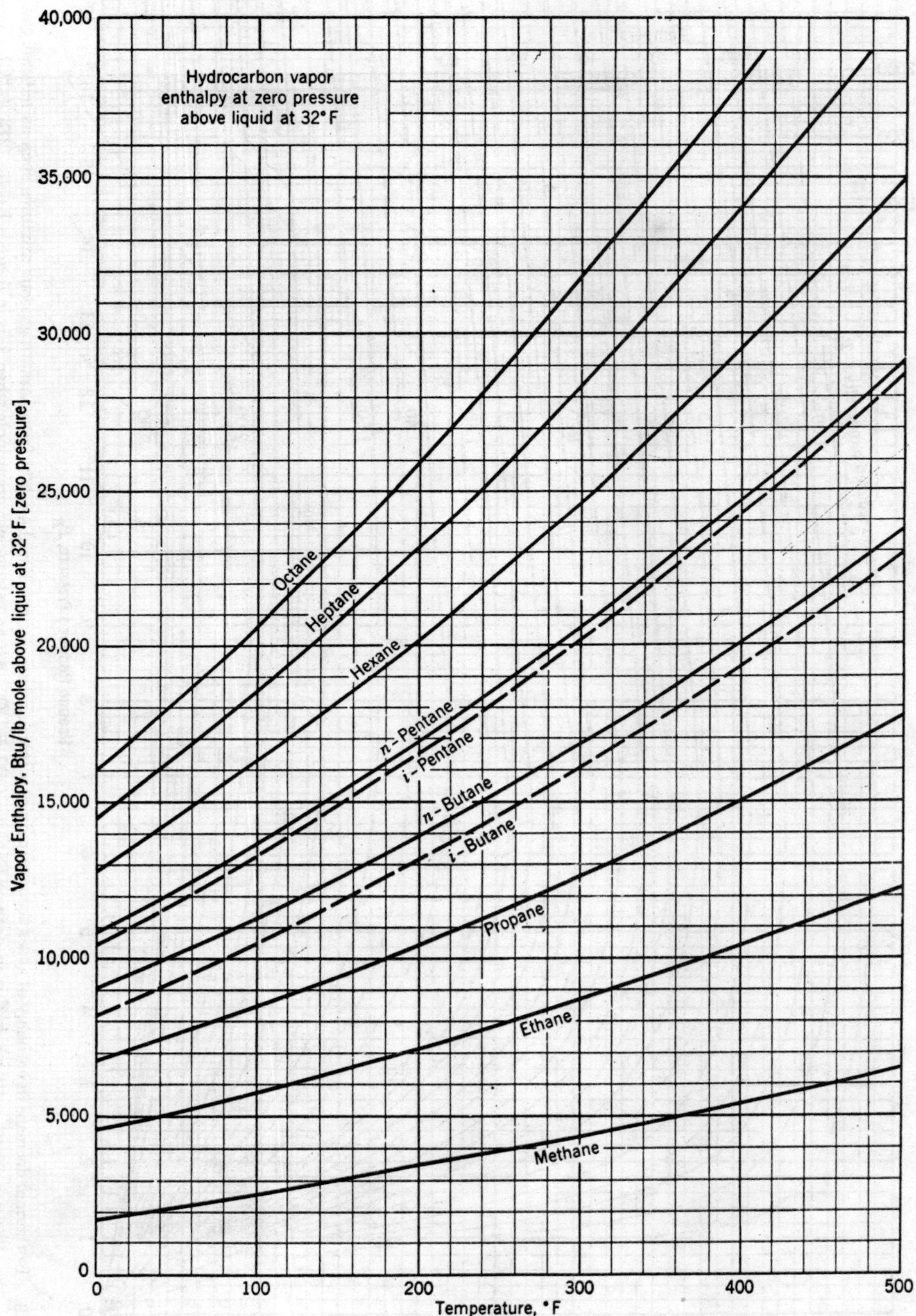

Fig. 548. Enthalpy of paraffin hydrocarbon gases as a function of temperature at constant low pressure.

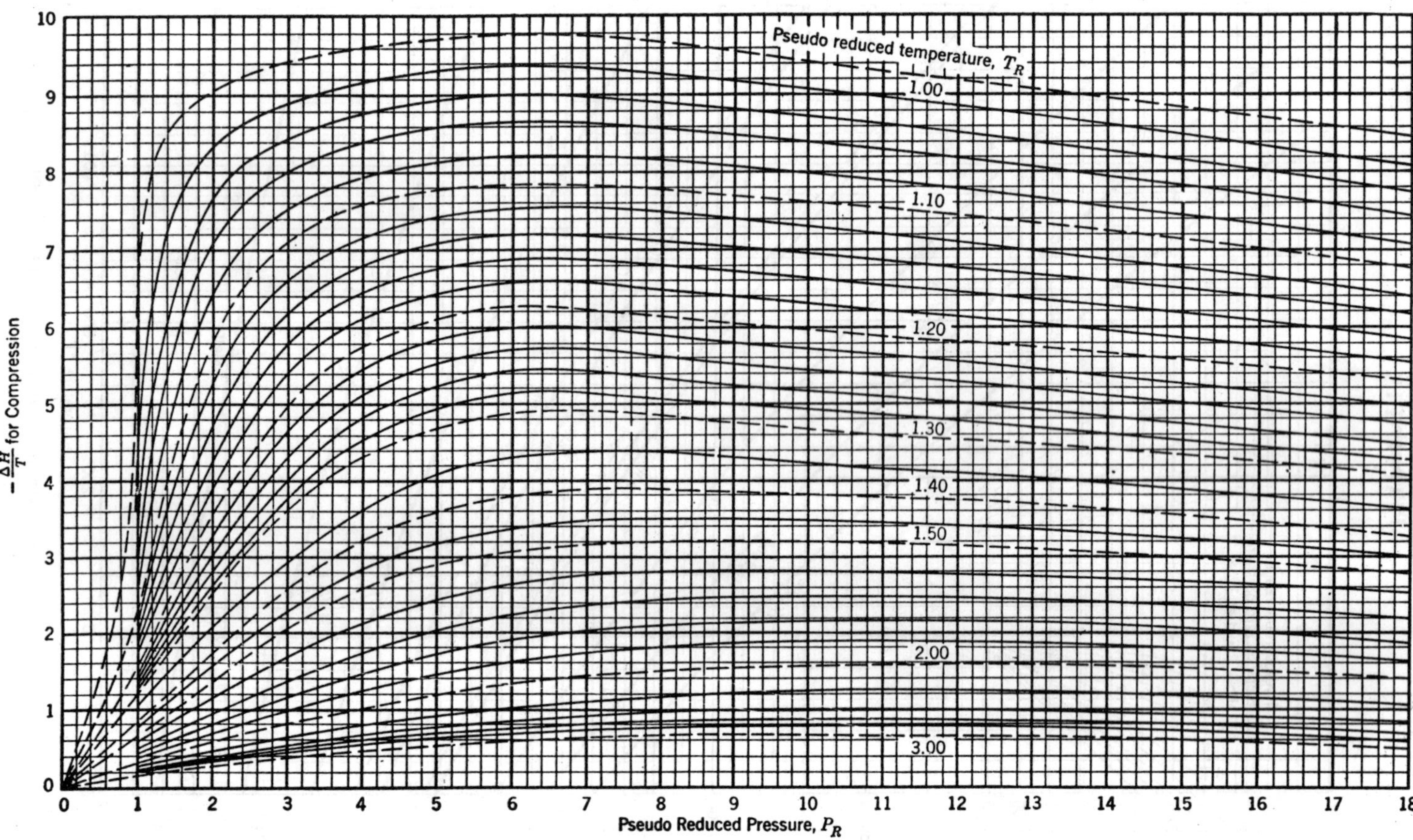

Fig. 549. Isothermal decrease in enthalpy of gases accompanying an increase in pressure. Based on data from paraffin gaseous mixtures such as natural gas. ΔH in Btu per pound mole when T is in degrees Rankine (460 + °F). ΔH in calories per gram mole when T is in degrees Kelvin (273 + °C).

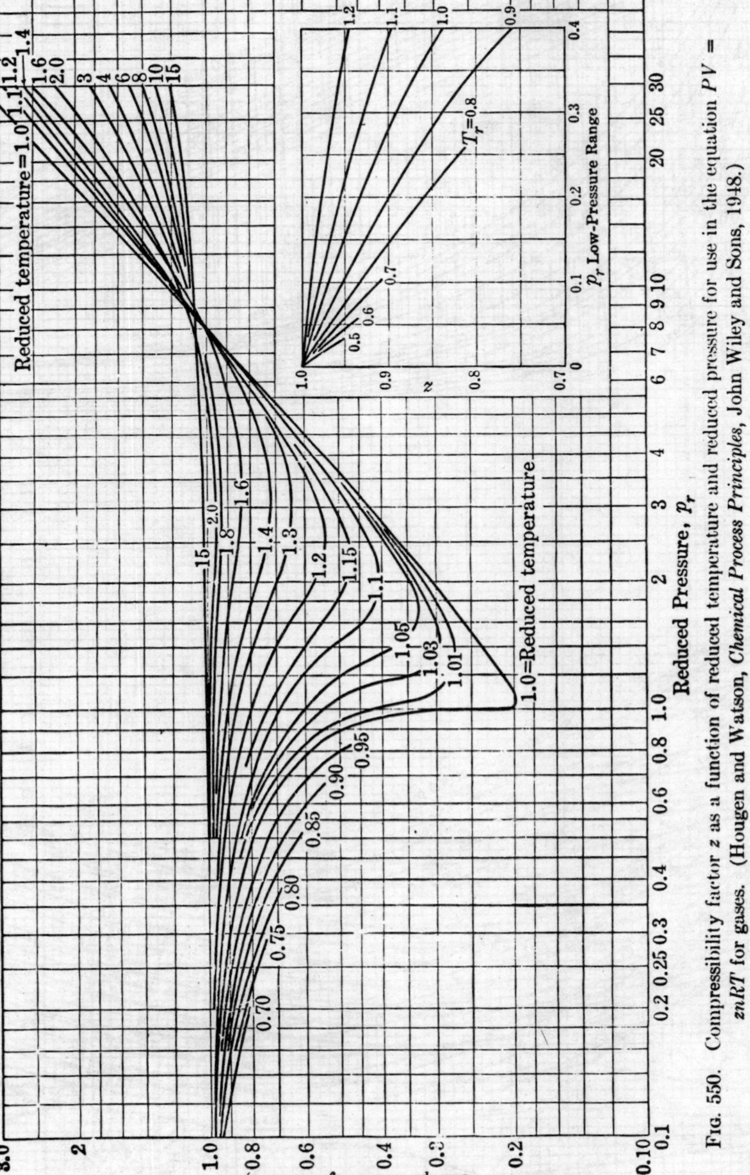

FIG. 550. Compressibility factor z as a function of reduced temperature and reduced pressure for use in the equation $PV = znRT$ for gases. (Hougen and Watson, *Chemical Process Principles*, John Wiley and Sons, 1948.)

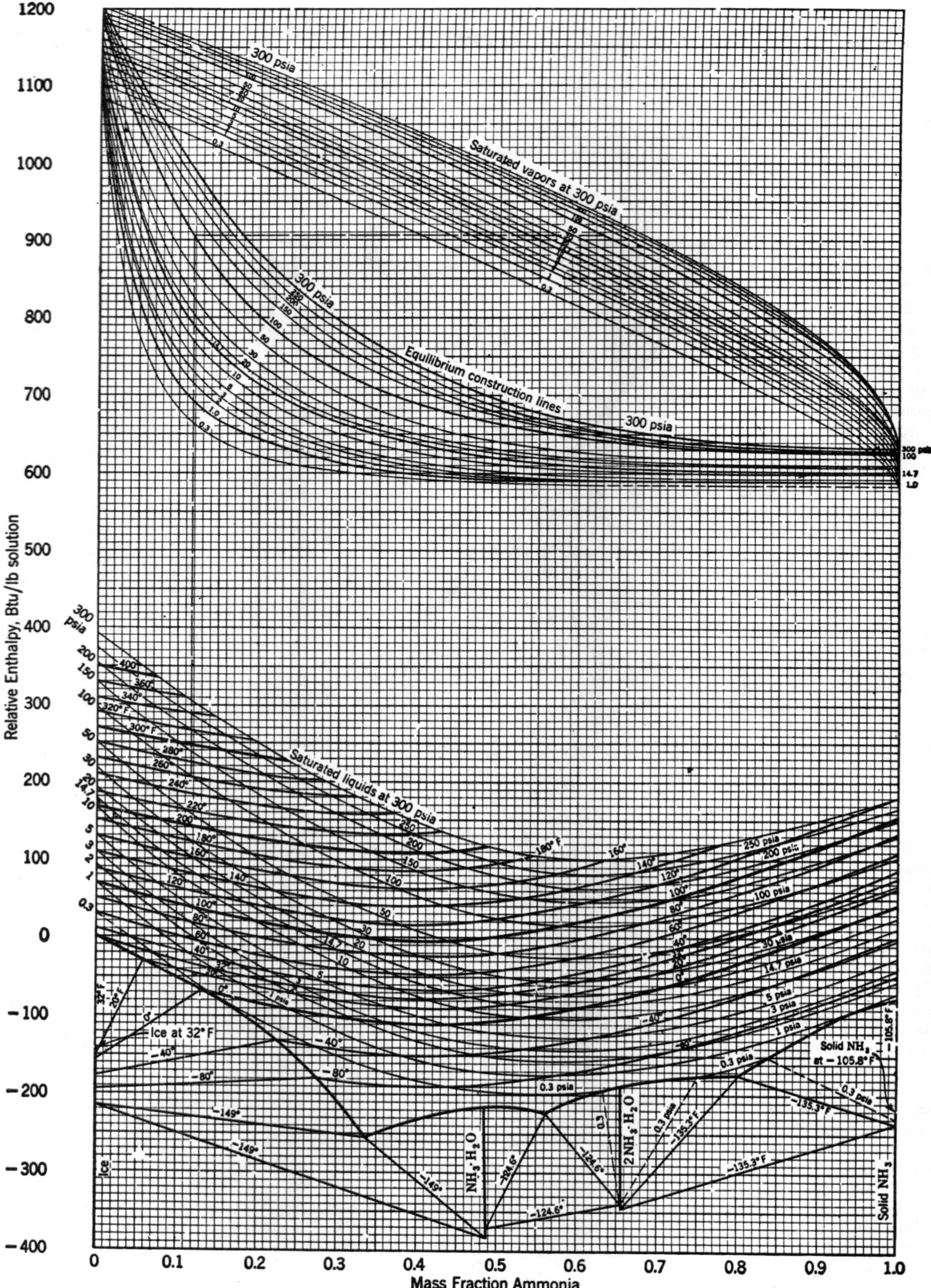

Fig. 551. Enthalpy-concentration diagram for the system NH_3—H_2O.

Reference states: water at 32° F and liquid ammonia at −40° F to conform with the usual steam and ammonia tables.

To determine equilibrium compositions a vertical line may be erected from any liquid composition at its saturation or boiling point and its intersection with the appropriate equilibrium construction line. A horizontal from this intersection will intersect the appropriate saturated vapor line at the desired equilibrium vapor composition. For example, liquid of 0.117 mass fraction NH_3 at 100 psia and 260° F is in equilibrium with va[illegible] of 0.625 mass fraction NH_3 as indicated by the dashed construction lines.

Index